T0319588

POLLINATION AND FLORAL ECOLOGY

POLLINATION AND FLORAL ECOLOGY

Pat Willmer

PRINCETON UNIVERSITY PRESS
PRINCETON AND OXFORD

Copyright © 2011 by Princeton University Press
Published by Princeton University Press, 41 William Street,
Princeton, New Jersey 08540
In the United Kingdom: Princeton University Press, 6 Oxford Street,
Woodstock, Oxfordshire OX20 1TW
press.princeton.edu
Jacket photograph: The bee *Eucera longicornis* foraging on the sweet pea
Lathyrus. (c) Andrej Gogala

All Rights Reserved

Library of Congress Cataloging-in-Publication Data

Willmer, Pat, 1953-
   Pollination and floral ecology / Pat Willmer.
      p. cm.
   Includes bibliographical references.
   ISBN 978-0-691-12861-0 (cloth : alk. paper) 1. Pollination. 2. Polli-
nation by insects. 3. Pollination by animals. 4. Plant ecology. I. Title.
QK926.W55 2011
571.8'642—dc22   2010049061

British Library Cataloging-in-Publication Data is available

This book has been composed in Times

Printed on acid-free paper ∞

Printed in the United States of America

10 9 8 7 6 5 4

CONTENTS

PART IV FLORAL ECOLOGY

PREFACE

When I first became interested in pollination biology around thirty years ago, there was essentially just one book to consult for core information, the classic by Faegri and van der Pijl, *The Principles of Pollination Ecology*, which has been the inspiration to generations, its last edition appearing in 1979 but still structured around the 1966 original. But twenty years on I became frustrated that there had been no single volume since that date which covered the field. This is particularly extraordinary given that the subject itself has expanded and blossomed prodigiously, both in its own right and as a testing ground for wider theories, as well as having assumed greater importance in both scientific circles and as a wider public concern.

It has not been all quiet on the publishing front since the 1970s of course. There have been excellent books with more of a natural history approach (M. Proctor et al. 1996; Buchmann and Nabhan 1996) and botanical books concentrating on floral structures (Barth 1985; Endress 1994) or on floral development (Glover 2007). Over the same period many extremely useful edited volumes have also appeared, each with chapters contributed by experts updating readers on particular themes, such as nectar and nectaries (Bentley and Elias 1983; Nicolson et al. 2007), anthers and pollen (D'Arcy and Keating 1996; Dafni et al. 2000), and floral scents (Dudareva and Pichersky 2006). Several books have compiled useful reviews on practical techniques (Dafni 1992; Kearns and Inouye 1993; Dafni et al. 2005) or on applied aspects of crop pollination (Free 1993), while Chittka and Thomson (2001) took a fresh approach by assembling chapters on aspects of pollinator cognition and learning.

But perhaps most important have been a select few edited volumes with broader pollination coverage and sometimes a more eclectic assemblage of contributions. Here the editors have been very much the lead-ers in the field: Jones and Little (1983) began the trend with *Handbook of Experimental Pollination Ecology*, while Lloyd and Barrett (1996) produced *Floral Biology: Studies on Floral Evolution in Animal Pollinated Systems*, opening up a broader and more ecological and evolutionary perspective on the whole subject. Then two collected contributions came along in 2006, picking up the themes that had been emerging in the preceding decade: Waser and Ollerton in *Plant-Pollinator Interactions: From Specialization to Generalization*, and Harder and Barrett in *Ecology and Evolution of Flowers*. These have been seen as especially important in expounding new approaches and exposing older assumptions to a more critical analysis. The debates about generalization and specialization, and the usefulness of the syndrome concept or of alternative modeling approaches, have been particularly thought-provoking, so that anyone starting out as a researcher in pollination ecology will find much to engage them from the authors represented in those two books.

Those readers, however, would not get the background information, the core material of pollination and floral biology. A beginner would still have to explore very widely across the mass of scientific journals, hoping to light upon a reasonably up-to-date review article. And in the last decade the primary literature has undergone an extraordinary explosion in volume, almost more than in any other field, partly because of the practical and applied environmental problems now seen to be emerging with pollination services, and partly because around twenty-five years ago interest began to spread much more widely beyond the European and North American temperate pollination systems, with expertise emerging especially in the New and Old World Tropics, in southern Africa and Australia, and in Japan and China.

For all these reasons, it seemed to me timely to attempt a book that would cover the subject more thoroughly, providing chapters on every key topic, and serve both as an introduction for anyone coming into the field as advanced undergraduate or postgraduate and as a source book for professionals. Begun half way through the decade, it has inevitably burgeoned beyond the original intentions as the expanding literature burst around my ears, to the distress of my publishers. But the resulting book will, I hope, achieve at least part of its aims and will not too greatly annoy those highly respected authors whose work I draw heavily on. Some, seeing pollination biology as pioneering a new era where ecological modelers have more to tell us than old-style field workers, might find my approach too traditional. But at least those coming in to the field hereafter may gain a better understanding of the foundations on which models can be built and so will better see where the key issues may lie.

Pat Willmer, November 2009

ACKNOWLEDGMENTS

It is a pleasure to begin by thanking the many generations of students and postgraduates who have let me know when I taught them in ways that inspired their interest and when I merely confused them. From them, I realized where the real fascinations of pollination lay and where my own interests were too biased or quirky for a wider audience—I have tried, not always successfully, to play down those quirks. Many former postgraduates and postdocs, now running research groups of their own and blossoming into major players in the world pollination and mutualism debates, have been especially important to me, and I thank them all for discussions, arguments, and the chance to read and learn from their own works; in particular I must pay tribute to Graham Stone (GS) and Simon Potts (SP), and more recently to Clive Nuttman (CN), Betsy Vulliamy, Gavin Ballantyne (GB), Caroline King, and Jonathan Pattrick, and to those whose names are followed by initials for use of their photos.

On a wider front, I confess to being a very poor networker, so that I know the pollination biologists of the world far less well on a personal level than I should, which is my loss. My communications with them have nevertheless been wonderfully invigorating, and I thank so many of them for their generosity in exchanging ideas. In particular, I have benefitted at various times from conversations about bees or pollination or wider mutualisms, whether prolonged or brief (and some so brief they will scarcely remember me!), with Jane Memmott, Steve Johnson, Judith Bronstein, Sally Corbet, Florian Schiestl, Dick Southwood, Naomi Pierce, Amots Dafni, Nigel Raine, Ingrid Williams, Cathy Kennedy, Francis Gilbert, Dino Martens, John Free, Jette Knudsen, James Cresswell, Steve Bullock, James Cook, Peter Yeo, Scott Armbruster, Mo Stanton, Spencer Barrett, Dini Eisikowitch, Barbara Gemmill, Peter Gibbs, Aubrey Manning, Gidi Ne'eman, Sue Nicolson, Samy Zalat, Theodora Petanidou, Chris O'Toole, and Rob Raguso.

Several of these colleagues have read and commented wisely on outlines, drafts, or chapters for me, and to them I owe enormous gratitude and an apology where I have insufficiently taken on their criticisms; the misconceptions or infelicities that remain are entirely my fault. Particular thanks to Clive Nuttman on color and scent issues, Peter Gibbs on incompatibility and on general issues of specialization and syndromes, Francis Gilbert (especially for updating my dipteran taxonomy), and Jonathan Pattrick for taking a broad beginner's view of the book and being even more persnickety about semicolons than I am!

I am deeply grateful to many who have allowed me to use their photographs, which usually put my own efforts to shame. All are acknowledged by initials in the plates, and all of course retain their copyrights in the photos reproduced here. Susie Whiten (SW), Alison Fernie (AF), Ian Johnston (IJ), and Sandy Edwards (SE) were generous with their own photographs from the UK and abroad. Special thanks to Steve Johnson (SJ), Dino Martens (DM), Karen Sarkisyan (KS), and Andrej Gogala (AG) for supplying superb photographs of the more difficult animal visitors. I also owe a massive debt of gratitude to my illustrators, especially Dawn Toshack for most of the pictures of flowers and visitors and their detailed structures, and Caroline King for reproducing most of the graphical figures from other published works, as well as to Dimitri Karetnikov for advice on figures in general. I must also thank others at Princeton University Press who have patiently supported my efforts to complete this work; for the initial enthusiasm of Robert Kirk especially, and the subsequent diligence of Stefani Wexler, Beth Clevenger, Gail Schmitt, and Jennifer Slater, as well as the anonymous reviewers who offered comments at an early stage.

Finally a thank-you to many colleagues and friends who have helped to keep me sane through this whole process, especially Clare and Iain and Morven, Sandra who was so often a rock at work, Peter and Elisabeth, and neighbor Pam; all of these have been invaluable.

PART I

ESSENTIALS OF FLOWER
DESIGN AND FUNCTION

Chapter 1

WHY POLLINATION IS INTERESTING

The flowering plants (**angiosperms**) account for about one in six of all described species on earth and provide the most obvious visual feature of life on this planet. In the terrestrial environment, their interactions with other living organisms are dominant factors in community structure and function; they underpin all nutrient and energy cycles by providing food for a vast range of animal herbivores, and the majority of them use animal **pollinators** to achieve reproduction. Most of the routine "work" of a plant is carried out by roots and leaves, but it is the **flowers** that take on the crucial role of reproduction.

A flower is usually **hermaphrodite**, with both male and female roles. Hence it is essentially a structure that produces and dispenses the male **gametophytes** (**pollen**), organizes the receipt of incoming pollen from another plant onto its own receptive surfaces on the **stigma**, and then appropriately guides the pollen's genetic material to the female **ovules**. The flower also protects the delicate male and female tissues (stamens and pistils) and has a role in controlling the balance between **inbreeding** and **outbreeding**, hence influencing the genetic structure and ultimately the evolution-

ary trajectory of the plant. But the plant itself is immobile, so that incoming pollen has to be borne on some motile carrier, sometimes wind or water but much more commonly on a visiting animal. To quote one source (Rothrock 1867), "among plants, the nuptials cannot be celebrated without the intervention of a third party to act as a marriage priest"! A pictorial overview of the stages is shown in figure 1.1, covering the processes of pollination that are the focus of this book.

A flower also serves to protect the pollen as it germinates and as the male nucleus locates the egg and then to protect the ovules as they are fertilized and begin their development into mature **seeds**. However, these later events (**germination** of the pollen and **fertilization** of the ovule) are technically not part of **pollination**, and they are covered here only as needed to understand the characteristics and effects of pollen transfer.

Since flowers bring about and control plant reproduction, they are central to much of what goes on in the terrestrial world, and pollination is a key **mutualism** between two kingdoms of organisms, perhaps the most basic type of exchange of sex for food; the plant gains reproductive success, and the animal—usually—gains a food reward as it visits the plant. But the visitor does not "want" to be a good pollinator and has to be manipulated by the plant to move on and to carry pollen to another plant. In practice, only about 1% of all pollen successfully reaches a stigma (Harder 2000).

Nevertheless, pollination by animals (**biotic pollination**) is both more common (Renner 1998) and usually more effective than alternative modes of abiotic pollen movements using wind or water, and animal pollination is usually also associated with more rapid **speciation** of plants (Dodd et al. 1999; K. Kay et al.

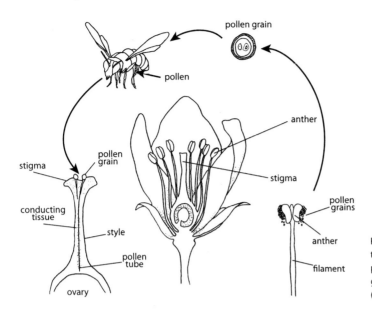

Figure 1.1 The central processes of pollination in a typical angiosperm flower, with the route taken by pollen from anther to stigma (followed by pollen tube growth into the style) in an animal-pollinated species. (Modified from Barth 1985.)

2006). Discussion of animal pollination therefore dominates this book, and around 90% of all flowering plants are animal pollinated (Linder 1998; Renner 1998). Furthermore, plants are, of course, the foundation of all food chains on the planet, and their efficient pollination *by* animals to generate further generations is vital to ensure food supplies *for* animals. Natural ecosystems therefore depend on pollinator diversity to maintain overall biodiversity. That dependence naturally extends to humans and their agricultural systems too; about one-third of all the food we eat relies directly on animal pollination of our food **crops** (and the carnivorous proportion of our diet has some further indirect dependence on animal pollination of forage crops). Pollination and factors that contribute to the maintenance of pollination services are vital components to take into account in terms of the future health of the planet and the food security and sustainability of the human populations it supports.

Beyond its practical significance, the flower-animal mutualism has been a focus of attention for naturalists and ecologists for at least two hundred years and provides almost ideal arenas for understanding some of the fundamental aspects of biology, from evolution and ecology to behavior and reproduction. It is perhaps more amenable than any other area to providing insights into the balance and interaction of ecological and evolutionary effects (Mitchell et al. 2009). Flowers are complex structures, and their complexity admirably reveals the actions, both historical and contemporary, of the selective agents (mainly, but not solely, the pollinators) that we know have shaped them. These factors make floral biology an ideal resource for understanding biological adaptation at all levels, in contrast with many other systems, where there are multiple and often uncertain selective agents.

In this first chapter, some of these central themes are introduced to set the scene for more specialist chapters; it should be apparent from the outset that while each chapter might stand alone for some purposes, it cannot be taken in isolation from this whole picture.

1. Which Animals Visit Flowers?

At least 130,000 species of animal, and probably up to 300,000, are regular flower visitors and potential pollinators (Buchmann and Nabhan 1996; Kearns et al. 1998). There are at least 25,000 species of bees in this total, all of them **obligate** flower visitors and often the most important pollinators in a given habitat.

There are currently about 260,000 species of angiosperms (P. Soltis and Soltis 2004; former higher estimates were confounded by many duplicated namings), and it has been traditional to link particular kinds of flowers to particular groups of pollinators. About 500 genera contain species that are bird pollinated, about

250 genera contain bat-pollinated species, and about 875 genera predominantly use **abiotic pollination**; the remainder contain mostly insect-pollinated species, with a very small number of oddities using other kinds of animals (Renner and Ricklefs 1995).

The patterns of animal flower visitors differ regionally. In central Europe, flower visitors over a hundred years ago were recorded as 47% **hymenopterans** (mainly bees), 26% flies, 15% beetles, and 10% butterflies and moths; only 2% were insects outside these four orders (Knuth 1898). But in tropical Central America, the frequencies would be very different, with bird and bat pollination entering the picture and fewer fly visitors, while in high-latitude habitats the vertebrate pollinators are absent and flies tend to be more dominant. Some of these patterns will be discussed in chapter 27.

2. Why Do Animals Visit Flowers?

The majority of flower visitors go there simply for food, feeding on sugary **nectar** and sometimes also on the pollen itself. Chapters 7 and 8 will therefore deal in detail with these commodities, and chapter 9 will cover a few more unusual foodstuffs and rewards that can be gathered from flowers; chapter 10 will take an economic view of all these food-related interactions, in terms of costs and benefits to each participant. Major themes in other chapters include the ways that flower feeders can improve their efficiency: learning recognition cues to select between flowers intra- and interspecifically, learning handling procedures, learning to avoid emptied flowers, and avoiding some of the hazards of competing with other visitors.

Flowers are also sometimes visited just as a convenient habitat, often simply because they offer an equable sheltered **microclimate** for a small animal to rest in, a place that is somewhat protected against bad weather, predators, or **parasitoids**. Or flowers may offer a reliable meeting site for mates or hosts or prey, or for females an **oviposition** site providing shelter for eggs and larvae. More rarely they are used as a warming-up site by insects in cold climates, usually because the flowers trap some incoming solar radiation, which enhances their own ovule development, but occasionally because a few flowers can achieve some metabolic **thermogenesis** that warms their own tissues (chapter 9 will provide more details on this topic).

3. How Do Flowers Encourage Animal Visitors?

Many plant attributes contribute to attraction of visitors: J. Thomson (1983) usefully groups these as *plant presentation*. Some of these attributes are readily apparent to visitors, and these may be features of individual flowers (e.g., color, shape, scent, reward availability, or time of flowering) or features of whole plants or groups of plants (e.g., flower density, flower number, flower height, or spatial pattern). The more readily apparent plant presentation **traits** can be divided into **attractants** (advertising signals), dealt with mainly in chapters 5 and 6, which discuss visual and olfactory signals, and **rewards** (usually foodstuffs), dealt with in chapters 7–9. Aspects of the *timing* and *spacing* of flowers, and how these might be affected by competition between different flowering plants, are given more in-depth treatments in chapters 21 and 22.

Other floral attributes are more cryptic to the visitor and may only determine the reproductive success of the plant in the longer term; these might include pollen amounts, ovule numbers, the genetic structure of the plant population, the presence and type of **incompatibility** system, etc.

It is generally in the plant's interest to support and even improve its visitors' efficiency, encouraging them to go to more flowers of the same species (so ensuring that only **conspecific** pollen is taken and received) and to go to flowers with fresh pollen available and/or with receptive stigmas for pollen to be deposited upon. Many flowers therefore add signals of status to their repertoire, via color change, odor change, or even shape change. Visitors are thereby directed away from flowers that are too young or too old or already pollinated. Instead they will tend to concentrate their efforts on those (fewer) flowers per plant that are most in need of visitation, thus also being encouraged to move around between separate plants more often and to ensure **outcrossing** rather than **selfing**. Reasons for favoring breeding by outcrossing (i.e., with other plants) are covered more fully in chapter 3.

4. What Makes a Visitor into a Good Pollinator?

In many ways this is the crucial theme running through this book. It relates to what is probably the major current debate in pollination ecology, that is, to what extent pollination it is a *generalist* process and to what extent it is a *specialist* one. Pollination has in the past nearly always been categorized in terms of *syndromes*, with particular groups of flowers recognized as having particular sets of characteristics (of color and scent, shape, timing, reward, etc.) that suit them to be visited by particular kinds of animals; and it is implicit in this approach that these suites have often been arrived at and selected for by convergent evolution in plant families that are unrelated. Thus most authors have used terms such as **ornithophily** to describe the bird **pollination syndrome**, or **psychophily** for the butterfly pollination syndrome. Flower characteristics would be listed that fit each syndrome, and an unfamiliar flower's probable pollinators could therefore be predicted. Flowers in each category were seen as having a degree of specialization that suited them to their particular visitors, with some syndromes being more specialized than others. Nearly all earlier works on pollination were organized around this theme of syndromes, and it served as a useful structure for understanding animal-flower interactions for nearly two centuries. Without knowing this background it would be nearly impossible to follow the current debates that are a major focus for pollination ecologists, and it would also be very difficult to structure the information on flower attractants and flower rewards in chapters 5–9. This book thus retains a syndrome-based approach throughout its early chapters and explicitly considers the evidence in support of a syndrome approach in chapter 11; then it unashamedly covers each of the syndromes in turn in chapters 12–19, providing all the core materials on which later criticisms might be based.

The criticisms and debates focus around the reality of syndromes and how far they have been overplayed in the previous literature, perhaps blinkering or biasing our perceptions. Many authors now regard flower pollination as a much more generalized phenomenon, where most flowers get many different kinds of visitors and have *not* been and are *not* being heavily selected to specialize for the needs of one particular "best" visitor. This approach is specifically addressed in chapter 20. It will be a major argument there that the issue has been clouded by an as yet insufficient distinction between flower visitors and pollinators. So what does make a visitor into a good pollinator?

Physical Factors

Any animal that is to be an effective pollinator must have the ability to passively pick up pollen as its body moves past the **anthers** of a flower that it visits and carry that pollen to another flower. Normally this will be aided by the animal being a good *physical fit* in terms of size and shape, so that in alighting on the flower surface, or when inserting its tongue or beak of appropriate length, some specific part of its body touches the anther. Additionally pollen pickup and carriage will be aided if the animal has appropriate *surface structures*; pollen adheres well to feathers, fur, and hairy or scaly surfaces in insects but does not get transported well on shiny or waxy surfaces (and may even be damaged by certain surface secretions). Hence a small shiny beetle taking some nectar by crawling into the lower surface of a large tubular **corolla** where the anthers are in the corolla roof may well be a regular visitor to that flower but is unlikely to be an effective mover of its pollen; in effect, it is acting as a "cheat" as far as the flower is concerned and may be termed an **illegitimate visitor.**

Behavioral Factors

Different animals land on and forage at flowers in very different fashions. There are many aspects of behavior that will affect whether a given animal is going to be a good pollinator. Pollinators will seldom have a complete perception of all the aspects of plant presentation mentioned above, but they will respond to at least some of them in ways that are useful to the plant:

1. Their choices of *places and times* to visit, and exactly which flowers to visit, will be critical. Visits occurring before **dehiscence** (the splitting of the anthers to reveal the pollen) or after pollen depletion are normally of no value to the flower in fulfilling its male role, and visits before or after the stigma is receptive to incoming pollen are of no value to the flower in its female role.
2. Their *handling* of the flowers affect their pollen pickup and deposition characteristics; ideally they should receive pollen at a specific site on their bodies, and one that is also a good site for subsequent deposition of that pollen onto a stigma.

3. Their *handling time* per flower affects how many flowers are visited in a given time.

4. Their speed and directionality of *movement between plants* affect **pollen dispersal**.

5. They should not be too efficient at *grooming* off the pollen, or indeed at *eating* it.

6. Their *flower constancy*, that is, the likelihood that they will move to another flower of the same species, is perhaps most critical. If they innately or by learning prefer a particular flower **phenotype**, their high constancy will usually ensure that they move neatly and sequentially among conspecific flowers, not wasting pollen by depositing it in the wrong species. Constancy to a flower (considered in detail in chapter 11) gives economies to the visitor also; it may minimize travel distances, handling times, and learning effort and maximize pollen packing.

Behavioral factors such as these are often the key to being a good pollinator and of course are affected by the animals' abilities to learn as they become more experienced as foragers. The ability to form a **search image** or to respond consistently to other cues, associating particular signals with the presence of food, hones their foraging ability and can cement their relationships with particular flowers. Hence later chapters of this book, in considering particular groups of animals that visit flowers, include careful consideration of their sensory and learning capacities.

Physiological Factors

Animals have differing physiological strategies and constraints, and these too can affect their energetic needs and thence their flower-visiting patterns, as will be discussed in chapter 10. Most animals (including nearly all invertebrates, and therefore many of the insect flower visitors) lack elaborate internal physiological regulatory systems, and their *thermal balance* and *water balance* are strongly influenced by environmental conditions. They cannot function if they are too hot or too cold or are short of water, and must forage in times and places that provide suitable microclimates, using the sun's radiation to warm up by basking, or shady places to cool down again, and seeking (usually) sites that are relatively humid. However, birds and bats are physiologically more sophisticated and can regulate their body temperature and water balance more precisely; they generate heat internally through their own metabolism (**endothermy**) and regulate their own body fluids with efficient skin exchanges, respiratory controls, and excretory organs. They can in principle forage at almost any time and in any habitat, though they may still conserve their own energy by picking more equable sites.

The distinction does not lie exactly between the vulnerable invertebrates and the highly regulated birds and mammals, however. It is now clear that a rather small proportion of insects can also show endothermy, at least some of the time when they need to warm up in the absence of solar inputs (chapter 10); this applies especially to most bees, a few hoverflies, some large moths, and some beetles, occurring more sporadically in other groups. It is perhaps no coincidence that endothermic abilities in insects are most common in the flower-visiting groups, which have access to ready fuel supplies in the form of nectar but which may also need to compete for that nectar in the cool of early mornings or at dusk.

Given the list of factors that can turn a visitor into a good pollinator, two obvious points should emerge:

1. A great many of the animals that go to flowers for a short drink of nectar may be rather poor at pollinating that flower. Those with a poor physical fit, those that cheat, and those with little or no flower constancy are likely to be especially ineffective.

2. Of all the visitors, bees are likely to be especially good as pollinators. They rely solely on flowers for food, both as adults and as larvae, and so must visit more flowers than any other animals. Their sizes, hairy surfaces, and variably long tongues, their excellent learning abilities, communication systems and floral constancy, and their endothermic capacities all equip them to visit flowers efficiently (from their own perspective) and effectively (from the plant's point of view). Although they are sometimes described as **pollen wasters** (because they, or rather their offspring back at the nest, eat so much of the pollen that they pick up), they are by far the most important pollinators in most ecosystems, and they do achieve high pollen export from visited flowers (Harder and Wilson 1997); plants have adapted over evolutionary time to make

best use of them by providing more than enough pollen to cater for their needs *and* ensure that sufficient pollen still commonly reaches other flowers.

5. Costs, Benefits, and Conflicts in Animal Pollination

Plants with hermaphrodite flowers benefit greatly because a single animal visit can allow both pickup of pollen and its deposition on a stigma, fulfilling both the **male** and **female functions** of that flower at the same time. Animals benefit greatly by finding easily acquired foods, both sugary (nectar) and often also proteinaceous (pollen). Pollination by animals may therefore be a mutualism, of benefit to both participants, but it is *not* **altruistic**; for the animals, pollination of the flowers they forage at is almost always just an irrelevant by-product. In fact the plant and animal have a conflict of interest, often with adaptation and counteradaptation going on from both sides through evolutionary time to try and get a bigger share of the benefits. So the situations that we see now are the end products of the long and sometimes quite duplicitous associations of flowers with animals.

The plant ideally wants a visitor that is cheap to feed, alighting only briefly, moving on rapidly to another plant, and being faithful to its chosen plant species; so ideally, the forager should be chronically underfed and continuously on the move. But (again ideally) the animal would prefer to be as well fed and lazy as possible, getting as much food as it can from one flower with minimum energy expenditure, being relatively sedentary, and then moving on to any other nearby flower with copious nectar, whatever its species (although we already noted that some degree of fidelity may improve its foraging efficiency).

Hence, although there are obvious benefits to both partners, there are potentially clear costs as well. The plant has to invest in attractants (its carbon and nitrogen resources are used to make flamboyant petals, pigments, and chemical scents), as well as mere rewards. If the plant reduces its rewards too far, the animal may not get enough food and will give up on that species. The plant generally also has to compete with other plant species for pollination, to obtain a share of the "good" pollinators, so it cannot afford to skimp on its offerings too much if it is growing within a reasonably diverse plant community. Many plants also have to

offset the additional costs of animal exploiters: those visitors who take rewards without pollinating (thereby **cheating**, chapter 24) and the flower eaters (florivores) or foliage herbivores also attracted by the pollination cues (chapters 25 and 26). For the animals, there may be costs linked to carrying the pollen they have inadvertently picked up (sometimes it is very unwieldy and can interfere with their wings or legs or sense organs), which may favor animals trying to cheat, and there may be costs also from the potential risks of predation or parasitism at flowers, since enemies can use them as a place to find prey or hosts reliably (chapter 24). There are also costs arising from the tendencies of the plants to cheat (chapter 23) by offering no real reward and sometimes by trapping the animal.

6. Why Is Pollination Worth Studying?

Pollination ecology can provide almost unparalleled insights into evolution, ecology, animal learning, and foraging behavior (fig. 1.2). It is perhaps the best of all areas to see and understand some basic biological processes and patterns; studies that deal exclusively with pollination biology have often had major impacts on general ecological and evolutionary theory.

Pollination interactions often show us *evolution* by natural **selection** in action almost before our eyes and provide some very clear-cut examples of **adaptive radiation** and, perhaps, of plant speciation. They are particularly useful for studying **coadaptations** (**coevolution**), because such interaction often involve relatively few organisms interacting with relatively high interdependence and incorporating the most fundamental of phenomena (reproduction for the plant, food supply for the animal). There are selection pressures on each side of the partnership, offering hopes that effects at the basic level of male and female **gene flow** can be quantified, sometimes (in crop pollination especially) on a time scale that can be detected within one scientist's period of study.

In terms of *ecology*, the study of pollination sheds light on how different organisms interact and affect each other, especially the competitive effects of plants upon each other, and on the various levels of interactions of plants with pollinators, including resource allocation, competition, exploitation, and simply cheating. In the last two decades there has been an increasing stress on community-level interactions in pollination

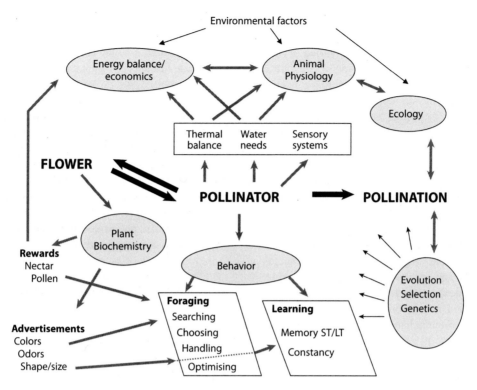

Figure 1.2 Key interactions of major biological topics and themes promoting interest in the study of pollination.

biology, now seen as an especially useful (because highly quantifiable) arena for more general work on community structures (J. Thomson 1983); so this book inevitably considers the community ecology of pollination, especially in later chapters.

Pollination biology also gives exceptional insight into the ecology of reproductive strategies and the complexities of sex and reproduction. Flowers usually are hermaphrodites, but they have many ways of organizing their sex life sequentially or spatially to maximize their reproductive output and **fitness**. This book contains rather little coverage of plant reproductive strategies beyond the basics, because the field has now become dominated by theory and modeling, and the topic has also been extensively and recently reviewed in other works (e.g., Harder and Barrett 2006).

In the realm of animal *behavior* some key influences can be especially easily measured and manipulated with flower visitors, and it is no accident that much of the key work on **visual discrimination**, learning behaviors, and above all **optimal foraging** has used pollinators, especially bees. **Optimal diet theory** can model how animals should behave in an environ-

ment offering different proportions of alternative prey as potential food items of differing value (also taking into account factors such as conspicuousness and different variances). The theory predicts that a range of outcomes from complete specialization on one kind of prey item to complete generalization on all possible items is to be expected, even from the same animal, as the prey parameters are varied. Substituting "flower" for "prey item" (and with the immense advantage of very easily quantified **caloric rewards** from nectar), it is not unexpected that pollinators similarly turn out to show almost the full range of possible foraging behaviors. Furthermore, they have proved ideal subjects with which to develop foraging models that can take into account different constraints on the foraging animals, whether from different physiological limits or from different cognitive skills. Learning ability is especially needed where resources are of intermediate predictability (Stephens 1993): that is, too unpredictable over one or a few generations for fixed behavior patterns to be favored, but not so greatly unpredictable that an individual cannot track the changes. This exactly applies to floral resources, so that we should expect flower visitors from any **taxon** to be good learners. Fortunately this is also readily tested with real or model flowers where just one trait at a time can be varied or associations of traits compared; again, social bees are ideal animals to work with, reliably emerging from their nests and traveling straight to the flowers provided, then displaying clear choices between alternative flowers.

For all these reasons, and with the added concern over human-induced effects on pollinator services in relation to biodiversity and to crops, interest in pollination ecology has burgeoned in the last ten to fifteen years, and the subject is attracting strong interest beyond the traditional academic centers of the developed world, giving us valuable insights into more varied habitats in Asia, Africa, and South America and into a greater diversity of interactions. Increasingly these systems are being modeled, and our preconceptions (of these and of other kinds of mutualisms) are being challenged. But the models are sometimes hampered by reliance on inadequate records, and understanding, of flower-visitor behaviors, and one of the most important issues for the immediate future is ensuring that the new generation of pollination ecologists understand the core subject material of floral biology and can measure and categorize *pollination* as distinct from mere *visitation* to feed into their models. We are in need of many and better quantitative studies of the *effectiveness* of visitors (for example, the average number of conspecific outcrossing **pollen grains** deposited on a stigma at an appropriate time by a given visitor in a single visit; chapter 11). Then we can properly understand plant and pollinator communities and pollination **networks**, and the effects of potential extinctions of flower visitors/pollinators on the communities of which they are a part. This book therefore hopes to provide in a single source a useful reference for all the aspects of floral biology and pollination interactions that need to be considered to give a real appreciation of these fascinating mutualisms.

Chapter 2

FLORAL DESIGN AND FUNCTION

Outline

Flowers are essentially the containers for a plant's sex organs, but they must perform several interrelated functions. Most obviously, and taking center stage here, they make and mature the **gametes** and then dispense the male gametes in such a way that they will be transported to another appropriate flower in the process of pollination, leading to fusion with female gametes. However, the flower must also be structured so that it protects the crucial sex organs through pollination and seed set, both from the environment (excessive rain or drought, freezing, heat load in full sun or during flash fires, physical damage in high winds) and from herbivorous animals (whether these be tiny sucking insects or large browsing and trampling vertebrates). The relative importance of the protective function in flowers varies enormously, and for many plants the issue of protection of the living tissues is perhaps less crucial than for most animals; plants are usually modular, with many sex organs in many flowers, so that they can afford to lose some flowers and regrow a

new supply (whereas animals normally are not modular and have a very limited and nonrenewable supply of gonads). Investment in flower protection is therefore often quite limited, but it may become paramount for species with only intermittent flowering, or with just a few large and expensively constructed flowers, or with an extremely limited supply of pollinators; the various ways in which flowers can be protected are considered in more detail in chapter 24. Here we will concentrate primarily on the design features related to **sexual** function in pollination.

Flowers come in an astonishing variety, whether viewed across the entire spectrum of angiosperm families (appendix) or merely one **family** at a time. A family such as the Ranunculaceae has flowers ranging from simple bowls (e.g., *Ranunculus* species such as buttercups) to complex **bilateral pendant** tubular designs with elaborate internal nectaries accessible only to long-tongued bees (e.g., many *Aconitum*) and to tiny fluffy flowers where only the anthers are conspicuous (e.g., *Thalictrum*). Other families show similar variation but often produce the same three broad kinds of flowers, suggestive of convergent evolution.

Not only are individual flowers exceptionally varied in shape, size, and appearance, but the story is further complicated by the tendency of flowers in many families to mass together as **inflorescences**. Here the overall **floral display** is determined by the combined appearance of a cluster of small flowers (sometimes then termed **florets**) into one seemingly whole structure, which may in practice be the unit that is perceived, both by humans and by flower visitors, as the functional flower. We tend to refer to the complex multiple inflorescence of hydrangeas, or of many umbellifers, as their flowers; and we almost always speak of the floral displays in the composites (Asteraceae—

TABLE 2.1
Basic Floral Anatomy

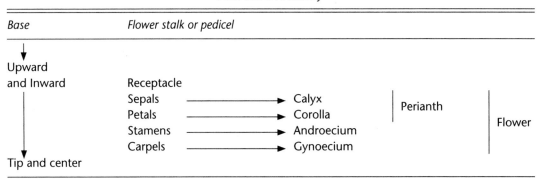

Base	Flower stalk or pedicel			
↓ Upward and Inward	Receptacle			
	Sepals ——————→	Calyx	Perianth	Flower
	Petals ——————→	Corolla		
	Stamens ——————→	Androecium		
↓ Tip and center	Carpels ——————→	Gynoecium		

daisies, asters, chrysanthemums, and their kin) as flowers, whereas in fact each is made up of many different florets, one kind making up the central disk and another kind forming the outer rays, which appear as petals.

To consider flower design, then, we need to understand both the key components of an individual flower, to analyze their juxtaposition within one flower, and to review how the separate flowers can be assembled together, in many flower families, into complicated inflorescences that become functional units in terms of display and attraction.

1. Essential Flower Morphology

Flowers are complicated assemblages of plant parts, modified originally from leaves. Repetitive patterns of units occur in series, growing **centripetally** and sitting upon the flower stalk. Table 2.1 shows the general patterns of floral parts, with the four main series of structures from base to tip, or outer to inner: **sepals**, **petals**, **stamens**, **carpels**. This fundamental structure can be detected in most flowers from the basic arrangement in figure 2.1A, though the details vary enormously; these details can often be conveniently expressed in terms of a simple flower diagram of the type shown in figure 2.1B. In particular, the numbers of each part are highly varied: primitive magnoliids have very large numbers of petals, as do cacti, but in many **dicot** plants there are just five petals and five sepals, whereas in most **monocots** these structures come in threes. Even here many exceptions occur, and some important groups, such as the crucifers (Brassicaceae) and the poppies (Papaveraceae), have their parts in multiples of two or four. These petal and sepal numbers may or may not be re-

flected in the inner series of stamens and carpels: in the geranium family all the parts are in fives, and in the lily and iris families all are in threes, but in some other families this numerical consistency is lost, often with fewer carpels and more stamens than there are petals and sepals.

During flower development each part or series forms in sequence, and is in turn influenced by the initiation and growth of neighboring structures. The original **apical meristem** that will form the flower is protected in a **bract**, within which floral organs are initiated; then as floral **morphogenesis** progresses, the first series of structures, the sepals, grows and surrounds the **bud**, within which the petals and the **androecium** (male) and **gynoecium** (female) form sequentially. To add further variety, the fundamental structures may fuse into a ring as they arise or during **ontogeny** to form tubular organs. And sometimes parts of *different* floral organs may fuse, to form organ complexes. However, the basic underlying anatomy can still usually be described by a simple floral diagram as in figure 2.1B.

The primary functions of flowers are always performed by these basic and genuinely floral structures, but as we will see, some of the secondary functions (protecting, guiding, advertising, and rewarding) may be taken over by organs outside the flowers.

2. The Perianth

The outermost structures of a flower collectively form the **perianth**; this sits upon the **receptacle**, which is really just the expanded often cone-shaped end of the flower stalk, or **pedicel**. The perianth itself may be

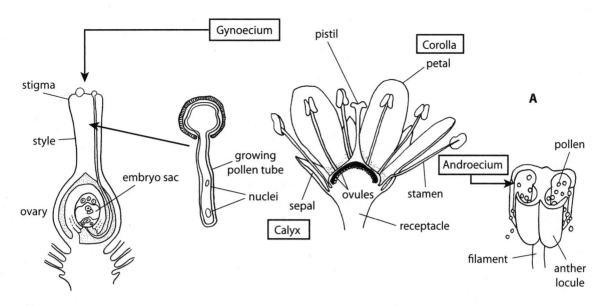

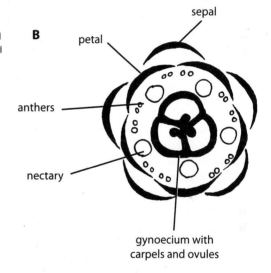

Figure 2.1 (A) The basic structure of a flower, with both female (gynoecium) and male (androecium) reproductive parts (modified from Faegri and van der Pijl 1979 and later sources). (B) A classical flower diagram showing an idealized structure in transverse section and the relation of the four major whorls of floral parts.

protected by extrafloral bracts, which in some cases are brightly colored and serve both protective and advertising functions.

The perianth is made up of rather thin and flat structures that enclose and usually protect the rest of the flower, and its tissues are sterile (nonreproductive). At **anthesis** (floral opening) the perianth commonly becomes a major part of the floral display, attracting the pollinators.

There may be either one or two series of perianth parts. Where there is only one series (or occasionally two but with no structural differentiation), the component parts are usually called **tepals**. However, in the majority of plants there are two series of distinctly different structures—the sepals and petals—forming the perianth (as shown in table 2.1).

Calyx and Sepals

The outer series of floral parts is mainly protective at all times in most plants and is termed the **calyx**, made up of sepals, often the same green color as the foliage but rather more robust. The calyx makes a cup (in which the rest of the flower sits), either by sepals overlapping each other or by their fusion (producing the condition called **gamosepaly**). Sepals can be formed from a thickened **epidermis** that is relatively solid and free of airspaces or can be toughened by **sclerenchyma** to give a semiwoody protective base to the flower; in extreme cases, such as *Costus*, the sepals form a lignified coating over the whole inflorescence through which small cohorts of individual corollas emerge (plate 34F, G). The external surface of a calyx may be further protected with glandular hairs (for example, in many Solanaceae) or with a waxy layer (for example, in eucalypts). Either of these additions may help to deter possible folivores, and some of the waxy calyces may also serve to reflect solar radiation, keeping the base of the corolla relatively cool.

There are also quite a number of plant genera in which the calyx sepals take on all or part of the role of **advertisement** and pollinator attraction, both visual and olfactory (with scents from the glandular hairs). Sometimes the sepals form the main ring of colored visual attractants (and tend to be referred to incorrectly as the petals); this is common in the Ranunculaceae, including *Helleborus* (where petals may become nectaries; chapter 8), and many *Anemone* and *Pulsatilla* species.

Alternatively just one or a few of the sepals are enlarged and highly colored, giving the main visual signal in some Rubiaceae and Verbenaceae. Something similar occurs in many euphorbs, such as *Dalechampia* (plate 11F), although there it is bracts rather than sepals that produce the show (the "flowers" are in fact compound inflorescences contained within the brightly colored bracts as a group of one female and several male tiny flowers). Or attractive sepals may act in concert with petals, whether being of the same color (e.g., the *Salvia* in plate 13G) or contrasting, as in some *Abutilon* (plate 14A) and some *Clerodendrum*, or enlarged, highly modified, and very differently colored (plate 13F). In yet other cases, the use of sepals as advertising structures occurs in only some flowers of a complex inflorescence; the central flowers are of usual type, whereas outer ones have one or more enlarged colored sepals, increasing the showiness of the whole floral display. Many *Hydrangea* species and hybrids are familiar **temperate** examples (plate 11E), as is coriander (plate 11D).

A further specialization occurs with the production of a protective fluid by internal sepal surfaces, producing a **water calyx,** or *water jacket* (also achievable by a cup-like calyx or leaf morphology that traps rainwater, as in many **bromeliads**). This moat of water around the flower base (plate 34D) may help to exclude ants and other small visitors with little role in pollination (e.g., Carlson and Harms 2007; and chapter 24) and may also have a cooling function. Such watery calyces are well known in various tropical plants, including many Bignoniaceae (they were first described in *Spathodea* from this family). Other genera produce

viscous or mucilaginous material; for example, individual *Commelina* (Commelinaceae) flowers (plate 12A) emerge through a copious layer of thick **mucilage** held in the calyx, although the function of this substance is rather uncertain.

Finally, there are cases of rewards deriving from sepals; by far the commonest examples are extrafloral nectaries, where glandular areas of the sepals produce nectar that is predominantly collected by ants, in relation to their plant-guarding activities (chapter 25). Other examples are rare, although **food bodies** arising from sepals and fed on by beetles are not uncommon in the Winteraceae and in some Annonaceae.

Corolla and Petals

The inner series of the perianth structures is the corolla, which is made up of petals and takes on the familiar main display/advertisement function in most angiosperms. The petals tend to be arranged alternately with the sepals (fig. 2.1B and examples in fig. 2.15). As with sepals, the petals may be fused for all or part of their length (**gamopetaly**) to produce a tubular corolla. The color(s), size, shape, position, or orientation of petals may all contribute to the attractiveness of the display.

The petals are usually colored differently from the foliage and are often more delicate in texture, with a thin epidermis and a mesodermal layer that has plentiful air spaces. However, thickened or fleshy petals are not uncommon, and thick, white or cream-colored corollas are often found in heavily scented dusk- or night-flowering species that are visited by bats or moths using fragrance as an important cue. Petals may release scent through their surfaces or bear specific scent-producing **trichomes**, or **papillae** (chapter 6). In some cases, particularly among the Magnoliaceae and Liliaceae, they bear nectar-producing tissues, and occasionally they provide a surface for **secondary pollen presentation** (chapter 7).

Petals may also present complex microtextured surfaces, which provide easier **landing platforms** for visitors and may also play a role in close-up flower recognition (Kevan and Lane 1985). Textures involve ridges, overlapping plates, or patterns of pimples, often differing between the upper and lower petal surfaces, or between the upper petals and the lower landing-platform petals in bilateral flowers. For example, through direct tactile perception, **bumblebees** could discriminate among flowers of *Antirrhinum* mutants that lacked special conical epidermal cells on petal surfaces (Whitney, Chittka, et al. 2009).

Occasionally the petal tissues, or parts of them, combine with parts of the next inner ring of tissues (the stamens) to form a *corona*. This appears as an "inner corolla" in flowers such as daffodils, where it forms the protruding trumpet (plate 6E), and in passionflowers (plates 5A,B and 26G), where the corona is filamentous and becomes an elaborate inner circle of contrasting display.

Both series of perianth structures are usually wilted or shed after anthesis, and in some plants, pollination per se, or the fertilization of the ovules that follows pollination, may act as a trigger for the **wilting** or **senescence** of the corolla. However, sepals are sometimes retained to assist in **fruit** development (for example becoming the wings of the **nuts** in the large Southeast Asian **dipterocarps**) or to provide a contrasting background for fruit advertisement. Petals may wilt on the calyx or may be shed by a specific **abscission** mechanism involving **ethylene** (a common within-plant signaling system). But even petals are occasionally retained and thickened to form part of the fruiting structure.

3. The Androecium: Male Structures

The male functional organs of a flower, collectively termed the androecium, are a series of stamens. Each is formed from a thin **filament**, which may be erect, curved or pendant and which supports the terminal anther. The most common type of stamen has an anther with four pollen sacs (or **locules**), arranged in two pairs, each pair termed a **theca**. The thecae open at maturity in the process termed dehiscence, so exposing the pollen (dehiscence and its control will be covered in chapter 7).

Relative to other flower parts, stamens are rather constant in size and shape and design (see the review by Bernhardt 1996). In the phylogenetically basal family Magnoliaceae, the distinction between the anther and the filament is poor, but in nearly all other flowers, a two-lobed anther sits on the end of a much thinner and elongate filament. Stamens are rarely more than 20 mm long, although in some plants the filaments are fused basally and can then be strong enough to support

lengths of up to 50 mm. This length is often achieved quite late in the floral development process, and elongation can be very fast; in some grasses the filament can grow several millimeters in just a few minutes as dehiscence approaches. On average, filaments are somewhat longer in grasses and other anemophilous (wind-pollinated) flowers than in zoophilous (animal-pollinated) species. However in many of the zoophilous flowers with fused petals, the filaments are inserted onto the inside of the resulting corolla tube (adnate filaments; e.g., fig. 2.2F–H) rather than arising basally, so they are technically short but still function as if they were elongate. Rarely, the filament is elaborated with fine hairs (easily seen in many *Verbascum* species) or with nodular thickenings (e.g., in *Sparmannia*), which may enhance their visual signaling effect (fig. 2.2K).

The filament is usually attached to the base of the anther, but sometimes anthers are suspended close to their midpoint by a thin and flexible junction; this is seen in many Liliaceae (fig. 2.2J; plate 12C), where the anthers hang down and move freely in seesaw fashion about this joint.

Within the anthers, the main design variation is in the location and form of the furrows that split apart at dehiscence (fig. 2.3). Most commonly the furrows (or *thecal slits*) open toward the center of the flower (**introrse anthers**), so that pollen is presented inward in the direction of the female organs; in flowers visited by larger animals, such as bats or birds, the introrse anthers may open broadly to produce a flat pollen-dispensing surface that can press against fur or feathers, as seen in many passionflowers (plate 5A,B). Externally or laterally opening anthers (**extrorse** or **latrorse**, respectively) are also found, but less commonly; they tend to be smaller and are borne on shorter filaments. Occasionally, there is only one furrow and the locules are united internally to open as a single chamber; *Pavonia* is a good example. At the opposite extreme, the locules may subdivide internally (either transversely or longitudinally) to give a **polysporangiate** anther. Sometimes dehiscence slits may lose the standard elongate pattern and become curved or almost circular on the tip of a bulbous anther head, giving a **valvate** stamen (fig. 2.2E).

In some plant groups, especially zoophilous taxa, stamens are modified by becoming united into a more complex structure (fig. 2.4), a phenomenon known as **synandry**. This is quite common in the large orders

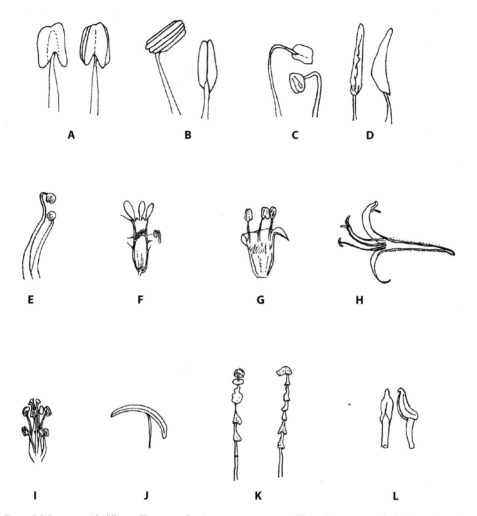

Figure 2.2 Stamens with different filament and anther arrangements: (A–C) simple patterns with dehiscent slits on anthers; (D) elongate with elaborate dehiscence furrow; (E) anther rounded and valvate; (F–H) adnate filaments arising from petals, internal or protruding, in various Boraginaceae and Caprifoliaceae; (I) dimorphic stamens set at two heights in *Oxalis*; (J) seesaw anther head on filament, in *Lilium*; (K) filament with nodes in *Sparmannia*; (L) anther with terminal pore in caesalpinoid Fabaceae. Not to scale.

Malvales and Fabales, and it may involve just the filaments or the more distal anthers as well, resulting in a tubular androecium (or even a solid cylindrical structure if there are no female organs in that flower). This greater rigidity in the central floral organs may provide a stronger area to grip or to maneuver around for insect flower visitors. A further possibility is for the anthers to unite at the tips but without the filaments doing so; this is often seen in **composite flowers** (see *Composite Flowers* below). It also occurs in many **buzz-pollinated** plants (chapters 7 and 18), such as *Solanum* or *Cassia*, giving a single pollen-dispensing structure that a visiting bee can grasp and vibrate its body against, and in these kinds of plants, it is commonly associated with loss of the anther's **dehiscence furrows** and their replacement with one or more terminal or lateral pores through which pollen emerges (**poricidal anthers**). Single-pored anthers are also found in other situations: in some bellows flowers, such as *Cyphomandra*, that are squeezed by a bee to release the pollen, and in some of the **carrion flowers** that attract flies, where the anther is polysporangiate internally

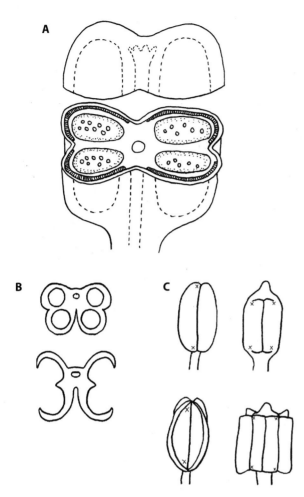

Figure 2.3 Anther heads: (A) transverse section showing internal structure; (B) transverse section of typical tetrasporangiate anther with two locules, closed and then open at dehiscence; (C) longitudinal dehiscence (left) and valvate dehiscence (right) seen in frontal view, x–x marking the dehiscence furrows. (Redrawn from Endress 1994.)

but where aggregated pollen emerges from the single pore in a slimy thread or droplet. One other specialization to be noted is the phenomenon of sensitivity in stamens: in a number of plant families the filament is somewhat contractile and may shorten or curve in one direction when stimulated by a visitor (chapter 7 and *Compsite Flowers* below).

As well as dispensing pollen, whether partly as a reward or solely for reproductive purposes, stamens may serve some other important roles. They are often used in advertisement and attraction, via visual or ol-

factory signalling. Where they are long and protruding, they markedly increase the apparent size of the flower or inflorescence and may result in a **brush blossom** (see *Brush Blossoms* below). Where the rest of the perianth is very small, the anthers then become the main visual signal. While anthers are commonly brown or dull in color when closed and take on the color of the pollen when dehiscing (usually yellow; but see chapter 7), the filaments are often white or almost colorless and so may provide a visual contrast to enhance visibility of the pollen. In a few cases the filaments change color very strikingly as a flower ages (or perhaps as it is pollinated); one case is *Koehneria* from Madagascar (plate 16E). Scented stamens are known in some Chloranthaceae and Solanaceae (including *Solanum*), again enhancing the flower's attraction. Additional functions include guidance for the tongues of pollinating visitors, especially where the corolla is tubular but quite wide, and exclusion of the tongues of inappropriate/illegitimate visitors, by such features as hairiness (sometimes as a band of hairs on the filament at a particular depth) or by partial fusion, where the stamens insert onto the sides of the corolla, as in many composite flowers (see *Composite Flowers*).

The most obvious variation in androecium structures comes with **heteranthy** (also termed heteranthery), the phenomenon of having two kinds of anther (fig. 2.5). In its less dramatic forms, heteranthy involves two whorls of stamens within a flower that are slightly different in developmental rate and so appear morphologically distinct almost until they dehisce; in these cases there may be little or no functional difference between the two sets, although where the two groups are set at different heights, they may exploit two different kinds of flower visitor. The more extreme versions, seen in fig. 2.5 and in examples on plate 12, occur where one set of anthers is relatively cryptic and bears the reproductively important pollen, while the other is showier and contains **fodder pollen** (chapter 7). Several genera in the Lecythidaceae have hundreds of stamens per flower: in the cannonball tree *Couroupita*, there are prominent central feeding stamens, plus a semicircle of stout pollinating stamens set on an *androecial hood* around the lower half of the flower, depositing pollen on the feeder's undersurface. Even more extreme are some of the blue-flowered *Commelina* species (plate 12A), with three kinds of stamen and anther: two cryptically blue and lateral, one central and either blue or yellow, and three shorter and bright

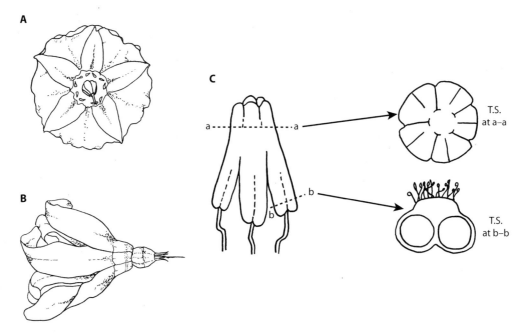

Figure 2.4 Examples of synandry (full or partial fusion of anthers centrally) in various angiosperms: (A) *Solanum*; (B) *Dodecatheon* (shooting star); (C) *Sprengelia*, transverse section showing basal separation and terminal fusion of anthers. (Parts (A) and (B) redrawn from Buchmann 1983; part (C) drawn from flowers.)

yellow (usually termed **staminodes**; see below). All produce viable pollen, but that from the staminodes is much less viable; the lateral stamens achieve most of the cross-pollination, the central one provides rewards to pollinators (and some delayed selfing possibilities), and the yellow staminodes serve as bright attractive signals mimicking copious pollen on offer (Hrycan and Davis 2005). Other species of *Commelina* have rewarding yellow anthers, **unrewarding** yellow anthers, and brown reproductive anthers, and removal of either of the first two types reduces pollen export from the third (Ushimaru et al. 2007).

The design and size of stamens may usually be relatively constant, but variability in *stamen number* is extreme. Flowers may contain just one stamen, as in most asclepiads and most orchids, or may have several hundred, as in some Cistaceae and Myrtaceae. Occasionally there are in excess of two thousand stamens in a flower. Indeed, almost this full range can occur within just one order of plants: *Ascarina* has one stamen and *Tambourissa* has about two thousand, both being in the Magnoliidae.

Staminodes are structures formed from the same series of embryonic tissues as the functional stamens but are sterile and have no reproductive function. They may augment the visual display (*Theobroma* or *Jacaranda*, fig. 2.5) or have scent-emitting tissue (**osmophores**) or secrete some nectar, and in some cases they can move during the life of the flower, bending over to protect the female structures.

4. The Gynoecium: Female Structures

The female organ system, technically the gynoecium, is the most central part of the angiosperm flower, and while superficially rather simple, it is internally very complicated, having interconnecting chambers and canals. The gynoecium is made up of one or several pistils, each of which contains one or more carpels. Most flowers have several carpels, two to five being the most common, though these may be laterally fused together more or less completely (**syncarpy**) and thus contribute to only one pistil (the advantage of this being that received pollen can be distributed among all the carpels after successful pollination). Rarely the number of carpels is huge, up to about two thousand in *Tambourissa*, which as noted above also has around two

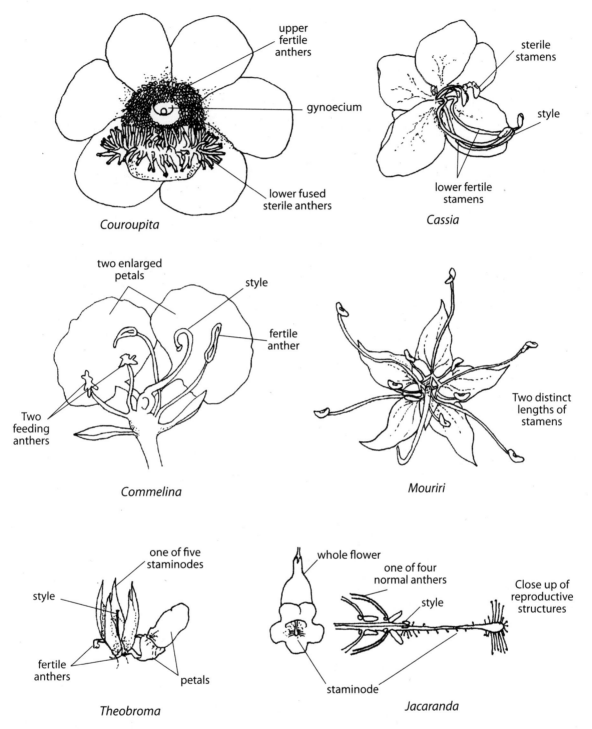

Figure 2.5 Heteranthy in various flowers, where stamens come in two different forms: different lengths, as in *Mouriri*, or modified for feeding and reproductive purposes, as in *Couroupita*, *Commelina*, and *Cassia*; or with one or more modified as staminodes in *Theobroma and Jacaranda*. (*Couroupita and Commelina* modified from Endress 1994; *Mouriri* and *Cassia* modified from Buchmann 1983; *Jacaranda* and *Theobroma* drawn from photographs and flowers.)

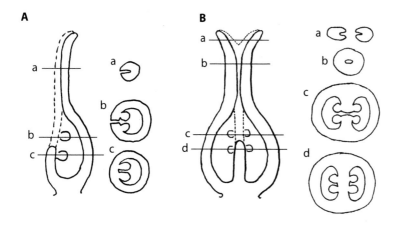

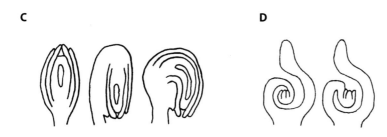

Figure 2.6 Basic gynoecium and ovule structures in medial long section and the associated terminology. (A) Unicarpellate (i.e., gynoecium with a single carpel containing two or more ovules); (B) syncarpous, here with two carpels; (C) patterns of curvature as the ovule develops: (from left to right) orthotropous, anatropous, and campylotropous; (D) ovule curvature relative to carpel curvature, (left) syntropous and (right) antitropous. (Redrawn from Endress 1994.)

thousand stamens. Only in the legumes is a **unicarpellate** state commonly found (though legumes are so numerous that this may amount to around 10% of all angiosperms).

Pistils and carpels can be rather complicated, and their terminology is confusing for nonbotanists. For pollination biologists, the term "pistil" is the most useful to remember, because **unisexual flowers** (or phases of flowers) are often referred to as **pistillate** and **staminate** to indicate female and male, respectively. In practice, for most pollination studies it is only necessary to consider two key zones of each pistil or carpel: the apical surface, which forms the stigma on the tip of the more or less elongate **style**, and the underlying **ovary,** comprising one or several ovules (fig. 2.6). Each ovule consists of an undifferentiated multicellular mass termed the **nucellus**, which is surrounded by an integument that leaves open only a narrow apical channel, the **micropyle.** The basal end of the ovule is attached to the rest of the carpel via a short stalk. As the flower matures, one cell within the nucellus undergoes **meiosis** to give four **haploid** cells, and one of these enlarges as the **embryo sac**. The normal further

divisions of the embryo sac are shown in figure 2.7; the end result is eight nuclei, one of which at the micropylar end of the ovule becomes the egg itself, which after fertilization from a pollen grain will become the plant's seed (chapter 3).

The number of ovules in a flower is highly variable: many plants have only one ovule per carpel, a state seen in grasses and most of the Ranunculaceae (in which case, the term "ovary" is redundant); but at the other extreme, orchids may have thousands of ovules per carpel.

The style is a rather diverse organ both in size and in shape, with a more or less distinct stigmatic surface at its tip (fig. 2.8). Larger styles are common in wind-pollinated flowers and in those with many ovules. Whatever the size, styles are sometimes of a compact and cone-like shape, but more commonly they are elongate with stigma and style distinct from each other. The style often forms a fairly rigid central column in the flower in insect-pollinated species, its main function being to get the stigmatic surface into the "right" position, but it may also be providing grip for some visitors and perhaps preventing too much damage

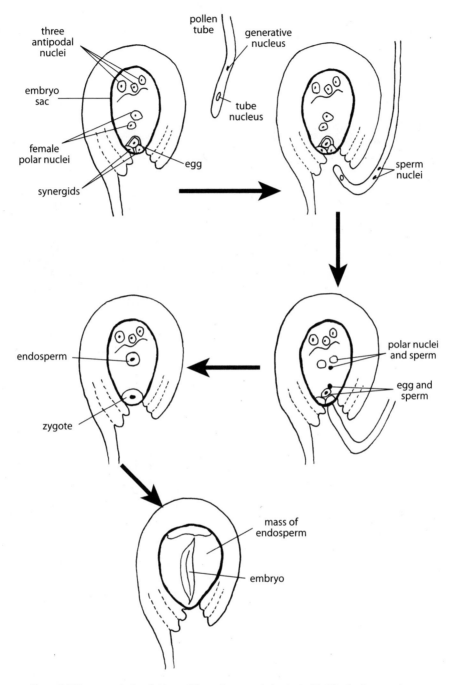

Figure 2.7 Five stages in the divisions of the embryo sac during typical fertilization in an angiosperm.

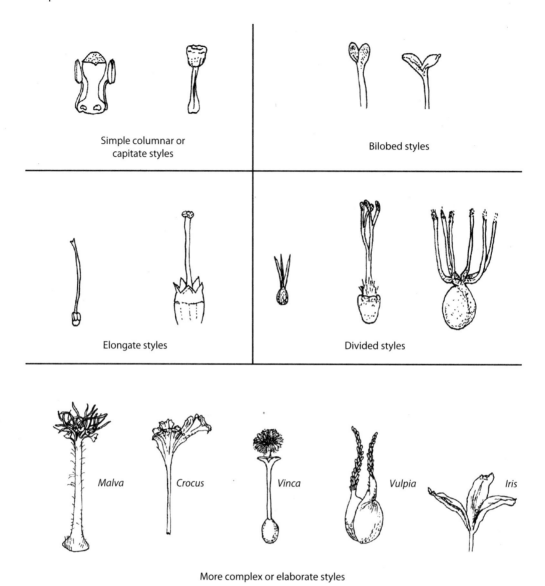

Simple columnar or capitate styles

Bilobed styles

Elongate styles

Divided styles

Malva   *Crocus*   *Vinca*   *Vulpia*   *Iris*

More complex or elaborate styles

Figure 2.8 Varieties of styles and stigmas in angiosperms, ranging from typical short columnar structures to elongate and highly divided forms and including the crested styles of *Crocus* and the enlarged petaloid styles seen in *Iris*. Not to scale. (*Malva, Crocus, Vinca, Vulpia, Iris* redrawn from Barth 1985; others redrawn from photographs and flowers.)

during nectar-seeking behaviors. Quite commonly the style elongates throughout the life of the flower, so that it starts shorter than the anthers and later grows to match or exceed them in length, thus facilitating foreign pollen pickup (chapter 3). Occasionally the style recurves as it grows, and this can be influenced by pollen deposition to promote outcrossing while allowing selfing as a backup or safety net (e.g., Yu and Huang 2006). In some flowers the style is strongly asymmetrical, bent to one side of the flower (plate 12F–H), while in irises the style can become petaloid, and its partly tubular form helps to restrict and control pollinator access (plates 9C and 15C).

Many stigmas have a somewhat lobed tip, with three or five **stigmatic lobes** (plate 12G) being the most widespread patterns, reflecting the presence of three or five carpels in the major plant families. Bilobed, forked styles are also common (plate 12F). The surfaces of the lobes are often strongly textured, with many papillae or with a feathery (**plumose**) appearance: in some *Crocus* species the stigma has showy lobes and crests (plate 3G). This surface may also be described as wet (sticky) or dry. The wet/dry categorization is probably not very helpful in relation to pollination; most stigmas have some form of **stigmatic exudate**, usually **lipid** rich and markedly hydrophobic, so that an appearance of stickiness merely indicates how abundant this exudate is at a particular time. However this wet/dry distinction is often important in relation to the mechanisms of **self-incompatibility** in the plant (covered in chapter 3): wet stigmas are relatively free of protein material and are not selective for pollen adhesion, whereas dry stigmas have thin proteinaceous exudates and only the conspecific pollen is retained.

Several other stigmatic phenomena are important for pollination, and most of these relate to temporal patterning. The timing of the secretion of the stigmatic exudate may be important and is often carefully controlled, occurring in one or more waves; it may itself be triggered by pollination. Likewise, the timing of the opening (or closing, or both) of the stigmatic lobes is crucial in many flowers in relation to determining when pollen can be received; in some flowers, the stigma exhibits sensitivity and can close when touched. In bladderwort (*Utricularia*), the two stigmatic lobes rapidly fold together after being touched by a bee entering the flower. In monkey flowers (*Mimulus*), physical contact produces a faster closure than pollen arrival, and touched stigmas with no pollen will normally reopen within 2–5 hours (Fetscher and Kohn 1999), whereas pollen receipt produces delayed reopening or permanent closure (fig. 2.9).

The surface of the stigmatic lobes is primarily responsible for determining whether a pollen grain that has landed there will germinate or not. This surface has a limited period of **receptivity** to pollen, which may vary from just a few minutes in some grasses to around 4 to 5 days in tomato (*Lycopersicon*, now included in *Solanum*) and up to 3 weeks in some orchids. The surface is structurally and chemically complex, with receptor sites concerned with several phases of pollination and beyond:

1. Pollen hydration (which may take only a matter of minutes), then
2. Pollen recognition, then
3. Pollen adhesion and **lectin** binding, and finally
4. Pollen germination, for which there are also zones with enzyme-secreting activity, often particularly associated with high levels of peroxidase during peak receptivity.

Pollen grains commonly imbibe water from the stigma and germinate within a few hours of their initial deposition. Germination does not happen in water alone, but for some species it can be induced in simple sugar solutions, suggesting a fairly uncomplicated osmotic effect. The **pollen tube** emerges into a mucilaginous secretion, which may provide some of the nutrients required for later stages of pollen tube elongation.

Within the style is a pollen tube transmitting tract leading down to the ovules. This tract may be epidermal or may involve deeper tissues, depending on the taxon; in groups with very numerous ovules (especially the Orchidaceae, but also including some Ericaceae and Rafflesiaceae), the tract is particularly well developed and secretes substantial mucilage internally. Pollen tubes can cross between carpels readily in this tract, so that their final destination in terms of meeting an ovule is not predetermined by where they landed on the stigma. However, as the tract approaches the ovules, it splits into separate parts, one for each carpel or ovule. Growth in the tracts is normally unidirectional, from stigma to ovule, but if pollen is introduced experimentally at a midpoint, it can often grow equally well in either direction. The act of pollination may cause the ovary to release a growth stimulus for pollen tube elongation, providing a pulsed gradient of chemical signals that the pollen tube follows down to the ovules; ions and protons are implicated, and calcium/calmodulin signaling is also involved (Holdaway-Clarke and Hepler 2003; Shi et al. 2009). It also seems likely that the style tissues help to transport the pollen tube, since tiny latex beads introduced into the style can be seen to move toward the ovules at a speed similar to that of tube elongation (Sanders and Lord 1992).

Only rarely is this pollen tube tract absent in angiosperms. However, in a few self-fertilizing **cleistogamous** flowers (chapter 3) pollen can germinate within the anthers where it was produced and grow down the filament or through the anther wall to the stigma.

The ovary of the flower is the most crucial organ in reproductive terms, and its relationship to the receptacle on which the whole flower sits is important. In

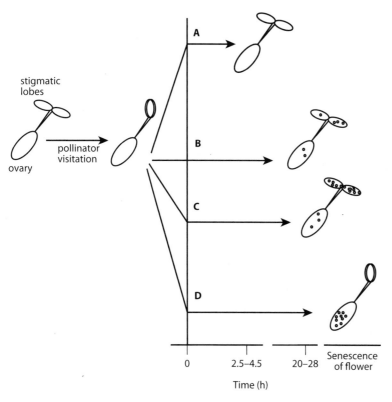

Figure 2.9 Stigma behavior in *Mimulus aurantiacus*, with initial closure in response to touch or visitation, followed by (A) quick reopening if there has been no pollen receipt, or (B) somewhat delayed reopening with some pollen receipt, or (C) a slow reopening with more pollen but few fertilized ovules, or (D) permanent closure if fertilization levels are high. (Modified from Fetscher and Kohn 1999.)

many flowers the receptacle forms a simple (often nectar-secreting) dish on which all the true floral parts sit, with the carpels at the center; but in other cases the receptacle has become deeper and sturdier, with the carpels more or less embedded within it, and the other floral structures then appear to be borne above the ovary. Ovaries in the latter case are termed **inferior** (seemingly lying below the point of origin of the sepals, petals and stamens) as opposed to **superior** ovaries, where the outer floral structures clearly arise below the base of the ovary (fig. 2.10). There are many possible intermediate states, but in some of the major families this character is fairly fixed and is a useful aid to identification (table 2.2). Inferior ovaries are rare in more phylogenetically basal plant families and are generally assumed to be an evolutionarily more advanced state resulting from selection for increasing protection by the receptacle. However, changes and reversions in this ovarian position character are fairly common; for example, ecological considerations may intervene, and bird-pollinated plants even from more basal taxa are often found to have inferior ovaries,

where the receptacle can give some much-needed protection against damage from a bird's bill.

Arrival of a pollen tube within the ovary and close to an ovule may occur some hours after pollination, this interval depending in part on the length of the style. Pollen tube arrival initiates the series of events that lead to fertilization and the successful reproduction of the plant, which will be discussed in chapter 3.

5. Flower and Inflorescence Features

Single Flowers and Inflorescences: Functional Units and Blossoms

Basic floral anatomy not only can yield the full range of familiar single flowers but also allows flowers to be arranged en masse to form a composite inflorescence with very different appearance and handling needs (plates 1 and 11). The latter may be a loose open structure, with individual flowers set well apart, or it may form a compact mass, with flowers almost or actually

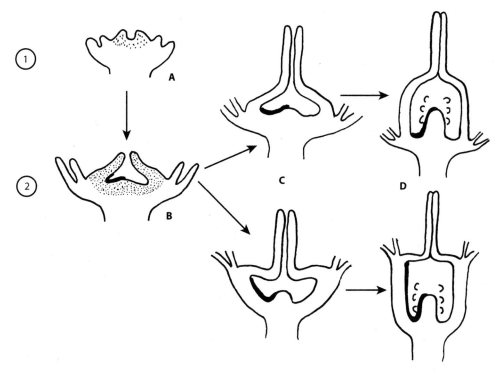

Figure 2.10 Superior (1) and inferior (2) ovaries, resulting from different rates of expansion of the enclosing and adjacent tissues. Meristem tissue is stippled in A and B. The thick line represents corresponding tissues in (B), (C), and (D) to clarify relative expansions. (Redrawn from Endress 1994.)

touching. An inflorescence may be perceived as a single display unit—effectively as one large flower—by most visitors and is commonly referred to as "a flower"; a single daisy or thistle or scabious is almost self-evidently "one flower" in all practical senses. A useful convention is the use of the term **blossom** for a functional unit. Thus a single buttercup or foxglove or orchid is a blossom, a daisy or a willow catkin is a blossom, and at the opposite extreme, just one of the three parts of a single iris flower may serve functionally as a blossom (see below).

The various tight packing arrangements seen in the inflorescences (sometimes called "flower heads") of umbellifers and composites may have the benefit of producing a large landing platform for the many kinds of visitor that cannot hover while feeding, while simultaneously keeping individual flowers very small, matching the size of the visitor's mouthparts (rather than accommodating the visitor's whole head or body, as is usually appropriate for a single flower). It is also

likely that dense packing of a flower head can be a useful strategy for avoiding, or at least reducing, illegitimate visits to flowers, preventing access by those visitors who might otherwise bite into the bases of the corolla tubes and steal the nectar (chapter 23).

The form and structure of inflorescences are summarized in figure 2.11 and table 2.3. Although the terminologies given there seem fairly clear-cut, in practice many inflorescences defy easy classification and have mixed characters (e.g., many labiates are **racemose**, but the branches within the **raceme** are **dichasial cymes**). For practical purposes, it is usually easier just to refer to an inflorescence; botanical taxonomists need the more complicated approach, but pollination biologists usually do not.

The differing structures of inflorescences can nevertheless have marked effects on visitor behaviors. Jordan and Harder (2006) compared bumblebee visits to racemes, **panicles**, and **umbels** each with 12 flowers and found that the bees visited slightly more flowers

TABLE 2.2
The Position of Ovaries within the Flowers of Some Major Plant Families

	Always superior	Always inferior	Variable
Monocots	Alliaceae Commelinaceae	Orchidaceae Iridaceae Musaceae Zingiberaceae	Agavaceae Araceae Bromeliaceae
Dicots	Acanthaceae Apocynaceae Boraginaceae Caryophyllaceae Cistaceae Convolvulaceae Fabaceae Gentianaceae Geraniaceae Magnoliaceae Malvaceae Papaveraceae Polemoniaceae Scrophulariaceae Solanaceae Verbenaceae Violaceae Primulaceae *	Apiaceae Araliaceae Asteraceae Caprifoliaceae Cornaceae Hydrangeaceae Myrtaceae Onagraceae Rubiaceae Begoniaceae * Cactaceae *	Ericaceae Lythraceae Nymphaeaceae Rosaceae Saxifragaceae

* usually rather than always.

on the umbels but had far more consistent foraging paths on the raceme, which (combined with different temporal regimes of pollen presentation, discussed in chapter 3) should lead to less selfing on the racemes than on umbels.

Likewise, the position of flowers on inflorescences can markedly affect their characteristics, as shown in table 2.4 for a typical umbel (the carrot, *Daucus*; Perez-Banon et al. 2007). Here the lower orders of umbel have fewer flowers each and a higher proportion of male-only flowers, so that the overall sex of the plant changes as it ages; so too does the **sex ratio** on an individual umbel (fig. 2.12B), and thus the plant as a whole has male phases alternating with female/hermaphrodite phases (fig. 2.12C).

There is often a sequential decline in resource allocation among flowers on an inflorescence as well.

Vallius (2000) explored this effect in the orchid *Dactylorhiza* and found upper flowers to be smaller, to have lighter **pollinia**, and to produce smaller seed capsules; but these effects could be ameliorated if lower flowers were removed artificially, which suggests some influence from resource allocation. Kliber and Eckert (2004) tested the underlying causes of this in *Aquilegia canadensis* and found that detailed aspects of floral morphology (**nectar-spur** length, sepal size) showed very little sequential change. However, ovule and pollen production declined by 9% and 18%, respectively, with flower sequence up the **spike**, an effect that could be modified in either direction by resource availability. In this plant the main adaptive reason for sequential decline in allocation was probably **herbivory**, which was greater on the higher and later flowers.

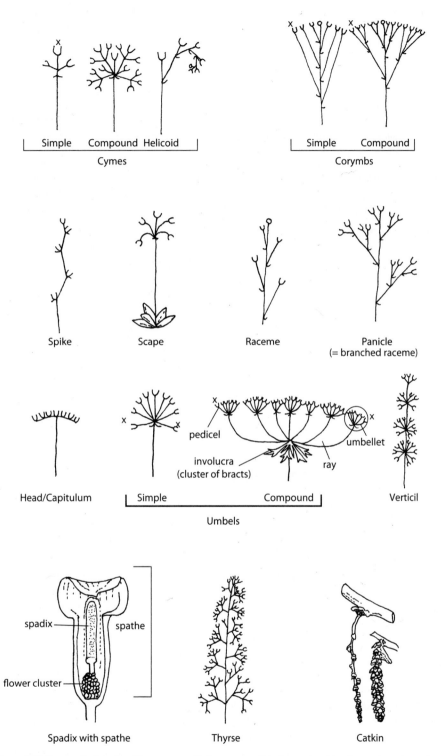

Figure 2.11 The terminology for different types of inflorescence. (X indicates the oldest flowers in each case.)

TABLE 2.3
The Structures and Terminology of Inflorescences

Structure	Term
Oldest flowers at base	Racemose inflorescence
Flowers stalkless	Spike
Flowers stalked	Raceme
Stalks all or mostly branched	Panicle
Varied stalk length to give flat top	Corymb
Oldest flowers at apex	Cymose inflorescence
Two branches paired below apex	Dichasial cyme
Single branch, to one side	Scorpioid cyme
All flowers radiating from common point	Umbel
All flowers stalkless on top of stem	Head or capitulum

TABLE 2.4
Variation in Characteristics on the Umbels of *Daucus carota*

Umbel position	No. of umbels	Total flowers per umbel	Hermaphrodite flowers* (%)
Central	1	4,829	61.0
Primary	17.6	4,895	33.6
Secondary	84.4	3,006	16.1
Tertiary	55.7	1,223	15.6

Source: Modified from Perez-Banon et al. 2007.
*Hermaphrodites are peripheral and males central.

Symmetry

As shown in figure 2.13, there are essentially two main types of symmetry in flowers: **radial symmetry** (also termed **actinomorphic**, or **regular**) and **bilateral symmetry** (or **zygomorphic**, or **dorsiventral**, or lateral; also sometimes termed **irregular**) (Endress 1999). Neal et al. (1998) gave a detailed review of the terminology and confusion that can arise, as summarized in table 2.5. Radial symmetry is clearly the ancestral or primitive state, being present in the more ancient plant families, and it remains very common; but bilaterality has evolved and been lost repeatedly (Ree and Donoghue 1999). This has happened at least partly because the switch from radial to bilateral symmetry can occur very easily, often involving just one or two genes (*cycloidea, dichotoma*) (Coen and Nugent 1994; Cubas et al. 1999; Glover 2007), and is organized via a *floral symmetry gene network* that may be linked to plant breeding systems (Kalisz et al. 2006). In modern plants, radial symmetry occurs in around 80% of all families, and bilateral symmetry in 33% of dicot families and 45% of monocot families (Neal et al. 1998); the evolution to bilaterality probably occurred repeatedly by parallel recruitments of *cycloidea* homologs (Preston and Hilernan 2009).

Radial flower shape offers similarity of approach and landing from any direction, simplifying the pattern recognition task for the visitor. It is generally also associated with a radial offering of rewards: either a complete ring of **nectary** tissue around the base of the ovary or a series of separate nectaries related to the number of petals present, plus a mass of central anthers. A radial flower can therefore be accessed and

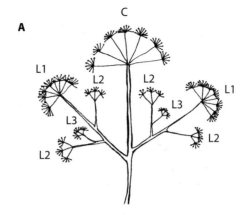

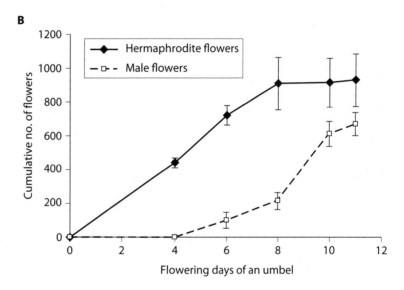

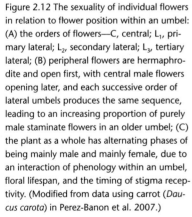

Figure 2.12 The sexuality of individual flowers in relation to flower position within an umbel: (A) the orders of flowers—C, central; L_1, primary lateral; L_2, secondary lateral; L_3, tertiary lateral; (B) peripheral flowers are hermaphrodite and open first, with central male flowers opening later, and each successive order of lateral umbels produces the same sequence, leading to an increasing proportion of purely male staminate flowers in an older umbel; (C) the plant as a whole has alternating phases of being mainly male and mainly female, due to an interaction of phenology within an umbel, floral lifespan, and the timing of stigma receptivity. (Modified from data using carrot (*Daucus carota*) in Perez-Banon et al. 2007.)

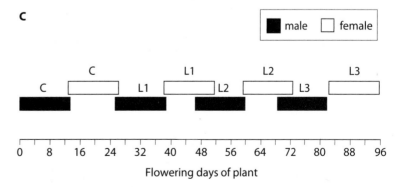

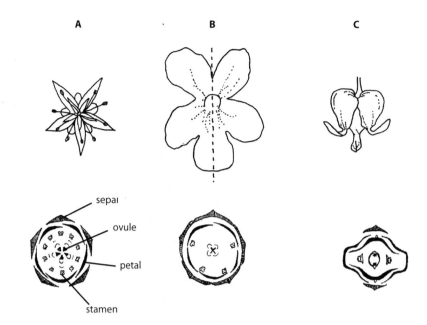

Figure 2.13 Types of flower symmetry and the associated flower diagrams: (A) typical actinomorphy, or radial symmetry; (B) typical zygomorphy, as in a labiate flower; (C) the curious symmetry of *Dicentra* flowers, with up/down, rather than right/left, symmetry. There are five axes of symmetry on (A), and dotted line shows single axis of symmetry on (B). (Modified from Barth 1985.)

TABLE 2.5
Interpretations of Floral Symmetry

Planes of symmetry	Symmetry type	Synonyms	Generic examples
Many	Radial	Actinomorphic, regular, multisymmetrical	*Primula, Narcissus*
Two	Disymmetrical	Biradial, bisymmetrical	*Dicentra*
One	Bilateral	Zygomorphic, irregular, bilabiate	
	Right-left symmetry	Bilabiate, medial zygomorphic	*Salvia, Corydalis*
	Upper-lower symmetry	(No term in use)	
None	Asymmetrical		
	Floral organs in spirals	Radial, actinomorphic, regular	*Nymphaea, Magnolia*
	Left or right handed	Enantiomorphic, enantiostylic	*Cassia, Solanum*

Source: Modified from Neal et al. 1998.

fed upon by a whole range of visitors, and it has often been noted that beetles, flies, and lepidopterans particularly favor radial flowers. Some of these land and forage in a relatively disorganized or haphazard fashion, but others, also including bees, show a more controlled foraging, often working methodically around the ring nectary or the series of nectaries, sometimes with an appearance of counting a full circuit of petals or nectaries and leaving the flower after one full circuit (fig. 2.14).

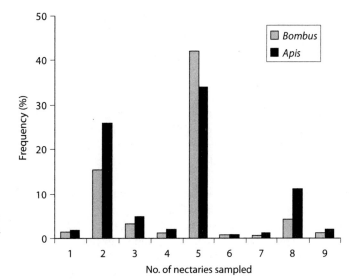

Figure 2.14 Apparent "counting" by bumblebees and honeybees feeding on strawberry flowers (*Fragaria*) with a radial design. Many bees sample just one or two nectaries, but the majority sample all five and very few do more than one full circuit (i.e., few sample more than five times). (Data collected by Emily Clark and author.)

In contrast, bilateral shape allows the manipulation of visitor approach so that the incomer is forced to enter the corolla in a specific manner, which can in turn ensure appropriate passage past the anthers. Bilateral flowers are nearly always oriented to open in a horizontal plane, and artificially changing their orientation leads to reduced visits or a higher proportion of illegitimate visits or both, as Ushimaru et al. (2009) demonstrated with **syrphid flies** and *Commelina* flowers, attributing the effects to both poorer recognition and poorer landing behavior. (However, it might be noted that something rather similar can happen in radial flowers: **hummingbirds** consistently approached from a face-on direction for naturally horizontal *Silene* flowers, but when the birds visited artificially fixed vertical flowers, they arrived from the side and from any compass direction, so contacting the sex organs in random fashion [Fenster et al. 2009]).

Bilaterality is often combined with concealment of the nectar, again ensuring that the visitor has to approach appropriately and insert its head or tongue "correctly." There are some indications that in bilateral flowers a more perfectly symmetrical bilaterality is also favored by visitors, perhaps because such symmetry indicates superior quality and strong developmental stability (Müller 2000). For example, bumblebees preferred symmetrical *Epilobium* flowers over somewhat lopsided asymmetric forms (Müller 1995a). But Midgley and Johnson (1998) give examples of visitors showing no such preferences, and it is always

difficult to disentangle **innate preference** from previous visual experience here; West and Laverty (1998) certainly found no preference expressed in naive bumblebees.

There are some indications that bilaterality can accelerate diversification and speciation in angiosperms (Sargent 2004), although this finding was only partly supported in a later analysis (K. Kay et al. 2006).

Inflorescences often serve to turn bilateral individual florets, which can gain the benefits of zygomorphic anatomy and visitor directionality, into an appearance of radiality at least from a distance; this may give the plant some of the benefits of both floral types. Examples can be seen in several of the categories dealt with in section 2, *Particular Floral Shapes*, below.

Nectar Spurs

A wide range of flowers from most of the architectural categories discussed in the next section have tubular extensions termed nectar spurs (plate 19 shows examples). Spurs may be formed from various different tissues, but most commonly they originate from petals; they may project from the back of the flower or may be tucked within it. They provide a very convenient way of offering a larger nectar reward without at the same time filling up the corolla (which would make this nectar available to inappropriately small or short-tongued visitors). Spurs have clearly evolved repeatedly across most plant families (Hodges 1997), and they can be up to 400 mm long in a Madagascan orchid (Nilsson 1988). There is good evidence that their appearance in a particular taxon can give a major boost to floral radiation and speciation (Hodges 1997; K. Kay et al. 2006; and chapter 11). There may be a punctuated, fast change in spur length during speciation episodes, since in columbines (*Aquilegia*) the spurs seem to have evolved rapidly at times to fit the adaptive peaks predefined by pollinators (Whittall and Hodges 2007).

In *Aquilegia* each of the petals normally has a visible backward-pointing spur (fig 8.2), producing a **revolver flower** effect (see below). In contrast, *Delphinium* species (larkspurs), in the buttercup family, have petaloid sepals, just one of which is elongated at the underside of the flower into a long spur. Very long-spurred species occur in the orchids, for example in *Platanthera*, where pollinating moths only approach as close as they need for their tongue to reach to the end of spur; hence they effect much less pollination in shorter-spurred individuals, giving ongoing selection for increasing spur length.

The presence of nectar spurs can lead to **convergence** within flowering communities. For example, the South African flora has species from at least 10 different plant families, including *Ixia*, *Geissorhiza*, and *Pelargonium*, all with spurs around 90–100 mm in length and all visited by long-tongued nemestrinid and tabanid flies (chapter 13). In Kenya a number of orchids converge on a white-flowered long-spurred pattern (Martins and Johnson 2007), and the **hawkmoth** visitors have matching **tongue lengths**; in these cases the spurs contain a nectar concentration gradient that may have adaptive significance (chapter 8).

Pendant Habit

This is another common feature that may occur in any flower design, but it is particularly common in bowl- and bell-shaped flowers (plate 4). A pendant habit excludes many visitor types because of the agility in flight and landing that is required to visit effectively (either an ability to feed while clinging upside down or an ability to feed while hovering). A simple hanging bell is a common structure, but more complex variants do occur. For example, in Turk's cap lilies (*Lilium martagon*) each petal of the bell has a narrow flanged groove at its base leading to a nectary, so that visitors (mainly hawkmoths in this case) must hover and probe almost vertically upward with their tongues, repeating the feat at each petal in turn.

Pendant flowers may also help to protect the internal organs from rain and weather-related damage, and an interesting example here is *Pulsatilla cernua*, whose flower stalk bends from erect to pendulous and back again during anthesis (Huang et al. 2002). During the key period of pollen presentation, the pollen is protected from rain in the pendant flower; thereafter, as the petals elongate, they develop unwettable hairs so that the flower, once erect again, still protects its remaining pollen.

Floral Shape Change

Some flowers show **diurnal** changes in shape and accessibility, for example, opening wide in the daylight and closing at night or during rain (e.g., *Nemophila*, *Hepatica*, *Crocus*) or more specifically, opening in response to sun and closing fully or partly when it becomes cloudy (e.g., *Adonis*, some *Mesembryanthemum*, some *Gentiana*). *Bergeranthus* flowers in South Africa open daily at about 1530 and close after about 3 hours, closely tied in to ambient temperatures above 23°C (Peter et al. 2004), and they are almost exclusively visited by bees during their opening hours. Gentians are particularly interesting because they may show two kinds of flower closure, one in response to environmental conditions (e.g., cooling temperatures or approaching thunderstorms [Bynum and Smith 2001]) and a second, more permanently in direct response to being pollinated (He et al. 2005). In the latter case, the closed flowers are often retained on the plant for some time to add to the display effects. Fireweed flowers (*Chamerion angustifolium*; syn *Epilobium*) show a similar pollination-induced closure mechanism within 4 hours of pollen receipt (Clark and Husband 2007); the time to close is reduced when pollen loads are larger and perhaps increased when **self-pollen** is deposited, indicating a consistently adaptive response.

Less drastic shape changes, such as increasing reflexion of petals or sepals, are also a common occurrence, and such changes often act as a signal of the age of an individual flower and hence perhaps the likelihood of pollen availability; an example of this effect was described for *Anemonopsis* visited by bumblebees (Pellmyr 1988).

Smaller shape changes may also occur within one part of a flower, including elongation of the anthers or stigma linked to **protandry** or **protogyny** (chapter 3) or altered curvature of these structures. Or there may be subtle and reversible movements of parts of the flower, seen for example in the lateral (inward) movement of anthers in *Mahonia* when the flower is stimulated by physical contact from a visitor (plate 12B), increasing the chances of pollen contact with animal surfaces.

6. Particular Floral Shapes

The listings below approximately follow the categorizing of flower types by Kugler (1970) and by Faegri and van der Pijl (1979), although with some simplifications. Many schemes have been produced over the years, some more strictly botanically correct, but the main requirement here is to distinguish types in relation to possible or actual visitors.

Open Disk or Bowl Flowers

This simplest of flower types is extremely common across many plant families (plates 2 and 3), with the petals forming a relatively flattened disk when fully open. The petals may be touching or overlapping laterally for much of the flower's life but can become quite separate and well spaced as the flower becomes fully mature, giving a more star-shaped appearance (fig. 2.15). In other cases the petals persist in a more raised orientation forming a deeper bowl, but still with very free access to most visitors.

This kind of floral arrangement most commonly occurs with a single layer of four to eight petals, as in buttercups (*Ranunculus*), *Anemone*, *Helleborus,* poppies (*Papaver*), and many rosaceous plants, such as bramble (*Rubus*), hawthorn (*Crataegus*), cherry (*Prunus*), or *Potentilla*. More rarely in wild plants (but rather commonly in garden hybrids) there are multiple petal arrays, as seen in many *Rosa* species. These flowers have a mass of anthers positioned centrally within the cup or bowl and have shallow exposed nectaries, so that both pollen and nectar are freely available to visitors.

Some examples within this type are more complicated. For example, hellebores have tubular petal nectaries within the bowl, concealing the nectar from short-tongued visitors (plate 19B). Other genera contain pollen-only flowers, with no nectar on offer; *Papaver* is a classic example, but this situation is also common in *Helianthemum, Adonis*, and some species of *Cistus*.

Tubular Flowers with Radial Symmetry

Here the perianth parts are elongated and arranged at their base to form the tube, while in the outer corolla the petal tips are **reflexed** somewhat to form a flattened landing area around and above the tube (plates 4 and 6 show examples). The flowers typically have fewer reproductive parts that are more fixed in number compared with more primitive disk and bowl flowers, and they have regular and small numbers of petals or calyx lobes. This design leads to visual concealment of the nectary, and as a result, the nectar can only be reached by a visitor with a suitably elongate tongue. Such flowers occur in many plant families, often where the "typical" flower form is quite different.

This functional state can be achieved in several different ways (fig. 2.16). Elongation by uniting both sepals and petals into separate tubes is the commonest form, and the corolla then usually has a narrow tubular basal region, into which only visitor mouthparts can be intruded, and a flat outer disk of petal tips, to offer landing or gripping space. This design (plate 6) is very familiar in flowers such as gentians (*Gentiana*), periwinkles (*Vinca*), forget-me-nots (*Myosotis*), and primroses (*Primula*), each of these examples being from a different family. The outer tube formed from the sepals acts as support and protection to the base of the corolla and may help protect against illegitimate nectar robbery from short-tongued visitors.

Uniting the petals into a tube that has little or no support from sepals is also possible, as seen in honeysuckles (*Lonicera*, plate 16A) and many **bellflowers** (Campanulaceae, e.g., plate 4D); these examples generally have protected inferior ovaries and so need less physical protection from the sepals. Tropical versions are perhaps commoner and include *Stephanotis* (plate 6H), *Dipladenia*, and *Plumbago*, the latter having a very narrow tube of fused blue petals and a calyx bearing fine sticky hairs externally (plate 6G), perhaps

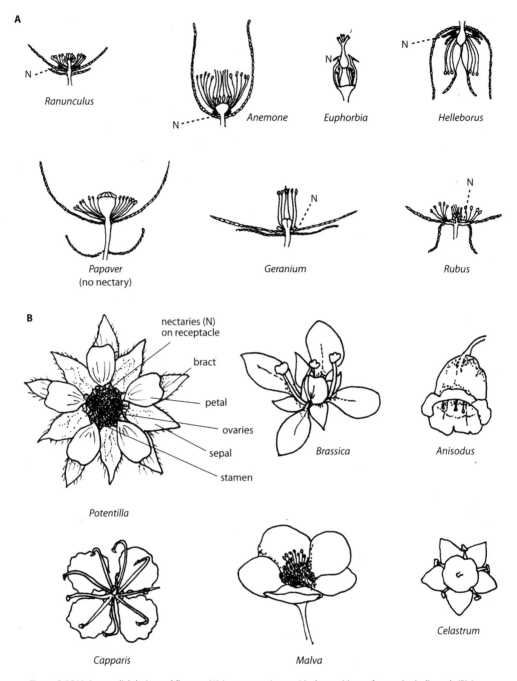

Figure 2.15 Various radial designs of flowers: (A) in cross sections, with the positions of nectaries indicated; (B) in frontal view. Note that this design includes star shapes (often with petals and sepals alternating) and bowls or bells. (Redrawn from Proctor and Yeo 1973.)

deterring ants (chapter 25). Red tube-shaped flowers are also common in bird-pollinated plants such as *Hamelia* or *Aloe* (plates 26 and 27).

Elongation by forming a sepal tube is an alternative approach and is characteristic of many plants in the Caryophyllaceae, especially in the **genus** *Dianthus* (pinks, campions, etc.). The sepals form a distinctive tubular cup from which thin dissected petals emerge and spread out to give a frilled effect.

In the Brassicaceae (a taxon usually with small and

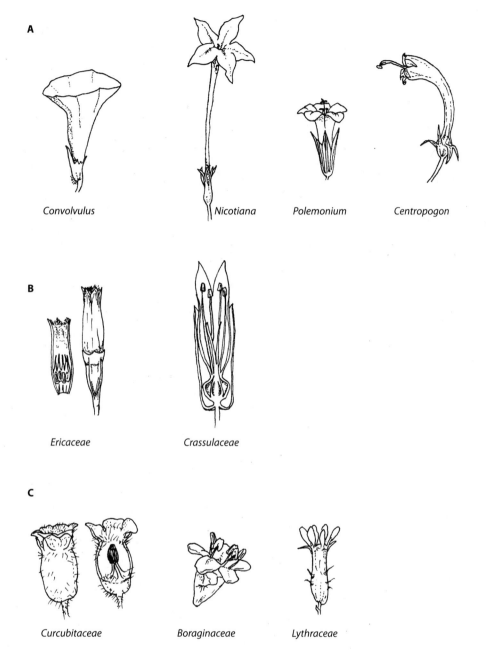

A

Convolvulus   *Nicotiana*   *Polemonium*   *Centropogon*

B

Ericaceae   *Crassulaceae*

C

Curcubitaceae   *Boraginaceae*   *Lythraceae*

Figure 2.16 Types of tubular design and how they are achieved, including elongate separate but overlapping, or fused, petals (and sometimes sepals).

shallow cross-shaped flowers), some of the larger-flowered genera, such as cuckoo flowers (*Cardamine*) and wallflowers (*Erysimum*), have the four sepals elongated and pressed tightly together (without actual fusion) to form the tube, while the four petals are separate and very narrow basally but expand apically above the calyx tube. In some Geraniaceae species (e.g., herb robert, *Geranium robertianum*) and in some mallows (*Lavatera*, plate 3E) and *Hibiscus* (plate 3D), a similar design occurs, although with five or six sepals and petals. In some of these examples the inner surface of the corolla effectively becomes ridged, producing a series of channels leading down to the nectar and foreshadowing the revolver flower, which will be discussed in *Trumpet-Shaped and Bell-Shaped Flowers*, below.

The risk with arrangements that lack petal or sepal fusion is that illegitimate visitation can occur, often when honeybees force their way into the tube between perianth segments that are merely pressed together. Hence a less risky way of achieving the same effect is to deepen the receptacle substantially; this can be seen in purple loosestrife (*Lythrum salicaria*) and in various species of *Daphne* (Thymelaceae) (plate 6A), where several millimeters of depth are provided by the receptacle, above which lie short petaloid sepals. And even with only slight receptacle deepening, an effective functional flower depth can still be achieved by adding a dense ring of anthers above the receptacle, in effect forming a tube into which a visitor's tongue must be inserted; some species of *Prunus* use this system.

For all these flowers, which converge by various means on a common functional design, the critical feature in terms of attracting a particular kind of visitor is the match between corolla depth (from the tube rim to the nectar or pollen) and tongue length. The same is true for flowers in the next section, although in their case, added complexity and more precise matching is required.

Tubular Flowers with Bilateral Symmetry

These are typical zygomorphic tube- or funnel-shaped flowers (plate 7), for which Westerkamp and Classen-Bockhoff (2007) used the term **bilabiate** (fig. 2.17A). Nectar is again concealed and, as a result, can normally only be reached by a visitor with a suitably elongate tongue, but the flower form is more complex relative to the preceding group. The flowers are normally arranged in a consistent orientation on the plant—more or less horizontally—and a particular direction of approach is needed to land and enter. Hence the visitor needs good perception of three-dimensional shape and reasonable learning ability to gradually improve its handling of the flower, fitting its own bilaterally symmetrical body into the bilateral flower.

These flowers are again typified by a small and fixed number of reproductive parts (stamens, styles, ovules) and regular small numbers of petals or sepals (usually three to six), these perianth parts being united to form the tube. Furthermore, the uppermost petal is often expanded apically as a hood, protecting anthers and/or stigma, and one or more lower petals are modified as a landing platform.

By far, the majority of these flowers are **nototribic**, that is, they have dorsal anthers and styles that deposit and pick up pollen on the back of the head or **thorax** of a visitor; they are also sometimes termed **gullet flowers**. Examples in temperate floras include a host of flowers in the two large families Lamiaceae and Scrophulariaceae, as well as many orchids. In the tropical floras many additional examples are provided by two additional very large families, Acanthaceae and Gesneriaceae, and of course by a great many more orchids. In fact, under the term "bilabiate," Westerkamp and Classen-Bockhoff (2007) found examples in 38 angiosperm families, and these include the great majority of blooms that would be conventionally described as specialist bee-visited flowers.

Many of the labiate flowers (Lamiaceae) have corollas that are essentially just two-lipped. The upper lobe forms a roof or hood over the stamens (protecting the pollen both against external conditions and against being too easily gathered by bees) and then protecting the later-developing style (since protandry is the norm; chapter 3). The lower lobe forms a landing platform for visitors (fig. 2.17C), although this can be much

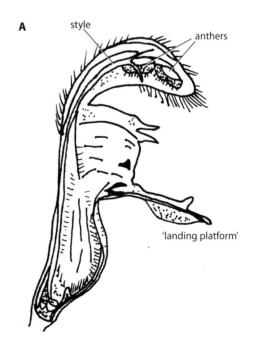

A

style

anthers

'landing platform'

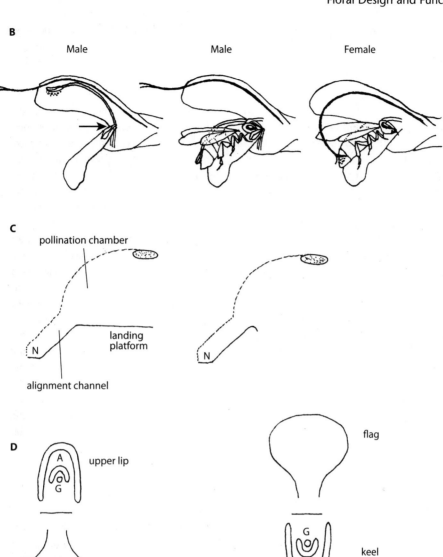

B

Male   Male   Female

C

pollination chamber

landing
platform

N

alignment channel

N

D

flag

A
G
upper lip

keel
G
A

lower lip

A = androecium
G = gynoecium

Figure 2.17 The typical bilabiate floral design: (A) as seen especially in Lamiaceae and Scrophulariaceae (modified from Proctor et al. 1996); (B) changing form in a protandrous species such as *Salvia*, where the stigma is out of reach in the male phase but an insect pushing on the lever-like anther base (arrow) causes the fused anthers to bend down and contact its back, whereas in the female phase the style is elongated and curved to contact a visitor in a similar place (modified from Heinrich 1979a); (C) in diagrammatic cross section showing the landing platform, the pollination chamber and the narrower alignment chamber, and an alternative version with little or no landing platform, commonly bird-pollinated (redrawn from Westerkamp and Classen-Bockhoff 2007); (D) labiate-type lip flower and legume-type flag flower variants compared (modified from Endress 1994).

reduced in many of the larger bilabiate flowers. The two lobes are separate apically (forming the "pollination chamber," in Westerkamp and Classen-Bockhoff's (2007) terminology, fig. 2.17C) but united into a tube basally for about half of their length (the "alignment channel"). Nectar is secreted at the base of this tube, often accumulating as a substantial column within the tube (and can be visible from the outside in some thinner-walled corollas).

Both the flowers and their visitors come in a very wide range of sizes and with different degrees of accessibility. Large temperate examples include archangel (*Lamiastrum*, plate 31E) and some *Salvia* species, with flowers often 20 mm or more in length that can be accessed only by longer-tongued bumblebees; *Phlomis* (plate 7A) is also in this category, though with its roof curved over it has a rather closed-access corolla that a bee must force its way into. Medium-sized forms, such as deadnettles (*Lamium*), skull caps (*Scutellaria*) and selfheals (*Prunella*), have tubes 8–15 mm deep and are visited by a wider range of bees, whose bodies fit neatly between the hood and the landing platform as their tongues probe for nectar. Smaller-flowered varieties include many of the familiar hedgerow plants, such as woundworts (*Stachys*), bugles (*Ajuga*), and woodsages (*Teucrium*). Also in this category are many of the classic Mediterranean herbs, including mints (*Mentha*), thyme (*Thymus*), and marjoram (*Origanum*), which often occur as flower spikes (plate 10). In most of these flowers, the fully enclosed tubular part of the corolla is only 2–6 mm long, so that they can be visited by a substantially wider range of insects, including medium- to long-tongued flies and short- to medium-tongued bees (honeybees and solitary species). Even large bumblebees still visit these smaller labiates but insert only the head under the hood rather than the whole body. A given *Bombus* species may get the pollen of *Ajuga* on its head but the pollen of *Lamium* on its dorsal thorax and so remain an effective **cross-pollinator** of each even when visiting both on the same **trip**. Where these flowers occur on spiked inflorescences, the stamens often project beyond the upper lip of the flower, but they get some protection from the bract or calyx of the next flower upward on the spike (e.g., plate 10A–D).

Labiates also come in larger varieties in the tropics suited to bird, and occasionally bat, visitation, but many of these examples have a reduced lower lip because a landing platform is no longer needed. *Salvia* provides good examples in those species showing large and highly elongate red corollas. This genus has an unusually large variety of form and visitors and is also unusual in having a kind of lever action in its stamens (fig. 2.17B) to deposit pollen on the visitor (Reith et al. 2007; and chapter 7); these levers have evolved at least three times independently (Walker and Sytsma 2007). They are still present in the bird-pollinated species but have often become immovable, assumed to be so because the lever effect is not necessary with a large bird visitor (Wester and Classen-Bockhoff 2006).

In the original family Scrophulariaceae (now somewhat realigned taxonomically) similar kinds of flower and a similar range of sizes occur. The large foxgloves (*Digitalis*–Plantaginaceae) belong here, though these are very open, with little trace of upper and lower lobes; their downward-pointing orientation and hairs in the corolla tend to exclude many "inappropriate" visitors, the main successful foragers being bumblebees. In the medium-sized range come yellow rattles (*Rhinanthus*–Orobanchaceae) and snapdragons (*Antirrhinum*–Plantaginaceae), and then the smaller eyebrights (*Euphrasia*–Orobanchaceae) and toadflax (*Linaria*–Plantaginaceae; plate 32C). Louseworts (*Pedicularis*–Orobanchaceae) are particularly common examples throughout the northern hemisphere and exemplify the type. The corollas are two-lipped as in labiates, but here the upper hooded lip is laterally compressed, protecting the stamens within, which occur in two inward-facing pairs that are protectively pressed together until a visitor lands on the flower (fig. 2.18). The structure of the upper lip requires a visiting bee to enter with its head at a slant, so levering the hood open and forcing the anthers apart, and this action releases pollen onto the bee's head. In North America the larger early flowering species, such as wood betony (*P. canadensis*), have a style that is unusually long and protruding from the yellow corollas, and Macior (1968) showed that when visitors (mostly queen bumblebees) alight, the long style is exactly sited to pass into the bees' neck region (the groove between the head and the thorax), in which previously deposited pollen can rarely be completely groomed off by the bee, so ensuring pickup of **nonself-pollen**. Some "scrophulareous" types have developed a more sophisticated form of protection, with the opening much restricted by a raised lower lip to give a "door" effect, where the upper and lower lips are closely pressed together, such that only a relatively strong and

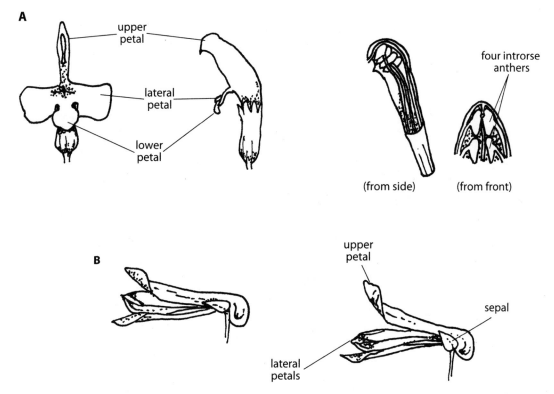

Figure 2.18 Tubular flowers in Scrophulariaceae, typically with lateral compression of the upper petal and inward facing anthers: (A) *Pedicularis* flower, frontal and lateral views, and cut away to show the anthers (redrawn from Faegri and van der Pijl 1979); (B) *Corydalis* flower before and after being opened by visitation (modified from Proctor et al. 1996).

"clever" visitor can lever them apart and force a way in to the rewards. This can be seen in nemesias, toadflaxes, and snapdragons (plate 30F), where additional constraints are imposed by the presence of a nectar spur, which keeps most of the nectar out of reach of all but the long-tongued visitors. In such flowers, bees seeking only pollen may often work the flowers upside down, scraping pollen from the anthers that lie in the roof of the upper corolla lobe.

Some other plant families can produce similar and rather typical zygomorphic tubular flowers. For example, the *Viola* flower is essentially a shallow version of this floral type (plate 19C), with five separate petals, of which the lowest is prolonged backward into a spur, and with a complex central cone incorporating the anthers and a hollow style (chapter 8). The main visitors to most wild violets are bees, both bumblebees and a range of solitary species, though some hoverflies also visit.

An open zygomorphic design also occurs in many hummingbird-pollinated flowers, often the Gesneriaceae, where the upper lip is pronounced and overhanging but the lower lip is almost absent or strongly **recurved** (e.g., plate 26F), removing any landing platform effect and requiring the bird visitor to hover in front of the flower as it feeds. The term "gullet flower" is particularly applied to these flowers.

There are also some **sternotribic** tubular flowers that should be mentioned, where the anthers and style are more ventral and come up from the base of the flower, usually depositing pollen on the underside of a visitor. Many of these are less obviously tubular, instead being *keel*, or *flag*, flowers (dealt with in *Keel Flowers* below), but there are some strictly tubular sternotribic flowers that should be discussed here. The monkshoods (*Aconitum*) provide a good if somewhat elaborate example (plate 7D). Although part of the Ranunculaceae, a family that is regarded as primitive and

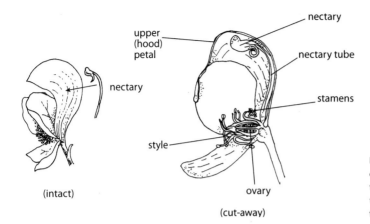

upper (hood) petal

nectary

nectary

nectary tube

stamens

style

ovary

(intact)

(cut-away)

Figure 2.19 Monkshood flowers (*Aconitum* sp.), with complex sternotribic morphology and an elongate internalized petal nectary only accessible to a long-tongued bee working upward from below. (Redrawn from Proctor et al. 1996.)

normally has simple radial flowers, the monkshoods have evolved a complex bilateral morphology (fig. 2.19) that is almost exclusively shaped for bumblebee visitation; no other visitors are effective. The sepals form the principal colored attractant structures (purple, blue, or cream in different species), the uppermost of the five sepals forming a substantial hood shaped rather like a monk's cowl or a rounded helmet. Under this hood are two enlarged tubular nectaries (formed from petals), the other nectaries being small or entirely absent. The stamens develop before the style is mature (i.e., the flowers are protandrous; see chapter 3) and are reflexed downward initially but stand out horizontally as they begin to dehisce, so that they dust pollen on the **abdomen** of a visiting bee that alights on the lower sepals and moves upward to reach the nectaries. Only after the stamens are all dehisced and withering do the maturing stigmas become exposed. In the blue *A. napellus* the nectaries are fairly long with a small reservoir at the tip and can be visited by medium- and long-tongued bees, whereas in the alpine species *A. vulparia*, which is predominantly creamy yellow, the nectaries are extraordinary, up to 22 mm long and spirally coiled, so that only the very longest-tongued bumblebees are able to access the nectar.

Perhaps not surprisingly, some scrophulareous flowers have also produced almost sternotribic flowers. In North America the summer-flowering pink lousewort (*Pedicularis groenlandica*) has a very elongated curved style protected by a lobe of the upper corolla lip, and **worker** bumblebees alight on top of this beak-like structure in a way that would suggest sternotribic pollen deposition. However, the style's curvature is such that its tip comes under the bee and then up the side of the thorax to lie close to the front surface of the abdomen, pollen therefore alighting on a surface that is difficult to groom clean (cf. *P. canadensis* discussed earlier).

Trumpet-Shaped and Bell-Shaped Flowers

Although these designs could be seen merely as versions of radial tubular flowers, here the tube is more open, and visitation by insects involves the whole body entering the flower. Hence the crucial factor in coevolution with a preferred visitor is matching flower tube diameter to body size, and there is no necessity to match corolla depth to tongue length (although there are flowers of this type mainly visited by vertebrates, where the visitors may merely insert their tongue).

These flowers are perhaps best seen as intermediate between open bowl flowers and radial tubular flowers, and of course there is in practice a continuum between all three designs. However, the dicot trumpet-shaped flowers usually have some petal fusion to form the base of the trumpet, with the petals becoming separate only as the corolla flares outward. And whereas open bowls can be (and are) visited by almost any insect, the trumpet-shaped flowers are commonly only visited by those with a closer match of body size to trumpet diameter. Some of the more sedentary smaller insects (pollen beetles, for example) may crawl inside, but small active fliers are less likely to be regular visitors. The stigma and stamens tend to be central, more or less strongly united into a single structure, with the anthers dehiscing outward to deposit pollen on insects entering the trumpet.

A classic example is *Campanula* and many related genera in the Campanulaceae, all commonly known as bellflowers. Many are pendant, some projecting sideways from the flower stalks and a minority pointing upward, but all are essentially bell shaped with a slight flare at the outer rim of the corolla (plate 4D). Nectaries lie at the base of the bell, but the five central stamens have rather expanded bases, so that five short channels lie between their bases and lead to the nectar. Looked at head on (fig. 2.20, plate 3B), there is thus a ring of separate narrow tubes in the flower base, and a visitor must probe each in turn to get the full reward; this design is sometimes termed a revolver flower, having the appearance of the barrels of a revolver and also conveniently reflecting the fact that a visitor must revolve around within the flower to sample each channel in sequence, taking up a series of feeding positions. In fact the shortest-tongued visitors cannot reach the liquid reward at all, and revolver flowers of this kind are usually only visited by insects (mainly bees) having a moderate tongue length. Simple revolver flowers are also found in some *Abutilon* (plate 3C) and in a more complex form in the genus *Nasa* (Loasaceae) where they can be divided into two types, *tilt revolver* (spreading white-yellow petals with brightly colored scales concealing the nectar entrances) and *funnel revolver* (more erect orange-red petals, nectar freely accessed through noncontrasting funnel-shaped scales). The former condition is ancestral and the flowers are visited by bees, especially **colletids**; the latter state has arisen at least twice and the flowers are mainly visited by hummingbirds (Weigend and Gottschling 2006).

Another well-known trumpet flower is bindweed (*Convolvulus* and related genera such as *Ipomoea, Calystegia,* etc.), opening to an almost perfect cone shape (plate 3A). The very common *Convovulus arvensis* has flowers about 20 mm deep and up to 30 mm in diameter, while other species can be somewhat smaller or substantially larger. The flowers have a rather narrow tube at the extreme base where the nectaries lie, again with a restricted revolver flower form with access via five channels between the stamen bases. Most of these flowers last only one day.

Similar designs, though slightly more elongate, are seen in some alpine gentians ("trumpet gentians," such as *Gentiana asclepiadea* and *G. acaulis*, plate 6B) that are typically visited by bumblebees. They stay open for several days and are protandrous, again with the revolver flower effect internally. Some of these species have "windows"—patches of pale translucent walls near the base of the corolla, so that an insect may seem to move toward the light as it enters. *Soldanella* is another alpine example of this type, although pendant and with divided petals (plate 4F). A larger version of the design is seen in *Datura* (syn. *Brugmannsia*, plate 3H), visited a little by bees in daylight and intensively by hawkmoths, and sometimes bats, around dusk.

Monocot versions of the trumpet-type flower generally lack petal fusion and tend to have the stigma and stamens further apart. In this group protogyny is common, with the styles longer than the stamens. Typical examples occur in the Liliaceae, including lilies

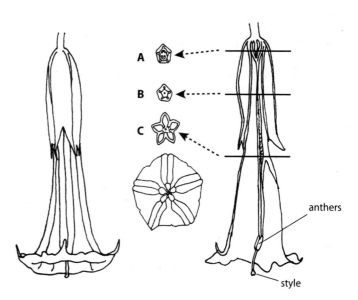

Figure 2.20 Revolver flower, especially seen in face-on views of *Brugmansia*; note the five channels within at transverse section (c), formed by the stamens and corolla. See plate 3 for a similar effect in *Campanula* and *Abutilon*. (Redrawn from Endress 1994.)

anthers

style

(*Lilium*), tulips (*Tulipa*), and fritillaries (*Fritillaria*), where nectar is secreted in grooves at the bases of the petals (these nectary grooves being enlarged into cuplike structures in some of the larger fritillaries such as crown imperial, *F. imperialis*, plate 19A). Some Iridaceae also have bell-shaped flowers, notably *Crocus*, with the stamens and styles more united as a central column and the bell upturned. Trumpet design also occurs in the monocot bluebells (*Hyacinthoides*) and Solomon's seals (*Polygonatum*). Daffodils (*Narcissus*) are essentially of this type too, although they are unusual in having the "extra" corona tube projecting forward beyond the normal ring of spreading petals. Most of the genera mentioned are visited by bees, especially bumblebees and some solitary *Anthophora*, though longer-tongued syrphid flies may also be useful pollinators. However, it is worth noting that many of the cultivars of these bulbous plants grown in gardens receive very little visitation, having been bred for unusual colors and forms that are less attractive or less rewarding to insects.

Spherical Flowers

In a rather small range of flowers, we find corollas that widen out to a broad central space before contracting to a narrow mouth, giving a roughly spherical overall shape. The petals are generally fused along most of their length to create this shape. The family Ericaceae provides many examples (plate 5), such as heather and heath (*Erica, Calluna*), bilberry and cranberry (*Vaccinium*). Snowberry flowers (*Symphoricarpos*, from the honeysuckle family) are also rather spherical, but with a flared opening. The monocot grape hyacinths (*Muscari,* from the lily family) are also of this general form, again with flared "teeth" at the mouth, though in this case they are packed together as a tight inflorescence spike (plate 5C). Some tropical flowers also produce a spherical form, especially in the family Rubiaceae, and are very often white in color (plate 5E), though some larger red versions occur and tend to be bird visited.

Keel Flowers

Keel, or *flag*, flowers (plate 9D–H) are in some respects just a modification of the bilabiate or tubular zygomorphic flower type, but they are very distinctive and commonly involve specializations for triggered pollen release, and so deserve a category of their own. They are particularly characteristic of the legumes (Fabaceae), familiar as pea and bean flowers, often collectively termed **papilionate** flowers. However, a remarkably similar design has also arisen convergently in the Polygalaceae, with very detailed congruences of structures (Westerkamp 1999).

The flowers have a short tubular green calyx, usually clearly five-lobed, at least at the tip. Within this the corolla is formed from five petals, but with the central uppermost petal being enlarged as the **standard**. The two petals just below this to either side are termed the **wings**, forming lateral projections to which a visitor can cling; and the lowest pair are folded tightly together to form a somewhat boat-shaped structure termed the **keel**, which encloses the stamens and stigma and is only open toward its tip (fig. 2.21). All five petals usually remain free in the sense that they do not fuse to each other, but a variety of complex folds and projections tend to occur that effectively lock them together, at least basally. Within the keel the stamens do fuse together; there are normally ten, nine of which unite along a substantial part of their length to form a tube in which nectar accumulates from a basal nectary, while the tenth and uppermost stamen remains free to allow access to this nectar.

In all these flowers the stamens and stigma are very close together within the keel, and self-incompatibility (chapter 3) is therefore commonplace. In some species there is a degree of **self-compatibility**, but the stigma does not become receptive until after its surface has been abraded during the course of at least one visit, such that a second visit (hopefully from a cross-pollen-bearing visitor) is needed to effect pollination.

When a visitor alights on such a flower, it will normally grip onto the wing petals and insert its tongue between the more or less erect standard petal and the upper edges of the keel, sliding the tongue inward to reach nectar in the **staminal tube**. A variety of events can then occur to cause pollen deposition on its body:

1. In the simplest forms, such as white melilot (*Melilotus alba*) and the clovers (*Trifolium*), the wings and keel are pressed down and apart by the visitor's body and the rather rigid stamens and style are thereby exposed so that they will contact the abdomen of the probing animal (usually a bee). Once the visitor leaves,

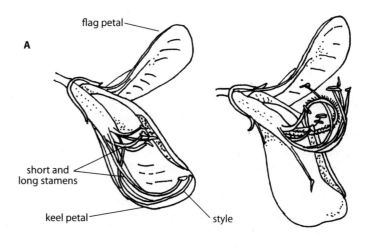

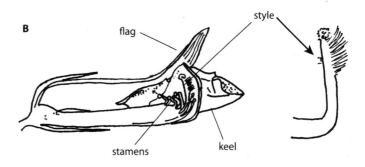

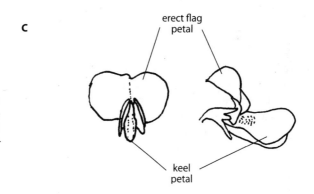

Figure 2.21 Structure of typical keel blossoms in the Fabaceae: (A) untripped and tripped flowers of broom (*Cytisus*), with half of calyx and corolla removed to show stamens and style (held in place by the keel, then explosively released by a visitor); (B) cross section of vetch (*Vicia*), showing the pretripped position of the brush-bearing style; (C) frontal and side view of typical "pea flower" undisturbed, showing the normal appearance of the floral display. (Part (A) modified from Faegri and van der Pijl 1979; (B) and (C) modified from Proctor et al. 1996.)

the keel springs back upward to protect the reproductive organs.

2. In the vetch genera *Vicia* and *Lathyrus* (including some beans and sweet peas; plate 9H) the basic mechanism is similar but there is a secondary pollen presentation system where the style is sharply bent upward terminally, with a brush of fine hairs near the tip and just below the stigmatic surfaces. Normally the stamens dehisce inside the closed bud, and therefore shed their pollen onto this brush or onto the inside of the end of the keel, where the brush picks it up. When the

flower opens, the stamens are already withered, but the stigma brush is exposed by the first visitor and carries the pollen onto the visitor's underside. In these flowers only quite large visitors with long tongues (bees, some hoverflies) can reach the nectar, and their weight is enough to depress the keel and ensure that the brush is exposed. In some *Lathyrus* species, the keel structures are asymmetric, and visitors must approach from the left and will receive contact from both style and pollen on their right side.

3. In genera such as *Lotus* (trefoils), *Anthyllis* and *Hippocrepis* (vetches), and *Lupinus* (lupins), the keel petals do not part on the midline to expose the stamens and style but instead are united except for a small slit at the tip. When the wings and keel are pressed down by the weight of a visitor, the stamens below force a ribbon of pollen out of this slit, acting rather like a piston. Later on in the life of the flower the stigma visibly protrudes through this same slit.

4. Taking this mechanism one stage further, some legumes have a form of **explosive pollen presentation**. In the common forage crop lucerne (*Medicago sativa*), the stamen tube is initially held down by projections on the keel petals, but the weight of a visitor dislodges these projections and allows the stamen tube to spring upward (often termed **tripping** of the flower). In common European gorse (*Ulex,* plate 9D), which is nectarless but extensively probed in early spring by both *Bombus* and *Apis*, the two keel petals are stuck together initially but the weight of the visitor causes them to break apart thus uncovering the stamens. The broom flower (*Cytisus*) is even more elaborate, with two sets of stamens that spring out once the keel petals are disunited, one set hitting the bee's underside and the other set springing outward and upward to hit the sides or back of the bee. Once visited, these "explosive" flowers tend to become limp and are rarely visited again. Plate 9, E and F, shows examples of flowers during and after tripping.

All the cases mentioned so far have been **leguminous**, but as so often happens in floral biology, a useful floral design giving some control of visitors has been invented convergently, and a remarkably similar design is found in genera such as *Corydalis* and *Fumaria* (plate 7E; Papaveraceae). Here there are just four petals, the upper one forming a small upright **flag** and also bearing a nectar spur at its base. The two lateral petals curl inward and fuse at their tips, so forming a sheath that encloses a thick and rigid style, which becomes covered by its own pollen in the bud stage as in *Vicia* (paragraph 2 above). Bumblebees and larger **solitary bees** depress the sheath as they visit and probe for nectar and so receive pollen on their undersides. One species of yellow corydalis from alpine Europe (*Pseudofumaria lutea*) even has an explosive pollen release system like that of gorse.

Iris-Type Flowers

In the case of iris-type flowers, there are six petals modified into two sets of three: an upper set expanded as the standard petals that project upward and provide the main floral display, and a lower set that projects downward and are termed **fall petals** (fig. 2.22, plates 9C and 15C). The falls are modified at their inner end as large grooves that are roofed over by an expanded petal-like style to form a tube leading to nectar and housing a single anther. The distal region of the fall is also expanded to enhance the floral display and to give a landing platform; its upper surface often has a central elaborately ridged and furrowed area (the *beard*) that is differentially colored and gives a good grip to the bees that normally pollinate the flower by entering between the standard and the fall. In effect, the single iris flower has become three separate functional units, each of which can be pollinated separately and each of which is essentially similar to a single zygomorphic tubular flower as described in the section *Tubular Flowers with Bilateral Symmetry* above, so that Westerkamp and Classen-Bockhoff (2007) include this category in their bilabiate grouping.

Iris flowers are unusual (perhaps only matched by orchids) in showing a great diversity of stamen-mimicking structures, including hairs with yellow tips, comblike structures, velvety cushions, and yellow lobes.

Orchid Flowers

Orchid flowers appear complex and very different from most other flowers (plate 10F–H) but are essentially formed from a tripartite basal plan not unlike that of the Liliaceae. They have six perianth segments, three inner and three outer (fig. 2.23), with the lower

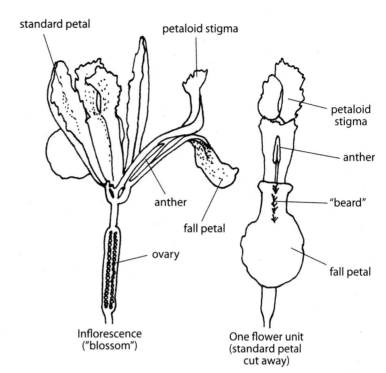

standard petal

petaloid stigma

petaloid stigma

anther

"beard"

anther

fall petal

fall petal

ovary

Figure 2.22 Basic structure of an *Iris* flower, essentially formed from three separate functional units acting as one blossom.

Inflorescence ("blossom")

One flower unit (standard petal cut away)

inner petal expanded and termed the *lip*, or **labellum**, which provides a landing platform and sometimes bears a nectary (often with a nectar spur) at its base. The greatest degree of modification is in the central reproductive structures, where many stamens have been lost and the remainder are fused with the stigmas into a central **column**, which projects above the labellum. Of the original stamens, only one is normally fertile (though there are two in the more primitive lady's slipper orchids). The stamen forms a stalk at the back of the column, with its anther housed under a small hood that protects the two large pollinia, which are formed from a mass of tiny pollen grains held together by viscid elastic threads (chapter 7). In all orchids, only two stigmas are functional (though often fused together), the third becoming a small projection (the **rostellum**) on the top of the column, producing a sticky exudate conveniently close to the anthers (fig. 2.23B), and thus helping to stick the pollinia to the bodies of passing insects.

Within this basic anatomical framework, enormous variation in shape and size occurs, and many orchids have sophisticated mechanisms to ensure pollinia removal and deposition in precise places and at carefully timed intervals, often involving movements of parts of

the column. Some of the most striking of these involve **pseudocopulation**, a topic covered in chapter 23.

Perhaps not surprisingly, orchid diversity (with more than twenty thousand species) is linked with high pollinator diversity, but in fact over 60% of orchids are bee pollinated. They are especially associated with nonsocial or only primitively **eusocial bees** (bumblebees in northern habitats and **euglossine** bees in the tropics; chapter 18); these animals are adept at visiting widely spaced plants and populations and have very precise learned behaviors ensuring strong fidelity and precise **pollen placement**.

Brush Blossoms

"Brush blossom" is a useful general term for densely headed inflorescences that function as a single unit of attraction, and for practical purposes these can be regarded as a single pollination unit. Many tens or hundreds of tiny individual flowers are packed together side by side to give a highly visible flat or domed landing surface, from which separate corollas can be probed for nectar. Anthers protrude to form the surface "brush," and may be the most conspicuous part of the

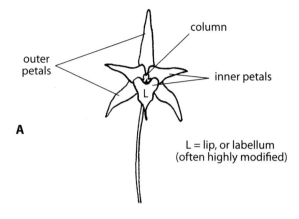

A

L = lip, or labellum
(often highly modified)

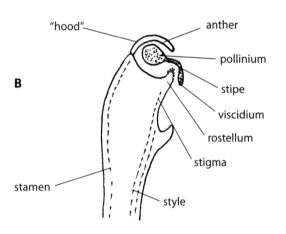

B

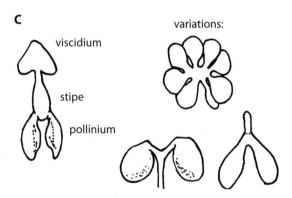

C

Figure 2.23 (A) Basic orchid floral anatomy. (B) Close up of column and rostellum. (C) Various pollinia and pollinaria.

flower, so that pollen is freely available at the surface of the blossom, from which it can be collected (by bees and hoverflies) or can be brushed passively onto the underside of other visitors as they pass over. The brush blossoms offer a reward to a very wide range of visitors and typify **generalist flowers**, those usually lacking specific adaptations to particular pollinators.

Several plant families produce a classic version of this flower type, in which a broad inflorescence is made up of numerous small white or cream-colored flowers, from which a massed froth of anthers emerge. Plate 1 includes some examples. The herbaceous umbellifers, such as hogweed (*Heracleum*), cow parsley (*Anthriscus*), and pignut (*Conopodium*), are the most obvious cases; they are abundant in most temperate hedgerows, along with their kin that provide important vegetables (carrot [*Daucus*], parsnip [*Pastinaca*], fennel [*Foeniculum*], etc.). But the same design can be found in other families, such as the scabious (*Scabiosa*, Dipsacaceae), rampions (*Phyteuma*, Campanulaceae), and various members of the monocot Liliaceae family (onion relatives such as *Allium*). It also occurs in much larger and woodier hedgerow shrubs, such as many *Viburnum* species (guelder rose, snowball tree, etc., Caprifoliaceae) and *Sambucus* (elder, Adoxaceae); in *Cornus* species (dogwoods, family Cornaceae), which are common throughout the northern hemisphere; and in *Hydrangea* species (Hydrangeaceae). On a much smaller scale it is also seen in many low-growing plants such as saxifrages and sedums. Since many of the families mentioned are regarded as fairly advanced, it is clearly not appropriate to regard these generalist flowers as in any sense more primitive than flowers specialized for particular pollinators.

Brush blossoms of all these kinds are visited by almost any kind of insect: flies are nearly always common, along with beetles, wasps, honeybees, and short-tongued solitary bees. The prevalence of this design may even be an indication of Müllerian **mimicry rings**, in which plants exploit the evolutionary advantage of a common advertising system. Their visitor spectrums do vary somewhat, probably largely due to differences in scent, which range from rather musty to extremely sweet and which can even vary within a species in different populations (e.g., Tollsten and Ovstedal 1994, working with *Conopodium*).

An alternative kind of brush blossom occurs in many genera of the Fabaceae in the mimosoid subgrouping. Plate 8 shows examples in various size ranges. Here masses of tiny florets are assembled into a

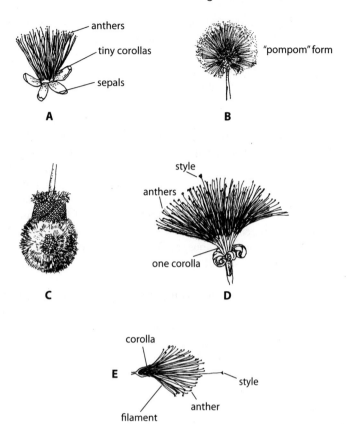

Figure 2.24 A range of brush blossoms: (A) basic form; (B) mimosoid flower of acacias; (C) and (D) two forms seen in Bombacaceae, *Parkia* and *Pachira*; (E) typical structure of one flower from within the brush, with much-reduced corolla. (Parts (C) and (D) redrawn from Endress 1994.)

spherical or elongate (spicate) inflorescence (fig. 2.24), often yellow or white but sometimes purple, pink, or red. The classic example is the acacia, which is extremely common as a shrub or small tree across all the southern continents (though recently undergoing reclassification such that only the Australian genera may now be strictly termed *Acacia*). Here the corollas are tiny and the blossom is in effect formed by the massed anthers emerging as a pom-pom, offering pollen freely to all animals alighting on the surface. In other genera the whole structure is much larger and may be visited by hawkmoths, birds, and bats; examples include *Albizzia* and *Newtonia*. Similar examples occur in the Myrtaceae and among eucalypts, and some of the bottlebrush flowers, such as *Callistemon*, also fit here (plate 8D).

Composite Flowers

The family Asteraceae is probably the largest worldwide, exceeding even the orchids, with some 25,000 to 30,000 species, and has to be recognized as one of the most advanced and most successful of all the angiosperm taxa. Its flowers, which are usually instantly recognizable composite "daisy" or "thistle" designs (plate 8F), must have contributed in large measure to this success. The composite inflorescence is made up of many individual tiny flowers, each with one ovule, that open successively over a period of days. Different pollination events may take place over the surface of the flower head on different days and times of day and may be effected by different visitors. The advantage may well be that different (and often neighboring) flowers get different sexual partners, and the inflorescence as a whole ends up with seeds sired by a range of different fathers, opening up far more opportunities for variation and adaptation to new environments.

A typical composite inflorescence (**capitulum**, fig. 2.11) has florets of two separate types, most easily seen in daisies (*Aster*), in ragworts (*Senecio*), or on a grand scale in sunflowers (*Helianthus annuus*). Plate 8G shows a particularly clear example. The outer part of the blossom is made up of **ray florets**, which generally look just like petals and have asymmetric, strap-like corollas, while the center is made up of small tubular **disk florets,** which are radially symmetrical. The ray florets, commonly white or pink/purple, provide most of the advertisement, although in many species they are aided in this role by the contrasting color (almost always yellow) of the central disk florets.

The structure and function of the two kinds of floret from a typical composite flower are shown in figure 2.25.

1. Ray florets are purely female in function, with a short bilobed stigma, and they open before any of the central disk florets are mature, so making the capitulum effectively protogynous (even though the individual disk florets are usually protandrous).

2. Disk florets, which contain both male and female organs, are packed tightly together in the center of a blossom. They have the calyx reduced to a simple ring of long hairs (the **pappus**) around the inferior ovary and a simple tubular corolla emerging above this, widening just at its tip into five short lobes. The five stamens are separate within the corolla, but the introrse anthers are fused apically into a long tube that surrounds the style; pollen is therefore shed to the interior of this tube. The style (with its bilobed stigmatic tip closed) then begins to grow up through the anther tube as the floret opens (usually in the morning), pushing out the pollen as it advances. In many species, including the common knapweeds (*Centaurea*), this emission of pollen is not continuous but is triggered by visitation, which causes the stamen filaments to contract a little, so extruding a small dose of pollen onto the visitor's underside; successive small doses of pollen are dispensed onto each succeeding visitor, the filament length slowly being restored between each visit. Only when all the pollen has been extruded (usually not until day 2 of the floret's life) does the style tip emerge, spreading open its stigmatic lobes. Nectar-gathering from these flowers is difficult since the anther insertions effectively block the corolla tube; it is likely that most insects foraging on composites can gather only nectar that has risen above the anther insertions (Corbet 2000).

The relative proportions of disk and ray florets, and the sexuality of each, provide the main sources of variation in the pollination characteristics of Asteraceae. In coltsfoot (*Tussilago*) both floret types are present

but the central disk florets are purely male. In some genera, particularly those known loosely as thistles (*Carduus, Cirsium, Centaurea*, etc.), only the tubular (disk) florets are present, and the mechanism of pollen presentation is similar to that already described. However, in somewhat fewer genera, only the strappy ray florets are present, the most obvious example being the ubiquitous dandelion (*Taraxacum*). Here the ray florets are of necessity hermaphrodite and function rather like the ray florets of a daisy, although in most cases the style, rather than acting as a simple piston in pushing the pollen out, is more like a bottle brush, having hairs on its sides that carry the pollen upward off the anthers, again before the style tip spreads open to reveal stigmatic surfaces. Occasionally species occur with the male and female functions on different plants; the thistle *Cirsium acaule*, for example, has some all-female blossoms and some (larger) hermaphrodite blossoms.

The other important variable that affects visitor patterns is the depth of the florets, especially the disk type. Asteraceae do show a considerable variation in corolla depth, but this shows little correlation with visitor diversity and is instead linked strongly to phylogeny (Torres and Galetto 2002). The later-evolving taxa have shorter corollas and greater visitor diversity, probably reflecting a tendency toward generalist pollination in the family.

Unsurprisingly, then, nearly all composite flowers receive many kinds of visitor; for *Senecio jacobaea* a list of 178 insect species was compiled by Harper and Wood (1957), many of them also visiting other composites and umbellifers. Clearly much pollen will be lost to the wrong species of flower by such generalist visitors, but composite flowers commonly seem to work by flowering in great abundance in any one area, so that at least some conspecific pollen is received. As with white umbellifers, a case could be made for Müllerian mimicry rings such that visitors need only one search image at any one time; hence all the white and yellow daisies form one grouping, all the pure yellow dandelions, hawkbits, hawk's-beards, and cats-ears form another, and perhaps all the purple/blue thistle types form a third, with these three groups recurring across habitats, and indeed across continents.

Some of the many visitors to composite capitula may be unwelcome, especially those insects that lay their eggs in flower heads (chapter 24); a composite blossom provides a particularly good food store and

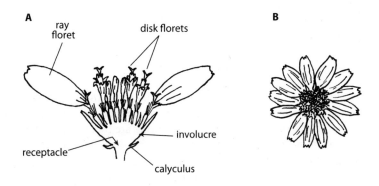

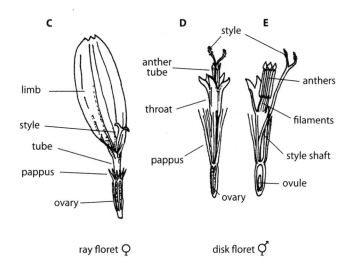

Figure 2.25 Ray and disc florets in Asteraceae: (A) the basic anatomy, and some of the special terminology used; (B) an overhead view of a typical radiate version; (C) a single ray floret; (D) intact single disk floret; (E) single disk floret cut away to show arrangement of anther and filaments and style.

protected environment for an insect larva to develop in. Many composites are indeed targeted by ovipositing female flies, beetles, and moths, leading to infestations of eggs and then larvae that put all the seeds at risk. This may be one constraint on the success of the Asteraceae and on the size of capitula that they have produced (M. Proctor et al. 1996).

Trap Flowers

Trap flowers represent very specialized forms of floral architecture. They are highly variable as a category and involved in particularly unusual kinds of pollinator interactions where a visitor is detained within a chamber of the flower for a more or less fixed period and only allowed out in a manner that ensures pollen

pickup. They are therefore dealt with separately in chapter 23.

7. Flower Size and Size Range

As with shape, flower size can be considered either as a function of one flower or of an inflorescence or flower head. It is helpful to deal briefly with inflorescences first, since evidently the formation of an inflorescence can be an important way of achieving an effective large display size while retaining individually small flowers. Indeed, apparent size can be enhanced by what is effectively cheating—having showy peripheral sterile flowers (plate 11D,E) around a core of functional fertile flowers, as in some *Viburnum* or *Hydrangea*, or sterile ray flowers around fertile true flowers in the aster and daisy family (plates 8F–H and 9A,B).

There are some obvious physical constraints on individual flower size. Small **annuals** cannot support very large flowers, but trees and large shrubs can do so. Plants that can allow their flowers to lie on the ground, or to float on water, can also support much larger floral size. Note also that large flowers are more costly in terms of plant metabolites and water, and the balance of pollinator preference and plant constraint can often affect flower size differently in varying habitats (Galen 1999a). For example, rosemary flowers are smaller in the drought conditions of Spanish coasts but larger at higher elevations in moister mountain sites (J. Herrera 2005), though flower visitors at both sites prefer larger flowers. Insolation also affects flower size or overall floral display directly (e.g., S. Cunningham 1997).

Within the constraints of physical support and shape, flower sizes nevertheless come in an enormous range, and there may be four or five orders of magnitude difference in flower size within quite small taxa. Barkman et al. (2008) found that rapid change in floral size was possible over evolutionary time, with the largest flowers often increasing in size the fastest (*Rafflesia* by about 20 cm in diameter every million years). But design may have to change markedly with size, so that closely related flowers of very different sizes may appear unexpectedly divergent. This issue of **allometry** (how shape may change with size and scaling effects) is well known in animals but has largely been ignored for flowers. J. Herrera (2009) reported almost isometric corolla size relationships in a restricted range of **entomophilous** Mediterranean species, though with a higher allocation to corolla tissue in bilateral forms compared with radial forms. Across a larger size range, mechanical stability does require different kinds of support in different size ranges, and the largest flowers therefore tend to have fused perianth organs to provide basal stability, whereas the free (unfused) upper parts are kept relatively short (less so if pendulous). Large petals may also be strengthened by ribs of thicker tissue to keep them spread open, a feature noticeable in the *Convolvulus* group, where larger versions of the flowers have distinctive radiating ribs (plate 3A).

But exactly what we measure as flower size needs some consideration. Corolla *diameter* is the most commonly assessed feature, and certainly it is most often what matters in terms of advertisement to an incoming visitor. But corolla diameter does not take into account contour or edge effects, which often have more effect on insects (chapter 5) and may not reflect the true area that is being responded to, especially if used comparatively for different designs of flower. Here some kind of planar projection score is needed, but it is hard to use in field studies, and for tubular flowers with complex frontal views some composite of flower size tends to be used (e.g., width by breadth). Tube *length* is another aspect of size; it is rarely important for advertisement but is crucial in terms of access in relation to tongue length. A composite measure of flower *mass*, or total flower **biomass**, for a plant is another alternative; it is easy to measure though not very clearly related to either advertisement or access, especially where petals are thick and waxy. There is no single easy answer here, but corolla diameter and tube length remain the most commonly used measures.

The smallest flowers are barely visible to the human eye, reduced to a tiny group of sex organs with almost no display from petals or sepals (many of these being wind or water pollinated); the largest (species of *Rafflesia*, which are always found at ground level; plate 22G,H) may be very nearly a meter in diameter, while *Amorphophallus* blooms (plate 22E) can reach 3 meters in height. Davis et al. (2008) in reviewing floral gigantism pointed out that it is most commonly associated with small beetle- and carrionfly-pollinated species.

Clearly in these cases there is also some link with pollinator type or syndrome. A larger flower is essential if a bird or bat is the main visitor since the flower must accommodate a big enough reward to pay the animal and must be strong enough to withstand its activities; although how large the corolla (or the corolla opening) must be will depend on whether the animal then inserts its whole head or just its tongue. Smaller flowers containing smaller rewards will potentially suit most invertebrate visitors, but the apparent flower or inflorescence size (the front-on display) may be influenced by how far the main visitors are visual foragers. Relative to other plant parts, flower size might be more strongly constrained (by the need for a fit with particular pollinators), and it does indeed show lower variability within species than does leaf size, especially in zygomorphic species as compared with radial species (van Kleunen et al. 2008).

The size of individual flowers is likely to be strongly interactive with the size of a plant and with how many flowers are present at a time (again linked to strength and support issues); and these two factors together determine the overall display. Flower size is therefore strongly tied in with flowering patterns and with aspects of visitor type and preferences, and it is discussed more specifically in chapter 21.

8. Flower Sex and Flower Design

Thus far we have mostly considered the design of sexually "typical" hermaphrodite flowers, those with both male and female parts. While these are the most common form encountered, they pose some problems to pollination biologists looking for selective pressures on design, because selection through male function may operate in different directions from that through female function. For example, color and shape can have different consequences for each gamete type; selection through male function (pollen dispersal) may favor wide corollas, while female function (pollen receipt) may be better served with narrower more restrictive corollas (D. Campbell 1989a,b), selecting for more specialist visitors. It is therefore often easier to investigate selection effects on floral features using species with unisexual flowers, and they have consequently received more than their fair share of attention.

Single-Sex Flowers

Many flowering plants have separate staminate (male) and pistillate (female) flowers. These may occur either on the same plant, giving a **monoecious** condition, or on completely separate plants, in which case the species is strictly **dioecious** (like most animals). Further details and terminology for plant sexual systems are considered in the next chapter; for now it can be noted that complexity arises because there are also plants with both male and hermaphrodite (**androdioecious**), or both female and hermaphrodite (**gynodioecious**), flowers.

In some plant families unisexual flowers are the norm, and then the two types are usually structurally very distinct, showing **sexual dimorphism** (Delph 1996; fig. 2.26). Classic examples are found in many temperate trees, such as the dioecious willows and the monoecious hazels, which have showy male catkins

and very nondescript, small female flowers (chapter 19). In other cases only a few members of a taxon are unisexual; this includes some familiar *Silene* and *Potentilla* species, where the two sexes of flower are not very distinctively different, and there may be vestiges of the stamens in a female flower and of the pistils in a male flower.

It has usually been reported that male flowers are routinely larger and showier than female flowers, an observation first noted explicitly by Darwin (1877) and one suggesting that selection has been driven more strongly by pollen dispersal rather than pollen receipt. This would be in accordance with **Bateman's principle** that male success is usually limited by access to mates, and female success by access to resources (Bateman 1948; G. Bell 1985). Traditional functional adaptive explanations for large males needed some revision after the findings of Plack (1957, 1958) with gynodioecious *Glechoma* plants, where removing anthers from hermaphrodites resulted in decreased corolla size, fully **emasculated** (all anthers removed) flowers being similar in size to the naturally occurring female flowers, though they could be restored to their normal, larger size when treated with the hormone **gibberellic acid**. This result suggests a proximate cause for a correlation between pollen presence and larger flower size, with anther-derived hormones affecting corolla size. Hormonal effects on phenotype could of course still be adaptive; the link between anther number and corolla size may indicate some inbuilt constraint, but the strength of this link could itself be subject to selection (Stanton and Galloway 1990).

G. Bell (1985) drew together data on 79 plant species and found that in 74 cases the female corollas were smaller and, in most cases, were also fewer in number. However a review by Delph (1996) covered 552 monoecious and 367 dioecious species from a total of 102 families and indicated that the "males are larger" view is substantially oversimplified, sexual dimorphism being in reality much more variable. Using a simple form of **comparative analysis** to correct for phylogenetic effects, she found **size dimorphism** in 85% of the cases, but only slightly more male-larger-than-female situations than the reverse, and with no differences between animal- and wind-pollinated species (fig. 2.27). Once habitat was factored in, though, it became clear that males were larger than females more often in temperate flowers (fig. 2.27B), with no such effect in tropical species. Bawa and Opler (1975)

A

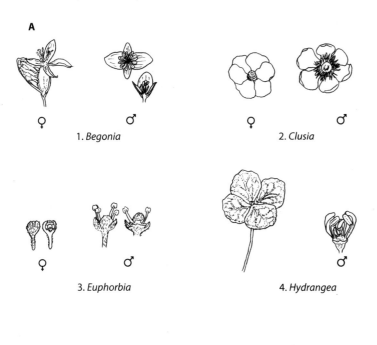

1. *Begonia*

2. *Clusia*

3. *Euphorbia*

4. *Hydrangea*

B

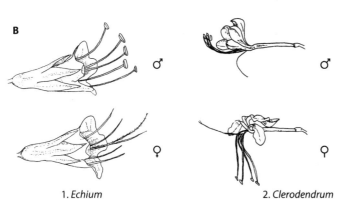

1. *Echium*

2. *Clerodendrum*

Figure 2.26 Unisexual flowers or flower stages: (A) structures of unisexual flowers, with separate pistillate or staminate versions, or with showy outer sterile flowers and inner male flowers in *Hydrangea*; (B) effectively unisexual stages in protandrous flowers of (1) *Echium*, in which the anther heads are lost and the style elongates in the later female phase and (2) *Clerodendrum*, in which the anthers droop down and the style elongates and becomes erect to give the female phase. (*Clerodendrum* redrawn from Endress 1994.)

had suggested that in tropical trees, large female flowers were favored because the larger petals were needed to enclose and protect the relatively larger ovaries and/or nectaries. This has not been explicitly tested, and other explanations may be found. For example, it might be that larger ovaries are expensive to produce, which might select for fewer flowers and then require those few flowers to be more attractive (Delph 1996); this would also fit with observations about the anomalous temperate *Silene latifolia*, which has about 15 times as many small male flowers as large females, with a size-to-number trade-off (Meagher 1992). Another possibility is that there is a male-female reward skew in tropical species, making females more unre-warding and thereby selected for greater attractiveness; this might fit with the observation that about one-third of tropical species have unrewarding female flowers, which are pollinated by deceit (Renner and Feil 1993).

For gynodioecious species, where both hermaphrodite and female flowers exist on one plant, Delph (1996) found that in 98% of cases the hermaphrodite flowers were larger, and pollinators generally showed a preference for the hermaphrodite flowers (e.g., Eckhart 1991, and Delph and Lively 1992). Again, though, other differences also exist, including higher nectar rewards in the hermaphrodite flowers; and compensatory factors may occur in the female flowers, which often

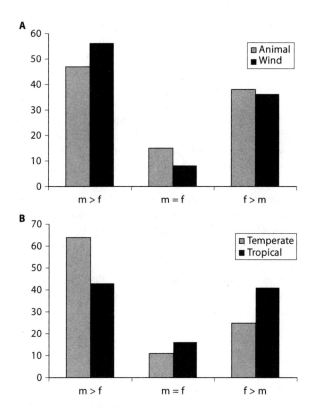

Figure 2.27 Sexual size dimorphisms in flowers: (A) biotic versus abiotic pollination types, with no significant differences; (B) temperate versus tropical types, where temperate flowers more commonly have significantly larger male flowers. (Modified from data in Delph 1996.)

persist longer on the plant (e.g., Ashman and Stanton 1991, with *Sidalcea*), or are more accessible (e.g., Atsatt and Rundel 1982, with a hummingbird-pollinated *Fuchsia*).

With both monoecism and dioecism, it is clear that flowers of the two sexes may differ either in resources or in morphology or both, so they may attract either different pollinators or the same ones at different rates. Overall, female flowers may be visited at lower frequencies than males, because females must be nectar-only (or have no reward), whereas males may have both nectar and pollen and be more attractive to visitors; but this will be confounded because some visitors only collect nectar and may not discriminate.

Floral Design in Selfing Flowers

For most plants, selfing happens accidentally when self-pollen falls or is knocked onto the stigma in the same flower or is moved in from another flower on the same plant by a visitor. This will often be blocked by self-incompatibility mechanisms or overridden by cross-pollen arriving and out-competing the self-pollen. However a few flowers in unusual circumstances are obligate self-pollinators, and others have **facultative** mechanisms to achieve or allow selfing when no cross-pollen has been received, giving a form of **reproductive assurance**. Obligate systems can rely on keeping a flower closed (so that no external pollen can touch the stigma), whereas facultative systems normally operate in open flowers and require suitable arrangements of male and female parts, both in space and time, often involving the relative movement of either anthers or style later in the flowers' life. Since all these systems are intimately connected with plant reproductive strategies, the details will be discussed in the next chapter in connection with ways of ensuring selfing.

9. Overview: Essentials of Floral Design in Relation to Cross-Pollination

A flower's design will affect not only which animals *can* feed there but also which ones *do* visit and which return. Hence the morphological design features discussed in this chapter are only part of the story; other flower traits are related to distant and local recruitment by advertisement and to management by payment of reward to ensure that a visitor gets enough benefit to encourage further visitation. Chapters 5 to 10 cover these aspects in detail, and these same themes underlie many other parts of the book. Advertisement can be visual (flower shape and size and color—chapter 5) or olfactory (flower scent—chapter 6) or occasionally tactile or auditory. Just as flower morphology has to fit a visitor, so the floral signals need to be matched to the sensory capacities of a visitor. Flowers can be seen as surfaces that act as sensory billboards (Raguso 2004a) of interacting messages.

As a brief foretaste of what is to come, we can split flowers and their type of pollination into four broad categories.

1. For flowers that use wind and water pollination, advertisements and rewards are irrelevant, so that most aspects of floral size, shape, and color can be disregarded and no specific reward need be offered. Flowers are

often green or greenish yellow, very small with no real perianth, and thus cryptic. The crucial design features include large exposed anthers that release copious pollen, and stigmas (maturing at a different time, usually later) that also are large and protruding, often feathery, and frequently very sticky with a visible glistening exudate on the surface.

2. For flowers that exploit a range of fairly generalist animal visitors, disk and bowl designs are common, as are inflorescences made up of many small open flowers, in either case with rather exposed nectaries and abundant pollen freely accessible on numerous anthers. Visitors scrabble over the surface of the flower or inflorescence, and the result is often described as **mess pollination**.

3. Flowers that use more specialized animal pollination, with most of their pollen transfer effected by just one kind of visitor (and we will meet much debate in later chapters as to how common this is!), tend to recruit animals that show high floral constancy. Here the shape and size of the flower can be matched to visitor shape and behavior, so that an incoming animal fits the flower and has appropriate "sticky" (often furry, feathery, or scaly) surfaces in the right place to receive pollen from the anthers and to then deposit it reliably on the stigma of the next flower visited. Nevertheless, the range of possible designs covers the entire spectrum described in *Particular Floral Shapes*. Tubular corollas with basal nectaries are particularly common, because the length and width of the tube can be matched to visitor size or mouthpart length or both. But some radial, cup-shaped flowers are also fairly narrow in the range of visitors they receive; for example, hellebores, where the corolla cup may contain somewhat tubular nectaries, or passionflowers, which look "open" but often have nectaries protected by lids that only forceful visitors can breach.

4. Even more specialized flowers have complex designs that exclude all but a tiny range of visitors; often these have morphologies that control the animal's movement through a flower and may even control the timing of its activities within a flower. Traps, light windows, and glue-dispensing tissues are parts of the flowers' armory in achieving these controls; in later chapters we will meet examples in the flies that visit arums, bees that visit Dutchman's pipes, and euglossine bees that visit bucket orchids.

It is worth closing this chapter on floral morphology with some consideration of the constraints that may operate. There is a considerable literature (reviewed by Conner 2006) on the extent to which floral traits are genetically correlated or are correlated with other adaptive traits in the plant, and thus are not free to diverge to an "ideal" phenotype; Conner concludes that traits can evolve under selection with surprising independence even though constrained by **genetic correlation**. There is also increasing recognition that floral traits must balance the conflicting needs of attracting pollinators and *avoiding* attracting extra florivores or herbivores (Irwin et al. 2004; and chapters 20–24), while also being affected by abiotic factors, so that flower features are often strongly constrained by multiple interacting selective forces. Hence floral morphologies need to be viewed in a more holistic framework than was customary in earlier studies.

Chapter 3

POLLINATION, MATING, AND REPRODUCTION IN PLANTS

Outline

1. Plant Fertilization
2. Plant Sex and Plant Mating Systems
3. The Benefits of Crossing and Selfing in Plants
4. Methods for Avoiding Selfing within a Flower
5. Methods for Avoiding Selfing between Flowers within a Plant
6. Methods for Ensuring Selfing

———————

Reproduction in plants, as in most organisms, can be either sexual or **asexual**, but the generation of new variants (which is the underlying necessity for adaptation to new or changing conditions and for evolutionary change) requires that at some point in the life cycle sexual reproduction occurs. **Diploid** cells must undergo **meiotic division** to produce haploid cells (gametes) with half the normal **chromosome number**, and a gamete from one individual must then interact with another gamete (preferably but not necessarily from another individual) to effect fertilization and so reinstate the diploid state, resulting in an embryo that grows into the mature individual.

For plants, which are not in themselves motile, the essential problem is that of moving immotile male gametes to an equally immotile female gamete; this is the process of pollination, with the essential end result of fertilization and then seed production. **Ephemeral** and annual **plants** must be pollinated each year if they are to adapt and survive. In contrast, some trees and shrubs live for decades, or even centuries, and may only flower very occasionally, while others spread as **vegetative clones** over large areas and very long periods; in these cases pollination is just an occasional and seemingly tiny part of the life cycle, but even so, it must at some point occur if new sites are to be colonized before the adult plant eventually dies. Pollination is therefore a crucial feature, on some timescale, of all flowering plants.

As described in the previous chapter, in angiosperms the pollen grain is the male gamete, the equivalent of a **spore** in simpler plants. The ovule (egg) contains the female gamete. Pollination is the process whereby pollen grains from one flower get to the stigmatic surface of another flower, so that a pollen tube can germinate, grow, and penetrate the ovary, delivering a male gamete that effects fertilization and so gives rise to a **zygote** (the diploid embryonic seed). If the pollen lands within the same flower that produced it, or on a different flower from the same plant, then **self-pollination** occurs and **self-fertilization** (or **autogamy**) ensues; only if the pollen is transported to a flower on an entirely separate (genetically distinct) plant do we see *cross-pollination* and **cross-fertilization** (or **allogamy**, also known as *xenogamy*), potentially followed by the full benefits of sexual reproduction, with viable and variable seeds. Cross-fertilization requires that the pollen comes from another individual, but of course from the same species (i.e., is conspecific rather than **heterospecific**). To achieve this, the plant needs to employ some mobile agent, either a moving abiotic agent (wind or water currents) or a mobile animal. Animals have the advantage of directed movement, sometimes to the extent of considerable fidelity to a particular plant species, so that wasteful heterospecific pollen transport onto stigmas may be much reduced.

Pollination biology is often seen as largely concerned with how plants ensure that conspecific cross-fertilization, rather than selfing, occurs. However it is important to remember two other points. First, successful cross-fertilization is often mainly due to a plant's self-incompatibility to its own pollen rather than to pollination specialization, so that post-pollination events alter the apparent mating patterns resulting from pollen transfer. Second, selfing can serve as a very useful strategy for many plants and especially as a fail-safe when crossing has failed or has become impossible at particular times or in difficult environments. Thus there is a whole range of strategies in plants, from a complete barrier to selfing in some species to an almost inevitable self-fertilization in flowers that dehisce pollen onto their own receptive stigmas even before the flower bud opens.

1. Plant Fertilization

Pollen is the plant's male gametophyte, and within a pollen grain there are two haploid nuclei, one vegetative and one generative, the latter acting as the male gamete. The ovule is the small female gametophyte, primitively made up of eight cells, one of which acts as the haploid female gamete. Pollination requires that pollen grains land on the stigmatic surface, which connects via the style to the basal ovary. Most of this book is concerned only with pollination itself, the process of getting the right pollen to the stigma. The events that follow, leading to fertilization, are reviewed here only briefly (while recognizing that post-pollination events are often a major key to plant reproduction).

Pollen interactions with the stigma and style are the crucial intermediaries between pollination and fertilization and were reviewed by Lord and Russell (2002). After a pollen tube lands on a conspecific stigma with which it is compatible, it will take up water from the surface (the hydration stage) and then germinate. Germination results in the emergence of a pollen tube from one of the pores on the pollen grain (occasionally, in just a few plant families, more than one tube emerges, and all but one then serve solely in anchorage). The pollen tube penetrates and grows down into the style, and the **vegetative nucleus** of the pollen grain passes into the tube, followed by two gametes formed by the splitting of the **generative nucleus** just before, or just after, the grain germinates.

The tube elongates until its tip penetrates the ovary at its apical micropyle (chapter 2); the vegetative nucleus then disintegrates, but the two male gametes are delivered, one of which effects fertilization to give a diploid zygote, which can then develop into the embryonic form of the future seed. The other gamete fuses with a diploid nucleus within the ovary, and the resulting cell divides repeatedly to form the (triploid) **endosperm** tissue, which provides the nutrition for the developing embryo (often a substantial food resource, as in many cereal and legume seeds). This *double fertilization* system, where two male gametes fuse with two different nuclei in the female, is typical of angiosperms but rare in other plant groups.

The control of germination is complex and largely determined by compatibility effects, which are discussed in the section *Genetic Separation by Self-Incompatibility*, below.

2. Plant Sex and Plant Mating Systems

Sexual Systems

The great majority of modern plants are **bisexual** or hermaphroditic, that is, they have stamens (with pollen) and carpels (with ovules) all present in the same flower; this is the condition in at least 80% of all angiosperms, and the benefits in terms of single-visit **pollination effectiveness** are obvious. But in the remaining 15%–20%, a great range of other unisexual or polygamous conditions are found, with about 6% of angiosperms (Renner and Ricklefs 1995) having two quite separate sexes (dioecy—the familiar condition seen in most animals) and a further 10%–15% having a range of more "mixed" strategies. Table 3.1 summarizes the possibilities, with the many variations on monoecy and dioecy, and gives some examples. In dioecious species the male plants tend to start flowering earlier, whereas in monoecious species it is common for female flowers to open before male flowers (Stephenson and Bertin 1983). The advantages of dioecy include easier avoidance of inbreeding and higher **fruit set** compared with monoecious or hermaphrodite plants. The disadvantages are that the animal visitors need to move between individuals of both sexes, so the plants become even more sensitive to changing

Table 3.1
Patterns of Sexuality in Flowers and Plants

Sexual type	Flower types present on any one plant*			Examples
	Male-only (staminate)	Female-only (pistillate)	Hermaphrodite	
Bisexual flowers (hermaphrodite)	-	-	+	80% of all angiosperms
Unisexual flowers				
Monoecious	+	+	-	Many trees (Fagaceae, Betulaceae); Cucurbita; Euphorbia
Gynomonoecious	-	+	+	Some Asteraceae (female ray and bisexual disc florets)
Andromonoecious	+	-	+	Many Apiaceae (umbellifers); some Asteraceae; Solanum
Dioecious	+	-	-	Many Salicaceae; Urtica; Ilex; some Silene
	-	+	-	
Gynodioecious	-	+	-	Many Lamiaceae; Plantago, some Saxifraga
	-	-	+	
Androdioecious	+	-	-	Rare: Phillyrea, Datisca
	-	-	+	
Polygamous				
Polygamomonoecious (on same plant)	+	+	+	Rare: Cocos nucifera (coconut palm)
Polygamodioecious (on separate plants)	+	+	+	Rare: Sanguisorba (salad burnet)

*Some plants also have a proportion of a fourth category of sterile flowers, usually at the edges of inflorescences, adding to the floral display but contributing little to resource costs: some Asteraceae, also Viburnum, Hydrangea. (See Pseudoflowers in chapter 23.)

pollinator abundance, and pollinator and/or pollen limitation may be common (see below).

These variations inevitably have marked effects on pollination systems when compared with the norm of hermaphrodite flowers, and dioecy and its variants are more common in abiotically pollinated plants and (perhaps more surprisingly) in climbing plants (Renner and Ricklefs 1995; J. Thomson and Brunet 1990). Furthermore, selection for type of mating system in a given plant species can be brought about by pollinator activity but may also be influenced by the activities of other biotic agents; for example, herbivores often damage male or hermaphrodite plants more than female plants (Ågren et al. 1999; Asikainen and Mutikainen 2005).

Some plants can change sex, and doing so may depend not only on their genetic makeup but also on environment, including factors such as day length, water availability, soil nutrient levels, and **plant growth substances** (Meagher 1988). It is therefore not uncommon for a normally male plant to produce a few female flowers, or vice versa. The **perennial** jack-in-the-pulpit, *Arisaema triphyllum*, can change sex as it grows, from male to female, but it can reverse this change if the plant becomes smaller again (Bierzychudek 1982). Recent evidence indicates that sex expression is primarily controlled by a small group of genes (**MADS**) that clearly diversified within plant taxa well before the origin of the angiosperms (Lawton-Rauh et al. 2000; Glover 2007). Flowering plants thus inherited an ability to switch sexual systems rather readily, which probably facilitated the evolution of diverse selfing and crossing strategies for multiplication. Sex expression in plants is clearly highly plastic, a fact that can be useful in greenhouse crop husbandry to control the reproductive system, but one that requires pollination biologists to keep a careful eye on sexual behavior of their study plants.

It should be noted that many plants also have asexual reproductive options, varying from simple **vegetative** spread (e.g., using **rhizomes** or runners or **budding** from **bulbs**) to the production of **bulbils** (new **viviparous** plantlets produced in place of an inflorescence, as in some *Allium, Lilium,* and *Saxifraga*) to **agamospermy** or **apomixis** (producing seeds without fertilization, as can occur in *Ranunculus, Citrus,* and some Rosaceae and Asteraceae). Such systems are not especially relevant to pollination biology, though they may provide an alternative where pollination is limited by weather or by low pollinator availability.

Because most flowers are hermaphrodites, they carry out both the male and the female roles, and so contribute the plant's genes to the next generation in two different ways. These two different roles have rather different costs and are subject to different selection pressures. The male function is usually cheaper, involving spreading the pollen successfully to as many other conspecifics as possible; the costs are those of attractants for pollinators and of making the pollen. The female function, proceeding through the setting, growth, maturation, and then dispersal of seed, is usually considerably more expensive in terms of both energy investment and time; and where separate sex flowers occur, the females are often smaller and fewer, as discussed above, perhaps reflecting this greater expense (G. Allen and Antos 1993). But these issues are rarely that simple; all studies to date that have measured selective forces for both male and female fitness have found them to have separate and often contrasting effects on flowers and often to vary substantially between sites and between years (Conner 2006).

Plant Mating Systems

Plants tend to employ animal **vectors** to move their pollen around and can exploit the animals' behavioral flexibility by their use of advertising signals and rewards to improve an individual plant's own mating system. Animals tend to move both within and among neighboring plants, and their efficiency of pollen carriage varies markedly, so that most plants end up with a "mixed" mating strategy at the point of being pollinated (Harder and Barrett 1996; Goodwillie et al. 2005), with some uniparental or biparental inbreeding. In self-incompatible plants this can then be "improved" to become a system dominated by outcrossing, although this should not be viewed as a common endpoint; Vogler and Kalisz (2001) reviewed available data to show that most plants achieve a mixture of selfing and outcrossing, with 49% of all animal-pollinated species lying between 20% and 80% outcrossing.

To some extent, plants can control their own mating opportunities by the way they manipulate their visitors, and in this sense can indulge in **mate choice**, just as many animals do. Attracting a larger number of each of a range of pollinators is one option and is mainly determined by the advertising traits and the rewards offered. Restricting pollen removal at any one visit can have similar effects and is determined by another set of floral traits: anther positioning, the mechanisms and timing of dehiscence, and again the patterns of reward offering, which affect visitor behavior and visit length. Pollen acceptance in the maternal plant also exerts an influence, varying according to the duration and area of stigma receptivity and of stigma-visitor contact. Hence the relation between visitation, pollination, and resultant fertilization is rarely as simple as may at first appear. Many of these issues are addressed more fully in later chapters.

Female Functions in Flowers

The female function and female success of plants have traditionally been much easier to measure, most simply by counting (and often weighing) the seeds on the parent plant; this led in the early literature to an overemphasis on female success as a measure of pollination success. In many studies, assessments of the important of pollination and pollinators concentrated on how or whether pollen receipt was limiting to fruit or seed set. Even there, it is easy to oversimplify: one confounding variable is the widespread abortion and shedding of "surplus" seeds on many plants where space or nutrients are limiting, since in trees and shrubs with big floral displays it can be impossible for the parent to support all the fruits that result, with as many as 90% **aborted** in some years (e.g., *Banksia*, Vaughton 1991). Hence even groups with characteristically low seed set, such as the asclepiads (Wyatt and Broyles 1994), can be successful and even invasive. In many **herbaceous plants**, though, there is good evidence that additional experimental **hand-pollination** of flowers does lead to higher fruit set than in open-pollinated plants, implying that pollination is often limiting on female function. A review by Burd (1994) estimated that of 258 species, at least 62% had their seed set limited by pollination at some stage. Again though, caution is needed: it is crucial to ensure that resources are not the real constraint across both time and space, because some plants may have low seed set one year as a result of high seed set in previous years (J. Zimmerman and Aide 1989; Primack and Hall 1990), and some can divert resources from other parts of the plant into seed set under certain conditions. These issues were taken into account in models generated by Calvo and Horvitz (1990), and they also concluded that pollination would

be limiting for many species. Nevertheless, seed set in one or a few seasons is only part of the story; **seed dispersal**, germination, and establishment are at least as important, and many plants suffer high mortality rates at all these stages. Thus, in an ideal world all are agreed that it is the female **lifetime reproductive success** of the whole plant that really needs to be measured, and this is often exceedingly difficult.

Male Functions in Flowers: Implications for Gene Flow

Male function depends on dispatching pollen into the world, and most of the attractive and rewarding features of flowers are selected to enhance the male role as a **pollen donor** (Stanton et al. 1986). However male function and male fitness are rather hard to assess, requiring estimates of pollen movement through a floral community. Except with the large and easily tracked pollinia of orchids or asclepiads, this was difficult to measure in the past with any certainty (Snow and Lewis 1993; Klinkhamer et al. 1994), until the advent of automated particle counters that could handle normal pollen grains. Pollen dispersal and gene flow are not the same thing, of course (D. Campbell 1991). Dispersal measured as a visitor's flight distance between plants ignores **pollen carryover** between successive plants and **pollen competition** within styles after arrival in a flower; all are key aspects of male success. Looked at another way, pollen produced in and dispensed from a flower can have a range of possible fates (fig. 3.1). Some may never be removed from the flower (though potentially the remaining pollen could be involved in autonomous or facilitated selfing, which will be discussed in *The Benefits of Crossing and Selfing in Plants*), some may be lost to wind or rain, and some may be eaten at the flower. Thus only a small proportion may be "properly" exported from the flower and dispersed. S. Johnson et al. (2005) measured these components for *Disa* orchids (fig. 3.2) and found that only 3% of the pollen was properly exported. Even then some may be lost during transport (dislodged by the airstream or by pollinator movements, including specific grooming), and if carried on a bee, much may end up as food in a larval cell. Harder (2000) found that only 1% of typical pollens reached conspecific stigmas, the result of high removal and transport losses; in fact it is a rather common finding for many plants

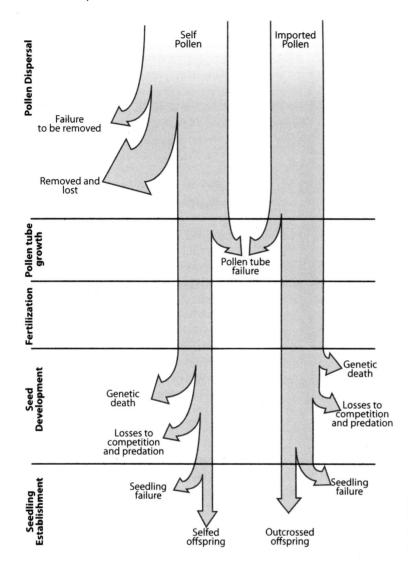

Figure 3.1 The possible range of self- and cross-pollen grain fates from fertilization to seedling establishment. (Modified from Harder and Routley 2006.)

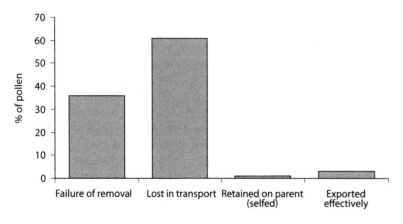

Figure 3.2 Analysis of pollen fate in *Disa cooperi* orchids; only about 3% of pollen is exported, and 1% achieves facilitated self-pollination. (Modified from data in S. Johnson et al. 2005.)

that less than 1% of a plant's pollen ever reaches a stigma (D. Levin and Berube 1972; Galen and Stanton 1989; Harder and Thomson 1989; Wolfe and Barrett 1989; H. Young and Stanton 1990a). And quite high proportions of pollen grains that are effectively dispersed to another conspecific flower may still be lost (in an evolutionary sense) in various other ways at a post-pollination stage.

Measuring Pollen Dispersal

Pollen dispersal can be seen as a composite of three components:

1. Distances travelled between plants by visitors
2. Directions traveled
3. Pollen carryover from one flower to another on the visitor's body (Lertzman and Gass 1983)

Each of these has often been studied separately, but they have rarely been analyzed together in a given system. Most of the emphasis until recently was on comparative assessments of the amounts and diversity of pollen carried on animal bodies, coupled with estimates of the animals' movements within and between flowers. But there are a great many variables that can complicate such estimates of animal behaviors, many of which will be discussed in later chapters (especially chapters 21 and 22). Furthermore, direct measurement of pollen dispersal or flow has been rather difficult, relying on techniques that are known to be oversimplistic. In particular, flowers were often emasculated to ensure that self-pollen did not interfere with estimates of incoming pollen, but it is known that emasculation can itself alter visitor behavior and pollen dispersal (M. Price and Waser 1982; J. Thomson 1986; Morris et al. 1994). In practice such studies have rarely extended to the analysis of real pollination effectiveness of a given visitor or real pollen flow for a given plant. However, there are a few laudable exceptions, such as J. Thomson's work (1986) on *Erythronium*, where pollen deposition on successive stigmas was directly measured and shown to vary positively with bumblebee size, with bee visit duration, and with nectar volume but to vary negatively with bee grooming. (Much more on the definition and measurement of pollination effectiveness can be found in chapters 11 and 20.)

A common alternative for analysis of pollen dispersal has been the use of artificially applied powders as pollen mimics, which can be placed on the anthers of known flowers and then seen when deposited in other flowers. Fluorescent dye powders were traditionally favored because they showed up clearly when the flowers were lit with an **ultraviolet** (UV) source (e.g., Waser and Price 1982, 1984; Hessing 1988; Morris et al. 1994). There are reasonable or even very good correlations between dye and pollen movements in some studies, especially where passive pollen collection is involved (D. Campbell et al. 1991; Fenster et al. 1996; Rademaker et al. 1997; Adler and Irwin 2006). However, such powders sometimes do not really behave like pollen (e.g., Waser 1988), as shown in figure 3.3A, and for bee-pollinated plants they may be more or less like pollen depending on which bee is doing the collecting (fig. 3.3B).

Attempts to measure pollen flow have also been made using genetic markers that leave a visible cue: Griffiths (1950) used rye grass where the pollen-donating parent had distinctive red shoot bases and showed pollen movement of up to 30 m (by wind pollination in this case).

A different approach has been the mathematical modeling of pollen movement, which relies on the kind of estimates of pollen carriage on animals and animal movements mentioned above. Lertzman and Gass (1983) specifically drew attention to the deficiencies of early models; in particular, these were flawed by assumptions that all or nearly all pollen was deposited on the next flower visited (i.e., with no carryover), therefore greatly underestimating pollen flow. Lertzman and Gass pointed out that work with both hummingbirds and bumblebees (e.g., J. Thomson and Plowright 1980) had indicated pollen carryover occurring both significantly further and much more variably. Thus models explicitly incorporating measurements of pollen carryover between successive flowers (discussed further below) were advocated. A simple model is shown in figure 3.4A. Harder and Barrett (1996) improved matters further with a model that took account of post-flower-visit behaviors, especially

1. grooming by bees and flies,
2. **proboscis** coiling by lepidopterans which can dislodge pollen, and
3. older pollen burial beneath newer deposits on larger vertebrate visitors.

Hence in this model the pollen on the body was effectively seen as being in two compartments, one portion exposed and one portion (after grooming, etc.)

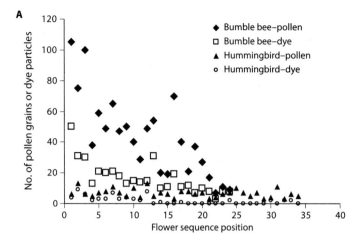

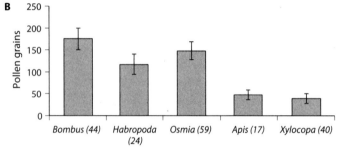

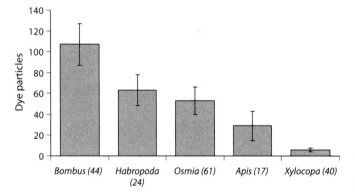

Figure 3.3 Evidence that pollen does not necessarily behave in the same manner as dyes applied to flowers: (A) in terms of movements to successive flowers of *Delphinium nelsonii*, by bees and birds, with pollen in solid symbols and dye as open symbols (modified from data in Waser 1988); (B) in terms of different genera of visiting bees with pollen above and dye below (SEs also shown, and with values of *n* in parentheses (modified from Adler and Irwin 2006).

that was in a "safe" site, now being unavailable for deposition on a stigma (fig. 3.4B). The pollen that accumulated in layers could also be modeled as in figure 3.4C (Harder and Wilson 1998). Hoyle and Cresswell (2006) pointed out that there was also scope for some secondary pollen dispersal, where grains were deposited in one flower (on the stigma or even on the petals) and were then picked up and moved on again by a subsequent visitor; although in practice they found this effect to be very small, at least in *Brassica* species.

With the advent of improved molecular techniques that provide genetic markers, there are options now for detailed genetic analysis of pollen origin, of **paternity analysis** for plants, and of source variation in the pollen loads on animals, allowing real estimations of pollen flow and thus male-function gene flow. Our understanding has been particularly enhanced by genetic paternity analysis with more discriminating tools such as amplified fragment length polymorphism (AFLP) and microsatellites (Adams et al. 1992; Austerlitz et al. 2004; Burczyk et al. 2006), and more recently still, by analytical tools such as TwoGener and developments

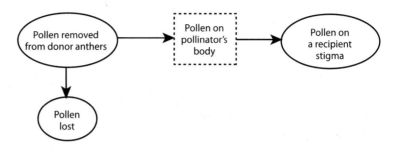

A Single compartment model

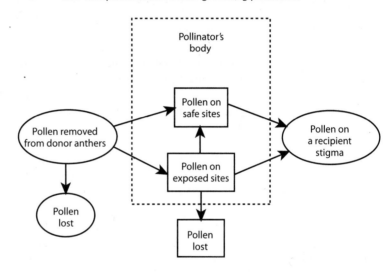

B Two-compartment model for grooming pollinators

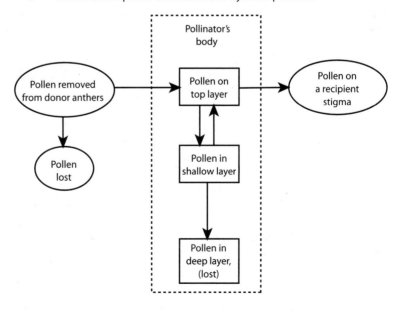

C Model where pollen accumulates in layers on pollinator

Figure 3.4 Models for pollen transport: (A) simple passive transport; (B) two-compartment model, in which some pollen acquired is transferred to a safe site and unavailable for deposition on flowers; (C) model with pollen accumulating in layers on a visiting animal's body. (Modified from Harder and Barrett 1996; and Harder and Wilson 1998.)

thereof (Robledo-Arnuncio et al. 2006). These kinds of study have also stressed the importance not merely of citing mean pollen flow measures but also of considering real and effective pollen flow, because it may only take one pollen grain going a much greater-than-usual distance to sire an offspring. To give some perspective, Ellstrand and Marshall (1985) measured paternity in radish seeds and found that pollen flow measured genetically was more than twice as great as that estimated from pollinator foraging or from fluorescent dyes. But Ellstrand (1992) and his various collaborators further showed that while most pollen might move only very locally, some pollen could travel between populations more than 150 m apart via insect vectors (Marshall and Ellstrand 1988; Devlin and Ellstrand 1990) and up to 5 km when bird pollinators were involved (e.g., Byrne et al. 2007). Krauss (2000) pointed out the implications for conservation strategies of these new inputs toward understanding mating patterns and paternity.

Cresswell (2005, 2006; Cresswell and Hoyle 2006) modeled the problem, reviewed the available data, and concluded that gene dispersal via pollen shows a leptokurtic decay from a point source, though the scale of this may be meters or kilometers (fig. 3.5), a finding that endorsed those reported in previous reviews (e.g., Morris et al. 1995). The number of flowers in the plant population also affects the outcome, with a less sharp decay pattern as flower number increases, so that (as expected) pollen from external populations has much less chance of siring offspring in larger flower patches. And superimposed on these patterns is the inevitable variability in male success, with certain individuals flowering far more of the **progeny** than others (fig. 3.5C).

Most studies have also found higher levels of multiple paternity than might have been expected. For example, D. Campbell (1998) recorded a 68–100% occurrence of multiple siring in *Ipomopsis* seeds, with 4–9 sires per seed, which is well matched with observed hummingbird movement patterns. Mitchell et al. (2005) found that 95% of *Mimulus* fruits in their experimental arrays had multiple sires, and the mean value per fruit was 4.6, in this case probably resulting from multiple probes to flowers by bumblebees, each delivering pollen from 1 to 3 fathers. Michaelson-Yeates et al. (1997) compared the effect of *Apis* and *Bombus* visits on paternity for clover (*Trifolium repens*), finding the numbers of fathers per pod to be 1, 2 or 3–4 for 0%, 50%, and 25%, respectively, after *Apis* visits, and 10%, 30%, and 50% after *Bombus* visits.

More recently still, Matsuki et al. (2008) used direct measures of pollen grain genetics with *Magnolia obovata* and were able to show that much of the pollen on bees and on small beetles was self-pollen, whereas that on flower beetles (subfamily Cetoniinae) was largely cross-pollen, making the latter probably far more efficient as outcrossing pollinators (although these authors did not take the further step of assessing pollen actually deposited on stigmas). These techniques also open up new possibilities for measuring the pollination effectiveness of different visitors (chapters 11 and 20).

Note, though, that molecular genetic tests may underestimate self-pollination, because self-pollen may germinate slower, tubes may grow slower, or selfed embryos may abort; established sirings of eventual seeds may therefore show less selfing than the level of self-pollen deposited. Conversely, if self- and cross-pollen behave (germinate, grow, and fertilize) similarly within a species, there may be a confounding factor of timing, since self-pollen may be deposited from early within-plant visits when anthers are freshly dehisced, giving a higher proportion of selfed seeds, compared with later between-plant visits, when anthers are relatively depleted.

Measuring Pollen Carryover

Analysis of pollen carryover was discussed briefly above, and Lertzman and Gass (1983) gave four different models incorporating ever more of the expected variables (e.g., pollen amount on anther, stigma capacity, pollen pick-up patterns and amounts, pollen deposition patterns and amounts, and pollen layering on the animal's body). Most commonly, carryover has been estimated in the field using one "source" flower, with all others in the vicinity emasculated. Waser (1988) tried this with *Delphinium* at sites in the Rockies and found the expected decrease in pollen carryover to successive flowers visited, but also found variation with visitor. Thus for bumblebees, pollen was still detectable on the 20th flower, and for hummingbirds on the 35th, as shown in figure 3.3A. Robertson (1992) collated data from several studies and found pollen carryover values (pollen carriage beyond the first flower visited) between 50% and 99%, his own work on *Myosotis* giving a value of 90% and significant pollen

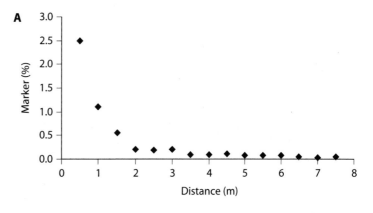

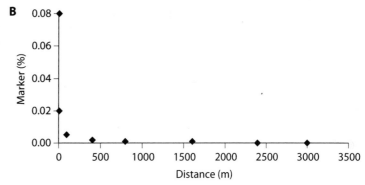

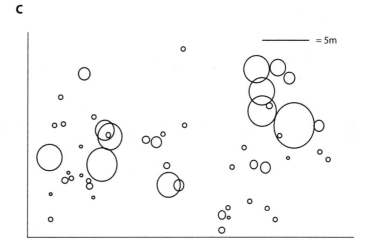

Figure 3.5 Gene dispersal measured using genetic markers and their occurrence in seed set at different distances from the source, showing a leptokurtic distribution: (A) jute plants; (B) rape plants; (C) circle diameters represent numbers of progeny attributable to individual males of *Chamaelirium*, showing high variability of paternal gene flow. (Parts (A) and (B) modified from data in Damgaard and Kjellsson 2005; part (C) modified from Meagher 1991.)

deposited (by a tachinid fly) on at least 13 further flowers.

Direct observation of natural pollen carryover, as distinct from models or tests with manipulated emasculated flowers, is rarely achieved because it is inherently difficult. It can normally only be done straightforwardly using **heteromorphic** flowers, where some charac-

teristic of the pollen differs between the flower morphs. Mixtures of pollen grains can then be counted on stigmas, and the decay curve for outcrossed pollen can be constructed for visits to previously unvisited flowers. An example with a hummingbird-visited **distylous** plant is shown in figure 3.6A (Feinsinger and Busby 1987), showing the general decline after the first visit

A

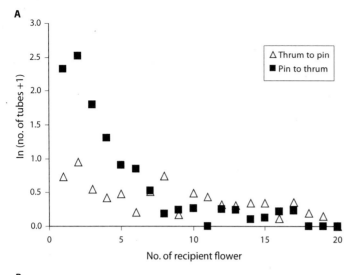

B

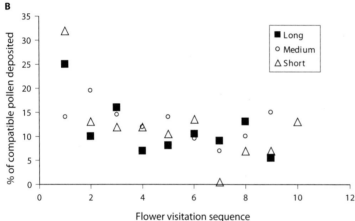

Figure 3.6 Estimates of pollen carryover: (A) with distylous *Palicourea* visited by hummingbirds (modified from Feinsinger and Busby 1987); (B) with tristylous *Pontederia* visited by bees, showing effects for long-, medium-, and short-styled flowers separately (modified from Wolfe and Barrett 1989). Note that figure 3.3A gives an example for normal (homostylous) *Delphinium* plants.

made; and figure 3.6B shows the more gradual decline in a bumblebee-visited **tristylous** plant (Wolfe and Barrett 1989). Such tests sometimes involve pollen color polymorphism, as in *Erythronium grandiflorum* (J. Thomson 1986), but in most cases this color difference rapidly disappears as the pollen germinates (chapter 7), so the work is very difficult in practice. More commonly, studies have involved flowers with **heterostyly** (see *Spatial Separation of Male and Female Parts* below), where the styles are set at different heights and dispense slightly different pollens. With such techniques, it has been shown that **geitonogamy** (self-pollen carried over between flowers but within the same plant) is quite commonplace in many species of plant and that it does increase as expected with the number of open flowers present on a plant (Hessing 1988; Dudash 1991; de Jong et al. 1992; chapter 21).

However, with genetic markers now available it is becoming possible to measure not just pollen dispersal but also pollen carryover, although for the latter any use of markers must be accompanied by detailed observations of visit patterns. Karron et al. (2009) used arrays of *Mimulus* plants, each with a unique genetic marker, to analyze pollen movement and selfing following a sequence of visits by a single bumblebee and found that the mean selfing rate increased from 21% for the initially probed flowers to 78% for the fourth flower that was probed (fig. 3.7).

Influences on Pollen Dispersal and Carryover

Obvious influences on pollen dispersal include the amount and timing of pollen available in flowers and the number of flowers per plant and per plant patch. All of these factors will interact with the attraction and

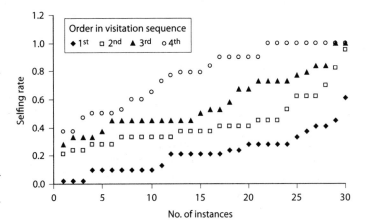

Figure 3.7 The proportion of selfing recorded in successively visited flowers of *Mimulus ringens*, each point representing a single fruit assessed for paternity. (Modified from Karron et al. 2009.)

behavior of different pollinator types: how much pollen they carry and where on the body, whether and when they groom, when they visit in relation to dehiscence, how much short-term flower constancy they show, and how far they travel between plants. A clear example of pollinator effects was shown with *Streptocarpus* species by Hughes et al. (2007), where one species in a woodland visited by long-tongued flies had a genetic structure indicating infrequent gene dispersal and limited between-population foraging, whereas another species visited by sunbirds showed much greater pollinator-mediated gene flow and belonged to a **clade** with much longer phylogenetic branching.

There are some more specific behaviors that will influence pollen dispersal, such as **territoriality** (common in several kinds of pollinator) or **trap-lining** (visiting the same sites or flowers repeatedly in specific spatiotemporal sequences, often over many successive days; again, not uncommon, and documented for various bees, butterflies, hummingbirds, and bats). Ohashi and Thomson (2009) specifically analyzed the effects of trap-lining on pollen movements and suggested that it increases variation among populations in both genetic diversity and the level of **inbreeding depression**, thus affecting rates of floral evolution.

For any one plant, the overall pollen dispersal and carryover will depend on the composition of the visiting fauna at any one time, and this will in turn depend on factors such as the weather (which matters for both wind and insect pollination). Hence any modeling of pollen dispersal and the resulting *paternity shadow* has to be remarkably complex (e.g., Cresswell 2005, 2006) if it is to be at all realistic.

Pollen Discounting

Pollen discounting is the extent to which ineffective self-pollination occurs and thereby reduces the number of pollen grains left to be involved in cross-pollination (Holsinger et al. 1984). It is controversial as to what degree pollen discounting really matters, given that pollen grains are usually so much more numerous than ovules that some selfing could normally be achieved with very limited effects on the pollen still available for outcrossing (D. Lloyd 1992; Holsinger and Thomson 1994; Harder and Barrett 1996, 2006). The latter authors point out that it is not pollen grain number that matters, but rather pollen mating opportunities. If less than 1% of all pollen that is produced ever reaches an appropriate stigma, the key question is, how much of the pollen that would otherwise reach other plants is involved (discounted) in selfing? In this light, discounting may be low for plants that bear single flowers but much more important for multiflowered plants (Ritland 1991; Kohn and Barrett 1994; Harder and Barrett 1996).

Pollen Limitation

It is often assumed that limited reproductive success in a given zoophilous plant species is the result of poor pollination or seed set in plants, which in turn is due to pollen limitation; that is, insufficient pollen receipt resulting from insufficient visits by pollen-bearing animals. Of course this limitation on reproductive output could arise from many other prezygotic and postzygotic factors; and even where it is due to inadequate pollen receipt, this could mean the receipt of poor-quality pollen (too much self-pollen or heterospecific pollen), rather than just low pollen quantity. Aizen and

Harder (2007) modeled these aspects to show that quantity limitation is likely only at low ranges of pollen receipt, whereas quality limitation can be important at all ranges. Most measurements of pollen limitation are made by supplementing cross-pollen receipt by hand-pollination, a method that does not adequately distinguish between these causes. Furthermore, much of the literature does not really distinguish between pollen limitation and **pollinator limitation** in practice. Thus, while we noted earlier that at least 60% of over 250 plants investigated have been reported to show limitation of fruit set as a result of a less-than-ideal pollination interaction (Burd 1994; reviews by Ashman et al. 2004; Knight, Steets, et al. 2005), this figure is somewhat suspect for the reasons given above. Knight et al. (2006) also pointed out the biases in estimates due to timing and type of sampling regimes, as well as the bias against reporting negative results.

Larson and Barrett (2000) attempted an overview of pollen limitation that took account of many of the problems, with phylogenetically corrected analyses, and were able to extract some patterns from the available data (table 3.2). Pollen limitation was less intense in self-compatible and autogamous species than in outcrossing species (as might be expected). Temperate species, species from open habitats, herbaceous species, and species offering nectar as a reward to visitors were less pollen limited, but specialized and generalized flower types showed little difference from each other. Phylogeny was found to have only a minor influence, though there were differences between families in mean scores (fig. 3.8). However, no single trait was a good predictor of the extent of pollen limitation. This is perhaps not surprising given the clear demonstrations that pollen limitation varies widely between adjacent populations and between years for any one population (see, e.g., Buide 2006).

Although the extent of pollen limitation may be variable and may have been overestimated, it is undoubtedly common, and its prevalence underlies many of the current debates in pollination ecology. The major consequences of this bottleneck to fitness are somewhat controversial; a reversion to selfing is one option (Eckert et al. 2006), and the use of delayed facultative selfing as a backup option is also reasonably common (Fenster and Martén-Rodriguez 2007), but the most frequent outcome seems to be a shift of floral traits leading to a selective shift of pollinator type (S. Johnson 2006). The latter view is supported by considering the orchids, a taxon where flowers are nearly always pollen limited but where self-pollination is rare and pollinator shifts are very common. Thus pollen limitation is a key issue at the root of the supposition of a strong selection for floral traits that confer more efficient pollination, those traits often being strongly correlated and giving rise to the suites of convergent traits that are perceived as floral syndromes.

3. The Benefits of Crossing and Selfing in Plants

Why is cross-fertilization (allogamy) usually best? This is a very similar question to that of why sex evolved, and the answer to that one has provoked enormous controversy and debate (Maynard Smith 1978; G. Bell 1982; Agrawal 2006; deVisser and Elena 2007). Restricting discussion to plants and pollination, there is no doubt that cross-fertilization provides some major benefits. First, it leads to each new individual plant inevitably having a new combination of genes, which allows for variation and adaptation to a changing environment—not just to abiotic changes such as climate or soil characteristics, but also to changes in other similarly varying and evolving organisms, especially the herbivores, parasites, and diseases against which the plant must defend itself. Second, cross-fertilization provides some defense against natural **mutations** in the genetic material: if one parent carries deleterious **recessive genes**, they can be masked by the **dominant genes** from the other partner, or if one of the two parental **genomes** becomes damaged, the effects of nonfunctional **alleles** can be masked by the correct functioning of the equivalent alleles on the **chromosome** inherited from the other parent. This **masking** in turn can occasionally allow the potential beneficial effects of a mutation to be realized, since a damaged but masked allele may code for another **polypeptide** that persists with slightly different properties, and that in time may come to serve as an **isozyme** of the original enzyme coded at that site, and may work better in slightly different conditions, so extending the environmental range of the plant. Over an even longer time scale this polypeptide could potentially take on quite new functions. A third major benefit is the avoidance of inbreeding depression, which can result from increasing homozygosity as the mean values of fitness-related parameters begin to decline. An allele promoting outcrossing will then increase in frequency if the fitness

TABLE 3.2
Factors Affecting Pollen Limitation

Factor	Mean pollen limitation	SE	Sample size
Self compatible	0.31	0.03	102
Self incompatible	0.59	0.04	66
Selfing	0.16	0.04	35
Crossing	0.38	0.03	97
Temperate	0.30	0.02	139
Tropical	0.56	0.04	85
Open habitat	0.33	0.03	82
Forest	0.42	0.03	108
Herbaceous	0.32	0.03	102
Woody	0.52	0.04	82
Nectar present	0.35	0.03	91
Nectar absent	0.47	0.05	42
Unspecialized flowers	0.38	0.03	130
Specialized flowers	0.42	0.03	94

Source: Modified from Larson and Barrett 2000.

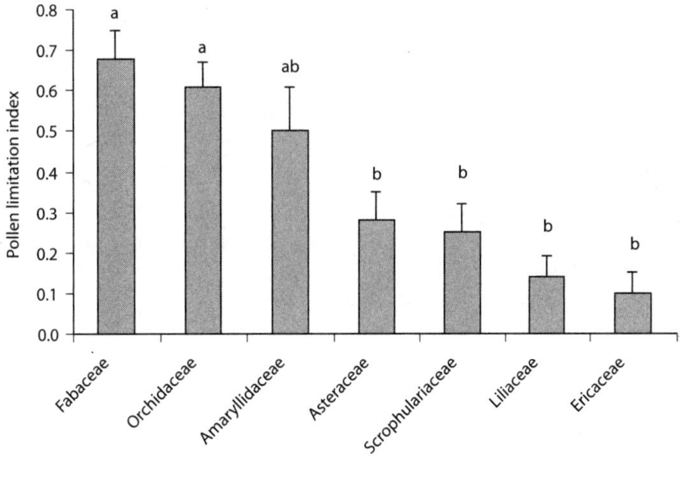

Figure 3.8 Mean values of pollen limitation estimated for seven important angiosperm families. SEs are shown, and statistically homogeneous groups carry the same letter above the bars. (Modified from Larson and Barrett 2000.)

of the outcrossed progeny is markedly greater than that of selfed progeny (D. Lloyd 1992). The result of these benefits is that outcrossed hybrid **heterozygous** plants generally do better than more **homozygous** relatives; thus, many populations over time come to contain more heterozygotes, and more of the surviving adults are heterozygotes than are the initial set of seedlings (e.g., Schmitt and Ehrhardt 1990 with *Impatiens*).

Given the benefits of outcrossing, how common is selfing (autogamy) and why does it still occur? Selfing is quite common in many species and is sometimes the

TABLE 3.3
Modes of Self-Pollination for Flowers

Timing relative to cross pollination	Mode	Using pollen vector?	Reduced pollen export?	Provides reproductive assurance?
Before	Prior	No	Some	Yes
Simultaneous →	Competing	No	No or some	Yes
Simultaneous ——→	Facilitated	Yes	Yes	Yes
→	Geitonogamous	Yes	Yes	No*
After	Delayed	No	No	Yes

Source: Modified from Lloyd 1979, 1992.
*unless it induces more pollinator visits.

norm; there is effectively a continuum between plants that cannot self-fertilize at all, through those that do it very rarely, those that use it sometimes, those that have it as part of their routine life cycle, and on to the other extreme of plants that nearly always self-pollinate and self-fertilize (D. Lloyd 1979, 1992). Table 3.3, based on Lloyd's reviews, summarizes the possible kinds of selfing that may occur and some of their consequences. Selfing that occurs at the same time as possible outcrossing can be either intrafloral without any influence from pollen vectors (competing), intrafloral but aided by the activities of a vector (facilitated), or extrafloral (geitonogamous), with pollen export by a vector to another flower on the same plant.

In temperate ecosystems about two-thirds of all plants can self if they have to, the proportion being highest in small and annual species (Aarssen 2000) but markedly lower in large woody perennial plants. In tropical ecosystems, the levels of potential selfing are probably lower for all plants (Bawa et al. 1985), and Ward et al. (2005) found more than 90% outcrossing for 45 species from tropical rain forests.

Where a vector is involved, selfing may be an inescapable by-product of adaptations that aid outcrossing, but where there is no vector, some kind of benefit to the plant must be involved. The observation that a loss of mechanisms to prevent selfing is one of the commonest of all trends in flowering plants (Schoen and Busch 2008) reinforces this point. Probably the single most important benefit is that self-fertilization gives guaranteed reproduction, even in the absence of any potential mates. And remember that self-fertilization is still a sexual event, with independent assortment of

chromosomes at meiosis and therefore the possibility of recombination events, producing at least some genetic variability in progeny (unlike truly asexual reproduction by runners, for example).

For annual plants, then, with short lives and only one flowering season, the obvious argument in favor of selfing is that it does provide the reliability of a fall-back or fail-safe mechanism (often termed reproductive assurance) when crossing has not occurred, which may be the case in years or in localities when the normal external agent of pollen transfer, whether an animal or the wind, has proved inadequate. This issue was thoroughly (and critically) reviewed by Eckert et al. (2006). Pollinator limitation is often a serious factor in **marginal habitats** and in poor seasons, and in those circumstances producing any seed at all, even if it is selfed seed with some disadvantageous mutations perpetuated, is better than nothing. Annual weedy plants with selfing habits are typically associated with open and unpredictable environments and with early spring or late autumn flowering in seasonal habitats, where cold or wet weather may limit the options for crossing. In some genera, there has been repeated reversion to selfing and to smaller, inconspicuous flowers in species living at the margins of their outcrossing ancestors' range, where environmental conditions are extreme (e.g., in *Clarkia*; Runions and Geber 2000; Mazer et al. 2004). A reproductive assurance case can also be made for selfing in some highly specialized flowers (such as some orchids and the elaborate *Tacca chantrieri*), where delayed selfing provides a fail-safe at the end of the flower's normal lifespan; Fenster and Martén-Rodriguez (2007) reviewed this point thoroughly, showing that the occurrence of traits for both pollination specialization and self-pollination in the same species is not paradoxical.

Another argument in favor of selfing is that it is cheaper. Most plants that are habitually self-pollinating have flowers that are smaller and fewer in number than those of close relatives that require cross-pollination. Good examples are provided by some of the common *Geranium* species: the large and brightly colored meadow and wood cranesbills (*G. pratense, G. sylvaticum*) are pollinated by insects, especially bees, while dovesfoot cranesbill (*G. molle*) and cut-leaved cranesbill (*G. dissectum*) are both self-pollinators and have flowers with much smaller petals that are dull pink in color. Selfing plants clearly do not need the expensive investment in flower color, size, and number to put on an attractive floral display. Similarly, they tend to have little or no nectar, saving on calories offered to visitors. Perhaps even more important, the individual flowers commonly have fewer anthers or fewer pollen grains or both, and fewer ovules. Cruden (1977) recorded average figures of less than 100 pollen grains per ovule for selfers, and of 700 or more for outcrossing species.

A third argument for selfing is that it is potentially quicker, and in that case pollinator limitation is not a necessary part of the explanation. Snell and Aarssen (2005) reviewed the associations of traits in selfing annuals systematically and found them to have shorter plants, smaller and shorter-lived flowers, shorter development times, and smaller seeds compared with related outcrossing species. They noted that these traits can all be explained by a simple time-limitation hypothesis that does not invoke any particular limitation of mates or of pollinators. Speed of reproduction may be critical for ephemeral plants with short growing seasons or for individual plants from late-germinating seed faced with the onset of winter in a seasonal habitat. A selfing flower can be fertilized as soon as its anthers and stigma are ripe, with no time wasted enduring windless conditions for an anemophilous plant or in waiting for appropriate visitors if zoophilous. A selfing plant may even be able to hasten this ripening relative to an outcrossing species, since many self-pollinating plants are notable for simple and "juvenile" flower structures that resemble those of related species when at a younger developmental age. In effect, the developmental program is slightly adjusted so that the maturation of anthers is speeded up relative to the maturation of petals or other showy structures (Guerrant 1989; Diggle 1992). Thus in *Mimulus*, selfing species have smaller and short-lived flowers with more rapid bud development than those of related outcrossers, suggesting that there has been specific selection for rapid development (Fenster et al. 1995). For plants that use selfing as a fallback, this adjustment may have other useful effects. In a juvenile flower, any differences between anther and stigma height are likely to be reduced; again geraniums provide examples, with the anthers and stigma of *Geranium molle* maturing simultaneously and close together, while the anthers and stigma of *G. sylvaticum*, which is protandrous, grow to different heights. Furthermore, in juvenile flowers, any intrinsic self-incompatibility mechanisms (which will be discussed in *Genetic Separation by Self-Incompatibility* below) are less likely to be fully operational.

The three advantages mentioned so far largely apply to short-lived annual plants. Perennials, of course, have the option of reproducing in subsequent years, and so have less need of selfing as a safeguard against the possibility of not being able to reproduce at all. Classic examples of longer-lived temperate plants that do habitually show a kind of selfing are *Rubus* (bramble, or blackberry, and its relatives) and *Sorbus* species such as whitebeam; both use agamospermy extensively, effectively dispersing by cloned seeds. So too do some common weeds, including many dandelions (*Taraxacum* and the related composite genera *Hieracium* and *Crepis*) and the common meadow grass (*Poa pratensis*) (Mogie 1992). This raises the fourth argument in favor of selfing, which is also the one widely quoted for animals: that it can be seen as adaptive for invaders of unpredictable and short-lived habitats, where establishment needs to be rapid. In those circumstances, producing a larger number of offspring that are very similar to the successful invader to take advantage of the transient resources is more important than producing variable offspring, some of which may not be as well suited as the parent to that particular transient habitat. Put simply, selfing is a good way of rapidly expanding a plant population. Of course the argument then also works for weedy, annual, self-fertilizing species, which epitomize good colonizers of new habitats and efficient users of transient resources following major disturbances such as fire, flood, or **anthropogenic** interventions, achieving assured seed-set from just one or two initial invaders. Clearly, this argument is connected to the preceding one concerning the speed of reproduction for annuals, although on a slightly different time scale.

A fifth advantage, or set of advantages, is sometimes invoked and is related to genetic effects. Where seed-set is low in an outcrossing plant, it ought to be vulnerable to invasion of its habitat by a mutant form that is self-fertilizing, so that selfing should spread rather easily in some circumstances. However, this argument is contentious: the issue here may be that seed-set (the traditional measure of plant success in laboratory trials) is not a good measure of fitness in field situations, since seedling establishment and parental longevity also have enormous effects in maintaining or increasing population size. It is frequently reported that, in this broader sense, fitness is maximized at moderate, or even at low, levels of seed-set (Calvo and Horvitz 1990; Calvo 1993), which require only moderate levels of pollination success and are achievable by moderate animal-mediated cross-pollination. On that basis, there may be less selective pressure for plants to evolve toward selfing than might sometimes be supposed. However, there are additional genetic reasons for selfing being favored, in that any genetic mating-system modifiers that reduce the rate of outcrossing should, at least in theory, be able to bias their own transmission; this is the "automatic selection" hypothesis for the evolution of selfing (Schoen et al. 1996). For example, a gene promoting outcrossing is transmitted only via ovules and pollen that are themselves both the result of outcrossing, whereas a gene promoting selfing will benefit from both of those scenarios but also from progeny produced by selfing, and so should be able to spread (at least whenever inbreeding depression is not too strong, see below). Fisher (1941) first pointed out this effect and noted that a partial selfer has an inbuilt advantage over an exclusive outcrosser because it can transmit its genes in three ways rather than in just two, giving, in theory, a 50% advantage.

Even without these genetic issues, but given the fourfold ecological benefits for selfing of being reliable, cheaper, quicker, and good for colonizing, it would seem that the countering benefits of crossing—generating variability and masking deleterious genes—must be rather powerful. This is evident also from the pattern of occurrence of selfing (mainly at the margins in both time and space) and from the ability of almost all habitually self-pollinating plants to achieve cross-pollination when possible. Thus even the small, dull, low-pollen and low-ovule flowers of plants that are associated with self-fertilization are readily able to be cross-pollinated when they receive a rare visit from an insect. Levels of 99% selfed seed are not uncommonly recorded in some weedy plant species, but populations with no crossing at all seem to be exceedingly rare, and Stebbins (albeit in 1957) specifically asserted that there were no examples.

This discussion perhaps leads back conveniently to the disadvantages of selfing. Where it is very common, it has consequences for the levels of variation within plant populations, for the reasons already given as selecting for cross-fertilization. Plants become much less varied than equivalent outcrossers and may become genetically fixed into particular subpopulations (a problem for taxonomists in the past, who may have identified as separate species what were really just local morphological types of a single species). Lack of variation in selfed plants can potentially lead to inbreeding depression (effectively the opposite of **hybrid vigor**, so that recessive deleterious alleles are expressed), with many possible manifestations: slower growth, smaller seed size, lower seed numbers, and slower germination have all been recorded, the precise response varying with species but always leading to an overall reduction in establishment and survival. Morgan et al. (1997) suggested that the costs of potential adult inbreeding depression are the major reason why perennials avoid selfing, and this may be broadly true for all angiosperms (Barrett 2002), because they could in theory suffer a gradual irreversible decline in fitness (although S. I. Wright et al. (2008) suggested several ways in which the worst genomic consequences might be avoided). Plants that do habitually self-fertilize may have survived the initial hurdles of inbreeding depression over a few generations by eliminating the most disadvantageous genes through high mortality, and perhaps preserving particularly useful gene complexes, so that the burden of inbreeding depression was quite rapidly reduced; but in almost all cases, the survivors still achieve better performance if crossed (A. Richards 1986; D. Charlesworth and Charlesworth 1987). The only exceptions may be some **polyploid** plant hybrids, where selfing is more favored (Barringer 2007), and the increased chromosome number reduces or masks the effects of deleterious mutations (Husband et al. 2008).

The lack of variation can also, of course, be a major and potentially lethal disadvantage if the habitat changes. And local fixation of the **genotype** and phenotype

TABLE 3.4
Self-Fertilization versus Cross-Fertilization

Crossing	Advantages	Generates variation
		Masks some mutations and damage in genome
		Permits evolution of new functions in masked alleles
	Disadvantages	Requires higher investment in flowers
		May generate some poor quality offspring
Selfing	Advantages	Reliable (as back-up when crossing fails)
		Cheaper
		Potentially faster
		Allows rapid exploitation of transient resources
	Disadvantages	Inbreeding depression
		No masking of disadvantageous alleles
		No variation that allows adaptation to changed conditions

could presumably become an additional handicap over evolutionary time, with separate subpopulations eventually so different that they cannot interbreed and restore variation.

A summary of the advantages and disadvantages of selfing and crossing is presented in table 3.4. The balance of costs and benefits will vary with the availability of the prevailing pollinator, as specifically demonstrated by Kalisz et al. (2004) and by Moeller and Geber (2005). But it is evident that cross-fertilization has important advantages for most plant species most of the time in most localities and that self-fertilization is usually better avoided or resorted to only as a last resort strategy. Self-fertilization in those plants that normally use outcrossing is therefore prevented, as far as possible, with a range of mechanisms within a flower and within a plant, as documented in the next two sections. Where these tactics fail and selfing occurs, embryo abortion is common, along with inbreeding depression and unmasked deleterious genes. Indeed, plants that have selfed extensively often show almost total abortion, as described in *Ipomopsis* by Waser and Price (1991).

4. Methods for Avoiding Selfing within a Flower

Avoidance of selfing can be achieved either by separating the male and female parts of the flower, so that there is a **prestigmatic** block to self-pollen ever being received on the style (and these methods will often have a very marked effect on floral design), or by a physiological system within the floral tissues, so that the flower recognizes and blocks self-pollen at a **poststigmatic** stage. An overview of these mechanisms is given in table 3.5. Many plants use both systems, probably because each has pros and cons. Prestigmatic blocking is cheap to achieve but somewhat risky. Poststigmatic blocking is highly effective in avoiding fertilization but still allows self-pollen to land on stigmas, where it may block the receptive surface (**pollen clogging**) and interfere with subsequent germination of cross-pollen; for example, in self-incompatible *Polemonium viscosum*, Galen et al. (1989) recorded that self-pollen on the stigmas, although blocked from germinating itself, could reduce germination of cross-pollen by up to 32% and seed set by up to 40%.

Temporal Separation of Male and Female Parts

Also known as **dichogamy**, temporal separation occurs widely in flowers that have sexual phases ("sex with a temporal offset"), which comprises nearly 90% of all angiosperms. This may be either protandry—having the male phase first, with anthers dehiscing before the flower's own stigma is receptive—or protogyny—the opposite, with the stigma already shriveling by the time pollen is offered. In some plants the

TABLE 3.5
Mechanisms for Avoidance of Selfing in Hermaphrodite Flowers or on Monoecious Plants

Temporal separation (dichogamy)

Protandry	Male phase first: pollen released from anthers before stigmas are receptive. Very common, especially in flowers pollinated by bees and flies, with higher reward in lower flowers and pollinator working upward from base to tip of inflorescence
Protogyny	Female phase first: stigmas receptive before anthers mature and release pollen. Less common, except in generalist flowers and monoecious flowers, especially those pollinated by beetles and wasps, and in some specialist trap flowers

Spatial separation (herkogamy): — Male and female phases are simultaneous, but style and stamens are kept spatially apart within each flower

Ordered herkogamy	Pollen and stigma positioned sequentially in relation to pollinator access
	Stigma at mouth of flower and pollen deeper within corolla; bell or tube-shaped flowers; stigma often one
	Pollen at mouth of flower and stigma deeper within corolla; narrow tubular flowers
Temporal herkogamy	Pollen or stigma or both change position through life of flower, occupying similar positions at different times.
	Pollen near mouth of flower initially; style grows into same position after pollen depletion
Movement herkogamy	Floral parts move in response to pollinator
	Stigmas close when contacted or after receiving pollen
	Anthers move inward as pollinator probes flower
Reciprocal herkogamy	
Heterostyly	Polymorphic styles; usually in tubular flowers and accompanied by SSI systems (see below)
Distyly	Different flowers have either long style and short stamen, or vice versa (pin and thrum flowers, e.g., *Primula*)
Tristyly	Similar to distyly, but long, medium, and short lengths occur
Enantiostyly	Style deflected to right or left of the main floral axis; usually pollen-only flowers with no nectar

Genetic separation by self incompatability (SI): — Controlled by chemical recognition between pollen and stigma, due to specific genetic SI loci. When there is the same SI allele at this locus, whether pollen is from the same plant or from another plant within the same population, pollination cannot occur

Sporophytic (SSI)	Parent genotype determines compatability; stigma expresses the incompatability *S*-allele (usually only two alleles and only one locus) Pollen arrests at stigma surface
Gametophytic (GSI)	Progeny determines compatability; pollen grain contains an incompatability *S*-allele (usually multiple alleles, only one locus) Pollen growth begins but arrests within style

switch from one phase to the other has a fixed timing, normally from one to a few days, but in others it can be triggered by external events. Dichogamy not only reduces selfing by preventing "sexual interference" within the hermaphrodite flower (D. Lloyd and Webb 1986), but if it is synchronized across all the open flowers on one plant, it can also reduce the cost of pollen wastage, limiting geitonogamous interaction between those flowers and thus leaving more pollen available for outcrossing (Harder and Barrett 1996).

Protandry is particularly widespread, especially in rather specialized zygomorphic flowers (Bertin and Newman 1993). **Phylogenetic analysis** has revealed that protandrous species tend to be bee or fly pollinated, whereas protogynous species are more likely to be wind or beetle pollinated (Sargent and Otto 2004). Protandry can be more or less relative in its timing, so that there are varying degrees of overlap of the male and female phases. Occasionally, the gender phases are absolutely distinct, as in some *Impatiens* (balsam) species where the androecium completely covers the gynoecium in young flowers; the gynoecium swells later on and pushes the androecium away (Schemske 1978). Protandry can also occur in a protracted form with a sequential ripening of anthers; for example, in *Saxifraga aizoides*, a yellow-flowered alpine species commonly grown in rock-gardens, the filaments curl into the center of the flower one at a time and dehisce there, with the style maturing only once all the anthers have dehisced.

Protandry is almost the rule in certain families (Lamiaceae, Caryophyllaceae, and Asteraceae), and is common in others (Campanulaceae, Apiaceae), although the degree of protandry differs between species within these taxa. It may vary even between populations, as Schoen (1982) showed with the annual *Gilia achillefolia* (Polemoniaceae), where communities with the largest protandry index (the temporal separation between the sexual phases) had the highest levels of cross-pollination.

In contrast, protogyny is rather uncommon, perhaps because it reverses the "normal" order of development of successive whorls of floral organs. It occurs in some very generalist flowers (especially the small open flowers pollinated by beetles and wasps), notably in the families Ranunculaceae (e.g., the pasque flowers, *Pulsatilla*) and Brassicaceae. It is also found in some wind-pollinated plants and in many plants with unisexual flowers. Many of the protogynous flowers are self-compatible. They tend to be spring-flowering and are often associated with polar and **boreal** regions (chapter 27). These points may be linked to one major advantage of protogyny: the anthers can persist into old age if the flower has not been pollinated (rather than shrivelling at a given age to allow stigma maturation), and therefore the flower can still self-fertilize as a last resort, as is often necessary in habitats where pollinators are scarce and crossing is unreliable.

Protogyny also occurs almost of necessity in some of the specialist trap flowers that will be described in chapter 23, where the pollen-bearing visitors are held in a chamber around the mature stigmas and only released once the flower's own anthers (sited on the exit route) are ripe. An even more extreme case of structural complexity occurs in the tiny protogynous flowers of the tree *Guazuma ulmifolia* (allied to cocoa, in the Malvaceae; fig. 3.9), which initially presents a visitor with a central female chamber bounded by five strap-shaped petals; these will then wither and intermingle and thus prevent access to this female chamber at the same time as the wilting sepals expose lower entrances to five male chambers containing nectar, each with a "window" effect to lure the visitor into position for pollen receipt (Westerkamp et al. 2006).

In some species, such as the avocado pear (*Persea*, family Lauraceae), all the plants are protogynous, but subpopulations open at different times of day, with a closed phase in between, therefore facilitating outcrossing (Ish-Am and Eisikowitch 1991). A few plant species can have individuals that are either protandrous or protogynous (e.g., the tropical tree *Trochodendron*; Endress 1994), again reducing selfing and promoting outcrossing.

Spatial Separation of Male and Female Parts

Spatial separation of the male and female structures within a flower is also known as one form of **herkogamy**. It may be functionally important in preventing unwanted physical interference between the sexual structures (stamens restricting access to stigmas, or styles obstructing exit routes for pollen-bearing visitors), but it is often also linked with flower morphologies that manipulate a visitor's movements in the flower to ensure that it touches the sexual structures appropriately. Somewhat complex measures may be needed for relatively **specialist flowers** that rely on

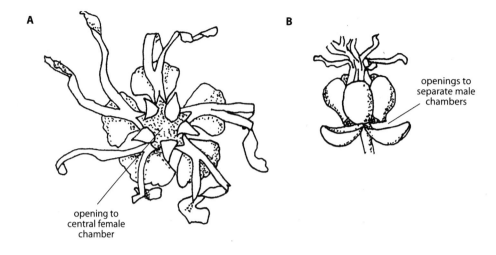

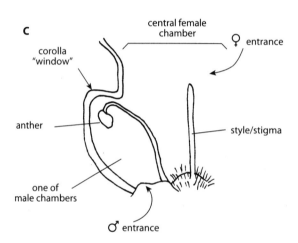

Figure 3.9 The complex flowers of *Guazuma ulmifolia*: (A) seen from above (with entrance to female chamber open); (B) seen from the side (female entrance now closed by withering tissues, and male entrances from below now open; (C) a cut-away detail of one male chamber and the female chamber. (Modified from illustrations in Westerkamp et al. 2006.)

precise pollen placement on a visitor, since outcrossing then often depends on the stigma being in a similar position to the anthers to receive that pollen from the same part of the animal's body. There is often a conflict between avoiding self-interference and achieving precision in pollination.

In **ordered herkogamy**, a visitor meets male and female parts sequentially. In its simplest form, in bowl-shaped flowers, the stigmas are commonly central and the anthers peripheral; most visitors will land centrally, so depositing any pollen they carry onto the stigmas, and (usually after feeding) they then move to the

edge of the flower to take off, so picking up fresh pollen. In tubular corollas, ordered herkogamy can be achieved by unequal protrusion of the stigma and stamens; most often the stigma is at the mouth of the flower and anthers bearing pollen are deeper within the corolla. In brush-blossom inflorescences with many small short-tubed flowers, the anthers and stigma of any one flower can be kept apart thus, but since visitors tend to walk all over the inflorescence, they will usually move pollen successfully between florets. In larger flowers, longer stigmas are particularly associated with bell-shaped or widely tube-shaped flowers, and where these are zygomorphic, the stigma is often dorsally sited and may be one-sided. Thus an incoming visitor deposits foreign pollen from its mouthparts or head as it enters and picks up fresh pollen on its back. However, in very narrow corollas, the situation is often reversed, and pollen may be dispensed by long stamens with anthers at the mouth of the flower, while the stigma lies deeper within the corolla; here a typical visitor (bee, moth, or bird) normally only inserts its tongue, which carries pollen on its brush-like tip, so picking up fresh pollen as it withdraws the tongue from one flower and then depositing it as it probes for nectar deep into the center of the next flower.

Ordered herkogamy can be very efficient in promoting outcrossing. There can be great variation within a species in relative stigma-anther lengths, and this is usually directly linked to variation in the rates of selfing (Motten and Stone 2000; Takebayashi et al. 2006). Relative lengths of the stigma and anthers can also be under strong selection; for example, in the Australian orchid-like genus *Stylidium* (trigger plants), pollen placement is very precise, and populations in different areas within different communities have differences in anther and style morphology to ensure placement on the pollinator's body in sites that reduce overlap with **coflowering** plants (Armbruster et al. 1994).

A second option is to change the relative positions of anthers and stigma as the flower ages (clearly this combines an element of temporal separation, especially protandry, and so might be termed **temporal herkogamy,** though the term "heterodichogamy" is also used in special cases). This effect can be achieved in bowl-shaped flowers if the stamens move either inward to the center after shedding their pollen (e.g., *Geranium, Hypericum*), or outward (e.g., *Malva*), in both cases taking the anthers away from the maturing stigmas. In zygomorphic flowers the stamens are commonly long and dorsal, so that pollen is deposited on a visitor's back, but the style is either lateral or ventral and often short; then later in the flower's life, the style either moves or grows so that it ends up in a similar site to the anthers, thus solving the dilemma created by the need for precise pollen placement on visitors. This is what happens in many familiar tubular flowers, such as *Penstemon*, and so allows the pickup of imported pollen from the same site on the body of the same pollinator species. Plate 26H shows a bird-pollinated example of this kind. Another example can be seen in

Ribes (plate 14G), where younger flowers have exserted white anthers and a short style, whereas in older flowers, the elongate yellowish style emerges beyond the anthers.

A third variant is **movement herkogamy,** where floral parts move only in direct response to the presence and touch of the pollinator. This may occur where the male and female parts are normally separated, and then either the stigmas close when contacted or after receiving pollen (e.g., *Mimulus, Utricularia*; chapter 2) or the anthers move inward as a pollinator probes the flower (e.g., *Mahonia*, plate 12B). Alternatively, in some rosaceous plants, such as rock-rose (*Helianthemum*, plate 2E), the anthers are initially clustered centrally but splay apart quickly after visitation, so exposing the central maturing style to later visitors.

A fourth and particularly interesting version is *reciprocal* herkogamy, also known as heterostyly. It involves plants that have **polymorphic** styles, and it occurs in at least 28 families (though often has been lost again in certain lineages within groups that acquired it; Sakai and Wright 2008). It is best known in the form of **distyly,** where a particular species has two different flower forms with different style lengths (Barrett 1990; Barrett et al. 2000). The primroses (*Primula*) with their *pin* and *thrum* varieties are the most familiar case (plate 6C). Here, the **pin flowers** have an elongate stigma reaching the end of the corolla tube and anthers sited half way down that tube, while the **thrum flowers** have anthers set at the top of the tube and a stigma only half the length (fig. 3.10). Furthermore, the pin flowers have much more strongly papillate stigma surfaces and less variably sized pollen grains. Similar arrangements occur in a range of common plants, shown in table 3.6, including some tristylous plants with three, rather than two, variants (i.e., having long, medium, or short styles). In all distylous cases, pin pollen pollinates thrum stigmas and vice versa. In part this is due to the position of the anthers inducing quite different placements of pollen on the body of a visiting insect (Wolfe and Barrett 1989). Usually this involves different positions on the visitor's proboscis, so that thrum pollen is placed high up the tongue, near the face, of a bee or **hoverfly** or **bombyliid** fly, and thus is deposited on the pin stigma of another flower. In most cases, however, there is also self-incompatibility, so that even if pin pollen does land on a pin stigma, it will not fertilize the flower. Thus, the physical placement by the visitors can be seen as an "extra" that helps

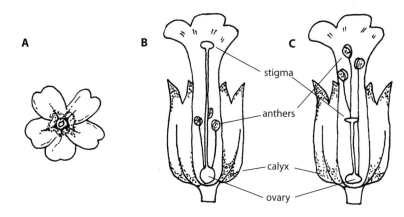

A   **B**   **C**

stigma

anthers

calyx

ovary

Figure 3.10 Pin and thrum variation in *Primula* flowers: (A) face view of thrum flower; (B) cut-away of thrum flower; (C) cut-away of pin flower. (See also plate 6.)

TABLE 3.6
Occurrence of Heterostyly in Well-Studied Plants

Family	Genus	Common name
Distylous		
Primulaceae	*Primula*	Primrose, cowslip, oxlip
	Hottonia	Water violet
Boraginaceae	*Pulmonaria*	Lungworts
	Cordia	Manjack
	Lithospermum	Gromwell
Oleaceae	*Forsythia*	Forsythia
	Jasminum	Winter jasmine
Menyanthaceae	*Menyanthes*	Bogbean
	Nymphoides	Fringed water lily
Rubiaceae	*Galium*	Bedstraw
	Arcytophyllum	
	Coussarea	
	Psychotria	
	Mussaenda	
	Palicourea	
Erythroxylaceae	*Erythroxlum*	
Linaceae	*Linum*	Flax
Lamiaceae	*Chloanthes*	
Tristylous		
Oxalidaceae	*Oxalis*	Sorrel
Lythraceae	*Lythrum*	Purple loosestrife
Pontederiaceae	*Eichhornia*	Water hyacinth
	Pontederia	Pickerel weed

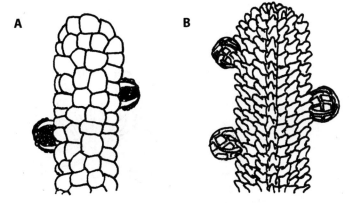

Figure 3.11 Two distinct pollen types and associated style types in *Armeria*: (A) papillate pollen on cob stigma; (B) cob pollen on papillate stigma. (Drawn from photographs in Proctor et al. 1996.)

reduce the clogging of stigmas with unsuitable pollen, but it carries the cost that most plants can at best fertilize only half (or one-third) of the population. It also affects visitor handling time, for example in *Jasminum* (J. Thomson 2001a), where bees and bee-flies spent less time handling short-styled flowers compared with long-style flowers.

Further variants also occur, where there is heteromorphy without heterostyly. A number of plants, such as sea lavenders (*Limonium*) and thrifts (*Armeria*), have two flower types but with similar style lengths, the differences residing only in the sculpturing of the stigmatic surface and of the pollen grains. There are **cob stigmas** (resembling corn-on-the-cob) and **papillate stigmas**, with smaller cone-like protrusions, and the two types of pollen are even more distinctive, the "cob" pollen having larger scale patterning (fig. 3.11).

The genetic controls of these systems are rather simple. There are two alleles at a single gene in *Primula* and other pin-and-thrum plants, with thrum plants having one of each allele and pin plants having two copies of the same allele. In the tristylous plants there are two genes, each with two alleles.

The evolutionary origins of heterostyly are not entirely clear, with alternative theories from D. Charlesworth and Charlesworth (1979) and from D. Lloyd and Webb (1992) that vary in the sequence in which characteristics were acquired; the former authors viewed inbreeding avoidance as the main selective force, the latter proposed that selection for efficient crossing is preeminent, so that incompatibility systems come into play only later in the evolutionary sequence. Barrett and Harder (2006) summarized this argument and the evidence in favor of Lloyd and Webb's view. Hetero-

styly occurs in diverse taxa, with no obvious phylogenetic pattern except that it is absent in the more phylogenetically basal angiosperm groups and absent in the most advanced groups such as orchids and composites. Most commonly it occurs in tubular and radially symmetrical flowers pollinated by insects, which have relatively low numbers of stamens, and it seems highly likely that heterostyly has evolved convergently several times, aided by its very simple genetic control. This control system may also explain its tendency to break down rather easily: a number of heterostylous plants have reverted to **homostyly** and lost any self-incompatibility, either within certain populations (e.g., some primroses [Piper et al. 1984]) or across most populations of a species. In the case of *Eichhornia crassipes*, reversion has helped a species to become a pernicious invasive weed (Barrett et al. 1989). Simplicity of control and easy reversions might also partly explain reported cases of almost continuous variation in style length, notably in *Narcissus*. However, in Spain (the center of diversity for this genus) Barrett et al. (1996) found that different populations of several *Narcissus* species were always either mono-, di-, or trimorphic for style length and never continuous.

A final kind of herkogamy that is relevant here is **enantiostyly**. Here the style is sharply deflected to right or left of the main floral axis, usually in pollen-only flowers with no nectar, such as *Cassia* (fig. 2.5), or in large pendant flowers, such as *Hibiscus* and *Gloriosa* (fig. 3.12A). There may be one type of flower (left or right), or more commonly two with mirror-image organizations (fig. 3.12B). In the latter case these may be present within the same plant (**monomorphic** enantiostyly) or on separate plants (dimorphic),

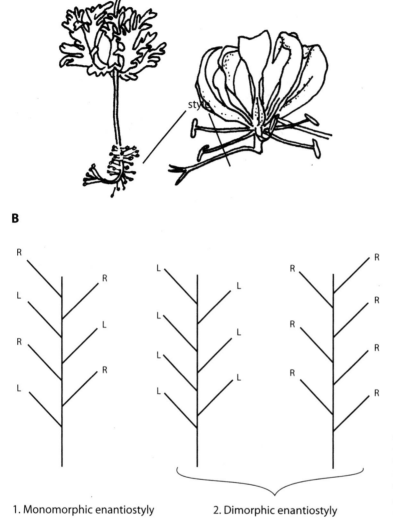

A

Hibiscus　　　　　*Gloriosa*

style

B

1. Monomorphic enantiostyly　　　2. Dimorphic enantiostyly

Figure 3.12 (A) Enantiostyly in *Hibiscus schizopetalus* and in *Gloriosa superba* with the stigma sharply curved to one side (redrawn from Endress 1994; see also plate 12). (B) Diagrammatic view of different types of enantiostyly, with L and R referring to the left and right side displacement of the styles (redrawn from Barrett et al. 2000).

only the second of these being a true genetic polymorphism (Lin and Tan 2007) seemingly controlled by a single **locus**, with the right-styled orientation often the dominant one. Neither type is common, and dimorphic enantiostyly has arisen independently in just three or four taxa, all monocots (Jesson and Barrett 2002). Sometimes the style deflection is accompanied by anther deflections (reciprocal enantiostyly). The probable advantages of enantiostyly are protection of the style (from forceful hovering visitors and from interference by male organs) and also an increased

chance of delayed selfing as a reproductive assurance strategy. Jesson and Barrett (2005) further showed that enantiostyly increased the precision of cross-pollination in bee-pollinated *Solanum* and thus reduced geitonogamy.

These systems of heterostyly and enantiostyly are compared schematically in figure 3.13, together with a third possibility in which stamens are constant but styles vary in position (stigma-height dimorphism); the figure also shows the intermorph matings that result.

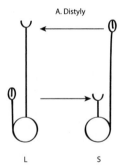
A. Distyly

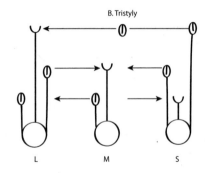
B. Tristyly

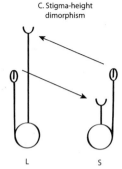
C. Stigma-height
dimorphism

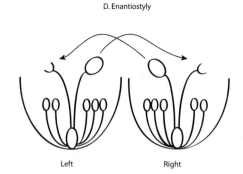
D. Enantiostyly

Figure 3.13 A diagrammatic comparison of the four main kinds of style variations and of the intermorph pollen movements that occur and that tend to maintain the polymorphisms with equal frequencies. (Redrawn from Barrett et al. 2000.)

Genetic Separation by Self-Incompatibility

An alternative way to avoid selfing in angiosperms involves self-incompatibility (SI) mechanisms, whereby a cellular physiological mechanism operates at or beyond the stigmatic surface and prevents self-pollen from delivering gametes to fertilize the ovules. This process requires a system for recognizing and discriminating between self and nonself, an ability that is not uncommon in living organisms across all the phylogenetic kingdoms. However, in most taxa, it involves the rejection of nonself (as, for example, in animal defenses against diseases), whereas plants with SI systems most unusually discriminate against self, rejecting self-pollen and encouraging nonself-pollen.

SI is controlled by a single genetic S locus with many alleles and is mediated by chemical recognition systems operating between the pollen and stigma. The presence in the haploid generative cell of the pollen grain of one allele identical to either of the alleles in the recipient stigma will block pollen tube growth. The main SI mechanisms (GSI and SSI, see below) are prezygotic in function, that is, they operate to prevent fertilization; when the same allele is present in pollen and stigma, regardless of whether the pollen is from the same plant or from another plant within the same population, fertilization cannot occur.

SI is now reported in nearly 40% of all angiosperm species (probably including up to two-thirds of grasses [Grime et al. 1988]) across more than a hundred families. However, it is also frequently lost in plant lineages, because it is chiefly controlled at the single S locus, where any variation collapses when there is a transition to self-compatibility.

At present, several different systems of self-incompatibility are recognized, with variants also occurring, and classification systems differ between authors. However, for our purposes it is appropriate to consider two main systems, gametophytic and sporophytic, along with a third that is more problematic. The topic has been usefully reviewed by Barrett (1988), Takayama and Isogai (2005), A. Allen and Hiscock (2008), and Igic et al. (2008), all drawing attention to the occurrence of several different mechanisms at the genetic, molecular, and cell-signaling level. In the future, as more species are categorized, the classification of SI types is likely to be revised, and the distinctions made below may be abandoned.

Gametophytic Self-Incompatibility

Included within gametophytic self-incompatibility (GSI) is a range of mechanisms by which the pollen tube is recognized and rejected as it grows, most commonly while it is within the style and before it reaches the ovule. It is only the haploid genome of the progeny that determines compatibility, hence the term "gametophytic SI." The pollen tube, which contains the gametophyte nuclei, either bursts, ceases to grow, or just grows very slowly. In species of poppy (*Papaver*) and in the evening primrose *Oenothera organensis*, both of which have been especially well studied, the *S* locus seems to be truly polymorphic, with some 25–45 alleles in most natural populations. O'Donnell and Lawrence (1984) estimated that in a wild *Papaver* population, any one plant was fully compatible with about 80% of the others but entirely unable to breed with perhaps 5%, most of those being its near relatives and presumably sharing the same *S* allele genotype.

In the grasses (Poaceae) a slightly different gametophytic SI system operates, with two interacting multiallelic loci, *S* and *Z*, and here self-pollen is recognized and blocked at the stigmatic surface. A few other multilocus systems have also been identified, involving three (buttercup, *Ranunculus acris*) or four (sugarbeet, *Beta vulgaris*) loci, each with several alleles. In all these cases the alleles at each locus must be the same in the pollen and in the stigma for rejection to occur, so the number of plants with which any one individual can breed is increased.

McClure and Franklin-Tong (2006) reviewed GSI, identifying two different mechanisms operating at the cellular level. In the Solanaceae, Rosaceae, and Scrophulariaceae, there is a ribonuclease (*S*-RNAase) as the style *S*-component and an F-box protein as the pollen *S*-component, so that the pollen tube RNA gets degraded; whereas in the Papaveraceae, there is a small peptide secreted as the style *S*-component, and the pollen component may be a transmembrane receptor that triggers a calcium-ion cascade leading to cell death. The Poaceae system is probably different again, and in this rapidly advancing field yet more mechanisms are certain to be described.

Sporophytic Self-Incompatibility

In sporophytic self-incompatibility (SSI) the pollen growth is always arrested at the stigma surface, and the system therefore involves rejection of the pollen grain rather than of the growing pollen tube. The mechanism is termed "sporophytic" because the pollen grain exhibits not only the potential incompatibility of the *S* allele in its nucleus, but also the incompatibility reaction of the other *S* allele present in its parent (i.e., in the **sporophyte**), apparently by having these *S*-gene protein products embedded on or in the pollen wall. Hence the parent diploid genotype determines overall compatibility. The recipient (female) stigmatic surface recognizes and responds (at least in some cases via a receptor kinase) to this specific protein on the pollen wall, so that the grain either fails to germinate or is blocked just as the incipient pollen tube starts to penetrate the stigma. Sporophytic SI is therefore very fast-acting and efficient, and it is particularly characteristic of two large and successful families, the crucifers (Brassicaceae) and the composites (Asteraceae). Again a single genetic locus with multiple alleles is implicated (Hiscock and McInnis 2003), although there may be perhaps 17 genes within the *Brassica* SSI locus. There are usually only two alleles, but occasionally far more (e.g., at least 24 alleles have been suggested for *Sinapis* species; Ford and Kay 1985).

Late-Acting Self-Incompatibility

Late-acting self-incompatibility (LSI) incorporates a range of examples in which incompatibility acts at a late stage: as the pollen tube enters the embryo sac or even after fertilization has occurred, in which case the embryo is aborted. It was first observed in cocoa (*Theobroma*) (Cope 1962), and Seavey and Bawa (1986) drew attention to many cases in the literature where selfed flowers did not yield fruits even though the self-pollen tubes grew freely down the style and penetrated ovules; they called the phenomenon late-acting self-incompatibility. LSI mainly operates postzygotically, unlike GSI and SSI. It is now known to occur in a broad spectrum of plants, especially some legumes and a range of tropical woody plants in the Bignoniaceae and Bombacaceae (Gibbs and Bianchi 1993). These trees can afford the apparently expensive habit of waiting until fertilization to initiate fruit abortion, since like many longer-lived plants they produce far more flowers than could ever mature into fruits, and they naturally undergo abortions for many reasons in different times and places.

It is not entirely clear whether LSI is different in kind from the "natural" abortion of many embryos already mentioned as a common phenomenon in plants undergoing excessive selfing and thus with deleterious recessive alleles (*genetic load*). However, A. Allen and Hiscock (2008) hypothesize that LSI may be a phylogenetically basal kind of SI mechanism in flowering plants.

In reviewing SI systems, it is evident that relatively few plants have been examined in sufficient detail, and it is impossible to be sure of the patterns of occurrence or of evolutionary origins for the various mechanisms. However, it is clear that various SI systems are very efficient as regulators of the breeding system, allowing a plant to cross with most of its near neighbors while avoiding being fertilized by gametes from its own pollen or that of a near relative. SI systems are therefore particularly suited to plants normally pollinated by external agents, such that the plant cannot choose its own mates. But the boundaries between **antiselfing systems** are actually rather blurred. Certainly some plants with an identified SI system successfully avoid any selfing, but others appear to be only partially self-incompatible, especially in the borage family (Karoly 1994). Furthermore, in a few examples from the Boraginaceae and Onagraceae and in wallflowers (*Cheiranthus*), there is an additional form of SI, *cryptic*, where selfing may often occur but is rarely successful because any cross-pollen that lands on the stigma will germinate and grow more quickly (R. Bowman 1987) and so outcompete the self-pollen. This is very probably a quite common way of avoiding the drawbacks of selfing, and underlines again the rather fuzzy borders between physiological SI and some of the built-in side effects that result from the disadvantageous effects of selfing.

5. Methods for Avoiding Selfing between Flowers within a Plant

If a zoophilous plant is multiflowered, a trait that may be essential to get a big enough display to attract any pollinators, there is a substantial risk that a visiting animal will go from one flower to another on the same individual, and there would then effectively be no outcrossing between plants even though pollen is moving between flowers. As far as the plant is concerned, this particular form of selfing (geitonogamy) is genetically no different from, and no better than, selfing within a flower (autogamy).

Of course one advantage of SI mechanisms to control selfing is that they work just as well for the whole plant as they do for individual flowers, so for about a third of plant species, geitonogamy is avoided using the poststigmatic internal mechanisms described in the previous section. However, SI systems do not prevent reception of self-pollen, which can still be detrimental to the plant, potentially clogging up the stigma and preventing outcross pollen tubes from growing. Self-pollination in a self-incompatible plant also undermines the male function of the flower, since that pollen could have been better used to sire seeds on conspecific plants. Hence measures to avoid self-pollen deposition beyond relying on SI may still be useful and appropriate (Bertin 1993; Snow et al. 1996).

Just how important geitonogamous selfing is as a selective agent remains controversial. Some authors argue that it does indeed markedly reduce reproductive success in plants, producing high selfing rates in many self-compatible species and reduced seed-set in many SI species. However, the likelihood of geitonogamous pollen transfer may often be rather low and should clearly vary between species in relation to certain key variables:

1. The number of open flowers on the plant (most plants have few flowers at a time)
2. The behavior of the flower visitors (most pollinators visit even fewer flowers per plant than are available, but the ordering of their flower visits within a plant may be crucial [Karron et al. 2009])
3. The level of pollen deposition on each visit (with significant pollen carryover from previously visited individuals (see above and Robertson 1992), which also varies with visitor type
4. Local plant density.

While the "risk" of geitonogamous selfing can therefore be reduced by adjusting these factors, it can also be almost entirely avoided in several additional ways by appropriate flower design, covered briefly below. Chapter 21 will explore these issues in more depth, since they can have a major influence on flowering behavior in plants.

Temporal Separation of Male and Female Parts

Temporal separation, either protandry or protogyny, as described in *Methods for Avoiding Selfing within a Flower*, could be effective for a whole plant as well as for individual flowers, but this would require that all the flowers that were open at any one time were in the same sexual phase. In practice this is rather unusual except in some composites, umbellifers, and a few tropical plants (Webb and Lloyd 1986). Where synchrony does occur, and particularly where there are multiple flowers in an inflorescence, the switch from one phase to the other usually requires a trigger rather than occurring automatically as the flower matures. For example, in *Aster* complete removal of pollen is necessary before the florets expose their stigmas. In umbellifers too, stigmas in one umbel generally only become receptive when *all* the flowers have passed the staminate phase (T. Richardson and Stephenson 1989; M. Proctor et al. 1996) and there is synchronous development within each successive order of umbels (i.e., from the early-opening central primary umbel to the subsequent secondary and tertiary outer rings of umbels; Cruden and Hermann-Parker 1977; table 2.4); hence visitors may in principle choose between male (pollen and nectar) and female (nectar only) umbels (e.g., Davila and Wardle 2007). Another way to achieve temporal separation occurs in longer-lived plants that are unisexual at any one time but change sex within a year or between years, giving rise to the **sequential hermaphrodite** condition (McDade 1986; Cruden 1988).

For most plants with solitary flowers, or with less tightly organized inflorescences, and where flowering periods are longer than a few days, it is disadvantageous to have all the flowers at the same age and same sexual stage at once. It is then that spatial mechanisms come into play.

Spatial Separation of Male and Female Parts

Even if the flowers on a plant are not all synchronized in the same sexual phase, selfing may still be partly avoided by spatial patterning of flower age combined with manipulation of a visitor's behavior. A common strategy is to have younger male flowers at the top of a spike-shaped inflorescence, and older female ones lower down, or (less commonly) vice versa. Many familiar spiked garden plants, such as *Delphinium, Aconitum,* and *Corydalis*, are strongly protandrous and also open their flowers sequentially from the base upward. Thus for most of their flowering period they have young male flowers at the top of the spike and older, female-phase flowers basally. This very neatly exploits the tendency of many visitors, especially bees, to work upward on spikes (Corbet et al. 1981; Kudo et al. 2001). A bee leaves a plant at the top, carrying pollen from the fresh male flowers it has just visited and lands at the base of another plant, where it deposits that pollen on fresh stigmas in the older female-phase flowers (fig. 3.14). The same phenomenon can occur with other pollinators, and Jersakova and Johnson (2007) specifically tested the effects with a moth-pollinated orchid, manipulating levels of protandry and showing much greater geitonogamy on upper flowers of the artificially nonprotandrous inflorescences.

Plants using this strategy may also control visitor movements more explicitly by having differential nectar rewards in each flower phase (chapter 8), that is, higher rewards in the basal flowers, with progressively lower nectar volumes higher up the spike that are still acceptable to the bee since it does not have to move far to reach successive flowers.

Specific Mechanical Blocking Systems

Although rare, there are occasional examples where plants can physically block self-pollen from landing on a stigma on the same plant. One particularly neat case occurs in the orchid *Eulophia foliosa*, which is pollinated by elaterid beetles (Peter and Johnson 2006). Here there is an anther "cap" that remains on the **pollinarium** after it has been picked up by a beetle, the retention time being on average about 5 minutes. The residence time of a beetle on an inflorescence averages around 6.5 minutes (though it is somewhat dependent on humidity), so that the cap does not usually fall off (and thereby allow the pollinarium to fit on to a stigma) until after the beetle has left the plant where it first picked up the pollen.

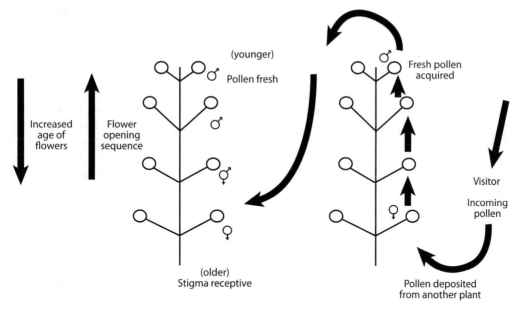

Figure 3.14 The principle underlying the advantages of upward movement by animals during visits to flower spikes that open from the bottom sequentially and that have protandrous flowers so that the upper younger flowers are male-phase.

Reversion to Single-Sex Flowers

For a plant in which geitonogamy is too much of a problem, the ultimate solution is to switch to separate-sex flowers. Having separate male and female flowers on one plant (monoecy) clearly only partly solves the problem of selfing, and geitonogamy can persist. Intermediate conditions, with separate sex flowers occurring on the same plant as hermaphrodite flowers (table 3.1), can in some cases improve matters, but having separate male and female plants (dioecy) is the only way to completely eliminate geitonogamous selfing.

Monoecy characterizes about 10% of all plant species; it is most widespread in the familiar temperate tree families and genera, including hazel, birch, beech, oak, sallow, poplar, hornbeam, and witch-hazel, nearly all of which are also protogynous. Most **conifers** are also monoecious, so that in terms of biomass (with most of the temperate and boreal forests contributing), it is evident that monoecy is extremely important. Dioecy is present in quite high proportions in a few families, such as Euphorbiaceae and Cucurbitaceae, but overall in only about 5%–7% of all angiosperm spe-

cies. A rather smaller percentage of the temperate floras are dioecious (notably including willows, hollies, and nettles). Habitat-related effects are therefore apparent for both kinds of unisexuality: higher levels of dioecy occur in Mediterranean shrubs, and the occurrence is even higher in tropical forest trees (up to 11%–30% in tropical habitats generally, with particularly high levels in the **Neotropics**) and also higher on isolated islands (table 3.7 and chapter 27), whereas monoecy is more common in temperate than in tropical habitats. These values come mainly from Bawa (1980a), who conducted a detailed review. Renner and Ricklefs (1995) produced a survey with phylogenetic corrections, revealing some further patterns, which are summarized in table 3.8.

Among herbaceous taxa, sexuality is more varied than in trees and shrubs, with monoecious and dioecious species occurring sporadically, even in the same subfamily and genus as hermaphrodite species.

An advantage of monoecy is the ability to vary the proportions of male and female flowers and, in some cases, to change this ratio facultatively with age or environment. Older plants may more easily support fruits and so have more female flowers, while young plants

TABLE 3.7

Occurrence of Dioecy in Plants in Relation to Habitat

Location	% dioecy
Tropical forest	
Costa Rica, lowland	23
Panama, lowland	9
Venezuela, montane	31
Temperate, Mediterranean, and semi-arid	
USA	4
UK	3
California	3
South Australia	4
India	7
Island	
Jamaica, montane forest	21
Hawaii	28
New Zealand	13

Source: Modified from Bawa 1980a, 1990, and others.

TABLE 3.8

Dioecy in Relation to Plant Group, Mode of Pollination, and Geographical Occurrence

Plant Group (percentage distribution)	
Monocots	5%
Dicots	8%
Mode of pollination (percentage distribution)	
Insect pollinated	5%
Vertebrate pollinated	<1%
Abiotically pollinated	21%
Geographical occurrence (number of genera)	
Old World tropics	402
New World tropics	217
Pantropical	86
Temperate	149
Other (mainly Australasia)	75

Source: From Renner and Ricklefs, 1995.

probably do better to invest in pollen as a contribution to the next generation. In general, male flowers are larger and showier than female flowers (Bawa 1980a; Givnish 1982; Q. Kay and Stevens 1986; Vamosi and Otto 2002; but see *Flower Sex and Flower Design* in

chapter 2), though both male and female flowers tend to be markedly smaller and less conspicuous than hermaphrodite flowers in close relatives. The greater attraction of male flowers is crucial for unisexual plants (often being enhanced by early flowering and longer flowering period), since pollinators *must* visit males before they go to females, and male flowers normally benefit from many visits whereas a female flower may be fully pollinated by just one visit. It is often difficult to see why visitors go to the female flowers at all, as these have no pollen reward and often have much less nectar (chapter 8); so one solution for the plant is **intersexual mimicry**, whereby females resemble males as much as possible so that visitors cannot discriminate between them (chapter 23). Vamosi and Otto (2002) pointed out that species with showy male flowers may be at greater risk of extinction in years with poor pollinator availability and that this may be another reason why dioecious plants are uncommon relative to hermaphrodites.

Reversion to monoecy may be rather rare; the distribution of the condition in modern plants suggests that it is often ancestral (in most trees, sedges, and perhaps **aroids**) rather than secondary. In contrast, evolution of dioecy from hermaphrodite flowers has evidently happened rather often, even though it requires each plant to reduce its reproductive potential by half. Removing the possibility of selfing may be part of the reason, but avoidance of stigmatic blocking by "clogging" with self-pollen may be even more important. Possibly, plants with small flowers are particularly likely to revert, since they are less able to avoid selfing by spatial arrangements (Proctor et al. 1996). Reversion also seems to occur more often in heterostylous plants, perhaps because many insect visitors become unable to reach the pin anthers or the thrum stigmas (Beach and Bawa 1980; Muenchow and Grebus 1989). However, D. Charlesworth (1993) argued that dioecious flowers could readily arise in species whenever pollen is not itself an attractant. Hence they might be selected for in almost any plant that is not normally bee pollinated; as soon as a plant is freed from the constraint of attracting bees, it can revert to smaller and cheaper and eventually unisexual flowers, perhaps passing through a gynodioecious stage with both female and hermaphrodite flowers on some plants (B. Charlesworth and Charlesworth 1978). This interpretation remains controversial because most of the examples where it might apply are close relatives of hermaphrodites; furthermore, it cannot easily explain dioecy in large taxa such as willows, hollies, some aquatic plants, and many tropical-forest tree taxa.

Summarizing this section, it should be clear that avoidance of selfing can be seen as a key factor in the evolution of plant life-cycles (monoecy and dioecy), of self-incompatibility, of flowering patterns (male and female phases, often spatially organized on the plant, and few open flowers per day), and of floral design (separation of male and female organs). These kinds of selective pressure have been invoked repeatedly since Darwin's insights into floral biology (1876) and represent major themes over the following 130 years in the pollination literature, as reviewed by D. Charlesworth and Charlesworth (1987), de Jong et al. (1992, 1993), and Snow et al. (1996).

6. Methods for Ensuring Selfing

Many species of plant that normally outcross are self-fertile and will use selfing as a backup (reproductive assurance) if cross-pollination fails. A common way of doing so is via the differential growth rates of self- and cross-pollen tubes (Stephenson and Bertin 1983), so that any cross-pollen received will "win" against competing self-pollen by reaching the ovules first; but if no cross-pollen is received, the self-pollen will achieve fertilization. Another method is for the stigma lobes to begin their period of open receptivity very distant from (and usually above) the flower's anthers, but then to bend down later in their life if they have not received any pollen, so arriving in a position to be selfed as a last resort. This process is well documented in bellflowers (*Campanula*, Faegri and van der Pijl 1979).

A more extreme version of facultative selfing when crossing has failed occurs at the plant, rather than the single-flower, level. Some plants have two different kinds of flowers (i.e., the flowers are dimorphic): those in the early season being open and attractive to visitors and incapable of selfing, and those maturing later being closed and relatively cryptic. The latter are termed cleistogamous flowers (normal flowers being *chasmogamous*), and they are reviewed by Culley and Klooster (2007). Classic examples are the woodland violets (*Viola*) of Eurasia, which have attractive sweet-smelling flowers in early spring and are visited by the relatively few pollinators active at that time but which in later months produce small, closed-up flowers that are entirely self-pollinating and mainly hidden beneath the foliage (Beattie 1969), thus ensuring at least some seed set. (Note, however, that Rocky Mountain *Viola* species showed no relation between allocation to cleistogamy and flowering time [Forrest and Thomson 2008], although there was more cleistogamy in the larger plants). A number of other plants, such as the wood-sorrels (*Oxalis*) and some short-lived legumes and grasses, behave rather like European violets, and some form of cleistogamy is now known in nearly 700 species across 50 families, particularly in association with heterogeneous environments.

A few plants are unusual in having "deliberate" selfing mechanisms, either as their sole pollination mechanism as they age or as mechanisms that come into play only when visitation has not occurred. An example of the former is the orchid *Holcoglossum amesianum*, where the anther rotates through 360° to place its own pollinia into the stigma. In this species no visitation apparently occurs, and the plant grows in windless, insect-scarce habitats; it probably has no other options than selfing (K. Liu et al. 2006). Another case is *Caulokaempferia coenobialis* (Zingiberaceae), where a pollen "drop" is automatically released from the freshly dehiscing anther on to the horizontal style below, forming an oily film with entangling threads; this film then gradually slides toward the stigma over the next 9–24 h and provides a delayed self-pollination mechanism as backup, visits by pollinators again being extremely rare in this plant (Y. Wang et al. 2005). Examples of facultative deliberate selfing include *Incarvillea sinensis*, which is normally bee pollinated, where the abscising corolla with adherent stamens from an unvisited ageing flower is dragged by wind as it droops, so allowing pollen to contact the stigma (Qu et al. 2007); and *Mimulus verbenaceus*, where the senescing corolla drags anthers down over the style by its own weight as it abscises (Vickery 2008). In the orchid *Cyrtopodium polyphyllum*, accumulating rainwater on the stigma dissolves the adhesive material to form a droplet of sticky fluid that contacts the pollinia, and subsequent evaporation moves the pollen grains back onto the stigma (Pansarin et al. 2008).

Chapter 4

EVOLUTION OF FLOWERS, POLLINATION, AND PLANT DIVERSITY

Outline

1. The Origin and Early Evolution of Flowers
2. The Diversification of the Angiosperms
3. The Advantages of Animal Pollination
4. Pollination, Floral Variation, Plant Speciation, and Plant Diversity

---

1. The Origin and Early Evolution of Flowers

Animals of various kinds must have been interacting with green plants—both for shelter and to find food—almost as soon as land plants evolved. Any foliage would provide some shelter, both in terms of alleviating the microclimatic conditions and by offering some possibility of hiding from predators. It would also offer decomposing plant material as potential food, where **microorganisms** had already softened and partially digested the plant walls. However, eating fresh, green plant tissue was a later development, as true herbivory is a difficult skill that requires modifications to the gut to extract the maximum nutrition from rather poor-quality food and the specialization of mouthparts to pierce or chew the tough outer layers of **cellulose** or **lignin** that protect and support terrestrial plants.

Coming into close association with plants would provide new opportunities for some animals, and from the late **Silurian** period some taxa of insects were eating, and probably inadvertently dispersing, various kinds of plant spores, a habit that was commonplace by the late **Carboniferous** period, 300 million years ago (MYA). For example, some **seed ferns** had very large spores, which were probably too big to have been wind-borne and would have been spread via animal fur or in feces. Animal-mediated "pollination" in the very loosest sense may therefore have occurred well before "modern" flowers appeared, in groups of seed plants that are now extinct. Some insects still feed primarily on fungal and moss spores today, inevitably also helping to distribute them indiscriminately and so occasionally to move gametes between individuals.

Gymnosperms, a taxonomically loose term for all the seed plants other than the flowering plants, dominated the land for at least 200 million years and evolved a range of reproductive systems that could provide the starting point for more specific and targeted modern floral pollination. Most groups of nonangiosperms today use abiotic pollination to move male gametes between plants (as did most of those that are now extinct). The conifers, by far the largest surviving order of gymnosperms, have reproductive structures organized in unisexual **cones**, either on the same tree or separate trees, and all are essentially wind pollinated, producing copious pollen that is captured on a sticky **pollination fluid** near the micropyle of the ovule in female cones (chapter 19).

So where and when did the transition from wind to animal vector occur? Culley et al. (2002) reviewed the patterns of evolution of wind pollination in angiosperms and found **anemophily** to be associated with small simple flowers and rather dry pollen, as well as with certain environmental conditions (see table 4.1

TABLE 4.1

The Traits and Environments Associated with Wind Pollination and Biotic Pollination

	Wind pollination	Biotic pollination
Floral morphology		
Flower location	Held apart from vegetation	Variable
Inflorescence structure	Pendant, often condensed	Highly variable
Flower structure	Simple, small	Showy, often large
Flower sex	Usually unisexual	Usually hermaphrodite
Perianth	Absent	Showy
Stigmas	Feathery	Simple
Pollen	10–50 µm diam.; usually dry	Very variable size; not dry
Stamens	Long	Variable
Nectaries	Absent	Present
Fragrance	Absent	Usually present
Habitat		
Plant density	Moderate to high	Low to high
Plant diversity	Low	Low to high
Windspeed	Low to moderate	None to low
Rainfall	Infrequent	Rare to common
Humidity	Low	Low to high

for an overview), including a moderate density of plants with low diversity, and moderate wind speeds in a fairly dry climate. These conditions probably did exist during the early phases of flowering-plant evolution, and subsequent changes in these conditions may have favored biotic pollination or **zoophily**. Changes between the two pollination types may still occur, often via an intermediate or transitional state, when both wind and insects have a pollen-transferring effect (**ambophily**, perhaps a more common condition than is usually recognized). But exactly where and when biotic pollination arose is still uncertain. As figure 4.1 shows, there are actually three groups of land plants that use animal pollination, and two of them are "older" and rather unfamiliar taxa—the **cycads**, or **tree ferns**, and a tiny group of gymnosperms known as the **gnetaleans**, both of these groups normally being dioecious. It is therefore quite likely that animal pollination arose at least twice, each time in a different group (Norstog 1987; D. Lloyd and Wells 1992; Labandeira et al. 2007).

Within the tree ferns, wind may play the major pollinator role in the family Cycadaceae, whereas in the family Zamiaceae wind just moves pollen to the edges of the cone-like flowers (fig. 4.2A; plate 36B), but insects, especially small beetles and occasionally thrips, are attracted to these cones by the sticky pollination fluid (Norstog 1987), which contains sugar and **amino acids,** and the animals thus help to move the pollen in to the ovules. Animal-pollinated cycads certainly also exist (e.g., Kono and Tobe 2007) and have different scent profiles from anemophilous species (Pellmyr et al. 1991), and cycad toxins, such as cycasin, may also play some role in attracting the pollinating animals (Schneider et al. 2002). Terry et al. (2007) reported a role for cone warmth in attracting pollinators to cycads, the heat from the plant tissue interacting with odor signals, although this thermogenic effect is absent in the isolated African cycad genus *Stangeria*, where the cones appear to mimic the scent of fermenting fruits and thereby attract pollinating sap beetles (Proches and Johnson 2009).

In the Gnetales the three known genera are *Gnetum* (from the tropics), *Welwitschia* (only from the Namib Desert), and *Ephedra* (from various Mediterranean climate zones). In the first two genera, wind pollination

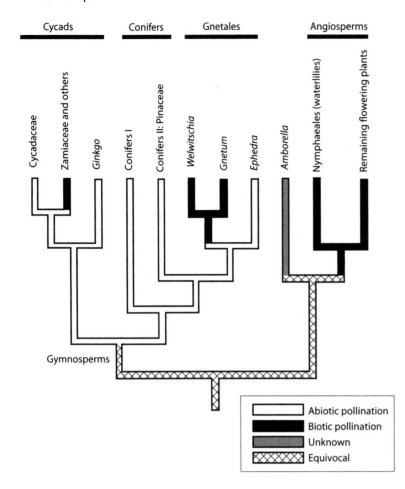

Figure 4.1 A modern phylogeny of seed plants, indicating the occurrence of abiotic and biotic pollination modes; on this view, biotic pollination has arisen at least three times. (Redrawn from Pellmyr 2002, based on earlier sources.)

is probably negligible, and pollen movement is certainly aided by nonspecialist insect visitors. In fact in *Gnetum,* insect pollination is probably the norm (Kato and Inoue 1994; Momose, Ishii et al. 1998), given the lack of wind in tropical forests; various animals, but especially small moths, visit the flowers (again somewhat cone-like, fig. 4.2B) to lick the sugary droplets of pollination fluid off the ovules, and this liquid also helps to trap and rehydrate the pollen. In *Ephedra,* wind pollination does occur in some species, but fly pollination has also been reported (Bino et al. 1984). We should note in passing that the Gnetales are also important as the potential ancestors of angiosperms (Doyle and Donohue 1987; P. Bell 1992) in that they too have the double fertilization system (chapter 3). Other now extinct gymnosperm groups may also have had more flowerlike structures (e.g., the **Bennettitales**) and have attracted insects, so that very unspe-

cialized insect pollination was probably occasional even as early as the Carboniferous (A. Scott and Taylor 1983) and was likely to be fairly common in various **Cretaceous** gymnosperms (P. Crane et al. 1995; Labandeira et al. 2007).

Angiosperms have flowers by definition but are also defined by particular vessel types in their stems, by pollen grains with columnar structures in their protective layers (chapter 7), and by the double fertilization system. Traces of true angiosperms (Anthophyta) first appear in the fossil beds of the lower Cretaceous, around 136–130 MYA. Deposits from that period reveal the dominance of ferns and gymnosperms, but among these are the first signs of angiosperm pollen, and fossils of angiosperm leaves and fruits of similar date also occur. However, an origin somewhat earlier than this is probable (D. Lloyd and Wells 1992), and from molecular studies, the earliest common ancestor

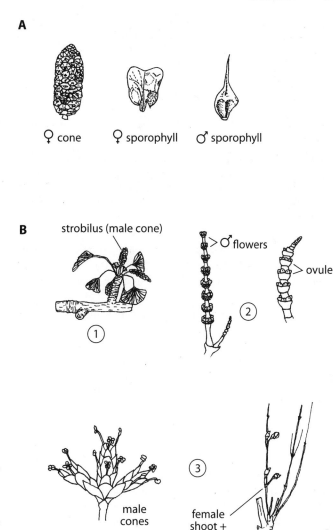

Figure 4.2 Nonangiosperm flowers: (A) cone-like flowers of Zamiaceae, showing a female cone and a single female sporophyll, and a single male sphorophyll; (B) simple flowers—(1) male cone of *Ginkgo*, (2) male and female cones of *Gnetum*, and (3) male cones and female shoots of *Ephedra*. (Redrawn from Proctor et al. 1996, partly based on earlier sources and on Bell 1992.)

may indeed have to be traced back at least 190–200 million years (D. Soltis et al. 2007).

These first angiosperms were probably small, weedy herbaceous plants that appeared in tropical habitats in the understory of forests, only moving to higher latitudes after about another 20 million years, but then doing so very rapidly. They would probably have had their pollen moved mainly by wind, requiring only rather simple "flower" structures perhaps similar to the cones of the cycads and Gnetales with large extended stamens dispensing abundant pollen and sticky stigmatic surfaces. Molecular analysis has shown that the

Amborellaceae is probably the most primitive living angiosperm taxon, with the Nymphaeaceae, Chloranthaceae, and other minor groups as basal sister lineages from which the magnoliids, monocots, and dicots arose (appendix). The tiny flowers of the Chloranthaceae (fig. 4.3) provide a likely model for those of the earliest angiosperms (P. Crane et al. 1995; Weberling 2007), being just a few millimeters in size and borne in dense inflorescences. These early bisexual flowers most probably originated by the appearance of "ectopic" ovules in the center of a male cone-like structure; this is the "mostly male" hypothesis for flower

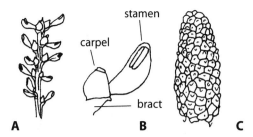

Figure 4.3 Perhaps the earliest angiosperm flowers, as seen in modern Chloranthaceae: (A) an inflorescence of *Sarcandra* and (B) a single flower; (C) the all-male inflorescence of *Hedyosum*, with simple flowers consisting of one stamen. (Redrawn from Endress 1994.)

origins, supported by genetic and developmental evidence (Frohlich and Parker 2000). Glover (2007) provided a review of developmental processes in relation to floral evolution.

Even in these earliest forms, which were somewhat like a modern magnoliid in their reproductive tissue arrangements but without petals, some insect transport of pollen may well have occurred, and an element of insect pollination could be regarded as almost ancestral for the angiosperms as a whole (P. Crane et al. 1995; Friis et al. 2001; Hu et al. 2008).

2. The Diversification of the Angiosperms

From modern genetic knowledge, it is clear that early angiosperms were equipped with a high degree of developmental flexibility based on a "tool kit" of floral genes that has been narrowed down over time. The early forms possessed a range of MADS genes that control floral organ identity, and duplications of these have been important in generating diverse floral forms (D. Soltis et al. 2007). There was clearly a rapid diversification among the basal angiosperms, between about 143 and 140 MYA (M. Moore et al. 2007). And within about 20 million years of the initial appearance of floral traits, much larger and more complex flowers in the Magnoliales are evident as fossils (fig. 4.4), rapidly followed by forms such as saxifrages, planes, and lilies (Crepet et al. 1991 or P. Soltis and Soltis 2004 for reviews). Within another 40 million years or so, the angiosperm pollens become hugely abundant in nearly all fossil beds (fig. 4.5). This suggests that the angiosperms had strong evolutionary advantages over their ancestors and contemporary rivals, and it seems highly

likely that this advantage was linked to full adoption and improvement of biotic pollination. A major reason may be that whereas structures that could reasonably be termed flowers appeared in both abiotically and biotically pollinated earlier plant groups, only the angiosperms show *hermaphrodite* flowers. It is not entirely clear whether this is their primitive state, because the Amborellaceae family that is probably most basal is unisexual and the Chloranthaceae show both versions of sexuality; D. Lloyd and Wells (1992) assumed a bisexual ancestor. But hermaphrodite flowers certainly arose very early in angiosperm evolution and must have given a major spur to the development of pollination systems, because a single flower could then achieve visitor contact with both male and female parts. Hence a single visit could effect both male-function and female-function pollination, and manipulation of visitor behavior could begin to pay off; not surprisingly, 80% of all modern angiosperms have hermaphrodite flowers.

The first angiosperm flower visitors would rarely have achieved single-visit success, though. They could have included collembolans, beetles, short-tongued midge-like flies, and small wasps, as well as some extinct groups (Endress 1987; Willemstein 1987; Gottsberger 1988; Kato and Inoue 1994), though Thien et al. (2009) have argued that dipterans (flies) were probably the first really important pollinators. Many of these early visitors presumably operated in a rather clumsy fashion, chewing at the softer parts of the flowers, eating the pollen, or licking at stigmatic exudates, as still exhibited by the pollinators in groups such as the Winteraceae (D. Lloyd and Wells 1992). Note that this may have been a significant selective pressure on development of protective carpel tissues around the previously exposed ovules and the seeds they then produce—this development being the defining difference between gymnosperms ("exposed seeds") and angiosperms ("hidden seeds"). The carpels in turn would enable the development of a protruding style above the ovules, which may have been a key factor in promoting pollen tube competition and in due course allowing the development of stylar self-incompatibility systems (chapter 3), such that hermaphrodite flowers could largely avoid self-fertilization and begin to select for genetically superior pollen grains (Mulcahy and Mulcahy 1987).

Thus, as long as a particular plant species was abundant in a given habitat, and the generalist clumsy

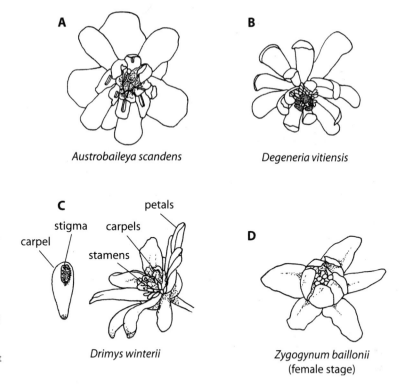

A *Austrobaileya scandens*

B *Degeneria vitiensis*

C *Drimys winterii*

D *Zygogynum baillonii* (female stage)

Figure 4.4 Flowers of some primitive groups within the Magnoliales and Winteraceae. In (A) and (B) the stippled areas are substantial staminodes. (Parts (A), (B), and (D) redrawn from Endress 1994; part (C) redrawn from Proctor et al. 1996.)

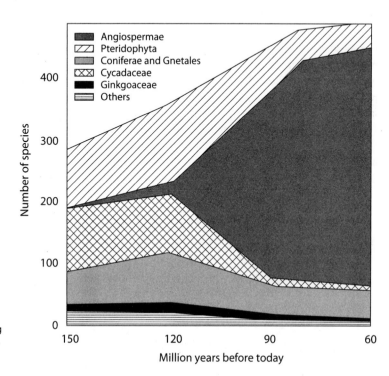

Figure 4.5 The timing of major radiations in the land plants, with angiosperm radiation becoming dominant from about 120 million years ago. (Redrawn from Lunau 2004.)

visitors were likewise abundant, this kind of nonspecific animal-plant relationship centering on angiosperm flowers would still confer an advantage over wind pollination (although there were probably several independent reversions to abiotic pollination when these conditions were not met [P. Crane et al. 1995; Friis et al. 2001]). But all these flower-eating (florivorous) behaviors are deleterious to the flower and would select for further floral modifications to achieve more efficient take-up and transport of pollen and to limit the amount of flower damage and the amount of pollen eaten. Pollen is expensive both in the reproductive sense of loss of gametes and in a strictly nutritional sense because it uses up both the nitrogen and the phosphorus reserves of the plant (Petanidou and Vokou 1990; chapter 7); it is also nonrenewable. The provision of an alternative food that was almost purely carbohydrate (admittedly using up carbon reserves, but plants are rarely short of carbon) would be a major economic advantage. Hence the first nectar-producing flowers would get significant benefits, turning a visitor's attention away from the pollen as food and offering instead an ideal fuel in the form of sugars, which the plant could "top up" if needed. Here the early dominance of insects as flower-feeders would be crucial, since adult winged insects are usually short-lived and often require little else but carbohydrate as food, with only the larval stages undergoing significant growth that is dependent on protein or fat intake. Many plants can produce sugary exudates, and phylogenetically basal terrestrial groups often secrete nectar at the bases of the petals or on the anthers. Early angiosperms were presumably under strong pressure to develop more organized nectaries and to deliver an appropriate nectar supply for their insect customers (A. Meeuse 1978), while increasingly making their pollen rather sparse and cryptic, delivering small deposits of it onto some part of a visitor's body well away from the mouthparts. As soon as the fitness gain to the plant resulting from its pollen being moved exceeded the losses of tissue eaten, then selection on the plant would change from deterring to attracting animals, and the latter in turn would, in effect, switch from being herbivores to pollinators. And while a nonspecific generalist flower design probably remained a very good solution for reasonably abundant plants in communities that were not too diverse, there would also be pressure to be more specialized for some plants. This issue of the relative merits and relative importance of generalization and

specialization in flowers is one that will recur repeatedly in later chapters.

Throughout the Cretaceous period, then, there were probably two main lines of angiosperm evolution. One involved the successful insect-pollinated flowers, which had large, radially symmetrical, bowl-shaped forms that offered a relatively protected site for easy landing and feeding, increasingly showy petals and simple fruity scents, and a tendency to fix the numbers of floral parts (especially in whorls of five); hence the Rosidae are well represented by the mid-Cretaceous, and floral nectaries are evident in late Cretaceous rosid fossils. Paralleling these were the wind-pollinated plants, particularly larger trees, which were developing tiny flowers on catkin-like inflorescences that were highly suited to the dry breezy conditions of the time.

Between about 80 and 50 MYA, straddling the time of the great Cretaceous extinction at 65 MYA, the biotically pollinated flowers began to show much more variation, which was at least in part linked with the evolution of more varied insect taxa (fig. 4.6). Longer-tongued flies appeared in this time period, soon followed by butterflies and moths, all providing major catalysts to floral evolution. The spread of humid tropical forests would also have boosted floral radiation, with more than half the total angiosperm diversity in the world now being located in those habitats. Zygomorphic (bilaterally symmetrical) flowers were present by the end of the Cretaceous, with their petals beginning to fuse basally forming tubular corollas. Crucially, by about this time one group of originally predatory sphecid wasps had evolved into the specialist flower-visiting bees (Apoidea). The oldest definite bee fossils date from about 80 MYA, and these specimens are very like modern stingless *Trigona* bees, which are highly specialized and eusocial, suggesting that bee evolution had begun somewhat earlier (Michener and Grimaldi 1988).

Once bees were present, the selective pressure on flower foods changed again. Bees are (almost) unique in feeding their larvae on pollen, and they therefore require far more pollen to be offered in the flowers that they favor than do other pollinators; but in turn they "repay" the plant for this supply by visiting far more flowers than any other kind of visitor as they industriously stock every one of their larval cells full of a nutritious pollen-nectar mixture. With bees around and becoming very effective as pollinators, the demand for more generous pollen-producing flowers would be

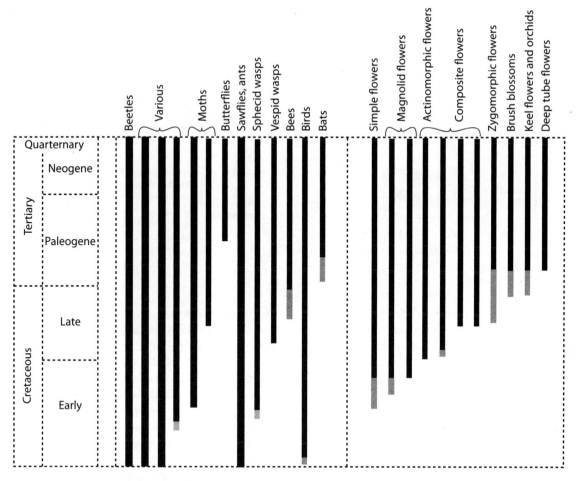

Figure 4.6 Timings of radiations in insect evolution, with major pollinating taxa appearing between 85 and 50 million years ago, at similar times to the appearance of zygomorphic flowers and brush blossoms and then keeled papilionate flowers. Origins of vertebrate groups containing pollinators are also shown; functional pollinators within these groups appeared rather later. (Modified from Schoonhoven et al. 2005, based on earlier authors.)

substantially increased again. Today a high proportion of plants in temperate ecosystems produce relatively large volumes of pollen, sometimes with no nectar reward. They can cope with the loss of much of this pollen to feed succeeding generations of bees, because bee behavior ensures that enough of the pollen from the bees' bodies will reach an appropriate stigma in a conspecific flower and achieve effective pollination for the pollen-donating plant.

Bees may also have provided the spur for evolution of ever more elaborate floral forms, with three-dimensional complexity arising from fusion or differential growth of petals and other structures; the bees' ability to discriminate between, and to learn to handle,

varied flower types would promote increased flower constancy and hence favor specialization in both partners—and eventual speciation in the plants. Certainly today, bee-plant genera are more **speciose** than genera with more generalist visitor spectra: in California, for example, one survey concluded that bee-plants averaged 5.9 species per genus, while generalists averaged only 3.4 species per genus (V. Grant 1949). Thus rosid flower types diversified widely, and by the early Tertiary the more derived asterid flowers were also appearing and diversifying.

The most recent major development in biotic pollination was the advent of vertebrate pollination, with birds and bats both getting involved in the early to

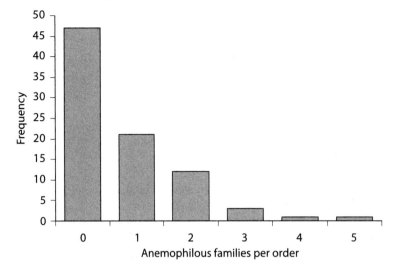

Figure 4.7 Numerical occurrence of abiotic (anemophilous) pollination: most orders of plants have no families showing anemophily, although two orders (Caryophyllales and Sapindales) do each contain four to five wind-pollinated families. (Redrawn from Ackerman 2000.)

mid-Tertiary (approximately 60–50MYA). Very long-tubed flowers associated with hummingbirds appeared at around these times, and other vertebrate-flower associations, with bats, perching birds, and nonflying mammals, all date within the last 40–50 MY (fig. 4.6). From here onward, though, it becomes much harder to see broad trends in pollination evolution, because plants changed to more or to less specialized systems, or switched to more or less selfing, or even reverted to wind pollination, according to the particular demands of their changing habitats, pollinator abundance, and competitive interactions with other plant species. Some of the trends already mentioned clearly went into reverse for a few plant taxa, notably the grasses, which are now mainly wind pollinated, and the composite and umbellifer subgroups that now have tiny rather nonshowy flowers aggregated into attractive inflorescences.

Overall, it is clear from figure 4.7 that as a means of cross-fertilization in plants, biotic (zoophilous) pollination, by whatever animal group, is technically uncommon when assessed in terms of the numbers of broad taxonomic groups that use it. In practice it is *numerically* dominant, among species, because the angiosperms adopted it so rapidly and fairly comprehensively and have come to dominate all other plant taxa on land in all habitats other than the boreal forests. Note, however, that it is not possible to say what proportion of modern angiosperms is biotically pollinated, since large areas of the tropics are not adequately

investigated. We do know that the radiation of angiosperms has very closely paralleled the radiation of pollinators (Pellmyr 1992a; Grimaldi, 1999). The Coleoptera, Diptera, and Lepidoptera had all evolved earlier, but they did not radiate greatly until the angiosperms appeared, and the diversification of flower-visiting groups of Hymenoptera clearly began as the angiosperms began to radiate. Furthermore, taxa with animal pollination are now far more speciose than families that have abiotic pollination (Dodd et al. 1999). Hughes (1976) recognized only about 500 plant types, roughly equivalent to species, in the late Carboniferous (about 290–280 MYA), increasing to 3,000 types by the early Cretaceous (150–140 MYA) and to 22,000 types by its end (65 MYA); in the modern world there are perhaps 250,000 plant species. All of this indicates rather clearly that the evolution and exploitation of animal pollinators has had a strong influence on plant speciation, and overall patterns of floral change through evolutionary time can readily be assembled (fig. 4.8).

But whether the link between plant diversification, on the one hand, and biotic pollination and floral specialization, on the other, is strictly causal is still open to debate (fig. 4.9). Arguments against a key causative role for pollinators cite the point that angiosperm diversification was not particularly rapid at the time of origin of the angiosperms (when at least some animal pollinators were present) but instead took off somewhat later (Sanderson and Donoghue 1994). Midgley

Monocots   Dicots

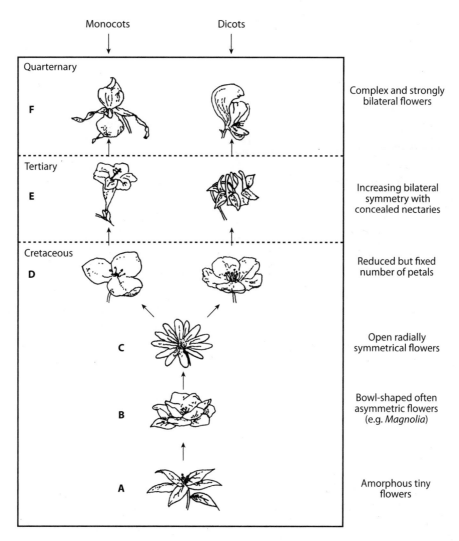

Quarternary	**F**	Complex and strongly bilateral flowers
Tertiary	**E**	Increasing bilateral symmetry with concealed nectaries
Cretaceous	**D**	Reduced but fixed number of petals
	C	Open radially symmetrical flowers
	B	Bowl-shaped often asymmetric flowers (e.g. *Magnolia*)
	A	Amorphous tiny flowers

Figure 4.8 Overall patterns of floral change within the angiosperms through about 100 million years of evolutionary time. (Modified from Leppik 1971 and later authors.)

and Bond (1991) suggested that biotic gene dispersal was not as important in promoting plant diversification as some innovations within plants themselves that affected their growth and regeneration rates. More specifically, Armbruster and Muchhala (2009) argued that the link between specialized pollination and plant species diversity worked the other way around: that is, where there was high diversity within plant clades, the increase in numbers of similar plants within a community tended to promote floral **character displacement**, which was then evident as increasing specialization. This is the *character displacement model* seen in figure 4.9, as contrasted with the three alternative models that work in the reverse direction.

Abiotic pollination merits consideration in its own right (chapter 19), but here it is more useful to move on to the reasons for successful evolution of zoophilous pollination.

3. The Advantages of Animal Pollination

The intervention of animals as a means to move pollen around has proved to be the best way for angiosperms

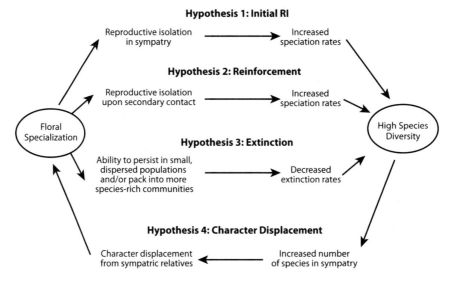

Figure 4.9 Four hypotheses relating floral specialization to plant species richness; the character-displacement hypothesis postulates a causal relationship in the reverse direction from the other three, such that higher plant diversity promotes specialization in flowers. (Redrawn from Armbruster and Muchhala 2009.)

to achieve reproductive success. In a world formerly dominated by gymnosperms and wind pollination, any plant visited by insects that could pick up pollen on their bodies and move it around more directionally than unpredictably moving air would have a significant advantage. There are two key reasons for this:

1. Above all, perhaps, zoophilous pollination is very much cheaper in terms of the pollen amounts needed (table 4.2), often by several orders of magnitude. A single birch catkin may release over five million pollen grains, where an animal-pollinated flower will rarely exceed a few thousand.
2. It can occur in small isolated populations with scattered individual plants. For example, in tropical forests there may be only one tree of a species per square kilometer, with dozens of other species intermingled. Animals can get around this problem because
   a. they can and do form search images for the right plants and seek them out, producing flower constancy (chapter 11) and reducing wasteful heterospecific pollen deposition;
   b. they can move over long distances—especially the flying animals that are dominant as the three main pollinating groups (insects, birds, and

bats), so increasing the *dispersal efficiency* of pollen (Inouye et al. 1994);
   c. they can specifically help in preventing selfing and promoting crossing, if the plant has its stamens and stigma placed carefully and attracts an animal with the right shape and behavior to ensure the precise pick up and subsequent precise placement of pollen.

Hence the need to attract and retain the services of specific animals may become a powerful selection pressure on flowers, and it was this insight that led to the idea of specialist pollination syndromes, a theme that is dominant in the pollination literature. Pollination also, inevitably, becomes a coevolved and *mutualistic* arrangement, in which both participants can benefit.

However, it is crucial to reiterate that pollination by animals, while commonly mutualistic, is *not* altruistic: the animal visitors want food, and pollination is just an unintentional by-product of their activities, one from which they very rarely get any fitness benefit whether they move pollen or not. In practice the plant and animal have a conflict of interest (which will be explored more fully in later chapters), with adaptation and counter-adaptation occurring on both sides as each

TABLE 4.2
Pollen Amounts Required for Different Modes of Pollination

Plant type	Agent	Plant species	Pollen grains per ovule
Trees	Wind	*Corylus avellana* (hazel)	2,550,000
		Fagus sylvatica (beech)	637,000
	Insect	*Acer pseudoplatanus* (sycamore)	94,000
		Tilia cordata (lime)	44,000
Herbs	Wind	*Plantago lanceolata* (plantain)	15,000
	Insect	*Senecio jacobaea* (ragwort)	5,000
		Polygonum bistorta (bistort, adderwort)	6,000

Source: Based on data cited in Faegri and van der Pijl 1979.

tries to get a bigger share of the benefit. The relationship can best be described as *reciprocal exploitation.*

4. Pollination, Floral Variation, Plant Speciation, and Plant Diversity

The extent to which pollinators may have contributed to plant diversification and speciation may be debatable, but there is little doubt that pollinators can both produce **reproductive isolation** (when they are reasonably specialist, at least for the duration of a given foraging trip) and contribute to ongoing gene flow between plants. The end products, in this simplified scenario, would be pollinator-mediated selection on flower features, pollinator-mediated reproductive isolation for the plant, and hence, at least potentially, **cospeciation** of plants and pollinators.

Indications from models of speciation are that even moderate pollinator specialization could initiate and maintain divergence in the form of flowers, by producing **assortative mating** in the plants. Assortative mating is nonrandom fertilization occurring between two individuals that are either much more (+) or much less (−) like each other than the average in one or more characters (color, shape, etc.); positive assortative mating leads to the alleles with similar effects becoming coupled in the population, and so alters the variance in the associated characters.

But beyond models and speculations, is there experimental and observational evidence for any such effects?

Do Pollinators Select for Floral Divergence?

In later chapters of this book we will examine the evidence that small changes in flower morphology are genetically "easy," often relying on changes in single genes (Stuurman et al. 2004 on nectar control; Hoballah et al. 2007 on color control, and Galliot, Stuurman, et al. 2006 for a review of the field). It will also become clear that small changes in flower morphology can alter pollinator behavior and floral choices (chapter 11). But the crucial question in terms of evolution is how far floral form is heritable, and this is surprisingly hard to establish. In greenhouse conditions heritability can be more easily measured and has often been recorded as quite high (e.g., Schoen 1982; Holtsford and Ellstrand 1992; Mitchell and Shaw 1993; Andersson and Widén 1993), but this ignores many possible environmental effects, particularly temperature, which itself can influence flower development and thus features such as flower size (e.g., Kinet et al. 1985). In field conditions it is hard to assess the heritability of flower features since these only become apparent quite late in a plant's life cycle, after many environmental influences will already have operated. However, there are a few careful analyses that do give reasonably controlled values. To pick out three on particularly well-studied plants, *Phlox drummondii* was reported to have a heritability of 0.15 for corolla length and width (Schwaegerle and Levin 1991), *Ipomopsis aggregata* showed strong genetic control of stigma-anther separation (D. Campbell et al. 1994), and *Polemonium viscosum* showed heritabilities close to 1 for corolla length and width but very low heritability for height at flowering and for the number of flowers in the flower spike (Galen 1996a).

Given this kind of information, we then need to be able to demonstrate pollinator-mediated selection. Correlations of flower features with reproductive success are often cited as evidence for such selection, but they can be misleading, because other environmental factors can affect outcomes. For example, in *Polemonium viscosum* low altitude populations are mainly small bee and fly pollinated, but bumblebees (which pollinate higher-altitude populations) prefer larger flowers with broader corollas; an increase of corolla width from 14 to 16 mm gave a 10% increase in female function when measured as seed production (Galen and Newport 1987) and also a 14% increase in male function via increased pollen export per bee visit (Galen and Stanton 1989). But flower size also correlated with the nectar production rate, so it is hard to know which factor was the driver here. Ideally we need not only observed correlations, but also experimental manipulation of the pollinators or of the flower traits. When *Polemonium* flower size and flower reward were manipulated independently, Cresswell and Galen (1991) found that bumblebees were influenced by both; they chose plants by corolla size, but then the number of visits made to flowers on a plant was almost entirely the result of reward size. Galen set up caged populations of plants and introduced bumblebee queens, recording all visits to known flowers on known plants, from which she calculated coefficients of bumblebee-mediated selection (Galen 1989, 1996b). For a single generation, the response of corolla width to bumblebees was estimated as up to17%, and for corolla length, the value was up to 18%. These data suggest that selection over just one to three generations could yield the larger and broader flowers found in the high-altitude populations of this plant, where bumblebee-pollination dominates, thus driving divergence away from the lower-altitude populations visited mainly by smaller bees and flies.

Even where features are manipulated, we still need to be wary of direct and indirect effects. The key study mentioned above by D. Campbell et al. (1994) with *Ipomopsis aggregata* is instructive: hummingbird visitors were assumed to be selecting for greater stigma exsertion from the flowers, with the more exserted stigmas receiving more pollen. But the authors also showed (using independent manipulations of stigma exsertion and of the timing of the staminate and pistil-late phases of the flowers) that increased pollen receipt was in fact largely due to the longer time of stigma presentation, with no direct effect of stigma exsertion on pollinator efficiency. Pollinator-mediated selection was operating but was effectively indirect: through the genetic correlation of stigma exsertion and length of the pistillate phase.

This is an important theme, which will be revisited in greater depth in the specific context of pollination syndromes in chapter 11, section 2, *Why Pollination Syndromes Can Be Defended*.

How Do Pollinators Drive Floral Divergence?

P. Wilson and Thomson (1996) suggested five ways in which floral divergence could arise by selection:

1. ADAPTATION TO DISTINCT NICHES: Pollinators are seen as opportunities that plants can take advantage of and try to fit with, their characters shifting locally to suit whatever animals are available.

2. CHARACTER DISPLACEMENT: Species diverge to reduce local competition with other plants present (a response that is classically invoked for flowering time, see chapter 21). A good example comes from studies of *Phlox drummondii*, whose flowers are pink in most habitats but red where the species coexists with the pink congener *P. cuspidata* (Levin 1985), and in those same habitats hybrids are selected against by the assortative behavior of butterfly visitors. This effectively reinforces the isolating barriers.

3. ADAPTIVE "WANDERING": The direction and strength of selection varies over short timescales, so that floral morphology might "wander" in different directions. For example, pollinators can produce a selective force leading to floral divergence and speciation in several ways in *Ipomopsis* (e.g., D. Campbell 1989a,b) or in *Lavandula* (C. Herrera 1988), where selection varies between years and between sites just a few kilometers apart.

4. CHARACTER CORRELATIONS: There might be selection on characters that are essentially physiological or developmental, with floral traits "dragged along." Many floral traits are likely to be genetically correlated, either by **pleiotropy** (where two traits are affected by a single locus) or due to **linkage disequilibrium**

(either where the two traits are on the same chromosome, or are on different chromosomes but are nonrandomly associated during chromosome assortment). In Neotropical *Dalechampia*, where colored bracts are the main attractive feature and **resin glands** provide the main reward, Armbruster (1991, 1996) showed that bract length covaried with the area of the resin gland and with the distances between the gland and stigma or between the gland and anthers. Such effects could originate by selection, but they may just represent nonadaptive effects of **genetic constraints**. There is ongoing debate about how far genetic correlation is a constraint in floral evolution on a broader scale, but clearly there is scope here for 'apparent' pollinator-mediated selection on flowers.

5. GENETIC DRIFT: Characters that are neutral in small populations could lead to divergence. **Genetic drift** might be most relevant in small populations of very specialized flowers (such as orchids that interact with euglossine bees, chapter 9), or in heterostylous plants (chapter 3).

K. Jones (2001) offered a somewhat different (although overlapping) breakdown of selection types that might operate:

1. DISRUPTIVE SELECTION, in which intermediate phenotypes are ignored by the flower visitors and so have low relative fitness. Eventually there is a good chance of reproductive isolation occurring between the two extreme phenotypes. This kind of selection could operate in either of two ways. First, it could exert pressure on reproductive traits, where selection will happen especially quickly; so, for example, red alleles are more likely to become coupled with other red alleles, and white with white, with the intermediates becoming rarer, and these intermediates are in turn likely to become more selected against by pollinator choice, as may occur in *Polemonium* (Galen et al. 1987) and in *Ipomopsis* (D. Campbell et al. 1997). Second, selection could operate on ecological characters to produce increased variance but without necessarily leading to evolutionary branching, unless the ecological trait is genetically correlated with another trait that produces assortative mating via effects on pollinator choice. Selection against the ecological intermediates would allow genetic associations to persist and avoid being broken up by repeated recombination events, and gene complexes could build up leading to genetically isolated groups within populations (genetic neighborhoods). Thus assortative pollinator behavior could reinforce the divergence of the neighborhoods and can drive speciation.

2. "SELECTIVE" SELECTION, in which foraging animals faced with mixed floral communities tend to focus their visits on nonrandom subsets of the available flowers and to show flower constancy (chapter 11). Different pollinators will have different effects. Bees and hummingbirds are often described as relatively specialist, but in both cases they are opportunistic and prone to override any innate preferences by experience and learning. Specialist "choosy" visitors will obviously affect floral evolution more strongly and more quickly; if specialism is dominant in a community, it would tend to become more speciose (an example might be the Cape flora, with many long tubular corollas and specialist flies; S. Johnson and Steiner 2000, 2003). In this kind of situation, assortative mating could result from partitioning of floral resources, both between taxa and between individuals within a taxon. In fact, individual preferences, changing within a trip or day or season, must have very strong effects at local level and may often lead to establishment of, and preference for, hybrids, so that the pollinator pool is partitioned much more finely than by conventional taxa. Hybrids and polyploids are of course common in plants (natural hybrids making up 5%–22% of all species in natural communities [Ellstrand et al. 1996]), and polyploidy itself can be a major factor in establishment of new species.

There is, then, good evidence for pollinator-mediated selection and appropriate trait heritability in flowers, and there are well-established mechanisms by which this could bring about floral change, reproductive isolation, and evolutionary divergence or specialization. It would be easy to produce several books just on these topics, but for our purposes, taken together with the known evolutionary history of plants and pollinators and of interrelated speciations (not one-to-one, but at least matching reasonably through time), we may accept that it can happen. How far and how often it does happen is a more complicated issue, relating back once again to the issue of how far floral specialization and associated visitor specialization are important factors in pollination ecology. We will return to that theme in chapter 11, after a thorough look at all the floral advertisements and rewards on which selection might act.

PART II

FLORAL ADVERTISEMENTS
AND FLORAL REWARDS

Chapter 5

ADVERTISEMENTS 1: VISUAL SIGNALS
AND FLORAL COLOR

Outline

Visual attraction by flowers is substantially related to flower shape and size, and the basic aspects of these were covered in chapter 2. But above all, for most visitors, color and color patterns are attractive. **Trichromatic** color vision occurs in many terrestrial animals, and in the evolution of insects it certainly predates the flower-visiting habit. Undoubtedly most of today's key pollinating taxa have good color vision, and flowers should have been selected to interact with their visitors' visual abilities. So it is not surprising that plants use color signals to make their flowers conspicuous in a mainly "green" world. Since flowers comprise successive whorls of sepals, petals, stamens, and carpels, in principle any part of that complex structure (or all parts) can be colored, and the production of colored floral displays therefore exhibits extraordinary variety.

It is important to know something about how colors are made in biological systems, about how they are perceived by animals' eyes, and thus about their ecological significance. Then we can see how color interacts with the other visual signals from flower size, shape, and outline in enhancing floral attraction to visitors.

1. Floral Pigments and Floral Color

The underlying physiological basis of flower color is floral pigments (table 5.1 and fig. 5.1). There are three main types, all of them secondary plant products found either dissolved in the aqueous **cell sap** (flavonoids and betalains) or retained within **plastids** (carotenoids); Grotewold (2006) provides a good review. Flowers mostly reflect light from the near UV (350 nm) to deep red (700 nm) using combinations of these pigments.

Cell-Sap Pigments

The **flavonoid** pigments, the commonest type, are water soluble and thus readily dissolve in the cell sap. Most of them are built around a core **anthocyanidin** molecular structure of coupled six-carbon rings (sometimes alternatively termed aglycones). There are three particularly common anthocyanidins (fig. 5.2): **pelargonidin**, **cyanidin**, and **delphinidin**, of which cyanidin is probably the primitive form (Harborne 1993).

TABLE 5.1
Floral Pigments

Major groups	Types	Examples	Color	Additions	Effects	Distribution
Flavonoids (in sap)	Anthocyanidins/ anthocyanins	Cyanidin Pelargonidin Delphinidin	Red, maroon Scarlet, orange Purple, blue	Sugars	Adds blueness	Very common and wide-spread
				Metal ions (Fe, Al)	Stronger color, color change	
	Anthoxanthins Flavones Flavonols Chalcones Anthochlors	Gossypetin	White, cream, yellow with strong UV component Intense yellow			
	Anthocyanins and anthoxanthins together ("co-pig-mentation")		Red-browns, orange, and some pink colors			
Carotenoids (in plastids)	Carotenes Xanthophylls	β-carotene Lycopene Auroxanthin	Orange-yellow Orange-red Yellow, lemon			Very common
Betalains (in sap)	Betaxanthins Betacyanins		Bright yellow Bright red, purple			Rare

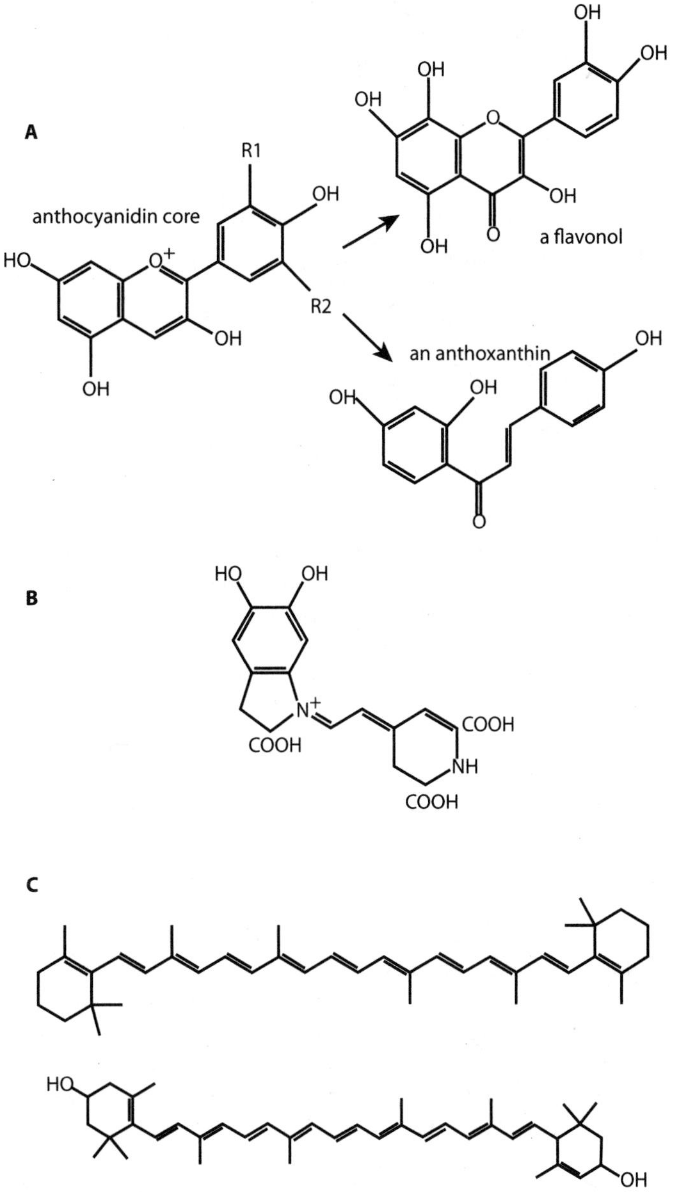

Figure 5.1 The three main types of floral pigment: (A) the anthocyanidin structure underlying flavonoids, together with a flavonol and a typical anthoxanthin at left; (B) a betalain; (C) carotenoids, with a simple carotene above and a xanthophyll below.

These three core molecules differ markedly in their light absorption properties, and hence colors, as shown in table 5.1. Whichever anthocyanidin is present, it is always linked with at least one sugar molecule to give an **anthocyanin**. Other flavonoids, with somewhat more variable core structures, are collectively termed **anthoxanthins** and give the deep cream through to deep yellow colors, as seen in primroses (*Primula*, plate 6C), and European gorse (*Ulex,* plate 9D), or ap-

proach orange shades in cases where additional pigments, **anthochlors**, are present, as in many Asteraceae and some snapdragons (*Antirrhinum*).

Betalain pigments are also dissolved in the cell sap; they incorporate a five-carbon and a six-carbon ring and are chemically related to the **alkaloids** that provide the defensive secondary chemicals in many plants (including poisons such as nicotine, atropine, etc.). In plants they are always linked with sugar molecules.

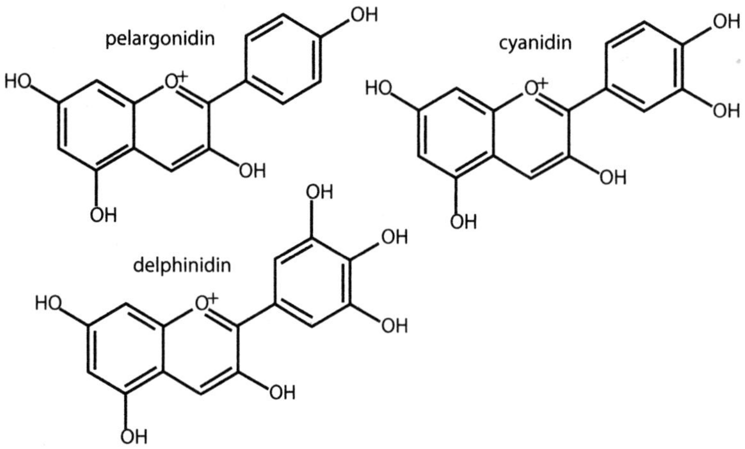

Figure 5.2 The three principle anthocyanidins occurring in flowers.

They occur in relatively few plant families as floral pigments but produce particularly intense colors in the deep crimson and magenta range, as seen in *Bougainvillea, Portulaca*, and some of the more spectacular cacti (plate 13B), or deep yellow, as in *Opuntia*. They may also occur sporadically in fruits and in other parts of plants, a notable example being beetroot (*Beta vulgaris*).

Plastid Pigments

Plastids are intracellular organelles, of which **chloroplasts** are the most familiar type. Chloroplasts normally contain two types of chlorophyll pigment (both green and imparting the primary coloration to foliage and stems) and smaller amounts of orange **carotenes** and yellow **xanthophylls**, which have an accessory role in photosynthesis. These last two occur in flowers as well as foliage and are collectively in the group termed **carotenoid** pigments. They are simple **hydrocarbon** tetraterpenes (four five-carbon units arranged linearly), insoluble in water, and either oxygenated (xanthophylls) or lacking oxygen (carotenes).

Carotenoids are important as protective and coloring pigments in many parts of plants and give the traditional coloration to carrots, tomatoes, and some fruits. In flowers they similarly produce yellow, orange, and red colors, as seen in daffodils (*Narcissus*, plate 6E), marigolds, (*Calendula, Tagetes*), and some lilies.

Production of Colored Flowers Using Pigments

Merely listing pigment types and examples is not sufficient to define the subtleties or range of flower colors. In practice, flowers achieve complex shadings and intensities of color in a number of different ways, and they can produce elaborate color patterns even within one floral part. What follows largely uses examples from petals, but the same principles may be found in other structures (bracts, sepals, stamens, etc.) in a smaller number of plants.

1. Pigment concentration is perhaps the key variable, and concentrations in petals varying from less than 0.01% to more than 15% (dry weight) produce very different effects. This is largely under simple genetic control, with expression of just one or two genes often determining pigment amounts (Galliot, Hoballah, et al. 2006).

2. Both plastid and cell-sap pigments may occur in the same cell of a petal (copigmentation), giving a resulting shade that could not be achieved by a single pigment. In particular, flavones and flavonols, though rather colorless by themselves, will bind with anthocyanins in stacked molecular complexes to produce more subtle shades.

3. Multiple pigments may be present in different cell layers of the petal, combining in varied proportions according to the thickness of each cell layer to

give variations of tones and patterning across the length or breadth of the petal. Such a petal may then appear as different colors when viewed from above or below, and it may appear quite different depending on how incident light strikes the petal surface.

4. Different combinations of pigments may be spatially varied in different cells across a petal to produce color shading, often darkening toward the center.

5. Spatial variation may give rise to distinct color patterns within a petal. Common types are spots in the ventral throat of a tubular corolla (as in foxgloves [*Digitalis*], plate 15G) or stripes radiating out from the center of a radial flower (as in many geranium, pansy, or mallow flowers) or from the throat of a zygomorphic flower (as in African violets [*Saintpaulia*] and Cape primrose [*Streptocarpus*], plate 15A). In such cases the patterns may serve to direct visitors into the center of the flower, where the nectar is situated, ensuring that they pass the anthers en route, and the patterns are often described as *nectar guides* (or **honey guides**), which will be covered in more detail in section 6, *Nectar Guides*. These guide lines or spots are often purer colors with strong pigmentation against a more diffuse and less intense background petal color.

6. Different structures within the whole flower may be of different colors, often giving better contrast and better visibility from a distance. In radially symmetrical flowers this arrangement commonly occurs where the central mass of anthers and stigmas are yellow and contrast strongly with white, pink, or purple petals (e.g., plates 2B,C, 8F, 9A, and 13E) or where a darker colored rim surrounds a paler center (e.g., plate 14B). In zygomorphic flowers, the upper (banner, or flag) petals may be strikingly different in color from the lower petals (e.g., plates 19F and 32E). Moreover, the pigmentation of the sexual organs is often particularly pure and that of the petals rather diffuse (e.g., *Trillium*, plate 14F).

7. Different parts of an inflorescence may be differently colored (plate 14); for example, the yellow disk florets of daisies are very distinct from the white peripheral ray florets, and the outer sterile flowers of some umbellifers are strong white colors while the central functional flowers are yellow-green. The brush flower *Dicrostachys* is a striking example in which the unopened yellow florets lie below the open lilac ones (plate 14H). In other cases the bracts (plates 13F and 14A,E) or the stamens or styles (plate 14B,D) provide the contrast. Sometimes petal color is augmented by an apparent window effect, where the paler tube is seen within (plate 14C).

Other Ways of Producing Floral Color

The perceived brightness and coloration of a flower may be affected not only by its pigmentation but also by reflection effects. This is particularly common for intensely white flowers: white is produced in part by an absence of pigment and also in part by refraction and reflection of light at the cell surfaces and at the air spaces within the petals. This phenomenon is often seen in flowers that open at dusk, when spectral differences are not easy to perceive.

A classic example is found in the common buttercups (certain *Ranunculus* spp.). Here the upper epidermal cells are very thin with a smooth surface, and they contain a dense oily solution of yellow carotenoid pigment (unusually not enclosed in plastids, perhaps linked to the cells being nonnucleate). These cells lie immediately above an air layer and a deep layer of starch-filled cells that appear densely white. The back of the petal is matte, and its cells contain yellow pigment in plastids. Together these layers produce a petal with an intensely bright and glossy yellow appearance from above (traditionally attributed to reflection, but probably also with refraction at interfaces of the oily layer and the cell or air layers; B. Glover, pers. comm.) and a much less glossy yellow surface below (plate 13A).

Cell structure may also influence color intensity, for example in *Antirrhinum* flowers, where conically shaped cells in the petal epidermis have a major effect on the perceived color and thus affect insect visitation (Noda et al. 1994; Glover and Martin 1998). More specifically, Whitney, Kolle, et al. (2009) showed that surface microstructures on petals, including long cuticular striations, can act as diffraction gratings and result in iridescence, lending another potential variable to visual perceptions; they recorded such iridescence in ten angiosperm families. In *Hibiscus* and *Tulipa*, they showed that petals change their hue depending on the angle of observation, and they also demonstrated that bumblebees could disentangle color cues from irides-cent cues when approaching such flowers.

Environmental Effects on Floral Color

Chelation with metal ions has a major effect on blue coloration in flowers, with iron, magnesium, and alu-minum all potentially interacting with anthocyanins. This reaction can give the very intense colors found in blue cornflowers (*Centaurea* species) and in *Commelina* (plate 12A), or in some blue-purple *Hydrangea* species. The amount of these metal ions in the soil can affect flower color, as is well known with hydrangeas, where the same plant can produce either intense blue flowers or rather washed-out pink flowers depending on the soil it grows in. It is not only the soil's metal content that matters though, since metal-ion uptake via the roots is affected by the plant's nutritional status (especially by how much nitrogen, phosphorus, and potassium are available) and also by soil acidity.

Pigment intensity within the plant is also affected by other environmental variables, especially tempera-ture and water stress. In particular, the biosynthesis of flavonoids is sensitive to light levels, temperature, and mineral availability; and flavonoids are in turn precur-sors for the synthesis of anthocyanins, so that most flower pigments will be affected by these environmen-tal factors. Hot, dry conditions favor more pink and purple morphs over white morphs in a range of species, an observation clearly documented in *Polygonum, Vicia, Cirsium*, and *Digitalis* by Warren and Mackenzie (2001).

2. The Problems of Defining and Measuring Color

There are real problems relating to the apparently straightforward concept of color. Color is not simply the property of an object, for it results from the visual and sensory processing capacities of the viewing organ-ism; the same surface may be seen as different colors by different sensors and hence by different animals. In other words, color as perceived by any given animal is not a direct indication of the spectrum of radiation giv-en off by an object but instead is an abstract phenome-non, often species specific or potentially even individu-al specific. Furthermore, color perception also depends

upon the light environment in which a surface is seen (Kevan 1983), that is, the reflected/absorbed spectrum from that surface is modified by differing ambient light conditions. And color perception also depends on the contrast with the background against which it is perceived. Defining and describing color is really difficult—what you see is not exactly what you get!

Humans also tend not only to define a color but also to qualify this with other issues about purity, intensity, or brightness of that color. *Hue* is often used infor-mally to describe such qualities, though the word can also be used more precisely to define the relative (rath-er than absolute) outputs of **cones** in an eye. We do not know whether (or if so, then how) pollinators perceive qualities such as these, although we do know that they crucially depend on the contrast of colors against background.

Many problems stem from these differing percep-tions of color. Advances in studying the subject have involved devising systems for measuring or describing color, together with greater understanding of the visual capacities of animals, and correspondingly more so-phisticated measures of behavioral discrimination.

Various methods for measuring color are given in the literature (Kevan 1978; Chittka and Kevan 2005), although in practice they tend to produce complemen-tary results. Using the human eye to match colors to standard color charts, which are ordinarily produced for the printing industry or for horticultural purposes, is a fairly widespread practice. In the field, nondigital photography with **monochromatic** filters is possible but requires great care with the exposure and needs quartz, rather than glass, lenses to capture UV light; pictures taken against uniform black backgrounds are advisable. Color video can be used and then converted into "insect view" format, where UV becomes blue, blue becomes green, and green becomes red (see be-low). If a flower can be taken back to a laboratory, its color can be determined more rigorously by spectro-photometry, which has built-in corrections for contrast and other variables, and the color can be represented as a locus on a standard diagram, although color pattern-ing and features such as nectar guides are then lost.

At the most sophisticated level, the concept of a **color space** (also known as a chromaticity diagram) can be developed for a given animal from the known interactions of its various **color receptors**. Thus for a human, the color of an object can be defined as a point in a specific three-dimensional color space, where

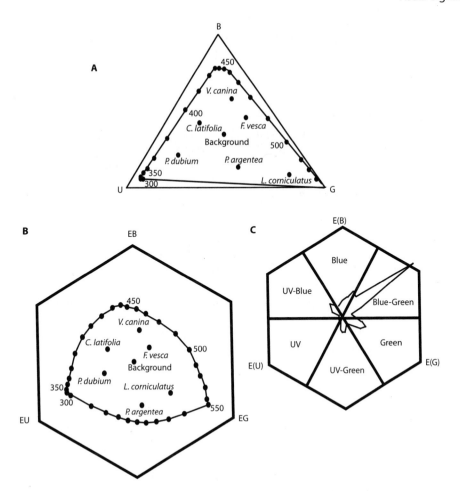

Figure 5.3 (A) Color triangle for a honeybee, with the three points representing the three photoreceptors (green, blue, and UV) and with the reflectance from various flowers plotted within the triangle (thus *Lotus corniculatus* stimulates mainly the green receptor, for example). (B) Color hexagon, again for a honeybee, where E represents receptor excitation, showing the intermediate colors and with the same flowers plotted. (C) A similar color hexagon, where the dotted line shows the distribution of floral colors from a large sample of real flowers, indicating the predominance of "bee blue-green" flowers. (Redrawn from Chittka et al. 1994.)

each of the three axes measures the photon catch of one of the three receptor types in the eye. The perceived hue of the same object is then represented by a *color triangle* within this space, with the three corners mapping onto the relative blue-, green-, and red-cone stimulations (Chittka and Kevan 2005). However, even for humans this is extremely controversial. Extending it to any or all flower-visiting insects is proving very difficult, although a reasonably secure color triangle for the honeybee is shown in figure 5.3A. A *color hexagon* can also be developed by converting a three-dimensional interaction plot for the three receptors

into two dimensions, where a given point in the hexagon is defined by the constant difference between the various receptor signals, which is probably a good representation of the algorithmic processes happening in the bee brain; an example is shown in figure 5.3B.

3. How Animals Perceive Flower Color

Color vision requires an eye with several types of **photoreceptor** that differ in their spectral sensitivities. Most flower-visiting insects are trichromatic, with

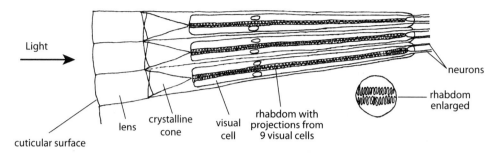

Light

neurons

rhabdom
enlarged

cuticular surface

lens

crystalline
cone

visual
cell

rhabdom with
projections from
9 visual cells

Figure 5.4 Structure of three adjacent ommatidia, which are the functional units of the insect compound eye, also showing the central rhabdom where projections from each of the visual cells converge and overlap and where the photoreceptors are sited. (Redrawn from Barth 1985.)

three different **photopigments** in their compound eyes. The ancestral condition in vertebrates appears to be tetrachromy, and this state persists in birds, although these also have oil droplets with high carotenoid content in the **retinal** cones that act as color filters and modify the spectral sensitivities of each cone. Hence flower-visiting hummingbirds and sunbirds potentially have more sophisticated color vision than do insects. In contrast, most mammals, including flower-visiting bats and rodents, are effectively dichromatic, having lost two of their ancestral receptor types and retaining just red/green and blue retinal cones, although humans and most other primates are secondarily trichromatic, with distinct red, green, and blue cones.

Effective color vision depends not only on the receptors and their pigments but also requires the neural processing ability to translate incoming signals from the photoreceptors, and in this area we as yet know rather little about taxa other than vertebrates and insects (mainly bees).

Insects

The basic element of an insect compound eye is the structural unit known as an **ommatidium**, which is made up of from seven to nine radially arranged receptor cells, each containing within it an area of highly folded membrane termed the **rhabdomere**. The rhabdomeres of each of the sensory cells are stacked together at the center of the ommatidium to form the **rhabdom** (fig. 5.4), which is the site of photoreception and uses membrane-bound **rhodopsin** as the key visual receptor pigment (as in virtually all eyes). Each cell's rhabdomere may have its own **spectral sensitivity**, and in the honeybee there are usually four green-sensitive cells, two blue-sensitive cells, and three UV-sensitive cells in a given ommatidium (fig. 5.5), although the third UV cell is often only apparent basally. In fact in bumblebees the situation is even more complex, with three types of rhabdomeres: one having green and blue receptors; one having green and UV receptors; and one having all three types of receptor (Spaethe and Briscoe 2005), each type potentially seeing a flower rather differently.

When spectral sensitivities are calculated, it is apparent that an insect's visual system is different from the human visual system (fig. 5.6 and 5.7). In general, the insect spectra are derived from three pigments peaking in the UV, blue, and yellow/green zones, in effect "shifted down" by about 100 nm to shorter wavelengths relative to humans' spectra. This may be especially useful in a light environment where UV is relatively scarce, and where detecting its presence may give additional useful information (Kevan 1983), particularly in shade, where UV light is less attenuated than the rest of the daylight spectrum. However, notice in figure 5.6 that there are also marked differences *between* insects; indeed, some flies are only dichromatic, some butterflies have up to six receptor types and a visual sensitivity extending well into the red wavelengths (Goldsmith and Bernard 1974), and many nocturnal insects are effectively color-blind. Table 5.2 shows further examples.

Among the hymenopterans, visual spectra vary at the genus and species levels, the best known being that of the honeybee *Apis mellifera*, where the receptors

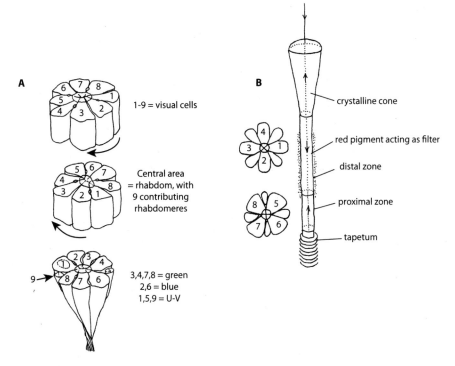

Figure 5.5 (A) Distribution of the nine different color cells in cross-sections of the ommatidium of the honeybee, with different cells contributing more strongly to the rhabdom at different levels. (B) This distribution is seen more clearly in the ommatidium of the butterfly *Pieris rapae*, where cells 1–4 make up the rhabdom distally, cells 5–8 are located in the middle section (and just cell 9 basally, not shown). (Redrawn from Barth 1985.)

peak at 344 nm (UV), 436 nm (mid-blue), and 544 nm (green, but overlapping into the orange/red zone); this is compared with the human spectrum in figure 5.7. If the spectrum is regarded as a circle rather than a line, then the circle can be closed by "mixing" the two ends; in humans this means mixing reds and blues, resulting in purple, but in insects such as bees it involves mixing yellow and ultraviolet, giving a color that is unimaginable to humans, but commonly labeled as **bee-purple**, and evidently easily discriminated by bees. Many other attempts have been made to label "insect colors" with descriptive terms suited to human perceptions, but these are perhaps not especially useful here.

Since most insects have good vision in the UV range, it was common in the past to see this as the key difference from humans and therefore to photograph flowers as seen under UV (following pioneering photographic work by Knuth in the 1890s). This UV picture was seen as a good guide to what flower-visiting bees really saw, and there is much in the literature about flower patterns as seen in UV (e.g., Barth 1985). Guldberg and Atsatt (1975) reported the UV reflectance of 300 flower species, looking for patterns; they found a positive correlation of UV patterning with flower size, and an increase in such patterns in yellow and purple flowers, with less patterning on bird-pollinated flowers and on anemophilous flowers. They also found more patterning in certain families (Fabaceae, Geraniaceae, and Ranunculaceae, perhaps also Onagraceae and Amaryllidaceae) and much less in others (Ericaceae, Polemoniaceae).

But apparent flower color and color pattern are influenced by differing illuminations, which in practice will mean that the light and shade patches experienced while foraging have a substantial effect on what is perceived. Hence, visitors such as bees above all require good color contrast, particularly with green backgrounds, and to this end are likely to be using integrated input from all their receptor types. Thus UV reflectance alone is not a good indicator of detectability

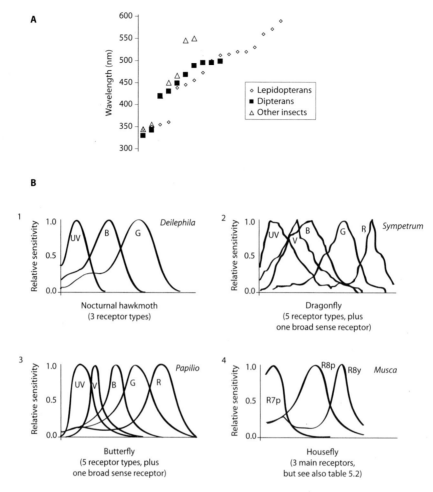

Figure 5.6. Spectral sensitivities of various insect eyes: (A) peak wavelengths for insect rhodopsins from various taxa (scored simply by occurrence in no particular order) ranging from 330 to 600 nm (modified from data in Stavenga 2006); (B) spectral curves for the three to six receptors in specific insects (modified from data compiled in Kelber 2006).

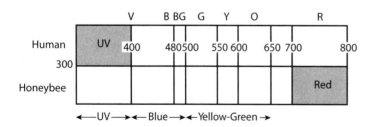

Figure 5.7 Comparison of the spectral range of human and honeybee eyes. (Redrawn from Barth 1985.)

in the field. Furthermore, the evolution of bee UV vision cannot have been driven by flowers, as used to be thought, since the same receptors occur in nearly all insects and crustaceans, implying that the visual receptor system was set quite early in invertebrate evolu-tion. Kevan et al. (2001) therefore regarded UV as no more important for flower visitors than the blue, green, or red wavelengths.

In fact the *absence* of UV reflection in most flowers that appear white to humans is often critical for bees;

TABLE 5.2
Receptor Sensitivities in the Eyes of Some Flower-Visiting and Other Insects

Genus	Common name	Receptor peaks (nm)
Sympetrum	Dragonfly	490, 520, 620
Gryllus	Cricket	332, 445, 515
Periplaneta	Cockroach	365, 510
Coccinella	Beetle	360, 420, 520
Deilephila	Hawkmoth	345, 440, 520
Lycaena	Butterfly	360, 437, 500, 568
Vanessa	Butterfly	360, 470, 530
Papilio	Butterfly	360, 400, 440, 520, 600
Musca	Housefly	335, 355, 460, 490, 530
Tenthredo	Sawfly	328, 464, 540, 596
Cataglyphis	Ant	345, 440, 520
Callonychium	Solitary bee	360, 404, 536, 600
Bombus	Bumblebee	345, 440, 575
Apis	Honeybee	344, 436, 556

Source: Various, including Kelber 2006.

this is because if UV, blue, and green were all reflected in similar proportions from some traditional "bee flowers," they would occupy the same color space as most of the surrounding vegetation (which to insects may seem rather gray or gray/yellow (Kevan 1972, 1978). Insect-attracting flowers need pigmentation to stand out against this background in as wide a range of light conditions as possible. Dyer and Chittka (2004) highlighted the point that bumblebees do indeed make errors in color discrimination in differing light conditions, although they can compensate in part through recognizing shape variation.

Taking such factors into account, table 5.3 shows how some common flowers may appear to humans and to bees. Note that two yellow, or blue, or red flowers, both appearing very similar to us, may appear quite different to a bee, and for many insects flower color diversity must appear much greater than it does to humans. Kevan (1983) expanded this point for particular floral communities, showing how much more varied the trichromatic plots of flower assemblages from the Arctic and north temperate ecosystems are for insects (while to humans they are dominated by white and yellow variants). Figure 5.8 shows one of his examples.

Although the insect compound eye has many advantages, its **visual acuity**, or resolution, is really quite poor; in fact, resolution is necessarily traded for keeping the eyes small. The ability to distinguish shape and form at a distance is therefore decidedly limited, and most flowers will probably appear just as blotches of color (again emphasizing the importance of contrast effects) until an insect is quite close. This may explain why insects seem to need a visual pigment system that can see flowers in a much greater range of color types than humans can achieve. Their greater color discrimination is probably also aided by neural interactions between the receptors (Kelber 2001), so that in fruit flies the detection of green fruits is aided by lateral inhibition between green and blue receptor cells, and similar effects may explain strong preferences for yellow over green in many herbivorous insects.

Vorobyev et al. (1997) tried to take these additional issues into account to give a more sophisticated understanding of what is going on when an insect sees a flower, using *Apis* and the orchid *Ophrys*. They analyzed video recordings through selected filters and with a UV-sensitive camera and then built up the reflectance spectra of the flower linked to a profile of receptor sensitivity in the bee eye. This gave the effects of different color sensitivity, as well as poor **spatial resolution**, and suggested that what flower-visiting insects see is a rather imprecise coarse spatial array of patches that are

TABLE 5.3
Floral Colors for Humans and for Bees

Flower	Reflectance range	Appearance	
		To humans	To bees
Primula auricula	99% yellow, 1% UV	Yellow	Yellow
Caltha palustris	95% yellow, 5% UV	Yellow	Bee purple
Verbascum nigrum	60–80% yellow, 20–40% UV	Yellow	Bee purple
Gentiana clusii	90% blue	Blue	Blue
Veronica chamaedrys	70–80% blue, 20–30% UV	Blue	Violet
Dryas octopetala	High blue and yellow, little UV	White	Bee blue-green
Prunus avium	High blue and yellow, little UV	White/pink	Bee blue-green
Raphanus raphinastrum			
Morph A	All colors reflect	White	White
Morph B	Yellow, some UV	Yellow	Bee purple
Penstemon barbatus	High red, some blue, no UV	Red/purple	Bee black
Papaver rhoeas	High red, some blue, some UV	Red	Bee UV

Source: Data from Barth 1985 and others.

Note: All the color appearances attributed to bees are suspect because we cannot tell what a bee really perceives.

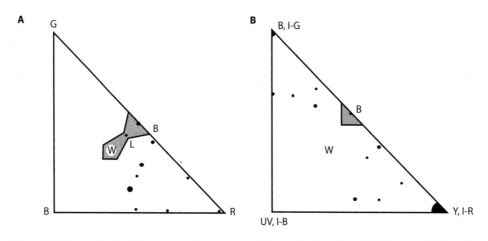

Figure 5.8 Trichromatic plots of flower colors from a community in the Canadian Arctic, with the center (W) in each case representing an achromatic white flower where all color receptors are stimulated equally: (A) against the human visual spectrum (blue-green-red); the shaded area represents 48 white or white and yellow flowers with the dots representing just 12 flowers outside this range as seen by a human; (B) against the bee spectrum (green/ yellow, blue, UV); the shaded area represents 13 insect-yellow flowers and the dark shading represents insect-green flowers (14); clearly the overall range of colors perceived by a bee in this community is greater. (Modified from Kevan 1983.)

colored differently from what we would see. Evidently it is unwise just to record insect recognition of flower color patterns in human perception terms, although in practice nearly all pollination biologists still do so, for lack of any comparably easy method.

Birds

Most vertebrates have inherited a **tetrachromatic** visual system, and one of their four ancestral cone types is primarily a UV sensor. Many birds retain this system, with four types of single cones and an additional double-cone type, where two cones are tightly juxtaposed. For a bird, the color space is therefore four-dimensional and almost impossible to map directly onto human perceptions; we cannot in any meaningful way describe just what a bird will see.

The four classes of cones have sensitivities peaking at 543–571 nm (LW), 497–509 nm (MW), 430–463 nm (SW), and then either 402–426 nm (VS, or very short wavelength) or 355–376 nm (UVS, or extreme short wavelength). Either of these last two types confers sensitivity at short wavelengths beyond the range of human perception. Birds can therefore, in principle, detect UV radiation, and many do have lenses and corneas that are transparent to UV light. Huth and Burkhardt (1972) specifically demonstrated UV sensitivity in hummingbirds behaviorally, long before their retinas had been physiologically analyzed. Cuthill et al. (2000) reported definite evidence for UV vision in at least 35 bird species but also noted that a role for UV vision in flower visitors had not really been established. UV patterns certainly appear to be rather rare in typical red-colored hummingbird flowers, although a few examples do occur (Lunau and Maier 1995).

Birds also have oil droplets in their cone cells that filter out light over part of the receptor's range, narrowing the waveband response of each cone and decreasing the overlap between cone responses. The oils probably also improve **color constancy**, whereby objects appear the same color even when the illumination changes.

General Effects

Of course for most flower visitors, distinguishing between colors also depends on the size of the flower, that is, of the colored target. Hence the term "mini-

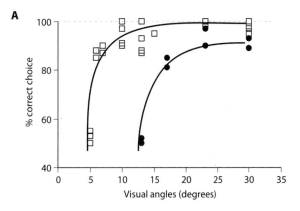

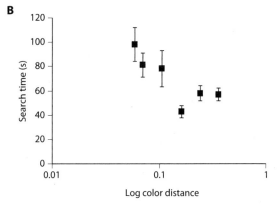

Figure 5.9 (A) Minimum visual angles for detection of different-colored discs presented against a gray background, for bees. Open squares show results using stimuli with good green contrast, and closed circles are for stimuli lacking green contrast and needing to subtend a much larger visual angle before they were detected. Hence (B) shows that there is a longer search time for targets with poorer color contrast (here expressed as log color distance between two stimuli). (Modified from Giurfa and Lehrer 2001.)

mum visual angle" is often quoted in the literature (e.g., Giurfa, Vorobyev, et al. 1996, and fig. 5.9A). A honeybee approaching at a distance can detect a flower of any color (if it has good contrast with the background) at a visual angle of about 5°; but they need to be closer, with this angle increased to around 15°, if there is no good contrast against a green background, and search times then become markedly longer.

From this discussion, it should be apparent that within any one taxon, even closely related species can have eyes and pigments that differ markedly in spectral sensitivities and in their ability to resolve differences, suggesting that visual systems are highly plastic. For this reason, strong selection on floral color may be unlikely, except where the visitor spectrum to a particular plant species is small and consistent (i.e., for a rather specialist plant).

4. Color Preferences in Animals

Flower color is important in pollination ecology because the main pollinating taxa do have good color vision, and it is well documented that **color preferences** occur in different animals. In more sophisticated flower visitors, including bees and vertebrates, it is widely assumed that there may be an initial innate preference (Lunau and Maier 1995) and a subsequent learned preference associated with rewards received. At least in hawkmoths (Goyret, Pfaff, et al. 2008) and bumblebees (Gumbert 2000), this dichotomy has been experimentally demonstrated to be broadly true, although an experienced bee will revert to its innate preference if no color similar to its acquired or trained preference is on offer. Thus long-range attraction by display of an appropriate flower color can be a simple and effective way to attract the desired visitors and potential pollinators. It can also be fast: butterflies, flies, and bees can all learn a color-reward association in a single trial (fig. 5.10A), although precision of choice varies with the exact color pairing offered (fig. 5.10B).

Table 5.4 shows a simplified summary of supposed (usually innate) color preferences for flower-visiting animals. A table such as this raises some major problems, partly because it probably confuses innate and learned preferences, and also because versions of it have been in circulation for well over a hundred years that were at least initially principally based on observation and correlation, without experimental testing, thus leading to some rather circular arguments. Even now, rigorous experimental tests of animal color preferences are often missing. An additional problem is that colors of petals and other floral structures are sometimes correlated with additional plant traits that exert contrasting or complementary effects, and these have rarely been tested for. One example occurs with the genus *Raphanus* (wild radish), where different color morphs vary substantially in their inducible glucosinolate concentrations (glucosinolates being one of the main chemical defenses in the crucifer family). The darker pink/purple- and bronze-flowered morphs produced more defenses when fed on by *Pieris* caterpillars than did the white and yellow morphs (fig. 5.11), so the herbivores preferred the yellow morphs, coun-

teracting the flower-color-selection from pollinators because these (especially bees) also preferred the yellow and white flowers (S. Strauss et al. 2004; Irwin and Strauss 2005; S. Strauss and Whittall 2006).

However, based on reasonably careful observation and experiment we do now have good data on the color preferences of some groups that supplement the broad generalizations found in table 5.4. (Details and references on these points are given in the chapters dealing with specific groups).

1. Beetles usually have preferences for white, cream, and yellow/green flowers, but in a few cases make more specific choices; for example, southern African monkey beetles will select orange flower models over yellow, blue, or red.
2. Many generalist flies that visit flowers have a preference for yellow, or yellow and white, flowers. More specialist flower-visiting flies, such as syrphids and bombyliids, prefer pink, purple, or blue flowers, although syrphids also like yellow centers.
3. Butterflies in general like white, yellow, orange, pink, and red; more primitive ones, such as **nymphalids**, often prefer yellow, and others commonly frequent pink or red blossoms. Some advanced butterflies with good red sensitivity may respond particularly strongly to red non-UV-reflecting flowers.
4. Diurnal hawkmoths, such as *Macroglossum* and *Manduca*, have an innate preference for blues but will learn a preference for almost any color.
5. Nocturnal moths and bats usually prefer strong whites, creams, or pastel shades that are conspicuous in poor light; but they always require scent cues in addition.
6. Bees vary in their choices; honeybees prefer flowers toward the blue end of the spectrum when naive, but this is partly dependent on background contrast and is readily overridden by learned preferences.
7. Birds have no innate preference for red colors but learn this as a choice (see below).

However, preferences cannot be taken too far when comparing insect taxa. Figure 5.12 shows an example of the range of flower colors visited by different insects expressed in terms of the bee color spectrum, with most groups predominantly selecting "blue-green" flowers within the choices offered. It might be noted here that these data were taken as evidence against *color syndromes* (Waser et al. 1996; chapter 21).

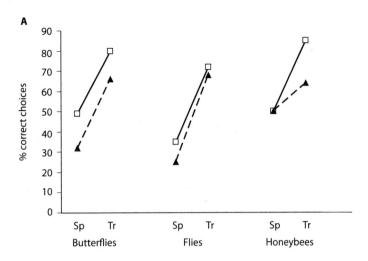

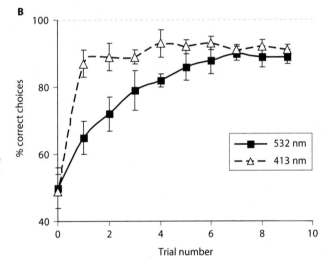

Figure 5.10 (A) Color-learning performance compared in butterflies (*Papilio*), flies (*Lucilia*), and bees (*Apis*) with spontaneous color choices in naive individuals (Sp) contrasted with trained choices (Tr, after one sucrose reward). Correct decisions made after training on different colors (red and yellow for butterflies, blue and yellow for flies) are similar, except for bees, whose training with blue discs is less successful than with violet (redrawn from Weiss 2001, based on various authors). (B) Percentage of correct choices through successive trials for bees trained on two different-colored targets; learning is faster with a 413 nm target (blue) than with the 532 nm target (green) (modified from Menzel et al. 1993).

TABLE 5.4
Flower-Visiting Animals' Supposed Color Preferences

Vertebrates	
Bats	White, cream, dull green, dull purple
Non-flying mammals	Red, brown, dull shades
Birds	Red, orange
Insects	
Bees	Blue, purple, pink, white; yellow in less advanced and/or short-tongued types
Beetles	Cream, dull light green; red or orange in a few species
Butterflies, diurnal moths	Red, pink, purple; yellow in less advanced types
Moths	White, cream
Flies (most)	White, cream, yellow/green
Carrion flies	Brown, deep red, purple
Wasps	Brown, green

Source: Modified from Kevan 1983, Scogin 1983, and various sources.

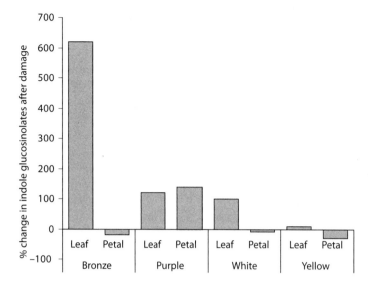

Figure 5.11 Evidence that flower color and floral defences interact. Here the percent change in glucosinolate concentrations in leaves and petals varies for different morphs of *Raphanus* after damage by *Pieris* caterpillars, with the bronze morph showing a much greater induced defense in its leaves and the purple morph showing the greatest induction in its flowers but with yellow flowers showing almost no change. (Redrawn from Strauss and Whittall 2006.)

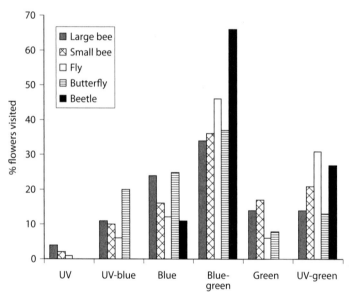

Figure 5.12 Color preferences of pollinating insects at a site in Germany with 154 flower species, showing the proportions of flowers of different colors (scored for the insect visual system) visited by different taxa, with blue-green (usually human white) preferred by most groups (see fig. 5.3), although with large bees and butterflies also visiting high proportions of UV-blue and bee-blue flowers (blue, purple, and pink to humans). (Modified from data in Waser et al. 1996.)

Where color preferences do exist in particular flower-visiting animals, whether innate or learned, the resultant floral color constancy is often critically important for foraging insects, in turn ensuring the visitor's floral constancy that is so important for the plant. In principle, then, colored flowers are a good way to ensure constancy for most insects, and also for bats, which learn color well, although this is probably less true for birds, which have good color vision but learn colors and color associations less easily. Fairly precise color constancy has been demonstrated experimentally in several bee species and in at least one moth (Kelber and Pfaff 1997). In fact for bees, constancy is highly correlated with color differences as tested experimentally (P. Hill et al. 2001; chapter 18).

If there are flower-color preferences, a further question arises as to what degree flower color is adaptive and associated with particular pollination syndromes? This is hard to test rigorously except with a genus naturally exhibiting very large color variation and visitor variation. Armbruster (1996) investigated it with 35 species of *Dalechampia*, using the colors of bracts, sepals, resin glands, and stigmas. With phylogeny controlled for, he found only one rather weak association, that between bract color and pollination mode, with species pollinated by female euglossines preferring white or pink bracts and those visited by smaller **megachilid** bees preferring green bracts. There were no significant links between pollinator type and the colors of other features. Overall, he suggested that selection for white or pink bracts might relate to existing pigment-synthesizing abilities in particular clades, the pink bracts tending to occur in species that also had red pigments (anthocyanins) in the leaves and stems (thus pleiotropy lies at the root of any apparent preferences). Rausher (2008) analyzed evidence for color transitions in flowers as related to pollination and found it to be equivocal, quite often being related to nonpollinator agents acting on the pleiotropic effects of flower-color genes.

However, there are a good many examples, from work on *Penstemon, Mimulus*, and other large genera, that reveal color as a key component in delineating particular syndromes; some of the work on *Penstemon* is covered in *Specific Flower Colors and Associated Visitation Patterns* below, and other examples will be met in more detail, in context, in chapters 11 and 20.

5. The Ecology and Evolution of Flower Color and Color Preferences

Chittka et al. (2001) considered whether there was coevolution of flower color and pollinator color vision. The latter came first in evolutionary terms, so presumably flowers would have had to adapt to it to set off a coevolutionary process. Some indications that this may have occurred come from the recurrence across many families of common color patterns, such as yellow central structures and yellow pollen with high color purity (Lunau 1992a, 1996a, 2000), and increased peripheral shortwave reflectance (Kevan 1983). Observations that **sympatric** heterospecific flowers tend to be as different as possible (e.g., in *Phlox*), with hybrids disfavored, and that flower colors do often have sharp steps in their spectral properties just at the wavelengths where pollinators are most sensitive (Chittka

1997) may also support the hypothesis of coevolutionary influences.

Specific Flower Colors and Associated Visitation Patterns

Yellow Flowers

Yellow flowers and white and yellow combinations are very common in most floras. In general, many of the less specialized flower visitors have an apparent preference for yellow, as seen, for example, with many flies and some butterflies. More specifically, Lunau and Wacht (1994) showed that the hoverfly *Eristalis* had an innate response (the proboscis extension reflex, PER, which is present in completely naive insects) to monochromatic light at a narrow wavelengths in the yellow part of the spectrum (520–600 nm). This response corresponds to the known spectral sensitivity of the longest wavelength photoreceptor present in most insects. Most pollens are yellow and reflect light in that wavelength. In fact, yellow pollen is ancestral (chapter 7), the pigments that produce yellow colors being protective against UV damage, and perhaps yellow pollen entrained in flower visitors a preference for the color yellow. This possibility suggests that a flower of similar color to pollen may have been a very good ancestral state to attract a good range of generalist insects, and it would also fit neatly with the continuing abundance of yellow-centered flowers.

Many yellow flowers also have strong UV reflection, which will give particularly high contrast against a pure yellow background, so that guide marks can readily be added to an ancestrally yellow flower.

Red Flowers

Red flowers are reasonably common, but more so in some habitats (and some seasons) than others. Why should this be? There appears to be a correlation with bird visitors, hence red flowers are abundant in habitats where bird pollination is common. Crosswhite and Crosswhite (1981) suggested that red had been selected for because it gives the best signal strength to birds at a distance. Alternatively it has often been assumed that many insects, including bees, cannot detect red wavelengths (600–650 nm), so that red flowers would be a way of attracting animals other than bees—mainly hummingbirds—so explaining the greater frequency of red flowers in Neotropical communities.

We now know that bees can distinguish red flowers reasonably well. However, most of the red flowers with reflectance around 650 nm do present a problem for detection in that they produce poor contrast against a green foliage background. Hence insects do find and visit red flowers but will usually locate both red and UV flowers more slowly than blue and yellow flowers when presented against a green foliage background. Hence red flowers are rarely "'preferred" by bees, hoverflies, and other good insect pollinators.

Why then do birds visit red flowers? Hummingbirds probably *do not* have an innate preference for red. They may detect red more easily than bees do, but they will visit flowers of all colors largely according to the nectar quality that they receive. That is, their reported preference for red is almost certainly a learned association. It arises not so much because the birds "like" red, but because red flowers are missed by other visitors and thus when visited by the birds are still full of nectar (K. Grant 1966; J. Brown and Kodric-Brown 1979; Rodriguez-Girones and Santamaria 2004). Perhaps for that reason, within a genus hummingbird-pollinated plants are indeed more likely to be red (J. Thomson et al. 2000), and increased numbers of red flowers do indeed occur in the Neotropics. In a multi-dimensional study of *Penstemon* species by P. Wilson et al. (2004, 2006), color was the single best predictor of hummingbird visitation, and species attracting birds (rather than bees) showed an evolutionary shift from blue/violet colors to pink/magenta colors; then with a further shift to red/orange, they not only attracted birds but also deterred bees.

There may also be another (associated) reason for red bird-visited flowers: if hummingbird flowers within a given Neotropical community converge on red, they can then form a mimicry ring supplying the birds with nectar through a whole season, by which all of them gain (Bleiweiss 2001; and chapter 21).

In these scenarios, birds will learn a preference for red flowers. Then even when the flower itself is relatively cryptic, some hummingbird-visited flowers may attract the birds by having red colors elsewhere—on the bracts or, even more spectacularly, as red spots on the leaf tips in the Central American *Columnea florida* appearing only when the (relatively hidden) flowers are mature.

This does not mean that a Neotropical red flower is purely associated with birds of course; the classic and well-studied red hummingbird flower *Ipomopsis aggregata* has a reflectance from 300 to 600 nm and does attract many hummingbirds, but in practice it is also visited by bumblebees, solitary bees, syrphids, hawkmoths, and butterflies (Mayfield et al. 2001).

Finally we should note that red flowers are also common in an entirely different habitat, around the Mediterranean basin in early spring (anemones, tulips, poppies etc.). In this case, bird visitation is not the issue, and in the Eastern Mediterranean the red flowers are mainly visited by beetles (Dafni et al. 1990). Here the flowers are coming into bloom before there is much foliage, and perhaps red colors stand out well against barren sandy backgrounds.

Blue Flowers

Blue flowers, and often mauve and purple flowers, commonly have bees as their main visitors (Kevan 1983), although hoverflies and bombyliid flies are also common. A preference for blue flowers may be linked to an association of color with reward: blue and bee-UV/blue flowers have higher rewards, at least where this has been tested—in Israeli (Menzel and Shmida 1993) and German (Giurfa et al. 1995) floras. Species of *Bombus* from Europe, Japan and North America, when tested in a naive state, all preferred blue and violet flower to any others offered (Chittka et al. 2001). However the preference could be strongly dependent on background in bumblebees, which were indifferent to color tested against a uniform green background but preferred blue against a background of real foliage (Forrest and Thomson 2009).

Are There Temporal or Spatial Color Patterns in Flowering Communities?

The concept of temporal or spatial color patterns is a persistent idea in the literature, although it is expressed in different forms. One version suggests that within one habitat there are waves of particular colors progressively through the season, conveying the idea that at any one time many flowers converge on a particular color, perhaps to form a mimicry ring. Gumbert et al. (1999) tested this general idea in flowering communities in Germany, assessing flower color as bees would see it, and found that in practice most communities were not significantly different from random in the colors of simultaneously blooming species; the exceptions were some of the rarer flowers, which might be under stronger selection to secure pollination.

Alternatively, it may appear that within species earlier and later flowering populations may have different colorations. For example, *Lupinus nanus* has less blue reflectance and fewer nectar guides in seasonally early strains, but it is a stronger blue and has more obvious guide patterns later in the season, when pollinators are more abundant and competition is greater (Horovitz and Harding 1972). Similarly, different color morphs of *Epacris impressa* are related to pollinator abundance (Stace and Fripp 1977): red morphs flower in winter, when there are no insects but birds do visit, whereas white morphs occur in spring, when insect numbers are increasing. This again supports the case for a selective effect on flower color exerted by pollinators.

Shading also has an effect. Yellow flowers dominate many grasslands and deserts, but whites and pinks are commoner in nearby woodlands (H. Baker and Hurd 1968), or paler flowers occur in dense woodlands compared with open woodland (del Moral and Standley 1979). Similar trends can sometimes be seen within a species, for example *Ipomopsis aggregata* tends to be red within forests but white or pink in open sites (Elam and Linhart 1988). Some such effects are very likely to be temperature related, and there is little direct evidence (there has indeed been little testing) of effects relating to pollination ecology, but it is nevertheless easy to speculate on links with pollinator abundance: flies and moths are commoner in shaded forest floors and bees more abundant in open habitats.

There are also some suggestions of biogeographical patterns in flower colors (e.g., Kevan 1972; Scogin 1983). Temperate regions have the widest range of flower colors. The high Arctic has an unusually large proportion of white and yellow flowers (Kevan 1972) and a low incidence of UV-reflecting flowers (Kevan et al. 2001), perhaps relating to the rarity of bees and prevalence of flies. High-elevation habitats often have many blue flowers, whether in Europe, the Rockies, or Southeast Asia (Weevers 1952), an effect that is produced by an unusually high proportion (>60%) of delphinidin-containing flowers while only 2% contain red pelargonidin (Acheson 1956). For some specific plants there is a trend to paler colors with altitude (species of *Trifolium*, *Campanula*, *Cypripedium*), whereas for others the trend is reversed, with *Primula*, *Daphne*, and *Crocus* species (all rather early flowering) often

more intensely colored at altitude (Rafinski 1979; Kevan 1983). The tropics have more bright red and orange flowers and fewer blue ones; Beale et al. (1941) linked this finding to high proportions (38%) of pelargonidin-containing plants. Note that many of these studies date from a period when the chemical analysis of flower color was in vogue and have not really been followed up by ecological analyses; but these trends have traditionally been linked to trends in pollinator availability with more insects (especially flies) at high altitudes, more bee diversity in temperate zones, and more flower-visiting birds in the tropics.

Plant species composition may also have an influence. Ostler and Harper (1978) reported a specific trend for more blue flowers in high-diversity communities and more yellow flowers in less specoise zones; again perhaps this could be linked to the greater numbers of specialist bees in very diverse communities, but strict evidence is lacking.

Effects of Color Polymorphisms

There are many recorded examples of pollinators discriminating between flower color morphs and perhaps thereby exerting **directional selection**. This kind of study is useful because the morph variation does not affect all visitors equally; for example, with *Raphanus* morphs *Pieris* butterflies and *Eristalis* flies both preferred the "insect-purple" morph, while *Bombus* preferred the white flowers (Q. Kay 1976, 1978). For *Claytonia*, preferences differed between flower visitors, herbivores, and pathogens (probably helping to maintain the polymorphism), and flower color seemed to be selected largely independently of other floral traits (Frey 2004, 2007). With computer simulations and using both virtual and real hummingbirds and bumblebees, Gegear and Burns (2007) showed that color preferences alone could drive ecological speciation in polymorphic flower populations.

Selection can also operate in different directions for different pollinators. Positive frequency-dependent foraging is normally expected, but there may be negative frequency dependency where competitive effects interact with reward patterns. For example, in *Clarkia xantiana*, Eckhart et al. (2006) found no overall preference for morphs with and without strong-colored petal spots when testing pollinator assemblages as a whole, but there was a strong positive frequency dependence in the most widespread bee *Hesperapis* and a preference for whichever morph was locally in a minority for two other specialist bees, so that a mixture of positive and negative effects probably maintained this polymorphism and may also have contributed to pollinator coexistence.

But such effects are not always present. In *Polemonium viscosum* in the Rockies, most pollinators did not discriminate between white and blue/purple morphs (Galen and Kevan 1980), even though these should appear to the visitors as very distinctive "insect yellow" versus "insect green or blue-green" (although bumblebees did choose between them almost entirely on the basis of different scents, described as sweet and skunky, respectively, as discussed in the next chapter). It is likely that some of these morph color differences may just be due to **edaphic** or climatic differences and have no real selective effect.

Other Selective Influences on Flower Color

Protective coloration, **crypsis**, and **aposematism** are well known in other organisms; do they occur in flowers? There is perhaps some correlation between toxicity and flower color (Hinton 1973), and it could be argued that the ancestral flowers were originally brightly colored as an aposematic signal, to warn off contemporary reptiles and other herbivores. Floral color can still also be under selection for repellency to herbivores; for example, Irwin, Adler, and Brody (2003) with *Raphanus* and Frey (2004) with *Claytonia* showed that the outcome of conflict between pollinator attraction and herbivore repellence was the maintenance of a floral color polymorphism.

Crypsis is presumably counterproductive for animal-pollinated flowers, for which the main requirement must be to attract attention. But some organisms do manage to be cryptic *on* flowers, thus becoming better predators (e.g., crab spiders; chapter 24). Furthermore, many of the very small animals that feed on or in flowers, such as thrips and bugs, are pale and cryptic against generalist cream or yellow-green flowers. Flowers may also benefit from having color patterns that suggest the presence of a potential mate to an incoming insect, which may explain the presence of darker, often pink/purple, florets at the center of carrot umbels (Goulson et al. 2009), the plants with more such dark florets attracting more beetle visitors.

One other aspect of flower color to be considered is the relation of color to thermal biology; pale-colored, highly reflective flowers may remain cooler than more absorptive, dark-colored flowers. This point is taken up again in chapter 9 in relation to floral heat as an alternative form of reward.

6. Nectar Guides

Nectar guides, which are distinct in color from the rest of the corolla (at least as interpreted by humans), are common in flowers and most often get their coloration from flavonol pigments (W. Thompson et al. 1972). Early studies (e.g., Daumer 1958), in which flowers were observed with UV filters, indicated that nectar guides, whether as spots or stripes or radiating patterns, were differentially distributed in flower types (table 5.5). These data could be taken to indicate a greater likelihood of guide marks when the flower is more complicated and the route to the nectar less obvious, a point specifically made by Kugler (1963) and Mulligan and Kevan (1973) and strongly implying adaptive significance. Furthermore, where guide marks were apparently absent, Kugler (1963) reported that the anthers or stigmas were usually strongly UV absorbing, so marking the floral approach in a similar fashion.

To many insects, then, these color-based guides may help to highlight the architecture of the flower during the approach, making foraging more efficient (Heinrich 1975a). However, direct evidence for their importance in guiding a visitor's approach has been rather mixed, perhaps because of an early overemphasis on their UV components (e.g., Knoll [1925] with hawkmoths, A. Manning [1956a] with bumblebees, and Free [1970a] with honeybees). Daumer (1958) did produce neat evidence for guided responses in honeybees with *Helianthus*, in that when he artificially reversed the petals, the bees reversed their normal behavior, landing centrally and then walking out to the rim and only extending their proboscis on reaching the UV-absorbing guide marks that were now present at the flower edge. Hoverflies (*Eristalis*) found the center of a dummy flower faster when guidelines were present (Dinkel and Lunau 2001) and also showed specific responses to colored-dot guides on saxifrage flowers, where there is a centripetal transition from red to orange and then to yellow dots (Lunau et al. 2005). More recently, Medel et al. (2003) showed that varied guide shapes in *Mimulus luteus* differentially affected insects (which preferred guides pointing to the center)

TABLE 5.5
Occurrence of Nectar Guides in Different Flower Types

By flower form	% with guides	By visitor type	% with guides
Bowl, bell and tubular	~50%	Butterfly	83%
Gullet and composite	~70%	Bee	76%
Leguminous flag	88%	Generalist	50%

Source: Modified from Kugler 1963 and Kevan 1983.

and birds (which preferred heart-shaped guides; fig. 5.13C). And where pigments are lost and such guides are missing, bees appear to take significantly longer to find the flower center and to probe correctly.

Although the guides are often visible to us (plate 15), some are only revealed to the human eye with UV photography (Guldberg and Atsatt 1975; Harborne 1993); being strongly UV reflective would make the guides conspicuous to bees, or if strongly UV absorbing, they would produce a contrasting dark mark to an insect (fig. 5.13A shows some examples). Very commonly nectar guides do appear (especially to bees) as strikingly contrasted with the background color of the petals, whether as spots or stripes. Like anthers, the marks tend to have strong color purity (Lunau et al. 1996).

The guide patterns take on many different forms. Concentric circles of different colors are common in radial flowers (e.g., Asteraceae, Primulaceae, and *Myosotis* and other flowers in Boraginaceae; plate 16F), commonly with shorter wavelengths peripherally (blue, purple) and white or yellow centrally (plate 21A). A circle of spots is common (e.g., plates 5F and 15H). Radiating lines also occur in many radial flowers (e.g., Geraniaceae, plate 21B; and fig. 5.13B). In zygomorphic blooms a single dark dorsal blotch (plate 11G) may help orient a visitor, or there may be contrasting blotches near the flower center (fig. 5.13C, also showing different sizes and shapes linked with bees and birds); where many smaller spots occur, they again tend to become denser toward the entrance (as in foxglove, plate 15G). Lines converging toward the corolla entrance are common in orchids, as well as in Fabaceae and Violaceae (plate 15), effectively pointing to the rewards. In laboratory studies, A. Anderson (1977) showed that the stripe widths and spacings are often such that they closely match the preferred patterns of honeybees.

For nectar guide spots, and for contrasting central colored areas in flowers, there is some evidence that they may serve as a mimic of pollen and anthers; they are often white or yellow and are more frequent on flowers where the pollen itself is hidden (e.g., in foxgloves, white or purple spots occur on pink and purple flowers in which the pollen is concealed, but no spots occur on the yellow foxglove, where the anthers visibly protrude). Heuschen et al. (2005) showed a specific and innate preference in bumblebees for two-colored flowers, where the inner color areas were similar to the color of pollen. Indeed most flower pattern preferences in insects have been interpreted as an innate response to anther-like and pollen-like markings (Osche 1983; Lunau 2000), which would normally guide an incoming visitor to the food-dispensing (and conveniently past the pollen-depositing) part of a flower. Hence bees, which show an antennal response to scented anthers by touching the pollen with the tip of each **antenna**, show the same response most strongly (even in naive untrained bees) to flowers bearing marks—such as two erect ovals—that resemble anthers (Lunau 1991). Pohl et al. (2008) found that in dichogamous plants, the sexual phases of flowers with bright yellow, anther-like floral guides were less easily discriminated by bumblebees; in effect, the spectrally appropriate guides ensured that bees still visited (and potentially pollinated) female-phase flowers even though these yielded no pollen.

7. Floral Color Change and How It Can Control Pollinators

Many flowers change color in their lifetimes just due to the loss of pigmentation that accompanies gradual senescence and wilting. But others show a more

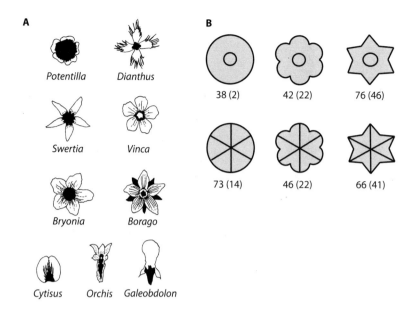

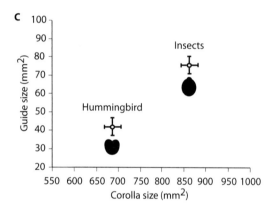

Figure 5.13 Nectar guides in flowers: (A) guides on apparently uniformly colored flowers as seen in UV light (though these are now known not to be as especially significant as once thought; see text) (modified from Kugler 1966); (B) honeybee landings on six pure yellow models and on the same models with blue nectar-guide lines added; the model with nectar guide is always preferred, the number of landings with (and without) guides being shown below each model (modified from Free 1970a); (C) mean size (face on) of corollas and of guides on flowers visited by hummingbirds and insects (also showing some difference in shapes of the guides (redrawn from Medel et al. 2003).

controlled color shift, with underlying specific pigment changes, unrelated to any loss of **viability** or **turgor** and followed by a significant period of flower retention. Color change that is potentially adaptive and not merely related to ageing is known in more than 80 angiosperm families. Familiar examples occur in lungworts (*Pulmonaria*), which change from pink/red at the bud-opening stage to purple or blue when mature, a change that is commonplace in other Boraginaceae as well as in Polemoniaceae, Convolvulaceae, and some of the common gullet flowers in Fabaceae and Scrophulariaceae. This pattern of occurrence of adaptive color change provides indications of multiple convergent evolutionary changes.

The color change can involve all or just part of the flower. Partial changes may affect almost any tissues. For example, in *Lantana camara* all the corolla petals change, originally from yellow through orange to red but now with other combinations in hybrids (plate 16B); and in *Raphiolepis umbellata* (Rosaceae), the stamens change from yellow to red. In the horse chestnuts (*Aesculus*) nectar-guide spots on the lower petals change from yellow through orange to pink (plate 16C), and nectar is present only in the yellow or orange phases. In some species, the floral scent also changes (Lex 1954). And in *Catalpa bignonioides* there are conspicuous corolla marks interpreted as **pseudostamens** (Lunau 2000) that change from bright yellow to dark purple (Lunau 1996b), again accompanied by a marked reduction in nectar reward.

Very often, the changes occur in response to visitation or some aspect of pollination and can be interpreted as an adaptive response by the flower that serves to direct visitors to those flowers that are not yet pollinated. The common linkage between color change and reduced reward status again indicates an adaptive response, the plant ceasing to put valuable sugar resources into flowers that are already fertilized while maintaining the reward in the unchanged flowers that still signal their need for a visitor.

Appropriate responses to color change are known in many pollinators. Examples include different species of butterflies that show different color preferences when responding to *Lantana* (Dronamraju 1960; Ram and Mathur 1984), and hoverflies responding to a change in the **coronal scale** color (plate 16F) of *Myosotis* (Nuttman and Willmer 2008). Honeybees and bumblebees visit only lupins with prechange banner petals (Schaal and Leverich 1980; Nuttman and Willmer 2003); bumblebees visit only young flowers of *Aecsculus* with yellow nectar guides (Barth 1985); *Anthophora* bees and bumblebees visit pink (but rarely blue) *Pulmonaria* (plate 16D; pers.obs.; Oberrath and Böhning-Gaese 1999) and also choose yellow (but not white) *Alkanna* (Willmer et al. 1994; Nuttman et al. 2006).

The direction of color change is perhaps most often toward a less conspicuous, and usually a darker, color. Some plants mainly visited by insects change from light pinks or oranges to a darker red, taking them out of the best visual sensitivity range; some nocturnal flowers change from pale to dark colors, becoming more or less invisible against the background foliage (e.g., *Cobaea*, Proctor and Proctor 1978). Changes in ultraviolet reflection are also known (C. Jones and Buchmann 1974), which should affect visibility and contrast for certain visitors. Some species have changes in petal markings that again reduce contrast and conspicuousness, as for example in lupins, where in one species a white banner spot changes to purple or in another species changes from yellow to blue, in each case making the spot much less distinct from the main petal color. There may also be cases of a change to a color that has greater attraction for a different kind of pollinator. For example in *Lantana camara*, flowers are normally yellow on day 1, orange on day 2, and thereafter pink, red, or purple. Nymphalid butterflies were found to prefer the yellow/orange flowers, while three other butterfly species in other families preferred

orange/pink flowers, and yet another was completely undiscriminating (Dronamraju 1960). A more spectacular effect occurs in *Quisqualis indica*, a climber whose long tubular flowers are white and horizontal on their first night of opening, when they are visited by hawkmoths, but which change over the next two days through pink and red and become pendulous with somewhat larger corolla diameters, still containing some nectar (which has now slid down toward the corolla mouth) and attracting daytime visits from bees, flies, and sunbirds (Eisikowitch and Rotem 1987).

Color change linked to changing reward status can potentially offer a useful resource signal to visitors, so that they only visit "prime" flowers. But if color change were also triggered by an aspect of visitation or pollination, then it would additionally serve as an ecological signal to visitors. As such, it would be one of a range of possible postpollination changes that could be adaptive (Gori 1983; and *Flower Longevity and Flowering Period* in chapter 21). These might include changes in morphology, orientation, or scent, as well as color, that either make the flower inconspicuous, unattractive, or inaccessible to a visitor or that signal that the flower is no longer worth visiting (after the animal has learned the association between signal change and reward change). Two examples of color changes that are triggered or accelerated by visitation or pollination are shown in figure 5.14. This triggering is by no means always present however, and (for example) it does not seem to occur in *Pulmonaria* (Oberrath et al. 1995).

Many flowers that change color, whether simply with age or as a postpollination response, nevertheless remain on the plant for some time after the change, even though they have little or no reward on offer. In this way they can continue to contribute to the overall floral display of the plant from a distance but can be discriminated and rejected by experienced visitors at close range (Weiss 1991, 1995a). There are some instances where the older color-changed flowers do not seem to add to long-range attraction (e.g., *Pedicularis monbeigana*, where the flower lip changes from white to purple [S. Sun, Liao, et al. 2005], or *Cryptantha* where yellow coronas change to white [Casper and LaPine 1984]). But more commonly, attraction increases. In *Alkanna*, for example, the flowers change from yellow to cream after pollination (Willmer, Gilbert, et al.1994; Nuttman and Willmer 2008), and the cream flowers remaining on the plant increase overall

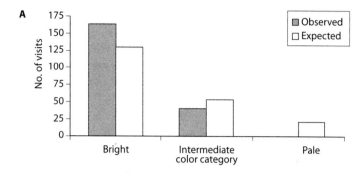

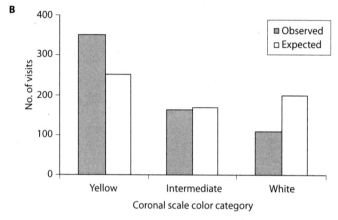

Figure 5.14 Effects of floral color changes on visitation: (A) in *Alkanna*, *Anthophora* bees are more likely to visit bright yellow flowers and hardly ever visit the pale postchange flowers (expected frequencies based on availability of color phases in the population) (from data in Nuttman et al. 2006); (B) in *Myosotis*, hoverflies choose the flowers with yellow coronal scales and largely ignore those where the scales have changed to white (see plate 16F) (redrawn from Nuttman and Willmer 2008).

attractiveness to bees from a distance (Stone et al. 1999). The same is true in *Erysimum scoparium*, which changes from white to purple (Ollerton, Grace, et al. 2007). *Myosotis sylvatica* is particularly interesting because the corolla changes with age from pink to blue, but this shift has no effect on visitors; the adaptive change is restricted to the coronal scales around the neck of the corolla turning from yellow to white (as in *Cryptantha*), an effect to which visiting hoverflies are very responsive, showing a PER with 94% frequency to a yellow corona but only 14% frequency to a white corona (Nuttman and Willmer 2008). The corolla color change is automatic and simply age related, while the coronal change is accelerated by pollination (fig. 5.14B). The retained postchange flowers may increase long-range attraction and also improve the landing platform for visiting hoverflies.

Where they do occur, these kinds of postpollination changes could be useful to both partners. The postchange flowers persisting on the plant enhance overall attractiveness at a distance so they benefit the plant; at close range they also dissuade visitors from wasteful visits to already-pollinated flowers and direct

them toward higher rewards, so benefitting the animal, and thus also to more "useful" (possibly pollinating) activity on the prechange flowers that will again benefit the plant. However, Kudo et al. (2007) demonstrated with artificial inflorescences that the effect can vary with overall display size and flowering pattern and is not always beneficial to the plant.

Mechanisms of changes have been investigated in *Lupinus pilosus*, where color change occurs in the banner petal spot only, which goes from white to pink. Bees strongly prefer flowers with white spots (fig. 5.15). Flowers that were not visited (because they were bagged) did not have any growing pollen tubes and did not undergo much color change. In contrast, exposed visited flowers with growing pollen tubes (i.e., already pollinated and probably fertilized) did change color, over the normal time course (fig. 5.15B). These observations suggest that color change is indeed functional, is triggered by pollination, and can direct visitors to other as yet unpollinated flowers (Nuttman and Willmer 2003). It also suggests that the growth of pollen tubes in the stigma is the probable trigger. Since this growth is, from the plant's perspective, akin to tissue

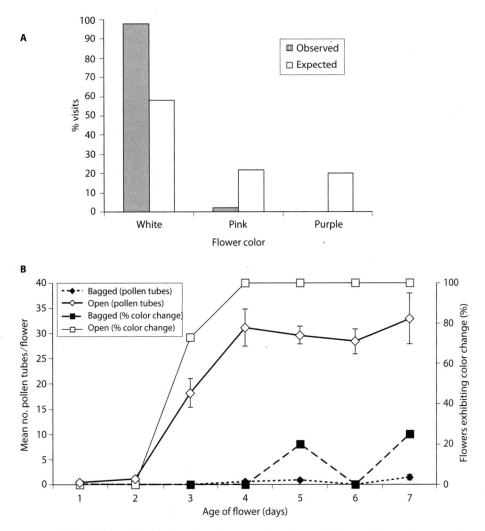

Figure 5.15 (A) Bumblebees visiting *Lupinus pilosus* show a strong preference for flowers with white banner petal spots, largely ignoring the pink or purple postchange flowers (plate 16). (B) The time course of color change in bagged versus open flowers: the bagged, unvisited flowers have almost no pollen tubes growing in their styles and a very slow color change only beginning on day 5, whereas open, visited flowers show a much faster color change, similar in time course to the growth of pollen tubes, both phenomena being effectively complete in four days. (Redrawn from Nuttman and Willmer 2003.)

damage, it is likely to be mediated by ethylene, because it is a major stress-signal molecule in many aspects of plant function that is also widely implicated in mechanisms of postpollination reductions in attractiveness, longevity, and color change (van Doorn 1997, 2002) and in wilting (Hilioti et al. 2000). Thus in *Viola cornuta*, where a color change from white to purple occurs after pollination, it has been shown that three genes responsible for anthocyanin production increase

their expression simultaneously, probably turned on via hormonal signals from ethylene and gibberellic acid (Farzad et al. 2003).

We now know that reversible color change is also possible. In the East African legume *Desmodium setigerum* (plate 9F and fig. 5.16), there is a unique ability to reverse the flowers' color and shape changes within their one day of life. A single morning visit by a bee "trips" the flower explosively to expose the anthers

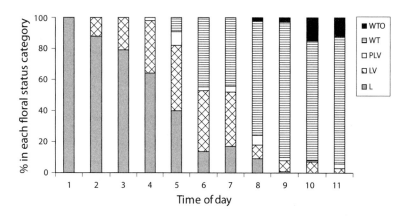

Figure 5.16 Reversible color and shape changes in *Desmodium setigerum* flowers in East Africa. Flowers are initially lilac (both keel and flag petals, L) and are irreversibly tripped by a single visit to expose the stamens and style (LV); thereafter they begin to pale (PLV) and fairly rapidly close and change color to have a white flag and turquoise keel (WT). However a small proportion reopen later in the day (WTO) and in some cases revert to a darker turquoise or recurrent lilac color; these flowers are ones which had received little or no pollen in their first visit (see text for more details). Note irregular time intervals. (Redrawn from Willmer et al. 2009.)

and also initiates rapid change (within about 2 hours) from lilac flag and keel petals to a white flag and turquoise keel, with the flag bending over to conceal the reproductive parts. But the flowers that receive little or no pollen on the stigma can then partially reopen to reexpose the stigma, at the same time showing a further color change to a deeper turquoise in keel, or flag, or both, sometimes with a recurrence of lilac coloration in the keel petal. Thus most of the flowers achieve full pollination from a single bee visit and undergo no further floral changes; but those that received insufficient pollen can reverse their signals to earn a second chance by eliciting further visits from other potential pollinators (Willmer, Stanley, et al. 2009). Hopefully further examples of this kind of adaptive reversal will emerge.

8. Other Visual Advertising Cues

Flower Shape and Outline

Flowers vary enormously in their form (chapter 2), and aspects of their size and shape are clearly part of the visual advertisement. Spatial flower features in relation to insect vision were reviewed by Dafni et al. (1997). Many visitors are particular about their angle of approach to flowers, which suggests they have a need for a standard shape image. There is substantial evidence for shape preferences in many bees and for at least some other flower visitors. For example, when visiting a crucifer flower (*Erysimum mediohispanicam*), bees preferred narrow, pointed petals while bee-flies preferred rounded, overlapping petals (Gómez et al. 2008). Bees also selected symmetrical flowers over those with moderate **asymmetry** (Müller 1995a; Giurfa et al. 1996), whereas Midgley and Johnson (1998) found no particular symmetry preferences in beetles and bee-flies, so a predilection for symmetry is not an inevitable condition in flower visitors.

Bees have an innate preference for flower-shaped objects and show a strong preference for somewhat dissected shapes, or shapes with a high **contour intensity** or "edginess." Hence they prefer flowers with frilled edges to flowers with simple outlines and will generally prefer six petals to three, and multiple-petaled, daisylike outlines to simple six-petaled outlines (Lehrer et al. 1995). Honeybees are poor at discriminating among simple solid shapes or among highly edgy shapes (fig. 5.17A) but easily sort the latter from the former, preferring any of the edgy shapes. However, they also discriminate between edgy shapes set at different angles, detecting as little as 4° difference, and preferring whichever angle they have been trained to (Wehner 1971, 1981). They learn complex shapes—those with a high ratio of contour length to enclosed area—more quickly than simple shapes (fig. 5.17C).

However, working on lavender (*Lavandula stoechas*, plate 11H), Herrera (2001a) found that he could remove the large purple bracts acting as flags above the corollas, making the overall inflorescence appearance quite different, with no effect on male or female reproductive success (i.e., pollen grains removed or pollen tubes growing), and he suggested that the concept of visual cues as key factors in insect visitation has sometimes been overplayed in the literature. But Higginson et al. (2007) found that removing the purple bracts on lavender did reduce the inspection rate

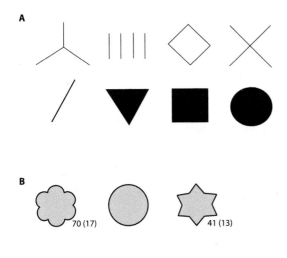

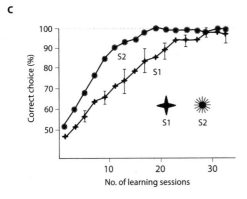

Figure 5.17 Edge effects in bees: (A) honeybees show poor discrimination within the shapes on the upper and the lower lines but easily discriminate any of the upper line from any of the lower line (redrawn from Barth 1985, from data given by von Frisch); (B) bee landings on different shapes; trained on the middle shape, they then prefer to land on either of the more dissected shapes (numbers are landings on alternative models, with landings on original model in parentheses) (redrawn from Free 1970a); (C) learning curves for honeybees, with faster learning when the training object is the more dissected, "edgy" shape (redrawn from data in Schnetter 1972).

by *Apis*, suggesting that there are more subtle responses to the visual display than those revealed by measuring the final reproductive display. Keasar et al. (2007) tried a similar manipulation with the colored bracts of *Salvia* flowers and found that honeybees significantly reduced visitation to manipulated flowers; these researchers thought the key difference was that *Salvia* bract sizes are strongly correlated with the number of open flowers on the inflorescence, so pro-

viding an honest signal of reward, whereas this is not always true in lavender (for bract size or even bract presence).

Apart from all the work on bees and some flies, Kelber (2002) examined pattern discrimination in a hawkmoth and found innate preferences for pattern over uniform color, and for particular patterns: radial patterns were preferred over tangential, which were preferred over striped. These latter pattern preferences were fixed and could not be overridden by experience, perhaps an important factor in a hovering pollinator where patterns guide the feeder to the reward.

The different spatial preferences of bees, bee-flies, and moths, as outlined here, suggest that corolla shape can be strongly selected for by visitor frequency, and indeed the study by Gómez et al. (2008) noted geographical variation in *Erysimum* flower shape linked to local pollinator frequencies.

Flower Texture and Microstructure

Texture in petals and other floral parts may offer both visual and tactile cues to visitors (Kevan and Lane 1985; Comba et al. 2000), and sometimes this feature has significance for pollinators. Petals may be highly reflective because their surfaces are very smooth or have a somewhat matte appearance where the epidermis is finely structured. Glover and Martin (1998) showed a specific role of petal cell shape, which is related to color and pigmentation, in the pollination of *Antirrhinum* flowers, where conical cells in the epidermis affected the perceived color for bees and may also have affected their tactile perceptions (Whitney, Chittka, et al. 2009). These kinds of "texturing" cells may turn out to be quite common. Comba et al. (2000) showed that the presence of conical petal cells raised intrafloral temperatures, which in turn increased the pollinator preference for these flowers (chapter 9).

In a few cases nectar guides also offer mechanosensory stimuli, taking the form of grooves or ridges, and these can help guide visitors to the nectar (e.g., moths: Goyret and Raguso 2006).

Flower Movement

The effect of flower movement has been explored only minimally, but Warren and James (2008) found that

Silene flowers that "waved" (i.e., moved in a light breeze because their stalks were thin and flexible) attracted more insect visits, although of shorter durations, and had a high heritability of stem mobility and an optimum median mobility overall. More exploration is needed of floral waving as a means of attracting pollinator attention.

Combining the Visual Cues

In practice, there can be little doubt that most flower visitors will be taking into account many aspects of a flower's visual signals: color(s), size, shape, outline, and patterns, including nectar guides. At close range they may also potentially respond to flower textures and surface sculpturing. The overall message is that visual floral advertisement is complex and that the cognitive abilities of the pollinators are rather sophisticated and quite varied. Nevertheless, several authors have pointed to a suite of fairly consistent features found in a great many flowers that seem to indicate a degree of coevolution with the cognition of their main pollinators. These would include

1. the presence of strong UV patterns,
2. the color contrast between corolla and androecium,
3. the predominance of yellow color in pollen,
4. yellow colors in flower centers,
5. the greater reflection of shorter wavelengths more peripherally, and
6. gradients of color purity increasing toward the center.

9. Why Do Flower Colors Diverge? Selection and Floral Color

It is evident from much of the preceding discussion that flower color really matters and may be under strong selective pressure. In the multidimensional study of *Penstemon* species by P. Wilson et al. (2004, 2006), color was the single best predictor of hummingbird visitation, so that as we saw above, blue species attracted bees, pink species attracted birds, and red/orange species attracted birds and deterred bees. In *Mimulus* hybrids (Sutherland and Vickery 1993),

bees responded primarily to color and less to shape, whereas birds were little affected by color differences but responded to petal shape and flexion. Later work on this genus involved genetic manipulations to produce the color of one species in flowers that were in other respects another species (H. Bradshaw and Schemske 2003), and here both birds and bees used color as their main cue when other characters were kept roughly equal.

Selection for flower color is not necessarily operating to produce particular pollination syndromes but is likely to be acting more directly on visitor constancy (Gegear and Laverty 2001), so that floral constancy is often highly correlated with color difference above a threshold (fig. 5.18). Selection can operate quite rapidly, partly because the divergence of color is quite easy to achieve at a genetic level: in *Ipomoea*, single insertions of the highly mobile **transposons** in the genome are known to effect most changes (Fry and Rauscher 1997; Clegg and Durbin 2003; Galliot, Stuurman, et al. 2006).

However, selection for *optimum* flower color in any given population may be prevented by gene flow from nearby populations, by genetic drift, or by positive **frequency-dependent selection** by pollinators; and it may also be limited by phylogenetic constraints. It has already been noted how easy it can be to get drawn into **adaptationist** reasoning about flower colors, and much of the literature about pollination syndromes rather blithely assumes strong selection for particular flower colors to suit particular flower visitors, a point that will become clearer in later chapters. But it is useful at this point to reiterate some further notes of caution:

1. There can be substantial effects of pleiotropy in operation. A high concentration of anthocyanins in flowers tends to be correlated with a high concentration of anthocyanins in the leaves and stems of the same plant; in other words, prior selection on **vegetative traits** may have favored the expression of floral pigments initially, so that flower color is an **exaptation** (Armbruster et al. 1997; Armbruster 2002; and S. Strauss and Whittall 2006).
2. Plant pigments have many other functions besides coloration and so may mislead the unwary. Carotenoids are accessory pigments to chlorophyll in photosynthetic processes, and many pigments are

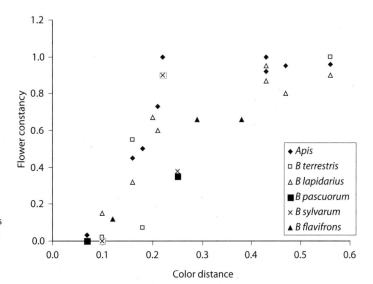

Figure 5.18 Flower constancy depends on the detection of difference; thus it increases as color differences between pairs of flower types increases, although the rate of improvement of constancy varies between the honeybee and five species of bumblebee. (Redrawn from Giurfa and Lehrer 2001.)

also involved in antiherbivore effects, in protecting against UV radiation, or in protecting against frost and heat damage. Anthocyanins confer better tolerance of heat and drought stress, so that, for example, blue morphs of *Linaria* do better in drought years but a polymorphism is maintained because white flowers do better in years of good spring rain (Schemske and Bierzychudek 2001).

3. There are substantial direct environmental effects on flower color, which can affect visitation patterns and influence selection within a species (see the section *Environmental Effects on Floral Color* above).

4. Pigment levels are sometimes correlated with other plant traits; for example, opium poppy plants with higher morphine alkaloid levels also have differently colored petals (Gyulane et al. 1980).

5. There is good evidence that pigments can be expensive to produce, so that investing in strong coloration can affect plant vigor in complex fashions. Thus there are always trade-offs for a given plant between producing optimal flower color and maintaining other key functions.

6. Attempts to correlate floral color, floral pigmentation, and pollinator classes have undoubtedly suffered on occasion from uncorrected phylogenetic influences.

Despite all those cautions, there are some clear examples of selection for gradual color change in relation to pollination, often involving major shifts in the type of pigment used (e.g., from red cyanidin-based colors to blue and purple delphinidin-based colors). Harborne and Smith (1978) largely avoided taxonomic confusions by choosing a single family (Polemoniaceae) as a case study, this being a family containing a large range of color types and pollinator types. They found a clear relationship between the type of anthocyanin pigment present, the perceived color, and the visitors recorded: red-flowered species contained pelargonidin and mostly attracted hummingbirds, blue and purple flowers contained delphinidin and were bee-visited, and pink flowers contained cyanidin-delphinidin mixtures and were preferred by lepidopterans. We will meet and revisit further examples of this type in chapters 11–18.

Chapter 6

ADVERTISEMENTS 2: OLFACTORY SIGNALS

Outline

The previous chapter concentrated on visual signaling by flowers, but it is usually not appropriate to look only at flower color and shape as attractants, given the abilities of most flower visitors, and insects in particular, to detect and respond to scents or odors as well; the latter are often the major component of floral attraction to visiting animals.

Floral scents mostly result from the production of small amounts of simple **volatile organic compounds**. The molecular size of these components largely determines their volatility, and hence the distance they will travel from the plant over a given time span. In practice, the range of small carbon-based volatiles that are reasonably cheap to produce is limited, so to avoid excessive overlap most plants almost inevitably use mixtures of compounds to form their scents, with specific ratios of components characteristic of each species. They can thus achieve a novel and often unique floral bouquet, so that scent can become an isolating mechanism between plant species, each odor potentially attracting different visitors. Virtually all flowers therefore produce complex bouquets with many components (as few as 1–8, but often in excess of 100, and with a norm of around 20–60). However, not all of these may be important as signals to visitors. Quite commonly, the most abundant compounds are not particularly strong elicitors of responses in visitors, whereas a few key components (often the rarer ones) are crucial signals. In sunflowers, for example, there may be as many as 130–150 constituents, of which only around 30 are necessary for a honeybee to perceive the mix as being "sunflower odor" (Laloi et al. 2000).

Plant volatiles emitted as scents are typically lipophilic compounds, all of them of low molecular weight (< 300 atomic mass units, or daltons), able to cross biological membranes, and evaporate into the atmosphere readily. Around 1,700 are known, and more await identification. They are secondary plant products resulting from standard plant biosynthetic processes (Pichersky et al. 2006), and some may even be waste products that are excreted following various metabolic processes. Their occurrence shows rather little phylogenetic pattern, although some compounds are restricted to particular lineages (Raguso 2004a).

1. Types of Scents

Individual flower odor compounds can be classified by scent type, as defined by the human nose, or by chemistry, as shown in table 6.1. Representative chemical structures are shown in figure 6.1A, and the basic biosynthetic pathways that produce the main groups of compounds are shown in figure 6.1B.

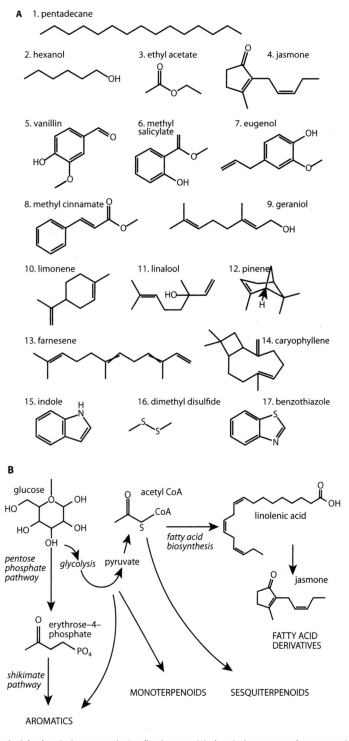

Figure 6.1 The principle chemical types producing floral scents: (A) chemical structures of representative examples—1–4 are aliphatic (hydrocarbon, alcohol, ester, and ketone, respectively, related to fatty acids); 5–8 are benzenoids; 9–14 are terpenoids, 9–12 being monoterpenes and 13–14 sesquiterpenes; 15–17 are unusual nitrogen or sulphur containing scents; (B) main synthetic pathways to the key groups of volatiles.

TABLE 6.1
Flower Scents

Chemical classification	Examples
Terpenoids	
Monoterpenes (10 carbon)	Geraniol, citronellol, linalool, α-pinene, limonene, camphor, menthol, cineole
Sesquiterpenes (~15 carbon)	α-farnesene, caryophyllene, ionone
Benzenoids	Vanillin, eugenol, methyl salicylate, methyl cinnamate
Aliphatic compounds	
Hydrocarbons	Pentadecane
Esters	Ethyl acetate
Ketones	Jasmone
Alcohols	Hexanol
Scent-type classification	
Flowery	Terpenes; benzenoids; some alcohols, ketones, esters
Dung or carrion	Amines, ammonia, indoles, skatole
Pheromonal	Alcohols, acids, esters; some terpenoids; (mimicking insect pheromones)

Numerical occurrences of known compound types are shown in table 6.2, derived from analyses of nearly a thousand taxa. Flower scents are most commonly **terpenoids**; these are assembled from five-carbon base units termed **isoprenoids** and can be divided into **monoterpenes** (C_{10}) and **sesquiterpenes** with various other multiples of the five-carbon base units as their backbone (usually C_{15}). In this group, the several isomeric and chiral variants of linalool, ocimene, myrcene, and pinene are the most frequently occurring. Only slightly less common are the simple **aliphatic compounds**, such as hydrocarbons, **esters**, **ketones**, and **alcohols** (especially C_4–C_{18}). Less abundant are the **benzenoids**, based on the aromatic six-carbon benzene ring.

The terminology of all these groups is somewhat muddled, because many of the terpenes and ketones are informally named using the common term (or sometimes the Latin name) for the plant from which they were identified; hence pinenes, geraniol, and limonene are terpenes from pine trees, geraniums (*Pelargonium*), and lemon plants, respectively, while jasmone is a ketone found in jasmine plants. However all these compounds are also found much more widely, across a range of taxa, and indeed may be more dominant in plants that are very distant evolutionarily from their namesake; for example, geraniol and citronellol are both more common in *Rosa* than in geraniums or citrus flowers. Other common flower volatiles retain their more correct chemical names in the literature, having first been isolated in other sources; methyl salicylate, for example, is closely related to aspirin, famously derived from willow plants, and gives the well-known scent of wintergreen.

Most of these examples can broadly be described as having a flowery smell, but generally speaking, human observers are poor at discriminating or describing flower scents beyond a relatively crude level, often merely making comparisons to a few well-known basics (lavender, rose, citrus, etc.). Detection and discrimination by the human nose can be improved by sealing particular flowers (or parts of flowers) in a container for a few hours, so that the volatiles released become concentrated in the headspace of contained air, and this procedure can be of some use in field studies where the more advanced technologies for scent analysis (discussed in section 2, *Collection and Measurement of Flower Scents*, below) are not available. There are, of course, rare individual humans who have much more highly developed discrimination and well-trained noses, and they are highly prized in the perfume industry!

TABLE 6.2
Numerical Distribution of Known Floral Scent Compounds

Type of compound	Structure	Number	Totals
Aliphatics (straight chain)*	$C_1–C_2$	64	
	$C_3–C_4$	74	
	$C_5–C_6$	123	
	$C_7–C_8$	72	
	$C_9–C_{10}$	68	
	$C_{11}–C_{15}$	76	
	$C_{16}–C_{20}$	44	
	$C_{21}–C_{25}$	7	528
Benzenoids and phenylpropanoids	$C_6–C_0$	23	
	$C_6–C_1$	133	
	$C_6–C_2$	78	
	$C_6–C_3$	83	
	$C_6–>C_3$†	12	329
C_5 branched-chain compounds	Saturated	40	
	Unsaturated	53	93
Terpenoids	Mono, acyclic	147	
	Mono, cyclic	148	
	Sesqui, acyclic	48	
	Sesqui, cyclic	116	
	Irregular	97	556
Nitrogen compounds	Acyclic	42	
	Cyclic	19	61
Sulfur compounds	Acyclic	37	
	Cyclic	4	41
Miscellaneous	Cyclic	111	111

Source: Modified from Knudsen and Gershenzon 2006.
*$C_1–C_2$ through $C_{21}–C_{25}$ represent ranges; $C_6–C_0$ through $C_6–>C_3$ represent bonds.
†C_6 linked to chains longer than C_3.

Less commonly, and usually in relation to attraction of carrion flies and beetles, some flowers do not have pleasant flowery bouquets but instead emit odors that are moderately or extremely unpleasant to the human nose. In the first group are the yeasty and fermenting smells, often found in the phylogenetically basal angiosperm groups such as Annonaceae, and comprising the classic fermentation products, such as ethanol, ethyl acetate, and butanol derivatives (e.g., Goodrich et al. 2006). In the second (really unpleasant) group, an odor of dung, rotting meat, or carrion is common, mimicking the kinds of small $C_2–C_5$ molecules that result from the decay of proteins in dead flesh (ammonia, **amines**, or **indoles**). Many of the Araceae have these odors, often picturesquely described in the literature (e.g., "dead horse," "wet dog," or "wet decaying hide"). So too do several flowers famous for producing cloying, long-distance stenches: *Stapelia*, *Rafflesia*, and giant *Amorphophallus* plants of Indo-Malaysia and Madagascar, whose putrid emissions can cause humans to faint (plate 22). Analysis of various stapeliads (Jürgens et al. 2006) revealed four types of volatile profile, shown in table 6.3, roughly paralleling the categories also found in Araceae. Many of

TABLE 6.3
Fetid Floral Odors in Stapeliads

Mimic of	Main volatile contents
Vertebrate herbivore feces	High *p*-cresol, low in polysulfides
Carnivore feces, or carcass	High polysulfides, low *p*-cresol
Carnivore feces, or carcass	Heptanal, octanal
Urine	Hexanoic acid

Source: Based mainly on Jürgens et al. 2006.

these flowers are also thermogenic (chapter 10), their heat helping to volatilize the scents and probably also increasing the perceived resemblance to recently dead flesh. However, some of these carrion flowers also have a rather sweet terpenoid odor beneath their stench (e.g., Kite and Hetterschield 1997), which may play a role in close-range attraction. In at least a few cases, including the voodoo lily (*Sauromatum*, Araceae), this sweeter odor emanates from within the internal trap chamber of the flower (chapter 23), with the putrid odors given off by the protruding **spathe** as a longer-range attractant (Hadacek and Weber 2002).

There is also a third group of floral scents, which mimic the visiting animals' own species-specific attractants (known as **pheromones**), and in insect-pollinated flowers many such compounds occur. Orchids are especially adept at this pheromonal **mimicry**, and closely related species of *Ophrys* produce different bouquets, each attuned to their own preferred visitor for whom they may have an **aphrodisiac** effect: these comprise mainly hydrocarbons in *O. insectifera*, which is fly pollinated, and mainly terpenoids and alcohols in bee-pollinated *O. fusca* and *O. lutea* (Bergström 1978; Borg-Karlson 1990). Taken to a further extreme, one orchid (*Dendrobium sinense*) appears to mimic a component of the alarm pheromone of *Apis* bees—not as a way of attracting these bees but rather to attract pollinating hornets that normally prey on honeybees (Brodmann, Twele, et al., "Orchid mimics honey bee," 2009). In brood-site mutualisms (chapter 26), where flowers harbor their own pollinators as larvae in situ, volatiles that may be pheromone-like again

help to attract the required animals, often with high specificity (especially in the fig-wasp mutualisms). In fact, this area of pheromone-mimicking scents becomes very complex, since some of the insects involved are known to have derived their own pheromones from the host plants on which they feed in the first place; for example, many beetles that are damaging pests in pine forests (such as *Dendroctonus* and *Ips*) use pinene or derivatives of pinene as their own sex pheromone. Hence it is often tricky to disentangle the evolutionary history of such chemical signals and to be sure just who is mimicking whom.

There is an additional and nonfloral source of scents that may be important in pollinator attraction which occurs in the form of odors derived from leaves (Caissard et al. 2004). It is not uncommon for leaves to contain some of the same scents as their flowers, but in the Mediterranean dwarf palm (*Chamaerops humilis*), the volatiles emitted by the leaves are distinctive and have become the main attractant for the pollinating weevil *Derelomis* (Dufay and Anstett 2004). A close relationship exists between the production of leaf odor and floral **phenology**, with maximum foliage odor coinciding with anthesis and thus beneficial to the plant in terms of pollinator attraction.

2. The Collection and Measurement of Flower Scents

The analysis of floral scent is considerably less problematic and potentially less subjective than the analysis of flower color. Recent advances in collection and measurement techniques allow good qualitative and quantitative evaluation of tiny samples of organic substances. Raguso and Pellmyr (1998) and Tholl and Rose (2006) give reviews of the standard techniques and apparatus used, and the volatiles detected from current techniques are reviewed in detail by Knudsen et al. (1993), Dobson and Bergström (2000), and Knudsen and Gershenzon (2006). Kearns and Inouye (1993) provide a useful overview of alternative techniques for those seeking practical advice on odor analysis. However, all these authors have highlighted the problem that differing results accrue with different methods of collection and analysis (see also Agelopoulos and Pickett 1998).

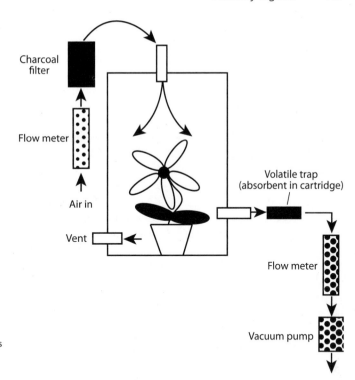

Figure 6.2 A schematic overview of collection techniques for floral volatiles using headspace sampling from an enclosed flower. (Modified from Tholl et al. 2006.)

Collection

An overview of the standard collection and measurement system is shown in figure 6.2. *Headspace collection* is the first step: it allows the gathering of scents emitted from living plants and individual flowers in a highly sensitive and noninvasive manner. Essentially the flower is enclosed in a lightweight bag, into which the volatiles diffuse, and the air from this bag is then pumped out (with a simple battery-operated pump for field use) such that it passes through a small tube (a few ml in volume) containing an appropriate adsorbent material. If the bag is left in place for some time, the volatiles from even delicately scented flowers can be enriched, although there are possible risks of raised humidity and compromised gas exchange affecting the plant's normal physiological processes. Alternatively, the time course of emissions can be followed by taking separate adsorbent aliquots at shorter intervals. For even more detailed analyses, dynamic headspace sampling can be used with a continuous airflow through the bag (the inlet being protected by a charcoal filter to remove external scents), or more elaborate "push-pull"

techniques can be applied in laboratory settings (Tholl and Rose 2006). Scents with very low molecular weight and high volatility may have to be sampled and analyzed with refrigeration, and some very large scent molecules (C_{21}–C_{29}) may have to be extracted directly with solvents rather than being collected from the headspace.

The adsorbents used during headspace collection are usually commercial standards such as Tenax or Porapak, with small, fixed amounts in the collecting tubes. After a known period of collection, the adsorbent cartridges are sealed and preserved under refrigeration until they can be returned to a laboratory for analysis. Some polar compounds are collected with only low efficiency in these adsorbents, so that more than one collection with different adsorbents is often advisable.

Measurement

The adsorbents are eluted into glass vials with organic solvents (pentane, hexane, dichloromethane, etc.), and one or more standards are added as quantitative comparators. Chemical analysis with GC-MS is then the standard procedure. With capillary gas **chromatography** (GC) samples are injected into the GC column, and each component elutes at a rate determined both by its volatility (boiling point) and by its affinity to the column packing material used. This is combined with *mass spectrometry* (MS) to analyze the extracted compounds as they emerge sequentially from the chromatography column, making it possible to separate and identify hundreds of floral odor components by comparison with extensive standard mass spectral libraries. Different solvents can be used, again with varying effectiveness for different compounds; hexane, for example, is usually the choice, but it is a poor eluent for polar compounds. Dobson et al. (2005) give very detailed practical guidance.

In recent years a fast and transportable miniature GC system has become available, known as the zNose, which allows sampling intervals as short as three minutes. This instrument gives results directly in the field that are comparable with laboratory analyses and often with even better temporal resolution.

For laboratory analysis of very small volatile amounts, or for finely divided temporal emission sequences, the techniques of *solid-phase microextraction* (SPME) have also come into use, although these do not allow re-elution, and thus re-analysis, of the original sample.

Ideally, the detected compounds can then be bought or synthesized and tested separately—either neurologically or behaviorally—on known flower visitors to decide which are the effective attractants (although mixtures, rather than single components, may be crucial for some visitors). Detection complemented by neural confirmations of the effect is especially suitable for testing insect reactions to flower scents, and is accomplished with **electroantennogram** (EAG) techniques to measure the insects' responses; when the insect is exposed to particular scents, the EAG directly records responses occurring in the neurons that exit from the antennal **chemoreceptors**. At a further level of refinement, the responses of single olfactory neurons can be obtained with specific electrodes.

At best, GC can be combined directly with EAG or other neural recordings by testing each compound on an insect or other recipient as it elutes from the GC column; an example is shown in figure 6.3 for the hawkmoth *Manduca sexta* when tested with different flowers (Fraser et al. 2003). Behavioral checks on the effectiveness of volatiles with bees or butterflies that show innate PERs to particular odors (and/or colors) are also useful, and checks on flight responses in wind tunnels charged with particular scents can also be used.

Amounts Detected

The amounts of volatiles collected from flowers are hugely variable, usually between picogram and microgram quantities. In general, beetle- and moth-pollinated flowers have the largest amounts of fragrance, while hummingbird flowers often have no detectable scents.

3. Variation in Floral Scents

Generic, Specific, and Individual Variation

Although the techniques available give readily quantified and reasonably unambiguous results, characterizing the scent of a flower is nevertheless somewhat difficult. Scents are highly variable within a genus, suggesting that there are specific adaptations, as for example in *Narcissus* (Dobson, Arroyo, et al. 1996), *Primula* (Gaskett et al. 2005), or *Silene* (Wälti et al. 2008). Even within a species, there can be dimorphic flower scents; for example, *Polemonium viscosum* has two scent-types that are readily distinguished by bumblebees (Galen and Kevan 1980; Galen and Newport 1988). Intriguingly, another plant formerly described as occurring with two scent morphs, the orchid *Chiloglottis valida*, has now been shown to be two species with two different pairs of thynnine wasp pollinators, the two occurring with different altitude ranges (Bower 2006); a neat case of scent as the key variable in speciation.

There is also marked individual variation in scent, which can again be quantitative or qualitative, involving varied ratios of components in different individuals that can in turn affect visitation. The individual differences can be spatially organized through a community; for example, Knudsen (2002) reported a geographical cline of scent similarity within a population of a palm (*Geonoma*) in Amazonia. An extreme case of individual variation was documented by Patt et al.

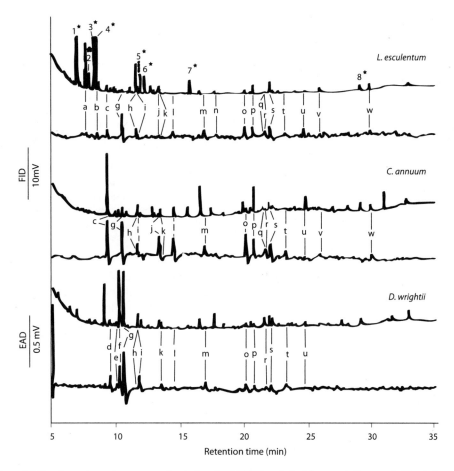

Figure 6.3 Gas chromatography–mass spectrophotometry (GCMS) analysis of flower volatiles, coupled to electroantennogram (EAG) recording; here the response of the moth *Manduca sexta* to the volatiles from three plants: tomato (*Lycopersicon esculentum*), bell pepper (*Capsicum annuum*), and *Datura wrightii*, with GCMS traces set above EAG traces in each case (redrawn from Fraser et al. 2003). Major components are labelled with lowercase letters: greatest responses are from (c), (*E*)-β-ocimene; (g), (*Z*)-3-hexenyl-acetate; (m), phenylacetaldehyde; (o), methyl salicylate; (r), geranyl acetone; and (s), benzyl alcohol. Note that several components in tomato scent elicited no response (*).

(1988), who found a thousandfold variability in floral emission strength within individual plants of the orchid *Platanthera stricta.* Hence simplistic views of fragrance as a common isolating mechanism in speciation have been somewhat undermined by observed natural variation and by the absence of simple inheritance patterns (Raguso and Pichersky 1999, Raguso 2001), as well as by the complex effects of temperature, humidity, and photoperiod that cause fragrance emissions to show high plasticity.

Scents of flowers can also vary with sex, either with the sexual phase in protandrous or protogynous hermaphrodites, or with flower sex in dioecious species. In gynodioecious *Fragaria virginiana,* for example, the hermaphrodite flowers are more strongly scented than the female flowers, and their scent is preferred by visitors, contributing to the twofold increased visitation on the hermaphrodite flowers (Ashman et al. 2005). Thus scent can drive pollinator behavior even in a weakly scented and small-flowered plant. Ashman (2009) reviewed differences in scents across 33 sexually dimorphic species and suggested that males usually had stronger scents (often accompanied by some composition differences) but that

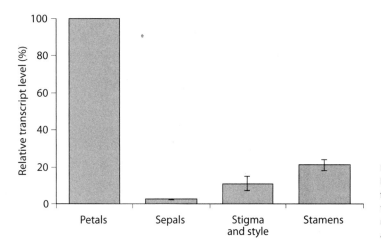

Figure 6.4 Localized gene expression of synthases, here shown for benzoic/salicylic acid methyltransferase (bsmt) in *Nicotiana suaveolens*; the relative transcript level (using mRNA as a control, ± SE) is much higher in petals. (Redrawn from Dudareva and Pichersky 2006.)

differences were less common where females were rewardless.

Spatial Variation in a Flower

Scent production in many flowers is tissue specific, and the main floral odor may be released from several sources within the flower; but emission is particularly commonly from petals or from the androecium (stamens, anthers, and pollen). Sepal scents occur more rarely, and these are often dominated by sesquiterpenes.

While scent from petals is usually emitted from the entire corolla (e.g., in roses), it can sometimes be more localized, for example, just from the corona in *Narcissus*. Less commonly it can be even more local, for example, arising from specific **scent marks**, which are often aligned with the visual nectar-guide markings (as in buttercups, *Ranunculus*, Bergström et al. 1995). Alternatively there can be different scents for the petal as a whole and for the nectar guides (e.g., *Papaver, Tropaeolum, Viola,* some *Narcissus*). Occasionally, scented areas may act as nectar guides without any corresponding visual markings (e.g., *Silene latifolia*, where there are scent marks at the base of each petal). Lex (1954) and von Aufsess (1960) sampled many flowers with the human nose as detector, and their data suggested that scent marks were present in at least 88% of all flowers.

Some plants have localized scent glands, or *osmophores* (Vogel 1990). These are especially notable as discrete entities in dung- and carrion-scented flowers and in some orchids, where the odor emanates from differentiated glandular regions and where the scents may take a liquid/crystalline form rather than being diffused as gases. However, the distinction between osmophores and scent-emitting epidermal zones, which may also have specialized cellular structures, is no longer clear-cut, and the term "osmophore" is now often used for any scent-producing cells (Effmert et al. 2006). The mechanisms of release of volatiles from epidermal cells (always across the plasma membrane and cuticle rather than through pores or stomata) are reviewed by Jetter (2006).

Where scent production is localized, this is now easier to detect because of our increasing understanding of the genetic control of floral scent, with key synthase and transferase enzyme-controlling genes now well characterized (Dudareva and Pichersky 2006). Most commonly the biosynthetic enzymes that control scent volatiles (for example LIS, a linalool synthase, or BAMT, which produces methyl benzoate) are most highly concentrated in the osmophore regions of petals. This concentration potentially allows for the localized and more precise detection of relevant gene expression; figure 6.4 shows an example.

Adding to this complexity, anthers and pollen are often differently scented from the rest of the flower (Bergström et al. 1995; Dobsonk Groth, et al. 1996; Jürgens and Dotterl 2004), and for pollen, the source is usually the oily, sticky outer layer of **pollenkitt** (chapter 7). Nectar can also carry its own scent (chapter 8). Again, these local intrafloral areas of different scent or of stronger scent could be providing a scent guide to visitors; the nectar guides, or the pollen, may smell the strongest or may be qualitatively different, thus cueing

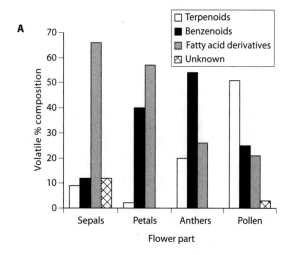

Figure 6.5 (A) Percentages of four classes of volatiles produced in different parts of *Rosa* flowers (redrawn from data in Dobson et al. 1990). (B) The key volatiles concerned.

a visitor's orientation to the rewards within a flower. A good illustrative example is the rose flower, with many hybrids derived from *Rosa canina* and *Rosa rugosa*. In both these species, the petals dominate the overall bouquet with terpenoids and benzenoids, including strong doses of geraniol and citronellol; the sepals add sesquiterpenes; and the anthers and pollen contribute a suite of additional odors to the whole, few of which occur in the perianth tissues (Dobson et al. 1990). This variation is illustrated in figure 6.5.

Temporal Variation

Problems in defining flower scent arise not only from individual variation and spatial variation, as outlined above, but also because flower odor may change through the life of a flower (Patt et al. 1991), simply with ageing or, more specifically, in relation to the changing conditions of the flower (visitation and/or pollination; see section 6, *Change of Scent and Control of the Pollinators*, below). Scent may also change rhythmically in quantity or quality on a diurnal basis with an underlying and often complex circadian control (Dudareva and Pichersky 2006; Hendel-Rahmanim et al. 2009); many flowers are more strongly scented in the early morning (e.g., *Antirrhinum*, some *Rhododendron*, many orchids), and conversely, some are noted for strong scent at dusk (e.g., *Nicotiana, Daphne, Lonicera, Stephanotis*). A few emit scents that are quite different in kind by day and by night (Matile and Altenburger 1988). This kind of variation can also occur within a genus of course; for example, some *Petunia* species are scented at midday, and others, such as *P. axillaris*, only produce their scent at dusk as their hawkmoth visitors become active (Hoballah et al. 2005).

Some flowers may vary their scent in relation to the competing needs of attracting pollinators while not attracting florivores, so that the thistle *Cirsium arvense* has a strong floral fragrance when pollinators are active but low emissions at times of florivore activity (Theis et al. 2007).

Any of these temporal patterns may also be modified by the plant's environment and by changing weather. For example, Hansted et al. (1994) showed strong effects of temperature on odor emissions from flowers of *Ribes nigrum* (blackcurrants).

Variation with Pollinator Type

For all the reasons covered in the last three sections, convincing studies of floral odor for a given species must include a large number of plants and good temporal resolution of the odor-emission patterns; and if seeking comparative patterns, one must also avoid generalizing at higher taxonomic levels until a wide range of species have been sampled. Some older studies are therefore of limited use in trying to discern ecological or evolutionary patterns in flower scents.

Despite these qualifications, there are clear possibilities of convergence in floral scents for particular kinds of visitor, with certain kinds of scents being attractive to a particular suite of flower-visiting animals.

This phenomenon is well documented and has been rigorously reviewed by Dobson (2006). Some examples are as follows (see table 6.4 for details):

1. GENERALIST FLOWERS: For open-bowl flowers (especially in the Apiaceae, Rosaceae, and Ranunculaceae), no particular patterns emerge, except that all three main types of scent are usually present, and usually the common types of each (which may in itself link with generalist attraction: Tollsten et al. 1994). The most frequently encountered components are β-ocimene and pinene.

2. FLOWERS VISITED BY DYNASTINE SCARAB BEETLES: These flowers are mainly in the Annonaceae, having chamber morphology and often thermogenesis (Gottsberger 1999), and the emitted volatiles commonly include methoxylated benzenoids.

3. FLY FLOWERS: There is no overall convergence of scent types in fly flowers, perhaps as expected. The odors are extremely varied: floral, musty, or carrion-like, depending on the particular syndrome involved (chapter 13). The most general fly flowers, such as many Apiaceae and larger flowers like the hawthorn, have an essentially sweet smell but with undertones of sourness akin to sweat or urine, which arise from various aliphatic acids and some nitrogenous compounds. The same is true of midge-pollinated flowers, such as *Theobroma cacao*. Flowers visited by hoverflies smell sweeter, similar to those visited by bees, and have a higher terpenoid content. **Carrion-fly** flowers have the typical fleshy smells referred to earlier. Fruit flies in the Dacinae subfamily that collect floral scents for use as a male pheromone (chapter 9) are almost exclusively attracted to flowers releasing either methyl eugenol or raspberry ketone (Metcalf 1990).

4. BUTTERFLY FLOWERS: Andersson et al. (2002) and Andersson (2006) proposed convergence for butterfly flowers, where the scents are delicate and the common components are benzenoids, such as phenylacetaldehyde and 2-phenyl ethanol, and terpenoids, such as linalool and its oxides and the rarer oxoisophorone (a particularly attractive component in *Buddleja* and *Centranthus* flowers). There may be higher levels of the benzenoids in European plants and of linalool and its derivatives in New World plants.

Table 6.4

Typical Scent Components Associated with Particular Pollinator Groups

Pollinator group	Commonest floral volatiles
Generalist	Variable: aliphatics, terpenoids, benzenoids all present
Beetle	
Temperate generalists	Variable; often some N compounds
Tropical scarabs	Methoxylated benzenoids
Other tropical beetles	Esters and terpenoids
Weevils	Aliphatics
Carrion beetles	Heptanal, octanal, hexanoic acid, ρ-cresol, polysulfides
Flies	
Temperate generalists	Aliphatics and N compounds
Carrion flies	Heptanal, octanal, hexanoic acid, ρ-cresol, polysulfides
Thrips	Variable, too few data; β-myrcene in cycads
Bees	
Nectar/pollen foragers	Variable, complex, terpenoids often abundant
Fragrance-seeking euglossines	Simple, benzenoids and monoterpenes
Fig wasps	Simple, terpenoids dominant
Lepidopterans	
Small moths	Aliphatic esters
Yucca moths	Aliphatic hydrocarbons, alcohols, sesquiterpenes
Butterflies	Benzenoids, terpenoids
Nocturnal perching moths	Benzenoids, terpenoids
Sphingid moths	Methyl benzoate, linalool
Birds	Weak or no scent
Bats	
Microchiroptera	S compounds prevalent
Megachiroptera	Weaker scents, fewer S compounds

5. MOTH FLOWERS: Convergence in flowers visited by nonhovering moths was specifically noted by Knudsen and Tollsten (1993). Flowers attractive to these settling moths are said to have a "quite floral" scent, with acyclic terpenoids, such as linalool and farnesol; simple aliphatics, including phenylacetaldehyde; and sometimes the rosy components, such as citronellol and geraniol (Dobson 2006). Small amounts of indoles and oximes are also common.

6. HAWKMOTH FLOWERS: Miyake et al. (1998) reported strong fruity or flowery scents in the evening,

with oxygenated terpenoids and nitrogenous oximes prominent, in flowers visited by these hovering **sphingid** moths. This pattern was confirmed by Dobson (2006), who found more benzenoid esters and less of the aldehydes than in settling-moth flowers, and often rather more of the nitrogenous components. Linalool is often a dominant element in the well-known highly scented moth flowers such as *Lonicera, Stephanotis, Crinum*, and some *Oenothera*.

7. BEE FLOWERS: Dobson (1987) reported that bee flowers usually have the typical floral scents (also

variously divided up as rosy, spicy, aromatic, and ion-onic [Dobson 2006]), which are released at various times of day (chapter 18). The scents tend to be dominated by terpenoids (especially ocimene, myrcene, pinene, linalool, limonene, caryophyllene, and far-nesene), with low amounts of benzenoids and aliphatics (although some rosaceous plants have moderate levels of benzenoids). It is not yet clear if there are patterns of variation correlated with the main visiting genera. Particularly strong convergence is, however, evident in euglossine bee flowers, which often possess carvone oxide as a scent component (Whitten et al. 1986; Williams and Whitten 1999; Schiestl and Roubik 2003; chapter 9), and perhaps also evident in those flowers that provide **oils** for bees, where benzenoid ethers predominate.

8. BAT FLOWERS: Many authors concur in seeing convergence (e.g., Knudsen and Tollsten 1995; Bestman et al. 1997; von Helversen et al. 2000; Winter and von Helversen 2001) in bat flowers. These have strong scents, often fruity or musky, that are especially intense at dusk (when visual cues are less useful). The bouquets tend to be dominated by monoterpenes and often also include unusual sulphur-containing compounds (e.g., dimethyl sulfide), especially in the Neotropics (S. Pettersson et al. 2004), where glossophagine bats have particularly poor visual searching abilities (chapter 16).

9. HUMMINGBIRD AND PERCHING-BIRD FLOWERS: In these flowers, the convergence is very clearly to a lack of any complex scents (Knudsen et al. 2004; Dobson 2006), matching the poor sense of smell in birds. Where scents have been detected they are often of the vegetative scents also associated with foliage.

As a corollary of convergent scent patterns in relation to pollinators, it follows that families or genera whose species have different pollinators show very varied scent profiles. A good example was given by Knudsen et al. (2001) with the Neotropical palms, in which beetle-, fly-, and bee-pollinated species showed quite different odor compositions, with the beetle-visited examples showing particularly highly specific attractants for their pollinators. Even more spectacularly, Jürgens (2009), using new data from Goodrich and Raguso (2009), could separate 9 genera and 21 species of the phylogenetically basal family Annonaceae into non-

overlapping fragrance clusters (fig. 6.6) that corresponded to species groups visited by (a) small nonspecialist beetles; (b) saprophilous flies and beetles (the genus *Asimina*, which has scents of decay and also maroon coloration); and (c) nonsaprophilous beetles and moths. Two other individual species from this family, one of which is visited by chrysomelid beetles and the other by euglossine bees, had their own distinctive odor patterns. In similar (but even more precise) fashion, three sympatric euglossine-pollinated species of Araceae in South America have their own distinctive floral-scent profiles and attract different bees, while there is a fig wasp with a single chemical attractant that confers specificity (chapter 26). From these data it has been suggested that changes in floral fragrance were key innovations in promoting pollinator shifts.

4. The Discrimination and Detection Ranges of Flower Odors

In insects, the antennae and often the mouthparts bear the olfactory sensillae, each normally housing two to five olfactory neurons that connect directly to the brain. In bats the nasal epithelium is large and projects to the olfactory bulb centers of the brain, which are proportionately much larger than in humans. Birds, however, have rather small olfactory centers. In all cases, though, the immediate detection system is provided by *odorant binding proteins* (OBPs) in the sensillae of the antenna or nose; these proteins confer the specificity required and trigger a second-messenger signaling cascade that eventually depolarizes the neuron and sends a message to the central processor in the brain (Dryer and Berghard 1999; Carlsson and Hanson 2006). Thus, a flower-visiting insect, such as the hawkmoth *Manduca sexta*, may have neurons that respond individually to separate molecules (e.g., geraniol) or to a small group of related molecules (e.g., geraniol and the various forms of linalool). The latter type of neurons, with somewhat broader "tuning," are probably more useful for responses to flower odors and can be used in concert with each other to give extraordinary discrimination. Using this kind of detection system with overlapping neuron tunings, humans with about 300 different OBPs can recognize more than 400,000 different scent molecules. But human noses are poor

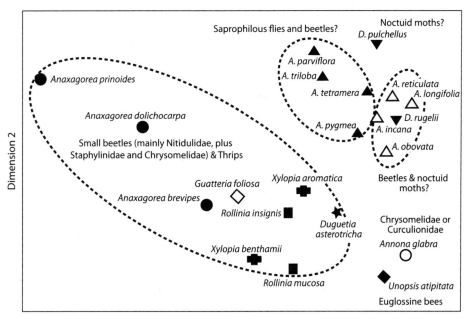

Figure 6.6 Floral odor clusters (generated via nonmetric multidimensional scaling) among the Annonaceae, in relation to known or suggested pollinators (redrawn from Jürgens 2009). In particular, the saprophilous *Asimina* species (closed triangles) are all maroon in color and have a particular scent profile distinct from moth and beetle pollinated *Asimina* (open triangles) and *Deeringothamnus* species; flowers visited by small beetles have a more varied but still distinct odor.

guides to floral scents, and most flower-visiting animals have far greater discriminatory powers for odors than we do.

Bees are astonishingly good at discriminating scents; for example, when 1,816 odor pairs were tested with honeybees, they could distinguish between 1,729 of these (Vareschi 1971). They can also distinguish very small-scale odor patterns much more clearly than humans, being able to discriminate between two different scents, or two concentrations of one scent, when applied to each of the two antenna; this means they can also use subtle spatial scent differences within flowers as cues. They could even discriminate between four different cultivars of snapdragon plants (*Antirrhinum*) that varied only in the *ratios* of constituent odors (G. Wright et al. 2005). However, bees and other insects are not greatly different from humans in terms of scent acuity (i.e., the concentrations needed to elicit a response).

Different kinds of scents will operate at different ranges, being emitted from their source as an *odor plume* that spreads downwind and is affected in its shape and duration by the local wind and air movements as well as by the inherent volatility of the components. Attraction and orientation of the target animals requires that they detect the signal provided by the scented plume of air, then differentiate the signal from background scents, and then code and interpret the information. Long-range effects are particularly common in moth and bat flowers that open toward dusk (Haynes et al. 1991), and floral scent plumes can travel over hundreds of meters in the still air that often occurs around sunset. As with pheromones, detection of scent may trigger upwind flight and then activate a visual search; in moths, for example, the search begins for white objects resembling flowers.

At shorter ranges, scent may be a cue for the correct directional approach, entry to a flower, and orientation within a flower and may even prolong visits in butterflies (Pellmyr 1986). However, scent usually acts in concert with visual cues at this range (and occasionally with sound cues perceived by echolocation in bats; chapter 16).

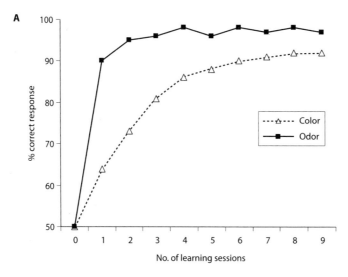

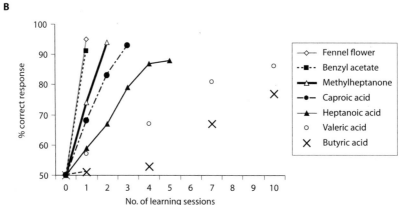

Figure 6.7 (A) Bees can learn floral odors in a single visit and learn odor faster than they learn color (redrawn from Menzel et al. 1974). (B) Fennel-flower bouquet is learned more quickly than almost any single chemical component of flower volatiles (redrawn from Schoonhoven et al. 2005, based on data in J. Gould 1983 and Kriston 1971).

5. Learning Floral Scents

Scent is more conducive to associative learning when linked to floral reward than is any other cue, at least in most insects (Menzel and Müller 1996; Raguso 2001, 2006), so that scent learning significantly enhances flower fidelity in many pollinators and thus the chances of correct (conspecific) pollen transfer (G. Wright and Schiestl 2009). Hence artificial addition of phenylacetaldehyde to two species of *Silene* flowers made their odors more similar and immediately increased the transfer of a pollen-analogue between them (Wälti et al. 2008), showing the importance of odor learning to visitors' ability to discriminate and to maintaining reproductive isolation for the plant.

Honeybees can usually learn a complex floral odor reliably in a single visit (fig. 6.7), although they take longer to learn single volatile compounds. They learn faster when the scents are floral and can readily learn to associate floral scent with (in hierarchical order) the size, shape, color, and texture of flowers and with floral rewards, especially nectar (so that olfactory signals may be ideal as honest indicators of nectar reward). Overall, honeybees can learn to distinguish among at least 7,000 different flower bouquets, and to "logically" associate these with other cues, such as visual expectation and positional effects. Thus, for example, they may ignore a favored flower scent if it is offered at an inappropriate height above the ground. Scents may also induce recall of navigational information and visual memory in honeybees (Reinhard et al. 2004) and in *Apis* and other social bees may be learned from nest mates (Dornhaus and Chittka 1999).

Because of the potentially complex interactions of scent, vision, and behavioral cues, it is always important to consider cues from all sensory modalities in concert and in particular to avoid the excessive concentration on visual signals that tended to dominate the early pollination literature (Raguso 2006).

6. Scent Change and Control of Visitors

Changes in the Flower with Age or with Visitation

Scent from a flower appears to be easy to change in an evolutionary sense. For example, within the genus *Clarkia* most species are unscented and attract mainly bees, but *C. breweri* is scented and hawkmoth pollinated, and it differs by just one gene for one compound (with small changes in the biosynthesis of another) (Raguso and Pichersky 1999). Effects can be seen even within a species, as in *Polemonium viscosum*, which at higher altitudes (alpine tundra) is bumblebee pollinated and has larger corollas and a sweet scent, while the flowers below the tree line are mainly visited by flies and small bees, being smaller and with a "skunky" odor from their calyx, which serves to repel abundant ants (Galen et al. 1987).

But what about scent change in contemporary time? Postpollination odor does change, in at least some orchids, ranging from an effect within minutes in *Catasetum* to days in *Platanthera* (Tollsten and Bergström 1989; Tollsten 1993). Flowers of *Nicotiana attenuata* also change odor within a few hours after pollination (Euler and Baldwin 1996). *Cirsium* thistles reduce scent emissions by about 90% within 48 hours of being pollinated, and bees are then three times less likely to visit the flower (Theis and Raguso 2005). *Lantana* flowers also change odor as they change color with age, losing their most attractive components (Raguso 2004a). In both *Antirrhinum* and *Petunia* there is a reduction in methylbenzoate emission after pollination, and the key trigger is the growth of pollen tubes to the ovary (Negre et al. 2003), the BAMT gene being directly suppressed by ethylene that is released as the tubes grow.

Postpollination odors may sometimes be maintained as part of an overall distance attraction to the plants (e.g., Eisikowitch and Rotem 1987, with *Quisqualis* flowers), but in other cases, they may be reduced or altered to provide a signal that a particular flower is no longer worth visiting (Lex 1954; Tollsten 1993), as we saw for color change in chapter 5.

A further possibility is a change of scent contingent upon plant damage by herbivores, as demonstrated in *Cucurbita pepo* (Theis et al. 2009), where male flowers had higher fragrance levels after leaf damage while female flowers (inherently more strongly scented) were unaffected. This observation may indicate yet another mechanism whereby floral odors mediate the interactions of plants, pollinators, and other animals, a theme reviewed by A. Kessler and Halitschke (2009) in the context of the variable tissue-specific regulation of odor chemistry.

Changes of Scent Added by Visitors

Many eusocial bees have been shown to scent-mark individual flowers after visiting them, essentially by leaving a "footprint" of volatile chemicals, from glands in the tarsi, so that subsequent visitors avoid the recently depleted flowers. Scent-marking is known to occur in bumblebees (U. Schmitt and Bertsch 1990; Stout et al. 1998; Goulson et al. 2001; Witjes and Eltz 2009), in honeybees (Free and Williams 1983; Wetherwax 1986; Giurfa and Nuñez 1992; Giurfa 1993; Williams 1998) and in certain **stingless bees** (V. Schmidt et al. 2003; Nieh et al. 2003). A bee may avoid flowers visited only by conspecifics or may also avoid those marked by **congenerics**; in practice honeybees and bumblebees both avoid flowers visited by each other and seem to share the volatile 2-heptanone as a component of the repellent mark. Scent marking of flowers may operate even across orders, because bees will avoid flowers recently visited by a hoverfly (Reader et al. 2005).

However in some circumstances both *Apis* and *Bombus* seem to *increase* their visitation to recently probed and highly rewarding artificial flower models (Cameron 1981; U. Schmitt and Bertsch 1990; U. Schmitt et al. 1991), indicating that a positive scent signal can also be deposited. Stingless bees can also deposit attractant odor marks on good food sources to aid recruitment (Jarau et al. 2004). Thus two different odors may be involved in scent-marking, and honeybees, at least, may have an attractant marker that is longer-lasting than their repellent marker (Stout and Goulson 2001).

However, this scent marking of flowers is not automatic, and so is not "merely" a footprint in a mechanistic sense. Goulson et al. (2001) found that many social bees make their decision to visit a flower purely on direct detection of the reward when it is visible and will leave a scent-mark only on species where the rewards are hidden. Nor is the mark simply predetermined. Stout and Goulson (2002) showed that the duration of marker repellency differed among flower species for the same bumblebee visitors, and this period varied inversely with nectar secretion rates. Repellency lasted just a few minutes on comfrey flowers (*Symphytum*), which replenishes its nectar rapidly, but could persist for 2–24 hours on the smaller and less nectariferous vetch-type flowers of *Melilotus* and *Lotus*. This might mean that signaling bees can alter the strength of the mark they deposit, or that receiving bees can alter the duration of their response to repellent marks (or of course both may occur). Additionally, bumblebees have been shown to be more than twice as likely to reject scent-marked complex flowers as scent-marked simple flowers (Saleh et al. 2006). Evidently there is considerable sophistication here that is still to be explored, not least since different bumblebee species appear to deposit very different scent components (Bergman and Bergström 1997). However, Wilms and Eltz (2008) showed that the marks left could be regarded as "merely" footprints (rather than specific pheromonal cues) in the sense that bees did not normally distinguish between flowers that had been fed at and those that had just been passively walked over by another visitor.

It is easy to see why marking a visited flower with scent would be useful for social bees, who are geared to maximizing foraging returns, and so making a colony's food gathering efforts as efficient as possible. However, flower foot-printing is not unique to eusocial bee species. It also operates in species of *Xylocopa,* a genus of large bees with varying levels of sociality (e.g., Frankie and Vinson 1977), and even in the entirely solitary genera *Centris* (Raw 1975), *Colletes, Andrena, Tetralonia,* and *Osmia* (Yokoi and Fujisaki

2009). In fact, in solitary *Anthophora* (F. Gilbert et al. 2001) the marks left are complex multicomponent volatiles, individually recognizable and varying with foraging needs, and perhaps used as dominance or exclusion markers. It is possible that flower marking originated as a side effect of **territorial marking** behaviors in solitary bees and has merely taken on new roles in directing feeding activity, so enhancing colony foraging success, in eusocial species.

7. Special Cases

Fragrance Cues in Deceptive Systems

Many flowers achieve pollination by forms of deception, offering apparent rewards (food or egg-laying sites or mating opportunities) that are misleading and where there is no real benefit to the forager (these are covered in detail in chapter 23). But we should note here that a high proportion of these deceptions rely on odor cues, exploiting the dependence of so many flower visitors on fragrance signals. Most famous examples come from among the orchids, and figure 6.8 illustrates one case that gives a sense of the sophisticated deceptions involved. Here the orchid *Ophrys speculum* provides an illusory copulation opportunity to male scoliid wasps (*Camposcolia ciliata*), mimicking both the appearance and scent (pheromone) of the female wasp. But the system also incorporates a change of scent after pollination, in particular a decrease in the emission of 9-hydroxydecanoic acid, that parallels a substantial reduction in this compound from mated female wasps, so that males are not fooled into visiting an already serviced flower (Ayasse 2006). For an even more spectacular example, Schiestl et al. (1997, 1999) described the case in which a flower, *Ophrys sphegodes*, practices sexual deception of visiting male bees, *Andrena nigroaeana*, by its shape and color but above all by its bouquet, which mimics the female bee's pheromone. There are 27 compounds in the orchid's floral bouquet, and a decrease in total volatiles follows pollination, but Schiestl and Ayasse (2001) also found that after pollinia deposition there was a specific increase in farnesyl hexanoate, which was a repellent to the bee. More details of this and other odor-related deceptions are given in Chapter 23.

Another intriguing example is that of *Epipactis helleborine*, an orchid pollinated by chemical deception

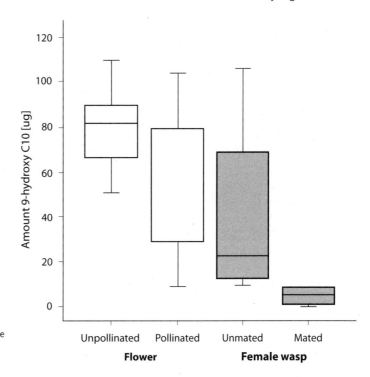

Figure 6.8 *Ophrys speculum* scents pre- and post-pollination, showing the larger amounts of 9-hydroxydecanoic acid before pollination and the larger amounts of this same chemical in unmated as opposed to mated females of the main pollinator, the wasp *Camposcolia ciliata*. (Redrawn from Ayasse 2006.)

of social vespid wasps (Brodmann, Twele, et al., "Orchids mimic green-leaf volatiles," 2008). Here the odors from the flowers are classic green-leaf volatiles, which are normally emitted from damaged leaves and therefore indicate to the wasps the apparent presence of herbivores such as caterpillars, the usual prey of these wasps.

However, odor mimicry is not always used. Galizia et al. (2005) analyzed both odor and visual signals in *Orchis israelitica*, which mimics a lily (*Bellevalia*), and found a shift of the mimic's visual signals toward the model but could detect no olfactory mimicry, suggesting that this orchid relies on long-range attraction from neighboring model plants.

"Unscented" Volatiles as Advertizing Signals

Some flower visitors can respond to carbon dioxide given off from flowers. In *Datura wrightii,* visiting *Manduca* moths detect a substantial pulse of CO_2 given off by the flowers as they open in the evening, at the same time as nectar production is peaking (Guerenstein et al. 2004). Possibly the gas production is related to the high metabolic activity associated with nectar

secretion and so could be an honest signal for assessing nectar content to an incoming moth. However the effect is context dependent for females (Goyret, Markwell, et al. 2008), so that CO_2 may be mainly used as an indicator of host-plant quality for bouts of mixed feeding and egg-laying.

Other Functions of Floral Scents

Some anemophilous conifers produce a spectrum of odors from male flowers that differ from those from female flowers, commonly producing more α-methyl ketones and alcohols. Indeed, there may be a general link between α-methyl ketones and anemophily (Dobson and Bergström 2000), suggesting that these compounds, which can be deterrents to insects, constitute an ancient pollen-defense mechanism. But angiosperm flower scents can also be used to repel unwanted visitors. Well-documented cases involving potential pollinators are rather rare, but one neat demonstration involved *Pieris* butterflies, which were seeking sites for oviposition but would not visit *Osmanthus* flowers because of the deterrent floral volatiles (Omura et al. 2000).

The best known cases involve ants, which are generally unwelcome as flower visitors (chapters 11 and 24). Flower scents that repel ants were initially described in the unusual system involving aggressive **ant-guards** in *Acacia* trees (Willmer and Stone 1997a). Here volatiles released from the pollen briefly repelled the ants from young dehiscing flowers, which therefore became accessible for visitation by pollinators, and the diurnal patterns of scent release thereby differentially influenced the flowers' interaction with pollinators and with guards. Examples of ant repellence are now known in other acacias, and more generally in a range of flowers (Junker et al. 2007; Willmer, Nuttmann, et al. 2009; chapter 24 for details).

But floral scents may also influence floral enemies (Raguso 2004a,b), since many of the components of floral scents are the same as those that defend against herbivores. These may remain effective against generalist feeders but frequently become cues to specialist feeders on leaves or (it follows) on flowers. Galen (1983, 1999b) showed that ants preferred to feed at the more sweet-scented of the two *Polemonium* morphs already referred to, causing damage to the corolla and style and so exerting negative selection. In *Cucurbita*, different volatile components from the blossoms may attract squash bees (*Peponapis*), or attract both these bees and the specialist herbivore cucumber beetles, or in the case of indole, attract only the herbivore (Andrews et al. 2007), so that plants can affect their interactions with mutualists and antagonists by varying the proportions of three key volatiles. In the case of scents that mimic carrion or dung, there may be a specific additional benefit of herbivore deterrence (Lev-Yadun et al. 2009), since herbivores may associate such smells with the likely presence of carnivores.

And floral scents can also attract the enemies of other flower visitors, which may or may not benefit the plant. For example, **ichneumonid** wasps that parasitize pollen beetles are attracted to oilseed rape (where the beetles are feeding) by the flower volatiles (Jonsson et al. 2005) and thus may help to reduce pollen losses to these relatively destructive visitors.

New possibilities for understanding the multiple roles of floral odors are opening up with the use of genetically transformed plants. For example, D. Kessler et al. (2008) knocked out genes to show that both repellent and attractant elements in *Nicotiana* scents were needed for maximum seed-capsule production; attractants served to bring in bees, and repellent nicotine served to reduce **florivory** and nectar robbing and to enforce only short drinking bouts on visiting hummingbirds. M. Whitehead and Peakall (2009) indicated other ways in which molecular tools applied to flower scents should assist analyses of plant evolution and population ecology.

Floral Fragrance Collection: Scents as Rewards

This is an important aspect of floral fragrance, where the scent itself is gathered by a flower-visiting animal (especially by euglossine bees) and thus acts as a payoff to the bee. It is therefore dealt with separately in chapter 9, as a special case of floral rewards.

8. Overview: Interactions of Scents with Other Floral Signals

Flowers can be seen as sensory billboards (Raguso 2004a), where all the signals that they emit across different sensory modalities can interact to attract, guide, or repel particular visitors. Classic examples of such interactions include flower scent and height in orchids, scent and color and temperature in carrion flowers, or scent, ultraviolet and color visual signals, and auditory echo print in some bat-visited flowers (see chapter 16).

In general it is true that among signals there is visual precedence in diurnally flowering species and olfactory precedence in nocturnal species. This applies to both the flowers and their visitors and obviously reflects the relative importance of different senses in the animals involved. Thus, Balkenius et al. (2006) found nocturnal moths mainly responding to odors but related diurnal species strongly favoring visual stimuli. There may also be range effects, so that insects and bats foraging by day or around dusk often use odor cues at longer range and visual signals when they are closer. This is not invariable though, since Andersson and Dobson (2003) showed that *Heliconius* butterflies mainly used visual cues to select flowers, but olfactory signals could then initiate and maintain foraging at closer range. And in some cases different modalities are clearly synergistic, so that a full response requires two or more signals to be present; for example, *Manduca* hawkmoths will not feed normally on *Datura* unless both visual and olfactory cues are presented

(Raguso and Willis 2004). Similarly, bumblebees are now known to learn faster, and to achieve a higher accuracy in their decisions, when they are given differences in both modalities simultaneously in model flowers (Kulahci et al. 2008).

Whatever the precedence or the balance, for the more specialist flowers there is seemingly always some matching between the signal strength from the flowers and the sensory strengths of preferred visitors, so that odor cues are especially obvious in bat- and moth-pollinated flowers and common also in the broad swathes of bee-pollinated species. But an increasing appreciation of the role of nonpollinators (thieves, florivores, and others) in shaping selection on flower signals and of environmental factors that also affect signal strengths undoubtedly means that there are further levels of sophistication yet to be uncovered in the details of floral odors.

Chapter 7

REWARDS 1: THE BIOLOGY OF POLLEN

Outline

A pollen grain contains the male gamete of the angiosperm plant and is thus the equivalent of a spore in many other plants. In essence, the structure of a pollen grain is adapted to protect and nourish the male gamete during its maturation and subsequent transit between plants, and then to ensure that one of its nuclei reaches and fertilizes an ovule within another plant. The grain must alight on an appropriate stigma and germinate, then develop a pollen tube down through the style along which the nucleus can travel so that the nucleus passes down to the ovary at the base.

As discussed in chapter 4, pollen is the primary reward for flowers in an evolutionary sense and probably the resource for which animals first went to flowers. Its inherent characteristics—small easily managed nutrient-containing packages—make it a useful resource to exploit as food, potentially collectable by almost any animal. It remains a crucial reward for pollen-eating and pollen-gathering visitors, such as some flies, some beetles, and virtually all bees. But for many contemporary visitors it is *functionally* secondary, an extra reward for animals that primarily track nectar resources and pick up pollen incidentally. Furthermore, since pollen is formed and dispensed in the male organs of flowers, it occurs only in about half the flowers in dioecious species; hence cross-pollination by mainly pollen-seeking visitors is potentially very difficult for dioecious plants.

Pollen thus has two functions—reproduction, and reward of visitors—that are mutually incompatible. As an offering to visitors, pollen is expensive for a plant, both in terms of the nutrients invested in it and in terms of the direct loss of genetic material every time it is consumed. Whether pollen is the primary or the secondary reward, then, it must usually be produced in considerable excess over the needs of reproduction, and a flower should be expected to have adaptations to reduce pollen grain losses to inefficient or inappropriate visitors.

Pollen quantity and quality have been extensively reviewed (e.g., Stanley and Linskens 1974; Bernhardt 1996; D'Arcy and Keating 1996; Cruden 2000; Roulston and Cane 2000), often in an attempt to understand pollinator preferences for particular kinds of pollen-dispensing flowers. But it should be said at the outset that pollen characteristics have proved to be primarily related to the plant's phylogenetic relations and its reproduction, and there is relatively little evidence that aspects of pollen shape, size, or chemistry have been shaped by selection from pollinators.

1. Pollen Grain Characteristics

Pollen Grain Structure

As an anther forms in the developing flower bud, it can be seen to be four-lobed (normally) and composed of

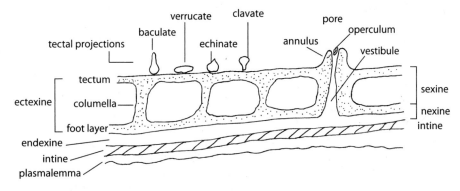

Figure 7.1 The structure of the four main pollen grain layers, showing some alternative terminologies. (Modified from Nepi and Franchi 2000.)

four layers of rather large cells with unusually large nuclei. These cells divide repeatedly lengthwise as the anther elongates, and also divide inwards once, to give an outer cell layer that forms the *tapetum* (part of the anther wall), and an inner cell layer whose cells become the *pollen mother cells*. At this stage meiosis occurs (cell division from the normal diploid to the reproductive haploid state), every mother cell thereby producing a quartet of cells each with half the parental chromosome number. These individual cells separate slightly from each other and develop their own surface structures to become four distinct pollen grains. Within each grain the initially single haploid nucleus divides again to form a vegetative nucleus and a generative nucleus. The latter divides again to give two nuclei, which are the true gametes, and thus each mature pollen grain normally has three nuclei present.

Structurally, the resultant pollen grains have four principle layers (fig. 7.1), which protect the pollen against desiccation, overheating, genetic damage from excessive UV radiation in strong sun, and pathogenic attack from bacteria or fungi. The form and functions of these layers have been reviewed in detail by Heslop-Harrison (1979) and Hesse (2000).

The most superficial layer is a thin oily material called pollenkitt, which is secreted by the tapetum tissue within the anther and consists primarily of lipids (especially linolic acid, linoleic acid, and arachidonic acid; Solberg and Remedios 1980) that are the components of broken-down tapetal cells and cell organelles. It also contains some pigments, mainly flavonoids in gymnosperms but mainly carotenoids in zoophilous angiosperms. These pigments have the advantage of absorbing UV strongly but reflecting longer wavelengths that could initiate overheating. Pollenkitt is particularly associated with insect pollination and can be entirely absent in some anemophilous plants; furthermore, where it occurs, it sometimes dries out after a few hours, losing its stickiness, and the pollination mode may then shift toward anemophily (e.g., in *Cyclamen, Calluna, Erica*, and *Tilia*; Hesse 2000). Parts of the pollenkitt layer, especially within the grooves of the surface sculpturing of the grain, also include small amounts of proteins, again derived from the tapetum, that give adhesive properties and are involved in self-incompatibility effects (section 3 in chapter 3, *Methods for Avoiding Selfing within a Flower*). These are also the proteins (*sporophytic proteins*, sometimes also termed *pollen loadings*) that produce allergic reactions to pollen (hay fever) in many humans. The proteins reveal their main functions after landing on the correct stigma: they flow off the grain and form a "foot" that penetrates the microscopic channels on the stigmatic surface, and are thus the essential recognition molecules involved in incompatibility.

Two other surface components may occur. In brassicas and some other groups, *tryphine* is found, while in the Ericaceae, Onagraceae, and some caesalpinioid Fabaceae, strands of material known as *viscin threads* are found (with the even stickier *elastoviscin* occurring in orchids). These probably all have similar functions to pollenkitt, providing increased adhesion both between the pollen grains and between pollen and vector. Skvarla et al. (1978) reported that viscin threads were beaded in moth- and bird-pollinated taxa, but smooth in bee-pollinated groups, such as evening primroses, but there is no further clear evidence of such functional correlates. Hesse et al. (2000) indicated that in different plants they could function either to facilitate pollen presentation or to limit and regulate pollen dispensing.

Beneath the pollenkitt, the main outer wall of the pollen grain is termed the *exine*, and this is again primarily assembled from materials supplied by the tapetum lining the anther. It consists largely of a polymeric terpenoid molecule termed *sporopollenin* (which also forms the viscin threads on the surface). The exine usually has several sublayers, with suitably complex associated terminology shown in figure 7.1. In some families (such as Asteraceae) it is exceptionally thick, while in others (such as Strelitziaceae) it is merely a thin skin over the pollen grain. Whatever its substructure, though, the exine as a whole is extremely resistant to decay, to physical damage, and to chemical attack.

Inside the pollenkitt and exine lies a second wall, the *intine*, essentially a normal plant cell wall made up mainly of cellulose, hemicelluloses and pectin, with a little protein (Heslop-Harrison and Heslop-Harrison 1991), and with some apertures through which the pollen tubes can grow at the time of germination. Inside the intine is the *plasmalemma* (*cell membrane*), where calcium fluxes occur that are linked with pollen hydration and germination (Malho 1999). Finally there is the **cytoplasm** of the cell, although this may be less than half of the total volume of the grain, especially in smaller pollens. The cytoplasm is usually rich in carbohydrates, especially **polysaccharide** starch and the simple **hexose** sugars, which form the fuel reserve for development and germination of the gamete. Some lipid may also be present as an additional energy store, especially in smaller grains.

Much of the diversity of pollen grains (fig. 7.2 and plate 18) arises from the exine layer's *surface features*, which can be extremely variable in terms of pattern and sculpturing, and form the main identification features of the specific pollen. Especially important are the *pores* (or apertures) and *furrows* that demarcate the site of origin of the pollen tube after germination and that also act as exchange sites for water and for recognition substances. The terminology of pollen grain features is reviewed by Thanikaimoni (1986). Dicot plants often have pollen of an elongate shape with three pores (triporate) or three longitudinal furrows (tricolpate), each of which may contain a central pore;

in fact all eudicots (see appendix) have the **synapomorphy** of triporate pollen grains, although some have spherical or tetrahedral shapes with additional regularly spaced pits and corrugations. In contrast, monocot pollens usually have just one pore or groove (monoporate or monocolpate). Furness and Rudall (2004) suggested that an increase to three pores per grain gave dicots a significant advantage, ensuring contact between at least one aperture and the stigmatic surface, which may have contributed to dicot radiation and success. However, there are no other obvious correlations of pore type or distribution with pollination mode or other aspects of ecology, except perhaps that aporate (inaperturate) or monoporate pollens occur rather often in plants from wet or moist habitats and are accompanied by unusually thin exine layers (Furness and Rudall 1999), so that pollen hydration is probably easily achieved. Aporate pollen is also found in sterile pollens that are provided only for visitors to feed on (see below), where the pollen is itself dimorphic (Furness 2007).

The more general surface *sculpturing* of pollen is remarkable, and the associated terminology was clarified by Punt et al. (1994); a range of pollen types is shown in the scanning micrographs in figure 7.2. It is not really clear why there is so much variation in surface structure, although there has been much speculation here. There is no obvious physical fit with stigmatic surface features, so that the stigma/pollen interaction is certainly not a lock-and-key relationship, as has sometimes been implied. Probably some correspondence occurs between pollen surface features and the general dimensions of patterning on the stigmatic surfaces, since pollen grains need to make close contact there while taking up water for germination, but it is not a very specific relationship. Nor is there any obvious fit of pollen grain structure with any structural features of visitors, such as bees. Certainly having a rough surface and general dimensions of 20–60 μm (plate 18) will help pollen to be picked up onto bees' plumed hairs (or onto lepidopteran scales, which have their surface divisions in a similar size range), but the detailed surface sculpturing is on a scale at least ten times smaller than this and cannot be matched with any animal's pollen-collecting structures (Simpson and Neff 1983).

Most authors therefore find no links of pollen sculpturing to pollinator type, although such studies are always hampered by the lack of data across enough

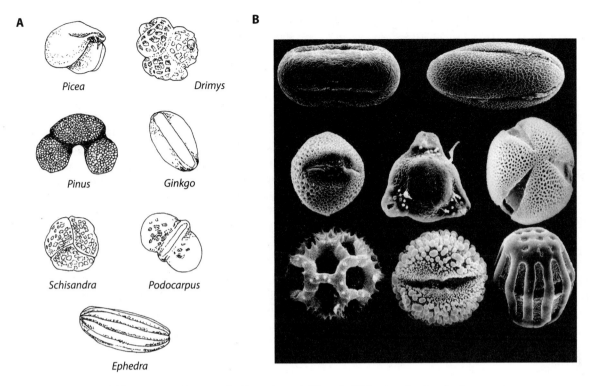

Figure 7.2 (A) Appearance of a range of pollen grains, not all to scale. (B) A range of surface structures seen in scanning electron micrographs of pollen grains, showing pores, furrows, reticulations, and protrusions. Top row: *Anthriscus, Lamium*; middle row: *Hedera, Chamerion, Verbascum*; bottom row: *Taraxacum, Ilex, Asclepias*. (Photographs kindly supplied by the Food and Environment Research Agency of the UK).

genera and species. Ferguson and Skvarla (1982), in a review of pollen morphology in leguminous plants, did point to several features that appeared to vary with pollination mode. Hemsley and Ferguson (1985) subsequently showed that within the genus *Erythrina*, the pollens are less delicate when they are from species pollinated by perching birds compared with those from species pollinated by hummingbirds, the latter also having less pollenkitt present, while Ferguson and Pearce (1986) also found differences related to pollination mode within *Bauhinia*. Outside the legumes, Grayum (1986) suggested that the pollens of Araceae were variously correlated with large beetles (large grains or **polyads**), with small beetles (polyplicate pollen, having multiple parallel folds along the main axis), and with flies (more spinose pollen, having tiny spines across the surface). However, most of these comparisons lacked a good phylogenetic framework and so could not use a strict comparative method to eliminate the effects of phylogenetic constraints. In contrast, Linder (1998) attempted a phylogenetically

controlled analysis for windborne pollens and showed that an apparent tendency to have smoother surfaces was probably a taxonomic artifact. Stroo (2000) took a similar approach with bat-pollinated plants and found no correlations for surface structures or sculpturing, the only significant effect being larger grains in chiropterophilous species than in other plants (chapter 16).

The point made by P. Crane (1986), and echoed by many others, is apposite here: pollen grains are evidently *not* optimally designed for a single very specific function in a very specific habitat, but instead are balancing the needs of at least two functions and are suited to work reasonably well in varying ecological contexts and environmental conditions.

Pollen Grain Size and Shape

Pollen grains vary from 4 to 350 microns (μm) in diameter, but most are in the range 15–60 μm, with an average around 35 μm. Table 7.1 gives further examples.

TABLE 7.1
Pollen Grain Sizes

Plant	Pollen grain diameter (μm)
Myosotis	5
Echium	12
Lecythis	20–30
Acacia (polyad)	30–50
Brassica oleracea	50 × 30 (ovate)
Bauhinia	100
Typha (tetrad)	50
Cleome	17–30
Adansonia	60–67
Bombax	43–49
Musa	50
Heliconia	60–80
Pseudobombax	68–75
Mirabilis	80
Cucurbita pepo	130–160
Lavatera	100
Durio	50–70
Kigelia	40
Agave	70–80
Cucurbitaceae	100–200
Malvaceae	150–200

This range in diameter leads to a greater than 10^5-fold difference in grain volume. Grain size tends to have a moderate phylogenetic component (e.g., the borage family has small pollen, with *Myosotis* grains at just 5 μm, while the mallow and **cucurbit** families often have large pollen, cucumber pollen commonly around 200 μm in diameter). However, there is no particular relationship between pollen size and pollinator (e.g., Harder 1998), at least for bee- and bird-pollinated plants. There is a fairly crude match between large flower size, long styles, and large pollen grains, as might be expected simply from allometric considerations; and this correlation does suggest that larger-bodied pollinators (such as birds and bats) that visit larger flowers will usually get larger grains even without specific selection for this association. There are also some distinctions between abiotically transferred pollens (chapter 19), with anemophilous pollen usually small and spherical, epihydrophilous pollen large and spherical or slightly elongate, and hypohydrophilous pollen essentially filiform (Ackerman 2000).

Pollen grain size can be variable within a species in relation to environmental factors. The nutritional status of the plant has some effect: higher nutrient levels increase the grain size in *Clarkia* (Vasek et al. 1987), and both nitrogen and phosphorus specifically promote larger pollen in cucurbits (Lau and Stephenson 1993, 1994). Altering the plant's nutrient balance through herbivory or artificial defoliation can decrease pollen size, as well as decreasing pollen grain number, in *Chamaecrista* (Frazee and Marquis 1994) and *Cucurbita* (Quesada et al. 1995). A similar reduction in grain size has been recorded following removal of the leaves in *Raphanus* (Lehtilä and Strauss 1999), although soil nutrient availability had no such effect in this plant (H. Young and Stanton 1990a).

Grain shape, as with surface patterning, seems largely unrelated to issues of fit with the stigma or pollinator, but it does give crucial information for pollen identification, a range of possibilities and terminologies being shown in table 7.2.

Pollen Advertisement: Color and Odor

Through visual or olfactory signals, pollen itself may be part of the attraction apparatus of a flower, especially where there are particularly large anthers or where many anthers form a brush effect (as in many Rosaceae or in mimosoid Fabaceae, where the anthers form the entire outer surface of the inflorescence). Examples can be seen in plates 2, 8, 13, and 14. Indeed, Lunau (2000) argued that pollen was the original floral visual signal and that much of the color patterning and morphology of zygomorphic flowers is based on mimicry—*signal copies*—of anthers or pollen (chapter 5).

Pollen grains are most commonly yellow in color, extending to cream or orange shades. More rarely they can be white, brown, black, red, or even blue or purple (plates 13H and 27E). The color may sometimes be adaptive in itself; for example, in bird-pollinated orchids (Dressler 1971) and cacti (Rose and Barthlott 1994), dull-brown pollens occur, even though related insect-pollinated species have the usual yellow pollen. Such coloration might indicate a lesser signaling function for pollen in ornithophilous plants, but it has also

TABLE 7.2
Pollen Terminology

Class	Clypeate: grooves between circular or polygonal exine areas Fenestrate: ridges between circular or polygonal exine areas Plicate: parallel folds following long axis of grain Saccate: one or more expended sacs on main grain (only gymnosperms) Spiraperturate: one or more spiral apertures
Shape	Irregular Oblate: polar axis shorter than equatorial diameter Prolate: polar axis longer than equatorial diameter Spheroidal: polar axis and equatorial diameter roughly equal, hence a sphere
Cross section	Circular Elliptic Lobate Polygonal Quadrangular Triangular
Aperture type	Colpate: apertures elongated, at equator or all over grain Sulcate: aperture elongate and at distal pole Porate: apertures circular or elliptical Poroid: apertures as above but with indistinct margins Colporate: porate aperture with a raised area (*colpus*) within it Ulcerate: single rounded aperture at distal pole Annulate: aperture sharply differentiated from rest of grain surface Operculate: aperture surface with marked exine sculpturing

Ornamentation	Baccate: rod-like protrusions
	Clavate: club-shaped protrusions
	Echinate: spine-shaped protrusions
	    Microechinate: spines less than 1μm tall
	Fossulate: elongate irregular grooves
	Foveolate: rounded depressions
	Gemmate: globular protrusions
	Granulate: tiny granular protrusions
	Perforate: small, scattered perforations
	Psilate: more or less smooth
	Reticulate: irregular depressions in network-like surface,
	    Microreticulate: irregular depressions less than 1μm deep
	    Reticul-cristate: reticulate with network formed by rows of protrusions
	Rugulate: flattened, elongated exine structures
	Striate: flattened, elongate, parallel exine structures
	Verrucate: wart-like protrusions

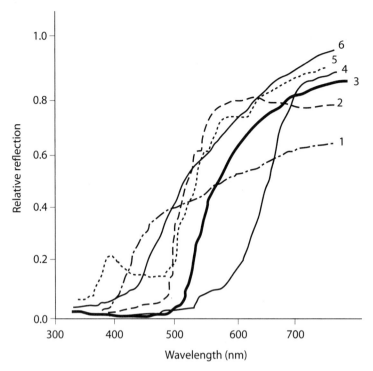

Figure 7.3 Examples of the spectral reflectance of pollen grain surfaces: (1) *Calystegia* (whitish); (2) *Mimulus guttatus* (yellow/orange); (3) *Chrysanthemum vulgare* (bright yellow); (4) *Lilium henryi* (deep orange); (5) *Linaria vulgaris* (hidden and strongly UV reflecting); (6) *Pinus cedris* (yellowish). (Redrawn from Lunau 2000.)

been interpreted as a matching to beak or face color to reduce the risk of conspicuous pollen being groomed off. Many pollens change color as they dry out, often appearing whitish as they desiccate with age, which could again serve as a useful signal from the pollen itself, independently of flower color.

Most of the color in fresh pollen derives from pigments in the pollenkitt layer. Bright yellow colors are mainly due to carotenoids such as β-carotene, lutein, and antheraxanthine, which occur in most zoophilous pollens (Dobson 1988), although often less intensely in pollens that are concealed within the corolla (Lunau 2000). The carotenoids in modern pollens are probably secondary products, since most nonangiosperm pollens and most fern and moss spores contain paler flavonoid pigments, such as quercetin and chalcones (Brouillard and Dangles 1993), which have greater UV absorption. Figure 7.3 shows examples of the spectral reflection of some pollens, including anemophilous examples, based on Lunau (2000).

In a few taxa the pollen color is polymorphic; we met examples in *Commelina* species in chapter 2, which have yellow, blue, or brown anthers and pollens.

The dog's tooth violet, *Erythronium grandiflorum*, may have either red or yellow pollen, which has proved useful in looking at pollen dispersal and gene flow (J. Thomson and Thomson 1989). Hairy flax (*Linum pubescens*) has white or blue pollen, associated loosely with yellow and blue corollas, respectively, but with some opposite combinations occurring (Wolfe 2001). Some species of *Nigella* have two pollen-color morphs, yellow and violet, and their frequency correlates with habitat, the darker morphs being commoner on north- and east-facing slopes and also being more susceptible to drought (Jorgensen and Andersson 2005), perhaps indicating an effect of, or reaction to, insolation levels. However, in this genus there seem to be rather inconsistent site-specific effects of pollen color on visitors (Jorgensen et al. 2006). Some plants with dimorphic anthers (heteranthy) also have two kinds of pollen: a notable example is *Lagerstroemia*, in which the fertile reproductive pollen is borne on six large anthers and is blue, whereas the food (fodder) pollen (which can germinate but never achieves fertilization) is provided by 30–40 much smaller anthers and is yellow (Pacini and Bellani 1986); these two pollens also differ in exine patterning and pore number. Dimorphic pollen is also well documented in the legume genus *Cassia* (fig. 2.5). This again shows heteranthy (plate 12D), with protruding, showy yellow, brown, or red anthers that are buzz pollinated by bees and yield food pollen, whereas the larger, dull-colored, functional pollen is in the inconspicuous dorsal anthers (chapter 23). In *C. bauhinoides* the pollen yield is 21% large grains (about 80 µm) and 79% small grains (about 27 µm), a finding interpreted as reflecting the need for fodder pollen to be more abundant (Linsley and Cazier 1972). Other species in this genus show similar dimorphy, with at least a twofold size difference between the two pollen types. Allocation of resources to two different anther types has also been shown in *Monochoria* (Pontederiaceae) (Tang and Huang 2007), where mirror-image flowers (enantiostyly, see chapter 3) occur, each with one large purple anther mainly contributing to pollination, and five smaller, yellow feeding anthers; the feeding anthers have much less pollen in them in a selfing species (*M. vaginalis*) than in an outcrossing species (*M. korsakowii*).

Even more remarkably, some heterostylous plants (chapter 2) have different pollen morphs on their differently positioned (but otherwise very similar) anthers. We saw earlier a case of morphological differences for *Armeria* (fig. 3.11), but color differences are rather more common. For example, in the tristylous *Lythrum salicaria*, the short- and middle-length stamens in or near the corolla mouth bear bright yellow pollen, whereas the long stamens that project well out of the corolla have inconspicuous green pollen (Lunau 1995); but since all morphs have a bright yellow pseudostamen mark at the corolla entrance, it may be hard for visitors to select between the morphs.

Pollen grains frequently carry their own odor, which originates in the pollenkitt and is therefore significantly more common in zoophilous plants than in anemophilous plants. The odors involve the same kinds of volatiles that produce other floral scents (Dobson and Bergström 2000; chapter 6) but are generally not identical to the odor components from the rest of the flower (Dobson, Groth, et al. 1996), although the pollen may adsorb some additional odor components from the surrounding floral tissues after dehiscence. Pollen odors may originally have been functional in pollen defense against pathogens, such as fungi or bacteria (Char and Bhat 1975), and against flower feeders, only later taking on a role in attraction (Pellmyr

and Thien 1986). In modern plants, pollen odor may serve both attractive and defensive functions, and the volatiles that produce the scents may also have some role in pollen-stigma interactions. It could be argued that the presence of specific pollen odors, which can be discriminated and preferred by visitors such as bees, is evidence that the pollen is "deliberately" provided as a reward to pollinators in many plants. Certainly honeybees can choose between pollens on the basis of their **phenolic** odors, preferring low phenolic content, even though this has no nutritional significance (F. Liu et al. 2006); and they can also deactivate some of these phenolics within the hive (F. Liu et al. 2005).

In the last chapter in *Other Functions of Floral Scents*, we noted that some conifers produce specific odors from male flowers and also that pollen volatiles can be repellent to some flower visitors in *Acacia* and other plants. However, where defensive volatiles occur in zoophilous flower pollens, they might need to be overlaid by stronger attractant scents for the desirable visitors, at least while the pollen is freshly available (Dobson and Bergström 2000). This should be particularly important in entomophilous plants, which tend to have stronger pollen odors than ornithophilous species (Dobson 1988). There are indications that beetle-pollinated species have especially strong pollen scents and that beetles are particularly attracted by pollen volatiles (e.g., S. Cook et al. 2004), as perhaps are thrips (Kirk 1985) and some moths (LeMetayer et al. 1993). A few specialist bees are also specifically responsive to pollen odors when seeking their preferred plants (Dobson and Bergström 2000). Thus olfactory cues from pollen could be involved both in animals choosing between floral species and in selecting the most rewarding flowers within a species, at least where the anthers are not visually exposed.

One other possible role for pollen volatiles, unrelated to advertisement, is that of pollen allelopathy, whereby pollen of one species can exert negative effects on another species' pollen, thus gaining a competitive advantage (Murphy 1999; Dobson and Bergström 2000; Roshchina 2001). Usually conspecific pollen on a stigma acts to inhibit germination of heterospecific grains (Gaur et al. 2007), but sometimes foreign pollen inhibits the correct pollen's growth, worsening the physical effects of stigmatic clogging. This has been documented for rather few plants however.

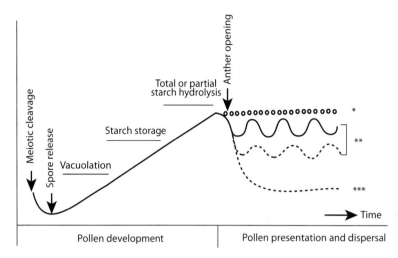

Figure 7.4 Stages in pollen development up to dehiscence, and subsequent volume reductions as desiccation occurs at variable rates in relation to environmental conditions: (*) aquatic pollen; (**) zoophilous or anemophilous pollen than completes its dehydration after presentation, varying with humidity; (***) usually anemophilous pollen, with no mechanisms to prevent dehydration. (Redrawn from Dafni and Firmage 2000.)

2. Storage and Delivery of Pollen in the Plant

Storage in Anthers

Anther design was discussed in chapter 3. Varying from one to several hundred per flower, most anthers are ovate or elongate and commonly contain four separate pollen-filled locules arranged in two pairs, giving the anther a two-lobed appearance. Within each locule the pollen grains are bathed in a nutritive locular fluid, and at maturity this fluid reduces in volume (fig. 7.4) either by evaporation or resorption (Pacini 2000). It is the change in locular fluid that triggers anther opening, and also leads to the pollen drying out somewhat, to a water content of less than 20%, so that it enters a relatively dormant state immediately prior to dispersal.

The amount of pollen within an anther reasonably matches how much of it is required for effective reproduction, so there is substantially more in species that only produce pollen as a reward (e.g., poppies, where a single flower may produce more than 2 million pollen grains) when compared to a similar-sized and similarly visited flower that also produces nectar. Table 7.6 includes some further information on grain number as related to ovular number.

Some anthers are nonfunctional (chapter 2), being devoid of pollen and purely serving for attraction of visitors. In most such cases there are both normal and empty anthers within the same flower.

Delivery by Anther Dehiscence

Anthers are initially sealed structures and must be unsealed in the process of dehiscence, often only undergoing this process once the corolla itself opens (although in some flowers the anthers open before the petals separate). Anthers usually open by longitudinal splitting along the visible furrows of the locules; the slit generally runs down the whole length of the anther, and it may be directed to the inside or the outside of the corolla (introrse or extrorse). Sometimes the slit is not elongate but curved, to give a valvate anther (fig. 2.3), and in *Acacia* and related genera, this type of opening results in the anther heads appearing to have a stalked lid (termed *an anther gland*) that lifts up at dehiscence (fig. 7.5), the stalk serving almost like a zip-fastener around the locules prior to this.

A locule usually opens gradually, over the course of a few minutes or hours, depending on species and environmental conditions. Its time course may vary adaptively, at least in *Penstemon*; here opening is more rapid and synchronous in bird-adapted flowers than in bee-adapted species (J. Thomson et al. 2000).

Once an anther is open, the pollen can exit in one of several ways. In the simplest cases it may be passively released directly into the air, either dropping from pendant anthers or being moved off by air currents; this is of course characteristic of anemophilous plants (chapter 19) but also occurs in quite a range of mainly entomophilous flowers as they age, and the pollen not collected by visitors progressively dries out.

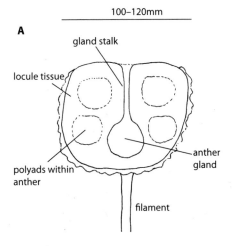

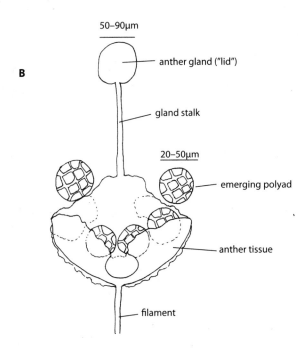

Figure 7.5 Acacia anther heads (A) with "lids" (anther glands) that lift on a stalk (B) to reveal the composite polyads.

surfaces to adhere to are offered, with pollen then moving directly from anther to vector.

Alternatively, in a minority of flowers the pollen grains may be "launched" dynamically in several ways, either into the air or onto a visitor. One possibility uses *filament movements*; in *Mahonia* the anthers curve inward when the flower is touched, thus converging on the sides of a visitor's tongue or face (plate 12B). In some composite flowers the filaments are sensitive and contract slightly when the flower is visited, so that the closed tip of the style in effect moves up within the "tube" of stamens and brushes some pollen out onto the visitor (chapter 2; and fig. 7.6). Such movement allows for the controlled, progressive delivery of pollen, about 20–25% of the total being emitted at the first contraction for *Centaurea*, with lower amounts for each succeeding visit (Percival 1965).

Similar effects can occur in groups other than the composites, with anthers held down until triggered or with anther appendages acting as trips or levers to dispense pollen. In *Kalmia* the ten anthers are held folded outward by pockets in the petals and then spring upward and inward when a flower is probed. In *Torenia*, J. E. Armstrong (1992) described a flanged outgrowth from the pollen-sac wall, which when pressed produces a lever action in the anthers, forcibly shedding pollen grains. Somewhat similarly, in many species of *Salvia* there are just two functional anthers, in each of which one of the two lobes is abortive and expands downward to form a flap that almost blocks the entrance to the tubular flower. When a bee visits, it pushes against these two sterile lobes or inserts its tongue into a small hole in the staminal lower lever, which makes the two fertile anther lobes swing down about a see-saw pivot point and so contact the bee's abdomen (fig. 2.17B; Reith et al. 2007). In some salvias this mechanism is less well developed and the lower anther lobes still produce some pollen, while in many scarlet-hummingbird-pollinated salvias of the Neotropics, the see-saw mechanism has been lost, although there is still only one anther lobe and pollen merely rubs onto a visiting bird's head. Han et al. (2008) most clearly demonstrated dose-controlling lever effects in *Incarvillea arguta*, where downward-pointing appendages on the anthers act as levers to dispense pollen, one set initiating pollen release as a large-enough visitor (e.g., a bumblebee) enters the flower and a second set producing a further aliquot of pollen as a bee exits; here

In animal-visited plants the pollen is usually not allowed to fall after dehiscence but instead is retained within or just at the surface of the anther chambers (*primary pollen presentation*), sometimes with progressively more pollen exposed as the anther slit elongates over time. The pollen grains are restrained by their own sticky surface properties or by viscous threads (or both), until the flower is visited and new

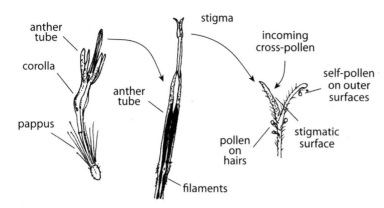

Figure 7.6 The pollen-dispensing brush of a typical composite, where the style emerges from within an anther tube and has hairs on its outer surface that sweep pollen out and upward as it grows; the style therefore bears self-pollen externally before its own stigmatic surfaces spread apart to receive incoming cross-pollen.

proportions of pollen remaining in the flower declined from 27% to 10% and 7% in successive visits.

These anther movements are all relatively slow and contrast with the fast *"trigger" effects* that are seen in a smaller range of flowers and that sometimes even produce a relatively explosive pollen release. The triggered ejection of pollen at the time of dehiscence can result from the sudden release of mechanical stresses set up in the anther as it dehydrates (see below), for example, in the castor oil plant *Ricinus* (Bianchini and Pacini 1996), where launching is weaker at high humidities. Or triggered release can occur as the anthers move, as in bunchberry (*Cornus canadensis*), where the pollen is catapulted vertically from the anthers either onto a visiting insect or into the wind (Whitaker et al. 2007). Ejection may sometimes result from electrostatic forces as, for example, in some anemophilous plants where the inside of the anther bears *orbicules*. The latter are made of sporopollenin similar to that on the pollen surfaces, so that the pollen may be ejected from the anther as soon as it opens because of the repelling similarity of electrostatic charges on these two surfaces (Keijzer 1987). Release can also be triggered well after the time of initial dehiscence, for example, in legumes such as beans, broom, sweet peas, etc., where the keel formed from the two lower petals (chapter 2) acts as a landing platform for visitors. The visitor's weight depresses the keel and allows the stamens to spring up and strike pollen onto the animal's underside.

Triggering can also be used to achieve precise placement of pollen onto different visitors. In Australian *Stylidium* trigger-plants, there are rapid movements of the combined column (style plus stamens), which at rest is arched back to one side of the petals but which, being sensitive at its base, will strike forward when a bee or fly visitor alights, thus depositing or picking up pollinia from a precise point in space. Of 31 species examined (Armbruster et al. 1994), there was little overlap in the placement of the pollinia on a visitor, so ensuring specificity of pollen transfer to a conspecific flower.

All of the above examples involve primary presentation. However *secondary pollen presentation* (G. Howell et al. 1993; Yeo 1993) occurs in a relatively small proportion of plants (though across at least 25 families). Here the pollen is dispensed from the anthers onto some other part of the flower, where it adheres because of the pollenkitt, and is subsequently picked up by visitors from that secondary site. For example, in *Campanula* species, the anthers dehisce in the closed-bud stage when they are closely adjacent to the style, so the pollen is dispensed onto the rather hairy outer surfaces of the style tip, then the anthers subsequently shrivel (Leins and Erbar 1994). The pollen must then be picked up secondarily from the style by incoming insects, and only later (after a few hours or days) does the style itself open at the tip to reveal receptive stigmatic surfaces; if these are not dusted by cross-pollen over the next day or two, they may curve downward to contact any remaining self-pollen as a last resort. Something similar occurs in those genera of the Asteraceae that have only ray florets, such as dandelions (chapter 2), and also in *Anthurium* inflorescences (Araceae), where a piston-like mechanism forces pollen from the hidden male flowers out of an opening below the **spadix** (Westerkamp 1989). In some flowers pollinated by nonflying mammals, pollen is

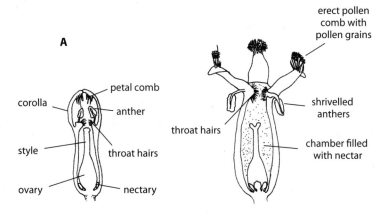

A

erect pollen
comb with
pollen grains

corolla

petal comb

anther

style

throat hairs

ovary

nectary

throat hairs

shrivelled
anthers

chamber filled
with nectar

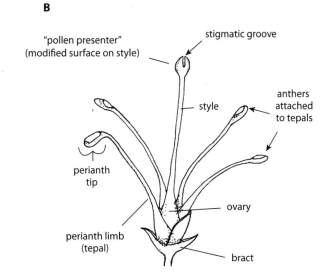

B

"pollen presenter"
(modified surface on style)

stigmatic groove

style

anthers
attached
to tepals

perianth
tip

ovary

perianth limb
(tepal)

bract

Figure 7.7 (A) *Acrotriche* pollen combs, picking up self-pollen while reflexed within the bud and then held erect and pollen-presenting on the open flower, with the anthers now withered (modified from McConchie et al. 1986). (B) Pollen presenters in *Protea*, where the anthers are sited directly on the perianths (strictly these are tepals), and their pollen is transferred to the emerging style surface within the bud; the main display features of a mature *Protea* inflorescence is often a mass of these "pollen presenters" (see also plate 27).

dispensed onto pollen combs on the petals, as in *Acrotriche* (fig. 7.7A), or is delivered from *pollen presenters*, as in many *Protea* species (fig. 7.7B).

Rate, Timing, and Control of Dehiscence

Timing of male functionality in flowers can be affected by two main factors: the elongation and maturation of the stamens as a whole, which is usually characteristic for a given species and presumably developmentally preprogrammed, and also the dehiscence and pollen-dispensing pattern of the anthers.

The *rate* at which dehiscence occurs is highly vari-

able. It can be simultaneous in all the anthers of one flower, even to the extent of being explosive, as described above. It may even be simultaneous over a whole plant, or a whole population of plants—for example, in some *Acacia* species, dehiscence occurred within 1–2 hours in all the plants of one species at a given locality (Stone et al. 1998, and fig. 7.8). Alternatively it can be sequential between flowers on the same plant, or sequential for different anthers within one flower; in many hellebore flowers the anthers ripen and open one at a time. In some plant families it is common for stamens to occur in two or more whorls that mature at different rates. For example, a ring of five stamens may become erect and dehiscent as the

A

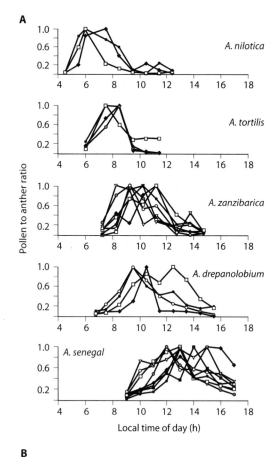

B

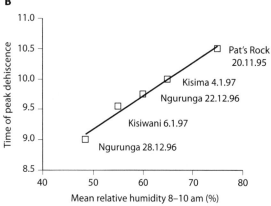

Figure 7.8 (A) Patterns of daily dehiscence through time for five East African acacia species. (B) An example of relative humidity effects on dehiscence timing, for *Acacia zanzibarica* from African sites (redrawn from Stone et al.1998).

bud opens, then shrink back against the petals as a further ring matures some hours (or even days) later, as in some Geraniaceae, including the common cranesbills, and in stitchworts (*Stellaria*) and some other Caryophyllaceae.

Further variation comes from the behavior of the dehiscence slits on the anther locules. Normally these slits open synchronously along their length, with all the pollen in that anther becoming available. But in some plants (e.g., the rock rose *Cistus*, a major bee-plant in Mediterranean habitats) the slits progressively split along their length so that pollen is gradually revealed and dehiscence is effectively staggered within one anther (**pollen dosing**). In a few genera where polysporangiate anthers occur (chapter 2), they are sectioned internally along their length, with each successive section dehiscing separately through its own pore; this again gives a useful way to partition pollen presentation through time. Such systems make obvious sense: effective pollen export and seed-siring are almost always improved if multiple visitors each remove just a portion of any one flower's pollen (Harder and Thomson 1989; LeBuhn and Holsinger 1998), especially where visitors are messy or wasteful, or are groomers who take large amounts of pollen that gets packed into **scopae** and that can never be deposited on stigmas. It is also particularly beneficial for flowers that occur in inflorescences, where extensive geitonogamous selfing would lead to large amounts of pollen being "discounted" for any useful purpose (chapter 3). In fact, any system of controlled pollen delivery, such as gradual anther opening or triggered release, can improve male function for the plant and may therefore be strongly selected for (Harder and Wilson 1994). Castellanos et al. (2006) showed a relation of pollen-dosing strategy to pollinator type by analyzing *Penstemon* and *Keckiella* species and found that (after controlling for phylogeny) the hymenopteran-adapted species did present pollen more gradually than hummingbird-adapted species (birds being on average less common but more efficient at pollen delivery).

The *daily timing* of anther opening in most plants shows a general correlation with pollinator activity. Broadly speaking, and in line with expectations, flowers visited mainly by diurnal bees or butterflies tend to dehisce in the early morning, while those pollinated mainly by moths or bats tend to dehisce in the later afternoon or even after dark. The timing of dehiscence can be tight and confined to a very brief temporal window in some species, where relatively specialist visitors occur. For example, *Alkanna* showed very precise daily dehiscence in the arid wadis of the Sinai, with associated tight patterning of visitation by female *Anthophora* bees (Stone et al. 1999). Nansen and Korie (2000) showed a close temporal relation between the time of honeybee activity and the time of pollen release on *Cistus salvifolius*, with a time lag of just 28–60 minutes over a 10-day period. There was also a good match between the timing of visits, handling times, and pollen availability in *Solanum* flowers for three sonicating bee species studied by Shelly and Villalobos (2000); here, early-morning dehiscence gave a high pollen reward in new flowers around 0700–0900h. Examples from later in the day have also been reported: in *Ludwigia elegans* (which is related to evening primroses) pollen was not available until late afternoon, and two dusk-active Brazilian bees (*Tetraglossula* and *Heterosarellus*) arrived precisely on cue to gather its pollen (Gimenes et al.1996). The same was true for *Perdita* bees that specialize in collecting pollen from *Mentzelia*, whose anthers open in the late afternoon (Michener 2000).

But perhaps the clearest evidence comes from *Acacia* flowers (Stone et al. 1996, 1998). Many species that are all very similar in flower morphology co-occur in African savanna habitats and have no barriers to heterospecific pollen deposition. All of them attract mass visitation, since they are almost the only plants in flower for parts of the season; and most species have no nectar, so that pollen is clearly the visit-structuring agent. Flowers generally last only one day, but crucially within that day, different species dehisce at different times at a particular site. Figure 7.8 displays data for five species in Tanzania showing that pollen availability sharply peaked for any one species. Likewise, the visitors (especially *Apis* and small megachilid bees) had peaks of activity on a particular species of tree, and within a particular daily "pollen window" their visits were all or mainly to one *Acacia* species; hence, by the time an individual bee moved to another species it would have deposited and groomed off most of its pollen so was fairly free of heterospecific grains. Individual bees were clearly tracking pollen availability closely, and their daily round of activities was largely dictated by the patterns of *Acacia* dehiscence.

There is not as yet enough work on what factors *control* dehiscence: to what extent it is preprogrammed, is dependent on other changes within a flower, or is

linked to the environment. Many factors are implicated, as summarized in table 7.3 (based on Pacini 2000). Most of these factors are in part programmed by the development of the flower. But in some flowers the usual timing of dehiscence can be substantially affected by the weather; anthers dehisce quicker in dry than in humid air, confirming that drying-out processes are involved (see above and chapter 2). For many plant species the daily cycles of declining relative humidity and rising ambient temperature enhance evaporative water loss from the locular fluid and accelerate the structural changes in plant tissue that lead to dehiscence (Pacini 1994, 2000; Keijzer 1999), also allowing the pollen to partially dehydrate just before it is dispersed. In the *Acacia* communities just mentioned, which have diurnal species-specific peaks of dehiscence, the daily changes in relative humidity acted as a specific cue, and anther opening was delayed on more humid mornings, albeit to different extents in different species (Stone et al. 1998); figure 7.8B shows an example of humidity effects on dehiscence timing in *Acacia* species. The effects of weather on timing can be striking in other plants: for example, in the woodland dog's mercury (*Mercurialis annua*), which has an unusually long flowering period, anthers may open around noon in February but as early as 0700 in July when the temperature is high and the humidity low (Lisci et al. 1994). Similarly, with hellebores, which were mentioned above for having anthers opening one after another, the number opening on any one day is temperature dependent (Vesprini and Pacini 2005).

Of course many flowers dehisce in the evening, when the humidity is normally increasing. While it is possible that the lower humidity through the morning could take time to induce sufficient drying for later dehiscence, it is more likely that simple abiotic cues are not the whole story when it comes to determining the timing of dehiscence. Opening at dusk may be the best strategy for a plant that is "hedging its bets" on visits by several kinds of pollinators; the pollen becomes available to **crepuscular** and nocturnal pollinators, usually bats and moths, but many plants will be far from depleted overnight, and their remaining pollen becomes a major resource for bees through the following day. The alternative of opening and dehiscing in the morning could induce a surge of bee visits, potentially leaving nothing for nocturnal visitors. For plants that have floral morphologies inviting visits from a

TABLE 7.3
Factors that Affect Anther Dehiscence

Mechanical
  Number and placement of layers thickened with lignin in walls of locules, causing shape change during dehydration, so exerting stress on slit or furrow
  Fusion of the two locules as septum degrades, increasing the volume of the cavity so pollen no longer entirely bathed in fluid and can start to dehydrate

Physiological
  Anther stomata: regulating water loss after flower opening and before anther opening
  Cuticle thickness and permeability in different parts of anther
  Increasing osmotic pressure as starch is hydrolyzed
  Water translocation out of locules

Biochemical
  Enzymes degrade the septum between the locules
  Calcium oxalate particles help to weaken the septum

Environmental
  Temperature: incoming radiation affecting anther evaporative losses and accelerating dehiscence
  Humidity: lower relative humidity increasing evaporation and accelerating dehiscence

Source: Modified from Pacini 2000.

range of long-tongued animals, then, selection for a dusk anthesis and pollen presentation may require the uncoupling of any relationship between dehiscence and humidity and more dependence on internal cues.

It is worth mentioning here that pollen exposure by anther opening is not always an irreversible event, and some plants can effectively re-close their anthers. Valvate stamens, in particular, are often capable of closing, often with a **diel** patterning. Or closure may occur in response to rain, so that the pollen is not wetted (which can irreversibly damage pollen grains and/or reduce its adherence to visiting bees [Dafni 1996]) or wastefully washed from the flower. Examples of closure in plants with normal longitudinal dehiscence slits are less common, although *Mirabilis* (Cruden 1973) and some lilies (Edwards and Jordan 1992) can probably do it.

3. Pollen Packaging

Usually pollen occurs as individual grains, but in some cases it gets packaged into larger units. This most often occurs in flowers that are strong nectar providers, where nectar is evidently the main reward, or alternatively at

the other extreme, in those with no nectar reward, where pollen is clearly the main offering to visitors.

Pacini and Franchi (1998) list thirteen pollen aggregation systems, but essentially the phenomenon can be achieved in four different ways.

Inherently Sticky Pollen

Here, pollen grains have a particularly sticky coating that comes from the anther tapetum. One example is the pollenkitt that is found in many families, including composites and labiates, and another is tryphine, which is found in the crucifers. This phenomenon is common in many animal-pollinated plants.

Gluing Processes

Pollen grains are sometimes bound together with a glue that is secreted from some other floral part. This method is rather rare and is found mostly in tropical plants. A good example occurs in *Tylosema* (De Frey et al. 1992), where more distant parts of the anther secrete a sticky mucilage.

Tangling

Rather than being glued, pollen grains can be held together by threads that enmesh the grains, binding them somewhat more loosely than a glue. Commonly the threads are viscin or elastoviscin threads that originate inside the pollen sacs, but other parts of the anther may also produce filaments that trap and entangle pollen grains. The system appears to be most frequent where the main pollinator is a large and messy feeder, such as a bat or bird or large moth; a good example is *Heliconia*, where pliable, sticky threads entangle pollen grains and form them into aggregates that stick readily to bird feathers and beaks (Rose and Barthlott 1995). Butterfly-pollinated *Caesalpinia* also have pollen that is clumped by viscin and becomes attached to the forager's wings (Cruden and Hermann-Parker 1979). However, pollen tangling also occurs in aquatic angiosperms, where the pollen grains are highly elongate and tangle with each other to form floating pollen mats (chapter 19).

A related system has been reported in *Montrichardia* (Araceae), where pollen grains explode when wetted, and threads of intine material emerge and entangle the grains to improve adherence to shiny-surfaced pollinators (Weber and Halbritter 2007).

Tetrads and Other Polyads

Yet another system involves the coherence of the pollen grains by means of common walls. The basal form is the **tetrad**, where four grains derived from the same ancestral cell are loosely held together, due to lipid deposits on the pollen wall, or are more tightly linked, via exine bridges. Tetrads occur quite frequently in the Annonaceae, Winteraceae, Ericaceae, and Asclepiadaceae; a common temperate example is *Chamerion* (fireweed, or rosebay willowherb). Sometimes tetrads can themselves be additionally aggregated by viscin threads, as in many *Rhododendron* species.

A smaller range of flowers have composite polyads, notably the Orchidaceae, mimosoid Fabaceae, and Annonaceae. The clumps of pollen are always in multiples of 4, giving 8, 16, 32, or 64 grains per unit. Polyads are particularly well known in *Acacia*, a massive genus within the Leguminosae. Here the corollas are reduced and there is a brush inflorescence of 50–150 flowers, with large numbers of stamens forming a

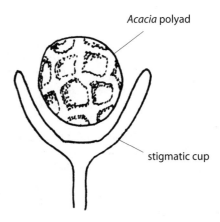

30–80 µm in

different species

Acacia polyad

stigmatic cup

Figure 7.9 Acacia flowers have cup-like stigmas, each of which can accommodate just one of the complex pollen grains (polyads).

spherical or spicate surface (plates 8A and 20G). Each stigma is a cup-like structure (fig. 7.9) into which a single polyad fits. This method ensures that all ovules in the same ovary have the same father. Kenrick and Knox (1982) studied the number of ovules in the ovary and of pollen grains in the polyad in 45 *Acacia* species: 40 species with polyads of 16 grains always had 5–16 ovules per ovary; three species with 8-grain polyads had 2–10 ovules; one species with 32-grain polyads had 24 ovules, and one with simple tetrads had 1–3 ovules. Thus there appears to be good matching, with always more than one (but rarely more than two) pollen grains per ovule, perhaps giving the most efficient system. Cruden (1977) had proposed that this was usually the case with compound pollen grain systems. However, in various other taxa (Winteraceae, Lloyd and Wells 1992; Pyrolaceae, now in the Ericaceae, Knudsen and Olesen 1993; Orchidaceae, Neiland and Wilcock 1995), the ovule number substantially exceeds the number of pollen grains. Even in *Acacia*, fruits carrying multiple paternity seeds are not uncommon when traced with molecular marking (Muona et al. 1991).

More complex are the pollinia found in two families, the Orchidaceae and the asclepiads within the Apocynaceae. In these taxa all of the pollen grains from one anther are bound together as a single mass, embedded in elastoviscin, and usually separated only by the intine walls, with exine confined to the outer surface of the whole pollinium. In orchids the pollinia have accessory structures, often including long stalks and a sticky area (the *viscidium*) used to give adhesion to a visitor, and have suitable breakage points built in; the whole complex is then more correctly termed a pollinarium. Orchid pollinaria are morphologically highly diverse (S. Johnson and Edwards 2000), and the pollen mass itself may be soft (about 7% of species) or hard (80%), or even segmented (11%, in a few genera such as *Satyrium*, where the sections are then termed *massulae*, and a pollinator—often a moth for this genus—can deposit pollen on to several successive floral recipients as the segments break off in turn). One orchid pollinium may contain between 5,000 and 4 million pollen grains (Schill et al. 1992), and a given orchid flower generally contains a number of ovules roughly similar to the number of grains deposited per visit (Neiland and Wilcock 1995; Nazarov and Gerlach 1997).

Orchid pollinia are normally held dorsally on the flower in a small hood on top of the column formed from a single stamen (chapter 2). From here a pollinium will usually get placed onto the tongue, the dorsal surface of the head, or the back of an animal visitor, depending on visitor behavior. The pollinium is then carried around and is resistant to grooming, the viscidium glue having set so that it resists normal scraping. During this phase the pollinarium stalk often bends progressively with a time course of a few seconds to several hours (S. Johnson and Edwards 2000) so that the pollen mass ends up correctly oriented to meet the stigma on another flower. The pollinarium can then be picked up at the next visit made by its carrier to a new flower, often being inserted into a specifically shaped slit on the stigma so that force is exerted to pull it off the moving animal. In some orchid species the pollinia can be as big as an insect's head, and if placed on the dorsal thorax it may impede full wing-movement.

In the asclepiads there are diad pollinaria, the products of two anthers that fuse at their tips due to a sticky glandular secretion. Such flowers tend to be complex, with hairs or coronal structures that guide visitors past the pollinaria (Kunze 1991). In some asclepiads the pollinaria are glued to the feet of the visitor (often a butterfly or moth, sometimes a bee) and then come off when it alights on a new flower; for example, in *A. curassavica* there is a slippery surface at the landing point and a visiting butterfly slides down, its feet touching against the stigmatic slit as it regains its balance. Morse (1981a) reported bumblebee feet so laden with asclepiad pollinaria that their foraging efficiency markedly declined. In other asclepiads it is the tongue rather than the feet that transfer pollinaria; for example, *Metaplexis* pollinaria clip on to the tips of the tongues of various moths, being pulled out of the flowers along a "guide rail" (formed by an anther slit) as the moth feeds on nectar (Suguira and Yamazaki 2005). Stigmatic surfaces in asclepiads are nearly always on the underside of the style, and various devices serve to fix the pollinaria in the right place (Kunze 1991). But despite these specialized strategies to ensure correct pollinaria receipt, seed set in asclepiads is notoriously low, often 1%–5% (Wyatt and Broyles 1994).

Why do plants bother with polyads or pollinia? Aggregated pollen has arisen independently at least 39 times, but the adaptive benefits have rarely been explored (Harder and Johnson 2008). A major cost is that transfer is quite tricky and often fails. But the pollen is very efficiently removed from an anther, with minimal pollen wastage in transit, and there will potentially be highly efficient deposition on a conspecific stigma. Hence, when the system works properly, several thousand or tens of thousands of pollen grains are transferred in one visit; and the species that use it tend to have thousands of ovules in the ovary. In fact, the number of ovules is usually lower than the number of incoming grains per pollen package, thus allowing gamete competition (Pacini and Franchi 1999).

Orchids, in which pollinia are best studied, tend to be minority flowers in any given habitat, and it could be difficult to produce enough of the usual type of pollen to distribute it onto enough shared pollinators to ensure effective cross-pollination. So it may be a better strategy to concentrate on rare pollinators and develop specific relationships with them. In that situation a reliable pollen dispensing and attachment system is favored, with unusually durable long-lasting pollen that cannot readily be eaten or discarded. Hence pollinia are an ideal solution. Harder (2000) found that whereas only 1% of typical pollens reached conspecific stigmas due to high removal and transport losses, for orchid pollinia removal failure was high (about 50%), but around 8% of the pollinia did eventually reach a stigma.

But flowers with pollinia *cannot* have pollen as their main food reward, because they would lose all their reproductive potential when visited. Asclepiads use profuse nectar as a reward instead; orchids sometimes offer nectar, and sometimes offer scents to specialist bees (chapter 9), but a great many use deceit tactics in the form of pseudocopulation (chapter 23), leading Harder and Johnson (2008) to speculate that there is an explanatory link between pollen aggregation and **deceit pollination**.

4. Pollen Gathering by Animals

Passive Gathering

It is generally assumed that most pollen is transferred entirely passively onto a visiting animal's body when physical contact is made. Pollen grains adhere to the surfaces of flower visitors because of their inherently sticky properties (pollenkitt, tryphine, viscin threads). Their adhesion is often enhanced by the characteristics of the animal's surfaces: branched hairs for bees, scaled surfaces for lepidopterans, fur or feathers in vertebrates. Pollen acquired in this passive fashion may then be groomed off, either because it becomes a nuisance, is to be eaten immediately (in syrphid flies, masarine wasps), or is to be transferred to a storage site (in bees: initially either the **crop** or a specific scopa on legs, thorax, or ventral abdomen, and then eventually the nest). The grooming step can be a substantial and time-consuming process, contributing markedly to the overall cost-benefit analysis of flower feeding.

There are some indications that what is termed *passive gathering* can be augmented by *electrostatic attraction* between pollen grains and animals' bodies (Corbet et al. 1982; Erickson and Buchmann 1983; Vaknin et al. 2000, 2001; Armbruster 2001). Plants are inherently slightly negatively charged in warm still air, the magnitude of the charge field depending on the plant's composition and architecture and on weather conditions. Charge is greater at sharp tips, and therefore often higher in flowers than elsewhere (Dai and Law 1995). Hence it is possible that pollen grains would acquire negative charges within the flower. In contrast, insects, and especially hairy bees, tend to acquire a positive surface charge as they fly (Erickson 1975; Erickson and Buchmann 1983; Gan-Mor et al. 1995). This charge difference might aid the transfer of pollen onto visiting insects; even without direct contact, pollen grains could be attracted onto a close-up bee. Pollen might also be unloaded from bees onto dry stigmas in reverse fashion (Woittiez and Willemse 1979), especially where stigmas protrude from flowers and provide foci for charge accumulation (Corbet et al. 1982; Bechar et al. 1999). Small electrostatic charges could also lead a visiting insect to leave an electrical "footprint" on flowers, detectable to subsequent visitors for at least several minutes and thus potentially allowing avoidance of recently depleted flowers.

Direct evidence for electrostatic involvement in pollen pick-up and deposition remains somewhat sparse (reviewed by Vaknin et al. 2000), and measurements of the required charges and distances involved are still equivocal (Dai and Law 1995, Gan-Mor et al. 1995), but some careful modeling has indicated that charges of the right order do result in increased deposition of pollen grains onto stigmas, and the theory is intriguing and worthy of more thorough investigation. There are already some indications that the use of electrostatically charged pollen can improve fruit set in commercial pollination applications (chapter 28).

Active Targeted Gathering

Rather few kinds of pollinators eat pollen and so gather it "deliberately," but those that do are important and numerically dominant kinds. Syrphids may eat pollen directly from the anthers. A few bees (very young adults in many taxa, and a few genera of primitive bees throughout their foraging lives) also gather pollen directly with their mouthparts, eating some immediately and carrying the rest back to the nest in their crop. In such cases only the pollen that is accidentally scattered during consumption (or that passively adheres to other parts of the insect) is of any use for pollinating.

Most bees normally do not eat pollen at the flower but instead collect it—with their legs and mouthparts, by rubbing the ventral body surface across a flower, or by using various comblike or rake-like structures (chapter 18). In these cases, the pollen is then packed away on or into specific collecting areas, often with the addition of some moistening nectar. Hence there may be substantial interruptions to foraging while a bee lands, grooms, and packages its pollen, although more advanced bees require only a brief hover between flower probings to achieve grooming and packaging.

Although active pollen gathering may be highly

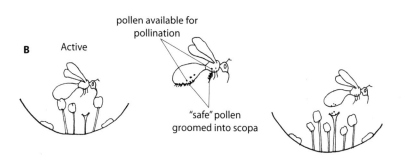

Figure 7.10 Principles of passive versus active pollen gathering (see also fig. 3.4).

beneficial to a flower-visiting animal, there are problems from the plant's point of view:

1. Much of the pollen is sequestered into sites on the visitor where it becomes useless and will never contact a stigma. Therefore, selection on the plant might act to place as much pollen as possible in sites on the visitor that are hard to groom, examples being the mid-dorsum, or the extreme front of the thorax and "neck." With suitable arrangement of floral parts, there is then a good chance that another conspecific flower will receive some pollen from this same site.

2. Many Hymenoptera (especially ants) have **metapleural glands**, which secrete small quantities of compounds that protect their nests against fungal and bacterial growth but that can also kill pollen (Beattie et al. 1985; F. Harris and Beattie 1991; chapter 24).

3. When saliva and nectar are added to the pollen gathered by bees, the pollen may suffer biochemical change, with sugars being inverted and the pollen's water balance being disrupted, since bees do not "want" the pollen to germinate once deposited in the nest cell.

For such reasons, usually the longer pollen adheres to a visitor the more its viability is compromised (Stanley and Linskens 1974). However, orchid pollinia may be exempt from this (Nilsson 1992) because they are deposited rather than gathered and usually adhere to their pollen vector on a short stalk, keeping the grains away from damaging secretions.

Figure 7.10 summarizes the benefits of passive and active pollen gathering and shows how pollen is thereby separated into "compartments" with different values to flower and to visitor. From the plant's point of view, those visitors who gather pollen passively are generally more useful. Grains moved into the scopa by a targeted feeder are normally no longer useful for pollination, being in the wrong place and rapidly rendered nonviable anyway; only the pollen that gets missed during grooming is of use to the plant. The plant therefore benefits more from a visitor that is "messy" and cleans itself only occasionally. Fortunately, completely efficient grooming is almost impossible; one honeybee can carry up to 20 mg of pollen in its scopa but will always have some left on its body hairs no matter how long it grooms itself.

There is one other type of active targeted gathering: the pollen collecting by fig wasps and by yucca moths, whereby pollen is gathered in the mouthparts or in specific pouches on the body and then "deliberately" deposited onto a stigma (**active pollination**, chapter 26). Active gathering is clearly of direct benefit to the plant, although this has to be offset against losses of ovules to the feeding activities of the visitors' offspring.

Gathering by Sonication: Buzz Pollination

The special system of **buzz pollination** is used almost exclusively by bees and can greatly increase their foraging efficiency. It is probably a relatively ancient phenomenon in angiosperms (Proença 1992), and Buchmann's (1983) analyses showed how widespread it now is in taxonomic terms, with 6%–8% of all angiosperms (15–20,000 species) being "buzzed" during pollination. It occurs in at least 72 families of plants, being particularly common biogeographically in Australia and taxonomically in various families allied to the lilies; table 7.4 shows examples. Many of these are nectarless plants, with pollen as the only reward (see section 8, *Pollen-Only Flowers*). In colder climates it is often the summer-flowering species that are nectarless and buzz pollinated, with earlier spring congeners offering nectar and showing the usual type of pollen presentation (e.g., *Pedicularis*, Macior 1983).

Buzz pollination, or **sonication**, is particularly found in small- to medium-sized flowers, mostly radial or only slightly bilateral, often pendant in habit, commonly blue, purple, pink, or white in color, and often with contrasting bright yellow anthers (plate 17). The flowers commonly open and dehisce in the early morning and are relatively short-lived. All these features could be said to be typical of bee flowers (chapter 18). Crucially, though, buzz pollination is almost always associated with anthers that have a pore at the tip or side (*porose* or *poricidal* anthers), rather than dehiscing lengthwise, so that pollen can only emerge from a small exit. The anthers come in a great variety of shapes, often unusually long with elaborate tips (which may influence pollen dispersal speeds and directions; Marazzi et al. 2007), but they all show this unusual feature of a single or double pore or short slit, sometimes partly covered with a lip or valve (fig. 7.11). Furthermore, a high proportion of buzzed flowers have a central cone of tissue (the *hypanthial cup*) comprising several such anthers fused together with the central style (Buchmann 1983; Harder and Barclay 1994).

During sonication, a visiting bee grasps this central cone with its legs or mouthparts and vibrates her body, using the indirect thoracic wing muscles acting simultaneously and antagonistically but in a partially uncoupled state and producing an audible buzz. The resulting vibration may be sufficiently fierce that the bee has to grip hard with her **mandibles** to avoid being shaken off the flower, which leaves visible wounds on the anther cone. The number of marks on this cone is a good indication of how effective pollination has been, and in tomato greenhouses this "bee kiss" is a useful sign to commercial growers (Buchmann and Nabhan 1996).

In just a few unusual cases, the anthers are hidden within the corolla and not even accessible to a bee. In *Conostephium,* bees such as *Leioproctus* grasp the top of the corolla tube, and their vibration is transmitted to anthers via the stiff filaments (T. Houston and Ladd 2002). This example is also unusual in that the anthers are not porose, opening progressively via a longitudinal slit.

Pollen in buzz-pollinated flowers is usually only 30–50 µm in diameter (as little as 13–18 µm in *Solanum* species), and bee species that use sonication extensively tend to have scopae with short, fine hairs spaced about 10–20 µm apart (Buchmann 1983). The pollen is generally very dry and powdery rather than sticky, with reduced pollenkitt but high lipid and low starch content. The grains may have less sculpted surfaces than the pollen of nonsonicated plants (e.g., Buchmann 1983), although though this may be confounded by phylogenetic effects.

When a porose anther is shaken at an appropriate frequency by a visiting bee, pollen emerges from the pore in puffs so that it is dusted over the animal's body. In *Dodecatheon* tested with different frequencies, Harder and Barclay (1994) found optimal pollen removal at 450–1000 Hz, suggesting that the system is tuned to a higher frequency than a bee can normally produce, which may help to limit the pollen taken on any one visit. Sonication at up to 400 Hz for one second would release about 10% of the pollen to a visitor.

Once again, pollen gathering may perhaps be aided by electrostatic effects; Buchmann and Hurley (1978) suggested that electrostatic attraction from pollen to bee (and then from a bee to a dryish stigma) was particularly likely in buzz-pollinated flowers such as *Cassia* or *Dodecatheon*, where the pollen grains are small

TABLE 7.4
Typical Buzz-Pollinated Flowers

Family	Genus	Common name	Geographical occurrence	Reference
Boraginaceae	*Borago*	Borage	Europe	
Solanaceae	*Solanum*	Bittersweet	Widely distributed	Buchmann 1983
	Lycopersicon	Tomato	S America	
Primulaceae	*Dodecatheon*	Shooting star	North America	Harder and Barclay 1994
	Cyclamen	Cyclamen	Europe	
	Ardisia	Coralberry	Widely distributed	
	Lysimachia	Loosestrife	Widely distributed	
Liliaceae	*Dianella*	Flax lily	Australia	
Amaryllidaceae	*Galanthus*	Snowdrop	Widely distributed	
	Leucojum	Snowflake	Central Europe	
Papaveraceae	*Papaver*	Poppy	North temperate	
Gesneriaceae	*Ramonda*	Phoenix flower	Southern Europe	
Ericaceae	*Vaccinium*	Cranberry	North American species	
Myrtaceae				Proença 1992
Fabaceae	*Cassia*	Senna	Widely distributed	Gottsberger and Silberbauer-Gottsberger 1988
	Lupinus	Lupin	Widely distributed	Harder 1990a
	Chamaechrista	Partridge pea	Eastern USA	Neff and Simpson 1988
Scrophulariaceae	*Pedicularis*	Lousewort	Widely distributed	Macior 1983
	Actinidia	Chinese gooseberry	Eastern Asia	
Dilleniaceae	*Hibbertia*	Guinea flower	Australasia, Madagascar	Bernhardt 1984, 1986
	Schumacheria			
Epacridaceae	*Conostephium*	Pearl flower	Australia	Houston and Ladd 2002
Melastomataceae	*Rhexia*	Meadow beauty	N America	Larson and Barrett 1999

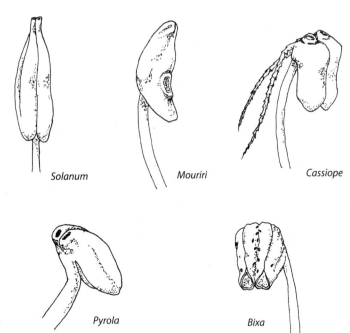

Figure 7.11 Examples of structures of poricidal anthers, in flowers that are "buzzed," from five different families. Usually one to two pores or slits are present, often partly covered. (Redrawn from Buchmann 1983.)

and dry. Buchmann (1983) pointed out that since many sonicated flowers are pendant, most of the pollen (probably far more than is observed) would be lost to gravity after buzzing unless it were electrostatically attracted to the bee's surface.

Buzz pollination is highly suited to bees, since they have well-developed abilities to vibrate their indirect wing muscles without flight for thermoregulatory reasons (chapters 10 and 18). King and Buchmann (2003) showed that for *Bombus* and *Xylocopa* the vibration seen on flowers occurred with a lower thoracic displacement but a higher frequency than that used in flight (and often at the second harmonic of the flight frequency). The frequency used by bumblebees can be varied within a species according to the flower being visited and can also vary between *Bombus* species foraging on a common flower (King 1993). In some **andrenids** the frequency changes as a bee works steadily up the cone-like anthers (this has been termed *buzz-milking*; Cane and Buchmann 1989). Buzzing behavior may also change through the course of a day as pollen supplies alter (King and Buchmann 1996). Bumblebees that are active soon after dawn may buzz for less than a second at one flower, but the buzz periods lengthen to 5–10 seconds later in the morning, sometimes with multiple buzzes on each flower, and

the bees work many more flowers to get a full load. In afternoon visits each bee may buzz one flower many times, often rotating around the anthers or the cone of anthers as she does so. Presumably the tendency of buzzed flowers to have quite large and plump-looking yellow anthers that do not shrivel as they gradually empty helps to keep bees coming even when the pollen is almost all gone. (This trait may explain the prevalence in Australia of orchids such as *Caladenia* or *Thelymitra* that appear to mimic buzz-pollinated flowers with unusually plump attractive anthers, although they cannot themselves be buzzed). Most sonicating bees must, of course, interrupt their foraging occasionally to make nectar-gathering trips, and the frequency of these may also vary through the day in relation to the weather and to the filled status of cells within an individual bee's nest.

Vibratory buzzing is often performed by bees on some flowers that are not especially adapted for sonication, especially a range of rosaceous plants including wild roses and some ericaceous species. Buzzing presumably facilitates pollen dislodgement even from those anthers and can therefore allow very fast working of the flowers. For example, on *Potentilla erecta*, which has normal anthers, *Bombus* commonly produced buzzing visits and worked more than 30 flowers per minute, while *Apis* (which cannot buzz—see below) only visited 11–12 flowers per minute with a similar net pollen gain per flower (Buchmann 1985).

But buzz pollination poses a problem, because one visitor could potentially take nearly all the pollen from one flower as food and transfer very little. Hence a fairly high proportion of sonication-adapted flowers find ways around this via heteranthy (chapter 2), presenting dimorphic pollen through a functional division of labor between feeding anthers and fertilizing anthers (as originally proposed by Darwin). The best-studied example is *Cassia* (fig. 2.5 and plate 12D), which has showy, protruding anthers that are buzzed to yield food pollen and also has dorsal, cryptic anthers in which the functional, fertile pollen is. The fertile anthers release small amounts of pollen precisely onto the visitor in a safe (ungroomed) site while the visiting bee is feeding (van der Pijl 1954). Similar situations are common in the family Melastomaceae, where corollas and ventral pollination anthers are generally pink/purple whereas the dorsal feeding anthers are bright yellow; bee visitors partly support their weight on the pollination anthers while sonicating the upper anthers and become covered in fertile pollen released by the bodily vibrations. In *Melastoma malabathricum*, Luo et al. (2008) found that although all ten of the dimorphic stamens were buzzed simultaneously, pollen from the fertile, purple set was far more likely to land on stigmas than pollen from the yellow feeding anthers, again because large proportions of the fertile pollen are deposited in ungroomable areas. The same was shown for *Solanum rostratum* flowers, where bumblebees preferred to manipulate the feeding anthers, but more pollen was exported from the pollination anthers (Vallejo-Marin et al. 2009).

Bumblebees are the best-known sonicators, but examples are now known among other bees—andrenids, halictids (Shelly and Villalobos 2000), stenotritids, megachilids (Neff and Simpson 1988), **anthophorines** (Batra 1994; Stone et al. 1999), and euglossines. However, sonication is not available to all bees, and the necessary higher frequencies conspicuously do not occur for *Apis* (King and Buchmann 2003), which cannot dislodge pollen in buzz-pollinated flowers. This may be critical in considering just how effective introduced honeybees are as pollinators (chapters 28 and 29).

Finally, although buzz pollination is almost entirely restricted to bees, it is not *quite* exclusive to them. A syrphid, *Volucella mexicana*, can do it, this hoverfly being a good visual mimic of small *Xylocopa* bees that themselves sonicate *Solanum* flowers (Buchmann 1983). The fly visits hundreds of *Solanum* in turn, grasping and vibrating the anthers just like a bee. King and Buchmann (2003) also recorded sonication in *Eristalis*, another bee-mimicking hoverfly (plate 23A).

The Energetics of Gathering Pollen

Passive gatherers of pollen incur few costs unless the grains deposited on their bodies become so abundant as to interfere with their normal behavior and necessitate grooming. However, animals that gather pollen by sonication, and those that gather it for consumption (by themselves or by their offspring), must forage according to some balance of benefits and costs, the costs being measured either in time spent or in energy used (Petanidou and Vokou 1990). Pollen feeders generally seem to forage to maximize their gross efficiency, that is, the ratio of benefits gained to energy used, with little influence from time considerations (whereas time constraints are often very critical to nectar-feeders; chapter 8).

A. Müller et al. (2006) estimated that bees gain only about 40% of the pollen in a flower on average, and that solitary bees need to visit a very large number of flowers to gather enough pollen to raise one offspring (values ranging from around 20 to over 2,500 flowers, depending on the bee and the plant). The energy expended in this activity must usually be offset by gains from nectar, either from the same plants or (less often for solitary bees) from other species visited.

Pollen Transfer

Individual flower visitors to postdehiscent flowers normally carry away large numbers of pollen grains, and they transfer a proportion of these to multiple floral recipients, since each recipient stigma will receive only a fraction of the pollen being carried by the visitor. Usually each successive stigma receives progressively less pollen from a given donor, the amount decaying very quickly for the first few visits and then somewhat more slowly (the process of pollen carryover; chapter 3 and fig. 3.6). The rate of decay of carryover varies according to the various sites onto which pollen is deposited from the donor flower, and the sites (with or without redistribution by grooming) from which it can be donated to recipients.

Plants have significant control over their own patterns of pollen transfer. There should usually be selection for strategies that counter any visitor's tendency to take too much pollen at a single visit, especially in relation to bee visitors (Westerkamp 1996, 1997a,b), where most of that pollen will become food for larvae. Plants can control the availability of pollen to some extent (Harder and Thomson 1989) by

1. managing the packaging and delivery systems (the dehiscence processes) in time and space, as described above and also mentioned in chapter 2 for particular groups, and

2. controlling the collection system (the deposition of pollen onto visitors) by manipulating the behavior of their visitors. This may be most obviously achieved via flower morphology, which can govern the spatial interaction of visitor and anthers. But it can also be influenced by offering other rewards, so that nectar production (chapter 8) can be strongly tied to dispensing and controlling pollen. Nectar volume and concentration both positively affect visit duration (longer visits generally being linked to greater pollen pick-up) and also influence the numbers of visits per inflorescence or per plant (which will in turn affect delivery of cross-pollen). Furthermore, because nectar tends to accumulate in unvisited flowers through time, individual visitors will usually remove more pollen per flower when pollinators are rare, and less pollen when they are common and very active (because they will then be depleting flowers more regularly and keeping the **standing crop** of nectar lower). Pollen transfer can thus differ markedly even between populations of the same plant in relation to pollinator abundance and the interactions of pollen and nectar.

Pollen Rejection

The phenomenon of pollen rejection is not very often reported, but it does occur. In the Malvaceae, pollen rejection is well known in cotton (chapter 30) and also occurs in *Callirhoe* and *Kallstroemia* (Estes et al. 1983). It has also been documented in *Passiflora* by Corbet and Willmer (1981), who speculated that for a plant to make its pollen repellent would be a useful way of reducing pollen wastage. Given the increasing evidence for scent-based repellence in flowers, and more specifically in pollen (chapter 6), this is a subject that deserves greater attention.

5. Pollen as Food

Pollen Composition

Pollen analysis requires rather large volumes of material, so that its composition has mainly been determined from anemophilous species, such as pine or oak, or using pollen pellets recovered from honeybees (which are probably contaminated by sugary additions). The available data are therefore derived from rather biased and imperfect samples. Hence, the commonly reported values, which have been repeated across the literature (Stanley and Linskens 1974; Colin and Jones 1980), are open to question (Solberg and Remedios 1980; Roulston and Cane 2000). The best data come from the painstaking hand-collection of samples, although even here the results obtained are dependent on the methodology. The more reliable updated values from Roulston and Cane (2000), Roulston and Buchmann (2000), and Roulston et al. (2000) are as follows:

Protein	2.5%–61%	lowest in anemophilous species, highest in buzz-pollinated species; but effects of phylogeny are overriding
Starch	0%–22%	highest in anemophilous species, many others < 1%; but no significant effects after correcting for phylogeny
Lipid	1%–18%	no patterns observable
Energy	16–28 J g^{-1}	no patterns related to pollination mode observable.

These data suggest that there is very little selective effect of the pollination system on pollen nutrition, and the authors found no links between pollen content and pollination syndromes; in particular, the needs of the

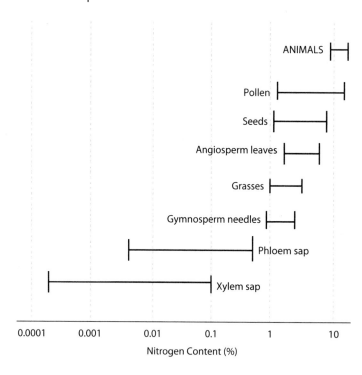

Figure 7.12 Pollen nutritional value compared with that of other plant tissues and with animal flesh; for nitrogen content (directly linked with protein), pollen can be 0.5%–12.5% dry weight, higher than for other plant parts and overlapping with the range in animals. (Modified from Strong et al. 1984.)

growing pollen tube are probably more important in determining pollen protein.

Pollen also contains useful amounts of phosphorus and other minerals and has a higher content of water-soluble vitamins and of sterols (up to 1%) than other plant materials. It has a lower water content than most leaves and fruits, although this is highly variable, from 5%–50% at different ages but generally falling below 20% at dehiscence (in all except grass pollens, which can remain close to 50% water). Pollen therefore provides a reasonably balanced diet for animals, at least when compared with most other plant tissues (fig. 7.12). It is routinely eaten by some beetles, thrips and springtails, a few nonspecialist flies and many hoverflies, a few butterflies, some flower-visiting birds and mammals, one sphecid wasp and the masarid wasps, and by virtually all bees.

Pollen Digestion

The outer exine layer of a pollen grain is totally indigestible for almost all animals (McLellan 1977) and is therefore discarded with the feces (incidentally mak-ing pollen a useful tool for palynologists deciphering fossil deposits, since exine features persist intact for millions of years). The intine layer, being mainly cellulose and pectin, is also hard to digest and relatively slow to decay. However, pollen grains if they are to have any food value must be split open by animals. There is little evidence for mechanical splitting by mouthparts in most animals that utilize pollen in the diet, with the possible exception of some beetles (Samuelson 1994), a few ceratopogonid midges (de Meillon and Wirth 1989), and thrips that do pierce pollen grains and suck out the contents (Kirk 1984). For most other feeders, pollen grains enter the digestive tract intact, and the gut must be either inducing the pollen to germinate or burst it (perhaps with osmotic shock) or penetrating it with digestive enzymes, these three alternatives being difficult to distinguish in practice (Roulston and Cane 2000). Germination, or more aptly "pseudo-germination," is quite plausible, since this happens fairly readily when pollen is kept in slightly acidic fluids containing sugars; and there are observations of cytoplasm extruding through pollen grain apertures in hoverfly midguts (Haslett 1983) and in honey possum stomachs (Turner 1984).

TABLE 7.5

Percentages of Pollen Grains Emptied in the Guts of Various Pollen Feeders

Taxon	Examples	% empty grains	Pollen type	Reference
Mammals	Rodents	50–83	*Protea*	Van Tets 1997
	Bats	47–86*	*Cacti*	L. Herrera and del Rio 1998
		27–82*	*Pseudobombax*	L. Herrera and del Rio 1998
		53–57	*Banksia, Callistemon*	Law 1992
	Marsupials	60–78	*Banksia*	Turner 1984
Birds	Passerines	~40	*Banksia*	Wooller et al. 1988
		0–18	*Eucalyptus, Zauschneria*	Brice et al. 1989
		>90	*Opuntia*	B. Grant 1996
Bees	*Apis mellifera*	50–98 (decreasing with age)	*Castanea, Trifolium*	Crailsheim et al. 1992

*Large variation because three different species of bat were assessed.

However, germination may well be supplemented—or replaced—by osmotic shocks occurring as the pollen moves into different gut areas with varying fluid concentrations (Dobson and Peng 1997; Roulston and Cane 2000). Whatever the mechanisms, and they may differ between taxa, the exine and intine are somehow split usually at the grain apertures, and the internal cytoplasm contents can be then be used as nutrition.

But the exine may make up at least 30% of pollen volume, and this value is even higher for smaller grains, so it follows that larger grains will be "better" food because a higher proportion of their volume will be nonexine. Larger grains (>80 μm) may also have a higher lipid content than small grains (Endress 1994). However, with up to 50% of each grain's volume being useless, pollen is hardly the "ideal" food, as sometimes portrayed, and very few animals rely totally on it. In fact only bees, most syrphid flies, and the uncommon masarine wasps get their protein solely from pollen. Not surprisingly, bees are also particularly efficient in the percentage of pollen grains that are emptied in the course of gut transit (table 7.5), and they are particularly efficient at assimilating protein from pollen, with recorded values of 77%–83% usage (J. Schmidt and Buchmann 1985); this is much higher than other invertebrates achieve, although it is almost matched by the assimilation efficiencies recorded for some pollen-feeding bats (Law 1992) and marsupials (Turner 1984; A. A. Smith and Green 1987) and even for a flower-visiting Galapagos finch (B. Grant 1996).

Adaptive Features?

Roulston and Cane (2000), in common with earlier authors, concluded that there was no good evidence for specifically adaptive relations between pollen food value and the type of pollinator; and there was equally little or no evidence that flower visitors could assess or respond to the pollen's nutritional quality except perhaps in the crude sense of grain size.

In search of possible adaptive nutritional features, the focus has inevitably been on bees, whose main dietary input is pollen. Franchi et al. (1996) presented evidence that bees preferentially took pollens with low or negligible starch content, selecting floral species in which the stored starch had already been hydrolyzed to sugars; but this apparent preference may just reflect pollen longevity (see section 7, *The Longevity and Viability of Pollen*, below). There is only limited evidence in the older literature that bees can, or do deliberately, select more protein-rich pollens (Schmidt and Johnson 1984; van der Moezel et al. 1987). For example, honeybees with artificially reduced pollen quality in their hives could not make a switch to gathering more protein-rich pollens (Pernal and Currie 2001). However, S. Cook et al. (2003) showed that *Apis* preferentially took higher protein pollens when given artificial choices, and Robertson et al. (1999) recorded *Bombus* as choosing between patches of *Mimulus* flowers on the basis of pollen protein quality. Hanley et al. (2008) reported that in British herbaceous plants, the obligate insect-pollinated flowers, and especially bumblebee flowers, had a significantly higher pollen protein content; curiously, they found no relation between protein content and grain size, although their samples did not include the very large or very small pollens.

Bee growth and performance are certainly affected by the quality of their pollen diet, however. For *Osmia* eggs grown on ten different pollen types, only those pollens richest in protein supported development through to adulthood (M. Levin and Haydak 1957). Likewise, for bumblebees growth was faster on higher-protein pollens (Regali and Rasmont 1995), while for honeybees, longevity increased with pollen protein content (Schmidt et al. 1987). There is also evidence that bees may produce smaller offspring when feeding on protein-poor pollen (Roulston and Cane 2000).

Aside from bees, some cases are known where pollen content *appears* to be specifically coadapted with a particular pollinator. Most strikingly, there can be very different overall protein contents within genera that have different vectors for different species. For example, in the genus *Agave*, *A. americana* (pollinated nocturnally by bats) has 43% protein in pollen, whereas *A. schottiana* and *A. parviflorum* (both insect pollinated) have 8% and 10% protein, respectively. Taking this a stage further, in the bat-pollinated saguaro cactus (*Carnegiea*), protein is abundant in the pollen and is particularly rich in the two amino acids tyrosine and proline (Howell 1974). Bats not only have high energy demands per se (chapter 10) but also need unusually large amounts of tyrosine to make the collagen that forms their wing membranes (collagen is 80% tyrosine repeats), and proline is an important component of bat milk. At first sight, this observation indicates adaptive nutrition provided by pollen. However, proline has important functions for the plant too, being crucial in cell-wall formation during pollen tube growth; thus high proline and high overall protein levels are needed in larger flowers with longer styles, and for this reason alone these features can seem (spuriously) to be associated with large bat-pollinated flowers. In fact, some anemophilous pollens such as *Zea*, which have particularly rapid pollen tube growth, are substantially richer in proline than any of the *Agave* pollens (Solberg and Remedios 1980).

Does pollen ever have specific nutritional attractants for specific visitors? Although early examples have been discounted, the answer may still be yes; there is some evidence that bees like particular fatty acids found in some pollens (C.Y. Hopkins et al. 1969), some having an antibacterial effect and others improving larval nutrition (R. Manning 2001). Adaptive nutritive traits in pollen might be more likely where pollen is offered specifically for feeding, as in heteranthous plants. In some Lecythidaceae (fig. 2.5), the food pollen grains are known to be larger than the reproductive pollen (Mori et al. 1980). However, in *Commelina* some of the "food anthers" are deceptive and have little or no food reward, and some contain a milky fluid instead of pollen. Curiously, sterile pollen is reasonably common even in nonheteranthous but functionally female flowers (Cane 1993a), for example, *Actinidia*, *Rosa setigera*, and some *Solanum* spp. This case has usually been regarded as one of deception, but the sterile pollen can still be nutritionally useful to bees (especially its lipid-rich pollenkitt layer; e.g., Cane 1993b) and may even have as much amino acid content as the fertile pollen (Davies and Turner 2004).

Conversely, does pollen ever have repellents or toxins? We noted that some volatiles present may be deterrent to ants, but these are not really connected with pollen "nutrition." However, *Zigadenus* pollen can kill bees, due to certain alkaloids present (Hitchcock 1959, Tepedino 1981), yet the plant is bee pollinated. Other recorded pollen repellents or toxins are mannose (E. Crane 1977, 1978) and polyphenolics (Carisey and Bruce 1997), generally also found in the foliage of the same species (Detzel and Wink 1993).

6. Pollen Preferences?

The majority of pollen feeders are **polylectic**, that is, they take pollen from several different plants opportunistically. Only rarely are flower visitors classified as pollen specialists, that is, narrowly **oligolectic**, or even **monolectic** (elaborations on this terminology are considered in chapter 18, since it is primarily relevant to the pollen-gathering bees). Polylecty makes sense not least because so many of the relevant animals have an active adult life cycle longer than that of any one plant's flowering season. But at any one time, specialists will concentrate on a few related plant species, ignoring others of equal or greater value (note: they cannot reasonably be described as oligolectic *unless* alternative pollen sources are present and are being ignored). Some of these specialists have pollen-gathering structures that are seemingly linked with harvesting grains of particular size or shape.

While most flower visitors are described as nonspecialist, they may still exercise a good deal of *preference*. Many bees do behave in a narrowly oligolectic fashion in practice and can often show almost complete floral (hence pollen) constancy on any one day or at least on any one trip (Westerkamp 1996). Many, perhaps most, bees also have a degree of polylecty that is in part imposed by nectar accessibility rather than by pollen, given that it is quicker to collect both resources from the same flower. Even the somewhat specialist bees may take some novel pollens when they are offered in feeding trials (e.g., N. Williams 2003, using *Osmia* species). However, a broader study with specialist bees found that they failed to develop on nonhost pollens; pollen from Asteraceae and Ranunculaceae was unacceptable except to bees that were specialists on those groups, and some *Heriades* bees proved so fussy that they ceased nesting if offered only nonhost pollens (Praz et al. 2008a,b). These authors suggested that some pollens may possess protective properties that hamper digestion by bees, undermining the general view that pollen is "easy protein" for flower visitors.

We know relatively little about flower visitors' assessment of pollen quantity or quality or about their decision making on continuing to gather particular pollen rewards; but since these issues relate almost entirely to bees they are again reserved for chapter 18.

7. The Longevity and Viability of Pollen

In general terms, pollen can be a long-lived material under the right circumstances, since it can be gathered, stored, and sent around the world (in careful packaging) to plant breeders and gardeners (e.g., Shivanna and Johri 1985; Hanna and Towill 1995). Most pollens, if kept fairly dry and cold and in the dark (or more specifically without exposure to UV radiation), do persist for weeks or months. But grass pollens (i.e., most cereals) are notoriously short-lived, often remaining viable for only a few hours, and sometimes only minutes; for reasons explained above, they must be kept at higher humidities. Evidently there is much variation in artificial pollen management that is still to be explained.

However, in more natural conditions, pollen longevity and pollen viability are rather difficult terms to define or to measure (Dafni and Firmage 2000) and are used differently in different contexts. In some cases, viability is taken as merely the capacity to grow, but that of course depends on the conditions provided. Growth on a stigma or in vivo are more precisely definable, but ultimately, in terms of pollination biology, it is the ability to grow *and* to result in effective seed set that is critical, which in turn may depend upon a pollen type's competitive ability. Any or all of these terms may or may not be linked to such laboratory-assessed factors as germinability and stainability, so the whole field is fraught with uncertainty in assessing reported results from diverse sources. Nevertheless, in the applied contexts of horticulture and crop breeding, it is often crucial to assess the viability of pollen samples, and many methods have been developed for doing so (reviewed by Dafni 1992; Dafni and Firmage 2000). Some involve stains for the cytoplasm (since pollen lacking cytoplasm is inevitably dead) or tests of respiration, but these methods tend to overestimate viability since pollen with stainable respiring cytoplasm is not necessarily viable. Enzymic methods, where applied stains change color if subjected to breakdown by active pollen enzymes, are usually more reliable. Of these, the fluorochromatic reaction (FCR, using the dye fluorescein diacetate) is still the most widely used technique, in which treated pollen grains fluoresce in UV light if still functional; but here again there may be only limited correlation with real seed-setting viability. Better reliability comes from testing germination directly in a culture medium and detecting growing pollen tubes with suitable microscopy; or testing germination on natural stigmas, then digesting the stigma and observing pollen tubes in dyed preparations. However, these methods take significantly longer and even then cannot take account of possible interactions *between* pollen grains, both facilitatory and competitive (see below). Hence in practice, accurate assessment of pollen viability remains extremely difficult.

A great many factors affect pollen viability in natural conditions, as summarized in figure 7.13 (Dafni and Firmage 2000). These can be loosely divided into factors acting before dehiscence and factors acting after the pollen becomes exposed in the environment.

Some aspects of pollen longevity and viability are determined within the anther long before dehiscence. When forming within the locules, pollen grains tend to

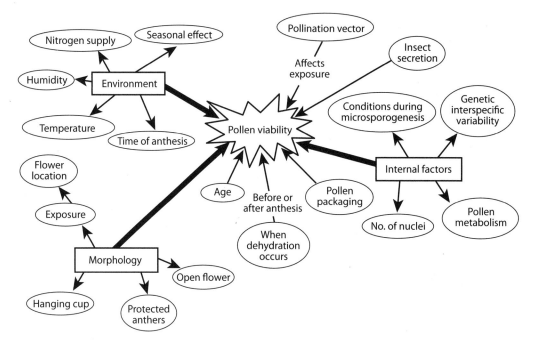

Figure 7.13 The many factors, extrinsic and intrinsic, affecting pollen viability. (Redrawn from Dafni and Firmage 2000.)

store carbohydrates gained from the anther tapetum layer as starch grains; but as the pollen matures, these are usually partially or wholly hydrolyzed into sugars prior to anther splitting. This hydrolysis can be crucial for pollen function, because it affects longevity and environmental resistance via partial dormancy (Pacini 2000). Where the starch is hydrolyzed and the sugar levels increased, the pollen grains become more resistant to gain and loss of water, since they can polymerize the sugars to change their own internal osmotic concentration. Furthermore, the resultant sucrose-rich pollen survives better than very starchy pollen, owing to the well-known and ubiquitous effects of sucrose, trehalose, and other sugars in stabilizing biological membranes during thermal or osmotic stress. Pollen that is produced in water-stressed plants tends to have lower sugar levels, poorer survival and viability, and a lower seed-siring capacity (H. Young and Stanton 1990a).

Pollen size and packaging, both of which are determined prior to dehiscence, also affect the dehydration rate, since this will always be lower where the surface-area-to-volume ratio is low. Large individual grains will dry out more slowly than tiny grains, and tetrads, polyads, and pollen masses will dry out even more slowly, prolonging their life. This is one reason why

orchid pollinia are among the longest-lived of all pollens, with records of 30 to 50 days not uncommon (Neiland and Wilcock 1995).

Once released from the anther, pollen is inevitably subject to environmental stresses, and its viability will decrease (J. Thomson and Thomson 1992; Dafni and Firmage 2000). This decline occurs more or less rapidly depending on intrinsic factors: size, sugar content, and perhaps partly the presence of certain protective proteins (S. Campbell and Close 1997). Pollens that are partially dehydrated and can regulate their water content, such as many Mediterranean species (e.g., *Cistus, Olea*) and xeric palms (e.g., *Chamaerops*) and some temperate plants (e.g., *Mercurialis*), are all reasonably long-lived. Those that remain starchy and cannot regulate their water loss or uptake are fragile, and their release is usually restricted to times and/or places where the humidity is high; examples include *Cucurbita, Zea,* and many common grasses, which as already noted may remain viable for only a few minutes even in greenhouse conditions. However, the rate of decline will also depend on external factors. Some of these still relate to the plant itself, including the degree of exposure or protection of the pollen in the anthers and the microclimatic amelioration afforded by other aspects of the plant's architecture that keep the humidity higher and reduce the temperature extremes. Other extrinsic factors relate more broadly to the environment: diel temperature regimes, radiation inputs, low humidity and drying winds. But there will also be contact with and collection by flower visitors to contend with, imposing physical and potentially chemical stresses (since animals may produce secretions that can damage pollen). Even without these influences, the most resistant pollen types do continue to decline in viability after release, the result of natural ageing that is under the influence of several different enzymatic pathways. Hence it is almost impossible to define a sharp cutoff between viable and useless pollen.

To what degree pollen longevity varies in a systematic or adaptive fashion is unclear, but a few interesting points have emerged:

1. There are apparent links between low pollen longevity and high pollen competitive ability (Harder and Wilson 1994; J. Thomson et al. 1994), which could select for a rapid dispensing schedule from the plant's point of view.

2. There are also associations between high pollen longevity and low or unreliable visitor activity (e.g., Beardsell et al. 1993); longer stigma receptivity is also involved, which again makes adaptive sense for the plant. Winter-flowering *Cyclamen* species in the Mediterranean have long flowering periods, long stigma receptivity periods, and high pollen longevity, all of these parameters recorded as 16–20 days (Swartz-Tachor 1999). Some evidence from *Banksia* species also supports such a link; in *B. menziesii* most of the pollen lasts less than 24 h, but pollen removal efficiency is high (Ramsey 1988), whereas in *B. spinula* pollen can often last 8–12 days, but its removal from the anthers is only partial even after several days (Vaughton and Ramsey 1991).

3. Short-lived pollen is often associated with greater selfing capacity (Franchi et al. 1996), whereas outcrossing is favored by longer-lived pollens. Again the dispensing pattern is relevant (Thomson and Thomson 1992), with the gradual release of successive small doses of pollen (either from individual anthers opening progressively or from many sequentially opening flowers in an inflorescence) being matched with shorter-lived pollen.

4. The most obviously useful association would be to have longer-lived pollen where the transit time between donor and recipient plants was high, as in highly dispersed populations. Here there are few direct tests, but the extremes may be exemplified by orchids, most of which are rare and widely dispersed and have very long-lived pollinia, and by anemophilous grasses, which commonly occur in dense swards and have among the shortest-lived of all pollens. In pursuit of this same idea, Dafni and Firmage (2000) calculated the mean pollen longevity for insect-pollinated plants as 8.5 days, and for windborne pollens as 21.5 h, although each had a very wide range.

Thus, the longevity of pollen could be adaptive for the plant. In principle it could also be an aspect of quality that is of importance to some pollen-feeders. However, even pollen that is effectively dead (nonviable) from the plant's point of view can still serve as useful food for a hoverfly, a beetle, or a bee and its larvae. There is, perhaps not surprisingly, little evidence that flower visitors can detect whether pollen is viable. But they may well be able to detect when it is very desiccated and therefore perhaps less valuable to them; in that sense, sucrose-rich pollens may have some benefits, explaining the findings of Franchi et al. (1996) mentioned earlier.

8. Pollen-Only Flowers

Some flowers—around 20,000 species—have become specialized as pollen-only providers, perhaps because of the selective pressures to meet the demand for extra pollen from bee visitors, as outlined in chapter 4. It appears that only predominantly bee-pollinated flowers have taken this route, and some of them combine the strategy with having buzz-pollinated anthers to effectively exclude all other pollen seekers. Familiar examples include the poppies (*Papaver* spp.), the rock-roses (some *Cistus* and most *Potentilla* and *Helianthemum*), and some of the flowers with feathery inflorescences that are dominated by attractive colored anthers, such as some *Solanum* and *Hypericum* or, in a more extreme form, meadow rue (*Thalictrum*; although this can also be wind pollinated). One problem, of course, is that this approach works only in a hermaphrodite flower—it would be impossible to attract visitors to separate female flowers if pollen were the only reward, except by some careful intersexual mimicry or similar subterfuge (chapter 23). Another problem is that the pollen has to be protected somehow from being entirely used as bee food; this may be partly solved by having two different kinds of anther (heteranthy), as discussed above in relation to buzz pollination.

Do pollen-only flowers genuinely produce more pollen? Vogel (1978a) suggested that this occurred, and Pellmyr (1986) found some evidence for it in *Cimicifuga*. Mione and Anderson (1992) compared pollen amounts relative to the ovule number (P/O ratios—see below) in various Solanaceae and found no link, but subsequently Cruden's larger analysis (2000) showed a clear and significant effect across a range of taxa with nectarless flowers.

9. Pollen Competition

It was an assumption of most early pollination biologists that there are usually more pollen grains deposited on a stigma than are needed to produce maximum seed set and that therefore pollen grains compete with each other once they have landed and begun to produce pollen tubes. Hence the first arrival could get a head start in the race to fertilize ovules, with differential growth rates of the tubes partly depending on the pollen's genetic constitution (Mulcahy et al. 1992), and potentially leading to differential fertilization success (Snow and Spira 1991, 1993). Some pollen sources do always grow faster in styles, whereas others have different speeds in different styles, so the growth of pollen tubes must involve the interaction of the style and the pollen tissues, probably at the level of allelic combinations, with good combinations being favored (as opposed to self-incompatibility, where some allelic combinations get blocked, as mentioned in chapter 3). Pollen allelopathy, which was referred to earlier, may also be involved in pollen competition.

An explicit test of the key assumptions about pollen competition was presented by Mulcahy et al. (1983) using *Geranium maculatum* flowers (these have large pollen grains and produce a convenient maximum of five seeds per flower). Figure 7.14A shows the number of grains per stigma through 2 hours of observation immediately after the stigmas became receptive and visitation began, and it reveals a rapid accumulation, such that a stigma will on average receive 50–60 grains and may have well over 100. The authors showed that

only approximately 30 grains were needed for maximum seed set (which would take only about one hour to be deposited), and that pollen tube growth from the stigma to the ovule required 30–120 minutes (fig. 7.14B). In this situation, given the highly variable growth rates for the pollen tubes, slow-growing grains from a first visit could readily be overtaken by faster growers from a second visit, and competition could therefore result both between grains deposited simultaneously (and quite probably from the same donor) and between those deposited at different times by different visitors (and probably from different donors). Arriving early and growing fast would therefore both be advantageous, especially because it is also known that growth rate is strongly heritable: gametes resulting from fast-growing grains give rise to fast-growing sporophytes (Ottaviano et al. 1980) and taller seedlings (Mulcahy 1974).

Several other factors might affect this competitive interaction. The amount of pollen present is probably critical in some plants, since a high pollen dose may delay or inhibit germination of grains (Cruzan 1990), and dispersed pollen grows faster than clumped pollen. There might also be an advantage for those grains that landed in the most favorable places on a stigma (perhaps landing in crevices rather than on top of the papillae). The environments in which the grains were formed could also matter, and pollen grains from plants with poor nutrient status might do less well (H. Young and Stanton 1990a; Stephenson et al. 1992), as might those that were formed at lower temperatures. A specific effect of plant age was shown in *Raphanus* (D. Marshall et al. 2007) whereby some pollen donors did better in young plants and others in older plants; these differences also correlated with the flowering time of the donors, a finding that raises the possibility of "temporal" specialization in the pollen donors.

Pollen competition generally leads to more heterozygosity and less selfing, although this may change with the age of the plant, and there may be more selfing in an old plant that has not yet had much fertilization success, perhaps because its recognition systems change subtly with age (Holsinger 1992).

However, *competition enhancement* can also occur, in that pollen growth may be stimulated by the presence of other grains on the stigma, especially if those grains came from different parents (Mulcahy et al. 1992). A very low dose of pollen may be too low to stimulate germination at all—in other words, pollen

A

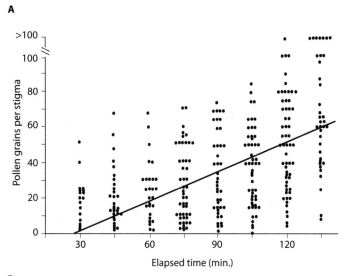

B

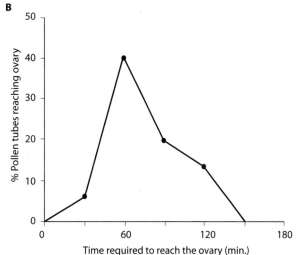

Figure 7.14 Aspects of pollen competition at stigmas: (A) number of grains per stigma accumulating through time as visitors arrive; (B) variation within pollen grains of the timing of tube growth, averaging around 60 min. (Redrawn from Mulcahy et al. 1983.)

grains could have a "mentoring effect" on each other (Lord and Russell 2002). This would imply that quite complex recognition systems are operating.

10. Conclusion: How Much Pollen Is Needed?

The efficiency of pollination is bound to be affected by the number of pollen grains released from a plant, whether wind- or animal-pollinated. In fact there is a key difference between pollen as a reward and the alternative of nectar as a reward in that pollen is nonre-

newable, whereas nectar can be (and usually is) topped up regularly. A flower has a fixed number of pollen grains, which is established before it even opens, but it can exert only limited control over how this number is portioned out to visitors. The number of grains a flower (and in turn a plant) produces can therefore be one of the most critical factors in its pollination strategy.

How many grains a plant "should" produce is inevitably a complex function of many selective forces, which have been reviewed by Cruden (2000). Key morphological variables include pollen size, the stigma surface area (SA), the pollen-bearing surface areas (PBA) of the visitor(s), and the ratio of these last two

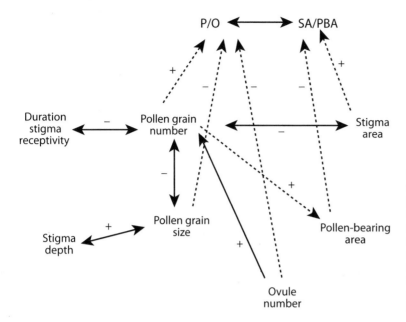

Figure 7.15 Interrelationships of pollen traits with other features of animal-pollinated plants. Note that the pollen-bearing area of the visitor, and the ovule number of the plant, should have positive relationships with grain number, whilst stigma area, stigma receptivity duration and pollen grain size have negative effects. The pollen/ovule ratio (P/O) can be a useful value, differing for various pollination modes (see fig 7.16). (Redrawn from Cruden 2000.)

factors (SA/PBA). Physiological factors, such as the duration of stigma receptivity, the longevity of pollen, and the nature and strength of incompatibility reactions, also come into play. The environment has an additional effect, since insolation can change pollen production within and between plants (e.g., Etterson and Galloway 2002). However, population parameters are perhaps the most critical, especially the average interplant distance, which is even more crucial for wind-pollinated plants.

Cruden reviewed many of these influences in terms of *pollen/ovule ratios* (P/O), which he proposed (1977) as a useful indicator of the likelihood of pollen effectively reaching a stigma. Pollen grain number (P) is generally inversely related to pollen size, reflecting an obvious trade-off in terms of investment (e.g., Vonhof and Harder 1995), although restricted pollen outlets on anthers or flowers may also have an effect (e.g., from buzz-pollinated porate anthers or from legume flowers that are triggered to release pollen from a narrow slit between the petals).

P and O (ovule number) are positively correlated across plant species, both within families (Small 1988, for Leguminosae; Lopez et al. 1999, for Fabaceae; Kirk 1993) and between families (Cruden and Miller-Ward 1981). P/O ratios can vary between populations within a species, but generally in a direction consistent with ecological constraints (e.g., in *Blandfordia*, Ramsey et al. 1994). P and O may also vary between indi-

viduals and between flowers on an individual, but usually there is either no relationship or a positive relationship between them at this level. Both P and O may change through a flowering season, although P/O often remains constant (e.g., in *Raphanus*, H. Young and Stanton 1990b). Hence, use of the ratio (P/O) removes some of the variation and allows more useful and meaningful comparisons between different pollination modes and different breeding systems (Cruden and Jensen 1979; Cruden and Lyon 1989; Cruden 1997), though as yet, phylogenetic effects are not fully taken into account, an omission that probably confounds some of the patterns seen.

Some of the interactions of floral traits found by Cruden's group are shown in figure 7.15. There are positive interactions between pollen size and the depth of the stigma, and negative links with pollen number, the latter also being negatively influenced by stigmatic area and the duration of receptivity. The authors also showed that the P/O ratio was negatively correlated with the SA/PBA ratio, so that more pollen is needed per ovule when the surface of the stigma is relatively small. For population-level factors, they found that the P/O ratios were

1. *higher in outcrossing plants* (usually 1,000–10,000, rarely 100,000) than in selfing plants (<1,000),
2. *lower in hermaphrodite plants* than for any other sexual system (unisexual, polygamous, etc.),

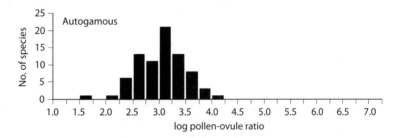

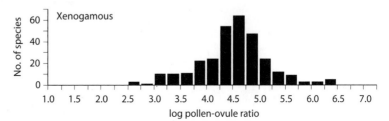

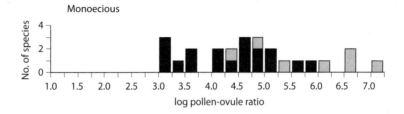

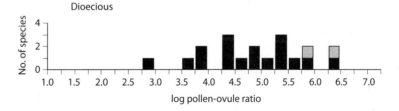

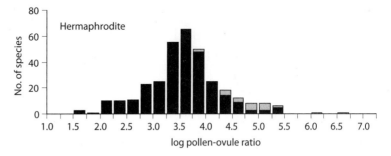

Figure 7.16 A summary of pollen/ovule ratios (P/O) compared between various plant breeding systems and pollination modes. The differences between the breeding systems are significant, with more pollen grains per ovule in xenogamous plants. Animal-pollinated hermaphrodite species (solid bars) have lower P/O ratios than hermaphrodite wind-pollinated species (hatched bars), a difference also reflected in monoecious and dioecious species. (Modified from Cruden 2000.)

3. lower in plants with packaged pollen,
4. higher where pollen is the only reward offered, and
5. *much higher in anemophilous plants* than in zoophilous plants.

They also found little effect of whether pollen presentation is primary or secondary. And perhaps most important, their analyses also failed to find any broad differences in the P/O ratio between plants pollinated by different kinds of animal (bird, bee, fly, etc.).

TABLE 7.6
Pollen/Ovule Ratios

Pollination type Genus or family	Pollen grain number	Pollen size	Pollen/ovule ratio
Animal-pollinated plants			
Medicago	small	large	low
Trigonella	larger	smaller	high
Amsinckia, Cryptantha, Cassia, Isopogon	large	tiny	high, > 25000
Cactaceae, Cucurbitaceae, Onagraceae	small	large	low, < 1000
With compound polyads or pollinia			
Orchidaceae	very high	tiny	very low, 1–20
Acacia	low	medium	very low, 1(–3)
Wind-pollinated plants			
Generally	large	small	very high, $> 10^6$
Ambrosia			highest, $\sim 10^{15}$
Water-pollinated plants			
Generally, freshwater	variable	variable	low, 100–1000
Potamogeton spp.			medium, 2–40,000
Generally, marine			low, 1,000–10,000

Source: Based on data and sources in Cruden 2000.

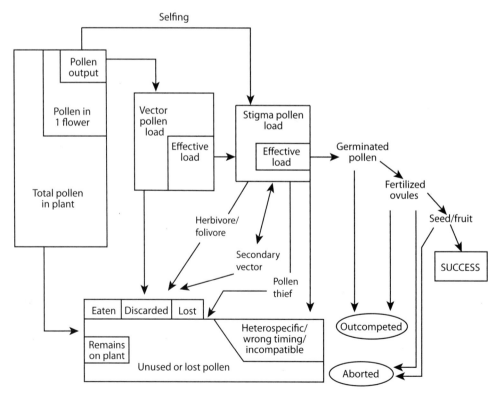

Figure 7.17 A schematic overview of all possible pollen fates.

Some of these broader findings, which are summarized in figure 7.16, are based on rather small samples from rather few families and have also been criticized as oversimplifications that are highly subject to local differences in mate availability and competition. But they do provide a useful overview of potential selective effects and a broad clue to how much pollen a flower should produce. Some examples of approximate P/O ratios are therefore given in table 7.6, together with other key morphological and ecological features, for a range of plant taxa. This information broadly supports the generalizations that Cruden (2000) set out, although plants with compound pollen or pollinia inevitably do not easily fit the other trends. A later analysis by Michalski and Durka (2009) using 107 species provided further support for point 1 above, although the relation was stronger for wind-pollinated plants and only marginal for zoophilous species.

Narrowing down our view from ecological processes, how much pollen a plant should produce is also influenced at a much finer scale by the processes taking place on and in the style of an individual flower. Pollen grains must first germinate and grow functional pollen tubes, and since grains may sometimes facilitate each other's growth, the number of pollen grains required on the stigma may be much more than one grain per available ovule (as in the *Geranium* example above). Pollen grains must also be sufficiently numerous to overcome interference from incompatible pollens. Pollen must then be sufficient to fertilize the available egg cells and produce seed; and there may be a minimal number of fertilized seeds required to produce a fruit that is worth retaining and investing in (e.g., Stephenson 1981). Several studies indicate a minimum requirement of four to six pollen grains per ovule to give the best seed set (e.g., Snow 1986; Murcia 1990). This also allows the higher-quality pollen grains to outcompete poor-quality pollen (providing a degree of **sexual selection**), since larger pollen loads and mixed paternity pollen loads tend to produce higher-quality progeny (e.g. Niesenbaum 1999).

As an overall conclusion from all these effects, then, the number of pollen grains carried as a result of a single visit to a donor flower should—from the plant's point of view—be enough to deposit sufficient grains to compete successfully on the first stigma encountered and to allow sufficient pollen carryover to compete for fertilization with a few more potential mates. A summary of all the possible outcomes for a pollen grain (*pollen fate*), which is shown in figure 7.17, indicates just how complicated any estimate of requisite grain number may be.

Chapter 8

REWARDS 2: THE BIOLOGY OF NECTAR

Nectar is the main secondary floral reward in an evolutionary sense, appearing on the scene probably in the late Cretaceous in angiosperms, later than pollen. But it has very often become the primary offering of a flower (chapter 4), thereby protecting the plant's investment in the reproductively useful pollen. As a commodity, nectar is easy for plants to produce and easy for animals to handle, its sugars being simple to metabolize and thus to use as a readily available fuel for an animal's activities. The general properties of nectar are usefully reviewed by de la Barrera and Nobel (2004) and in an edited volume (Nicolson, Nepi, and Pacini 2007).

Nectar supply is somewhat unpredictable in time and space, however, and dedicated nectar feeders therefore benefit from having a storage region in their gut (the crop, in most nectarivorous insects and in birds) that can hold large amounts of nectar when it is available and then release the stored fluid gradually to the absorptive gut beyond, so buffering the animal's body fluids from the osmotic shock of a sudden sugar load.

Nectar is not necessarily associated with flowers, since it also occurs in ferns, where it attracts honeybees, and can commonly be secreted extraflorally on the leaves, **petioles**, and stems of a substantial proportion of terrestrial plants. Thus nectar—in the broadest sense of a sweet sugary secretion from a plant—must have existed before angiosperms evolved. A range of animals would have become accustomed to search around early land plants for sugary exudates (perhaps also finding stigmatic exudates and the **honeydew** released by sap-sucking insects). Hence modern flowers in a sense have just taken over nectar production for a more specific function that enhances their reproductive success.

Two models for the evolution of nectar secretion have been proposed by de la Barrera and Nobel (2004). They note that a flower requires a substantial water input as it grows and expands its large surfaces, which would leave an excess of sugars from the **phloem** that could simply be excreted into the flower base to form the beginning of a nectary. Alternatively, a simpler "leaky phloem" model may be at work, relying mainly on the high hydrostatic pressure in the phloem to force a leak out into the incipient nectary across relatively weak and still expanding tissues. The authors propose that both models may now operate together as underlying factors in nectar production in modern flowers.

Nectar is extremely convenient for pollination biologists to study; it can be extracted from most flowers

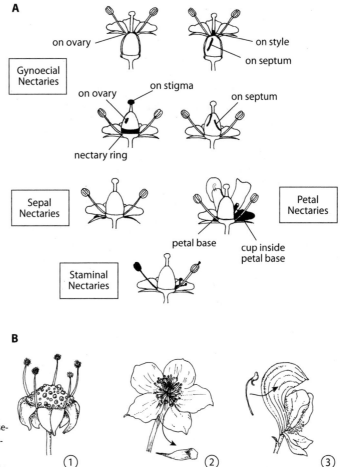

Figure 8.1 (A) Nectary types and sites in flowers (redrawn from Bernardello 2007). (B) Specific examples: (1) surface secretion in ivy (*Hedera*); (2) conical petal nectaries in *Helleborus*; (3) concealed petal nectary within the arching upper petal of monkshood (*Aconitum*) (redrawn from Barth 1985).

rather easily and can be measured, both in terms of volume (V) and concentration (C), rather easily, thus providing a direct assessment of the calories that a forager can gain from a single visit to a given flower. In this lies much of the attraction of flower-visiting behaviors for biologists studying optimal foraging and of pollination ecology as a focus for analysis of more general ecological principles.

1. Nectaries

Floral nectar is produced by nectaries, which tend to appear as yellowish-green surfaces in a flower and are often shining or glistening, sometimes with visible nectar droplets on their surface. They can be derived from various different floral tissues and therefore have no fixed shape or size or position in the flower (Pacini et al. 2003 and Bernardello 2007 provide reviews). Nectaries are usually more or less basal within the flower, so that a visitor must insert its tongue or head or whole body into the corolla and will reliably contact the anthers in passing and so pick up pollen. But essentially a nectary can be positioned wherever is appropriate in terms of manipulating the visitor's position, and there can be little doubt that nectaries have arisen convergently many times within the angiosperms. They certainly show no particular patterns associated with taxa or habitat or pollinator type.

There have been many different classification systems, but a simple scheme based on position is probably most useful. Some common types, as defined by location, are shown in figure 8.1, and include the following:

1. SEPAL NECTARY, which is on a sepal (or the equivalent on a calyx)

2. PETAL NECTARY, which is formed from part or all of a petal. This type is common in the buttercup family (Ranunculaceae), where a fold of petal tissue secretes the nectar; sometimes whole petals are modified as nectaries, and may be termed *nectar leaves*. For example, in many *Helleborus* species the petals become conspicuous curved tubular cups containing nectar, with the sepals forming the main visual attracting surfaces (plate 19B). In *Trollius* there are special spoon-shaped petals within the flower that are light yellow in color and contain the nectar. In monkshood flowers (*Aconitum*), which are visited almost exclusively by bumblebees, there are two petals curled into tubes internally to form nectaries, so the nectar is held right at the top inner part of the flower, where it cannot be reached by other visitors (fig. 8.1B).

Petal nectaries occur in other families too; in many *Lilium* species the nectaries are in grooves formed by the base of the petals, and in *L. martagon* (the Turk's cap lily) these grooves are covered in fine hairs. In a few orchids parts of the petals are enlarged into 'lollipop hairs' that secrete nectar fed on by short-tongued beetles and wasps (S. Johnson et al. 2007).

3. STAMINAL NECTARY, where all or part of a stamen is nectariferous, either the filament or the anthers or both being involved. Species of *Pulsatilla* (pasque flower) have nectaries at the base of their stamens. In pansies and violets (*Viola* spp.) there are fleshy "tails" on the base of the stamens that secrete the nectar, which is then stored in a short nectar spur. The pistil in violets is a hollow structure filled with mucilage, which oozes out when a bee moves the somewhat flexible style while feeding at the nectary; the mucilage helps stick some pollen to the bee. The style straightens again when the bee leaves, releasing the pressure, and excess mucilage and pollen are then drawn back in to the hollow.

4. GYNOECIAL NECTARY, where fluid is secreted from some part of the ovary tissue; this situation very often gives rise to a *disk nectary*, which is situated by the base of the stamens and ovary and often appears as a fleshy ring right around the inside of the flower in more advanced families of dicots and in most flowers with an elongate corolla tube. Alternatively there may be a *septal nectary*, which occurs on septa that separate the carpels of the ovary. This type is common in many monocot families. Less commonly there is a stigmatic or stylar nectary, or a small ring or disk nectary situated on top of the ovary.

It is also quite common to find that a nectary (especially a sepal or petal nectary) discharges into a special nectar spur (chapter 2), an elongate tube from the base of the flower that acts as a storage site for secreted nectar. Nectar spurs are recorded in at least 15 families and are ubiquitous in certain taxa. They can help to keep the secreted nectar at a stable composition that is unaffected by evaporation or by rainfall (but see section 4, *Nectar Volume*, below) and can protect against unwanted or ineffective visitors and invasive microorganisms. They are common in moth and hummingbird flowers, where a particularly long tongue is used to extract the reward, as well as in flowers visited by longer-tongued bees. They can be up to an extraordinary 40 cm long in a Madagascan orchid (fig. 8.2 and chapter 14).

Finally, although always produced in a nectary, nectar can sometimes be transferred via ducts to another position on the flower (*secondary nectar presentation*), several examples being given by Bernardello (2007).

2. Nectar Secretion

The tissue of a nectary has a typical ultrastructure (Fahn 1979; Nepi 2007) made up of three key components: a thin epidermis, often with trichomes or papillae or obvious glands, that overlies one or several layers of **parenchyma**, and a **vascular bundle** supplying phloem fluid (with or without **xylem**). All nectaries are ultimately supplied from phloem vessels, and nectar can be viewed as modified phloem sap. But the sap is probably altered into "pre-nectar" as it moves across the phloem **sieve plates** into the nectary tissue and modified again as it moves from parenchyma across to the epidermis and through epidermal membrane, cell wall, and cuticle. Figure 8.3 shows a diagrammatic summary.

Parenchymal Processes

The parenchymal cells of a nectary may form a single layer, or they may constitute a more complex structure

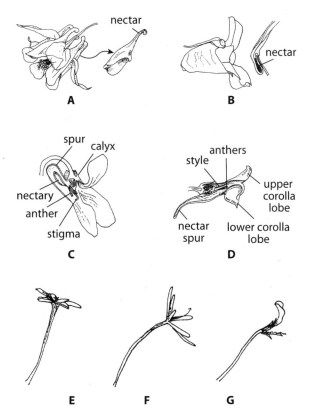

Figure 8.2 Elongate nectar spurs in (A) columbine (*Aquilegia*), also showing a single petal and its spur; (B) *Impatiens glandulifera*, with a cut-away view showing the spur; (C) *Viola*, with a large spur normally hidden at the rear (modified from Proctor and Yeo 1973); (D) toadflax (*Linaria*) with a spur below (modified from Proctor et al 1996); (E–G) three southern African examples, *Ixia, Geissorhiza*, and *Pelargonium* (parts (E–G) redrawn from Goldblatt and Manning 2006).

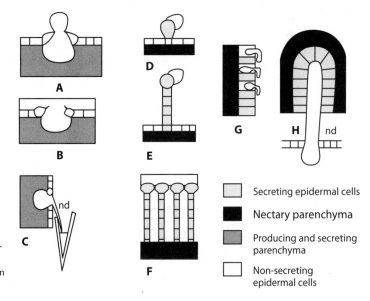

Figure 8.3 Common types of nectar secretion processes: (A–C) parenchyma cells producing nectar; (D–F) production by trichomes; (G, H) epidermal/epithelial cells. Processes in (A), (D), and (E) produce droplets; those in (B), (F), and (G) produce a continuous layer; process in (C) secretes into a nectary spur, and process in (H) into a nectary duct. (Modified from Pacini and Nepi 2007.)

with one layer of obviously specialized cells overlying a subnectary parenchyma of larger and less specialized cells. The cells of the secretory parenchyma are often multinucleate, and when actively secreting they have large vacuoles. In many plants they contain starch granules, which are broken down as nectar is secreted; but in other cases the supply of carbohydrate for the nectar comes directly from photosynthesis via the phloem, and starch is absent (Pacini et al. 2003). This clearly reflects the two distinct modes of nectar production in plants:

1. Sugars produced from *stored starch* within the nectary
2. Sugars produced directly from *photosynthesis*

The relative advantages of these two systems were compared by Pacini and Nepi (2007), as shown in table 8.1. However, there seem to be no systematic patterns determining which mode occurs in a given plant, and sometimes both may be present; any plant secreting nectar by night, or before the plants are in leaf, must be using stored starch reserves, but could use a photosynthetic source by day or later in the season. The only clear pattern is that where secretion is fast and copious, stored starch granules are involved and nectary tissue tends to be more substantial.

Nectar secretion from the parenchyma cells may be either *granulocrine* (vesicles from the cellular endo-plasmic reticulum fusing with the plasma membrane) or *eccrine* (individual molecules transported across the membrane). Where fast production and starch granules occur, transport out of the cell is normally eccrine (Nepi 2007).

A single clear model for how nectaries function is still lacking (Pacini and Nepi 2007). There is no doubt that secretion from the nectary tissue involves active (i.e., ATP-requiring) processes, and the secretion of sugar across the epidermis is probably linked to carrier molecules with subsequent osmosis, although the details remain unclear (Pais and Figueredo 1994; Nicolson 1998; Vesprini et al. 1999; Pacini and Nepi 2007) and processes may indeed vary substantially between plant species.

Epidermal Exudation

The epidermal layers of nectaries are highly variable, within and between species and even with environmental conditions. They may be smooth and flattened, but often they have trichomes on their surface where the secretion of nectar is concentrated; these trichomes may be tiny and unicellular or substantial, easily visible, multicellular structures. A number of ways for nectar to be exuded by the epidermis have been recorded:

TABLE 8.1
Advantages and Problems Associated with Two Different Nectary Types

	Advantages	Disadvantages
Sugars from photosynthate	Nectar can be produced for days or weeks.	Nectar cannot be produced at night or on leafless plants.
	Nectar removal by visitors may lead to further production.	Nectar production is slow and dependent on nectary size.
	Nectary may benefit plant even after pollination.	Nectar concentration cannot be very high.
Sugars from stored starch	Nectar can be made at any time.	No further nectar can be made after consumption.
	Production rate can be high and concentration can be high.	Less immediately responsive to the local environment.
	Resorption may be easier.	

Source: Modified from Pacini and Nepi 2007.

1. Leakage through the plasma membrane and cell wall and accumulation beneath the cuticle, which breaks or leaks under the pressure (a common route where trichomes are present)
2. Through microchannels or pores in the cuticle
3. Facilitated transport by the cell walls, which sometimes have special wall growths
4. Gradual death of the epidermal cells leaving nectar exposed
5. Via modified stomata that can no longer open and close.

The last of these seems to be the commonest mode, and modified stomata have been reported in the nectaries of many dicot families, although perhaps not in monocots. They may be held within deep depressions but are normally raised slightly above the surface when the nectary is actively secreting. There seems to be little control of their aperture (which does not respond to turgor or to ionic signals in the ways characteristic of leaf stomatal apertures), although nectary stomata are sometimes occluded by solid plugs of unknown material.

The initial outcome of epidermal processes is a series of microdrops of nectar on the epidermal surface, but the final result may be a thin layer of nectar spread across the surface, or a series of discrete droplets (potentially coalescing over time) above each stomata or each trichome, or even a clear nectary surface with the nectar rapidly moving off by gravity or capillarity into a spur or other storage site. Lopes et al. (2002) reported large nectary trichomes on the inner surfaces of petals of the hummingbird-pollinated *Lundia cordata*, from which nectar trickles down to the base of the corolla for storage.

Secretion and exudation of nectar normally continue for just a few hours but can persist for as long as a few days. The petal nectaries in early spring flowering species of *Helleborus* are unusual in that they produce nectar for more than a week, and in some species up to 20 days; in this genus secretion is *holocrine*, the entire epidermal cell discharging its contents into the nectary (Vesprini et al. 2008). But environmental effects can complicate the picture. Temperature and light intensity are bound to affect nectar production, since they affect the photosynthetic rate, which in turn will translate into stored starch. Nectar production may decrease at both very low and very high temperatures, and the limits will be set differently for different plants and in different habitats. Light effects (e.g., Boose 1997) are particularly problematic in forest understory sites, where reduced investment in nectar may be required. Water availability is also critical (M. Zimmerman and Pyke 1988a; Wyatt et al. 1992; Mitchell 2004), and plants tend to produce their highest nectar crop after a good rainfall. Severely water-stressed *Chamerion* plants dramatically reduced their nectar volume (Carroll et al. 2001), although without changing the nectar concentration. These kinds of effects, of course, make it more difficult to estimate the selective influences on nectar from potential pollinators.

3. Nectar Chemical Composition

Whatever its location in the flower, nectar is essentially a mixture of sugar and water, with concentrations of sugar normally ranging between about 10% and 75% (although lower values are often recorded, perhaps especially after rain or dew have affected the nectar). This composition provides a highly suitable fuel supply for almost any flower visitors.

Sugar Composition

The sugars in nectar are predominantly **glucose**, **fructose**, and **sucrose** in various proportions (at least eleven other carbohydrates have been identified on occasion, but these are inconsistent and always minor). How far the types of the three main sugars present in a given nectar are adaptive in relation to pollinator type or are determined by phylogeny or are a secondary consequence of flower morphology has been much debated.

Sucrose is a **disaccharide** formed by combining one glucose and one fructose; and since the **monosaccharide** (hexose) sugars glucose and fructose are not normally found in plant phloem, they must either be made from phloem-derived sucrose within the nectary cells or by sucrose breakdown (**inversion**) after the nectar has been secreted. The invertase enzymes transglucosidase and transfructosidase have been detected in nectars (e.g., Frey-Wissling et al. 1954), and invertases certainly also occur within nectar parenchyma. However, glucose and fructose do not necessarily occur in similar amounts (Percival 1961; H. Baker and Baker 1983), so that simple inversion cannot be the whole story; there is evidence that the hexoses are partly and differentially recycled through other bio-chemical pathways before being secreted into the nectary (Wenzler et al. 2008). In practice, out of 765 nectars tested by Baker and Baker (1983), 649 contained all three sugars and 78 contained only the hexoses (all other combinations occurring at low frequency except that no nectars contained only fructose or only sucrose and fructose).

To aid comparative assessments, Baker and Baker (1983) suggested a four-point scale that is produced by calculating the sucrose to hexose ratio, r, so that where equal amounts of sucrose (S), fructose (F), and glucose (G) are present the ratio is 0.33/0.66 (i.e., $r = 0.5$) and where sucrose is present in the same total amount as F + G, then $r = 1.0$. Hence they proposed the following:

$r > 0.99$ sucrose dominant

$r = 0.5$–0.99   sucrose rich

$r = 0.1$–0.5   hexose rich

$r < 0.1$ hexose dominant

From a review of sugar types in large numbers of plants, they suggested that the ratio r has adaptive value, and they drew several conclusions (table 8.2):

1. Certain families typically had sucrose-rich nectars (Lamiaceae, Ranunculaceae), and others typically had hexose-rich nectars (Brassicaceae, Asteraceae). (They noted that this finding raised some phylogenetic issues that might be interfering with their analysis of animal preferences, for example many of their butterfly flowers were in the family Asteraceae.)
2. Hummingbirds, hawkmoths, and long-tongued bees appeared to prefer sucrose-rich resources, while perching birds and short-tongued bees and flies appeared to prefer hexose-rich nectar.
3. New World microchiropteran bats preferred hexose-rich nectar, while Old World megachiropterans preferred more balanced sugar mixtures.

They also compared nectars within families and within genera in which more than one pollinator-type was recorded. Taking *Penstemon* as an example, they found that 7 hummingbird-pollinated species had sucrose-rich nectar, while 17 insect-pollinated species had hexose-rich nectar (and just one had sucrose-rich nectar); later work by J. Wilson et al. (2006) largely confirmed this, with bee-pollinated *Penstemon* scoring a mean r value of 0.315, and bird-pollinated species scoring 0.950.

This aspect of nectar composition could potentially act as an important structuring agent for visitor activities and might reflect coevolution between visitor needs or preferences and floral nectar output. Baker and Baker produced some interesting, and apparently adaptive, explanations from their findings:

1. SUCROSE-RICH NECTARS IN HUMMINGBIRD-FLOWERS occurred because they matched the birds' recorded preference in taste tests (H. Baker 1975). Sucrose preference might be related to the lower viscosity of a sucrose solution compared to an inverted hexose solution (of equivalent caloric value), making a sucrose-rich solution more manageable for a longer-tongued flower visitor. Later work has also indicated that hummingbirds tolerate sucrose particularly well because their intestinal walls contain a sucrase enzyme (del Rio 1990). However, the data on hummingbird preferences are often conflicting (e.g., Hainsworth and Wolf 1976; Stiles 1976; del Rio 1990; del Rio et al.1992), and when other options are not available, hummingbirds will freely take hexose-rich nectar from plants and also from nectar-rich introduced trees, such as eucalypts. The distinction between sucrose-rich hummingbird flowers and hexose-rich passerine flowers is not clear-cut, since many South African passerine flowers have sucrose-dominant nectar and the birds are readily able to digest this. In fact, birds' sugar-type preferences are probably strongly linked to the concentration used in the tests (Lotz and Schondube 2006), with many birds switching from hexose preference at low C to sucrose preference at higher C.

2. SUCROSE PREFERENCE IN LONGER-TONGUED BEES perhaps evolved in relation to viscosity. Bees were represented mainly by honeybees in early experimental work and did generally prefer sucrose, although subsequent work indicated little or no preference with other bees (Wells et al. 1992). Studies with Mediterranean floras indicated that bees and wasps of *all* tongue lengths slightly preferred high-sucrose nectars, while flies, beetles, and butterflies preferred lower-sucrose nectars (Petanidou 2005), results that are somewhat in conflict with those in table 8.2.

TABLE 8.2
Nectar Sugar Composition

| Main pollinator type | % with particular sugar ratios [sucrose / (glucose + fructose)] | | | | No. species tested |
	<0.1	0.1–0.5	0.5–0.99	>1.0	
Numbers	195	231	149	190	765
Percentages	25	30	19	25	
By family					
Brassicaceae	90	10	0	0	10
Asteraceae	51	42	6	2	52
Scrophulariaceae	9	43	21	26	53
Lamiaceae	0	33	24	43	21
Ranunculaceae	0	10	19	71	21
By pollinator type					
Insects					
Short-tongued bees	44	40	10	6	310
Flies	40	38	10	12	72
Wasps	11	39	22	28	18
Beetles	11	33	22	33	9
Perching moths	7	32	26	35	43
Butterflies	7	22	32	39	75
Long-tongued bees	6	37	24	33	203
Hawkmoths	3	13	31	53	61
Birds					
New World passerines	92	8	0	0	12
Other passerines	73	24	3	0	66
Hummingbird	0	13	32	55	140
Mammals					
New World bats	33	67	0	0	27
Old World bats	14	43	29	14	7
Nonflying mammals	0	40	40	20	5

Source: From H. Baker and Baker 1983.

3. BATS' PREFERENCE FOR HEXOSE-RICH NECTARS might have been developed from eating fleshy fruits that are also hexose rich; and H. Baker et al. (1998) further reported marked sugar composition differences for both nectars and fruits for tropical bird and bat flowers that supported this theory.

However, there are serious criticisms to be leveled at these findings. Several authors, but most cogently Nicolson and Thornburg (2007), pointed out that the ratios are misleading in themselves, because "sucrose-rich" nectar means any value above 33% sucrose rather than above 50% (using simple percent sucrose values would be better). Furthermore, these pollinator matches and sugar preferences may be somewhat illusory and have more to do with the type and degree of exposure of a nectary than with an adaptive match to pollinator type. Sucrose appears to remain dominant in deep, hidden nectaries (as preferred by longer-tongued bees, hummingbirds, and moths), while glucose and fructose levels become progressively higher in nectars that occur in shallow, exposed sites in flowers, inevitably the flowers that are visited by shorter-tongued visitors. This point was made implicitly by Percival (1961), based on her numerous analyses of nectars. Several related points can be made here:

1. First, all water-based solutions progressively lose water by evaporation in drier air, and a more exposed nectar will be evaporating faster because it is more exposed to dry air (and perhaps also because it is more easily irradiated and warmed up). Nectar containing only sucrose would reach a higher concentration by evaporative equilibration than would a hexose-rich nectar at the same humidity (Corbet, Willmer, et al. 1979). Thus hexose-rich nectars might be selected for (to remain more dilute and more manageable) in exposed flowers. Their presence in bat flowers may be particularly necessary to raise the osmolarity and so help draw water out of the nectary to allow a more dilute solution to be offered. In practice, bats offered choices showed no inherent preference between sugar types (Rodriguez-Pena et al. 2007).

2. The sucrose-to-hexose ratio is probably partly a reflection of easier postsecretion inversion in exposed nectar in full sunlight. Again, Petanidou (2005) confirmed differences across families in her own analysis of Mediterranean plants, with high-sucrose ratios in Lamiaceae, and low sucrose levels in Liliaceae and Apiaceae, findings that might support the inversion theory because the latter two families generally have flowers of more open/exposed architecture. Galetto and Bernardello (2003) found similar phylogenetic effects in Argentinian plants but no clear pollinator-related effects.

3. There is a possible link to the action of microbes in the nectar. Exposed nectar will be more prone to invasion by airborne fungal spores and bacteria, which could introduce invertase activity and give rise to hexose sugars. Interestingly Baker and Baker noted this effect themselves, reporting that in Costa Rica the nectar of *Inga vera* changed from sucrose-dominated just after anthesis in the afternoon ($r = 1.78$ at 1600h), when hummingbirds and bees visited, to a rather sour-smelling product some hours later with much more hexose ($r = 0.50$ at 1800h), when the flowers were visited by bats, the effect being "possibly under the influence of microorganisms," an effect also suggested by Salas Duran (unpublished, cited by H. Baker and Baker 1983) working with this species. C. Herrera et al. (2008) showed a strong link between microbial abun-

dance and nectar composition in three bumblebee-pollinated plants, with increasing fructose levels, and thus potential deleterious effects on pollinators, as the density of the yeast increased. The same group of researchers (C. Herrera et al. 2009) found that around 30%–40% of all nectars tested across three different plant communities contained yeasts, those with high yeast content being favored by bumblebees but negatively associated with solitary-bee visits. When interpreting such findings it must be borne in mind that visiting pollinators can themselves add microbes to nectar and so influence its subsequent composition (e.g., Canto et al. 2008).

4. Nectars vary more in field conditions than in greenhouses, again indicating that there are biotic influences from exposure and/or visitation, perhaps especially involving yeasts (Canto et al. 2007).

For all these reasons, the type of sugar present in nectar may be largely irrelevant, primarily a consequence of phylogeny and exposure of the nectar to the environment, and there may be rather little actual causal or adaptive link between sugar composition and pollinator type. That conclusion is also supported by observations that as some nectars age, they have a decreasing proportion of sucrose; and that there is extreme variation in nectar concentration and sugar composition even within a plant patch, or more dramatically, even within separate nectaries of a single plant, as is the case for hellebores (C. Herrera, Perez, et al. 2006). The counterobservation that in some plant species the sugar ratios are surprisingly constant between plants and across flower ages may be explained by the existence of defenses against microbial action in some nectars, as discussed in the section *Other Nectar Constituents and Properties*, below.

Data on this issue, although still widely quoted in the literature (and often supported by later workers where the same problems could have occurred), are therefore not generally included in later discussions of particular syndromes in chapters 12–18.

Amino Acids

Most nectars contain small but measurable quantities of amino acids, which often constituting the second most abundant constituent after sugars, and these have again been proposed to have adaptive significance for

TABLE 8.3
Amino Acid Levels in Nectars

Main pollinator type	Mean amino acid concentration ($\mu mol\ ml^{-1}$)
Carrion and dung flies	12.50
Butterflies	1.50
Most moths	1.06
Wasps	0.91
Bees	0.62
Most flies	0.56
Hawkmoths	0.54
Hummingbirds	0.45
Bats	0.31
Perching birds	0.26

Source: Modified from H. Baker and Baker 1973, 1986.

some pollinators (H. Baker and Baker 1973, 1986, 1990). The data shown in table 8.3 for particular pollinator types suggest high values of amino acids for carrion flies, moderate levels for most lepidopterans and bees, and very low levels for all birds and bats. These levels seem to match the particular animals' need for nitrogen, which is usually obtained from protein in the diet. Thus, within the insects, butterflies (who are normally taking *only* nectar as adults) generally obtain nectars with more amino acids than do bees (who can meet their nitrogen needs from pollen). For the vertebrates, both birds and bats take some insects in their diet as well as floral nectar, and bats also tend to eat some pollen, so neither taxon has much need for amino acids in the nectars they imbibe.

However, the significance of this apparent correlation of amino acid levels with pollinator needs has been doubted (e.g., Willmer 1980, Gottsberger et al. 1984, 1990), partly because of technical issues relating to collection and measurement. There are two key arguments, both linked to the underlying correlation of nectar volumes and concentrations (V and C) found in particular kinds of flowers with particular visitor types:

1. Flowers are rather easy to damage with the glass microcapillaries used when collecting nectar, which can leach amino acids from the plant tissues into the nectar; this damage is particularly likely when sampling low volumes of sticky nectar from flowers. Both natural visitation and repeated sampling by capillary can lead to reports of raised amino acid levels (Willmer 1980). Likewise, it is rather easy to displace pollen grains into nectar in a narrow tube when sampling, and just a few grains of *Hibiscus* pollen can greatly increase the recorded amino acid level in the flower's nectar (Gottsberger et al.1990).

2. Nectars with low sugar concentrations have lower viscosities and tend to produce relatively larger diffuse spots when deposited onto filter papers for **colorimetric** amino acid testing, even if constant volumes are collected and applied. Thus, the more dilute nectars from the long, tubular moth, bird, and bat flowers automatically tend to produce an apparently lower concentration of amino acids. Where sample volume and spot size were controlled for during collection, some of the differences reported in table 8.3 tended to disappear.

On this basis, there is little evidence that amino acid levels in nectar are important to, or even detected by, most flower visitors (e.g., Gardener et al. [2003] for tropical stingless bees, Leseigneur et al. [2007] and Nicolson [2007] for sunbirds). It should also be noted that the amino acid concentration in any given species is fairly variable anyway, markedly more so than amino acid composition (Gardener and Gillman 2001), and can change significantly with the age of the flower (Gottsberger et al. 1990). Moreover, recent analyses using reliable high-pressure liquid chromatography (HPLC) techniques (rather than colorimetry) have largely confirmed the absence of a relationship between amino acid levels and life form or flowering season, and very little effect of the taxon (e.g., Petanidou et al. 2006). These authors, working with Mediterranean plants, did find a relationship between phenylalanine and γ-aminobutyric acid (GABA) contents and pollinator preference, especially for longer-tongued bees; they proposed that in hot climates, where nectar is often very concentrated, amino acids may be more effective as phagostimulants than sugars, and phenylalanine is a known stimulant for honeybee feeding.

Are there any adaptive effects relating to amino acids in nectar? There are indeed some indications that certain butterflies do prefer higher amino acid contents, as provided by Erhardt and Rusterholz (1998) with *Inachis* and by Mevi-Schutz et al. (2003) with *Coenonympha*; in the latter case the level of preference varied with the quality of the larval food plant, and hence the need for nitrogen. A preference for

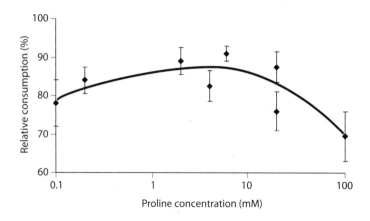

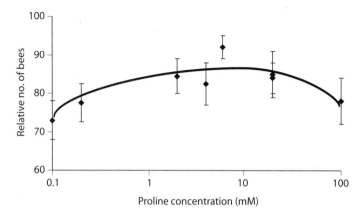

Figure 8.4 Preference for proline in nectar by bees, shown as relative consumption and as number of bees attracted, both peaking at around 4–10 mM proline. (Redrawn from Carter et al. 2006.)

amino-acid-rich nectars may enhance the butterflies' fecundity (Mevi-Schütz and Erhardt 2005) and support the longer lifespans in tropical species (Beck 2007) or both. Carter et al. (2006) also reported that honeybees exhibited a specific preference for proline in nectar (fig. 8.4), having identified proline as an especially dominant amino acid in *Nicotiana* nectar; they proposed that this may be related to insects' use of proline as a fuel during flight initiation. One other potentially adaptive effect occurs in extrafloral nectaries, where ant-defended plants sometimes offer nectar with higher amino acid levels and/or reduced sucrose, both of which are known to be preferred by ants (Heil et al. 2005; and chapter 25), thus helping to keep the ants away from the floral supplies.

Other Nectar Constituents and Properties

Nectar may contain very small amounts of other solutes in addition to sugars (Kearns and Inouye 1993),

and until quite recently these were thought to be unimportant in structuring visitor activities. Lipids, **antioxidants**, and traces of some deterrents found in plants—alkaloids, phenolics, **glycosides**, etc.—have all been detected and may occur in more than half of all floral nectars. In at least some cases, these defensive compounds may be induced in the nectar (as in other plant tissues, including corollas) by the actions of herbivores damaging the foliage (Euler and Baldwin 1996; S. Strauss et al. 1999; Adler et al. 2006), and the nectar can thus become toxic. However, toxicity can also be constitutive rather than induced (Adler 2000), and bees often die quite rapidly after visiting certain flowers (e.g., species of *Caesalpinia*, del Lama and Peruquetti 2006). Quite probably some such cases are merely coincidental, since the secondary components of nectar may just be an unavoidable side effect of leaf defense, being contaminants released as the plant delivers an extract of its own phloem across a leaky surface.

More recently, however, attention has focused on

the possible adaptive significance of some of these *defensive constituents* in nectar. Rhoades and Bergdahl (1981) had earlier suggested that secondary compounds in nectar could have two different roles:

1. A restricted role in deterring illegitimate floral visitors, following from an early suggestion by Janzen (1977) that nectar deterrents to certain animals (especially ants, which might otherwise interfere with pollination) could be common.
2. A broader role in helping to sort out pollinator types. Here an example may be the pollination of the milkweed *Pachycarpus grandiflorus* by *Hemipepsis* pompilid wasps (plate 20F); the open flower is visually cryptic with copious exposed nectar that appears to make specialized pollination unlikely, but Shuttleworth and Johnson (2006) found that bees were deterred by the nectar while the wasps were not.

The general issue of ant-deterrent nectar has largely been disproved (Feinsinger and Swarm 1978; Haber et al. 1981), because most nectars are highly attractive to ants; in practice, flowers have alternative mechanisms to keep ants out of or away from flowers (Guerrant and Fiedler 1981; chapters 24 and 25). However, some nectar constituents have proved to have a repellent or defensive role. For example, avocado nectar is disliked by honeybees perhaps because of its high mineral concentrations (Afik et al. 2007). Adler and Irwin (2005) manipulated the levels of the alkaloid gelsemine in wild jessamine plants (*Gelsemium sempervivens*), and thus in the nectar, and found reduced visitation by both pollinators and nectar robbers when alkaloid levels were raised, with pollen export reduced by up to a half, although pollen receipt and fruit set were unaffected. They proposed that secondary compounds in nectar have more costs than benefits (Irwin and Adler 2008). However, working with the same plant, Gegear et al. (2007) noted that the level of deterrence to pollinators was strongly dependent on the context and the availability of alternative nectar sources, so that the costs could be highly variable. Liu et al. (2007) similarly pointed out that the sugar concentration of the nectar served as part of the context, so that phenolics in nectar could be attractive to bees when C was 15%–35% but was deterrent outside this range. Intriguing results from Singaravelan et al. (2005) indicated that some secondary compounds (including caffeine and nicotine) had no deterrent effect for *Apis* when added to

nectars and might even have acted as stimulants that produced dependence or addiction.

Many of the defensive compounds known in nectars are volatiles and therefore overlap with the issue of *nectar scents*. These are rather unexplored, but some nectars do take up scent from adjacent floral tissues and some have novel scents of their own (Raguso 2004b). Defensive volatile scents in *Nicotiana* include nicotine (Kessler and Baldwin 2007), which repels moths in particular, thus reducing their foraging time and nectar removal, but because it also increases the number of visits, it may favor outcrossing and permit a lower nectar reward to be offered; and the nicotine also has a benefit in deterring nectar-thieving ants. It seems likely that attractively scented nectars could be helpful to both plant and pollinator, perhaps providing an honest signal of reward status to incoming visitors.

Nectar taste may be important to some bees and to butterflies and is likely to be largely due to amino acids (Gardener and Gillman 2002); this area is not well explored, although we noted above that phenylalanine is favored by bees, for which it is a known phagostimulant. Proline may also have a specific role in taste, since it can trigger the normal insect salt-receptor neurons and so initiate feeding. In certain aloes, floral nectar with a high phenolic content also produces a taste effect, being very bitter, and this appears to deter honeybees and sunbirds while being attractive (either by its taste or by its brown color, or both) to the bulbul birds that are the most effective pollinators (Johnson et al. 2006). Bitter nectar in *Nicotiana* plants apparently enforces modest drinking behavior on visiting hummingbirds, so that they move around more, whereas it has little effect on hawkmoths (Kessler et al. 2008). The presence of ammonia in *Lathraea* nectar could also be cited here as a case of deterrence by taste, keeping birds away from the flowers but being acceptable to bumblebees (Prŷs-Jones and Willmer 1992).

A further aspect of nectar composition that has only recently received scientific attention is that of *nectar color* and the pigments that produce it. Colored nectar is rare, but it can be found in at least 67 taxa worldwide, with a broad taxonomic spread, indicating at least 15 separate evolutionary origins (Hansen, Olesen, et al. 2007). Yellow, orange, and red are the most common colors found, but green and blue also occur, as do brown and almost black. Colored nectar appears rather often in island plants and with species growing at altitude, but its most obvious association is with vertebrate pollination. In the most famous case, colored nectar occurs in the Mauritian plant *Trochetia blackburniana*, where the brightly colored red nectar contrasts with the central white coronal color, and this coloring has now been firmly linked with lizard pollination (chapter 17; Hansen et al. 2006). The *Phelsuma* geckos of the island strongly prefer the colored nectar to clear nectars. The nectar color may act as an honest attractant signal of nectar availability, or alternatively as a deterrent signal to thieves (or of course both of these); it could also be that the pigments responsible are useful antimicrobials.

This brings us neatly back to the fact that some nectars do invert or ferment rather easily, so changing their sugar composition when fungal spores or other microbial contaminants enter the nectar. The nectar accumulates ethanol and may become intoxicating; it can make bees drunk (e.g., when foraging at some rhododendrons). Ehlers and Olesen (1997) reported fermentation in *Epipactis* orchids and noted it made wasps more sluggish, helpful from the plant's point of view as it prolongs their visits and decreases their chances of avoiding pollinia attachment. Since the wasps import some fungal spores and bacteria as they visit, they may in practice be responsible for making their own nectar resource alcoholic. Nectar yeast may also warm some winter flowers (Herrera and Pozo 2010). Strongly alcoholic nectar produced by yeast fermentation also occurs in the palm *Eugeissona tristis*, which emits a strong brewery odor. The fluid is drunk in substantial quantities by tree shrews and other small **arboreal** mammals, although they show no signs of intoxication (Wiens et al. 2008) and may well be effective pollinators (see chapter 17).

Microbial inhibitors in nectar may have an important role in preventing inversion or fermentation in many plants, and there is increasing evidence of such nectar components serving as part of the floral defense system (Carter and Thornburg 2004; Nicolson and Thornburg 2007). Proteins identified in nectar (*nectarins*), particularly from leeks (*Allium*) and tobacco (*Nicotiana*), have antibiotic effects. In tobacco they form part of a novel biochemical pathway that produces hydrogen peroxide. In the nectary tissue this compound is a potent microbial inhibitor, derived from an NADPH oxidase (Carter et al. 2007), and the whole system is termed the *nectar redox cycle*, which is now thought to be quite widespread in angiosperms.

A summary of possible effects of nectar constitu-

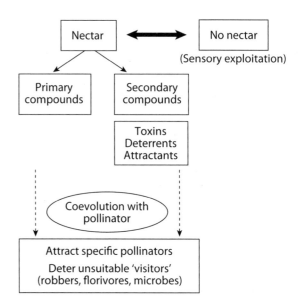

Figure 8.5 A summary of possible effects of nectar constituents on pollinators and on unsuitable flower visitors. (Modified from Brandenburg et al. 2009.)

ents on floral visitors is given in figure 8.5, which shows the interactions of primary and secondary compounds on attraction and deterrence of different visitors (Brandenburg et al. 2009).

4. Nectar Volume

The volume of nectar (V) is usually easy to measure from the length of a column of the fluid when it is drawn up into a volumetric microcapillary (e.g., before the same sample is dispensed onto a **refractometer** for concentration measurement, as discussed in the section *Nectar Concentration and Viscosity,* below). A single microcapillary will serve for most flowers (1, 5 or 10 µl), although for bird and bat flowers it is often necessary to use larger collection tubes or to resample several times until the flower is emptied. For very narrow flowers, such as composites, or for very small volumes or where the concentration is so high that the fluid will not rise into the capillary, volume measurements are much more difficult; some estimate can be obtained from centrifuging composite flowers (e.g., Tepedino and Parker 1982; Nuttman and Willmer 2008) or by simply counting the visible droplets on an open inflorescence (e.g., Apiaceae or Proteaceae), which can give a reasonably reliable value (S. Lloyd et al. 2002). Where volumes are very small, samples can also be collected with filter-paper wicks to allow total sugar estimates, but concentration values are then highly inaccurate. Washing out the flowers with water may be the best alternative (Morrant et al. 2009), but this method probably underestimates the amount of sugar present.

The nectar volume per flower is a major determinant of the interaction between a flower and any visitor, because of the essentially conflicting selective forces operating on plant and animal. The plant will benefit from keeping V as low as possible, minimizing its own costs (e.g., Harder et al. 2001) and ensuring that the visitor moves on to more flowers (taking with it the essential pollen that can ensure male success in the next plant generation). But V must not be so low that the visitor "gives up" on that floral species. In contrast, the visiting animal will benefit from visiting as few flowers as possible, each with as large a V as possible; ideally it would fill up all its gut capacity from one flower, remaining relatively sedentary, and incidentally transferring minimum pollen around the system.

Nectar volume should therefore be under strong selection and amenable to genetic dissection, although research in this area has only just begun (Mitchell 2004). Boose (1997) estimated a heritability of 0.64 for nectar production in a hummingbird-visited plant (*Epilobium canum*), although that number was not based on detailed genetic analysis. Work with *Petunia* species in controlled conditions (Galliot, Hoballah, et al. 2006) confirmed the presence of syndrome-related volume effects, with consistent tenfold differences (means of 1.2 μl in bee-pollinated *P. integrifolia* and of 13–23 μl in hawkmoth-pollinated *P. axillaris*, which has much longer corollas); this could be traced to four minor quantitative trait loci (QTL), which (using an intermediate cross) summatively accounted for approximately 30% of the differences in V. In *Mimulus*, however, control was linked to a single major QTL, whereby about 41% of the difference in mean volumes between two related species could be explained (Bradshaw et al. 1995), and in this genus hummingbirds (but not bees) were readily able to distinguish the nectar volume effects of artificial allele substitutions in hybrids (Schemske and Bradshaw 1999).

The "correct" volume of nectar dispensed per flower should therefore vary greatly according to its principle visitor/pollinator type (e.g., Faegri and van der Pijl 1979; Cruden et al. 1983). A typically bat-pollinated plant may need to produce several millilters per flower; quoted average figures include *Ochroma* 7.5 ml/flower, *Parkia* 5.7 ml/flower, and *Agave* 2.0 ml/flower, although only 100–200 μl was recorded in various bat flowers within the Gentianales (Wolff 2006), whereas Tschapka (2004) extracted samples varying from 100 μl up to 21 ml from various coflowering bat-visited species in Costa Rica. Hummingbird flowers also contain high volumes (though rarely as high as the upper end of the bat-related scale). Both these visitor types are relatively large and endothermic, so they have high energy demands (chapter 10). In contrast, a fly-pollinated generalist flower may dispense almost unmeasurable amounts, less than 0.05 μl. Bee flowers should come somewhere in the lower-middle range of this spectrum, commonly with a mean reward of perhaps 0.1–10 μl, reflecting both the fact that bees are generally somewhat larger and more expensive to fuel (given the demands of their partial endothermy) than are flies, and that bees are also gathering resources for their progeny as well as for their own individual needs. Figure 8.6 shows values for maximum volumes dispensed by flowers as collated by Opler (1983) for tropical plant species.

However, generalizations about nectar volume in the field lead immediately to two important problems, which bedevil the literature on nectar availability in flowers:

1. Nectar *secretion* patterns and *replenishment* (top-up) patterns vary greatly both between and within species, so that there is massive temporal component of variation in nectar rewards, which will be considered further in section 7, *Nectar as a Water Reward*, below.
2. At any given time, there is very considerable *variation* in nectar reward between flowers within a plant (e.g., Pleasants and Zimmerman 1983; Possingham 1988; Real and Rathcke 1988; Boose 1997; and see below and chapter 23 on "empty" flowers) and between different adjacent plants of the same species (e.g., Leiss and Klinkhamer 2005a; Leiss et al. 2009).

In fact, the distribution of recorded nectar volumes in a given species is usually left-skewed (fig. 8.7) and approximates to a negative exponential (G. Bell 1986; and references in F. Gilbert et al. 1991); some flowers are very rewarding and others nearby are almost empty.

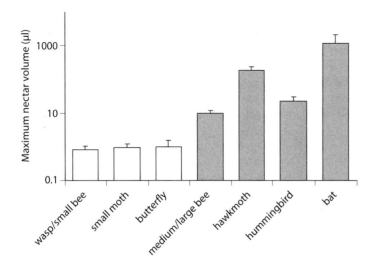

Figure 8.6 Pooled estimates of pollinator preferences for nectar volumes across many taxa, with SEs. Shaded bars represent endothermic groups. (Redrawn from Opler 1983.)

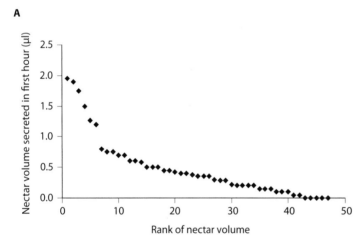

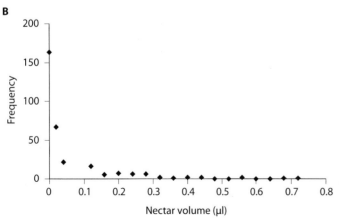

Figure 8.7 Volumes per flower in nectar production are generally strongly left-skewed: (A) across species, volumes in first hour of opening, for 47 temperate species ordered by rank; (B) within a species, volumes for standing crop of nectar in *Polemonium foliosissimum*. (Redrawn from F. Gilbert et al. 1991.)

This remains true even when visitation has been precluded, so it is not merely that some flowers have just been emptied by another forager. Nor is it necessarily the result of genetic differences, although there may be some spatial aggregation of higher-reward plants affecting genetic structures (Leiss et al. 2009). And it is not just environmentally determined, for while it is true that insolation can enhance nectar secretion (e.g., Rathcke 1992), the patterns of variation often show little relation to sun-shade patterns on a plant. Rather, extensive variation appears to be a common pattern, particularly where flowers are organized into complex inflorescences, indicating that it pays to have "blanks" and "bonanzas" among the flower population (Brink 1982; Thakar et al. 2003).

Why should this be so? It may be that it keeps a visitor somewhat uncertain but prepared to persist because it might find the occasional bonanza flower. However, there are also alternative explanations that rely on different visitor responses: Biernaskie et al. (2002) showed that high variance in nectar volume (and hence reward) could lead to foragers making fewer flower visits when working on a variable inflorescence, so again promoting outcrossing. Whatever the cause, this spatial variation markedly affects the temporal aspects of flower visitation. Biernaskie and Cartar (2004) also reported that the variability in nectar per flower was positively correlated with the size of the floral display, so that "risk-averse" foragers would often leave a highly floriferous patch after just a few visits, again ensuring that there was not too much intraplant (potentially geitonogamy-inducing) movement. Moreover, plants with low overall nectar production may have a selective advantage within a patch by benefitting from the enhanced local-pollinator attention; although at a population level, the higher-reward plants would have the advantage (Leiss and Klinkhamer 2005a). This issue of "empty" flowers is considered further in chapter 23 as a form of cheating by an individual plant.

There are further sources of variation in V that might be mentioned here. One is that some flowers can in effect *partition* their nectar, having an easily accessed nectar pool as well as a more difficult-to-access, somewhat hidden supply. A good example is the desert willow *Chilopsis* (Whitham 1977), where there is a pool of up to 8 µl of nectar overlying a further supply (about 1 µl) contained within five radiating grooves and taking up to ten times longer to extract. Whitham showed that bumblebees would just take the easily reached pool nectar when supplies were good but would stay and drain the dregs when nectar became scarce. Many flowers were visited twice, once by well-fed bumblebees and then later either by hungry, stressed bumblebees or by smaller, less energy-demanding solitary bees, for which 1 µl of nectar was a valuable reward. This kind of nectar partitioning may be more widespread than currently realized.

Another source of variation is that some flowers can at least partially *resorb* nectar that has not been gathered by the end of a day or a flower's lifetime or once the flower is pollinated. This aspect of nectar supply has been somewhat neglected but is now well-documented in many genera across both monocots and dicots (Nicolson 1995; Nepi et al. 2001; Pacini and Nepi 2007; Nepi and Stpiczynska 2008). Resorption can occur even from storage sites in nectar spurs (Stpiczynska 2003). **Nectar resorption** would most obviously pay off when a plant is unvisited for a substantial period or is reaching its natural age limit; but it also occurs after pollination in the case of at least a few cucurbits and orchids (Koopowitz and Marchant 1998; Luyt and Johnson 2002; Nepi and Stpiczynska 2007), and the sugars that are resorbed can then be diverted to the growing ovules, giving the plant a substantial resource recovery. And of course a resorption system also gives many plant species an ability to regulate the values of V and C on offer even more precisely.

A third kind of variation, as yet largely unexplored, is the effect on V from the supply side as the plant's cellular structure changes in responses to edaphic or environmental variation. For example, Petanidou et al. (2000) found that the stomatal opening on nectaries decreased in the summer, when water was scarce.

5. Nectar Concentration and Viscosity

Nectar *concentration* (C) is almost invariably measured via the related property of refractive index with handheld refractometers, giving a reading in percent (this is % w/w, or percent of sucrose equivalents, or how many grams of sucrose are present per 100 g of solution). The common range of C encountered in temperate flowers is 20%–50%, rising toward 70% in warm conditions on sunny days. Mean C values are usually rather less in flowers from the moist tropics (sometimes only 5%–10%), and they may be higher in hot arid conditions, with a maximum at about 75%.

Nectar can be taken up by animals either by capillary adhesion or by suction (Kingsolver and Daniel 1995; Krenn et al. 2005), and many nectarivores use a combination of both. Mouthparts that use adhesion and capillarity tend to be wettable apically and to perform visible licking or lapping motions; those that use suction tend to be much longer and remain motionless while feeding. These differences have important consequences in relation to nectar concentration.

In earlier literature, specific "mean" figures for preferred concentrations were sometimes quoted—for example, bats were said to prefer 17%, hummingbirds 21%, hawkmoths 27%, and bees 46%—but these kinds of precise averages are unhelpful, because they mask enormous variation. However, table 8.4 shows that different visitors *do* prefer particular concentrations, or at least concentration ranges, when they select among flowers. Note that this table gives values for nectar commonly found in flowers visited by particular animal types that reasonably match the so-called preferences, but it also gives values for the optimal concentration for fast intake as recorded in laboratory work, and these do not necessarily match up with apparent preferences; we will return to this discrepancy below. Nevertheless, generalizations as in the table do reflect one simple reality: animals with longer tongues cannot drink more concentrated nectars. This arises primarily as a consequence of *viscosity* effects—a concentrated sugar solution becomes too sticky and viscous to rise up or be sucked up a long thin tube. And the effect is not linear, since viscosity rises exponentially with C, so that a 60% sugar solution is roughly 28 times as viscous as a 20% solution. Concentrated nectars cannot travel fast enough up the long hollow tongue of a butterfly or bee, so the rate of gaining food is too low to be profitable. There is therefore an optimum concentration for a given tongue length and type to achieve the best rate for acquiring calories. At low concentrations this rate is limited by the low energy content of the nectar, and at high concentrations, it is limited by viscosity. The optimum will vary primarily with mouthpart morphology, but also with aspects of behavior (for example, a visitor that can hover will make only short individual-flower visits and so may need

Table 8.4
Visitor Preferences for Nectar and Optima for Nectar Uptake

	Mean in flowers (%)	Preferred concentration (%)	Optimum concentration (%)	Reference
Butterflies	17–40		30–40	May 1985, 1988, 1992; Hainsworth et al. 1991; Boggs 1988
Hawkmoths			34	Josens and Farina 2001
Short-tongued bees			45–60	Roubik and Buchmann 1984
Euglossine bees	38	34–42		Roubik et al. 1995; Borrell 2007
Stingless bees		35–45	60	Roubik and Buchmann 1984
Bumblebees	36	30–55*	50–65	Pyke and Waser 1981; Harder 1986
Honeybees		30–50*	60	Roubik and Buchmann 1984
Hummingbirds	20–25	20–25	40–45	Hainsworth and Wolf 1976; Cruden et al. 1983; Nicolson and Fleming 2003
Sunbirds	21		30	Beuchat and Nicolson, cited by Nicolson 2007
Honeyeaters			30–50	Mitchell and Paton 1990
Bats	15–20	60		H. Baker et al. 1998; Winter and Von Helversen 2001

*exceptionally variable, with colony status and environmental conditions.

quicker uptake than another visitor that is foraging by crawling over the flowers).

Hence there is an extensive literature on the tendency of long-tongued hummingbirds, bats, and lepidopterans to seek dilute sources, often 15%–30%, and on the preferences for higher nectar concentrations in other insects (usually >30%). Bees vary considerably in tongue length (chapter 18), but very roughly speaking, the longer-tongued species may feed better with 30%–50% nectars, and short-tongued ones may be able to cope with 45%–60% solutions. At the far extreme, flies (mostly with quite short lapping probosces rather than sucking tongues) can take nectar concentrations up to 65%–70% (and even beyond, to the point where the nectar is effectively crystalline, since they can then regurgitate some saliva into it and lap up the resultant fluid). Not surprisingly, then, there is a reasonably good relationship between flower type and nectar concentration, with lower values of C in the long tubular flowers preferred by long-tongued insects, and more concentrated nectars in the open flowers with exposed nectaries that are more often visited by flies and short-tongued bees.

These relationships, however, lead us into a classic "chicken and egg" dilemma, since nectar in open flowers will naturally tend to evaporate and become more concentrated through time (section 9, *Postsecretory Changes in Nectar: How Flowers Control Their Nectar and Their Pollinators*, below). We will later look more carefully at the concentration and timing of secretions and at methods to protect nectar from postsecretion effects to determine if the nectar is really produced or maintained at preferred and hence "adaptive" concentrations.

It should also be noted that nectar concentrations probably bear some phylogenetic imprint, so that within one taxon, such as the Gentianales (from a tropical site), the mean C values for bat, bird, and moth flowers were 14%–17% as expected, but the mean C for bee and fly flowers only rose to 25%–31% (Wolff 2006).

Occasionally flowers are able to offer different concentrations of nectar at the same time. This situation could readily arise where some was stored in a protective nectar spur while the rest was more exposed within the corolla. In fact, spurs themselves can provide a way of varying the nectar concentration, since Martins and Johnson (2007) reported gradients of C along their length in angraecoid orchids, from 1% at the mouth to around 20% at the tip; they proposed that this sets up a "sugar trail" that entices hawkmoths to probe deeper in to the spur, while at the same time deterring inappropriate short-tongued visitors. Other spurred flowers may have similar gradients (Martins, pers. comm.), but these are largely unexplored. More rarely, variation in the offered concentrations is achieved where a droplet of fluid hangs out of a flower and becomes more concentrated than the bulk nectar within (e.g., in *Anagyris* [Valtuena et al. 2007], where the main pollinators are perching birds requiring the more dilute supply).

In passing, note that sugar concentration is not the only determinant of viscosity in nectar, since there are several reports of unusually viscous and almost gel-like nectars in flowers with otherwise dilute nectar, possibly due to the incorporation of polymers. The rodent-pollinated lily *Massonia* (plate 28F) in southern Africa has nectar with a sugar concentration of around 20% but high viscosity (S. Johnson et al. 2001), and the authors speculated that this discourages robbing by insects while facilitating lapping by rodents. The Mediterranean *Phlomis fruticosa* also has an unusually high viscosity in one of its floral morphs (Petanidou and Smets 1995), perhaps helping to delay sucrose breakdown in these long-lived flowers. Some species of nocturnal bat-pollinated Musaceae, including *Musa*, produce gelatinous nectar too, with a mucilaginous secretion added to otherwise normal nectar (Fahn and Benouaiche 1979). Most strikingly, some *Combretum* species in Brazil secrete a ring of "nectar" as a sugary gum surrounding the style, which is plucked off and swallowed by passerine birds, and at the same time they also acquire some pollen on their heads (M. Sazima et al. 2001).

Returning to the question of optimal nectar concentrations for maximum ingestion rates, which are also shown in table 8.4, it is noticeable that these occur rather often at higher ranges of C than those of the nectar normally chosen in the field (e.g., Roubik and Buchmann 1984 for bees). The calculated optima depend on the feeding mechanism (capillary action or suction); details are considered for each pollinator group in later chapters. But most studies modeling nectarivore tongue actions (Josens and Farina 2001; Borrell 2007) have found that intake rate is determined largely by viscosity (fig. 8.8), and caloric intake is highest at unexpectedly high C values. There are a few caveats here: for example, hawkmoths and honeybees

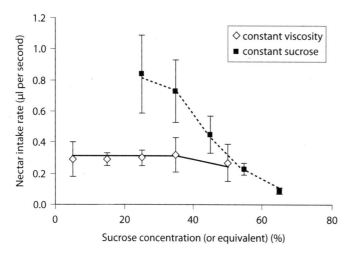

Figure 8.8 The relation between intake rate and nectar viscosity (here tested with *Euglossa* bees): when viscosity is kept constant but sugar varies, intake rate is constant, but when sugar is constant and viscosity varies the intake rate slows dramatically at high viscosities. (Redrawn from Borrell 2006.)

can vary their feeding rate in relation to perceived sweetness as well as viscosity. There is also a marked viscosity-temperature interaction (with an inverse proportionality), so that an animal feeding on warmer nectar, in tropical or semi-arid climates, will have fewer problems with overly viscous nectar (Heyneman 1983), and both the warmer nectar and the warmer body temperature can lead to significantly greater ingestion rates. But why are the optima recorded in laboratory studies higher anyway? Some hints will be found in later chapters on particular groups, but one factor is probably the provision of a separate plentiful water supply in most laboratory work, whereas in nature most foragers have to get their water supply from the flowers as well, and this may well lead them to select more dilute sources than optimal caloric input would dictate (section 7, *Nectar as a Water Reward*, will pick up on this point). But the discrepancies are also at least partly because, like so many other floral traits, nectar characteristics may be under opposing selective pressures from other agents, such as nonpollinating flower visitors or nectar thieves or even herbivores. In particular, nectar robbing (chapter 24) can vary with nectar concentration (Irwin, Adler, and Agrawal 2004), robbers having either similar or different preferences to pollinators (e.g., Gardener and Gillman 2002), so that some flowers may keep nectar dilute to deter robbers such as ants which tend to like nectar at 40%–50% (e.g., Paul and Roces 2003).

Few studies have definitively tested whether V or C is the key element in preferences. However, Cnaani et al. (2006) altered the nectar concentration and volume separately in artificial flowers and found that bumblebees responded more rapidly to changes in C, even when profitabilities were matched. Given choices of 40% and 13% the bees almost abandoned the dilute source regardless of its volume; when the choice was between 7 µl and 0.85 µl of the same concentration, the low-volume flowers still received about a quarter of the visits. The authors suggest that the bees may be able to assess C more quickly than V, or may need a smaller sample size for this assessment.

6. Nectar as a Sugar and Energy Reward

It is relatively straightforward to combine measurements of nectar V and C (again taking into account temporal and climatic effects) to get the mean total *sugar amount per flower* or to convert this to caloric reward for a particular species. There are three precautions that affect the value obtained: first, the % w/w values obtained from a refractometer should be converted to a w/v or molarity basis, either with tables or using an equation (Kearns and Inouye 1993; Dafni et al. 2005); second, a nectar dominated by hexoses has only 95% of the caloric value of a sucrose nectar (table 8.5); and third, there are possible corrections relating to the minor nonsugar constituents (Inouye et al. 1980). The corrections are never very great (Nicolson and Thornburg 2007) but are unwisely ignored in simplistic studies. For temperate plant species, the final

TABLE 8.5
Energetic Rewards of Nectar and Other Floral Rewards

Reward	Concentration (g sugar 100 g^{-1} solution)	Energy content (J g^{-1})	Reference
Glucose		15.7	
Sucrose		16.5	
Nectar 20%	20	3.3	
Nectar 40%	40	6.6	
Nectar 60%	60	9.9	
Pollen		21–25	Simpson and Neff 1983
		16–28	Roulston and Buchmann 2000
Oils		30–40	

calculation generally gives values in the range of milligrams of sugar per flower. Calculated sugar amounts can be particularly useful for comparisons between available floral resources in a habitat and for estimating foraging budgets for particular visitors.

As table 8.5 also shows, the energy values of nectars are low relative to the other main floral reward (pollen), although this is partially offset by the rather high percentage of indigestible material in pollen grains (chapter 7), so that weight for weight, a 60% sugar nectar may be about as rewarding as pollen in pure energy terms.

The use of nectar V and C converted to energy intake, as calories or joules, can give a direct handle on the "currency" that a particular flower-foraging visitor might be seeking to optimize, whether by maximizing its net rate of energy gain, by maximizing its energy efficiency (rewards minus costs), or by minimizing its risk by selecting the floral resource with the lowest variance. These themes will recur in particular for birds, bats, and bees in later chapters.

A complication arises in flowers that are dioecious or protandrous or protogynous and have different nectar offerings in each sexual state or phase. Female flowers generally have lower nectar production and lower energy reward than male flowers; early examples were reviewed by Dafni (1984), and details for *Rubus* were given by Ågren et al. (1986). In protandrous flowers, the male phases may produce more nectar than the female phases, with a difference of 65% being

found in a gesneriad (Carlson 2007), and with an associated lengthening of hummingbird visits in the male phase, which is indicative of strong sexual selection. But this male bias is not always true; for example, in squash plants (*Cucurbita*), female flowers had higher C and hence higher overall sugar reward (Nepi et al. 2001), and the same was true (with a threefold difference in production) in *Alstroemeria* flowers (Aizen and Basilio 1998). In some cases the male and female flowers (or floral phases) on a plant are in particular spatial relationships, which may be related to nectar rewards. For example, in *Delphinium* Waser (1983a) showed that the older female flowers at the base of the flower spike had twice the sugar (0.4 mg/flower) of the younger male ones with dehiscing anthers in the upper spike (0.2 mg/flower). Waser suggested this was selected for by the usual pattern of foraging of bees, which tend to start at the bottom of a spike and work upward, (fig. 3.14). By offering the "best" nectar flowers early on, the plant ensured that bees tended to keep working upward in hope of finding more of the bonanza flowers. But this pattern is not universal; for example, Arista et al. (1999) showed that flies tended to move downward on a spike, and hummingbirds seem to start at the top or the bottom with roughly equal frequency and move much less predictably (chapter 15). Furthermore, Kliber and Eckert (2004) showed a general trend of decreasing sugar reward in later flowers opening on spikes (usually those toward the top) and suggested that this was partly because the later (higher) flowers were more likely to get eaten by herbivores and were therefore less worthy of investing scarce resources in. Whatever the explanations of the patterns, though, they argue once again for a necessary caution in generalizing about floral rewards available to visitors from just a few flower measurements.

7. Nectar as a Water Reward

One further issue here, already hinted at, is that floral nectar, although commonly regarded merely as a sugar reward, can also serve as a "water" reward for some visitors, and perhaps especially for bees. This situation particularly arises in dry climates because of the potential water balance problems for a forager, which may affect its activity and flower choices. Many bees do not collect free water (*Apis* is an exception) but instead rely on inputs from nectar; in deserts there is unlikely to be standing water anyway.

Perhaps the clearest example of water as a constraint comes from work on *Chalicodoma*, a mason bee, which in Israel visits *Lotus* flowers. Willmer (1986) showed that the plant had to provide nectar that was sufficiently dilute to keep the bee in positive water balance while foraging in the hot dry climate. Measurements of the timing and duration of visits to different flowers, together with the assessment of body fluid concentrations before and after foraging trips, showed that the bees foraged to get a constant water input per trip from the nectar with relatively little regard to the sugar intake. Important and size-dependent water constraints were also reported for *Xylocopa* bees in desert conditions (Willmer 1988), while Olesen (1988) speculated that *Echium wildpretii* flowers in the arid Canary Islands served as a water resource for bees, providing nectar at 11%–12% concentration and being increasingly visited by honeybees when other water sources were withdrawn. Nectar that is more dilute than predicted from caloric optima may quite often be traced back to the forager's need for water.

However, in tropical climates there is much more likely to be a problem for any animal with a high energy throughput of getting *too much* water from their flower visits (since they also generate very significant amounts of metabolic water in flight) and so having to unload some of this excess. This is especially true for bigger tropical and subtropical bee species (Willmer and Stone 1997b), and even for large bumblebees in temperate habitats (Bertsch 1984); the need to eliminate water may provide a nonthermoregulatory explanation for their habits of regurgitating nectar onto the tongue to allow evaporation (**tongue-lashing**), as well as causing the constant dribble of dilute **excreta** that sometimes occurs in flight. To illustrate these conflicting water balance problems, carefully controlled studies by S. Roberts et al. (1998) showed that for flying *Centris pallida*, there was strongly positive water balance at lower ambient temperatures but that this switched into water deficit above 31°C as evaporative losses persisted and metabolic heat production (and therefore metabolic water production) decreased. Thus, for many larger bees in warmer climates, water balance must be assessed in some detail before predicting how it might impact on flower choice and nectar requirements. The same may apply to larger flower-visiting sphingid moths.

Nectarivorous birds and bats may have even bigger problems and must also manage their water balance in relation to nectar intake. Fleming and Nicolson (2003) showed that excretory regulation is the main associated control route in *Nectarinia* sunbirds, which almost shut off water excretion when feeding on high-concentration nectars and have extremely dilute cloacal fluid after the intake of nectars below 10%; on the latter diets the birds could not maintain their energy balance, with solute recovery becoming too expensive and heat loss being compromised. Sunbirds with relatively low water intakes appear to be able to regulate the water flux from gut to body and never really absorb some of the water, the fraction ingested being only around a third at the highest intake rates (McWhorter et al. 2003; Nicolson 2006). However, hummingbirds, often taking in several times their own body mass of nectar per day, do appear to absorb and somehow cope with all of the ingested water (McWhorter and del Rio 1999).

For these larger animals, some other explanation is needed as to why nectar in their chosen flowers seems to be more dilute than would be optimal. One suggestion, which will be pursued in chapter 15, is again that the dilute nectar is partly used to deter illegitimate visits from insects, who would find nectar at 15%–20% concentrations quite unmanageable in terms of their energy budgets.

8. Daily, Seasonal, and Phylogenetic Patterns of Nectar Production

Temporal patterns of nectar secretion vary widely, both in terms of the timing of onset and the rate of flow, between species and between individual flowers. The patterns may be partly determined genetically (Mitchell 2004; Leiss and Klinkhamer 2005b), with evidence of a circadian clock in some species (e.g., *Hoya carnosa* always secretes around midnight, Matile 2006), but they are certainly also modified by environmental factors in many species. Different species secrete fresh nectar at different times of day, at different times in relation to the opening of the flower, or both; often the peak flow coincides with maximum pollen availability, but sometimes with stigma receptivity (de la Barrera and Noble 2004). This variability also interacts with **floral longevity**, since short-lived flowers tend to produce nectar once only and often rather rapidly, but in longer-lived flowers there are commonly regular diurnal patterns of somewhat slower nectar production.

In fact a flower could offer its nectar in various different fashions:

1. Secrete a relatively fixed volume once only, on offer as the flower opens
2. Secrete at a steady rate for a fixed period, regardless of whether or not there is any depletion
3. Secrete continuously during the life of a flower, regardless of whether or not there is any depletion
4. Secrete with a fixed diurnal pattern, if a flower lasts more than one day; sometimes also linked with sexual phases (e.g., secrete early in the male phase, then cease, and secrete again one or two days later in the female phase)
5. Secrete in relation to visitation, to keep a standard volume available by regular "topping up" (sometimes termed **nectar homeostasis**); either for a fixed period or throughout the life of the flower. However, replenishment on this scale is far from cost-free because it can have negative effects on seed set (Ornelas and Lara 2009) by draining plant energy reserves.

Which of these options is occurring can be hard to assess, since secretion patterns may vary somewhat in relation to environmental conditions (section 1, *Nectaries*, above) and may also interact with the experimenter's chosen sampling regime. **Bagging** with fine nylon or cotton mesh is often used to prevent visitation in the hope that this removes a key variable and reveals what the flower does "naturally"; a single sample taken from an unvisited flower that has been bagged for some time (often 24 hours) is commonly quoted as being representative of the "maximum nectar reward" for that species. However, this can often be a much lower V than the summed volumes from the same flower sampled repeatedly at intervals over the same period, since the flower will refill after it has been artificially emptied and can in reality produce much more nectar than is seen in a single bagged sample. In other words, at least for longer-lived flowers, replenishment and nectar homeostasis are commonly occurring. This was shown for example in *Penstemon* flowers by Castellanos et al. (2002), in which full replenishment was achieved in 2–3 hours; and in *Moussonia* by Ornelas et al. (2007), where repeat-sampled flowers produced two to four times as much nectar as single samples had suggested.

Problems can also operate in reverse though: for example, Carpenter (1983) sampled a range of hummingbird-visited plants and showed that accumulated nectar from a flower bagged for 24 hours was, in some species, of similar volume to the totals summed from repeat samplings through a 24-hour period (in *Ipomopsis, Penstemon, Castilleja*) but for *Aquilegia* the 24-hour accumulated volume was much *higher* than the summed interval samples, perhaps due to corolla damage resulting from the sampling itself. Repeated sampling is always a tricky issue, even in larger bird-visited flowers. And bagging a flower carries its own problems, which will be discussed in the next section.

Even with these provisos, it is probable that all of the secretion patterns listed do occur and with no immediately obvious phylogenetic or biogeographic patterns. However, there may be some clear adaptive responses, at least for when secretion *starts*, that are related to habitat or to the principle flower visitors. For example, most tropical bat-visited flowers begin to secrete nectar at dusk, as do some moth-visited plants, presumably avoiding losses to diurnally active foragers (which was also discussed in relation to dehiscence and pollen availability in the last chapter). Many desert species secrete nectar at dawn and dusk (a crepuscular pattern, although this term is sometimes used just for dusk), which would allow these plants to invest water into nectar production at times when the ambient relative humidity is higher and the plant is less water stressed (Bertsch 1983; Nicolson 1993). It is also true, however, that many nondesert plants secrete nectar mainly at night, perhaps because doing so allows them to accumulate significant volumes in readiness for dawn and early-morning visitors. Many temperate flowers have **unimodal** patterns of secretion, where the main production of nectar is overnight or just predawn, and even plants that attract both nocturnal and diurnal visitors (such as the common milkweed *Asclepias syriaca*, which is visited by both bumblebees and moths) tend to secrete their nectar mostly at night, despite the fact that, for the milkweed, the nocturnal visitors are less effective at moving pollinaria between flowers (Morse and Fritz 1983).

It remains unclear to what extent temperature and humidity are direct cues for the timing of secretion of nectar; the observation that most plants show fairly specific diel peaks may indicate that **photoperiodic triggers** are more important. However, temperature can affect overall production in different fashions that are generally assumed to be adaptive. For example, in the typical Mediterranean genus *Thymus*, nectar V, C, and sugar content all increased with temperatures up to 38°C (as long as there was no water stress), whereas in the closely related Mediterranean but shade-loving *Ballota*, neither light nor temperature had significant effects (Petanidou and Smets 1996); and in cool-temperate flowers, such high temperatures often lead to effectively nectarless flowers. Jakobsen and Kristjansson (1994) showed adaptive intraspecific effects for *Trifolium*, where clones from Iceland had optimum nectar secretion at 10°C compared with Danish clones, which peaked at 18°C, the differences being strongly heritable. Such fine-tuning of the timing of secretion peaks by daily climatic variation may be more common than has yet been realized.

Bearing all this in mind, it would be instructive to see some examples of nectar availability patterns, and a chosen few are shown in figure 8.9. While taking actual values with some degree of skepticism, it should be very obvious that the time and pattern of production does vary greatly between species. But it also varies

1. for different plants within a species (fig. 8.9B),
2. for different morphs within a species,
3. for different populations,
4. within a plant measured on different days (fig. 8.9C),
5. for flowers in different positions on a plant or inflorescence,
6. with flower age,
7. with sexual phase,
8. with time through the flowering season, and
9. in relation to visitation.

Note that as yet we have only considered daily and seasonal variation; taking things to a larger time scale, it is also evident that there can be substantial differences in nectar production for the same species or floral community between years, with Carpenter (1983) reporting a tenfold difference at the same site in successive years for some tropical trees (fig. 8.9D).

9. Postsecretory Changes in Nectar: How Flowers Control Their Nectar and Their Pollinators

The discussion so far has implied that, despite many sources of variation, flowers merely have to supply nectar at a suitable C value for a particular visitor and could simply "select" their nectar concentration through the course of evolution according to the size and tongue-length of their main or preferred visitors. However, this is difficult to achieve in practice because most plants cannot merely secrete nectar at the "right" C and then leave it to be gathered, because of another major factor influencing nectar in varying habitats: that is, the **postsecretory changes** that occur with respect to nectar concentration (Corbet, Willmer, et al. 1979; Willmer 1983; Búrquez and Corbet 1991). Any fluid exposed to dry air will tend to equilibrate with it, essentially by losing water, and the loss will be more marked in hot arid habitats but less evident in the humid tropics. Thus, after secretion from the nectary tissues, nectar will almost always tend to dry out and gradually become more concentrated and more viscous over time. This was well known more than two hundred years ago, as evidenced by the writings of Krünitz in his massive encyclopedia (published 1773–1858) and those of Sprengel (1793). Krünitz believed such desiccation was a real problem for flowers because the concentrated thickened nectar would "cover and block the exits and impede fruit development."

It could be argued that the plant loses control of the situation once the nectar is released into the corolla and the environmental influences take over. However,

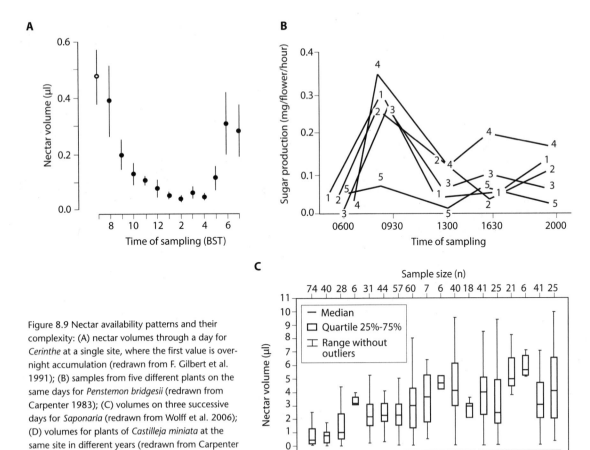

Figure 8.9 Nectar availability patterns and their complexity: (A) nectar volumes through a day for *Cerinthe* at a single site, where the first value is overnight accumulation (redrawn from F. Gilbert et al. 1991); (B) samples from five different plants on the same days for *Penstemon bridgesii* (redrawn from Carpenter 1983); (C) volumes on three successive days for *Saponaria* (redrawn from Wolff et al. 2006); (D) volumes for plants of *Castilleja miniata* at the same site in different years (redrawn from Carpenter 1983); (E) volumes from sites at different altitudes for *Penstemon bridgesii* (redrawn from Carpenter 1983); (F) volumes from flowers of *Satureja* growing on adjacent unburnt and recently burnt sites (redrawn from Potts et al. 2001).

matters are still not that simple. For example, in temperate habitats nectar is commonly secreted at perhaps 20%–40% sugar, but although it does then get more concentrated, especially on warmer and drier days, it does *not* change as much as expected if it were merely a free sugar solution in the open air (fig. 8.10). Furthermore, whereas nectar in open cup-shaped flowers equilibrates fairly rapidly and becomes viscous, concentrated, or even crystalline (>75% sugar), rendering it useless to many flower visitors, nectar that is initially similar but in an enclosed elongate corolla may remain relatively dilute throughout the day so that it is readily drawn up by an animal with a longer tongue.

The explanations of these effects were provided by

Corbet, Unwin, et al. (1979) and Corbet, Willmer, et al. (1979), who showed that tubular flowers have their own internal microclimate of more humid air, which slows the rate of equilibration with ambient conditions and so helps the flowers control their own nectar. Long, enclosed flowers have higher internal relative humidity and may also be cooler or warmer than the ambient air at different times of day—overall, they have a more stable microclimate (fig. 8.11). These authors specified a number of floral features that contribute to these effects: elongate corollas, relatively enclosed corollas, hairs within the corolla just above the nectary, constrictions of the corolla wall beyond the nectary, and even lipid layers over the nectar as

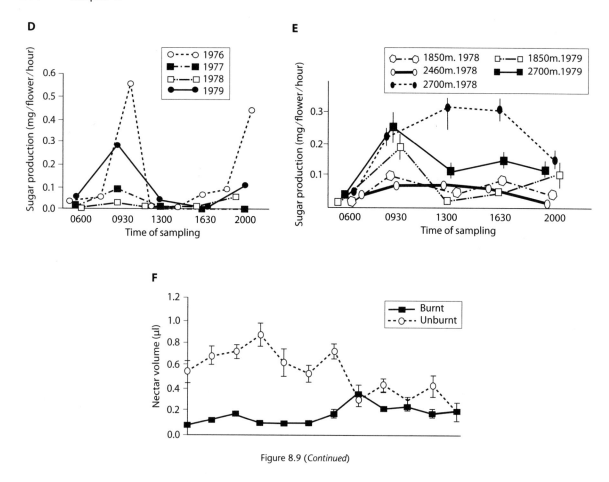

Figure 8.9 (*Continued*)

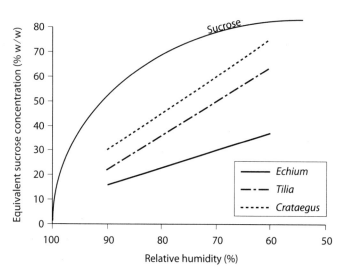

Figure 8.10 Postsecretory nectar changes in relation to environmental relative humidity: the curve represents the expected concentrations for solutions of pure sucrose, exposed freely to air of the given humidity. But nectar within flowers changes much less, and the relative constancy of the nectar depends on the protection afforded by the floral shape—nectar is highly protected in tubular *Echium*, where the concentration rises only from about 15% to about 35%, but nectar is relatively exposed in the open, bowl-shaped flowers of *Crataegus,* where concentrations rises to more than 65% in the drier air. (Modified from Corbet et al. 1979.)

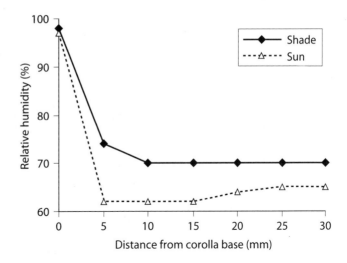

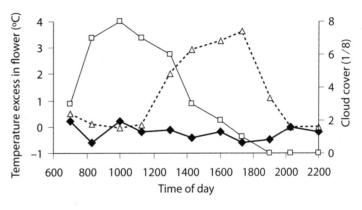

Figure 8.11 Nectar and microclimate within temperate flowers: gradients in humidity within *Echium vulgare* in sun and shade show that these tubular flowers retain their own more humid microclimate, which in turn helps keep their nectar more stable (see fig. 8.10); and temperature differentials within *Crataegus*, varying with cloud cover (open squares) such that even in these more open flowers, a sunlit corolla can be nearly 4°C above ambient. (Modified from Corbet et al. 1979.)

protective "waterproofing" to delay evaporative loss. Hence a classic tubular bee flower, such as *Echium vulgare*, can keep its nectar below 50% C (suitable for most bee visitors) through most of a temperate summer day, while nectar in an adjacent cup-shaped flower, such as *Crataegus*, may reach greater than 60% between 1100h and 1500h, and the nectar on an open umbelliferous flower, such as *Heracleum*, evaporates fairly freely and can become almost crystalline (hence only suitable for flies and a few beetles) for much of that same day (Corbet, Unwin, et al. 1979; Willmer 1983). Of course on cool, overcast, or wet days, these differential effects will be very much reduced.

Even for nontubular flowers there are features that can help protect the nectar from postsecretory change. Pendant flowers can trap rising warm moist air from the ground below and can act as umbrellas safeguarding against dilution by rain. Many brush blossoms

probably benefit from a microclimate within and below the mass of anthers that helps to keep nectar concentration more stable. Hairy or furry calyces or sepals may protect inflorescences from rain, as shown for some *Protea* flowers (Nicolson and Thornburg 2007).

Above all, then, it is the structure and *shape* of flowers that is crucial, not just for reasons of morphological fit with visitors but also in this extra respect of helping to determine the flowers' temporal pattern of nectar concentration. Concealed nectar may require particular behaviors from the visitor, but it also tends to remain at a C best suited to the preferred visitor. This will apply irrespective of initial nectar C at the time of secretion and will be to some degree coadapted with the climatic regime. It follows that the timing of initial secretion in a flower will in turn affect the concentration of the nectar encountered by a visiting bee at different times of day and to differing extents in different kinds of flower.

To complete this story, it should be noted that sometimes in tropical habitats the ambient humidity is so high that secreted nectar would actually tend to become more *dilute* by equilibration, potentially making it too dilute to be profitable even for hummingbirds. Here the flower can keep a *drier* microclimate inside itself in equilibrium with its nectar, which therefore does stay more concentrated than the external humidity would predict, as shown for *Justicia* (Corbet and Willmer 1981). This effect is probably especially useful for hummingbird-pollinated flowers, because these animals strongly prefer nectar at around 15%–20% sugar (chapter 15).

10. Problems in Measuring and Quantifying Nectar

From the above discussion, it will be evident that although measurements of nectar V and C are technically simple, there are many complicating factors in knowing what in practice a visitor will receive when it visits a single flower:

1. The timing of nectar secretion, its possible modification by temperature and drought stress, and its almost inevitable interaction with postsecretory changes related to the weather will have to be taken into account when trying to assess the average nectar reward that a flower offers. This requires careful sampling at different times and in different weather conditions.

2. There may be substantial variation in nectar reward even at the level of flowers on different parts of the same plant; nectar profiles from the sunlit and shady sides, or the outer flowers and middle flowers, may differ quite considerably (e.g., Stone 1995).

3. Whatever their secretion concentration, their initial secretion rate or volume, or their time of peak secretion, most flowers will have a different nectar reward on offer depending on their recent visitation history (i.e., how recently or how often their nectar has been entirely removed or partially depleted by a visitor) and on their ability to refill themselves after being visited.

4. Sampling from bagged flowers would in theory remove this visitation-related problem, but it raises other difficulties that are often unacknowledged. For one thing, bagging in itself can affect the floral microclimate, perhaps making it warmer and more humid, and thus can alter the nectar. Hence nectar V or C measured from a bagged flower does not reflect the reality of what a visitor will typically get. And sampling from bagged unvisited flowers may not even tell us what the plant can do "at best" without any visitors, because a flower may stop secreting once it is "full" even though it could go on producing more nectar if it were being regularly depleted. Furthermore, visitation is a natural part of the plant's experience, and what any new visitor receives is always and naturally affected by the plant's previous visitation history. Hence nectar readings taken from bagged flowers have little place in pollination ecology, although they may help plant physiologists understand what is possible.

5. Knowledge of refilling patterns and recorded visit rates are still not enough to predict available nectar volumes; there must also be information on the mean volume a given visitor can or will extract, which is likely to be different for different visitors. These volumes may themselves vary with nectar C (hence with time of day and weather conditions) and with the proportions of other rewarding flowers available nearby, etc.; so they must really be measured on site and not extrapolated from previous studies.

6. Scientific sampling of nectar for experimental assessment may in itself promote subsequent further production (using the homeostatic mechanism); but removal of nectar could at least in theory also trigger reduced secretion in plants where just one or two visits normally effect pollination. Given the evidence of postpollination resorption cited above, it would seem possible that the two main floral rewards can interact, such that pollen removal or deposition could affect nectar production, probably via the various plant signaling systems that mediate postpollination abscission or color change. These kinds of effect have rarely been looked for (e.g., Aizen and Basilio 1998) but would be appropriate adaptive responses for plants where resources were scarce and could constitute an important factor in structuring animal visits.

7. The sugar and amino acid composition of nectar can be directly affected by visitor behavior and

frequency as we have seen in earlier sections, via tissue damage and/or by introduction of microbes.

For all these reasons, what has been measured from a given flower, or a few flowers, at any one time can be seriously misleading in terms of assessing the natural standing crop of nectar, (the average reward actually encountered by a visitor from flowers unprotected by bags and undergoing normal visitation [Corbet 2003]). The standing crop will be highly correlated with the plant's nectar production when visitation is very low, but when visitors are plentiful, it may be only very slightly related to the supply side, that is, to the V and C of nectar that the plant is secreting. Thus a hummingbird plant being continuously harvested by visitors can frustratingly yield almost no nectar from most flowers to a probing microcapillary.

Not surprisingly, there is a considerable literature of inadequate data on the supposed "typical" nectar V and C values, and diurnal secretion patterns, for particular plant species; one-off measurements of a flower's nectar can never be trusted. Even the best published data cannot be relied upon when seeking understanding of pollinator activity patterns in relation to floral nectar reward; each study must include its own inventory of the rewards available at the appropriate time(s) and place(s). Many flowers must be sampled from many plants throughout the diurnal cycle and in different weather conditions to give a reasonably accurate picture of what a visiting animal can expect to find in a typical flower at a particular foraging time. And given that foraging animals moderate their behavior and flower choices in relation to what they experience (which will be discussed at length in chapter 21), it is also likely that the standing crop encountered by a foraging animal that is selecting its flowers (on whatever criteria) will be different from that measured by a careful pollination ecologist who is randomly selecting flowers.

In summary, it is exceptionally hard to give average figures for nectar reward for any particular flower species or to say what the natural standing crop of nectar for that species is, because

1. nectar changes with the environment, to an extent that depends on floral morphology,
2. nectar changes with the rate and type of visitation, and
3. sampling the nectar and the frequency of repeat sampling affect its subsequent production (and/or

resorption)—in effect, a typical "uncertainty principle" of measurement is in operation.

11. The Costs of Gathering Nectar

Flower visitors have to suck or lap up liquid nectar, and the main cost to any visitor is the handling time associated with this (actual digestion being very easy). Handling costs depend on

1. the volume of nectar ingested,
2. its viscosity and hence the time needed for capillary rise or the force needed to suck, which is sometimes altered by the time spent adding saliva to concentrated nectars,
3. its depth within the flower, in relation to the animal's tongue length, and
4. the individual visitors' experience and learning ability.

Most assessments of nectar-feeding animals have shown that their behavior appears to maximize the net intake rate, although slightly different rules may operate for bees (see later chapters). Thus, unlike pollen feeders, most nectar feeders are critically affected by time costs in their overall budget (chapter 10).

A flower visitor could be structuring visitation on the basis of nectar either before a flower is visited (by evaluation of what is on offer) or afterward (by evaluation of what has been received). Can flower visitors assess nectar reward at a distance? If they could, this ability should affect their activity patterns by making foraging more efficient and potentially reducing the ratio of feeding time to trip length. The evidence on this point has been seen as somewhat equivocal (Corbet et al. 1984). However, F. Gilbert et al. (1991) demonstrated that *Anthophora* bees could distinguish between individual *Cerinthe* flowers on the basis of their nectar secretion rate, perhaps using flower age as a cue; and some flowers do signal their age and altered reward status rather clearly by such cues as color change (chapter 5). Direct assessment may also be possible, either where a particular nectar is strongly scented so that its abundance could be assessed at some distance from the flower, or where the corolla tube is relatively transparent and the nectar meniscus can be detected visually; there is some indication of this occurring in honeybees (Goulson et al. 2001).

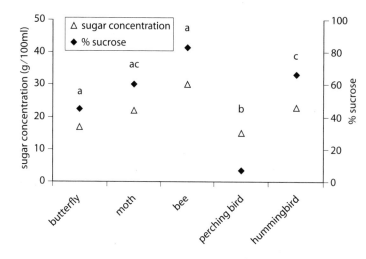

Figure 8.12 Mean values for nectar concentrations and compositions (% sucrose) in Acanthaceae visited by different pollinator groups, where letters designate significantly different groups. Flowers visited by perching birds have low overall concentrations and particularly low sucrose levels. (Redrawn from Schmidt-Lebuhn et al. 2007.)

TABLE 8.6
Nectar Parameters Correlated with Floral Traits and Pollinators in Four Groups of the Loasaceae

	Nectar volume per flower (µl)	Nectar concentration (%)	Morphology	Main pollinator
Group I	1.5–3.5	40–80	small, white, radial	short-tongued bees
Group II	9–14	40–60	orange, bowl or bell	long-tongued bees, hummingbirds
Group III	40–100	30–40	orange, bowl or bell	hummingbirds
Group IV	80–90	10–15	various, geoflorous	small non-flying mammals

Source: Modified from Ackermann and Weigend 2006.

12. Overview

Nectar is a crucial factor in determining the interactions of flowers and their visitors, and not surprisingly, many studies show good, apparently adaptive, links between nectar characteristics and visitor or pollinator needs. These connections persist even when taxonomy is factored out; for example, a recent analysis with the Acanthaceae showed that the link between nectar and pollinator was much more important than phylogenetic constraint (Schmidt-Lebuhn et al. 2007), and figure 8.12 shows the composition and concentration effects for sugars that these authors found compared with pollinator classes. Similarly Ackermann and Weigend (2006) could readily separate the flowers of the Loasaceae into four groups on the basis of their nectar measurements, which corresponded with the main pollinators and associated floral traits (table 8.6). The fact that a water-stressed plant reduced its nectar V markedly but maintained its nectar C might indicate that the latter can be most crucial for attracting the "right type" of visitor, in accordance with the constraints imposed by viscosity and tongue length. The match of nectar with visitor is also very clearly reflected in several analyses of pollinator shifts within given taxa, where the nectar reward can readily change from low to high, or vice versa (e.g., in studies on bee and bird Penstemon referred to above, or on Fuchsia which may switch from bird to bee pollination); and adaptive shifts can even occur from ancestrally no reward to offering a nectar reward (in Disa orchids this has occurred at least three times; Johnson et al. 1998). Many

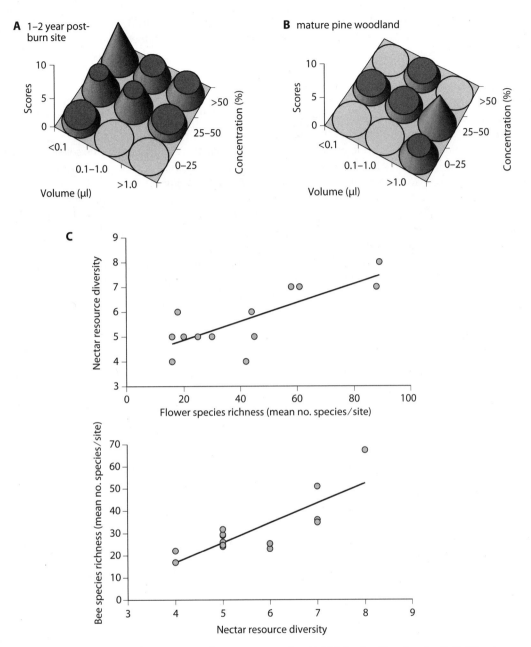

Figure 8.13 Nectar niche diversity for a Mediterranean community with high floral and bee diversity: (A, B) different flowers produce different combinations of mean nectar volume and concentrations, and the relative filling of nine resultant nectar niches can then be compared between different sites; here the mature site (B) has fewer nectar niches, whereas the site recovering from a recent burn (A) has more niches more fully filled (especially the lower volume and higher concentration niches attributable to rapidly growing and weedy annual plants); (C) the strong relationships between nectar resource diversity and both flower and bee species richness measures. (Redrawn from Potts et al. 2004.)

more examples of such shifts will be noted in later chapters, and there is much scope for further work here given that we now have some understanding of the genes (two QTL in *Petunia*, for example) that exert control over nectar volume (Stuurman et al. 2004).

At a community level, the organizing and predictive effects of nectar can therefore also be substantial. Thus Potts et al. (2004) showed that the diversity of available nectar resources across a mixed Mediterranean plant community, in terms of C and V, described as *nectar niche diversity*, was the most important determinant of the flower-visiting community's structure and complexity, as summarized in figure 8.13.

However, another overriding message from this chapter is that nectar as a commodity is easy to measure on a one-off basis from one flower but annoyingly hard to characterize fully in time and space for a given species, so that simplistic average measures of C or V

should always be viewed with some suspicion. Nevertheless a given plant does have a good measure of control over its nectar and sugar dispensing systems, since

1. nectar secretion is an active process with a potential "on-off switch," and
2. postproduction changes can also be controlled, or at least limited, by morphology.

Each plant therefore does have reasonable scope to get exactly the "right" nectar reward in its flowers at a given time to pay off its visitors and still encourage movement of those visitors onward to other similar flowers. Thus flowers can manipulate animal behaviors quite precisely in space and time to get the best possible pollen movement out of one flower and into another, and so can use their nectar to exert control over their own reproductive success and the gene flow within their population.

Chapter 9

OTHER FLORAL REWARDS

Occasionally flowers offer neither pollen nor nectar as a foodstuff to their visitors but instead yield other rewards; or they may offer these as "extras" in addition to some pollen. This chapter reviews these possibilities, considering a range of oils, waxes, scents, and resins (on which topics Simpson and Neff [1981, 1983] provided earlier reviews), as well as some less tangible rewards that can be obtained from flower visits.

1. Oils

First described in detail by Vogel (1969), fatty oils as an offering in flowers are now known from at least 80 genera across several families (table 9.1) and from nearly 1% of flowering plant species (Buchmann 1987; Steiner and Whitehead 1991). The *oil flowers* occur in four distinct domains of the world, those in the New World being unrelated to those of the Old World but convergently similar. In each case the oils are produced by secretory hairs termed **elaiophores**, which are usually in paired glands sited at the base of a flower (fig. 9.1); these have evolved independently several times, and once evolved, they have rarely been lost. They may be derived from either epithelia or trichomes (fig. 9.1A,B). In the orchid genus *Grobya*, there may be three different elaiophores, the one on the lip providing an "oil guide" that entices bees onward to the deeper-sited glands (Pansarin et al. 2009).

The oils produced in flowers are mainly **diglycerides**, with C_{14}–C_{18} (rarely C_{20}) backbones (fig. 9.2), often with smaller amounts of hydrocarbons, esters, and aldehydes; they differ in character from the lipids that are sometimes found as trace components of nectar. They tend to be colorless or pale yellow and are usually odorless.

All known examples of oil flowers are essentially solitary-bee pollinated, and the bees are mainly from the taxa Melittidae, Ctenoplectridae, and especially Anthophorinae. Most harvest the oil using their feet rather than their mouthparts (Roberts and Vallespir 1978; Buchmann 1987; Steiner and Whitehead 2002). There may be an association between the oil-collecting habit and pollen gathering by sonication, since most of the anthophorines that gather floral oils also use buzz-pollination at least some of the time (Simpson and Neff 1983), thereby acquiring rather dry pollen that lacks much lipid content (chapter 7). The oils the bees obtain could then act as lubricants, aiding pollen packaging and perhaps reducing pollen desiccation, although they may also be used as a cement to help line, and thereby waterproof, the nest cells. There is no doubt, however, that the collected oils also serve directly as a liquid foodstuff for the bee larvae, with or without added nectar or pollen. Oils as a food are particularly nutritious, two- to threefold more valuable weight for weight than a reasonably concentrated sugary nectar (table 8.5). When they occur in flowers, oils

TABLE 9.1
Plant Families Interacting with Oil-Gathering Bees

Oil produced; oil bees visit	Deceptive, no oil production; oil bees visit without reward, or take nectar
Cucurbitaceae	Caesalpiniaceae
Iridaceae	Gesneriaceae
Krameriaceae	
Malpighiaceae	
Melstomataceae	
Orchidaceae	
Primulaceae	
Scrophulariaceae	
Solanaceae	

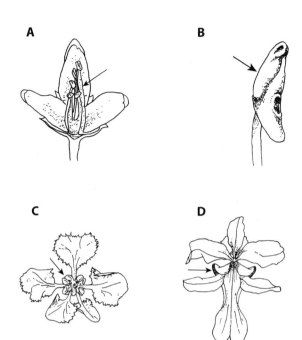

Figure 9.1 Different types of oil glands: (A) trichome elaiophores, glandular hairs on petal bases in *Lysimachia*; epithelial elaiophores: (B) on anther of *Mouriri*; (C) at base of *Malpighia* flower (seen from below); (D) on *Oncidium* orchid, seen from underside. (Parts (A) and (B) redrawn from Buchmann 1987; parts (C) and (D) redrawn from Endress 1994.)

are often the only collectable reward visible, the pollen being hidden, although a few species do offer both oils and nectar.

Examples of oil-producing flowers are particularly abundant from the Neotropics, most notably from savanna and moist-forest habitats. Three plant families—the Iridaceae, Orchidaceae, and Malpighiaceae—contribute particularly large numbers of examples, with nearly all of the species in this last family producing oil rewards. However there are also oil-producing flowers in the Cucurbitaceae, which are associated with *Ctenoplectra* bees. Some species of the genus *Angelonia* (Scrophulariaceae) also produce oils, the flowers having pockets at the back where the elaiophores are located (Machado et al. 2002); these are visited by long-legged *Centris* bees. In South America a range of unusual dwarf epiphytic orchids also offer oil (Dressler 1993), although these are not particularly specialist since various bees visit several of them and probably also gather a range of their pollens (Schlindwein 1998). Orchids with oil rewards also occur quite commonly in southern Africa and include species of *Disperis* and some related terrestrial genera, which are pollinated by *Rediviva* bees (Steiner 1989) in a much more specialist interaction. These bees also visit oil-bearing *Diascia* and *Ixianthes* flowers, which are related to snapdragons. Steiner and Whitehead (1996, 2002) reported asymmetries of specialization: one *Ixianthes* species (*I. retzioides*) is pollinated only by *Rediviva gigas*, but that bee in turn visits several other oil-

secreting plants and some of its populations may not visit *I. retzioides* at all, and similarly, *Colpias* flowers are entirely dependent on *R. albifasciata*, but that bee also visits other Scrophulariaceae.

Specializations for gathering oils do occur in some bees (fig. 9.3). Notably, the Neotropical *Centris* and *Epicharis* species have forelegs bearing unusual hairs that are like flexible combs or scoop-like blades (*elaiospathes*), which can rupture oil-bearing glands and then scrape up and transfer the oils to patches of stout bristles on the bee's hind legs for carriage to the nest. **Melittid** bees have "hair felt" patches on their legs that take up floral oils by capillarity, while the unusual ctenoplectrids have abdominal oil-mopping hairs. The southern African *Rediviva* bees have exceptionally long front legs (fig. 9.3E) that can gather oil by probing into the long spurs borne on their orchid flowers.

Symphonia species (Clusiaceae) have a more unusual use for their flower oils; they produce pollen mixed in an oily fluid (*anther oil*), which helps the pollen to be absorbed by capillarity into the stigmatic

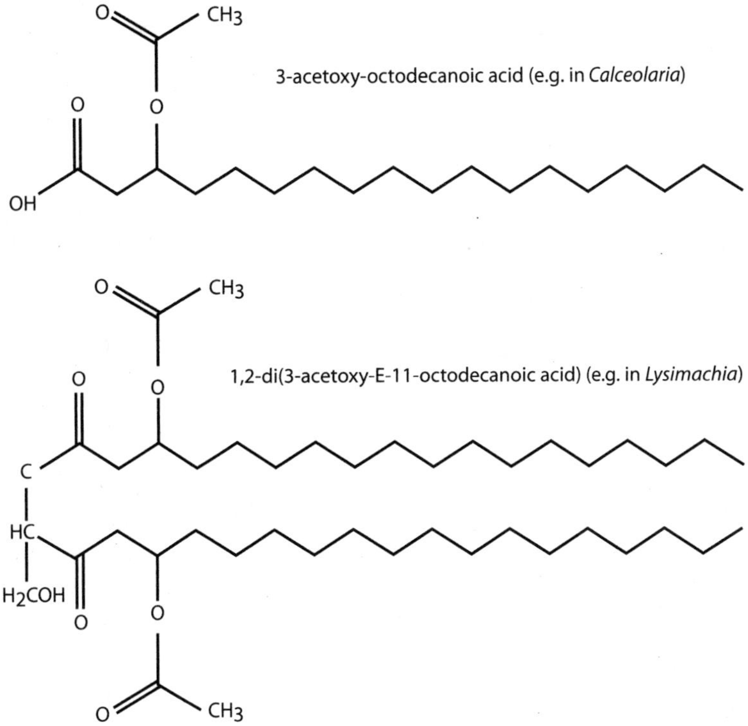

3-acetoxy-octodecanoic acid (e.g. in *Calceolaria*)

1,2-di(3-acetoxy-E-11-octodecanoic acid) (e.g. in *Lysimachia*)

Figure 9.2 Floral oils are normally based on free fatty acids with chain length C14–20; (above) a typical acetoxy-substituted version; (below) a diglyceride version.

pores after deposition (usually by a hummingbird; Bittrich and Amaral 1996).

The most familiar examples of oil-producing flowers in temperate habitats all have trichome oil glands and include *Calceolaria* from the figwort family (Scrophulariaceae) and yellow loosestrife, *Lysimachia* (Myrsinaceae), a common weed in damp places in most northern continents that is almost exclusively visited (as in plate 19H) by *Macropis* bees (Cane et al. 1983). The flowers of *Lysimachia* perhaps attract bees initially by specific scents in their oil (Dotterl and Schaffler 2007), but experienced bees also learn and respond to the visual cues from the flowers. Oil alone (for cell lining) and oil and pollen together (for larval food) are apparently collected in separate foraging trips.

It is an oddity of oil flowers that they very frequently act as models for deceptive orchids that match their shapes and colors and produce shiny surfaces that appear (quite falsely) to be offering oil. This is particularly common in the *Oncidium* orchids, which mimic many of the Malpighiaceae and also *Calceolaria* and attract the same bees that normally visit their models.

Oncidium cosymbephorum explicitly benefits (by higher seed set) from its resemblance to the rewarding shrub *Malpighia glauca* (Carmona-Diaz and Garcia-Franco 2009).

2. Resins and Gums

Floral **resins** have been reported in occasional genera that are abundant in the tropics. *Clusia* and *Clusiella* (Clusiaceae) (Skutch 1971) and *Dalechampia* (Euphorbiaceae; plate 11F) are best known, although resin is also reported from *Mouriri* flowers (Melastomataceae; Buchmann and Buchmann 1981). The floral resins in *Clusia* and *Dalechampia* are terpenoids (usually triterpenes in *Dalechampia*).

There is evidence that *Clusia* resin rewards are specifically antibacterial within the nest (Lovkam and Braddock 1999), and the same may be true for the resins of other genera. Quite a range of bees, especially megachilids, will collect resinous plant secretions oozing from vegetative structures (stems and leaves) where the resins have a protective antiherbivore effect

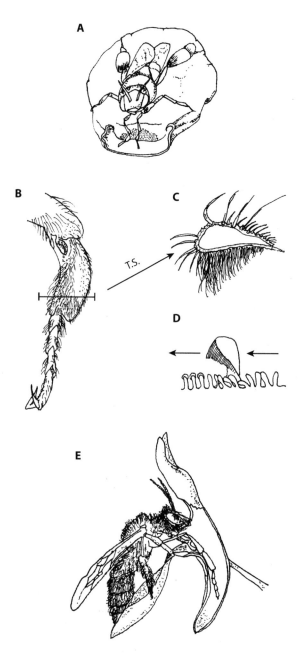

for the plant; the bees take them back to their nests again to serve as a cell-lining and so help prevent fungal growth (e.g., dipterocarp resins, Messer 1985). This may give a clue as to how the habit of offering resins arose in flowers; floral resins may be more attractive to bees than stem-oozes, being more easily gathered or worked with, and they would also be more predictably located once a floral search image had been acquired.

The flowers of *Dalechampia* normally have a pair of conspicuous and highly colored bracts around otherwise very reduced corollas (fig. 9.4), and it is the lip of the bract that forms a resin gland (Armbruster and Webster 1979; Armbruster 1984, 1997). Female bees (euglossines, **meliponines**, and megachilids, plus occasionally *Apis*) have been recorded as visitors, gathering the resin as a nestlining material. Armbruster et al. (2005) showed that visiting *Heriades* bees selected flowers by the size of their bracts, rather than by gland area, even when the latter was easily visible; the bracts appear to serve as an "honest signal" for the size of the reward.

Some aroids are also known to have floral resins.

Figure 9.3 Oil-collecting apparatus in bees: (A) anthophorid bee collecting oil with its front-leg mops (the oil then transferred to the hind legs); (B) the collecting mop on the foreleg, and (C) in transverse section; (D) the mop being drawn as a scraper across the oily hairs of the oil gland; (E) a *Rediviva* bee extracting oil from a *Diascia* flower with its elongated spurred front legs. (Parts (A–D) modified from Barth 1985, based on earlier sources; part (E) redrawn from Whitehead et al. 1984.)

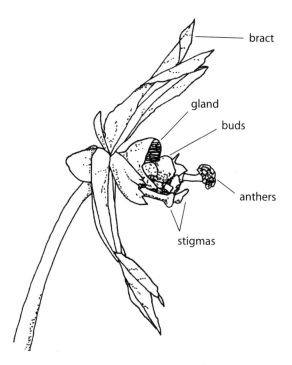

Figure 9.4 A *Dalechampia* inflorescence with oil gland sited above a small group of flowers. (Redrawn from Armbruster 1996.)

The functioning of these deceit-pollinated trap flowers is covered in chapter 23, but it may be noted here that the upper part of the spathe has male flowers with pollen, whereas the lower part (which forms the trap for visiting flies) has female flowers that in at least some species produce resinous secretions.

Once again, orchids are mimics and have the occasional habit of mimicking resin plants, a notable example being some *Maxillaria* species that have a patch of shiny apparent resin on their lower lip, which in reality is nothing more than callus tissue.

3. Stigmatic Exudates

Stigmatic exudates are common in angiosperms, where their main functions relate to pollen capture, adhesion, and germination (chapter 2), together with protection of the stigma against damage or desiccation. However, in many cases these exudates also provide a good oily food source, and sometimes this can be the primary reward, attracting beetles, flies, and some other insects. Generally the material exuded is composed of lipids and amino acids, with some phenolics, alkaloids, or antioxidants also present (F. Martin 1969; H. Baker et al.1973; Lord and Webster 1979). In a few cases the reward is targeted rather specifically, as for example in certain *Aristolochia* that are primarily pollinated by flies which become trapped in a floral chamber where they feed on these amino-acid-rich exudates (Baker et al. 1973). Exudates are also abundant in some palms, where again it is mainly flies that feed on them (Simpson and Neff 1981).

Nearly all orchids also have stigmatic exudates, which are primarily used to stick the pollinia to visitors; in these cases the exudate is rarely fed on and is sometimes secreted almost explosively. But some *Dactylorhiza* orchids provide an exudate that is a food source (whereas most of the genus is food-deceptive, having no nectar and no other reward); the common spotted orchid *D. fuchsii* has an oily exudate that is also rich in glucose and amino acids and is fed on by both honeybees and bumblebees (Dafni and Woodell 1986).

Whereas the above examples have oily stigmatic secretions, there are a few cases—for example, species of *Anthurium*—where the stigmatic fluid may be up to 8% sugar (Croat 1980). In some *Asclepias* species this same kind of sugary exudate flows away from the stigma and collects in a nectar reservoir where it is visited

by pollinating insects, so producing a reward in a group that is normally rewardless.

In plants where exudates are used as food, the styles tend to be short or rudimentary and have a broad stigma, so that the pollen-receptive function and the growth of pollen tubes are both relatively resistant to damage.

4. Scents

Fragrances as rewards to flower visitors are found in at least seven different plant families, all from the Neotropics. Orchids are by far the most common examples, up to seven hundred species having been recorded as dispensing scent.

Orchids and Euglossine Bees

Many genera of orchids use scent as their main lure to flower visitors and offer no other rewards; perhaps 10% of all species of Orchidaceae fall into this category (van der Pijl and Dodson 1966). For most of these, their main visitors are bees from the Euglossinae subfamily, an exclusively Neotropical taxon common in the canopy of rain forests (Dressler 1982). These bees (about two hundred species) live in small primitively social groups or in solitary nests (chapter 18). Females gather nest materials and collect pollen and nectar in the normal bee fashion. However, the males, which are not associated with the nests and must forage for nectar for themselves through their unusually long lifespan of three to six months, will also gather flower scents, and in a fashion unique to this group.

Since their scent gathering is almost always from orchids (Lunau 1992b; Eltz et al. 1999; N. Williams and Whitten 1999), these animals are sometimes termed **orchid bees**. The orchids they visit are strongly fragrant but have no nectar and no accessible edible pollen. Instead, the male bees scrape up oily and waxy compounds from the scent organs on flower (usually on the orchid labellum). For this they use brush-like structures on their front tarsi that act like mops (fig. 9.5), taking up by capillarity the fragrances that seep out of the flower. The droplets of scent are transferred to a rake of hairs on the middle legs and then passed to a small opening on either of the very swollen hind **tibiae**. These openings bear fine feathery hairs that

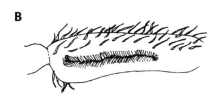

Figure 9.5 Brushes on feet of euglossine bees: (A) setae from front tarsal brush; (B) hind tibia, with slit surrounded by setae; (C) detail of hind tibial setae. (Redrawn from photographs in Williams 1983.)

receive the scented material and pass it through a channel to a large hollow chamber within the leg. Here the perfumes are finally stored; the chamber has many internal ridges, each bearing masses of feathered hairs that take up the scented liquid (fig. 9.6). This tibial chamber functions as a perfume bottle, and each male euglossine is effectively carrying around its own built-in lure. A single male bee may carry up to 60 μl of scent material (Vogel 1966).

The scents gathered contain some benzenoids but are predominantly monoterpenes, including classic floral fragrances such as cineole, myrcene, ocimene, pinene, eugenol, limonene, and linalool; any one flower may offer a mixture of 6–10 compounds (Dodson et al. 1969) from this range. In quite a few of the euglossine-visited flowers, the relatively unusual monoter-

pene trans-carvone oxide is present, and it may be a key attractant for these bees (Whitten et al. 1986; Teichert et al. 2008). There are indications that the bees also add compounds of their own making to the mix from glands within the tibia (fig. 9.6), perhaps making lipid materials that reduce the volatility of the scent mix and so prolong its useful life. From their labial glands, which act as extractors and carriers for the fragrances, they also add straight-chain lipid materials, which are then recycled back from the hind-tibial pockets for reuse, forming a "lipid conveyor belt" (Eltz et al. 2007).

The floral fragrances are so attractive to the males that with a few drops of eugenol or cineole on a filter paper, it is easy to attract large numbers of euglossines down from the canopy in most central and southern American rainforests.

It was initially assumed that the scents are subsequently released by the males for reproductive functions, perhaps specifically as pheromonal cues for reproduction (N. Williams 1982; Dressler 1982). However, females are not attracted directly to the scents in the flowers, and a simple sex-pheromone function therefore seems unlikely. Male behavior involves trap-lining between flowers, and the same route may be used by more than one male at different time intervals. At particular sites on their own route, males will rest on the vegetation and flutter their wings before flying up and circling the site for a while. Eltz, Sager, and Lunau (2005) showed that the males perform specific and intricate leg movements during this circling flight, transferring perfume from the tibial pouches to a tuft of hairs on the mid-tibia, where they are ventilated and wafted into the air by the wing movements, confirming that the scents are in use at this stage and that the males are emitting a puff of perfume into the air. But it is still unclear whether they are a signal to females, to other males, or to both. Often a few more males will arrive, presumably attracted by the scent. Females are attracted to the patrol routes, and particularly to these sites, but whether the attraction is the scent itself or the presence of a small **lek** of fluttering, circling, and conspicuous males, is debatable.

Issues of specificity are interesting in these particular associations. The bees depend heavily on the orchid fragrances they collect for their reproductive success, and the orchids likewise depend on the euglossine bees for pollination. However, each bee species needs to assemble its own unique and species-specific fragrance

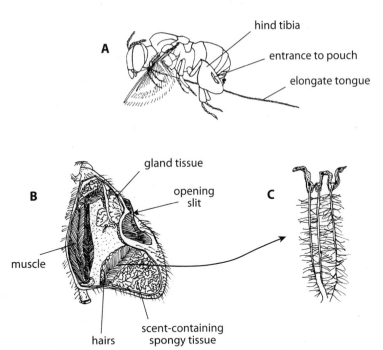

Figure 9.6 The tibial sponge-like glands in a male euglossine bee: (A) in situ, with slitlike entrance; (B) a cutaway view, including the gland tissue and the spongy tissues that store the scent; (C) a close up of the hairs making up the spongy storage tissue. (Redrawn from photographs in Williams 1983.)

mixture; for example, three species of *Euglossa* have quite distinctive mixtures (table 9.2). To acquire their mixes they must normally visit more than one orchid species; so there is a constraint acting against full specialization. Eltz, Roubik, and Lunau (2005) showed that individual bees are *less* attracted to the orchid species they have previously visited, so they must learn and remember the fragrances they have already collected, and so are showing a limited kind of negative floral constancy. However, Dressler (1968) showed that 11 different species of orchid deposited and picked up their pollinia from 11 distinct sites on a euglossine's body (fig. 18.14), so that effective specificity for the orchid can still be considerable.

How exactly do male bees foraging for fragrances benefit the plant? Scent gathering takes a significant time, and on at least some of their preferred orchid species, the male bees appear to suffer an intoxicating effect from the odors, becoming clumsy and rather sluggish, with a loss of full motor control (Evoy and Jones 1971). This may help the flowers to "manage" their visiting bees quite precisely and direct them into places they could otherwise avoid. For example, *Stanhopea* and *Gongora* orchids proffer their scents in tissues sited where the male bee must hang upside down, and when intoxication sets in and his control lessens,

he loses his grip on the slippery surface and falls off onto the cushioning anthers below. Some *Stanhopea* species additionally have a chute-like arrangement to ensure that the bee lands correctly with his dorsal thorax aligned on the sticky pollinia. Even more specialist examples occur in the orchid genera *Catasetum* and *Cycnoches*, both with unisexual flowers. *Catasetum* orchids are pendant, the male flowers showy and the females drab, but both have the same scent (H. Hill et al. 1972); the males have a large viscid disk below the pollinia, and the rostellum projects forward into two antenna-like structures, such that a touch on these by a scent-gathering bee (working at the waxy scent-bearing tissues in the hooded labellum) causes the disk and the pollinia to be shot forward onto the bee's back. The same bee visiting a female flower (which is inverted) has to hang upside down to gather scent, and the pollinia then swing off his back under their own weight, on an elastic thread, and are picked up on the grooved stigma below. The swan orchid *Cycnoches* operates similarly, requiring a visitor to hang upside down to gather scent and to let go with his hind legs, so that his body swings down and touches a cover that conceals the anthers, again causing an explosive reaction that shoots the pollinia onto the underside of his abdomen.

TABLE 9.2

Fragrance Mixtures in Three Species of *Euglossa* Bee

Component	E. cognata	E. imperialis	E. tridentata
Cineole	0	22	3
Unknown sesquiterpene ketone	21	0	0
Methoxy cinnamic alcohol	0	0	11
Unknown A	0	18	0
Unknown B	0	11	0
Unknown C	11	0	6
α-Pinene	0	10	<1
Methoxy cinnamaldehyde	0	0	10
E-Nerolidol	9	6	3
Benzyl benzoate	0	0	9
Methoxy cinnamyl acetate	0	0	8
Dimethoxy benzene	8	0	1
Benzyl cinnamate	7	0	1
Geranyl acetate	0	0	7
E,E-Farnesene	7	0	0
Hedycaryol	0	0	7
β-Pinene	0	5	<1

Source: From Eltz, Roubik, and Lunau 2005.

Note: Values are percentages for any component making up at least 5% in any one species.

Perhaps even more extraordinary are the large and elaborate bucket orchids in the genus *Coryanthes* (Dodson 1965), which combine scent offerings with trapping (fig. 9.7). Male bees (usually *Eulaema* or *Euglossa*) are attracted by scent to the pendant flowers, where they find a textured zone at the base of the lip at which they scrape vigorously to gather scent in the normal way. In the process, some fall off and land in the "bucket," a large cup formed by the labellum, into which drops of fluid drip and collect. The bees cannot climb out of this bucket except via a narrow hole near the base of the column; this first leads them beneath the grooved stigma (which extracts any pollinia the bee was carrying) and then, as they approach daylight and "freedom," past the anthers and pollinia. It may take up to thirty minutes for a trapped bee to get free of the flower, because the last part of the route out is very narrow and the rostellum restrains the bee while pollinia are glued to his back. Each flower needs only one bee to pass through its exit route to fulfill both its male and female functions, and once this has occurred the flower ceases scent production and wilts within a few hours.

Other Examples of Fragrance as a Reward

Scents are also gathered from a solanaceous genus *Cyphomandra* (M. Sazima et al. 1993) by euglossine bees; three different species in this plant genus have very different scents and attract three different euglossines. As the male bees gather scent droplets from the flower, they push against the anthers and trigger a pneumatic bellows-arrangement such that the poricidal anthers eject jets of pollen and cover the bees' sternum with pollen grains. Curiously, postpollination events here include a substantial enlargement of the corolla with an accompanying color change.

Some Neotropical species of *Dalechampia* vines have volatile scented oils instead of resins, which again are collected by male euglossines. In chapter 6 we also met examples of aroids, such as *Sauromatum*, that release pleasant scents within their trap chambers, along with the rather putrid odors from the spathe. Another example is *Anthurium*, where in some species the lower part of the spathe has perfume sacs, from which scents are gathered by euglossines (Croat 1980).

Something rather similar to the euglossine behavior

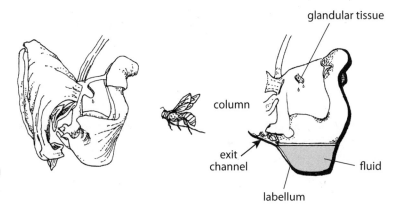

Figure 9.7 *Coryanthes* (bucket orchid) structures: whole flower and cut-away view showing labellum and a bee emerging past the column. (Modified from Barth 1985.)

is now known to occur in at least two groups of flies. Tephritid fruit flies pollinate *Bulbophyllum* orchids (Tan and Nishida 2005; Tan et al. 2006), and here the male flies land on the petals and climb to the orchid lip, forcing it open and lapping at the oily scented compounds secreted there. Then the fly is catapulted into the orchid's column as the lip springs back, so picking up the pollinia. These flies subsequently use the phenylpropanoids from the flower, probably to make their own pheromone. Fruit flies in the subfamily Dacinae, particularly the genus *Bactrocera*, also collect floral scents, mainly from orchids, that they store in their rectal glands and use as a pheromone (Nishida et al. 2009).

Although not strictly volatile scents, **pyrrolizidine alkaloids** are acquired from various parts of their host plants (including flowers) by some **danaid** butterflies, and these can become part of their pheromones as well as of their defenses (Boppré 1978; Reddy and Guerrero 2004).

5. Floral Tissues

In a few instances, particularly in the more basal plant families, special areas of a flower are modified as food bodies, which have high contents of carbohydrates, lipids, or proteins, and flower visitors will nibble or scrape away some of this tissue to eat. The modified areas are often at the base of petals or the tips of stamens and are derived from epidermal or parenchymal cells.

Consumption of such tissues requires mouthparts that can chew, so this phenomenon is most commonly found in beetles, but also in some birds and a few bats.

Floral food bodies are primarily provided with translocated carbohydrate reserves (sugars and starch) from the phloem; much more rarely they are enriched with protein, as reported in *Calycanthus*, which is fed on by beetles (Rickson 1979). Some bat-visited flowers, rather surprisingly, offer sugary bracts that are eaten (e.g., *Freycinetia*, Cox 1982; and chapter 16). However, cytoplasm is also a possible reward, and one *Epidendrum* orchid has modified liquid-filled cells lining the corolla base that yield a cytoplasmic reward to visiting moths, whose proboscis tip is covered in tiny spines (chapter 14) that can scrape at the tissues.

Whenever a significant part of a flower is eaten by chewing or scraping, it is essential that the flower as a whole is relatively sturdy and relatively long-lived. It is also necessary that visitors are attracted away from the ovules and anthers and toward the expendable parts, and this is probably achieved by localized odor cues.

6. Other Possible Nonfood Rewards

Brood Sites

Flowers functioning as a brood place could be considered as a further possible reward offered by particular plants. This scenario occurs where a flower-eating or seed-eating animal lays its eggs in flowers (which would normally make it a floral parasite), but then gets exploited in turn by the flower as a pollen vector. Simple examples of this occur in large-headed flowers, such as thistles, a favorite brood site for many flies.

As an alternative to (or an extension of) this, the

progeny of the flower visitor, once hatched from the eggs, may consume the seeds of their host plant. Probably the plant then gets a net loss overall, and the mutualism breaks down, although some examples of this nursery pollination do provide a net benefit to the plant (chapter 26). In a very few spectacular examples these brood-site mutualisms have been elaborated into active pollination, as in *Yucca* and in *Ficus*, where the seed parasites of a plant have become the key pollinator. These cases are again dealt with fully in their own right in chapter 26.

Microclimatic Protection and Warmth

Flowers can offer suitable shelter from rain and wind for many small insects, and a number of bees are particularly noted for taking shelter in flowers or using them as sleeping places, at which time a certain amount of pollen is then exported on the body, a concept called *shelter pollination*. This practice occurs especially in the males of solitary bee species, which lack their own nest; a well-known temperate example is *Chelostoma florisomne*, which can often be found sleeping in the pendant flowers of harebells. *Serapias* orchids are also quite common sleeping sites for bees. In the eastern Mediterranean, clusters of scarab beetles commonly occur overnight in bowl-shaped flowers (chapter 12). Certain beetles in the genus *Pria* may have more specific relations with female *Leucadendron* flowers, which are more cup-shaped and enclosed than the male flowers and are used apparently specifically and solely as shelter from frequent rain (Hemborg and Bond 2005), being visited 90% more often on wet than on sunny days.

More than simple shelter may be on offer however. Flowers quite often exhibit their own internal microclimate (fig. 9.8, and other examples in fig. 8.10) by virtue of relatively elongate, enclosed corollas, and thus they offer significant amelioration of climatic conditions to their visitors. Within a moderately large and sturdy corolla tube there may be substantial temperature increments, giving a microclimate that allows a visiting insect to warm up, although the primary benefit to the plant may be that of potentially speeding up ovule maturation, pollen tube growth, fertilization processes, and seed development (Kjellberg et al. 1982). Bumblebees will choose to forage at warmer flowers, given a choice (Rands and Whitney 2008).

There are more specific cases where flowers offer *warmth* to insects. This reward to a visitor may be a side effect, in that the main benefit accruing to the plant may again be accelerated reproductive development, or a faster volatilization of plant scents, or even a better resemblance to a warm piece of dung or carrion. But in cooler habitats it may play a significant secondary role as a means of attracting flying insects to a flower (Hocking and Sharplin 1965; Cooley 1995), thus improving their energy budgets by reducing thermoregulatory costs (which will be discussed in the next chapter). Heat from cones is known to attract pollinators to cycads (Terry et al. 2007) and seed-feeding insects to conifers (Takács et al. 2008). Whitney et al. (2008) showed that warmth and sugar concentrations in angiosperm flowers were assessed independently by bumblebees, but sucrose levels usually took precedence over warmth (or at least over nectar temperature).

Additional warming in flowers can be achieved in at least four different ways:

1. THERMOGENESIS: Thermogenic tissues occur in at least ten plant families, mainly among the basal angiosperms (Thien et al. 2000), and are especially well studied in aroids and palms. Floral tissues in various aroids can generate heat from the spadix, often at specific times of day (e.g., Bay 1995; Seymour and Schultze-Motel 1998). In *Syngonium* the warming may occur each night during a three-day lifespan, and in some species there are two periods of warming on particular nights (Chouteau et al. 2007). In *Homalomena*, thermogenesis increased the floral odor emissions from the spadix which specifically attracted *Parastasia* scarab beetles that served as effective pollinators, whereas other beetles were unaffected by the heat or scent and visited the flowers indiscriminately at all stages (Kumano-Nomura and Yamaoka 2009).

Some thermogenesis also occurs in members of the Annonaceae, where the closed floral chamber of *Xylopia* has been recorded as 8°C above the ambient temperature (Ratnayake et al. 2007). Night-flowering water lilies in both South America and West Africa are moderately thermogenic too and have specific associations with scarab beetles, suggesting a coevolved specialized relationship going back at least to the early Cretaceous (Ervik and Knudsen 2003). Seymour and Matthews (2006) observed beetles spending long periods in these flowers, in the absence of any nutritional reward, and suggested that the temperatures main-

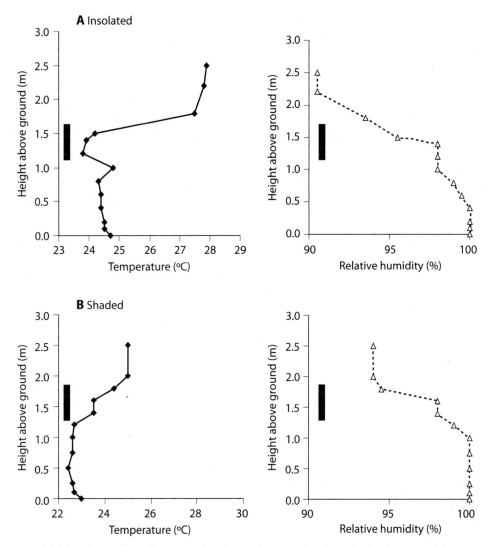

Figure 9.8 Microclimate within a flower patch, here for *Justicia aurea* in Costa Rica, showing the higher relative humidity and lower temperatures within the patch, especially in insolated areas (A). The solid bars show flower height. (Modified from Willmer and Corbet 1981.)

tained in water lilies are precisely adjusted to reward the captive beetles by allowing their active mating behaviors to continue.

2. HIGHLY ABSORPTIVE SURFACES: Flowers that are dark in color are likely to be particularly absorptive of solar radiation and may therefore warm up well above ambient temperatures; this warmth would be especially useful at dawn when other possible sites are still cold. The Mediterranean iris *Oncocyclus* has large, dark-colored flowers, with hidden pollen and no nectar, and in Israel it is not visited in the daytime but does provide a night shelter to solitary male bees. These

bees emerge from the iris flowers earlier than from any other site at dawn, when the irises are around 2.5°C above air temperature (Sapir et al. 2006); it appears that floral heat is the only reward on offer here.

High absorptivity may also be achieved by specific cell structures, for example, the poppy *Papaver radicatum* from arctic regions, in which the inner layer of the petal epidermis is papillate and traps light reflected from the underlying mesophyll cells (Q. Kay et al. 1981).

3. PARABOLIC REFLECTION: Some bowl-shaped flowers have unexpectedly raised internal temperatures

at their centers. This phenomenon is generally associated with highly reflective white or yellow internal surfaces and an effectively parabolic shape (fig. 9.9), giving maximum reflection of the incoming radiation onto the center of the flower (Kevan 1975c; Heinrich 1993). Insects basking in this central zone can have a body temperature 5–15°C above the ambient air temperature. Cup-shaped flowers of this kind may therefore be especially useful at high altitude or latitude (chapter 27) or in the early spring in temperate habitats. There are also instances of nodding flower heads in some high-altitude plants whose intrafloral temperature is kept above ambient by reflection from the substratum.

4. SOLAR TRACKING (HELIOTROPISM): **Solar tracking** occurs in at least 18 plant families, most familiarly in the sunflower (*Helianthus*) where it can be easily observed occurring en masse in a field crop. In effect the stalk rotates so that the corolla is always pointing directly at the sun through the course of the day (fig. 9.10), thus achieving maximum warming (Kevan 1975c, 1989; Ehleringer and Forseth 1980; Kjellberg et al. 1982; Luzar and Gottsberger 2001). Heliotropism tends to occur in flowers from warmer climates that flower early in the season when pollinators are scarce. For example, the Mediterranean marigold (*Calendula arvensis*) traverses an angle of about 20° east to west each day and can thus maintain a 20°C temperature excess, enough to attract bombyliid flies, which warm up sufficiently to fly on to another flower (Orueta 2002). At least some of the reflective bowl-

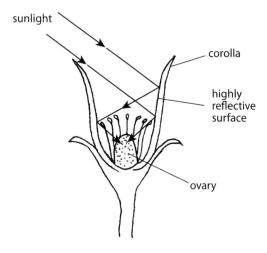

Figure 9.9 Parabolic reflection, with Arctic flowers shaped to act as internal reflectors, warming the ovules. (Modified from Kevan 1975a.)

shaped flowers from northern tundra or polar habitats (e.g., some *Papaver, Dryas, Pulsatilla,* and *Ranunculus* species) also use this strategy.

In heliotropic flowers the stalk undergoes continual slow movements, which in *Ranunculus adoneus* involve an eastward-bending of the peduncle (by differential growth on the shaded side) in the morning, with gradual unbending through the day (Sherry and Galen 1998). This is controlled by light at the blue end of the spectrum (Stanton and Galen 1993). It has clear benefits to the plants, so that *Ranunculus* flowers that were artificially restrained (Galen and Stanton 2003) had about 40% fewer pollen grains germinating than those undergoing their natural solar tracking.

Some flowers are known to combine several of the above effects and act as "micro-greenhouses," their enclosed structure and reflectance properties allowing noticeable internal warming. *Crocus* flowers, for example, can be 2–3°C above ambient air in cold weather, which will stimulate opening of the corolla, access by pollinators, and pollen tube growth (McKee and Richards 1998); white and purple varieties warm up more than yellow ones. *Narcissus* flowers can have an internal temperature excess of 8°C in early spring, especially around the stigma and anthers (fig. 9.11); *Andrena* bees caught inside the flowers had markedly raised temperatures, and their visitation rate was positively related to the floral temperature (C. Herrera 1995b). Even more precisely, Jewell et al. (1994) noted that keel temperatures in *Lotus corniculatus* were higher in morphs with dark keels than in those with light keels and that the darker morphs tended to occur in the cooler microsites within a population. Given that bees can choose flowers on the basis of their warmth and can learn to associate flower color with flower temperature (Dyer et al. 2006; chapter 18), further observations of floral temperature increments in natural situations are much needed.

On the other hand, some flowers in hot climates may have a problem keeping the gynoecium cool and thus avoiding damage to the carpels. Some tropical Convolvulaceae flowers were shown to exhibit strong evaporative cooling effects, and if greased to prevent such evaporation, their internal organs could overheat (Patino and Grace 2002). The flowers also showed nonrandom orientation, pointing toward, but not directly at, the sun, which may have facilitated transpiration (Patino et al. 2002). Whether floral cooling could also

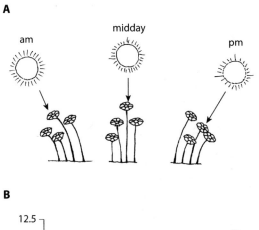

A

am

midday

pm

B

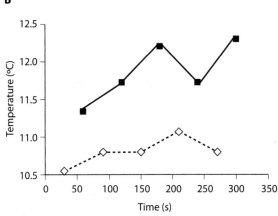

C

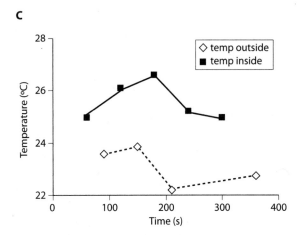

Figure 9.10 Solar tracking, or heliotropism (A), and the resultant temperatures inside (open circles) and outside (closed circles) in (B) *Papaver* and (C) *Dryas* flowers. (Parts (B) and (C) drawn from data in Hocking 1968.)

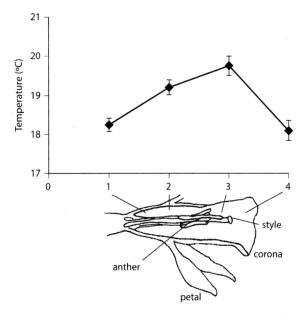

Figure 9.11 Variation in temperature within *Narcissus longispathus* at four points along the corolla, showing the higher temperature close to the anthers and style. (Modified from data in C. Herrera 1995b.)

be of benefit to pollinators remains unclear, but it would not be unexpected in very hot climates, where flying insects can easily overheat (chapter 10).

Meeting Places

Flowers can also offer a kind of reward to visitors in that they provide a reliable place to encounter other flower visitors. This may be particularly useful for mating opportunities, or for predatory opportunities. This theme will be discussed in more detail in chapter 24.

7. Overview

Most of the rewards discussed here are very minor in a numerical sense, although they may be the key to some particularly fascinating pollination systems. They do open up possibilities for new dimensions in animal-flower interactions, many of them currently underexplored. In fact these rather unusual kinds of rewards are always worth bearing in mind when investigating new interactions, and may prove to be rather more commonplace than is presently appreciated.

Chapter 10

REWARDS AND COSTS:
THE ENVIRONMENTAL
ECONOMICS OF POLLINATION

Outline

Pollination is usually thought of as a mutualism, of benefit to both partners, each of which gains in fitness. In such a relationship, both should be trying to maximize their survival and ultimately their reproductive success, which will require balancing their costs against the rewards and hence assessing the net benefits gained. Disentangling the economic aspects of the interaction for each participant has become a major area of study in pollination ecology, and this chapter explicitly considers the costs that may be incurred and that have to be offset against the rewards gathered, as outlined in the last three chapters. This becomes a particularly intriguing approach when we remember that, although pollination may be a mutualism, there is always a conflict of interest between the participants, for whom the cost-benefit analysis may work in very different ways.

1. The Conflicting Requirements
of a Plant and Its Pollinators

A given plant essentially requires effective intraspecific pollen transfer gained as cheaply as possible. Viewed broadly, for this result it needs the "right" visitors; animals that are the right body size or tongue-length or physical fit, that are suitably "sticky" to pollen grains, that will forage at the right time of season or day, and that will show the right behaviors and suitable flower constancy. These traits should all be acquired for a minimum payoff, that is, the ideal visitors should be cheap to pay for their services, in nectar or pollen or both. As the preceding chapters have shown, plants may be selected to achieve these aims by having suitable morphology, using suitable attractants (color, shape, nectar-guides, scent, etc.), and by offering a suitable reward per flower. They may adjust their rewards to keep visitors close to their limits of getting a net profit, thus ensuring that the animal moves around as many flowers as possible; Voigt et al. (2006) gave a good example of this for a bat pollinator.

Flower visitors, on the other hand, are likely to want maximum foraging profit for minimum effort in extracting it and in moving between plants to get it. A given visitor should pick flowers where it can easily make an energetic profit; essentially, for a nectar forager, this will be where the energy content of the nectar accessible to it exceeds the energetic cost of making the visit. So the visitor will choose carefully on which plant and where and when to focus its foraging. A well-adapted visitor essentially wants to minimize the cost of foraging—which is mainly measured as time spent, a composite of time in transit between flowers and the handling time on the flowers—in relation to how much energy is extracted from those flowers.

Very large visitors, which have high energy costs (birds, bats, some bees and moths) may need so much nectar (or pollen) that it is not worthwhile for the plant to supply them. But very small visitors, while cheap to feed, may not move enough pollen around effectively. Furthermore, in any size range, if the reward in any one flower is too big, then a visitor may gain so much payoff that it does not need to go on to other flowers on other plants anyway; whereas if the reward is too small, the visitor may give up on that plant species entirely. In both cases there will be little or no pollen transfer.

Calories gained and lost can be seen as the currency in what is essentially an economic payroll transaction, so that pollination studies can be handled as a classic cost-benefit analysis, following the initial insights of Heinrich and Raven (1972). Supply and demand should be interactive and roughly balanced in an ideal marketplace transaction.

So who controls what? The most obvious general idea might be that the plant determines the reward, and the animal therefore has to follow that reward as best it can. This is made more likely since it is almost always more vital for the plant to get the balance right, because it may go extinct if not cross-pollinated, whereas the animal can just go elsewhere for food; this is equivalent to the *life-dinner principle* operating in predator-prey interactions (the predator is working for its dinner, but the prey should be under even stronger selection since failure is at the cost of its life). In subsequent sections we will explore the reality of how far the plant is in control and the animal is merely the follower.

2. The Costs Incurred by the Plant

Nectar and Pollen Production

There is no doubt that nectar is fairly expensive to produce, as reflected by demonstrations that it can be resorbed by plants, thus conserving their resources (chapter 8). This expense is also revealed in the consequences of extra nectar production. For example, Pyke (1991) found that if nectar was artificially removed from *Blandfordia* flowers, more was usually produced

by the plants, but the final seed output would then be reduced. Ashman and Schoen (1997) determined that *Clarkia* flowers that lasted 35% longer also invested about 30% extra into nectar production, but they then sustained reduced seed production.

There have been rather few direct analyses of the costs of making nectar. Pyke (1991) provided examples of nectar production that used up to 37% of a plant's available energy; this was one of several studies to show that removing nectar from a flower can cause it to raise its net nectar production, so that at least the costs incurred are dependent on usage. Southwick (1984) found that in alfalfa crops about 15% of above-ground production went into nectar, while for *Asclepias syriaca*, up to a third of the daily **photosynthate** went into nectar production (equivalent to 3% averaged over the whole year). In general, values of around 30% per day during the flowering season for moderately large flowers may not be unreasonable. But since flowers themselves are expensive (see below), this nectar cost may be only a small part of their total price; for *Agave* a value for nectar of 4% of the costs of the flower spike was recorded (Howell and Roth 1981), while for *Pontederia*, Harder and Barrett (1992) estimated that 3% of a flower's energy content was in nectar at any one time, and that values were of about this order for a range of other plants. It would seem that nectar production, while highly variable, can be a substantial (though not exorbitant) cost to the plant (de la Barrera and Nobel 2004). Such expense might help explain the prevalence of **automimic** cheats with empty flowers (chapter 23). Interestingly, though, some flowers go on producing nectar after they are fully and efficiently pollinated, so the costs are perhaps not too restrictive. For example, Harder and Barrett (1992) recorded that *Pontaderia* produced 45% more nectar than was necessary, perhaps as a **bethedging** mechanism in case of poor weather and poor pollination on some days.

The amounts can be looked at in another way. An herb-rich grassland may produce about 25 g sugar/m², over a year (Heinrich 1975a; Proctor et al. 1996), which is about enough to support 1 *Apis*/m² through the summer (whereas one *Bombus* might need 3–4 m² to break even). A productive alfalfa field at its peak might therefore support around 50 *Apis*/m² for the short flowering and nectar-producing season.

Pollen production is perhaps only 5%–20% as costly. Pollen is tiny in terms of mass, but it is still

A

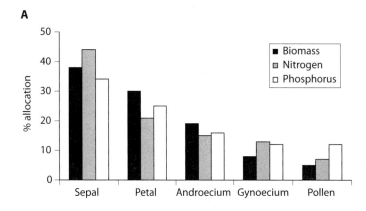

B

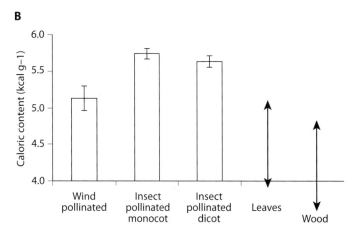

Figure 10.1 (A) Proportional costs of different parts of flowers, in terms of biomass, nitrogen and phosphorus (redrawn from Ashman and Baker 1992). (B) Caloric content of pollens compared with other plant parts, bars showing mean and SE, and arrows indicating range (drawn from data in Petanidou and Vokou 1990).

expensive because it uses up both nitrogen and phosphorus from the plants' reserves (Petanidou and Vokou 1990; and fig. 10.1), and it is nonrenewable. Estimates of its cost are generally indirect and unsatisfactory, although Howell and Roth (1981) measured pollen cost as just 1.2% of the total cost of an *Agave* inflorescence. Rameau and Gouyon (1991) confirmed a negative correlation between pollen grains per flower and seed production using *Gladiolus*, indicating a trade-off of costs between two expensive items.

The Cost of Making Flowers

As well as financing the rewards within its flowers, a plant also invests in the physical structures and chemical components of each of its flowers. Most studies have assessed the cost of a flower by treating the flower as a whole physical unit; but the accessory advertising costs (notably scents and pigments) cannot be ignored, although they often involve chemicals that the plant is making in its vegetative parts anyway.

That flowers are costly is obvious from the greatly enhanced flowering that can be produced by providing additional resources through watering and feeding, as is evident to any gardener or farmer. The effects, however, can vary greatly with life history. Hence the **monocarpic** *Ipomopsis aggregata* quickly produced more, larger, and longer-lasting flowers after soil fertilization, along with more nectar and better pollen receipt, whereas the perennial *Linum lewisii* showed no effects until the year after fertilization, when overall biomass and seed production were enhanced (Burkle and Irwin 2009).

Primack and Hall (1990) tracked lady's slipper orchids (*Cypripedium acaule*), which produce a single large flower per year that amounts to about 18% of the plant's dry weight, but which has no nectar reward. A single successful pollination (by a bumblebee) produced a seed capsule with a few thousand seeds, and

TABLE 10.1
Percentage Dry Weights of Floral Components in Plants with Different Mating Systems

	No. species recorded	Calyx (%)	Corolla (%)	Stamens (%)	Pistils (%)
Outcrossers	12	20.5	40.8	27.8	12.3
Facultative outcrossers	11	41.9	27.8	23.1	10.3
Facultative selfers	13	48.5	23.3	14.7	23.7

Source: From Cruden and Lyon 1985.

this investment in flower and seed reduced leaf area the following year by 10%–13%, also reducing the plant's probability of flowering at all; overall, the costly effects of a single successful flowering event could persist for at least four years. This is rather an extreme case, but evidently the costs of making flowers themselves are not trivial.

Ashman and Baker (1992) assessed the costs of different floral parts for *Sidalcea* (fig. 10.1): sepals and petals dominated the budget, and the androecium required greater biomass (but not very much more nitrogen and phosphorus) than the gynoecium. However, it is important to use dynamic estimates of costs here (Ashman 1994), since some constituents of the flower (especially nitrogen and phosphorus) can be resorbed and recycled during floral senescence. The large petal and sepal structures also have high surface-area-to-volume ratios and so may incur a cost in water loss that could be significant in more arid climates.

Some studies have specifically addressed the relative costs of the male and female parts. C.A. Smith and Evenson (1978) investigated the energy contents of various structures in *Amaryllis* (Amaryllidaceae) and found that 30% of the total energy of the flower went into the ovary and anthers and only 9% into the gametes (ovules and pollen). Furthermore, investment in ovules was only 30% that of investment in pollen, although once the full protective structures linked to the ovaries and anthers were taken into account this pattern was reversed, with 17% allocation to female structures and 11% to male structures. Similarly, in *Smyrnium* (Apiaceae) Lovett Doust and Harper (1980) reported that the male gametes were costlier than the female gametes, but that maternal care from the protective structures reversed the total costs per sex.

The mating system in the plant will certainly influence flower costs in relation to male and female func-

tions, though. In the dioecious shrub *Oemleria*, Antos and Allen (1994) showed that the attraction features were markedly heavier in the male flowers (petals and hypanthium constituted 63% of the reproductive biomass, compared with 50% in females), whereas in *Solanum carolinense*, which has both perfect and female-sterile flowers, the latter were significantly cheaper to produce (Vallejo-Marin and Rausher 2007). Cruden and Lyon (1985) found differential allocations to floral parts in selfing and outcrossing plants (table 10.1), with proportionately more investment in the corollas and stamens in outcrossers and in the calyx and pistils in selfers. Pollinators may also affect the costs to the plant by their discrimination of male- and female-phase flowers, being especially deleterious to the plant's costs if they avoid the female flowers; for example, Bierzychudek (1987) found only 21% of all visits were made to female *Antennaria* flowers although these made up 44% of the population, and female flower visits were also of much shorter duration (fig. 10.2).

Galen et al. (1993), however, pointed out that in some plants the flowers themselves contribute significant photosynthate to the plant, meeting the carbon costs of sexual reproduction; in the snow buttercup, the female organs (carpels) had particularly high photosynthetic rates, which balanced the costs of attracting pollinators, perhaps making the female organs less costly overall than male organs.

The Cost of Pollination-Dependent Reproduction

How expensive is sexual reproduction itself? Assessing this requires careful microcalorimetry of plant parts, but it is not easy to give meaningful values for a species because the balance of sexual and asexual reproduction in so many plants depends on the conditions,

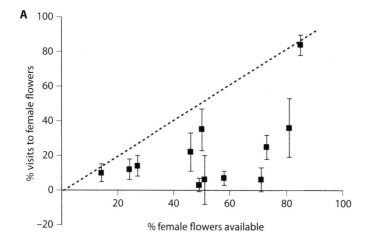

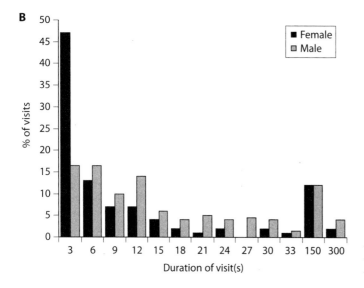

Figure 10.2 Visits to male and female flowers of *Antennaria*: (A) the proportion of visits to females is much lower than their percentage availability (with SDs shown); (B) frequency distribution of visit durations, usually longer to male flowers (medians are 4 s for females and 11 s for males). (Modified from Bierzychudek 1987.)

with sexual effort often increasing with poor soils, stressful climatic conditions, competition from other plants, etc. Reproduction generally does not occur until the plant reaches a certain size, so its timing and its relative cost can be massively affected by these abiotic and biotic factors.

Data from several studies are shown in table 10.2; most of these found a trade-off between sexual and vegetative reproduction (Evenson 1983; Obeso 2002), and many found the cost of sexual reproduction to be less than 10% of the whole life budget, although not infrequently up to 30%. At the extreme, in a **semelparous** century plant (*Agave palmeri*) pollinated by bats, Howell and Roth (1981) recorded a massive 60%

investment in sexual reproduction in the single flowering episode occurring in the mature plant (one flower stalk, with 1,265 flowers, and with an energy content of 70,000 kJ).

Where plants are unisexual, females tend to be smaller than males in woody species, an effect attributed to the costs of reproduction, but oddly, females tend to be larger in herbaceous plants. M. Harris and Pannell (2008) suggested that this results from the relatively higher nitrogen content of herbs, so that male growth and nitrogenous pollen production are more compromised than female growth. In *Mercurialis annua* they showed that males devoted more of their biomass to nitrogen-harvesting roots to offset the

TABLE 10.2
Costs of Reproduction in Plants as Percent Dry Weight Allocation

Family	Species	Total reproductive effort (%)	Sexual (%)	Asexual (%)
Onagraceae	*Circaea quadrisulcata*		3–6	12
Liliaceae	*Allium* spp.		27–55	1–3
	Trillium erectum		8–11	
Ranunculaceae	*Ranunculus repens*	48–60		
Fabaceae	*Trifolium repens*	20	2	18
	T. pratense	12	0	
	Lupinus nanus	61	29	32
	L. variicolor	18	5	13
	L. arboreus	20	6	14
Asteraceae	*Achillea millefolium*	29	2–3	26
	Artemisia vulgaris	11	2–3	9
	Cirsium arvense	22	7	15
	Taraxacum officinale	38	25	13
	Tussilago farfara	46	26	20
	Solidago canadensis	12	1–2	
Scrophulariaceae	*Mimulus primuloides*	41–56	2–14	
Palmae	*Astrocaryum mexicanum*	35	35	0
Typhaceae	*Typha latifolia*	14	6	8

Source: Based on data and sources in Evenson 1983.

costs of pollen production. Females, in contrast, paid for their more carbon-rich seeds by investing in photosynthesizing tissues.

Other Possible Costs

Although the previous sections have outlined the main costs incurred by a plant in achieving flowering and sexual reproduction, there may be some additional minor expenditures to consider that will vary with habitat and circumstances. In hotter climates there may be a cost to keeping flowers cool (perhaps by evaporative water loss or flower movements to improve shading). In arctic and alpine zones, the costs incurred in keeping a flower warmer than ambient may be considerable: less so if achieved passively by shape and reflec-

tivity, but more so if active heliotropic movements or thermogenesis are involved (chapter 9). At least in this case, though, the costs will be partly offset by the resulting greater attraction of the flowers to pollinators.

There may also be significant costs of defending flowers, since they normally make the plant more conspicuous and may bring in both florivores and herbivores. Chapter 25 will consider these aspects more fully.

3. The Costs Incurred by a Flower-Visiting Animal

For a foraging animal, the costs depend on its energy expenditure, which must not exceed its energy gain from the flowers visited. The *gain* can be assessed as

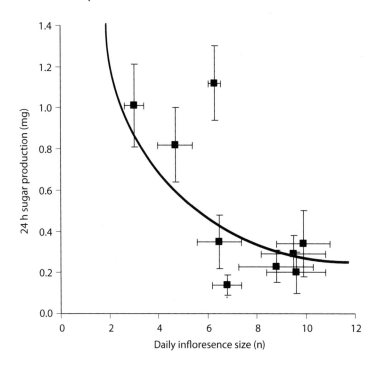

Figure 10.3 The relatively poor fit of sugar production with inflorescence size, across nine species of legume (with means and 95% CL), against a predicted line. (Redrawn from Harder and Cruzan 1990.)

the product of the mean energy obtained per flower and the number of flowers visited per unit time, both relatively easily measured factors, although there are complications in determining the mean nectar reward because it fluctuates spatially and temporally and in response to visitation (chapter 8). The net energy uptake rate will depend on the nectar available and the probing time of the visitor per flower. Probing time comprises both access time and ingestion time, the former varying with depth or other aspects of corolla morphology and the latter varying with nectar concentration and viscosity. Harder and Cruzan (1990) tried to model the relationship between nectar production and certain floral features, such as corolla depth and number of flowers per inflorescence, and found no good single predictors that could be used across a variety of plants. Sugar production per flower was expected to decrease with inflorescence size, but in fact, it often increased; its relation with flower size was variable, sometimes but not always increasing in large flowers; and flower size showed no clear relation with inflorescence size. Figure 10.3 shows the poor fit between the observed and predicted relations of sugar production to daily inflorescence size across nine legume species that they tested. In general there seems to be no easy way to predict nectar availability and the

associated costs for a particular plant; rather, the relevant traits have to be measured.

The energy *expenditure* for a given animal will be determined by both inherent physiological factors and by the patterns and timings of its movements on and between flowers.

Physiology-Dependent Costs

Some costs depend upon metabolic rate and will vary with body size and with thermal strategy. Whereas most pollinators are essentially **ectothermic**, deriving any extra body heat from the sun by basking (which produces a raised body temperature, T_b), some larger insect flower visitors and all the common vertebrate flower visitors are **endothermic**, with a metabolic rate that is roughly ten times that for a comparably sized ectotherm. For birds and mammals endothermy is nearly always obligate, the raised metabolic activity is a permanent feature of their lives, and T_b values are normally 36°C–40°C at all times. The metabolic rates of nectarivorous vertebrates do not diverge from the expected value for their body sizes, although temporary exceptions do occur in small hummingbirds and a few very small bats and marsupials, which cannot get enough fuel to maintain a fully raised T_b through the night and become torpid to save energy, lowering their metabolic rate and T_b many degrees below their normal active setting (temporal heterothermy).

For insects such as bees and many large moths, together with a few flies and beetles, all of which are stoutly built and insulated by hairs or scales, their endothermic ability is facultative, and they are often therefore called **heterotherms**. These animals warm up their flight muscles metabolically via a "shivering" mechanism (muscle activity uncoupled from muscle shortening), which raises the T_b prior to take-off in cooler ambient temperatures. This ability is present in many well-insulated insects above about 30 mg body weight (Stone and Willmer 1989a); below this size, heat loss by convective cooling is too great to allow significant warm-up. Among the bees, bumblebees have particularly sophisticated thermoregulatory mechanisms. As the temperature rises, their successively employed mechanisms include: endothermic warm-up by shivering before flight with heat retained in the thorax; cessation of flying to allow additional shivering; shunting heat from the thorax to the abdomen for excess heat dissipation; cessation of flying to cool down; and evaporative cooling from regurgitated nectar. Using these strategies they can partition their body heat between the head, thorax, and abdomen as needed (Heinrich 1979a, 1993). Some desert bees also have particularly strong endothermic warm-up abilities, which may facilitate flight in the cool early mornings (Stone and Willmer 1989a; Willmer and Stone 2004). Honeybees, however, are only moderately endothermic and have less elaborate regulatory mechanisms than *Bombus* (Heinrich 1980, 1993; Coelho 1991; Heinrich and Esch 1997), although the cavity-nesters (*A. mellifera, cerana*) have a better ability to warm up than do the less high-powered, open-nesting species (*A. dorsata, florea* and *laboriosa*) (Underwood 1991).

Given that endotherms have tenfold higher metabolic rates, any level of endothermy absolutely depends on high-calorie food inputs, so that the demands placed on received floral rewards are much greater for birds and bats and rodents, and are especially high for hummingbirds and some bats that can hover at flowers. For both sunbirds and bats, the daily energy expenditures are among the highest recorded for any animals, with a daily calorie turnover that is about 65% of the total body caloric content. This value is higher still

for hummingbirds (Winter and von Helversen 2001), partly because their flight is a little more expensive than that of bats. Scale also becomes very important: larger species of bird have larger energy requirements and may stay longer at any one plant, because the costs of flying scale up faster ($W^{0.97}$, Tucker 1974) than the costs of sitting and feeding ($W^{0.73}$, L. Wolf et al. 1975). So from simple energetic considerations, larger birds may probe more flowers per plant, and the same probably applies to bats.

But remember that the heterothermic bees and hawkmoths also have very high energy demands: a bumblebee foraging in early spring at 5°C requires two to three times more energy than the same bee foraging in summer at 25°C (Heinrich 1979a). Larger moths and bees can adjust both their thermal strategy and their foraging style to take account of variable nectar rewards and changing ambient temperatures, as will be discussed in section 4, *Environmental Effects on Plants and Animals*, below; and several species show higher body temperatures with higher nectar sugar content. In fact, Mapalad et al. (2008) showed that *Bombus* adjusted their thoracic temperature not just in relation to nectar but also according to the quality of pollen they fed upon, being warmer both while feeding and on return to the nest when pollen protein levels were high. In all these cases it is not entirely clear whether a raised body temperature was adaptive via increased flight speeds or reduced return times or both.

Locomotory Costs

The costs of moving to flower patches and moving between flowers are a crucial part of a flower-visiting animal's energy budget, and the incurred costs will depend on the distance between flowers or flower patches and on the distances the animal has to travel from its nest site or resting place to the flowers (inter-patch and between-patch costs). These costs are very different depending on whether the visitor is flying or walking and also on whether it lands or hovers while feeding. Table 10.3 summarizes some of the comparative costs of these styles of locomotion. Note that hovering in front of flowers is especially costly; for example, a 100-mg bee that lands on a flower for a minute may expend just 0.3 J, while a similarly sized hawkmoth hovering for the same period expends at least 100 times more energy. But the costs of hovering are not that

TABLE 10.3

Energy Expenditures of Different Kinds of Pollinator

Pollinator	Activity	Body Mass (g)	Energy use (J g^{-1} h^{-1})	Power consumption (W)	Mass specific power* (W kg^{-1})	Sugar used (mg h^{-1})	Reference(s)
Butterfly	Flying	0.2	58				Data in May 1988
	Feeding	0.2	1.6				Data in May 1988
Hawkmoth	Resting		2–10				
	Hovering	1	1,250	0.35	350–500	75	Heinrich 1975b, 1979a
Honeybee	Flying	0.1	~800				Heinrich 1980, 1993
	Resting	0.1	1–3				
Bumblebee	Flying	0.3	~1,000				Heinrich 1975b, 1993
	Resting	0.3	2–10				
Small hummingbird	Hovering	5	900	1.25	261	270	Heinrich 1975a
	Flying	5	760	1.06		230	L. Wolf et al. 1975
	Resting	5	250	0.35		75	L. Wolf and Hainsworth 1971
Large hummingbird	Hovering	13.5	900	3.39		730	L. Wolf et al. 1975
Sunbird	Flying	13.5	836	3.14		680	L. Wolf et al. 1975
	Feeding/flying	13.5	322	1.21		260	L. Wolf et al. 1975
Bat	Hovering	7–18		1.1–2.6	159		Voigt and Winter 1999
	Foraging	8–9	220				Voigt et al. 2006
Marsupial (*Tarsipes*)	Foraging	5–12	140				Nagy et al. 1995
	Resting	5–12	52				

Note: Some data are recalculated from the source given.

*These data are means for the groups (from Voigt and Winter 1999).

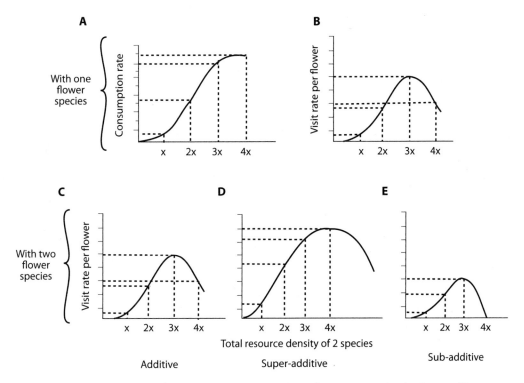

Figure 10.4 Functional responses and visit rate for pollinators with one floral species (A, B) and various possible effects when there are two floral species interacting (C–E). (Redrawn from Geber and Moeller 2006.)

much greater than those of forward flight, as is sometimes assumed; and since the mass-specific power used in hovering declines with body size, a bat expends the least energy per unit of body weight of all the hoverers (Voigt and Winter 1999; and fig. 16.9B), although this effect is probably limited to a maximum hovering size of about 30–35 g.

Within a given taxon, costs also vary in relation to wing loading (the ratio of body mass to wing area), so that energy consumption in flight is higher for species with higher wing loading (e.g., Heinrich 1993), and flower visitors with relatively smaller wings may need to select more rewarding flowers or more clumped flowers if they are to make a profit. The ability to carry fuel reserves may also come into play, since most insects have rather low lipid or carbohydrate reserves (e.g., May 1992; Heinrich 1993) relative to larger vertebrate flower visitors. Wing loading also varies with the amount of pollen and nectar carried, and in honeybees, pollen foragers were shown to have a metabolic rate 10% higher than nectar foragers, along with higher thoracic temperatures and a more horizontal body

position while flying (Feuerbacher et al. 2003). However, pollen loading (at a typical 18% of body mass) did not affect wingbeat frequency or wing action in that study, although the wing power output increased by 16%–18%.

Within a patch, the locomotory costs relate to the visitor's *functional response* (how its consumption rate changes with resource density, as shown in fig. 10.4A). In general, the consumption rate increases with the flower density up to some asymptotic value, where handling time is limiting. For more specialist, "choosy" flower visitors the functional response is likely to be saturating (Type II); whereas for generalists that switch between available flower types freely, or for foragers on particularly high-density patches, or for animals that improve handling markedly by learning, it is more likely to be sigmoidal (Type III). At the same time, the visitation rate per flower will normally rise to a peak and then start to decline as resource density increases (fig 10.4B). Analyzing the functional responses and visit rates in mixed species communities is more complicated (Geber and Moeller 2006), with flower preference and flower constancy coming into play, and overall visitation rates (fig 10.4C–E) may then increase or decrease (depending additionally on whether the floral species are very similar to handle and whether their rewards are equivalent). If visitation rates increase, though, the relative contribution of locomotory costs within the patch will be reduced.

Locomotory costs between patches, or between patch and nest, are often the greater part of the budget. For some more generalist flower visitors the distances travelled on a given day may be relatively small, and where there is no home base, an animal can simply move through the landscape in any direction favored by weather and flower abundance. With bees, birds, and bats all having a home or roost that is regularly returned to, we instead see **central-place foraging**. Here the distances travelled on any one foraging trip can be many meters or kilometers; for example, honeybees routinely travel 600–800 m, but some individuals fly several kilometers from their nest. With the estimate that 95% of the foraging occurs within 6 km of the nest, the total foraging area for honeybees would still be more than 100 km^2 (Beekman and Ratniecks 2000). Bumblebees, flying at 11–20 km h^{-1} (Heinrich 1979a) may range even further afield (Osborne et al. 1999; Walther-Hellwig and Frankl 2000). However, for all but the social insects the actual time spent flying to feeding sites can be quite a small proportion of the total daily time budget: for example, most hummingbird species spent more than 80% of their time resting at the nest or in trees and only around 5%–15% of their time foraging (L. Wolf and Hainsworth 1971).

Feeding Costs

Feeding in itself is probably not vastly expensive for any flower feeders, but the costs of handling and processing the rewards must be factored in to any overall budget. For nectar these costs vary with the concentration and viscosity of nectar and the tongue action of the forager (a topic introduced in chapter 8). For an endotherm there is also a cost incurred in warming the nectar to body temperature, which for a hummingbird feeding on cold nectar in North American sites can require a significant increase in metabolic rate (Lotz et al. 2003). For an active pollen feeder, there will also be costs relating to the difficulty of extracting pollen from the anthers, whether by scraping and scrabbling or by

sonication (chapter 7). Roughly speaking, the energy contents of individual flowers are highly correlated with mean foraging profits at those flowers, the link being clearest for nectarivores (e.g., May 1988 for butterflies). Any kind of food gathering may also show dependence on the temperature or humidity (both affect nectar, and humidity can affect the pollen's stickiness).

The major cost not yet covered is the time and locomotion invested, after arrival, in searching out the rewards within a flower, whether nectar and pollen or specialities such as oils or fragrances. Where a preferred visitor is a high-energy forager (hawkmoths, large bees, birds, and bats), it may be economically preferable for the plant to reduce these costs by investing heavily in easy detectability (e.g., by large size or conspicuous position or gaudy advertisement) and thereby save a little on the nectar reward it must provide to offset the visitors' costs.

Other Costs

Although rarely considered, it is important not to forget other costs that may come into play on some or all of an animal's foraging trips. These would include the potential risks (and therefore costs) of being attacked on or near flowers by predators or the costs of being infected by parasites. Even if not caught by these attackers, there may be costs of simply avoiding or evading undue harassing attention by predators or parasites, or indeed of potential mates, any of which may use the flowers as useful meeting places. These factors can lead to foragers moving around and between flowers more quickly and more often, potentially improving the outcrossing of the plant but adding to the time and energy used up by the animal.

Overall Costs

Foraging is costly. A 500-mg bee spends around 600 J h^{-1} in flight, which would require an intake of around 40 mg of glucose just to cover the average costs of arriving at the flowers, before any profit could be made. Collecting food from more distant flowers is therefore only worthwhile if they are offering unusually high rewards, and the threshold reward at which making the trip becomes economically feasible can rise exponen-

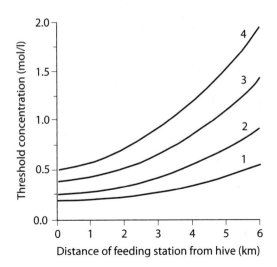

Figure 10.5 Nectar concentration threshold for recruitment in *Apis* increases with distance from the nest (and on days 1–4, when conditions varied) and hence with profit. (Redrawn from Schoonhoven et al. 2005, based on Boch 1956.)

tially with distance from the nest. Within certain limits (bearing in mind the problems with unduly high nectar concentration discussed in chapter 8), either the concentration of the nectar supply needs to be higher (fig. 10.5) or, alternatively, the volume per flower must be much higher. Nevertheless, for bees (which are highly efficient flyers) the travel costs of foraging at 4–5 km from the nest may still be only around 10% of the nectar yield of such a trip (T. Wolf et al. 1999), given the very high loads that they can return with (perhaps 20% body weight of pollen, or up to 90% body weight of nectar, for a bumblebee; Heinrich 1979a).

Energy intake needs will scale with body mass (M) both intra- and interspecifically, so that smaller individuals should forage and respond somewhat differently from large conspecifics. Using orchid bees (euglossines), Borrell (2007) calculated that the energy intake rate should scale proportionately with $M^{1.0}$, and the optimal sugar concentration should be independent of body size, but he found that in reality, energy intake rate scaled with $M^{0.54}$ partly because proboscis length was *not* directly dependent on body mass. Thus increasing tongue length tended to reduce both the energy intake rate and preferred sugar concentration, as expected (chapter 8), and most euglossines preferred nectars at 34–42% concentration regardless of body size.

For a given flower visitor, it is relatively straightforward to measure the rate of nectar intake (and hence of

calories) from measured rates of visitation and knowledge of the mean sugar content of the flowers visited (at that time, and in that place). We can also estimate the energy used in foraging from the observed activity schedules, knowing the time spent per flower, the time flying, the time resting and/or grooming, and the overall trip length. This calculation is where pollination biology starts to overlap very substantially with foraging ecology, and it is no accident that flower visitors have often been the animals of choice for theorists working on foraging behavior. It is rather easy to work out exactly how many calories a visitor will get per flower and the variance between flowers and between plants; to calculate how many calories it costs them to sit drinking, to move between flowers by flight or by walking, and to move between plants; and to estimate costs of moving between flower patches or from home (nest) to flower patch. Hence the entire foraging transaction can be modeled rather neatly, and it is a small step to begin testing possible foraging models. For example, are the visitors

1. maximizing their net rate of energy gain (rewards), or
2. maximizing energy efficiency (rewards minus costs), or
3. minimizing risk by selecting the lowest variance source?

Many assessments of nectar-feeding animals have shown that they maximize the net intake rate, so that (unlike pollen feeders), nectarivores tend to be strongly affected by time costs. However, in the case of social bees (where the literature is vast, because they are so much easier to work with than most other visitors), they may instead maximize their net foraging efficiency or minimize their risk. Maximizing efficiency reflects their need to provide the greatest daily delivery of resources to the nest, and it requires these social pollinators to visit more flowers per patch or per inflorescence, and to work longer at each flower. A preference for minimum-variance flowers leads to **risk-sensitive, or risk-averse foraging**, and is commonplace in *Apis* and *Bombus* species in many temperate studies (chapter 18; and reviews by Cartar and Abrahams 1996; Perez and Waddington 1996). Learning alters foraging tactics as well, since relatively naive workers behave very differently from an experienced bee. In particular, handling time can be reduced by learning (fig. 10.6), often by as much as tenfold after an appropriate

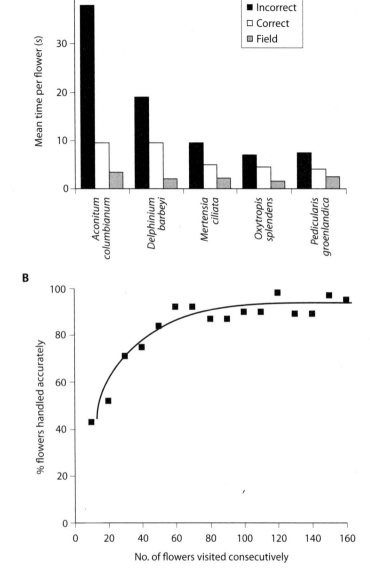

Figure 10.6 (A) Mean visit duration for correct and incorrect flower visits for inexperienced bumblebees and the much faster visits of experienced bees in the field (drawn from data in Laverty 1980). (B) The improvement of accurate handling with time and experience as more *Impatiens* flowers are visited consecutively (redrawn from Heinrich 1979a).

learning period, although even an experienced bee on a very complex flower may get a lower reward than if feeding at a simpler flower.

Studies on other kinds of flower visitor are less common, but Wolf et al. (1972) compared three species of hummingbirds foraging at two species of *Heliconia* and calculated the nectar taken per flower per bird per visit. They found that nectar extraction efficiency was highest for the plant species with more concentrated nectar, but that the smallest of the three birds performed the best because while it took nectar at the

same volumetric rate as other birds, it used less energy in collecting it. Mendonca and Dos Anjos (2006) calculated the activity costs for Brazilian hummingbirds at up to 30 kJ per day, where the flowers they visited (mainly *Palicourea*, Rubiaceae) yielded just 66.5 J each through much of the day, so that birds were visiting several hundred flowers per day to make a profit.

Tactics and costs for any one visitor will vary not just in relation to its own behavior, but also with regard to that of others present on the flowers, introducing competition into the calculations. Where a mixed population of visitors is at work on a flower patch, which will be steadily depleted of nectar, each species is likely to have different thresholds at which they quit as the nectar reward lessens toward or below their particular break-even point. If foraging costs scale in accordance with body mass (e.g., Heinrich and Raven 1972; Heinrich 1993), then all else being equal, the largest visitors would be predicted to leave first, whereas the smaller ones should be the last to leave. However, body size and tongue length are sometimes in conflict: on more tubular flowers, where a long tongue is needed to reach the receding nectar levels, a larger insect with a longer tongue may be able to work the flowers successfully when the smaller insects can no longer reach any nectar. In essence, there is an economic trade-off here between cost thresholds and depth thresholds (Corbet 2006). This is particularly evident in highly specialist relationships such as the trap-lining hummingbirds and large euglossine bees, which move predictably between widely spaced tropical canopy trees.

4. Environmental Effects on Plants and Animals

Thus far, we have considered the direct costs to the plant and to the animal of flowering and of flower visiting. However, it should be evident that both sides of this balance are in turn affected by variation in the environment, and it is worth exploring these effects on the economics of pollination more explicitly.

Environmental Effects on the Plant

Overall Plant Growth
Plant growth is affected by many aspects of the environment, most obviously light levels, soil nutrients and water, and temperature. Hence the resources available for investment in producing above-ground tissues and a large attractive plant will be strongly affected by environmental variation, and the potential for investing in flowers rather than vegetative tissues will be similarly prone to environmental controls (see *Floral Display*, below).

Floral Rewards
Nectar secretion patterns are influenced by insolation, and the resultant concentration and volume available

in a given flower depend critically on temperature and humidity (chapter 8). There will also be effects of precipitation (rain) and of wind, since both of these will alter the microclimate and humidity around flowers. Figure 10.7 summarizes the overall effects of environmental variation on nectar.

Pollen production may also be affected by environmental factors, for example, by insolation or light levels (e.g., Etterson and Galloway 2002, working with *Campanula*). The time of dehiscence is almost certainly influenced by temperature and humidity, in at least some plants; the example of dog's mercury given in chapter 7 is an extreme case. The stickiness of pollen surfaces also seems to be modified by humidity, which will potentially affect handling time. Hence bees, which require particularly large amounts of both pollen and nectar, may partition their efforts throughout a day or across several days as needed; they tend to forage more for pollen in warm and windy weather, and especially at low humidities, avoiding pollen with dew or rainwater on it (Peat and Goulson 2005), whereas in more humid conditions (often in the early morning) they will gather nectar, because it will often be less concentrated and more manageable.

Floral Display
Investment in sexual reproductive organs (flowers) is bound to vary substantially with the environment. Overall floral display on a given plant is a composite function of flower size, flower numbers, flower spacing, and floral longevity (chapter 21); and the summed local display of many plants can also be important in attracting visitors. There have been surprisingly few direct tests of environmental effects on individual flowers. Work on *Datura* (Elle and Hare 2002) showed that irrigation produced longer showier corollas, although it did not directly affect outcrossing rates. Sánchez-Lafuente et al. (2005) explored a range of environmental effects with *Helleborus* in Spain and recorded changes across habitats in all the plants traits they measured: plant and flower size, floral display, corolla length, the number of stamens, and the number of nectaries. However, only the overall floral display had a significant positive effect on visitation.

Another major influence, as yet somewhat unexplored, comes from edaphic factors, particularly the interactions of a plant with its underground mutualists (notably the mycorrhizal fungi that are associated with plant roots and that affect nutrient uptake). When the

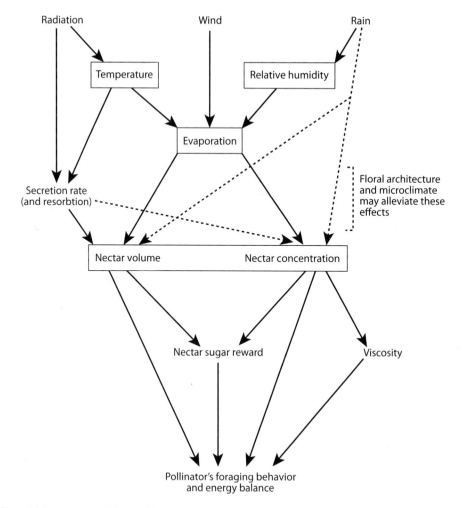

Figure 10.7 An overview of the possible environmental effects on nectar parameters, with floral architecture and microclimate as intermediaries.

soil fungi in a flower-rich grassland were suppressed by the application of fungicide, the pollinator community shifted from larger bees to smaller bees and flies, with an average 67% reduction in visits per flower stem across 23 plant species, which was attributed to reduced floral display (Cahill et al. 2008). There clearly needs to be increased attention paid to the effects on flowering and pollination of this below-ground trophic level.

Structures that protect the more delicate parts of flowers from the environment should also be factored in, including enlarged bracts beneath the corolla that act as supports and a defense against thieves, or bracts suspended above and around corollas to shield against excessive radiation or rain damage to pollen (as, for example, in the handkerchief tree *Davidia;* J. Sun et al. 2008). Investment in such structures could be part of the economic considerations for the plant.

The timing of flowering can be especially crucial (chapter 21). Having many flowers at once (**big bang flowering**) may attract more pollinators, but it also drains the plant's energy reserves rapidly, and it increases the risk of selfing, so there is an economic trade-off for the plant that must be considered (Harder and Barrett 1995; Schaffer and Schaffer 1979; Klinkhamer and de Jong 1993; Klinkhamer et al. 1994). And optimal floral longevity is also part of the economics (Ashman and Schoen 1994): a plant must balance the cost of maintaining an existing flower (increasing the overall display) relative to the cost of making a new one. Here the balance depends on the rate at which pollen is received, the rate at which pollen is exported, how much pollen is needed for full fertilization, and how quickly the flowers become visually unattractive (wilting, shrivelling, or generally changing their appearance). In practice, it may be economically helpful to keep old (functionless) flowers on a plant as long-range attractants, so long as a visitor can then detect their poor status on closer inspection and not waste time visiting them. Some plants do retain old flowers, which provide additional showiness, and couple this with signals (subtle color or scent change) that direct the visitor's probings to new flowers (chapters 5 and 6).

Note that any or all of the factors affecting optimal flower longevity may again depend on temperature and humidity (directly or via effects on the visitors). There may also be much less subtle, weather-related effects on display: in more fragile flowers the effects of prolonged rain (soggy brown petal clumps in place of attractive flowers), of especially intense exposure to sun (bleaching and rapid desiccation of petals), and even of strong winds that dislodge flowers from their petioles are all going to influence pollination outcomes.

Environmental factors can also affect the allocation of resources to sex within a plant. For example, resource supplementation (fertilizer and water) with *Asclepias* increased its fruit set, largely through effects on female fitness components, which indicates a degree of gender-specific selection (Caruso et al. 2005). Even the sex of flowers on a plant may be affected by the environment, especially with some hybrid crops (e.g., cucumbers) where growers can manipulate conditions to alter the sex ratios of plants or of flowers and so achieve desirably seed-free fruits.

All of this suggests, again, that the plant and the environment together may be "controlling" the situation and the rewards, and the visitors just have to follow. The plant is in control of factors that affect the handling time for a visitor, at the level of single flowers (nectar volume and concentration, rate of replenishment), and also controls factors that affect the travel time for a visitor (spatial effects within plants—such as the location of rewards with height or on the inside or outside of a bush—and between plants). Additionally, the plant controls floral longevity, which in com-

bination with the rate of flower production governs overall attraction to visitors via the daily and seasonal offering of rewards. For all these reasons, it might be thought that if a plant grows in the right sort of environment and if appropriate visitors are about, then all will be well in terms of pollination.

Environmental Effects on Animal Visitors

Overall Animal Distribution

At a very general level, geographical and local distributions of animal taxa and populations impose extremely obvious limitations on plant-pollinator interactions. A few examples will suffice: hummingbirds only occur in the New World tropics, bumblebees are mainly cool temperate species, carpenter bees are mainly warm temperate, and stingless bees mainly tropical, whereas functionally equivalent beetles, moths, and many kinds of flies occur in almost all habitats.

Indirect Effects

Any potential pollinator will be affected by the environment in terms of the availability of its other needs beyond access to flowers. Above all, such necessities may include other food resources, because many of the pollinating taxa have additional (nonfloral) needs; for example, birds and bats usually eat some insects as well as nectar. Even groups that rely almost entirely on flower foods will require an environment that provides enough floral diversity; many bees need more than one flower type either simultaneously or sequentially to get adequate quantities or qualities of nectar and pollen to raise a brood. Beyond the demands of food, many pollinators also need an environment that provides shelter: birds and bees need nesting sites (wild bees nest in the ground, in sandy banks, in dead trees, in snail shells, etc.), and bats need safe daytime roosting sites.

Direct Effects of Temperature and Radiation

Thermal effects on pollinators are often crucial determinants of their behavior and are now reasonably well known and useful as predictors of flower-visiting activities (Willmer 1983; C. Herrera 1995a,b). Clues as to the effects of these thermal constraints come readily from looking at patterns of insect visits to one plant, where different visitors are active at different times according to their own thermal costs. Figure 10.8 gives an

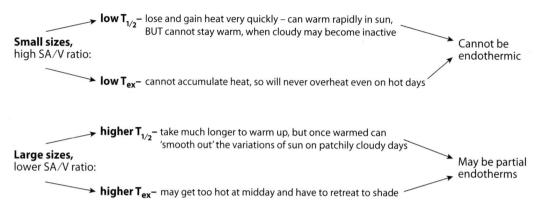

Figure 10.8 An overview of the thermal relationships and needs of insects in different size ranges; the partially endothermic insects are in the minority but include important pollinators among the larger bees, beetles, and moths.

overview of the effects of temperature on thermal strategies in animals of varying sizes and thermal properties. For flower visitors, this means that small, brightly colored, highly reflective species can forage in full sun at midday, but larger, darker-colored, or well-insulated hairy species may get too hot at such times. Conversely the latter types may be able to absorb enough radiation to warm up in the early morning and late afternoon, and thus visit flowers at times when the smaller foragers are too cold for activity. Any species with an endothermic capacity can circumvent these restrictions and be active at any time even without insolation but may still run the risk of being too hot if flying in full sun.

These effects of size, color, and physiology lead to the kinds of activity distributions shown in figure 10.9 for lime trees (*Tilia*). The flowers produced abundant nectar throughout a day, so there were few reward constraints, and the temporal patterns of insect types and frequencies were correlated primarily with radiation. Heterothermic bumblebees visited early in the morning in cool air but avoided the middle of the day, when they would risk overheating. Ectotherms, including small bees and most flies, foraged only in the warmer midday conditions. Since thermal exchanges and costs depend critically on body mass (which affects conduction, convection, and radiation) and on the reflectance of body surfaces (mainly affecting radiative exchanges), there were consistent patterns of mean body size and color (reflectance) through a day for visitors to single flower species (*Tilia* and *Heracleum*), as shown in figure 10.10.

For these reasons, the activities of flower visitors

are *not* always very well correlated with nectar reward. Instead, for many smaller insects, the timing of activities is strongly linked with the weather and with the thermal costs they experience while foraging (and remember also that flowers may offer a microclimate that can facilitate warming and reduce the energy costs of insect visitors). From this it follows that a plant may be following the visitors' needs rather than the other way round. A given insect may pick a less rewarding flower because it is in sunlight or may choose to visit it later in the day when it is cooler but also when the flower has fewer rewards available.

The exceptions to these constraints are the heterotherms and endotherms, which can be active independent of the environment by making their own metabolic heat, but which then need much larger rewards from flowers to make a profit and tend to be more strongly dependent on the nectar reward, seeking out the most rewarding flowers irrespective of where and when they occur. Such considerations may in part explain why rewards in temperate flowers visited by larger bees are often highest in the early morning (when bees can be active but most other insects are still torpid), and why flowers growing at higher (and therefore cooler) latitudes secrete more nectar than the same species at lower latitudes (Heinrich and Raven 1972). At least for bees, foraging behavior may also be directly affected by the warmth of the flower (Dyer et al. 2006), with bumblebees preferring a warmer foraging reward in the absence of any nutritional advantage and warmth itself thus becoming an additional reward, as was described in chapter 9.

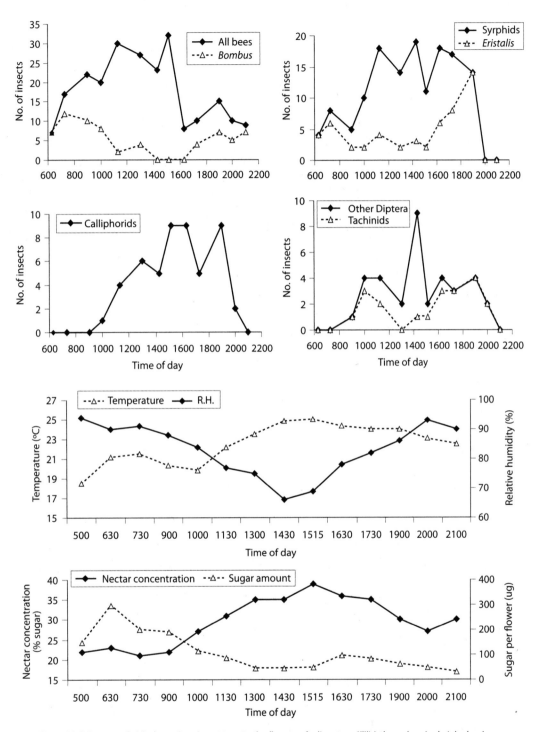

Figure 10.9 Patterns of visits by various insect taxa to the flowers of a lime tree (*Tilia*) through a single July day, in relation to the microclimate and nectar availability. Taxa with open triangles are those with a degree of endothermic warm-up ability. (Redrawn from Willmer 1983.)

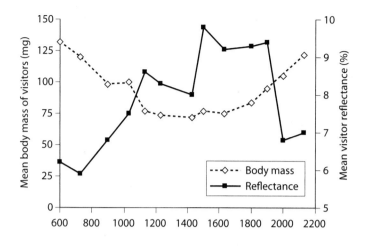

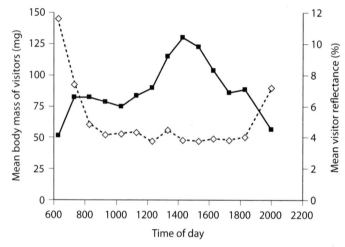

Figure 10.10 Consistent patterns of mean body mass and reflectance (color) patterns through a diurnal cycle, for two floral types: lime (*Tilia*) (above) and hogweed (*Heracleum*) (below), with larger, darker species present in the morning and evening, and smaller and brighter species foraging in higher temperatures around midday. (Redrawn from Willmer 1983.)

On the basis of these kinds of results, we can make sensible predictions for all types of foragers (knowing their physiological thermal strategy and their size and color) as to when and where they will visit flowers, which flowers they will prefer, what situations they will favor, and how (or whether) they are likely to fit in to the overall pollination strategy of a given plant. We can also predict that some plants may bet-hedge their nectar offering to attract some partially or fully endothermic pollinators through a cool early dawn period yet still have nectar left that can attract other nonspecialist visitors later in the day, an especially useful trait when the weather is poor or unpredictable, or where the visitors are scarce in the early morning (e.g., Heinrich and Raven 1972).

Physiological constraints may also affect where, in a broader sense, animals visit flowers. For example,

Roubik (1993) worked on the stratification of bee visits in tropical canopies, showing that larger euglossine bees tended to forage high up and more in the open, where it was windier, thus facilitating convective cooling in these large dark hairy bees; whereas other smaller and shinier bees were more active in sites at low strata in the forest, where there were consistently high ambient temperatures and little air movement. Nocturnal bees also tended to prefer the upper canopy.

Thus it should be clear that the environment affects both the visitors and the plants but that perhaps the visitors and their foraging costs are often the more crucial factor. Animals can *only* visit flowers when and where they do not get too warm (or too cold), and if they have no other water source, then also when and where they get enough (but preferably not too much)

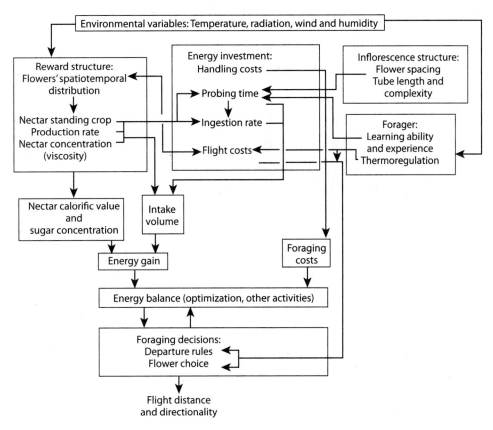

Figure 10.11 A summary of energy budgets for a nectarivorous flower visitor. (Redrawn from Dafni 1992.)

water from the nectar. Here, then, we have reasons why the animals, in concert with the environment, are often "in control" of flower-visiting patterns. They can choose flowers within and between plants according to their own needs; according to the age, stage, and position of the flower on the plant; and in relation to the microclimate, perhaps using some distant reward assessment and perhaps also leaving pheromone signals to facilitate locating flowers again or avoiding revisiting emptied flowers. Thus they have control over their own rate of visiting, their handling time (hovering versus landing, flying versus crawling, etc.), and their flower constancy. The plant, therefore, rather than setting the agenda, may just have to *respond* to the visitor's choices with an appropriate reward to keep it happy. From this perspective, flowers might be expected to organize their rewards, their time and place of flowering, etc., to give the greatest benefit (economic pay off) to the most efficient of their potential visitors.

5. Resulting Interactions

An obvious message from this chapter is that flower-visitor interactions are economic transactions, but that components of the environment control both of the partners, directly or indirectly, and so affect the balance point of the system and the outcome of the transaction in any given time and place. The resulting patterns can clearly be very complex, and figure 10.11 summarizes all the parameters that may feed into pollinator energy budgets. Both plants and insects are differentially affected by environmental factors, and patterning through time and space is the result. Hence communities of pollinators do differ very considerably on spatiotemporal axes. A good example was described by Eckhart (1992) for flower visitors working on *Phacelia*, where the commonest foragers varied greatly across three populations (fig. 10.12) and between different parts of the season. On a smaller scale, Willmer and Corbet (1981) illustrated some of these

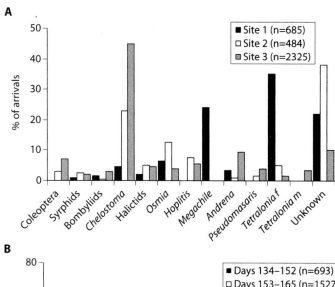

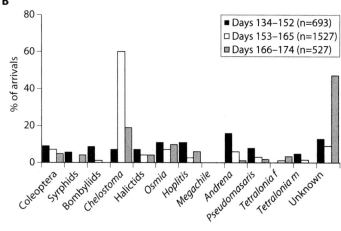

Figure 10.12 Visitors to *Phacelia* flowers, varying substantially (A) for three different sites, and (B) for three successive seasonal periods. (Redrawn from Eckhart 1992.)

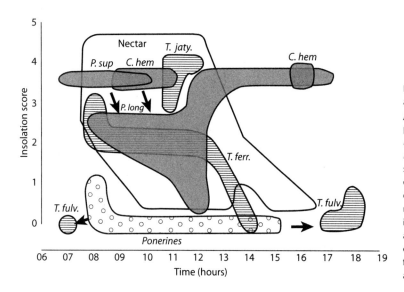

Figure 10.13 Patterns of visitors through time and against climatic conditions (insolation) to *Justicia aurea* flowers in Costa Rica. Contour peaks are shown for birds (solid gray: *Phaethornis longuemareus*, *P. superciliosus*, and *Campylopterus hemileucurus*, all preferring more sunlit sites); for bees (horizontal stripes: larger and darker *Trigona fulviventris* active in shaded sites early and late, *T. ferricauda* in moderately sunny sites, and the very small *T. jaty* active only in sunny midday periods); and for ponerine ants (circles), persistent in shaded areas. The contour for maximum nectar availability in the flowers is also shown. (Modified from Willmer and Corbet 1981.)

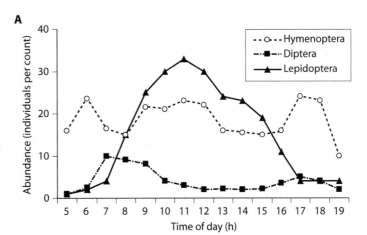

Figure 10.14 (A) Diel patterns of activity for three major taxa of insects on lavender. (B) Detailed patterns for the four main bee visitors, differing in size, with the larger bees active morning and evening and the smaller bees around midday. (C) Resultant pollen transfer rates, as grains per stigma per minute, through the day. (Redrawn from data in C. Herrera 1990.)

issues with a hummingbird-pollinated plant, *Justicia aurea,* where other visitors (stingless bees and ants) were partitioned in time and microclimatic space in ways that could be predicted from their sizes and reflectances (fig. 10.13). Similar effects of color and size on the activities of stingless bees were documented by Pereboom and Biesmeijer (2003).

C. Herrera (1990) also addressed this complexity with *Lavandula* and showed that daily cycles in the plants and their various pollinating visitors did not match up neatly. Pollen availability peaked in the late afternoon, nectar around midday, and various animal visitors showed different diel patterns (fig. 10.14A,B) that related to their own needs. Hence both the pollen transfer effectiveness for the plant and the average distance pollen was moved varied markedly throughout a day (fig. 10.14C), a point to bear in mind when we come back to the issue of pollinator effectiveness in the next chapter.

The idea that any particular plant has coevolved with a particular pollinator, so that just cataloguing its key floral traits will predict who its pollinator is, is often an oversimplification. In reality, in a diurnally and seasonally changing environment, any one plant is likely to be visited by, and potentially may therefore be pollinated by, a range of different visitors at different times of day, in different weathers, and through the period of flowering; and the efficiency of pollination may likewise vary through time and space. Early in its season, a plant may primarily attract a large endothermic bee, which visits in the early mornings, but as the season progresses and warms, the same plant may also

be attractive to smaller visitors with higher ambient-temperature thresholds, who visit in the sunnier hours of each day. Some of these visitors—although by no means necessarily all of them—may also be effective as pollinators. Hence it should be no surprise that a classic hummingbird flower, *Ipomopsis,* was often visited by bumblebees in a year when birds were rare and the nectar abundant enough in the long corolla for some of it to be reached and to give a profit (Pleasants and Waser 1985); put simply, economics can sometimes override syndrome matching.

Environmental interactions may also explain why a plant species can vary in its pollination in different parts of its natural range. Work on *Aconitum* (monkshood) in the Californian Sierra by Brink (1980) afforded a good example. This flower is almost exclusively pollinated by bumblebees, and at high altitudes where it is cooler, the flowers have particularly long nectar spurs, which are especially suited to larger bees with longer tongues, thus achieving good cross-pollination and seed set with little need for asexual reproduction. However, at lower altitudes there are fewer visits by *Bombus,* which are outcompeted in these milder climes by *Apis* and various solitary bees; here the same monkshood species has much shorter spurs, suited to these shorter-tongued bees, but pollination is rather poor, and the plant can make numerous asexual bulbils as a safety net.

Again, such instances stress the need to look carefully at energy budgets—gains and losses—in individual pollination systems, not simplistically or in a generalized way, but in terms of a specific plant in its specific

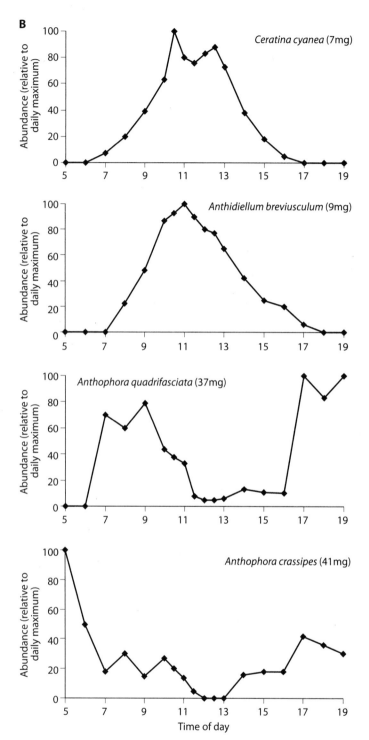

Figure 10.14 *(Continued)*

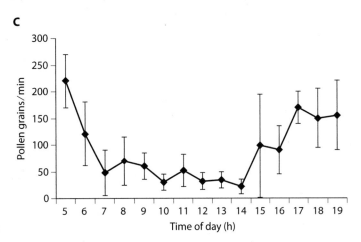

Figure 10.14 *(Continued)*

environment and as part of the local economic community. In most situations the many possible visitors have both different relative costs and different relative effectiveness as pollinators to the plants. There may be quite variable problems and solutions to the issues of achieving pollination even for the same species, due to local environmental effects on both flowers and their visitors; so it is crucial to study specific plants, in defined sites with good environmental measures, and to avoid the easy generalizations.

PART III

POLLINATION SYNDROMES?

Chapter 11

TYPES OF FLOWER VISITORS: SYNDROMES, CONSTANCY, AND EFFECTIVENESS

Outline

In the next few chapters, different kinds of flower visitor are reviewed in some depth, using the literature accumulated for over a century documenting their flower visits and floral selection, their color and scent preferences, their food and energy requirements, and aspects of their behavior on flowers. Here, this is all set in context with an explicit introduction to the idea of pollination syndromes, and a review of evidence that syndromes are both real and useful. Without this background, a newcomer to the field might have difficulty understanding or disentangling the arguments. Once the syndromes themselves have been covered, chapter 20 offers a critical analysis of the syndrome concept in theory and in practice.

The key issue to emerge is the balance of specialization and generalization—whether pollinators are commonly specialist for particular kinds of flower and/or flowers are specifically adapted to attract and be pollinated by particular animals. If either or both of these propositions are supported, then the use of the syndrome framework in pollination makes good sense. But if most flower visitors are very generalist in their flower choices, and most plants receive visits from and are pollinated by a whole range of different animals, then we would probably be better off leaving the syndrome concept behind us. Hence specialization and generalization are issues at the forefront of modern pollination studies, and will be the focus of concern in chapter 20.

1. Pollination Syndromes

It is evident that flowers show enormous adaptive radiation, but also that the same kind of flower reappears by convergent evolution in many different families. Thus many families produce rather similar, simple bowl-shaped flowers like buttercups; many produce similar zygomorphic tubular lipped flowers; and many produce fluffy flower heads of massed (often white) florets. These broad flower types are the basis of the idea of pollination syndromes—the flowers have converged on certain morphologies and reward patterns because they are exploiting the abilities and preferences of particular kinds of visitor. Thus we can speak of hummingbird-type flowers, long-tongued-bee flowers, and so on. In this view, most angiosperms are seen as sufficiently specialized to produce recognizable suites of convergent traits that recur in flowers of very different evolutionary origin but that share similar pollinators. The underlying message is that different kinds of flowers attract different visitors, cementing the mutualisms and by implication tending to make flowers increasingly specialist, and visitors increasingly selective.

This idea of floral types and associated pollination syndromes is far from new: it appeared in papers of the

TABLE 11.1
Classical Pollination Syndromes

Syndrome	Pollinator	Timing of anthesis	Main colors	Nectar guides	Scent
Cantharophily	Beetles	Day or night	Cream, green, usually dull	No	Strong, fruity or fermenting
Myophily	Flies	Usually day	White, yellow, greenish	No	Usually mild, not sweet
Sapromyophily	Carrion and dung flies	Day or night	Purple/red/brown, mottled	No	Strong, decaying meat or feces
Psychophily	Butterflies	Day	Red, orange, yellow, mauve	Maybe	Slight to moderate, sweet
Phalaenophily	Most moths	Dusk, night	Cream, yellow, greenish	No	Fairly strong, sweet
Sphingophily	Hawkmoths	Dusk, night	White, cream, pale green	No	Strong and sweet
Melittophily	Bees	Dawn, day	Pink/purple/blue, white, yellow	Yes	Moderate, usually sweet
Ornithophily	Birds	Day	Red, orange	No	Usually none
Chiropterophily	Bats	Dusk, night 1 night only	Dull white, dull beige/green	No	Strong, fruity or fermenting

1860s and 1870s by Delpino and was taken up in the early twentieth century by Knuth. Vogel (1954) published an explicit summary of these ideas, which were further elaborated by van der Pijl (1961), Baker and Hurd (1968), and Faegri and van der Pijl (1979). The study of syndromes has been of enormous interest, because they serve as good models for adaptive evolution more generally and give us a focus for understanding key processes of natural selection.

Waser (2006) offered a good historical review of the syndrome idea but suggested that it arose initially out of an expectation of harmony and an intrinsic order in nature and has been insufficiently held up to critical analysis. He offered a substantial critique of the pervasive effects of syndrome-based thinking in the literature. These perceived problems with the pollination syndrome orthodoxy are explored in depth in chapter 20; but since the concept has been so dominant and is built in to much of the literature, any student of pollination needs to know about it in some detail, as a background for reviewing its strengths and weaknesses and assessing its current utility. Table 11.1 therefore shows

Shape	Nectar site	Nectar volume and concentration	Pollen amount	Pollen deposited
Radial, flat or bowl-shaped	Exposed	Low V—mid C	Mid/high	Face, legs, underside
Radial, flat; or flat inflorescence	Usually exposed	Low V—mid/ high C	Low/mid	Legs, face, thorax
Radial or bilateral ± deep with trap	None		Mid	Most of body
Small, long tube, often *en masse*	Concealed	Low V—low C	Low	Face, tongue, (± legs)
Usually radial, moderate tube	Concealed	Low/mid V—low C	Low	Face, tongue
Usually radial, long tube or spur	Concealed	Mid V—low/ mid C	Low	Face, tongue
Bilateral or radial, Exposed or short/ medium tube	Exposed or Concealed	Mid V—mid C	Mid	Head, dorsal, or ventral body
Bilateral or radial, short/medium tube	Concealed	High V—low C	Low	Forehead, beak, throat
Bilateral or radial, bowl or brush	Usually exposed	High V—low/ mid C	High	Face, head

the features of the basic pollination syndromes, as set out in many texts since the 1960s, and this table underlies the next seven chapters of this book.

Few would doubt that the syndrome idea has had its uses as a convenient pattern recognition system. Pollination biologists have become accustomed to speaking in syndrome terms, and find little difficulty in doing so, recognizing that there are certain "nodes" in a multidimensional phenotypic space that occur over and over again across taxonomic boundaries. It is often very useful that an observer with no prior knowledge of a particular flower's visitors can make an intelligent guess as to the likely pollinators, especially if it is possible to measure the nectar and observe the timing of flower opening and dehiscence. Identification of syndromes and broad flower types has also permitted a rough analysis of how these vary in frequency in different habitats, or in parts of one habitat: for example, it is possible to document that there are more fly-type flowers at high latitude, or more large bee and moth flowers in a tropical forest canopy and more small bee, hummingbird, butterfly, and beetle flowers on the forest floor (see chapter 27). Recognition of such patterns can be the first step in seeking the reasons for them and in understanding the community as a whole. Whether the patterns that observers have thought they perceived are really supported by the evidence is the theme for chapter 20.

If syndromes are real and important, they must be dependent on key features of the visiting animals. Some of these features are simply *anatomical*, concerning the physical matching of an animal body (or body parts) to flower structure. Others are *behavioral* and cognitive: the learning abilities and foraging and flower handling patterns of visitors. Others again are *physiological*: how thermal control systems affect visitors' ability to forage at particular times and places. A fourth suite of features are broadly *ecological*, relating to life history patterns and matching phenologies. Each of these themes is taken up elsewhere, but some factors that underlie an appreciation of syndromes—in particular, aspects of floral constancy and of visitor effectiveness as pollinators—are put in context below.

2. Why Pollination Syndromes Can Be Defended

A point that needs to be stressed at the outset is that pollination syndromes have always been recognized (often quite explicitly) as statistical rather than absolute constructs, suggesting that particular sets of floral characters were likely to be over-represented in plant species that were visited by particular types of visitor. The idea of a syndrome was never intended to be diagnostic, such that mere observation of a flower's characters could indicate unequivocally what its pollinators would be. The main authors of the syndrome idea were fully aware that pollinator communities and guilds vary enormously, with some flowers getting just one visitor type and others at the far end of the continuum getting many different visitors, varying in time and space. For example, van der Pijl (1961) described his flower-type groupings as "classes with bad boundaries but a clear center." Hence terms like "fly flower" and "bee flower" were intended only as approximations, to be tested by empirical study. It is unfortunate that the syndrome idea has come under fire for being something much more than it was ever meant to be—there is some element of attacking a straw man here.

Convergent Character Suites Do Exist

There are clear and irrefutable arguments, set out in chapters 12–18, for real convergence of traits across unrelated flowers—in morphology and color and perhaps most clearly and compellingly in scent—for such categories as bird, or bee, or bat flowers. Particular floral traits do act as attractants for particular kinds of visiting animal and as filters against other kinds of visitor. Thomson and Wilson (2008) used the term "evolutionary attractors" in considering the convergent states toward which phenotypes are drawn, with intermediate phenotypes rare, and discussed how evolutionary transitions between such states might occur (notably through different pollen-transfer efficiencies of visitors, loss of function in pigment pathways, or mutations with large effects on flower phenotypes). A few examples of suites of floral traits as phenotypic nodes within a genus should suffice here:

1. In the genus *Passiflora*, despite very complex flower morphology (fig. 11.1), species occur with distinctive suites of features that correspond with bee, bird, and bat pollination (Varassin et al. 2001); table 11.2 shows details. The quantities and type of nectar, the color of the flowers, and above all the scents and volatile components all match with the traditional syndrome suites for these visitors; bee and bat flowers show markedly stronger scents with more complex fragrance compositions. In at least one species, the Andean *P. mixta*, there is a single effective pollinator (the sword-billed hummingbird, *Ensifera ensifera*), which is declining due to forest fragmentation and may lead the plant into extinction (Lindberg and Olesen 2001).

2. In the genus *Penstemon*, Wilson et al. (2004, 2006) reported that the great majority of about 270 described species are predominantly bee pollinated, and these show the characteristic colors, corolla shapes, and anther exsertions of bee flowers. But they noted that about 40 species show the suite of adaptations suited to hummingbirds (red colors, longer, narrower corolla tubes with less pronounced lower lips, and markedly exserted and accessible anthers), and that these species also show a high frequency of hummingbird visits. This genus therefore yields clear examples of characteristic flower types that can readily be quantified and mapped in multidimensional ordinations, which in turn map on to pollinator change.

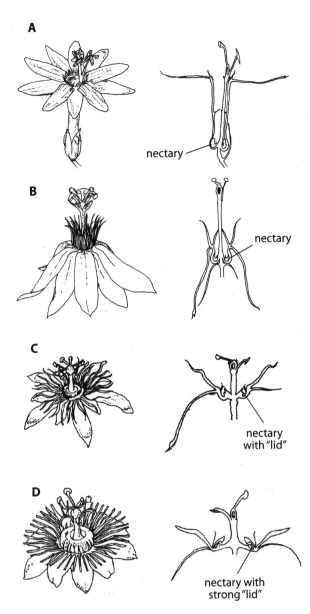

Figure 1.11 *Passiflora* species with different pollinators: (A) and (B) bird pollinated, usually red, with a moderate to long tube and anthers held high above (Part (B) is *P. coccinea*); (C) and (D) bee pollinated, usually white or purple, with a flat corolla, a protected lidded nectary, and anthers set only a short distance above and held horizontally when dehiscing (Part (C) is *P. rubra* and (D) is *P. holoserica*). (Redrawn from Endress 1994.)

3. Even within a species, there may be morphs across the distributional range that correspond with different visitors and visit frequencies. In southern California, *Mimulus aurantiacus* has red morphs coastally and yellow morphs at inland sites, the red morphs being favored by hummingbirds (95% preference in artificial arrays) and the yellow morphs showing greater than 99% hawkmoth visitation (Streisfeld and Kohn 2007).

It follows logically from this that pollinators can often be predicted from floral traits. Throughout the pollination literature, there are cases where investigators have predicted the likely pollinator of a flower from its visible and measurable characteristics and have then been able to prove the point. Examples can be found in many chapters of this book, but here is one recent case. Four sympatric species of morning glory (*Ipomoea*) were compared with regard to their traits and then to their visitors (Wolfe and Sowell 2006). Two (*I. hedereacea* and *I. trichocarpa*) were clearly bee types, with blue or purple flowers, large corolla openings, and small nectar volumes, while the other two (*I. quamoclit* and *I. hederifolia*) were bird types with red flowers, narrow openings, and large volumes of dilute nectar. The total pollinator fauna consisted of 11 species of hummingbird, bee, and lepidopteran, but the first two plant species were the only ones visited by bees (getting 75% of their visits from these), while the last two were the only ones visited by birds (although 80% of their visits were from sulfur butterflies). The authors conclude unequivocally that "pollination syndromes do aid in partitioning the pollinator fauna."

There are also many elegant examples of pollination biology detective work that make no sense except in the light of an appreciation of convergent pollination syndromes. The most famous is Darwin's successful prediction of a very long-tongued moth pollinator to match the exceptionally long-spurred corollas of a Madagascan orchid, the moth being discovered decades later (Nilsson 1988; and see chapter 14). Perhaps the neatest example, though, is the ieie plant, *Freycinetia arborea*, in Hawaii, anecdotally reported as pollinated by rats but evidently with few appropriate characters (see chapter 17). Cox (1982, 1983) watched the flowers much more carefully and saw visits only by an introduced bird, the Japanese white-eye; and characters of the flowers did fit with the perching bird syndrome (chapter 15). On checking historical records, Cox found at least three perching bird species

TABLE 11.2
Passiflora Species Exhibiting Different Syndromes

	Habit	Colors	Scent	Time of anthesis	Nectar concentration	Pollinators	Pollen placement
P. mucronata	Erect or inclined	White	Sweet	0300	19–32	Bats	Head
P. alata	Pendant	Red purple	Strong, sweet	0430	~ 45	Anthophorid and euglossine bees	Dorsal thorax
P. speciosa	Erect	Red	Faint, acrid	0530	35–40	Hummingbird (Phaethornis)	Head
P. galbana	Erect	White	Strong, fishy	1730	22–32	Bats	Head

Source: From Varassin et al. 2001.

recorded as visitors, all now extinct or very rare; and museum specimens of the extinct Kona crossbill and of the extinct oue both showed plentiful pollen matching that of the ieie flowers on their head feathers.

Character Suites Can Be Statistically Supported

Syndromes as statistical or morphometric constructs can themselves be tested statistically, and some efforts along those lines have begun. Jürgens (2006) tested the characteristics of 53 species of Caryophyllaceae (diurnal and nocturnal, some self-pollinating) against their pollination mode with a canonical discriminant analysis (CDA) and found significant correlations between floral morphometry and mode, the key characters being factors that defined nectar accessibility, relative positions of anthers and stigmas, and visual attractiveness of petals. The best overall diagnostic trait was the distance between the nectar source and the tip of styles, where contact with pollen was made, a measure that neatly aligns with the physical fit of the pollinator; and this measure also differed significantly between the pollination modes.

In a similar fashion, Smith et al. (2008) tested the pollinator shifts occurring in 15 species of *Iochroma* (Solanaceae) with various Bayesian tree models using

different patterns of trait evolution. They found that species with high nectar reward and large floral display were statistically more likely to be hummingbird pollinated, although corolla length and color were not significantly linked with particular groups. Tripp and Manos (2008) used the large genus *Ruellia* (Acanthaceae) for a similar rigorous analysis, finding significant floral differences among species with different groups of visitors, and with various reverse transitions apparent between hummingbird- and bee-visited species. Lazaro et al. (2008) assessed the associations between floral traits and pollinator identity in three Norwegian flowering communities and found significant relationships, although hoverflies and butterflies visited different plants in the three areas. Ecological generalization was more a property of the plants and pollinators present than of specific floral traits.

As a final example here, Marten-Rodriguez et al. (2009) mapped 11 floral traits across 19 species from the Antilles and 4 from Costa Rica from the family Gesneriaceae, to give clustering patterns in phenotypic space. They found two convincing clusters explicitly equating to hummingbird- and bat-pollination syndromes, and discriminant analysis correctly classified 19 of the 23 species into the predicted categories.

However, one large-scale study has refuted any statistical robustness across the whole range of pollination syndromes (Ollerton et al. 2009); this analysis will be considered again at some length in chapter 20.

Spatiotemporal Discrimination Is Critical

There are many existing data sets showing multiple visitors to particular plants, and these are widely used in support of generalization and incorporated in most published analytical models. But these studies commonly lack good spatiotemporal disaggregation of the data, rarely considering the varying patterns through time and space of either abundance or visitation, and so giving serious sampling artifacts.

For example, a long-lived (vertebrate) visitor may be recorded as visiting ten different plant species during a study, but may in practice have visited only one or two at any one time, switching between species through a season as their flowering peaked in sequence (see chapter 21). Hence a bat or bird can be termed a generalist if seen in overview but is effectively specialist at any one time, and this will be missed unless there is a good phenological data set. Thus studies over longer periods have tended to indicate more generalization in visitors than have shorter studies (see Ollerton and Cranmer 2002).

Aggregation of data on visitation from a broad spatial range carries similar problems, since a widespread species may interact with many visitors across its whole range but with only one or a few in any one locality (e.g., Fox and Morrow 1981). Some plant species vary in their pollination in different parts of their range, or at different times of flowering, and there may be local selective effects toward specialization in each zone (see Herrera, Castellanos et al. 2006; Johnson 2006). In practice, unless there is strong gene flow among populations, each relatively isolated population becomes adapted to its local pollinators, so that we may record generalization at the species level but obvious specialization at a regional or population level. There is some suggestion that where plant ranges are more affected by human activities they tend to be more generalist (see chapter 29), whereas natural ecosystems are more likely to show higher degrees of specialization (Johnson and Steiner 2000).

Thus what might seem to be a problem for a syndrome approach, where one plant with two flowering seasons has only one of these overlapping with its spe-cialist visitor, can be overcome without the plant necessarily being regarded as a generalist. As a temporal example, the globe mallow has visits from one species of *Diadasia* bee when it flowers in spring, and from three species of *Perdita* bees in autumn (see Minckley and Roulston 2006); but this can still be reasonably called a very specialist plant. Examples of variation across a spatial range include the *Mimulus* example mentioned above, and Brink's (1980) work on *Aconitum* at varying altitudes in California, described in chapter 10.

Plants Are Often "Bet-Hedgers"

Along similar lines, chapter 10 made it obvious that even quite specialized plants should "bet-hedge" the timing and patterning of their reward (typically their nectar offering), to try to attract the best (coadapted, coevolved, specialist) pollinators during the key early hours after anthesis and dehiscence, but potentially to still have enough reward left to attract other nonspecialist visitors later in the day (or over a few days). This will be particularly useful in habitats where weather can limit activity, such that conditions may be too poor for morning visitation (e.g., Harder and Barrett 1992). Some of the environmental effects on visitor timings (and on nectar) that would influence such effects were explicitly addressed in the last chapter.

This kind of bet-hedging of reward may set a limit on specialization, but certainly does not argue against a plant being as specialist as it can. And again, if timing of visitors is not taken into account, the plant will appear more generalist than is really the case in terms of its effective pollination.

There Are Advantages to Being a Specialist

One obvious advantage, to both plant and animal, lies with improved foraging efficiency. Examples occur in many of the following chapters, but three classic studies will make the point. First, Strickler (1979) evaluated different bee visitors to *Echium vulgare* and found that oligolectic bees could fill nest cells (and so produce offspring) more quickly than polylectic bees foraging on the same plant; one specialist bee (*Hoplitis anthocopoides*) visiting little else was much more

TABLE 11.3
Differing Efficiencies of Bee Visitors to *Echium vulgare*

Bee	Dry body mass (mg)	Flowers visited per stem	Anthers depleted of pollen	Handling time (s)	Pollen removed (mg)	Pollen collection rate (mg min⁻¹)	Pollen needed per cell (mg)	Time on flowers per offspring (min)
Specialist*								
Hoplitis anthocopoides	16.6	2.5	4.0	5.7	0.20	2.1	17.8	8.4
Generalists†								
Megachile	16.8	2.1	3.9	10.4	0.14	0.9	35.2	37.8
Osmia	14.3	–	3.7	15.7	0.19	0.8	14.4	17.4
Hoplitis producta	4.4	2.2	3.1	29.3	0.14	0.3	6.4	19.8
Ceratina	3.6	1.9	2.8	31.7	0.12	0.3	4.1	13.3

Source: Based on Strickler 1979.

* Taking pollen from this plant only; and acquiring pollen and nectar simultaneously during a visit.

† Taking pollen from several plant species; usually taking pollen and nectar separately, and/or from different species.

effective, on all measures (table 11.3). Second, specialist *Habropoda* bees were more efficient visitors than polylectic bees on blueberries (*Vaccinium ashei*; Cane and Payne 1988). And third, specialist long-tongued bumblebees such as *Bombus consobrinus* and *B. vagans* (or *B. hortorum* in Europe) handle the complex monkshood (*Aconitum*) flowers more efficiently than generalists (Laverty and Plowright 1988), rapidly becoming highly effective foragers for nectar, pollen, or both (fig. 11.2), and thus gaining this plant as an almost exclusive resource.

This leads to another potential advantage, that of reducing interspecific competition. For animals, this may be an unlikely driving force for specialization, if sister species in practice tend to have similar flower preferences; and, for bees at least, oligolectic species tend to specialize on hosts that are widely attractive to many generalist bees, where competition tends to be high (Minckley and Roulston 2006; and see chapter 18). But for plants, selection for efficient despatch and receipt of pollen when pollinator availability is limiting may be a key force in driving floral shift and eventual speciation (Johnson 2006).

Neither of these points means that specialist relationships will inevitably evolve; but they enhance the likelihood of specialization, at least in certain circumstances.

Specialization Is Not Necessarily a Risky Strategy

Although some one-to-one species interactions do carry extinction hazards, as in the passionflower example already mentioned, many or perhaps most specialists turn out to be not as much at risk as might have been thought, since most of them have fallback mechanisms of clonality, longevity, facultative selfing, and so on, that give a measure of reproductive assurance. Fenster and Marten-Rodriguez (2007) showed that facultative delayed autonomous selfing was in fact rather common in plants with specialized floral traits, especially those that were protandrous; and statistical analysis within the genus *Schizanthus* likewise found higher levels of delayed autonomous selfing in the more specialized species (Perez et al. 2009), serving as a safeguard against pollinator failure. In other words, there is no inconsistency to being both specialized and able to facultatively self-fertilize (which has sometimes been proposed as a paradox, acting against the evolution of specialization).

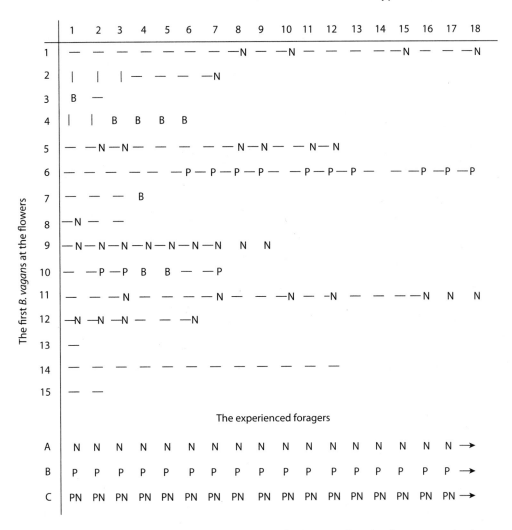

Figure 11.2 Bumblebees (*Bombus vagans*) foraging on *Aconitum* flowers; inexperienced workers represented at the top make incorrect approaches (vertical bars), or probe for nectar among the anthers (–), or visit buds (B), only occasionally achieving a correct N or P visit; whereas experienced foragers (three examples shown below) make repeated correct visits for N or P or both. (Modified from Laverty and Plowright 1988.)

One good specific example of effective fallback systems is *Ixianthes retizioides*, a South African shrub, which seems to be surviving very well in the long term despite the loss of its specialist oil-collecting bee (Steiner and Whitehead 1996). A further example comes from many of the agaves in North America, which are bat plants in all classical respects, but which are surviving with bee pollination where bat migration corridors or bat roosts have been disrupted— a scenario perhaps inevitably used by some (e.g., Slauson 2000) to argue against the existence of specialization.

Pollinator-Mediated Selection Exists for Particular Floral Traits

Selection on flower traits can be shown to operate unequivocally, and over relatively short time spans, often leading to more specialist floral designs. For example, there has been rapid evolution and speciation in *Aquilegia* (Hodges 1997), with greater species richness in the taxa with longer corollas, and especially rapid radiation following the acquisition of nectar spurs, a finding also supported by a more substantial phylogenetic analysis across the angiosperms (K. Kay et al.

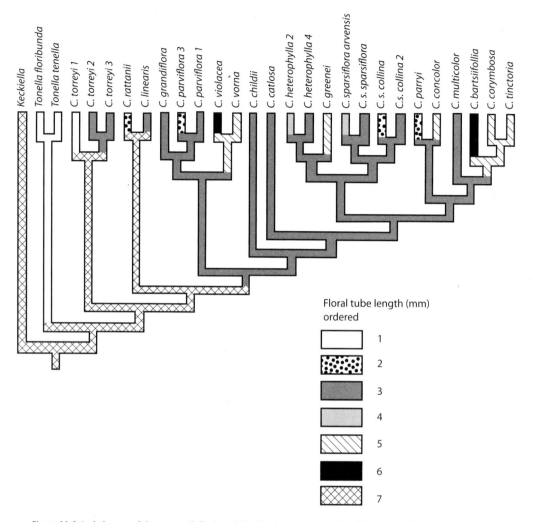

Figure 11.3 A phylogeny of the genera *Collinsia* and *Tonella*, showing the rapid radiation into relatively specialized species with different tube lengths in *Collinsia*, with *Tonella* staying generalized and with only two species. (Redrawn from Armbruster 2006.)

2006). This would indicate that increasing specialization can rather quickly lead to increased success within a clade (see also Dodd et al. 1999).

Similar analyses of shifts between specialist pollination syndromes have been documented in many other genera. In the African orchid genus *Disa*, Johnson et al. (1998) showed 27 species with multiple shifts, with 19 different pollination syndromes, including convergent origins of butterfly pollination twice, carpenter bee pollination twice, and nocturnal moth pollination four times. Effects were even more spectacular across the Iridaceae in southern Africa (Goldblatt and Manning 2006; see chapter 27 and details in table 27.1). Elongation of corolla tubes was clearly

specifically related to selection from hawkmoths within the genus *Gladiolus*, the species *G. longicollis* having tubes ranging from 56 to129 mm long and showing a positive relation between tube length and seed set from visiting moths with 85–135 mm tongues (Alexandersson and Johnson 2002). In the *Collinsia* and *Tonella* group (Scrophulariaceae; Armbruster 2006), the former has proved to be specialized and has 20 species, the latter unspecialized and with only two (fig. 11.3). The evolution at least twice of "funnel revolver" bird-pollinated flowers in *Nasa* is also of note here (Weigend and Gottschling 2006; see chapter 2); and a great many other examples will be seen in other chapters where a genus or family has repeatedly produced two or more flower types, each associated with particular pollinator types. In fact, much of the pollination literature from experienced field workers would make little or no sense if this obvious linkage between flower type and selection by principal pollinator type is removed.

Herrera's analysis (2001b) of variation in traits of various papilionate legumes in southern Europe is especially helpful in showing much more variation in traits unrelated to pollination (e.g., pedicel length, stem diameter) than in those features of petal and calyx that would affect visitors' flower choices. Based on this and many other examples from Mediterranean floras, Herrera, Castellanos et al. (2006) advocated the need for analysis of selective shifts in flowers at an intraspecific geographical level (raising the issue of spatial discrimination again), and their results with *Lavandula* do indicate local differentiation in floral traits driven by variable selection from local pollinators (fig. 11.4). Johnson (2006) likewise stressed and demonstrated the need for more intraspecific studies of plants mapped against the geographical pollinator mosaic, to detect local specialization; and, pursuing this theme, Anderson and Johnson (2008) provided a compelling example of local matching of flower depth and bee-fly proboscis length in *Zaluzianskya*, the fit of fly and flower materially affecting plant fitness across a substantial geographic range. Finally, several examples reveal that pollinator-mediated selection works strongly at the population level and not merely the species level. The same authors (2009) showed local convergent variations of corolla depth and the proboscis lengths of pollinating nemestrinid flies (*Prosoeca*) across guilds of up to 20 plant species, with the fly tongues varying between 20 and 50 mm in different sites. Again, Medel et al. (2007) documented geographic effects in *Mimulus luteus* in Chile: corolla size increased with the proportion of bees in the pollinator assemblage of a given locality, and nectar-guide size increased with the proportion of hummingbirds present, again indicating pollinator-mediated divergence. Even more strikingly, an Andean cactus (*Echinopsis ancistrophora*) showed extreme variation across 11 populations studied (Schlumpberger et al. 2009), with corolla lengths varying between 4.5 and 24.1 cm, the shortest flowers having morning anthesis and low nectar production (0–15 μl) while the longer flowers opened at dusk with prolific nectar (up to 170 μl). Sphingid moths were prevalent only in the four populations with the longest corollas, and all other populations were mainly pollinated by solitary bees.

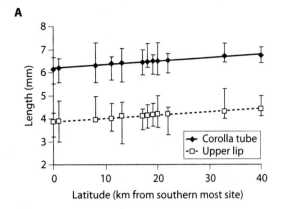

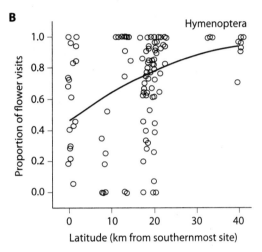

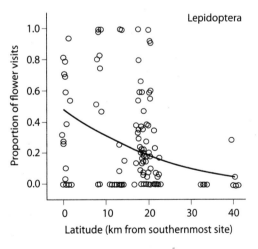

Figure 11.4 Local scale variation in *Lavandula latifolia* and its pollinators, in southern Mediterranean sites: (A) increases in corolla tube length and upper lip length, from south to north across 40 km, with vertical bars showing range; (B) proportions of flower visits due to Hymenoptera increasing (above) and Lepidoptera decreasing (below) across the same latitudinal range. (Redrawn from Herrera et al. 2006.)

A

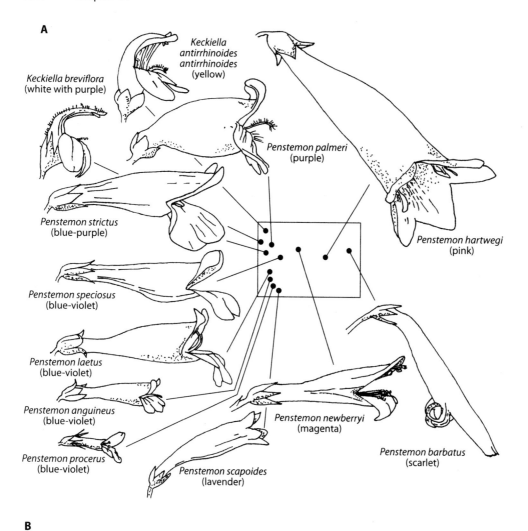

B

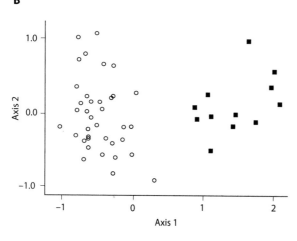

Figure 11.5 Bird- and bee-pollinated *Penstemon* species, and their characteristic floral traits: (A) selected species illustrated against an informal ordination (redrawn from Wilson et al. 2004), and (B) 49 species with known visitation patterns in a multidimensional scaling ordination where axis 1 correlates well with color and major shape traits, and with bird versus bee or wasp visitation, while axis 2 relates more to display size and anther to base distance; open circles are bee-pollinated species, and closed squares are bird-pollinated species; (C) a summary of the occurrence of bee (open circles) and hummingbird (closed circles) pollination against a parsimonious phylogeny of 194 species of *Penstemon* and *Keckiella*, with 23 shifts of syndrome required to produce the 29 bird-pollinated species (Parts (B) and (C) redrawn from Wilson et al. 2006).

C

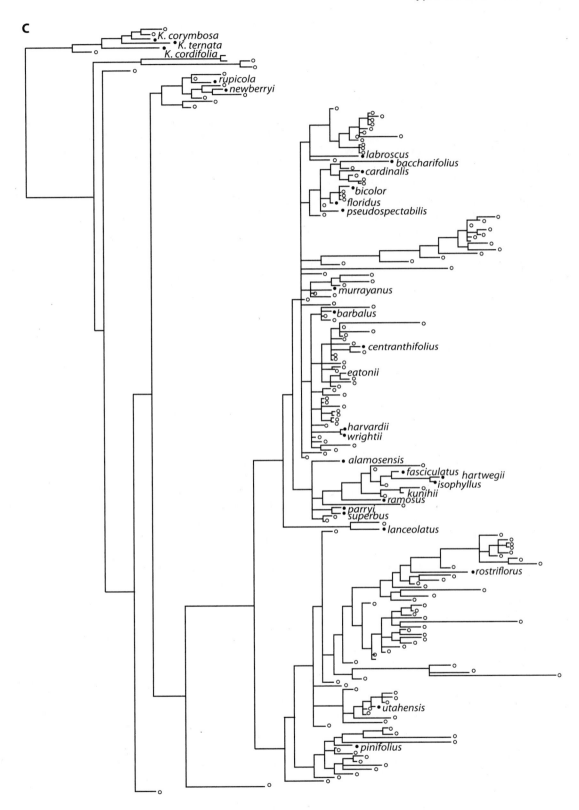

A particularly striking and persuasive example is the substantial work by Wilson et al. (2004, 2006) on the genus *Penstemon*, referred to under point (1) above; the association of hummingbird traits and visitation is shown in figure 11.5, A and B. These data have been mapped onto a phylogeny of the group, revealing multiple (at least 13, perhaps up to 24) separate and convergent shifts from ancestral bee traits to derived hummingbird-related adaptations; Wilson et al. (2007) used updated analyses to locate 21 separate origins of ornithophily. About half of these evolutionary shifts have produced species that exclude bee visitors seeking nectar, with the other half producing varying degrees of intermediates. There have also been single shifts to long-tongued fly flowers and to butterfly flowers, and a few shifts to specific visitation by large bees (*Xylocopa*—flowers with more open vestibules and shorter corollas) or small bees (*Osmia*—long narrow corollas). Figure 11.5C summarizes these findings. Many radiations to new species do not entail pollinator shift, but there are alternative stable character suites (i.e., syndromes), with relatively few species exhibiting the intermediate character suites. These authors detected fixed patterns and sequences for the evolution of hummingbird-related floral traits, interestingly with nectar modification as the likely first stage. They used the term "syndrome gradient" to describe the situation in this plant genus, with pollination shifts (from bee to bird) as evolution along this gradient.

Another study including careful phylogenetic mapping was that of *Schizanthus* by Perez et al. (2007), who showed that integration of corolla traits was clearly shaped by pollinator-mediated selection. Here, integrated character sets differed between species, and when plotted against a molecular phylogeny showed no relatedness patterns, indicating separate selection on different species acting on a palate of possible traits present in the genus.

Work on *Mimulus* (Bradshaw and Schemske 2003) took the evidence to the genetic level and showed that a single allele substitution at the *yellow upper* (*YUP*) *QTL* locus alters the color of *M. lewisii* flowers from pink to yellow-orange, and that of *M. cardinalis* from red to dark pink. Where these two species normally had overlapping ranges they were more than 99% reproductively isolated, attracting bumblebees and hummingbirds, respectively. After manipulating color changes by substituting the alternative YUP allele, *M. lewisii* received a 68-fold increase in hummingbird

visits (so was only slightly more likely to be visited by a bee than by a bird), and *M. cardinalis* received a 74-fold increase in bee visits. Thus an adaptive shift in pollinator preference and a resultant major change of specialization were initiated by a single major mutation. Very similar effects are also known in *Petunia*, where a single gene can cause color change and a major shift in the pollination syndrome (Hoballah et al. 2007).

To conclude this section, the widespread convergent evolution of suites of floral traits across often very distantly related plant families is, in itself, unambiguous evidence that pollinators have a major (perhaps the predominant) role in shaping floral adaptations (e.g., Fenster et al. 2004; Johnson 2006). The fact that we are now more aware of other interacting and sometimes opposing selective forces on flowers (see chapters 20 and 25) does not alter this key issue.

Constraints on Specialization Are Hard to Find

Despite the enormous literature on pollination biology, it is rather hard to pinpoint specific cases of developmental or genetic constraints that limit specialization, although the idea of such constraints has often been promulgated. It might be noted that several studies on *Raphanus* by Conner (reviewed in 2006) indicated that genetically correlated traits in flowers, which might be expected to be constrained, could in fact evolve independently rather easily; and other studies have usually found a lack of correlated constraint with other traits on key floral features such as color (e.g., Frey 2007). An analysis of floral and mating traits in *Clarkia* (Dudley et al. 2007) specifically concluded that there was no evidence of genetic constraints governing joint distributions of particular floral and plant traits.

Specialization and Multiple Distinct Syndromes Are Common in Tropical Ecosystems

There are now a number of good studies in previously underexamined areas of the tropics that do show the prevalence of specialization. One example comes from Sarawak, where out of 270 species studied across an entire forest system only 37 were considered to be diverse or generalist in their pollination system (Momose, Yumoto et al. 1998). Another case is that of high-altitude tropical grassland in Brazil, where careful observations showed that of 106 plant species one-third were monophilous and one-third oligophilous (defined as visited by one or two functional groups of pollinators), with the remaining third generalist (Freitas and Sazima 2006). (These authors regarded their ecosystem as unusually generalist relative to other Brazilian and Venezuelan zones, and linked that to the fragmentation of the high grasslands by repeated glaciations.)

The issue of specialization versus ecosystem type will be examined properly in chapter 20.

The Existence of a Generalist Syndrome

There is an idea in parts of the literature that the existence of very obviously generalist flowers—umbellifers, daisies, hawthorn, or lime blossom—inevitably undermines the concept of syndromes. This is extremely debatable. Such flowers were neatly described by Proctor et al. (1996) as "catering for the mass market"; but of course their market is that of the small and the short tongued, and they do not generally receive visits from long-tongued insects, or from vertebrates. They can reasonably be described as a syndrome in their own right, as in chapters 12 and 13, characterized as open, radial, bowl shaped, or flat; with accessible pollen; with exposed nectar at relatively high concentration and low volume; commonly white, cream, or yellow-green in color; and often very small and grouped together as inflorescences. Such flowers receive many visits from flies, beetles, smaller wasps, and short-tongued bees. The often repeated view that the syndrome classification fails to include most of these flowers is just misleading. A more balanced view is that this group of flowers are morphologically specialized to be ecological generalists (Corbet 2006). Even among the most classically generalist flowers such as *Heracleum*, the numerous visitors are not by any means all of equal effectiveness as pollinators (chapter 12; Zych 2002, 2007).

Increased Pollinator Diversity Promotes Plant Diversity

Pollination is primarily an animal-mediated phenomenon, but there seems no intrinsic reason why more diverse pollinator assemblages should increase plant diversity unless specialization is a rather important component of the animal-plant interactions. We saw in chapter 4 some evidence that plant diversity and pollinator group diversities went hand-in-hand through evolutionary time. On a more contemporary time scale, Fontaine et al. (2006) manipulated the functional diversity of pollinator assemblages (bees and hoverflies) in caged areas of an undisturbed meadow in France, and within two years found a 50% increase in plant species diversity in areas with high pollinator diversity, as well as a striking complementarity between functional groups of pollinators and plants. They did not use the currently out-of-favor term "syndromes" in their study, but it was clearly the number of syndromes present in the system that was increasing.

Many Flower Visitors Show Strong Floral Constancy

Most of the animal groups that visit flowers regularly do so in nonrandom patterns. They choose particular types of flower, and do so repeatedly. Moreover, they very commonly return to the same species of flower, or even the same patch of that species, trip after trip and day after day. This is a crucial component of the argument in favor of pollination syndromes and is therefore taken up below in more detail.

3. Flower Constancy

Floral syndromes must in part be dependent on floral constancy, which implies that a visitor moves sequentially and reliably among conspecific flowers, whether for pollen, nectar, or occasionally more unusual rewards. There may be strong selection for flower constancy in visitors, as it is potentially beneficial for both partners of the mutualism. Constancy to a flower species, and/or to a site, can improve the foraging efficiency and economy of the animal by minimizing travel distances, handling times, and overall foraging times, while also maximizing pollen packing; but it should be favored only where savings on handling time on the chosen plant exceed the costs of increased travel time between patches while other flowers are ignored. Flower constancy (and hence pollen constancy) is also crucial for the plant, helping to ensure reproductive isolation and maintain species differences.

All the advertisement signals and control of rewards in flowers could be selected to achieve both attraction and constancy.

The concept of flower constancy is not completely straightforward, though. First, it may be usefully divided into *passive constancy* (Thomson 1982), where only one plant species is in flower, or where flowering species are strongly aggregated and effectively enforce intraspecific movement, and *active constancy*, where several plant species are intermingled but only one is visited; only the latter is of real concern here. Second, it has to be distinguished from a general preference for one kind of flower phenotype. Flower constancy is sometimes termed flower fidelity, but this latter term has also been used in a broader sense to encompass both floral constancy (as seen in the temporal behavior of foragers) and oligolecty, referring to a more fixed feeding specialization relevant to a species or taxon rather than to individuals (Cane and Sipes 2006). In other words, specialization and active flower constancy may partly be imposed by floral characters, but sometimes are dictated by innate characters of the visitors. Thus the terms "innate fidelity" or "fixed preference" have also been used to refer to these latter cases, with "'learned fidelity" or "labile preference" as the alternative terms for constancy related to ongoing selection of flower features. In this book, the term "constancy" is used to avoid confusion, as we are mainly concerned with forager behavior. Even with this usage, though, flower constancy as a feature of an individual flower visitor can be used to refer to successive trips, or even trips on successive days (e.g., Free 1970b), but it is more commonly used to refer to behavior within a single foraging trip, a kind of "temporary oligolecty" in a given flower visitor. Flower constancy is then manifested as a tendency to specialize on flowers of a particular plant species, or even more specifically as a tendency to visit the same flower type as the one last visited.

For animals, floral constancy may be based upon flower shape, or color, or upon scents, or a combination of these. There may even be constancy to foraging height that can be exploited by a plant, for example in butterflies (Levin and Kerster 1973), honeybees (Faulkner 1976), and solitary bees (Frankie et al. 1983). In any of these cases, constancy requires some degree of learning based on recognition of features, and it may also require learned flower handling skills. However, it also requires a *limitation* on learning, pre-

cluding the visitor from remembering how to deal with too many flower types at any one time. A sensory system and behavioral motor patterns that can be temporarily fixed in one mode might be ideal, and this underpins the search image idea whereby flower visitors have a perceptual bias for a particular flower at a given time (Goulson 2000; Dukas and Kamil 2001; and see the review by Chittka et al. 1999).

But absolute flower constancy would be counterproductive to the animal, as the forager would never discover other more rewarding sources. Hence most flower-foraging animals regularly check other flower species. Among the bees, solitary species do it more than social ones; in the latter, the checking can be left to the scout bees, who monitor and pool information about other flowers available to the nest, and can then instruct the foragers on the best flowers to visit.

Measuring and Categorizing Flower Constancy

Using the simplest form of recording, there are innumerable direct observations of foraging, spanning nearly 150 years of work by naturalists and field biologists, to confirm that in natural settings many flower-visiting animals will regularly or repeatedly visit just one flower species, whether for nectar, or pollen, or both.

Records of pollen load composition are another useful indicator when assessing flower constancy, especially for bees where the pollen from many successive flowers is loaded into a scopa, and such records generally indicate quite strong floral constancy (table 11.4). But the table may mask even greater constancy in practice. In honeybees, Grant (1950) and Free (1963) recorded 90%–98% constancy to one plant species on a single trip (see also Hill et al. 1997), and in bumblebees Heinrich (1976a) found 55% of individuals with just one pollen type but another 32% with just two types (see also Heinrich et al. 1977; and Chittka et al. 1997). So even where pollen samples on the body are not pure they rarely contain more than two or three types and are strongly dominated by just one pollen form. An alternative assessment can be gained from the contents of individual cells from bee nests, and this again often indicates constancy over the many trips required to fill one cell (examples are given in chapter 18). These kinds of data might perhaps be biased by different ease of pickup of pollen types but still are good indications of flower constancy, especially in so-

TABLE 11.4
Bee Pollen Load Compositions (Relative Frequency)

Genus	% pure loads	
Andrena	44–68	
Halictus	75–84	
Megachile	65–75	
Anthophora	20	
Bombus	49–69	
B. lucorum	66	Free 1970b
B. muscorum	37	Free 1970b
Apis	62–94	Multiple sources
Trigona	88	White et al. 2001

Source: Data modified from sources in Grant (1950) except where shown.

cial bee species. (Note that they are not necessarily good indicators of pollen-specialization—oligolecty versus polylecty—for reasons discussed by Cane and Sipes 2006 and pursued in chapter 18). Records of interplant constancy or switching also give good constancy values for other bees, for example 77% in *Trigona* (Slaa, Cevaal, and Sommerheijer 1998) and around 43% in *Trichocolletes* (Gross 1992).

Bees are much easier to gather data from, by any or all of the techniques listed above, but a good degree of floral constancy has also been reported for butterflies (Lewis 1986, 1989; Goulson and Cory 1993; Goulson et al. 1997), for beetles (Pellmyr 1985; de Los Mozas Pascual and Domingo 1991), and for certain flies (Goulson and Wright 1998); details are given in the next few chapters as appropriate.

As a rough overview, constancy is lower in visitors that combine foraging with other tasks, that is, most nonsocial animals. Social bees only do flower visiting once they become foragers, but butterflies, beetles, and flies might be combining feeding, mating, and oviposition all in one trip. For example, egg-laying *Pieris* butterflies switch plant species more often than do males or nonovipositing females (Lewis 1989). Thus in groups other than bees constancy is improved if tasks can be kept separate, and this is often achieved by innate sensory recognition patterns, so that *Pieris* individuals will extend their tongues to blue or red target

colors but will tap their tarsi on yellow-green targets, indicating innate systems for distinguishing flowers from leaves. Another neat example comes from dunglies (Sarcophagidae), which prefer yellow colors linked with sweet scents and brown colors linked with carrion scents (here mistaking the flower for an egg-laying site), and therefore tend to visit *either* flowers *or* dung at different times and not to mix their visits.

All of the above could be criticized as somewhat anecdotal or indirect, or both. Several more precise indices of flower constancy have therefore been utilized for comparative purposes (see Dafni et al. 2005). They rely essentially on measures of floral visitation frequency or of floral visitation sequence. Some of the indices proposed are more suitable for laboratory choice tests than fieldwork. Waddington (1983) used probability matrices for all the possible transitions between flower types, but this gets complicated with more than three or four flower types. The Bateman index uses a calculation analogous to genetic analysis, where constancy can range from −1 to +1, but can produce strange results (e.g., where many visits in turn to the same plant are preceded and succeeded by single visits to another, which registers as random visitation). Levin (1970) instead used a fidelity index that divided constant transitions (between the same flower types) by all transitions made, and this has been more widely adopted and termed the *constancy index*, with a range of 0–1.

However, constancy also depends on plant density, and on season, so that no one measure is necessarily definitive for a particular foraging species.

Causes of Floral Constancy

Constancy probably has multiple causes (Chittka et al. 1999), but it is commonly attributed to limited cognition on the part of flower visitors, making them unable to store or process or recall information adequately about multiple flower types (Waser 1986; Lewis 1993; Dukas 1998; Goulson 2000; Menzel 2001). This theory can perhaps be subdivided into two versions:

1. Darwin ascribed constancy mainly to limitations on pollinators' motor learning, constraining their handling skills on particular flowers. Waser (1983b, 1986) formalized this as pollinators being constant because they cannot remember too many handling skills

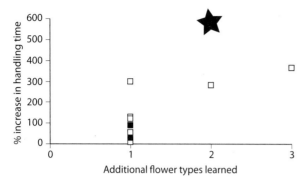

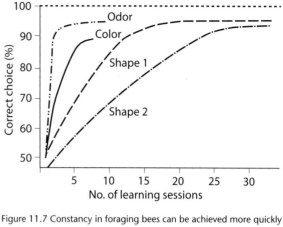

Figure 11.6 Testing Darwin's explanation of flower constancy; data from six different studies as summarized by and redrawn from Gegear and Laverty (2001), showing the increase in handling time for a visitor that switched to learning new flowers, compared with an experienced constant visitor. The increase in handling time for one extra flower is usually only 0–2 s, or up to ~100%, although switching through three or more flowers can produce ~300% increases, and up to 600% in the only study (*) involving butterflies rather than bees. (Open squares show complex flowers, closed squares represent simple flowers.)

Figure 11.7 Constancy in foraging bees can be achieved more quickly (and more accurately) with scent cues than with color (see also fig. 6.6A), while learning based on shape is even slower, requiring up to 20–30 learning sessions for the simpler forms. (Redrawn from Schoonhoven et al. 2005, from data in Menzel et al. 1974 and Schnetter 1972.)

simultaneously, while Lewis (1993) more specifically proposed that learning to handle new flowers disrupts the memory of old handling skills for other flowers. Collectively, these views can be termed "Darwin's interference hypothesis"; in effect, constancy arises because it minimizes the costs of learning and relearning flower handling skills. If this approach is correct, then an experimentally forced switch of an animal to a new flower should increase its handling time on the old one that it was accustomed to. A flower visitor could be trained on flower species A, switched to species B, and then retested on A. If it were then slower than before in its handling of A, this could be due to the direct negative effects predicted by the Darwinian hypothesis, or simply due to forgetting, but the latter could be controlled for by testing A twice with a similar time gap between trials and no intervention of species B. Tests of this kind are shown in figure 11.6, based on several studies (pooled by Gegear and Laverty 2001), and indicate that there is only a limited effect on handling speed for bees although a much bigger effect for butterflies (see also Goulson, Stout et al. 1997). The reduced performance on the original flower was markedly more pronounced when there was more than one new flower species, then giving a roughly 50% decrement in handling time for bumblebees. Overall, Gegear and Laverty concluded that there was only a weak link between switching cost and the strength of con-

stancy, suggesting only limited support for this version of constancy theory.

2. An alternative is the search image hypothesis, which views the limitation as being at the sensory level rather than lying with motor skills. The examples given above for *Pieris* and for dung-flies indicate the importance of innate recognition skills in flower handling. There is also evidence that bumblebees do stick to one color of flowers requiring different skill types, but avoid flowers that require the same skill in different colors (P. Wilson and Stine 1996). Gegear and Laverty (2005) tested bumblebees with an increasing number of trait differences between available artificial flowers and found that the bees showed higher constancy when flowers differed in more traits, suggesting a limitation in their ability to seek out or remember too many combinations of flower traits at the same time. So sensory learning clearly does enter into the constancy debate. Moreover, the sensory modality involved will affect how fast animals learn to revisit a flower and how readily they give up on it. For example, *Apis* learn floral scents faster than colors, and can choose a flower scent correctly after a single visit around 95%–100% of the time, especially where the odor is relatively complex (chapter 6). In contrast, colors need 3–6 visits to get to 90% accuracy, and shapes need at least 20 visits (fig. 11.7).

Are the two ideas mutually exclusive? Clearly not, but it is very hard to test between them, requiring the use of artificial flowers with careful independent control of variables. With such tests, it seems to be generally true that constancy is greater when perceived floral characters are more distinct, and greater again when the flowers are made to differ in more than one trait. Bumblebees do appear to search for the same image just after leaving one flower, but this effect decays within 4–5 s, so it cannot fully explain their normal levels of constancy. However, honeybees are much more constant when tested on mixed artificial arrays and in the field, and they may be more reliant on search images than are *Bombus*.

It may be fair to assume that both sensory and motor limitations affect constancy. This suggests that the constraints on pollinator behavior have over time selected for divergence of floral traits in mixed communities of outcrossing plant species. Where plants compete with each other for visitors, we might expect selection for more isolated floral trait phenotypes that are defined by the sensory or motor limitations of the predominant local visitors. Interestingly, Ostler and Harper (1978) analyzed floral features in 25 communities, and found that flower diversity (for color, tube shape, and bilaterality) was correlated with the number of co-occurring species, that is, that plants were indeed showing isolation in sensorimotor space. However, this does not necessarily mean that they inevitably become reproductively isolated or that in the longer term constancy from visitors can drive the plants to speciate (Chittka et al. 1999).

4. Pollinators and Visitors

Many animals are flower visitors, but they are by no means all effective flower pollinators. There is a crucial need to distinguish between these, and much of the literature is bedeviled by this problem; a failure to make the distinction between all visitors and effective pollinators is a key issue in the current debates about generalization and specialization in pollination biology (Waser and Ollerton 2006). As we shall see in chapter 20, recent studies have often meticulously recorded all visitors to particular plant species and then constructed **pollination webs** from the data, when in practice only tiny proportions of the visits of some animals might be involved in effective pollination. As a conse-

quence, an impression of broad generalization is perhaps mistakenly presented.

In other words, visitors may or may not provide a pollinator service, and the nature of their visits can be limiting to plant success. This limitation can be either by being *insufficient* (too few visits received) or by being *inferior* (visits are frequent, but anthers and/or stigmas are rarely contacted and pollen dispersal is poor, so that mate choice for the plant is low and any offspring that do result may be of poor quality). These two kinds of unsatisfactory visitation have different selective effects on floral characters:

1. Insufficient pollination results from visitors being inherently rare or being more attracted to other plants in the community, and it tends to select for selfing with assured (though potentially lower quality) seed output
2. Inferior pollination may be intraspecific, where there is poor pollen dispersal within a plant species (poor carryover, or excessive intrafloral or geitonogamous pollen discounting); or it may be interspecific, usually due to inappropriate size or behavior, or poor flower constancy, in the visitors. Either of these may tend to select for greater specificity in the interactions.

The goal, if we seek to understand pollination ecology properly, has to be to measure the pollinator effectiveness of each type of visitor, and the pollination efficiency as received by the plant, so that we know how useful both the apparently ideal (coadapted, coevolved?) visitors and the apparently "wrong" visitors really are. In fact the issue can be broken down into several key components.

First, of all the recorded visitors, which of them

1. visit at the right time, when fresh pollen is available? (any predehiscence, or postdepletion visits are useless for the plant's male function; and visits before or after stigma receptivity have no value for female function);
2. effectively pick up and carry pollen on their bodies? (dependent on being a suitable size and having suitable surfaces, as well as on appropriate behaviors for picking up pollen and not grooming it all off too quickly or eating it all);
3. show enough floral constancy to visit the same species again, frequently and quickly?
4. transfer some viable pollen to the next conspecific stigma they encounter?

And then, for the plants, how much transferred pollen is needed for effective fertilization?

Defining and Measuring Pollinator Effectiveness

In principle, this might appear straightforward, and both pollinator effectiveness and pollination efficiency are widely used terms, but they mean different things to different people. The concepts have rarely been taken the same way in any two studies; Inouye et al. (1994) identified 18 different usages of the first term and 7 of the second, and proposed clarifications which regrettably have not been widely taken up. Table 11.5 presents some of this complexity.

In the past some authors have assessed efficiency or effectiveness as a pollinator simply in terms of pollen load analyses, or the average percentage of conspecific pollen on visitors captured at flowers or at nests; but these measures do not distinguish flower constancy from pollen constancy and cannot give real estimates of the probability of pollination (for example, on some plant species bees may actively gather pollen from anthers and have a high pollen constancy, but never in practice contact a stigma in the visited flowers). Other authors have used relatively simple measures of seed or fruit set per visit received, but this is complicated because different plants require different amounts of pollen for full fruit set (and therefore different visit numbers), and because time spent per visit can affect visit efficiency. Ne'eman et al. (1999) proposed a measure based on pollen loads (the pollination probability index PPI) as the product of the mean proportion of conspecific pollen on a given visitor species and the proportion of those visitors carrying that pollen at all, but although useful as a rule of thumb in given contexts (and especially with bees, which can be assessed at the nest) this measure too has limitations. Other measures have been used that are particularly suited to crop assessments, but those are rarely useful to a pollination ecologist.

A measure of effectiveness, to be most useful, should incorporate both

1. *per visit effectiveness* for a given visitor species making a visit of normal length, estimated from conspecific pollen deposited on stigma, or pollen removed, or seed set, but with the first of these be-

ing most reliable and most relevant (and also easiest) for a pollination ecology study and
2. visit *frequency* for that visitor.

Both of these may have to be measured against a temporal axis, in relation to dehiscence and stigma receptivity timing in the plant concerned. The overall effectiveness for fair comparisons between visitors should then ideally be the product of these two measures, and that product is sometimes termed "pollinator importance" (e.g., Bloch et al. 2006; Reynolds and Fenster 2008) although others would still call it pollinator effectiveness.

Those are normally the crucial variables that need to be measured, but in addition it may sometimes be important to factor in *visitor mobility*, that is, the distances traveled between visits by particular animals. A visitor that forages regularly and with good floral constancy and carries large pollen loads may still be a poor pollinator if its mobility is low and it mainly moves pollen within a plant or within a clone of a plant. For example, birds and bats both visited *Syzygium* trees in Australian rainforests, and while bats carried six times more pollen than birds they visited flowers much less frequently; but bats were far more mobile and often moved between forest fragments, so were likely to carry pollen of higher value to the trees (Law and Lean 1999). Mobility may be very important for larger vertebrate flower visitors, but for most insects this factor seems to matter very little in determining pollination effectiveness (e.g., Pellmyr and Thompson 1996).

As yet, records of any or all of these values are available for very few systems (cf. Sahli and Conner 2006), and accruing such data is—or should be—a major imperative for current workers. There are good signs recently that this has begun to be appreciated, with more studies appearing. But it is critically important to record exactly what is meant by effectiveness or efficiency in each study; it is to be hoped that some consensus will emerge in line with the ideas of Inouye et al. (1994) and the discussion here, which largely accords with that work. Furthermore, the data will ideally need to be statistically assessed for significant differences between pollinator importance values (Larsson 2005); an important step in this direction was taken by Reynolds and Fenster (2008) working with *Silene*, with a simulation approach as the recommended methodology for attaching SE values, allowing

Table 11.5

Definitions and Measures for Quantifying a Good Pollinator or Pollination Interaction

Term used	Reference	Measured
Pollination intensity	Primack and Silander 1975	Number of pollen grains deposited on a stigma in a single visit*
Pollination effectiveness	Motten et al. 1981	Seed set by a single visit of a given pollinator
Pollination efficiency	Spears 1983	Seed set resulting from a single visit relative to seed set with unlimited visits
Pollination efficiency	Richards 1986	Proportion of pollen grains produced that reach stigma of same species
	Or	conspecific grains reaching a stigma relative to number of ovules to be fertilized
	Or	reciprocal of pollen/ovule ratio
Pollination effectiveness	Montalvo and Ackerman 1986	Correlation between visitation rate and average seed set
Pollination efficiency	Dafni et al. 1987	Proportion of stigmas touched by visitor
Pollinator efficiency	Dafni et al. 1987	Proportion of fruit set in a given period of flower opening (e.g., morning, afternoon, night)
Absolute pollination efficiency	Galen and Stanton 1989	Pollen grains (or dye) removed divided by pollen grains delivered to compatible stigmas
Pollen transfer effectiveness	Herrera 1990	Product of flower visitation rate and pollen grains deposited on stigma in single visit†
Pollination efficiency	Willmott and Búrquez 1996	Number of pollen grains deposited on a stigma in a single visit*
Pollination probability index	Ne'eman et al. 1999	Product of % of conspecific pollen on a visitor and % of visitors carrying that pollen at all
Pollination effectiveness	Potts et al. 2001	Product of hours of stigma receptivity and calculated mean visits per 1000 flowers per hour
Pollinator importance	Bloch et al. 2006	Product of per visit effectiveness and visit frequency for given visitor†

* Per visit effectiveness as used in text.

†Overall effectiveness, as (per visit effectiveness × visit frequency), as used in text.

TABLE 11.6
Pollination Effectiveness of Single Visits to Spring Wildflowers

Plant	Visitor	Percent effectiveness	Seeds per fruit
Percent styles with pollen tubes			
Erythronium	Apis	93	10.3
	Andrena	88	6.4
Thalictrum	Halictid bees	100	–
	Bombylius	44	–
Percent fruit set			
Claytonia	Andrena	69	4.0
	Bombylius	64	3.4
	Gonia	83	3.4
Stellaria	Nomada	100	2.8
	Bombylius	55	2.6

Source: From Motten 1986.

comparisons between pollinators, sites, and seasons. Recalling Herrera's (1990) work with *Lavandula*, and the variation of effectiveness for different visitors and for the plants overall through a day (see chapter 10, and fig. 10.9), it is also important to record the key data in a time-sliced fashion and avoid too much premature "lumping" of the observations made. Finally, with recent molecular genetic developments it will also be possible to factor in the proportions of self- and cross-pollen transported and/or deposited by each visitor (see Matsuki et al. 2008).

Measures of Pollinator Effectiveness Show Crucial Differences between Visitors

Early Studies

Some of the early attempts to address this issue are instructive, and highlight key points about pollinator/visitor comparisons and their relevance to specialization. Strickler's data (1979) showed the specialist bee *Hoplitis* to be much more effective than other visitors to *Echium* (see table 11.3) and therefore to fill its cells with pollen more quickly. Another example was given by Motten (1986), for spring wildflowers; he painstakingly measured single-visit effectiveness for many different visitors to many species of plant, as styles with pollen tubes growing, or as fruit set. Table 11.6 shows

that bees on various plants could get close to 100% effectiveness (although this varied between plant species), while bombyliid flies were usually markedly less effective on all flowers.

Bee Examples

An example with precise indications as to why different effectiveness values arise was given by Wilson and Thomson (1991) for the effects of single visits by different animals to flowers of *Impatiens pallida* (balsam) (fig. 11.8). This is on all standard criteria a typical bee flower, but different bee species have different values; hence it has often been described more explicitly as a classic bumblebee flower. The authors showed that *Bombus* only collected nectar, crawling into the vestibule formed from the sepals to drink from a curved spur; each bee largely ignored the anthers, but as it fitted the flower rather snugly some pollen was deposited as a long dorsal stripe on the body, and multiple grains were then moved on to a subsequent stigma. In contrast, *Apis* specifically collected pollen, so left little on the anthers, but it moved very little pollen onto the next stigma (most being groomed off to the scopae in between visits). *Dialictus* (a smaller, solitary bee) also collected pollen, and so depleted the anthers, but as it was interested only in the pollen it never went to flowers in their female receptive phase, so deposited nothing on stigmas. Thus the three bee genera had very

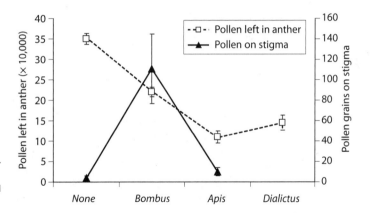

Figure 11.8 The effectiveness of single visits by bees to *Impatiens pallida*, for *Bombus*, *Apis*, and *Dialictus* (a smaller halictid bee) and for control flowers without visits. *Bombus* is by far the most effective pollinator; see text for more details. (Redrawn from data in Wilson and Thomson 1991.)

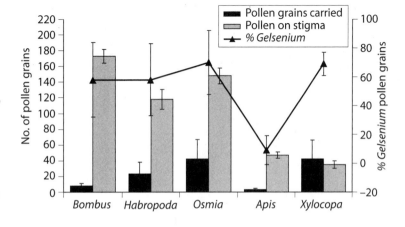

Figure 11.9 Different bee visitors to *Gelsemium sempervirens*. Mean number of pollen grains carried is highest in *Xylocopa* and *Osmia*; proportion of pollen that is *Gelsemium* indicates good constancy in all but *Apis*; but mean pollen grains transferred to stigmas per visit indicate best performance by *Bombus* and *Osmia*. Bars show SEs. (Redrawn from Adler and Irwin 2006.)

different effectiveness values on this flower. However, subsequent work showed that honeybees do sometimes forage for *I. pallida* nectar, and are then more effective in moving pollen and producing seed set (Young et al. 2007), so stressing the importance of recording precise foraging behavior when assessing effectiveness.

Adler and Irwin (2006) worked with *Gelsemium* plants, visited by species of *Xylocopa, Osmia,* and *Habropoda*, which carried the most pollen on their bodies (overall rather than per visit; fig. 11.9), and by *Bombus* and *Apis*, which carried less. All except *Apis* were reasonably flower constant, but *Bombus* scored highly on pollen deposition along with *Osmia* and *Habropoda*, and *Xylocopa* was the most frequent visitor. The authors took this broadly as evidence of nonspecialization for the flower, but it also underscores the necessity of measuring the right thing(s) in any study of pollination effectiveness. Bee size does not predict pollen carriage, and pollen carriage does not predict pollen

deposition; visit frequency was crucial to the assessment here.

Kwak (1993) showed that bumblebees deposited about twice as much pollen per unit time as did syrphids when visiting *Succisa pratensis*, while Kwak and Bekker (2006) recorded *Scabiosa* flowers receiving 145 pollen grains per inflorescence in three hours of visits by *Eristalis* flies, but just 0.7 grains from apparently equivalent visits by other small fly species. Stout (2007) reported only one-third of insect visits to *Rhododendron* flowers (mainly by bumblebees) resulting in pollen on stigmas; moreover, there were significant differences in effectiveness in relation to individual bee size, even within species. Galloni et al. (2008) looked at seven Mediterranean legumes and found very different visitor profiles and pollinator importance (PI) scores for each, with the former a poor predictor of the latter. Maximum PI values varied for each plant (>0.7 for *Bombus* on *Cytisus scoparius*, >0.8 for *Apis* on *Hedysarum coronarium*, but <0.2 for *Xylocopa*

A

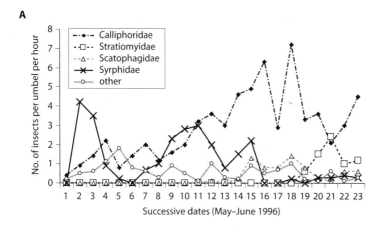

B

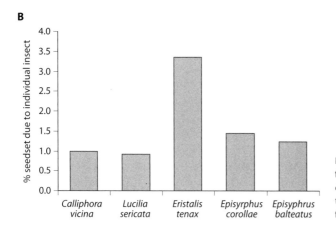

Figure 11.10 (A) Visit frequencies to the umbels of carrot by different taxa of flies. Syrphids including *Eristalis tenax* are not the commonest visitors, but (B) prove to be much the most effective when scored as seed set per umbel per visitor on caged flowers. (Redrawn from Perez-Banon et al. 2007.)

on *Spartium junceum*); furthermore, visitor diversity was negatively correlated with fruit set, while maximum PI value was positively correlated (hence those species with one highly important pollinator achieved the highest fruit production scores).

Even in architecturally specialized flowers such as orchids, different bees may have different effectiveness. In *Cypripedium parviflorum*, ten species of bees visited and were successfully temporarily trapped in the pouch of the flower, but only two species of *Andrena* successfully removed the pollinaria (Case and Bradford 2009).

Fly Examples

Visitation to the umbelliferous carrot (*Daucus carota*, Apiaceae), traditionally regarded as a fly flower with somewhat generalist characteristics (chapters 12 and 13), was assessed by Perez-Banon et al. (2007). Flies were indeed the major visitors, and calliphorid flies (especially the greenbottle *Lucilia*) made up about 67% of all visits, with syrphids at 16%, stratiomyids at 7%, and other flies around 10% (fig. 11.10). But the large syrphid *Eristalis tenax* was by far the single most effective pollinator in terms of deposition of pollen on stigmas, even though other visitors were also carrying good pollen loads and showing good floral constancy.

Lepidopteran Examples

Bloch et al. (2006) found different effectiveness and visit frequency values for five butterfly species visiting *Dianthus* flowers, which could not readily be predicted a priori. Likewise, Rocca and Sazima (2006) showed that *Citharexylum* flowers received many visitors, in-

TABLE 11.7
Visitors to *Lavandula latifolia* and Components of Their Effectiveness

	Percent flowers receiving pollen	Percent female stage flowers visited	Pollen grains on stigma, per visit	Mean interflower flight distance (cm)
Bees				
Halictus sp.	54	73	38	37
Anthidium sp.	70	54	17	61
Megachile sp.	45	55	21	25
Anthophora sp.	39	46	27	17
Ceratina sp.	18	63	8	49
Xylocopa violacea	50	53	18	65
Apis mellifera	45	49	27	21
Bombus lucorum	67	50	26	23
Flies				
Bombyliid	0	15	0	18
Tachinid	8	31	NA	NA
Calliphorid	16	67	4	NA
Hoverflies:				
Eristalis tenax	24	62	16	32
Volucella sp	19	58	11	72
Sphaerophoria sp	11	49	12	NA
Lepidopterans				
Macroglossum stellatarum	33	41	7	28
Pierid	0	18	0	74
Satyrid	12	46	7	110
Nymphalid	24	41	2	168
Hesperid	11	39	14	88
Lycaenid	6	35	NA	NA

Source: From Herrera 1987.

cluding hummingbirds, bees, and moths; the plant would have been a generalist by any standards if judged on visitation, but was in fact pollinated only by five species of sphingid hawkmoth. Another case of hawkmoths, foraging on *Caesalpinia gilliesii,* is described in detail in chapter 14 but is also useful here; several moths of highly variable tongue lengths occurred as visitors, but only the species with tongues around 55 mm long were good pollen carriers (More´ et al. 2006). Coming close to the ideal of carefully dissected pollination effectiveness studies are the various works by C. M. Herrera (1987, 1988, 1990) on *Lavandula,* which documented pollen loads per single visit for more than 8 species of lepidopteran plus around 15 bee species, and 5 fly species. Some examples from these works are given in table 11.7, showing in particular the greater pollen deposition by bees and hoverflies, and the greater interflight movement of some butterflies.

Vertebrate and Multiple Taxa Examples

Hargreaves et al. (2004) studied the apparently ornithophilous *Protea roupelliae* and found that its visitors included sunbirds and many different insect types; but cages that excluded only the birds resulted in a massive reduction in seed set, confirming that only the birds were effective pollinators. Fumero-Caban and Melendez-Ackerman (2007) measured the effectiveness of long- and short-billed hummingbirds, the perching bananaquit, and introduced honeybees foraging at *Pitcairnia angustifolia* (a bromeliad). They predicted the first of these to be the coadapted pollinators from the floral traits, and the long-billed hummingbirds did indeed deposit pollen most efficiently. However, these were rather rare as visitors, so that the honeybee and the nectar-robbing bananaquit also had some role to play. In a study providing a comparison of lepidopterans and birds, Boyd (2004) combined visit rates and pollen deposition measures for *Macromeria* and found that hummingbirds were most effective, since hawkmoth visits were less frequent and they tended to deposit less pollen per visit.

Another example comes from work on *Penstemon* species (Wilson et al. 2004, 2006). This is discussed as a multispecies study in more detail in chapter 20, but here just two species may be contrasted. *P. barbatus* has red flowers and is classically adapted for hummingbird visitation; birds removed on average 9684 pollen grains per visit, and deposited 12 grains per stigma. When the same birds visited the purple/blue bee-adapted *P. strictus* they were much poorer quality visitors, removing on average 3148 grains and depositing just 2 per stigma. Bumblebees operated very differently, taking 4508 grains from *P. strictus* and depositing an average of 5 grains on each stigma (although this does mean that they produced lower pollen carryover than hummingbirds), but failing to contact anthers at all in *P. barbatus*. Other bees such as *Osmia* and *Anthophora* species performed even better on *P. strictus*, where they specifically gathered pollen from the anthers.

Two outstanding cases with multiple visitor types provide a suitable climax here. The first is the pollination of *Ceiba* flowers (Gribel et al. 1999). This mass-flowering plant attracts bats, marsupials, monkeys, and hawkmoths after opening its flowers at night, and then bees, wasps, and hummingbirds early the following day. But only the phyllostomid bats were effective pollinators, and in particular all the morning visitors were useless because any pollen tubes initiated by their activities did not traverse the style to the ovary before stylar abscission occurred. *Aphelandra acanthus* is somewhat similar, attracting mainly bats and some moths by night and then hummingbirds by day

(Muchhala et al. 2009), but in this case, although bats carried by far the most pollen and were responsible for at least 70% of the effective pollination, they also deposited a high proportion of foreign pollen, perhaps reducing their "quality" as pollinators and potentially explaining why the flower shows somewhat generalist characters (narrow corolla base, long period of anthesis, prolonged stigma receptivity) in addition to its more obvious bat-related traits.

Generalist Examples

Most of the above examples involve plants that are fairly specialist in the traditional senses of morphology, attractants, and rewards. It is only fair to add an example of pollination effectiveness measured in a plant that was found to be, and was from a syndrome perspective expected to be, a real generalist. Kandori (2002) worked with *Geranium thunbergii*, with dish-shaped flowers, and found 45 insect species as visitors, of which 11 species (including Hymenoptera, Diptera, and Lepidoptera) were efficient pollinators. Bees, especially larger bees, were the best pollinators overall, but the most important species was not consistent through time or across years. The author explicitly noted that casual observations of visitation would be quite insufficient to identify important pollinators.

Other Variables to Consider

It should be evident from the examples given so far that pollinators providing the most effective service are often *not* the most abundant (Montalvo and Ackerman 1986; Schemske and Horvitz 1989; Pettersson 1991). For example, on *Asclepias syriaca* nocturnal moths were more effective in terms of seed set per visit, but diurnal visitors (especially bumblebees) were more abundant and therefore overall did account for more of the effective pollination visits (Bertin and Willson 1980; Jennersten and Morse 1991). Likewise, Larsson (2005) recorded visit efficiency for foragers at *Knautia* flowers, and showed that a specialist bee (*Andrena hattorfiana*) both picked up and deposited more pollen than other, more generalist, visitors, although its comparative rarity meant it accounted for only about 6% of the total pollination. In practice, the effectiveness of very common visitors is often undermined simply by their size; thus for *Eucalyptus regnans* trees small dipterans were extremely abundant flower visitors but only larger dipterans were able to carry substantial pollen between flowers (Griffin et al. 2009).

The behavior and "choosiness" of visitors also matter. Brunet and Sweet (2006) measured the outcrossing rate of *Aquilegia coerulea* plants in relation to different visitors' activities, and found significantly increased rates with hawkmoth visitation compared with all other flower visitors (bumblebees, solitary bees, syrphids, muscids), partly because hawkmoths were particularly good at selecting female-phase flowers for their first visit to a plant.

This last example offers a reminder that quality is not just the amount of pollen moved appropriately per visit, but also the selection slope for a particular visitor—how it affects selection on a given trait (Wilson and Thomson 1996), also seen as the interaction term between visitor species and floral character. Specialization can occur when better-quality visitors have a strong effect on selection while inferior visitors have little impact; and different pollinators can impose differential selection on the same plant species in different places, depending on the mixture of those pollinators present.

Fully convincing and complete measurements of pollinator effectiveness or importance have rarely been achieved as yet. Some of the cases cited may support retention of the "most effective pollinator principle" advocated as a useful approach by Stebbins (1970), but for other plant species this seems unlikely (e.g., Mayfield et al. 2001; Wolff et al. 2003). Quite where the overall balance lies is still to be decided.

5. Overview

A simple view of pollination biology would be that

1. plants tend to evolve toward specialization in attracting specific visitors, and

2. visitors develop constancy or fidelity to that plant.

Hence flower visitors could contribute to speciation of plants

1. through specialization, as agents of selection fostering further divergence of floral characters and
2. through their constancy, as agents of restricted gene flow.

The end product, in this simplified scenario, would be visitor-mediated (or more strictly pollinator-mediated) reproductive isolation for the plant, and potentially cospeciation of plants and pollinators.

There is some evidence for this (e.g., Powell 1992, for the particularly specialist interaction of yuccas and yucca moths), but as an evolutionary pathway it is far from universal. Many would assert that research in the last three or four decades has largely shown that pollination does *not* tend to become more specialized through time, and that most flower visitors are not representative of any particular syndrome. In fact, Waser et al. (1996) crucially maintain that floral *generalization* is to be predicted as the best strategy for a flower visitor whenever the abundances of preferred species are low, or where those abundances are fluctuating in time and space; and they suggest that such conditions are probably the norm. Specialization and constancy are therefore not to be expected except in very stable ecological contexts and should not be seen as an indicator of an advanced pollinator (Waser 2001). Indeed, the causal link between specialized pollination, increasing reproductive isolation and speciation rates, and hence increasing plant diversity has been challenged (Armbruster and Muchhala 2009). Such themes will be taken up again in detail in chapter 20, after we have reviewed the main groups of flower visitors and the features that are held to characterize more or less specialized pollination syndromes.

Chapter 12

GENERALIST FLOWERS AND GENERALIST VISITORS

Many insects that have other core diets (especially **entomophagous** species, and a range of herbivores) will top up on some floral nectar at times for an easy energy boost, and thus become potential occasional generalist pollinators. However, alongside these there are some rather more regular flower visitors, spending some part of most of their adult lives feeding in flowers, and they can be seen as a generalist flower visitor cohort, constituting a predictable part of the visitor spectrum of some kinds of flowers. These regular visitors include a range of beetles, and some social and solitary wasps, together with a few more unusual taxa. Inevitably they generate extremely varied visitation patterns. Nevertheless, some shared characteristics—both of the animals themselves and of the flowers that they frequent—can be identified.

1. Coleoptera: The Beetles

Coleoptera is by far the largest order of insects as currently described, and the beetles began to diversify early in insect history, becoming abundant quite early in the Mesozoic. These primitive beetles probably formed a main early group of pollinators, given their widespread associations with more basal plant groups (magnoliids, palms, aroids, and also some gymnosperms); although the common view that beetle flowers are primitive, and that the very first flowers were visited and pollinated by beetles, is disputed by several authors (see Gottsberger 1974; Meeuse 1978; Bernhardt and Thien 1987; Bernhardt 2000). Thien et al. (2009) in particular argue that flies were the more likely earliest visitors, and that beetle-related flower traits (especially odors) were a later acquisition.

While about 30 families of beetles contain at least a few examples of flower visitors, the main **anthophilous** groups today—those that are of real importance and abundance as flower pollinators—are soldier beetles (Cantharidae) and longhorn beetles (Cerambycidae). Both these families rely almost exclusively on floral rewards at some point in their life cycle. A few kinds of scarab (Scarabeidae) are also strongly flower associated, notably the monkey beetles of southern Africa (Mayer et al. 2006). Families such as the scarabs and cantharids did not diversify until the massive radiation of angiosperms took off in the Tertiary. Some of the flower types associated with these and other particular beetle families are shown in table 12.1.

Overall, beetles are quite an important cohort of pollinators, often overlooked in the literature, perhaps because they are more obvious in the southern hemisphere and in arid or tropical climates (Momose, Yumoto et al. 1998) than in temperate Europe and North America. Bernhardt (2000) suggested that at least 184 angiosperm species are pollinated almost exclusively by beetles, and roughly another 100 species have beetles as an important part of their pollinator spectrum. Collectively, the flower types that the beetles mainly

TABLE 12.1
Beetles as Flower Visitors

Family	Common names	Flower families and effects
Chrysomelidae	Leaf beetles	Eating pollen on Ranunculaceae, Asteraceae, some Nymphaceae, Dipterocarpaceae
Cantharidae	Soldier beetles	Eating pollen and nectar on Apiaceae, Asteraceae; occasionally associated with special nutritive tissues. (Especially *Rhagonycha* spp. in Europe)
Nitidulidae	Pollen beetles	Eating pollen on Ranunculaceae, Brassicaceae, Leguminosae. (Especially *Meligethes* and *Pria* spp.)
Staphylinidae	Rove beetles	Eating flower parts, pollen and nectar on various families
Scarabaeidae	Scarabs, chafers	Eating many types of flowers, usually destructively; common on Araceae and monocots
	Monkey beetles	Eating pollen and flower parts; especially on Asteraceae and Aizoaceae
Elateridae	Click beetles	Eating many types of flowers, usually destructively
Cerambycidae	Longhorn beetles	Eating pollen and/or nectar, sometimes flower parts, sometimes special nutritive tissues; especially open flower types (Ranunculaceae, Rosaceae, etc.)
Curculionidae	Weevils	Eating flowers or stigmatic secretions, often on Annonaceae, Dipterocarpaceae

visit can be seen as exhibiting a beetle pollination syndrome, termed **cantharophily**.

Beetle Feeding Characteristics

Beetles come in a wide range of sizes and highly diverse appearance as adults, but they share the key features of a protective hardened elytra formed by the forewings, with strong flight mainly using the hind wings. They can therefore easily alight on flowers and are rarely deterred by other predators or parasitoids (less protected than themselves) who might already be there. They have essentially chewing mouthparts, using the mandibles that characterize most of the more unspecialized insect feeders, and arranged parallel to the body axis, making complex manipulations quite difficult. They can therefore be rather destructive as feeders, many of the larger species consuming whole flowers, including the petals and the ovule tissues. They also tend to mate quite boisterously in the flowers and commonly defecate there too; they are therefore sometimes termed "mess-and-soil pollinators."

Nevertheless, quite a number of beetle types, whether or not they are destructive, do disperse moderate amounts of pollen reasonably well, moving several meters and sometimes tens of meters between successive flowers. Some of the best beetle pollinators have decidedly hairy bodies that make them effective pollen carriers. Regular flower visitors tend to behave differently on different flowers, eating some species but nondestructively pollinating others (e.g., Hawkeswood 1989). In a few cases they even combine predation and effective pollination, as with a meloid beetle that eats the entire perianth of *Calydorea* flowers (Iridaceae) but leaves the ovary undamaged and setting seed (O'Hara, cited in Bernhardt 2000).

A small percentage of flower-visiting beetles have specializations to gather pollen, albeit with a view to eating it; some chafer beetles, and a few cerambycids, have dense pads of long hairs on the tips of their **maxillae** that act as pollen brushes, while beetles in the genus *Malachius* have unusual bristles with spoon-like tips, used specifically to scoop up pollen (mainly from primarily anemophilous plants). Nectar-gathering specializations also occur, as an elongate rostrum in some *Lycus* species and most unusually in the form of an elongate proboscis in *Nemognatha*. Figure 12.1 shows some of these specializations.

A few beetles show reasonable floral constancy as a consequence of their relative feeding specialization, sometimes linked to a habit of ovipositing in preferred

A

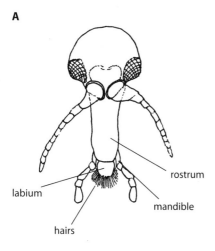

labium

rostrum

mandible

hairs

B

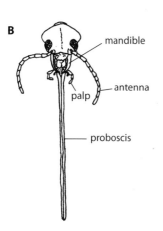

mandible

antenna

palp

proboscis

C

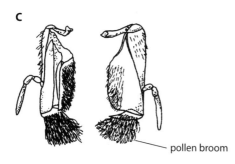

pollen broom

Figure 12.1 Beetle flower-feeding specializations: (A) elongate head in nectar-feeding *Lycus* with rudimentary mandibles and long hair tufts on maxillae and labium (redrawn from Stamhuis 1992); (B) the beetle *Nemognatha*, with elongate proboscis; (C) "pollen broom" on the maxillae of a rose chafer beetle, *Cetonia*, seen from either side (B and C redrawn from Barth 1985 based on earlier sources).

flowers. *Cetonia* beetles (Scarabeidae) showed good levels of constancy to *Viburnum opulus* flowers (Englund 1993), flying on average 18 m between plants and thus acting as good long-range pollen dispersers. De los Mozas Pascual and Domingo (1991) documented constancy in an alleculid beetle, while Pellmyr (1985) showed constancy in a byturid beetle; a similar specialization also occurs in *Byturus tomentosus*, which visits almost nothing but raspberry flowers, where it also oviposits (Willmer et al. 1996). Several beetles also show good constancy to palm flowers (e.g., Eriksson 1994; Listabarth 1996), and some other examples of reasonably constancy to particular beetle flowers in Mediterranean-type habitats are given in the section *Specific Beetle-Flower Interactions* below.

Beetle Sensory and Learning Abilities

Weiss (2001) reviewed the sensory systems of flower-visiting beetles. Many studies have indicated odor reception as the main cue for this group, with color coming into play only at closer range (e.g., Eriksson 1994; Pellmyr and Patt 1986), although for some of the more specialized interactions mentioned in the section *Specific Beetle-Flower Interactions* below, color seems to play a more essential role. Weiss (2001) gave several instances of trichromatic systems in beetles, with UV and green receptors particularly common and occasional examples of red sensitivity, the latter often linked to visitation at the more specialized beetle flowers.

Weiss also reported an absence of studies on flower-learning behaviors in any beetles, and this largely remains the case.

Beetle Flowers

Cantharophilous flowers are particularly well represented in the magnoliids (especially the Annonaceae; see the appendix), and in some phylogenetically basal monocot families such as Araceae. Bernhardt (2000) estimated that this kind of flower may have evolved convergently at least 14 times. He also noted that these flowers are understudied, since so many occur in the canopies of tropical forests and so few in accessible well-documented habitats (see table 12.2).

The generalist flowers visited by a diverse range of

Table 12.2
Occurrence of Cantharophily in Different Habitats

Habitat type	Location	Percent taxa showing cantharophily	References
Mediterranean	Israel phrygana shrubs	0.0025	Dafni et al. 1994
	South Africa fynbos shrubs	15	Collins and Rebelo 1987
	South Africa karroo shrubs	34	Picker and Midgeley 1996
Tropical rainforest	Malaysia dipterocarp forest	20	Momose et al. 1998
	Neotropics, lowland canopy	0	Bawa et al. 1985
	lowland subcanopy	11	Bawa et al. 1985
	cloud forest	45	Seres and Ramirez 1995

beetles are typically white, cream, or green in color, and rather dull as opposed to shiny. Color change is rare, perhaps indicating that flower-visiting beetles are not very receptive to detailed color cues. Morphologically, the flowers are radially symmetrical, and come in two main designs; the first group are usually quite large (able to house several beetles at once), either flat or bowl shaped (e.g., *Magnolia, Annona*) and often borne on woody rather than herbaceous plants (plate 2A), while the second group includes tiny flowers grouped as a terminal inflorescence and providing a platform to walk on (e.g., *Viburnum,* some *Acacia*, plus various euphorbids, palms, and aroids; plates 1 and 21 include examples). The flowers commonly have rather strong fruity or fermenting or slightly spicy smells, although in some species this crosses over into quite unpleasant odors reminiscent of the early stages of decay or of sweaty mammalian bodies. Odor emission may be combined with a degree of thermogenesis (see chapters 6 and 9), both in aroids and in some *Magnolia* species (Dieringer et al. 1999). While many of these flowers may seem to be generalist, and attract various flies, beetles, and even bees, some are apparently pollinated by just one genus of beetle (e.g., *Talauma ovata*, pollinated by *Augoderia* beetles: Gibbs 1977). More explicit evidence from genetic analysis of *Magnolia obovata* pollen indicated that only the cetoniine flower beetle visitors were effectively moving cross-pollen between flowers, the grains on bees and on other smaller beetles being nearly all self-pollen and with relatively few total grains being moved (fig. 12.2; Matsuki et al. 2008).

Some beetle-pollinated flowers partially circumvent the problem of beetles' destructive feeding habits by offering "food bodies" within the flower, appearing as white or cream spherical or tubular tissues on the tips of petals or sepals. *Calycanthus* flowers have such tissues on tepals, stamens, and staminodes (Rickson 1979), which produce a protein-rich diet for a diverse range of visiting beetles. It might be expected that flowers pollinated by beetles would also have some protection, at least for their style and ovules, against the worst of the potential feeding damage. They do tend to have sunken inferior ovaries, or in the case of composite flowers show rather robust basal corolla walls such that the ovules cannot readily be reached. Their pollen is also quite abundant (another way of counteracting messy or destructive feeding habits), often presented on exposed or exserted anthers, and the grains are often unusually oily in appearance and to the touch. When nectar is present, it is likely to be of relatively low volume and exposed, so it will become quite concentrated on warmer days.

Bernhardt (2000) sought to distinguish four different types of flowers solely or mainly visited by beetles. His categories were bilabiate, brush, chamber, and painted bowl (table 12.3). The first two are rather uncommon with respect to beetle visitors, mainly incorporating orchids and palms, respectively. However, the other two are quite common, the first corresponding to more traditional beetle-flower categories.

Chamber blossoms include all bowl-shaped forms offering a cave-like interior, either where multiple inflorescences are enclosed within stout bracts, or where a large cup-shaped perianth with multiple whorls of petals protects the reproductive parts (hence including

A

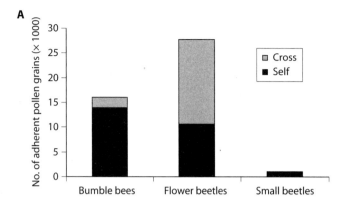

B

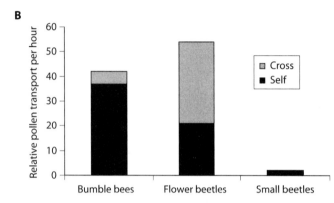

Figure 12.2 The amount of pollen grains (A) and the pollen per hour (B) transported by various flower visitors to *Magnolia obovata* flowers, separated into self- and cross-pollen as determined by genetic analysis; flower beetles are the most effective movers of useful cross pollen. (Redrawn from Matsuki et al. 2008.)

magnolias, many aroids, and Annonaceae). Few such flowers offer nectar, their edible rewards being either pollen or tissue or stigmatic secretions; many offer rewards at night as well as in the daylight hours. Chamber flowers are particularly common in the wet tropics, where their form gives some physical protection.

Painted-bowl blossoms include flowers with an erect, deep cup shape, unprotected from above, and with simple perianths usually consisting of just five or six petals. The corolla may enclose a mass of anthers (e.g., *Ranunculus*, some peonies), or just a few anthers located centrally (tulips, many Iridaceae). Flowers are brightly colored, often red, orange, pink, or yellow, and many have dark or black markings centrally, or dark pollen (Dafni et al. 1990; Steiner 1998a). They open mainly by day and are relatively odorless; nectar is absent or very sparse, and pollen the only reward. Painted-bowl flowers are reasonably frequent in warm temperate and Mediterranean zones (especially southeastern Europe and southern Africa). This categorization does usefully draw attention to a type of beetle-

visited flower not really included in older definitions of cantharophily.

Specific Beetle-Flower Interactions

There are a few distinctive flower-beetle associations worthy of particular mention, where a particular group of beetles is associated with more unusual flower types.

1. Mediterranean habitats often show distinctive examples of the painted bowl bulbous flowers highlighted above. For example, unusually hairy scarabaeid beetles (*Amphicoma*, and other genera) in the Eastern Mediterranean are conspicuously associated with various red-colored bowl-shaped flowers with dark centers, including the tulips, poppies, and anemones that flower somewhat sequentially and produce a distinctive phase of bright red early spring flowers. *Amphicoma* beetles prefer models that are at the

TABLE 12.3
Beetle Flower Types

Morphological type	Occurrence in plant taxa	Abundance	Examples	Visitors
Highly generalist				
Large, flat or bowl shape	Widespread	Common	Many families	Diverse beetle taxa
Tiny, massed as inflorescence	Widespread	Common	Umbellifers, mimosoids	Diverse beetle taxa
Moderately specialist				
Bilabiate or gullet	Orchids	Rare, mainly Mediterranean		Scarabeids, elaterids
Brush	Palms	Uncommon, mainly wet tropics		Curculionids, nitidulids
Chamber	Some monocots magnoliids, dipterocarps	Common, mainly tropics	*Annona, Clusia, Shorea, Magnolia, Nuphar, Sterculia, Philodendron*	Mainly curculionids, nitidulids, some chrysomelids
Painted bowl	Bulbous monocots, rare in dicots	Common, especially Mediterranean	*Tulipa, Anemone, Papaver, Ixia, Moraea, Homeria, Ranunculus*	Mainly scarabeids, some chrysomelids

Source: Modified from Bernhardt 2000.

normal orientation and height of the flowers, and of similar shallow depth (Dafni and Potts 2004). These beetles are also known to have red photoreceptors (Briscoe and Chittka 2001). Clusters of the beetles can be observed feeding and often sleeping overnight in these flowers in large numbers. The same is true for false blister beetles such as *Oedemera*, which feed and rest on various Asteraceae, Papaveraceae, and Cistaceae.

2. Somewhat similarly, in southern Africa beetles show quite specialized visitation to iridaceous plants (Steiner 1998a) and to certain orchids (Steiner 1998b), and up to a third of all sand-plain fynbos species are described as scarab pollinated (Picker and Midgley 1996). Goldblatt et al. (1998) reported convergence of herbaceous perennials from several families into a typical form suited to hopliine monkey beetles: bowl-shaped red, orange, or cream flowers, with colored

"beetle marks" at petal or tepal bases, little scent, and rarely any nectar (i.e., rather similar to the Mediterranean flora described in point 1). They collected 26 species of beetles from 40 plant species, of which 40% were visited by just one beetle type. But specific relationships were rare, only 28% of beetles carrying single pollen types and many beetles carrying up to five different kinds, indicating relatively unspecific foraging. Van Kleunen et al. (2007) showed that monkey beetles were attracted explicitly to the visual signals including the dark beetle marks, and relied very little on scents, as described for other cantharophilous flowers.

In comparable habitats in Australia, there are also rather high numbers of somewhat similar flowers pollinated mainly by beetles (Irvine and Armstrong 1988, 1990).

3. The tiny black pollen beetles in the genus *Meligethes* are very obvious flower visitors in temperate

habitats, seeking food in the base of a wide range of flowers (see plate 21F); they can often be found in large numbers on *Brassica* flowers, or on *Ranunculus*, and are a common presence within the corollas of edible peas and beans (and more irritatingly of sweet pea cultivars) in gardens. S. M. Cook et al. (2004) showed that they probably located the flowers mainly by pollen odor, and they survived better when rape pollen was included in their diet. They may stay for several hours within one flower, so probably move relatively little pollen and over no great distance. In the same family (Nitidulidae), the genus *Pria* sometimes have specific relations with flowers, and *P. cinerascens* is associated with sexually dimorphic *Leucadendron* shrubs (Proteaceae) where it mates, lays eggs, and consumes nectar and pollen in male flowers; the female flowers are more cup shaped and enclosed and are used as shelter from frequent rain (Hemborg and Bond 2005).

4. *Donacia* beetles (Chrysomelidae) may have a special relationship with American *Nuphar* water lilies (e.g., Lippok et al. 2000), which are female on day 1 and then male for a further 1–3 days (see plate 4H), and which trap their visitors overnight. However, in European species the lily flowers are more often pollinated by syrphid flies, honeybees, or sweat bees (Lippok and Renner 1997). In other water lilies (*Nymphaea* and *Victoria,* including some nocturnally flowering species), *Cyclocephala* scarab beetle pollination has been reported (Prance and Arias 1975; Prance 1980). However, Hirthe and Porembski (2003) recorded more efficient pollination by early-morning bees than by nocturnal dynastid beetles for *Nymphaea lotus* in West Africa.

5. The staphylinid beetle *Eusphalerum scribae* visits the flowers of the carnivorous plant *Pinguicula*, although interestingly it does so only in fairly shaded sites, the main visitors to sunlit flowers being thrips (Zamora 1999).

6. A variety of dung- and carrion-associated beetles are exploited by carrion flowers (mainly members of the Araceae, and loosely included in the chamber blossom group), in a similar manner to their exploitation of dungflies and carrion flies. Examples include *Amorphophallus* (Beath 1996), *Philodendron* (Gibernau et al. 1999), and *Lysichiton* (Pellmyr and Patt 1986). This is technically termed coprocantharophily. Young (1986, 1988) studied **carrion beetle** pollination of *Dieffenbachia* and pointed out that pollen translocations of 20–80 m were not uncommon. Some weevils are also associated with trapping flowers in the Annonaceae, for example *Xylopia* flowers, which attract *Endaeus* weevils with fruity scents and trap them (using petal closure) in a chamber whose temperature may then be strongly elevated (Ratnayake et al. 2007). Flowers that trap insects for pollination purposes are covered more fully in chapter 23. But one more unusual case of dung-beetle pollination worth a mention occurs in *Orchidantha*, a member of the ginger family, where the flowers grow at ground level and smell strongly of dung; here there is no reward for the beetles and no trapping is involved (Sakai and Inoue 1999).

Effectiveness as Pollinators

There are few direct measurements of pollination effectiveness for particular beetles, and specific determinations of pollen transfer are lacking for the cantharophilous syndrome as a whole (see Gottsberger 1988, 1989a,b; Goldblatt et al. 1998). One of the few clear examples concerns *Cetonia* beetles, which are more effective pollinators of the strongly scented *Viburnum opulus* than any of that plant's other visitors, moving pollen up to 18 m between plants and flying consistently between plants (Englund 1993). Some other beetles can also travel moderate distances between plants (perhaps up to 20 m in temperate beetles, with records over 80 m in tropical species), so they probably achieve a reasonably good pollen transfer range. Kwak and Bekker (2006) reported cantharid beetles making up 14% of all visits to *Anthriscus*, carrying an average of 5600 of its pollen grains, and moving an average of 80 cm between its flower heads.

However, as a general rule, beetles are less mobile than flies or bees, and being highly protected by the elytra are inclined to sit passively in flowers for long periods, ignoring provocations that would disturb and disperse other insects. Thus many of the larger beetles probably move only small amounts of pollen between rather few flowers. Since some do show reasonable fidelity, they at least will take the pollen to an appropriate place when they do finally move on from a flower.

Overview

Many flower-visiting beetles are indisputably destructive and may damage or completely devour quite large flowers over the course of minutes or hours (plates 1A, 21A, and 21H). Even those identified as reasonable pollen carriers and pollen movers are rarely sufficiently specific in their visits to the generalist flower types described above. Likewise, none of the features given as being part of a cantharophilous syndrome are unique to that syndrome. It is no surprise, then, that flowers visited by beetles are almost invariably also visited by at least one other order of animals, and often by several different orders, becoming part of a generalist flower syndrome rather than just existing as a cantharophilous syndrome.

2. Hymenoptera: The Wasps

"Wasps" is a rather broad term for an assemblage of hymenopterans, ranging from tiny gall wasps and parasitic wasps to the more familiar stinging yellow jackets and hornets. Technically, both ants and bees are also forms of true wasps, being derived from the main ancestral wasp lineage. However, it is obviously more sensible in this book to deal with ants and bees separately, as each has very different interactions with flowers. What remain as wasps can be divided into those without a waist (confusingly not termed wasps at all, but **sawflies**, in the suborder Symphyta), and those with the classical wasp waist (suborder Apocrita). The latter can be divided again into those without stings, where the ovipositor is used to lay eggs in hosts and larvae develop parasitically (termed the Parasitica), and those with the ovipositor modified as a sting (the Aculeata, including the familiar larger wasps, plus the ants and bees).

Types with Nonstinging Ovipositors

Symphyta
Some sawflies use flower nectar as fuel, and they may also consume sap, pollen, and honeydew. Female sawflies in particular tend to visit flowers, and will often eat the stamens and petals as well as the intended floral rewards; thus, although they can be common in some warmer months and may well move significant amounts of pollen, their potential benefits as pollinators may be markedly offset by the floral damage they cause. However, the family Xyelidae has mouthparts more specifically adapted to eating pollen, while some other families have longer probosces that can extract semiconcealed nectar, and all of these are less damaging to flowers (Jervis and Vilhemsen 2000).

Most other sawflies have short mouthparts, only 2–4 mm long, and are therefore most commonly seen on open exposed flowers. Unlike most wasps they are not red-blind, but apparently cannot see ultraviolet wavelengths, which may have some effect on their flower preferences. They are sometimes seen on various Rosaceae, some Asteraceae, and especially Apiaceae, being locally abundant on forms such as hogweeds (*Heracleum*) and parsnip (*Pastinaca*). Certain sawflies are also very common visitors to buttercups (*Ranunculus*) in spring. Hence their general flower preferences are essentially similar to the types favored by short-tongued flies (chapter 13). However, in a few **sawfly** species the flower preference can be quite restricted and thus appear as specialized, often focused on the flowers of the plants whose foliage is used as larval food. This is especially notable in willow sawflies such as *Euura* where the adults feed mainly on *Salix* catkins, and often just on one species of *Salix*. Another case of specialization occurs in an Australian duck orchid, *Caleana major*, pollinated by a sawfly (*Lophyrotoma*) that is fooled into pseudocopulation with the flowers (see chapter 23).

Parasitica
Some ichneumon and braconid wasps (all parasitoids of other insects) are common flower visitors, taking nectar (both floral and extrafloral), sap, or honeydew. Most have extremely short mouthparts (effectively less than 1 mm) so they mainly visit open flowers with exposed rewards, being common on similar plants to those mentioned above for sawflies, but perhaps even more restricted to the Apiaceae and short-corolla Asteraceae. Tooker and Hanks (2000) reported that many of these parasitic wasps were nevertheless quite restricted in their visits, to just a few plant species.

There are again a few more specific interactions. The twayblade orchids (*Listera*) are often described as "ichneumon pollinated," commonly visited by ichneumon wasps such as *Pimpla* and *Ichneumon*, although the flowers also attract small beetles, flies, and some sawflies. These orchids are a startling case of rather specific and complex pollination mechanisms in a relatively generalist flower; a visiting ichneumon or some other similarly sized insect crawls up the narrow labellum (the lowest petal) feeding on nectar that lies in a central groove, until it reaches the central column of the flower, whereupon the tip of the column (the rostellum) rapidly exudes a drop of sticky liquid and cements the two pollinia onto the insect's head. Curiously, any small bees that visit the twayblade do not elicit the same rapid response from the orchid and are rather ineffective as pollinators.

Matching this complexity, at least one Australian orchid, *Cryptostylis*, appears to have a pseudocopulatory deceit pollination relationship with an ichneumon (*Lissopimpla*), again described more fully in chapter 23.

Chalcids, Cynipids, and Other Gall Wasps

These are all very tiny insects and endowed with very short mouthparts, so they are not usually common as flower visitors. Some species do use exposed flowers occasionally as a food source, but there is no good evidence of their effectiveness as pollinators. However, the fig wasps, from the related family Agaonidae, are a famous case of highly specialized flower visiting, described in chapter 26.

Types with Stings: True Wasps

Chrysids and Scolioids

The chrysids (rubytail wasps) are brightly metallic insects, parasitic on true wasps and bees. They are mostly short tongued and once again fairly common on Apiacae. But a few European genera (e.g., *Panorpes*) have longer mouthparts (6–7 mm) and visit a wider range of flower types, including bowl-shaped flowers such as *Tilia* and *Rubus*, and sometimes *Ranunculus*. They are unlikely to effect much pollen transfer however, as they have very shiny surfaces.

Scolioids (velvet ants and their kin) are also parasitic on other hymenopteran larvae, and the females are often wingless, making them largely ineffective as pollinators. Most of the winged forms visit open shallow flowers occasionally, but some larger hairier forms have special pseudocopulatory relationships with orchids (for example, *Campsoscolia* pollinates *Ophrys speculum*, and several other examples are known from Australia; Stoutamire 1974, 1983). The hairy flower wasp of central Europe (*Scolia hirta*) is more unusual in being a regular visitor to a wide range of white, pink, purple, and blue flowers, taking nectar from short corollas (Landeck 2002).

Sphecids, Pompilids, Tiphids, and Eumenids

Members of all these groups of solitary wasps (especially the females) are not uncommonly seen gathering some nectar as fuel, between their more protracted prey-gathering foraging trips (they capture other insects or spiders, store them in their nests, and lay their eggs upon the paralyzed prey). While some of these wasps feed only upon the juices of their prey, others regularly take liquid food from flowers, in the form of honeydew, sap, or nectar. Indeed, some appear to need significant sugary inputs from flowers at the start of their adult lives, while nest building, and can therefore be transiently common on flowers in spring. They will often visit flowers with fully exposed nectar, especially the generalist umbellifer flowers such as *Heracleum* already mentioned in this chapter. But they are also found on plants that have partly concealed nectar, with more bowl-shaped flowers (e.g., *Ranunculus*, *Geranium*, *Rubus*, or the snowberry *Symphoricarpos*). Some also visit short tubular flowers, either single tubes (as in some of the classic herbs such as *Thymus* and *Mentha*, or *Calluna* heathers, on which eumenids—mason wasps—can be abundant), or composite inflorescences with multiple corolla tubes such as ragwort (*Senecio*), goldenrod (*Solidago*), or *Cirsium* thistles. A few tiphid wasps even gather nectar as a nuptial gift passed from male to female.

The choice of flowers by solitary wasps may sometimes be linked to water balance issues, so that on hot summer days they are more likely to choose flowers with somewhat protected nectar that stays more dilute and thus gives them useful water as well as sugar; this was the case with *Cerceris arenaria*, feeding preferentially on nectar from white melilot (*Melilotus alba*) in hot conditions (Willmer 1985). The same constraint may also apply in arid and desert habitats, where digger wasps and spider-hunting wasps can be quite abundant: the very large tarantula hawk wasp is a common visitor to asclepiads in southwestern US deserts (Punzo 2006).

Again, there are a few examples of more specialized interactions of flowers and solitary wasps. Some species in the genus *Bembex*, endowed with unusually long tongues for a sphecid, can pollinate smaller leguminous flowers, including the important forage-crop plant alfalfa (*Medicago sativa*). Some sphecid wasps show quite high levels of flower constancy when visiting asclepiads (Theiss et al. 2007). The sphecid wasp *Argogorytes* can pollinate the fly orchid *Ophrys insectifera* by pseudocopulation, and pompilid wasps pollinate some of the *Disa* orchids in southern Africa in similar fashion (Johnson 2005).

One genus of pompilids (*Hemipepsis*) includes specific pollinators of diverse African plants. The orchid *Satyrium microrrhynchum* is one example, and produces exposed nectar on erect "lollipop hairs" so that the wasp receives pollinaria on its head (Johnson et al. 2007). Other species in the genus are pollinators of some milkweeds, apparently due to nectar scents that deter other visitors (see chapter 8; Shuttleworth and Johnson (2006)) and also in some milkweeds due to the dull color and flower odor that attract the pompilids (Shuttleworth and Johnson 2009a). Finally, yet other *Hemipepsis* wasps are specialist pollinators of some *Eucomis* species in similar habitats, these again having complex odors but rather cryptic coloration (Shuttleworth and Johnson 2009b).

The small family Tiphidae includes the thynnine wasps, which are particularly famous for their pseudocopulatory interactions with the hammer orchids (*Drakaea*) of western Australia (Peakall 1990; see chapter 23).

There is also one sphecid (*Krombeinictus*) that has vegetarian larvae fed solely on flower produce (nectar and pollen), thus paralleling the bees (Krombein and Norden 1997).

Vespids

These are the true social wasps, with colonies made from chewed-up wood fibers, each colony containing a queen, many workers, and (intermittently) some male drones. In temperate habitats the nests are normally seasonal, with only the new generation of queens surviving through the winter to found a new colony in spring; hence numbers tend to peak in late spring through to early autumn. In tropical habitats nests can survive all year round.

Adult vespid wasps feed their larvae on captured insects, but also often take some liquid food for themselves and for distribution in the nest, and nectar is usually the preferred option (although their fondness for jams and other sweet treats at picnic sites is legendary!). Sugar-seeking habits are especially pronounced in autumn, and some of the common wasps (*Vespa, Vespula, Dolichovespula*) cannot complete full development without an input of sugars to supplement the main animal diet.

Vespids again like similar flowers to the generalist flies, and as with many of those they are commonly red-blind but UV sensitive, so they are frequently seen on white and yellowish Apiaceae, notably *Heracleum* and *Angelica*, and are often also abundant on *Eupatorium*. Since these wasps (having nests as refugia) often persist longer into the cold weather of autumn than other insects the exposed flowers of ivy (*Hedera*) also attract their attention right through to October or November in north temperate habitats, and they are more effective pollinators of ivy flowers than the calliphorid or syrphid flies that also visit (Ollerton et al. 2007).

The common yellowjacket wasps have somewhat longer tongues than most of the solitary wasps covered in earlier sections. Thus they also frequent a range of more specialized flowers than the truly generalist flies. Some genera of inconspicuous rounded flowers have on occasions been termed "wasp flowers" (fig. 12.3). These tend to be small and globular (around 3–8 mm in diameter), either upright or pendant on the plant, and often dull pink, red, purple, or brown in color. They generally lack strong scents, and they have relatively little pollen. Classic examples are figworts (*Scrophularia*, plate 20D), snowberry (*Symphoricarpos*), and species of *Berberis* and *Cotoneaster* each with small globe-shaped flowers, both of which (as gardeners will know) can be thronged by wasps in early autumn. Heathers can be attractive too, although their floral globes have too small an opening to admit the wasps' heads and therefore they tend to suffer robbing, with nectar taken through holes in the petal bases (often made by short-tongued bees and then exploited secondarily by the wasps—see chapter 23).

Species of the orchid *Epipactis* may also be specifically adapted to wasp visitation, and other more precise wasp-orchid relationships exist with the wasps exploited for deceit pollination; the most spectacular is perhaps the pollination of a Chinese *Dendrobium* by hornets (*Vespa bicolor*) attracted by a scent that mimics honeybee alarm pheromone (see chapter 6).

Masarids

Masarids are a unique group of pollen-collecting wasps, closely related to the vespids (perhaps better included within them as the subfamily Masarinae), but

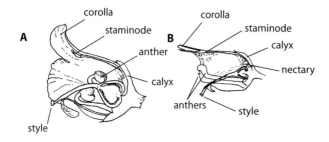

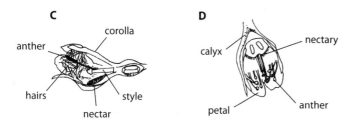

Figure 12.3 Wasp flowers with typical small spherical or bell-shaped corollas: (A) and (B) figworts, *Scrophularia nodosa* in female phase with anthers unextended, and *S. auriculata* in later male phase (modified from Faegri and van der Pijl 1979); (C) snowberry, *Symphoricarpos*; (D) *Cotoneaster*. (Parts (C) and (D) drawn from photographs.)

having developed the pollen-feeding habit similarly to but independently of the bees (Gess 1996). Thus the trait of vegetarianism in aculeate larvae has arisen three times, in a single sphecid genus, in masarids, and in the bees.

The masarids occur in warm temperate areas and can be abundant on flowers, collecting pollen on their faces where they bear substantial hair. From here the pollen is transferred to the crop, mixed with nectar for onward transit to the nest. Flower preferences are generally similar to those of smaller shorter-tongued bees: small zygomorphic tubular flowers such as *Teucrium, Penstemon*, and many of the common labiate herbs. However, a few species are more specialist, and should not really be included in a chapter on generalist visitors; for example, *Ceramius bureschi* in Greece is a monolectic pollen-gathering visitor to *Nigella arvensis* flowers, where it also mates (Mauss et al. 2007).

3. Hymenoptera, Formicoidea: The Ants

Ants have evolved from wasps and are closely related to both wasps and bees. They have highly complex societies where workers can be recruited and organized by pheromonal signals. They are extremely abundant in nearly all habitats, their biomass probably at least equaling that of humans and far in excess of that of any

other animal group. Thus they could potentially be very useful to plants from sheer force of numbers. But they are very unlike both of their taxonomic near neighbors in possessing some key characteristics that greatly reduce their potential as flower pollinators.

1. Ants are small, and the great majority—the workers—are wingless. This makes them a poor physical fit for most flowers and too immobile to be good pollen transfer agents.

2. They have shiny and hairless surfaces, suited to life in cramped underground nest tunnels but offering poor opportunities for pollen adhesion.

3. They have elongate mandibles, useful for various kinds of manipulation as well as aggression, but poor for sucking up nectar or handling pollen—they are mostly carnivorous.

4. They have metapleural glands that produce antibacterial and antifungal agents, a necessary adaptation to life in humid litter and within nests; but these metapleural secretions are also severely damaging to pollen longevity and fertility (e.g., Beattie et al. 1984; Hull and Beattie 1998; Galen and Butchart 2003).

Yet ants are evidently highly attracted to anything sugary and regularly seek out honeydews and other sugary

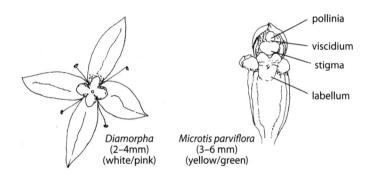

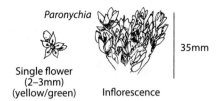

Figure 12.4 Examples of ant flowers. (Drawn from photographs.)

secretions on flowers; it is at first sight surprising that they do not more often take the nectar contents of flowers. While they can be quite common on the white or cream platform-like flower heads of umbellifers and euphorbs (plate 20), and sometimes on larger bowl flowers (where they tend to enter around the floral margins and fail to contact anthers or pick up any pollen grains), they are very rarely effecting any cross-pollination in these situations. Overall, ants are potentially deleterious for flowers, effectively acting as nectar thieves.

There is, however, a small group of plants that could be said to be adapted for ant pollination (Hickman 1974; Wyatt 1981; Peakall et al. 1991). They are generally prostrate and often twining among other plants, with inconspicuous small open flowers close to the stems with almost no stalk (fig. 12.4). Ants rove around among the stems, moving between several flowers on each foraging trip because the rewards from each flower are so low. The flowers are self-incompatible, so that any movement between flowers of different plants that are intertwined can potentially achieve crossing. Examples that fit this pattern include rupturewort (*Herniaria*), *Paronychia* (Proctor et al. 1996), and honewort (*Trinia glauca*; Carvalheiro et al. 2008) in Europe; while *Diamorpha smallii* (Wyatt 1981) and

Polygonum cacadense (Hickmann 1974) are of similar form in North America. Figure 12.5 shows examples of the movement of ants, or of pollen transfers achieved by ants, in some of these plants.

Occasional examples of ant pollination occur in rather different flower forms. One example here is *Cytinus hypocistis*, a parasitic plant whose flowers briefly emerge from the host tissues (*Cistus* or *Halimium* plants). In Spain this plant received more than 97% of its visits from various ants (deVega et al. 2009), which acted as true pollinators by reliably contacting the anthers and stigmas and carrying quite large pollen loads (although with different effectiveness in different ants, particular species of *Pheidole, Plagiolepis,* and *Crematogaster* being most effective). For their size, the flowers of this *Cytinus* have large reproductive organs, plentiful pollen, and a long life, which together with the low growth habit and a sweet scent (close relatives all having musty mammal-pollinated or sapromyophilous odors) may enhance the attraction for and effectiveness of ant visitors.

Most recorded examples of ant pollination in the earlier literature were from warm dry habitats, where ants are especially abundant, so that ant pollination was often described as most frequent in deserts and arid zones. However, there are recent studies from

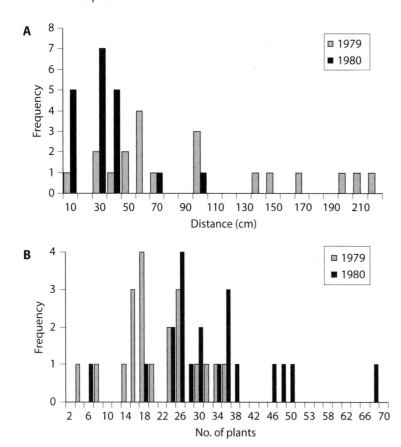

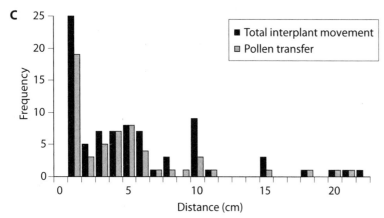

Figure 12.5 Ant foraging patterns on various flowers: (A) distances moved in 5 minutes by *Formica* ants on *Diamorpha*, mostly less than 50 cm but occasionally longer (two additional instances up to 5–10 m were recorded), and usually involving (B) 20–40 of the tiny plants (Parts (A) and (B) redrawn from Wyatt and Stoneburner 1981); (C) total interplant movements, and pollen transfer, for *Iridomyrmex* ants visiting *Microtis* orchids, usually involving distances of less than 10 cm (with two instances at 46 and 76 cm not shown) (modified from Peakall and Beattie 1989).

montane habitats. Puterbaugh (1998) explored ants as flower visitors in three alpine plant species, showing effective pollination in gynodioecious *Paronychia pulvinata*, especially on hermaphrodite plants. Gomez et al. (1996) studied pollination in some Mediterranean mountain habitats (comparing these with lowland arid examples) and found ant pollination in five species, the ants' role being heavily dependent on their abundance, so that they only became the main pollinator when they greatly outnumbered all other visitors. Garcia et al. (1995) reported ant pollination of *Bordera pyrenaica* in the Pyrenees, where ants moved more pollen than ladybirds and flies visiting the same flowers. For *Trinia glauca*, a rare plant in need of conservation in parts of Europe, ants were also found to be major pollinators but their visits were reduced if alternative alien plants (such as *Cotoneaster*) were present (Carvalheiro et al. 2008).

Ants may sometimes supply a fallback pollination mechanism for plants whose other visitors are rare in particular times and places. For example, Ramsey (1995) reported ant-induced self-pollination as a reproductive assurance mechanism when bees or birds had failed to pollinate flowers of *Blandfordia grandiflora* (a perennial herb from montane tablelands) although each ant carried on average only 28 pollen grains, half of these on the legs and unable to contact stigmas. He also noted only very small effects from integument contact on pollen viability in this system. Gómez (2000) recorded ants as pollinators of *Lobularia maritima* in summer, when they made more than 80% of all visits and were at least as effective as winged insects, again with no apparent detriment to the pollen. Gómez and Zamora (1992) discussed ant pollination of mass-flowering *Hormathophylla*, where the sheer abundance of ants made them the commonest pollinators, effective so long as the pollen did not remain on their bodies for long.

There are some more specialist examples of ant pollination too. Peakall and Beattie (1989) described worker ant pollination of an Australian orchid *Microtis parviflora* (shown in fig. 12.5C), albeit with a high degree of geitonogamous selfing occurring, as ants moved only about 120 mm on average between flowers. More recently Sugiura et al. (2006) reported partial ant pollination of the orchid *Epipactis thunbergii*, mainly pollinated by hoverflies but potentially serviced by ants in cooler weather, the pollinia again apparently unaffected by contact with the ants' integument. Peakall (1989) and Peakall et al. (1990) gave the most unusual example of pollination of *Leporella* orchids by pseudocopulating male *Myrmecia* ants (see chapter 23).

Given that so many ants are attracted to a wide range of flowers that they do not pollinate, it is not surprising that many plants have adaptations to prevent these ants from doing too much damage by removing valuable floral resources. There may be three options for excluding ants from flowers, all covered in more detail in chapters 24 and 25:

1. Physical barriers
2. Chemical deterrents
3. Decoys, offered in the form of food some way away from the flowers. This can lead on to the phenomenon of "ant guards" as a biotic defense against herbivory and **folivory**.

4. Thysanoptera: The Thrips

Thrips (also known as thunder-flies) are tiny insects, usually only 1–2 mm long, and with curious feathery wings unlike those of any other insect group. They have piercing and sucking mouthparts, and are pests on some crops not because they do immense physical damage but because they can transmit viruses. They are often very abundant within flowers (several hundred may occur within a single *Acacia* inflorescence, for example), and they take both nectar and pollen, the latter by sucking the contents out of individual grains. Thus they have been noted as pollen destroyers and accidental pollinators (Kirk 1984, 1985, 1987). They can carry small but significant numbers of pollen grains on their rather bristly bodies, but until recently had received little attention as possibly important pollinators.

However, the increasing research on ecosystems beyond temperate Europe and North America in the last twenty years has drawn attention to their role in pollinating some important plant taxa, giving rise to the term *thripophily*. They are active on some of the irregularly mass-flowering dipterocarp trees that dominate south-east Asian tropical forests (Appanah and Chan 1981), where the thrips' ability to multiply rapidly within the flowers as the trees suddenly and unpredictably produce enormous numbers of flowers is probably crucial. In the same region they are also noted as pollinators of some Annonaceae (including *Popowia*, where the entrance to the floral chamber is too small for other insects to enter; Momose, Nagamitsu et al. 1998), of some ant plants in the forests (Moog et al. 2002), and of endemic moraceous plants (a sister group to figs) in New Guinea (Zerega et al. 2004). Thrips are also reasonably common as pollinators in lowland New Zealand forests (Norton 1984), and in Australia they are recorded as important pollinators of some cycads (Mound and Terry 2001; Terry et al. 2005) and of the unisexual tree *Wilkiea*, where both male and female flowers serve as brood sites (G. A. Williams et al. 2001). Thrips are also known to visit and pollinate the ubiquitous tropical weed *Lantana* in its yellow phase (Mathur and Mohan Ram 1978); this is in accord with their apparently enhanced responses to white, yellow, and blue colors in preference to greens, reds, or ultraviolet (Matteson and Terry 1992). However, the anthophilic genus *Frankliniella* (blossom thrips) can also demonstrate a preference for red colors, consistent with its preferred host plants, which include *Malvaviscus* and *Hibiscus* (Yaku et al. 2007).

In colder climates, where bees and butterflies are rare, thrips may also come into their own, especially on Ericaceae. Hagerup (1951) noted them as pollinators of heathers (*Calluna* and *Erica*) in the Faeroe Islands, and at least one species (*Ceratothrips ericae*) goes through its entire life cycle in heather flowers. Garcia-Fayos and Goldarazena (2008) estimated that they also contributed around 20% of the pollination of bearberry (*Arctostaphylos uva-ursi*) across various European populations. In wetland plant species, thrips may also mediate the self-pollination of certain plants (Baker and Cruden 1991).

Thrips may even have some commercial significance as pollinators; earlier records noted their ability to transfer pollen in onion (*Allium*), bean (*Phaseolus vulgaris*), sugar beet (*Beta vulgaris*), plum (*Prunus*), and probably also cacao (*Theobroma cacao*, primarily pollinated by small flies; see chapter 28). Relatively recently they have been promoted as potential commercial pollinators of certain chilli plants (Saxena et al. 1996).

5. Other Insects: Cockroaches, Grasshoppers, Bugs, etc.

Here we are limited to some very sparse and usually brief records. Nagamitsu and Inoue (1997a) demonstrated cockroach pollination in *Uvaria elmeri* (Annonaceae) in a dipterocarp forest in Sarawak; the plant is a climber and its flowers cauliflorous, giving access to crawling cockroaches, which visited mainly at night and fed on stigmatic exudates and some pollen. Flowers are cream or brown, with a smell described as decaying wood or mushroomlike. A second example of cockroach (*Amazonina platystylata*) pollination was reported for a *Clusia* species on inselbergs (isolated mountain tops) in French Guiana (Vlásaková et al.

2008), this cockroach having a rather rough surface to which pollen adheres, and feeding on a tissue exudate. The flowers are unusually strongly scented with acetoin, which is a known component of cockroach communication systems.

Termites (and perhaps gnats) apparently pollinate the most unusual orchid *Rhizanthella gardneri* in western Australia, which is almost entirely subterranean, growing on the roots of *Melaleuca* bushes. It produces bowl-shaped dark purplish flowers that only just protrude among the leaf litter under these shrubs, giving ready access to wingless termite workers.

In the Galapagos Islands, with a very sparse pollinator community (chapter 27), there are regular flower visits by a short-horned grasshopper (*Halmenus*), which Philipp et al. (2006) reported as carrying pollen between flowers of 5 out of 12 species investigated. On Reunion Island a more specific interaction with an orthopteran occurs, with the orchid *Angraecum cadetii* being pollinated by a cricket, which took pollinia from around half of all the flowers visited and deposited them on stigmas in 27% of flowers (Micheneau et al. 2010).

The spoon-winged lacewing, a neuropteran, has been reported as having specialist pollen- and nectar-feeding mouthparts (Krenn et al. 2008) and to be a proficient flower visitor, selecting Brassicaceae and Asteraceae, but especially *Achillea*.

The only clear report of a pollinating hemipteran bug involves *Macaranga tanarius* plants, where the flowers interact with anthocorid and mirid bugs in a brood-site mutualism (see chapter 26); the bugs pierce a ball-like **extrafloral nectary** to gain food, breed in the chambers formed by floral bracts (Ishida et al. 2009), and in the process move significant pollen onto stigmas. This system may have evolved from a predatory interaction of bugs with flower thrips, which have a similar brood-site mutualism with other *Macaranga* species, and on which these bugs often feed.

6. Overview

From all the descriptions given in this chapter, a *generalist pollination syndrome* can readily be established that encompasses most of the flowers visited by a range of beetles, various different taxa of wasps, and a sprinkling of other insects. The traits included would be

1. open accessible flowers, bowl shaped or flat;
2. small or medium-sized flowers (very small examples often grouped together as inflorescences);
3. commonly white, cream, or yellow-green in color;
4. mild to moderate scents, fruity or musty but not unattractive;
5. highly accessible pollen, often in large quantities;
6. exposed nectar, occurring at low volume and usually high concentration.

These same flowers, in accordance with their generalist appeal, may also of course be visited by a range of short-tongued flies and bees; but it is worth remembering that they are not *completely* generalist, in the sense that they do not provide adequate reward for larger or endothermic pollinators and are visited only by the small and the ectothermic insects.

However, within this grouping there are some phenotypically generalist flowers (e.g., *Rubus, Tilia*) that can be termed cornucopia species, having a higher nectar reward and attracting a greater visitor diversity including some longer-tongued bees. It may be useful to have a subdivision *cornucopia generalist* syndrome to accommodate these unusual cases, which could also include some more morphologically specialist flowers like the thistles *Cirsium* and *Centaurea* (Ellis and Ellis-Adam 1993; Corbet 2006), where the nectar production is so high that the nectar entirely fills the corolla tube, suiting them to long-tongued visitors, but also able to be accessed by flies and some beetles.

Another important point to make is that even the most generalist of flowers are not necessarily equally served by all their visitors. Many small composite flowers are assumed to be generalist, but this can often be based on a lack of careful studies. For example *Hypochaeris salzmanii* is a duneland species in Spain, a typical composite that one might expect a priori to be visited by a wide range of small insects; but in fact the capitula open in the late morning, are visited only by a very few species of solitary bee, and close by mid afternoon (reopening with similar timing over the following days). Without carefully analysis over several hours, the fact that a seemingly generalist flower (inflorescence) has a highly specialized floral biology would be missed (P. E. Gibbs, pers. comm., from observations by S. Talavera). *Heracleum sphondylium* is even more useful here; it is a widely cited and exemplary generalist in every respect, and can attract at least 40 insect taxa, yet Zych (2002) reported that in Polish populations only 53% of these carried significant pollen, and the only consistently important pollinators were syrphid flies and greenbottles (Zych 2007). Likewise for the carrot (*Daucus*), studies have recorded in excess of 250 species of visitor, the composition and number varying geographically, but with solitary bees, honeybees, and some sphecids and hoverflies variously recorded as most efficient (see Koul et al. 1989; Perez-Banon et al. 2007 for details).

Chapter 13

POLLINATION BY FLIES

The flies (order Diptera) constitute a very diverse group of insects, all characterized by just one pair of wings, the ancestral rear pair being modified as flight- and balance-control organs termed halteres. Hence flies are often very agile fliers, able to take off and land in any direction and often to hover (rare in other insects). Fly mouthparts are essentially suctorial, but can be either piercing and sucking (using either plant or animal fluids) or merely sucking and lapping without the ability to pierce tissues. Many types of fly have the ability to regurgitate saliva onto potential foodstuffs, making the material more liquid and manageable, and some use "bubbling" behavior to speed evaporation of excessively dilute fluids. The feeding habits of flies are therefore highly varied, and different taxa are able to suck, lap, chew, or bite, so that flies can be found taking advantage of almost all possible foodstuffs. A great many groups (from at least 45 fly families) have a strong preference for sugary fluids, and therefore commonly take some nectar as part of their adult diet, by sucking or lapping at flowers; quite a number also take pollen.

Some flies after visiting a flower will carry moderate amounts of pollen on their bodies, making them potentially useful as pollinators; but many of them do not move large distances between plants (the mean interplant distance is usually less than 1 m), reducing their value for effecting cross-pollen movements. However, a few taxa are more inclined to make long-range movements and thus are more useful. One family in particular, the hoverflies (Syrphidae), are specifically equipped for pollen feeding and rely almost entirely on flowers for their adult food intake; they also move much more regularly and systematically through flower patches and are well known as efficient and important pollen vectors in temperate zones.

Before going into further details on any particular flower-visiting flies, some basic fly taxonomy is necessary to make sense of this group as possible pollinators. An overview is shown in table 13.1. Formerly the flies were subdivided into three main parts, but molecular evidence has clarified matters and there are now just two suborders, the second divided into four main infra-orders (see Yeates and Wiegmann 1999; Wiegmann et al. 2003). The most primitive flies are in the suborder Nematocera (meaning "threadlike antenna"), mostly very small and with very short mouthparts, and commonly with aquatic larval stages. These are the midges, mosquitoes, gnats, and craneflies, all of which are poorly endowed with attributes that might make them good pollinators but which can nevertheless be quite common on flowers; and they may represent the earliest of all flower-visiting animals (see chapter 4 and Thien et al. 2009). They are especially abundant in arctic and montane habitats (see chapter 27) where other insects are rare. They take mainly nectar from flowers, although some bibionids, mycetophilids, and scatopsids also eat pollen according to

TABLE 13.1

Dipteran Taxonomy, Showing Some of the Main Groups of Flies Potentially Found on Flowers;
Those with Significant Flower-Visiting Habits Are Shown in Bold

Suborder	Infra-order	Families	Common name
Nematocera		Tipulidae	Crane flies
		Psychodidae	Owl midges
		Culicidae	Mosquitoes
		Ceratopogonidae	Biting midges
		Chironomidae	Nonbiting midges
		Simuliidae	Black flies
		Bibionidae	Fever flies, St Mark's flies
		Mycetophilidae	Fungus gnats
		Cecidomyiidae	Gall midges
		Scatopsidae	
Brachycera	Xylophagomorpha	Xylophagidae	
	Stratiomyomorpha	**Stratiomyidae**	Soldier flies
	Tabanomorpha	**Tabanidae**	Horse flies
		Rhagionidae	Snipe flies
	Muscomorpha		
	Heterodactyla	**Nemestrinidae**	Long-tongued flies
		Apioceridae	Flower-loving flies
		Bombyliidae	Bee-flies
		Asilidae	Robber flies
		Therevidae	
	Eremoneura	Empididae	Empids
		Dolichopodidae	Dance flies, long-headed flies
	Cyclorrhapha	Platypezidae	
		Phoridae	Scuttle flies
		Syrphidae	Hoverflies
	Schizophora	**Conopidae**	
		Tephritidae	
		Trypetidae	
		Sepsidae	
		Sciomyzidae	
		Coelopidae	
		Sphaeroceridae	
		Ephydridae	
		Drosophilidae	Fruit flies
		Chloropidae	
		Scathophagidae	Dungflies
		Anthomyiidae	Lesser house flies
		Fanniidae	(House flies)
		Muscidae	House flies
		Calliphoridae	Blow flies, bluebottles
		Sarcophagidae	Flesh flies
		Tachinidae	

Source: Taxonomy based on B. M. Wiegmann et al. 2003.

early records (Willis and Burkill, 1895–1908, UK flower-visiting records). Gnats, midges and mosquitoes of both sexes may take floral nectar as fuel, although for the females it is blood meals that provide the main fluid intake. Many of these insects, of either sex, are particularly active on flowers toward dusk, when their flights in search of their vertebrate hosts tend to reach a peak.

The second and much larger suborder is the Brachycera, usually stouter and with shorter antennae. The more ancestral parts of this taxon (see table 13.1) include a few moderately important flower-visiting groups, notably soldier flies (stratiomyids), which are brightly or metallically colored flies that occur reasonably commonly on flowers with exposed nectar, such as umbellifers. Rhagionids can also be quite common on such flowers; relatively large, with elongate pointed abdomens, they take some nectar but use flowers more commonly as encounter sites for prey or for mates. The same could be said of apiocerids, therevids, and asilids, in the slightly more advanced Muscomorpha/Heterodactyla grouping; also here are the acrocerids, where the genus *Eulonchus* contains some flower visitors selecting geraniums and similar forms with good floral constancy (Borkent and Schilinger 2008). But this taxon above all contains two key flower-visiting families, the bee-flies (Bombyliidae) and the nemestrinids. Bombyliids are long tongued (commonly 8–12 mm, but occasionally much longer than this) and are reasonably common in temperate habitats but especially abundant in Mediterranean and semiarid habitats. The Nemestrinidae are also long tongued, some species having exceptionally long proboscis up to 70 mm, and are important in southern Africa, sometimes exclusive to particular long-corolla flowers. These two families originated around 210–180 MYA (Wiegmann et al. 2003), before the evolution of angiosperm flowers, so the earliest forms may have fed on Jurassic gnetaleans and other preflower structures.

The cyclorrhaphan and schizophoran subgroupings are enormous and within the former the key group is the specialist flower-feeding hoverflies (Syrphidae), which are rather like bees in their reward needs and floral preferences; they commonly have tongue lengths of 2–8 mm. The scuttle flies (Phoridae) are also potential pollinators of certain flowers, including Araceae and some of the elaborate Aristolochiaceae (Rulik et al. 2008). The Schizophora encompasses the higher flies, taxonomically difficult and with a host of small families. The drosophilids—fruit flies—are sufficiently

familiar to have a common name; beyond them, the familiar house flies, blowflies, bluebottles, and botflies, as well as most of the dung- and carrion-visiting fly types, are the most recently evolved families (Calyptratae), many with flower-visiting propensities.

1. Feeding Apparatus

Flies are endowed with a proboscis (fig. 13.1) made up from a basal rostrum, plus the ancestral unpaired labrum and **hypopharynx**, which together normally form a short tube that lies on the **labium**. The labium forms the ventral wall of the food canal, and its tip is expanded into a pair of conspicuous flat pads termed the labella, each labellum bearing grooves known as **pseudotracheae** (Gilbert and Jervis 1998; see also Krenn et al. 2005). Where the proboscis is relatively short, flies can exploit many exposed fluids with a dabbing or lapping action, drawing fluid onto the pads (the pseudotracheal grooves often have hydrophilic inner linings to aid fluid uptake) and then upward into the mouth; but they can also use quite solid materials by first suspending the particles in saliva regurgitated through the hypopharynx.

In a few of the families, where there are flies with a more elongate proboscis, feeding from long tubular corollas becomes possible, as well as from the generalist open flowers repeatedly mentioned in the previous chapter. For example, in bombyliid flies the ventral part of the proboscis base is extended, and the suctorial mechanism is also more powerful, giving a tongue that can penetrate and suck fluid from quite deep corolla tubes. The labellar musculature is also altered, and these flies can feed from laterally opening flowers as well as those with frontal and dorsal openings (Szusich and Krenn 2002).

Where substantial pollen is taken in (suspended in nectar, especially in syrphids) the mouthparts tend to be shorter and the labella are broader with more pseudotracheae, the width of the furrows perhaps reflecting the preferred sizes of pollen grains (Gilbert and Jervis 1998).

2. Sensory and Behavioral Capacities

It has usually been noted that flies use visual cues for longer-range flower-seeking and then cue in to olfactory signals at closer range. The eyes of day-flying

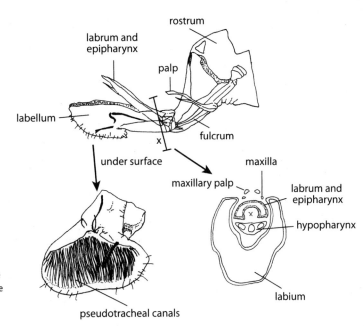

Figure 13.1 The basic fly proboscis, with views of the labellar surface and pseudotracheae, and a transverse section showing the food channel at X. Dark areas are underlying sclerites. (Largely modified from Gilbert and Jervis 1998.)

flies are large and sophisticated by insect standards, and all groups of flies so far tested have good trichromatic color vision. Flower foragers in at least some families (syrphids, calliphorids, anthomyids, tephritids) have an apparently innate preference for yellow, closely aligned with the wavelengths reflected from pollen in the centers of most of the generalist flowers; an innate response was clearly demonstrated in a syrphid by Sutherland et al. (1999). However, some of the more specialized flower-visiting bombyliid bee-flies prefer pinks, mauves, and blues, associated with particular radial flower forms (Johnson and Dafni 1998).

Flies have long-range chemosensors (sited on the antennae) that are abundant and receptive to many floral odors, plus taste receptors (contact chemosensors, on the mouthparts and on the feet) that are highly attuned to detection of sucrose and/or glucose.

Various taxa show associative learning, linking color or odor to reward, although some have innate preferences that persist despite training (e.g., syrphids retain their strong preference for yellow; Lunau 1992c). For *Lucilia* (the common greenbottle calliphorid) this kind of associative learning can occur with just one trial (Fukushi 1989), as shown in figure 13.2. Some flies can learn and use spatial landmarks, and various male syrphids use this ability to become territorial, managing to hover in one site reliably over successive days using visual cues, and chasing off intruders with

considerable targeted accuracy (Collett and Land 1975). With sugar-seeking habits plus sophisticated sensory abilities and behavioral repertoires, a whole range of flies are therefore important components of the flower-visiting fauna in many habitats.

There is little evidence for different kinds of foraging behaviors in different fly groups; for example, on *Clematis ligustifolia* Borkent and Harder (2007) recorded similar visit and revisit frequencies across a range of culicids and small to large muscids, and on *Shepherdia* they found no differences between syrphid and empid behaviors.

3. Generalist Fly Flowers

A wide range of small open flowers with exposed nectaries, occurring either singly or in clustered umbels or capitula (see fig. 2.11) are accessible to, and favored by, a multitude of flies. This includes both the more primitive midges and mosquitoes and the advanced syrphid and muscid types. The flies operate using the "spit and lap" technique that allows feeding on nectar of almost any concentration, but which is especially useful when exposed nectar becomes more concentrated on warmer and drier days. Muscids in particular can take nectar that is effectively crystalline, in excess of 75% concentration, when virtually all other flower

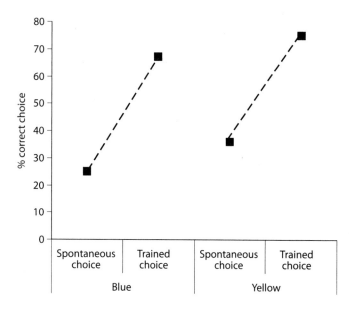

Figure 13.2 Substantial associative learning occurs in a single trial in the calliphorid fly *Lucilia cuprina*, trained on either blue or yellow paper circles mimicking flowers. (Drawn from data in Fukushi 1989.)

visitors are unable to use this; they can be seen actively feeding on a range of familiar umbellifers—such as hogweed (*Heracleum*), carrot (*Daucus*), and parsnip (*Pastinaca*)—on the hottest summer days.

These kinds of flowers are traditionally referred to as *myophilous*, and the generalist fly flower syndrome as **myophily**. The typical flowers have the following properties:

1. They are small and open, flat or shallow bowl shaped, radially symmetrical
2. They are often clustered into inflorescences
3. They are white or cream, or sometimes greenish-yellow, in color
4. They have mild, sweet or musty, but usually not unpleasant smells
5. They open in the daytime, often producing nectar throughout the middle of the day
6. The nectar is exposed, high concentration and low volume.

There is an evident overlap here with the generalist syndrome described in the last chapter, and it is quite reasonable to merge these all into one syndrome, encompassing flowers visited by all the beetles, wasps, and short-tongued flies.

Similar floral designs can be found on a larger plant framework on many familiar temperate trees that have open accessible flowers—lime (*Tilia*), sycamore (*Acer*), rowan (*Sorbus*), and hawthorn (*Crataegus*), as well as some fruit trees (*Malus* and *Prunus*). Not surprisingly, all of these too are frequented by many of the fly groups, and the plants could be included as exemplifying the myophily syndrome. In fact the reality of "generalist fly flowers" can readily be tested by enthusiastic observers in their own gardens: a single sunny day spent watching the flowers of *Tilia, Heracleum*, or *Crataegus* can easily yield records of ten or more fly families, and hundreds of individual dipteran visits (although visits from beetles and wasps may also occur, and activity from at least a few bees is also likely). Examples of such records can be found in Corbet, Unwin et al. (1979), Willmer (1983), and Zych (2002, 2007).

Small flies from the lower taxonomic groups also visit plants that are low growing and so produce their flowers close to the ground, where the microclimate may be more equable and humid, suited to a small flying ectothermic animal. Some alpine plants with tiny clustered flowers exemplify this scenario—saxifrages and sedums, among others. Climbers with flowers growing close to surfaces have similar properties, and flies are common at ivy flowers in autumn. Flies also visit a range of somewhat taller fairly generalist plants with bowl flowers, such as meadowsweet (*Filipendula*), rock-rose (*Potentilla*), and other Rosaceae, or certain kinds of *Clematis* (Ranunculaceae); or those with very short tubular corollas growing in masses, for example some asters and scabious, valerians (*Valeriana*), mints (*Mentha*), and some spurges (*Euphorbia*). Again, white and rather pale pinks or yellows tend to dominate, and the individual flowers are radial or nearly so, usually close enough together for a fly to walk between them across the inflorescence surface (see plate 1B,G).

A few of the lower Brachycera fly families should also be picked out for mention here. The stratiomyids or soldier flies, with conspicuously colored and somewhat flattened bodies, are reasonably frequently seen on flowers, especially in waterside habitats; their tongues are quite short but spread out at the tip into large labella that lap up medium-concentrated nectars very effectively. They visit Apiaceae regularly, and some of the shorter-tubed Asteraceae. Empids are also regular flower visitors, although primarily predatory; their mouthparts look relatively well suited for nectar feeding, being designed for piercing, but the proboscis is rather rigid and inflexible (suited to its role of stabbing into prey) so in practice works best when probing into shorter corollas than one might predict on tongue length alone. Empids occur quite commonly on daisy-type flowers with medium corolla lengths, such as knapweeds and thistles (*Centaurea* and *Cirsium*), which hold nectar at the base of tubes several millimeters in length.

Within the Schizophora, flies from several families are generalist flower visitors, all having short tongues that merely mop up nectar. Sepsids are small ant-like flies, often common on umbels. Some of the tephritids and trypetids are brood-site parasites of flowers, laying their eggs in flower heads of various composites; they may transfer some pollen between plants, but any benefit to the plant is probably outweighed by the subsequent loss of flowers or seeds to the feeding larvae. The same may be true of the fruit flies (Drosophilidae), which are often attracted to nectar and sometimes lay eggs in flower heads (Miyake and Yafuso 2003, 2005), particularly of flowers in the family Araceae with rather fruity scents. Chloropids are small, often yellow-colored flies that again visit Apiaceae and Asteraceae but are additionally regularly recorded on forget-me-nots (*Myosotis*); the family also includes the frit flies (*Oscinella*) which are serious pests of cereals where they lay their eggs.

Higher schizophorans include most of the flies familiar to the layman and irritatingly present in urban situations. They are generally quite large and round bodied, and many are rather bristly in appearance (es-

pecially the tachinids). A number of the families are important, albeit rather generalist, flower visitors, and they are probably more use as pollinators than the lower muscomorphs, primarily because of their larger size. All the groups commonly termed "house flies" (muscids in the broad sense, including anthomyiids and fanniids) are generalists flower foragers (plate 1D), and because they are so numerous and so enamored of sugars they are probably the second most important kind of flower-visiting flies after the hoverflies. They can be found intermittently on almost any open flower offering access to sugary fluids. They therefore fit fully into the generalist myophily syndrome, although there are indications that they are more attracted to sweet-smelling flowers (such as meadowsweet (*Filipendula*), rowan (*Sorbus*), and mignonette (*Reseda*)) than some other flies. *Lucilia* (greenbottles) and *Calliphora* (bluebottles; see plate 1G) have similar habits to the muscids and can be particularly effective as pollinators of various *Allium* species such as leeks and onions (Clement et al. 2007), although the bluebottles also like to visit carrion (and so can be particularly problematic in spreading unwanted bacteria onto sugary foodstuffs). The same is true of the common yellow dungfly *Scathophaga*, which obviously frequents various kinds of dung but also visits flowers to find prey and to feed on both nectar and pollen; it can be common in late spring on buttercups and hawthorn, and in autumn on brambles and various umbels. The broad grouping of higher flies together contributes a substantial generalist flower-visiting cohort in most habitats. In northern parts of Europe and North America, and in mountain habitats, they may be more numerous than either bees or beetles; and with the generalist nematoceran flies added on they strongly dominate in the really cold alpine, taiga, and tundra zones.

Figure 13.3 shows distributions of various generalist myophilous flower visitors along a humidity or rainfall gradient in Patagonia (Devoto et al. 2005), with flies common in the wetter habitats and decreasing in areas of low rainfall; this may again be partly associated with temperature differences.

4. Specialist Nectar-Feeding Fly Flower Types

Those families of flies endowed with a longer slender proboscis may visit all the flower types referred to

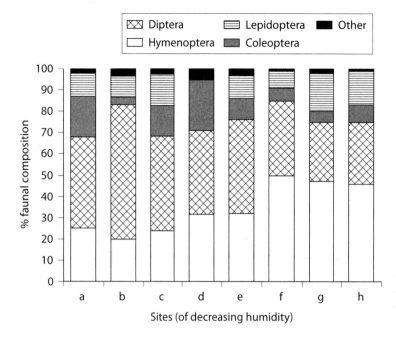

Figure 13.3 Relative composition of flower-visiting fauna at eight sites in Patagonia, ordered from left to right along a decreasing humidity gradient; flies decrease in abundance in the drier sites. (Redrawn from Devoto et al. 2005.)

above, but can in addition visit a substantially broader spectrum of floral morphologies, inserting their mouthparts into elongate corollas. Such flies tend to be more specifically seeking nectar and may show quite specialist relationships with particular flowers. The two classic examples are the bombyliids (bee-flies), common in many parts of the world, and the nemestrinids, which are largely restricted to southern Africa.

Bombyliids are unusual in having a long forward-pointing tongue that is not retractable, so they are easily recognized in flight and when hovering above flowers. However, in other respects they are confusing to an observer because (as the name implies) they appear very bee-like, having unusually rounded and furry bodies (plate 23H). When sitting on or hovering over flowers probing for nectar, they are very easily mistaken for small bees. In Europe, species of *Bombylius* reach the size of smaller bumble bees, and have tongues (fig. 13.4) that can be 10–12 mm in length; in Africa their diversity is greater and their tongues can be even longer.

The bombyliids are highly specialized nectar feeders, working systematically through floral communities, while selecting their preferred species carefully primarily by sight; and since they can carry substantial pollen on their body hairs they are clearly effective pollinators (although some may also eat a little pollen).

Kastinger and Weber (2001) indicated that their importance has usually been underestimated. They visit moderately small tubular flowers, which in temperate habitats are mainly white, blue, and purple (more rarely yellow). They are especially active in spring in temperate habitats, preferring sunlit sites, but visit flowers all year round in the tropics. Familiar preferred temperate flowers include violets (*Viola*), grape hyacinths (*Muscari*), stitchworts (*Cerastium*), various small-flowered Labiatae such as bugle (*Ajuga*) and ground ivy (*Glechoma*), and Boraginaceae such as forget-me-nots (*Myosotis*) and lungwort (*Pulmonaria*). Some also visit primroses and cowslips (*Primula*).

The bee-flies have a semiparasitic relationship with bees in their reproductive habits, since they lay eggs at or in the entrances of solitary bee nests, and their larvae develop by feeding on the stored provisions within; this habit has presumably led to selection for resemblance to bees.

The extreme case of long-tongued flower-visiting hovering flies is the family Nemestrinidae. All representatives have tongues more than 15 mm long and sometimes up to 70 mm (more than four times the body length; see fig 13.4C). In the same southern African habitats, and exploiting many of the same flowers, there are some very long-tongued bombyliids and (oddly) also some members of the family Tabanidae

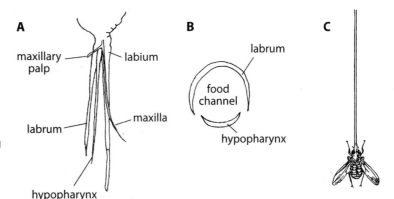

Figure 13.4 (A) and (B) *Bombylius* tongue structure, and transverse section of the food channel formed by labrum and hypopharynx (modified from Gilbert 1981). (C) The extremely long tongue of a southern African nemestrinid fly, *Megistorhynchus* (redrawn from Barth 1985).

with tongues of up to 47 mm (plate 23G), although tabanids in other habitats are typically endowed with short stabbing mouthparts that can inflict painful bites on vertebrates. This unusual dipteran community interacts with a whole range of apparently specialized floral species (Johnson and Steiner 1995, 1997; Goldblatt and Manning 1999, 2000), some of the flowers perhaps forming mimicry rings (Johnson et al. 2003). These flowers are from several plant families, including Geraniaceae (pelargoniums), Iridaceae (iris/gladioli), and Orchidaceae; and they have many shared features, especially zygomorphic deep tubular morphology, often with additional nectar spurs and conspicuous nectar guides, and copious nectar at 20%–30% concentration. The various authors familiar with these interactions propose the recognition of a specific *long-tongued fly* syndrome to reflect both their importance in these ecosystems (see chapter 27), and the close match of morphological traits of insects and flowers. Johnson (2006) showed the functional synergy of fly and plant types here, and the convergent evolution in both partners, although he also noted that the flies and plants occur in "guilds," lacking obligate one-to-one relationships, a particular plant species potentially being visited by more than one type of fly that in turn visits several plant species.

As well as the bombyliids and nemestrinids, there are other families of fly with less extreme elongate tongues that link to flower-visiting behaviours.

1. Conopidae often have moderately long tongues that are extended further by elongated labella. Various species can reach nectar in corollas 4–7 mm deep, but in practice they tend to visit open exposed nectaries in the Apiaceae, Asteraceae, and Rosaceae, and in the autumn they are fairly common on ivy flowers. However, they spend rather longer just "sitting around" on a wider variety of flowers than is required for their feeding, because they use flowers as encounter sites for their prey (bees and other hymenopterans). Hence they may be recorded as visitors on flowers where they are not feeding, and are making almost no contact with anthers.

2. A few species of tachinid fly (beyond those already mentioned in southern Africa) are moderately long tongued. For example, *Siphonia* and *Prosena* both visit small tubular flowers such as *Mentha* in Europe. Some exceptionally bristly tachinids such as *Dejeania* also occur on *Acacia* inflorescences in Africa, apparently probing for both nectar and pollen (pers. obs.).

5. Hoverfly Flower Types

Syrphidae, the hoverflies (or sometimes simply called "flower flies"), are by far the most important flies to be properly equipped for pollen feeding and specifically deriving all or nearly all of their food (as adults and as larvae) from flowers. As such they often rival the bees in importance as pollinators in particular habitats or for particular crops, and they have been better studied than most other flower-feeding flies. They are particularly important in north temperate habitats and on relatively nonspecialist flowers (once again the Apiaceae, the Rosaceae, and many Asteraceae and Brassicaceae);

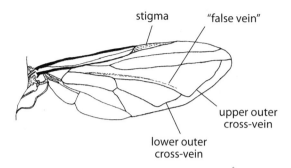

stigma   "false vein"

upper outer
cross-vein

lower outer
cross-vein

Figure 13.5 Syrphid venation, easily recognized by the false vein and the apparent outer margin formed by the outer cross veins.

thus they are very good pollinators of crops such as rape (Jauker and Wolters 2008). They may take nectar or pollen or both (Gilbert 1981, 1985), in part depending on the flower type and their needs at the time. They show good floral constancy in mixed-array tests and in the field (for example, *Melanostoma* had on average just 2.7 pollen types in its gut when caught; Hickman et al. 1995), so they have all the key attributes of good pollinators.

Syrphids come in a wide range of sizes, many of the tiny ones being elongate and shiny black or metallic, while the medium to large species are often black and yellow (plate 23B,D,E) or black and red and so are regularly mistaken for wasps. A few are specific mimics of bees: the drone flies in the genus *Eristalis* (plate 23C) can be remarkably like male honeybees, while the genus *Volucella* contains excellent bumblebee mimics and can even have morphs within one species, each morph having color bands appropriate to mimic a particular bumblebee type. All syrphids have some covering of hair, which can be branched or even plumose, but these bee-mimic species are especially hairy (in part no doubt to keep up the resemblance) and therefore become particularly good pollen carriers. The mimicry is assumed to be effective in deterring predators, and several hoverflies are readily able to deceive birds into ignoring them as potential prey (Bain et al. 2007). But for human observers it is relatively easy to distinguish hoverflies once they are at rest, as they have just one pair of wings like all dipterans and have extremely standard and rather unusual wing venation (fig. 13.5; plate 23D).

The mouthparts of hoverflies are variable in length (usually 2–4 mm, but up to 5–8 mm in *Eristalis* and *Volucella* and 12 mm in *Rhingia*), but they are reason-

ably consistent in design, with a more or less elongate rostrum from which emerges a more or less elongate proboscis (made up from the labrum and the hypopharynx) (fig. 13.6). In species with very long tongues, such as *Rhingia* in Europe (fig. 13.6C), both rostrum and labrum are elongated (unlike most other longer-tongued flies, where the rostrum remains short). The tip of the proboscis expands into a pair of bristly labella, which are used to dab at a flower and pick up superficial pollen (Gilbert 1981). Alternatively, pollen is gathered either by rubbing anthers between the labella, or by inserting the labella into larger anthers and scraping out the pollen. The face and tongue are regularly cleaned with the legs, allowing pollen to be transferred to the mouth and eaten. The labella can be closed together almost as a tube when feeding on nectar in tubular flowers, or spread out over a wider liquid surface in more open flowers. This allows the syrphids to feed on a particularly wide range of flower types.

Gilbert (1981; see also Gilbert and Jervis 1998) analyzed feeding behavior in common European hoverflies. All species took pollen as their main protein source (more so in females, as they have to provision their eggs with substantial protein), and the proportion of pollen taken was roughly reflected in the recorded density of the ridges (pseudotracheae) on the labella surfaces. The more polyphagous genera visiting several sources for pollen (such as *Platycheirus, Episyrphus, Sphaerophoria*) were shown to be commoner in open habitats, compared with more forest-loving oligophagous types (Branquart and Hemptinne 2000).

Not all hoverflies take in nectar. The smaller types, often with shorter tongues, are mainly pollen feeders (e.g., *Syrphus, Episyrphus,* and *Melanostoma*), although others of similar body size have longer tongues with smaller labella and take both nectar and pollen (e.g., *Metasyrphus, Platycheirus,* and *Syritta*). Varying amounts of nectar were used in the diets of some medium and all larger species tested (*Sphaerophoria, Rhingia,* and *Eristalis*), presumably reflecting their higher energy needs. The largest hoverflies such as *Eristalis* could fill their crop with nectar in 75–220 minutes depending on the flowers chosen (Gilbert 1983).

As might be expected, hoverfly flowers include a multitude of designs. There is a good correlation between proboscis length and flower depth (Gilbert 1981; fig. 13.7), even though flies with longer tongues can also exploit shallower open flowers. The most favored

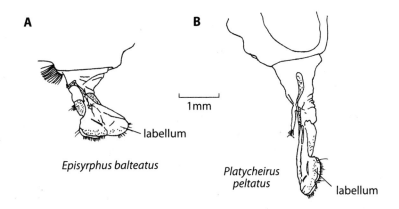

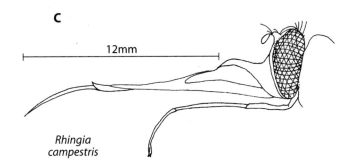

Figure 13.6 (A) and (B) General syrphid proboscis types in *Episyrphus* and *Platycheirus,* and (C) elongated pointed proboscis in the genus *Rhingia.* (All modified from Gilbert 1981.)

families are Apiaceae, Asteraceae, Ranunculaceae, Brassicaceae, Caryophyllaceae, and Rosaceae, many of these corresponding with the generalist fly flowers discussed in section 2, *Sensory and Behavioral Capacities*, above. The range of Asteraceae used is certainly extended for the hoverflies, with more of the longer-corolla flower species coming into the frame; syrphids regularly visit hawkbit (*Leontodon*), hawkweed (*Hieracium*), and dandelion (*Taraxacum*). Gilbert and others have also reported a strong propensity to visit the (normally wind-pollinated) flowers of grasses (Poaceae) and of plantains (*Plantago*), especially for two very common genera, *Melanostoma* and *Platycheirus*. On open flowers such as these, the timing of hoverfly foraging depends in part on size and coloration as discussed in chapter 10 (figs. 10.9 and 10.10), with a particular tendency to activity in cooler weather in the large and hairy drone flies (*Eristalis*), as these have some endothermic ability.

For the medium- and long-tongued syrphids, a further range of plants beyond these generalist families are exploited. Some of the smaller zygomorphic flowers from the families Lamiaceae and Scrophulariaceae (such as *Stachys, Glechoma,* and *Ajuga*) are especially valued by such flies, which feed on them much more efficiently than can other flies; they approach in an appropriate direction and handle the relatively complex morphology of the flower rather easily, unlike muscids and calliphorids, which seem to visit more randomly. Other preferred flowers include smaller legumes such as melilot (*Melilotus*) and clover (*Trifolium*), and varieties of scabious (*Knautia* and *Scabiosa*). Extraordinarily, members of the genera *Volucella* and *Eristalis* that mimic bees also manage to mimic their habits of buzz pollination on some of the classically sonicated flowers (see chapter 7).

A few plants are sometimes regarded as having more specialist hoverfly flowers (Kugler 1938). Many of these share a particular arrangement of paired sideways-spreading stamens set at a slightly higher level than the central and slightly downwardly directed stigma (fig. 13.8), as seen in some species of *Veronica* (speedwell), and in enchanter's nightshade (*Circaea*). When a hoverfly grasps the stamens to feed, they droop slightly under the weight, and the underside of the insect body contacts the stigma. Both these plants are

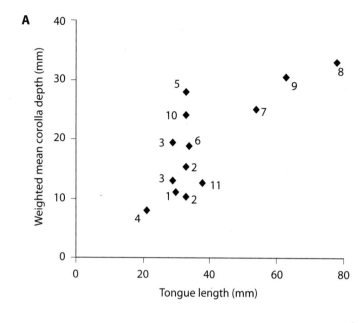

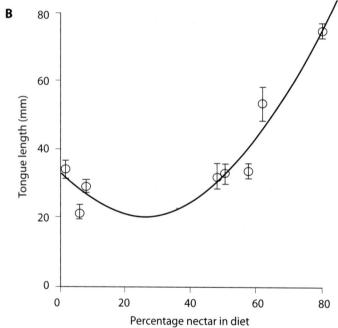

Figure 13.7 Correlations of syrphid tongue lengths with average corolla depth of flowers visited for nectar (A), and with percentage of nectar taken in the diet (B). 1, *Syrphus ribesii*; 2, *Metasyrphus corollae*; 3, *Episyrphus balteatus*; 4, *Melanostoma*; 5, *Platycheirus*; 6, *Syritta*; 7–9, *Eristalis* spp; 10–11, *Sphaerophoria* spp. (Redrawn from Gilbert 1981.)

common in the shady moist habitats of woodland, where appropriate smaller-bodied hoverflies such as *Melanostoma, Baccha,* and *Syritta* also occur (and where similarly sized bees that could also work the flowers are rather uncommon). Another rather specialist example is the slipper orchid *Paphiopedilum villo-*

sum, which is almost exclusively visited by syrphids, lured in by glistening staminodes and an apparent perch to land on, as well as a urine-like scent; this is pollination by deceit, as in attempting to land the flies slide off into a trough and can only escape (unrewarded) by squeezing out past the pollinia (Bänzinger 1996).

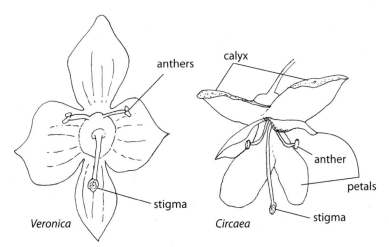

Figure 13.8 Flower forms said to be typical for hoverflies, with paired high stamens and a downward-pointing stigma. (Drawn from photographs.)

For the really long-tongued hoverflies, favored flowers include red campion (*Silene*), some geraniums, common bluebells (*Hyacinthoides*), the bindweeds (*Convolvulus* spp.), and some of the small to medium balsams (*Impatiens*). *Rhingia* is particularly adept at feeding on nectar from common bindweed in hedges and extending its attention to ornamental forms growing in gardens and is also a regular visitor to forget-me-nots and lungworts.

It should be evident from this discussion that hoverfly flowers overlap very substantially with any listing of bee flowers. It is reasonably common to find hoverflies as main pollinators on bee flowers in regions or at times where bees are relatively scarce, as in the moist woodlands mentioned above, or in winter in grasslands of the subtropics, where hoverflies have been reported to take over as pollinators of flowers such as *Sisyrinchium* (Freitas and Sazima 2003). In recent years, with a scarcity of honeybees in many areas (see chapter 29), there have been marked increases in hoverfly numbers, especially of the ubiquitous and almost worldwide *Episyrphus balteatus*.

Hoverflies also have sensory and behavioral attributes that are rather bee-like and accentuate their importance as pollinators. Their color and shape preferences are attuned to their being flower visitors; in laboratory trials several showed a clear preference for yellow flowers, or yellow centers, and smaller types such as *Episyrphus* chose smaller rather than larger flower models (Sutherland et al. 1999). A strong bias for yellow is accompanied by poor discrimination of either blues or deep reds from grays. Additionally they are "clever" flower visitors, with good associative learning. Many have an excellent ability to learn spatial landmarks, which links to male territoriality, one individual returning to hover in the same sunny patch or woodland edge day after day. Hoverflies can handle complex flower morphologies with considerably more facility than most other insects and can deal with zygomorphic corollas quickly and easily. In addition, they can work systematically around the florets of composite flowers (fig. 13.9), "counting off" a full circuit and then leaving (Gilbert 1983). Crucially, the hoverflies also tend to show a high degree of flower constancy. Studies with mixed floral communities showed very different but highly consistent flower choices made by *Eristalis* and *Helophilus* (Parmenter 1958), while Kugler (1950) revealed that individual constancy occurred, different individuals of *Eristalis tenax* being faithful to quite different flower species on a given day.

6. Carrion-Fly Flower Types

These flowers practise deceit pollination, and flies are very commonly involved. In essence this is a form of brood-site mimicry, covered more generally in chapter 23. The flowers mimic animal carcases and carrion, the preferred egg-laying site of various kinds of fly whose larvae require dead or decaying flesh as food; they attract in the adult flies but usually offer no rewards. The technical term for this syndrome is **sapromyophily**

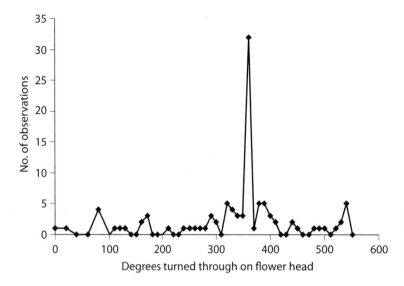

Figure 13.9 The hoverfly *Eristalis tenax* appears able to "count" a circuit when foraging on *Aster* flower heads, with the majority of individuals completing a 360° turn and then leaving the flowers. (Redrawn from Gilbert 1983.)

(although something similar also occurs with dung beetles, when it is termed coprocantharophily; chapter 12). A closely related brood-site mimicry is found in some flowers that resemble the gills of fungi, attracting visits by egg-laying flies, and termed **mycetophily**; here the best-known examples involve fungus gnats (very small nematoceran flies in the family Mycetophilidae). Within the Saxifragaceae, a fungus-gnat pollination syndrome with unusual saucer-shaped flowers has evolved repeatedly (Okuyama et al. 2008).

Members of the family Araceae (the aroids) are best known as carrion flowers, but examples also occur in Aristolochiaceae, Apocynaceae/Ascelpiadaceae, Taccaceae, and Orchidaceae. Some examples are relatively simple, while others involve complex trapping mechanisms. Various asclepiads use simple brood-site mimicry, a well-known example being *Stapelia*: the large flowers have reddish or purplish corollas and distinctive patterns of hair on the petals, increasing the resemblance (both visual and tactile) to dead animal surfaces. Most strikingly, the flowers are strongly and unpleasantly scented; volatile profiles were shown in table 6.2 (Jürgens et al. 2006). They attract a range of muscid and calliphorid flies, which lay eggs on the flowers and in the process get the typical asclepiad pollinaria (chapter 7) stuck on their feet. Some other asclepiads (e.g., *Tavaresia*) enhance their attraction to flies by incorporating motile structures inside (or at the mouth of) their flowers, such as hairs or filamentous tissues that constantly move or vibrate in the slightest breeze; similar oscillating structures occur in some *Ceropegia* species and in a few orchids.

Rafflesia, a genus including the world's largest flower, also fits in here. Like its close relative *Rhizanthes*, the *Rafflesia* plant occurs in the shaded understory of Asian forests, growing as a parasite on tree roots and with only its enormous reddish-brown flowers visible above ground level (plate 22G,H). Both genera emit volatiles and some carbon dioxide and have a degree of thermogenesis (Patino et al. 2002), all helping to attract the blowflies such as *Lucilia* that act as pollinators (Beaman et al.1988). Flies seek to access the enticing "flesh" by entering anther grooves with hair-lined ridges that guide them precisely, so that the viscous pollen matrix is deposited on their backs; then the flies squeeze into female flowers under a ring of stigmatic tissue where pollen is rubbed off. Figure 13.10 shows the architectural features involved.

The genus *Tacca* is also worth mentioning; again it occurs in Asian forests. Widely referred to as the "bat flower" from its shape, but nectarless and with elaborate filiform appendages in dark purple or almost black colors and strong decaying odors, this is again fly visited (albeit at low frequency, with selfing being common; Zhang et al. 2005; and see Fenster and Martén-Rodriguez 2007).

Examples of trapping systems in carrion flowers are described in chapter 23, so here we concentrate only on the floral features linked to fly attraction. Examples include *Aristolochia*, where the corollas are

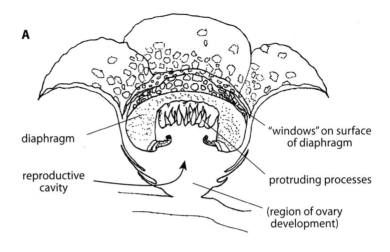

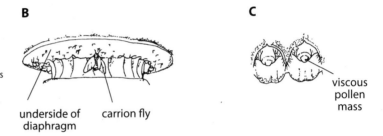

Figure 13.10 *Rafflesia* structures: (A) general cut-away view; (B) blowfly feeding on male structures under the rim of the diaphragm; (C) closeup of male structures, with channels leading to the anthers dispensing a viscous pollen mix. (Modified from Endress 1994, based on earlier sources.)

typically tubular but strongly protogynous so that they depend on attracting flies already covered in pollen (Brantjes 1980). The flowers have expanded petal lobes, often greenish or mottled with purple and white, and with a long tail (or sometimes several shorter tails around the edge of the petals), all providing visual attraction (fig. 13.11 and plate 22A). They emit strong carrion smells, often intensely over just a few hours, and these are sometimes localized into the petal tails which act as osmophores. The base of the corolla (containing a central cone of stout styles and anthers) secretes a little nectar with a high amino acid content and is also transparent so it appears as a window, attracting the flies down to the reproductive organs in pursuit of food and a way out. For the small Eurasian species, the pollinators are mainly small nematoceran flies, especially biting midges or, in the case of *A. pallida*, male phorids (scuttle flies) (Rulik et al. 2008). For the larger and more complex South American *Aristolochia* flowers, a whole variety of fly genera are attracted, although only a small range (usually sepsids, muscids, and calliphorids) are actually trapped by any

one species. There is rather good specificity (and little overlap) between flower species and fly genera. These New World *Aristolochia* species often have a U-shaped corolla, and are called "Dutchman's pipes" because of their form (fig. 13.11B).

In the Asclepiadaceae the highly variable genus *Ceropegia*, occurring throughout the southern continents, provides all the major examples of carrion fly flowers (Vogel 1961). Here again the corolla is long and tubular, commonly green/grey, with the petal lobes converging and usually uniting at their tips to produce a lanternlike structure (fig. 13.12 and plate 22B,C). The flower may also have a paler window area basally, around the sexual organs and nectaries. *Ceropegia* species usually lack the purple fleshy external appearance and the strong decaying scents of other carrion flowers, instead having quite delicate scents. They attract a range of small flies with fairly good specificity. The flies pick up pollinia on the underside of their heads or mouthparts while drinking nectar and may deposit them into grooves adjacent to the nectaries in a subsequent flower.

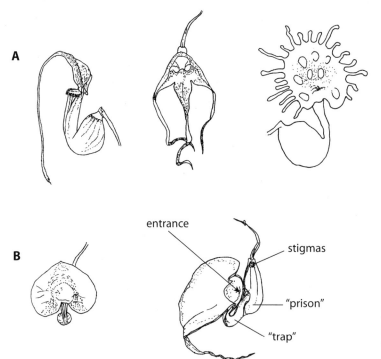

Figure 13.11 (A) Various *Aristolochia* flower structures (redrawn from Endress 1994). (B) New World Dutchman's pipe type, in front view and transverse slice (redrawn from Proctor et al. 1996 and from flowers).

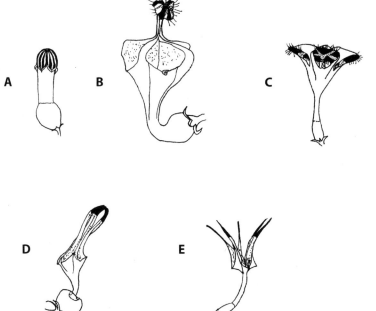

Figure 13.12 A variety of floral forms in *Ceropegia* species: (A) *C. ampliata*; (B) *C. distincta haygarthii*; (C) *C. sandersonii*; (D) *C. robynsiana*; (E) *C. stapeliiformis*. Scent-producing areas are black, and waxy or slippery surfaces are stippled. (Redrawn, not to scale, from Vogel 1961.)

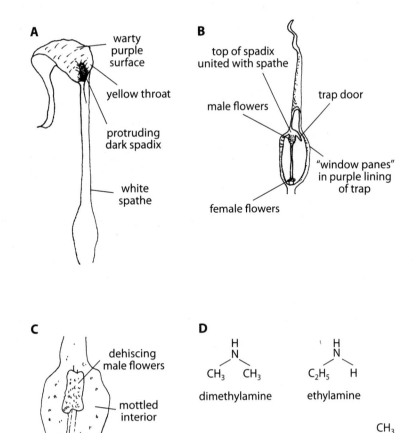

Figure 13.13 Structures of typical saprophytic aroid inflorescences: (A) and (B) external view and transverse section of *Cryptocoryne*; (C) internal view of spadix of *Arum* (A–C redrawn from Proctor et al. 1996); (D) the main volatiles emitted by common *Arum* species.

In Europe it is primarily the Araceae, and specifically the genus *Arum*, that provide carrion fly-trap examples (Meeuse and Morris 1984). Many of these aroids smell strongly of decaying meat, with dark purplish-brown and often hairy surfaces. They have unusual inflorescences with a central club-shaped *spadix* and an outer leafy *spathe* (fig. 13.13) that together take over all the attractive functions, and either of these structures may be expanded into long tails. Inside the resulting upright inflorescence tube, the flowers exist in single-sex arrays, the females (little more than an ovary with a flattened stigma on top) being at the base of the spadix and the male flowers (merely paired stamens) in a ring just above this, both protected deep within the spathe cavity. In the familiar European *Arum maculatum* (lords and ladies or cuckoo pint),

and the similar Mediterranean *Arum nigrum*, the spathe opens during the night, the stigmas mature, and the spadix emits a strong fecal odor from dawn onward, composed of ammonia, ethylamine and diethylamine, putrescine, indoles, and skatole (fig. 13.13D; and see chapter 6). Dung and flesh flies and varied small nematoceran flies may arrive in considerable numbers over the next few hours, drawn in from a distance by the scent and at closer range by the colors and sometimes by the waving tails. In *A. maculatum* a high proportion of the insects that effect pollination are small flies in the owl-midge family (Psychodidae) (Lack and Diaz 1991), while in Asian species a range of biting midges (Ceratopogonidae) and blackflies (Simuliidae) have been recorded.

One odd feature of most kinds of aroid is their

substantial heat generation during flowering. This was at one time thought to be an attractant, inducing flies to enter the spadix voluntarily; but in practice there is no doubt that the flies fall in accidentally, and are not attracted to warmth per se. It seems likely that the heat produced in the spadix by chemical thermogenesis is mainly helpful in rapidly vaporizing the volatile scents as they are secreted from the plant's tissues (Meeuse 1966). Similar high levels of heat production have been reported in various *Ceropegia* and *Aristolochia* species.

From these descriptions of examples from three key families, it is not difficult to identify key convergent features of these flowers.

The Sapromyophilous Syndrome

1. Large tubular flowers with spreading upper corolla, often growing close to the ground
2. Dull red, purple, brown, or sometimes greenish coloration, often with mottling
3. Petal surfaces with hairs and often with longer tail-like structures
4. Strong unpleasant scents, mimicking dead or decaying flesh or excreta, with vaporization sometimes aided by thermogenesis
5. Window effects from translucent corolla bases
6. Abundant pollen, little or no nectar (amino-acid-rich where present, perhaps to sustain the fly until it meets and is trapped by its next flower)
7. May have trapping mechanisms (downward-pointing hairs, slippery papillate surfaces, rings of protective bristles, constrictions, overhangs; chapter 23 gives details)
8. May have release mechanisms (withering of hairs, relaxation of constrictions, or change of floral orientation from pendant to horizontal)
9. Strongly protogynous
10. Relatively short lived, often 2–4 days

Orchids can readily be found that fit within this same syndrome, in both Old and New World floras. Red or brown flowers and foul odors occur in various *Bulbophyllum* species, on which flies alight and crawl toward the middle, finding themselves tipped rapidly toward the central column when they try to grip the orchid labellum, which is a finely balanced spring mechanism. Different species of *Bulbophyllum* exploit different flies, each being "sprung" by a different weight of visitor. Similar mechanisms occur in some Neotropical *Masdevallia*, which lure in flesh flies by color and odor and offer no reward (Dodson 1962).

7. Some Other Specialist Cases

Various gall midges are known to have relatively specialist interactions with flowers. For example, *Schisandra* is pollinated specifically by female *Megommata* midges (Cecidomyidae), the females unusually eating the pollen as a major part of their diet (Yuan et al. 2007). Cocoa flowers (*Theobroma*) are also routinely pollinated by cecidomyid midges (Young 1985; chapter 28). More unusually, the monoecious tree *Artocarpus integer* is pollinated by gall midges, but here a fungus that infects the male flowers is crucial to the interaction (Sakai et al. 2000), as the midges eat the fungal mycelium while ovipositing on the flower and picking up pollen, then transfer this to (uninfected) females presumably because of an odor-based sexual mimicry.

Fungus gnats, as well as engaging in deceit pollination with flowers that resemble their oviposition sites, have rather specialist interactions with a range of spring-flowering woodland plants such as *Tolmeia* (Saxifragaceae), where larger flies and bees are ineffective and act merely as pollen robbers on male-phase plants (Goldblatt et al. 2004).

Orchid pollination is also sometimes effected by flies other than carrion types. Male mosquitoes visit *Habenaria* in North America (Thien 1969), and various nematoceran flies work the flowers of the genus *Pterostylis*. Tachinid flies pollinate an Andean orchid, *Trichoceros antennifera*.

8. Overview

From a worldwide perspective, flies are second only to bees in their importance as flower visitors, often making up for their relatively poor pollen-carrying capacities by their sheer abundance (Larson et al. 2001). They may also be less rigorous visitors and make visits to fewer flowers per plant, so producing less geitonogamy, and they show reduced grooming so that less pollen may be wasted from the plants' point of view. Anthophilous flies probably do not get their fair share of attention in the pollination literature. They may be at least as important as bees in some tropical and semiarid zones and are often more important on some islands where bees are uncommon (including large islands such as New Zealand), and especially in the cold high-latitude and high-altitude habitats considered in chapter 27.

However, the details given in this chapter should make it evident that there is not one fly pollination syndrome (myophily) but several. Most flies could best be included within the generalist syndrome described in chapter 12, and it is these for which the term "myophily" is most often used (and see table 11.1).

But there could also be grounds for regarding the long-tongued flies of southern Africa, and perhaps even more so the carrion flies, as representing additional syndromes in their own right. Furthermore, the hoverflies (and a few other groups) are distinctly more specialist than the majority of flies, and in many respects their flower-visiting choices are allied closely with the characteristics of bee pollination discussed in chapter 18. In these last three syndromes—long-tongued flies, carrion flies, and hoverflies—quite specialist and efficient pollination relationships have developed, with good constancy, good pollen carriage, and potentially rather high pollination efficiencies.

Chapter 14

POLLINATION BY BUTTERFLIES AND MOTHS

Outline

1. Feeding Apparatus
2. Sensory and Behavioral Capacities
3. Psychophily: Butterfly Flowers
4. Phalaenophily: General Moth Flowers
5. Sphingophily: Hawkmoth Flowers
6. Overview

The order Lepidoptera contains the butterflies and moths and represents around 10%–11% of all described insect species. The types that are relevant here are mostly rather large insects (although there are also many thousands of species of small micromoths), but they are usually not particularly strong fliers, flying effectively only over short ranges. Their larval stages (caterpillars) are herbivorous, feeding on plant leaves or occasionally woody material or flowers; but as adults all are liquid feeders, sucking up fluids using a long, coiled, and elastic proboscis. This unusual tongue can be used to feed on many possible liquids, including plant sap, fruit juices, and excrements, but it is mainly employed to drink nectar. Since all lepidopterans have sculpted scales and hairs on their body surface (their name means "scaled wings"), it is inevitable that some pollen sticks to them, especially their tongues and faces, while they are flower visiting. Thus they have built-in adaptations as potential pollinators.

Lepidopterans are split into four suborders, but all the flower visitors occur in about 16 families within the largest of these, Ditrysia (the other three groupings include only micromoths). Ditrysia incorporates many superfamilies, mostly consisting of large moths and yet more micromoths, but two contain the evolutionarily more recent butterflies (thus neither "butterfly" nor "microlepidopteran" is a strictly acceptable taxonomic term).

Butterflies are loosely defined by their relatively thin bodies, colorful wings, and the thickened club-like tips of their antennae; in the superfamily Papilionoidea are the nymphalids, pierids, lycaenids, papilionids, and riodinids, while Hesperoidea contains the skipper types. All butterflies are diurnal and associated with warm and relatively still summer weather. However, the vast majority of *moths* (from the other superfamilies) are crepuscular, or active at night time. Hence the two groups have different constraints and different floral needs, and it has been traditional to link them to two differing kinds of flowers, opening by day and in the evening, respectively, and to specify two distinct lepidopteran pollination syndromes. However, it is in fact more useful to have three groupings, with moths split again, the family Sphingidae (sphinxmoths or *hawkmoths*) being distinct from all the other moth families. This is partly because very few butterflies and just a handful of species from most moth families can hover, and usually only weakly, whereas the sphingid moths are able to fly rapidly and strongly over a long range and can also hover expertly. This has the important outcome that they do not need a landing platform on their flowers. Sphingids also generally have rather good endothermic abilities (see chapter 10) and so can be active in much colder conditions, after dusk and into the night, when all other (ectothermic) lepidopterans are unable to fly. In many ways, the hawkmoths are the most effective pollinators in this order of insects, and they are very important in many warmer habitats. Here then we will treat the settling moths and the hovering hawkmoths separately.

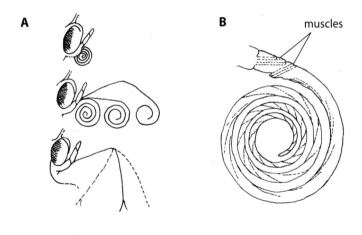

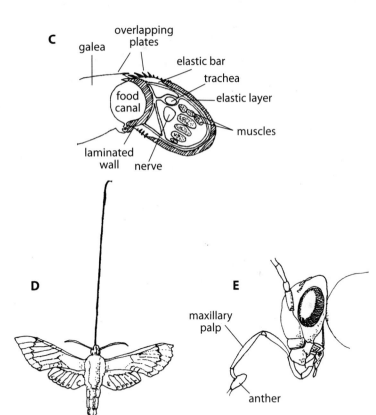

Figure 14.1 Anatomy of the lepidopteran proboscis: (A) schematic view with the tongue in various states of protrusion and during sucking; (B) in longitudinal section showing the main internal muscles; (C) transverse section of the double feeding tube; (D) the fully extended highly elongate tongue to be found in certain hawkmoths such as *Cocytius*; (E) the more unusual feeding apparatus of the relatively primitive moth *Micropteryx*, with pollen-scraping modified maxillary palps. (Parts (A), (B), (D), and (E) redrawn from Barth 1985; (C) redrawn from Proctor et al. 1996.)

1. Feeding Apparatus

The lepidopteran proboscis (fig. 14.1) is derived from the paired maxillae rather than from the labium and hypopharynx as in flies. The two maxillae are held together by hooks and teeth, giving a watertight seal around the central tube (Hepburn 1971). The maxillae are themselves hollow, filled with **hemolymph** (the term for insect blood or body fluid), so the structure can be uncoiled partly by increased pressure of this fluid, aided by small muscles that change the tube's cross section. There are also lateral bars ensuring that the tongue unrolls in an organized manner in the plane of the body axis, and a flexible layer containing the highly elastic protein resilin, to ensure that it coils up neatly again into an almost perfect spiral when pressure is released by opening a valve at the upper end. Suction within the tube is created by a cibarial pump in the head, and as in other nectarivores fluid viscosity has a major influence on the food intake rate. However, hawkmoths can vary their pumping in relation to both viscosity and concentration and achieve peak intake at around 34% sucrose (Josens and Farina 2001). In general, lepidopteran optimal feeding rates have been recorded at 35%–45% sugar concentrations (May 1985, 1988; Boggs 1988; Kingsolver and Daniel 1995; see table 8.4). But for butterflies the actual feeding rate is often rather slow, given the need to probe many tiny flowers in succession; Hainsworth et al. (1991) recorded *Vanessa* butterflies taking nearly 50 minutes to gain an average of 28 µl from a *Lantana* plant.

With this feeding apparatus, any fluid is potential food, and many lepidopterans do have minor dietary inputs from decaying fruits, oozing saps, or animal fluids. Indeed, for a few, such as the white admiral butterfly (*Limenitis camilla*), saps and excrements are the major food, while for the two-tailed pasha (*Charaxes jasius*) saps and fruits are the sole foods. However, the vast majority of lepidopteran species use the proboscis to exploit nectar as the main nutritional intake.

The tongue is always elongate, up to 16–18 mm in European and American butterflies and often much longer in flower-visiting moths (fig. 14.1D and table 14.1). Tongue length is roughly correlated with body mass and with wing area in European butterfly species (fig. 14.2; Corbet 2000); but nymphalids have relatively shorter tongues and smaller wings, while hesperids have unusually long tongues but small wings. Since wing loading affects foraging costs (chapter 10, and see May 1988), it follows that lepidopterans with unusually small wings prefer to forage on tightly massed flowers or those with higher than average nectar rewards; hence nymphalids and hesperids do not commonly visit the solitary flowers that attract many other butterflies. In tropical species, body length and proboscis length are again correlated allometrically, and Kunte (2007) proposed that outliers were rare because disproportionately long tongues were constrained by the longer handling times that such tongues imposed.

Tongue length certainly sets a direct limit on the range of flowers from which nectar can be extracted, and butterflies do not normally visit corollas deeper than their tongue length. Because the mouthparts are elongate, most lepidopterans require quite dilute nonviscous nectars (chapter 8); concentrations above 30%–40% become exceedingly difficult to suck up. However, some **noctuid** moths (perhaps rather more species than currently recorded) can spit saliva to give dilution of nectars that are at first too concentrated for them. The silver Y-moth (*Autographa gamma*), common in Europe, can take quite concentrated nectars in this fashion. In fact, a few moths can take much more dilute fluids and regurgitate them for evaporation, as well as taking the more concentrated fluids by the salivating technique (Wei et al. 1998).

In addition, the tip of the lepidopteran proboscis is covered in tiny spines and can scrape at tissues in the base of a flower to get some sap released. This is probably why many butterflies can be seen apparently successfully feeding suctorially at flowers that seem to produce no nectar. Chapter 9 discussed at least one case where this results in a specific relation between an orchid and a moth. More generally, this strategy may extend the range of visited plants beyond that of bees and flies, and it could be an economic solution of direct benefit to certain nectarless plants, ensuring that they invest in rewards only when they are actually being visited.

Some lepidopterans with short adult lives take little food after emergence and spend most of their time mate seeking, using up lipid reserves they accumulated as larvae. Those species with limited lipid reserve may have to be more selective about the flowers they visit so that they achieve a greater foraging profit, whereas species with large lipid reserves (for example, up to 16% dry mass in *Agraulis vanillae*) can afford to be unselective and visit flowers with small nectar volumes (May 1992). On a similar theme, some species rely solely on protein they accumulated as larvae to mature their eggs, but many do require some food input other than nectar to give at least a minimal input of extra nitrogen. Recent evidence indicates that some prefer nectars with higher amino acid levels, presumably for this reason (chapter 8). Alternatively, tropical butterflies in particular will often take urine and feces, such as bird droppings, and many will show "puddling" behaviors on muds and small water patches, gaining salts and probably some nitrogen from the microbial growths in such fluids.

TABLE 14.1
Lepidopteran Tongue Lengths

Family	Species	Common name	Tongue length (mm)
Butterflies			
Lycaenidae	Polyommatus icarus	Common blue	~8
	Lycaena phlaeas	Small copper	7–8
Papilionidae	Parnassius apollo	Apollo	12–13
	Papilio machaon	Swallowtail	18–20
Nymphalidae	Boloria pales	Shepherd's fritillary	9–10
	Vanessa atalanta	Red admiral	13–15
	Aglais urticae	Small tortoiseshell	14–15
	Inachis io	Peacock	16–18
	Agraulis vanilllae	Gulf fritillary	17–21
Satyrinae	Coenonympha pamphilus	Small heath	~7
	Maniola jurtina	Meadow brown	10
Pieridae	Pieris brassicae	Large white	15–16
	Pieris rapae	Small white	12–15
	Phoebis sennae	Cloudless sulphur	27–32
Hesperidae	Thymelicus sylvestris	Small skipper	14–15
Moths			
Pyralidae	Pyrausta		4–9
Noctuidae	Autographa gamma	Silver-Y	15–16
	Most temperate species		10–20
Sphingidae	Macroglossum stellatarum	Hummingbird hawkmoth	24–28
	Hyles lineata	Striped hawkmoth	25–33
	Sphinx ligustri	Privet hawkmoth	36–42
	Agrius convolvuli	Convolvulus hawkmoth	65–80
	Manduca sexta	Tobacco hawkmoth	85–95
	Manduca spp (C America)		45–135
	Deilephila elpenor	Elephant hawkmoth	250–280
	Cocytius cluentis (S America)		250
	Xanthopan morgani (Madagascar)		200–250+
	Mean for C American sp		49

Further feeding specialization is seen in the long-lived Neotropical *Heliconius* and *Laparus* nymphalid butterflies, which feed on pollen to acquire amino acids and so meet their nitrogen requirement (Gilbert 1972; de Vries 1979; Penz and Krenn 2000). Here the butterflies use a behavior perhaps derived from their proboscis-cleaning technique; they gather a small ball of pollen and add a drop of regurgitated nectar and/or saliva to it, kneading the resultant mass for some time (by rolling and unrolling the proboscis) before sucking up the fluid, now enriched with amino acids since the saliva is known to contain protease enzymes. Eberhard et al. (2007) showed that the more pollen these butterflies ate the more eggs they produced. A few moths from the primitive family Micropterygidae also feed on pollen, using **maxillary palps** (fig. 14.1E) to scratch

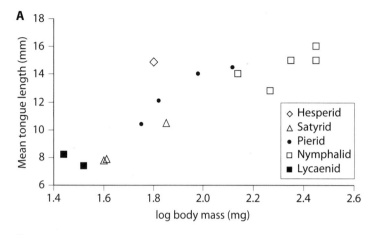

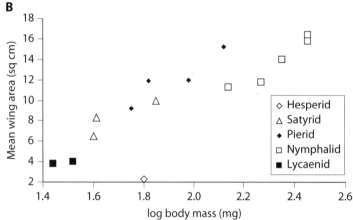

Figure 14.2 (A) Tongue length of butterflies in relation to their log body mass, and (B) wing area in relation to log body mass, for five families. (Modified from Corbet 2000.)

pollen out of anthers on various open bowl flowers, particularly *Caltha* (marsh marigold) and related forms.

2. Sensory and Behavioral Capacities

Visual Systems

Lepidopterans generally have very good color vision with a wide spectral range of 300–700 nm, normally involving three (but sometimes four, five, or six) visual pigments, and particularly acute in the UV range. Unlike most insects, around half of the butterflies (including *Heliconius, Papilio,*and *Pieris*) also have sensitivity at the red end of the spectrum. This is probably lacking in moths, and the common hummingbird hawkmoth (*Macroglossum stellatarum*) is known to be red-blind (Kelber et al. 2003) with its three receptors responding to green, blue, and UV wavelengths.

Butterflies tend to search visually for flowers, and their color preferences have been studied in some detail. Early work commonly found innate preferences for blues and purples, but it seems that color choices vary at specific and even intersexual levels (Weiss 2001). Studies with some tropical *Heliconius* butterflies recorded preferences for blues and oranges (Swihart and Swihart 1970), while swallowtails showed innate preference for yellow, blue, and purple (Weiss 1997), and the nymphalid *Vanessa indica* preferred yellow and blue colors (which also took precedence over odor cues; Omura and Honda 2005). Yet other butterflies have an innate preference for red coloration (Weiss 2001), although in most laboratory studies learned preferences are readily superimposed on the innate choices. Variation is obvious, so that a study of butterfly preferences in a mixed alpine meadow flora found no overall color-related patterns; two out of six species preferred red flowers, one strongly preferred yellow flowers, and the remaining three showed no particular patterns (Neumayer and Spaethe 2007).

For the moths, color preference tests are rare, though an anecdotal preference for pastel shades and white and cream colors is fairly widely documented. Long-range visual searching is usually not required as odors are more useful in dim light. However, fully nocturnal genera such as *Hyles* and *Deilephila* can still discriminate flower colors, even at starlight intensities (Kelber et al. 2003). *Macroglossum* is unusual among hawkmoths in being diurnal, and when naive this moth has an innate preference for paler blue flowers, with a radial pattern and some central contrast; but it rapidly learns to visit almost any flower color or form (Kelber and Balkenius 2007). Likewise, *Manduca sexta* moths have an innate preference for blue, which rapidly turns into a learned preference for white flowers with experience of natural foraging (Goyret, Pfaff et al. 2008).

Olfactory Systems

Lepidopterans mostly have good long-range olfactory sense, although this is less obvious in butterflies, which often respond to odor as an attractant only at close range. Naive *Heliconius* butterflies preferred floral scents (*Lantana*) to vegetative scents (Andersson and Dobson 2003). Naive nymphalids such as *Inachis* and *Aglais* preferred a typical butterfly flower odor (*Cirsium*) to the odor of nonbutterfly flowers, whatever the color association offered (Andersson 2003), and once experienced they could also learn to switch their preferences to the most rewarding species. But in general butterflies' discrimination of odors is poor relative to other pollinating groups.

In contrast, olfactory discrimination is exceptional in some of the nocturnal moths with very large and plumose antennae bearing huge numbers of distance chemoreceptors (often in excess of 50,000), and these moths can often detect the pheromones emitted from the opposite sex at a range of several kilometers. Likewise, they may use the strong scents of some of their preferred night-flowering blooms as long-range cues. Some may have innate odor preferences; for example, *Manduca* sphinxmoths switched from readily available (bat-adapted) *Agave* flowers to their host plant *Datura* as soon as the latter became abundant and even

though its flowers had much less nectar, apparently due to an innate preference (present in naive moths) for the *Datura* odor (Riffell et al. 2008).

Chapter 6 mentioned common volatiles in lepidopteran flowers, and although there were distinctions between the three taxonomic groups covered here there were also some overlaps in components between butterfly and noctuid moth flowers, and between noctuid and sphingid flowers. But it is noteworthy that some flowers pollinated by butterflies in the daytime and by moths nocturnally can switch their scents; for example the orchid *Gymnadenia* has a scent dominated by benzenoids in daylight but lower levels of benzyl alcohol and methyl eugenol at night (Huber et al. 2005).

Lepidopterans also have good contact chemoreception via their tongues and their feet, used in both mating and host-plant finding for oviposition, but perhaps less involved in flower discrimination.

Learning

Butterflies and moths can readily achieve associative learning, linking shape and/or olfactory cues to other features of flowers such as color (Weiss 1995b), or to features of nearby leaves. Indeed some butterflies and many hawkmoths learn color- or odor-to-reward association in just one trial (see fig. 5.10A for an example with *Papilio*). They can also reverse-learn rapidly when given aversion stimuli, and can learn negative associations (i.e., to avoid unrewarding stimuli) (Kelber 1996; Weiss 1997). The papilionid butterfly *Battus philenor* achieved a learned preference for unrewarded green colors with 96% accuracy, and for red, yellow, or blue with about 75% accuracy (Weiss and Papaj 2003), and could learn a different color-sucrose reward association at the same time (dual conditioning). Butterflies can also readily learn an association between color phase and reward within a flower species that changes color.

There are clear links between learning abilities and normal foraging experiences, especially in relation to day or night activity. *Macroglossum* moths, perhaps in common with other more diurnally active moth species, would learn odor cues only when the color itself was not attractive (Balkenius and Kelber 2006), whereas nocturnal hawkmoths such as *Deilephila* responded primarily to odor stimuli (Balkenius et al. 2006). Odor learning in night flyers can be very precise, with

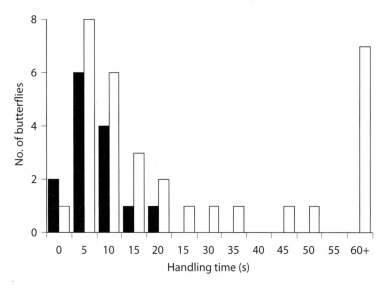

Figure 14.3 Learning of flower handling by checkerspot (*Euphydryas editha*) butterflies: time to locate nectar in *Arctostaphylos* for butterflies caught feeding on either *Arctostaphylos* flowers (black bars), where techniques had been learned, or on *Calyptridium umbellatum* flowers requiring different handling (white bars). (Redrawn from McNeely and Singer 2001.)

Helicoverpa moths readily learning to prefer flowers with added phenylacetaldehyde and discriminating among floral bouquets with and without this single odor change (Cunningham et al. 2004).

Many lepidopterans also learn associations between egg laying on host plants and nectar foraging at flowers, with one host plant serving both as fuel stop and nursery site. Longer-lived species may also achieve associative learning for navigation purposes, allowing them to learn complex foraging routes; for example, *Heliconius* butterflies learned visual landmarks that allowed them to return each evening to the same nocturnal roost sites and to trap-line each day between remembered flowers (Mallet et al. 1987).

In keeping with this behavioral sophistication, at least some butterflies and moths can learn to handle flowers more efficiently with practice. For example, the common cabbage butterfly *Pieris rapae* significantly decreased its handling time with experience (Kandori and Ohsaki 1996), learning to avoid nectarless outer florets on *Erigeron* and go straight to the central, rewarding florets; they could also remember rewarding flower colors for at least three days. The checkerspot *Euphydryas editha* when caught already feeding on the "difficult" flower *Arctostaphylos* took only about 10 s per flower (fig. 14.3), whereas an individual caught feeding on a simpler flower took on average 43 s at first on *Arctostaphylos*, although it speeded up with later trials (McNeely and Singer

2001). The same butterfly species, however, was unable to learn and improve its efficiency for finding oviposition sites. The small skipper *Thymelicus flavus* increased its handling time by on average 0.85 s per flower whenever it switched to a new species (Goulson, Ollerton et al. 1997). Hence these butterflies showed a reasonably degree of floral constancy, with 85% of visits being to the last species visited (Goulson, Ollerton et al. 1997). Other butterflies and moths can almost certainly be similarly constant; Lewis (1989) reported a nearly threefold increase for pierids in the likelihood of a visit when the flower encountered was the same species as the previously visited one, and Wiklund et al. (1979) reported that 90% of all visits by the wood white (*Leptidia sinapis*) were to *Viola* and *Lathyrus* flowers, with the latter being the *only* flower visited once it was reasonably abundant (although visits were mainly illegitimate). For hawkmoths, Willmott and Búrquez (1996) reported nothing but pure pollen loads for various species captured when visiting the convolvulaceous *Merremia*, where a single moth visit could give sufficient fertilization.

Generally speaking, where the corolla shape is appropriate for visitation, butterflies prefer larger individual flowers to smaller ones. They were also recorded showing a preference for larger displays when visiting inflorescences, although they then probed smaller proportions of the open flowers (Arroyo et al. 2007; and see chapter 21).

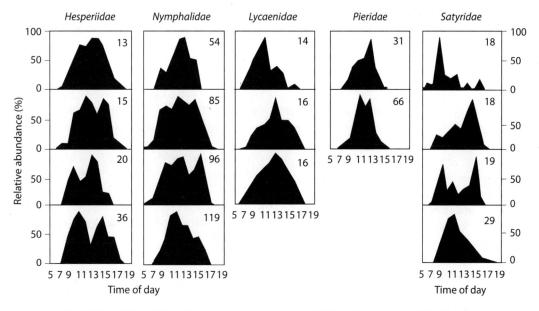

Figure 14.4 Butterfly daily activity patterns on lavender (*Lavandula latifolia*) for 17 species from 5 families, shown as abundance relative to the specific daily peak, and with body mass in milligrams shown for each species. (Redrawn from Herrera 1990.)

Foraging Behavior

Most butterflies are active by day, often for a particular part of the day related to their body size and color (see chapter 10); examples of their activity patterns on lavender are shown in figure 14.4, the patterns being essentially unimodal and centered on the hottest part of the day as expected for strictly ectothermic insects. They feed from flowers only after landing, walking between adjacent flowers (or the florets of an inflorescence) wherever possible. In contrast, most settling moths are active at dusk and into the night, mainly feeding from a resting position, and operating at lower body temperatures than butterflies.

The hawkmoths are again primarily crepuscular and nocturnal, but feed mainly while hovering. Although they usually have a reasonably strong endothermic ability, hawkmoths tend to function only on warm evenings and in warmer climates (Herrera 1992). For example, on *Mirabilis* flowers they operated only above 13°C ambient temperature, so on cooler evenings pollen transfer was nil and the flowers selfed (del Rio and Burquez 1986). The temperature dependence of butterfly and moth pollination may account for observations that their foraging decreases at alti-

tude (e.g., Cruden et al. 1976); although it may be showing a tendency to spread to both higher latitudes and altitudes as warming climates occur, with *Macroglossum* now reported throughout the United Kingdom, where it was formerly a southern species, and similar effects documented for several butterfly species in Europe and North America (chapter 29).

3. Psychophily: Butterfly Flowers

Butterflies must land and settle to feed and so prefer either relatively large solitary flowers with platforms to hold on to, or, more commonly, flat-topped inflorescences (see plate 24), although their detailed preferences do vary with species (Feber and Smith 1995; Corbet 2000). Commonly visited inflorescences comprise either moderately sized flat-topped flowers or many small vertical tubular florets set side by side, providing a horizontal landing. Because of the effects of wing loading on foraging costs (see above and chapter 10), butterflies (especially those with relatively smaller wings) tend to forage on massed flowers, avoiding solitary flowers unless these are particularly rewarding (Corbet 2000). In general, their tongue structure restricts butterflies to relatively dilute nectar sources; May (1988) recorded foraging at 11 flower species in Florida with 17%–40% nectar and with volumes ranging from below 0.03 μl to around 1 μl per flower.

The butterfly flower syndrome is termed *psychophily* (see also table 11.1), and this should probably also include the few but quite important examples of day-flying moths such as the burnet moth, *Zygaena*. It should involve the following:

1. Massed flowers forming a flat or gently curved inflorescence, each flower having a long corolla (or a medium corolla tube but with a nectar spur) to be probed by the uncoiled tongue

2. Stamens and style commonly somewhat protruded beyond the upper lip of the flower, so that pollen is deposited on the face or head of the visitor

3. Petals often intensely colored and attractive; deep pink or blue, cream, yellow, orange, or red, and often with yellow center

4. Petals without nectar guides, or with only simple patterning of this kind

5. Usually a mild but pleasantly sweet scent

6. Day flowering

7. Offering small to medium volumes of dilute nectar suited to the long tongue and the relatively low energetic expenditure of a foraging butterfly

8. At least sometimes, nectar with higher than normal amino acid content; it may be that tropical species have a higher need for amino acids in nectar to support their greater longevity (Beck 2007).

Perhaps the most familiar and typical examples are *Buddleja* (often called butterfly bush) and the tropical genus *Lantana*, both these plants being widely used in butterfly farms. Buddlejas have an unusual scent mix (ocimene, oxoisophorone, and farnesene; Tholl and Rose 2006) that may be particularly attractive, although this has not really been explored, while *Lantana* scents are dominated by terpenoids (Andersson and Dobson 2003). Some common native plants preferred by butterflies in Europe and North America are given in table 14.2. It is noteworthy that nearly all the plants mentioned are perennials, few annuals providing appropriate resources; in agricultural ploughed areas, only a few persistent annuals such as radish (*Raphanus*) and mustard (*Sinapis*) are of any use to butterflies (Dover 1989).

In warmer and more tropical floras, many of the milkweeds (*Asclepias*) and the Rubiaceae serve as butterfly plants, as do various mimosoid Fabaceae such as the nectar-producing species of *Acacia*. Red and yellow flowers of *Caesalpinia* and its relatives (fig. 14.5) are also pollinated by butterflies, especially large papilionids, with the pollen from the elongate anthers becoming deposited on their wings as they flutter in front of the flowers (Cruden and Hermann-Parker 1979). Orchids visited by butterflies include the alpine species of *Nigritella*, with small dark reddish flowers and a vanilla-like scent. Pyramidal orchids (*Anacamptis pyramidalis*) perhaps match the traditional syndrome of psychophily better, having dense flower heads of small long-spurred pink flowers; they are visited by various fritillaries, day-flying burnet moths, and nocturnal noctuid moths, all of which may accumulate many pollinia on their tongues (but apparently without any nectar being present to reward them).

Generally butterflies do not carry very much pollen on their bodies, but sometimes they become visibly heavily dusted with grains, especially about the face, and are then efficient pollen dispersers (e.g., Levin and Berube 1972; Linhart and Mendenhall 1977; Murphy 1984; Pettersson 1991). For example, on *Lavandula* Herrera (1987) recorded various bees visiting flowers regularly but producing a high level of geitonogamy with short interflower flights; whereas butterflies visited less frequently but tended to do so bearing cross-pollen, making longer flights between individual flower visits, and so effecting better outcrossing. In fact butterflies can fly substantial distances between plants (Waser 1982); usually only 1–10 m for temperate butterflies, but up to 25–75 m for larger tropical species and with a maximum recorded at 350 m for *Heliconius* (Murawski and Gilbert 1986), although at relatively slow overall flight speeds of 25–30 m/h (Shreeve 1981). Effective pollen transfer is probably highest in tropical butterflies, where the interplant movements are so much longer. However, since many butterflies do carry only low pollen amounts and may move only a meter or less between plants, a proportion of studies have recorded poor pollination efficiency and ascribed parasitic or nectar-thieving status to their studied butterfly species (e.g., Wiklund et al. 1979; Venables and Barrows 1985). Some debate on these issues was provided by Tepedino (1983) and Courtney and Hill (1983).

TABLE 14.2
Common Butterfly Flowers in Temperate Habitats

Single flowers
    Ragged robin   *Lychnis flos-cuculi*
    Catchfly   *Silene* spp.
    Wallflower   *Erysimum cheiri*

Multiple inflorescences
    Pinks and carnations   *Dianthus* spp.
    Valerians   *Valeriana, Centranthus* spp.
    Scabious   *Scabiosa, Knautia* spp.

Asteraceae with good landing surfaces
    Hemp agrimony   *Eupatorium cannabinum*
    Fleabane   *Pulicaria dysenterica*
    Knapweeds, thistles   *Centaurea, Cirsium* spp.
    Goldenrod   *Solidago virgaurea*
    Daisies   *Aster* spp.

Smaller labiate flowers without constricting upper lips
    Mint   *Mentha* spp.
    Woundwort   *Stachys* spp.
    Bugle   *Ajuga* spp.
    Germander   *Teucrium* spp.

"Cornucopia generalist" flowers with particularly abundant nectar
    Bramble   *Rubus fruticosus* agg
    Lime   *Tilia* x *europea*

Additional flowers for smaller butterflies (e.g. lycaenids)
    Forget-me-nots   *Myosotis* spp.
    Also sedums, phlox, hebes, verbenas and escallonias in
       gardens.

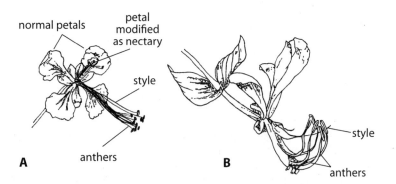

Figure 14.5 Butterfly-pollinated flowers in the Fabaceae: (A) *Caesalpinia*; (B) *Amherstia*. (Redrawn from Endress 1994.)

Quite high levels of floral constancy have been recorded, for example, in checkerspots (Murphy 1984) and in white pierids (Lewis 1986, 1989; Goulson and Cory 1993; Kandori and Ohsaki 1996). However, on some occasions these same butterflies are rather random and nonconstant in their flower-visiting behaviors (e.g., Ellis and Ellis-Adam 1993 with *Pieris*); perhaps in these cases a long association with horticultural practices where the host plant cabbages and ornamental plants grow in close proximity has something to answer for. A substantial study over 12 years of butterfly generalization and specialization in northeast Spain analyzed 100 butterfly species and showed that the degree of generalization in flower use was rather similar across a range of biotopes for a given species (Stefanescu and Traveset 2009). Generalization was more common in species with long flight periods, as expected, and was also commoner in butterfly species from open habitats as compared with forest species which tended to be more flower constant. However, neither rarity nor body size was an important determining factor.

4. Phalaenophily: General Moth Flowers

The larger flower-visiting moths, mostly from the families Noctuidae but also including some Geometridae and Pyralidae, are moderate flyers but generally cannot hover. They operate particularly at dawn and dusk, but can be active into the night in warm summer weather; otherwise, the absence of radiation from the sun leaves them too cold to fly. Hence they tend to increase in abundance and importance in warmer climates. For example, there is a marked increase in their frequency in the south-western United States in the warm deserts and semiarid zones, where many canyons also provide shelter sites. In the seasonal cerrado regions of Brazil, settling moths are common pollinators, and moths as a whole pollinate 21% of the common trees (Oliveira et al. 2004).

The general moth syndrome, **phalaenophily**, involves tubular flowers again, often with strongly dissected outlines that provide a preferred visual display (chapter 5). They show a broad range of possible colors but are usually in pastel shades of rather subdued intensity, more often at the white/cream end of the broader range discussed for butterflies. For example, of 21 moth-visited species studied in Brazil, Oliveira et al. (2004) found 11 white, 7 cream, and 3 yellow. Certainly the rather strong creams and whites in some of the flowers that open at very low light levels make them easier to discern visually. Schremmer (1941) noted a tendency for moths to visit pale pink and yellow flowers in the early evening but to ignore these and visit only white flowers as it became darker. But location of flowers by moths is nearly always aided by strong and often particularly delicious scents, detected by the often massive antennae.

The flowers tend to have the same general morphology as butterfly flowers (since the tongues to be inserted are rather similar), commonly with 5–15 mm tubes and very small apertures. But moth flowers rarely occur as massed inflorescences, often instead presenting in small groups to provide some kind of landing site, or at least a surface that can be gripped by the forelegs while the moth is feeding. They provide fairly dilute nectar in low to moderate volumes; larger amounts than in butterfly flowers on average, reflecting the larger body sizes and the cooler conditions in which the animals operate. Their longevities are variable, mostly 1–3 days but sometimes considerably longer in temperate and cold-habitat species.

Common choices in Europe are campions and catchflies (*Silene*), pinks (*Dianthus*), soapwort (*Saponaria*), and *Phlox*, most of these being low-growing plants in accordance with many moths' tendencies to fly low, close to foliage. Thus on spikes of the orchid *Platanthera* (Inoue 1986) noctuid moths were more likely to visit shorter spikes and pick the flowers on their lower regions, visiting them systematically (whereas sphingid moths preferred large spikes and visited more sporadically). Honeysuckle (*Lonicera*) and jimsonweed (*Datura*) are also classic moth flowers, their horizontal or pendant flowers usually being gripped by the forelegs (whereas hawkmoths, which also frequent these plants, tend to hover at the flowers). Four-o'-clock plants (*Mirabilis*) that open in late afternoon are also frequented in the wild in the United States and in many European gardens, and the same is true of *Gaura* and *Calliandra* flowers.

In the tropics, typical moth flowers include many from the family Rubiaceae, and some from Apocynaceae, Fabaceae, and Solanaceae. Quite a number of these are plants much favored as scented ornamental plants for conservatories in more temperate climes—*Jasminum* and *Gardenia* for example. It is rather common to be able to predict moth pollination a priori with plants whose visitors have not been recorded; for example, *Struthiola ciliata* in southern Africa has creamy white elongate (~20 mm) corollas, opening around 1800 h and emitting a strong sweet scent, with 0.02–0.2 μl of 20%–34% nectar, and it did indeed turn out to be visited exclusively by two species of noctuid moth (Makholela and Manning 2006).

Clearly there is some overlap between butterfly and moth flowers, and it is fairly common for such flowers to be visited by butterflies in daylight and then by moths at dusk. Moth flowers also overlap with sphingid flowers covered in the next section. In some cases, evidence for the relative effectiveness of each has been gathered:

1. We met the example of *Asclepias syriaca* in chapter 11, where nocturnal moths were more effective per visit, but diurnal visitors (mainly bumblebees) were more abundant and therefore more effective overall (Bertin and Willson 1980; Jennersten and Morse 1991)
2. *Saponaria* is visited both by *Autographa* moths and by sphingid hawkmoths, the latter producing markedly higher seed set (Wolff et al. 2006)
3. Miyake and Yahara (1999) suggested that on honeysuckle (*Lonicera*) the nocturnal hawkmoth pollinators removed rather little pollen per visit but still showed significantly greater pollen deposition efficiency than the later diurnal visitors (here mostly long-tongued bees)

However, some flowers that might be used by both nocturnal moths and later diurnal butterflies or bees have ways of avoiding what for them are the less desirable visitors. For example, some *Silene* species can roll up their petals during the day, so appearing withered and unattractive, but unroll them again in the evening when moths are active, maturing two separate sets of stamens on nights 1 and 2, and then the styles on night 3 (Kephart et al. 2006).

Orchids pollinated by noctuids are not uncommon. *Gymnadenia* has smallish pink flowers that are fragrant, and unusual among orchids in producing copious nectar. The so-called butterfly orchids in the genus *Platanthera* with narrow-petaled and strongly fragrant white flowers are in fact mainly visited by noctuids, although some hawkmoths also visit in the colder more northerly parts of Europe (Nilsson 1983a).

Having described the normal patterns of phalaenophily, it is only fair to record some exceptions. Most notably, many smaller moths with shorter tongues should probably be regarded as closer to being generalist, since they will commonly visit quite open flowers such as buttercups, brambles, and even willow catkins. At the other extreme, there are two moths that have particularly specialist interactions with flowers, in the form of active pollination; these extraordinary cases, involving yucca moths and senita moths, are covered in chapter 26.

5. Sphingophily: Hawkmoth Flowers

Sphingids are distinguished from other moths in several ways. They have very strong forward flight, with a fast wing beat. They commonly feed at flowers while hovering, without gripping the flowers with their legs (although some will reduce the costs of hovering by resting the front legs on a flower; Heinrich 1983a). Wasserthal (1993) showed that some use a particular technique called "swing-hovering" when feeding at corollas shorter than their own tongues, where their body is positioned some distance away from the flower and can sway from side to side while the tongue stays in place within the flower, this perhaps being an antipredator adaptation (avoiding or confusing the spiders and mantids that may lie in wait in the flower; see chapter 24). Sprayberry and Daniel (2007) reported rather precise tracking between a hovering sphingid and the movements of the flowers they fed at (whether caused by winds or by the moths own wing beats), allowing the moths to continue to feed efficiently.

Compared with butterflies and settling moths, sphingids often have exceptionally long tongues (see table 14.1), and tongue lengths correlate well with wing lengths (e.g., Bullock and Pescador 1983). They have stout bodies, weighing from about 100 mg (similar to a honeybee) up to 6.5 g (about three times the familiar large bees such as some *Bombus* queens). Their bodies are often decidedly furry, providing the good insulation that is an essential accompaniment to endothermy. Hawkmoth metabolic rates are high compared with other insects (see table 10.3). For an animal of this size, the metabolic cost of endothermic warm-up is relatively insignificant but the costs of hovering are high (see chapter 10 and table 10.3). A medium-sized hawkmoth should require about 1.3 mg of sugar per minute to support hovering and would normally need to visit many flowers per minute to provide this.

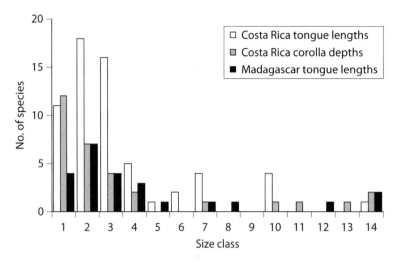

Figure 14.6 Hawkmoth tongue lengths and associated corolla lengths in the flowers they visit in Costa Rican dry forest (showing a good match, though with some long tongues used in shorter flowers); and tongue lengths from a Madagascan sphingid community, showing a similar range. "Size classes" are at intervals of 1.5 cm. (Redrawn from Agosta and Janzen 2005.)

However, the nectar from just a single large hawkmoth orchid (such as the Madagascan angraecoid orchids; see below) would provide enough energy for at least a minute of hovering (Nilsson et al. 1985).

Usually sphingids will hover for only a second or two in front of, or above, or below, their selected flower, with only the tongue inserted. The flowers are therefore generally quite large, with long or very long tubular corollas (not uncommonly 40–100 mm) with rather larger apertures than in settling moth flowers, and they are arranged singly rather than bunched. They may be erect or horizontal or pendant, and here different hawkmoths have particular preferences, also varying with flower shape. For example, flower orientation in *Aquilegia* dramatically affected hawkmoth visitation: manipulation of the upright flowers of *A. pubescens* to be pendant, so resembling those of *A. formosa* (the two species occurring together in a hybrid zone in California) resulted in hawkmoths visiting the upright flowers at more than ten times the frequency with which they visited the pendant flowers (Fulton and Hodges 1999).

Vogel (1954) reported that sphingid flowers commonly have a star-shaped radially symmetrical appearance as the insect approaches, and a narrower corolla than similarly sized flowers visited by day-flying butterflies; both of these features could be seen as aids to initial location and then proper tongue insertion in low-light conditions. When species of *Mandevilla* (Apocynaceae), which has a specific triggering mechanism for pollination, were visited, Moré et al. (2007) showed that both tongue length and effective tongue width of sphingids were important in constraining

visitation and thus bringing about specialized interactions on flower species with different corolla lengths. Likewise, in Costa Rican hawkmoths there was a strong positive correlation between body size and tongue length, and between both of these and flower corolla depths (fig. 14.6) in the flowers visited (Agosta and Janzen 2005).

However, some hawkmoths will avoid expensive hovering at narrow tubular flowers when possible and may for example land on and clamber inside some large trumpet-shaped flowers, such as amaryllids, *Datura*, and some *Convolvulus*. Or they may land on the blossom top when working at inflorescences with brush-type corollas much shorter than their own tongues, such as *Capparis* or *Cleome*. When they do visit brush blossoms (like the *Albizia* species and acacias, plate 8A,B,E) they may show more specificity than has previously been supposed; for example, on *Caesalpinia gilliesii*, with long brush-type flowers, moths of four different tongue lengths occurred but only the species with long tongues (~55 mm) were really good pollen carriers, the short- and medium-tongued species foraging from below and not touching anthers, and the very-long-tongued species hovering where they were out of reach of the anthers (Moré et al. 2006).

Most of the sphingid flowers are dusk or even night flowering and tend to be intensely white or cream in color. Their scent is a very strong trigger of feeding behavior and will on its own elicit endothermic warm-up in preparation for foraging activity (Brantjes 1973). Many sphingid flowers are much more strongly scented than butterfly or settling moth flowers—for example, honeysuckle species (*Lonicera*), tobacco plant (*Nicotiana*), oleander (*Nerium*), four o'clock plants (*Mirabilis*), *Pittosporum, Stephanotis,* and *Plumeria*—with scent being the major attractant from a distance although relatively unimportant once the moth is close to a flower. *Lonicera japonica* can be visited in the day by bees, but as noted above the nocturnal hawkmoth visitors move pollen longer distances between flowers, so it pays the flowers to open initially at dusk (Miyake and Yahara 1998) and to produce their strong linalool-dominated scent in the early night hours (Miyake et al. 1998).

Typically, hawkmoth flowers also have more copious nectar than other lepidopteran flowers, supporting the larger and endothermic body of the visitor; although the nectar is still usually dilute in order to flow easily through the thin proboscis. But in some of the shorter flowers from which the hawkmoths feed, the nectar can be moderately concentrated; and Wolff et al. (2006) reported higher seed set on *Saponaria* flowers that had been doctored with abnormally concentrated nectar, suggesting that the moths preferred this when it was on offer (see chapter 8).

Hawkmoths can carry large pollen loads on their copious insulating scales and may move up to 3–400 m between the plants visited. They are also less likely to visit many flowers on the same plant than some other groups; for example, on *Ipomopsis aggregata* they visited 4.7 flowers per plant, where hummingbirds visited 7.4 (Waser and Price 1983). Good flower constancy has been recorded (Knoll 1922); and hawkmoths will become odor constant quite readily and retain this learned preference until at least the next day (Brantjes 1973). Thus they have both physical and behavioral attributes that enhance their effectiveness as pollinators.

In certain habitats hawkmoths become a really substantial component of the pollinator fauna. They are especially important as flower visitors in regions where dusk temperatures are higher, hence their abundance in many Mediterranean-type and savanna habitats (see chapter 27); some will fly by day in such regions and may then visit a range of more typically butterfly flowers. They are also abundant in the moist tropics where guilds of at least 100 species can occur, visiting 30 or more sphingophilous plants. But perhaps the most famous case of locally enhanced importance is the large and isolated island of Madagascar, where a relatively

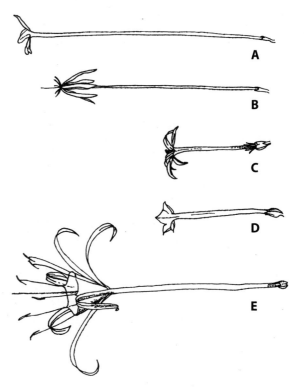

Figure 14.7 Hawkmoth-pollinated flowers (other than orchids) with long corollas or long nectar spurs: (A) *Posoqueria latifolia* and (B) *Oxyanthus formosus*, both Rubiaceae; (C) *Isotoma longifolia*, Campanulaceae; (D) *Nicotiana sylvestris*, Solanaceae; (E) *Hymenocallis littoralis*, Liliaceae. Approximately to scale, and (A) is ~ 100 mm long. (Redrawn from Endress 1994.)

primitive group of hawkmoths have established some highly specialist relationships with many very long tubular flowers (plate 25 shows two examples), especially a range of angraecoid orchids (Nilsson et al. 1985, 1987). Figure 14.7 shows some of these.

The most famous example is *Angraecum sesquipedale*, a large white orchid with a nectar spur measured by Darwin (1862) at 11.5 inches (290 mm). Any visitor, he predicted, would require a tongue just a little shorter than this to effect pollination, and forty years later Madagascan hawkmoths (*Xanthopan morgani*) with tongues 200—250 mm in length were indeed discovered (and later still a further species, *Coelonia solani*, of comparable tongue length was also found on the island). Several other long-spurred angraecoid orchids are pollinated by the relatively common hawkmoth *Panogena lingens* with a tongue of about 120 mm, in each case placing their pollinia in slightly different places on the proboscis and face (Nilsson et al. 1987). These are long-standing and unusually specialized interactions, and although they are asymmetrical (the moths often visiting a range of flowers, including those with shorter corollas where they need not insert the whole tongue), it is perhaps no coincidence that they have arisen in a very isolated and stable tropical habitat, where many other unusual species have evolved.

6. Overview

How good are lepidopterans as pollinators? A fair answer here would be that butterflies and moths are reasonably effective as flower pollinators, with some more specialist and probably more effective examples among the larger and partially endothermic sphingid species. The lepidopterans have moderate to good flower constancy and moderate levels of specificity in flower choice, dictated mainly by the characteristics of their tongues. With butterflies, for example, Bloch et al. (2006) found that five different species visiting *Dianthus* flowers were all reasonably specialist, and all had different effectiveness, only two of the species being of major importance.

For most lepidopteran species the flight strength and range are not high, however, so that pollen movement may be less than ideal. In fact, in terms of pollen dispersal, the lepidopterans probably act as two compartment pollen carriers (see chapter 7; Harder and Barrett 1996). This is because, after feeding, when the tongue is coiled up, pollen tends to get redistributed, so that only a proportion remains available for deposition on stigmas (Levin and Berube 1972). Pollen carryover is therefore not a process of simple sequential decay from flower to flower. However, levels of pollen deposition in single flowers can be quite high, especially (again) for sphingids on their preferred flowers such as *Lonicera, Datura*, and some mimosoids.

Chapter 15

POLLINATION BY BIRDS

Outline

1. Feeding Apparatus
2. Sensory Capacity
3. Behavior and Learning Capabilities
4. Bird Flowers and Ornithophily
5. Overview

Bird pollination or *ornithophily* is a widespread phenomenon, underappreciated in early literature as it is absent in Europe; but it is familiar through much of the United States and as far north as Alaska and occurs throughout the tropics, in the eastern Mediterranean and Middle East (although not in the more northern parts of Asia), as well as in most of Australasia and Africa. Flower-visiting birds are also surprisingly widespread taxonomically, the habit being recorded in at least 50 families, although many of these are only occasional visitors and do more damage to flowers than they do good in transferring pollen. Many common birds do visit flowers by biting through or piercing their corollas, notably tits and warblers. As table 15.1 shows, there are perhaps only ten separate groups that have developed flower-visiting behavior to the point that they have become significant pollinators for at least some flowers. Together these ten families make up about 10% of all birds.

Primarily birds take nectar from flowers, although some may also eat pollen and occasionally take solid floral tissues. Nearly all bird-pollinated plants produce copious nectar that is dilute and sometimes rather slimy in character. It is remarkably consistent in concentration within a plant species, and across all species shows a range normally between 15% and 30% sugar,

remarkably often averaging around 20% (see chapter 8). In some species the nectar is so abundant that it drips from the flowers or can be shaken out in showers from a tree (e.g., *Erythrina, Grevillea*) or drunk by a human from a flower (e.g., some *Banksia* in Australia). These nectar characteristics match birds' requirements for low-viscosity fluid that can be rapidly drawn up by the tongue and yet for large caloric rewards to offset the very high costs of flight and of endothermy (chapter 10).

All flower-visiting birds also require other food sources, and most get their protein (nitrogenous) input from elsewhere, mainly by eating insects. Possibly they were led to exploit nectar from flowers while searching around those flowers for the many insects that visit them. Modern nectarivorous birds appear to have a lower nitrogen requirement than omnivorous birds anyway, reducing their reliance on other foods, although it is not yet fully known how they conserve their nitrogen supply (Tsahar 2006).

Although ornithophily is usually listed as just one syndrome (as in table 11.1), in practice it is helpful to divide the flower-visiting birds into two categories and thus (in section 4, *Bird Flowers and Ornithophily*) to recognize two somewhat different floral syndromes.

Hummingbirds

Hummingbirds constitute the family Trochilidae from the New World tropics, and the syndrome associated with them is sometimes termed *trochilophily*. They evolved in South America and colonized North America later, probably around the mid-Tertiary (Grant 1994). Some species extend far up into the United States and even Canada, but these are all migratory, their movements closely following the flowering peri-

TABLE 15.1
Bird Families Involved in Pollination

Family	Common names	No. of flower-visiting species	Distribution	Main plants visited
Trochilidae	Hummingbirds	300–350	Neotropics, parts of North America	Very varied
Nectariniidae	{ Sunbirds { Flower-peckers	{~110 {	Africa, southwest Asia Australasia, southeast Asia	Very varied
Promeropidae	Sugarbirds	2	Southern Africa	Proteaceae
Meliphagidae	Honeyeaters, spinebills, bellbirds	160–180	Australasia	Varied, including Proteaceae
Zosteropidae	White-eyes	~90	All southern continents	Varied, including mistletoes
Psitaccidae	Brushtongued lorikeets (parrots)	60–70	Australasia, southeast Asia	Eucalyptus
Fringillidae	Hawaiian honeycreepers	23	Hawaii	Tree lobelias
Dicaeidae	Flower-peckers	55–60	Australasia, southern Asia	Mistletoes
Thraupidae	Honeycreepers Bananaquit	15 1	Neotropics and Hawaii North and South America	
Icteridae	Orioles, caciques	90*	North and South America	Varied

Source: Modified from Stiles 1981; Proctor et al. 1996.
* Only a few of these are flower-visitors.

ods of their favored plant species (Grant and Grant 1968), so that they move south at breeding time and then return northward (or to higher altitudes) after breeding. These more northern populations may require particularly large nectar supplies from their flowers, as they expend more energy in keeping their body temperature high and in warming up substantial volumes of cold nectar as it enters their digestive system (chapter 10).

Hummingbirds are most abundant in the north and west of South America, although they have been most intensively studied in Costa Rica and California. They are mostly very small birds, commonly from 3 to 10 g in body weight and from 5 to 20 cm in body length but with at least half this length often comprised of beak and tail.

Perching Birds

This category includes all the bird groups shown in table 15.1 other than the hummingbirds. None of these other taxa can hover efficiently, so they must perch to feed (although some sunbirds do in practice hover effectively for up to 30 s when visiting alien New World species introduced into their habitats: Geerts and Pauw 2009). The pollinating perching birds come from a range of families from various parts of the Old World tropics but also include rare examples (**honeycreepers**, bananaquits (plate 27H)) from the New World. The most important and abundant examples are the sunbirds of Africa (e.g., *Nectarinia, Cynniris*) and of southern Asia (e.g., *Arachnothera*, an unusual example in being very long beaked). The Australasian honeyeaters and wattlebirds (Meliphagidae) are important pollinators of some of the key shrubby families in Australia (including many *Banksia* and *Protea* species, 27D), and are represented by 67 species, many of which are strongly dependent on nectar and often move seasonally between flowering species and across flowering areas (Paton and Ford 1983). This continent also has the greatest range of other bird flower feeders, with chats, wood swallows, tree creepers, bower birds, and butcher birds also recorded as pollen or nectar

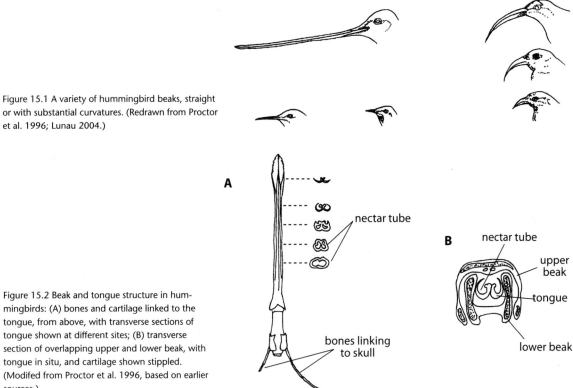

Figure 15.1 A variety of hummingbird beaks, straight or with substantial curvatures. (Redrawn from Proctor et al. 1996; Lunau 2004.)

Figure 15.2 Beak and tongue structure in hummingbirds: (A) bones and cartilage linked to the tongue, from above, with transverse sections of tongue shown at different sites; (B) transverse section of overlapping upper and lower beak, with tongue in situ, and cartilage shown stippled. (Modifed from Proctor et al. 1996, based on earlier sources.)

feeders (well over 100 Australasian species being involved in total). In New Zealand, bellbirds (*Anthornis*) are particularly important as pollinators.

These perching birds are usually a little larger in body size than hummingbirds, although most are still among the smaller bird types (sunbirds 5–17 g, white-eyes about 11 g, honeyeaters usually 10–20 g); only the Australasian groups seem to get larger, with some wattlebirds weighing in at over 100 g.

1. Feeding Apparatus

Hummingbirds

Hummingbirds have very varied beak lengths often closely matched to the length of the corollas that they visit (although they may also fly right in to some larger trumpet-shaped flowers). Longer beaks probably give access to a wider range of flowers (Bleiweiss 1999; Temeles and Kress 2003). Most have straight and very thin beaks, but some (notably the hermit humming-birds) show beak curvature, generally downward but occasionally slightly upward (fig. 15.1); again, this may match the shape of preferred flowers. However, the North American species are of rather uniform beak length and shape, so that in this region there is limited specialization of particular flowers to particular species; in more southerly countries beaks are much more varied and the flowers visited are also less uniform in shape, size, and even color.

The hummingbird beak itself is unusual among birds in that the upper and lower **mandibles** overlap very considerably laterally, forming a tube (fig. 15.2), and when the bird is feeding the long tongue works rapidly in and out within this tube. The tongue itself is deeply forked, each half folding up lengthways to form a double tube, in which nectar probably rises mainly by capillarity aided by lapping (Kingsolver and Daniel 1983); suction has only a minor role when the tongue is withdrawn back into the mouth and nectar is squeezed off it. The tip of each tongue tube is usually decidedly frayed in appearance, giving a brush effect that may aid initial capillary uptake. Heyneman (1983) predicted optimal feeding rates with nectar at around 22%–26% concentration, although this may vary upward a little, perhaps even to 40%–45%, when flower handling costs and the unloading phase (from tongue to mouth) are added in (table 8.4; Kingsolver and Daniel 1983; Gass and Roberts 1992).

Some hummingbird beaks have markedly serrated edges; this appears to be most developed in the shorter-beaked species that are often nectar thieves (Ornelas 1994).

Perching Birds

Here, across a wide range of families, beak lengths are commonly 10–20 mm, with little or no curvature. However, there are convergences across all these groups, and also across to the hummingbirds, in beak and tongue characters. Sunbirds parallel the **trochilids** very closely in their range of short and long, curved and straight beaks, and show a similar level of matching of preferred flower lengths to beak lengths. The South American swordbill, *Ensifera,* has a beak about 100 mm long, longer than its own body. The large honeyeaters of Australasia have more conventional beaks but share with the sunbirds a convergence on bifurcated, longitudinally folded, and brush-tipped tongues, which can extend 10–40 mm beyond the beak and flicks in and out at high frequency. The exception to these recurring patterns comes with the lorikeets, which have typical if rather blunted parrot-style beaks and short tongues, and feed extensively on pollen, mostly from *Eucalyptus* flowers.

The genera *Diglossa* and *Diglossopis* (tanagers known as flower piercers) are unusual in being flower-visiting specialists but with specializations for cheating, as their name suggests. The sharp beak bears a downward-curved hook which pierces the sides of corollas to gain access to nectar, and these birds are responsible indirectly for a great deal of secondary theft by subsequent visitors of all kinds (see chapter 23). Sympatric species often coexist, each with different hook sizes (Mauck and Burns 2009).

2. Sensory Capacities

Here we can deal with all the flower-visiting birds together, as the sensory systems of trochilids and of perching birds show little or no differentiation.

Birds have good color vision and excellent visual discrimination of color and form. They most often have a tetrachromatic system (see chapter 5), with four types of cones having peak sensitivities around 550 nm (LW), 500 nm (MW), 445 nm (SW), and either 410 nm (VS, very short) or 365 nm (ES, extremely short). They also have oil droplets that filter and narrow the waveband response of each cone, probably also improving color constancy. The VS and ES receptors should detect UV radiation, and Huth and Burkhardt (1972) specifically demonstrated UV sensitivity in hummingbirds, although UV patterns appear to be rather rare in typical red-colored hummingbird flowers (see Lunau and Maier 1995).

The apparent preference for red flowers is not "real," as discussed in chapter 5; it arises not from unusual red sensitivity of the birds' eyes, or from an innate bird preference for red, but from learned associations of high nectar rewards and fast handling speed with red flowers that are unvisited by other groups and therefore unusually rewarding. Hence birds will learn to visit any color equally well in an experimental situation. It may be that red is used by bird flowers specifically because it is the least likely color to attract bees, but even that is an oversimplification as bee choices also depend strongly on contrast (chapters 5 and 18). It is certainly true that in many communities virtually all hummingbird-preferred and hummingbird-adapted flowers are red; for example, in California at least 17 plant families produce tubular red flowers with high-volume and low-concentration nectar, and with little scent. But here it is likely that strong convergent evolution has occurred (and perhaps a degree of mimicry, with red flowers forming *mimicry rings* that offer easily located resources for birds in any one habitat throughout the year).

Clear evidence for a red flower association comes from the multidimensional study of *Penstemon* species by Wilson et al. (2004, 2006), where red color was the single best predictor of hummingbird visitation. Within this genus, species attracting birds (rather than bees) shifted from blue/violet colors to pink/magenta colors, and shifting to red/orange additionally deterred bees.

In comparison with other vertebrates, birds are reported to have very good acuity (the ability to resolve very small objects visually); but they have rather poor spatial contrast sensitivity (Ghim and Hodos 2006), perhaps a necessary trade-off for the excellent acuity, UV detection, and color vision.

Most birds have rather small olfactory bulbs at the front of the brain and poor odor discrimination. Unlike other terrestrial vertebrates they rely very little on chemical signals (pheromones) for intraspecific communication (although some colonial nesters can readily detect self- and nest odors), and they forage mainly by using acoustic and/or visual signals. Hence the lack of strong (or in many cases of any) scent in flowers that seem adapted for bird visitation (chapter 6) is entirely to be expected.

3. Behavior and Learning Capacities

Hummingbird Foraging Behaviors

The hummingbirds are incredibly agile fliers, including the abilities to fly backward and to hover, so they have no need to perch while feeding and can visit flowers of any orientation or placement on a plant. Some flowers (e.g., certain *Heliconia* species) force them to visit with their heads almost upside down, with a more than 90° twisting of the neck. Aizen (2003a) recorded higher proportions of down-facing flowers in bird flowers than in insect flowers, perhaps because birds still pollinate during rain and the nectar must be protected from flooding. Varied flower architectures and consequent bird behaviors ensure that pollen is placed in specific locations on the beak, face, head, or back of these birds, and thus potentially correctly transferred only to the right species at subsequent visits.

Coupled with their agility, hummingbirds can show considerable flight speeds and fast flower handling times and may visit hundreds or even thousands of flowers in one day. Given that continuous flight to seek out flowers, and especially hovering while feeding at them, are energetically expensive (chapter 10 and table 10.3), these very high rates of visitation are essential to support both the flying style and the endothermic physiology. Activity costs for Brazilian hummingbirds were calculated at up to 30 kJ per day (Mendonca and dos Anjos 2006), directly fueled by the newly ingested sugars (Welch et al. 2006). However, the length of their foraging day may be curtailed by physiological constraints. They are in many cases so small, with such high surface area to volume (SA/V) ratios, that maintaining their body temperature (T_b) at the normal avian levels of 38–42 °C throughout a 24 h cycle is extremely expensive. Many therefore allow their T_b to drop markedly at night (i.e., they are temporal hetero-

therms rather than complete endotherms); for several hours in each daily cycle they are effectively torpid, with a T_b regulated at a much lower value (often only 20–25 °C), and with no flower-visiting activity possible. They tread a delicate tightrope, requiring intensive flower visits in the daylight hours to support their high energy demand and then keep them "ticking over" through the night-time torpidity. This perhaps explains why they appear to forage as net rate maximizers (Houston and Krakauer 1993), rather than being concerned with risk or variance of the nectar sources (see chapter 10).

Learning in Hummingbirds

Hummingbirds very easily learn to feed at artificial sugar sources, even when these lack any of the typical floral cues. This makes it easy to work with them and to test preferences. Most studies have indicated a lack of innate preferences for particular colors, in accordance with the previous discussion. Individual birds may exhibit rather different flower preferences within a particular habitat, again indicating that temporary preferences are learned and not innate. Such preferences can also be modified by conditioning, especially in relation to reward quality; and the birds learn very quickly that red flowers are usually unvisited by other animals and thus full of appropriate nectar.

Healy and Hurly (2001) reviewed much of the work to date on hummingbird memory and learning tasks, pointing out that simple rules and patterns proposed in earlier work (e.g., Pyke 1978, 1981) are not particularly helpful as the birds show flexible behaviors strongly influenced by their experience and learning. Experimentally, it is much easier to teach hummingbirds "shift" tasks than "stay" tasks, which is in accordance with their normal habit of emptying a flower at one visit and then moving on (although they *can* learn stay tasks, as they effectively do when working at commercial feeders). They appear to have good memory for locations, so can remember point locations of flowers. When trained on artificial flower arrays they can learn specific spatial information fairly readily, remembering complex distributions of full and empty flowers, especially if these are laid out in clear patterns. They can also remember previous contents of individual flowers to some extent, so will avoid water-filled flowers and in test conditions will choose nectar-filled ones that were visited but not completely depleted. They even appear to be able to remember morph-specific (long-style versus short-style) differences in daily schedules of nectar production in *Palicourea* (Ornelas et al. 2004).

Since hummingbird flowers normally dispense quite large nectar volumes, but of dilute fluid, each offers just a few milligrams of sugar; and since they appear to refill only slowly, the birds need to visit a great many flowers each day, but should avoid sampling previously visited flowers. With some preferred flower species having relatively transparent corolla walls, the birds may be able to detect empty flowers by visual cues at a distance, but with many plants this is impossible, so they could in theory adopt one of two options or rules to avoid resampling empty flowers:

1. where plants are clumped, always move to new flowers in the patch based on immediate experience; or

2. where flowers are more dispersed, remember individual locations of flowers and trap-line between them.

There is little or no evidence that hummingbirds do have simple movement rules in quite the way that many bees do. For example Pyke (1978), Waser and Price (1983) and other authors noted that hummingbirds do not consistently work either up or down a spike of flowers, and they show no consistent directional movements either. Waser and Price also recorded visits to more flowers per plant than for other visitor types; on *Delphinium nelsonii* hummingbirds visited a mean of 2.5 flowers per plant (mean for bumblebees is 1.5), and on *Ipomopsis* they visited a mean of 7.4 flowers per plant (hawkmoths 4.7). However, in both cases, the birds made less frequent flights between very-near-neighbor plants.

Overall, they appear to choose flowers just by their proximity and their size, visiting the nearest flower so long as rewards are good, but flying further away (in a random direction) as soon as the last flower visited gives a poor reward. Learning is based mainly on spatial locations, with color or pattern visual cues of flowers less important. In fact Sutherland and Gass (1995) found that their decisions on revisitation are made above the level of the flower, and probably above the level of individual plants, but rather on a broader spatial scale.

In accordance with their good spatial learning, *territoriality* around feeding sites is common, especially among male hummingbirds. Single males may defend hundreds or even thousands of flowers, returning to the same patch each day; they tend to feed at the edges of their territory early in the day and then move inward later on. This again implies reasonable spatial recall of where they have already fed, by remembering their own movements, or patches of flowers within the area, or even specific flowers (Healy and Hurly 2001). The first of these is the likely main mechanism, since it is matched by skills shown by the same birds in their known lekking behaviors and in trap-lining behaviors. But there may also be contributions from the second type of spatial recall, in view of the Sutherland and Gass data mentioned above, although it is currently unclear whether this is by vector sequences (perhaps unlikely given the lack of consistent movement patterns) or by using landmarks.

Territorial behavior, especially when conducted around just one tree or part of a tree as often happens, must impose serious limitations on outcrossing for the plant. Fortunately, perhaps, territories are not held for long (compared with those of passerine birds, for example) and only occur in a proportion of the hummingbirds. Territoriality is rare or absent in the hermit hummingbirds (the subfamily with particularly curved beaks) but relatively common in the much showier (exhibitionist) and usually straight-beaked nonhermits. For these reasons the hummingbirds have been further subdivided into four feeding groups on the basis of size, beak, and behavior (Stiles 1981):

1. The hermits have curved beaks, are relatively drab, live at low levels within the shade of forests, do not hold territories, and are coadapted in beak length and curvature with the flowers they prefer to visit, which tend to be herbaceous.

2. The largest nonhermit exhibitionist species have some beak curvature and tend to visit only flowers that are obviously hummingbird adapted, again showing a good match of beak length to corolla length. They feed at large red flowers on both herbs and trees, out in the open rather than in shade, and are commonly territorial.

3. Medium-sized exhibitionists have straight beaks and tend to visit smaller flowers, often pink or white in color as well as red, many of these being more typical insect-visited flowers, although they visit more of the

classical bird flowers when nectar levels are high in the morning, and may then be territorial.

4. The smallest exhibitionists have very short beaks and do much of their visiting illegally, piercing flower bases.

In Central America there is a particular group of plants, in the genus *Heliconia* (plate 26E), that provides a range of flower sizes and growth habits, and which has proved particularly useful for investigating hummingbird behaviors, following pioneering works by Linhart (1973) and Stiles (1975). Some species grow in shady sites within forests, having small plants with few flowers, while others colonize margins and river banks and can produce large stands of hundreds of flowers. The first group are visited by hermits, the second group by exhibitionist territory holders. Using powdered dyes to mimic pollen movements, Linhart showed that the forest species received fewer visits but their pollen moved over longer distances, supporting the idea that visitor territoriality would not be very helpful to these plants. Stiles additionally showed that the forest species had lower nectar energy output, while the plant clumps at forest margins could supply the entire daily needs of their hummingbird residents. All the indications were that the hermits were functioning as trap-liners, moving between many small known clumps with good floral constancy, while the exhibitionists showed similar constancy but almost no interpatch movement. Arizmendi and Ornelas (1990) showed that hermit hummingbirds were also much more mobile wanderers in the longer term, following nectar-rich resources around seasonally over quite large areas of forest, while the larger exhibitionist birds were almost always residents within a small forest zone. Forest plants with smaller rewards therefore offer enough to be worth a daily visit over the few days or weeks that they are in flower, but not enough to be worth defending. In practice, it turns out that the majority of *Heliconia* species use the former strategy; those growing in large clumps at margins are the exception, and may be exploiting the greater light to get fast vegetative growth there while being forced into the compromise of reduced outcrossing (but not, since they are mostly self-compatible, of reduced seed set). This is one of the clearest cases where concentrated work on a particular plant genus has allowed pollination biologists to dissect the alternative strategies that

plants can adopt to balance growth and reproduction (mediated by pollinators) according to habitat and resources available.

Foraging and Learning in Perching Birds

Perching birds are all generally much less agile in flight than hummingbirds. Although most sunbirds and the spinebill meliphagids can hover briefly, and occasionally do feed while hovering, they all prefer to feed from a perch; and the other families have no capacity for hovering at all and thus no alternative but to feed while standing. They all need rather strongly built flowers to withstand their probing activities, and they move much more slowly between flowers than do hummingbirds.

However, sunbirds parallel hummingbirds in their range of short, long, curved, and straight beaks and show a similar matching of preferred flower lengths to beak lengths. Larger species with more curved beaks tend to work flowers most efficiently, while the smallest species again show a strong tendency to cheat and pierce corollas, even on flowers where they could visit legitimately. Medium species usually with straight bills can get a net gain from much more dispersed flowers than the large sunbirds and so tend to do better on lower flower densities.

In Australian species of honeyeaters, Paton and Ford (1983) showed preferences for plants with more flowers and for flowers with more nectar. Nectar standing crop was highest when flowering was at its peak, because visitation rates per flower were then lowest, and pollination levels were also therefore lowest. Birds have sometimes been shown to be "inappropriate" in their foraging behavior from the plant's point of view though; honeyeaters foraging at *Telopea* flowers (Proteaceae) in New South Wales produced 60% of the fruit set in the upper third of inflorescences (even though anthesis was from the base upward), probably because they stood at the top of the inflorescence and probed downward (Goldingay and Whelan 1993); there are likely to be many other examples of differential fruiting resulting from perching birds' relatively immobile habits.

Like hummingbirds, passerine flower visitors perform better on tasks requiring them to avoid previously rewarding locations (shift tasks) than to return to them (stay tasks). This may partly be an adaptation to the generally depleting nature of nectar supplies, since omnivorous honeyeaters abandon their shift tendency when feeding on other types of food (Sulikowski and Burke 2007).

As discussed in chapter 10, larger species of bird have higher energy requirements and may stay longer at any one plant as the costs of flying scale up faster than the costs of sitting and feeding, so that larger birds should in theory feed from more flowers per plant. In Australia, for example, the eastern spinebill, with a body mass of 11 mg, probed 26–69 flowers per minute on *Eucalyptus* and other single flowers, while the much larger wattlebird (110 g) probed 30–91 flowers per minute (Paton and Ford 1983); although on the large clustered inflorescences of *Banksia* the distinction disappeared and both probed only 2 flowers per minute. Australian spinebills show a good ability to discriminate the most rewarding flowers within a plant, for example on *Correa*, where they picked out the developmental flower stage with highest nectar reward and tended to avoid flowers that showed visual signs of having been robbed (although they were not able to avoid those containing flower mites) (Scoble and Clarke 2006).

Many flower-feeding perching birds are territorial at least some of the time (around a quarter of all honeyeaters, for example; Pyke et al. 1996). Territoriality may occur where nectar is in short supply, on a patch of flowers, or on part of a whole tree (for example on *Eucalyptus*). When defending territories the smaller species are often better pollinators, as they visit individual flowers more frequently (Paton and Ford 1983). But territorial behavior tends to be seasonal, restricted to peak flowering season and to the densest flower patches, and larger species may exclude smaller species at such times. For example flowering *Banksia* were visited by New Holland honeyeaters in early June (fig. 15.3); then some of these were displaced by little wattlebirds in later June; then as flower numbers increased the wattlebird territories enlarged and were more strongly defended so that all honeyeaters were excluded through late July and August. Finally the wattlebirds left and the honeyeaters returned in early September (Paton and Ford 1983).

A few perching birds have more unusual flower-feeding behaviors. One example is the cacique (Icteridae), which feeds on the explosive flowers of certain *Mucuna* species that do not normally open unless the wings or keel are subject to pressure (Agostini et al. 2006). The bird inserts its closed beak and then opens its mandibles, forcing the flower to open, subsequently feeding on the nectar as well as on caterpillars of *Astrapetes* butterflies that are often hidden within the flower.

4. Bird Flowers and Ornithophily

Hummingbird Flowers

These are typically day flowering and commonly possess red (or less commonly orange, yellow, or pink) tubular corollas, usually with little color marking or patterning. They thus lack nectar guides in the usual sense, although the shape of the orifice may act to orient the birds correctly into the flowers (Smith et al. 1996). The flowers are rarely scented, and odor is not a foraging cue. They normally have no landing platforms, typifying the gullet flower design referred to in chapter 2. Corollas are routinely described as long, although the specific length is roughly matched to the beak length of the main visitors, as discussed above; but in a few cases such as *Aphelandra* longer corollas can be visited by shorter-beaked birds because they undergo a reversible compression when the flying bird moves in to the flower (Rengifo et al. 2006). The flowers tend to have strong, thickened basal corolla walls to prevent damage, but the corolla beyond the base may be quite delicate and translucent enough for the nectar column to be visible through it, so that a visitor can detect "empties" from a distance. The anthers are almost always dorsal, often pressed up against the inside of the upper corolla or exserting somewhat from it, thus dispensing pollen onto the top of the bird's head. Figure 15.4 shows some typical examples of bird-visited flowers, including some convergent forms; see also plate 26.

The nectar is copious and of low concentration (high *V*, low *C*), giving a low viscosity suited to the long capillary-action bird tongue, and it is generally secreted into the base of the corolla tube. In most species studied the nectar is dominated by sucrose (but see chapter 8, as the reasons for this have been controversial). Ornelas et al. (2007) showed a strong positive correlation between corolla length and nectar volume per day across 289 species of hummingbird-pollinated flowers, and Fenster et al. (2006) also found nectar reward to be highly correlated with flower size within a species (*Silene virginica*), so giving a suite of associ-

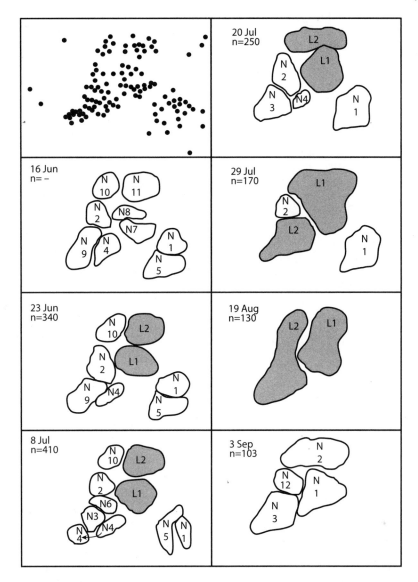

Figure 15.3 Seasonal changes in territory sizes for honeyeaters (N) and wattlebirds (L, shaded) feeding on *Banksia marginata* in Australia. Top left, the position of each *Banksia* bush; then territories for seven days from June to September, with numbers of inflorescences open shown below the date. Wattlebirds move in as flowering increases, and their territory sizes increase until honeyeaters are entirely excluded in August. (Redrawn from Paton and Ford 1983.)

ated traits (attractants and rewards) implying strong pollinator-mediated selection.

In any one habitat where hummingbirds are present (or through which they migrate) several species of bird-pollinated plant are present, reflecting the birds' need for resources over a longer time period than is normally provided by any one plant species. Figure 15.5 shows an example of flowering peaks for key hummingbird species in the cloud forests of Costa Rica (Feinsinger 1978). Some widespread or well-studied examples of hummingbird flowers are listed within table 15.2, and can be seen in plate 26. This flower type is particularly common in the Neotropical families Gesneriaceae and Acanthaceae. Sometimes the flowers are organized in spikes above the plant's foliage, each flower pointing up, down, or roughly sideways (suited to the birds'

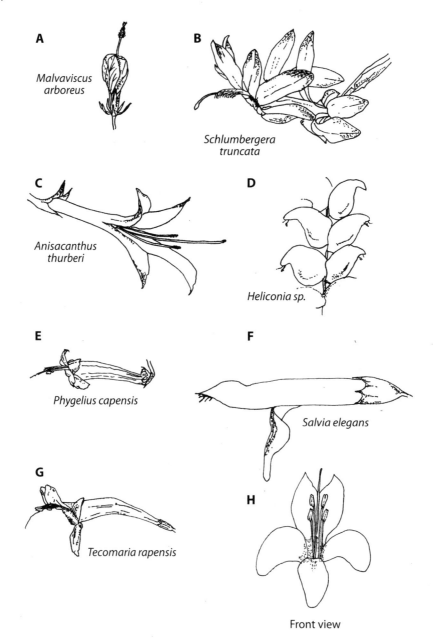

A *Malvaviscus arboreus*

B *Schlumbergera truncata*

C *Anisacanthus thurberi*

D *Heliconia sp.*

E *Phygelius capensis*

F *Salvia elegans*

G *Tecomaria rapensis*

H

Front view

Figure 15.4 Some classic bird-pollinated flower types, all from different families. (A–D) forms visited mainly by hummingbirds: (A) Malvaceae, (B) Cactaceae, (C) Acanthaceae, (D) Heliconiaceae; (E–G) convergent flower forms used by both hummingbirds and perching birds, in Scrophulariaceae, Lamiaceae, and Bignoniaceae. (H) indicates how these last three forms appear from the front. All those shown are red, orange, or deep pink. (A, B, E, G, redrawn from Endress 1994; others drawn from photographs.)

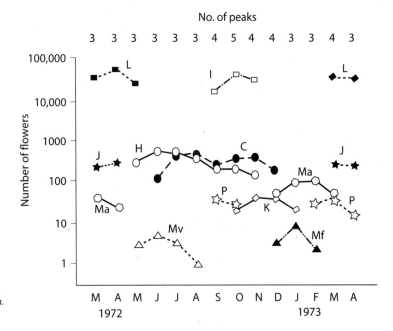

Figure 15.5 The staggered peak flowering times for hummingbird flowers in cloud forest in Costa Rica, with at least three species available in each month of the year: L, *Lobelia*; I, *Inga*; J, *Justicia*; H, *Hamelia*; C, *Cuphea*; Ma, *Malvaviscus*; P, *Psittacanthus*; K, *Kohleria*; Mf, *Manettia*; Mv, *Mandevilla*. (Redrawn from Feinsinger 1978.)

spectacular abilities to hover in almost any position); but they can also be pendant, held singly well away from a twig or branch, or may emerge from a robust tightly packed inflorescence (as in many plants of the Zingiberaceae and Costaceae and in bromeliads, where the bracts may provide the main color signal). The abundant and numerous *Heliconia* species have very large bracts supporting the flowers and providing the main display (commonly red, or red and yellow) while the flowers themselves are much smaller, often white or green, just protruding from the bract.

Hummingbird preferences for pendant flowers are anecdotally reported to be very common though not well tested experimentally; this apparent preference may just reflect covariance with some other trait (such as pedicel flexibility to prevent breakage during visits; Hurlbert et al. 1996).

However, there is an important caveat to be made here. We noted earlier that the North American hummingbird species are of rather uniform beak length and shape, so in these zones there is limited specialization of particular flowers to particular bird species, and the flowers visited do show strong uniformity. James (1948) listed the commonest species visited in the United States east of the Mississippi (table 15.3), and though derived from eight different families all are predominantly red and of similar size and morphology

(corolla length). But in Central and South America the trochilid beaks are much more varied, and the flowers visited are concomitantly also much less uniform in shape, size, and even color; here it is less easy to define a single hummingbird flower type, and there is more overlap with the other types of bird flower described in the next section. Hence Neotropical hummingbirds appear repeatedly as the main visitors in table 15.2 where some other bird is also visiting; the plants visited commonly have radial tubular or spurred flowers, but they may also sometimes be keel flowers or brush blossoms, and are strongly represented under "other designs."

Perching Bird Flowers

As a broad generalization, these flowers tend to be tubular gullets, but wider and more open at the mouth than hummingbird flowers. They are often sited directly on stout twigs where the birds can land and probe, whether upward or downward or laterally. They offer moderate to large volumes of fairly low-concentration nectars, often hexose dominated in Neotropical species but with sucrose common in African examples (again, see chapter 8). Since the great majority of these birds do not eat pollen, the flowers have relatively stiff

TABLE 15.2
Typical Bird-Pollinated Flower Types

Floral design	Genus and species	Family	Visitors	Distribution
Gullet flowers, zygomorphic, tubular	*Columnea* spp	Gesneriaceae	Hummingbirds	Central America
	Justicia spp	Gesneriaceae	Hummingbirds	Neotropics
	Mimulus cardinalis	Scrophulariaceae	Hummingbirds	Neotropics
	Salvia splendens	Lamiaceae	Hummingbirds	South America
	Antholyza spp	Iridaceae	Sunbirds	Africa
	various	Bromeliaceae	Hummingbirds	Neotropics
	Costus spp	Costaceae	Hummingbirds (and bees)	Pan-tropical
Other zygomorphic tubular flowers	*Penstemon* spp	Scrophulariaceae	Hummingbirds	North America
	Kniphofia	Liliaceae	Sunbirds	Africa
	Aloe spp	Liliaceae	Sunbirds	South Africa
Radial tubular flowers	*Abutilon megapotamicum*	Malvaceae	Hummingbirds	South America
	Fuchsia spp	Onagraceae	Hummingbirds	Central and S America
	Ipomopsis aggregata	Polemoniaceae	Hummingbirds	North America
	Manettia inflata	Rubiaceae	Hummingbirds	Neotropics
	Aeschynanthus	Gesneriaceae	Sunbirds and others	Indonesia
	Erica spp	Ericaceae	Sunbirds	South Africa
Keel flowers	*Erythrina* spp	Fabaceae	Hummingbirds, sunbirds	Pan-tropical
	Chadsia spp	Fabaceae	Sunbirds	Madagascar
Brush flowers	*Callistemon* spp	Myrtaceae	Flower-peckers, honeyeaters	Australia
	Eucalyptus spp	Myrtaceae	Honeyeaters, lorikeets	Australia
	Acacia spp	Fabaceae	Honeyeaters	Australia
	Banksia spp	Proteaceae	Honeyeaters, spinebills	Australia
	Calliandra spp	Fabaceae	Hummingbirds	Central America
Spurred flowers	*Delphinium cardinale*	Ranunculaceae	Hummingbirds	North America
	Tropaeolum pentaphyllum	Tropaeolaceae	Hummingbirds	South America
	Impatiens niamniamensis	Balsaminaceae	Sunbirds	Central Africa
Capitulate inflorescences	*Protea* spp	Proteaceae	Sunbirds	South Africa
	Mutisia spp	Asteraceae	Hummingbirds	South America
Other designs	*Strelitzia* spp	Strelitziaceae	Sunbirds	South Africa
	Heliconia spp	Heliconiaceae	Hummingbirds	Neotropics
	Passiflora coccinea	Passifloraceae	Hummingbirds	Neotropics
	Marcgravia spp	Marcgraviaceae	Hummingbirds	Neotropics
	Dalbergaria spp	Gesneriaceae	Hummingbirds	Neotropics

TABLE 15.3
The Most Important Hummingbird Flowers in the Eastern United States

Common name	Genus and species	Family
Cardinal flower	*Lobelia cardinalis*	Campanulaceae
Columbine	*Aquilegia canadensis*	Ranunculaceae
Jewelweed	*Impatiens capensis*	Balsaminaceae
Oswego tea	*Monarda didyma*	Lamiaceae
Red buckeye	*Aesculus pavia*	Hippocastanaceae
Trumpet creeper	*Campsis radicans*	Bignoniaceae
Trumpet honeysuckle	*Lonicera sempervivens*	Caprifoliaceae
Flame flower	*Macranthera flammea*	Scrophulariaceae

Source: Based on James 1948.

stamens freely exposed to brush pollen onto the visitors' feathers.

However, it is possible to recognize several other distinct types of flower that could be said to be associated with bird pollination (table 15.2 and fig. 15.6). Zygomorphic tubular flowers are less often of the gullet type, instead retaining their lower corolla lobe, and with the upper lobe generally less elongated, again commonly in the color range red, orange, and yellow. Aloes provide a classic example in southern Africa, their drooping tubular flowers arising from a stout central stem to which a bird can readily cling (plate 27F). Some pendant *Datura* species in South America are also visited by perching birds, including the swordbill, working the flowers from a central stem or from a perch on branches below. Elongate radial flowers with slightly flared mouths are also abundant, often in the families Gesneriaceae, Acanthaceae, and Scrophulariaceae, so that the normal zygomorphic character of those families is lost in adapting to bird visitors. Here the stamens and style may protrude noticeably from the corolla. Brush blossom flowers with densely clustered spherical or cylindrical flower heads are particularly common as bird flowers in Australia, where *Banksia* and *Callistemon* (bottlebrush flowers) are very well represented. Again the stamens are protruding and fully exposed, and the center of the blossom often houses a visible supply of shiny nectar pooled from the separate flowers (plate 8D). However, some of the capitulate *Banksia* species have clustered smaller flowers, resembling thistles; they have their corollas facing upward and provide a landing platform for the birds.

Bird flowers with nectar spurs can have various corolla designs but merit a grouping of their own here because the spur seems to serve not only to store the nectar but also to direct the rather sharp bird beak away from the ovaries and off to one side of the flower (or sequentially to several sides if there are several spurs). Interestingly, the angraecoid orchids on Reunion Island, from a group pollinated by hawkmoths in nearby Madagascar, have developed much reduced spurs and have very little scent, being pollinated by the endemic white-eye (Micheneau et al. 2006); this is a most unusual case of pollinator shift, as birds very rarely pollinate orchids.

The genera given in table 15.2 as "other designs" are particularly interesting in revealing both the great morphological variation in flowers visited by birds and the convergences that appear across a great range of families. The first two somewhat resemble each other and are both from families within the Zingiberales (see the appendix), closely related to the Musaceae (the banana family). *Strelitzia*, commonly known as bird-of-paradise flower, is pollinated by perching birds and has large horizontal flowers on stout stems, supported in large bracts (plate 27A). Each has a large six-section perianth, but three petals are erect and orange, providing the main display, while the other three are prone and blue, two of them (on which the bird must stand) being larger and concealing the stamens, the third one smaller and holding the nectary. The weight of the bird standing on the two blue petals causes their lobes to spread apart, and sticky strings of pollen are deposited on the bird's underside; these petals are also rather

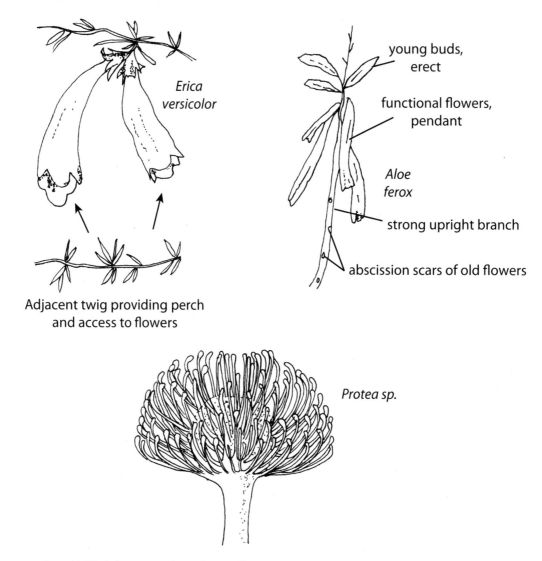

Figure 15.6 Typical access types for perching bird flowers; either from an adjacent lower twig to pendant flowers, or to successive flowers on the same vertical stem, or where the whole inflorescence provides a sturdy platform to stand on as in *Protea*. (Drawn from photographs.)

slippery and hard to hold on to, ensuring that the bird grips on tightly and must work the flower for some time.

The last three "other design" bird-pollinated species listed in the table fit no category. Some species of *Passiflora* (e.g., *coccinea, vitifolia;* plate 26G) are visited by hummingbirds and have the typical passionflower morphology modified to allow easier access to nectaries, with petals reflexed so that there is no landing platform, and they are bright red in color. The re-

lated genus *Tacsonia* has flowers with very elongated corollas and is visited by the swordbill in South America. Even more curious are the pendant *Marcgravia* inflorescences of the Neotropics, with flowers whose appearance is dominated by large nectar pouches formed from modified bracts, shaped so that the head of a visiting hummingbird must contact pollen on the whorls of flowers that lie above (fig. 15.7). Here the nectaries most unusually provide the main display, being yellow when young and turning red with age. In

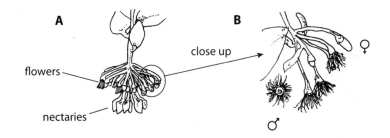

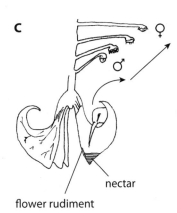

Figure 15.7 The unusual structure of hummingbird-pollinated *Marcgravia* inflorescences: (A) whole intact inflorescence, with nectaries protruding below; (B) closer view of male and female phase flowers (also shown on left and right of (A)); (C) section of the inflorescence with the progression from lower male to higher female stages of flowers, and the nectaries below, each with a flower rudiment. Arrows show the movement of a bird's head as it leaves. (Redrawn from Proctor et al. 1996.)

yet another curious genus, *Dalbergaria*, the leaves often provide the color signal bearing large red spots, with small yellow tubular flowers concealed behind them.

Finally it is worth mentioning a few oddities. One of these is the mistletoe family (Loranthaceae), which is especially associated with perching birds, and particularly sunbirds in most zones. Here the attraction for the birds is the seeds rather than the flowers, but in some cases the birds also passively pollinate the flowers, and in up to half of the species (including the New Zealand genus *Peraxilla*) the flowers are explosive and birds' activities trigger the removal of a cap on the bud (Ladley et al. 1997). A second curiosity is the behavior of some honeyeaters in Australia that feed not at the tiny rewardless flowers of the tree *Acacia terminalis*, but at the conspicuous red extrafloral nectaries, transferring pollen as they do so (Knox et al. 1985). And a third strange case is a *Calceolaria* species that offers a fleshy food body to a non-nectarivorous bird (*Thinocorus*) (Sersic and Cocucci 1996), perhaps because in the tip of South America there are no oil-bees that serve as the usual pollinators for this genus (chapter 9).

The Ornithophilous Syndrome

From the extremely varied flower designs mentioned in the last two sections, it is still possible to extract a series of characters that define bird flowers (Faegri and van der Pijl 1979). Cronk and Ojeda (2008) pointed out the strong convergent evolution involved here, and the potential for tracking the gene systems involved in the near future.

The flowers show all or most of the following:

1. Day flowering, opening early morning
2. Dehiscence and nectar secretion mainly in the morning
3. Vivid colors, with red dominant and some contrast (yellow, white, sometimes blue) also commonly present
4. Coloration often from longer-lasting structures such as bracts, stems, leaves
5. Absence of nectar guides (although the orifice shape may act as guide)
6. Flower tubular, and/or pendant or nodding, often with nectar spurs

7. Lower lip absent or curved back
8. Basal corolla walls thickened, or protected by thickened sepals or bracts
9. Stamen filaments stiff, often dorsal and united; anthers often exserted at maturity
10. Ovary protected, usually inferior
11. Large volumes of low-concentration nectar
12. Absence of scent
13. Large spatial separation between nectar supply and the anthers (when dehiscing) and the stigmas (when receptive)

Most of this is self-explanatory, and clearly a few of these points—(4) and (5) in particular— apply more to hummingbird flowers than to perching bird flowers; but most do occur in all the flowers in table 15.2.

However, one character worthy of further mention is the nectar. Nearly all bird-pollinated plants produce copious nectar that is dilute and remarkably consistent in concentration, averaging around 20%–25%. Bird flower nectar is thus consistently more watery than bee flower nectar (Pyke and Waser 1981). These nectar characteristics would seem very clearly matched to birds' morphological and physiological constraints; they need low-viscosity fluid (Kingsolver and Daniel 1983; Roberts 1995), and yet they require large caloric rewards. Gutierrez et al. (2004) noted a good correlation between the nectar supplied in 29 Andean hummingbird-pollinated plants and the nectar requirements of the associated 13 bird species, varying in energetic demand as they molted, reproduced, and moved around the system. Likewise Stiles and Freeman (1993) found they could reasonably well predict whether a flower was visited by hermit or nonhermit hummingbirds from measurements of secretion rate and concentrations of sugars (taking into account altitude effects where secreted volumes and percent of sucrose declined with elevation and percent of fructose increased). In fact, hummingbirds have even been divided into lowland and highland species on this basis, the former preferring somewhat more concentrated and lower-volume supplies; temperatures are generally higher than at altitude, and the nectar stays manageable at somewhat higher concentrations, while slightly smaller overall rewards are required as keeping warm is cheaper (Forcone et al. 1997).

There is some controversy here, however (e.g., Bolten and Feinsinger 1978). Hummingbirds in experimental situations generally prefer higher concentrations of 30%–45% (Roberts 1996; and see table 8.4); and we noted earlier that when flower and nectar handling costs are added in to calculations, Kingsolver and Daniel (1983) suggested that optimal concentrations would rise to perhaps above 40%. Why then do birds seem to prefer lower concentrations when foraging in the wild? Some have argued that the low concentrations occurring naturally in the birds' preferred flowers are partly to deter bees from taking the nectar, as 20% concentrations would normally be unprofitable for a bee. Hummingbirds tolerate dilute nectar better, perhaps because they have extremely efficient kidney tubules that rapidly excrete excess water (McWhorter and Rio 1999). This may make good sense, because certain smaller bees can potentially visit some typical hummingbird flowers, although larger bees may be precluded where the corolla tube is particularly narrow (e.g., some *Penstemon*; Wilson et al. 2006). But when small or medium bees do visit for nectar they may be very poor pollinators—they often fail to contact anthers as these are dorsal and/or exserted, and they may not touch stigmas in wider flowers. Such considerations would strongly favor floral changes (such as very dilute nectar) to discourage bee visits. It might therefore be argued that changes to the nectar (larger volumes and lower concentration) may be the necessary first stage for a plant in switching to bird pollination. This would also fit with the observation of Johnson and Nicolson (2008) that flowers adapted for hummingbirds and sunbirds (both specialized nectarivores) have nectar around 15%–25% concentration, while flowers attracting generalist bird visitors often have extremely dilute nectar (8%–12%).

That may be a sufficient explanation for bird flower nectars. However, it has also been noted that birds drinking at artificial nectar feeders have abnormally high volumes on offer and feed very much more rapidly than they do at flowers (Gass and Roberts 1992), suggesting a different balance between the tongue loading and unloading phases; and that they normally also have access to water separately in such situations, which may alter their preferences for sugar solutions.

5. Overview

Extreme specialization is precluded in ornithophilous relationships for a number of reasons. Above all, birds

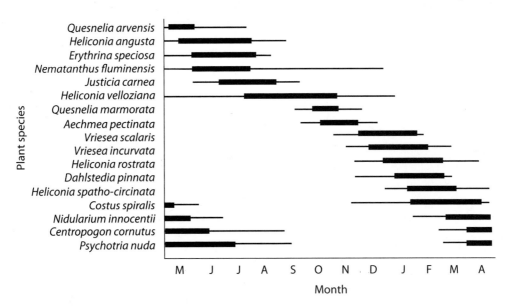

Figure 15.8 Floral usage succession through a year for the hummingbird *Ramphodon naevius* in southeast Brazil. (Redrawn from Sazima et al. 1995.)

are long-lived animals, and most will need to feed for much longer periods than can be provided by any one plant's flowering season. Thus Sazima et al. (1995) showed that hummingbirds may need up to 20 species in succession through a year (fig. 15.8). Second, there is only limited scope for precise pollen placement on the feathery surface of a bird to preclude heterospecific pollen pickup and deposition, so that Murcia and Feinsinger (1996) found considerable interference between different pollens, with intervening flower visits reducing the pollen transfer from donor to correct recipient; the offsetting bonus being that heterospecific grains took up little space on stigmas and did not produce serious interference for pollen tube growth.

Hence one-to-one relationships are extremely rare. Any given Australian bird flower species may get visits by a range of honeyeaters and other birds, and species of Proteaceae and Myrtaceae routinely received visits from at least 8, and up to 24, bird species (Paton and Ford 1983); rather lower numbers have usually been recorded in Central America for hummingbirds.

However, we have already noted that some degree of specialization is induced by matches of corolla shape and beak curvature, and this may also be linked to a general size fit between bird and flower. Larger species tend to have curved bills, in both hummingbird and nonhummingbird groups, perhaps because no pollinators could go on evolving longer beaks or tongues linearly, especially for those that land to feed, as they could not then easily probe into flowers (Westerkamp 1990). Larger bird-pollinated flowers do tend to show similar degrees of curvature, to accommodate their birds' beak shapes. Where there appears to be a mismatch of beak length and corolla length, Temeles et al. (2002) showed that corolla width also played a part in flower choices; and one case where a flower shortens slightly from the pressure of a short-beaked bird arriving was mentioned earlier.

Perhaps the best example of specialized curvature fit comes from the sexually dimorphic Caribbean purple-throated hummingbird (*Eulampis jugularis*), the sole pollinator of two *Heliconia* species (Temeles and Kress 2003). *H. caribaea* has flowers that fit the short straight beaks of the male birds, and *H. bihai* has flowers that fit the longer curved beaks of the female birds. Across several islands, the males promote inbreeding in *H. caribaea* while the females promote outcrossing in *H. bihai*. However, on the island of St Lucia *H. caribea* is rare, and *H. bihai* compensates by having a second morph with flowers matching male bills, while on Dominica it is *H. caribaea* that has a second morph with flowers that suit the female birds. In each case the nectar provision in the plant is also matched to the birds' needs, the male hummingbirds being substantially larger than the females. Further work on hummingbird communities in the Antilles islands using a pollination network approach (see chapter 20) again found that specialization levels were broadly size related: larger bird species were more functionally specialized on particular plants, whereas smaller hummingbirds had broader floral niches and began to overlap with insect-visited flowers (Dalsgaard et al. 2008).

A level of specialization and increased efficiency as pollinators can also be inferred from the movement behaviors of birds referred to above, including territoriality around a plant patch or tree, and is likewise favored by the community structures of some pollinating birds. Feinsinger and Colwell (1978) found there were usually just 2–3 species of hummingbirds on Caribbean islands, each with different characters and different flower choices: commonly one territorial short beak, one long-beak nonterritorial trap-liner, and one less specialist. In larger communities there were often these same three types (maybe two of each) and also some "marauders" and "filchers," plus often, in denser forests, a very long-beak understory high-reward trap-lining species. Community structure is less clear-cut in perching birds, which are usually less dependent on floral food, although there may be some pattern in Australian honeyeaters (Carpenter 1978a) and in South African sunbirds and sugar birds, where coexisting species do differ in morphology, beak length, territoriality, and aggression.

Following this theme of only moderate specialization, it is interesting that the same plant genus can appear to be coevolved with different birds in different places. A classic example is the legume *Erythrina*, whose flowers are always bright red and make a spectacular showy display when in flower on the tree; but one species in Indonesia has large and widely open keel blossoms, facing inward on strong stalks and readily reached by nonspecialist birds; another has small flowers with a greatly reduced keel and stamens freely exposed, depositing pollen on sunbirds and white-eyes; and *E. umbrosa* and other New World species are medium sized but with long enclosed tubular keels dispensing pollen only at the tip, onto the throats or bellies of hovering hummingbirds. Yet other New World species have their nectaries protected by a petal flap, only visited legitimately by perching birds such as parrots and icterids (Cotton 2001).

Another neat example of substantial specialization

toward a particular bird comes with sunbird pollination of certain Asclepiadaceae (milkweeds), described by Pauw (1998). Milkweed pollination has usually been regarded as relatively nonspecialist, although it involves a particular mechanism for pollen transfer since the pollen is packaged as pollinaria (chapter 7). Each of these comprises two pollen-filled sacs plus a mechanical clip (the corpusculum) that usually gets attached to various parts (feet, or proboscis, or some bristles) of an insect visitor. But Pauw noted that in South Africa *Microloma sagittatum* flowers were visited mainly by *Nectarinia chalybea*, and the corpusculum attached to the frayed edges of the bird's tongue, allowing very efficient pollinaria transfer to conspecific flowers. This is unusually specialized given that for most sunbirds pollen transfer is via feathers. Oddly, this was presented as a "blow" against specialized syndrome approaches (Ollerton 1998) because the flower's appearance suggests insect pollination; but it could equally be seen as a nice example of specialization (within constraints) to exploit locally available visitors.

There is very little in this chapter about bird interactions with pollen, or measures of their effectiveness in moving pollen. Perhaps that is fair; the interaction is somewhat limited. During a flower visit, pollen is likely to be deposited on a bird's body in a relatively specific place, but little grooming off occurs except where pollen grains are obtrusive on eyes, beak, etc. Thus the pollen tends to accumulate, and in many cases birds can be seen as single-compartment pollen donors (fig 7.12). However, where visitation is intensive and flowers produce large amounts of pollen, a bird's body may acquire multiple layers of pollen (Price and Waser 1982; Lertzmann and Gass 1983; Feinsinger and Busby 1987) such that the innermost layers are effectively useless for deposition on stigmas, giving a two-compartment pollen dispersal pattern. In either situation, the pollen carryover from birds is probably rather simple to model, and in chapter 3 we noted that work with hummingbirds had indicated carryover occurring both significantly further and more variably than was once supposed, so that Waser (1988) reported carryover up to the 35th flower visited (fig. 3.6). Several estimates of pollination effectiveness for avian species were given in chapter 11, and although birds often carried low pollen amounts their frequency of visit tended to improve their ratings compared with other possible vectors. Thus both hummingbirds and other perching species can be moderately specialist and highly effective pollinators of those species where they are the dominant visitors.

It is worth ending with a reminder that birds which are not specialized for flower visiting at all can nevertheless effect pollination on some plant species. On islands this may be particularly common, so that on Tenerife opportunistic nectar-feeding tits, warblers, blackcaps, and finches produced 93% of effective visits to native *Isoplexis* flowers (Rodriguez-Rodriguez and Valido 2008). In Europe the common blue-tit can pollinate willow catkins (Q. Kay 1985). Not infrequently, blue-tits and great-tits (*Parus* species) also pollinate the large-flowered crown imperial fritillary (*Fritillaria imperialis*) widely grown in gardens (Búrquez 1989), perhaps attracted because these flowers have extremely copious nectar (plate 19A). And in south-eastern Asian urban settings some species of *Erythrina* (a typical red-colored keel blossom, referred to above) are regularly pollinated by a range of medium-sized ordinary passerines such as starlings, thrushes, and drongos.

Chapter 16

POLLINATION BY BATS

Outline

1. Feeding Apparatus
2. Sensory Capacities
3. Foraging Behavior and Learning
4. Bat Flowers
5. The Bat Pollination Syndrome: Chiropterophily
6. Overview

Flowers visited and pollinated by bats constitute the syndrome termed **chiropterophily**, as bats were traditionally united in the mammalian order Chiroptera although they are now classified as two distinct and separately evolved orders, Megachiroptera and Microchiroptera. Bats are primarily nocturnal, and as flying endothermic mammals have extremely high energy demands (chapter 10). Furthermore, their flower visits often require hovering for short periods (with wing beats of 12–16 per second and involving an upward tilt of the wings to generate lift); this increases their energy demand further, although—somewhat surprisingly—not by an enormous amount (see Winter and van Helversen 2001); in fact glossophagine bats have a lower mass-specific hovering cost than hummingbirds or hawkmoths (table 10.1). However, having rather large body masses, in practice the bat species that visit blossoms for a major part of their diet are linked with some unusual and very high-reward flowers (Kunz and Fenton 2003).

The Megachiroptera constitutes an Old World grouping of generally larger species. The flower-feeding species come from Africa, southeastern Asia, and Australasia (although almost absent from the extreme southern zones). They lie within two subfamilies, Pteropinae (strongly dependent on fruits for food; thus generally known as fruit bats, and including the so-called flying foxes in the genus *Pteropus*) and Macroglossinae (predominantly nectar feeding, and sometimes called flower bats). A few of the Macroglossinae rely totally on flowers for their food, taking protein from the large amounts of pollen groomed from their fur.

The Microchiroptera has a worldwide distribution but its members are primarily insectivorous; flower feeders occur only in the New World, in the family Phyllostomidae. Several subfamilies include flower visitors, but often of a rather unspecialized type, feeding mainly on larger brush-blossom plants. However, in the subfamily Glossophaginae, which has about 29 species currently identified, there are more highly adapted flower visitors, with long slender snouts and long tongues and particularly dense fur where the individual hairs are (unusually for bats) almost plumose like the hairs of bees, giving excellent pollen-holding capacity. Many glossophagine bats are therefore specialist visitors to gullet-type flowers, being able to hover for a second or two while drinking nectar. Indeed, some flowers are visited by both frugivorous and nectarivorous glossophagines; the palm *Calyptrogyne* has flowers attracting both perching frugivores (taking fruitlike tissue from the flowers) and hovering nectarivores, with fruit set being higher after the former have visited (Tschapka 2003).

Bats occur in most habitats, but the flower-visiting species from either order occur predominantly in the tropics, extending about 30°N and 30°S of the equator. As well as humid tropical zones, this range includes hot arid areas in the southern United States, where bats pollinate many desert plants (especially the large-flowered cacti), although these bats tend to be migratory (McGregor et al. 1962). And while bat pollination is usually a phenomenon of tropical latitudes, this does not always mean hot habitats, since bats also occur at high altitude (especially in the Andes), where they encounter cool and even freezing temperatures at night.

Flower visiting is by no means uncommon in bats. Cox et al. (1991) reported 92 different plant genera in 50 families being visited by flying foxes (*Pteropus*) around the Indian Ocean and South Pacific islands. Mexico alone has 11 nectar-feeding bat species, visiting cacti, *Ipomoea*, *Agave*, bananas, and several other commercially important species. Thus bat pollination is not an uncommon phenomenon, with up to 4% of plants in some Neotropical forests being chiropterophilous (see table 27.3).

Flower-visiting bats from both orders are generally larger than the flower-visiting birds discussed in chapter 15. In the New World, flower bats range from 10 to 35 g (rarely to 50 g) body mass, and in the Old World usually from 15 to 300 g (although flying foxes can be in excess of 1.5 kg with a wingspan of 1.8 m). Their chosen flowers are correspondingly larger and more robust than those visited by birds, although often in the same genera. Indeed, as bats are large relative to *most* other pollinators, they commonly forage on larger woody plants—shrubs and trees—rather than on herbs. The exceptions to this are some smaller partly or wholly bat-pollinated herbs that are epiphytic or scramble over open rocky ground.

With large plants and large flowers and high energy demands, it is no surprise that floral rewards are also substantial, among the largest in all plants. Neotropical flowers supplying glossophagines commonly contain more than 100 µl of nectar, while some flowers visited by the largest bats may contain several milliliters each.

Since the action of a bat pollinating a plant takes place mainly in the dark and at some distance from the ground, bat pollination was understandably less well known until quite recently, and even now studies tend to be more common for the accessible desert cacti than for tropical forest trees. Some of the evidence regarding bat pollination is still indirect, relying on analysis of pollen loads on netted bats, or of pollen grain remains in fecal sample from bat roosts; but increasing use of night viewing equipment and photography, and radio tracking of tagged bats, are adding to the direct observational data.

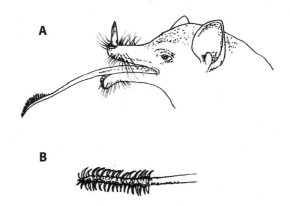

Figure 16.1 Megachiropteran tongue (A) from the side and (B) the tip from above. (Drawn from photographs and museum specimens of *Glossophaga soricina*.)

1. Feeding Apparatus

With fruit dominating their diet, the Megachiroptera have pronounced snouts and a long slender tongue (fig. 16.1); some of the species that feed extensively on flowers have tongues that are 60%–70% of their own body length. A few of the Macroglossinae, while retaining the standard bat mouth and tongue design, take no fruit in their diet and rely totally on flowers into which the snout and tongue can be inserted, the Queensland blossom bat (*Synocteris australis*) being a well-studied example. This species and some near relatives are among the relatively rare bats that take large amounts of pollen in their diet, grooming it from fur with their claws, teeth, and tongue.

The Microchiroptera have moderately short snouts and tongues and are relatively awkward when handling flowers, clinging on mainly upside down and often consuming parts of the flower tissues. However, in the more highly adapted Glossophaginae, there are long slender snouts and longer tongues (often 50%–95% body length). The exceptional case is *Anoura fistulata* from Ecuador, with a tongue up to 85 mm, 150% of the body length, and stored back in the thoracic cavity when not in use (Muchhala 2006). This bat is the sole pollinator of *Centropogon nigricans*, which has 80–90 mm corollas; the only known case of a 1:1 relationship for bat flowers. Remarkably, it has also been shown that (artificially manipulated) even longer-tubed *Centropogon* male-phase flowers deliver more pollen onto visiting bats, and also receive more pollen per bat in the subsequent female-phase flowers (Muchhala and Thomson 2009).

The tongues of nectarivorous bats have features in common across the orders and families, having tips with a brush of fine bristles or narrow scales, often pointing slightly backward, and allowing nectar uptake by capillarity. Flower-feeding bats tend to have fewer teeth than their insectivorous relatives, although a few species do use their teeth to aid in the consumption of floral tissues (e.g., the bracts of *Freycinetia*, and some sugar-rich parts of *Madhuca* and *Bassia* flowers).

Bats show reasonably good digestion of pollen with higher extraction rates in the flower feeders than in frugivores (Mancina et al. 2005), apparently related to different levels of the appropriate enzymes.

2. Sensory Capacities

All the members of the Megachiroptera have good eyesight and excellent olfactory senses, but they have poor navigational abilities in the dark as most of them lack any echo-location system (the genus *Rousettus* is

an exception, able to echo-sound using clicks made by the tongue). In contrast, all the Microchiroptera bats have relatively poor eyesight but excellent olfaction and **echo location**, along with good binaural sound localization acuity, and they can navigate in complete darkness.

Olfaction

This is the sensory modality for flower location common to all bats, more important than visual attraction in the dim light of dusk, so that bat flowers are almost always highly scented. They often have rather "odd" smells, fruity or fermenting to our noses; terpenes are commonly present, often with long-, medium-, and short-range variants. In fact scents in bat flowers are noteworthy for their similarity across a very wide taxonomic spectrum of bat flowers (Knudsen and Tollsten 1995; Bestman et al. 1997); figure 16.2 shows the relative preference of *Glossophaga* bats for some of the characteristic components (Winter and von Helversen 2001). Many but not all bat flowers contain the sulfurous volatiles that are evidently highly preferred.

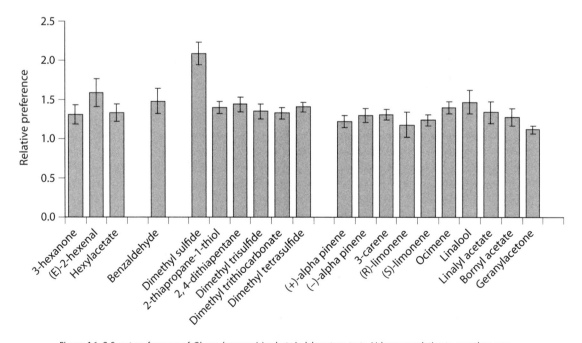

Figure 16. 2 Scent preferences of *Glossophaga soricina* bats in laboratory tests. Values are relative to scentless controls and given as means ± SE. (Redrawn from von Helversen et al. 2000.)

Van der Pijl (1936) suggested that the flowers are mimicking the scents from the glandular secretions of bats, used as attractant cues for flocking and roosting together.

Bats' noses appear to house a rather standard vertebrate olfactory epithelium and may also contain mechanoreceptors and cold/warm receptors. Their whiskers, sited around the nose, also give mechanosensory input that may help guide the snout or tongue correctly into the flower.

Vision

Bats are generally recorded as color-blind, having no cones, although some *Glossophaga* species are sensitive to UV wavelengths (using the rhodopsin in their rods), which may help them detect pale flowers against a dark background when there is a faint light available. In addition, several bats tested so far do have functional color vision opsin genes (Wang et al. 2004), although their expression may be limited.

Megachiroptera have good pattern-recognition abilities, again useful in detecting the contrast of flower outlines against backgrounds, and have a good spatial memory to recall locations once seen (Holland et al. 2005). Even some Microchiroptera probably use vision for longer-range flower detection in all but the darkest nights, with echo location coming into play at closer range.

Most typically the flowers that bats choose are therefore white, cream, dull beige, or sometimes drab greenish in color, with low color saturation. Those that open and are visited at dusk tend to be white, providing the best possible visual signal, whereas night-opening species show the duller colors, perhaps helping to keep them unnoticed by potentially inappropriate diurnal visitors or robbers during the day before they open. Some species have darker colors (deep reds, purples, or browns) centrally, perhaps helping to disguise the corolla opening from any visually searching and "inappropriate" visitors; see plate 28A–E.

Ears, Sound, and Echo Location

It is now clear that nectar-feeding glossophagine bats often find their flowers in dense tropical forests using echo location, which is their primary system for orientation in flight. Some flowers have evolved features that make their echo more conspicuous or more easily discriminated. The famous example is the flower of *Mucuna holtonii*, a chiropterophilous high-canopy vine (fig. 16.3), which has an unusual concave upper petal (vexillum) that provides a distinctive echo signal (von Helversen and von Helversen 1999). Figure 16.3 also shows the echo-location traces recorded from a flower as a (simulated) bat approached from different angles, compared with the echoes from a closed bud and from an adult flower where the concave petal was filled with cotton wool. The flower has an explosive nectar-exposing mechanism triggered by the first bat visit, so that later visits are relatively unrewarding, and bats are able to distinguish the virgin flowers from the already exploded flowers (von Helversen and von Helversen 2003). Other *Mucuna* species without bat pollinators do not have the concave vexillum.

Glossophaga bats have been trained with various geometric objects to test their echo-based discrimination and can readily pick out hemispherical surfaces of different diameter (Simon et al. 2006). They may also be able to use object-specific binaural echoes to determine floral orientation and even identity (Holderied and von Helversen 2006). Close to a flower, they may use echo acoustics to guide themselves to a nectar chamber (von Helversen and von Helversen 2003).

Many other bat flowers have rather concave bell shapes and may prove to be acoustically conspicuous (von Helversen et al. 2003). Some of the bat-pollinated cactuses house their flowers within a hairy socket which may be a sound-absorbing system to enhance the signal from the flower (Winter and von Helversen 2001).

3. Foraging Behavior and Learning

Many bats feed by hovering at a flower, each hovering bout lasting typically only 0.3–0.6 s, so that most of their time is spent on forward flight, commuting between flowers. There are relatively few calculations or observations on numbers or frequency of flower visits per trip (or per night) for bats, or on how many visits each flower might expect to receive. Vogel (1958) suggested that each flower of the sausage tree *Kigelia* received around 50 visits from bats in its single night of existence, but this is a little unsatisfactory as the plant is an introduced alien at the site studied in South

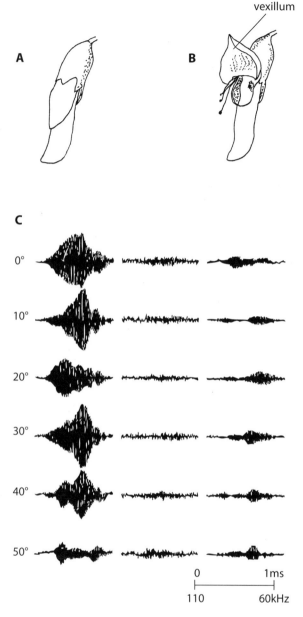

A

B

vexillum

C

0°

10°

20°

30°

40°

50°

0   1ms

110   60kHz

Figure 16.3 The structure of *Mucuna holtonii* flowers: (A) an un-opened bud, (B) an intact open but unvisited flower with the vexillum erect. (C) The ultrasound records from an open flower on the right, from a bud in the middle, and from a flower with the erect vexillum padded with cotton wool on the right, where the echo is largely suppressed. The echo also decreases as the angle of approach starts to exceed about 30°. (All redrawn from von Helversen and von Helversen 1999.)

America. For more reliable studies with native bromeliads, visits from 1–3 bats per flower were recorded, each bat going to about 40–50 plants and any one bat revisiting a given flower 3–30 times per night (Winter and von Helversen 2001; and see below on revisitation patterns). For *Werauhia* flowers in Costa Rica, Tschapka and von Helversen (2007) recorded *Glossophaga* bats hovering for a mean period of just 320 ms at a flower, this period correlating with recorded nectar secretion rates; hence there were very high flower visitation rates.

Bats have excellent spatial memory (Thiele and Winter 2001), which for nectarivorous species must help them to visit known and reliable high-value nectar sources repeatedly. They often live in very large roosts that occupy fixed locations for years or decades; a single roost site in some Asian cave-dwelling species such as *Eonycteris spelea* can contain tens of thousands of individuals. The roosts may be a very long distance from some of the trees that the bats pollinate. Hence they have to travel long distances on each night's foraging trip, with some records of regular flights of up to 40 km, taking perhaps 1.5–2 hours in each direction and flying together in large flocks. In contrast, some smaller tree-roosting Asian species roost alone or in very small groups, in or close to their preferred feeding trees. For Central American species, Heithaus et al. (1975) found a rough proportionality between size of bat and flight distances. Hence the foraging patterns and behaviors of bats are extremely variable, depending on body size and on the spatial relationship between roost and feed sites in any given habitat. Indeed the same species may show quite different patterns when studied in different places and even in the same place in different seasons as flower density changes. For example, the medium-sized bat *Artibeus jamaicensis* has variously been recorded as a group forager, and as a solitary trap-lining forager (Sazima and Sazima 1977); the same authors reported *Phyllostomus* bats varying the size of their foraging group according to numbers of open flowers on *Lafoensia* and *Bauhinia* trees.

Despite the variability in foraging patterns, a few broad generalizations can be made. With regard to diel patterning of activity, most bats do not become active until around sunset, and therefore do not arrive at flowers until it is almost dark, in some cases not being present at feeding sites until at least an hour after dark. Infrared sensors increasingly indicate that peak feeding may be well after midnight (e.g., Tschapka and von Helversen 2007). Many species then continue foraging almost until dawn, although the larger bats often have shorter feeding periods of around 3 hours (all else being equal), reflecting both their relatively lower metabolic costs and their often greater return flight time to the roost site. Howell (1979) reported that the 3 hour feeding bout of *Leptonycteris* bats cost about 63% of their total daily energy expenditure.

New World flower-visiting species are almost always encountered feeding in small flocks, and the mass-roosting larger Asian species also tend to break up into small groups, rarely more than 30–50, on any one tree, perhaps with some degree of territory holding being involved (Gould 1978). These feeding flocks move from tree to tree together and may characteristically revisit the same tree more than once each night, with short bouts of resting in between. Heithaus et al. (1974) showed that bats in flocks working *Bauhinia* flowers would almost completely drain all nectar from a visited flower; and that gaps between the pulses of visitation (averaging 21 minutes) appeared to correspond with the time taken for emptied flowers to refill with nectar. It is possible that group foraging is an efficient way for large bats to utilize flowers that they can empty completely and that then refill reliably, especially on trees that are widely spaced within a forest (Heithaus et al. 1975). Group foraging may also improve search efficiency and maximize exploitation of plant patches (Howell 1979). In practice it also means that each flock visits just a few flowering species on any one trip, and individuals thus carry only slightly mixed pollen loads, many species showing reasonable flower constancy per trip.

As mentioned, smaller bats are more likely to forage alone and operate in less organized fashion, rarely emptying a flower completely and often revisiting a flower more than once, then moving on to any nearby flowering tree. Smaller species also have relatively smaller wings, with lower aspect ratio, which increases their cost of flight, so that they tend to take more frequent rests. These rest periods again often lead to a degree of pulsing in the visit pattern and may allow time for some pollen to be groomed from the fur. Some bat species, of both small and large body size, tend to cluster in small groups during the rests between feeding pulses, and this aggregation behavior will probably help to conserve each bat's body heat as the night air cools.

Body size also affects foraging site on a particular tree for many bats, with larger species not surprisingly visiting the outer flowers, often higher in the tree, and smaller more maneuverable species working at all heights and often penetrating to flowers deep within the tree.

Finally, some bats are alarmingly specific in their feeding; the endangered *Leptonycteris curasoae* apparently does not vary its home range from year to year but alters its foraging effort depending on the local abundance of *Agave palmeri* plants, so that its conservation depends on maintenance of adequate agave populations (Ober et al. 2005).

4. Bat Flowers

General Morphology

Bat-pollinated flowers are nearly always large and conspicuous, since easy detectability will decrease searching time for a foraging bat and so reduce the need for the plant to invest in nectar calories to offset its visitors' costs. Typically the flowers have an open bowl or bell morphology (fig. 16.4; and see plate 28A–E) giving easy access to the bat head and tongue, and with strong often fleshy corolla walls. The flowers are commonly set on a long and strong stalk to stand well clear of the foliage, and this stalk may be erect or may dangle well below the leaves (flagelliflory). Another option is for the flowers to grow directly on the trunk or branches (cauliflory) to give strong support, a pattern common in many upright bat-pollinated cacti such as *Carnegiea* and *Stenocereus*, and in the calabash tree (*Crescentia*, plate 28D). Alternatively, the flowers grow when the plant is leafless to ensure good access.

The stamens are generally set well forward or outward from the flower, giving a gap under the anthers where the bats' heads fit, and pollen deposition is therefore usually nototribic on the top or back of the head. Alternatively, there are very large numbers of more delicate stamens that the bat pushes into when seeking the deeper-sited nectar, and here the anthers deposit pollen all over a visitor's face and snout, or all over its underside as in the kapok tree (*Ceiba*) pollinated by bats in its native Neotropics and in most of the tropical countries it has now been spread to (Gribel et al. 1999).

Beyond these generalizations, table 16.1 gives a

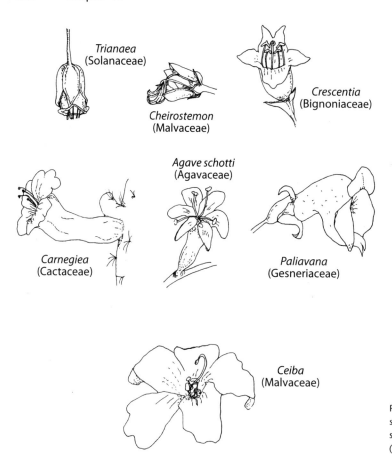

Figure 16.4 Typical bat flowers, open bowl or bell shaped, with long stalk or borne on trunk, stamens usually directed forward or outward. (Drawn from photographs.)

range of examples of bat-pollinated flowers, and they can be fairly readily divided into two main morphological types (Vogel 1958) spread across many plant families (although particularly well known from the Bombacaceae).

Brush-Blossom Types

These may occur as either single flowers or inflorescences. An example of a single brush flower occurs in the African baobab tree (*Adansonia*), with large creamy-white flowers (plate 28E) dangling below the foliage. Each flower has a ball or brush of anthers hanging just below the petals, and a very long stigma extending even lower (fig. 16.5).

Inflorescence brush blossoms are the commoner design, with many tiny flowers each contributing their long anthers to the overall effect. In most cases the blossom is upward facing. Nectar may be produced in each individual flower or in bracts separate from the

tiny flowers (e.g., *Marcgravia*). However, it is not uncommon for the blossom to be a mix of a few male flowers with abundant nectar and many nectarless hermaphrodite flowers; for example, in some *Albizia* species (plate 8B) where there is often just one large central male floret. Having such a mix helps to increase both the overall pollen amount (Heithaus et al. 1974) and the pollen to ovule ratio, as in *Pseudobombax* (which in some species offers no nectar at all).

Mimosa lewisii is a good example of an exclusively bat-pollinated brush-blossom flower from a genus that is normally entomophilous. It is unusual compared to most chiropterophilous plants in having rather small flowers without a strong scent, but is typical in other respects, the flowers being white and producing copious dilute nectar only by night, having unusually stiff filaments and styles, and inflorescences set well above the foliage (Vogel et al. 2005). Various members of the normally entomophilous Myrtaceae (e.g. some

TABLE 16.1
Types of Bat-Pollinated Flowers

Brush blossoms			
Single flowers			
Adansonia	Baobab	Bombacaceae	Madagascar, Africa, Australasia
Angophora		Myrtaceae	Australia
Sonneratia	Mangrove apple	Myrtaceae	Africa, Asia, Australia
Inflorescences			
Ceiba	Kapok tree	Bombacaceae	Pan-tropical
Mabea		Euphorbiaceae	Neotropics
Parkia		Fabaceae (mimosoid)	Pan-tropical
Agave (some)		Agavaceae	Neotropic deserts
Banksia (some)		Proteaceae	Australia
Barringtonia		Lecythidaceae	Madagascar
Planchonia		Lecythidaceae	Southeast Asia
Eucalyptus (some)	Gum tree	Myrtaceae	Australasia
Pseudobombax	Shavingbrush tree	Bombacaceae	Neotropics
Hymenaea	Flour tree	Fabaceae	Neotropics
Grevillea (some)		Proteaceae	Australasia
Melaleuca		Myrtaceae	Australasia
Gullet or bell flowers			
Bauhinia	Orchid tree	Fabaceae (caesalpinoid)	Tropical
Bombax	Silk cotton tree	Bombacaceae	Pantropical
Burmeisteria		Campanulaceae	Neotropics
Carnegiea	Saguaro	Cactaceae	Neotropic deserts
Durio	Durian	Bombacaceae	Southeast Asia
Cayaponia		Cucurbitaceae	South America
Cheirostemon		Malvaceae	South America
Cobaea	Cup and saucer	Polemoniaceae	South America
Epeura		Fabaceae	Pan-tropical
Freycinetia		Pandanaceae	Southeast Asia, Australasia
Hillia		Rubiaceae	Neotropics
Kigelia	Sausage tree	Bignoniaceae	Africa
Lafoensia		Lythraceae	Tropical
Louteridium		Acanthaceae	Neotropics
Mucuna (some)		Fabaceae	Pan-tropical
Ochroma	Balsa tree	Bombacaceae	Neotropics
Oroxylum	Tree of Damocles	Bignoniaceae	Southeast Asia
Parmenteria		Bignoniaceae	Neotropics
Symbolanthus		Gentianaceae	South America
Trianaea		Solanaceae	South America
Trichantha		Acanthaceae	Neotropics
Other designs			
Passiflora (some)	Passion flower	Passifloraceae	Tropical
Marcgravia		Marcgraviaceae	Neotropics
Musa (some)	Banana, plantain	Musaceae	Pantropical
Protea (some)	Sugarbush	Proteaceae	Southern tropics

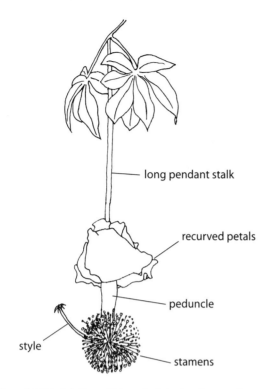

Figure 16.5 Baobab flower (*Adansonia digitata*) on long pendant stalk, and with asymmetrically directed style. See also plate 28. (Modified from Proctor et al. 1996, based on earlier sources.)

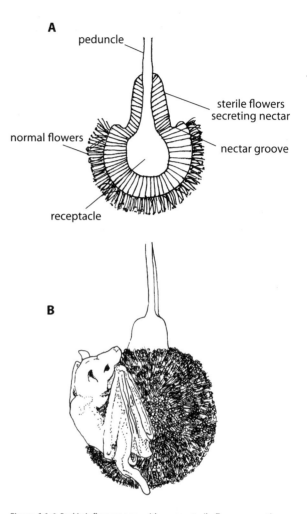

Figure 16.6 *Parkia* inflorescence, with upper sterile flowers secreting nectar into a common groove, and lower reproductive flowers; also showing the positioning of a visiting bat. (Redrawn from Faegri and van der Pijl 1979 and later sources, and from photographs.)

Eucalyptus and *Melaleuca* species) also have smallish brush blossoms that are visited by lighter-bodied species of *Pteropus* throughout Australasia (Beardsell et al. 1993).

Some *Parkia* species, such as *P. clappertoniana* from West Africa, are fairly specialist examples of this kind, with dangling inflorescences where hermaphrodite flowers form a ball on which a bat can cling with its hind legs (fig. 16.6). At the top of this ball are a smaller number of sterile flowers that secrete nectar, which accumulates in a groove at the top of the inflorescence. The flowers are initially purple but become rapidly paler and emit a fruity odor.

Gullet Types

These have a bell shape or a flared-trumpet shape, giving a substantial "front view" to the flower, but with the corolla narrowing at the base where the nectar is sited. Stamens and stigma may lie along either the upper or lower surface of the bell, but the anthers always protrude at least slightly, to deposit pollen on the bat's head or throat. Many examples come from genera that also have hummingbird-pollinated species. Some of the flowers are very large, where the bat enters and clings on, but many are quite small and only the face of a hovering bat enters the corolla. Century plants (*Agave*) in arid zones are classic examples, with 30–70 mm deep flared tubular flowers arranged in panicles. In *Agave palmeri* the timing of seed set as a function of time visited (fig. 16.7) reflects the efficacy of bat foraging (Howell and Roth 1981).

The sausage tree (elephant tree) *Kigelia* is a fine example of a large bat-pollinated gullet flower from the African savannas, with pale or dark red fleshy and

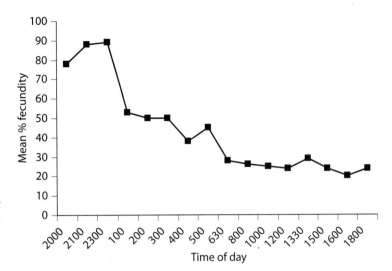

Figure 16.7 The relative seed set in *Agave palmeri* flowers in relation to the time (of night) at which they received a bat visit. (Redrawn from Howell and Roth 1981; note that timings are not regularly spaced.)

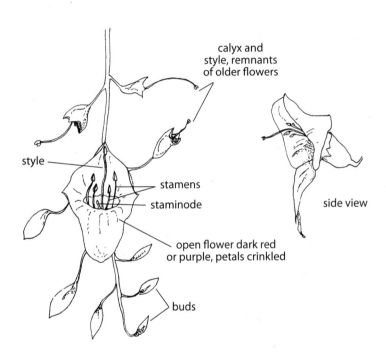

Figure 16.8 Pendant *Kigelia* inflorescence, with a single flower open. (Drawn from photographs.)

unpleasantly sour-smelling flowers (fig. 16.8, plate 28B) hanging on very long stalks. Each flower may be 70–80 mm deep and with a front-on diameter of 120 mm. The lower inside surface of the corolla has a wrinkled surface, presumably enhancing the visiting bat's grip. A similar kind of display—multiple large flowers on a long dangling stem—is seen in *Ochroma*, the balsawood tree, and in some *Mucuna*, which may be white or cream but also occasionally dark red.

However, most other gullet flowers are markedly smaller than these examples, and instead of dangling are held close to a branch for support.

One example familiar outside the tropics is the cup-and-saucer vine *Cobaea*, widely grown as an ornamental scrambling climber in temperate gardens. It has cup-shaped green flowers up to 50 mm across and with a flared rim, turning dull purple later (plate 28A), and with a strange rather sickly-sweet vegetable scent.

Other Designs

The banana genus *Musa* has different pollinators in different continents and for different species, but some Asian bananas are strongly associated with *Macroglossus* bats. The creamy flowers are tubular and downward pointing, aligned in a row beneath a large dark-red bract on which the bat hangs.

Bats are also common pollinators of certain mangrove species. Fruit bats and flying foxes use the trees as roosts as well as for food, and roosts of over 200,000 individuals of *Pteropus* have been recorded in Australian mangrove swamps. The mangrove genus *Sonnertia* in Malaysia is largely pollinated by the long-tongued fruit bat *Macroglossus minimus* that roosts in it, and by the cave fruit bat *Eonycteris spelaea*, whose roosts can be as much as 35–40 km distant (Hogarth 1999). The mangrove flowers last only one night, getting heavily damaged by visiting bats, and they have particularly long protruding stamens that deposit large quantities of pollen on the bats' fur. *M. minimus* seems to have a specialist relationship with this mangrove, and has never been recorded away from mangrove areas in Malaysia, where one species of *Sonnertia* flowers all year around and two others produce flowers in flushes for up to 9 months of each year.

The genus *Passiflora*, with its own unique morphology (fig. 11.1), includes some species such as *P. mucronata*, white and nocturnal and with extra large anthers sited very high above the other floral parts, that are associated with bat visitation (Sazima and Sazima 1978).

Timing

Chiropterophilous flowers typically open at dusk or in the early hours of the night and offer pollen and nectar rewards throughout most of the night. Flowers usually only last one night, during which they can become quite tattered and claw-marked by the bats' activities; shriveling after one night may help deter further visitors and protect the fertilized ovules, since these large flowers would represent a good-sized meal to many potential florivores. Rather few species last 2–3 nights, these often being protandrous (male on the first night, female thereafter; e.g., *Cobaea* and *Agave* species). Very rarely, flowers lasting more than one night continue to secrete nectar by day and become joint bat-and-bird flowers (some American desert cacti species, some *Puya* species in South America, and *Aphelandra acanthus* in Andean cloud forests; Muchhala et al. 2009).

Although each flower commonly lasts only one night, bat-pollinated flowers are usually produced over a long period, often much longer than nonbat flowers in the same genus. *Mucuna holtonii* inflorescences may persist for more than 6 weeks, and *Cleome moritziana* for up to 5 months, where others in these genera commonly last just a few days. This longevity of flowering at the level of the plants presumably exploits the good spatial memory of bats.

Pollen and Nectar Characteristics

Bat plant pollen has been studied by several authors, many finding large grain size to be characteristic (e.g., Chavez 1974; Taylor and Levin 1975; Stroo 2000). Ferguson and colleagues also supported a link between a particular form of net-like ornamentation (verrucate pollen) and bat pollination (Ferguson 1985, 1990; Ferguson and Skvarla 1982; Klitgaard and Ferguson 1992). The last study avoided some of the phylogenetic pitfalls by comparing bird-, moth-, and bat-pollinated species within just two caesalpinioid genera, again concluding that a verrucate surface was associated with bats. However, Dobat and Peikert-Holle (1985) refuted this, and the comparative study by Stroo (2000) came to the same conclusions using 130 species across 50 genera. Stroo found only increased grain size as a significant difference for bat-pollinated flowers, the mean diameter being 72 µm (compared to 64 µm for related but not bat-pollinated species); but he also showed a correlation between pollen size and style length, indicating a simple allometric effect (i.e., bat flowers are inevitably rather large, and so inevitably have longer styles and larger pollen grains). His study sample of bat plants included virtually every type of surface ornamentation found in pollens.

There are commonly very large amounts of pollen per flower, since so many bats will eat quantities of the grains. Bat tongues cannot readily transfer pollen, so pollen grain movement between flowers happens almost entirely via the fur. Bats caught during a trip may be very heavily dusted with pollen externally but normally have little or no pollen in their guts, indicating that those that eat it do so later when grooming; hence they can potentially transfer many pollen grains between plants within one trip. For the plant, providing unusually large pollen amounts is often achieved by having a proportion of purely male flowers among the normal hermaphrodites, as in some *Bauhinia, Cleome,* and *Albizia.* Alternatively, there may be larger numbers of stamens or larger thecae on each anther; and many bats will eat both pollen and anthers. The large amount of pollen may also be related to the large surface area of bats relative to the surface area of stigmas they must subsequently encounter (Heithaus et al. 1974).

From a nutritional perspective, pollen in bat flowers was reported to have higher protein content than normal (Howell 1974), so that captive *Leptonycteris* bats could survive in good health if given only bat flower cactus pollens as their nitrogen source. The night-flowering bat-visited *Agave americana* produces pollen with a remarkable 43% protein content. There have also been suggestions of specific bat-related amino acids in pollen; but the supposedly high levels of tyrosine and proline in *Carnegiea* and certain agaves are probably a reflection of large flower size and the requirements for rapid pollen tube growth, rather than a coevolutionary provision for bat feeders (see the discussion in chapter 7).

Most bat flowers have a very large nectar reward, in keeping with the large body size and high energy demand of the visitors; Voigt et al. (2006) showed that the field metabolic rate of nectarivorous bats was commonly twice that of similar-sized frugivorous species in similar habitats (fig. 16.9A) and suggested that nectar supplies have been selected to push pollinators to the limit and so ensure their multiple visitations. The nectar gathered is almost invariably dilute, recorded with a range of 3%–33%, but most commonly restricted to 15%–17% (von Helversen 1993). It is therefore nonviscous and allows fast uptake by the licking action of the bat tongue, though sometimes it is rather mucilaginous, which may help to deter sphingid moths active at the same times but with a capillary tongue. *Glossophaga* bats feeding on particularly dilute nectars increased their volumetric consumption but still gained less energy overall, and as a result they increased their feeding time and resting time; whereas bats taking their preferred higher concentrations had longer flying times (Ayala-Berdon et al. 2009).

Typical flowers for glossophagine bats secrete a minimum of 0.1 ml nectar each night but can go up to around 2 ml (Voigt et al. 2006), while the flowers fed on by large Old World bats sometimes exceed 20 ml per night (Dobat and Peikert-Holle 1985). In New World bat flowers the nectar is commonly glucose-rich (Baker et al. 1998), although this is not necessarily true of Old World plants, and different bats offered choices showed no preferences between hexoses and sucrose (Rodriguez-Pena et al. 2007).

In a few cases referred to earlier, bat-pollinated flowers offer alternative rewards in the form of sugary bracts or floral tissues that are eaten.

5. The Bat Pollination Syndrome: Chiropterophily

On the basis of the preceding sections, it is rather easy to put together a list of the characteristics that typify chiropterophilous flowers, expanding on the information given in table 11.1.

1. Anthesis, dehiscence, and nectar secretion occurring mainly at dusk and through the night
2. Flowers lasting only one night, but with a long flowering period for the plant
3. Flower usually large, bell or bowl shaped
4. Flowers set well away from foliage, often on a long stalk
5. Corolla walls often thickened, fleshy
6. Dull colors, white for dusk flowers or cream/beige/greenish for night flowers
7. Absence of visual or scent nectar guides (although the petal shape may act as an echolocation guide)
8. Strong fruity or fermenting scents, often with sulfide components
9. Stamens numerous, anthers protruding, plentiful pollen
10. Large to very large volumes of low-concentration nectar
11. Large spatial separation between nectar supply and the anthers when dehiscing and the stigmas when receptive

Bat flowers are therefore relatively easy to distinguish when they are fairly exclusively pollinated by these nocturnal mammals, although a proportion of them have less clear-cut traits when they are also visited by diurnal birds or insects (see below).

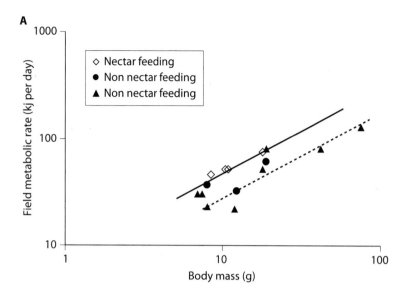

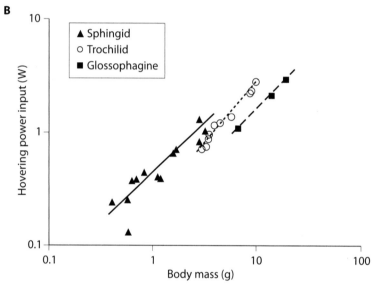

Figure 16.9 (A) Log-log plot of the field metabolic rate of different bats against body mass: the solid line is the regression for glossophagine bats, and the dotted line that for other eutherian mammals (redrawn from Voigt et al. 2006). (B) Hovering expenditure of bats, hummingbirds, and sphingid moths compared, showing the (log-log) regression lines for hovering power input against body mass: the three groups overlap, but bats have much the largest bodies (redrawn from Voigt and Winter 1999).

6. Overview

How specialized is chiropterophily? Bats are relatively recent arrivals in evolutionary terms, appearing in the Eocene around 50 MYA, with the vegetarian bats of both orders probably arising only some 15–25 MYA (Marshall 1983; Dobat and Peikert-Holle 1985). Some of the flower genera that they visit occur naturally in both Old and New World tropics, so are assumed to have evolved well before bats appeared, that is, before the breakup of the southern continent Pangaea, or at least before the separation of Africa and South America. Bats have then learned to exploit these flowers convergently in different continents, resulting in very similar visitation syndromes. And where bat-pollinated plants have been introduced to a new area by man, the local bat fauna has rapidly learned to visit them; examples include *Musa*, *Kigelia*, and *Durio*, which are aliens in the New World, and *Agave*, which has been widely introduced into the Old World. This again seems to indicate a rather specific bat pollination syndrome. The similar use of bat-like smells with volatile components in common may also be indicative of specific coevolution between bats and their flowers.

Many bat-pollinated plants occur in genera where other species are pollinated by birds, or by hawkmoths, or much less commonly by bees. Bird flowers are linked to bat flowers by their size and robust build and their high rewards, while hawkmoth flowers share the nocturnal habit and coloration, so it is quite feasible that many bat-pollinated plants had ancestors from one or other of these sources. In fact, some flowers show a mixed bat-and-bird syndrome; a good example is the cactus *Marginatocereus* from Mexico, with red flowers that open both by day and by night, pollinated diurnally by hummingbirds and nocturnally by bats.

As hummingbirds are rare, and there is strong competition from other synchronously flowering nocturnal bat-visited cacti, this is assumed to be a fail-safe strategy to ensure that some pollination does occur (Dar et al. 2006).

Bats also have some significance as crop pollinators (see chapter 28), pollinating up to 6% of all cultivated crops, and for example being the main visitors to agave in Mexico. They are also key pollinators of durian (*Durio zibethinus*), a commercially important and much-prized fruit crop in Malaysia (Start and Marshall 1976), where the visiting *Eonycteris* bats rely on mangroves for food in the 8–10 months of the year when durians are not in flower.

Chapter 17

POLLINATION BY NONFLYING VERTEBRATES
AND OTHER ODDITIES

Outline

1. Ectotherm Vertebrates Visiting Flowers
2. Pollination by Nonflying Mammals
3. Pollination by Unusual Invertebrates

Birds and bats are well known as pollinators, and each has merited its own chapter, but there are also some rather infrequent instances of pollination by other vertebrates, lacking any kind of true flight but able to access flowers either by climbing and gliding among trees or by seeking pendant flowers close to the ground. Most are mammals, but we deal first with a few cases of pollination from the lower vertebrates.

1. Ectotherm Vertebrates Visiting Flowers

Are the vertebrate ectotherms—fish, amphibians, and reptiles—of any value to flowers as pollinators? Fish are not recorded as flower visitors, for fairly obvious reasons. But they are at least occasionally facilitators of the pollination process for shoreline pond plants, where they prey on animals that compete with or reduce pollinator populations. A clear example was reported for fish consumption of dragonfly larvae, which otherwise as adults preyed on butterflies and bees (Knight. McCoy et al. 2005). Frogs have also not been clearly identified as pollinators, though a few species do take moderate amounts of pollen in their diets when wind-dispersed pollen from pines, oaks, and other trees falls onto pond surfaces and provides a short-term bonanza of protein (although this does not apparently enhance the animals' growth rates). Some tree frogs are regularly found resting within the protective corollas of Neotropical flowers, including large magnoliids, and they may also enter the chambers of certain carrion flowers such as *Amorphophallus*, where they could get the dual advantages of a warm humid site and a supply of flies; in either of these situations, some pollen transfer may occasionally occur.

These are disappointing negatives, but—although disregarded until recently—it is now clear that reptiles, and specifically lizards, do visit flowers and can sometimes be important as pollinators. This is especially the case on islands (Olesen and Valido 2003), perhaps because the reptiles achieve higher densities there due to lower predation risk.

1. Eifler (1995) reported that geckos in New Zealand (specifically in some small off-shore islands) are able to prise open flowers of New Zealand flax (*Phormium tenax*) and suck up the nectar. They also feed from *Metrosideros* flowers, and in both cases they become dusted with pollen on their throat areas. The scales in that site are unusually modified and carry substantial amounts of these pollens, so the relationships do appear to have a degree of coadaptedness.

2. On the Balearic Islands, lacertid lizards feed at and pollinate certain *Euphorbia* flowers (Traveset and Sáez 1997) and may carry a few hundred pollen grains on their bodies. Lizards also visit flowers on some other Mediterranean islands (Perez-Mellado and Casas 1997) and on Madeira (Elvers 1977).

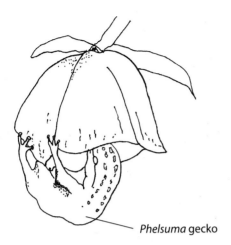

Phelsuma gecko

Figure 17.1 *Trochetia* flower with a feeding *Phelsuma* lizard, from Mauritius. (Drawn from photographs in Hansen et al. 2006.)

3. In Mauritius, the endemic gecko *Phelsuma cepediana* is a fairly generalist flower visitor (Nyhagen et al. 2001) but it also more specifically pollinates the endemic plant *Trochetia blackburniana* (fig. 17.1) while feeding on its nectar (Hansen et al. 2006; Hansen, Kiesbuey et al. 2007). This interaction is especially unusual in that the flower has brightly colored nectar (see chapter 8) which was an unsolved mystery for many years until Hansen et al. (2006) demonstrated that this served as a specific attractive visual signal to lizards.

4. Lizards are one of several effective pollinators of *Erythrina velutina* on an island off the coast of Brazil (Sazima et al. 2009); other pollinators are mainly perching birds.

5. Lizards also show some potentially pollinating interrelationships with a few of the carrion flowers discussed in earlier chapters. On Corsica particular species are regular visitors to the dead-horse arum, *Helicodiceros muscivorus,* where they bask on the warm dark red thermogenic tissue and feed on incoming flies, even inserting their heads fully into the chamber in pursuit of the flies, thus emerging with significant pollen on their snouts.

6. Perhaps even more surprising are the findings of Olsson et al. (2000) that lizards can act as pollinators' "hired help." The snow skink in Tasmania tears open the fused petals of flowers on the honey bush (*Richea*) to feed on nectar, and without this action seed set (via abiotic means or biotic pollination from insects that then have access) cannot occur.

2. Pollination by Nonflying Mammals

Although it has long been known that a range of nonflying mammals will visit flowers and perhaps pollinate them (e.g., Sussman and Raven 1978; Goldingay et al. 1991), it is only fairly recently that genuine instances of pollination by such animals have been properly documented, as reviewed by Carthew and Goldingay (1997). The term *eutherophily* has been employed by some, but is not entirely helpful since bats are also eutherians.

It is now clear that there are examples of mammalian pollination in all of the southern continents (Australia, Africa, and South America; see table 17.1), involving mammals as various as marsupials, rodents, shrews, lorises, monkeys, and lemurs (table 17.2). There are well-documented examples of pollination by 60 or more species of mammal across 20 families, with at least 17 species of marsupial well studied and 28 species of primate somewhat less well known. More of the nonmarsupials probably occur in Asia, but their nocturnal habits and the lack of studies there leave them under-represented in the literature; new examples emerge regularly, for example, the observations that in southeast Asian forests squirrels can also be effective pollinators (Yumoto et al. 1999). There are also less convincing records of pollination by large terrestrial grazing mammals in Africa, which will need some critical attention (see below).

Flowers that receive the attention of the varying taxa of nonflying mammals occur in at least 85 species, from 19 families (table 17.3) and are often clearly related to plants pollinated by bats or by birds, sharing some of the same characteristics such as large size and robust build. Proteaceae and Myrtaceae are the best represented families (tables 17.2 and 17.3). A proportion of the plants are in fact also pollinated by either birds or bats as well as by their nonflying visitors. Thus species within bat flower genera such as *Ochroma, Mabea,* and *Adansonia* are visited by typically small arboreal mammals such as marsupial opossums (Steiner 1981), or the small primates (lorises such as bushbabies, and occasional monkeys), although these can also

TABLE 17.1
Nonflying Mammal Pollinators: Habitats and Plant Families across Different Continents

Land mass	Habitat type	Main pollinators	Main families
South America	Rainforest	Monkeys	Bombacaceae
South Africa	Mediterranean fynbos	Rodents	Proteaceae
Madagascar	Wet and dry forest	Lemurs	Various
Australasia	Mediterranean and forest	Marsupials	Proteaceae, Myrtaceae

TABLE 17.2
Marsupial and Placental Mammal Pollinators

	Family	Genus	Common name	Body mass (g)
Australia				
Marsupials	Acrobatidae	Acrobates	Feathertail glider	10–15
	Burramyidae	Cercartetus	Pygmy possum	15–30
	Dasyuridae	Antechinus	Dibbler, antechinus	20–60
	Petauridae	Petaurus	Sugar glider	90–140
	Tarsipididae	Tarsipes	Honey possum	7–18
Placentals	Muridae	Mus	Mouse	90–130
		Rattus	Rat	
South and Central Africa				
Marsupials	Didelphidae	Didelphus	Opossum	up to 5 kg
Placentals	Muridae	Aethomys	Namaqua rock mouse	
		Praomys	Verreaux's mouse	
		Mus	Mouse	
		Rhabdomys	Striped field mouse	
	Gliridae	Graphiurus	African dormouse	18–30
	Soricidae	Crocidura	Musk shrew	
	Macroscelididae	Elephantulus	Elephant shrew	
	Primates		Monkeys, lorises, galagos, lemurs	up to 15 kg
	Carnivores		Viverrids	
South and Central America				
Placentals	Primates		Monkeys	
	Carnivores		Procyonids	
	Rodents			

be entirely destructive of flowers (Riba-Hernandez and Stoner 2005). Occasionally there are even visits by small carnivores such as genets (Viverridae) and kinkajous (Procyonidae), and there are good records of *Ochroma* pollination by a coati (Procyonidae, *Nasua*; Mora et al. 1999). Likewise, flowers classically associated with perching birds, such as *Telopea* (Goldingay and Whelan 1993) and *Combretum* in Africa, are also visited by small possums, and ornithophilous flowering trees may provide favorite foods for monkeys (plate 28H). On the other hand there are flowers that appear to be predominantly visited by the nonflying

TABLE 17.3
Flowers Regularly Visited by Nonflying Mammals

TABLE 17.3
Flowers Regularly Visited by Nonflying Mammals

Family	Genera	Species	Species of mammals recorded
Proteaceae	8	28	14
Myrtaceae	9	27	9
Bombacaceae	5	9	14
Capparidaceae	4	5	4
Leguminosae	2	2	6
Euphorbiaceae	2	2	5

Plus single species from 13 other families; see Carthew and Goldingay 1997.

animals but that also get a few visits from bats or birds. Thus within some Australian genera there are species mainly visited by birds, other visited mainly by small mammals, and others again that appear to be serviced almost equally by both groups; *Protea*, *Banksia*, and *Dryandra* serve as good examples (e.g., Paton and Turner 1985; Hackett and Goldingay 2001). Some of the differences between species of genera with mammals and other groups as pollinators are summarized in table 17.4. Two examples of mammal flower visitation are shown in plate 28, F and G.

It requires careful study to detect which of the visitors can be effective as pollinators and then to determine which are the most effective and genuinely important; Carthew and Goldingay (1997) gave a good summary of evidence. The balance of effectiveness of different pollinators may itself vary through time and space. Hence defining just where the syndrome of pollination by nonflying mammals begins is somewhat difficult, although the availability of closely related species with differing characteristics and different pollinators does provide useful comparative material.

Characteristics of Mammal-Pollinated Flowers

It is probably necessary to divide these into two morphological categories, with different kinds of visitor:

1. Ground-level (*geoflorous*) flowers, visited by rodents, or marsupial mice and rats. Flowers are either large and bell shaped, or smaller but borne in large clustered inflorescences, protruding sideways or downward from the foliage so that they are inconspicuous or concealed from above, and carried on short stout stems. There may be a guard of strong or even stinging hairs. Generally the corollas are dull in color, often green, brown, or dull purple, sometimes white. Nectar is profuse, accumulating in some part of the flower (often in troughs), occasionally even overflowing onto the ground below. The style is strong and protruding and sometimes hooked and wiry (at least in Proteaceae) to facilitate pollen pickup from fur (Carpenter 1978a). Anthesis is generally crepuscular or nocturnal, with flowers open over 1–2 nights. One unusual but consistent finding is the occurrence of the pentose sugar xylose in some species of rodent-pollinated Proteaceae, a sugar disliked by insects and birds but apparently favored by flower-visiting mice (Jackson and Nicolson 2002).

2. Arboreal flowers, visited by monkeys, opossums, lemurs, etc. Conspicuous upright blossoms are borne on clusters with short or no stalks, positioned a little below the tips of branches but clear of the leaves. A tough perianth of fused petals or sepals forms a cup for nectar which can be lapped up, requiring tongue or snout to be pushed among the brush of protruding stamens. Flowers often open *en masse*, giving a large display and large reward, so that even if some are damaged (especially by monkeys) there are enough to set seed. Colors vary from creams and greens (nocturnally visited) to bright reds especially where diurnal monkeys are active.

These two groups also share some common features. The nectar is of high volume and dilute or only moderately concentrated, dominated by sucrose, and occasionally also having a high alcohol content resulting from fermentation (making it attractive to some rodents and primates where alcohol tolerance can be high; Wiens et al. 2008). The pollen is abundant and readily released by slight disturbance to the flower; and the flowers are commonly relatively lightly scented, with yeasty or slightly musky odors.

Flowers with these suites of characteristics occur in rather different habitats in different continents as shown in table 17.1. In South America they are particularly associated with tropical rain forests (arboreal types, visited by monkeys; e.g., various Bombacaceae) and with higher-altitude cloud forests (shorter but dense vegetation, and high winds, where other pollinators are rare; e.g., *Blakea*; Lumer and Schoer 1986). In South Africa, mammal pollination is mainly recorded in the fynbos flora, a low scrubby Mediterranean-type region with hot dry summers, where ground-dwelling rodents are abundant and can readily pollinate low-growing Proteaceae as well as some bulbous plants, and where primates such as baboons may use floral nectars in aloes as a valuable water resource in the parched summer (Nicolson 2007). The rodent-pollinated *Protea* species are typically bushy plants having concealed dull flowers close to the ground; the concealment of flowers may help reduce nectar robbery by birds or insects, while also providing feeding places where the rodents are hidden from predatory birds. Johnson et al. (2001) documented pollination by four species of rodent in a *Massonia* lily (plate 28F), which they investigated largely because of its convergent resemblance to unrelated *Protea* species. In Australia, there are again *Protea* species of this habit, together with many *Banksia* and *Dryandra* species from the same family with bushy growth and low concealed flowers.

Characteristics of the Flower Visitors

For most of the small mammals that can pollinate flowers, nectar is only a part of their diet, sometimes available only for part of the year, so that the animals show little or no specialist adaptation to the flowers that they utilize. Most of them regularly take seeds, leaves, fruits, or insects as well as nectar, plus some pollen and flower tissues. For example, tamarin monkeys (*Saguinus*) in Brazil were shown to take nectar extensively in the dry months of July and August, concentrating on the flowers of *Symphonia* for about 25% of their feeding time, and apparently trap-lining between the trees of this species (on average more than 100 m apart); but they destroyed most of the flowers they fed at (Garber 1988) by lifting the corolla off the receptacle before licking the petals and discarding the bloom. Small arboreal New World opossums do show great agility in accessing flowers, aided by a long prehensile tail, but the same characters are equally required for feeding on fruits in the same localities. Few if any of these animals show any special characteristics of the mouth or tongue, and most cannot extend the tongue significantly beyond the snout tip.

The only exception to this lack of specialism may

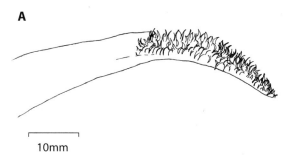

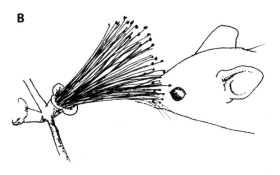

Figure 17.2 Tongue of honey possum (*Tarsipes rostratus*), and the approximate position of head and snout when feeding from a bottlebrush flower. (Drawn from photographs.)

occur in Australia, where some of the flower visitors (mainly but not solely marsupials) show specific morphological traits that are linked to their flower-visiting habits (Armstrong 1979; Turner 1982). The genera most commonly recorded as pollinators are shown in table 17.2; most of them visit the brush-blossom flowers of Proteaceae and Myrtaceae, which are available all year round in the warmer parts of the continent. Many have nectar and pollen as the largest part of their diets, with relatively small contributions from insects. Sugar gliders and squirrel gliders (*Petaurus*) take nectar and pollen predominantly in many habitats, but larger amounts of **homopteran** exudates (honeydew) in low-diversity marginal zones (Holland et al. 2007). The honey possum (*Tarsipes rostratus*) is the best known of all these Australian mammals (fig. 17.2), being an obligate nectarivore, smaller than most of its kin (and smaller than most of the flowers that it visits), and with an extended narrow snout that can probe into flowers for nectar. It has much reduced dentition, the small teeth helping to guide the tongue, which is long and brush tipped. The teeth also work together with ridges on the palate to scrape pollen and nectar off the tongue as it works in and out of the mouth. The honey possum is mainly nocturnal, but will feed on flowers by day in cool cloudy conditions. In common with the other marsupials listed in table 17.2, it has a modified gut morphology that is suited to digestion of large quantities of pollen. It also has an unusually low maintenance nitrogen requirement compared with other marsupials (Bradshaw and Bradshaw 2001), in keeping with the relatively poor nitrogen provision from pollen. It is a very common pollinator of Proteaceae, and in one area 97% of individuals captured had *Banksia nutans* pollen on their fur (Wooller and Wooller 2003).

Evidence for Effective Pollination

For Australian *Banksia* species that were accessible to varying kinds of visitors and that had visits throughout the 24 h cycle, Carpenter (1978b) provided evidence that both birds and rats caught near the plants carried significant pollen loads, and that pollen deposition was more effective onto the snouts of the rats than onto the (longer) bills of birds. In the same habitats, and again using *Banksia* species that attracted both kinds of visitor, Goldingay et al. (1991) found similar levels of seed set resulting from nocturnal pollination (by mammals) and diurnal pollination (predominantly by birds). They further showed that both the sugar glider and the antechinus (see table 17.2) could carry very substantial pollen loads after flower visits. Carthew (1993, 1994) went on to show that the behavior of these marsupials on the flowers was much more appropriate for effecting pollination than that of the eastern spinebill, one of the commonest avian visitors; and that the marsupials made at least as many trips between flowers as did the birds, often making large interplant movements using their terrestrial trails between the flowering *Banksia* plants.

Rodent visits to *Protea* species in South Africa are harder to observe as the flowers are hidden, but records from specimens caught in traps or photographically indicated substantial pollen carriage on snouts (Wiens et al. 1983). These authors also calculated that the nectar from the long-lasting *Protea* inflorescences would support all the local small mammal pollinators for many days. Hence when flower heads were bagged to exclude rodent visitation (but not insects) their seed set was reduced by 50% to 95% in different species.

Cocucci and Sersic (1998) found good evidence for a rare case of rodent pollination in South American *Cajophora* (Loasaceae) from pollen loads on the whiskers of mice, from pollen remains in their feces, and from their footprints on smoked plates set beneath the flowers.

The various genera of Bombacaceae in South America that are commonly visited by monkeys and opossums show mass flowering, and if the animals damage or destroy many flowers (as monkeys especially are prone to do) they may still give a net benefit to the plant in pollen transfer and seed set. Gribel (1988) observed the marsupial *Caluromys* visiting *Pseudobombax* flowers in Brazil, their faces fully inserted among the anthers to lap up nectar from the cup-like receptacle, spending around 3 minutes per flower and often moving between trees. These marsupials would sometimes use their feet to open up flowers that were slightly predehiscent. They have also been observed apparently making effective flower visits on species of *Inga, Ravenala, Ochroma,* and *Kigelia,* all of which have more chiropterophilous traits; but Gribel believed that *Pseudobombax* was more specifically a nonflying mammal plant.

In Madagascar, where there are no monkeys or possums, a particular phenomenon of lemur pollination has been recorded, for several different flower types. Lemurs can be observed eating some flowers whole and taking nectar from various trumpet-shaped flowers, probably effecting little pollination. But they are not always so damaging, since two species of *Eulemur* have quite different behaviors on the same flowers, one eating destructively and the other drinking nectar and potentially moving pollen (Overdorff 1992). Lemurs also visit some bat-pollinated plants (baobab and kapok trees) fairly nondestructively and may have some pollinating effect. But lemurs are particularly and specifically associated with pollination in at least two genera. In the legume *Strongylodon* (Nilsson et al. 1993) they can trigger the pollen removal mechanism (which is not the case for birds or insects). In *Ravenala madagascariensis* (Strelitziaceae; Kress et al. 1994) the largest species (the black-ruffed lemurs, *Varecia*) make regular visits, pulling apart the bracts and inserting their muzzles and then carrying pollen on their fur

TABLE 17.4
Comparative Characters within Genera of Plants Pollinated by Nonflying Mammals and by Other Pollinators

Australia, *Banksia*	*Bird-pollinated spp.*	*Possum/rat-pollinated spp.*
Flower color	Red, orange, yellow	Dull, brownish
Flower location	Exposed, high	Concealed, low
Scent	Scentless	Faint to strong
Flower opening	Diurnal, usually am	Nocturnal
Nectar quantity	Medium to high	Very high
Brazil, *Pseudobombax*	*Bat-pollinated spp.*	*Opossum/monkey-pollinated spp.*
Color	White, pale green	Various
Location	Branch tips on long stalks, often pendant	Conspicuous, upright, short stalk, not quite at branch tips
Scent	Strong	Faint
Flower opening	Few per night	Many simultaneously
Costa Rica, *Blakea*	*Insect-pollinated spp.*	*Rodent-pollinated spp.*
Flower shape	Wide open	Narrow bells
Color	White or Pink	Dull green/purple
Flower opening	Diurnal	Evening/night
Scent	Strong, sweet	None or faint
Nectar	None	Plentiful at night
South Africa, *Protea*	*Bird-pollinated spp.*	*Rodent-pollinated spp.*
Inflorescence shape	Cone shaped	Wider and flatter
Inflorescence position	Conspicuous, upright	Concealed, low, point up or down
Color	Red, bright colours	Dull brown/white
Flower opening	Diurnal	Evening/night
Nectar	Abundant, scentless	Superabundant, yeasty smell

between trees. This is perhaps the largest pollinator of all (Buchmann and Nabhan 1996).

There are yet other mammals that might be pollinators. General suggestions that large mammals such as giraffes may pollinate acacias in the African savannas have been made on occasions, and this was specifically reported by du Toit (1990). Granted that these animals do get substantial deposition of pollen on their faces when feeding on the lower and middle tree, where they eat 85% of all flowers within reach, it still seems most unlikely that they have a real benefit to the plant. Work by Fleming et al. (2006) showed that there was no greater seed set in parts of *Acacia nigrescens* trees that were grazed by giraffes than in other inaccessible

parts. The substantial florivory by giraffes is detrimental rather than favorable for the trees.

Birds, Bats, and Other Mammals: What Suits Who?

Table 17.4 gives four examples from different localities where comparisons can be made within one genus, to get some indication of the features that can reasonably be supposed to particularly suit nonflying mammals as pollinators. The fact that flowers from the same genus as bird- and bat-pollinated plants (and even insect-pollinated plants) can acquire traits that converge on a flower form and temporal patterning that match the behavioral abilities and preferences of small nonflying mammals could reasonably be added to the kinds of evidence for their effectiveness as pollinators listed in the previous section.

3. Pollination by Unusual Invertebrates

For completion, and for want of any other place to deal with them, this is a good moment to address the real oddities of pollination that are rare but still instructive. All the other examples of invertebrate pollination that we have covered have concerned insects, usually flying but occasionally nonflying (notably ants), but there are also occasional reports of pollination by noninsect invertebrates.

Pollination by snails—*malacophily*—is one possibility, largely ignored but recently supported by specific observations. Sarma et al. (2007) described the snail *Lamellaxis gracile* as a specific visitor to the prostrate morning glory *Volvulopsis nummularium* (fig. 17.3). As is typical of the taxon, the flowers opened in the morning and lasted only a few hours, and they were visited by both snails and honeybees (*Apis cerana*). But bees were not active on rainy days, when snails were demonstrably the sole pollinators. There are other anecdotal accounts of snails and slugs that crawl over flowers effecting some pollination, the pollen grains adhering to their mucus, but there must be some possibility of resultant nonviability of the pollen, which has not really been investigated.

It is also possible to imagine successful pollination by other terrestrial noninsect invertebrates such as woodlice, land crabs, or millipedes, but the only group among this assemblage that regularly do visit flowers are certain spiders. There are some species that may

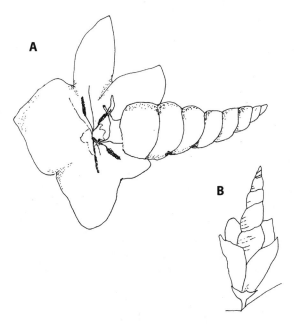

Figure 17.3 *Volvulopsis* flower with a visiting snail: (A) at an open flower, and (B) vertical with head deep into a partly opened flower on a rainy day. (Drawn from photographs in Sarma et al. 2007.)

imbibe nectar for both sugar and water rewards, and some crab spiders take nectar while waiting on flowers as ambush predators (chapter 24). More unusually, there are jumping spiders (Salticidae) that use nectar as a food (Jackson et al. 2001), and in laboratory tests they preferred 30% sugar solutions to water, so they were apparently utilizing the sugar. Tests with various spiders that do not build webs but wander over plant foliage, specifically on cotton, revealed that around 30% of females had ingested fructose from floral or extrafloral nectaries (Taylor and Pfannenstiel 2008). How far they are effective as pollinators is largely unexplored.

Chapter 18

POLLINATION BY BEES

Outline

1. Feeding Apparatus and Feeding Methods
2. Sensory Systems and Bee Perception of Flowers
3. Social and Sexual Effects in Bees
4. Behavior and Learning in Flower-Visiting Bees
5. Melittophily Types
6. Overview

Bees are very special as flower visitors, because almost uniquely they use both nectar and pollen as foods and rely totally on them for both adult and larval nutrition. Adults eat nectar and usually some pollen as well; larvae eat large quantities of both pollen and nectar (converted into honey). Thus any one bee is collecting not just for her own needs but for the offspring; her own sons and daughters if she is solitary, as most bees are, or the queen's offspring in the less common (but often much more noticeable) social species. The sheer number of visits made to flowers by bees is therefore much greater than for other taxa, and the distances moved (often tens of meters between plants, and sometimes several kilometers on a given foraging trip) are also larger than is the norm. Bee tongues (varying from short to very long in different species) can exploit many different floral designs. Bees are therefore far more varied in their interactions with flowers than any other taxon (Michener 1979, 2000; Roubik 1989). A vast range of flowers can be seen as primarily adapted for attracting bees, and these flowers tend to signal whether they are mainly offering pollen (such flowers are often yellow or white) or nectar (often blue, purple,

or pink). Bees commonly use a wider range of plants for nectar than for pollen and are rarely specialist in their choice of nectar sources, though with respect to pollen they can be strongly oligolectic or occasionally monolectic (chapter 7). This is all greatly oversimplified, of course, but provides a useful starting point for this chapter.

Bees have another important feature, in their common (but not universal) ability to produce and control extra internal metabolic heat (endothermy), so that they can warm up sufficiently to fly even when there is little or no sunlight. This gives them much greater independence of the weather than most insects can achieve; their abilities are matched only by some sphingid hawkmoths and a few beetles. So they can be active around dawn and dusk and often quite deep into the night; and some species can forage at flowers even when there is snow on the ground.

To give a very broad overview, then, the bee flower syndrome or *melittophily* involves flowers that have medium to long corolla tubes, often pendant, usually zygomorphic (i.e., bilaterally symmetrical rather than radial), commonly with a landing platform with complex texture or ridging so that a bee can hang on easily, and often arranged in spiked inflorescences. The flowers usually open in the early morning and offer their main rewards before midday, although a few are particularly rewarding in the evenings. The corollas are often in the blue-pink-purple-white color range, frequently having nectar guides on the petals, but may also be yellow or white. They have sweet typically floral scents, and they offer medium concentrated nectar in medium volumes, sited often quite deeply in the base of the flower. Pollen grains are in the range 15–60 μm, readily picked up on the feathery hairs of bees' bodies.

It is important to stress from the outset, though, that bees (and hence bee flowers) are highly variable. There may be in excess of 40,000 bee species worldwide (over 25,000 so far fully described), coming in highly varied sizes (not much over 1 mm body length in the tiniest *Perdita minima*, to around 50 mm in the giant *Chalicodoma pluto*), with varied tongue lengths, varied endothermic abilities, varied social structures, and extraordinarily varied flower visiting patterns and behaviors (Willmer and Stone 2004). To get a handle on all this diversity it is helpful to sort out some basic bee classification (table 18.1). Bees evolved from within the wasps, and many would argue that all the bees are so closely related to each other that they should be defined as just one family and seen as a small cohort of specialized wasps. However, the bees are sufficiently diverse in relation to their most important functions as flower visitors that it is more helpful to regard them as several families within one superfamily (Apoidea), with just the "higher" bees (including nearly all the social bees) within the family Apidae. Some additional (albeit artificial) subdivisions are useful:

1. The first four families shown in table 18.1 are often grouped together as short-tongued bees, the remaining three as long-tongued bees (see section 1, *Feeding Apparatus and Feeding Methods*).

2. Nearly all of the first six families, and the first two subfamilies of the Apidae, are entirely solitary; the only exceptions are some social halictids (see section 3, *Social and Sexual Effects in Bees*). Within the Apidae, all of the Bombini, Meliponini, and Apini are social, and some members of the Xylocopini and Euglossini also show degrees of sociality.

3. A range of bees from across all families are **cuckoo bees** (also known as *kleptoparasites*, nest parasites, or inquilines), laying their eggs on the provisions gathered by other bees, usually with specific hosts in the same family as themselves. These bees do visit flowers, but predominantly for nectar to fulfil their own adult food needs. The genera that act as cuckoos are asterisked in table 18.1 and are scattered throughout the last four families.

The tongue length, the size, and the strength of bees all affect flower access and thus flower choice. Their degree of communal living or sociality and their varied reproductive strategies also affect foraging behaviors.

Hence we will eventually need to consider several different types of "bee syndrome" to make sense of their interactions with flowers.

1. Feeding Apparatus and Feeding Methods

Mouthparts

Bee tongues essentially combine a lapping capillary tip with a suctorial tube, using a suction pump provided by a muscular chamber (the cibarium) in the front of the head that applies negative pressure to the channel within the tongue. The tongues can be of very varied length, but are of relatively uniform construction (fig. 18.1), differing mainly in the proportions of each section (see Krenn et al. 2005). The terminal **glossa** becomes highly developed compared with other insects, with a bristly tip region where the food is initially licked up via the hairy surface, leading to a rolled-up tubular section with the slit at the rear. Above the glossa lie the *maxillae*, which are modified into a proper tube sealed by tongue-and-groove junctions that permit some longitudinal sliding (Borrell and Krenn 2006). The suctorial action from the cibarium is applied via the maxillae, so drawing the liquid off the glossa. The **galeae** which surround the maxillae are projected forward in some bees (sometimes with the *labial palps* as well) to elongate the effective food channel.

In all bees the glossa is flexible and springy and can be drawn in and out in front of the main food channel to augment the lapping action. In short-tongued bees the glossa forms a terminal **flabellum**, something like a scoop rather than a long tube, being bifid (bilobed) in colletids and a single short probe in most andrenids and halictids (fig. 18.2).

In longer-tongued bees, the relative extension of different parts varies, some (mainly the less advanced bees) showing elongation of the maxillary section and others having a very long glossa and moderate maxillae and galeae (fig. 18.3). As tongue length increases, it is generally observed that sucking increasingly takes over from lapping, and in the euglossine orchid bees, with tongues that can be 30–35 mm long, feeding is probably entirely suctorial.

Apis is inevitably the best known bee, and here the glossa is rather specialized, with concentric rings of

Table 18.1
Families, Subfamilies, and Major Tribes and Genera Within the Superfamily Apoidea (Bees)

Family	Subfamily	Tribe	Genera	Common names
Stenotritidae				
Colletidae	Colletinae		Colletes, Leioproctus	
	Diphaglossinae		Ptiloglossa, Caupolicana	
	Hylaeinae		Hylaeus	Mask bees
Andrenidae	Andreninae		Andrena, Melittoides	
	Panurginae		Panurgus, Perdita, Melitturga	
Halictidae	Rophitinae		Rophites, Dufourea	
	Nomiinae		Nomia, Pseudapis	
	Halictinae		Halictus, Lasioglossum, Evylaeus, Sphecodes*, Augochlora, Agapostemon	Sweat bees
Melittidae	Dasypodainae		Dasypoda, Hesperapis	
	Melittinae		Macropis, Melitta, Rediviva	
Megachilidae	Megachilinae		Lithurgus, Chelostoma, Hoplitis, Heriades, Osmia, Proteriades, Anthidium, Stelis, Coelioxys*, Megachile, (+s.g. Chalicodoma, Creightonella)	Mason bees, Carder bees, Leaf-cutter bees
Apidae	Xylocopinae		Xylocopa, (+s.g Proxylocopa), Ceratina	Carpenter bees
			Allodape, Braunsapis, Exoneura	
	Nomadinae		Nomada*, Epeolus	
	Apinae	Exomalopsini:	Exomalopsis	
		Eucerini:	Eucera, Melissodes, Peponapis, Tetralonia, Martinapis, Gaesischia	
		Anthophorini:	Anthophora, Amegilla	
		Centridini:	Centris	
		Melectini:	Melecta*, Thyreus*	
		Euglossini:	Euglossa, Eulaema, Exaerete, Eufriesea	Orchid bees
		Bombini:	Bombus, (+s.g.Psithyrus*)	Bumblebees
		Meliponini:	Trigona, Scaptotrigona, Leiotrigona, Melipona, Plebeia	Stingless bees, sweat bees
		Apini:	Apis	Honeybees, hive bees

Source: After Michener 2000.

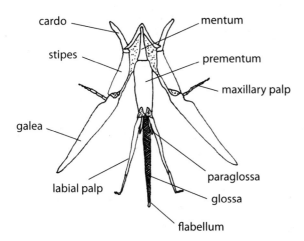

Figure 18.1 Tongue structural components in bees, spread apart. (Modified from Michener 2000.)

strong bristly hairs that can be erected or laid flat as the springy rod within the glossa retracts or extends. When the glossa is fully extended, the hairs are erect and fluid is absorbed between them; when the glossa retracts slightly, the hairs flatten and release the collected fluid into the maxillary food channel. The tip of the glossa also has some lateral movement and can be seen probing around when a bee is active in an open-bowl flower. Similar mechanisms operate in other long-tongued genera such as *Bombus* and *Anthophora*, although in both these the galeae are somewhat longer and the hairs of the glossa are rather flattened. In megachilid bees (*Megachile, Osmia*, and their various cuckoo associates) the galeae are longer again, sheathing the glossa for much of its length, perhaps appropriate for bees that regularly feed on leguminous plants with very restricted openings where the fragile glossa may need more protection to function properly.

It is evident from this discussion that bee tongues will vary greatly in overall length (table 18.2), and not just allometrically in relation to bee size (so that small bees may have long tongues and vice versa). Tongue length does have a strong phylogenetic component, with lengths largely following family lines, but there are cases where bees from long-tongued lineages have developed secondarily shorter tongues (e.g., in an anthophorine *Ancyloscelis* species; Alves-dos-Santos and Wittmann 1999) and a concomitant shallow flower preference.

There are complex relationships between tongue length and flower choice. In general, if a bee is seeking nectar from flowers it should preferentially visit species with corolla depths that approximately match its own tongue length (fig. 18.4A). Short-tongued bees (especially Colletidae, Halictidae, and Andrenidae) can take nectar from open shallow corollas, or from longer corollas at certain times of day when they are very full (although small bees may crawl right into the

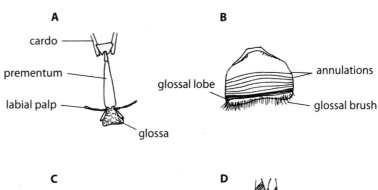

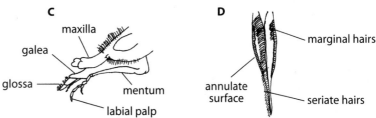

Figure 18.2 Variants of structure in shorter-tongued bees, with reduced glossae: (A) the standard type of structure, (B) the bifid and partially annulate glossa of a colletid, and (C) the mouthparts of same in side view; (D) somewhat longer annulate glossa of a halictid. (Modified from Michener 2000, and other sources.)

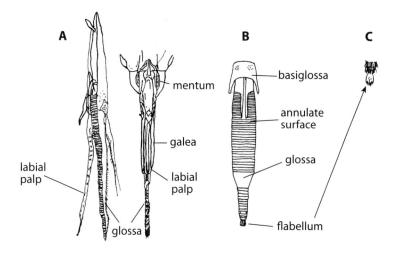

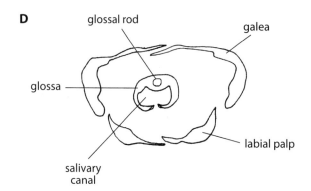

Figure 18.3 Longer-tongued bee mouthpart structures: (A) two different examples in resting position, (B) detail of the glossa and (C) of the flabellum at its tip; (D) the mouthparts in transverse slice in their extended position for feeding. (Modified from Michener 2000 and other sources.)

corolla tube, or may sometimes also visit illegitimately at long corollas, as discussed later). Long-tongued bees (especially most Megachilidae and Apidae) could theoretically access a wider range of flowers (Ranta and Lundberg 1980) but may operate best with fairly deep corollas (Inouye 1980a; Pleasants 1983), where they can harvest even the last dregs available, so gaining a reward over a much longer daily period than a shorter-tongued visitor. On an open flower with exposed nectaries, not only are their tongues unwieldy but also nectar in such flowers often becomes too concentrated (see chapter 8) and cannot be sucked into a long tubular tongue. Hence tongue length shows a clear interaction with time spent per flower (fig. 18.4B); medium-tongued species take longer than long-tongued species on a long-corolla flower (and

Ranta and Lundberg (1980) found that time per flower was in fact a U-shaped relation with tongue length for medium bees), but short-tongued species work faster than any others on open short-corolla flowers. Note that all bees would appear to do better on short-corolla flowers if time alone were considered, but of course longer-corolla flowers commonly have much larger nectar rewards, so that it is more than worthwhile to spend longer visiting them when tongue reach permits. Pleasants (1983) summarized this situation neatly (and see table 18.3). The situation is asymmetric and therefore subject to local conditions and bee abundances, because avoidance of longer corollas by short-tongued bees is probably mainly due to an inability to reach the nectar, whereas avoidance of shorter corollas by longer-tongued bees is substantially due to the actions

TABLE 18.2
Bee Tongue Lengths

Family	Genus and species	Body length (mm)	Tongue length (mm)
Stenotritidae			
Colletidae	Colletes	6–12	2–4
	Hylaeus	5–7	1–2
Andrenidae	Andrena	7–15	2–4
	Panurgus		
Halictidae	Halictus, Lasioglossum	5–12	2–6
	Sphecodes	6–8	2–3
Megachilidae	Megachile	6–9	9–12
	Osmia	6–11	7–9
	Coelioxys	12	4
Apidae	Nomada	9–12	4–5
	Anthophora, Amegilla	9–15	8–21
	Euglossa, Eulaema	8–30	15–35
	Apis (worker)	10–14	5–8
	Bombus	9–22	5–11
	Short-tongued	Worker	5–6
		Queen	8–9
	Medium-tongued	Worker	7–8
		Queen	10–11
	Long-tongued	Worker	9–10
		Queen	12–13
	Very long-tongued	Worker	14–16

of short-tongued bees (reducing nectar volumes below profitability, and also thereby speeding up the evaporative concentration of the remaining nectar into sticky, high-viscosity fluids that cannot be sucked into a long tongue). Figure 18.5 shows a model of length of tongue and the limits set.

From this discussion, it is clear that tongue length is strongly adaptive in relation to flower usage in bees. A particularly neat illustration of this comes from *Clematis stans* where the calyx tube determines access to nectar but changes in length as the flower ages (Dohzono et al. 2004). In young flowers the calyx is longer, and a long-tongued *Bombus* species is the main visitor, but in older flowers the calyx tube has shortened and a different *Bombus* with a short tongue visits; since a single visit cannot fertilize all the flower ovules,

multiple visits by two different bees, each of which has a specialist relation with the plant, is probably beneficial for seed set.

Some constraints on bee mouthpart structures are imposed because those same mouthparts must also function in nest construction for most bees (except the cuckoo bees). The tongue is used to smooth and consolidate the cell walls and to plaster saliva and other secretions onto these walls, helping to keep out rain and prevent fungal growth.

Gathering Nectar and Making Honey

Nectar is swallowed into the bee crop, a relatively large storage area lying in the abdomen. The crop permits

A

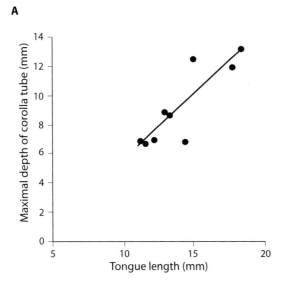

B

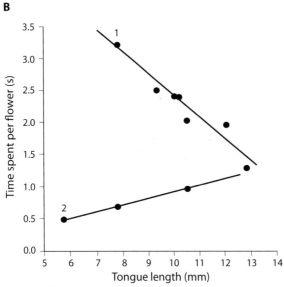

Figure 18.4 Bee tongue lengths in relation to (A) corolla length for queen bumblebees (redrawn from Ranta and Lundberg 1980) and (B) time spent per flower on long-corolla *Delphinium barbeyi* (1) and short-corolla *Aster bigelovii* (2), both for bumblebees (redrawn from Pleasants 1983).

fructose; see chapter 8). Most of the crop's nectar contents are not passed on to the absorptive parts of the gut to become fuel for the gatherer but instead are carried back to the nest site and regurgitated. In the nest this fluid gradually thickens as water evaporates from it, and may undergo further inversion, until eventually at around 82% concentration it qualifies as "honey." A very rough average of the concentration of nectar collected by social bees in temperate habitats might be around 40%–45%, so evidently a good deal of evaporation has occurred. In social colonies the nectar may be taken up and regurgitated several times by different bees until it is concentrated enough to be stored in the comb. In some bees and at some seasons it may also be mixed with substantial amounts of collected honeydew, the sugary excreta of homopterans such as aphids.

In social bees a large volume of honey can accumulate in a healthy nest, and honey from *Apis* (almost worldwide) and from stingless bees (in the tropics) is regularly gathered by other animals, including honey badgers, bears, and of course humans. It can be estimated (from figures given by Heinrich 1975b) that 1 kg of honey represents 35,000–50,000 foraging trips, and 13–22 million individual flower visits, which would represent up to 20,000 bee working hours.

Nectar preferences of bees have frequently been studied, although the conclusions of some of this work are suspect for reasons discussed in chapter 8. Frankie et al. (1983) reported nectar values for many species visited by larger bees in tropical dry forest, with average values of 24% and above but most commonly in the range 28%–42%. For stingless bees (*Melipona*), Roubik and Buchmann (1984) reported foraging on all concentrations from 20% to 75%, with a maximum imbibing rate at about 45% and maximum caloric intake at about 60%. In a later study on a variety of tropical bees, Roubik et al. (1995) recorded optimal nectars as being in the range 35%–65% overall, with means of 38% for euglossines, 44% for meliponines, and 48% for centridines. Only a minority of bees would take nectar at 10%–15%, and very few would take it at 65%–70%.

However, a cautionary note was added by Biesmeijer et al. (1999), who pointed out that apparently different nectar preferences in closely related stingless bees could in fact arise from thermally determined preferences for sunlit or shady sites in which nectar is inevitably differentially concentrated (see below and discussions in chapter 10).

storage of enough supplies to feed the young and by trophallaxis to feed nestmates, and (since it is impermeable) it also buffers the foraging animal from a sudden sugar shock. While in the crop nectar is usually inverted by the action of the enzyme invertase (i.e., any sucrose present is converted to glucose and

TABLE 18.3
Tongue Length and Corolla Choices

| Bee tongue length | Flower corolla length | | |
	Short (sc)	Medium (mc)	Long (lc)
Short (st)	**Best option**	Accessible IF not depleted below reach or profitability by mt or lt visitors; forage more slowly than on sc	Excluded, and/or nectar too dilute
Medium (mt)	Accessible IF not depleted below profitability by st visitors; may forage more slowly than on mc	**Best option**	Accessible IF not depleted below reach by lt visitors; forage more slowly than on mc
Long (lt)	Accessible but rarely profitable, especially after st or mt visits Nectar often too concentrated	Accessible IF not depleted below profitability by mt visitors Nectar sometimes too concentrated	Not much shared with other bees **Best option**

Source: Largely based on Pleasants 1983.

Gathering Pollen

The basic and probably primitive adaptation for gathering and carrying pollen in bees involves *passive* collection, through the presence of branched hairs on most parts of their bodies, often with particularly long or dense patches in appropriate places; on the legs, on the face, or on the top of the head where favored flowers are nototribic. Even male bees usually have some branched hair and therefore accumulate and transfer significant amounts of pollen while foraging for nectar. Only the **kleptoparasitic** genera are relatively hairless, with their remaining hairs reduced to a more scalelike form.

Passive collection does not always involve simple pollen adhesion to hairs; it also occurs very commonly with pollinia which become stuck by glue-like materials to flower-visiting bees (on their heads or backs from orchids, and often on their feet from milkweeds; in total, at least 13 different sites for pollinia deposition have been recorded). Such pollinia are rarely removed by grooming and are not fed to the larvae.

However, most female bees also use some form of *active* collection, specifically gathering pollen from anthers. Different bees exhibit a range of remarkable structural adaptations (mainly on their legs) to achieve this harvest (reviewed by Thorp 1979, 2000). The process often involves the mouthparts; several genera have modified hairs there for hooking pollen out of tubular flowers (especially Boraginaceae and Primulaceae; Müller 1995b, 1996), and some will bite the tips off anthers to get at deeper-sited pollen grains (Renner 1983). Alternatively, pollen gathering is done by the legs, usually the forelegs, since tarsi are generally endowed with small terminal claws suited to a probing or digging action, and hooked hairs may also be present (Müller 1995b). Thus bees are commonly described as "scrabbling" at the surface of fully open anthers, or "pumping" their forelegs in and out of more enclosed elongate anthers. Scrabbling actions are particularly noticeable where bees gather grass pollens, exposed on long feathery anthers; some bees are particularly adept at this (e.g., Adams et al. 1981; Tchuenguem Fohou et al. 2004) and take a fairly high proportion of this supposedly anemophilous pollen, especially in early mornings when dew is still abundant and powdery pollen clumps together. In a few bees with ventral pollen-collecting hairs, the gathering behavior may require rubbing the lower body surface across a flower or inflorescence with superficial pollen. And,

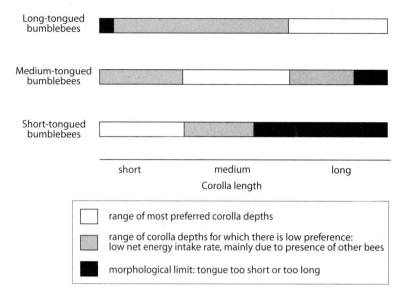

Figure 18.5 Explanatory model for the relation between bee tongue length and the limits on corolla lengths used. (Redrawn from Pleasants 1983.)

finally, many bees use sonication to vibrate pollen out of poricidal anthers (chapter 7). This process occurs in some species from all families but is absent in *Apis*; however, in a few cases honeybees are known to achieve a similar effect by drumming poricidal anthers with their forelegs to release pollen from the pores, as reported for cranberry flowers (*Vaccinium*) by Cane et al. (1993). Figure 18.6 gives a pictorial view of foraging on different flower types for a bumblebee.

We should also remember that bees occasionally gather different fluids such as oils or stigmatic secretions from flowers (see chapter 9), requiring yet other behaviors.

Carrying Pollen

Some of the more primitive bee taxa (within the family Colletidae) simply swallow the gathered pollen into their crops for transport to the nest. Most others collect together the pollen grains that have been both passively gained (grooming it off the branched hairs) and actively collected and then transfer it to a specific storage site. Pollen in the most inaccessible sites on the body ("safe" sites, especially along the dorsal midline; e.g., Kimsey 1984) cannot be groomed; this and a small proportion of the grains on the rest of the body that are missed even during thorough grooming remain available for pollination. In at least a few cases,

some pollen does just get rejected and discarded (see chapter 7).

Once gathered together, the pollen grains may again be carried in the crop (for some very young adults), but most commonly the grains are transferred by grooming to a specific *scopa* on legs, thorax, or ventral abdomen. Thorp (1979) reported that the density and plumosity of the scopal hairs on bees were related to the size and structural features of the pollen grains that the females carried, with more branched hairs when preferred pollens were small; and hairs may also be particularly plumose in oil-collecting bees (Roberts and Vallespir 1978). Where hairs are sparse and relatively nonplumose, larger grains can be accommodated: thus *Leioproctus* species that mainly collect grains of *Conospermum* pollen (large at 80–90 µm diameter) have much less dense scopae than related less specialist species (Houston 1989).

For some female bees, there are substantial interruptions to foraging while they find a resting site where pollen is packaged into the scopa; but many more advanced bees require only a brief hover between flower probings to achieve grooming and pollen packaging. Some bees carry the scopal pollen dry, others moisten it with nectar; in fact there is a general evolutionary trend from carrying pollen in the crop, to collecting it dry on simple scopae, to collecting it in a moistened form in specialized leg **corbiculae (pollen baskets)**, as in the family Apidae.

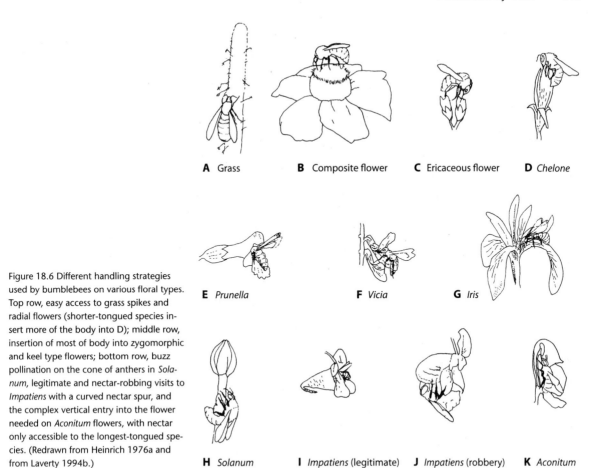

A Grass   **B** Composite flower   **C** Ericaceous flower   **D** *Chelone*

E *Prunella*   **F** *Vicia*   **G** *Iris*

H *Solanum*   **I** *Impatiens* (legitimate)   **J** *Impatiens* (robbery)   **K** *Aconitum*

Figure 18.6 Different handling strategies used by bumblebees on various floral types. Top row, easy access to grass spikes and radial flowers (shorter-tongued species insert more of the body into D); middle row, insertion of most of body into zygomorphic and keel type flowers; bottom row, buzz pollination on the cone of anthers in *Solanum*, legitimate and nectar-robbing visits to *Impatiens* with a curved nectar spur, and the complex vertical entry into the flower needed on *Aconitum* flowers, with nectar only accessible to the longest-tongued species. (Redrawn from Heinrich 1976a and from Laverty 1994b.)

Abdominal Pollen Collectors

These are all in the family Megachilidae, and the scopa is formed from a coating of fine curved and usually unbranched hairs all over the abdominal tergites. The legs have stiff brushes of bristles on the inner metatarsi and combs on the inner sides of the tibiae; all the legs can thus gather up pollen and transfer it back to the hind legs and from there onto the abdominal brush. Some pollen is probably gathered straight onto the abdomen from composite and brush-type flowers with dense superficial anthers; on *Acacia*, for example, small megachilid bees can be seen passing rapidly over the inflorescences with the legs operating a rowing action that propels the bee forward but has little function in pollen transfer, since that is occurring straight from the superficial anthers.

Megachilid bees are therefore readily distinguished from all others when carrying a pollen load because of the dense coating of (usually white or yellow) pollen on their undersides (see plate 29B). They also have somewhat flattened abdomens, and when pollen-laden they tend to fly with their abdomens more "cocked up" than other bees.

Leg Pollen Collectors

Bees with scopae on their legs occur in most of the other families, and again have stiff bristles on the metatarsi and often on other tarsal joints as well. However, the exact position of the scopa varies. Andrenids carry the pollen on diffuse areas of the hind leg surfaces (tibiae, femora, and trochanters; fig. 18.7A) and also on the thorax; but in halictids and colletids most of the pollen is carried in a pollen basket on the underside of the hind femora, and there is little or no pollen on the thorax although there may be some on the underside of the front abdomen. These bees tend to do

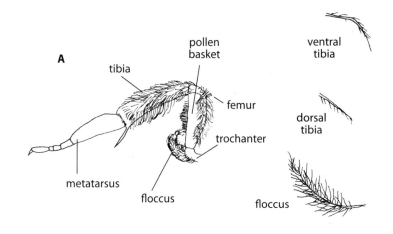

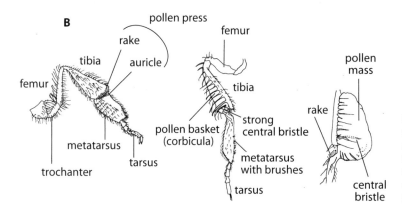

Figure 18.7 Scopae from bees' legs: (A) andrenid hind leg, with pollen brushes and femoral pollen basket, also showing hair structures from three sites; (B) apid corbicula on hind tibia, seen from the back and side, and with a view of the full basket. (Redrawn mainly from Proctor et al. 1996.)

most of their pollen grooming and packaging while on the flower, rather than in flight; each foreleg is rubbed on a middle leg, then each middle leg onto the hind leg, one side at a time. In melittids and in anthophorine bees the pollen collection system is focused on the hind metatarsi (which are enlarged) and the hind tibiae, all of which have elongated pollen-carrying hairs mainly on their outer surfaces. *Dasypoda* represents an extreme case of this arrangement, with exceptionally long hairs present, so that when fully loaded with pollen it appears to be wearing very large cowboy-style trousers (plate 29C).

The social bees *Apis* and *Bombus* also carry pollen on their legs, both gathering the pollen initially by careful grooming onto brushes on the middle metatarsi, where it is moistened with regurgitated nectar. From here it is scraped into the comb-like metatarsal brushes of the hind legs and then into the pollen baskets (corbiculae) on the outer surfaces of the hind tibiae (fig. 18.7B). It is compacted into the corbicula with a unique **pollen press**, present only in the social Apidae, between the top of the metatarsus and the base of the tibia, which forces a dense and sticky pollen mass upward and outward into the basket. The accumulating pollen load in the basket is shaped by the middle tibia into a neat kidney-shaped mass, held on a single central bristle, and its stickiness probably helps the rather sparse peripheral corbicular hairs to keep it in place. Displacement sensors give the bee information on how full its corbiculae are (Ford et al. 1981).

Hair structures vary between these bees, according to their preferred flowers. Some bees that spend much of their time buzz-pollinating have very thin hairs that can retain the small and rather dry pollen grains, whereas the squash and gourd bees in the southwestern United States, primarily visiting cucurbit flowers, have particularly coarse body and leg hair, suited to the large and oily pollen grains of their preferred plants. Bees that habitually gather very large *Convolvulus* pollen grains also have particularly broadly spaced hairs (Thorp 1979).

Unloading Pollen

Most of the pollen deposited passively on a bee's body is normally groomed for carriage into the scopa and so can be unloaded from there as a discrete pollen mass ready for packing into a nest cell. With *Apis*, it is not difficult for humans to extract the pollen pellets from each back leg of an incoming bee, these pellets then being sold in health food shops and the like (they would be better left to the bees!).

The only sites that are difficult for bees to groom are the middle of the back and a small area of the neck just beneath the head; if a plant can land its pollen in such safe areas, that pollen may have a better chance of evading being unloaded at the nest and hence of eventually reaching an appropriate stigma.

Selecting Pollen for Collection

The majority of bees are broadly oligolectic or moderately polylectic, taking pollen opportunistically from several different plants, but often restricting themselves to one family or sometimes one genus. Only rarely are bees narrow pollen specialists, not surprising when most (and especially the social bees) have an active adult life cycle longer than that of any one plant's flowering season. However, the time scale of recordings is a problem: Motten (1986) found that most individuals of andrenid, halictid, and *Osmia* bees that he sampled individually from a spring wildflower community had at least 50% of pollen from just one species on their bodies, with few bees carrying more than two species' pollen on any one trip; but these visitors shared many plant hosts and were rather polylectic when viewed on a daily or weekly scale.

Some more strictly specialist (narrowly oligolectic) bees do occur reasonably frequently though (Wcislo and Cane 1996; Michener 2000; Minckley and Roulston 2006), ranging from 60% oligoleges in Californian deserts (Moldenke 1976) to only 15%–20% in colder habitats such as Sweden (Pekkarinen 1997) and Nova Scotia (Sheffield et al. 2003). Taxonomically the pattern is unclear; some report it as seemingly random,

so that species with narrow oligolecty are found in the same genus, tribe, or subfamily as broad generalists. The clearest demonstration of this to date comes for the tribe Anthidiini, whose 72 western palearctic species were scored by Müller (1996) as 43% narrowly oligolectic, 18% moderately polylectic, and 35% strongly polylectic, with 4% unknown; the group had a strong preference overall for composites, with Fabaceae and Lamiaceae next most favored. However, Minckley and Roulston (2006) suggested that on the basis of modern phylogenies most bee clades are primarily all oligolectic or all polylectic, with the transitions that do occur being roughly equal in either direction. Narrow oligolecty is not necessarily fixed anyway; Williams (2003) found that the apparently specialist forager O*smia californica*, mainly feeding from Asteraceae, in fact retained reasonable flexibility to accept novel pollens, and that the apparent specialism did not extend to the larvae, which grew just as well on either novel or normal pollens. However, others have found poor larval growth on nonhost pollens (see chapter 7). Sedivy et al. (2008) reported a careful phylogenetic analysis of *Chelostoma* bees, finding 33 out of 35 species to be strict pollen specialists, on 8 plant genera with convergent floral features; the two generalist bees were clearly derived from this ancestral state. They suggested that evolution from oligolecty to polylecty was rather common, via a series of stages shown in figure 18.8, involving increasing physiological or sensory adaptation to new hosts.

While most bees are probably not strongly specialist in relation to pollen, they may still exercise a good deal of preference and limit their foraging to a subset of available resources. *Andrena vaga*, for example, feeds only from *Salix* catkins and responds specifically to the volatiles from *Salix* pollen (Dotterl et al. 2005); other cases of this kind undoubtedly await discovery. Even the highly polylectic honeybees seem to prefer some pollens over others (Schmidt 1982), perhaps basing choices on pollen odor or taste (Schmidt 1985), and the preference may be in part related to their hair characteristics and associated pollen sizes as discussed above. Thus, colonies of *Apis* in the Sonoran Desert gathered pollen from 35–55 plant species in a season, which represented about 25% of all available flowering plants (Buchmann and O'Rourke 1988). Many other bees also behave in an oligolectic fashion in practice, tending to show floral constancy on any one day or at least on any one trip when gathering pollen

Hypothetical stages of host-range evolution in bees   Mechanisms

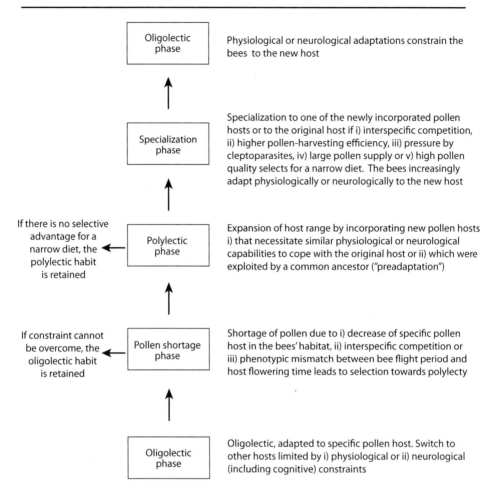

Oligolectic phase — Physiological or neurological adaptations constrain the bees to the new host

Specialization phase — Specialization to one of the newly incorporated pollen hosts or to the original host if i) interspecific competition, ii) higher pollen-harvesting efficiency, iii) pressure by cleptoparasites, iv) large pollen supply or v) high pollen quality selects for a narrow diet. The bees increasingly adapt physiologically or neurologically to the new host

If there is no selective advantage for a narrow diet, the polylectic habit is retained ← Polylectic phase — Expansion of host range by incorporating new pollen hosts i) that necessitate similar physiological or neurological capabilities to cope with the original host or ii) which were exploited by a common ancestor ("preadaptation")

If constraint cannot be overcome, the oligolectic habit is retained ← Pollen shortage phase — Shortage of pollen due to i) decrease of specific pollen host in the bees' habitat, ii) interspecific competition or iii) phenotypic mismatch between bee flight period and host flowering time leads to selection towards polylecty

Oligolectic phase — Oligolectic, adapted to specific pollen host. Switch to other hosts limited by i) physiological or ii) neurological (including cognitive) constraints

Figure 18.8 The constraint hypothesis for the evolution of floral host range, constancy, and specialization in bees. (Redrawn from Sedivy et al. 2008.)

(Westerkamp 1996; Thomson 1996). This is also broadly true in controlled "cafeteria" experiments where bees are offered many choices; for example *Megachile rotundata* visited only 21 species out of 209 offered (Small et al. 1997), those chosen being scattered across 14 genera and 7 families but sharing the key feature of a particular corolla tube length. Analysis of pollen loads carried by bees shows high levels of constancy within a trip (see chapter 11 and table 11.4). Analyses of pollen compositions from larval cells of bees also very commonly show that floral constancy for the day or two that might be required to

provision that cell is rather common, often with records of more than 95% of one pollen type per cell (Westrich and Schmidt 1987; and for examples see Corbet and Willmer 1981 with *Xylocopa*; Willmer 1986 with *Chalicodoma*; Willmer and Stone 1988 with *Creightonella*; Cripps and Rust 1989a,b with *Osmia*; Scott et al. 1993 with *Xylocopa*).

Cane and Sipes (2006) discussed some of the advantages and drawbacks of these various approaches in assessing floral constancy and degrees of "lecty" in bees, as well as presenting a modified terminology to define pollen specialization categories as shown in

Table 18.4
Definitions of Pollen Specialization and Constancy

Monolecty	*Probably a term best not used, implying total reliance on one species*
Narrow oligolecty	Pollen gathered from a single host clade, normally a genus.* Examples from the bees include *Macropis* using only *Lysimachia* (for pollen and for oils); *Andrena hattorfiana* using only *Knautia*; *Hoplitis anthocopoides* using only *Echium*; *Proteriades* using only *Phacelia*; *Evylaeus* using *Oenothera*.
Oligolecty	Pollen gathered from a few plant genera, closely taxonomically related.
Eclectic oligolecty	Pollen gathered from a few plant genera from widely separated clades. An example is *Osmia ribifloris*, using usually only one of *Cercis*, *Berberis* or *Arctostaphylos*, three unrelated genera with varied morphologies.
Mesolecty	Pollen gathered from a few plant genera, which are related but subsampled from one or more large plant families†; less predictable than the previous group. Examples include *Megachile brevis* (6-8 genera of Fabaceae); *Anthidium* (about six genera, from four families).
Polylecty	Pollen gathered from more than three plant families, but not from a high proportion of all available families. Examples include most eusocial and multivoltine bees, by necessity.
Broad polylecty	Pollen gathered from most genera and species, and from numerous families, taking supplies from at least 10% of all available resources. *Apis* is typical, and often samples 20%–40% of the available flora. Most bumblebees and stingless bees also belong here.

Source: Based on Cane and Sipes 2006.

*Use of a genus as the typical clade for definitions is not ideal, some genera attracting oligoleges being rather uniform (e.g., *Salix*, *Helianthus*), others being widely varied in phenotype (e.g., *Penstemon*, *Clarkia*, *Mimulus*)—genera typically used to look at selection for specialization, see chapter 11.

†Likewise a family with thousands of species has to be treated differently in terms of specialization from a small nonspeciose family, and a tribe of a large family such as Asteraceae or Fabaceae might be regarded as equivalent to a smaller family.

table 18.4. Note that many bees, seeking both nectar and pollen as efficiently as possible, may have polylecty in part imposed by nectar accessibility rather than by pollen. This raises an important issue in trying to define "bee flowers": since pollen gathering by bees is a major activity, correlations between tongue length and flower choice could be somewhat confounded, as a bee might spend significant time gathering pollen at flowers where the nectar is unreachable. However, many solitary bees do prefer to gather both nectar and pollen from the same flower and take back to the nest a balance of both resources, given that it is quicker to collect both resources from one flower all other things being equal; therefore a reasonable match of tongues

to flower choice persists. For social bees, though, there may be specialization for either nectar or pollen on a given foraging trip, and pollen-only visits may be to flower types apparently inappropriate to the tongue length. Of course some bees forage regularly on pollen-only plants, where their tongues are an irrelevance; often they then use their sonicating (buzz-pollinating) techniques. Pollen is also often gathered from "unsuitable" flowers, where there is an obvious mismatch between bee and flower morphology, and the process is then essentially a form of robbery with little or no possibility of pollen deposition onto distant stigmas. This typically involves small bees robbing anthers of much larger flowers, for example halictids and stingless bees working the exerted anthers of various hummingbird- or bat-pollinated flowers (plate 33E,F).

Little is known about bees' assessment of pollen quantity or quality, but there is evidence that they can detect quantities of available pollen on at least some plants (Harder 1990a; Rasheed and Harder 1997a; and Robertson, Mountjoy et al. 1999, who also found qualitative effects), either visually or using olfactory cues associated with dehiscence. *Agapostemon* bees could detect pollen levels visually in flowers with exposed anthers (Goulson et al. 2001), while *Bombus* could detect morphological change in *Anemonopsis* flowers as a cue to pollen availability (Pellmyr 1988). For *Alkanna* the anthers are visible only at very close range but nevertheless give adequate visual cues to *Anthophora* females (Stone et al. 1999). In *Acacia* detection may involve both vision and scent (Stone et al. 1998).

In addition to assessment of availability in advance of choosing flowers, bees must also make decisions on continuing to gather particular pollen rewards, and this is presumably based on volume or weight considerations (the degree of packing of the scopa, or perhaps rate of increase of body weight). Some bees have been shown to assess pollen reward per flower (Harder 1990a; Robertson, Mountjoy et al. 1999), perhaps using sensitive setae around the scopa. *Bombus* and *Ptiloglossa* could assess the pollen load they were receiving while sonicating *Solanum* flowers, again presumably mechanically (Buchmann and Cane 1989). Furthermore, Harder (1990a,b) recorded bumblebees making more visits per inflorescence when flowers were pollen-rich and returning to these inflorescences more frequently; and he showed that they abandoned pollen collection on a plant when the rate of gain was too low, even if the anthers were not fully depleted. There was also a good match between timing of visits, handling times, and pollen availability in *Solanum* flowers for three sonicating bee species (Shelly and Villalobos 2000).

How many flowers, and how much time, are required for a full pollen load? For *Apis*, Ribbands (1949) recorded full scopae deriving from 27 flowers of a *Papaver* cultivar, and from 60 to 180 flowers for *Nasturtium*, these requiring a minimum of 3 and a maximum of 18 minutes to complete. *Chalicodoma* bees working *Coffea* flowers required around 17 minutes to fill up with pollen (Willmer and Stone 1989). Obviously these timings are strongly influenced by time of day (relative to dehiscence), by weather conditions, and often by competition from other bees and other flower visitors.

In concluding this section, we should remember that, since bees may forage for nectar, for pollen, or for both at once on any one foraging trip, it is always important to note what they are gathering when assessing their activities. There may be pronounced temporal patterns for social bees, with bumblebees tending to take nectar early in the day and concentrate pollen collection into the drier midday hours (fig. 18.9A). Solitary species may have their flower visits patterned by the need to gather nest-cell building materials, as in fig. 18.9B where a **leaf-cutter bee** showed daily rhythms of leaf gathering and nectar or pollen flights. And even if a bee visits the same plant species for both nectar and pollen, it may be making choices as to which individual plant or flower it visits. For example, on lavender inflorescences, where female-phase flowers have more nectar, honeybees seeking only nectar chose inflorescences with more female flowers (generally the smaller inflorescences) and those seeking both pollen and nectar chose the larger inflorescences with greater numbers of flowers of both sexes (Gonzalez et al. 1995).

2. Sensory Systems and Bee Perception of Flowers

Vision and Color Perception

These sensory attributes have been intensively studied for bees, especially *Apis* and *Bombus*; in fact the compound eyes and visual system of the honeybee are by far the best known of any insect. The eyes have a high flicker fusion frequency, so can distinguish very rapidly moving or flickering objects. They are very sensitive to outlines, preferring complex dissected outlines to simple ones, and hence jagged toothed flowers to simple round ones (see fig. 5.17) and symmetric to asymmetric shapes (fig. 18.10), but they are not particularly good at discriminating shape categories. When tested with model flowers, bees commonly alighted at the jagged edge and then followed converging lines ("petal edges") into the center (Manning 1956a); this and other studies (e.g., Free 1970a) make it seem likely that flower nectar guides only have an effect at rather close range and do not strongly direct a

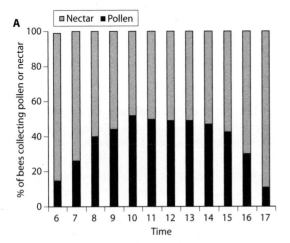

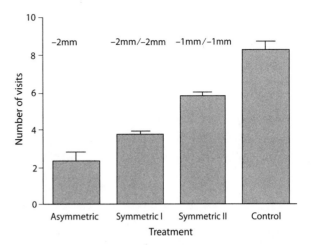

Figure 18.10 Honeybees show preference for larger and more symmetrical shapes; the control was an intact *Epilobium* flower, the treatments involved cutting off the amounts shown from the conspicuous lower petals. (Redrawn from Müller 1995.)

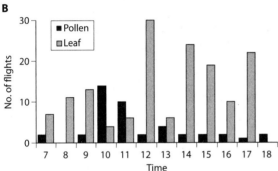

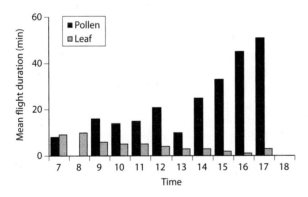

Figure 18.9 (A) Pollen- and nectar-gathering trips against time of day for *Bombus terrestris*, with nectar gathering dominant in the early morning and evening (redrawn from Peat and Goulson 2005); (B) pollen flights and leaf-gathering flights for the leaf-cutter bee *Creightonella* working on coffee flowers, with pollen flights most common and short 1000–1100 and lengthening later as the resource becomes scarcer (redrawn from Willmer and Stone 1988).

bee to the center of the flower as it approaches but only after it has landed. However, Free showed that bees could learn to follow the guides more quickly and directly to the center once they were experienced.

Within the eyes of *Apis* there are three receptor types, which have wavelength sensitivity peaks at 344 nm (UV), 436 nm (mid-blue), and 544 nm (green-yellow), the first of these being much the most numerous type. Other bees tested to date have rather similar spectral sensitivities, so it seems that relative red-blindness and a high ability to detect UV are common features. In the past, the sensitivity to UV led to much analysis of "what flowers look like to a bee," with many publications highlighting the nectar guides revealed in UV light (chapter 5). In the light of the last paragraph, these may not be so critical, at least to less experienced bees. However, known sensitivities must determine the bee perception of overall flower colors (Daumer 1956, 1958), especially for what humans perceive as white flowers. These are white to us because they reflect color in all parts of the spectrum that we can see, but to a bee many of them will appear in the blue-green range, because these flowers usually do *not* reflect UV wavelengths. In fact bees are extremely good at distinguishing between adjacent color tints, especially in this blue-green range (von Helversen 1972). But it follows that most white bee flowers are not white to their visitors—in fact, a true "bee white"

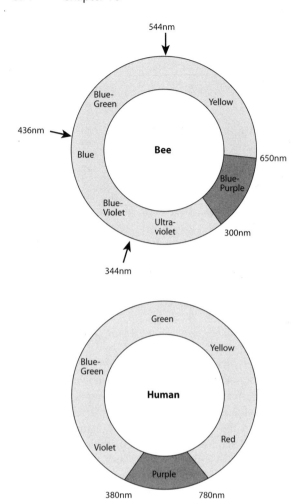

Figure 18.11 Honeybee receptor peaks and the bee color circle compared with that of humans; see also fig. 5.7. (Modified from Barth 1985.)

(mixing all three of the bee's visual wavelengths) is very rare in flowers. For similar reasons, flowers that we see as yellow, or blue, or purple, may appear to a bee as quite different shades (see table 5.3). Figure 18.11 shows the characteristics of the bee receptors and of the color circle for bees as compared with the more familiar one of humans, which helps to explain the results in table 5.3.

In recent years there has been increasing stress on the analysis of flower reflectance as a whole rather than at particular wavelengths (e.g., Kevan 1979) and a realization that what bees see must also take into account the effects of contrast (which for flowers will

usually be with green foliage or soil, both of which have little UV reflectance). Bees above all require this good color contrast, which uses input from *all* their receptor types, so that UV reflectance alone is not a good indicator of flower detectability. Hence honeybees offered blue versus white choices of artificial flowers behaved very differently in terms of interflower movements from bees offered blue versus yellow choices, which to them would have much higher contrast (Hill et al. 2001).

The visual capabilities of bees also depend critically on the light environment at a particular time: Dyer and Chittka (2004) found that bumblebees made errors in color discrimination in differing light conditions, although they could compensate in part for this by recognizing shape variation. Chittka et al. (2003) also showed that, under experimental conditions, bumblebees were remarkably pragmatic about their color preference during foraging tasks, so that flowers of normally preferred colors but offering no reward (or a *quinine penalty*) were very soon avoided in favor of rewarding flowers of slightly abnormal (normally unpreferred) coloration.

In general, it remains fair to say that bees prefer flowers that to us appear pink, purple, or blue, as well as many that appear to us as white; they also select some red flowers (if UV reflectance is also present, as it usually is), and (especially for short-tongued bees) many bright yellow flowers. But they have complicated preferences when faced with real flowers against real backgrounds (cf. Forrest and Thomson 2009). For example, bumblebees are affected more by peripheral than central colors, prefer flowers with at least two colors, and like the central area to be large and essentially pollen colored (Heuschen et al. 2005).

There are subtle differences among bees in their color vision, although again this has really only been explored for *Apis* and *Bombus*. Dyer et al. (2008) showed that bumblebees had poorer color discrimination than honeybees in behavioral tests but could detect certain contrasts at lower visual angles, indicating a trade-off between discrimination and detection, perhaps linked to habitat differences. However, no overall differences in color preferences have been found among a range of bees from different habitat types or from different continents. Perhaps this is in part because bee flowers come in almost all possible colors, and color contributes markedly to bee constancy on a day-to-day basis (see below). Indeed, Menzel and

Shmida (1993) showed that bee-pollinated flowers in Israel (one of the most bee-rich of all ecosystems) covered the entire range of colors that bees can perceive.

One additional aspect of bees' flower-color choice has recently been established. Bumblebees will choose warmer nectar feeders, and warmer flowers, in laboratory tests, and can learn to associate warmth with color, so that they will pick either pink or purple flowers depending on which has been warmed (Dyer et al. 2006). Such choices clearly fit with the aspects of environmental economics in floral biology discussed in chapter 10.

Olfaction and Taste

Bees have a strong olfactory sense, using chemoreceptors on the last eight segments of the antennae. These are on average one or two orders of magnitude more sensitive than human olfactory senses, and bees can clearly detect flowers that to us seem unscented and show preferences for some that to humans are only very slightly fragrant. Many classic bee flowers (hyacinth, primrose, pansy, etc.) have mild honeylike scents, or delicate floral fragrances much less intense than those of moth flowers. Some flowers attractive to bees smell quite musty to humans, for example some hellebores and asters.

Scent is more readily conducive to associative learning with floral reward than any other cue in bees (Menzel and Müller 1996), so that scent learning significantly enhances flower fidelity and thus the chances of correct pollen transfer. Honeybees can be conditioned with honey odors and will generalize to other honeys from their initial training (Bonod et al. 2003). Scents may also induce recall of navigational information and visual memory in honeybees (Reinhard et al. 2004). Furthermore, in social bees scents (von Frisch 1950) and scent associations (Dornhaus and Chittka 1999) may be learnt from nest mates, either from their body surfaces or from the nectar and pollen they carry.

Scent is often used by plants to manipulate bee behavior, especially in the deceptive pseudocopulatory relationships with orchids. The best-known example is *Ophrys sphegodes* with *Andrena nigroaeanea*, studied by Schiestl and his colleagues (Schiestl et al. 1997; Schiestl and Ayasse 2001), involving sexual deception of visiting male bees, which alight on flowers and at-tempt copulation, receiving pollinia in the process (chapter 23 gives details). Essentially, the flower shape and color help in the deception, but the orchid's odor bouquet is the key: it accurately mimics a female bee's pheromone.

The balance of visual and olfactory attraction modes in the relations of bees and flowers has been a subject of controversy for at least a century, many studies using honeybees or bumblebees with artificial flowers and applied scents to determine which sense is more critical (e.g., von Frisch 1954; Manning 1956a, 1957; Free 1970a). Most studies concluded that scent discrimination was important only at quite close range (whether using attractive scents or aversive scents such as clove oil), flower searching and approach behaviors being visual most of the time. Whether this is also true of the more crepuscular bees, searching at low light levels, is untested but perhaps less likely. Some effect of habitat type is also suggested (and is perhaps linked with light levels), since species of North American lousewort (*Pedicularis*) that grow in the open have less strong scents than those in shady wooded areas (Sprague 1962).

3. Social and Sexual Effects in Bees

The great majority of published studies on all aspects of bee behavior and flower visiting have focused on the common and familiar social bees, primarily *Apis* and *Bombus*, but in reality eusociality (despite having evolved separately at least eight times) is the minority condition for bees. This needs exploring in some detail as it has substantial effects on flower-visiting behaviors.

Levels of Sociality

There are various levels of sociality in animals as traditionally defined (e.g., Wilson 1971; Michener 1974, 1985; Hölldobler and Wilson 1990), although definitions differ (e.g., Crespi and Yanega 1995; Lacey and Sherman 2005). The bee superfamily includes representatives at all levels.

Solitary bees are the commonest, each female making and stocking her own nest. Nests are commonly tunnels in the ground, or in hollow stems, or crevices.

A single female chooses a nest site, constructs a nest tunnel, then makes separate cells and stocks each cell with food; she then lays an egg on the food, seals the cell, and starts another one. Thus she is only involved with her own progeny, and then only as eggs (Batra 1984). Females tend to alternate periods of cell construction and periods of flower foraging. In temperate habitats solitary bees have one generation a year, each emerging adult flying for only a few weeks and stocking perhaps 5–10 cells on average. But in warmer climates there may be several broods each year, and some tropical species breed continuously and aseasonally, so that individuals can be encountered foraging at any time of the year. It is very common for nests of one species to be grouped together as an aggregation (especially for soil-nesting forms and some carpenter bees, which are often strongly philopatric), but there is no real behavioral interaction between individual females.

Parasocial bees include those living in communal, quasi-social, and semisocial groups. Here the adult bee population is still a single generation at any one time, and no interaction occurs between a female and her offspring. *Communal* bees share a nest between several females but each makes and stocks her own cells; examples occur quite commonly in halictid and andrenid bees and in some euglossines (Paxton et al. 1999), and while there may be no benefit in increased offspring per capita there can be advantages in reduced nest parasitism (e.g., Soucy et al. 2003). In **quasi-social** and **semisocial bee** nests the females cooperate on nest construction and provisioning, but if quasi-social all adult females are egg layers while if semisocial only one or a few have active ovaries and the rest are worker-like and do not lay eggs, giving the first level of division of labor.

Subsocial bees are those where families of one adult female and several of her offspring, matured at least into larvae, are found in one nest, giving overlap of generations. The larvae are protected and fed by the mother, who then leaves or dies at about the time that they mature into adults themselves. There is no division of labor here, and the subsocial condition occurs only rarely and then often only for a part of the bee lifecycle; often this is for the initial breeding cycle of a more fully social species, before the first generation of offspring mature into workers.

Eusocial bees live in nests with two generations of mature bees simultaneously, a mother and her daughters. Bumblebees and honeybees are the most familiar here, but eusociality also occurs in tropical stingless bees and in some halictid bees. In *primitively eusocial* types, including *Bombus*, mothers and daughters are usually physically similar albeit different in size; a daughter (worker) can normally lay only unfertilized eggs, but occasionally functions as an egg-laying queen if necessary. In some *Lasioglossum* bees there are substantial size differences between queen and worker, but again workers can mate and function as queens if needed (Paxton et al. 2002). In all these primitively eusocial species, queens can forage normally when required to do so (especially in the spring); but in all of them communication concerning food sites is limited. In contrast, in *highly eusocial* bees (*Apis*, in colonies of up to 60,000–80,000, and the stingless bees such as *Trigona* with up to 180,000 individuals), the queens and workers are usually physically distinct and very different in size, giving a relatively rigid caste system. The workers are infertile, engaged on nest maintenance tasks, on feeding their nonforaging nest mates by trophallaxis (Michener 1974; Wainselboim et al. 2002), and on flower foraging. They often show polyethism, graduating to more complex intranest tasks as they age and eventually becoming foragers for the last few weeks of their lives. The queens are fertile but lack structures for gathering or manipulating pollen and cannot survive long outside their colonies, so that new colonies have to arise by swarming. Complex communication between nest mates concerning availability and location of food resources occurs, and this has major influences on flower-visiting patterns.

Levels of sociality are not always fixed. The carpenter bees *Xylocopa* and *Ceratina* (Watmough 1983; Dunn and Richards 2003) and some euglossine bees (Cameron and Ramirez 2001) can show facultative sociality, usually when a founding queen's nest is usurped by a second female and the foundress then becomes a guard and helper. Some **allodapine** and halictid bees are particularly labile in their degrees of sociality and may have two different social morphs within one species (Richards et al. 2003), for example being fully social in warmer climates and entirely solitary in cooler climates, with a mixture of behaviors in marginal zones (e.g., Potts and Willmer 1997; Soucy and Danforth 2002).

Effects of Sociality on Flower-Visiting Patterns

In most respects the flower-visiting patterns in parasocial and subsocial bees differ little from those of purely solitary species, since all females are stocking cells and most are laying their eggs on the resultant cell contents. However, there may be some interactions between foragers in relation to information about available flowers; bees pick up floral odors readily, from corollas and/or from pollen, and may therefore be signaling the resource on which they have been foraging as they return to a nest. This could lead to some targeting of foraging bouts and greater efficiency in the balance between travel plus searching time and handling time.

However, eusocial bees differ from other bees in several key features:

1. Some workers are not engaged in foraging at all, remaining nest bound for whole days or sometimes for their whole adult lives.

2. The foraging workers must gather much larger quantities of nectar and pollen than a similarly sized solitary bee; this is possible because they can spend all day foraging if weather and resource availability permit, having no other tasks to perform. Each foraging bee can carry relatively very large loads of nectar and/or pollen (figures of 92.5% body weight of nectar have been recorded for *Apis* workers). Foraging ranges may also be larger; in solitary bees distances traveled from the nest for 16 species were normally 150–600 m and correlated with body size (Gathmann and Tscharntke 2002; see also Beil et al. 2008 and Greenleaf et al. 2007, who present predictive equations for this effect with 62 species), whereas social honeybees routinely travel 600–800 m, with some individuals flying several kilometers from their nest. Carpenter bees can forage up to 6 km from their nests (Pasquet et al. 2008). Bumblebees routinely forage over 1.5 km (Osborne et al. 2008) and some can return home when taken at least 13 km from their nests.

3. The colony may be active for much longer periods, and all year round in some tropical species; hence a much longer seasonal sequence of forage plants is needed, giving a necessity for more generalization in

flower-visiting habits. In contrast, some solitary species may have an active flight period that is entirely contained within the flowering period of just one floral species and can therefore (at least potentially) be highly specialist.

4. Critically, nectar can be stored within the nest in open cells, for use by other nest-bound workers; thus both nectar and pollen can be collected very intensively on days of plenty and/or when weather conditions are favorable, but do not necessarily have to be gathered every day. Individual social bees do not need to gather nectar and pollen in the right proportions at any one time, so they are not forced into making compromises in collecting a balance of resources within one trip or within a few trips. The balance is resolved at colony level, by varying the proportions of nectar and pollen collectors.

5. Social colonies derive substantial benefit from the emergent physiological properties (often termed **colony homeostasis**) of a nest (e.g., Kronenberg and Heller 1982; Southwick 1991) or less often of a swarm (Heinrich 1981; Seeley et al. 2003); with this microclimatically stable refugia, physiological constraints on activity are somewhat relaxed, and individuals may be more inclined to take foraging risks and to visit flowers in poorer weather conditions.

6. There may be differences in behavioral and learning abilities in social bees that affect flower constancy and foraging efficiency, covered in section 4, *Behavior and Learning in Flower-Visiting Bees.*

7. **Social facilitation** may occur, when individuals influence and potentially enhance each others' foraging. This can operate at the levels both of recruitment to flower patches and of flower choice within patches, and is perhaps the major factor altering the flower-visiting patterns of social as compared with solitary bees. Social facilitation can occur just by chance because individuals move toward conspecifics irrespective of any flower associations (using inadvertent social information or ISI; see Baude et al. 2008; naive bumblebees were more efficient on patchy flowers when a demonstrator was present). Or it may be a precise response where conspecifics indicate the presence of a rewarding flower patch, which requires complex

interindividual communication systems but offers the huge advantage that bees may be recruited to and directed to the most profitable sources of nectar and/or pollen without having to search for themselves, and large numbers of individuals are recruited to newly opening high-reward flowers. More details are given under *Type 4: Bumblebees*, *Type 5: Stingless Bees*, and *Type 6: Honeybees*, below.

Social facilitation of choices between individual flowers can involve local enhancement when once again individuals prefer to feed close to a nest mate. However, more experienced bees may instead show local inhibition and space themselves away from existing foragers (thus presumably avoiding recently emptied flowers). Kawaguchi et al. (2007) showed that in bumblebees this effect varied with the familiarity of the foraging plant, bees choosing an occupied inflorescence of an unfamiliar plant but avoiding familiar plants when they were occupied by conspecifics. A familiarity effect also occurred in stingless bees (Slaa et al. 2003), although there the prevalence of enhancement or inhibition appeared to be species specific.

Many eusocial bees can also achieve social facilitation by scent-marking individual flowers after visiting them (see chapter 6), so that subsequent visitors avoid recently depleted flowers. However, this is not unique to eusocial bees, occurring also in some *Xylocopa*, *Centris*, and *Anthophora*.

Effects of Males and Females

In solitary bees the ratios of active males and females may be close to 1, although with variations through a season as males often emerge earlier. In social bees, the ratio is usually strongly female-biased, although males may become very abundant rather briefly, later on in the season. How far do these differences affect flower-visit patterns of particular bees?

By far the majority of bee foraging studies have concentrated on females, and male behaviors are relatively unknown, though it is widely assumed that they must visit fewer flowers (as they are only feeding themselves) and concentrate mainly on nectar gathering. Ne'eman et al. (2006) looked for male-female differences with three solitary bee species on five flowers in Israel and found that nectar amounts gathered were similar but males spent less time per flower. The female bees flew shorter distances between plants and

were likely to be more efficient pollinators through greater pollen movement, but males may have contributed more to long-distance pollen flow. It would be useful to know how generally applicable these findings may be for solitary bees.

4. Behavior and Learning in Flower-Visiting Bees

Innate Behaviors

Some aspects of **innate behavior**, shown in entirely naive bees, have substantial effects on flower handling. However, in the majority of studies it is unclear whether the bees tested were genuinely naive, so this section includes only the more unequivocal findings.

Naive bees landing on a flower show an innate probing response, usually toward the center of the bloom. Innate preferences occur in bumblebees for two-colored rather than single-color flowers, and for flowers with a large central area, especially where this is yellow and thus resembling pollen coloration (Heuschen et al. 2005). Naive bumblebees also unequivocally prefer bilaterally symmetrical shapes (Rodriguez et al. 2004).

At least some specialist bees also have innate preferences for their host pollen. For example, *Heriades truncorum* females would take only their host pollens (various composites) and entirely ceased nesting activity when offered nonhost pollens even if reared on those as larvae, showing an underlying genetic basis for host recognition (Praz et al. 2008a). It seems possible that pollens do possess properties that limit their use by some bees (Praz et al. 2008b).

Learned Behaviors

Beyond innate behaviors, all bees tested show good abilities to learn how to exploit more elaborate flowers, which enhances their speed at pollen or nectar gathering. Learning takes time; for example, a bumblebee encountering a new flower type could improve its handling accuracy on consecutive flowers (fig. 18.12A; Heinrich 1979b) but maximized its foraging efficiency only after around 30–100 visits, depending on flower complexity. Once a bee has invested this time it may pay for it to continue to forage on that spe-

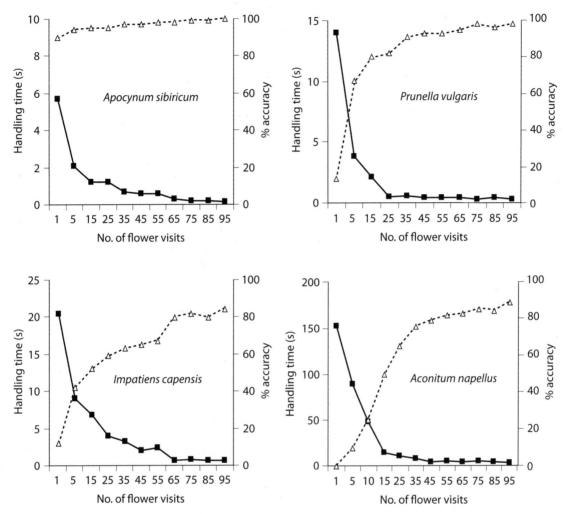

Figure 18.12 Flower handling times (squares) and handling accuracies (open triangles) for initially naive bumblebees at 100 consecutive flowers for a range of flower species. Bees learned the simple *Apocynum* and *Prunella* flowers quickly, but had longer handling time and longer learning times on the elaborate *Impatiens* and *Aconitum* flowers. (Redrawn from Heinrich 1979b.)

cies, leading to increased floral constancy (see chapter 11). Learned behavior is probably more marked and more complex in social bees than in solitary ones (Dukas and Real 1991; Campan and Lehrer 2002), and this may give the social species greater floral constancy and more efficient foraging. Learning in social bees is a topic of intense research activity (reviewed by Menzel and Müller 1996; Menzel 2001; Srinavasan and Zhang 2003), especially using *Apis* and *Bombus*, and what follows is largely based on work with those two genera.

Flower Features

Bees can learn various features of flowers, including shape and size, orientation and symmetry (e.g., Müller 1995a; Lehrer et al. 1995), and also color (e.g., Giurfa et al. 1995; Hill et al. 1997; Gumbert 2000). In effect this allows them to learn the quickest route to access the nectar or pollen (where to land, which direction to move in, where to probe) and so to learn a preference for a particular floral species at a given time (Menzel 2001). They learn simple odor or color reward-associated choices readily, but odor rather more rap-

TABLE 18.5

Flower Handling Time and Its Improvement with Experience for
Different Floral Types Visited by *Bombus* Workers

	Foraging success (%)	Handling time (s)	
	1st visit	1st visit	55th visit
Open bowl flowers			
Apocynum sibiricum	100	5.5	0.4
A. androsaemifolium	100	14.1	0.3
Short open tube flowers			
Prunella vulgaris	100	13.9	0.1
Vicia cracca	100	18.1	0.2
Impatiens capensis	70	20.4	1.7
Longer or constricted tube flowers			
Gentiana andrewsii	45	44.4	6.5
Chelone glabra	40	196.6	8.1
Elaborate closed flowers			
Aconitum henryi	35	134.7	13.6
A. napellus	29	153.5	3.0

Source: From Laverty 1994b.

idly (see fig. 6.6). In addition, honeybees show classical conditioning responses and can form complex associative links, for example between visual or olfactory floral advertisements and received floral profits. Multimodal choices improved their decision making even further, with faster learning and better accuracy (Kulahci et al. 2008).

Flower-Handling Behaviors

Figure 18.6 showed the variety of approaches that a bee might require for visiting different flower types. Table 18.5 gives examples of speeds of handling and of improving handling, for various kinds of flower, based on Laverty (1994b). Bumblebees needed almost no time to achieve appropriate foraging techniques on open-bowl flowers with freely accessible nectar, but could take at least 60–80 visits to become maximally efficient on complex zygomorphic corollas such as *Impatiens* (as in fig 18.12). Bumblebees also handled bilateral flowers about 45% faster than radial flowers and symmetrical bilateral flowers 20% faster than lopsided flowers (West and Laverty 1998). It follows that the caloric reward gained from the less easily handled

flowers must be significantly higher if the effort of foraging (even after learning) is to pay off.

The efficiency of memory for particular species is certainly reinforced by learned motor patterns (handling behaviors); bees that learned a single task in the laboratory were more efficient than those that learnt two (Chittka and Thomson 1997), and individual bees often learn their own particular ways of handling complex flowers. Hence they can apply learned flower features to distinguish between unfamiliar floral species and can use prior knowledge to detect poorly visible or cryptic flowers. These attributes impact upon their floral foraging efficiency and can substantially reduce the length of time spent foraging per unit reward collected. For example, pollen collection rate on poppy flowers increased substantially over four successive foraging bouts in bumblebees, giving significant time savings on the later bouts (Raine and Chittka 2007a). Learning and memory are not unlimited however; in that same study pollen collection was markedly poorer early the next day, indicating overnight forgetting. Limitations on learning are also indicated by the significant costs of switching to new flowers (Chittka et al. 2001; Dukas and Kamil 2001). This may be important in itself, as it ensures that a bee cannot easily remember how to handle too many flower types at once, again reinforcing constancy, to the potential benefit of both partners.

Flower-visiting behaviors that are learned will inevitably vary with bee age and experience. Newly emerged and very young worker honeybees could not achieve associative learning (for example, linking odors and rewards), this ability normally appearing about 5–9 days after emergence (Ichikawa and Sasaki 2003). However, olfactory information could be gained during the pupal phase which affected behavior in mature bees (Sandoz et al. 2000), so that adult bee foraging activity depended partly on the stored pollens encountered in the nest as juveniles, perhaps giving continuity in learned behavior from one season to another.

Foraging Routes; Spatial and Temporal Learning

Bees can also learn routes and pathways between flowers, whether in the long-range trap-lining of euglossines and some bumblebees or in a more local manner. Experienced bees evidently know where their previously visited plants are and approach them directly; and they can therefore be readily tricked into visiting plants distant from the nest that were rewarding and visited earlier in the day or on a previous day but that have had their flowers artificially removed. Similar effects occur on a much smaller scale, where a bee will on successive trips repeatedly alight first on a single highly rewarding blossom chosen from among many others in a patch. This must require a very exact awareness of flower position. It can be coupled with learned awareness of reward status through time, since many bees will travel directly to one species that dehisces early in the morning but then transfer to another (formerly ignored) roughly at the time when it starts to produce pollen or nectar some hours later. Clearly, new olfactory cues associated with the onset of reward availability could partly explain this, but there are many instances where the sequential change of preference appears without any apparent sampling of the alternatives. In fact, honeybees have an intrinsic clock mechanism, giving them a good sense of time (Moore 2001), allowing them to learn daily flower reward schedules. Foraging *Apis* can make precise associations between food presence and time of day, using this endogenous periodicity. How far inherent timing systems are present in other bees is uncertain, although circadian effects occur in the foraging of eusocial *Scaptotrigona* bees when they are kept under constant light (Bellusci and Marques 2001). Recent evidence indicates that in bumblebees even more precise learning of temporal patterns of reward status is possible, as bees can learn to time the interval at which higher-reward artificial flowers are presented, remembering both when a reward is produced and what type of reward it is (Boisvert et al. 2007).

Foraging Choices and Optimization

As a result of their substantial learning capacities, bees in general and eusocial bees in particular are noteworthy for their ability to make complex decisions about flower foraging. To summarize a great deal of research, in general eusocial bees seem not to respond to nectar concentration or sugar rewards per se, and not to maximize their rate of energy gain except perhaps in the very short term, but rather they often prefer minimum-variance flowers (risk-sensitive foraging), and will leave a high-variance inflorescence more quickly than one with constant resource (e.g., Waddington et al. 1981; Real et al. 1982; Harder and Real 1987; Possingham et al. 1990; Banschbach and Waddington 1994; Shafir et al. 1999; Biernaskie et al. 2002; and reviews by Cartar and Abrahams 1996; Perez and Waddington 1996). Plants should then evolve strategies that maximize their perceived profitability to a bee that is operating with this approach, resulting in *cognition-mediated coevolution* between plant and bee (Shafir et al. 2003).

However, this risk-averse effect commonly diminishes as the mean reward size increases, and the visitors become more inclined to take risks. In fact Taneyhill (2007) showed that with *Bombus* risk aversion diminished over time anyway, and his studies with a longer timespan showed rate maximization as the crucial currency (i.e., more like the situation in birds and bats). The bees' preferences may also change with colony state, so that a bee from a nest that is short of nectar may opt for gross intake of volume, while individuals from a richly stocked colony may make more subtle decisions about maximum profitability of particular flowers. And the flower preferences of social bees often vary intraspecifically, so that small workers may make quite different choices from large workers (see Harder 1988). Finally, preference may change with pollinator density, so that bumblebees became more generalist when competition for resources was higher in laboratory tests (Fontaine et al. 2008).

A

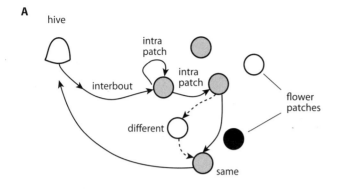

B

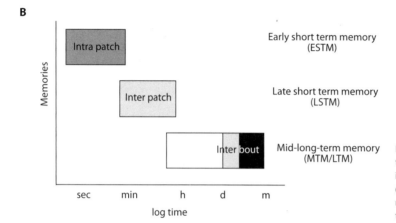

Figure 18.13 Menzel's model for different durations of bee memory patterns: (A) the nature of interbout, interpatch, and intrapatch decisions; (B) the hypothesized storage of relevant information in different memory levels with different time constants. (Redrawn from Menzel 2001.)

Bee Memory Patterns

Eusocial bees appear to have both long- and short-term memories (LTM and STM). LTM is affected by the experience of many flowers or many foraging trips, while STM may be related only to the last flower visited (Cresswell 1990; Real 1992). In fact Menzel (2001) distinguished three memory systems for honeybees (fig. 18.13), which may be more generally applicable to bees: the early short term memory (ESTM) works with a timespan of seconds, effective for movements within a patch of flowers, the normal STM lasts for minutes and can aid movements between patches, while spans of many minutes or hours in mid- and long-term memory should mediate behaviors across many foraging bouts and across days or weeks of foraging; and recall can even last for months with bees that overwinter.

In the laboratory, memories relating to flower characters once acquired can persist for three weeks or more (Chittka 1998), although they decay a little each night and improve again each day (Keasar et al. 1996). Foraging success in honeybees increases through the first week of foraging activity, which corresponds with median lifespan, so that these bees are spending a large part of their adult life in learning and improving their foraging skills (Dukas and Visscher 1994). They probably only achieve maximally efficient flower-feeding patterns as they approach the end of their lives.

5. Melittophily Types

Bees are unusually diverse as a pollinator group, and their diversity shows interesting global patterns, peaking at 40°–50° latitude (fig. 18.14A). But in relation to plant diversity the bees tend to show relatively low diversity in the tropics (Panama, Cuba, Philippines), and also in Australia and southern Africa, but rather high (ratios of 5:1 or less) in most of the north temperate habitats (fig. 18.14B).

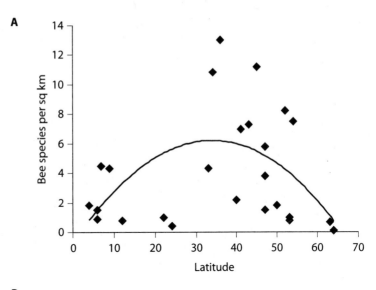

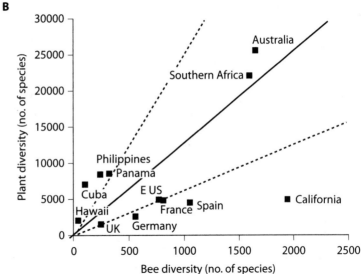

Figure 18.14 (A) Bee diversity patterns worldwide against latitude, corrected for landmass, and (B) the same, showing roughly proportional relationships with flower diversity, peaking in northern temperate and some Mediterranean habitats; solid line, ratio 10:1, dotted lines, ratios of 5:1 and 20:1. (Redrawn from Ollerton et al. 2006, compiled from various data sources.)

Given these complex patterns and the differences between various bee taxa, it is not possible to identify a single version of melittophily, and several different syndromes are therefore presented below.

Type 1: Solitary Bees

Solitary bees constitute by far the majority of all bee species, and it should be evident by now that it is almost impossible to generalize about them, as they are extremely variable in habit as well as in size and

tongue length (see table 18.2). There is really no justification for treating them together as one subsyndrome of melittophily. However, most of them have been relatively little studied so they must be considered together here for simplicity.

Solitary bees appear to be more common and speciose in savanna and desert habitats than in the more equable tropics (see references in Michener 2000). There are probably around 3000 species in Australia, 3000 in Africa, and about 3000 in the Mediterranean basin. The majority are ground-nesters, in drier and sandy or rocky patches of bare soil, linking to abundance in warmer and semiarid habitats where this kind of substratum is readily available. Their numbers and diversity can be reduced where site aspect allows inundation by seasonal rains or where there is heavy grazing to trample nests (e.g., Vulliamy et al. 2006). Most of them forage over relatively short ranges in practice, but even the small species can have flight ranges up to 1250 m on occasion (Beil et al. 2008).

The majority of the solitary bees encountered in Europe and North America are short to medium tongued, although some of those with medium-sized bodies (such as mason bees (*Osmia*), leaf-cutter bees (*Megachile*), and some of the anthophorines) can have tongue lengths equal to or slightly greater than their own body length. In the tropics, the tongues of euglossine bees may be disproportionately long, and these are dealt with separately below, as are the stingless bees.

Small solitary bees are commonly members of the halictid and the andrenid families in most palaearctic and nearctic habitats, although some colletids (with very short tongues) also occur in most communities and become dominant in Australasia. Medium and larger bees will commonly be megachilids (leaf-cutter bees) and anthophorines in Europe, but eucerines become common in warmer and semiarid habitats and larger numbers of allodapines occur in Australia and southeast Asia. There are differences in small- and large-bee behaviors (see Dafni et al. 1997), the smaller species tending to fly slower and lower and to hover at a flower before landing, while large bees fly higher and into forest canopies, landing directly on chosen flowers (see chapter 27). The larger bees are also more likely to select bilateral flowers with a "side-on" advertisement, while smaller bees select radial flowers with a particularly dissected rim.

Flower types that are visited by solitary bees tend to show a prevalence of Asteraceae and mimosoid Fabaceae, especially for those bees with an abdominal scopa as they can collect pollen simply by scraping their undersides across the dorsal surface of the flowers. In Europe and North America, the early spring species (often *Andrena* and some *Anthophora*) visit willow (*Salix*) catkins when they first emerge (plate 29H), then rapidly move on to some of the springtime yellow composites and to buttercups (*Ranunculus*), lungworts (*Pulmonaria*), bluebells (*Hyacinthoides*), fruit trees (*Malus, Prunus*), and speedwells (*Veronica*).

The nectar-rich flowers of hawthorn (*Crataegus*) are favored in late spring. In high summer a wide range of Asteraceae are visited: shorter-tongued species choose yarrows (*Achillea*) and tansy (*Tanacetum*), while the medium-tongued bees may prefer deeper-flowered knapweeds (*Centaurea*), bowl-shaped flowers such as *Campanula* and *Malva*, and small Fabaceae such as clovers (*Trifolium*). The long tongues of *Osmia, Megachile*, and later-emerging *Anthophora* visit larger Fabaceae, but also Boraginaceae, Scrophulariaceae, and Lamiaceae, including *Echium, Penstemon*, and many of the genera listed later for bumblebees (table 18.6). In late summer, some very short-tongued *Colletes* appear and frequent heathers (*Calluna*), while many of the medium- and longer-tongued solitary bees are abundant on bramble (*Rubus fructicosus*) and field bindweed (*Convolvulus*).

Similar listings for warmer and tropical climates could be compiled. For example, large-bee flowers in the relatively dry tropics have been treated in detail by Frankie et al. (1983), with anthophorine and euglossine bees the commonest visitors. They listed the flower characteristics as diurnal and generally lasting one day only, often opening rapidly and synchronously before 0530 h; large and colorful (yellow/orange or pink/purple dominating) and usually zygomorphic; usually hermaphrodite; and more often occurring on trees and vines than on other plant growth forms. In the dry season most bloomed en masse, attracting huge numbers of bees (up to 15,000), whereas the relatively few species that flowered into the wet season had more extended blooming periods (fig. 18.15, and see chapter 21). By contrast, flowers visited by small bees were more commonly those sited on trees, shrubs, climbers, and herbs, tending to be small and often white, and frequently radially symmetrical, each flower lasting one day but the overall flowering period lasting for several days or weeks. In warm savanna habitats the grass-visiting bees become noticeable in the early mornings, one example being the genus *Lipotriches* throughout the savanna regions of Africa (pers. obs. in East Africa; Tchuenguem et al. 2004 in West Africa).

Evidently, defining a flower type for solitary bees as a whole is an impossible task. However, a great many short or medium tubular and bell-shaped flowers are visited predominantly or exclusively by bees, and a high proportion of those visits will be from solitary species except where honeybee hives are abundant or,

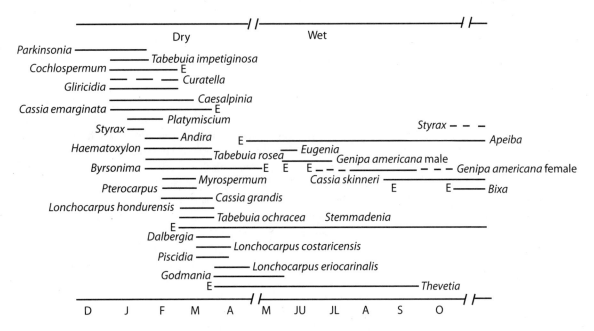

Figure 18.15 Flowering seasons of large-bee flowers in Costa Rican forest, with short mass-flowering seasons of mainly one-day flowers in the dry months; those labeled E had more extended flowering seasons and occurred mainly in the wetter months. (Modified from Frankie et al. 1983.)

more rarely, where bumblebees dominate in colder habitats.

Solitary bees often show flower constancy (Gross 1992) although it may be less developed overall than in social species for reasons already mentioned. European *Andrena* species showed strong constancy to various fruit-tree pollens or to sycamores (*Acer*) as indicated by their pollen loads (Chambers 1968). In North America, analysis of pollen loads from *Andrena, Anthophora, Lasioglossum,* and *Megachile* also showed good to very good levels of constancy (Grant 1950; table 11.4). Some tropical solitary bees are also extremely flower constant, for example, the *Creightonella* (*Megachile*) bees foraging in coffee plantations in Papua New Guinea (Willmer and Stone 1988). These examples perhaps give a somewhat unbalanced picture, merely indicating constancy when circumstances permit or dictate it (mass-flowering trees, or artificially concentrated resources in agricultural systems). However, there are also many more casual and unquantified records of solitary bees consistently preferring one flower among others available; see Proctor et al. (1996). Constancy in tropical euglossine bees that trap-line between flowers is also well known (see *Type*

3: Euglossine Bees below). Flower constancy is often limited by apparent rules for departure rather than rules for staying on a plant; for example, both *Anthophora* and *Eucera* showed patterns of departure from *Anchusa strigosa* that depended on the amount of nectar received from the last two flowers visited (Kadmon and Shmida 1992). In fact, decisions on foraging can be finely tuned to nectar rewards; Gilbert et al. (1991) demonstrated that *Anthophora* could distinguish between individual *Cerinthe* flowers on the basis of their nectar secretion rate (perhaps using flower age as a cue), and Taneyhill and Thomson (2007) showed that many bees did not obey the simple "depart after two empty flowers" rule, instead making complex decisions based on recent experience in a floral spatial array. Collevatti et al. (1997) likewise found that in practice the rule depended on patch density but also on body size for a range of Neotropical bees.

Specialist and monolectic solitary bees do occur. *Macropis* species are almost exclusively seen on flowers of yellow loosestrife (*Lysimachia*), from which they collect pollen and oils (see chapter 8; plate 19H), although some syrphid flies may also visit this plant on sunny days (Kwak and Bekker 2006). *Andrena hat-*

torfiana is specialist on the family Dipsacaceae, which in northern Europe largely restricts it to *Knautia* (Larsson and Franzen 2007), and *Andrena vaga* as mentioned above is a specialist on willows (*Salix*). The European species of *Melitta* are also fairly specialist, on certain legumes and campanulas; for *Campanula rotundifolia*, *Melitta* and *Bombus* seem to successfully exclude each other (Kwak and Bekker 2006). *Rediviva* bees are specialists on oil-producing *Diascia* species in southern Africa, and there is a precise matching between the lengths of oil-secreting spurs on the flowers and the lengths of the bees' forelegs that gather the oils, both across species and even between races within species (Steiner and Whitehead 1990). In North America, *Trachusa larreae* exclusively visits creosote bush, and *Hesperapis oraria* is a constant visitor to sunflowers on Gulf Coast dunes. *Perdita meconis* in Utah is an oligolege on just a few species of poppy and monolectic at any one site; for example, it is exclusive to the endangered bearclaw poppy *Arctomecon humilis* in some sites (which is visited earlier in the season by *Synhalonia* and later by *Apis*; Buchman and Nabhan 1996). Certain plant species are also fairly exclusively pollinated by particular solitary bees, even if those bees also visit other plants; for example, the orchid *Cyclopogon elatus* is pollinated only by *Augochlora* sweat bees (Benitez-Vieyra et al. 2006).

Type 2: Carpenter Bees

Xylocopa is the main genus of carpenter bees (named from their normal habit of nesting in holes dug into wood), and includes large, strongly built insects (plates 30C–E and 33A) resembling large *Bombus; Ceratina* and two other small genera also belong here, but are smaller and slenderer bees, often rather hairless and sometimes with metallic coloration, more like the solitary groups in their flower-visiting behaviors.

Xylocopa bees have somewhat shorter tongues than bumblebees, housed on stout beak-like mouthparts, highly sclerotized, which can be forced into quite enclosed nectaries. They have rather small scopae for their size and may carry some pollen in their crops. The genus shows varying degrees of sociality, from entirely solitary species to semisocial examples, variation occurring even within a species. These bees occur in most continents, although not in colder habitats, and the taxon includes some dusk and nocturnal species,

especially in the subgenus *Proxylocopa* (often foraging on *Capparis* around the Mediterranean and quite common visitors to night-blooming trees such as *Heterophragma* in India; Somanathan and Borges 2001).

Given their size and build, carpenter bees should need large and stoutly built flowers to visit; and they need moderate to high volumes of nectar to fuel their flight, at medium concentrations which pose no problem for their relatively short tongues. Like bumblebees they often pick flowers that occur on stout spikes that can bear their weight, and they tend to work upward on vertical flower spikes (Orth and Waddington 1997). They are commonly recorded on large individual flowers such as *Cassia, Hibiscus, Lathyrus, Virgilia, Thunbergia, Passiflora, Calotropis*, and *Fouquieria* but somewhat surprisingly also crop up regularly on much smaller aggregated flowers such as thistles, alliums, hydrangeas, lavenders, *Lantana*, and *Asclepias* (although they generally avoid the more open aggregated flowers of umbellifers, etc., where nectar rewards are just too small). Moreover, they are notorious even by bee standards for their propensity for robbing tubular flowers that are too narrow to allow their entry, by biting through the side of the corolla base. Perhaps more than most bees, then, they are prone to visit some flowers not obviously suited to them. For example, they will go to *Oenothera elata*, which opens toward dusk (it was always assumed to be a hawkmoth flower) and in doing so they can deposit pollen on up to 70% of stigmas (Barthell and Knops 1997). They also seem to be specialist pollinators of at least one orchid, the spectacular blue-flowered *Herschelianthe* genus in southern Africa (Johnson 1993).

Type 3: Euglossine Bees

Euglossines, or orchid bees, are medium to large bees of the Neotropics (Mexico to Argentina, but mostly in tropical forest). There are about 200 species; many (*Euglossa, Exaerete*, and *Aglae*) are brilliantly metallic blues, bronzes, and greens, and rather hairless, while the genera *Eulaema* and *Eufriesea* contain larger and hairy bees resembling stout bumblebees. They have rather weak endothermy and are only active in daylight, but all are fast flying and can cover long distances between flowers, often trap-lining along known routes each day. They predictably arrive at a site even if the flowers there have been removed (Janzen 1971),

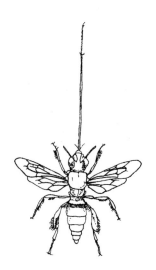

Figure 18.16 Proportions of tongue length to body length in a typical euglossine bee; see also fig. 9.6.

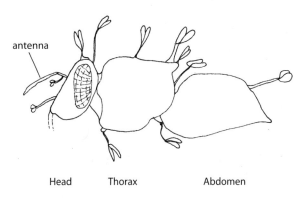

Head   Thorax   Abdomen

Figure 18.17 Precise placement of orchid pollinaria from 11 different species on bodies of euglossine bees. (Modified from Dressler 1968.)

so they are attracted to and sequentially constant to sites rather than to specific flowers. They have extremely long tongues, often substantially longer than their own bodies (see figs. 9.6 and 18.16, and plate 30B), so will take nectar from a wide range of long tubular flowers. They require fairly low-concentration nectar; in modeling their tongue action, which involves a particularly strong suction action, Borrell (2006) showed an optimum intake rate at 30%–38% concentration that fits well with recorded preferences of 34%–42% irrespective of body size (Borrell 2007).

The euglossines are also well known for their habit of taking scents as rewards from flowers to act as sexual perfumes (chapter 9). For this they use mainly orchids (occasionally some other flowers from Araceae, Gesneriaceae, and Solanaceae), with varying degrees of specificity. The chemicals, once scraped up, are stored in pouches on the rear legs and probably used as, or as precursors to, lek-related pheromones (chapter 9). In general, there are not precise one-to-one flower relationships for these bees, most species visiting several different orchids, and any one orchid may get visits from several species of euglossine. However, specificity for the plant is achieved by pollinia placement, which is very precise (on the head, thorax, or abdomen, and dorsally, laterally, or ventrally) as shown in figure 18.17. The architecture of each flower forces the pollinia-laden bee to twist about and maneuver in a way that ensures only the right pollinia are picked

off from the right site by carefully shaped stigmatic surfaces.

Euglossine bees are particularly renowned for traplining, traveling long distances along the same route each day between widely dispersed flowers. Janzen (1971) showed that they could find their way back to their nests when displaced at least 23 km away, doing so in a little over an hour. The tropical flowers they visited showed a rather typical pattern of flowering over a long period but with just a few flowers each day (see chapters 21 and 27). Each euglossine visited just one or a few species of plant and flew a set route each day between its chosen trees.

These bees are therefore almost uniquely important for the pollination of widely dispersed but obligately outcrossing tropical tree and orchid species in the New World.

Type 4: Bumblebees

Bombus species are stout, furry, and relatively large bees although they vary in size and tongue length (see table 18.2). Bumblebees are all primitively eusocial (except for a few kleptoparasitic species), living in small to medium seasonal nests, with only the queens normally being perennial.

Bumblebees were among the first invertebrates shown to exhibit partial endothermy (with warm-up rates up to 11°C min^{-1}), and their physiological and behavioral mechanisms for maintaining a constant T_b before and during foraging flights are particularly well known (reviewed by Heinrich 1979a, 1993). Nevertheless, temperature considerations remain important structuring agents for their diel activities. Some *Bombus* species are active in the very early spring and late autumn, flying at air temperatures only a little above freezing; they are also usually active earlier and later in the day than other insects and most other bees (including *Apis*) and tend to appear in lower numbers around midday (e.g., Willmer 1983; Stanghellini et al. 2002), with precise patterning related in part to specific body size. Larger species usually have lower thermal windows (i.e., they work in cooler conditions and avoid hotter ambient temperatures) than smaller ones, as simple physical principles would predict. Non-size-related physiological differences also occur between *Bombus* species, though; for example, of the common European bumblebees, the medium-sized *B. lapidarius* has a distinctly higher setting for its thermal window than most other species with which it is sympatric (Corbet et al. 1993).

Bumblebees are common in temperate and colder habitats, linking to their strong endothermic abilities and their very well-insulated furry bodies, and they are far more speciose in Eurasia than in North America. They are also important in some Neotropical ecosystems, but absent in most of Africa and occur only at altitude in India and Southeast Asia. They tend to be less common in woodland but abundant in pastureland, clearings, and forest edges. Overall, tubular flowers are often commoner than open flowers in temperate woods and occur throughout the season, especially in the semishaded parts accessible to *Bombus* but not to more sun-loving species. However, in large natural woodlands in North America there are usually more open flowers (many of them capable of self-pollinating) and fewer specialized bee flowers, and also fewer bumblebees, perhaps because of the lack of edges, hedges, and clearings.

Typical bumblebee flowers (table 18.6) are quite large, usually zygomorphic, and often long tubed or gullet shaped. They are very commonly from the families Lamiaceae, Scrophulariaceae, and Fabaceae; the majority of pollen collected often comes from fabaceous flowers (Goulson et al. 2008). Colors are often in the blue and purple range (sometimes with paler centers or even translucent lateral windows near the base, as in some gentians), and also including pinks, creams, and whites. However, there is also a range of yellow flowers much frequented by bumblebees, especially among the legumes (Fabaceae). Although there may be innate preferences in naive bees (see section 4, *Behavior and Learning in Flower-Visiting Bees*), in practice preferences that relate to the main flower-visiting periods are learned as the bees start to forage; they readily learn to associate reward size with flower size (Blarer et al. 2002) and maybe to associate either of these with floral colors and scents (Menzel and Müller 1996; Dornhaus and Chittka 1999, 2004). However, they do not transfer this learning to novel situations, thus limiting the overall efficiency of their foraging.

Bumblebee flowers often have features that make the nectar harder to gather, perhaps with barriers of hairs or constrictions in the corolla; this can be readily seen in many garden and hedgerow plants such as *Antirrhinum, Delphinium,* comfrey (*Symphytum*), and toadflax (*Linaria*). Many of the legumes visited have tripping mechanisms, and individual bee size (in effect body weight) therefore affects how well each bumblebee performs on flowers such as *Cytisus* (Stout 2000). On any of these more complex flowers, the first visit by a worker bee may be rather long and inefficient, as she struggles to locate a reward. However, handling time decreases sharply on subsequent visits and may be reduced tenfold after a reasonable learning period (Laverty 1994a; see fig. 18.12), greatly increasing the rate of intake of nectar. The most difficult bumblebee flowers tend to offer an unusually high reward, so that it pays a bee to take time to learn to handle them and then to show constancy to them.

There is a need for caution here, though, because flowers that are regarded as typical bumblebee types in practice make up only a moderate proportion of all the flowers that *Bombus* can be seen visiting. This is partly because our perceptions of bumblebee flowers are largely those chosen by the long-tongued species (e.g., *B. hortorum, B. consobrinus*), which take nectar that is out of reach of almost all other visitors in their habitats, and which always visit legally. In contrast, short-tongued species (e.g., *B. lucorum, B. bifarius*) visit many short tubular corollas and bowl-shaped flowers, and may also exploit nectar from longer tubes illegally, via holes made in the side with their mandibles (fig. 18.18 and chapter 24). Tongue length is therefore critical. Bumblebees on average visit flowers that are just a little shorter than their maximum tongue length (Brian 1957; Prŷs-Jones and Corbet 1991), so they can avoid the nectar dropping out of reach too

TABLE 18.6
Typical Bumblebee Flowers in Eurasian and North America

Family	Species	Common name	Colors
Zygomorphic or radial tubular flowers			
Scrophulariaceae	*Digitalis* spp.	Foxglove	Purple, white
	Pedicularis spp.	Lousewort, betony	Pink, red, yellow
	Antirrhinum majus	Snapdragon	various
	Linaria spp.	Toadflax	Purple, yellow
Boraginaceae	*Echium* spp	Bugloss	Blue/purple
	Symphytum spp.	Comfrey	Cream*
Ranunculaceae	*Aconitum napellus*	Monkshood	Blue, cream, yellow
	Delphinium spp.	Delphinium, larkspur	Blue
Lamiaceae	*Lamium* spp.	Dead-nettle	Purple, white
	Stachys spp.	Woundwort	Pink, purple
	Salvia spp.	Sage	Blue, purple, pink, cream
Gentianaceae	*Gentiana* spp.	Gentian	Blue, pale/white inside
Caryophyllaceae	*Silene dioica*	Red campion	Red/pink
Liliaceae	*Hyacinthoides* spp.	Bluebell	Blue*
Zygomorphic keeled flowers			
Fabaceae	*Trifolium* spp.	Clover	Purple, pink, white
	Vicia spp.	Vetch, broadbean	White*
	Lathyrus spp.	Vetch, sweet pea	Purple, pink, white*
	Lotus spp.	Trefoil	Yellow
	Medicago spp.	Medick, lucerne, alfalfa	Purple, yellow
	Ulex europaeus	Gorse	Yellow
Fumariaceae	*Corydalis* spp.	Corydalis, jack-by-the-wall	Purple, white, yellow
Brush-blossom flowers			
Dipsacaceae	*Scabiosa* spp.	Scabious	Blue, purple, pink
Asteraceae	*Centaurea* spp.	Knapweed	Purple
Pollen only, buzz-pollinated flowers			
Papaveraceae	*Papaver* spp.	Poppy	Red, pink, yellow
Solanaceae	*Solanum dulcamara*	Bittersweet	Purple
Primulaceae	*Dodecatheon* spp.	Shooting star, American cowslip	Purple
Ericaceae	*Vaccinium* spp.	Cranberry, bilberry	White, pink
	Arctostaphylos uva-ursi	Bearberry	White, pink
Other			
Balsaminaceae	*Impatiens* spp.	Balsam	Pink, cream
Violaceae	*Viola* spp.	Violet, pansy	Purple
Iridaceae	*Iris pseudacorus*	Yellow flag iris	Yellow

* Commonly visited illegitimately by shorter-tongued species.

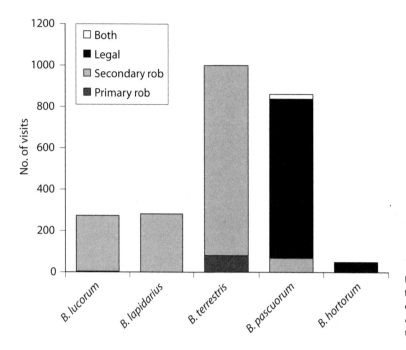

Figure 18.18 Robbing behavior patterns for five species of European bumblebees of different tongue lengths on *Linaria vulgaris*; *B. pascuorum* and *B. hortorum* have the longest tongues. (Modified from Stout et al. 2000.)

quickly as they forage, and they extract nearly all the nectar that is present.

Bombus species could be characterized as good generalists and rather opportunistic as a species; but in fact any one individual shows a good level of flower constancy, foraging primarily on a "major" flower and usually checking out one or two others ("minors") as well (Heinrich 1979b). Commonly, perhaps 90%+ of all visits are to one species out of many possibles, the other 10% being to just one or two minor choices. Figure 18.19 shows examples. An individual generally makes these decisions for herself and not just by following scout bees. The preferred flower is likely to switch readily within days or even hours, but on any one foraging trip a bee normally retains its strong major flower fidelity. Individuals are therefore specialists, with learned behaviors on their preferred plants; but different individuals from one nest may be operating with different preferences within the same flowering community (fig 18.19B). In this way, the nest may get the benefits of both generalization and specialization.

Bumblebee flower constancy has been particularly well documented, with both natural and artificial flowers, and Gegear and Thomson (2004) showed explicitly that bumblebees were economically better foragers when showing floral constancy. The bees may behave differently with very rewarding easy and conspicuous flowers and with less typically bee flowers that are less apparent; but they can be constant with both. For example, on houndstongue (*Cynogolossum officinale*), which has inconspicuous flowers that change from brown-red when young to purple when old, they often did not visit the plants at all for several days after opening (Manning 1956b), apparently unattracted by the strong mousy smell; then a few individuals did learn to visit and thereafter returned regularly, effectively trap-lining on the houndstongue and showing good constancy. They consistently preferred the younger flowers, soon searching out all examples in a particular area, while ignoring much showier and highly rewarding flowers such as foxglove (*Digitalis*). In contrast, other individuals of the same species started to visit *Digitalis* as soon as flowers appeared even slightly colored, often forcing their way in between still unopened petals and finding all the local plants very quickly.

However, constancy due to limitations from motor and sensory factors (as discussed in chapter 11) is not the whole story: persistence with the major plant may also depend on its continuing to match a reward expectation (Wiegmann et al. 2003), with a bumblebee sampling alternative flowers only when this expectation is

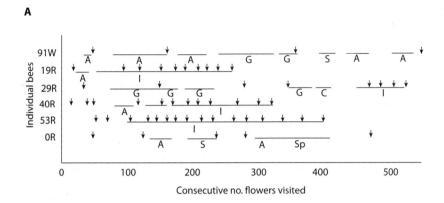

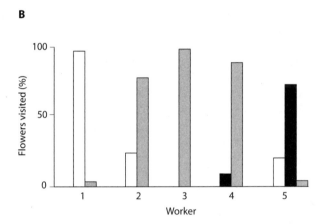

Figure 18.19 (A) Majoring and minoring in bumblebees; for 6 bees, each line represents at least 25 consecutive flowers of one species being visited, with arrow at end of trip, most bees visiting one species predominantly but with occasional visits to a second or third species (redrawn from Heinrich 1979b). A, *Aster*; C, *Campanula*; G, *Galeopsis*; I, *Impatiens*; S, *Solidago*; Sp, *Spiraea*. (B) The different flower preferences of five *Bombus vagans* workers working in the same field; white bars for dandelion, gray bars for red clover, black bars for carrot (modified from Heinrich 1976a).

no longer realized. Heinrich (1983b) showed that a bumblebee on a patch that is becoming unrewarding alters its behavior, increasing the flight distance between flowers and decreasing the angles between successive flights, so that it is progressively more likely to leave that patch. And Raine and Chittka (2007b) found that switching between flowers did not in itself reduce handling efficiency. Instead the bees lost efficiency by losing time flying between flowers; hence a new flower encountered within 0–2 s would be visited only if it was the same flower type as the previous one, but if encountered after an interval of more than 3 s it might be visited even if of a new type.

The result of this relative constancy is that for *Bom-*

bus 44%–65% of pollen loads have been recorded as monospecific (Free 1970b, and others). From the plant's point of view this means that bumblebees can be acting as highly specialized visitors. For example. *Alstroemeria aurea* flowers received more than 90% of all their visits from *Bombus dahlbomi* workers, even though the flower patches were changing from mainly male phase to mainly female phase as the season progressed; and this bumblebee also had by far the highest per visit effectiveness of all the insects that were recorded on the plant (Aizen 2001).

Bumblebees learn how to handle a very wide range of flowers, as summarized pictorially in figure 18.6., with examples in plate 31 including foraging on some apparently inappropriate flower types. Some examples of their averaged learning curves were shown in figure 18.12, although individual variation occurred with the age and size of individuals (Laverty 1994b). They are particularly likely to choose flowers that grow in spikes and to work upward on spikes, which correspondingly open from the bottom and are protandrous (chapter 3, fig. 3.13); specific examples for a bumblebee are shown in figure 18.20. Many such plants have stronger nectar production in their female flower phase, strongly promoting cross-pollination and outbreeding. However, this upward movement can occur even on plants with no nectar or age gradient in the flowers (Orth and Waddington 1997). Some bumblebee flowers may promote this behavior by having flower forms that keep the bee "upright" as it visits; for example, willowherb (*Chamerion angustifolium*) has drooping stamens and style that the bee clings onto and which it can really only feed from when in an upward-facing position (Benham 1969). Curiously, though, bumblebees preferred to forage on horizontal arrays of flowers rather than vertical arrays given the choice experimentally (Makino 2008), and foraging performance decreased with increasingly vertical arrays as travel time between flowers (whether upward or downward) increased.

Bumblebees also readily learn particular flight paths within a group of flowers (Thomson 1996), allowing them to use trap-lining foraging, and marked bumblebees have been shown to visit the same sites each day in specific sequences for up to a month (Heinrich 1976a). They use orientation flights (flying around in gradually increasing circles) to note the positions of flowers (as with nests). Such flights are performed less frequently with conspicuous plants such as foxglove and more with inconspicuous ones like houndstongue (Manning 1956b); the bees appear to estimate how much orientation is needed.

Once at a flower patch and feeding, bumblebees appear to have simple rules about whether to stay or to move on. On a poorly resourced area, they make longer interflower flights and have low turning angles, so tending automatically to leave the patch. On a good area, where most flowers are rewarding, they make shorter flights and turn a great deal more often, so tending to remain in the patch. For example, on *Trifolium* (white clover), Heinrich (1979b) found workers normally moving more than 75 mm on 67% of their flights, but this fell to 30% of flights when the nectar

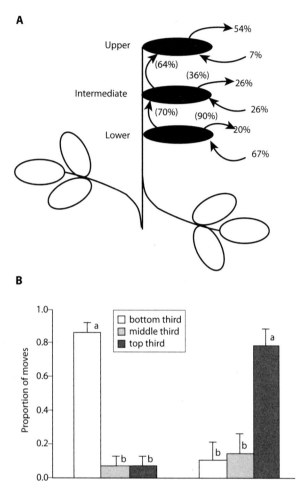

Figure 18.20 Bumblebee tendencies to move upward on spikes, showing (A) frequencies of first and last visits at different positions on *Corydalis ambigua* (redrawn from Kudo et al. 2001), with movements between spike positions in parentheses, and (B) proportions of arrivals and departures for each third of the vertical spike of *Pontederia cordata*, with SEs and different letters for significantly different proportions (redrawn from Orth and Waddington 1997).

was depleted, with 82% of those flights then being in the forward direction.

Bombus are also very good at avoiding flowers already visited by others, indicative of their ability to deposit and to detect "footprints" (chapter 6). In the same study by Heinrich, they veered away from up to 31% of approached flower heads after that patch had been heavily visited but from none in a patch that had had no *Bombus* visits for two days.

The arrival back at the nest of a *Bombus* well loaded with provisions increases the likelihood that another individual will leave on a foraging trip. This is perhaps due to sound communication by buzzing, but with additional effects from irregular runs within the nest by the incomer that seem to spread a chemical signal (Dornhaus and Chittka 2001), probably a specific pheromone from tergal glands (Dornhaus et al. 2003). However, bumblebees do not communicate any geographic details of food location. At best, they rely upon the odor of the provisions that they bring in, or odors temporarily carried on their bodies, which together with the general increase in activity prompted by their arrival can direct new recruits to the source the incoming forager has been exploiting. Thus bumblebees fed in the nest with artificial nectar laced with mint, carnation, or aniseed scents will thereafter forage preferentially from nectar sources with the right odor (Dornhaus and Chittka 1999).

Social facilitation is important for bumblebees, and they are attracted to inflorescences or flower patches where conspecifics are already foraging. Leadbetter and Chittka (2007, 2008) showed that they learned a foraging task (nectar robbing) more quickly in the company of holes in corollas and experienced conspecifics (i.e., social learning). There are even more specific indications of copied behavior (Worden and Papaj 2005), where naive bumblebees that had been allowed to observe trained (non-nestmate) bees selecting either orange or green flowers would subsequently prefer the same flower color themselves.

To give an overview of bumblebees, their constancy compares favorably with honeybees, and they often visit faster and more efficiently in terms of pollen pick-up and deposition, as well as being active earlier and later in the day and in poorer weather. *Bombus* may also learn flower-handling skills more quickly and distinguish flower age and reward status better. For all these reasons, they often provide better pollination services on crops than do honeybees (see chapter 28), and the presence of several species in any one habitat adds to their versatility overall. The relative abundance of different bumblebees is often related to competition between species of varying tongue lengths (see chapter 22), although in parts of Europe up to 16 species may co-occur (Goulson et al. 2008), the rarer bees tending to visit fewer flower species.

On the other hand, a typical bumblebee is "expen-sive," requiring about 10 mg h^{-1} of sugar to fly (Heinrich 1975a), representing 1610 joules (J) g^{-1} body mass h^{-1}. So a typical worker of 500 mg body mass would need 805 J to support an hour of activity at a power consumption of 225 mJ s^{-1} (mwatts, mW), a good match with measured values of 175–200 mW (Ellington et al. 1990). This generates some problems of "living on the edge"; Heinrich calculated that one colony of *B. vosnesenskii* needed 1.9 oz (around 50 g) of honey and 1.2 oz (35 g) of pollen every day. In practice, most nests when opened up normally have very little reserve, unlike an *Apis* hive. Honeybees are much more competitive, and one strong honeybee hive collects over 500 lbs (~220 kg) of honey, which would take away the resources of around 100 bumblebee colonies.

Type 5: Stingless Bees

The stingless bees constitute a single tribe (Meliponi-ni) of morphologically rather similar and easily identified bees (see plates 29E,F and 33E,F), with reduced wing venation and without the ovipositor modified as a sting as seen in all other aculeate hymenopterans. Common genera are *Trigona*, *Melipona*, and *Plebeia*, and they occur in the New World, Africa, southern Asia, and Australasia. They are common and important only in tropical and subtropical ecosystems, where their biomass can be really high, making up a large proportion of the pollinator community and sometimes up to a quarter of all bees (see chapter 27).

They are predominantly small, indeed sometimes tiny (2–13 mm long), although some *Melipona* are closer to medium sized by bee standards. All are fully eusocial, with perennial colonies and moderately sophisticated communication systems. *Melipona* species are relatively nonaggressive, whereas *Trigona* and related genera are usually highly aggressive to other bees. Most stingless bees have short to medium tongue lengths and so require quite small volumes of low- to medium-concentrated nectar, from small or short-tubed flowers. Nectar concentration can be critical to visiting patterns; for example, *Melipona* bees visited *Hybanthus* flowers only while nectar was below 60% sugar concentration (Roubik and Buchmann 1984), which can tend to produce diurnal bimodality with no visitation in warmer midday hours. However, the smaller species may commonly also act as nectar and pollen thieves on much larger flowers. Stingless bees on a pollen-gathering trip carry concentrated nectar in their crops when they leave the nest and probably use at least some of this to aid pollen adhesion to their corbiculae (Leonhardt et al. 2007).

Using individually marked *Melipona* bees, Biesmeijer and Toth (1998) found that about half the individuals from one nest specialized as pollen or nectar or resin collectors for their entire adult life; but other individuals switched daily, gathering pollen and/or resin in the morning and nectar in the afternoons. This difference also affected longevity, with pollen foragers only active for 1–3 h in the morning and surviving for 12 days on average, but nectar foragers sometimes active all day and lasting for only 3 days.

Recruitment occurs in some stingless bees and appears to be partly olfactory as individuals can be recruited to artificial syrups only if the syrup has a scent added. However, simple forms of alerting behavior on return to the nest, using zigzag runs that produce a general jostling effect, also occur (reviewed by Michener 1974). Some returning *Melipona* show even greater sophistication, making sounds (by wing vibration) whose duration is proportional to the distance from the food source. Aguilar and Briceno (2002) showed that the pulse pattern of the sound also varied with concentration of the located food, and information about the height above ground of the resource could also be conveyed (Nieh et al. 2003). However, floral odor on the body of returning foragers continues to play a role for most of these species. Individual experienced bees may also lead new foragers for a short distance in the right direction from the nest toward a good forage site; some species, especially when operating in the still air of tropical forests, can use pheromone trails outside the nest to guide new foragers, and the scent marks can be polarized with heavier scents nearer the goal and sometimes an additional unique mark at the feeding site (Schmidt et al. 2003). In addition, local enhancement at feeding sites has been recorded, where individuals prefer to feed close to a nestmate (e.g., Slaa et al. 2003). Some *Trigona* bees have been shown to have an intrinsic time sense, and so can show time-place learning with anticipatory visits to a known pollen or nectar source or to a daily food training site (Breed et al. 2002).

Stingless bees have been relatively unexplored as an ecosystem service in the past, but it is now known that at least 18 crops have stingless bees as major pollinators (Slaa et al. 2006).

Type 6: Honeybees

Honeybees (or hive bees) are all in the genus *Apis* (plates 1E,F and 30F). There are several species, native to Eurasia, but now introduced everywhere. Commonly only one species occurs in a given area, but where two do coexist (naturally, or after introduction) they are normally sufficiently different in size not to be direct competitors; for example, in Sudan *Apis mellifera* collects pollen in the early morning and late afternoon while the introduced *Apis florea* (the dwarf honeybee) forages undisturbed for pollen in the middle of the day, giving little overlap on the flowers (El Shafie et al. 2002).

Honeybees are in many respects average bees, average in size (*A. mellifera* around 90 mg, others a bit smaller or larger), in tongue length, and in endothermic abilities. Their thermoregulation is also only moderately sophisticated (Coelho 1991; Underwood 1991), lacking some of the elaboration of bumblebees. Above all, though, they are supremely eusocial, with large colonies in perennial nests that act as climatic refugia. Their unusual ability to gather water and use it to cool the hive evaporatively (e.g., Kuhnholz and Seeley 1997), and the habit of wing-fanning by the worker bees, which together can keep hive conditions relatively cool and comfortable even at air temperatures up to 39–40 °C, represent additions to the normal bee behavioral repertoire that impact on overall activity patterns for individual forager bees (see Heinrich 1993). *Apis* also have excellent communication systems that allow the foraging efforts of the workers to be very highly organized. As a result, trip lengths and durations can be exceptional, and a single hive might have a foraging area of 60 square miles. In consequence, one hive of *Apis mellifera* may take in around 25–40 kg of pollen per year and build up a resource many times this value over a few years.

In effect, honeybees will go to almost any flowers in most of the habitats where they occur and are often therefore termed "supergeneralists." Insofar as typical or preferred flowers exist for honeybees, they are not unlike those for short-tongued bumblebees. However, honeybees probably have less intrinsic preference for zygomorphic designs, and will often select a more open and radial flower, including the white/yellow/orange colors that are less often visited by other bees. It follows from this that honeybees as a species are collectively the most generalist and most polylectic of any pollinators; they are so unfussy that they sometimes collect other fine particulate matter such as coal dust, sawdust, and mold spores. Quite commonly they even visit classically anemophilous flowers, which may sometimes make up to 30% of their intake in savanna and desert sites.

However, as individuals on any one trip or any one day, honeybees can be at least as constant as *Bombus* in their behavior, with many successive visits to a single floral species where that species is dominant (and hence often on crops). There is a strong tendency for an individual worker to stick with one plant species even with better ones available nearby (von Frisch 1954; Free 1963). Ribbands (1949) recorded marked individuals sticking with the same plant for 12 successive days for a pollen crop and for 21 successive days for a nectar crop. Overall, though, they tend to be more constant when pollen gathering then when on nectar duty, at least in part because it is more difficult to package pollen from more than one source into the scopa. As a result, there can be a very high concentration of honeybees working a small area for pollen, which is potentially bad for the plants in terms of outcrossing.

Worker honeybees do tend to have temporary constancy to either pollen or nectar (Free 1970), although they often have a full load of one and an additional small load of the other. There is no clear developmental precedence of one activity over another as the bee ages (indeed, some bees gather pollen all their working lives and others gather only nectar), but switching between the two is common and switches from pollen to nectar are commoner than the reverse. Some honeybees persist as scout bees though, sampling all available flowers in the vicinity and relaying information back to the hive; these scouts obviously never become flower constant. They tend to be more abundant early and late in the season, when fewer flower types are available.

Experienced honeybees appear able to learn characteristics of nectar flow rate in a flower, achieving this within one foraging bout (Wainselboim et al. 2002), and thus improving foraging efficiency. Furthermore, where petal tissue is reasonably transparent, the level of nectar within a corolla tube may be assessed visually as a bee approaches (Goulson et al. 2001), al-

though such discrimination will not always be useful: where handling time is low and/or flowers are scarce, it may make economic sense for a bee to visit even the relatively depleted flowers that it encounters.

Apis mellifera workers make *orientation flights* when they first mature into foragers and *learning flights* when they first encounter new food sources; in both cases they are memorizing visual landmarks to guide their return. The orientation flights have a fixed duration but there is increased flight speed with increasing experience, so that later trips cover larger areas (Capaldi et al. 2000). In contrast, the learning flights increase in duration with landscape complexity but decline in duration as familiarity with the environment increases, and they also depend on the sugar concentration of the food source; all these features suggest that learning effort is adjustable (Wei et al. 2002). Both kinds of landmark memorizing can clearly affect subsequent activity patterns, potentially reducing trip durations and improving overall foraging efficiency. Honeybees can also store landmark information to locate sites they have previously visited, acquiring a series of stored visual images (*general landscape memory*; Menzel 2001). They navigate using directional (sun- and compass-related) cues, which feed into the landscape memory. Trip durations can be reduced—or at least the ratio of travel time to feeding time can be optimized—by such learning. Experiments with trained bees indicate an intrinsic time sense in *Apis*, where information is learned differently according to time of day (Gould 1987); thus a honeybee will return to the same patch of flowers at roughly the same time over several days.

As regards communication and recruitment within the hive, honeybees are of course well known for their "dancing" (fig. 18.21), but simple odor cues still play a role in information transfer, since a change in the stored food in a hive can redirect subsequent foragers (Free 1969). Honeybees may also augment the scent and hence attractiveness of floral sources once located, with secretions from their abdominal Nasonov glands (a habit more common in the subtropical *Apis cerana* and *A. florea* than in the familiar temperate *A. mellifera*). These simpler and probably ancestral recruitment behaviors are usually overlain or even overridden by the much more precise individual-to-individual communication, by both chemical and mechanical signals, that are contained in the returning honeybee's dances on the vertical walls within the nest. These dances

A

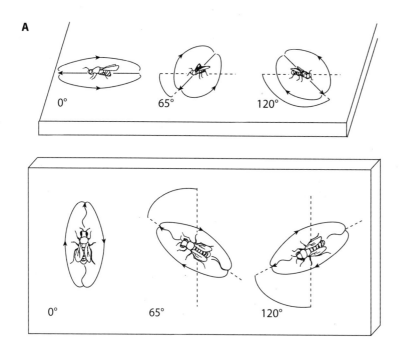

B

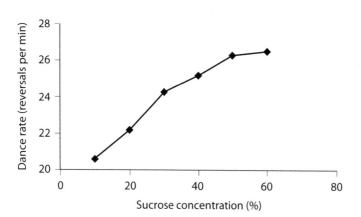

Figure 18.21 Honeybee waggle dance: (A) on a horizontal surface where the central waggle points to the food, and as translated onto the vertical comb (modified from von Frisch 1967, Barth 1985); (B) relation between dance rate and concentration of the sucrose source it refers to (redrawn from Waddington and Kirchner 1992).

(the round dance, the tremble dance, and the tail-wagging or waggle dance) can variously convey a great deal of information, ranging from vector parameters (i.e., range and direction to flowers), the identity of the flowers, and the nectar uniformity, concentration, and/or viscosity. Dance features also vary in complexity in relation to time of day and to environmental variables (von Frisch 1967; Steffan-Dewenter and Kuhn 2003; Thom 2003). Their effect can be modified by acoustic signals (piping) that may retard recruitment to low-quality food sources (Thom et al. 2003), and augmented by alkane and alkene scents emitted from the bee as she dances (Thom et al. 2007). In fact, odor signals borne on an incoming bee often override spatial information in the dance, and experienced foragers will commonly depart for a known source of that odor rather than seek out the new site encoded in the dance (Grüter et al. 2008).

Individual honeybee foraging decisions are not fixed responses to the recruitment signals, however. They vary with wing wear (Higginson and Barnard 2004), which is a reasonably good indicator of age and foraging experience; and also with individual assessments of predation risks on the flowers (see chapter 24). When a hive was rendered artificially low on pollen amount or pollen protein quality, for example, an increased proportion of the workers switched to pollen foraging, and the less experienced workers tended to collect heavier pollen loads on each trip (Pernal and Currie 2001), although none of the workers produced a quality response by changing their flower selection to gather more protein-rich pollens.

A honeybee requires about 10 mg h^{-1} of sugar to fly (Heinrich 1975a), representing 1610 J g^{-1} body mass h^{-1}, so that a typical worker bee of ~90–100 mg would need 160 J per hour of foraging. Power consumption is then 45 mJ s^{-1} (mW), about five times lower than in a big bumblebee. Many bee flowers have much less than 1 mg of sugar (chapter 8), so if we took a honeybee stomach filling as 50 mg of nectar, one complete trip could involve 100 to 1000 flowers of some typical *Apis*-visited plants; and it could require as much as 20 million flower visits to gather 1 kg of honey into the hive. A bee colony of 20,000 individuals will require about 15 kg of honey to survive through a typical temperate winter, but of course most hives produce more than this (the excess usually taken by humans). For reasons of diligence above all, but also range, constancy, and flower choices, honeybees are good (although not necessarily ideal) pollinators for a very wide range of flower types and for many introduced flowers and crops. However, they may outcompete and supplant native bee species which would if left undisturbed have provided a better service (see chapter 29); and their tendency to rob flowers also means that their specialization as flower visitors sometimes outruns what is ideal for the pollination of a given plant.

6. Overview

Bees are the dominant pollinators in many habitats across the world. In most temperate areas (except larger and denser woodlands) they predominate, and bumblebees are often the most important single native group, with social bees collectively dominating the pollination services. In Europe and much of Asia this pattern holds; the only exceptions are the northern boreal regions where flies predominate (although *Bombus* still occurs in all the boreal and arctic communities of the northern continents), and the Mediterranean fringes where solitary bees are more important and become extremely diverse, with flies and butterflies also more influential. In most of North America bees are again dominant, although a range of hummingbird-pollinated plants occur in open habitats. In tropical forests wind pollination and selfing are rare, so that animal pollination is critical; bats and birds come into play more obviously, but medium- to large-sized bees are often still the most numerous pollinators in the canopy and show some highly specialist relationships with flowers, especially with orchids. In open tropical habitats such as the savanna areas, wind pollination increases for the understory and grass layer, but bees are still important for the trees and herbs; and in arid zones and deserts bees are once again extremely important. Furthermore, in all these areas, even when they seem to be minority visitors in terms just of numbers, with more flies or butterflies present, the bees may still be the most important in terms of effective pollination.

Chapter 19

WIND AND WATER: ABIOTIC POLLINATION

Abiotic pollination involves the transmission and capture of pollen through a fluid medium, either air or water, and it occurs in at least 60 angiosperm families. Although it was once thought to be a somewhat random process, it is now clear that abiotic pollination is quite sophisticated, with significant adaptations in both the pollen releasing and pollen capturing processes that improve pollination efficiency (Cox 1991; Ackerman 2000). Phylogenetically controlled analyses show that, contrary to some earlier views, abiotic pollination gives rise to similar levels of species richness to those found in biotically pollinated sister groups, and that flowering periods are also of similar length (Bolmgren et al. 2003).

It is traditional to divide abiotic pollination into wind-mediated and water-mediated processes (anemophily and hydrophily, respectively), the two modes having some similarities; but, as will become clear, each of these categories contains substantial diversity and hydrophily in particular needs to be further split for clarity. Each mode has evolved repeatedly across the angiosperm taxa (Cox 1991), although anemophily is much commoner (about 98% of all abiotic pollination).

One point that should be stressed at the outset is that abiotic pollination gives no easy system for control over selfing, except by monoecy; and even then pollen is likely to be moving to very close neighbors that may be genetically very similar or clonal. Hence a high proportion of abiotically pollinated plants may have to use self-incompatibility systems.

1. Anemophily

Occurrence

Wind pollination is extensively used in modern plants, notably in most gymnosperms, in the catkin-bearing angiosperm trees, and in cereals (Poaceae and Cyperaceae). A few examples are shown in plate 36. It is predominantly a derived condition in angiosperms, associated with ecological conditions where zoophily is difficult; table 4.1 gave a simple overview of plant traits and environmental conditions associated with anemophily. Transition from zoophily to anemophily has occurred at least 65 times in such circumstances. Culley et al. (2002) reviewed the occurrence of anemophily, and Friedman and Barrett (2008) gave an analysis with added phylogenetic corrections, showing that it has evolved more often in taxa already having unisexual flowers or plants, with a loss of nectar and a reduction in ovule number occurring repeatedly.

Anemophily is an efficient and straightforward way of moving pollen between plants in open habitats where there are large stands of a particular species present—as in boreal forests dominated by just a few conifer species, or in prairie grasslands (although in both these situations some insect pollination may also occur). It is also effective when there are relatively few plant species, as in temperate woodlands where there may be a predominance of oak, ash, maple, beech, etc. It also dominates, perhaps somewhat surprisingly, in the temperate rainforests occurring in parts of Chile, the northwestern United States, and New Zealand, despite the

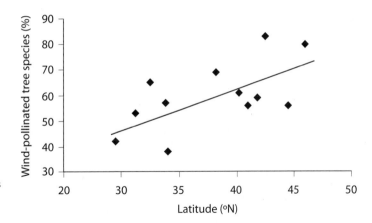

Figure 19.1 Percentage of wind-pollinated tree species with latitude in North America. (Modified from Regal 1982.)

high rainfall (see below). In fact globally the proportion of wind-pollinated plants increases steadily with latitude (fig. 19.1), reaching at least 80% in all northern forests (Regal 1982). Anemophily allows excellent long-range dispersal, unconstrained by the limited ranges of animal movements. Wherever plants of one species are reasonably gregarious, wind pollination may be not only sufficient but very possibly the most effective form of pollination, since cross-pollination of so many plants would require a larger insect population than a rather cool habitat could support through the main flowering season (Regal 1982; Whitehead 1983). Wind pollination may also be the only option in other situations where insects are relatively rare, such as salt-marshes and seashores, and some semiarid habitats where salty or alkaline soils dominate. Many emergent water plants are also anemophilous, since ponds and rivers (while plentifully endowed with insects) are not regular haunts of the main pollinating insect taxa; for example, Cook (1988) found that of 380 aquatic genera 119 contained anemophilous species.

But in more complex communities wind pollination becomes very wasteful. For one pollen grain to reach a stigma of perhaps 1 mm^2 surface area, the habitat might require up to one million pollen grains to land per square meter (Proctor et al. 1996). This is not an impossible target, given that single catkins of birch or hazel trees each produce about 3–5 million pollen grains, and single stalks of various grass cereals may produce 40,000–70,000 grains (only rarely do animal-pollinated plants begin to approach these pollen outputs, and then only in species such as poppies where pollen is their only reward). But the vast majority of anemophilous pollen grains will never reach a stigma of any sort, let alone a conspecific stigma. In the relatively simple forest habitats dominated by wind pollination, the pollen falling to the ground may amount to several tons per square kilometer every year. As table 4.2 showed, wind pollination therefore normally needs far more pollen grains to be produced per flower (or per ovule, per stigma, per plant) to succeed at all.

Habitat complexity is not the only issue determining pollination mode. In habitats with high rainfall, especially the cool temperate or tropical wet forests, much of the wind-blown pollen dispensed from flowers just gets washed out of the air. Wind speeds that are consistently too high, or much too low, are also a bar to success.

For all these reasons, anemophily is almost impossible in tropical rain forests; they are not only too diverse, but also too wet and too still, with little or no wind-produced air movement below the canopy. Estimates in tropical lowland forests are that only 1%–4% of all plant species are wind pollinated (Bawa 1990, 1994; and see Regal 1992), although higher rates occur in a few plant groups (e.g., Otegui and Cocucci 1999; Midgely and Bond 1991). Bullock (1994) also reported higher rates in Neotropical dioecious trees, where bee visitors occur but may be relatively ineffective.

A comprehensive listing of the taxonomic occurrence of anemophily was given by Ackerman (2000), indicating some examples in 60 families (16%–18% of all angiosperm families) spread across monocots and dicots. Most families have just a few examples; anemophily only tends to dominate in just a few taxa, such as the grasses (Poaceae), rushes (Juncaceae), and sedges (Cyperaceae); the major cool and temperate tree taxa (e.g., Betulaceae, Corylaceae, Fagaceae, Salicaceae); and some families with very weedy species such as nettles, docks, plantains, and goosefoots (Urticaceae, Polygonaceae, Plantaginaceae, Chenopodiaceae). It also occurs conspicuously in some composite weeds such as mugwort (*Artemisia*) in Europe or ragweed (*Ambrosia*) in North America. Hence in temperate sites the pollen count as sampled from the air is dominated by trees in spring (hazel, alder, willow, birch, yew), by grasses and trees in early summer (oak, beech, maple, pine), and then by plantains, sorrels, docks, nettles, and lime trees in later summer.

Anemophily can be quite common in a few taxa traditionally seen as entomophilous (Mahy et al. 1998) and may be regarded as an escape from the constraints of dependency on animals in certain conditions (Cox 1991). Numerous authors have described anemophily working effectively alongside insect pollination within a genus or species, the balance between the two being affected by environmental factors. Mahy et al. (1998) demonstrated sporadic anemophily in *Calluna* heathers; Berry and Calvo (1989) showed that its occurrence increased with altitude in the composite *Espeletia*, the same phenomenon occurring in the crucifer *Hormathophylla* (Gómez and Zamora 1996); Dafni and Dukas (1986) reported it in *Urginea* (a relative of hyacinths) in dry and sandy conditions; and it can occur as a backup mechanism (reproductive assurance) in common garden annuals such as *Linanthus* (Goodwillie 1999). While these mixed pollination systems are often seen as a switch to anemophily in a basically entomophilous taxon, remember that there can sometimes be considerable collection of wind pollen by insects, notably by certain bees using various grasses (see chapters 18 and 27). Where the two modes of pollination can occur simultaneously in one species, the term "ambophily" is sometimes used. Included in this would be some occasions when insect visitation triggers pollen release and some of that pollen is then transported effectively by wind (e.g., in the palm *Chamaedorea*; Listabarth 1992).

Wind-Pollinated Angiosperm Flowers

These share some features in common, as follows:

1. The flowers are commonly unisexual, as it would be nearly impossible to avoid a stigma becoming entirely covered with its own anthers' easily dislodged pollen, especially given the high output of pollen that is required. Sexual dimorphism can be unusually pronounced, since there are no constraints to site anthers and stigmas in similar positions (Weller et al. 2008). Where hermaphrodite flowers do occur, they are strongly dichogamous, with pollen produced at a time when the stigma is unreceptive.

2. Flowers are commonly grouped in inflorescences, and the inflorescences tend to be more condensed than in zoophilous relatives (e.g., Weller et al. 2006).

3. The perianth is always reduced or absent, having no attractive or protective functions.

4. Stigmas are distinctly larger than in most zoophilous plants, with finely divided, often plumose or papillate tips; and anthers are also large and exposed, often pendant on long stamen filaments.

5. The anthers tend to produce very numerous and fairly small pollen grains (20–40 µm, although a little larger in conifers) (Harder 1998).

6. Pollen grains are dry rather than sticky, due to differences in the pollenkitt (being thin or entirely lacking), which helps to reduce clumping.

7. Grains have relatively unsculptured surfaces (Crane 1986; Linder 1998) and a reduction in number or size of apertures, which probably helps to reduce water loss.

8. Grains often contain 1–3 air-filled bladders (**sacci**), especially in conifers.

9. Flowers normally have rather few ovules, very often just one, so producing just a single seed; hence they also have very high pollen/ovule (P/O) ratios.

10. Plants have a distinct flowering season, most often in early spring for angiosperm trees, when the foliage is still very sparse and does not obstruct mass air flow and pollen movement; and flowering onset is statistically earlier than for biotically pollinated plants (Bolmgren et al. 2003).

Examples can be usefully divided into four groups.

Catkins

Catkins (fig. 19.2) are typical of many wind-pollinated angiosperm trees, especially in the families containing

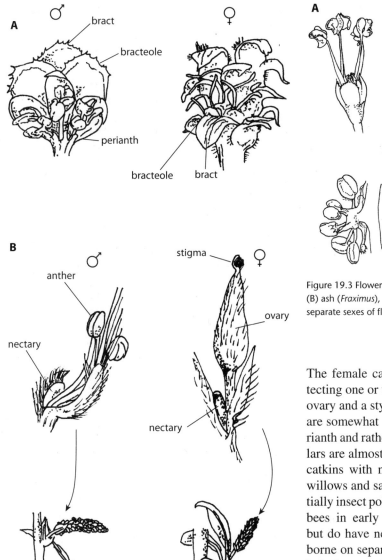

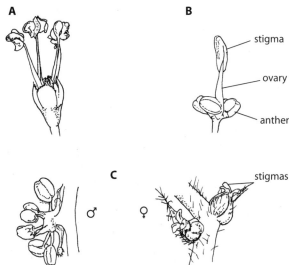

Figure 19.3 Flowers of other anemophilous trees: (A) elm (*Ulmus*), (B) ash (*Fraximus*), both hermaphrodite, and (C) oak (*Quercus*) with separate sexes of flower. (Redrawn from Proctor et al. 1996.)

Figure 19.2 Catkin structures from wind-pollinated trees: (A) groups of male and female flowers from catkins of silver birch, *Betula pendula*; (B) single flowers of sallow (*Salix*), and views of the male and female catkin structures. (Redrawn from Proctor et al. 1996.)

The female catkins again comprise large scales protecting one or two flowers, each of which is merely an ovary and a style. In wind-pollinated oaks, the flowers are somewhat less reduced, having a recognizable perianth and rather more stamens. In the Salicaceae, poplars are almost entirely wind pollinated and have long catkins with numerous stamens per flower, whereas willows and sallows still have catkins but are substantially insect pollinated, much favored by flies and small bees in early spring. The flowers are very simple, but do have nectaries, with male and female flowers borne on separate trees; anemophily still occurs regularly in some habitats and some seasons (Vroege and Stelleman 1990).

Other Wind-Pollinated Trees

The remaining important anemophilous tree taxa are the elms, ashes, and maples, usually with hermaphrodite and protandrous flowers (fig. 19.3). In elms (*Ulmus*) these are usually reddish in color, with a small upright cup-shaped four- or five-lobed corolla, four or five stamens protruding with rather large flat anthers at their tips, and two short papillate styles; the structure is evidently appropriate to wind pollination. In contrast, in the common ash (*Fraxinus*), the flowers (which are dark green and inconspicuous, with no corolla) have erect obvious stigmas and small round anthers at

birch (Betulaceae), hazel (Corylaceae), oak and beech (Fagaceae), and willow (Salicaceae). In birch and hazel the larger and more obvious catkins are male, formed from multiple tiny male flowers that comprise little more than a bract scale, a few bracteoles, a tiny perianth, and one or two pairs of long divided stamens.

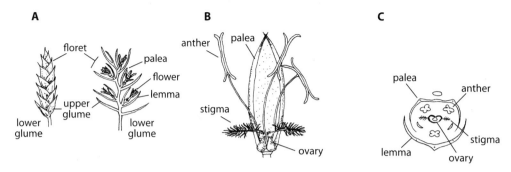

Figure 19.4 General structure of a grass spikelet (A) before and after opening; (B) a single flower; (C) flower in transverse slice. (Redrawn from Proctor et al. 1996, based on earlier sources.)

the base of the flower, appearing imperfectly suited to wind pollination. Since close relatives of ash are all entomophilous (e.g., lilac, jasmine, privet), it is likely that ash has rather recently switched to anemophily; indeed, some species of *Fraxinus* do have white fragrant corollas and are insect pollinated. An almost reverse situation is found in the maples (*Acer*), most of which are insect pollinated but where the common *A. saccharinum* is wind pollinated with small inconspicuous pinkish or green/yellow flowers.

Grasses and Sedges

The vast majority of these are anemophilous, the only exceptions being a few aberrant self-pollinating species. Grass flowers are small, greenish, and odd (see Friedman and Harder 2004); they are arranged as groups in spikelets, each enclosed by a pair of leaflike glumes (fig. 19.4; plate 36A). Each flower is also protected by a pair of small bract-like structures (the upper palea and lower lemma), within which is an ovary with two plumose stigmas and three long slender stamens with large pendant anthers. At the base there is a pair of scales that swell and force the flower to open and the anthers to dangle out in the breeze as the anthers mature. In some grass species, these typical hermaphrodite flowers are augmented by some separate male-only flowers, increasing the overall pollen output.

Herbaceous Plants

Perhaps the commonest wind-pollinated plants outside the trees and grasses are a group of unrelated "weeds," shown in figure 19.5. The most familiar are the stinging nettles (*Urtica*), with rather catkin-like greenish unisexual flowers, usually on separate plants. Female flowers are tiny and ovoid, with short tufted stigmas;

the male plants have more conspicuous flowers, due to the elongate pale anthers emerging from a four-lobed perianth. Unusually for a wind-pollinated plant, the stamens are under tension in the bud and emerge explosively to produce a substantial rain of pollen.

Plantains (*Plantago*) have spikes of protogynous flowers with small corollas; long stigmas emerge first, followed by even longer dangling stamens, and since the flowers mature at the tip of the inflorescence first, an inflorescence typically has female flowers at the top and male flowers with the conspicuous anthers further down. Some insects do visit plantains to gather pollen, and a few species have slight pleasant scents that may contribute to attracting insects.

In the dock family (Polygonaceae), many bistorts and knotweeds are mainly entomophilous with pink or red flowers that attract insects; but the true docks (*Rumex*) have large greenish inflorescences that are protandrous, with the typical wind-pollinated features of reduced perianth, large anthers, and large tufted stigmas. In goosefoots and beets (Chenopodiaceae), almost the entire family is either wind pollinated or selfing, and usually there is no corolla, with short stamens and stigma emerging directly from a five-lobed calyx.

Occasional examples of wind pollination can be found in other herbaceous families: the ragweed and mugworts in Asteraceae, the meadow rues (*Thalictrum*) in Ranunculaceae, and dog's mercury (*Mercurialis*) in Euphorbiaceae are well-known cases, all exhibiting the requisite enlarged protruding anthers and high-surface-area stigmas, giving a generally fluffier appearance than their zoophilous relatives.

It should not be forgotten that even many classic insect-pollinated herbs also experience a small degree of wind dispersal of their pollen, which may be useful

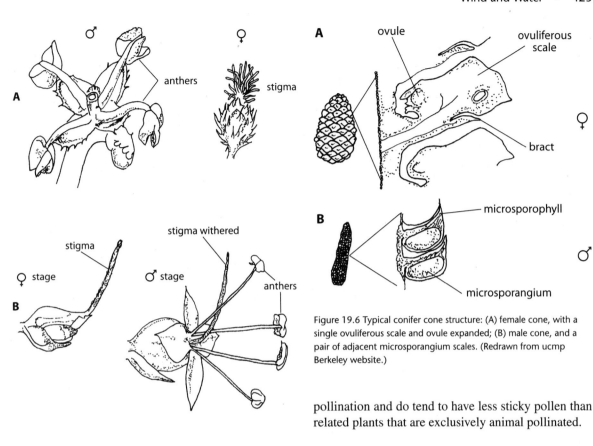

Figure 19.6 Typical conifer cone structure: (A) female cone, with a single ovuliferous scale and ovule expanded; (B) male cone, and a pair of adjacent microsporangium scales. (Redrawn from ucmp Berkeley website.)

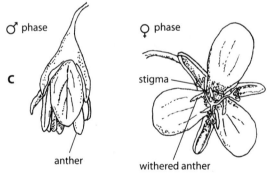

Figure 19.5 Flowers of some common anemophilous weeds; (A) nettle, *Urtica dioica*; (B) plantain, *Plantago lanceolata*, early female and later male stages; (C) dock, *Rumex obtusifolius*, male and later female stages. (Redrawn from Proctor et al. 1996.)

to them in poor weather when insects are scarce. Species with relatively open or bowl-shaped flowers, such as heathers (*Calluna*), lime trees (*Tilia*), and some rosaceous plants (*Helianthemum, Cistus*), as well as some willows, chestnuts, and maples already mentioned, may particularly benefit from some degree of wind

pollination and do tend to have less sticky pollen than related plants that are exclusively animal pollinated.

Wind-Pollinated Conifer Flowers

Traditionally known as cones, or more technically as **strobili**, these are rather different from angiosperm flowers and are always unisexual. Male cones mature earlier, each being made up from numerous stamens formed from a scalelike structure on the underside of which are two pollen sacs (fig. 19.6). Individual pollen grains bear 1–3 air-filled bladders or sacs (*sacci*) that probably reduce the rate of fall of the grains once released (Schwendemann et al. 2007), so increasing dispersal distance. They also have a crucial role in aiding flotation (Tomlinson 1994; Runions et al. 1999), including flotation on the pollination drop (see below).

Female cones mature slightly later, closer to the tips of the branches, and are somewhat larger, with thickened scales each bearing two ovules and with the scales set slightly apart from each other to give access to pollen via the micropyle; in some conifers (e.g., *Larix*, the larches) there are also colorful bract scales, making the female flowers relatively attractive.

The commonest situation is that any pollen grain

A

Non-saccate pollen, sinks into droplet on upright ovule

Saccate pollen, floats up into droplet

No droplet, pollen floats in on a raindrop

Pollen taken in by engulfment

Pollen grain germinates outside, tube grows inwards

B

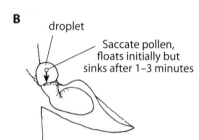 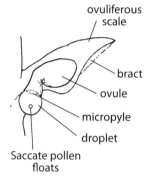

droplet

Saccate pollen, floats initially but sinks after 1–3 minutes

ovuliferous scale

bract

ovule

micropyle

droplet

Saccate pollen floats

Figure 19.7 (A) Schematic view of the pollen receipt options in conifer strobili, erect or pendant and with or without pollination droplets. (B) Illustrations of the first two options. (Modified from Owens et al. 1998.)

landing on a female cone near the micropylar opening into an ovule is drawn into that opening by a drop of sticky liquid, the *pollination drop*, a weak solution of sugars that helps to rehydrate the pollen and support its germination and that is then resorbed. However, details do vary in different taxa (Owens et al. 1998); figure 19.7 shows different possibilities for pollen receipt in conifer cones, depending on ovule orientation, presence or absence of a pollination drop exuding from the micropyle, and whether the pollen is buoyant (with sacci) or nonbuoyant so that it sinks.

After pollination, the scales thicken and close together, but fertilization often does not take place for weeks or months, even up to a year, and it may be a further year before the scales spread apart again into the typical dry cone structure, releasing the winged seeds.

Critical Factors Affecting Pollen Movement in Anemophilous Plants

The success of anemophilous pollen release and subsequent capture can be affected by a range of biotic and abiotic factors (e.g., Whitehead 1983; Okubo and Levin 1989), only some of which can be influenced by the plant itself.

Architectural Factors

1. *The position of release on the plant.* This is under the control of the plant, its growth and development leading to appropriate placement of flowers; usually at the periphery of the canopy for trees, or toward the apex for most grasses, with male flowers higher than female flowers.

2. *Spacing of conspecific and compatible flowers.* This is partly under the control of the plant, although only in the sense of exercising some indirect control of interplant spacing via successful germination or subsequent competition.

3. *Other vegetation structures.* Surrounding vegetation has filtering effects on the pollen emission and thus how far grains will be dispersed; a plant has little control over this except through competitive effects.

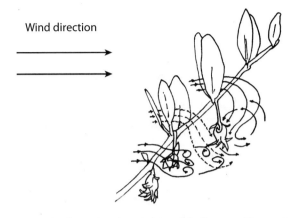

Wind direction

Figure 19.8 Pollen grain trajectories around the flowers and leaves of wind-pollinated jojoba (*Simmondsia chinensis*) showing the air turbulence that can lead to increased pollen contact with stigmas. (Redrawn from Niklas and Buchmann 1985.)

4. *The aerodynamics around the receptive surfaces.* This is affected by the geometry of surrounding structures and is not as random and inefficient as was once supposed. A flowing fluid (air or water) always has to divert around a solid object in its path, but the pattern of this divergence is complex. In practice, suspended particles within the flow tend to diverge less from a straight line due to their own momentum, so they are more likely to hit the obstructing surface, especially if the flow is roughly perpendicular to it. Increasing the local fluid turbulence helps to further increase the likelihood of impact, and flowers have many ways of achieving this and so improving their chances of pollen capture (fig. 19.8). Conifer cones are structurally complex, and Niklas (1984, 1985) interpreted this as being for good reason, their sculpted contours and often spiky projections helping to create elaborate vortices of moving air around them, slowing down pollen grains and keeping them in the vicinity of the cones for longer, with downstream eddies that recirculate the pollen back toward the receptive surfaces. However, more recent studies by Cresswell et al. (2007) could find no good evidence for such effects with *Pinus* and *Cedrus* cones.

Airflow round angiosperm wind-pollinated flowers, both trees and grasses (Niklas 1987), could also be "managed" by the plant's architecture. In the dioecious jojoba tree *Simmondsia chinensis*, from the arid southern United States, the small pendant female flowers lie at the tip of a branch underneath two unusually upright leaves, which were reported to create an almost laminar downward air flow directed at the flower, showering pollen most effectively onto the stigma (Niklas and Buchmann 1985; fig. 19.8). Likewise in the genus *Ephedra* Niklas and Buchmann (1987) demonstrated strikingly different behaviors of pollen from two species, largely due to very different pollen size, so that conspecific pollen was more readily trapped (Buchmann et al. 1989).

5. *The stigmatic collection efficiency.* This will be determined both by stigma architecture, relating to the aerodynamic flow as in (4), and by stickiness. Thus some plants appear to be able to select pollen of their own species from the air. For various *Pinus* species, Niklas and Paw (1983) found better capture of conspecific pollen than of other pine pollens. Linder and Midgley (1996) showed that for four genera of anemophilous plants the proportional capture of conspecific pollen was always above 40% and could be as high as 80%, when contemporaneous random collections on sticky glass slides were much more diverse (although the pollen selection was not as great as Niklas's models would predict). There are also suggestions that pollen capture at the stigma is augmented by electrostatic attraction (see chapter 7); pollen normally leaves the parent plant with a small negative charge, but carriage in the air could induce a moderately strong positive charge, and grains would then be attracted to the negative charge focused around the stigmas of recipient plants (Erickson and Buchmann 1983). Recent precision measurements of charges on anemophilous pollen grains indicated a good likelihood of electrostatic involvement in pollen capture at lower wind speeds (Bowker and Crenshaw 2007a,b).

Timing Factors

1. *Seasonal timing.* Most anemophilous flowers in seasonal habitats release their pollen early in the season, before there is much foliage, or more rarely in late summer when leaf cover is decreasing.

2. *The daily timing of release from the plant.* Again, this is substantially under the plant's control, not only by the typical time of anther dehiscence but also because many plants have mechanisms to prevent release at the normal time if there is no air movement. In catkins, pollen from one flower rests on the bract of

the flower below until shifted by air currents, and in most grasses the pollen is held in the shallow cups of the anthers even after dehiscence until some movement shakes the anther.

Atmospheric Factors

These are beyond any control of the plant.

1. *Air velocity.* An intermediate value is usually optimal, avoiding pollen release into still or sluggish air or into very high winds.

2. *Air turbulence.* Excess turbulence is unhelpful. Turbulence is affected both by the interplay of air currents and solid objects below (primarily the vegetation) and by convection currents of rising warm air.

3. *Air temperature and humidity.* Temperature alters the convection currents in the air, which are faster and more turbulent when there is a high radiant input from bright sunlight on clear days; and low humidity is favored, as it reduces the likelihood of imminent rain. Both temperature and humidity may also influence point 2 in the section *Timing Factors* above.

Overall Success of Anemophilous Pollen Movement and Deposition

Pollen that is successfully shed into moving air and not immediately washed to ground can travel impressive distances. Many records attest to pollen at altitude (at least 2000 m above ground) and in air far from land, both in the mid-Atlantic and well out into the Pacific Ocean. Pollen from Scots pine (*Pinus sylvestris*) has turned up in Spitzbergen 750 km to the north, and pollen from southern beech (*Nothofagus* spp.) occurs in the peat bogs of many Pacific islands at least 2000 km (and sometimes 5000 km) from the nearest growing trees. But in practice such long-distance movements are irrelevant for pollination, as most of this pollen will have been rendered nonviable by aging and by exposure to high UV loads (see chapter 7). The vast majority of wind-borne pollen is deposited close to its source, and within a few hours or days of leaving the parent tree. Figure 19.9 shows the general shape of measured pollen dispersal curves, and table 19.1 gives examples of estimates of the derived standard deviation measure (the root mean square of distances

moved) for common anemophilous plants. Figures are higher for trees, largely due to the greater elevation at which they shed their pollen, but for nearly all plants the mean figures are only a few meters or a few tens of meters. The shape of the curves is characteristically more highly peaked in the center and more spread out at the edges than a normal distribution would predict, and this curve is termed *leptokurtic*; interestingly, the same kinds of pollen dispersal curve occur for animal-pollinated plants (see chapter 3 and fig. 3.5).

Various key plant traits affect the outcome of wind pollination, summarized in figure 19.10, of which the most important is distance between potential mates (Cruden 2000). Nearly all anemophilous plants have abundant pollen and low ovule number, and hence very high P/O ratios (often in excess of 10^6; see table 7.5). Pollen grain number and the P/O ratio needed should theoretically be positively related to a plant's distance from a pollen source, so that in turn seed set will be inversely related to that nearest (suitable) neighbor distance. Allison (1990) confirmed these relationships for yew trees (*Taxus*), and seed set was also linked to plant spacing for at least three other wind-pollinated plants, *Espeletia* (Berry and Calvo 1989), *Staberoha* (Honig et al. 1992), and *Thalictrum* (Steven and Waller 2007), although in the last case only isolated plants in low-density populations were in practice pollen limited. For *Taxus*, there were also negative relationships between pollen amount and stigma surface area or stigma receptive period (Weis and Hermanutz 1993), just as in animal-pollinated plants (chapter 7). However, since pollen grains have to be small, the inverse link between grain size and grain number seen in zoophilous plants is often lacking, as is the relationship of grain size to stigma depth.

High P/O ratios are not necessarily only due to the uncertainty of pollen reaching a stigma, however, given the increasing evidence (summarized above) that this process is less random than once assumed. Midgley and Bond (1991a) suggested that intra-male competition could be at least as important in selecting for high pollen output and/or high P/O values, while Niklas (1992) proposed that pollen is more abundant largely because wind-pollinated plants can devote more of their metabolic reserves to it, not being constrained to invest in costly advertising materials.

On balance, wind pollination can be extremely effective in the right conditions, and some outcrossing anemophilous plants have the highest rates of gene

A

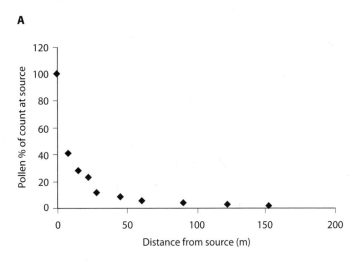

B

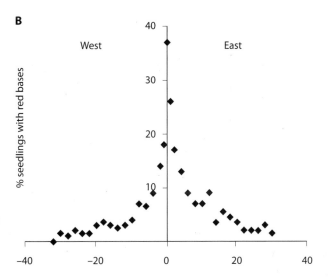

Figure 19.9 Pollen dispersal curves show a leptokurtic distribution, strongly concentrated near the source but also spreading out to longer distances than predicted by a normal distribution: (A) pollen from *Pinus elliotti*, captured on sticky slides; (B) seed set for a grass, *Lolium perenne*, to east and west of a population marked with visible red shoot bases. (Redrawn from Proctor et al. 1996.)

flow of any plants (Hamrick et al. 1995). This accords well with the fact that anemophily has been secondarily adopted by the relatively advanced and exceptionally successful family Poaceae (grasses) as the best solution for reproduction in open grass-dominated habitats.

2. Hydrophily

Whereas wind pollination is reasonably common and widespread across plant taxa, pollination by water is quite rare, occurring in several monocot families but just two dicot families, (collectively less than 3% of all angiosperm families). Hydrophily has a very patchy distribution indicative of multiple evolutionary origins. The great majority of aquatic angiosperms are in practice pollinated either by insects or by wind (Cook 1988), which is not surprising given the density of water (both freshwater and seawater) that produces substantial drag on any objects (such as pollen grains) moving within the flow, thus greatly reducing rates of mixing.

Dicot examples of hydrophily are found exclusively among the starworts (*Callitriche*) and the hornworts (*Ceratophyllum*). From the monocots there are examples from at least 29 genera in 9 families. Six of these families use hydrophily exclusively, and they tend to

TABLE 19.1
Pollen Dispersal Distances for Common
Anemophilous Plants

Common name	Genus	Standard deviation
Trees		
Ash	*Fraxinus*	15–45
Cedar	*Cedrus*	45–75
Elm	*Ulmus*	300+
Fir	*Pseudotsuga*	20
Pine	*Pinus*	15–65
Poplar	*Populus*	300+
Spruce	*Picea*	40
Herbaceous		
Sugar beet	*Beta*	2–4
Maize	*Zea*	6–9
Plantain	*Plantago*	50–60

Note: The measure used is "standard deviation," in meters, which is the root mean square of measured distances from the source plant. Modified from various sources.

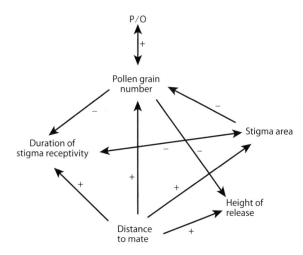

Figure 19.10 Factors affecting pollen grain number and pollen/ovule ratio in wind-pollinated plants. (Redrawn from Cruden 2000.)

be intimately associated with brackish or marine habitats, most notably the seagrasses (Cox 1988, 1993) which make up three separate but convergently rather similar families.

Hydrophily can take place underwater or at the water surface, and these two systems require rather dif-

ferent adaptations so are best treated below as two separate syndromes (Cook 1988; Cox 1988). However, they do share several crucial features:

1. The stigmas tend to be large, erect, and rigid, without the frilled or tufty forms of anemophilous plants (which would collapse in water)
2. The pollen is generally elongate rather than spherical (increasing the possibility of contact with a stigma) and often becomes clumped together in "rafts"
3. Pollen sizes are highly variable, since their transit in water is affected by their density rather than their mass
4. Pollen grains have a reduced exine and increased apertures and are much less rigid than terrestrial pollens (Pettitt 1984; Diez et al. 1988); they may also incorporate starch grains that enhance flotation (Mahabale 1968)
5. Pollen amounts are variable but generally quite low, so that P/O ratios also tend to be low (see table 7.5).

Hypohydrophily

Also known as hyphydrophly, **hypohydrophily** is underwater pollination, occurring in three dimensions. It is found in 18 genera, many of these being marine monocots; in fact all the marine angiosperms belong here. It may involve pollen moving through the water, but quite often it is whole anthers or even whole flowers that move. It requires some reasonable degree of water current flow to function at all (Pettitt 1984; Ackerman 1997a).

Underwater flowers are inevitably wet throughout the processes of pollen maturation and release and capture, imposing very different needs from those seen in terrestrial flowering systems. The flowers tend to have a much reduced corolla, usually white or greenish, with just one ovule and a large stigma; and they are usually unisexual. Pollen grains usually have an extremely thin exine, and when clumped together they tend to form loose gelatinous masses.

The hornwort *Ceratophyllum* presents probably the simplest case of underwater pollination and comes from a very ancient lineage pre-dating the main burst of angiosperm diversification (Dilcher 1995). In both sexes the flower (fig. 19.11A) is a tiny structure in a leaf axil, partly surrounded by a ring of 10–15 small perianth lobes. Female flowers are very simple, with a single ovary bearing an oblique and often bent style

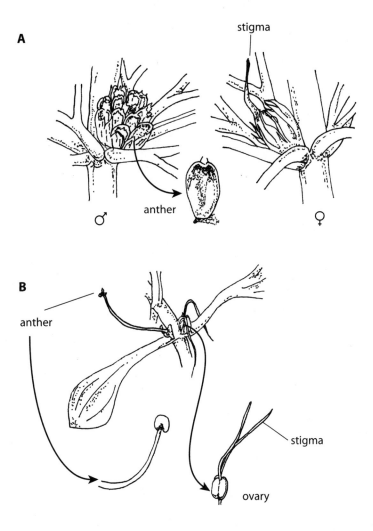

Figure 19.11 Aquatic flowers: (A) hornwort, *Cerato-phyllum* and (B) starwort (*Callitriche*). (Redrawn from Proctor et al. 1996.)

whose tip points upward. Male flowers consist of around 12–20 short stamens with stalkless anthers, bearing an expanded float. These stamens break away from the flower and float to the surface where they dehisce, but the pollen then sinks back down through the water column, much of it probably self-pollinating the parent plant unless currents are strong.

The only other dicot examples are the water starworts (*Callitriche*), which are less easy to characterize (Philbrick and Anderson1992). The plants vary from fully aquatic to amphibious and often produce flowers both above and below water. Flowers are extremely small and simple (fig. 19.11B), little more than a single stamen in males and a single ovary with two stigmas in females. In the submerged plants, pollination is primarily due to the most unusual phenomenon of growth

of pollen tubes through the water to an adjacent stigma, mainly resulting in selfing but occasionally achieving outcrossing. Even in emergent plants at least some fertilization is due to pollen tubes growing through the vegetative tissues to reach female flowers.

In the submerged monocots, extremely simple flower structures are again the norm (fig. 19.12). *Zannichellia* and *Najas* are pondweeds, and both have clusters of flowers, often one male and a few female. In *Najas* the pedicel of the male flower elongates and curves just before dehiscence, to bring anthers close to a branching stigma (Huang et al. 2001). Pollen grains are then released directly into the water, generally in small masses that adhere for a few minutes and then break up. Not infrequently the pollen grains begin to germinate while still floating freely (perhaps aided by

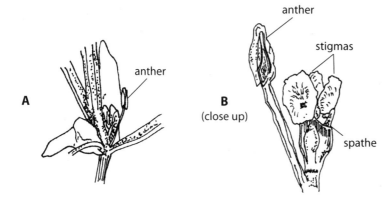

Figure 19.12 Flowers in the submerged monocot *Zanichellia*, with a close-up view of one male and three female flowers. (Redrawn from Proctor et al. 1996.)

the thinness of the exine), and it is likely that the resulting pollen tubes help to ensure catchment on a stigma.

In contrast, the marine seagrass monocots, distributed across three families but best known as eelgrass in the genus *Zostera*, are noted for their unusually long filamentous or spaghetti-like pollen, released as tangled floating pollen rafts easily visible to the naked eye. In most genera the individual pollen grains take this remarkable elongate form, probably convergently (Ackerman 1995), although in Caribbean turtle grass (*Thalassia*) a similar effect is achieved with more typical round pollen grains that are strung together in long strands. Most seagrasses have separate male and female flowers (fig. 19.13), held within narrow sheathed flowering spikes, where female flowers mature first. Once their similarly (although less extraordinarily) elongate stigmas have begun to wither, the anthers mature and emerge from the sheath, and the pollen rafts are released to drift through the eelgrass beds, where they will wrap around any elongate object they encounter, potentially thereby becoming trapped onto the stigmas of another plant. The elongation of grains gives a better change of encountering a stigma, especially as the female flowers produce local eddies around themselves (Ackerman 1997b) that make the pollen grains rotate and interact with the flow lines leading in to the flower. In species that are always submerged, the pollen grains have a similar density to that of seawater and move in three dimensions; but in intertidal species the anthers have many air spaces within them and the pollen is water repellent, forming rafts on the surface film and rising and falling with the tide. In the common *Zostera marina*, the pollen is neu-

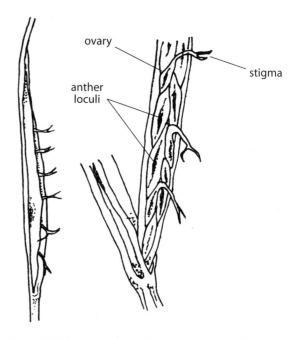

Figure 19.13 Flowers of seagrass (*Zostera*), showing a sheathed flowering shoot, and the same with sheath peeled back to show anthers and ovaries. (Redrawn from Proctor et al. 1996.)

trally buoyant and can function either in the surface film or when submerged (de Cock 1980; Cox et al. 1992).

Marine pollens provide a useful food source for fish, and in *Thalassia* there is a pronounced synchrony of anthesis and especially of pollen release from the males within one hour of dusk, perhaps ensuring maximum fertilization but also helping to avoid too much pollen predation (van Tussenbroek et al. 2008).

Epihydrophily

Sometimes termed ephydrophily, **epihydrophily** is pollination at the water surface and therefore in two dimensions. It can be either wet or dry; that is, the pollen may be released to float freely at the surface or it may remain adherent to the anthers, the whole flower moving across the water surface film to meet the stigmas just above the surface but effectively in air. Since pollination occurs at the surface, it is strongly affected by water currents, which can be much stronger at the air-water interface; therefore most plants using this system do not dehisce unless winds and current speeds are reasonably low. However, the fact that the pollen moves in one plane does then greatly increase the chances of hitting the target (stigmas) compared to three-dimensional underwater pollination.

In *Elodea*, one of the commonest waterweeds, it is the pollen that moves between plants in the normal manner. Both male and female flowers (fig. 19.14) occur on long stalks that reach to the water surface, where they open. In the male flowers, which tend to be rare, there are six small whitish petals so that they appear not unlike many terrestrial angiosperm flowers; but the sepals have water-repellent surfaces and so tend to push the rest of the floral structures out above the water film, acting as a cup on which the receptacle and stamens can sit erect. There are nine rather short and stout stamens, which dehisce very rapidly and scatter pollen grains over the surrounding water film, where they rest on tiny exine spines that hold back the water and so keep the pollen dry. In the female flowers, the three large water-repellent stigmas are somewhat recurved and protrude beyond the perianth, so that the flower sits in a small depression in the water film, on the tips of two or three of its stigmas. Any pollen grains, moved by air and water currents, that drift toward the female flowers are drawn into the surface dimple in the water meniscus and so inevitably contact these stigmatic surfaces.

Potamogeton pondweeds (fig. 19.15) often have systems rather similar to *Elodea* for pollination, but there is variation within the genus, some species having flowers emerging fully from the water surface and using wind to carry the pollen, and others such as *P. pectinatus* sometimes failing to reach the surface at all in deeper waters, their pollen being carried upward to the air/water interface in tiny air bubbles that form as the anthers dehisce (Philbrick and Anderson 1987).

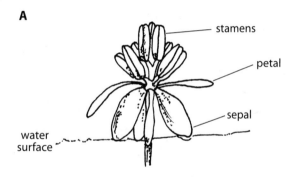

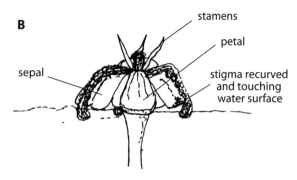

Figure 19.14 *Elodea* (Canadian waterweed): (A) male flower and (B) female flower with large stigmas, both emerging to effect pollination at the water surface. (Redrawn from Proctor et al. 1996.)

This may not be very efficient: Guo and Cook (1989) recorded seed set of 2%–7% in submerged inflorescences, compared with values of up to 40% in those that were emergent. Species of *Potamogeton* have been analyzed in terms of the predicted links between flower traits (Philbrick and Anderson 1987; Cruden 2000), showing that species with small inflorescences also tend to have small and short-lived stigmas and low P/O ratios (2000–10,000), and to exist in close proximity with neighbors, while as expected the species with larger numbers of flowers per inflorescence have large stigmas that are receptive for up to a week and have high P/O values (25,000–40,000), tending to live in highly dispersed populations where potential pollen donors are very distant.

One of the classic examples where whole flowers are the motile agents for pollination is the ribbonweed or tapegrass *Vallisneria*, originating in the tropics but widely naturalized elsewhere and familiar from its use in many aquaria. The plants grow in the bottom muds,

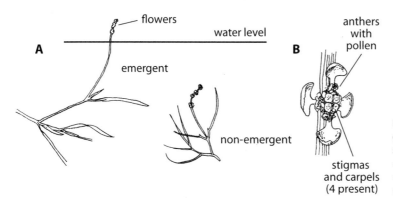

Figure 19.15 *Potamogeton* pondweeds: (A) growth habits of emergent and nonemergent species, and (B) a typical single flower of *P. berchtoldii*; usually protogynous so the stigmas are unreceptive once the pollen is dehiscing. (Redrawn from Proctor et al. 1996.)

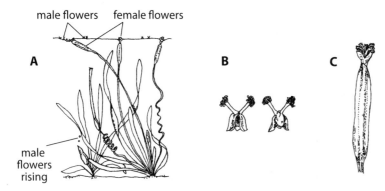

Figure 19.16 *Vallisneria*, tape-grass, showing (A) the growth form and male flowers breaking off to float to the surface; (B) male flowers at the surface; (C) a single female flower. (Redrawn from Proctor et al. 1996, partly based on earlier sources.)

with long ribbonlike leaves reaching almost to the surface. They are dioecious, with separate male and female flowers (fig. 19.16). Male flowers grow near the base of the plant but break free as they mature and then rise to the surface. Each is approximately globular, formed from three simple sepals, which once at the surface open to reveal two stout stamens, dehiscing to form two small balls of pollen. These tiny male flowers then drift about on the surface film, driven by wind and water currents. In contrast, the female flowers remain on the plant and are propelled to the surface on the tip of elongate spiraling stems. Once the tip of the flower reaches the surface it penetrates the surface film and its hydrophobic properties cause it to form and lie in a shallow dimple in that film. The three sepals then open to expose three large stigmatic lobes, covered in fine hairs that are strongly hydrophobic; the stigmas are therefore unwettable. When a male flower drifts close to a female, it tends to be drawn in by the dimple in the water surface and so hits the stigma.

Clearly the special properties of the water meniscus and surface tension phenomena play a role in bringing pollen to stigmas more reliably than chance alone, and the interaction of filamentous pollen grains with eddies around the open flowers play a part in increasing capture rates (just as for wind pollination). As with anemophily, the movement of pollen from donor to recipient is substantially better than would occur by chance alone.

We should also mention a further kind of hydrophily in the broadest sense of pollen movement by water, in that pollination can occasionally be effected by rain (*ombrogamy*; see Cox 1988). Cases of this could arise in particularly wet climates, and it has been specifically proposed for *Piper nigrum* flowers and perhaps for some cycads, but all of these have been disputed. A report of rain involvement in the pollination of a spruce (Runions and Owen 1995) seems more reliable; here the natural pollination drops secreted within the cone can scavenge passing pollen, but when the drops are enlarged after rainfall they do even better at floating pollen in to the micropyle. The case of rain-aided selfing in an orchid (Pansarin et al. 2008; chapter 3) should also be included here.

3. Overview

All modes of abiotic pollination share certain common characteristics, most notably the small drab flowers with reduced perianths, exposed anthers and stigmas, and usually reduced ovule number. However, they also all achieve markedly better pollen transmission and pollen capturing specificity than is usually realized, and are by no means random processes with only a tiny chance of success, as they are sometimes portrayed. In fact a diverse range of plants and their pollen grains have proved adept at exploiting the varied properties of fluid flow and at managing this flow to their own advantage.

Chapter 20

SYNDROMES AND WEBS: SPECIALISTS AND GENERALISTS

In recent years, many pollination biologists have embarked upon a reassessment of the classical approach to their subject, as established above all by Sprengel, Darwin, Vogel, and van der Pijl. For about 150 years the main emphasis was on showing the adaptive value of floral traits in relation to particular visitors, and then in showing how neatly plants can control these visitors to achieve cross-pollination. In turn this led to the establishment of particular relationships as pollination syndromes set out in chapter 11 and pursued in detail through chapters 12–19, where major groups of flower visitors were associated with certain combinations of structure, advertisement, and reward in flowers, which were seen as adaptive. Stebbins (1970) formalized aspects of this approach in terms of the *most effective pollinator principle* (MEPP), supposing that a given plant would evolve specializations suited to its more effective pollinators. In turn this leads to an assumption that specialization is desirable and promotes floral divergence and potential pollinator-mediated speciation through reproductive isolation, underpinning the modern diversity of flowering plants as was outlined in chapter 4.

Throughout the 1970s these views were somewhat amended as ecological principles of community structure and niche partitioning were assimilated into pollination studies, along with increasing appreciation of the complexities and convolutions of species interactions. By the 1990s many ecologists were increasingly pointing out that alongside seemingly quite specialist flowers there were many that were in practice visited by multiple kinds of visitors, and this was formalized by Ollerton (1996) as a "paradox": that flowers may appear to be phenotypically specialized but ecologically generalized.

The simple scenario of plant-pollinator evolution toward increased specialization, outlined at the end of chapter 11, is now seen as far from universal, and accumulating research work has stressed that pollination does *not* necessarily tend to become more specialized through time, and that plant-pollinator coevolution is far more complex and more varied than the original models implied. In the last two decades many workers have also turned their attention to a more precise analysis of the mechanisms and genetics of selection on floral traits and to modeling the evolution of specialization against a background of conflicting selection pressures. In addition, there has been much greater interest in intraspecific variation in floral characters; if a species is coevolved with a particular pollinator and all its features are selected to favor visits by that pollinator, why is there so much polymorphism within a flower species, so much variation in color, scent, and shape? And how does this variation, whether natural or artificially selected for by human intervention, affect the final outcome of flower visits? If floral variation within a plant species is substantial, how far are animals really exerting any selection on floral traits, how far are such traits adaptive, and how can this adaptation be quantified?

In short, are syndromes really a helpful way of approaching floral biology? Increasingly, ecologists have tended to stress that syndromes represent a serious oversimplification, that most flowers are generalists rather than specialists, that pollinator selection on floral traits is often rather weak and constrained by external ecological factors, and that "adaptationist storytelling" about flower-pollinator coevolution has been greatly overdone in the past. In this view, floral generalization is predicted as the best strategy for a flower visitor whenever the abundances of preferred species are low or are fluctuating in time and space (Waser et al. 1996). Specialization and constancy are not to be expected and should not be seen as indicators of an advanced pollinator (Waser 2001, 2006). This view has come to dominate the literature in the last decade, although it is worth noting that the paradigm shift did not begin from nowhere in the mid 1990s; in fact an earlier review by Thomson (1983) noted the rarity of obligate specialist interactions in pollination biology and pointed out that in practice pollination webs existed and (as with food webs more generally) were likely to be complex and cross-connected. By 1996 Herrera could assert that "there is now overwhelming evidence that syndromes are of little value in explaining interspecific variation in pollinator composition," while Ollerton and Watts (2000) claimed real difficulty in inferring the type of pollinator that visits a plant just from observing the floral traits. Various authors now start their data presentation from a standpoint of "generalization is the rule," accompanied by skepticism as to the reality of syndromes (e.g., Medan et al. 2006). This movement has proceeded to such an extent that some authors cannot bring themselves to use the word

syndrome at all, and speak (even then reluctantly) of "types of pollination systems."

In some respects, then, the world of pollination biologists has divided; one half focused on generalization and ways of dealing with it and predicting outcomes of it using increasingly sophisticated modeling approaches, and the other half continuing to document the specialist relationships they perceive in the field and to detect the pollinator-mediated selection that underlies this specialization.

This chapter follows on from the defense of syndromes mounted in chapter 11, and all the material in support of specific syndromes in chapters 12–19, by addressing the core issues raised by critiques of pollination syndromes and alternative approaches to characterizing pollination in communities, in the context of floral and pollinator specialization. Many will perceive it as covering the crucial issues for modern pollination ecology; although, in some sense, that perception is part of the problem!

1. Theoretical Arguments against Syndromes and Specialization

As outlined above, it is becoming commonplace to use a more ecological or community level approach to floral adaptation and divergence, considering flower communities as a whole rather than individual interactions. A good summary of this change of viewpoint was given by Wilson and Thomson (1996). The key paper to move things onward was by Waser et al. (1996); these authors stressed the distinction between laws and trends and noted that syndromes are not laws at all, but at best just fairly weak trends, with the links between floral traits and observed pollinators being much less strong and less simple than previous literature asserted. They asked that evolutionary pollination biologists should not let "the appeal of orderly pollination syndromes" obscure the richness and complexity of the relations between plants and flower visitors.

There has also been increased appreciation of the conflict at the heart of plant-pollinator interactions. Many mutualisms evolve out of essentially parasitic and exploitative relationships, and previous emphases in mutualism biology implied that coevolution should lead from this to ever-increasing specialization and stability, where the aims of both partners were best achieved; this is an attitude prevalent in all early

Darwinian literature. But, as earlier chapters have explicitly noted, the goals of plant and visitor are very different and often directly conflicting, so that a priori the outcome is often likely to be a more generalist compromise rather than increasing specialization. In a scenario of competing interests, instances where the two parties converge on a relationship of mutual dependence should be rather rare. Furthermore, the pollination syndromes discussed in earlier chapters are named and characterized by their supposed pollinator group, and hence they carry inferences about past coevolutionary processes having occurred that are untested and untestable.

It is also to be expected that there will be evolutionary and ecological risks to being a specialist; specialist associations are vulnerable to any perturbation, a classic ecological principle that is no less relevant in pollination ecology. A specialist plant would become vulnerable if the abundance of its key pollinator were reduced; similarly, a visitor adapted for a single type of plant would suffer upon a decline of that plant. Buchmann and Nabhan (1996) presented a number of good examples on this theme. High reciprocal specialization between a plant and a pollinator is especially risky where the abundance or the quality of interactions varies over time (Waser et al. 1996; Renner 1998; Vázquez and Simberloff 2002; Memmott et al. 2004); and in any obligate one-to-one relationship the consequences are potentially terminal.

Turning to the evolution of specialized floral types, these must presumably arise from a more generalized type (an issue considered more specifically below); and there must be a transition phase when adaptations to a particular pollinator are accruing but there is still exposure to other visitors. At this stage, any further gains in fitness in adapting to the specialist pollinator must more than offset losses from concomitantly becoming less adapted to the other possible pollinators. But pollination could be seen as occurring in a fine-grained environment (Aigner 2001, 2006), the plant experiencing a range of different environmental conditions in relation to the various visitors it may receive. Specialization is unlikely to proceed further in a heterogeneous environment, especially perhaps where each type of flower visitor itself experiences spatial and temporal population fluctuations.

Another crucial point is that many and perhaps most flower traits are undergoing selection from influences other than just their pollinators (e.g., Strauss and Whittall 2006; Irwin and Adler 2006; Gómez 2008), and thus their trait optimum is often a compromise (fig. 20.1). For example, many flower characteristics also influence herbivores or seed predators and are in turn influenced by them (e.g., Cariveau et al. 2004; and see chapter 25). Some floral traits are altered by infection from fungi etc., and others again are affected by environmental conditions. If the selective fitness value of a trait is affected in the same direction by factors other than pollinators, then the trait value seen is likely to reflect the preferences of a relatively specialized pollinator (fig. 20.1A); but if the selection from other interacting organisms is in the opposite direction then the trait seen in the natural population may be a trade-off, very unlike that preferred by a pollinator (fig. 20.1B), and a specialized interaction becomes more unlikely. Steffan-Dewenter et al. (2006) pointed out that a disregard for other biotic interactions (herbivory, seed predation, etc.) that might modulate plant-pollinator interactions could lead to a greater detection of "specialization" than really exists.

Syndromes are also seen to be undermined by the necessity to erect a category for the generalist flowers; for example Corbet (2006) reported this "absence of a category" as "a major shortcoming" for syndrome approaches. Such flowers get most of their visits from a diverse range of small and short-tongued insects (flies, beetles, small wasps, etc.), and they can be dominant in some communities. These generalist flowers are especially represented by the umbellifers, many Asteraceae, and some Rosaceae in temperate systems (as we have seen, but note also the cautionary message at the end of chapter 11); and by examples from a wider range of families in the tropics.

From all these theoretical standpoints, syndromes can be seen as far too "cosy" as a framework for understanding pollination ecology.

2. Practical Evidence against Syndromes and Specialization

Multiple Visitors to Single Plant Species

There have been repeated citations of examples where a classic plant from one syndrome gets visits from other kinds of pollinator; for example, Waser (1983a) mentioned a typical bumblebee flower (*Delphinium nelsonii*) visited heavily by hummingbirds and a typical

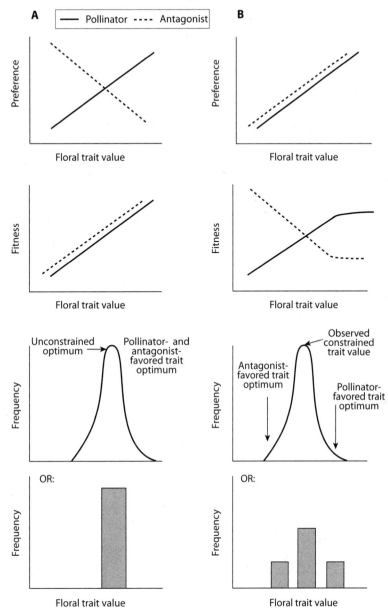

Figure 20.1 Floral trait evolution, with multiple agents—both pollinators and antagonists—acting: where preferences are opposing as on the left (A), they favor the same trait optimum; where their preferences are the same (B) the fitness effects will be opposing and there may be optima of traits as compromises, and a greater range of phenotypes. (Redrawn from Strauss and Whittall 2006.)

hummingbird flower (*Fouquieria splendens*) visited heavily by solitary bees, and this and many similar examples are regularly repeated.

More generally, it is an indisputable fact that any one plant may get flower visits from many different kinds of visitors, from different taxonomic groups, with visit patterns varying in space and time. These kinds of studies are presented at various taxonomic levels, and where insects are concerned they commonly involve orders or families of visitors. For example, Herrera (1996) gave records for seven different Mediterranean and southern European habitats (fig. 20.2), showing that it was very common for two or more orders of insect (and often all five common orders) to visit any flowers of any one plant species in a community, even where the flowers were tubular enclosed types and where (at least loosely speaking) most of the visitors were judged to be potential pollinators. At

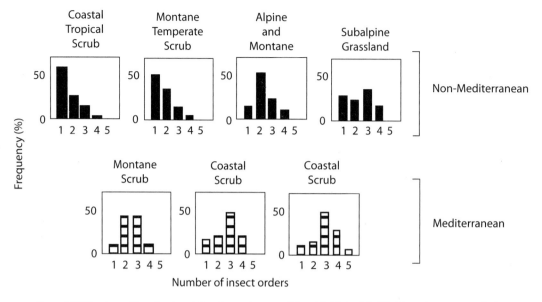

Figure 20.2 Numbers of insect orders visiting flowers in seven different habitats. (Modified from data sets compiled by Herrera 1996.)

these sites an average of 2.7 orders visited per plant species. Herrera's data have been taken to indicate that only a low degree of floral specialization is prevalent. Many other authors have come to a similar conclusion from data of this kind. In fact, it has generally been claimed that statistical analyses of observed visitation patterns give only very limited support for a classification of plant types in terms of traditional syndromes (e.g., Ellis and Ellis-Adam (1993) in Europe; Hingston and McQuillan (2000) in Tasmania; and more generally see Ollerton and Watts (2000)).

Particularly widely cited now are two analyses by Waser et al. (1996). The first was based on an early but large-scale study (Clements and Long 1923) of 94 plant species in Colorado, where the plant species each received visits from 1–62 animal species (mean 9.8), and 80% were visited by more than one animal species. Animals visited 1–37 plant species (mean 3.3), with 48% visiting more than one plant species. The second analysis was based on the work of Robertson (1929), who looked at 375 plant species in Illinois; here 33.5 animal species on average visited each plant species, and 91% of plants had more than one visitor type. On this basis, generalization was said to be predominant. However, several criticisms of these analyses can be made and will appear in later sections.

To pick out one more of many studies that have followed this lead, Minckley and Roulston (2006) worked in the Chihuahuan Desert with a bee-dominated community and found that each plant species hosted 0–13 visitor species, and there was a strong positive relation between number of visitors and number of types of visitor (fig. 20.3A). In terms of specialization, only *Penstemon parryi* was really specialist (fig. 20.3B), getting visits solely from *Osmia* bees, while oligolectic bees visited plants that also had many generalist polylectic bee visitors. Perhaps the most interesting finding was that oligolectic bees tended to visit rather simple flower types, not the more obviously specialized flowers. However, the simple flower types did not host significantly more visitor types (6.0 ± 2.6) than the more complex flowers (9.4 ± 1.7). These authors then reanalyzed the data from three other studies and again found that oligoleges and polyleges tended to converge on the same floral species (fig. 20.3C).

Spatial and Temporal Variation in Flower-Visiting Faunal Assemblages

Not only are flowers visited by many animal species, but the pattern of visits is highly variable in time and

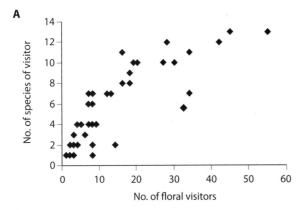

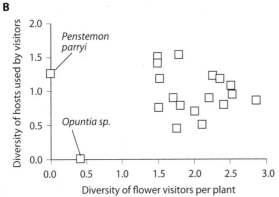

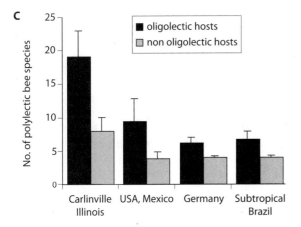

Figure 20.3 (A) Visitor number plotted against visitor type, in a Chihuahuan Desert bee-dominated community (redrawn from Minckley and Roulston 2006); (B) the Shannon-Weaver diversity of floral visitors per plant against the relative specialization of those visitors (~ the diversity of plants that they used), with *Penstemon* as an outlier visited by only *Osmia* bees (and *Opuntia* having a few but highly specialized visitors); (C) analysis of this and three other studies, by the same authors, showing an overview of the numbers of (generalist) polyleges hosted by plants that also do or do not host more specialist oligoleges, most of the generalists and specialists converging on the same hosts.

space (Herrera 1988, 1990, 1996; Waser et al. 1996; Armbruster et al. 2000; Price et al. 2005). Pollinator communities vary both across years (e.g., Herrera 1988; Traveset and Sáez 1997; Fenster and Dudash 2001) and within seasons (e.g., Ashman and Stanton 1991; Gómez 2000), and even in relation to local microclimate (Herrera 1995a; Zamora 1999). Price et al. (2005) gave a detailed analysis of a seven-year study on *Ipomopsis* which showed rates of visitation varying by more than an order of magnitude between sites and between years. Figure 20.4, A and B, shows variation in visitation to lavender flowers for 17 different sites in Spain and for sites close to or distant from streams (Herrera 1988), and figure 20.4C shows variation through the day at one site in terms of pollen deposition and distance flown between flowers (Herrera 1990).

Such variation is thus well documented and ubiquitous, occurring in most habitats, and critics of syndrome approaches point out that it is bound to lead to fluctuations in the magnitude and direction of phenotypic selection. Indeed, it has been alleged that an undue concentration on very small-scale studies in the past might also have led to a greater detection of "specialization" than really exists (Steffan-Dewenter et al. 2006).

There is a further component of temporal unpredictability in many pollination relationships, where a specialist visitor may be active before or after its preferred plant is in flower in a given year, or where the floral host may bloom twice a year but the specialist visitor overlaps with only one of these events. For example, *Larrea*, the creosote bush, commonly has a second late-summer bloom in southern United States deserts when only 2 of the 22 oligoleges that visit in the spring are present (Minckley et al. 1999, 2000). Still with the temporal theme, there are also obvious examples where long-lived pollinators are feeding over much longer periods than the flowering season of any one plant, so that the pollinator cannot be too specialist; this was mentioned in chapter 15, and a hummingbird example of the inevitable multiple plant usage was shown in figure 15.8.

There are likewise additional spatial issues, relating to range effects and the spatial unreliability of specialist visitors. Widespread and seemingly specialist plants are not overlapped across their whole range by any one or few visitor species. Again *Larrea* serves as a good example, with only 2 of the 22 oligoleges that visit it

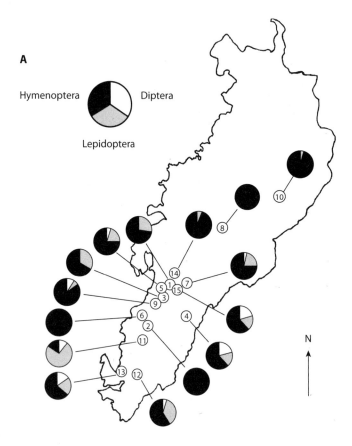

A

Hymenoptera   Diptera

Lepidoptera

N

having distributions covering more than half of the species range (Minckley et al. 1999, 2000). Such effects can be due to lack of suitable nesting sites or nesting material in some areas, but whatever the cause they should limit the plant's scope for specialization.

Finally, long-term climatic instability across time and space has repeatedly altered the communities in which organisms exist, thus acting against symmetrical specialist-specialist interactions (Ollerton et al. 2003).

Pollination of Plants without Their Natural Pollinators

There are well-known examples where a community of plants is successfully pollinated despite having no shared history with supposed natural pollinators. For example, there are at least 12 species of apparently ornithophilous plants on the Canary Islands, relict spe-

cies from preglaciation forests, but no classic bird pollinator species as visitors; the flowers are visited and effectively pollinated by various unspecialized warblers (Olesen 1985). On the scale of seminatural experiments brought about by human intervention, there are many cases where introduced plants are successfully pollinated by entirely new and unfamiliar pollinator assemblages. The syndrome approach also suffers in the light of experience with practical crop pollination, especially with introduced species, where it is commonly found that a plant is successfully pollinated by a new range of visitors when introduced to a new country (which after all is the reason why much of human cultivation worked at all).

Constraints on Specialization

In many areas of ecology, specialism is constrained by phylogenetic and morphological constraints, and this

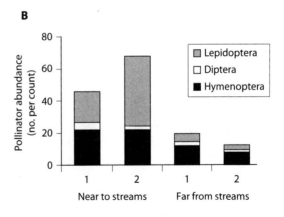

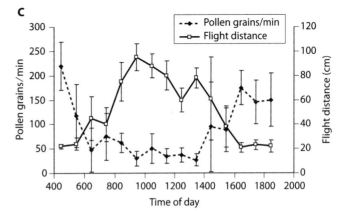

Figure 20.4 Visitor groups to lavender flowers varying (A) with site and (B) with proximity to streams, in Spain (Herrera 1988); and (C) the variations with time, in terms of pollen grains transferred to stigmas per minute and distance moved between flowers, showing greater efficiencies of pollen transfer in early and later daily periods (± SE) (modified from data in Herrera 1990).

is assumed to be true for pollination relationships as well. Intrinsic limitations on adaptedness may arise from genetic, life history, or developmental constraints (Kochmer and Handel 1986; J. Zimmerman et al. 1989; Delph et al. 2004). Other kinds of constraints are considered in section 5, *Refining Web-Based Approaches* below.

The Equivalence of Visitors as Pollinators

Somewhat counter to evidence given in chapter 11, it has been argued that many visitors to a flower will show functional equivalence in their ability to transfer pollen (e.g., Gómez and Zamora 2006), so that in practice they will act together to select for rather generalist floral adaptations and are not distinctive selective agents.

It could also be the case that many apparently specialist visitors are not necessarily good pollinators.

Minckley and Roulston (2006) asserted that "many are not," although in citing works on *Lantana* (Barrows 1976) and on *Heterotheca* (Olsen 1997) they picked examples where it is doubtful that most would call the visitors specialist anyway. They further noted that an apparently specialist visitor's efficacy can be diminished by morphological mismatch or inappropriate behavior such as robbing, although again it is doubtful that a pollination biologist would consider such an animal as a specialist.

Specific Criticisms

As specific evidence against the value of flower color syndromes, a study of German wildflower communities (Chittka, cited in Waser et al. 1996; see fig. 5.12) recorded that the classic color patterns seen as characteristic of particular syndromes were only weakly related to recorded visitors, and that floral color—often seen as the key element in the syndrome concept—was in fact not a good predictor of a flower's pollination system.

Another specific issue is the common occurrence of rewardless flowers, which is perhaps hard to explain if pollination syndromes and specialization are important, given the learning abilities of many foragers. Renner (2006) argued that the intermediate stage of transient rewardless flowers is crucial, preventing the learned discrimination and avoidance that would otherwise occur and so militating against too much specialization. In this context, Gómez and Zamora (2006) suggested that extreme specialization may *only* be found in permanently rewardless flowers, or in flowers producing unusual rewards such as oils, fragrances, or resins, where the visits of normal pollen and nectar seekers are discouraged.

The specific case of bees is also instructive here. Extensive work on how they forage and optimize their foraging has allowed detailed foraging models, especially for social species. Usually honeybees and bumblebees are generalist at the colony level but can be relatively specialist as individuals. Also they seem to prefer minimum-variance flowers, as reported in chapter 18. However, this is not always the case. First, preferences may change with colony state, so that whether the nest is short of nectar or already has good stores may make a difference to bees' preferences and "fussiness" (and hence to apparent specialization). Second, learning alters their preferences (in particular, handling time can be reduced by learning); but bees do not live long as foragers, so that at any one time many foragers will be naive, and even experienced bees may still end up with low foraging success on a complex (already learned) flower compared to an open, easy flower. Hence there could well be constraints against a flower visited by social bees (a dominant syndrome in many habitats) becoming too complicated or too specialized.

Orchids provide a further specific case, frequently appearing very specialized with low pollinator sharing and often using an animal's reproductive behavior (rather than foraging behavior) to achieve pollination (see chapter 23). According to Schiestl and Schlüter (2009), there is no association here between specialization and degree of pollinator sharing, although these authors did find that the degree of sharing was low, as expected, and that specialization did link to orchid species richness (cf. Sargent and Otto 2006).

3. Proposed Alternative Approaches

Given the currently perceived problems with a syndrome-based approach to pollination studies, it is not surprising that a range of other possible frameworks and approaches have been proposed in recent years. To some extent these are overlapping and complementary, but as a starting point they are summarized here as separate developments.

Plant-Pollinator Landscapes

Bronstein (1995) proposed that pollination ecologists should consider five types of community level landscape as a way of describing pollination systems. Her proposal accepts the idea of key traits in flowers that tend to cluster together but extends this to stress that these traits vary both within species and between species that may share an apparent syndrome. Different individual plants within a population may flower at slightly different times, have slightly different flower size or mean nectar volume, and attract a slightly different range of visitors. This landscape approach can be expressed as a contingency table (fig. 20.5), which also allows a useful focus on the plants' flowering strategies (synchronous or asynchronous flowering) and on the animals' foraging strategies (specialized, generalized, or migratory, also divisible into trap-lining, density-dependent, and sexually duped foragers).

Type 1: a landscape dominated by generalist pollinators, with the flowers blooming sequentially and in complementary fashion, with little overlap. Common in the tropics, but also in some highly seasonal systems including tundra. Little competition between plants, and the same plant may be visited by a sequence of visitors over a long flowering period. For example, lavender (*Lavandula stoechas*) in Spain receives up to 70 different kinds of visitor, each with a different activity peak over the three months of flowering. This can also lead to sequential mutualisms, for example of *Delphinium* and then *Ipomopsis* in the well-studied Rockies site.

Type 2: a landscape dominated by generalist pollinators but with plants all blooming at once. Common in deserts and in subtropical areas with a clear-cut rainy season, perhaps also in some temperate and alpine

POLLINATORS

		Specialized	Generalized	Migratory
PLANTS	**Synchronous**	**5** *Ophrys* orchids and male bees *Yucca* and yucca moths	**2** *Calathea* and *Euglossa* *Delphinium* and *Bombus*	*Fouquieria* and *Selasphorus* hummingbirds
	Asynchronous	**3** *Ficus* and figwasps *Dieffenbachia* and *Cyclocephala* beetles	**1** *Lavandula* and multiple insects *Psiguria* and *Heliconius* butterfflies	**4** *Agave* or *Pachycereus* and *Leptonycteris* bats

Figure 20.5 "Pollinator landscapes," expressed in a contingency table, with the five different landscapes described in the text identified by numbers. (Modified from Bronstein 1995.)

habitats. Obvious competition between plants, many with low reproductive success, visitors moving between species too much and wasting pollen.

Type 3: a landscape with specialist pollinators visiting plants with prolonged flowering periods, often in aseasonal habitats; uncommon, but an example is fig wasps, needing a sequence of figs all year round.

Type 4: a landscape of generalist migratory and trap-lining pollinators, for example bats or hummingbirds, switching through a variety of plants in the course of a year. Specialist at any one site and moving along nectar corridors of successive flowers. For example, Fleming et al. (1996) showed lesser long-nosed bats using localized areas of century plants (*Agave*) in autumn and again in early spring; the same bats foraged over wide areas on various *Agave* and columnar cacti (cardon, saguaro, etc.) in summer and various morning glory (*Ipomoea*) species in winter.

Type 5: specialist pollinators dominating, each visiting a small subset of the plant species available at any one time. Seen especially in deserts, tundra, and seasonal subtropics. Oligolectic solitary bees are the main players in this landscape, each linked to a widespread plant species such as creosote bush or mesquite, but also including some bees linked with orchids, and yucca moths linked to yuccas. Rather rare.

Bronstein also noted that types 1 and 5 are especially vulnerable to ecological disturbance and partner extinctions.

Ecological Models of Specialization and Generalization

Waser et al. (1996) generated a simple model to explain why generalization might be the norm in pollination ecology, with temporal variations in abundance, visitation rate, and visit effectiveness as the key factors. This was extended by Aigner (2001), who modeled a scenario with two pollinators where the predominance of specialization or generalization was largely explained by the strength of stabilizing selection on flower traits exerted by each, and by their abundance and visit rates. On his model, generalization was usually the expected result; but if there were two equally good pollinators A and B, where A exerted a stronger stabilizing selection effect, the flower phenotype would move toward the traits favored by pollinator A and could still therefore become specialized (fig. 20.6). A further model from Sargent and Otto (2006) more explicitly took account of local species abundance and of pollen import and pollen export so that male and female fitness functions were more easily separated. These authors found differing trade-off patterns in which generalization could either be favored or disfavored, and even one scenario of different pollinator attraction trade-off where initial evolution toward a generalist strategy could switch to favoring specialization once the population began to approach the generalist optimum; they also found that generalization could favor speciation. All these models have merit but as yet do not take into account the more dynamic role of pollinators in the whole interaction or

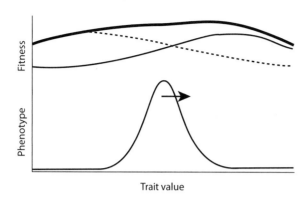

Figure 20.6 A model showing in the upper part two pollinators (thin and dashed lines) with different fitness contributions to a particular trait value, and the total fitness (thick line); the two pollinators make comparable contributions overall, but have very different selection "widths" so that one (the thin line) is exerting stronger stabilizing selection. Selection drives the phenotype in the arrowed direction which maximizes total fitness, so the trait value comes to match that from the pollinator exerting the stronger stabilizing selection, and a degree of specialization appears. (Redrawn from Aigner 2001.)

allow sufficiently for the variability in the pollinator environment (Morgan 2006). Morgan made an initial attempt to extend models in this direction and pointed to the need for much more sophisticated theoretical treatments in future.

Typological Approaches

Corbet (2006) drew attention to an existing but largely unregarded literature that had attempted to set up more functional groupings of flower types in relation to pollination. These groups are defined in terms of observable shared characters and corresponding functional groups (not necessarily taxonomic groups) of flower visitors. Ellis and Ellis-Adam (1993) compiled lists of visitors to 1300 species of flower in northwest Europe, and proposed a classification of floral types (largely based on early German classifications by Müller (1873) and by Loew (1895)), while also noting that anemophilous and pollen-only flowers did not form discrete clusters in their factor analysis. Corbet's study expanded on the resultant classifications of flower types and of the insects expected to associate with them, particularly adding details on floral microclimate and on insect energetics based in part on her own works. The result was a three-way functional classifi-

cation of plants, and a three-way largely taxonomically based grouping of associated insects (table 20.1); note that these groupings and the associated updated descriptors are based on statistical factor analysis from a large data set of recorded observations.

Corbet (2006) went on to dissect the insect groupings more carefully and pointed out that some insect taxa included species that visit flowers of more than one type. She noted that bees (superfamily Apoidea) could more usefully be divided into long- and short-tongued forms, rather than into family Apidae and other families as in the Ellis and Ellis-Adam (1993) analysis (and see chapter 18). Similarly with flies, she found the taxonomic divisions not always helpful; long-tongued hairy bombyliids are partial endotherms and visit *euphilous* flowers such as bluebells and primroses, as do a few long-tongued syrphids such as *Rhingia*, whereas most other bombyliids and syrphids select *hemiphilous* flowers. Beetles of different kinds also visit both *allophilous* and *hemiphilous* flowers. For all these reasons, Corbet proposed a more functional division of insects, related to their physical and physiological attributes. Here she concentrated on *depth thresholds* (the tongue must be long enough to reach concealed nectar in a tube), *cost thresholds* (accessible nectar energy content must be high enough to exceed costs of visiting), and *temperature thresholds* (body temperature must be high enough to permit flight for foraging bouts). A *cognitive threshold* should be added to this list, requiring sufficient learning ability to cope with complex floral structures. For allophilous flowers, none of these thresholds are very important; but for euphilous flowers they all come into play, and selection on the plant may act to favor specialist and/or exclude nonspecialist visitors by

1. lengthening the corolla to raise the depth threshold,
2. increasing nectar volume (but not concentration), to exceed an endotherm's cost threshold without preventing it from sucking up the fluid,
3. flowering at times (seasonally or daily) when temperatures are too low for ectotherm activity,
4. adding obstructions and complex features to raise the cognitive threshold.

In practice, this typological approach ends up looking not dissimilar from the syndrome system it set out to replace and improve on. However, Corbet (2006) also used it to offer practical suggestions in relation to both

TABLE 20.1
Typological Groupings of Flowers and Their Visitors

Flower type		Examples	Visitor type	
Allophilous	Flowers with fully ex-posed nectar, little or no intrafloral tempera-ture elevation Often white	Most Apiaceae	Allotropous	Short-tongued flies, non-bee hymenopterans, and beetles; poorly adapted as pollinators, broad diet other than flowers. Low T_{th}, no endothermy, small (<30 mg if hairy, <100 mg if not)
Hemiphilous	Flowers with moderate volumes of partly concealed nec-tar, often cup-shaped, with some intrafloral temperature elevation Often white or yellow	Some Rosaceae: *Prunus, Crataegus* *Rubus, Spiraea* Some Asteraceae: *Achillea,* *Leucanthemum* Also *Salix, Calluna,* *Ranunculus*	Hemitropous	Shorter-tongued bees, long-tongued flies, most lepidopterans; moder-ately good pollinators. Moderate T_{th}. Some ro-bust, hairy endothermic (bees, some syrphids); others slender but bask-ing (empid and syrphid flies, some butterflies)
Euphilous	Flowers with abundant deeply concealed nectar Often blue, pink, purple, yellow	Some Asteraceae: *Centaurea, Cirsium* *Taraxacum, Senecio* Many Fabaceae, Lamiaceae, Boraginaceae	Allotropous	Long-tongued, robust, with good insulation, high T_{th} and facultative endothermy; good polli-nators. Long-tongued bees, sphingid and noc-tuid moths, a few butterflies

Source: After Corbet 2006.

crop pollination and the management of invasive spe-cies and plant species under threat, in the hope that it could provide the baseline information appropriate to agricultural and conservation planners (see chapters 28 and 29).

Pollination Webs, Networks, and Matrices

The construction of pollination webs as a means to un-derstanding interactions follows on naturally from the study of ecological food webs in general, and requires that all visitors to all plants in a given community are recorded and mapped as an interacting weblike struc-ture (fig. 20.7A) which can then be statistically ana-lyzed. There has been a new emphasis on such pollina-tion webs since the late 1980s, tending to stress community processes and not individual associations. Jordano (1987) produced the first rigorous communi-ty-level analysis and in consequence suggested that pollination webs (or networks, as they are perhaps more often termed now) were much more generalized than had previously been assumed. Memmott (1999) plotted webs for British meadow communities and

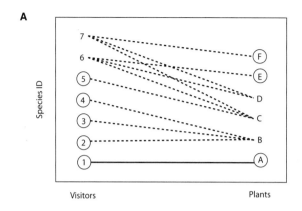

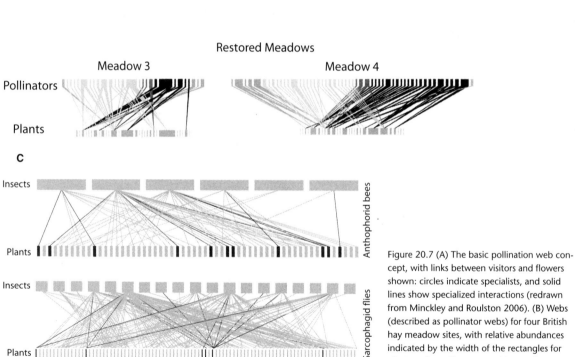

Figure 20.7 (A) The basic pollination web concept, with links between visitors and flowers shown: circles indicate specialists, and solid lines show specialized interactions (redrawn from Minckley and Roulston 2006). (B) Webs (described as pollinator webs) for four British hay meadow sites, with relative abundances indicated by the width of the rectangles for plants or for pollinators, and interaction strength by the line widths; darker lines are Hymenoptera (redrawn from Forup and Memmott 2005). (C) Three parts of the web derived from Robertson's Colorado data; here species with black rectangles and lines are aliens (redrawn from Memmott and Waser 2002).

found a highly connected interaction web (fig. 20.7B), where there were some specialist insects but with their preferred plants also being visited by many generalists that connected those plants to the rest of the web. Then Memmott and Waser (2002) took the discussion forward by plotting the pollination web for the entire Colorado data sets of Robertson used by Waser et al. (1996) and mentioned in section 2, *Practical Evidence against Syndromes*. Figure 20.7C shows three parts of this, for particular bees, flies, and moths, and it might be noted in passing that the flies appear a good deal more generalist than the bees, with the moths yet more restricted in their visit patterns.

In essence, the approach here is to record all visitors to all plants in a community, mapping the links between them, and then the **connectance** (*C*) of the web as a whole (the fraction of all possible interactions that actually occurred) can be calculated. As an example, the overall connectance of the pollination web from Waser et al. (1996) based on the Colorado data sets was 3.6%, a figure said by the authors to be quite high. However, it is at the lower end of values given for other webs (table 20.2; and see Olesen and Jordano 2002). Higher *C* values imply a more robust web (Dunne et al. 2002), but in practice connectances are of rather limited use for comparisons between webs, since their values depend quite heavily on web size and on sampling effort (see below). Olesen and Jordano (2002) corrected various networks for species richness and for network size and found some interesting geographic patterns: connectance values were lower at high altitude and lower in the tropics, with plants more generalized at higher latitudes and in lowlands but more specialized on islands.

Another approach is to calculate the **linkage** (the number of taxa with which a given species interacts), a term effectively the same as measures of "phily" or "tropy" used for plants and animals respectively in earlier literature (cf. Petanidou and Potts 2006). Linkage increases with network size (*S*, the sum of all plant and visitor species), so that linkage values for small island communities like those Lundgren and Olesen (2005) studied are low; more generally, this increase with network size follows a power-law relationship. From a plotted web the general strengths and importance of links within the structure can be assessed, and in more refined versions the *link strength* can also be factored quantitatively into the analysis (see below; Blüthgen et al. 2007).

Analysis of pollination webs or networks, and use of the associated terminology, now dominates significant parts of the pollination literature.

4. Some Problems with Pollination Webs

Visitation Is not the Same as Pollination

Flower visitors are not necessarily pollinators. This is perhaps the most crucial point, made in the very first chapter of this book, and it cannot be stressed too strongly: *visitation is not the same as pollination*, and simply scoring visitors to a plant can give grossly misleading results as to the degree of specialization in the interaction. It is certainly true that many plants get many kinds of visitors, as reviewed above, but the *value* of these visits can vary massively; visitors have different relative costs and different relative effectiveness. Evidence on this point was considered in detail in chapter 11, showing that in practice, for many kinds of flower with multiple visitors, most of the pollination may still be due to just one or a few kinds of visitor.

Most pollination web research to date is greatly oversimplified in this respect and should be described as visitation web research; it rarely takes enough (or any) account of the actual behavior and performance of different visitors. Put most simply, it is essential to look at the quality, not just the quantity, of visits. Most of the existing analyses use data based only on occurrence of pairwise interactions (which animals visit which plants) and rarely even incorporate data on how many times a particular animal visits a given plant. The Waser et al. (1996) analyses, referred to in *Multiple Visitors to a Single Species* in section 2, *Practical Evidence against Syndromes,* above, did not distinguish visitors from pollinators even though Robertson's database did include information on this. Vázquez and Aizen (2006) argued that grossly simplified approaches based only on visitation are still of value as measures of specialization when no better data exist, but recognized that this is only because very often no better measures of specialization currently *do* exist at a community level. But any attempt to measure *S* (the number of interaction partners recorded) for a species or *C* (connectance) for communities is bound to be inadequate unless it also attempts to incorporate some reasonable measure of pollinator effectiveness. In fact,

TABLE 20.2
Connectance and Linkage Values for Some Published Plant-Visitor Webs

	Connectance	Sources
Pre 1996 studies		
Japan, mixed forest	2.0	}
Japan, beech forest	1.9	}
Spain, scrub	8.9	} See Olesen and
US deciduous forest	28.1	} Jordano 2002
Venezuela forest	7.0	}
Andes, scrub	4.3	Arroyo et al. 1982
Greece, phrygana	3.4	Petanidou and Ellis 1993
New Zealand grassland	6.0, 6.6	Primack 1983
Canada, tundra	7.6, 10.4, 19.2	Hocking 1968; Kevan 1970;
Conifer forest/grassland, Colorado	3.6	Waser et al. 1996 (Robertson 1929)
Post 1996 studies		
United Kingdom, old hay meadows	35	Forup and Memmott 2005a
Restored hay meadows	27	Forup and Memmott 2005a
Arctic island	14	Lundgren and Olesen 2005
Arctic North Sweden	8.5	Elberling and Olesen 1999
Spain, scrub	9.1	} See Olesen and
Portugal, coastal scrub	25.0	} Jordano 2002
Average values for food webs	29.4	Jordano 1987

Note: Data from before 1996 have been reanalyzed as webs by later authors.

a reanalysis of the Robertson data set included in the review by Fenster et al. (2004) using only probable pollinators (and organizing them as functional groups rather than single species) reached an exactly opposite conclusion to that of Waser's group, that is, that specialization was widespread, with about 75% of the plants specializing on a single group such as long-tongued bees, pollen-gathering bees, or day-flying butterflies.

It is also very clear that quantified visitation data are explicitly misleading because the most abundant visitors are frequently not the best (or even good) pollinators. For example, Olesen (1997) showed that the best pollinators of a Texan composite plant, visited by ten different taxa, were among the least abundant of these. Ollerton (1996) and Johnson and Steiner (2000) stressed the same point, emphasizing that selection will be effected most strongly by the most important (most efficient) visitors, not the commonest visitors. Whether we talk in these terms of effectiveness, or more rigorously in terms of fitness functions (as

Gómez and Zamora (2006) require), this point cannot be overstated.

Sampling Effort Affects Outcomes

The connectance of webs provides some basis for comparisons between communities, but there are associated difficulties, largely related to sampling effort and how far (in a taxonomic sense) the web records are extended. To achieve a reasonably reliable web in all but the simplest communities certainly requires a major recording effort covering the whole flowering season, and over several seasons, since flower visitation may vary greatly between years (Herrera 1988; Williams et al. 2001). Olesen and Jordano (2002) showed that connectance decreases (roughly as a negative exponential) with increasing community size, so that the value obtained can be markedly biased by the sampling methodology and may be an unreliable indicator of specialization. Data from three major studies presented

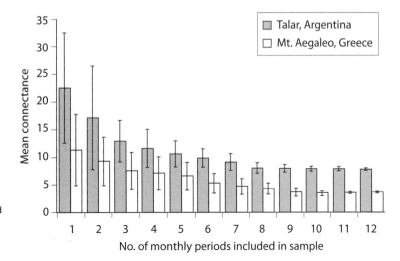

Figure 20.8 The decrease in values of connectance, C, as the number of months included in a web analysis rises, for both Greek and Argentinian sites. (Modified from Medan et al. 2006.)

by Petanidou and Potts (2006) supported that finding for Mediterranean sites (see also Medan et al. 2006).

However, on a more positive note, Nielsen and Bascompte (2007) showed that **nestedness** (see below) is less sensitive to sampling effort, so that some properties of ecological networks may be more resilient than previously thought.

Spatiotemporal Issues Are Unresolved

Most of the existing visitor or pollinator webs are certainly oversimplified, and two key aspects need to be explored and compared.

Spatial Webs

How large is the community or ecosystem for which a web should be constructed? If it is set too large, then many species recorded as present within it may not in practice overlap spatially at all simply because they *could* not do so. This can only be seriously addressed by constructing webs for different spatial scales and examining their similarities and differences, an approach that would be particularly appropriate in a community dominated by many pollinators with "homes" in fixed places around which they forage (most obviously solitary bees, perhaps).

Temporal Webs

Most studies to date have assumed that all partners coexist in the ecosystem at the same time and all are ac-

tive in pollen movement at that time. In practice this assumption is often simply wrong, with many species recorded as present through the study period but not temporally overlapping—they are unrecognized forbidden links (*sensu* Jordano et al. 2003). Medan et al. (2006) began to address this problem, using periodic sampling of an Argentinian xeric forest area (with year-long pollinating activity despite a moderately seasonal climate) over several seasons and comparing the networks recorded on a monthly basis. The composition of mutualist assemblages, and the network size and number of interactions, varied across months by more than one order of magnitude. Maximum activity in October and April (when many of the woody species flowered) was accompanied by maximum asymmetry of interactions with specialists interacting with generalists. Values of C decreased steadily as the number of months included in the analysis increased (fig. 20.8), and mean values for a 12-month sample ($C = 7.4\%$) were roughly three times lower than for separate one-month samples ($C = 22.4\%$). In that community, mean flowering periods for plant species were about five months and mean insect flight periods about fourth months. When this kind of analysis was repeated by Medan's group for the Greek data set compiled by Petanidou (see Petanidou and Ellis 1993; Petanidou et al. 1995), for which both flowering and insect activities (per species) were of much shorter duration (1–2 months), the decay of C with increased sample size was less marked. Medan et al. (2006) therefore concluded that the single values of C usually presented are often very misleading, especially when calculated for systems where the study duration greatly exceeded the average period of pollination activity for the plants and animals it contained. In fact, shorter sampling periods may give more meaningful values of C than very long and intensive sampling; C values might therefore best be calculated as means for separate short-duration analyses.

The time scale in other communities might need to be days or weeks rather than whole seasons; in other words, we need *time-sliced webs* within parts of a whole season. In certain situations the time scale may even have to be diurnal, and there would be merit in looking at webs in slices of just 2–3 h per day; an obvious example would be in communities where key species have flowers lasting only one day, and with very brief dehiscence periods within that day, as described for *Acacia* species in African savannas (Stone et al. 1998; and see chapter 21). A specific example of daily variation in web structure in this savanna ecosystem is shown in plate 37 (from Baldock 2007), which makes it very clear that time matters.

Other attempts to analyze webs temporally have begun. Alarcón et al. (2008) showed year-to-year variation in linkages in a Californian network, while Olesen et al. (2008) introduced a daily time base into an arctic network analysis and found linkage changes that helped to clarify how heterogeneous distributions of interaction numbers could arise. However, Petanidou et al. (2008) took a rather opposite approach with a long-term temporal analysis comparing four years of data, and used this to point out that specialization can be overestimated in short-term studies; their work certainly revealed extensive temporal plasticity in webs, with few interactions present in all four annual networks (which could also be taken to undermine confidence in conclusions drawn from many other network studies). An extended survey with six years of data (Dupont et al. 2009) reinforced the conclusion that species compositions varied markedly through the years, although the structural parameters of the network remained relatively constant.

5. Refining Web-Based Approaches

Webs may have various kinds of structure and pattern but these are rather difficult to see or analyze. An alternative is matrix analysis (reviewed by Jordano et al.

2006, and see Medan et al. 2007); this is a development of the web-based approach but uses two-way matrices that are easier to quantify and compare, making it easier to calculate structure within the web. Figure 20.9 clarifies terminology and examples of matrices, from idealized gradient structures (A) where all participants are relatively specialized, through compartmented structures (B) with separated zones of specialization, to nested structures (C) where a few species are highly generalized.

Lewinsohn et al. (2006) took a broad view of structure in plant-animal interactions and suggested that different models should always be probed, as they map onto different kinds of evolutionary processes. They found that simple *nested* patterns, generated by differences in abundance or dispersal of species, gave the best fit for plant-pollinator interactions, whereas plant-herbivore systems showed more complex structures with nested elements within *compartments*, more typical of a structure based on coevolutionary processes. Ollerton (2006) and Guimares et al. (2007) likewise addressed the whole range of mutualistic networks, adding in the concept of interaction intimacy, showing that more intimate (symbiotic) interactions would tend to form species-poor networks with compartments, while relatively nonintimate (nonsymbiotic) networks (such as pollinator-plant interactions) would be highly speciose and nested, with a core of generalists and with specialists often interacting with generalists.

Compartments within Webs

Compartments in webs imply recognizable subsets of interacting plants and animals, giving a blocked structure (fig. 20.9B) when viewed as a two-dimensional matrix. Compartments can be identified within webs by informal inspection or by statistical analysis, and in this area there has been an increasing interest to try and refine the utility of web-based approaches. Dicks et al. (2002) studying an English meadow showed in a formal manner that compartments were present, and they could separate two major groupings as

1. nectar-rich bee/butterfly flowers and
2. lower-reward fly (syrphid and brachyceran) flowers.

Others have recognized that compartments could be used to compute values of connectance separately for

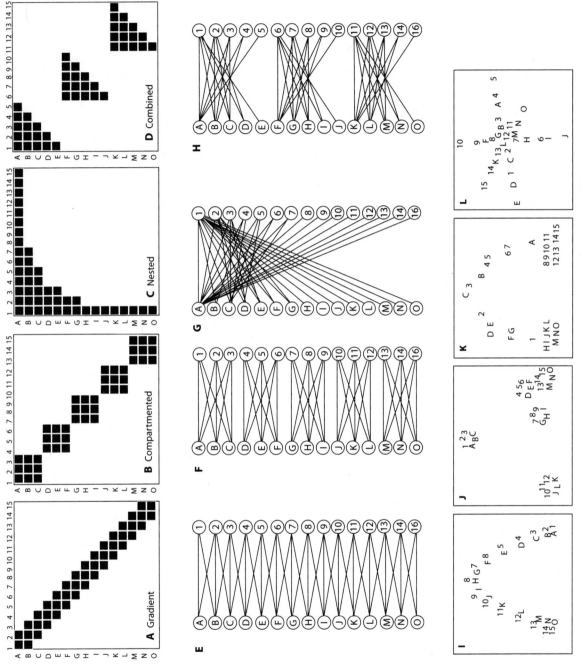

Figure 20.9 Different kinds of web or network and the associated terminology: (A) gradient, (B) compartment, (C) nested, and (D) combined, each expressed in different formats. (Redrawn from Lewinsohn et al. 2006.)

each part of the web and shown that each compartment can have a *C* value different from that for the web as a whole. Medan et al. (2006) attempted this with their Argentinian data set, finding higher *C* values for syrphid flies and lower values for muscoid flies and for bees, but with all values higher than for the data set as a whole. Using Petanidou's Greek data set, they identified compartments both as plant-centered subsets and as pollinator-type subsets of the data, and in both cases connectance was again higher for the groups alone than for the whole system.

Nestedness Approaches, and Use of Null Models

Bascompte *et al.* (2003) and Jordano et al. (2006) presented analyses of interactions where they found nested structures, meaning that there is a hierarchy or progression of included subsets (fig. 20.9C). Nested structures imply that plants with few interactions will only be interacting with generalist animals, and specialized species are interacting with a subset of the same partners of those that interact with the most generalized species. It is generally found that plant-pollinator mutualisms when analyzed as matrices center around a core of generalist species, with a high density of interactions, and with rather rare specialist relationships. However, the problems in sampling effort discussed in the last section apply equally in these cases and have not as yet been resolved.

Continuing the theme of matrix analysis, Vázquez and Aizen (2003, 2004, 2006) pointed out that it is difficult to assess the degree of specialization or generalization in any community without having some null model with which to compare the data; Bascompte et al. (2003) made a similar point for mutualisms more generally (and see Fortuna and Bascompte 2006; Medan et al. 2007; and Bascompte 2007 for a critique of progress to date). Essentially the starting point for these authors was that the existence of a pattern in flower visit frequencies per se does not necessarily mean that the pattern is unusual or unexpected. Their approach generated random data sets that shared key characteristics with the recorded data—essentially the number of species of plants and pollinators (by which they meant visitors, but see below), and the connectance or number of links between the interacting species. If the randomized data did not differ significantly from the observed data, then no mechanism other than random associations (complete generalization) needed to be invoked. Vázquez and Aizen (2006) assessed both interaction matrices and in a few cases interaction frequency matrices, based on 18 different published data sets from habitats ranging from subarctic plains to xeric scrub and including boreal and temperate forests (although not tropical forests). They used alternative null models with random interactions among species and random interactions among individuals. The first model showed significantly higher frequencies of extreme specialists and extreme generalists than would be expected at random (fig. 20.10); and when communities were compared in terms of their size (diversity) the more diverse communities supported an increasingly higher number of extreme specialists and generalists than implied by the null model. On this basis, even given the limitations of the data sets, high species richness appeared to favor the evolution of extreme specialization and extreme generalization. However, the authors also noted that there was a positive correlation between the number of individual visits received by a plant and the number of species that made those visits, and when this was factored out (by their null model 2) the model fitted the observations well; thus simple random interactions among individuals, assuming no difference between species except in their frequency of interaction, could explain the observed patterns. This is similar in principle to the *frequency of encounter* hypothesis used previously to explain species interaction patterns for herbivores on plants (e.g., Southwood 1961; Strong et al. 1984), linking to ideas of plant apparency (Feeny 1976) and invoking neutrality at the level of the individual as in other ecological models (e.g., Hubbell 2001).

Vázquez and Aizen (2006) went on to note that interactions were commonly asymmetric, with specialist animals often interacting with generalist plants and vice versa, and they found that the prevalence of such asymmetries increased as web size increased, so that larger, more diverse communities had not only more extreme specialists and generalists but also more examples of very asymmetric interactions. But as before they found that this could largely be explained away when the correlations between species' interaction frequency and their degree of specialization was factored out, in null model 2. In essence, this matches the find-

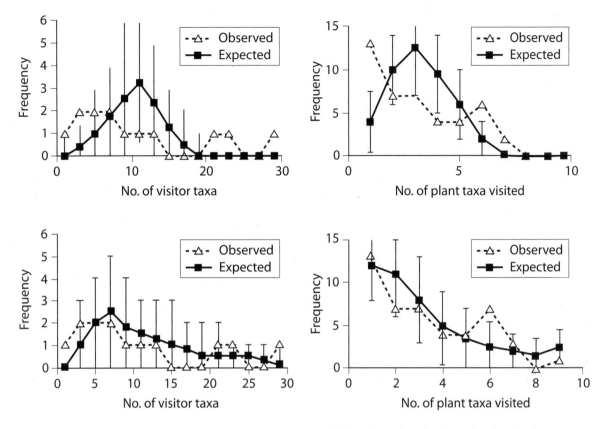

Figure 20.10 Two models comparing the observed and expected (with SE bars) values of *s*, the number of species of interaction partners (~ degree of specialization). Model 1 (above) assumes random interactions among species, and Model 2 (below) assumes random interactions among individuals. Model 1 shows more extreme specialists and generalists than expected at random; but this effect largely disappears using Model 2. (Redrawn from Vázquez and Aizen 2006.)

ings of Bascompte's and Jordano's analyses of nestedness, discussed above, where the nested structures showed specialized plant species interacting with a subset of the same animal partners that visited the most generalized species; nestedness inevitably implies high asymmetry levels.

The correlations between the number of interacting partners and the frequency of interactions may rest partly on visitor abundance, since rare pollinator species will be seen visiting more infrequently than common visitors. But other factors will also affect the fre-

quency of visits—in effect, all those factors that are traditionally viewed as aspects of pollinator syndromes, such as the animal's floral preferences, mobility, and food needs, or the plant's attractiveness and rewards. Vázquez and Aizen (2006) grouped together all these factors that are acting independently of species abundance, in a development of their model, and believed them to be relatively unimportant. However, they explicitly noted the simplifications inherent in their own model: that pollinator visits were independent of each other (fairly unlikely at any one time and place), and that sampling efforts were the same for all plant and visitor species (often unlikely).

Fortuna and Bascompte (2006) took the models one stage further by comparing different dynamical implications of network structure, one based on real interactions and two others on random networks. The real community models were found to decay sooner than random models when subjected to habitat loss, although they persisted for higher levels of fragmentation.

Neutral or random models have recently been challenged on several fronts. Santamaria and Rodriguez-Girones (2007) specifically tested neutral models against models incorporating complementary and conflicting biological traits and found optimal performance with a mixed model, moreover one where deterrence of floral enemies was at least as important as increasing pollinator efficiency. Likewise, Stang et al. (2007) put biological constraints back into a neutral model to explain the predominance of asymmetric interactions and found that asymmetry appeared in the simulations only if both "nectar holder depth" (usually corolla length) and abundance were factored in. They concluded that asymmetric specialization was the result of a morphological size threshold, and only above this did random interaction play a major part; and also that in lifelike asymmetric webs specialists had greater extinction risk than generalists, contradicting earlier web-based findings. Blüthgen et al. (2007) used a somewhat different analytical approach based on information theory, which takes interaction strength into account more explicitly, and found most of the networks they tested to be highly structured and nonrandom, with pollination webs significantly more specialized than seed-dispersal webs but not quite as specialized as ant-**myrmecophyte** interactions (fig. 20.11), and with rarely visited plants more specialized than frequently visited plants.

Accounting for the Visitor-Pollinator Problem

Short of really measuring pollination effectiveness, discussed in chapter 11, are there ways around this problem? Some authors have sought a way out by making assumptions about which visitors are going to be pollinators. One example is given by Petanidou and Potts (2006), who reported that they included only visitors that "contacted any part of a flower's reproductive organs"; but this is in practice nearly impossi-

ble to achieve in the field during transect surveys (and I was part of one of the study teams cited!). A second example comes from Memmott et al. (2004), who simply estimated that 80% of all visitors are effective pollinators, a figure that has been taken up by others (e.g., Vázquez and Aizen 2006). But the variance in effectiveness between visitors is bound to depend on the floral type; in open-bowl flowers pollinator placement and behavior is rarely crucial, whereas in tubular zygomorphic flowers it can be critical, and many visitors are likely to be ineffective pollinators. The correction value needed will vary with plant species and probably across communities; trying to get around the problem by using a standard (guessed at) value for the visitor/pollinator ratio is therefore inherently inappropriate.

Sahli and Conner (2006) tried another approach using Simpson's diversity index to estimate pollinator generalization across 17 species, as this feeds both richness and evenness into an estimate of "importance"; they found that pollinator richness explained only about 60% of the variation in diversity, indicating that importance was more affected by visit rate than by effectiveness. The same authors (2007) gave the most comprehensive survey to date of comparative pollinator effectiveness on a fairly generalist plant, using 15 genera visiting *Raphanus*. Single-visit pollen removal or seed set varied markedly, as did visitation rate, and the authors again concluded that in this species pollinator importance was most affected by visitation rate. This may well be the case for many more generalist plants (see also Engel and Irwin 2003). However, it might be expected that pollen deposition would be the more critical variable in more architecturally complex and/or more specialized flowers, and the authors stressed that effectiveness measures were still needed. As an example of the problems involved here, Ivey et al. (2003) suggested that, of all the factors they measured, time-dependent foraging behaviors were particularly important in determining forager effectiveness on a milkweed.

Some authors (e.g., Aigner 2001, 2006), have argued that minor (less abundant and/or less effective) pollinators can sometimes exert greater selective effects on flowers, so that "effectiveness" can be misleading. One of the best studied of all flowers, *Ipomopsis aggregata*, may appear to be a good example here, since Mayfield et al. (2001) found that long-tongued bumblebees deposited three times as much pollen per visit as the hummingbirds that are far commoner and

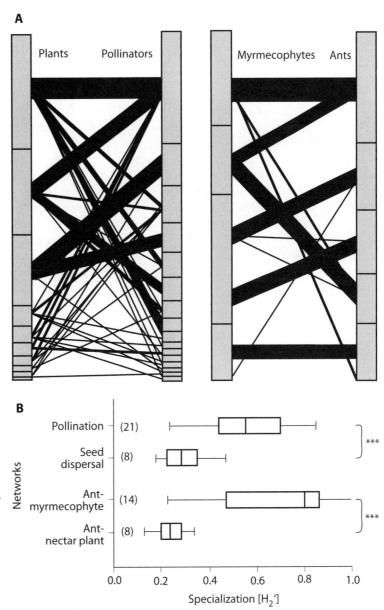

Figure 20.11 Specialization levels differ for different kinds of webs: (A) plant-"pollinator" webs are more complex and specialized and asymmetric than ant-myrmecophyte webs; (B) the numerical specialization parameter H'_2 is higher still in seed dispersal and ant-nectar interactions (numbers of webs compared are shown in parentheses). (Redrawn from Blüthgen et al. 2007.)

appear to be coevolved with the flowers. But this is a little misleading, as the bees visit the *Ipomopsis* flowers only when hummingbirds are rare and are not excluding them, so that the nectar can be reached and is at least briefly profitable for them (see also Pleasants and Waser 1985). And the *Penstemon* studies of Wilson et al. (2004, 2006) would not support Aigner's general contention, since these authors showed that species that attract both bees and hummingbirds are in

a transitional state toward hummingbird pollination (a conclusion based upon mapping traits onto a known molecular phylogeny for the group) and are predominantly under selection pressure to favor birds and to preclude bee visits.

Others have argued that very different pollinators could in fact be rather similar in their effectiveness, so that even if they have different phenotypic selective effects the net result will be generalization; this might apply to the findings of Motten et al. (1981) with *Claytonia*, Dieringer (1992) with *Agalinis*, and Vaughton (1992) with *Banksia*. Similarities of effectiveness may be especially likely where only one or a few pollen grains are required to fertilize one or a few ovules; but, as the earlier cited studies testify, such cases are likely to be the exception rather than the rule.

Morris (2003) attempted to build estimates of pollination effectiveness (variously using pollen removal or deposition, fruit set, or seed set as his measure) into analyses of possible effects of pollinator losses, using published data from 24 plant species. He found poor correlations between effectiveness and visit frequency and then (by testing effects of deleting species from the web models) concluded that visit frequency was in fact more crucial than effectiveness, so that frequent visitors with least between-year variability were much the most important regardless of their per visit effectiveness (see fig. 20.12).

Vázquez et al. (2005) specifically asserted that visit frequency was an acceptable surrogate for the total effects, since variation in frequency "overwhelms per visit effectiveness," and the mathematical modelling in their meta-analysis showed that interaction frequency *I* and total effect *T* were strongly positively correlated. However, Pellmyr and Thompson (1996) found relatively little effect of visit frequency or of pollen carryover, compared with single-visit effectiveness, in their work with *Lithophragma* when visited by bees, bee-flies, and moths, and many others have reported similar practical findings.

Forup and Memmott (2005a) complemented visitation webs with separate webs constructed in terms of pollen transport, derived from species of pollen grains recorded on particular insects active in their hay meadows. They found that bees were responsible for a higher proportion of the (potential) pollen transport than of the visits, rather as expected, and that generalization (here defined as carrying more than one kind of pollen) was greater than in the simple visitation webs. Gibson et al. (2006) also combined visitation webs with pollen transport webs in considering the pollination of three rare arable weeds, to predict which insects were key pollinators and which other plants they relied on. Lopezaraiza-Mikel et al. (2007) did the same, in examining the effects of an introduced alien (*Impatiens glandulifera*) on a native pollinator community; perhaps not surprisingly, they found that the visitor web would be a poor guide to pollination for the natives, as the pollen transport network was dominated by the invasive plant's pollen. In all these cases, the authors explicitly recognized that their data improved on simple visitation webs but that the ideal approach would be a "real" pollination web.

Where Are We with Webs or Networks?

Despite all the problem areas identified above (or perhaps because of them), general patterns seem to emerge from studies of pollination webs. Most of the results from web builders, with added matrix analysis and null models, have indicated that the frequency distribution of the numbers of links per plant or animal appears to decay according to a power law or a truncated power law, or sometimes with a faster exponential course. They also indicated that specialization in plant-pollinator interactions was, or was expected to be, less common than previously assumed, but that

1. in more diverse communities there could be a rather high occurrence of extreme specialists and extreme generalists, and
2. the specializations of plants and visitors were often not reciprocal, with a strong component of asymmetric interactions.

Vázquez and Aizen (2006) concluded that these inherent structural features could be accounted for by simple community processes and did not need a mechanistic explanation. They especially drew attention to the effects of species richness, differences among species, and the frequencies of interactions, and downplayed assumptions based on coadaptedness or coevolution. But the most recent network analyses have tended to put mechanistic coevolutionary aspects back into the picture, and the more they do so the better the fit to reality appears to be. A fine example of this is given by Stang et al. (2009), who inserted the match between proboscis lengths and flower depths, and the distributions of both these size parameters, into a Spanish plant-"pollinator" web, and found that they got a markedly better fit to the models and a better understanding of the interaction patterns. Those familiar with pollination syndromes will not be surprised by this.

There are now sufficient documented pollination networks that Olesen et al. (2007) could produce a meta-analysis of these, using 51 different published

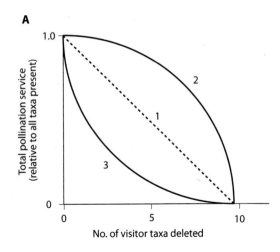

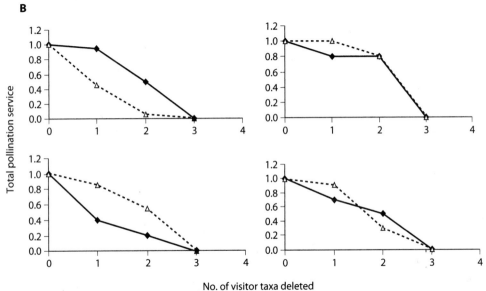

Figure 20.12 The effects of deleting species from constructed webs: (A) the principle, where random deletions (1) give a linearly declining service, or (2) where the earliest deletions are relatively nonessential (buffered pollination services) or (3) are most essential; (B) shows four examples, where solid lines are deletions made in order of increasing visit frequency and dotted lines are deletions made in order of per visit effectiveness, taken as meaning that visit frequency is more crucial than effectiveness. (Redrawn from Morris 2003.)

examples. Employing even newer analytical models, they found a prevalence of *modularity*, with weakly linked modules each of which contained strongly linked species within it that shared convergent traits (which sounds familiar!). They also reported that only about 15% of species were particularly important in most networks, often acting as *hubs* within modules or *connectors* between modules, such that their removal

could cause a network to fragment and initiate extinction cascades; these species would be especially important for conservation purposes. That is certainly a more realistic and helpful message than some earlier conclusions from network research. Aizen et al. (2008) also compared multiple networks, from forests and islands, to assess the effects of invasive aliens and found weaker mutualisms in the highly invaded networks with supergeneralist aliens taking over more of the links. Further analyses by Valdovinos et al. (2009) showed that well-established aliens could help networks to persist and stabilize.

However, it remains the case that so-called "pollination network" research to date is oversimplified at the most crucial level of the data it uses, and this underlines once again the gap between pollination ecologists and ecological modelers. That gap is only just starting to be bridged by incorporation of estimates of pollination efficiency into network analyses.

6. Specialization and Generalization: Core Issues

In many ways the relative importance of specialization and generalization has become the key question in modern pollination ecology, and it has been the focus of the most recent symposia and publications, notably a multiauthored book edited by Waser and Ollerton (2006). Many of the key arguments in that book are similar to those presented above. Of course specialization and generalization are extremes of an ecological continuum (Futuyma 2001), and all disagreements must be seen in this light, most plants and most pollinators probably falling somewhere in the middle; arguments are largely about the shape of this continuum and where the most common outcome lies. At the species level this could be measured as a value of S (the number of interaction partners recorded), while at the community level it could be measured as connectance C, as defined above; although in both cases with the proviso that the interactions used must be recorded with appropriate spatiotemporal discrimination and must be pollination events and not just visitations.

Defining the Terms

As various authors have pointed out (Renner 1998, 2006; Armbruster et al. 2000; Vázquez and Simberloff 2003; Minckley and Roulston 2006; Ollerton et al. 2007), the definition of the terms generalist and specialist is inherently open to confusion. There are several particular but interacting problems here:

1. Generalization and specialization as ecological terms usually involve factors such as niche breadth, in effect the number of resource items used by the organism being classified. But for a mutualism this can be viewed from the perspective of either participant, and for pollination that means either the plant or the visitor. A bee that visits many flower species is termed a generalist, while one of the flowers at which it forages may receive few or no other visitors and be termed a specialist.

2. Generalization and specialization refer both to the participants in an interaction (as in point 1) and to the interaction itself. Thus, for example, a bee *visitor* may become labeled as a specialist after observation of its foraging behavior, and its *interaction* with a given flower may also be described as specialist. Armbruster et al. (2000) distinguished these two approaches as *evolutionary specialization* (a process) and *ecological specialization* (a state), pointing out that the latter is relatively easily measured (at least in principle) while the process, especially that of evolution toward floral specialization, is more difficult to address. Other authors have used the term *functional specialization* as similar to ecological specialization, and Dalsgaard et al. (2008) devised a parameter that measures this.

3. The characterization of an interaction or of its participants may vary in time and space. Thus in practice the specialist bee referred to in (2) could be described as a generalist in another place or time, or in relation to the flowers of another plant species. Indeed, a bee may be completely specialist on one trip or through one day but generalist over its lifetime. And for social bees this problem is compounded, with individuals being more or less specialist through time but the colony appearing generalist at all times.

4. Even more confusingly, flowers can be described as specialist a priori largely by their morphology, scent, color, etc., following the syndromes approach, and this may or may not be confirmed by recording their visitors. Ollerton described this as *phenotypic specialization* (Ollerton et al. 2006, 2007), distinguishing it from both of Armbruster's ecological and evolutionary specializations in (2) above.

5. A practical problem arises where numbers of species visiting a plant are used to characterize plant specialization; Herrera (2005b) pointed out that when

relative abundances of each are included in an analysis, a diagnosis of generalization may turn out to be illusory. So we always need more than just species lists and must also record their abundance (and, as pointed out in chapter 20, section 3, *Ecological Models of Specialization and Generalization,* above, their visit quality).

6. There are issues concerning *absolute specialization* and *relative specialization*; plants in isolated small communities may be characterized as specialist just because there are few potential pollinators around, or taking a more absolute viewpoint they may be termed generalist because they use a high proportion of all the possible pollinators available to them.

7. There are also problems of *fundamental specialization* and *realized specialization*, in the same way that there are problems of fundamental and realized or achieved niche breadth for any organism. Fundamental specialization refers to all the potentially beneficial interactions for an organism in all possible ecological conditions (Vázquez and Aizen 2006), whereas realized specialization is the actual specialization recorded in a particular ecological context. In practice it is generally the latter that is used, but some authors impute aspects of fundamental specialization to their arguments as well.

All of these terms could be recast in the context of generalization rather than specialization, and to clarify that point the relations between some of the resultant terms are shown pictorially in figure 20.13.

Given all these problems with the terminology, some have argued that the term "specialist" should be descriptive only of flowers and of flower visitors rather than of interactions, and more precisely should mean

1. a flower species successfully pollinated by one or a few animal species (Renner and Feil 1993; Armbruster and Baldwin 1998; Armbruster et al. 2000; Fleming et al. 2001), or
2. an animal that harvests its resources from a narrow range of flowers.

This approach (albeit hampered by controversy about interpretations of "a few" or of "narrow") in turn means that generalization at the level of interactions can potentially still involve specialized flowers, or specialized visitors, or both. Taking this line, it is not possible to extrapolate from flower specialization to visitor or pollinator specialization (Renner 1998; Armbruster et al. 2000; and see Waser and Ollerton 2006).

Measuring Specialization, and Some Alternative Terms

In effect, most of the preceding discussions have involved ways of measuring specialization, above and beyond the efforts put into recording visits and recording visitors.

Petanidou and Potts (2006) argue that the term "specialization" had become unhelpful and should be replaced by the term "selectivity," designed to take into account the temporal variation within ecosystems. Selectivity (S) is in effect "relative tropy" for animals; the ratio of $(E-T)/T$, where E is the total number of plants that an animal could encounter and T the number that it actually interacts with, giving a value between 0 and 1. Exactly similar calculations can yield values of "relative phily" for plants. Some examples of the use of this parameter are considered in chapter 27, as these authors studied mainly Mediterranean systems.

In the same volume, Medan et al. (2006) used their temporally subdivided webs and their compartmentally divided webs to argue for a slightly different revision of terminology for web-based analyses (though this generates a slight anxiety: when adherents of a field feel the need to constantly revise exactly what they are measuring, then something may be amiss with the basic concepts). They proposed a *resource usage index* (RU), the ratio of effectively used mutualists to all available mutualists; specialists would have a value approaching 0 and supergeneralists a value close to 1. However, species with the same RU could still differ greatly in specialization; one species might visit 5 of the 20 available partners with equal frequency, while another might use the same 5 but concentrate strongly on just 1 in terms of visit frequency. Hence Medan et al. then had to apply an *evenness function* (E) to their RU parameter, and multiplied RU and E to get a single *generalization score* G, again ranging from 0 to 1. G is scale independent, while S and C are not. For their Argentinian data set, G and S were weakly correlated overall. The two components can be plotted against each other (fig. 20.14A); data points to the lower left are strongly specialist, and those to the upper right are supergeneralists. For three of the four seasons, the

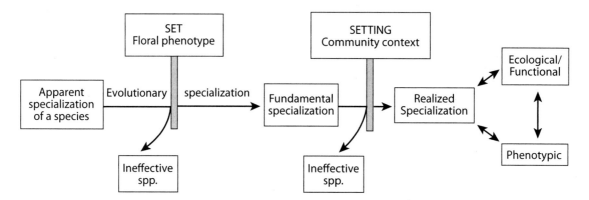

Figure 20.13 An approximate comparison of different types of generalization; the arrows could be seen as the *processes* of evolutionary specialization, whereas the boxes represent *states*. The floral phenotype and the community context act as filters. The apparent, fundamental, and realized specialization boxes reduce in size as these filters act. (Modified from ideas in Ollerton et al. 2007.)

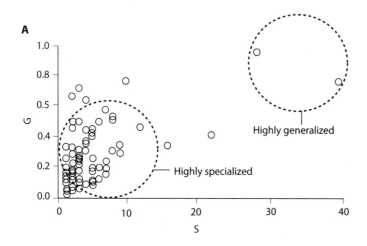

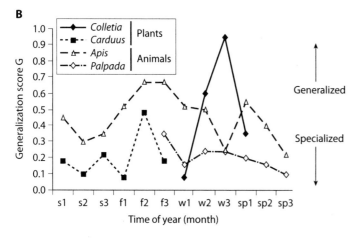

Figure 20.14 (A) A plot of generalization *G* score component against the selectivity *S* value, for plant-pollinator mutualists in an Argentinian system, where most species appear rather specialist (see text); and (B) a plot of *G* values through time, for four of the contributing species, showing marked variation through the year. (Modified from Medan et al. 2006.)

only marked variation they found was in the RU score: but in spring, values of evenness also varied, especially for animals. Following this up, they found that individual species varied in their generalist behavior across time (fig. 20.14B). For example, the shrub *Colletia* was relatively specialized at the start and end of its flowering period but became highly generalist during its peak flowering, when it could be characterized as a cornucopia plant (see chapter 21). A **syrphid fly** *Palpada* was fairly specialist at all times, with no more than three flower hosts at any one recording period. Interestingly, the introduced honeybee *Apis*, often labeled a supergeneralist, was in fact moderately specialist to moderately generalist in different months, but never highly generalized in its visiting, with a maximum of eight simultaneous hosts, so that its normal label could be seen as an artifact of summing all its hosts across time.

Why Might Generalization Be Common?

A simple answer might be, because it is usually better for the animals. In theory it should usually pay an animal to be a generalist, visiting any flower it meets that offers it a profit; it could easily include several plant species in its trips (especially if they have fairly similar floral structure so as to avoid the disruption of short-term learning of different handling). Hence the argument from the animal's point of view would be against the occurrence of too much specialization and in favor of at least moderate levels of generalization.

How then can generalization persist in plants, despite the very clear ability of visitors to choose between plant phenotypes? Gómez and Zamora (2006) suggested that generalization in itself becomes an adaptive strategy, where pollinators are strong agents of selection but have similar effectiveness, similar flower preferences, etc., and thus act together to generate floral adaptations. These authors suggested a need to distinguish between two phenomena. There is *nonadaptive generalization*, where spatiotemporal variability may mean that the selective regimes fluctuate through time and space, or where extrinsic factors throughout the plant's life cycle override selection imposed by flower visitors; and there is *adaptive generalization*, where different types of visitor impose similar selective effects on floral traits. Gómez et al. (2007) used *Erysimum* flowers as an example of a plant that has generalized flowers, visited by more than 100 species of insects, but where the degree of generalization differs between local and regional populations, the plants with intermediate generalization getting the best seed set. They suggested that there is an optimum level of generalization for any given generalized plant species.

Why Might Specialization Be Common?

Again, in simple terms, specialization is usually better for the plants. In theory, and in an ideal world, it should pay a plant to be a specialist, with one obligate visitor (i.e., with complete fidelity to one species) so that it only ever gets conspecific pollen deposited. For a plant, selection should be expected to favor characters that increase visitor efficiency, increase visitor fidelity and specialization, and avoid visitation by poor or parasitic visitors. Hence so many of the complex visual and olfactory cues—the attractants that bring in visitors—would be worth investing in heavily. Once the visitor is on the flower, selection should occur for traits that make rewards less accessible to all but the target visitor (nectar spurs, corolla barrier hairs, nectar lids, or partial/sequential dehiscence, etc.). Taken to extremes, a plant might evolve a complex physical path that a visiting animal must follow, to ensure correct pollen pickup and (in another place) deposition, with minimum profitable reward dispensed at the end of this path (i.e., the situation found in many orchids).

Following from the above, it is equally evident that there must be *some* reasonable level of specificity in flower visitation, or a plant would almost never get effective pollination. In other words, a degree of flower constancy by visitors has to occur, as discussed in chapter 11 and documented in chapters 12–18.

Theoretical considerations again would predict that a plant should be most likely to specialize in particular circumstances (see Waser et al. 1996, and many others):

1. Whenever pollinator availability, abundance, and behavior are reliable (but not when the pollinators are unpredictable from year to year)
2. When the plants are long lived and/or capable of vegetative reproduction if pollination fails, so they can afford to "risk it" as specialists

3. When the plants are rare or highly dispersed, so that specialization on high-fidelity pollinators will avoid the stigmas getting clogged up with useless pollen.

Conversely, annuals and weedy plants, colonizing unpredictable habitats with uncertain pollinator availability, should normally benefit from generalization. The same might be true for monoecious plants, where the risk of geitonogamy from too many small generalist visitors needs to be avoided.

How Much Specialization Does Occur?

This is the core question, and it cannot really be answered; the discussion here is confined to noting some trends and some examples to bear in mind. Most obviously, the number of plant species visited by any one animal visitor varies greatly. Some taxonomic groups are just very unfussy (e.g., many kinds of flies and beetles, although in both orders there are also examples of highly particular flower visitors). At the other extreme are the cases of specialization in fig wasps, which only ever visit fig flowers, and (at least in theory—but see chapter 26) only one species of fig each.

From the perspective of flower-visiting animals, examples of fairly extreme specialization are not too hard to find. Some solitary bees are highly specialist (Wcislo and Cane 1996), a good example being the genus *Macropis* in Europe, mentioned in chapter 18 along with many other cases of specialist solitary bees. Some visitors are quite generalist globally but highly specialized locally; for example, across an almost worldwide distribution the butterfly *Pyrameis carduii* visits a wide range of plants with concealed nectar, but within any one population its visitation is highly specialized. Migratory species (hummingbirds, butterflies) switch spatially and temporally as they move across their range but again are specialist locally; and we met in chapter 11 the case of a plant (*Passiflora mixta*) dependent on a single hummingbird species. Most of the long-lived vertebrate pollinators may visit a whole range of plants through an entire year but may be specialists visiting just one or a few species at any one time and switching temporally through their life cycle as flowering succession occurs (fig. 15.8 showed a hummingbird example). Solitary bees may be quite generalist at a species level in a particular locality but

rather specialist at the individual level. Social bees are generalist at the colony level but can be quite specialist at an individual level; the group may have a wide range of potential flower usage but any one bee will concentrate on one species at one time.

Minckley and Roulston (2006) offered a different perspective on animal specialization, however. They pointed out that most bees described as oligolectic (see chapters 7 and 18, and Cane and Stipes 2006) usually visit rather generalist plant types with good rewards, whereas more generalist bees tend to favor seemingly specialized low-reward plants. This again introduces an asymmetry into the concept of specialization (see also Jordano et al. 2006, Vázquez and Aizen 2004, 2006).

Examples of plant specialization are also readily available. The violet *Viola cazorlensis* is visited and pollinated only by the hawkmoth *Macroglossum stellatarum* (Herrera 1993), and different *Passiflora* species only by *Xylocopa* bees, or bats, or hummingbirds (and often only by one or two species of these visitors in each case). Many of the deceptive orchids are only pollinated by single species of bee, wasp, or fly (chapter 23). Most euglossine bees visit only a very few species of orchid, and many oil-collecting bees have highly specialized relations with particular plants (chapter 9), while some cucurbits are visited by a very narrow range of bees (see chapter 28 on crop pollination).

Olesen et al. (2007) concentrated on the morphological aspects of generalization and specialization in flowers with their large-scale meta-analysis of floral "openness" (using the Faegri and van der Pijl groupings where open flowers were dish, bell, and brush types, closed flowers were gullets, flags, and tubes; although the ascription of openness to each class was admittedly ill defined and potentially misleading). With over 1400 species of plants, they calculated a flower visitor generalization level (L) as the number of visiting animal species at a given site, and a relative value (L/A) as the proportion of the total visiting fauna at a given site that visited a given species. Neither L nor L/A was well correlated with the degree of openness of different blossom classes, although the authors allowed that correlation between L/A and openness is more likely within a blossom class anyway. It is noteworthy (and predictable!) that six out of their ten most generalized plants were in the dish-bowl category (and nine out of ten of the relatively most generalized); and figure 20.15 shows a key finding that the frequency distribution of L was heavily skewed to the left (20% of

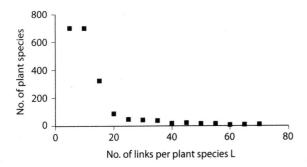

Figure 20.15 The frequency distribution of L values (number of links recorded per plant species, ~ generalization level) across a wide range of species, showing a strong skew to the left. (Redrawn from Olesen, Bascompte et al. 2007.)

all plants with $L = 1$, 11% with $L = 2$, a total of 46% with $L < 5$). Somewhat similar findings were reported in large-scale reviews by McCall and Primack (1992) and Fenster et al. (2004). Ramírez (2003) found a good relationship between blossom class and number of visitors; flags and gullets had just 1.1 and 1.2 types of "visitor," with dish, bell, and brush types having 1.6–2.0 types. Just how valuable any of these large-scale analyses are is somewhat open to doubt, as they admit sweeping generalizations about openness of flower types and lack any phylogenetic corrections; but it is interesting that they can come to such different conclusions.

It is appropriate here, once again, to refer back to the section in chapter 11 that documented existing data on pollination effectiveness of different animals operating on the same plants, and how often what appeared to be a generalist would in fact be rather specialized when recorded in terms of pollination rather than mere visitation.

Constraints on Specialization

In practice, the vast majority of plants are pollinated by more than one animal species, and often by two or more "kinds" of animal, so there must be ecological or genetic constraints that tend to limit specialization. There may be two main and obvious constraints operating.

Plant Abundance
Where a plant is common within a particular landscape, and the floral community is not particularly diverse, it may be more appropriate to be generalist. On the other hand, in species-rich communities it might be better for a plant to try and exploit visitor constancy, especially that of bees, and selection will then be for the variations of flower morphology, color, and scent that achieve this. Hence simpler plant communities are often dominated by Asteraceae and Apiaceae, and both of these can be seen as *secondarily* generalist families (see chapter 2). Where individuals of a plant are rare in general, or are widely scattered within a complex and diverse community, we tend to find flowers differentiated to ensure precise pollen carryover to a conspecific, as exemplified by orchids and asclepiads with aggregated precisely placed pollen; these increase in frequency in such communities, and bee flowers in general also tend to increase.

Pollinator Availability
Varied availability could be due to both the spatiotemporal unpredictability of pollinators and qualitative and/or quantitative differences in their effectiveness. Populations of any one kind of visitor may be very variable from year to year and between quite small adjacent areas, so that most plants have to hedge their bets by staying sufficiently attractive to at least two potential visitors in hope that at least one of them will be abundant enough at any one time and place. Insects are notoriously variable in number in space and time, particularly lepidopterans but also bees; they are affected by local extinctions, climate, migrations, etc. For example, bumblebees can vary at least tenfold, and sometimes up to 100-fold between seasons (e.g., *Bombus lapidarius* in Scotland (Willmer et al. 1994); *B. terrestris* in Israel (Dafni and Shmida 1996)). Even in the tropics insect populations can vary markedly (Wolda 1983; Roubik 1989). Hence different pollinators may be of differential importance in different years, the variation being stochastic, and this is often missed in short-term studies. But the effects are very real. For example, *Aquilegia caerulea* plants growing in the Rockies (Miller 1981) can be pollinated by bumblebees and by a hawkmoth (*Hyles*), these animals having different innate color preferences and hugely varying populations between years. The flowers are dimorphic for color, and seed set of each morph varies in relation to the relative abundances of the potential visiting pollen vectors. So over time the plant does better to maintain both morphs. A similar story can be seen for *Calochortus* in California, with different rather generalist

visitors in different parts of its range (Dilley et al. 2000); or for lavender in southern Europe, where there are large differences between years in the visitor spectrum (Herrera 1988).

Varied pollinator availability requires the plant-pollinator interaction to be flexible. In effect, selection may vary in direction over quite short timespans, and many plants cannot afford to get too specialized. Fontaine et al. (2008) pointed out that at high pollinator density an increased diet breadth (greater generalization) should be predicted from optimal foraging theory and demonstrated this effect with foraging bumblebees; so in this sense specialization is indeed a flexible trait influenced by competitive effects.

It is not surprising, then, that typically bee-pollinated plants can often also be pollinated by flies or by butterflies. Typical bird flowers may also be visited and pollinated by bees (as often noted for *Ipomopsis*), and sometimes by moths; and some of them may therefore show a mixed pollination syndrome. A rather exreme example is *Disterigma*, an ericaceous plant with short white corollas but copious nectar, preferentially attracting hummingbirds but also benefiting from small bee visitors and able to self-pollinate as a last resort (Navarro et al. 2007). Interchanges even between wind and insect pollination are possible, as seen in plantains and in some sallows/willows (see chapter 19).

On balance this probably means that most communities do tend to include a good many plants that would be scored as generalist, although this may be somewhat less true of long-term stable habitats, including tropical forests and some Mediterranean zones. "Generalist" is of course being used here in a rather different sense from the generalist flower syndrome discussed in this chapter and in chapter 12. And the presence of many generalists is not quite the same thing as generalization being the dominant phenomenon in pollination ecology.

7. Selection for Specialization in Flower-Pollinator Interactions

We still know surprisingly little about selection in pollination ecology; these issues have only really been studied in the last few years with suitable techniques available to give quantitative answers. We addressed this theme briefly in chapter 11, but it becomes essential to the debate pursued in this chapter.

Selection on Floral Traits

Effects of Specific Pollinators

Inevitably the best-known examples are derived from annual plants because these are easier to study over many generations. Primary case studies have been *Ipomopsis aggregata* in the field (with red tubular flowers mainly visited by hummingbirds) and *Raphanus raphanistrum* in the laboratory (with flat open flowers visited by various smaller insects, and usefully polymorphic for flower color). *Ipomopsis* shows phenotypic selection for long tubes with large diameter, for large nectar volume, and for an extended stigma set well apart from the stamens (Campbell et al. 1991), all consistent with efficient visitation and pollination by birds. For *Raphanus* the visitation effectiveness has been measured for many groups (butterflies, syrphids, *Apis*, small solitary bees) as pollen removal per visit (Conner et al. 1995), and effectiveness has been shown to be strongly related to anther exsertion for some taxa but not for others. Conner and Rush (1996) also found that large corolla size favored some *Raphanus* visitors but had little or no effect on others; and Lee and Snow (1998) found that *Raphanus* color preferences differed between flies and bumblebees. For this plant, selection pressures will presumably depend on the balance of visitors at any one time and place, and this will militate against specialization.

For a broader spectrum of plants the easiest aspect to analyze for any given species might be selection for increasing flower size. There are several cases reported where increased corolla size does enhance the efficiency of some visitors, but usually with little or no evidence that there is an efficiency cost for other visitor taxa; for example, Galen's work with *Polemonium* (Galen and Newport 1987; Galen 1989; Galen and Stanton 1989), or the Conner and Rush study of *Raphanus* mentioned above. Fenster (1991c) proposed corolla length (depth) as a better single measure, since greater length should be accompanied by a decreased range of visitors that would tend to be more specialist and effective pollinators. Hence the strength of selection should be greater where the tube is longer. He predicted that variance in corolla depth would be negatively correlated with mean corolla depth, and this was supported by his own work on ten hummingbird plants. But Herrera (1996) found less convincing relationships for insect-pollinated plants, with data for 58 species in southern Spain. He showed that longer-corolla species did have a narrower visitor spectrum (>20 mm, $n = 1$–1; <20 mm, n up to 90), but the coefficients of variation were usually only 10%–25%, with no significant relation between corolla length and variance, with or without the employment of phylogenetic comparisons.

Selection has also often been analyzed with hybrids, derived from plant species that use different pollinators. Early studies by Grant (1949, 1952) with two *Aquilegia* species were typical examples: he suggested that pollinators were agents of selection for floral form and also for reproductive isolation. Thus *A. formosa* with pendant red and yellow flowers and short nectar spurs attracts hummingbirds, while *A. pubescens* with erect pale yellow flowers and long spurs attracts hawkmoths; although these species are interfertile, in a natural overlap zone they persist because of visitor fidelity (although occasional hybridizations occur due to bees that will occasionally visit both species). Fulton and Hodges (1999) tested the basis of pollinator preference experimentally in this same system, finding that hawkmoths strongly preferred upright *A. pubescens* flowers over artificially inverted flowers.

A classic study by Schemske and Bradshaw (1999) also used hybrids to show disruptive selection on floral traits from different pollinator types, this time with *Mimulus*, where different species rely on different vectors: *M. lewisii* is mainly bumblebee visited and *M. cardinalis* is mainly attractive to hummingbirds. The species are interfertile, and they overlap and hybridize in parts of the Sierra Nevada. QTL mapping on an array of $F2$ hybrid offspring showed simple one-gene inheritance patterns for 9 out of 12 key traits, explaining more than 25% of the phenotype variation. This also fitted with visitor behavior (fig. 20.16), as both bees and hummingbirds preferred hybrids with big floral surfaces, but hummingbirds particularly liked a big nectar reward, and bees were particularly averse to hybrids with strong petal carotenoid content (making them deep red), preferring the larger and paler flowers offered. The effects were strong enough to give very little pollinator crossover even in the natural hybrid zone and thus presumably strong selection pressure on flower type, sufficient perhaps to eventually cause speciation. It should also be noted that both attraction and repulsion can be at work in this pressure for specialization. Aigner (2006) pointed out that the two most important traits (carotenoid concentration and nectar volume) were each under selection from only one of

the pollinators; and as they are controlled by genes in different linkage groups and probably on different chromosomes each could respond independently to selection. Thus, rather than necessarily producing disruptive selection, they may provide a strong drive to speciate.

This was the first major study to look at effects at the genetic level and pointed the way to much more work that is needed; Galliot et al. (2006) reviewed the field and its potential. The genes that determine color change are proving useful, with a single gene-mediated color shift documented to alter pollinator attraction in *Petunia* (Hoballah et al. 2007). There are some indications of genes that control nectar production in the same genus (Stuurman et al. 2004). It may be that scents will also be a particularly useful aspect to analyze given their apparent specificity, and that the Orchidaceae will be a good source here, as there are indications that in orchids pollinated by euglossine bees a single mutation can change the scent profile of a flower and hence the particular bee species attracted, giving the potential for rapid isolation and speciation.

Trade-offs for Different Pollinators

Trade-offs in floral traits for different pollinators are regularly proposed, and theory suggests that their presence will favor specialization (e.g., Muchhala 2007), but Aigner (2006) reported that good evidence has been hard to find. His own studies with the traditionally generalist genus *Dudleya* (Aigner 2005) involved a careful principal component analysis (PCA) of mixed species arrays of this crassulacean plant in California, where he found different assemblages dominated by hummingbirds, by larger bees (*Bombus, Anthophora*), or by small bees and flies. Long flowers with the reproductive parts inserted (PC2 in his analysis) had an advantage in all environments, but the slope of this relation was strong for hummingbirds, medium for large bees, and very slight for small bees and for flies. In experimental manipulations Aigner (2004) found that birds and bees both deposited pollen better in narrower (less flared) corollas, but more strongly so for birds (fig. 20.17); and birds also exported "pollen" (as a dye analog) more effectively from the less flared corollas. Aigner proposed that his results argued against generalists needing to trade off characters to support the "best" pollinators; for example, a population of *Dudleya* with wide corollas would be most effectively pollinated by bumblebees (i.e., they would be the MEPP),

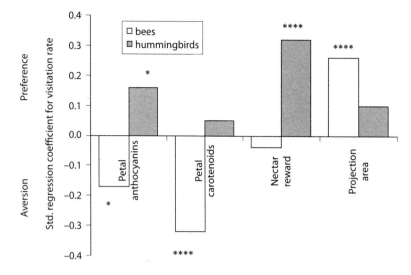

Figure 20.16 Bees and hummingbirds visit *Mimulus* flowers of different species, but when F2 hybrids are produced (*M. lewisii* x *M. cardinalis*) the bees show strong aversion to petal carotenoids, and strong selection of a larger head-on floral display ("projection area"), whereas hummingbirds select hybrids with particularly high nectar reward and have no obvious aversions. (Modified from Schemske and Bradshaw 1999.)

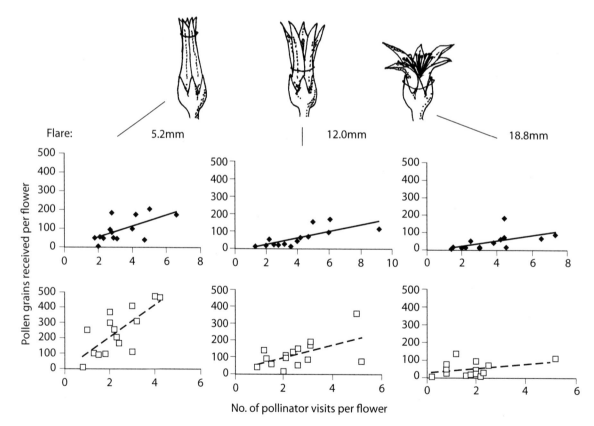

20.17 Effects of manipulating corolla "flare" in *Dudleya greenei*, for bumblebees (top row) and for hummingbirds (bottom row). Less flared corollas receive more pollen from both bird and bee visitors, but bees are slightly better pollen importers than birds on wide flowers and birds are much better than bees on narrow flowers. Unmanipulated flowers on average have a flare between 6 and 12 mm. (Redrawn from Aigner 2004.)

but nevertheless evolution would be largely driven by selection for specializations suiting hummingbirds as these specialist features pose no cost to the bees. This analysis indicates that evolutionary specialization in response to selection from one pollinator may not necessarily lead to increased ecological specialization, as measured by species richness of the visiting pollinators. In effect, for flowers that are rather generalist, different visitors will often act as selective forces on different traits, which is unsurprising given that different animals are attracted by different rewards and different visual and scent cues. The outcome may be a single floral phenotype with various features optimized for various visitors; the mariposa lilies studied by Dilley et al. (2000) may be a suitable example.

Counterarguments

There are some important caveats to the general argument that different pollinators might exert selection in different directions and so cancel each other out.

1. The limited evidence for trade-off adaptations to different pollinators in generalist plants may in part stem from studying isolated systems and omitting the effects of competition between pollinators. Where different visitors are competing for the resources of several plants, the potential for disruptive selection may be much greater, each animal group doing better if it preferentially visits the flowers that are least preferred by its competitors. It is very clear that competitive interactions of this kind do occur (see chapter 22), and do have marked effects on foraging behaviors that could in turn lead to stronger selection than is documented in the studies mentioned here.

2. Many of these studies also lack the key demonstration of pollination effectiveness and use visitation as a measure of pollination. Poor visitors, and indeed illegal visitors such as thieves and florivores (see chapters 23–25), may select for floral traits very different from those suited to pollinators. For example, in *Penstemon*, different species are bee or bird pollinated, and Castellanos et al. (2004) showed that selection was at least as much to avoid the wrong visitor as to attract the right one , so that antibee and antibird characters were favored.

3. There is a need to distinguish selection for floral traits from selection for reproductive isolation (Grant

1952; Waser 1998, 2001; Aigner 2006), the latter having more stringent requirements. Waser (1998) noted that specialization would rarely be so complete as to affect reproductive isolation directly, although Jones (2001) pointed out that it may affect speciation through assortative mating or by some other less direct route.

4. Selection in these kinds of studies may be constrained by covariation of characters (Mitchell et al. 1998). For example, Armbruster (2002) showed that floral color is partly determined by pleiotropic effects, linked to color of leaves, stems, etc., and so is nonadaptive in part. In some cases pigment pathways leading to flower color are also pleiotropically linked with production of secondary defensive compounds (see chapter 25).

5. Selection on flowers may also be overridden by selection for features unrelated to pollination, a point noted earlier. A commonly cited example is herbivore-mediated selection, taken up in more detail in chapter 25; as a single example here, *Paeonia broteroi* flowers were largely unaffected by pollinator-mediated selection because antiherbivory selection operated in the exactly reverse direction (Herrera 2000a).

6. Some recent studies have begun to document trade-offs explicitly, a good example being Muchhala's work (2007) with model flowers and captive birds and bats. He found flower "fit" to be crucial in determining pollination effectiveness, with wide corollas suiting the broad bat snout and narrow corollas suiting hummingbird bills, such that he could calculate selection for wider corollas when bats made more than 44% of all visit, and with intermediate corollas never selected for (i.e., generalization was always suboptimal).

Limitations of Existing Studies

Most studies to date have concentrated on the effects of selection for either female or male fitness. But pollinators can clearly influence both female success (seeds produced from cross-pollen from an appropriate vector, controlled by rewards and attractants) and male success (seeds sired from a flower's pollen landing on the stigmas of other conspecific plants, controlled largely by pollen dispensing and pollen stickiness, etc.). Some studies that have seemed to show little selection on floral traits have only looked at one side of this, usually the female side. A notable exception is the work of van Kleunen et al. (2008) with *Mimulus*. Their results were analyzed to avoid the hazards of covariance and showed that paternal selection gradients could be significant, notably in this case for smaller corollas with more red spot markings and for reduced anther-stigma separation.

Also, studies have generally considered just one generation of plants and rarely that plant's whole lifetime. Many authors have stressed the need for a lifetime, or at least life-cycle, approach to assessing effects on plant fitness (e.g., Gómez and Zamora 2006), but this is rarely achieved. It would allow for better estimation of pollinator effectiveness and also consideration of the extrinsic factors that constrain plant adaptive responses.

There is also considerable variation between the conclusions reached in laboratory and field studies, and in part for the same kinds of reasons; crucially, in field analyses edaphic factors and other interactions (such as herbivory levels) must also be operating, but are rarely assessed or controlled for. If such factors are ignored, pollinator-mediated selection on flowers may be substantially miscalculated.

For these reasons, individual variance in a given floral trait often accounts for only a small proportion of variance in fruit or seed production (Ehrlén 2002), and more general selection pressures may come into play. Thus for *Viola cazorlensis* in Spain, pollinated almost exclusively by *Macroglossum* hawkmoths, Herrera (1993) quantified selection rather precisely and showed that, while there was phenotypic selection on various flower characters, the additional ecological factors affecting plant size (soil type, herbivory, etc.) meant that only 2.1% of the variation in final fruit set (reproductive output) was explained by floral traits, and the one trait that seemed best to match the specialist pollinator (nectar spur length) was not subject to phenotypic selection over a four-year analysis. With similar approaches, Schemske and Horvitz (1988, 1989) found a value of 8% variation explained for *Calathea*, and Herrera (1996) found just 2% for *Lavandula*.

On a wider front, Crone (2001) showed that for various perennials fitness depended more on annual survivorship than on seed production (fecundity), again reinforcing the need for caution in assessing pollinator effects on perceived fitness in one part of the life cycle. It may be commonplace that selection on floral traits gets heavily diluted by influences from other factors.

Selection on the Pollinators

This is usually much harder to study, because lifetime fitness for an animal is more difficult to estimate, especially in the longer-lived and highly mobile pollinators. Furthermore, there are no direct effects on gene flow to be measured, whereas pollinators do affect gene flow directly for plants.

Pyke (1982) worked on bumblebees and showed that communities supporting three species of bees always included three with differing and largely nonoverlapping tongue lengths. A similar kind of result has been shown for hummingbirds (Feinsinger 1983). But this kind of evidence is rather inconclusive and indirect in relation to selection pressures.

However, there is one rather clearer hummingbird example concerning a sexual dimorphism, where Temeles et al. (2000) showed selection acting on the birds' beak lengths and shapes via competition for resources. Males and females of *Eulampis jugularis* visit different *Heliconia* species, and the males are aggressive and territorial around the more nectar-rich species, which have shorter straighter corollas, the male birds having shorter straighter beaks. But females of the same species frequent other (lower-reward) *Heliconia* species and have more curved beaks suited to accessing the flowers on these species. The end result for the plant is less interspecific pollen flow. Further work (Temeles and Kress 2003) neatly relating flower shapes to these bird beak characters on different Caribbean islands was discussed in chapter 15.

8. Generalization versus Specialization: Patterns in Different Ecosystems

A tendency for greater specialization in certain kinds of habitat has been widely discussed in the general ecological literature, and pollination ecology is no exception here. Working within the traditional syndromes framework, it is not hard to see that the more specialized syndromes occur far more frequently in warmer tropical areas. Most temperate habitats do not have bird or bat pollinators or plants with fragrances or resins as reward, while boreal, alpine, and arctic zones commonly lack the great majority of pollination syndromes. The number of taxonomic and functional groups declines with latitude; therefore the range of potential specializations is bound to decline too. Various authors who now assert that generalization is the norm for flowers nevertheless concede that specialization might perhaps occur in very stable ecological contexts, and thus that there might be habitat or geographical or very broad latitudinal trends.

At the habitat level, we will meet in chapter 27 some of the different patterns that occur in particular ecosystems (tropical, Mediterranean, high altitude or high latitude, and island). For now it is worth picking out one recent study specifically looking for effects of light levels on pollinator guilds (Sargent and Vamosi 2008). These authors assessed pollinator spectra at four different light levels from floor to canopy within various (mainly tropical) forests and found a general tendency for greater specialization at the lower light levels, perhaps because there is a greater need to avoid wastage of resources such as nectar that depend on photosynthate. They concluded that pollination syndromes may be relatively conserved, with specialized plants undergoing pollinator switches rather rarely; transitions to generalist pollination were mainly associated with beetle and fly pollinators.

Effects of Latitude

At the latitudinal level, Olesen and Jordano (2002) and Ollerton and Cranmer (2002) both reported major analyses of specialization patterns, using 35 and 29 published data sets, respectively, and reaching rather opposite conclusions. The first pair found tropical plants to be more ecologically specialized (with slightly lower connectance, after correcting for network size; fig. 20.18), while the second pair concluded that any such apparent trend was an artifact of sampling effort (see above). Neither group reported any significant trends in pollinator specialization.

Similar analyses were then presented by Bascompte et al. (2003) and Ollerton et al. (2006). The latter took the broadest view to date, using 32 published community-level studies. They categorized plants into one of 16 "broad functional pollinator systems . . . on the basis of their spectrum of pollinators (i.e., not using a pollination syndrome approach)." The plants were deemed to be specialized for a particular vector type only if more than 85% of their visitors were of that type, all others being listed as "insect-pollinated generalist" or "vertebrate/insect generalist." Communities were graded by latitude, altitude, and complexity of habitat structure. There were three major findings.

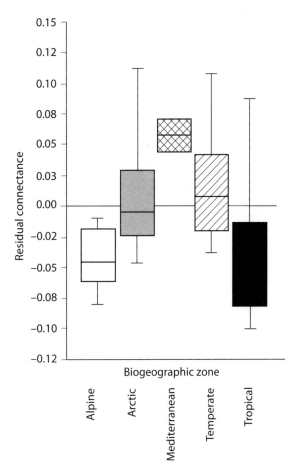

Figure 20.18 A comparison of networks from five geographic zones (corrected for different network sizes), showing lower connectance values in alpine and tropical habitats, with highest values (i.e., more generalization) in Mediterranean zones. The horizontal line is the median. (Redrawn from Olesen and Jordano 2002.)

1. As expected and usually assumed, there was a significant difference in the number of pollination systems at different latitudes, but the difference was essentially only between tropical systems and all others (fig. 20.19). Altitude had no overall effect, but there was an increase in pollination systems in complex habitats (i.e., woodland and forest).

2. There was some convergence across given latitudes and habitat complexities with the types of pollination systems that occurred there, and again no altitude effect. Tropical communities (and complex forest habitats) had higher proportions of plants specialized for small, large, and euglossine bees, for butterflies, for beetles, for wasps, and for thrips. They had lower

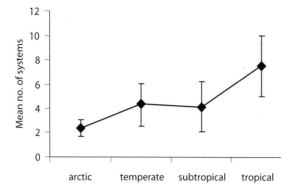

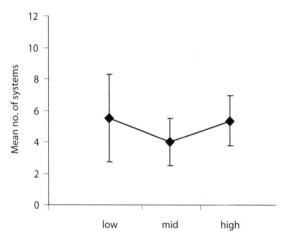

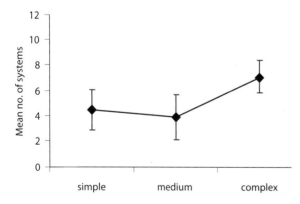

Figure 20.19 Comparisons of the number of pollination systems present in different biomes, different altitudinal zones, and habitats of differing structural complexity. See text for further commentary. (Redrawn from Ollerton et al. 2006.)

proportions specialized for flies and for birds. In contrast, temperate and subtropical communities had more plants pollinated by insects in general, by insects and birds, and by birds only.

3. Phylogenetic effects were important and had to be taken into account; some plant families may be inherently less prone to specialization and may be dominant in a particular community. Thus at the level of families, when compared across ecosystems, genuine evolved differences in pollination systems were more likely to be revealed, and Ollerton et al. (2006) found such effects for the two families they tested, the orchids and the asclepiads. For both families, species native to southern Africa were significantly more specialized than species from North America and Europe. Johnson and Steiner (2003) suggested that the high levels of specialization in the African flora are due to both phenotypic specialization and a rather depauperate pollinating fauna; there are relatively few bees overall (roughly 15 times as many plant species as bee species, compared with only 2–3 times as many in north temperate and subtropical systems), but unusually high numbers of long-tongued flies, large satyrid butterflies, hawkmoths, flower-visiting beetles, rodents, and oil-collecting bees.

Armbruster (2006) presented a taxonomically more restricted view of this same problem, but one that has the advantage of being based on his own observations across many habitats, with consistent methods of field analyses and an appreciation of the visitor/pollinator problem. His data on floral tube lengths certainly supported a latitudinal effect, with maximum values of 20 mm in arctic sites and 300–400 mm in tropical sites; although it is of course the variance that is increasing here, as the tropical sites also contain corollas much less than 20 mm long. Armbruster also compared four different taxa in detail and found evidence for specialization trends in three of them (see below); these resulted in reductions of niche overlap where species co-occurred and generally supported the idea of specialization as an escape from competition, especially in temperate and tropical habitats. The exception was the saxifrages, which retained open, radial flowers with little variation in reproductive structures, even where there were up to four congeners. This could be due to some inherent genetic or developmental constraint in the genus, which never shows any significant fusion of floral parts; but is perhaps more readily explained as selection for generalization in an arctic/alpine habitat suffering chronic pollinator scarcity.

A similar approach using a single taxon was taken by Goldblatt and Manning (2006), bringing together their extensive observations on the Iridaceae in sub-Saharan Africa. Within this one family they documented 17 distinct pollination systems, with recurrent shifts in mode across the various genera. Pollination by long-tongued bees (anthophorines) was thought to be ancestral in key groups, but the alternatives now include short-tongued bees, buzz-pollinating bees, oil-gathering bees, long-tongued flies (of four different types), large butterflies, settling and hovering moths, hopliine beetles, and sunbirds (see table 27.1 for more details). Only about 2% of the species could be scored as generalists. Direct observations of traits and visitors for 375 species allowed the authors to infer pollination mechanisms for another 610 species and to map pollination modes onto phylogenies to see the multiple shifts that must have occurred, especially in the bilaterally symmetrical Crocoideae.

But we should still recall that, even in the tropics, what were thought to have been good examples of specialization have come under fire. Lowland tropical forest communities in southeast Asia do contain many generalist and opportunistic flower visitors (Momose et al. 1998), but remember that this study was quoted earlier in numerical support of specialization as the norm. More specifically, various authors have pointed out that for bees the proportion of oligolectic species is higher in deserts, medium in temperate zones, and low in tropics; and in the Neotropics the hermit hummingbirds, with relatively long curved beaks, which were reported as significantly more specialist than straight-beaked species in the past, are now often shown to be just as generalist in their behavior (Cotton 1998). Levels of specialization were also reported to be similar along a rainfall/humidity gradient in Patagonia (Devoto et al. 2005), except that in wetter habitats flies became more abundant and bees less so (see fig. 13.3); this is probably linked to the association of flies with cooler temperatures.

Even some of the classic cases of one-to-one pollination syndromes from the tropics considered in chapter 26 are proving to be less than perfect; the obligate mutualism between senita moth and senita cactus turns out to be strongly supplemented by pollination from generalist small solitary bees (Fleming and Holland 1998), yucca moths visit sympatric yucca species such that hybrids result (Leebens-Mack et al. 1998), and many African fig species are now shown to have associations with several different fig wasp species.

Effect of Ecosystem Maturity

Beyond the issues of latitude and diversity, it should also perhaps be more common to find specialization in more mature habitats, so that patterns might vary with successional phase. Parrish and Bazzaz (1979) working in UK farmland, found these kinds of effects:

1. Short-lived ephemerals in the first successional stages were generalist open-bowl flowers, often selfing, and with relatively few visits
2. Mid-succession areas had **biennials** and perennials, including many composites, with long flowering seasons and high nectar rewards per flower or per inflorescence; mostly still fairly generalist, but important for bees and supporting the most diverse pollinator community
3. Later successional stages with tall grass, or mature woodland with glades and openings, had the most specialized plant types and the most distinct kinds of flora for different visitors.

The authors concluded that these effects arose at least partly from competition coming into play more strongly as the community aged. The same kinds of patterns, and perhaps similar explanations, apply in studies of Mediterranean communities undergoing successional changes after fire (Potts et al. 2003a,b; and see chapter 27).

Finally, it is generally reported that plant and pollinator specializations are reduced on islands, where "maturity" has an extra dimension of time of colonization, a theme also explored more fully in chapter 27.

9. Can Specialization Be Reversed?

From the evidence cited so far, it seems that there may be selection for specialization in plant-pollinator interactions in some situations and also selection against overspecialization, since there are costs incurred in departing from the global fitness optimum (see Waser et al. 1996). And it is generally assumed that specialization evolves from generalization, although relatively few studies have addressed this directly. Armbruster

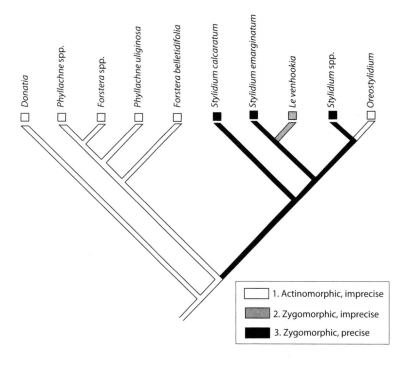

1. Actinomorphic, imprecise

2. Zygomorphic, imprecise

3. Zygomorphic, precise

Figure 20.20 A phylogeny of the Stylidiaceae based on ribosomal DNA, showing the progressive acquisition of more specialized pollination in the taxon. (Redrawn from Armbruster 2006.)

(2006) gave a clear example of the elongation of corolla tubes in the sister taxa *Collinsia* and *Tonella* (shown in fig. 11.3), where longer tubes occurred mainly after the most recent speciation events, generating species visited by long-tongued bees (*Osmia, Anthophora, Bombus*) and very distinct from the open-flowered *Tonella* species visited by small flies and bees. He gave a further example for the family Stylidiaceae, rather orchidlike in their complex fused (stamen + pistil) "column"; here the precision of pollen placement showed a clear phylogenetic trend toward specialization (fig. 20.20).

So can a trend to specialization be reversed? The second of Armbruster's examples shows clear indications that this can occur; most notably the genus *Oreostylidium* has reverted to a radial form with very imprecise pollen placement, perhaps linked to its dispersal to New Zealand where few specialist visitors were available. There are various other scattered instances in the literature: for example, honeycreeper beaks getting shorter in Hawaii (see chapter 27); the reversal of deeper more curved corolla tubes in *Aphelandra* (McDade 1992); and of course many reversions to self-pollination in angiosperms (Schoen et al. 1997). Tripp and Manos (2008) documented several reversals from

hummingbird to insect pollination in *Ruellia*, although they found that specializations to hawkmoth or bat pollination were more likely to be dead ends.

Probably the best examples come from the genus *Dalechampia*, with a well-mapped phylogeny and a series of highly specialized pollination scenarios in different species, basally involving resin-collecting bees but with a few fragrance-collecting bees (Armbruster and Baldwin 1988) and a few buzz-pollinated species (Armbruster 2006 provided a review). Several interesting trends appear here:

1. Within the resin-collecting bee-pollinated group (already rather specialized, as few bees have this habit), there have been repeated evolutionary trends to increase *or* decrease the resin production, linked to the size of bee attracted and usually correlated with changes in morphology affecting the gap between resin glands and anthers (fig. 20.21A); a clear *axis of specialization* can be identified.

2. At least three reversals to pollen collecting have occurred within this group. In particular, radiation in Madagascar has been accompanied by a switch to more general pollen-collecting bees (fig. 20.21B).

3. Whereas the first two cases are linked to changes in the pollinator species, a third example relates to the time of offering the rewards. Many species are open for 24 hours or more, but Armbruster identified at least four shifts to shorter diurnal opening times (assumed to reflect an increased level of specialization) (fig. 20.21C); however, there is one clear reversal in *D. schippii*.

Thus although there are rather few cases of reversal of specialization clearly documented against secure phylogenies as yet, the few studies with sufficient detail do give clear indications that reversals and parallelisms are not uncommon in plant groups; specialization is usually evolutionarily fairly labile, and trends to specialize or to generalize can go in both directions. This point may also be relevant to the debate (see chapter 4) as to whether specialized pollination promotes plant diversity, once widely accepted but explicitly questioned by Armbruster and Muchhala (2009).

10. Overview: Why Does the Argument over Generalization, Specialization, and Syndromes Matter?

This debate is currently at the forefront of pollination ecology and it matters greatly. This is largely because the currently favored network-based modeling approaches have in practice tended to find and to emphasize *an abundance and even dominance of generalists*, and a relative lack of specialized patterns across communities. They have found the interactions to be asymmetric, and most specialist plants to be rather often visited by generalist flower visitors. Furthermore, the resulting web models have been used to test the effects of removing one or more plants or animals from the modeled system and so to predict future threats and extinctions. Usually they have indicated a relative *lack of extinction threat* from simulated removal of one or even many pollinators (e.g., Morris 2003; Memmott et al. 2004) or from introduction of aliens (Memmott and Waser 2002; and see chapter 29). A common conclusion is that intense perturbation might lead to loss of specialist and rather sensitive pollinators, but that most plants would be buffered against this loss by the core generalist species. Memmott et al. (2004) explicitly recognized the problem that their webs had nothing to say about pollination effectiveness for visitors, but it is still only too easy for a message of community resil-

ience to become widespread when apparently comforting figures of predicted extinctions are produced. Following up the theme, Fontaine et al. (2006) began experimental tests of this with manipulations of simple biodiversity over two years, and found that as expected a community with the more functionally diverse pollinator assemblage had recruited more plant species than did one with an assemblage of less diverse pollinators. Pauw (2007) suggested that loss of a generalist oil-collecting bee (*Rediviva* sp.) in some conservation areas in the Cape had led to the failure of seed set in six of the species it visited, so confirming the network predictions. However, both of these studies perhaps presented rather odd perspectives on generalization. The view that webs and networks are resilient to species loss is particularly worrying given the limitations of methods that rely on visitation records rather than on pollination analysis; it could only too easily lead to inappropriate complacency.

It may be true that the syndrome concept has become too fixed and has tended to make field workers concentrate on the visitors that conform to expectation and not record or assess the effects of the nonconformist visitors seen at the flowers they study. The presence of 1:1 obligate relationships between plant and animal is rarer than perhaps once thought, and extreme specialization is certainly rare. Floral traits can sometimes be exaptations rather than adaptations (i.e., not due to current pollinators) and can often be partly due to selection from agents other than pollinators (Herrera 1996; Armbruster 1997). And, given the spatiotemporal unpredictability of some pollinator assemblages, it is probably true that there is unlikely to be consistent selection on floral traits, with heterogeneous assemblages also weakening directional selection at any one time. Stebbins's (1970) most effective pollinator principle implies that a flower's traits will be mainly selected by the pollinators that visit it most often and most effectively in a given region, but the visitors with highest effectiveness are often not abundant and may be especially unpredictable in time and space, thus again limiting the chances of very strong selection on floral traits (Herrera 1996).

For all these reasons, an *overemphasis* on specialized interactions may be unhelpful. It is rare to find flowering plants and flower visitors with tight, specific, obligate, and reciprocal pollination interactions. Perhaps a tendency to repeat the stories of orchids, figs, and yuccas too often has indeed conveyed too strong an

A

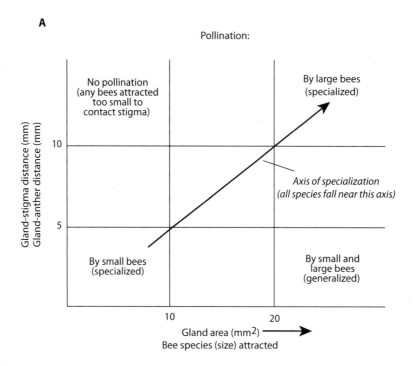

impression that strong specialization is the rule in pollination biology; and in that sense, the recent emphases on generalization have been thoroughly good for us.

But such emphasis may in turn risk the next generation of young pollination biologists losing sight of the reality of *convergent flower types* and the rather obvious degrees of specialization for attraction and retention of particular visitors covered in earlier chapters. Perception of patterns and trends is a very helpful way of finding one's way into a topic, and subsequently learning to dissect and criticize these patterns is an excellent way of getting to grips more deeply with the subject. In contrast, starting out with a perception of generalization, derived from models that may be difficult to grasp, can be discouraging to a beginner. It would be hard to understand the complexities of nectar variation discussed in chapter 8 without linking nectar rewards in particular flowers to particular visitors and their energetic needs, for example; indeed, it is hard to talk or write about pollination without using syndrome-like terms and referring to a bee-pollinated or a bird-pollinated plant, as most of the literature still does (beyond the world of the modelers). And it should be noted that even those who criticize or reject syndromes and propose alternatives nevertheless come very close

to using the syndrome traits and groups of traits in their analyses: Ollerton et al. (2006) divided up their visitors into remarkably syndrome-like groupings, and Corbet (2006) proposed more detailed typological groups that map very easily onto the syndromes discussed here in chapters 12–19.

At this point, though, it is only fair to deal with a critical major statistical analysis of syndrome robustness across 6 habitats and 482 plant species presented by Ollerton et al. (2009). This involved scoring presence or absence of 41 manifestations of 13 broadly defined floral traits. From this they derived three axes onto which classical syndromes could be plotted as sets of multiple alternative trait combinations in three-dimensional "phenotype space" (plate 38A), such that ten discrete syndromes were observable. However, the ordination of flowers from the six habitats studied only rarely lay within these syndrome groupings, and in about two-thirds of the cases the main pollinators could not be successfully predicted from the data. The stars in plate 38B show the actual phenotypic positions of flowers studied for one of the sites, in Peru.

The idealized syndrome plots show some interesting features that fit reasonably with expectations: for example, there is a proximity of bee and butterfly

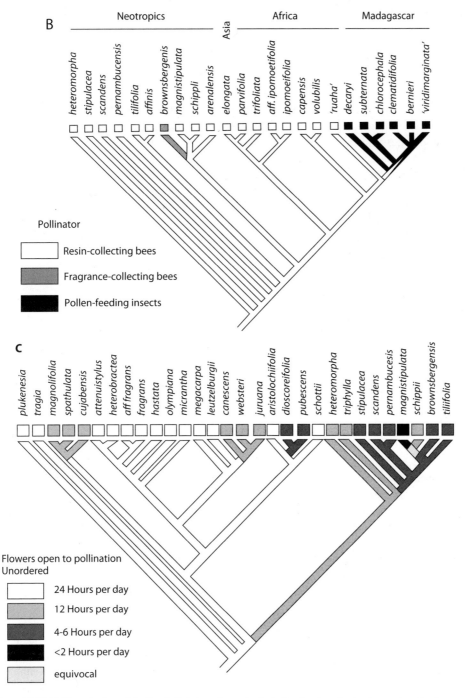

Figure 20.21 (A) The axis of specialization in *Dalechampia* species that attract resin-gathering bees, where the top left cell is empty and the bottom right cell nearly so (redrawn from Armbruster 2006). (B) An overall phylogeny of *Dalechampia*, showing a reversion to a more generalized pollen-gathering system using fairly unspecialized visitors in Madagascar (redrawn from Armbruster and Baldwin 1998). (C) A phylogeny of *Dalechampia* that indicates patterns in length of flower opening, with at least 4 shifts to shorter opening and more specialized interactions (redrawn from Armbruster 2006).

groupings, and of bat and nonflying mammal, although wasp flowers and moth flowers are particularly and perhaps unexpectedly distinct. The authors noted that for the real flowers some syndromes (bee and fly) were predicted more reliably than others (beetle, moth); and that some plant families were predicted well (zygomorphic taxa such as Lamiaceae, Fabaceae) while others were predicted poorly (e.g., Apiaceae, with many obvious generalists), although Asteraceae were surprisingly prone to successful predictions.

Taken at face value, this study should present a real problem for any defense of syndromes. However, the authors themselves take care to point out the difficulties: defining the syndromes quantitatively was extremely difficult, especially with a binary scoring, there was no weighting of traits, and samples often did not include the whole growing season so there may have been biases as a result. The "visitor versus pollinator" trap was only partly avoided: visitors were scored as pollinators if they were not obviously illegitimate, and if they were seen to contact the male and female reproductive organs.

Why then is there such apparently poor prediction? The distributions of points derived for real flowers are in fact rather weird upon closer inspection. Axis 1 was strongly positively influenced by bilateral symmetry, large flower size, diurnal anthesis, fresh scents, and nectar guides, but negatively by abundant nectar. Axis 2 was positively influenced by tubular shape, fresh scent, vivid colors, and nectar guides; negatively by strength of scents and exposed anthers. Axis 3 particularly indicated robust anthers, little or no scent, and flowers that were not white. Real flowers seem to be particularly likely to be more extreme in both senses in relation to axis 3 than expected, and to be more positive in relation to axis 2, and these patterns were consistent across all the habitats tested; but it is hard to see any obvious explanations, so that confidence in the utility of the axes is undermined. It seems very likely that the study simply does not capture the syndromes adequately, due to

1. the lack of weightings;
2. the inclusion of traits that are difficult to quantify (e.g., flower size, and especially nectar volume (chapter 7 discusses all the problems in deriving a single measure of this);
3. omission of other traits that are known to be extremely important (notably nectar concentration,

pollen amounts, corolla tube length, floral longevity).

Thus while the study will be taken by many who are thus inclined as evidence against syndromes, the authors are right to say "We do not take our results as evidence against convergent floral adaptation resulting from pollinator-mediated selection"; more conservatively, they merely warn against thinking solely in terms of a single most effective pollinator.

It remains the case that a reasonably experienced pollination biologist can in practice go out into the field, even in temperate sites, and say X is probably a bee flower, while Y is probably a hummingbird flower (and then test the predictions, or watch students do so). The flowers may indeed get other visitors as well, but these will often be of poorer quality and operating in less appropriate times or places. This aspect of a syndromes approach is exceptionally helpful as long as it is used with due care, and it can be properly tested (e.g., see Pauw (2006), who correctly predicted an oil-bee pollination in certain orchids; or Kleizen et al. (2008), who confirmed rodent pollination in two *Colchicum* species after predicting this from the floral features). In the tropics this approach as a starting point can be even more useful, as so many plants are unstudied and so many more specialist kinds of relationship are present, involving bats, hawkmoths, resin- and fragrance-collecting bees, long-tongued flies, fig wasps, etc. It is noteworthy that the main web work so far has been temperate, and specialization is probably lower in such habitats, where opportunist, clever, long-lived social bees are particularly abundant. And it may be worth noting also that these same temperate areas that are so much better studied are relatively young in a postglacial sense, so they may still be dominated by opportunist colonizing plants that are in an evolutionary sense not *yet* particularly specialized.

There is no doubt, though, that our *definitions* of generalization and specialization need to be reconsidered; as outlined in section 6, *Specialization and Generalization: Core Issues*, subsection *Defining the Terms*, the terms are often used too simplistically, without disentangling state and process, and indeed Armbruster (2006) believed that the current apparent contradictions stem at least partly from different emphases among pollination biologists on either ecological observations of state or evolutionary trends of specialization processes. For the future, specialization should not be seen only as a tight reciprocal specialization of plant and animal, exemplified in traditional syndromes, nor should generalization be seen as a lack of adaptation. Our perceptions of generalized or specialized communities or groups need to take into account some key points:

1. The *taxonomic level* used to characterize community components. Bees are an obvious case in point here, given the variety that allowed them to be subdivided into several separate syndromes in chapter 18. A flower may be described by one author as specialized for bee pollination and by another as rather generalist if it lacks features particularly associated with a particular bee genus (or indeed a particular bee tongue-length and body-size combination, occurring across various genera).

2. The *phyletic diversity* of the pollinators (a plant with multiple pollinators from very different taxa does not conform to the syndrome concept and is far more generalist than one with the same number of species of pollinator but all from the same genus or family; the latter should be termed a specialist).

3. The *number of available pollinators* in the region (a plant visited by a high proportion of the available fauna in a nonspeciose ecosystem is more generalist than one visited by a small fraction of the animals in a rainforest, even where the absolute number of pollinators is higher).

The new indices developed by Petanidou and Potts (2006) or Medan et al. (2006) could be helpful here. Pollination biologists need to accommodate ideas of adaptive generalization and of asymmetric specialization, incorporating a new appreciation that strong selection does not necessarily align only with strong specialization, since generalized flowers too can be under strong selection. We need also, and above all, to be able to incorporate genuine measures of pollinator effectiveness, and the web world needs to move from visitor data to real pollinator data if we want to see meaningful comparisons between communities. Careful analysis of pollinator effectiveness (i.e., both qualitative and quantitative surveys) can never be replaced by the compilation of impressive quantitative lists of the many different species, families, orders, or phyla of animals visiting a given plant species, the latter to be assembled into a web that inevitably appears to

stress generalization and that can then be used to show a (quite possibly illusory) resilience to extinctions.

Given the evolutionary conflict in flower-visitor associations, with both parties under selective pressure to exploit one another, it should follow that as a relationship becomes more closely integrated the success of the mutualism will inevitably vary, and the relationships will change in space and through time as selection leads each partner to maximize efficiency. Measuring visitation at one time and place gives a very poor overview and is rather likely to suggest generalization; but using a syndrome-based understanding and then measuring pollination over a longer time scale and larger area can show a different story.

11. In Conclusion: A Personal View

At present we are burdened with a situation where syndromes are out of fashion, and many authors go out of their way to avoid the term at all, while seeming intent on finding that communities are, broadly speaking, generalized. Petanidou and Potts (2006), good friends and experienced sensible colleagues of mine, presented data that documented quite high levels of specialization (high selectivity values) in their Mediterranean communities (fig. 20.22), yet by the end of their paper this had somehow turned into a conclusion that the communities were characterized by "very limited specialization." Remember also the quote given earlier from Ollerton et al. (2006): the authors categorized plants into one of 16 "broad functional pollinator systems . . . on the basis of their spectrum of pollinators (i.e., not using a pollination syndrome approach)." Here a syndrome is deliberately being presented as a mere listing of floral characters, whereas the idea has always been used in practice as a matching of observations of floral features and visitor frequency and behavior. The authors go on to worry that their categories lose "much of the subtlety of the distinction between . . . medium and large bee pollination, . . . or hovering- versus perching-bird pollination." Well, quite! Furthermore, in that same study the plants were deemed to be specialized on a particular animal group only if more than 85% of their visitors were from that group, all other plants being termed generalists—but surely few field workers would use the term "generalist" for a plant that received just 80% (or even 60%!) of its visits from one kind of animal and the rest from

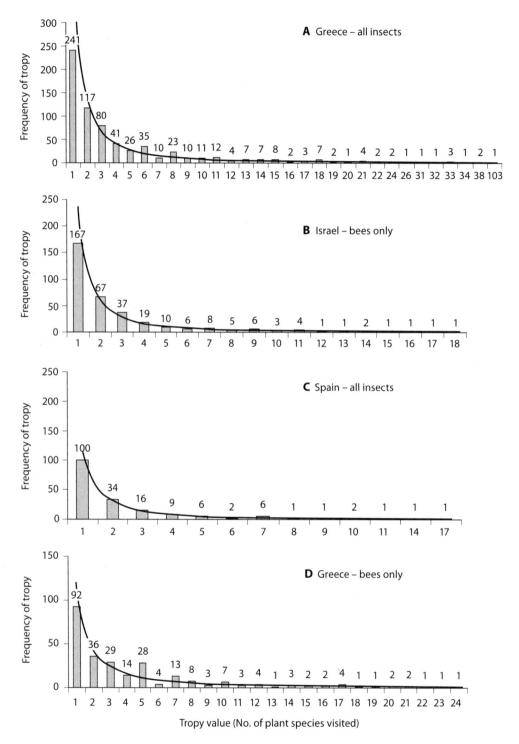

Figure 20.22 Frequency distributions of "tropy" values (numbers of plant species visited by a given animal) in three Mediterranean communities; in every case the largest number of animal species visit just one plant and there is a rapid fall-off to more generalized behaviors. (Redrawn from Petanidou and Potts 2006.)

a mixture of many others, even *without* properly observing (or testing for) visit efficiency in terms such as timing, pollen carriage and placement, etc. for each visitor. Even more disconcerting is the use of the term "generalized" for any plant having more than a single species of pollen carrier (for example, even where it is visited mainly by five different species of bumblebee, as in Waser et al. 1996); or for any insect visiting more than one species of plant (e.g., Forup and Memmott 2005a). This becomes rather absurd, and it would seem that we are bound to discover generalization when the boundaries are set in that way.

And there is another irony here; if generalization is indeed the norm, which would mean that most visitors have little flower constancy and forage rather at random, then visitation webs as currently constructed are almost bound to be fairly uninformative, and the effort that goes into constructing and modeling them might have been better directed elsewhere.

The situation has become unhelpful, and needs to be rethought with fewer preconceptions as to what is likely. My own bias—based on field experience and a complete lack of involvement with grand models—tells me that an appreciation of pollination syndromes and of the force of convergent evolution is wonderfully predictive. Observing a flower's features and phenology properly, and in particular taking the trouble to determine its rewards carefully on appropriate time scales (because as chapter 10 pointed out, economics really does matter), will tell me a great deal about its likely main pollinators. Watching that same flower will help me distinguish nonpollinating and pollinating visitors and very commonly (perhaps even 85% of the time!) confirm with which syndrome the flower is linked. If the grand models—once properly constructed on the basis of pollinators and not visitors—tell us something more, that will be wonderful; likewise, if new insights from molecular and whole-genomic analysis reveal new mechanisms underlying pollinator shifts as Cronk and Ojeda (2008) proposed. But in the meantime we should not disregard or throw out a century of accumulated expertise, or remove from pollination biology so much of the patterning and intricacy that makes it both intriguing and deeply exciting to those making their first (and then inevitably repeated) encounters with the subject.

PART IV

FLORAL ECOLOGY

Chapter 21

THE TIMING AND PATTERNING OF FLOWERING

———————————

Plants should flower in ways that maximize their own reproductive success. The "flowering pattern" is a composite of the timing and frequency of individual flowers opening, and also of flower longevity. These phenological factors vary between species but also within a species (and often between sexes for dioecious species). They may additionally be affected by abiotic phenomena, as well as by conflicting selection from pollinators, herbivores, and seed dispersers. Phenology can also in its turn affect pollinator attraction and, by altering pollinator behavior, can change pollinator efficiency. Flowering phenology can influence the plant's manipulation of its visitors in ways that should increase either or both of pollen transfer and pollen receipt. This chapter deals with the factors that affect flowering, and the effects of different flowering patterns on pollination outcomes. Table 21.1 introduces some of the parameters of flowering that might be reported for a particular flower, plant, species, or community.

1. Frequency of Flowering, and the Shape of the Flowering Period

Flowering patterns can of course be viewed at a species level; whether a plant is annual or biennial (both being *monocarpic*, with a single reproductive episode) or perennial. If perennial, a plant can flower each year (once or more than once), or it may flower supra-annually, perhaps for a few years irregularly (*episodic flowering*: Bullock et al. 1983), or at regular intervals varying from once every 2 years to once every 17 or 21 years in some forest trees. Both of these types of perennial are normally **polycarpic**. However, a few perennials flower just once at the end of their life, in which case they are again effectively monocarpic; this pattern is famously common in some palms and bamboos.

Alternatively, and perhaps more relevant from a pollination perspective, flowering can be viewed at the population level, where the terms **extended blooming** (a few flowers produced each day, over weeks or months) and **mass blooming** (many flowers each day, for just a few days) are commonly encountered.

Typical Flowering Patterns

The two basic categories mentioned may be further divided into several patterns as described below and in figure 21.1, originally modeled from the Neotropics (Gentry 1974; see also Poole and Rathcke 1979; Opler et al. 1980; Bawa 1983).

Cornucopia Flowering
Cornucopia flowering is one type of extended blooming: flowering is synchronous within a population, with moderate numbers of flowers produced each day

TABLE 21.1
Components of Flowering Phenology in Plants

Flower	Commencement	Date of first opening
	Longevity	Lifespan from anthesis to wilting
Plant	Commencement	Date first flower opens
or	Flowering period	Time from first to last flowers
population	Peak flowering time	Date of maximum flower number
or	(Mean) Number of flowers	On a given day, or at peak, or as a mean
species	Flowering time course	Plot number (or %) of flowers against time
	Relative flowering intensity	Number of flowers as % of highest number recorded
	Flowering synchrony	Degree of overlap of flowers on plant, or plants in population
	Flowering consistency	Across years

A

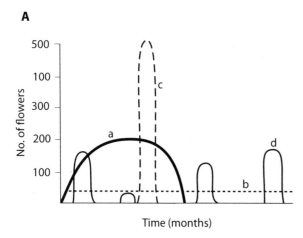

B

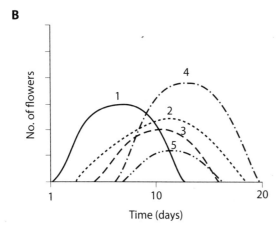

Figure 21.1 (A) Basic patterns of flowering phenology (modified from Gentry 1974): (a) cornucopia, (b) steady state, (c) big bang, and (d) multiple bang. (B) Patterns of cornucopia flowering in conspecific trees at the same site.

over a moderate period, up to several weeks, and usually at the same time in the annual cycle. This is the most common pattern, and was assumed by Gentry to be primitive and to be broadly characteristic of most temperate flora. Conspecific plants at a particular site may have somewhat different but overlapping curves (fig. 21.1B).

Steady-State Flowering

This is an extreme type of extended blooming: plants open a few flowers per day over a period of weeks to months. Several extreme cases of flowering for well in excess of 20 weeks in every year were reported for Neotropical trees, with a few flowering more or less all year round (Frankie et al. 1974). This kind of pattern tends to be linked with trap-lining pollinators, especially in tropical forest trees and shrubs; hummingbirds and bees learn a pollination route between widely spaced plants and go round this route every day to visit the few new flowers on each plant.

Big-Bang Flowering

Big-bang flowering is the typical form of mass blooming: all plants produce many flowers over just a few days. Examples include *Eugenia*, which flowers for just three days around Christmas Day each year, and *Casarea praecox*, where all plants in a population flower on a single day (Opler et al. 1976). Other plants produce their big bang at unpredictable times, where an unknown trigger appears to set off the whole population; this can be several times a year, or only once every few years (i.e., in superannual bloomers which are typically mass blooming). It relies on "intruding" on the regular visiting patterns of local pollinators with a brief, massive, and irresistible display. (It is also known as "mast flowering" or "masting" as a parallel to mast seed production; when it occurs synchronously in many species it may be termed "general flowering," particularly characteristic of tropical forests in Asia).

Multiple-Bang Flowering

In **multiple-bang flowering,** the flowers come out in small peaks (*episodic*), each usually lasting a few days, and not all the plants in a given population join in on every peak, so that peaks are of different heights. The peaks may occur at any time of year, and anything from two to five times a year.

Factors Influencing the Phenological Patterns

In general, the duration of the flowering period is related to the proportion of flowers and the number of flowers produced per day: sharply peaked patterns are achieved by high levels of flowering synchrony. Both *duration* and *synchrony* of the overall flowering period are therefore important, although obviously linked; in fact a highly asynchronous group of mass-blooming individuals could appear to an observer as a rather extended-blooming population or species. What determines these phenological parameters for any given plant species?

1. On a broad scale, environmental predictability and the temporal patterns of resources are obvious influences, with patterns therefore varying between broad habitat types. Cornucopia flowering is typical of highly seasonal ecosystems, with the tightness of the main flowering peak determined by the length of the spring/summer period; all other patterns are more likely in less seasonal habitats, although some plants do show big-bang peaks within a seasonal system.

2. Patterns of florivory, seed predation, and seedling survival are likely to be important (e.g., Kudo 2006; Elzinga et al. 2007). Mass blooming gives the possibility of temporal escape from these threats, or at least of "swamping the market" if blooming and floral predators (or subsequent seed set and seed predators) do coincide. Augspurger (1981) demonstrated that mass blooming in *Hybanthus* helped to satiate micro-lepidopteran seed predators, reinforcing the selective benefits of greater attractiveness to bees. Likewise Maycock et al. (2005) showed that masting in Bornean dipterocarp trees helped to increase seed and seedling survival.

3. Narrowed blooming time tending toward mass blooming should be favored where there is strong competition for pollinators, as it could help to limit interspecific pollen and gene flow. Discrete episodes of mass flowering among coexisting species might therefore be expected, and were described by Frankie (1975; and Frankie et al. 1983) for species of Fabaceae, mostly pollinated by medium and large bees, co-flowering in dry Costa Rican forests. For the relatively few species that flowered on into the wet season, blooming periods were much more extended as competition lessened.

4. Mass flowering should also be favored in plants that are rare, where just a few flowers would be unlikely to attract a pollinator at all. Habit may also be an influence here, with plants that bloom in dense forest canopies more likely to need mass blooming to produce a significant visual display, compared with plants that flower in the open or in a forest understory. In fact there are cases where forest trees of different species mass flower synchronously (general flowering), at irregular intervals of 3–10 years; for example, Sakai et al. (1999) documented that more than half of all flowering in a Borneo forest of more than 300 species occurred during one such flowering period.

5. Extended blooming patterns have some obvious advantages in terms of increased outcrossing with more partners, reduced geitonogamy, and lowered risk from sudden periods of bad weather and low pollinator activity. Potentially they give better control over investment in fruits in relation to resource availability.

6. Extended blooming may also be beneficial for colonizing species, as it will normally give longer periods of seed production and better chances of seedling establishment. Such species are also more likely to be generalist in relation to their pollinators, with fewer coevolved visitors available with which to synchronize. Opler et al. (1980) found that early successional species had longer blooming periods than species in adjacent mature communities, with the very long (even

year-long) flowering periods mostly occurring in the earliest successional communities.

7. Episodic multiple-bang flowering can in general be seen as a bet-hedging strategy (Gentry 1974), particularly appropriate in unreliable habitats; it is rather common in cloud forests (Haber, cited in Bawa 1983), and in shrubs and treelets rather than mature trees. However, examples also occur in the tropical and subtropical lianes and vines, climbing over other plants and flowering just after their "hosts"; they tend to have no nectar, and are only visited (mainly by bees) accidentally, so that this is effectively pollination by deception.

Note that some of the selective effects on flowering time can work in opposition to each other. For example, Marquis (1988) found that tightly peaked flowering in *Piper* led to better pollinator attraction and higher seed set, but that "off-peak" flowers suffered less seed predation, giving a degree of disruptive selection. In general, Elzinga et al. (2007) concluded that pollinators tend to favor peak flowering times (or earlier), whereas seed predators tend to favor off-peak and late flowering.

Constraints on Flowering Phenology

Within any one flowering pattern, there is of course much scope for variation. For a given species, a fundamental influence must be selection against interspecific pollen transfer and gene flow; this is linked to pollinator availability and hence competition for pollinators, as in (3) above, and in turn associated with the reward structure of the plant.

However, there are also more subtle but more wide-ranging effects, given that any particular plant species is also having to optimize other life history attributes apart from flowering: relative investment in vegetative growth to compete against neighbors for light and nutrients, or in flowers and/or seeds; investments in selfing (autogamy and geitonogamy) or outcrossing; investments in attracting pollinators or defending against folivores and herbivores; and in succeeding locally or colonizing new habitats.

Here we should raise the issue of *phenological displacement* that can occur in a community due to competitive effects between species; the phenomenon is dealt with fully in the next chapter as an aspect of escaping competition, but is an important potential constraint on flowering time for many species.

How Much Variation Is there within Any One Species, Population, or Individual?

Thus far we have looked at species-specific patterns of flowering, and the genetic basis of this flowering time control is beginning to be understood and explored: a cline in *Arabidopsis* flowering is modulated by the timing gene FRIGIDA (FRI) (Stinchcombe et al. 2004). But there is also some element of triggering in any one year that influences just when a population or an individual plant will begin to flower. Timing on this scale is well known to be influenced by day length (invariant between years) but also by temperature (Amasino 2005). Hence shifts in flowering time in response to recent climate change are already occurring (chapter 29).

Local and seasonal variations with climate (and hence also with altitude and latitude) are well known, with plants delaying flowering in a drought, or flowering later or more sporadically in a cold spring, or at higher elevations; clear examples of the latter came from communities at the fringes of alpine systems, where flowering depended directly on the timing of snowmelt (Kudo 1991, 2006). All these effects can readily be related to resource availability. Environment is also influential in several other ways (see chapter 10), affecting plant productivity, nectar and pollen characteristics, pollinator abundance, etc. Additionally there are some quite specific effects of shifts in flowering time within a species in relation to edaphic variation. For example, the grass *Agrostis* can shift its flowering time according to soil-based stresses; when growing on mine wastes (of which it is especially tolerant) the plant flowers slightly earlier, perhaps because it can thereby stay undiluted by pollen from the nontolerant grasses further away from the mine, so maintaining its own resistance to heavy metal pollution.

Variation should once again also be viewed in terms of synchrony within a species or population. In an outcrossing species, stabilizing selection for synchronous blooming across individuals might be expected, and was documented in the previously mentioned study on *Hybanthus* (Augspurger 1980). This author also reported (1983) that in five other species of shrub intrapopulation synchrony was greatest where the specific blooming period was shorter, as might be predicted. However, there are also instances of significant intraspecific variation within any of the four broad flowering categories defined above, with populations and individuals showing marked asynchrony. This can be quantified by defining full population synchrony as 1.0 and full asynchrony as 0; on this basis, Primack (1980) found populations with asynchrony levels varying from 0.34 to 0.74 in New Zealand plants, each population also varying from year to year.

Reasons for the favoring of asynchrony may vary in different environments and circumstances. Where populations are quite small and flowers are long lived, asynchrony could increase mating diversity by varying the neighbors available for a given plant to mate with across time. Where there is intense competition for pollinators within a species, as in some mass-blooming tropical canopy trees, moderate asynchrony may be a useful way to share out a limited pollinator community (Bawa 1983). Similar arguments would apply to phased flowering within a single canopy tree, where different parts of the crown may be preflowering, in full flower, or already in fruit (e.g., Medway 1972; Bawa 1983). Synchrony must also be allied to timing of events within individual flowers; for example, there is strong synchrony within species having very short periods of stigma receptivity, such as *Arum maculatum* (Ollerton and Diaz 1999).

Variation can also occur with sex, in dioecious plants. Males and females are under different selective pressures, and males may therefore flower earlier (both daily and seasonally) and for longer (Bawa 1983; Charlesworth and Morgan 1991). Male plants may also flower more frequently, with some female plants missing out a particular phenological peak due to over-investment in reproductive structures in the preceding episode (Bullock and Bawa 1981). Similarly, in gynodioecious species such as *Thymus*, hermaphrodites flower earlier and for longer periods than females (Ehlers and Thomson 2004).

Variation due to selective pressures from other biotic agents should again be stressed. Flowering may be altered locally to avoid herbivores, florivores, or seed predators, or to ensure synchrony with seed dispersers (and to avoid competition from other plants, i.e., phenological displacement as covered in chapter 22). Schemske (1984) showed an effect with herbivores in *Impatiens pallida*, where in mid-flowering periods

beetles destroyed many of the plants within a forest but not at the forest edge; but the former flowered 12 days earlier on average, which at least limited this effect. A clear example of florivore influences was found for *Helianthus*, where flower-feeding moths were much more damaging on early-flowering plants and so acted as the major selective agent on flowering time (Pilson 2000). Aizen (2003b) showed the contrasting effects of pollinators (hummingbirds, rare on early flowers) and of dispersers (marsupials, most effective when early flowers reached seed set) on selection for flowering times of an Andean mistletoe (*Tristerix*); the flowers were mainly produced in winter even though the pollinators were rare. And a comparative study across three valleys for *Astragulus scaphoides* showed that the benefits of synchrony could vary with site, although at least one economy of scale was evident in each valley (Crone and Lesica 2004).

Finally, it is noteworthy that real flowering patterns are rarely neatly symmetrical as shown in figure 21.1. In practice, most plants show a skew, either positive or negative, and Thomson (1980a) argued that this is often linked to their place within the flowering community. Those that flower early in a season show positive skew as they are unfamiliar to pollinators and must put on a good early show (see also O'Neil 1997). For example, early-flowering plants of *Phlox drummondii* are more showy, attract more pollinators, and achieve higher seed set than later-flowering plants, causing directional selection to flower even earlier (Kelly and Levin 2000). But when flowering density is especially high there may be a negative effect on flower visitation rates, which would favor a more asynchronous flowering (e.g., Gómez 1993).

2. When to Open a Flower

Daily

Most temperate ecologists will be so familiar with plants that open their new flowers in the early morning that listing examples would be spurious, although there are some equally well-known cases that first open around midday, and yet others that are inclined to open in the evening. The last of these in particular tend to be more specialized in morphology and to be linked to particular pollinators (moths, bats) that are active from dusk onward, so that the most effective pollinators encounter fresh flowers not depleted by day-time visitors. Time of opening may or may not be coincident with time of dehiscence (see chapter 7), although there is generally some relatively fixed relationship between these two events.

Seasonally

Seasonal timing of flowering on a broad scale certainly has a substantial phylogenetic component, with families differing in flowering time and with that difference preserved across floras as distinct as North America and Japan (Kochmer and Handel 1986). Figure 21.2 shows characteristic timings for the most important families in the North Carolina flora, with (for example) Violaceae and Ranunculaceae flowering early, Asteraceae and Convolvulaceae late, and other large families with intermediate timings. There are also some links with plant habit, trees and shrubs generally flowering earlier than herbaceous plants.

More precise timing is certainly linked to photoperiod, which has a direct inducing effect. This could be problematic near the equator, but Borchert et al. (2005) presented a model for photoperiodic control via detection of sunrise and sunset times that would work even in the equatorial forests of Amazonia. Triggering can also be due to rainfall (e.g., Dominguez and Dirzo 1995) especially in semiarid habitats, or to drought for the irregular mass flowering in Asian forests (Sakai et al. 2006). Links to temperature also occur (e.g., Appanah 1993), so that a given plant will start to flower later in a cooler spring for example. There are well-documented cases of wildflowers blooming earlier as global warming sets in (chapter 29; such effects could be serious if plants and their visitors are affected differentially, leading to an uncoupling of normal seasonal patterning). Mechanistic proximate effects on flower timing must be distinguished from ultimate causes that relate to selective advantages of flowering at a particular time. A major factor there has to be pollinator availability, which can be expected to have a strong seasonal component in all but the equatorial habitats.

Taking these low-latitude tropical habitats first, they can be almost aseasonal, although most do in fact show some climatic variation on an annual cycle. Several biotic and abiotic features of the environment could therefore operate as triggering factors. The wetter

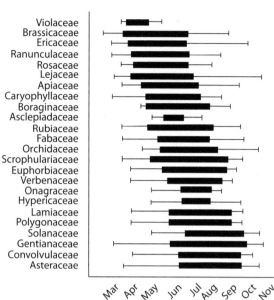

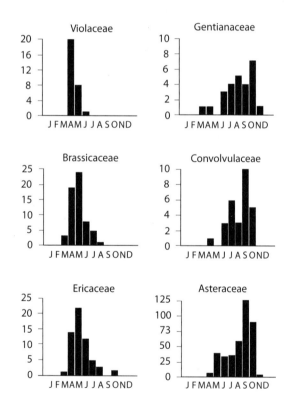

Figure 21.2 Timings for the 24 main plant families in the Carolinas, United States. A vertical bar marks the mean flowering time; below are frequency distributions for six representative early and late families. (Redrawn from Kochmer and Handel 1986.)

months may yield higher insect or fruit or foliage abundance that leads to higher population densities for bats or birds, or for the larvae of insects (especially caterpillars), so selecting for ideal flowering times that coincide with maximum adult pollinator numbers. Or the drier months in such habitats may (through reduced foliage) allow a more visible floral display, explaining a predominance of dry-season flowering in some parts of Costa Rica for example (Janzen 1967; Frankie et al. 1983). In rainforests of northern Australia a large-scale study (Boulter et al. 2006) could find no one predictor of flowering time (climatic, biotic, or phylogenetic), although floral seasonality did link to rainfall and temperature and there was some consistency within genera or families.

Temperate habitats show strong seasonality, and most plants here show a cornucopia flowering pattern, but usually with a strong seasonal bias. Basic patterns are readily identifiable, with spring-, summer-, autumn-, and rare winter-flowering examples. Some linkage with floral rewards probably occurs; species with low or no nectar may be selected to coflower with nectariferous plants that they resemble (see chapter 23), whereas the latter may be selected to "escape in time" from these deceitful mimics. This may in part explain the earlier blooming of particularly nectar-rich flowers in some temperate communities (Heinrich 1975a). Early blooming in temperate woodlands is also likely to reflect a need to grow and bloom before the canopy closes over with foliage. Another key issue in temperate habitats is the time required for seed development. Where a larger fleshy fruit is produced, most plants have to flower rather early (springtime) in order to have the fruits ready in autumn when the main migratory seed-dispersing birds are abundant.

In high latitude and polar communities, and also at altitude, flowering may be constrained to a rather short spring window, this being the only time when pollinators are reasonably reliably available; this issue is explored in chapter 27.

Beyond the simple climatic and latitudinal effects, competition within a community may further structure each species' flowering times, with *sequential* blooming of plants sharing main pollinator types commonly being seen as a potential method to reduce competition and reduce interspecific pollen movements. Alternatively, *convergence* of blooming times may occasionally occur, increasing the overall display and attraction; each species can then avoid some of the competition

by daily staggering of opening, or by varied spatial placements of pollen on the visitors. Chapter 22 covers these issues in depth.

Lifetime

For annual plants, flowering is generally the last phase of life, and may be accelerated or delayed by weather conditions and by resource availability. But for perennials, this translates into "at what age to flower for the first time," often inevitably linked with plant size at maturity, so that trees obviously flower first at a much older age than herbs. This may interact with flower number and size as well. For example, in *Polemonium viscosum* in the Rockies, the age of plants at flowering varied from 2 to 6 years, and had a strong effect on size of flowers (Galen 1993), mature plants producing larger displays; a 1-year delay in flowering gave a roughly 8% increase in corolla length and a 10% increase in flower number.

3. Flower Longevity and Flowering Period

Natural Lifespans and Floral Displays

All flowers have a natural aging process due to irreversible programmed cell death, largely independent of environmental factors (Rogers 2006), and at the end of their species-specific natural lifespan they more or less gradually become senescent, generally wilting, fading, and drooping at the petiole before abscissing from the plant. Abscission itself is usually controlled by ethylene acting as a growth regulator, although auxins may also act by influencing the sensitivity to ethylene; these processes were reviewed by Stead (1992). There may be some variations within a plant or a species due to temperature effects (e.g., Vesprini and Pacini (2005) with hellebores), but these produce only minor differences, with flowers lasting a little longer when the environmental temperature is lower. Most commonly flowers last for 1–3 days (fig. 21.3; Ashman and Schoen 1996), but longevity from anthesis to senescence can be as short as a few hours or may be up to weeks or several months (especially in orchids and asclepiads). The very large variation suggests that longevity in turn reflects adaptation to highly varied ecological conditions. Table 21.2 shows some of the observed patterns, with flower life lengthening

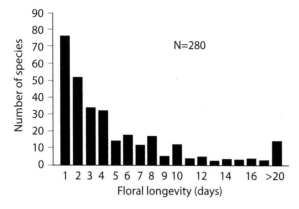

Figure 21.3 Flowering durations for 280 floral species. (Redrawn from data assembled by Ashman and Schoen 1996.)

away from the tropics (where a great many flowers last just one day or part of a day) and also with altitude. Similar patterns could be recorded within particular families such as Fabaceae and Rosaceae (Primack 1985a). Flower duration also varies with sex, female flowers lasting longer (sometimes several times longer) than male ones in most dioecious species.

Examples of very reduced lifespan include *Halimium* (a rock rose, Cistaceae) where the flowers open at 0530, may be pollinated by 0700, and show petal drooping from about 1000; or many species of *Oenothera* (evening primrose) which are open for just a few hours in the evening. In *Anagallis monelli* the flowers persist for about 8 days, but their effective life is shorter as they show **nastic** movement each day around 1800, drooping down into a nonfunctional state, then reviving and opening again around 0900 next day (and the arrival or germination of pollen does not alter this, the flowers still going on cycling for about 8 days).

At the other extreme, *Cymbidium* orchids have flowers that persist for at least 3 weeks, but this can extend to 8–12 weeks, attracting vectors throughout this period, and they normally only start to wilt once pollinated.

Longevity of an individual flower obviously affects the number of flowers that are open on the plant (display size), and the duration of the floral display, although this relation is complicated by the observation that removal of some buds on an orchid plant can increase the longevity of the remaining flowers (Parra-Tabla et al. 2009). Flower longevity can in turn influ-

ence the total number of flowers made per plant, because the cost of maintenance of an existing flower must be balanced against the cost of making new flowers now and in the future. Optimal floral longevity is therefore part of the economics of pollination (Ashman and Schoen 1994, 1996; Ashman 2004), and longevity has substantial effects on the quantity and quality of progeny; it should reflect the balance of maintenance costs and fitness consequences.

Determinants and Effects of Increased Longevity

What factors determine longevity? Primack (1985a,b) suggested that there would be a strong influence of habitat, some phylogenetic effects where particular taxa inherently have longer-lived flowers, and some relationship with the breeding system. But his data showed very little effect of pollinator type or syndrome, except for nocturnal bat- or hawkmoth-pollinated flowers which usually only last one night (see table 21.2 part B; the outlier is bird-pollinated Australian flowers, but the sample is strongly influenced by a few species of eucalypts). Ashman and Schoen added to the list of crucial influences the cost of making a flower, and how quickly a flower is pollinated and so fulfils its function. They developed an ESS model (an evolutionarily stable strategy) based on these key factors, in terms of pollen remaining in the flower and proportions of ovules fertilized against time. From this they could plot isoclines of optimal floral longevity for different values of m (the floral maintenance cost relative to its construction cost), as shown in figure 21.4. This reveals that short-lived flowers ($t = 1$–3 days) are favored when pollen and seed fitness accrue rapidly (top right-hand corner), and are more strongly favored when m is large (where maintenance of the flower is costly relative to making a new flower). The main implication is that floral longevity should be strongly related to rates of pollen removal and pollen deposition on stigmas, both of which should be determined largely by visit frequency. Ashman and Schoen showed that for 39 genera or species in previously published studies there was a strong negative correlation between visitation rate and floral lifespan (fig. 21.5), so that taxa receiving frequent visits were selected for short-lived flowers, while those with rare visits (often as low as 1 visit per 100 hours of flower life) tended to be much longer lived. The authors noted that selection

TABLE 21.2
Mean Flower Longevity against (A) Habitat and (B) Pollinator

		Mean (days)	*Range*
	Locality and season	*Mean (days)*	*Range*
(A) Habitat			
Tropical forests	Southeast Asia dipterocarp forest	1.0	1–1
	Neotropical dry forest	1.1	1–2
	Neotropical rainforest	1.3	1–3
	Mangrove forest	3.3	2–8
Temperate forests	Spring	6.9	2–12
	Early summer	5.7	1–14
	Late summer	2.5	1–-4
Grasslands	North American prairie	2.0	1–4
	New Zealand grassland	5.9	1–19
	African savanna*	1.8	1–-6
Montane	Chile	4.2	1–15
	New Zealand	7.8	3–-15
(B) Pollinator			
	Large bee	3.1	
	Small bee	4.7	
	Butterfly or moth	~6	
	Fly or other insect	~5	
	Hawkmoth	1	
	Bat	1	
	Bird (Australia)	12	
	(Neotropics)	1–2	

Source: Largely based on Primack 1985a.
* Author's own data.

will differ where flowers release pollen in a staggered fashion rather than all at once, or where stigmatic receptivity matures gradually so that pollen receipt is effectively staggered.

These kinds of relationship also explain why some flowers have particularly long lives in certain community settings. For example, *Kalmia latifolia* flowers lasted up to 21 days at a northeastern US site, whereas coflowering shrubs lasted on average 3.4 days; because *Kalmia* flowers produced rather little nectar they were poor competitors for bumblebee visits and needed to flower for longer as a form of reproductive assurance ((Rathcke 2003).

Differential effects can also be seen on inflorescences, where individual flowers vary in longevity. For example, in *Aquilegia buergeriana* longevity was greater in the first flowers than in later ones, especially through elongation of the male phase of the flowers (Itagaki and Sakai 2006).

What are the effects of changing flower longevity? From a pollination point of view, any increase should

1. increase the size and length of the floral display;
2. increase the number of visits received;
3. potentially increase the amount of pollen received and the pollen diversity, and hence the parentage and genetic diversity of offspring;
4. allow the plant to outlast coflowering competitors.

But it may also

5. potentially increase the heterospecific pollen receipt;
6. increase the within-plant pollen transfer.

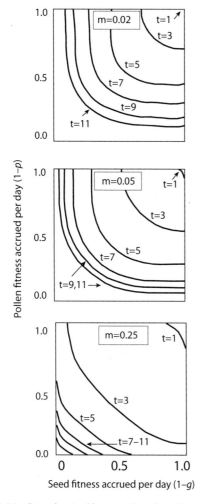

Figure 21.4 Isoclines of optimal longevity (*t*, in days) derived from a balance of pollen fitness and seed fitness, for three different values of maintenance costs of flowers (*m*). See text for further explanation. (Redrawn from Ashman and Schoen 1996.)

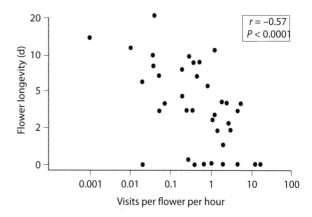

Figure 21.5 Negative correlation between log floral longevity and log visitation rate, each observation representing a mean for a genus. (Redrawn from Ashman and Schoen 1996.)

The balance of these pros and cons is likely to differ for different kinds of plant, but must always be carefully weighed up in terms of life history strategy.

Pollination Effects on Longevity

Pollination in itself can often be a trigger for floral changes and senescence (reviewed by van Doorn 1997). For example Weber and Goodwillie (2007) recorded *Leptosiphon* flowers that had achieved outcrossed pollination senescing after 1 day, whereas emasculated flowers lasted 2–5 days. In *Polygala vayredae* individual flower lifespan was negatively correlated with pollinator visitation rate, so that flowers could increase their longevity when pollinators were scarce, although fruit set and seed weight then suffered (Castro et al. 2008). The mechanisms involved in senescence after pollination are unclear, often (though not always) mediated by the plant growth regulator ethylene (Rogers 2006). In carnations, for example, ethylene produced from the pollinated stigma was translocated via the style to the petals, where it upregulated genes that biosynthesize further ethylene (tenHave and Woltering 1997), leading to petal wilt. But ethylene-independent pathways are also present in many flowers.

In some plants, effects are exactly opposite and pollination leads to flower retention and delayed senescence. This is potentially useful because, although flower lifespan is usually considered in terms of functional life (i.e., a flower still able to donate or receive pollen), it may often help to retain old and now functionless flowers on the plant. They then serve as long-range attractants contributing to the overall display, as long as visitors can then detect their poor status on closer inspection. Ideally, the effectively functionless flowers will undergo subtle changes as signals to visitors, as discussed in chapter 5 in relation to color change. In fact, flower longevity clearly has to interact with all the possibilities for postpollination changes, which can be of color, odor, shape, orientation, or reward status (Gori 1983; and see earlier chapters on color and odor change). All flowers tend to wilt and fade or turn brown before abscission, but many also have a separate pathway for change, involving induced effects that are not directly related to age but triggered by some aspect of visitation. These changes can be much more rapid (relative to the normal lifespan), and can affect one or more of the modalities in one or all parts of the flower. The issue of flower closure after pollen receipt was mentioned in chapter 2, for gentians and for fireweed (see Clark and Husband 2007), and many other examples probably exist, providing a clearly adaptive response to avoid wasteful pollen usage. In orchids, morphological changes are particularly common, some species showing petals folding over to cover the column, others undergoing a swelling of the column, and others again rotating the whole flower by 180° so that normal pollinators can no longer access the tube (e.g., *Angraecum* orchids normally visited by moths; Strauss and Koopowitz 1973). Alternatively, some flowers merely change color in one part of one petal (e.g., the alterations to the nectar guide color in *Aesculus*, described in chapter 5), while others undergo both color and odor change in the whole flower (e.g., *Parkinsonia*; Jones and Buchmann 1974).

In most cases, a change in advertising signal is also accompanied by a change in reward, usually a cessation of nectar production; and the change can clearly be detected by visitors, which then avoid the changed flowers in favor of as yet unchanged, unpollinated (and still rewarding) flowers. Whatever the change is, it is usually followed by a marked period of flower retention, giving a direct effect on longevity. Thus *Lupinus* flowers last for 5–6 days longer than "normal" if they have been pollinated and undergone the change in color of their banner petal markings (Gori 1989). Table 21.3 gives further examples, including a few where postpollination change has the reverse effect of reducing longevity as the flower wilts more rapidly (some *Petunia*) or even abscises (*Rhododendron*).

The mechanisms underlying these changes are complex, and probably vary between species, but can be broadly split into two categories:

1. Visitation-induced change. In some plants the physical effects of visitation are crucial, and floral change occurs when a visitor depresses the keel of a legume, or contacts anthers or stigma, or even just touches petals. Many such effects may work through a classical damage-response pathway in the plant, since even minor tissue damage can trigger an ethylene-based response leading to classical aging symptoms. This pathway is readily seen in action where the delicate petals of some flowers quickly discolor and wilt at the slightest touch.

2. Pollination-induced change. Here the change is specifically associated with effective pollination of the flower—whether determined by conspecific pollen deposition on the stigma, or by some later effect such as pollen tube growth or fertilization of ovules. Here again though there are suggestions that the ultimate pathway may involve ethylene-based signaling, since the penetration of style tissue by germinating pollen grains is in itself a form of tissue damage and can trigger ethylene production (e.g., de Martinis et al. 2002).

As a footnote here, a reverse effect should also be mentioned; that aging and wilting can of course affect a plant's pollination. Older flowers even before they wilt are more likely to self and also become less discriminatory among pollen donors and often set less seed (Marshall et al. 2010). Where a flower wilts naturally, but without having been pollinated, the collapse of the petals can also have a physical effect on the positioning of the anthers and ungathered pollen and can sometimes thereby induce self-pollination as a system of reproductive assurance. This is thought to happen, for example, in *Pedicularis dunniana* (Sun, Guo et al. 2005), where the wilting of the upper lip presses pollen onto the stigma. Other cases were given in chapter 3, section *Methods for Ensuring Selfing*, as examples of deliberate selfing mechanisms.

4. How Big Should Each Flower Be?

The general issue of flower size was considered in chapter 2, where effects of support and allometry were introduced, as well as the broad effects of visitor type. There are strong phylogenetic influences on floral dimensions, but within a given taxon substantial variation also occurs, much of the effective size of larger species being attributable to the expansion of substantial sheetlike structures (usually the petals) that form the advertisement of a particular flower. Larger sizes are thus achieved without too much extra mass or expenditure of valuable resources.

Not surprisingly, individual flower size is often a crucial factor in determining visitor frequencies and their effectiveness as pollinators. This may be a major reason why flower size is relatively constant within a species (e.g., Worley and Barrett 2000; Herrera 2001b).

TABLE 21.3
Longevity Effects in Flowers with and without Inducible Postpollination Changes

Species	Normal lifespan (days)	Time to change (hours)	Period of retention following change (days)	Change	Reference
Caesalpinia eriostachys	3–4	4	2–3	{Petal color, and { petal folding	Jones and Buchmann 1974
Caesalpinia pulcherrima	2		1–2		Cruden and Hermann-Parker 1979
Combretum farinosum	5		3		Schemske 1980
Cymbidium (various)	35–50	4	3–4	Color, and swelling of column	Arditti and Flick 1976 Harrison and Arditti 1976
Lantana camara	3		2–3	Petal color	
Lantana trifolia	2–3		1–2	Petal color	
Lupinus nanus	5–6		3	{ Banner petal	Dunn 1956
Lupinus bicolor	2		1	{	Dunn 1956, Nuttman and Willmer 2003
Lupinus sparsiflorus	4–5	4	3–4	{ Spot color	Wainwright 1978
Lupinus blumeri	10	24	8	{	Gibson, cited in Gori 1983
Parkinsonia aculeata				Color and odor	Jones and Buchmann 1974
Petunia (various)	?	<1	?	Corolla collapse	Gilissen 1977
Rhododendron (various)		<1	0	Corolla abscission	

Cresswell et al. (2001) found that *Brassica* flower size was conserved with varying plant density, the plants instead reducing their flower number as density increased. However, there is substantial evidence that within a species, all else being equal, pollinators will generally select larger available flowers (e.g., Bell 1985; Conner and Rush 1996; Andersson 1988) or will make more visits to plants with large flowers. Thomson (2001a) found this effect for lepidopterans but not for other visitors working jasmine flowers, and also showed that bees and bee-flies had longer handling times on larger flowers. A similar effect occurred with composite inflorescences (Andersson 1996) and with insect pollinators of *Madia sativa*, which chose blossoms with more ray florets (Celedon-Neghme et al.

2007) albeit with a trade-off for the plant of reduced autogamy in the larger blossoms. Most studies have shown that large flowers do increase male mating success, for example Martin (2004) with *Mimulus*; although a few plants with unusual mating systems do not match this expectation (e.g., Elle and Meagher (2000) with *Solanum*). Similarly it is generally true that more fully open flowers (with a bigger head-on display) are preferred over partly open flowers. If visitors do indeed exert selective pressure on flower size in this way, then outcrossing animal-pollinated flowers should show less variation in flower size than selfing species, and this was confirmed with a phylogenetically corrected comparison (Ushimaru et al. 2006).

However, the issue of flower size can be compli-

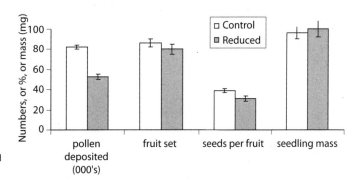

Figure 21.6 Manipulating floral sizes in *Cistus salvifolius* results in fewer dispersed pollen grains from the "reduced" smaller flowers, with little effect on seed set or seedling quality (means ± SEs). (Redrawn from Arista and Ortiz 2007.)

cated by the conflicting predictions of natural selection and sexual selection (Abraham 2005), since the former (incorporating optimality theory) suggests that *rewards* should largely determine visit frequency, irrespective of flower size. Many visitors may prefer larger flowers which may house large rewards; but if offered larger but artificially less rewarding flowers against small and high-rewarding ones Abraham found insect foragers switching to the smaller flowers in accordance with optimality predictions.

To reduce confounding variables, several studies have looked at flower size in relation to pollinators within a particular taxon and preferably in more equable habitats. Armbruster (1993, 1996) analyzed evolutionary trends in the genus *Dalechampia*, with comparative analysis of species across three distinct geographical areas, and showed a positive relationship between pollinator size and blossom size; for example, in the Neotropics smaller species were visited by *Trigona* bees, medium species by megachilid bees, and large species by *Eulaema* (euglossine) bees. He interpreted this size relationship as indicative of selection against losing pollen to sympatric congeners and thus against sharing of pollinators.

An alternative approach is to manipulate flower size experimentally, as Arista and Ortiz (2007) attempted with *Cistus salvifolius*. Here the smaller flowers suffered reduced pollen export (male function) but female success and offspring quality were unaffected, as might be expected if flower size is mainly related to attractants and male fertility (fig. 21.6). Ishii and Harder (2006) reduced flower sizes of *Delphinium bicolor* and found reduced visits per inflorescence but increased numbers of probes per visit by bees, so that the overall number of probes per flower was unchanged.

Given all this, is it better to have many small flowers

or a few large ones for roughly the same amount of flower tissue produced? Is there a trade-off between flower size and flower number? Sargent et al. (2007) tested this explicitly, looking at 251 angiosperm species with phylogenetic correction, and found a significant negative correlation as predicted by the trade-off idea (although this did not hold for all of the lower taxonomic levels tested). This issue of size versus number is clearly strongly interactive with the next topic, so we will pick it up again below.

5. How Many Flowers Should a Plant Have at One Time?

Do More Flowers Get More Visitors?

It is rather uncommon for plants to have just one flower at a time. Simply, more flowers per plant should be more attractive and receive more visitors. This could be because they are detected more easily by a passing potential visitor, or because they offer reduced interflower travel costs after arrival; both may apply, and the importance of each may vary with the economic costs of the particular visitor. There are many well-known cases of larger displays getting more visitors; for example, Augspurger (1980) with the mass-flowering tropical shrub *Hybanthus*; Andersson (1988) with bumblebees on *Anchusa*; Eckhart (1991) with *Phacelia*; and Kudo and Harder (2005) with several legumes. This generalization applies not only to separate flowers but also to aggregated ones: larger inflorescences tend to attract more visitors than smaller ones (e.g., Schmitt 1983 for *Senecio*; Cruzan et al. 1988 for *Phyla*; Thomson 1988 for *Aralia*; Makino et al. 2007 for *Cirsium*). And this applies not just to small insect visitors but also to large vertebrate visitors; for example, Pyke (1978) with hummingbirds on *Ipomopsis*, and Paton and Ford (1983) with honeyeaters on both *Correa* and *Eucalyptus*, showing more and longer visits at plants with more flowers.

In bees, at least, there is some evidence that there is an innate receiver bias for particularly strong floral display signals, so that honeybees chose a larger or more colorful floral display in laboratory tests even where there was no associated increase in reward quality (Naug and Arathi 2007). Such an effect could be more widely operative in floral display evolution than is currently recognized.

On the other hand, having more flowers on a plant increases the risks of geitonogamous selfing, as discussed in chapter 3; the plant thus has a real dilemma (Klinkhamer and de Jong 1993; Harder and Barrett 1995): whether to increase its flowering display but risk more selfing and greatly reduced pollen export, or to limit its attraction by having fewer flowers but where each is more likely to be successfully cross-pollinated.

This dilemma is partly resolved in some plants by an ability to adjust their display in relation to their recent visitation (*adaptive plasticity*). For example, pollinated inflorescences of the orchid *Satyrium longicauda* had fewer flowers than unpollinated ones, the effect being seen within a few days (Harder and Johnson 2005). The plant could put on a better display when visitors were rare but save resources with just a few flowers when visitors were abundant.

Do More Flowers Get More Visits?

Getting more visitors is one thing, but do individual flowers each then get more visits? Paton and Ford (1983) found that individual flowers on plants with more flowers received more visits from birds. However, this is probably not the norm; aggregated flowering was shown to result in similar or even *fewer* visits per flower in rather more cases, and this was modeled to be the optimum response for the plant by Iwasa et al. (1995). Lloyd and Schoen (1992) reported a very general tendency of flower visitors to visit only a fraction of the available flowers on any one plant before moving on. This could have many causes, both intrinsic (reward depletion, satiation, a need for other and varied resources) and extrinsic and arising from interactions with other animals (e.g., competition from other

visitors, mate searching allied to feeding, or predator avoidance). Robertson (1992) collated evidence to show that pollinators commonly visited fewer than 10–20 flowers per plant even when many more were available. Table 21.4 summarizes some findings on this point, for different kinds of visitor. The key question is then whether a visitor normally visits a fixed *proportion* of the flowers available to it.

It turns out that many animal flower foragers in practice visit either a nearly *fixed* or a *smaller* proportion of flowers as the number of available flowers increases. For example Kudo and Harder (2005) found more bumblebee visits to legume inflorescences with larger displays but no change in the proportion of flowers visited; whereas Ohashi and Yahara (2002) found that bumblebees visited a decreasing proportion of flowers on larger displays of *Cirsium*, as did Stout (2000) with bumblebees visiting *Cytisus*, and Feldman (2006) with syrphid flies and bees on *Brassica* plants. Fritz and Nilsson (1994) found rather fixed proportions visited in bee-pollinated *Orchis* but decreasing proportions in butterfly-pollinated *Anacamptis*. Arroyo et al. (2007) found fewer probes per flower head by butterflies on larger *Chaetanthera* inflorescences, and Harder and Johnson (2005) found a roughly linear relation of flower number to flowers visited by moths on *Satyrium*, although with indications of a reduced proportion at high flower number (fig. 21.7A). A similarly shaped relationship occurred with *Aconitum columbianum* visited by bumblebees (Pleasants and Zimmerman 1990), where around 20%–25% of flowers were normally visited, but with a slightly declining percentage when the number available was unusually high (in this case exceeding about 20 per plant) (fig. 21.7B). And in *Myosotis colensoi* plants, visited by tachinid flies (fig. 21.7C), the proportion of flowers visited decreased from almost 100% when only 1–2 flowers were present to a mean of around 20% when 100 flowers were present (Robertson 1992).

The particularly detailed study by Thomson (2001a) found that even on one plant the effects varied with visitor type; the number of visits was positively related to the number of open flowers on *Jasminum* for bee-flies and butterflies, but with hawkmoths visitation related more to flower size. To add further complication, for bees the response may change with experience, from choosing patches with more flowers initially to choosing patches with fewer flowers but higher rewards after a few hours of foraging (Makino and Sakai 2007).

Table 21.4

Numbers of Flowers Visited in Relation to Number of Flowers Present

Plant	Visitor	Mean flowers present	Mean flowers visited	Percent visited	References
Combretum farinosum	Hummingbird	>1000	12	~1	Schemske 1980
Delphinium nelsonii	Hummingbird	4.6 (2–8)	2.2	48	Waser 1982; Schulke, cited in
	Bumblebee		1.6	35	Snow et al. 1996
	Halictid bee		1.4	30	
Delphinium barbeyi	Hummingbird	15* (10–20)	3.4	23	Waser, cited in Snow
	Bumblebee		2.7	18	et al. 1996
Eichhornia paniculata*	Bee	3	2	67	Barrett et al. 1994
		6	2.5	42	
		9	3.3	37	
		12	3.5	29	
Geranium caespitosum	Bee	15* (10–20)		~50	Hessing 1988
Hybanthus prunifolius	Social bee	~200	10	~5	Augspurger 1980
Ipomopsis aggregata	Hummingbird	12.3 (10–20)	6.5	53	de Jong et al. 1992;
	Hawkmoth		4.7	38	Snow et al. 1996
	Bumblebee		2.8	23	
	Butterfly		2.5	20	
Malva moschata*	Bumblebees,	3	1.5	25	Snow et al. 1996
	anthophorid bees	6	3	50	
		12	3.5	29	
Myosotis colensoi	Tachinid fly	~100	19	~19	Robertson 1992
Sabatia angularis	Bee, wasp, hoverfly	35	2	6	Dudash 1991

* Experimental manipulations.

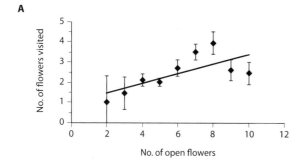

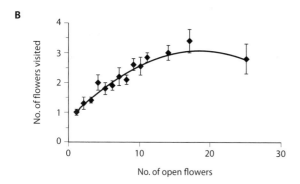

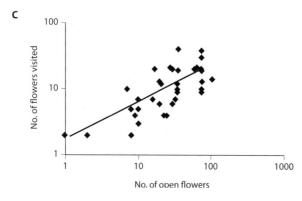

Figure 21.7 Examples of relationships between flowers available and flowers visited for (A) moths on *Satyrium* inflorescences (redrawn from Harder and Johnson 2005); (B) bumblebees on *Aconitum* (redrawn from Pleasants and Zimmerman 1990) (both means ± SEs); and (C) individual records (log scale) of tachinid flies on *Myosotis collensoi* (redrawn from Robertson 1992).

Such differences are likely to complicate the whole issue of selection on floral traits.

Where Does the Balance of Visitor Number and Visit Number Lie?

Figure 21.8 shows an overview: many flowers present at once may attract more pollinators per unit time

(fig. 21.8A); this can increase pollen receipt and removal, but it also tends to increase the number of flowers per plant visited by any one pollinator (fig. 21.8B), and so it can affect pollen dispersal and thus the mating system. Above all, it does increase the risk of selfing (Schoen and Dubuc 1990; de Jong et al. 1993; Barrett et al. 1994; Hodges 1995), so the trade-off for the plant is real. There should be a balance of the costs and benefits of having more flowers at one time, to achieve maximum plant fertility, roughly as in figure 21.8C. This issue of optimum floral display size was explicitly explored by de Jong et al. (1992), and Ohashi and Yahara (1999, 2001). It is partly a composite of the issues covered in sections **2,** *When to Open a Flower,* 3, *Flower Longevity and Flowering Period,* and 4, *How Big Should a Flower Be?* above, and it varies both within and between species, and with plant density. A fair summary of many findings would be that, as floral display increased,

1. visit rate per plant increased,
2. number of flowers probed per plant increased,
3. proportion of flowers probed per plant decreased, and
4. visit rates per flower could rise but were more often lower, although often only marginally affected.

As a result, several studies have shown a positive and roughly linear relation between overall floral display and pollination success for a plant, for example in *Asclepias* (Broyles and Wyatt 1990, 1995), for the orchids *Calopogon* (Firmage and Cole 1988), *Orchis,* and *Anacamptis* (Fritz and Nilsson 1994), and for bumblebee-visited *Delphinium* and *Aconitum* (Pleasants and Zimmerman 1990), with or without decreased visits per flower.

It is also possible that plants can "make more" of the benefits of increased display, and reduce the risks of geitonogamy, by their own behavior. One option is to offer highly variable nectar volumes in different flowers, so that risk-averse foragers move on to a new plant more frequently (see chapters 8 and 23; Biernaskie et al. 2002); or to have a gradient of nectar reward along the spike of a plant. Another option is the retention of older (postpollination) flowers to boost the display from a distance, as discussed above.

But there is one additional consideration to take into account, in that the effect recorded may depend on the spatial scale considered. Veddeler et al. (2006) noted that on a field scale bee density per coffee plant increased with a decreasing proportion of the plants

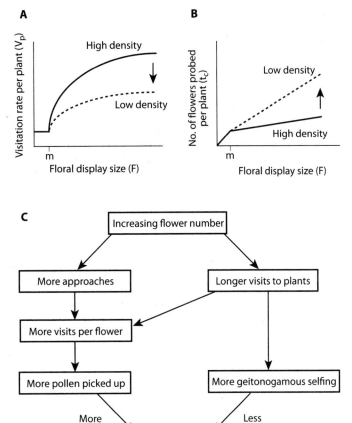

Figure 21.8 Models for visitation in relation to floral display size both for (A) the relative visit rate per plant and (B) the number of flowers probed per plant; (C) shows the likely causal links. (Redrawn from Ohashi and Yahara 1999.)

being in flower at the site (a dilution effect), whereas on a per plant or per branch scale bee density increased with increasing flower density (a concentration effect), with species richness also increasing with flower density only when viewed at the smaller scales. Patterns of visitation therefore need to specify rather carefully the scale on which they are recorded.

Effects on Pollen Carryover and Pollen Discounting

The extent of geitonogamy within a plant will depend on patterns of pollen dispersal for each kind of visitor (see fig 7.12). How much self-pollen and cross-pollen is deposited on a plant's stigmas during successive visits? Selfing rate is likely to be strongly correlated with flower number per plant, as shown by Schoen and Lloyd (1992) using *Impatiens pallida* flowers, where

there was a selfing rate of just 6% with one flower per plant, rising to 44% when there were three or more flowers, implying that selfing was mainly geitonogamous rather than autogamous. Snow *et al.* (1996) presented more extensive analyses using *Hibiscus moscheutos*, comparing plants with 3, 6, or 12 flowers present, and offered to pollinators (bumblebees and anthophorid bees) in fixed arrays. With 3 flowers present, selfing rate was 25%, while with 6 or 12 flowers this increased to ~50%. Similar tests with *Eichhornia* (Barrett et al. 1994) with 3, 6, 9, or 12 flowers again found a rising selfing rate, but the increase was only from 18% to 26%. Clearly geitonogamy is an important component of selfing, even with low numbers of flowers, but these studies indicate that the relationship between flower number and selfed progeny is usually asymptotic.

The pattern of geitonogamous pollen carryover between flowers and of outcrossing between plants can be readily modeled in a basic one-compartment pollen model (see chapter 7), with a simple decay of carryover from a single pool of mixed pollens, and several features then emerge (Harder and Barrett 1996):

1. Pollinator movement between multiple flowers on a single plant will inevitably produce mainly geitonogamy
2. Each recipient plant is likely to receive pollen from several outcross donors, but in proportions that vary with the number of flowers visited on the immediately preceding plants
3. Each donor plant contributes pollen to several recipients, but probably with a close match between the origins of male gametes contributed to the various flowers on any one plant.

In practice, geitonogamy will commonly decrease for the second and subsequent flowers on any one plant, as earlier pollen from outcross donors is revealed beneath the most recently acquired self-pollen.

However, where the visitors are bees geitonogamy may well be exacerbated by their tendency to groom pollen into the corbicula only when flying between plants (Harder 1990b; see chapter 18), so that while visiting flowers within one plant they will be moving self-pollen rather extensively. This of course raises the point that a single-compartment pollen system is probably quite uncommon, and the effects of flower number and visitor behavior often need to be modeled with a two-compartment model (fig. 7.12).

Pollen discounting becomes an issue here (chapter 7), as the pollen that is wasted in selfing is unavailable for crossing; it is probably not important as a cost in the rare case where there is a single flower per plant but does matter when there are multiple flowers (e.g., Ritland 1991; Kohn and Barrett 1994; Harder and Barrett 1996). One way to increase floral display without paying the costs of pollen discounting is to incorporate flowers that are not pollen donors or receivers—"pseudoflowers." Sterile flowers at the periphery of inflorescences are not uncommon, as in some *Hydrangea*, *Viburnum*, and many Asteraceae. The pseudoflowers contribute to long-range attraction but at close range can be detected as unrewarding and so are not wastefully visited (Gori 1983; Cruzan et al. 1988; Weiss 1991). Temporal dichogamy can also be important in reducing discounting costs, especially where visitors go to the female-phase flowers first.

6. What Determines Phenological Parameters for a Particular Plant Species?

Do the matters of flower size, number, and longevity vary in a consistent or patterned manner, for example with habitat, or with type of pollinator? And how much variation is there within any one species, or within an individual?

As a broad generalization, flower-visiting animals (as with any foraging animals) tend to feed in restricted areas, moving progressively from one flower to a near neighbor, which will increase the chances of geitonogamy and should therefore militate against large displays on any one plant. However, empirical data show that visitors do generally prefer larger patches, or larger inflorescences. Harder and Barrett (1995, 1996), with *Eichhornia*, found preference in bumblebees for different (manipulated) flower numbers per plant as in figure 21.9; the data there are for pairwise comparisons of different flower numbers, so that preference for 6 flowers over 3 flowers was almost the same as for 12 flowers over 6 flowers. An increased ratio of flowers in the pair led to increased preference, rather than flower number per se having an effect (and as shown in the last section the amount of selfing resulting in this plant was not very high anyway).

Preference for larger display could be either because this is more easily detectable from distance, or because the proximity of many flowers decreases flight costs, or both. But most studies have found that the proportion of flowers probed per patch tends to decline with increasing display size; that is, the number probed increases, but less sharply than the number available, an observation that does not fit easily with normal explanations for foragers preferring larger displays (increased detectability or reduced within-patch movement costs). This may be in part because of competition effects: in the Harder and Barrett work (1995, 1996) there was a decreasing proportion of flowers visited as flower number increased, but this proportion was affected by what other plants were available, being larger when competing plants were particularly low in flower number. Hence pollinator (at least bumblebee) behavior was context dependent, making the modeling more difficult. Ohashi and Yahara (2001) likewise modeled how many flowers an animal should visit, similarly finding that density dependence confused the issue, preference for large displays being lower with reduced plant density.

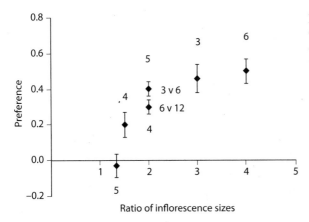

Figure 21.9 Mean preferences (± SE) of bees for larger numbers of flowers per plant in pairwise tests at different ratios in *Eichornia paniculata*. The zero axis indicates no preference; two different trials with a ratio of 2 between pairs were used, with 3 vs. 6 flowers and with 6 vs. 12; numbers above or below indicate the number of bees sampled. (Redrawn from Harder and Barrett 1995.)

For any visiting animal, the decision to move on should depend on the decrease in rate of energy gain (*patch depression*). This could arise from the nonrandom flower choice of the visitor (i.e., it chose the best ones first), especially where there was a spatial patterning to the reward (e.g., a gradient of reward up a spike). Or it could stem from the increased risk of revisiting already probed flowers, although most studies show this to be low; and the risk could be reduced anyway by high-capacity short-term memory, and/or by directional movement within the patch, and/or by footprinting, or even by detecting nectar levels before even probing the flower.

What could the plant do about all this to try to maximize visits but also increase movement between plants? Nonrandom flower choices could be held in check by reduced visual signals or reduced patterning of reward; or the plant could increase the risk of revisiting a flower, and therefore shorten a visit, by reducing "landmarks" on an inflorescence, with complex, close-packed arrangements of uniform small flowers so that visitors could not easily remember where they had already been. Alternatively, a plant could change the balance of within-plant to between-plant movements by making its adjacent flowers open asynchronously or by having more complex tubular flowers with longer handling times.

On balance, though, the benefits of attracting more

pollinators probably do not outweigh the costs of increased geitonogamy for large displays. A general conclusion in the literature has therefore been that where plants grow at low densities they should go for smaller displays and/or extended blooming, and *not* for larger displays (although other factors such as spatiotemporal sex separation etc., and abiotic constraints on flowering time, may influence this).

Given all the interactions between the open flowers on a plant, it is evident that the functional unit in terms of plant mating is not the individual flower but the overall floral display. The design of individual flowers has to be considered in this context, in terms of both ecology and evolution. Hence selection on flower number must occur in concert with selection on flower size and shape and on floral rewards, with larger inflorescences or more floriferous plants occurring where pollen removal is restricted and/or pollen carryover is substantial. As yet, major comparative analyses of the interactions between flower design and floral display, also taking into account plant density in the local area, are somewhat lacking.

7. Where Should the Flowers Be Placed?

How far can plants manipulate their visitors by spatial patterning? There are a number of possible ways in which this could happen, although each may also carry some costs.

Spike Inflorescences

Flowers can be arranged in spikes, opening sequentially from top to bottom or vice versa. Bees are well known for visiting spikes that open their flowers from base to tip from the bottom upward, and these flowers are commonly protandrous, so having older female flowers basally and young male flowers at the top (e.g., *Delphinium, Chamerion*; see fig. 3.13). They may also have a pattern of a gradient of nectar reward decreasing from bottom to top, which promotes shorter visits and therefore reduced geitonogamy; carpenter bees foraging on spikes of *Pontaderia* with no such nectar gradient did tend to stay longer and probe more flowers (Orth and Waddington 1997). But plants may allocate more nectar to female flowers when they have a particularly large floral display (Biernaskie and Elle 2005).

However, the top-down pattern of visiting behavior is common in flies, and hummingbirds seem to move up and down spiked inflorescences at random. Hence this kind of manipulation of the visitors only works well for certain kinds of plant, and there may be cases where offering too many spatial signals can increase visit length unhelpfully, as outlined in the last section.

Another aspect of spikes is their relative height, and taller spikes may generally get more visitors. In *Verbascum* taller plants achieved higher outcrossing rates, and the top part of shorter inflorescences did better than the lower parts (Carromero and Hamrick 2005). These effects are likely to be affected by the density and height of surrounding vegetation, and the height preferences of particular pollinators.

Radial Inflorescences

Flowers may be organized as radial inflorescences, such that visitor behavior is channeled into an orderly progression around a ring of florets. This occurs in hoverflies working on some kinds of Asteraceae, where there are indications of a fly detecting completion of a full circuit and so reducing wasteful visits for itself and excess crossing for the plant (Gilbert 1983; fig. 13.9). A similar kind of ability to "count" appeared to occur in bees visiting radial flowers with several nectaries spaced in a ring, observable, for example, with *Apis* and *Bombus* working *Fragaria* flowers (fig. 2.14). More information on diverse animals' behavior patterns on radial blooms would be very helpful.

Flower Position on Larger Plants

Flowers occurring at different heights on a large shrub or tree may attract different visitors, or different visitation rates from any one visitor. For example, the gum tree *Eucalyptus globulus* achieves higher outcrossing rates on its upper branches, and Hingston and Potts (2005) attributed this to the behavior of birds that foraged significantly more in the upper canopy and also commenced their foraging bouts more often near the tops of the trees.

Visitor behavior may also be partly controlled by positioning flowers on the periphery of a large plant or within its interior. Those on the inside will probably contribute less to the display, and will require more dextrous access flights through the foliage by visitors, but they will also experience a different microclimate, often having larger volumes of lower concentration nectar due to the reduced insolation they receive and/or the higher humidity within the foliage. Visitation of inner flowers can also be influenced by simple protection effects, since foraging there may afford an escape from possible predators or even from over-attentive mates (e.g., Stone 1995 with *Anthophora* bees; plate 29G). Some potential pollinators may visit surface flowers on a bush or tree early in the day (adequate reward, little disturbance) but move to the interior flowers later on (better reward, and predators or males more active).

8. Overview

There are clearly a number of good reasons why the timing and patterning of a flowering season may change from year to year or from place to place, and why the number of flowers per plant may change seasonally and even within a flowering season. Similar and additional reasons may affect the arrangement of those flowers and the nectar production per flower. Modeling the "ideal" setup for any one plant is therefore very complex and has only been partially successful as yet. And we have left out one aspect almost entirely, in that a plant's "decisions" about flowering must also have strong interactions with competitive effects—between flowers and between plants—for pollinator visits. These additional themes are dealt with in the next chapter.

Chapter 22

LIVING WITH OTHER FLOWERS: COMPETITION AND POLLINATION ECOLOGY

Plants rarely occur in isolation, but grow and flower as part of a community of mixed species; and they are rarely visited by just one kind of animal, but receive visits from several potential pollinator types, some of which will be shared with other plants. Thus plant-pollinator interactions have a strong community component, and both plants and animals are subject to potential competitive interactions. This was recognized very early on in pollination biology (e.g., Robertson 1895; and see review by Mitchell et al. 2009). However, many early studies largely side-stepped the issue, merely looking at a single plant and its visitor(s) or a single animal and the flower(s) that it visited, and thus often misrepresented the selective forces they exerted on each other.

What exactly is competition in the context of pollination? It is usually treated from the perspective of the plants, but competition is also likely to occur among and between the pollinators, and both these aspects must be taken into account. Competition can also occur at various levels—as a structuring factor in communities, as a selective force on an individual plant's phenology, morphology, or rewards, and also at a genetic level structuring pollen competition between males, and **female choice** between possible mates. Hence this chapter deals with a series of different but interactive issues.

1. Defining the Problems

Plants: Competition for Pollination

Types of Competition
Plant competition could be subdivided into the widely used ecological terms *interference competition* (one pollinator moving between two plant species, so reducing the effectiveness of its visits) and *exploitation competition* (one plant species receiving more visits, and so decreasing the visits to another).

However, Waser (1983a) proposed a specific definition of competition in pollination ecology as follows: any interaction in which co-occurring plant species suffer reduced reproductive success because they share pollinators. This can still be divided into the two different kinds of interaction:

1. Competition through interspecific pollen transfer, causing losses of pollen or of stigmatic surface, and hence of effective pollinator movements. This affects both male and female function and output. It can occur in both wind- and animal-pollinated systems, and even when pollinator numbers are not limiting.

2. Competition through pollinator preference, where one species attracts pollinators away from others and so reduces their reproductive output. This is only possible in animal pollination, and when pollinator numbers are limiting.

Competition between plants is therefore probably better referred to as competition *for pollination*, rather than *for pollinators*, although authors have differed as to which is more important (Silander and Primack 1978; Pleasants 1980, 1983; Waser 1983a).

The subdivisions of competition given above could also be seen as equivalent to:

1. Plants competing for pollen quality. If pollinators are abundant and indiscriminate or messy, plants may struggle to avoid clogging of their stigmas with self- or foreign pollen. Pollen quality is essentially pollen "purity," since heterospecific pollen directly wastes the male function and also wastes the female function by blocking up the stigmatic surfaces uselessly. Hence competition for pollen quality means that plants are competing for visits by high-fidelity visitors, and it can be either intraspecific or interspecific.

2. Plants competing for pollen quantity. If pollinators are limited, plants may have to compete to get enough visits with enough cross-pollen deposited, and intraspecific and interspecific competition will again both be operating. Essentially the plants compete to get as many visits as possible, by being as attractive as possible.

Competition for pollen quality may on balance be more important in most situations, and is well documented in various careful studies of interspecific interactions; for example Campbell and Motten (1985) manipulated the densities and mixing of two spring flowers *Claytonia virginica* and *Stellaria pubera*, and found that the latter suffered lowered seed set when forced to compete with *Claytonia*. This resulted from losses of *Stellaria* pollen to *Claytonia* stigmas, rather than any reduction in visitation (in this case by bee-flies). And when the rather similar *Mimulus ringens* and *Lobelia siphilitica* were artificially grown together (Bell et al. 2005) there was a 37% reduction in seeds per fruit in *Mimulus*, as those flowers received less conspecific pollen.

Mitchell et al. (2009) produced a useful overall pic-ture of the interactions involved in competitions for pollination, reproduced in figure 22.1, highlighting the effects of both visit number and visit quality.

Potential Outcomes of Competition

Competition can usually only be detected by its outcome, which may be a numerical change in one species, or a change in its resource use; Thomson (1980b) termed the latter a "niche shift." Most obviously, the immediate numerical outcome for a plant will be a reduction in seed and eventual progeny, although it may also sometimes be the production of maladapted hybrids or even local extinction. But it could additionally promote evolutionary divergence in resource use, where one aspect of resources is the pollinators. The former is more likely via pollinator preference effects, but the latter is more likely to lead to stable changes in floral traits and only rarely to extinction (Waser 1983a). However, in practice, the outcome as seen by a contemporary observer may well be quite different, as one or more of the plants have adapted some aspect of their morphology, behavior, or rewards to avoid the worst effects of competition; this issue is explored in more depth below. Thomson's paper stressed that both kinds of competitive effect are likely to affect community structure.

Interspecific or Intraspecific

Interspecific competition effects are usually the more obvious, but competition could occur between phenotypes within a plant species, and so could also be intraspecific. At this smaller scale, individuals of different phenotype compete in both their male and female functions, to maximize their genetic contribution to the progeny, which potentially produces competition to attract pollinators, to receive pollen, or to be pollen donors. Pollen grains themselves may compete at the stigma to reach the ovules, and fertilized ovules probably compete not to be aborted.

If coexisting plants overlap in flowering period and attract some or all of the same pollinators, they should compete for those pollinators (inter- or intraspecifically) whenever the plants occur above a certain density, and one or two plants (or species) in the community may then outcompete the others. For example, Mosquin (1971) found that cornucopian *Salix* and *Taraxacum* pulled pollinators away from earlier-blooming plants as soon as they started to flower. Likewise, in orchards, the presence of many dandelions in the grass

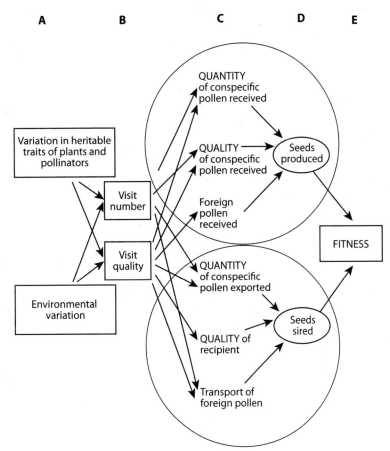

Figure 22.1 A conceptual framework showing the interactions that affect competition for pollination, through ecological and evolutionary variation (A) and its effects on visits (B), on pollen (C), and hence on seeds (D) and eventual fitness (E). The upper half of the figure involves female function, the lower half male function (seeds sired). (Redrawn from Mitchell et al. 2009.)

beneath trees affected the pollination of apples (*Malus*), which received fewer visits (Free 1968); here the apple flower's stigmas became clogged up with the foreign pollen, affecting the success of legitimate pollen, a case of classical interference competition.

Where there is fairly constantly competition between species, there is likely to be selective pressure on the species that are "losing out" to change some aspect of their morphology or behavior—perhaps to increase their reward, to enhance their attractants, or to change aspects of daily timing or of seasonal flowering period. The first two kinds of change might lead to receipt of a greater proportion of visits from the current pollinators, while any of these changes could lead to a more effective sharing of pollinators between competing plants (enhanced, for example, by changing stigma or anther positions so that pollen was placed and collected from a different part of the visitor's body,

avoiding mixing of pollens). Alternatively, a plant under competitive pressure could change some trait(s) that would result in different pollinators being attracted, or it could even to switch to wind pollination or selfing.

But where they occur at *low* densities two or more plant species may in practice *enhance* each other's success by attracting more pollinators overall and thus helping to maintain a pollinator community for each plant (e.g. Rathcke and Lacey 1985; Moeller 2004). That balance could be sensitive to small changes in pollinator and plant abundance, from week to week or from year to year. If the situation remained fairly constant, with mutual facilitation of the plants' pollination at low densities, the result could even be convergence of flowering times and of attractive features (the pollen clogging problems would remain, however, so that subtle stigma and/or anther changes could still be selected for). Over longer time periods, this kind of facilitation potentially leads to mutualistic mimicries, along with some interposed cheats, as discussed in chapter 23.

Animals: Competition between Pollinators

Here there is less potential for confusion. Pollinators are mainly competing for food resources, and the competition is usually rather direct and obvious, in the short term appearing as aggression, and as competitive exclusion in time or space or both. Over longer time periods, this should lead to structured communities, with coexisting species again diverging, so that they are just different enough from each other in their flower-visiting preferences, or behaviors, or phenology, or perhaps in their morphology (e.g., Heithaus 1974, 1979; and see Thomson 1980b). These kinds of effects are covered in section 4, *Reducing Competition among the Pollinators*.

However, there are also competitive effects to be seen where an alien bee or other pollinator has been introduced into a community, and these possibilities are dealt with in chapter 29 as one of the threats to pollination services.

Seeking Evidence for Competition

Competition is very commonly cited as a reason for character differences, leading to the observed general correlation between floral diversity and pollinator diversity. It seems intuitively obvious that there has been parallel adaptive radiation of the flowers and of the pollinators, shaped by competitive interactions. But clearly this is also linked to the debate over specialization in pollination systems (chapter 20), and whether specialization inevitably occurs as plants are selected to avoid competition and adverse hybridization.

In practice the evidence for competition is often far from clear. As always in ecology it is very hard to prove that any change has been due to competition, as all we usually see is the "ghost of competition past," often as limited niche overlaps (Rosenzweig 1981; Pritchard and Schluter 2001). It is particularly hard to get clear evidence from observational studies of single communities; rather easier where several communities are compared, and easier still where experimental manipulations are used (Connell 1975; Waser 1983a). It

is often unclear just how far pollinators are limiting to plant success; if they are not, then competition for their services is likely to be relatively unimportant anyway. While it is true that many plants fall well short of 100% possible seed set, this is very often at least partly due to limiting energy supplies (e.g., Sutherland 1980; Udovic 1981) rather than to limiting pollinator visits. However, there are good records of plants where seed set is significantly correlated with visit frequency (e.g., M. Zimmerman 1980), and very numerous cases where seed set is significantly higher with hand-pollination, implying pollen or pollinator limitation.

Because of the inherent problems of establishing proof of competition, the evidence here is compiled largely in terms of how competition might be, or might have been, reduced.

2. Reducing Interspecific Competition among Plants

In general, selection is supposed to work to reduce competition among coexisting species, so the expected outcome would be divergence along some axis. Differences between plants in a community could therefore be assumed to result, partly or wholly, from segregation due to competition for pollinators. This could involve

1. spatial separation; changing location to get a different subpopulation of pollinators, and so reducing interspecific visits;
2. temporal separation or *phenological displacement*; changing time of flowering to bloom at a time when the plant can get pollinators' undivided attention;
3. changing floral morphology or attractants; to attract more pollinators, or to attract a different pollinator;
4. changing floral morphology more subtly; so that the same pollinator gets pollen on a different part of its body;
5. Changing reward structure; which could reduce the need for phenological or structural change (although detecting this may be particularly difficult);
6. Switching pollination mode; changing pollinator type, or moving to wind pollination or selfing.

Note that methods that reduce exploitation competition also always reduce interference; but the reverse is not always true, since (for example) increased pollinator constancy, or spatial segregation, does not necessarily reduce exploitative interactions.

Spatial Displacement

The first possible response to competition involves biogeographic effects and is probably not uncommon. For example, Hurlbert (1970) showed that a greater temporal overlap between pairs of goldenrod (*Solidago*) species was accompanied by greater spatial separation. Pleasants (1980) found a similar set of patterns with five coexisting Asteraceae in the Rockies.

Phenological Displacement

Plants with substantial pollinator overlap might reduce or avoid competition by blooming at different times (mentioned in chapter 21). This could apply to both animal- and wind-pollinated plants, since both can suffer from competition through interspecific pollen transfer. The effect was noticed anecdotally in the late nineteenth century, when Robertson (1895) recorded that in northeastern America there was a succession of flowering from spring onward, with morphologically similar flowers coming in sequence rather than together. He also noted (1924) that introduced plants tended to have longer flowering periods than natives. Since then many other observational studies have reported similar effects, from arctic through to tropical zones (see Waser 1983a).

The effect is most clear-cut where there are just a few species in a community, or where only two species interact. For example, early-season flowering was favored in *Cynoglossum* only where it overlapped with more attractive *Echium*, even though this put the *Cynoglossum* flowers out of phase with the peak bumblebee activity (de Jong and Klinkhamer 1991).

Displacement effects are much more difficult to detect in complex temperate and tropical habitats, and flawed arguments are sometimes used. However, Pleasants (1980) showed a regularly spaced (nonrandom) pattern of flowering in Rocky Mountain meadow flowering communities, and Waser documented changes in flowering time for this Rocky Mountain flora that reflected competition for pollinators. He compared *Ipomopsis aggregata* flowering periods at two locations (Waser 1983a) and showed that the plant bloomed

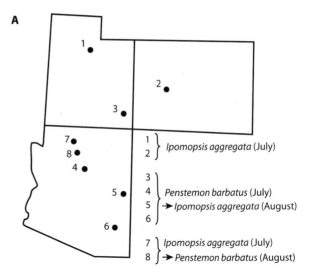

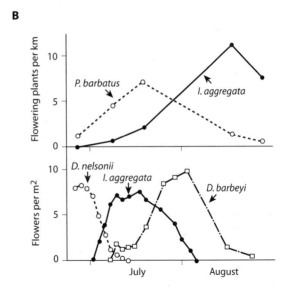

Figure 22.2 (A) *Ipomopsis aggregata* flowering times at sites in Arizona, Colorado, and Utah, with and without *Penstemon barbatus* that competes with it for hummingbird pollinators. (B) Direct comparison of *Ipomopsis* flowering at just two locations, with the additional presence of *Penstemon* or *Delphinium* species. *Ipomopsis* when on its own normally flowers in July, but may flower in July or in August depending on what else is in flower, and when in competition for pollinators it may switch to later flowering even at warmer, more southerly sites. (Redrawn from Waser 1983a.)

later in southern Arizona than in Colorado, explicable in terms of the flowering time of the main species competing with it for hummingbird visits at each site (fig. 22.2), whereas climatic effects alone would predict the opposite trend. He also compared this plant when flowering with and without *Penstemon barbatus*

TABLE 22.1

Percentage Seed Set in Floral Populations Coflowering in and out of the Period of Overlap, in Separated Control Populations, and in Competition Plots with Induced Overlapping Coflowering

| | Natural meadows | | | |
	Nonoverlap	Overlap	Control plots	Competition plots
Ipomopsis aggregata	10	14	9–14	6–9
Delphinium nelsonii	42–49	26–29	39–42	26–31

Source: Waser 1978b.

in the south-central United States; sometimes one flowered first and sometimes the other, suggesting real responses to competition rather than a natural fixed timing in either species.

However, pattern does not necessarily imply process (i.e., underlying competition) even if no other explanation can be thought up. Waser (1978a,b, 1983a) avoided this problem in part by setting up an experimental study using *Delphinium* and *Ipomopsis* species, again in the Rockies. Both are hummingbird pollinated, and in their natural habitats they flower sequentially, *D. nelsonii* first (see fig. 22.2). Waser transferred some plants of each up and down the mountains in pots. Normally those at low altitude flower earlier, but his manipulations allowed both plants to be flowering at the same time in certain mixed plots. They were then visited indiscriminately by the birds, and he showed a significant reduction in seed set in the mixed plots compared to the sequential sites (table 22.1).

For phenological displacements in natural situations to be firmly established, there must be evidence of reduced seed set when two species overlap compared to the nonoverlap periods; and visit rates per flower before, during, and after the overlap period should be recorded to establish the nature of the competition. Experimental evidence by bagging some flowers on one of the competitors during overlap and showing competitive release (as extra seed set) in the other is also useful. A substantial review of phenology patterns in plants by Rathcke and Lacey (1985) showed a range of reasonably well-supported examples of such displacement, and thereafter the idea became fairly

widely accepted as an explanation of how competition might act.

However, several critiques have pointed out that we may have got the whole idea of phenological displacement out of proportion, for various reasons:

1. Inadequate statistical testing. At the very least, there have to be good statistical analyses of flowering periods to establish that there is regular temporal spacing. For example Poole and Rathcke (1979) reanalyzed Stiles' data (1977) for ten hummingbird-pollinated plants and showed that peak flowerings were clumped mainly in relation to two dry seasons, and thus the patterns could be explained without invoking competition. However, earlier studies were not all deficient in this respect. Poole et al. (1979) used an analytical system that is conservative, depending only on peak flower date and ignoring the shape of the flowering curve; better statistical methods use computer simulations which can also retain the shape of the flowering curve. Thus Pleasants' study (1980) cited above used simulation models with his meadow communities and found that overlaps in flowering for at least half the pollination guilds were indeed significantly smaller than predicted on a random model. Results from such methods were reviewed by Pleasants (1983). Other suitable approaches were used by Stone et al. (1996, 1998) in assessing daily phenologies in acacias.

2. Most early studies looked at all plants in a community, although in practice subsets of plants in different pollinator guilds were probably not competing

with each other and no phenological spacing would be expected anyway. For bees, plants that are visited only for nectar are unlikely to be competing strongly with those visited only for pollen, and vice versa. Pleasants (1983) pointed to the need, ideally, to combine statistical phenological analysis with an understanding of guild structure and reward structure (cf. Sakai et al. 1999 in dipterocarp forests, or Lobo et al. 2003 with bombacaceous trees in the Neotropics).

3. The situation may be complicated further by likely asymmetries of underlying competition; an abundant species overlapping in flowering time with a rarer species is unlikely to lose too many visits, but the rare species may suffer severely, and is more likely to be the one to diverge in phenology (or in some other feature); but most models do not incorporate asymmetries, which may not predict "regular" blooming distributions.

4. Any of the other methods of reducing competition described here may also be operating, so that phenological spacing alone does not describe or detect the full competitive interaction; and phenological shifts may not occur at all if a competitor can invade and establish itself in a community just by being more attractive to the visitors.

5. Inadequate comparative methods may conceal phylogenetic constraints. Kochmer and Handel (1986) pointed out that each plant species and sometimes whole genera or families share similar flowering times; and when they examined similar areas of north temperate plant communities in North America, Europe, and Japan, with similar families and genera, they calculated that by far the biggest effect on flowering time was phylogenetic, suggesting that only limited divergence from ancestral flowering patterns was possible. So, for example, *Viola* species always flower in spring, *Erica* always in late summer, most of the temperate Fabaceae in early summer, and most of the Asteraceae in later summer (see fig. 21.2). Their review suggested that flowering time is a conservative character, and any competitive effect from pollinators may be fairly minor in comparison. A similar conclusion was reached by S. Wright and Calderon (1995) for a range of tropical species. However, others have assumed that flowering time constraints would be more relaxed in the tropics (see Bawa 1990), and certainly there are some

cases of related species flowering at different times; for example in southeast Asia, where *Shorea* species flower in succession in the canopy (Ashton et al. 1988), even though all tend to fruit at the same time. It seems that phylogenetic constraints can sometimes be overcome by strong selection. Lack (1982) offered the example of UK hay meadows, which had been mowed down in July over several centuries, and where there is now a massive flowering peak in June even in species that elsewhere flower later (e.g., knapweed, *Centaurea nigra*, and devil's-bit scabious, *Succisa pratensis*).

6. Morales et al. (2005) suggested that in seasonal habitats where the flowering period is restricted rather few species will be able to flower early or late and more will have to be concentrated in the middle (the *geometric mid-domain*), because of the time needed for flower development and fruit maturation. Hence regular distributions of timing are rather unlikely in such habitats.

Note that sequential phenological patterning of flowering is not necessarily just a case of reducing competition anyway; it can also help both plant species (or many species) by maintaining the pollinator community, since the animals have to be fed outside the flowering period of any one plant. In other words, as mentioned in the last chapter, sequential flowering has some other community-level advantages—the various plants are providing resources for an insect's or bird's complete life cycle. By building up a good population of pollinators all the plants benefit (although later flowerers benefit more from earlier flowerers than vice versa, at least within a season). Waser and Real (1979) proposed that this was important for *Delphinium nelsonii* and *Ipomopsis aggregata*, which may overlap a little and share the same visitors, but which help maintain pollinators for each other and thus in the long run for themselves. This effect may also be important for bat-pollinated plants; Lobo et al. (2003) found various species of Bombacaceae flowering in differently staggered patterns in different locations, the bats specializing on whichever tree was at its peak at any given location and thus taking only one pollen type at a time over several weeks.

In this scenario, it would then be a real advantage to be convergently similar in appearance to other species, but sequential in flowering. An effect of this kind is often proposed in western Europe for common hedgerow species of Rosaceae (e.g., Yeboah Gyan and Woodell 1987a,b), with successive flowering of hawthorn, then one or more species of wild rose, and then bramble; all have rather similar flowers, and they get a similar visitor spectrum in turn. This can be seen as a sequential mutualism when the situation is of benefit to more than one plant.

However, this whole story may sometimes work the other way, as briefly mentioned earlier: rather than spreading their flowering out sequentially to avoid competition and/or sustain the pollinator community, some plants stay flowering together and instead get *facilitation* effects, a form of Müllerian mimicry in flowering time. Schemske (1981) demonstrated floral convergence and pollinator sharing in two species of *Costus* (ginger family, Zingiberaceae) in the tropical forests of Panama. Here the two species had a large though not complete natural phenology overlap, and were shown to *benefit* from flowering synchronously. The flowers lasted 1–2 days, with very similar morphology and nectar availability (*C. allenii* 28 mm nectary depth, *C. laevis* 31 mm), and both used a euglossine bee as vector. This bee moved freely between the two species. Schemske used fluorescent powders to mimic pollen, and this turned up on both species' stigmas. Both achieved high fruit set (71% and 58%, respectively); and crucially there was no decrease in fruiting in the overlap period—in fact it was somewhat enhanced. Probably the presence of the other species gave double the nectar availability and increased the overall attraction to bees. Since the two species are intersterile, they could both still function; and there was little risk of pollen clogging, because neither plant was particularly abundant.

Rathcke (1988a,b) suggested a comparable temperate example, where the herbs on Rhode Island were converging on flowering time to get facilitation, although the evidence was not clear-cut. Thomson (1981) also recorded per flower visitation rates in several yellow and orange Asteraceae that were enhanced in the presence of other plants sharing the same visitors, suggesting a facilitating effect; he calculated that visitors were assessing overall flower density over a range of about 1000 m² for solitary bees and 500 m² for bumblebees. More recently Geber and Moeller (2006) showed data for three *Clarkia* congeners that indicated facilitation, with pollinator availability and visit rate increasing with the number of congeners present (fig. 22.3). This facilitation effect may even be

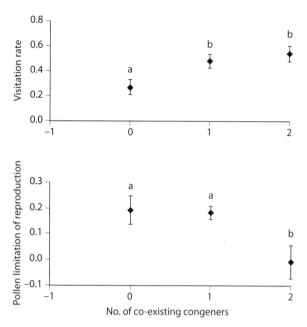

Figure 22.3 The facilitation effects of increasing numbers of congeners coflowering with *Clarkia*, resulting in increased pollinator visit rates and lowered pollen limitation (measured as the difference in seed set between the open-pollinated flowers and those receiving supplemented hand-pollination). (Redrawn from Geber and Moeller 2006.)

important for flowers that have no morphological similarity, since *Raphanus* has been shown to receive more visits when coflowering with typical arable weeds (*Cirsium, Solidago, Hypericum*), at least when these occur in relatively low density (Ghazoul 2006); this can perhaps be partly explained by a complementarity of rewards. On the assumption that such effects may well be more important than currently recognized, Buchmann and Nabhan (1996) suggested a widespread need among plants for other flowers to be around, supporting the required pollinator through (and more especially just before) their own flowering season. Feldman et al. (2004) tried modeling possible facilitations between two pollinators, finding that this could occur depending on the species' relative density, but was likely to be rare. However, Hegland, Grytnes et al. (2009) explored the occurrence of positive and negative interactions explicitly, with a community of 15 plant species in temperate grassland visited by bumblebees and/or flies. While most of the pairwise interactions (on visitation rates) were not significant, they found 17 statistically supported interspecific interactions, of which 14 were significantly positive, often between plants with similarly colored flowers. It seems from this that negative (competitive) interactions between coflowering plants may not after all be the dominant ecological community-level process, although data on the eventual reproductive output of species in such communities are still lacking.

Note that with synchrony and facilitation it might still be possible to reduce interspecific pollen movements and thus reduce competition, by having spatial differences in pollen placement or diurnal differences in flower opening or in the timing of pollen and nectar availability.

In fact, phenological displacement itself could operate diurnally, not just seasonally as in most of the classic cases examined; and there is now clear evidence that competitive diurnal temporal displacement does occur. Early reports included diurnal separation of visitation for *Erigeron* and *Viguera* in Rocky Mountain meadows (cited in Pleasants 1983), based on differing dehiscence times (morning for *Viguera*, after midday for *Erigeron*). A more impressive example was reported with East African acacias (Stone et al. 1998). Acacias are highly speciose but all very similar in flower morphology, and with no barriers to heterospecific pollen deposition. All attract mass visits, and in savannas they may be the only plants in flower for parts of the season. Flowers mostly only last one day, but crucially within that they dehisce at different times. Figure 7.9 showed the outcome for five species that coflowered at the start of the rainy season in Tanzania; anthesis and dehiscence were evenly (nonrandomly) spaced through the day, and the window of pollen availability was rather narrow for any one species. Visitors had similar peaks of foraging on particular species of tree; hence visits in a particular daily "window" were all or mainly to one species of acacia. And since visitors were mainly bees (grooming off body pollen at regular intervals), they were probably fairly free of heterospecific pollen by the time they went to the next dehiscing acacia. For these plants, dehiscing in a tight peak could also be relaxed when not needed: figure 22.4 shows that a sixth species in the same habitat, *A. thomasii*, had no daily peaks, flowering alone in the dry season when there was little or no competition. The precision of these patterns may link to the very long evolutionary stability of East African acacias and savannas, giving them enough time to have evolved such neat temporal divergence.

All these examples show that reducing or moving flowering seasons can help reduce competition. However, another possible solution to the competition problem is to *increase* floral longevity and flowering season, so that a plant can tolerate competition for part of its flowering period from some temporally overlapping species. There may be good adaptive reasons for both very short and very long flowering, depending on the plant community.

The other kind of evidence for the operation of such effects, in either direction, would be cases of the same plant in different communities having different flowering times. Here there is a need for caution, as the different communities may well vary in climatic and edaphic factors that could independently affect flowering time. Once this is taken into account, though, there are a few examples of species showing apparent character displacement in flowering times. Carpenter (1976) showed that in Hawaii two tree species (a myrtle *Metrosideros* and a legume *Sophora*), both pollinated by honeycreepers, coflowered with only a small temporal overlap at lower elevations, but at altitudes where only one of them was present it had a much longer flowering period, extending into the lowland overlap period.

Of course there may be other factors affecting flowering time (see Ollerton and Lack 1992). There will be a need to have seed ready for dispersal at the appropriate time, so flowering will be constrained by how long it takes for the fruit or seed to mature; and several factors may determine the "appropriate" time beyond just germination conditions; for example, keeping seed abundance out of phase with seed predators if possible, or in phase with fruit dispersers (perhaps less of an issue because these are usually long-lived vertebrates). A good example from the tropics was reported for *Hybanthus*, where there was higher seed predation on shrubs that were induced to flower outside their usual short highly synchronized season, when they perhaps normally saturated the seed predators (Augspurger 1981).

Changing Plant Morphology to Attract Different Pollinators

Here, diversity of flower shapes and colors could be seen as evidence in itself, as it must often have originated through competition in the past. But more

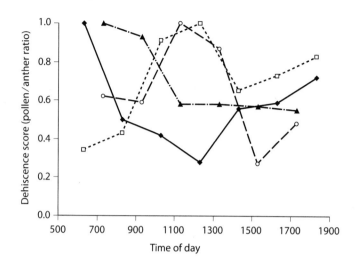

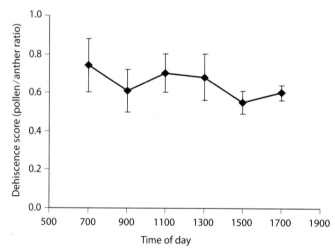

Figure 22.4 The dehiscence timing of *Acacia thomasii* on four individual trees in Kenyan savanna: unlike most species in the same habitat (see fig. 7.9) dehiscence is not synchronous (above), and mean pollen availability is almost constant through a day when averaged across trees (below). (Redrawn from Stone et al. 1998.)

specifically, plants should expect to promote pollinator constancy by differing in floral morphology on a local scale. Flowers with more unusual morphologies are most able to achieve floral constancy in visitors (Waddington 1983; chapter 11), and a negative correlation between morphological similarity and degree of temporal overlap (equated to degree of competition) was reported by Anderson and Schelfhout (1980) for a group of prairie composites (fig. 22.5). There is reasonably strong evidence that particular pollinators show constancy to shape, color, or scent, or to aspects of behavior such as foraging height, as covered in chapter 11, and any of these axes of constancy could be exploited by the plant. Examples for visual and ol-

factory signals were given in earlier chapters, and Waddington (1979) found that species sharing pollinators were more likely to be different in inflorescence height than those not sharing pollinators. To complete the list, the correlations of corolla length and tongue length are perhaps the clearest example of a shape effect, where different corolla tube lengths can promote resource partitioning among nectar feeders and so improve conspecific pollen transfer (Rodriguez-Girones and Santamaria 2007). How far tongue length and corolla length (or spur length) are strictly involved in a coevolutionary "race" is open to debate (see Ennos 2008; Hodges and Whittall 2008), and better phylogenetically corrected studies are needed here.

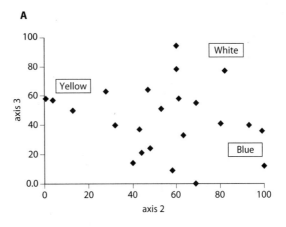

Figure 22.5 (A) PCA analysis for 31 prairie composites, based on floral features; of three PCA axes used, these two best separate the species, with PCA 2 strongly linked to blue flower colour, erect flowers, and large inflorescences and PCA 3 linked to white flowers with angled heads and large inflorescences; hence white flat-topped species are in the upper right region, yellow sunflower types in the left, and blue flowers in the lower right. From this, similarities between pairs of species (total nos. of pairs as open triangles) can yield the data shown in (B), indicating that the more similar two flowers are the less likely they are to overlap in their flowering periods (closed diamonds). (Redrawn from Anderson and Schelfhout 1980.)

It might be expected that for any two competing species the rarer one would be more likely to diverge, so there might be evidence of rare species being more attractive than their abundant relatives (Pleasants 1983). This effect has been noted by some (e.g., Beattie et al. 1973; Heinrich 1976b) but rarely tested for. However, Pleasants did record a negative correlation ($r = -0.82$, $p<0.01$) between species abundance and species attractiveness (although this was scored as number of visits per flower) in Rocky Mountain meadow flowers; he attributed this mainly to increased nectar production rate in the rarer plants (see the next section).

Reasonably clear evidence comes where one species differs in morphology according to the presence or absence of other interacting species. For example, two subspecies of *Polanisia dodecandra* were reported as sympatric in some zones, and where they co-occurred one subspecies had smaller flowers and shorter anthers (Iltis 1958), with some color distinc-

tion also appearing. In the genus *Solanum*, the degree of protrusion of both male and female structures differed substantially for species pairs where they were sympatric, the differences disappearing for species on their own; and the morphological differences in sympatric pairs appeared to constrain effective visitation to particular sizes of bee, so minimizing interspecific pollen movements (Whalen 1978). Changes in color in areas of sympatry were also classically reported for *Phlox* (Levin and Kerster 1967) and for *Clarkia* (Lewis and Lewis 1955).

Again, evidence from experimental manipulations may be more convincing, and early work by Levin was particularly clear. Levin and Kerster (1967) used *Phlox* species with different color morphs, and found that *P. pilosa* was strongly dominated by pink morphs when it grew alone but that when it coflowered with *P. glaberrima* (which is always pink) its populations were dominated by the white morph. When pink *P. pilosa* were transplanted into a mixed species population they were more frequently visited (by the main pollinating butterflies), and suffered much greater interspecific pollen transfer, than the naturally occurring white forms. Later, Levin (1978) exploited the differing inflorescence heights of two *Lythrum* species as a further test, constructing mixed arrays and observing foraging *Apis*; he found that interspecific flights decreased markedly when the height difference was greatest.

Changing Plant Morphology to Alter Pollen Placement

When visitor constancy is poor, it might still be possible for selection to manipulate floral morphologies rather more subtly to reduce interference competition, by altering pollen placements on the visitors' bodies. Sprague (1962) showed that two Californian louseworts, *Pedicularis groenlandica* and *P. attolens*, shared bumblebee pollinators but placed their pollen respectively on the underside and the head of the visiting bees. Likewise, Macior (1970) showed that coflowering species of *Pedicularis* in arctic and alpine regions of North America shared pollinators, but differed slightly in corolla length and in color, leading to differences in pollinator behavior and resulting pollen placement. For a very different flower type, Macior (1971) showed that pollinia from several co-occurring *Asclepias* species were located on different parts of the visiting bumblebees' bodies.

There are also well-known examples of differential pollen placements for hummingbirds in a range of plants (Stiles 1975; Carpenter 1978b; Brown and Kodric-Brown 1979), and for butterflies with *Phlox* (Levin and Berube 1972). A further neat example comes from flies visiting the Cape flora, where two taxonomically very distinct flowers, a *Pelargonium* (Geraniaceae, dicot) and a *Lapeirousia* (Iridiaceae, monocot), were visited by the same genus of long-tongued nemestrinid fly *Prosoeca*. Flowers of these genera are normally very different, but there they converged on a similar zygomorphic form (Goldblatt et al. 1995) (fig. 22.6), although the *Lapeirousia* deposited pollen on the head of the fly and the geranium deposited its pollen ventrally.

Another clear example came with bat-pollinated flowers in the genus *Burmeistera* (Muchhala and Potts

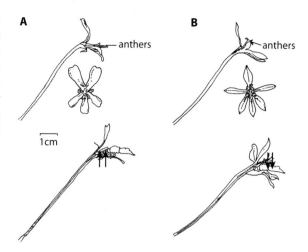

Figure 22.6 Frontal and side views of flowers in a southern African guild of long-proboscis fly-pollinated flowers, showing the convergence of form: (A) *Pelargonium*, (B) *Lapeirousia*, also showing the different pollen placement in each (ventral head and thorax and dorsal thorax and abdomen, respectively). (Modified from Goldblatt et al. 1995.)

2007). Different species (and different populations within species) showed varying degrees of exsertion of the reproductive parts from the corolla, and this feature determined where on the bats' heads the pollen was deposited. Coflowering species showed significant over-dispersion of exsertion length, suggesting local evolution to reduce interspecific pollen flow.

Finally, Smith and Rausher (2008) found convincing evidence of character displacement in a morning glory, *Ipomoea hederacea*, which when grown in the presence of *I. purpurea* showed a change in anther position toward clustering around the stigma; a coflowering congener was affecting the pattern of natural selection in a way that would contribute to reproductive isolation between the species pair.

Changing Reward

Here the evidence for divergence to avoid competition is unclear, not least because characterization of nectar rewards is so fraught with difficulties (chapter 8). Perhaps the clearest cases come where some nectarless populations may occur, for example in *Lobelia cardinalis* (Brown and Kodric-Brown 1979), or where some

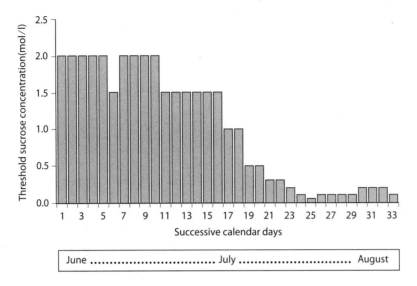

Figure 22.7 The threshold nectar concentration needed for honeybee recruitment; in early summer the highest thresholds coincide with the period of seasonal maximum nectar availability. (Redrawn from Lindauer 1948.)

shift in reward type has occurred, as in *Dalechampia*, where some coexisting species have shifted to resins rather than nectar, with less overlap in flowering than would be expected if no interactions were occurring (Armbruster 1986, 1992).

A related issue here may be changing the response to rewards from the pollinator's perspective. Honeybees have a markedly higher threshold for nectar concentrations to elicit recruitment behavior in the early summer, when flowers proliferate and nectar flow is abundant, than they have in midsummer, when flowers are still abundant but many more visitors are about and competition between flower visitors is rather high (Lindauer 1948). Thus, by late July and August, *Apis* will recruit to species with flowers where the nectar concentration is tenfold lower than the level they responded to a month earlier, and they are apparently adjusting their acceptance level to circumstances (fig. 22.7).

Changing Pollination Mode

This ought to be rather difficult to detect, but in fact sometimes the evidence is quite clear. One good example occurs with mountain laurel (*Kalmia latifolia*) in North America, which is self-fertilizing where it grows along with competing *Vaccinium erythrocarpum*, and bees ignore it; but it is outcrossing where the *Vaccinium* is not present, and there it gets sufficient bee visits (Rathcke and Real 1993).

Summary

From all the above examples, it is apparent that competition between plants for pollinators really does occur, and that the longevity and patterning of flowering or dehiscing are part of the solution from the plants' point of view. One recent example from an invasive plant may help to seal the case. *Impatiens glandulifera* (Himalayan balsam) is currently spreading widely and fast in Europe, especially on riverbanks, and it is also classed as one of the five worst aggressive invaders in North America. As a very fast-growing annual, it mainly seems to succeed by competitively "swamping the pollination market" (Chittka and Schurkens 2001); it is the most nectar-rich flower to be commonly found in Europe, with ten times the rate of sugar production of the next best species, and often literally overflowing with nectar (see plate 19G). Hence it takes visitors away from many native species; for example, *Stachys* received 50% fewer visitors when growing close to it. The natives thus suffer reduced seed set, while the balsam thrives and spreads and explosively strews around thousands of its own seeds.

3. Reducing Intraspecific Competition among Plants

Intraspecifically, an individual plant could perhaps succeed competitively by being just a little bigger, brighter, more rewarding, etc., than its neighboring conspecifics. But any one individual must not push this too far at any one time or place, or it will be at risk

1. with improved advertisements, of the visitor not recognizing other slightly less showy flowers as being of the same species;
2. with rewards, of the visitor becoming replete and not needing to go to other flowers.

Plants that outcompete conspecifics may severely compromise their own reproductive success by failing to receive or donate pollen adequately compared with other plants that stick to the norm for the species.

Intraspecifc effects have rarely been examined, perhaps for these reasons, but Caruso (1999) compared intra- and interspecific competitive effects for *Ipomopsis* and found the latter to be stronger.

4. Reducing Competition among the Pollinators

In searching for evidence here, the best source is most likely to be bees, since of all the pollinating taxa they are the most entirely dependent on flowers. Effects are indeed best known in multispecies communities of bumblebees.

Community-Level Effects

In the United States, Heinrich (1976b, 1979a) and Pleasants (1980, 1983) established that there were characteristically only four common *Bombus* species in any one area, one long-, one medium-, and one short-tongued, with a fourth fitting in at any tongue length but with a different seasonality (or sometimes with a nectar-robbing habit) (see also chapter 18). Each species was the primary visitor to a subset (guild)

of the available plants, although these guilds were somewhat overlapping. Segregations among the pollinator communities thus gave a niche separation axis for the plants, only those within one particular bumblebee species' guild being competitors with each other. Pleasants also reported a change in the reward structures for these different guilds, with 93% of the long-corolla flowers having both pollen and nectar, but only 50% and 16% of the medium and short corolla flowers, respectively, offering both rewards. Inouye (1978) calculated that within a community the bumblebees present differed in tongue length by a factor of about 1.3.

In the United Kingdom, the six most common *Bombus* species do differ in size and tongue length (table 22.2) (Prŷs-Jones and Corbet 1991), and the overlap where they co-occur is often quite small. These authors also noted that the longest-tongued bees tended to prefer horizontal flowers, and that the phenologically earlier bees (foraging up to June) were more likely to take both pollen and nectar on any one trip than the late-summer and autumn bees, both these factors potentially helping to preserve separation and reduce competition.

However, at more northern sites in Europe the overlap in flower-visiting between bumblebee species increases (Ranta et al. 1981; Ranta 1984), with overall larger body sizes and (perhaps allometrically linked) longer tongues. There is also a tendency for more species to co-occur (6–11 at a given locality, though only 5–6 for a given habitat type; although in Poland up to 16 species may co-occur; Goulson et al. 2008). This may be because higher-latitude habitats are generally more patchy with a highly mosaic structure (Ranta and Vapsalainen 1981), giving a less predictable resource base through a whole season, with smaller bee colonies and a lower ability to exclude other species.

Overall, these studies indicate that there is always some overlap in flowers used by bumblebees at all sites, reflecting the flexibility of the bees and their marked inter- and intraspecies variation in size and behavior; but that there are probably morphological displacements within most communities that would reduce or avoid competition.

Additional evidence that competition is a real structuring factor in bumblebee communities comes where competitive release has been demonstrated. Inouye (1978) showed that *Delphinium barbeyi* and *Aconitum columbianum* were both visited mainly by two bum-

TABLE 22.2
Characters and Differences of Common Worker Bumblebee Species in England

	pratorum	lucorum	lapidarius	terrestris	pascuorum	hortorum
Average tongue length (mm)	7.1	7.2	8.1	8.2	8.6	13.5
Main active period	June	August	August	July–August	August	June–July
Flower visits:						
Mean corolla length (mm)	7.4	5.1	5.1	6.3	8.3	8.7
% of visits						
for nectar only	23	71	75	80	60	38
for pollen only	10	28	12	11	7	5
for both P and N	67	1	13	9	33	57
to pendant flowers	39	1	0	7	13	16
to horizontal flowers	28	38	35	40	61	71
to upright flowers	33	61	65	53	26	13

Source: After Prŷs-Jones and Corbet 1991.

blebees, with *Bombus flavifrons* (medium-tongued) commoner on the monkshood (corolla length 8.4 mm) and *B. appositus* (long-tongued) commoner on the delphinium. However, when the primary visitor in a given monospecific patch was removed, the secondary visitor increased its flower visits significantly. In each case, Inouye showed that the primary visitor normally reduced the nectar availability to a point where it was not profitable to the other bee, so excluding it; but when the primary visitor was removed the secondary visitor experienced competitive release.

Diurnal competitive effects also occur (Heinrich 1976a; Pleasants 1983) and are perhaps more convincing than longer-term effects. For example, *B. bifarius* and *B. flavifrons* (short- and medium-tongued bumblebees, respectively) were both recorded visiting three plant species in meadows in the Rockies, but their daily abundances on each changed in opposite directions (fig. 22.8, from Pleasants1983). This arose because early in the day the short-tongued bee exploited newly opening rewards (pollen in *Polemonium*, nectar in *Helianthella*), leaving some shared plants to be exploited profitably by *B. flavifrons*. When *B. bifarius* returned to *Rudbeckia* and *Helianthella* later on, the medium-tongued bee was again excluded. (Note that temperature effects may be playing a part here, as outlined in chapter 10; and the study contained no detailed information on the plants' reward structures, so is in part inferential.)

Thus far the bumblebee examples have mainly concerned differences in tongue length, but bee body size also matters, and additional examples of this occur in the next section. The presence of honeybees probably also affects bumblebee community structures; for example there are very short-tongued *Bombus* species in North America, not present in Europe (Heinrich 1979a), perhaps because there is less competition from hive bees in most parts of the United States. But *Apis* presence can certainly affect bumblebee communities in the United States too; when *Apis* was naturally absent in some localities after a hard winter, Pleasants (1981) found local increases in short- and medium-tongued bumblebees, such that total bee numbers were almost exactly retained.

In the tropics more complex bee communities occur, and about half of all the bees in any one tropical zone are social (Roubik 1989, 1992), with very large, very long-lived, and highly flexible colonies. These social bees tend to visit over a huge area and to be highly generalist, especially in southeast Asia where four species of *Apis* are dominant, with a long flight range and an extensive plant list, taking advantage of very sporadic flowering that acts against specialization. However, where there is a degree of seasonality

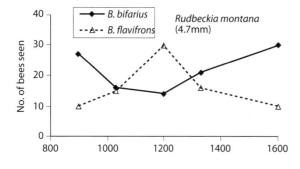

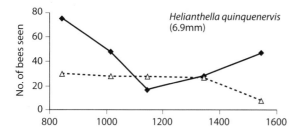

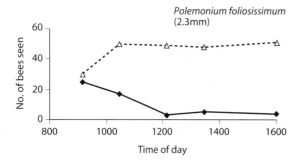

Figure 22.8 Diurnal patterns of two *Bombus* species activities in Rocky Mountain sites, foraging together on three different plants. Solid lines, short-tongued *Bombus bifarius*, ~5.8 mm; dashed lines, medium-tongued *Bombus flavifrons*, ~7.8 mm. See text for more details. (Redrawn from Pleasants 1983.)

in the tropics there is a trend to get more longer-tongued species toward the end of the flowering season, as diversity of flowers goes down and a higher proportion are of the large showy flower type with steady-state flowering. Roubik (1992) suggested that this and other patterns were indicative of a good deal of convergence in the bees, such that unrelated genera resemble each other in different continents, which he believed was suggestive of adaptive shaping by competitive interactions.

Competition between flower visitors might most

easily be seen at work in urban, disturbed settings where flowering plants can be sparse and often complicated by introduced exotics. However, Frankie et al. (2005) documented such effects in Californian cities and found less competition and less resource overlap than expected. First, the flower visitors (mainly bees) showed marked spatial and temporal sorting on the flowers (linked in part to sizes and colors and environmental tolerances, as in so many other examples). But additionally they found that plants appeared to be inducing separation by offering different reward patterns in different individuals.

A survey of the literature as a whole suggests that it may well be that these two mechanisms for reducing competition between flower visitors—spatiotemporal sorting and individual reward variation—are the two most commonly employed.

Competition on Particular Plants

The main evidence for such effects again comes from work with bees, and particularly with *Bombus*. Bumblebee size affects competition markedly, for example on goldenrod (*Solidago*) where Morse (1977) showed that smaller species such as *B. ternarius* would normally forage all over the plant, but retreated to the smaller outermost flowers on the inflorescence when larger bees such as *B. terricola* turned up and monopolized the flowers nearer the stalk, these larger bees being too heavy to land on the outer flowers.

Further examples can be seen for other bees. Willmer (1986) showed that two species of carpenter bee (*Xylocopa*) coexisted on *Calotropis* flowers as their main floral resource in an Israeli desert, but overlapped relatively little on a temporal scale, as the larger darker bee foraged at flowers in the morning and evening while the smaller species mainly visited in the warmer hours around midday (fig. 22.9). Willmer and Corbet (1981) reported several species of *Trigona* bee of differing sizes and coloration spaced out along microclimatic axes largely determined by shade in their visits to *Justicia* flowers in Costa Rica (fig. 22.10). Similar effects of size and color in stingless bees have been reported by Pereboom and Biesmeijer (2003), although in *Trigona* it is also common to see direct competition and fights, where the larger species usually wins (Johnson and Hubbell 1974). Four species competed on *Santiria* trees in Malaysia (Nagamitsu and

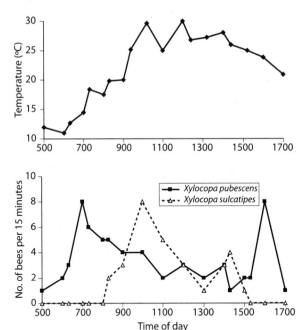

Figure 22.9 Two species of carpenter bee (*Xylocopa*) foraging on *Calotropis procera* in Israel, the larger and darker *X. pubescens* working early and later while the smaller *X. sulcatipes* forages in the warmer hours, avoiding only the very hottest midday period. (Redrawn from Willmer 1988.)

Inoue 1997b), with *Trigona canifrons* aggressively excluding the other three species during peak nectar flow; the others coped by arriving earlier at the trees, and in some cases returning later in the day as well (fig. 22.11).

Most flower visitors except for bees have other food sources, so there is rather little evidence of competitive interactions between butterflies (where larval food is far more critical) or hawkmoths (which even in Madagascar are fairly unselective between the various appropriate flowers); and beetles and flies are very opportunist, rarely competing with each other. Hoverflies might be an interesting exception, but there is in practice little evidence (Gilbert and Owen 1990) for community structures of the bumblebee type; specialist species do occur but appear to fluctuate with climate and other life-history factors, rather than with flower availability.

Moving away from insects, there is good evidence of community structure in hummingbirds (chapter 15), with coexisting groups of long- (curved), medium-, and short-beaked species partitioning the floral re-

sources, especially of *Heliconia* species in Central America and the Caribbean (Stiles 1975). Feinsinger and Colwell (1978) commonly found two or three species of hummingbird on islands, each with different characters and flower choices (one territorial short-beak, one long-beak nonterritorial trap-liner, one less specialist). In larger island communities these three types were joined by some "marauders" and "filchers" (see chapter 15), plus often a very long-beak understory high-reward trap-lining hummingbird.

Other birds, from the perching groups, are less dependent on floral food, and again tend to show less community structure, although there may be some pattern in Australian honeyeaters (Carpenter 1978b), and in South African sunbirds and sugar birds. The coexisting species in these examples again differ in morphology, beak length, territoriality, and aggression.

Bats, however, are largely structured by factors other than flowers, with considerable overlap in flower usage (usually rather more so than in the fruits that are the main diet at other times or for other species) (Heithaus et al. 1975).

5. Maximizing Paternity, Maternity, and Gene Flow in Coflowering Communities

Since males are the motile sex and females stationary, plants can be considered as showing male-male competition to "win" fertilization of females, and female choice to select between competing males, just as occurs in so many animals. Hence there is also a need to consider competition at this more genetic level.

Male Competition and Female Choice

The male components of flowers (or male plants in dioecious species) must compete by display and reward to get visitors and thus (potentially at least) become a pollen donor and sire some seeds. All the aspects of visual and olfactory display and advertisement covered in chapters 5 and 6, and of rewards covered in chapters 7–9, come into play here. However, there may be constraints on display due to the plant's architecture and spacing, while both advertisement and reward may be constrained by the resources available (set against other needs of plant growth).

Another aspect of male competition resides with

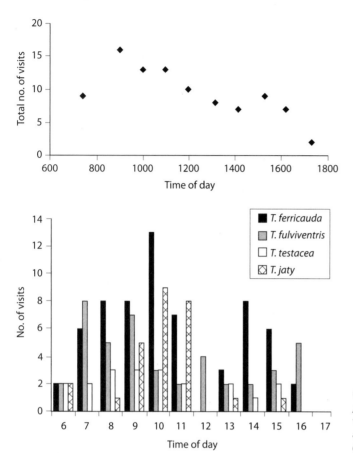

Figure 22.10 Four species of *Trigona* bees foraging on *Justicia aurea* in Costa Rica, with diurnal patterns related to their size and color resulting in reduced competition at any one time and place on the plants (see also fig. 10.13). (Redrawn from Willmer and Corbet 1981.)

the pollen itself, the male gametophyte. Chapter 7 described how each pollen grain must compete with others, to germinate, to grow its pollen tube, and to fertilize the ovules of the plant.

The female component of a plant can exercise choice by the interactions of the style with different pollens; some may be selectively blocked, at various levels in the stigma, style, or ovule, while others are favored. There may also be subsequent (often very extensive) abortion of excess seed, which can be selective, at the level of selfed seeds being unfavored relative to crossed seeds and even for different variants of crossed seeds (Lee 1988; Marshall and Ellstrand 1988; Rigney 1995).

In practice, though, it is hard to tell male-male competition from female choice (see Stanton 1994). It is rather likely that (as is well documented in other kinds of organisms) certain combinations of alleles coming together work particularly well, so that both

processes are probably occurring. Thus some pollen donors seem to do well in any female, but others will only do well in certain females and are worse performers in others (Marshall and Folsom 1991; Stanton 1994).

Gene Flow: Pollen Dispersal and Seed Dispersal

Gene flow in plant reproduction, where there is biotic pollination, depends above all on the visitors' behavior patterns, especially their interplant flight distances. But overall gene flow is not merely determined by pollination interactions. Pollen competition, discussed above and in chapter 7, may lead to somewhat increased gene flow and more heterozygosity in a population, with less selfing, and this sets some constraints on net gene flow, by determining the genetic constitution of an individual seed in a particular place.

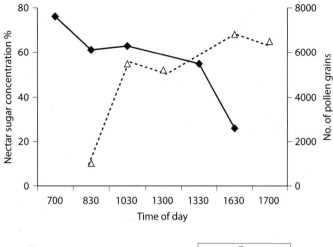

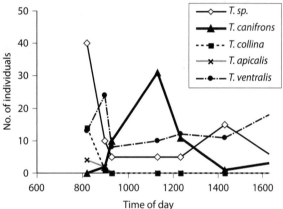

Figure 22.11 Nectar patterns and bee activities on *Santiria* trees, where *Trigona canifrons* excludes other bees at peak nectar times, so that the less aggressive bees only forage much earlier and sometimes later (when nectar concentration remains high but pollen levels per flower are declining). (Modified from data in Nagamitsu and Inoue 1997b.)

The quality of a particular cross may also be determined by the degree of genetic similarity between the parents: too similar could yield the familiar inbreeding depression, but too dissimilar could also be a problem, with outbreeding depression between closely related species or between relatively isolated populations of one species. There may be an intermediate degree of outbreeding that is optimal, and this too could affect competition. By extension, this could occur within a single population and lead to a preference for relatively small (physical) distance between mates, seen as an *optimal outcrossing distance.* For *Delphinium nelsonii* and *Ipomopsis aggregata* the actual outcrossing distances were shown to be shorter than the calculated optima in the Colorado Rockies (Waser and Price 1983). For example (fig. 22.12) the delphinium crosses generally gave better seed set with outcross distances of 3 m or 10 m than with 1 m or with any distances of 30 m or more. It is important though, that for birds, bees, and hawkmoths, the mean flower-to-flower flights were much smaller than the optimal outcrossing distances, with less than 10% of flights for either plant species carrying pollen the optimal distances. The authors speculated that the discrepancy between achieved and optimal outcrossing distances was a consequence of the conflict in plant-pollinator relationships; what is optimal for the plant conflicts with the inherently lazy and cost-cutting tendencies of the animal visitors (chapter 10).

Gene flow in practice proved to be highly localized in that study, at least as simply measured by pollinator behaviors and as assessed in the F1 generation. The same is true for some other temperate studies (e.g., Lertzmann with *Castilleja*, cited in Waser and Price 1983). However, Fenster (1991a,b) found some F1 fitness increases in *Chamaecrista fasciculata* at all

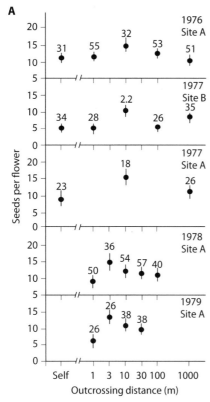

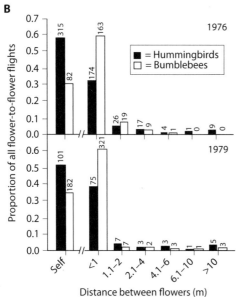

Figure 22.12 (A) Outcrossing distances versus seed set for hand-pollinated *Delphinium nelsonii*, for different years and sites, usually peaking at 3–10 m. (B) Actual interflower distances flown by bees and hummingbirds at the same sites, usually less than 1 m. See text for discussion. (Redrawn from Waser and Price 1983.)

crossing distances up to 1 km, although natural neighborhood sizes as determined by pollen flow were only 2–3 m in diameter. Fenster and Galloway (2000) with the same plant showed F1 superiority at distances of up to 2000 km, with relatively little loss of fitness even through to the F3 generation. Rather similar results have been reported for some tropical plants (e.g., *Inga*, Koptur 1984, where outcrossing at 1000 m was more effective than at 500 m or less), and for some desert plants (e.g., *Datura*, where the optimal outcrossing distance was several kilometers). Although the evidence is still patchy, it may be that optimal outcrossing range is often high and does show some correlation with habitat, perhaps covarying with seed dispersal range.

This naturally raises the more general and crucial point that after pollination the remainder of the gene flow in a plant population is due to seed dispersal. It is likely that (as with pollen dispersal) this has been seriously underestimated in the past, often due to an over-attention to primary dispersal and ignoring secondary movements (Levin and Kerster 1974). Seeds move away from their parents in many ways, often in a dormant state, long after their first transit from the seed pod to the ground; wind, flood, and motile animals ranging from ants to large mammals all play their part here. Survivorship of seedlings also matters and is rarely considered; in the Waser and Price study, for example, seeds surviving into their second year gave a rather different pattern, with the 1000 m crosses producing the best outcome. Meagher (1986, 1991) attempted to take all these processes into account and produced an intensive analysis of populations of the lily *Chamaelirium luteum*, which showed that genetic mixing was in practice quite limited, most plants being sires of seeds in a nearby female, and most seeds moving only short distances.

Far more detailed and longer-term studies are still needed, since the genetic constitution of a given plant population is evidently a composite resulting from many factors acting over long time scales.

6. Overview

Competitive effects between plants for pollination, and between pollinators for access to plant rewards, are almost certainly rather common though difficult to identify with absolute clarity. Available studies show effects in plant communities that can be attributed to escape from competition, but rather rarely do we have experimental evidence from manipulative studies to confirm this explanation. Structuring of pollinator communities, at least for bees and hummingbirds, is likewise strongly suggestive of competitive exclusion over varying time periods, with good correlative evi-dence from comparisons of communities at different latitudes or on islands. It may be that evidence for competitive effects is becoming clearer in the last few decades as pollinator communities increasingly come under threat and some members are displaced or lost, a theme taken up again in chapter 29.

Chapter 23

CHEATING BY FLOWERS: CHEATING THE VISITORS AND CHEATING OTHER FLOWERS

Outline

1. Mimicry of Flowers
2. Mimicry of Objects Other than Flowers
3. Overview

Since pollination is not an altruistic exercise, and there is a conflict of needs, both plants and pollinators are liable to cheat to their own benefit, and deception is very common in pollination biology (reviewed by Wiens 1978; Little 1983; Dafni 1984; Renner 2006). For a plant, this essentially means getting pollinated and hence fertilized without giving up any reward or resources. This can commonly be achieved by resembling a rewarding species, so attracting scouting animals by deceit, or less commonly by mimicry of objects other than flowers to which pollinators might be attracted for reproductive purposes. For a visiting animal, on the other hand, cheating will primarily mean extracting nectar or pollen in ways that do not carry any pollen to another flower; animal cheating is dealt with in the next chapter.

Mimicry (Batesian mimicry, in the terminology employed by zoologists) occurs where a flower pretends to offer a reward but does not in practice do so, so getting the benefits without paying the normal costs. The apparent reward is usually either nutritive (mimicking of another flower that would provide food) or reproductive (mimicking of a potential mate or a potential oviposition site). Nonrewarding plants constitute quite a large proportion of angiosperms, especially among

the orchids (Dressler 1993); but as with any such system the mimics must remain a small part of any one flower community, or the mimicry is likely to fail as visitors encounter too many cheats and too few models, and thus learn to avoid the area.

A flower that cheats by resembling other flowers may either mimic a rewarding flower with which it coflowers, or rely on innate or learned cues to attract a visitor, particularly in areas of low floral diversity. An example of a nonrewarding flower using innate cues rather than mimicry is the orchid *Calypso bulbosa*, flowering throughout the boreal forests of both America and Eurasia in early spring and visited by queen bumblebees, emerging from overwintering sites and prepared to visit any object looking somewhat like a flower (Boyden 1982). In such situations, some floral polymorphism may be beneficial to the mimic, naive visitors treating the different morphs as novel species each of which should be visited lest it offer a reward.

This, however, raises the point that true mimicry must be carefully defined (Vane-Wright 1980; Endler 1981), not just as a vague resemblance, but in terms of a measured increase of fitness; that is, the mimic must do better than it does in the absence of the floral model and its pollinators. In this context fitness means some parameter related to fecundity. We must also be wary of the overlap between the traditional *deceitful mimicry* and a phenomenon that has been termed *mutualistic mimicry* (or advertising mimicry, or floral mutualism, and equivalent to Müllerian mimicry), where several species convergently resemble each other, all are rewarding, and all share common pollinators, with each benefiting.

Floral mimicry is an oddity in that it serves to *at-*

tract the receiver of the deceitful signal; more usually the mimicking cheat's signal serves as a deterrent, a warning, or a camouflage.

1. Mimicry of Flowers

Interspecific Mimicry

This is the commonest scenario, where flowers of one species, lacking a reward, resemble the rewarding flowers of another species. Most of the examples documented to date involve bee or fly pollinators. Floral mimicry is especially common and clear-cut among orchids, which are frequently nectarless and achieve pollination in part by resembling rewarding species from other families; they are substantially less common than their models but flower close to them, where pollinators will be more abundant. This has been described as a magnet effect by Peter and Johnson (2008), who documented the spectral and morphological similarity of the orchid *Eulophia zeyheriana* to the blue-flowered *Wahlenbergia cuspidata* (Campanulaceae), thereby attracting visits from *Lipotriches* bees that were the sole pollinators of the *Wahlenbergia*. This kind of mimicry is very obvious in the common temperate genus *Orchis*, where flowers resemble those of sympatric unrelated genera flowering in the same locality at the same time. For example, *Orchis israelitica* mimics the liliaceous bulb *Bellevalia flexuosa* and shares its solitary bee pollinators, setting more seed when in the company of its model (Dafni and Ivri 1981a). *Cephalanthera rubra* (red helleborine) apparently mimics *Campanula* species with which it often grows, and again the orchid sets more seed when co-flowering occurs (here with shared visitation by *Chelostoma* bees; Nilsson 1983b).

Most of the examples come from more tropical climes, however, and are particularly striking where small and isolated habitats occur such that *varying* mimicry arises. For example, in the Caribbean, various species of yellow *Oncidium* orchids occur on different islands, each closely resembling flowers in two genera of Malpighiaceae, and all share *Centris* bees as pollinators (Nierenberg 1972). In Panama, *Epidendrum* orchids lacking any nectar resemble two models, *Lantana* and *Asclepias* (Boyden 1980; Bierzychudek 1981), both models having nectar. In this case the pollinia or pollen grains of each of the three species get deposited on different parts of the pollinating monarch butterfly (*Danaus plexippus*), so all may get successful specific cross-pollination.

Deceptive species may flower slightly before or after their models, and Internicola *et al.* (2008) found that flowering earlier could be beneficial as it reduced the need for strong similarity. It is also often useful for the deceptive species to have more than one model, making it harder for visitors to learn to avoid the mimics. Many nectarless *Orchis* and *Dactylorhiza* are variable intraspecifically, like the *Calypso* orchids mentioned in the introduction to this chapter. A classic and more specific example is *Disa ferruginea*, another orchid, occurring in South Africa, with two very distinctive color morphs and matching two different models with different colors in different parts of its range (Johnson 1994). This genus is adept at producing highly variable mimics: *Disa nivea* mimics the flowers of *Zaluzianskaya* (and classic loss of fitness in the mimic when it becomes too common relative to the model has been demonstrated experimentally; Anderson and Johnson 2006), while *Disa nervosa* mimics an iris-type flower (*Watsonia*; see plate 32B), and both are visited by long-tongued tabanid flies (Johnson and Morita 2006).

Most of these examples involve mimicry of simpler flower forms, but it has also proved possible for orchids to mimic some rather complex floral morphologies. For example, some Australian *Diuris* species mimic the leguminous shrubs *Daviesia* and *Dillwynia* (Beardsell and Bernhardt 1983), achieving a good likeness of the typical pea flower.

Examples that do not involve orchids are also reasonably common. Little (1983) documented mimicry between two desert annuals, where the model *Mentzelia* (Loasaceae), with abundant nectar in open actinomorphic flowers, was mimicked by *Mohavea* flowers (Plantaginaceae), with tiny amounts of nectar in more tubular corollas. Four species of solitary bee were constant to the model and were the only pollinators of the mimic, which they visited very briefly and apparently by mistake; they used the same behavior to gain access in both cases, although pollen from the mimic and from the model was deposited in different sites on the bees.

Interspecific mimicries are sometimes said to operate as part of a *floral mutualism*, where several similar species, all offering rewards, share pollinators through time or space, all being in a sense both "models" and "mimics" of each other, and all getting some benefit as mentioned in the context of phenology and competition in chapters 21 and 22. This is stretching the definition of Müllerian mimicry somewhat, and the term is probably best reserved for the rarer species within the community, or to communities of very few species where selective advantages can be established. The example of two coflowering *Costus* species (Schemske 1981) outlined in chapter 22 was one such case. Powell and Jones (1983) reported another, occurring between *Delphinium parryi* (model) and *Lupinus benthamii* (nectarless mimic) in California. Their leaves and general growth form were strongly convergent morphologically, both bearing blue/purple flowers (the lupin also with a pink banner spot darkening after pollination; see chapter 5). In the sites of overlap they flowered together (with the mimic surprisingly abundant), and both then received more pollinating visits by bumblebee workers.

Most other reports of floral mutualisms indicate sharing in time rather than in space, with sequential blooming periods reducing competition and improving the resource availability throughout the pollinator's life cycle. Examples of this kind were considered in chapter 21. However, in this situation, a cheating mimic can in theory insert itself into the flowering sequence of rewarding flowers and thus be a Batesian mimic. One classic example is a group of nine hummingbird-pollinated flowers in the western United States (data from Grant and Grant 1968; analyzed by Brown and Kodric-Brown 1979), where just one of the nine species (*Lobelia cardinalis*) is a rewardless cheat. The situation may also occur in European rosaceous hedgerow plants mentioned in chapter 22, where some wild roses in the sequential flowering pattern have no nectar but attract visitors because morphologically they resemble the other (pre- and postflowering) species that do offer rewards.

Intersexual (Conspecific) Mimicry

Where flowers are unisexual, the females are commonly rewardless (having of course no pollen, and often little or no nectar) and may achieve visitation because they resemble the rewarding and usually more abundant male flowers of the same species. Such mimicry seems to be quite common especially in tropical plant families; Renner and Feil (1993) found that about a third of the species from 29 genera in 21 families of tropical dioecious plants offered no reward in the female flowers. This kind of deceit pollination occurs, for example, in nutmegs (*Myristica*), with various beetles visiting the similar small yellow-green flowers on both male and female trees, although with a fourfold preference for the males; the two sexes have similar shapes and odors and the "preference" probably just reflects the greater abundance of male flowers in the community (Armstrong 1997). There is clearly again a problem of definition though (Willson and Ågren 1989): how similar does the female have to be to the male to constitute mimicry? After all, they are likely to be alike by reason of common genetics and developmental constraints. Mere resemblance is hardly surprising, and intersexual similarity can only really be termed mimicry if some features are unusually enhanced to improve the resemblance. In particular, the stigmas of the female flowers are often enlarged and yellow so that they look very much like anthers, as in *Begonia* (plate 32A) and some cucurbits (Ågren and Schemske 1991).

Strictly, though, there must be demonstrable selection for mimicry, that is, the success of female flowers will be a function of their degree of resemblance to male counterparts, and this has rarely been demonstrated. The clearest evidence for intersexual mimicry driven by selection comes where similarity arises from *nonhomologous* parts. For example, in *Jacaratia dolichaula* the female flowers lack corolla tubes but possess stigmatic lobes that markedly resemble the white corolla of the male flower (Bawa 1980b). Other examples include staminodes mimicking anthers in *Rubus chamaemorus* (Ågren et al. 1986), and corolla scales on the male flowers of red campion (*Silene dioica*) that mimic stigmas (Q. Kay et al. 1984, Willson and Ågren 1989). Note that in the latter case the male is modified to resemble the female, which is reasonable since both sexes benefit in the end.

It might be expected that such mimicries would involve olfactory as well as visual cues. In practice, it is more commonly reported that the male rewarding flowers are scented but the females lack strong odors, as in *Carica papaya* (Baker 1976); presumably the female flowers avoid wasting resources on scent because the pollinators (hawkmoths in this case) are drawn to the plant from a distance by the male flowers' scent and then make enough visual mistakes in visiting females at close range. However, in the date palm polli-

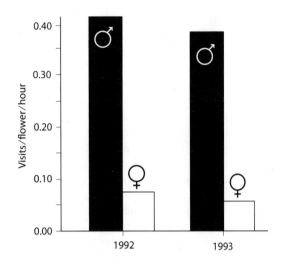

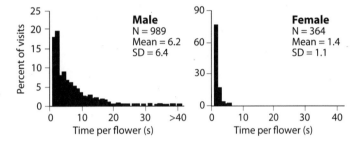

Figure 23.1 Preference for male flowers of *Begonia oaxacana* by bumblebees, showing mean visitation rates across two seasons, and longer visits per male flowers. (Redrawn from Schemske et al. 1996.)

nated by *Elaeiobius* weevils in Africa (see chapter 28), both male and female inflorescences attract many hundreds of visits using attractive scents, the volatile odors being produced continuously in males but only in short pulsed bursts in females; hence the beetles stay mainly in male flowers and only move on to females periodically, always well coated in pollen. Females have no reward, so this could be seen as effective deception by scent.

In these situations, pollinators that can detect the cheating and avoid the rewardless female flowers should be selected for, and there is some evidence of this, for example in *Thalictrum* (Kaplan and Mulcahy 1971), *Rubus* (Ågren et al. 1986), and *Ecballium* (Dukas 1987). It is also well known in *Begonia* species, for example *B. involucrata,* where *Trigona* bees visited males seven times more often and for ten times longer per visit than they did females (Ågren and Schemske 1991). That is a small part of a detailed study on tropical monoecious *Begonia* by Schemske and Ågren, much of it summarized by Schemske et al. (1996). *B. oaxacana* is a particularly good example, where inflo-

rescences usually contain both sexes of flower at the same time, with a female-biased sex ratio. The mature male flowers are displayed outward and upward, with two large white sepals, two smaller white petals, and dark yellow anthers; females are more pendant, still with two white sepals (smaller and less spreading than in males) but usually just one small petal, and a stigma that is green when young but becomes more yellow as the flower ages. Male flowers are therefore distinctly more showy and obvious, females more hidden and somewhat asymmetric. Visitors are mainly *Bombus*, and on average these showed a strong five- or sixfold male preference in their foraging, also visiting each male flower for about 4–5 times longer (fig. 23.1); thus there was clear discrimination against the rewardless females with no pollen. Females received a visit only when mistaken for males, and the bee then left again quickly. However, females could compensate for the rarity of visits: first by lasting longer, usually at least 14 days, within which time at recorded visitation rates they should have got at least one visit, and second probably by facing downward and thus resembling young male flowers, which are particularly attractive to bumblebees because likely still to be pollen-rich.

These last two phenomena—interspecific and intersexual mimicry—tend to involve resemblances exploiting innate behavior patterns rather than learned behaviors; so they might be expected to be relatively rare in bee flowers, since bees learn to distinguish readily between very similar flowers. But the floral mimics commonly exploit naive flower visitors which have not yet learned to detect the deceptive mimics either in time (e.g., newly emerged bees) or in a particular locality (e.g., migrating hummingbirds). For example, Nilsson (1980) reported that *Dactylorhiza* orchids in Sweden exploited the brief early emergence period of queen *Bombus*, before they established foraging routes and learned to distinguish the true resource qualities on offer. Hence most rewardless deceptive orchids are early-spring bloomers; those that do bloom later may rely on the attractions of coflowering species to bring in enough pollinators, as with *Traunsteinera* orchids in Europe that flower in late summer and benefit from pollinators attracted to coflowering similarly colored *Trifolium pratense* (Juillet et al. 2007).

Reliance on naive visitors raises the possibility that deception could occur even when no real model is present, relying purely on the inexperience of visitors and their innate flower preferences (Little 1983). The operation of such systems is open to debate, although in principle they would be similar to the pseudocopulation mimicries described in the section *Reproductive Mimicry of Potential Mates: Pseudocopulation* below, where visitor inexperience coupled with instinctive behavior is equally crucial. Whether a system that relies on a naive visitor should be termed mimicry at all is another question (discussed by Williamson 1982; Little 1983), since mimicry is generally held to rely on confusion between learned models and their deceptive mimics.

Aids to Mimicry: Pseudoflowers, Pseudonectar, and Pseudopollen

Pseudoflowers
Pseudoflowers represent a way of increasing the overall floral display, but where only some flowers are "real" and fully functional. It is a trick often exhibited by umbellifers, hydrangeas, etc., where outer flowers on the inflorescence are often more showy, with larger petals (plate 11D,E), but have no sexual function. In *Viburnum* the peripheral infertile flowers have stamens and pistils initially but these degenerate as the umbel becomes mature (Jin et al. 2007). In *Hydrangea macrophylla* the decorative flowers have larger receptacles and often one less petal (four instead of five), but they retain functional pollen (Uemachi *et al.* 2004).

Note that "pseudoflowers" can also sometimes result from fungal infestations of flower heads (e.g., in some *Euphorbia*), and the effects are then more likely to be negative for the host plant (Pfunder and Roy 2006).

Pseudonectaries
Some flowers have areas of apparently glistening moist and often yellowish-green surface that attract visitors but offer nothing, and this strategy can be particularly effective in bringing in Diptera. *Parnassia* is a famous case, with five staminodes forming semicircles of stalked pseudonectaries each tipped with a yellowish glistening knob (fig. 23.2 and plate 32F); these are in fact dry and rewardless, but they attract flies which probe at the tiny knobs. *Oncidium onustum* orchids in Ecuador and Peru have false nectaries at the base of the column that are attractive to *Xylocopa* bees, which apparently confuse these flowers with species of *Cassia* (Dodson, cited by Little 1983). Something similar occurs in many *Ophrys* orchids, covered in more detail below (although these examples are controversial because here the visitors are not seeking nectar anyway, and the glistening areas described as pseudonectaries may more plausibly be mimicking the eyes of the "pseudofemale").

Pseudopollen and Pseudoanthers
There are three possible scenarios here. The first is seen in dichogamous flowers, with male and female phases. In protogynous flowers in the early female phase there is no pollen present, while protandrous flowers are male first but may have lost all their pollen by the time their stigmas become receptive in the later female phase. In either case it may help to have bright anther-yellow guides or imitations of anthers or pollen present throughout the flowering period (see chapter 5, and Pohl et al. 2008). The stigmas can appear to mimic anthers, for example in many begonias where female flowers' stigmas closely resemble a group of stamens

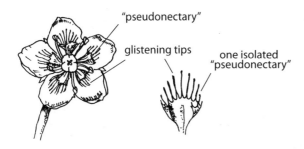

Figure 23.2 *Parnassia* flower with pseudonectaries; see also plate 32. (Redrawn from Barth 1985 and from photographs.)

(then known as "stylodia") or even the entire androecium of the male flowers (Vogel 1978b; Schemske and Ågren 1995; Schemske et al. 1996). In the kiwi plant (*Actinidia*), female flowers have pseudoanthers with sterile nutrient-poor fodder pollen (Cane 1993a). Other examples occur in *Parnassia* and in some *Saxifraga*. For protandrous species it is not uncommon to find the real anthers retaining turgor and bright yellow coloration even after all their pollen is gone, when anthers would normally be wilting. This occurs in African violets (*Saintpaulia*) and in *Exacum* (Gentianaceae), where the protandrous anthers remain fresh and rewarding, although usually empty by the time the stigmas become receptive (Faegri and van der Pijl 1979). The genus *Solanum* is particularly instructive here, as dioecy has re-evolved several times in the group (Knapp et al. 1998), and in all the dioecious species the male flowers have normal anthers and tiny stigmas, while the females have both stigmas and anthers but the pollen within is entirely sterile and nonfunctional.

A second strategy is the use of markings, usually on petals, that appear like anthers, and which add to the overall attractiveness of flower. Usually such marks occur as a central yellow mark or pair of marks, elongate or ovoid in shape, and bright yellow. This is especially clear in *Eichhornia*, where the real pollen is cryptic. Something similar occurs in many species of *Iris*, and is taken to extremes in *Iris germanica* and many hybrids (plate 32E) where the "beard" set in high relief on the landing petal strongly resembles a mass of anthers, with yellow tips on white filaments (the real pollen being concealed). Structural and textural anther mimics occur widely (fig. 23.3), notably in *Craterostigma, Mimulus guttata, Torenia*, and many Scrophulariaceae (Magin et al. 1989). In *Melampyrum* there are tiny nodules on the hairs of the flower's lower

lip that are of similar size to pollen grains. Pollinators are not usually deceived into making feeding attempts at these kinds of marks or structures, but the marks are exploiting the innate visual preferences of visitors (see chapter 5), who are thereby guided into the right part of the flower to find nectar, increasing their foraging efficiency. Imitation anther markings are generally sited above the true anthers in flowers with sternotribic pollination and below them where the pollination is nototribic.

The third possibility, often less clear-cut, is seen in the trend to supplement or replace functional anthers with deceptive structures such as staminodes. Plants where pollen is the major reward often show heteranthy, separating cryptic functional anthers from showy feeding anthers that become the advertising signal (see chapters 2 and 7); but often the feeding anthers become deceptive, with little or no functional pollen. Nepi et al. (2003) described the real and feeding pseudopollens in *Lagerstroemia indica,* the latter having more hexose sugars and more pores that perhaps made it more digestible. In some Theaceae, pseudopollen is produced in the anther connective and released into the pollen sacs where it mixes with true pollen and is gathered by bees (Tsou 1997); it perhaps provides some food value to the bees without wasting the reproductive capacity of the plant. Some nonheteranthous flowers achieve the same effect, with rather few real anthers but a few dummy ones to distract pollen foragers. The deception may be enhanced by deceptive structures on the anthers that look like additional anther tissues, including fluffy hairs on the filament (e.g., *Verbascum;* see fig. 2.2), or enlargement of part of the filament (e.g., *Cleome, Dianella*) or of the connective between filament and anther (e.g., *Blakea*). Some orchids go further, putting all their fertile pollen into pollinia and offering only some visually attractive but useless pseudopollen for visitors. *Paphiopedilum villosum* is particularly interesting here, having glistening staminodes that appear to be mimicking a food offering, and also a slippery wart-like area that apparently mimics a perch but in fact causes any landing insects (almost entirely hoverflies) to fall off into a trough from which exit is possible only up a tunnel that passes the column and pollinia (Bänzinger 1996). Simpler examples include *Maxillaria* orchids that produce yellow hair tufts shed from functional living cells (van der Pijl and Dodson 1966); *Cephalanthera* orchids which have yellow papillae on the lowest petal

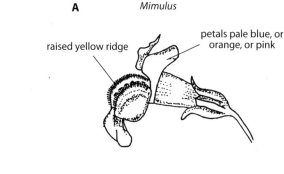

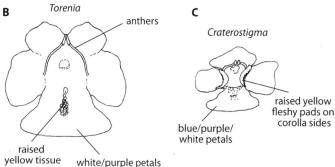

Figure 23.3 "Pseudopollen" embellishments, usually as raised or ridged yellow areas on the lower petals of flowers; see also plate 32. (Drawn from photographs.)

that seem to function as pseudopollen (Dafni and Ivri 1981b); and *Calopogon* orchids in North America, with stamen-like tufts. Here the cheating is often complete, and there is no mutualistic benefit to visitors; although in a few cases the pseudopollen of orchids does have some limited nutritive value (Davies and Turner 2004).

Empty Flowers as Mimics and Cheats

Emptiness can be a phenomenon of some flowers on a plant some of the time, of some flowers permanently, or of some species permanently. Each of these may have different benefits and effects, but each can be seen as a manifestation of mimicry or cheating at some level—a type of automimicry, also termed Browerian mimicry.

Many, perhaps most, plants will have some empty flowers on them on most occasions when a visitor arrives; for example, Thakar et al. (2003) detected them in 24 out of 28 species tested. They may have no pollen, or they may be devoid of nectar, or both. Frequently this will be due to recent visitation that has emptied a proportion of the flowers; this factor, coupled with inherent variability in nectar production and nectar replenishment rates within and between plants (chapter 8), makes flowers lacking any nectar reward a reversible but highly unpredictable fact of life for most flower visitors. Empty flowers may also arise because flowers only produce rewards in one phase of their life (nectar in the female phase, or pollen in the male phase), or because of transient environmental stress such as desiccation.

Because empty flowers are therefore commonplace, visitors must constantly encounter them, and aspects of their behavior are likely to be attuned to this (Gilbert et al. 1991). Or, looked at the other way around, empty flowers can become a way in which plants exploit the cognitive abilities and limitations of their visitors, so that the animal cannot distinguish between rewarding and nonrewarding flowers until it has begun its visit, by which time pollen uptake or deposition may already have happened. Having some empties, albeit by chance and by virtue of prior visits, is thus an effective way for the plant to save resources, and can become an evolutionarily stable strategy (Thakar et al. 2003; see also Smithson and Gigord 2003, Bailey et al. 2007).

But sometimes emptiness occurs because certain flowers do not produce pollen or nectar and are "deliberately" and permanently empty—they are genuine "blanks" rather than flowers that have become emptied. The occurrence of a proportion of empties among a population of rewarding flowers may then be regarded as a form of intraspecific Batesian mimicry (and the emptied flowers referred to in the last paragraph also serve this function, at least transiently). As always, the mimicry must rely on a low frequency of mimics and poor discrimination by the animal being cheated. The costs of visiting a flower with no reward should be at least balanced by rewards obtained from other flowers in the vicinity, or the animal will eventually give up its visits, and Bell (1986) modeled this as an ESS for the plant, where the proportion of cheats was D/H (where D is the discrimination time and H the handling time). Gilbert et al. (1991) found a good fit to this model in the field, with 75%–85% of flowers on *Cerinthe major* being empty or nearly so. Further modeling by Bailey et al. (2007) found that empty flowers enhanced pollination success for a plant, the optimal proportion of "empties" being lower in plants with low costs of selfing, and lower where pollinators were scarcer. Anand et al. (2007) pointed out that if empty flowers are really a cheating strategy then nectar volumes should be **bimodal** and demonstrated this to be the case in both *Lantana* and *Utricularia*, also showing that the proportion of empties was higher in particularly dense flowering patches for the latter species, supporting the gregariousness hypothesis where denser populations can support more cheats.

Empty deceptive flowers are signaling the presence of a reward where none is present. Taken to extremes, this can involve a species being entirely and permanently rewardless. Deception of this kind has evolved in many plant families (see table 23.1), but is especially well known in the orchids. Renner (2006) estimated that 3%–4% of all angiosperm species are permanently rewardless. This commonly involves not having nectar, and rather less often neither nectar nor pollen, but may also occasionally occur with rewards such as oils (e.g., in some Malpighiaceae; Sazima and Sazima 1989) or perfumes; Renner (2006) reviewed examples. Note that the offer of egg-laying sites (the section *Reproductive Mimicry of Brood Sites* below) is also a kind of empty flower deceit, as no such reward is really present. The same is true of many cases of sexual mimicry described in the section *Intersexual (Conspecific) Mimicry* above, where one sex is rewarding and the other (usually the female) has no reward.

The lack of a reward is particularly common in female flowers of unisexual species (Willson and Ågren 1989; Renner 2006), perhaps because females are under stronger selection to economize; this is not infrequently found in tropical trees and climbers (Ricklefs and Renner 1994). It may also be commoner in relatively specialized flowers with long fused corollas, where nectar would normally be hidden from sight anyway.

Orchids perhaps best exemplify the benefits of permanently rewardless flowers. They tend to occur in low numbers and highly spaced out, so that visit frequencies may be inherently low. Animal visitors cannot depend on any one species as a food supply, so will be making interspersed visits to other flower species that do offer rewards. This is acceptable for orchids (but not for many other plants) because their pollinia can be precisely placed on the visitors' bodies, thus ensuring correct intraspecific pollen deposition. Thus several attempts to alter orchid reproductive success by adding artificial nectar have found little effect (e.g., Smithson 2002). However, Jersakova and Johnson (2006) tested the role of empty flowers in a fly-pollinated orchid, *Disa pulchra*, showing that adding nectar to the nectar spurs increased the number of flowers probed, the time spent on a flower, the number of pollinia removed per plant, and the levels of self-pollen deposition, indicating that rewardless floral deception does help to increase interplant movement and reduce self-pollination.

Since the use of empty flowers as energy-saving mimics can benefit the plant, rewardless plants should often be able to invade pollination mutualisms. However, there is generally a good correlation between pollen export and a particular flower's investment in rewards; thus the extent of such invasions may be held in balance by visitors' (moderate) ability to discriminate empties. Given that many visitors will encounter empty flowers with great regularity, a rather limited ability to discriminate them may reflect that the cost of learning the relative reward status of individual plants or flowers (which include longer decision times, reduced flight speeds, and longer flights) is greater than the cost of making mistakes. Bees also seem to learn negative stimuli more slowly than positive stimuli (Dukas and Real 1993).

It is often assumed that empty flowers primarily rely on naive visitors, recently emerged as adults or recently immigrated into the area (e.g., Ackerman 1981; Smithson and Gigord 2003). However, reliance

TABLE 23.1
Examples of the Occurrence of Rewardless Flowers

	Family	Main visitors
A) Absence of nectar		
Begonia boliviensis	Begoniaceae	hummingbirds
Podophyllum	Berberidaceae	*Bombus* bees
Tabebuia	Bignoniaceae	bees
Iris pumila	Iridaceae	*Bombus* bees
Myristica	Myristicaceae	beetles
Mohavea	Scrophulariaceae	smaller bees
Solanum	Solanaceae	bees
Nymphaea (in female phase)	Nymphaeaceae	bees, flies, beetles
B) Absence of nectar and food pollen		
Plumeria	Apocynaceae	sphingid moths
Nerium	Apocynaceae	moths, bees
Many orchid species	Orchidaceae	flies, wasps, bees
C) Sexual mimicry or deception in monoecious species		
1. No nectar (or pollen) in female flowers		
Stelechocarpus	Annonaceae	beetles
Geonoma	Arecaceae	beetles
Antennaria	Asteraceae	pollen-foraging bees
Tussilago	Asteraceae	pollen-foraging bees
Begonia involucrata	Begoniaceae	*Trigona* bees
Sarcococca	Buxaceae	bees
Carica	Caricaceae	sphingid moths
Jacaratia	Caricaceae	sphingid moths
Clusia	Clusiaceae	bees or beetles
Ecballium	Cucurbitaceae	*Apis, Andrena* bees
Dalechampia	Euphorbiaceae	bees or beetles
Castanea	Fagaceae	bees
2. No nectar in male flowers		
Begonia ferruginea	Begoniaceae	hummingbirds
D) Brood site mimicry		
Aristolochia	Apocynaceae	flies
Asarum	Apocynaceae	flies
Ceropegia	Apocynaceae	flies
Stapelia	Apocynaceae	flies
Hydnora	Hydnoraceae	beetles
Ambroma	Malvaceae	flies
Rafflesia	Rafflesiaceae	flies

on naive visitors is unlikely in rewardless tropical flowers visited by long-lived bees or birds. It may perhaps occur for some spring-flowering temperate species, although even there most of the visiting bees learn so fast that they remain "naive" for only one visit. In fact, with *Dactylorhiza* orchids the later-blooming flowers higher on a spike (mainly visited by experienced bees) tended to have better pollinia export than those at the base which would have mainly encountered the earliest and most naive bees (Kropf and Renner 2005). Studies by Gumbert and Kunze (2001) on pairs of rewarding and rewardless orchids showed that bumblebees consistently visited whichever empty mimic most closely resembled the rewarding species they had just left, using their experience rather than behaving naively.

Finally, while this kind of mimicry by empty flowers will usually occur within a species and within a population, it could also work in mixed populations. For example, Brown and Kodric-Brown (1979) reported one population of empty-flowered *Lobelia cardinalis* (whose flowers usually have abundant nectar), and the aberrant population was reliant on deceiving birds because it resembled closely enough other tubular red flowers in the vicinity. There is a need for more studies on the varying frequency of empty flowers between populations.

2. Mimicry of Objects Other than Flowers

Mimicry of Other Animals: Aggressive Mimicry

Two species of *Oncidium* orchid are pollinated in a curious way that involves *Centris* bees, which appear to attack the flowers. The bees are territorial and will chase off other insects that invade their patch. They tend to perch near to a spike of *Oncidium* flowers, and when the flowers move in the breeze a bee will often dart in and buffet one of the flowers, seemingly reacting as if to an invading flying insect (Dodson and Frymire 1961; Dodson 1962). Flowers last about 3–4 weeks and so have a reasonable chance of being pollinated during this period, although fruit set is generally low.

Reproductive Mimicry of Brood Sites

Here flowers mimic the preferred egg-laying site of an insect, attracting them in but offering little or no re-

ward. The commonest examples involve mimicry of carcasses and carrion, attracting flying visitors whose larvae require dead or decaying flesh as food. This especially involves flies (sapromyophily, chapter 13), but also sometimes dung beetles (coprocantharophily, chapter 12). Additionally, some flowers mimic the gills of fungi, attracting visits by egg-laying fungus gnats (mycetophily).

Four plant families are particularly well known here: Araceae (the aroids), Aristolochiaceae, Apocynaceae (including the asclepiads), and Orchidaceae. In the less extreme cases, represented in all of these families, insects are lured to the flower by the scent and appearance of dead and decaying flesh; but in the rarer and more specialist cases the insects are trapped and detained by the flower, for periods ranging from a few minutes to more than 24 hours, often with many other insects also imprisoned, before being released suitably covered in pollen. It is helpful to deal with these two deceptive systems separately; basic descriptions of the flowers and their attractants were covered in chapter 13.

Brood Site Mimicry without Traps

Here some of the best-known examples are species of *Asarum* (Aristolochiaceae), which attract fungus gnats (nematoceran flies in the family Mycetophilidae; chapter 13). *Asarum* species mostly flower in the spring, when fungal fruiting bodies are rather uncommon, and they grow low to the ground with the flowers often hidden beneath foliage where they stay damp and shady. The perianth is dull purple or brown, slightly scented, and often with stripes, and at its base are translucent white patches of tissue that are kept damp by high transpiration. Female fungus gnats seek out the flowers and lay their eggs on these pale patches, their backs thus contacting the flower's sexual organs. In some species the patches of damp tissue are ridged, more specifically resembling mushroom gills. Similar features occur in some Araceae, in the genus *Arisarum*; here the best-known case is the mouse plant, *A. proboscideum*, again with flowers growing below the foliage, where the central spadix is modified into a fungus cap mimic, whose surface is again always moist and where the flies' eggs are laid. The surrounding spathe is translucent and probably produces a window effect that encourages flies to walk into the flower. A few orchids also achieve fungus-gnat attraction by mushroom gill mimicry, including some in the genera

Dracula and *Corybas* with modified semicircular ridged lips and rather fungoid or fishy scents, and a few low-growing *Cypripedium* species with structures resembling small mushrooms on their lips. *Dracula chestertonii* emits a scent recognizable as that of champignon, and 70% of its volatile emissions are typical mushroom constituents (Kaiser 2006).

Various asclepiad plants in the family Apocynaceae also use nontrapping brood site mimicry, but do so by resembling flesh or carcasses. A well-known example is *Stapelia*, but species of *Huernia* and *Caralluma* show similar features, all of these coming from southern Africa and southern Asia. Their large flowers are purplish with hairy petals, strongly and unpleasantly scented, thus resembling dead animal tissues. They attract a range of "higher" muscoid flies which lay many eggs in the flowers.

Brood Site Mimicry with Prolonged Trapping

Examples of trapping flowers occur in the same families mentioned in the last section, largely involving mimicry of flesh and enticing egg-laying flies. Usually the flowers have separate sexual phases or entirely separate sexes; the trapping happens in the female phase, after pollen has been delivered, and sometimes leads to killing by starvation or drowning. However, in most cases the trap area gives suitable microclimatic conditions for survival of the trapped insects (often small and delicate), with equable temperatures sometimes well above ambient, and relatively high humidity; some also provide light via translucent window areas.

The basic structures and mechanisms of *Aristolochia* flowers were described in chapter 13 (fig. 13.11). The trapping mechanism centers on a downwardly directed corolla tube lined with lubricated papillate cells, and embellished with downward-pointing hairs, so that a visiting fly slides down the corolla and is unable to climb out again. Once the flower is pollinated and its own pollen released (usually on the second day of its life), nectar and scent production cease, and the stigmas bend in on each other to avoid further pollen receipt; then the papillae and hairs shrivel (and in some species the whole corolla tilts up to a more horizontal position), so that the flies (often, though by no means always, after laying some eggs) can walk out. In some larger New World *Aristolochia* species with a U-shaped corolla (Dutchman's pipes) the flies slide from the petals down into the descending limb, then climb the ascending tube toward an apparent window, with the hairs in this part of the corolla helping their efforts. Specificity in this genus largely depends on size matching, not of the whole bloom but of the space around the reproductive organs in the trap end of the flower, since an effective pollinator must be able to fit snugly into this space if it is to deposit and then receive pollen; many flies are attracted, but only a small range are actually trapped by any one plant species, with sepsids, muscids, and calliphorids the commonest pollinators. In some cases the cheating aspect of this relation is perhaps suspended, as the flies do feed on a substantial stigmatic exudate within the sexual chamber.

A few *Aristolochia* species have diverged into fungus mimicry, with flowers borne near the ground and having a central mushroomlike structure at the flower entrance. The underside of this structure bears slippery lamellae, and flies attempting to lay eggs there tend to fall off into the trap below, subsequently ascending to the well-lit area around the stigmas (Vogel 1978b).

Ceropegia flowers were also described in chapter 13 (fig. 13.12 and plate 22B,C) and show the classic features of trapping by slippery internal papillate surfaces (on the "slide zone," which may be darkened and reddish), inward- and downward-pointing hairs which shrivel after one day as the corolla rises from pendant to horizontal, and a paler basal window area.

Arum and related genera (Araceae) exhibit these same features, albeit in an inflorescence rather than a single flower (Meeuse and Morris 1984). The visitors entering the mouth of the inflorescence get trapped for about 24 hours in a spacious reproductive cavity, between the club-shaped spadix and the outer leafy spathe (fig. 13.13 and plate 22D–F). They explore the spathe, perhaps additionally attracted by the appearance of light at the base from the window effect of its translucent walls; then they slip downward on the spadix or on the inner surface of the spathe, both of which are slippery and papillate. Near the base of the inflorescence they hit a ring of tangled bristles through which only small insects can pass, the larger ones having their fall arrested and then flying off. The small flies that pass all the way down then clamber about over the female flowers as they try to escape, depositing any *Arum* pollen that they were carrying. As pollen tubes grow and penetrate the ovaries, the stigmas wither and dehiscence begins, dusting all the trapped insects (which may be frenziedly laying eggs by this time) with large amounts of pollen. Early on the second day the ring of bristle withers, the papillae on the floral surfaces shrink, and all scent production ceases; the trapped flies escape, bearing pollen and leaving behind eggs that will hatch into doomed larvae with no possible food source.

In other aroid genera beetles commonly occur as pollinators. In *Amorphophallus* from Africa, Madagascar, and southeast Asia, these are large beetles, requiring a substantial overhanging ridge to prevent their premature escape, while in *Typhonium* it is tiny beetles that are trapped and the spathe actually contracts above them once they are in the female zone of the inflorescence, opening up again slightly once pollen is being shed. And one genus of aroids, *Colocasia*, has taken the trapping to a further extreme of visitor control, by having a one-way system. Flies are attracted by a strong unpleasant smell but can only enter the base of the spathe, through a small gap, being prevented from subsequently going upward beyond the female flowers by a constriction. As dusk arrives the entrance gap closes and scent production ends, so that flies are kept in overnight, but are gradually admitted to the higher (male) chamber to be dusted with pollen; only after dawn does the exit open, allowing them upward again, and out through the spathe tube.

Counts of insects in various plant traps have indicated that only two or three insect species tend to be present in any one spathe in substantial numbers, with just one species often strongly dominant. Another common finding is that roughly equal numbers of males and females are trapped (Knoll 1926; Drummond and Hammond 1991), perhaps because males are strongly attracted to sites where females are likely to be laying eggs. Seeds set in flowers are thus likely to have multiple paternity, although familial groups may occur in clumps with just a few of the trapped insects having the greatest effect on pollen movements (e.g., Nishizawa et al. 2005 for *Arisaema*).

The other family exhibiting sapromyophilous mimicry is the Orchidaceae, with examples from both the Old and New Worlds; again they usually trap the chosen insects for at least a short period. *Cirrhopetalum* is perhaps the best-known genus, having unpleasantly smelling flowers that attract flesh-flies. *Megaclinium* and some species of *Anguloa* and *Masdevallia* are also conventionally sapromyophilous, although in *Masdevallia* the trapping occurs by movement of the flower lip, forcing a visiting fly into the trumpet of the flower so that it can escape only via a narrow gap between lip

and column, passing the stigma and pollinia as it goes. *Bulbophyllum macranthum*, pollinated by large flies, is rather similar in mechanism; it is brown/purple in color, with thin (filiform) inconspicuous petals, and a smell of ammonia; when a fly touches the labellum this flips up and traps the fly inside against the column for several minutes (while the "glue" from the rostellum dries), so it emerges with pollinia on its back. This fly must then be deceived again by another flower, the glue by then being brittle, and the pollinia dislodged onto the stigma. A few species of the tropical slipper orchid *Paphiopedilum* also show all the classic sapromyophily traits but add a more specialized one-way trap system as seen in *Colocasia*, using it primarily to trap hoverflies that lay eggs on the staminodes (Attwood 1985).

We should mention some other cases of apparent trapping for pollination purposes that may or may not be related to brood site mimicry. Some *Bulbophyllum* orchids smell strongly of cloves and are attractive to just one egg-laying fly species, and *Pleurothallis* orchids likewise are somewhat aberrant, smelling rather fruity and attracting only *Drosophila* fruitflies, but oddly only the males (perhaps suggesting that the scent is mimicking female flies, and the syndrome is really pseudocopulatory). Further examples are found among the water lilies (Nymphaceae). Here the flowers open in the evening and are usually white and pleasantly scented, also generating significant metabolic heat and achieving temperatures several degrees above ambient. At this stage they are visited by various beetles or by flies. In the tropical giant water lily *Victoria amazonica*, the visitors are specifically scarab beetles (subfamily Dynastinae; Prance and Arias 1975; Seymour and Matthews 2006), while in temperate water lilies such as *Nymphaea* it is commonly hoverflies that visit, especially around dusk. In each case the visitors feed on stigmatic exudates; but as night falls the petals close up and often partially submerge, imprisoning the visitors for several hours. Next morning the anthers dehisce and the pollen-dusted visitors are released as the corolla reemerges above the water surface and opens up again. Visitors arriving on this second day and attempting to feed on the abundant pollen may avoid being trapped, but many suffer a worse fate as the inside of the flower becomes slippery and filled with fluid in which many insects drown. Clearly these flowers are manipulating their pollinators by trapping them, but it is unclear whether this is a brood site mimicry, or if they cheat by appearing to offer nectar or pollen on day 1 when neither is present. Interestingly, dynastine beetles are involved in a similar relation with some giant terrestrial aroids in the genus *Philodendron*, which open and produce strong scents in the evening and become exceptionally hot at the same time, offering exudates from a special area of sterile flowers. Here the beetles stay (although not trapped) until day 2, when they become liberally covered with pollen on their now sticky exudate-covered surfaces; late on day 2 the spathe starts to close and the beetles are driven out.

Whether or not these last few cases are included, brood site mimicry provides a particularly striking example of convergent evolution, with a suite of rather specialized characters recurring in quite unrelated genera and families. Thus it is easy to list the characteristics of sapromyophily, as in chapter 13. It may be a genuinely antagonistic interaction, rather than a mutualism, and thus unlike the mate mimicry scenarios dealt with in the next section. The visiting insect may receive no rewards and at the same time is induced to waste eggs. Given that the resources for egg-laying may be quite rare, carrion- and fungus-seeking visitors probably have to stay fairly broad in their host range (even when the flowers themselves are achieving quite high specificity). Hence the insects that are exploited by sapromyophilous plants cannot easily learn to avoid deceptive flowers in the future.

But there are also cases of genuine brood site mutualism, dealt with in chapter 26 (figs and fig wasps, yuccas and yucca moths) where the visitor may still be deceived by resemblance to a more traditional brood site, but in the process of pollinating does also get a suitable reward for its efforts, so that the overall interaction is mutualistic. There is a continuum here, not a clear dichotomy, and it is likely that the balance between antagonistic and mutualistic relations may vary with time and place even in the examples described in this chapter.

Reproductive Mimicry of Potential Mates: Pseudocopulation

In pseudocopulation syndromes, flowers resemble the females of the pollinating species (generally an insect), and are pollinated, usually exclusively, by the males during repeated attempts at "mating." At its simplest, this phenomenon is not very different from food-

deceptive mimicries described earlier, as it merely requires that the flowers attract mainly males, rather than being equally attractive to both sexes of the pollinator, and this can relatively easily be achieved by altered scent cues.

The orchids provide all the best-known examples of using pseudocopulation as a way to deposit their pollinia on a visitor (see Schlüter and Schiestl 2008), coupling this with a one-sided cheating exploitation as they are commonly lacking any food reward (about one-third of all orchids being rewardless and deceptive; Schiestl 2005). The two phenomena may be interactive; for example, the eastern Mediterranean *Orchis galilea* is pollinated exclusively by male *Halictus marginatus* bees, in the same habitats where many other rewardless orchids (including *Orchis israelitica*, mentioned above) mimic coflowering plants and attract both sexes of related solitary bees. Some of the most striking of orchid adaptations, especially in the genus *Ophrys* but also spread across many other genera, involve mechanisms to achieve pseudocopulation. The requisite specialized adaptations to particular pollinators are perhaps related to orchids' usually widely scattered and rather low-biomass distributions, where they may rather often be pollinator limited. Cozzolino and Widmer (2005) suggested that pollination by deceit was a major factor in promoting orchid diversity and success.

In the most sophisticated examples, the flower may mimic insect morphology very exactly, and in all known cases they then have a scent that precisely mimics the sex pheromones of the female insect (Borg Karlsson 1990; Schiestl 2005; Ayasse 2006). Often the two large orchid pollinia are also part of the lure, perhaps appearing as "superpollen." Pseudocopulation requires flowering that is synchronous with the adult emergence time of the specific insect, and is especially effective when the victim is a species where males emerge before females (protandry of this kind being common in bees and wasps) because the males then, at least for a brief time window, have nothing else to mate with. This may be a reason why deceptive orchid species tend to flower earlier than rewarding species and also show lower overall reproductive success (Kindlmann and Jersakova 2006). The phenomenon could again be said to rely on visitor naivete—it is mostly newly emerged males that visit, and once females are available the males, now experienced, are rarely deceived by flowers. In Australian *Caladenia* orchids, for example, only about 7% of wasps that approach the flower will attempt full copulation and so produce pollination, and the wasps never seem to be fooled more than once within a particular group of plants (Peakall and Beattie 1996). Partly for that reason, pseudocopulatory pollination is also expected to produce rather long range pollen movements, and in that *Caladenia* study the mean pollinia movement was 17 m, with some moving in excess of 50 m. This is probably of selective value for orchids occurring at low density.

The best-known examples, commonly known as insect orchids, are from the European and North African genus *Ophrys*, where single flowers are visually reminiscent of a variety of larger bees and wasps, or of flies, and with which male insects undergo all or part of their normal mating behaviors (up to and including sperm deposition). Similar phenomena, with greater diversity of floral form across genera, occur in the southern parts of Australia and Africa (see Dafni and Bernhardt 1990; Johnson et al. 1998). The principles and complexities of pseudocopulation can best be seen from a few specific examples.

The Genus *Ophrys* in the Northern Hemisphere

Kullenberg (1961) offered the classic review of this genus and of pseudocopulatory phenomena. *Ophrys* flowers have a highly modified labellum providing visual (and often tactile) mimicry, with appropriate colors, fringing hairs, and velvety textures. Scent glands along the rim of the labellum produce attractant odors that are moderately specific (to small subsets of orchid species rather than to individual species), variously containing terpenes such as cadinene, alcohols, aldehydes, ketones, and esters.

1. The mirror orchid *Ophrys speculum* is a common Mediterranean species where the flowers are structurally standard except for the thickened and enlarged labellum, which bears no nectary so the flower is rewardless. The orchid is visited from February to April solely by male scoliid wasps (*Campsoscolia ciliata*), somewhat larger than a honeybee; the females emerge several weeks after the males, so there is a substantial temporal window when the flowers can have undivided attention from the male wasps. The orchid's labellum is an oval of glistening metallic blue, bordered narrowly in yellow and fringed with long red-brown hairs, while the two upper petals are very narrow and dark red. The whole can be seen as mimicking a fe-

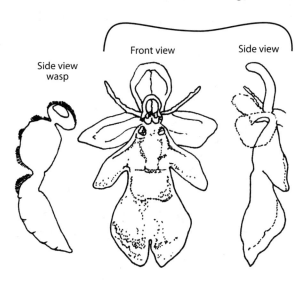

Figure 23.4 Physical mimicry by *Ophrys insectifera* of its pollinating wasp *Gorytes mystaceus*, shown in front and side views. The orchid also matches the wasp coloration on its labellum. (Modified from Kullenberg 1961.)

male, with the blue shiny surface representing the folded bluish wings of a wasp at rest, the fringe of hairs matching the reddish hairs around her abdomen, and the thin upper petals mimicking paired insect antennae. The male wasps readily visit bunches of picked flowers, and will seek out hidden flowers (indicating a strong olfactory cue), but once at closer range they ignore flowers with the lip removed as visual signals take over. On reaching a flower the male lands with his head near the column, and repeatedly thrusts his abdomen into the fringe of hairs on the labellum tip, picking up pollinia on his head in the process. After pollination, the odor of the flower changes (see fig. 6.7), matching the altered scent of the mated female wasp.

2. The fly orchid *Ophrys insectifera* is a more northerly European species, mimicking solitary sphecid *Argogorytes* wasps (fig. 23.4); it has the same kinds of features as *O. speculum*, but with a narrower and less hairy flower, as befits its need of visits from the thin, relatively hairless sphecid. Across its range from Scandinavia to southern Spain, its flowering period varies to coincide with male *Argogorytes* emergence. Furthermore, the scent of *O. insectifera* and the pheromone of *Argogorytes* are both unusually well endowed with aliphatic hydrocarbons, when compared with other *Ophrys* and other solitary hymenopterans.

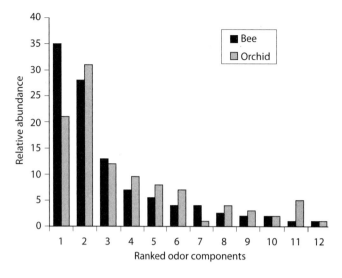

Figure 23.5 Floral scent mimicry in the orchid *Ophrys sphegodes*; here showing 12 compounds whose proportions in the sex pheromone of the pollinating bee *Andrena nigroaenea* and the scent of the orchid are closely matched. (Modified from Schiestl et al. 1999.)

3. *Ophrys lutea* is a Mediterranean species, with a bright yellow labellum bordered by dark bluish patches and a raised central darker area. It is visited by male *Andrena* bees, but the males land head down and pick up pollinia on the tips of their abdomens; presumably the bees see the orchid labellum as a female bee sitting on a yellow flower. The scents for this pairwise interaction are dominated, on both sides, more by alcohols, ketones, esters, and terpenes.

4. *Ophrys sphegodes,* the early spider orchid, is widely distributed across much of Europe. Here there is remarkable precision and specificity of chemical signals: the flowers produce the same set of compounds, and in similar relative proportions, as those in the cuticle-derived sex pheromone of their pollinator *Andrena nigroaenea* (Schiestl et al. 1999). There are 15 compounds from female bees active in attracting male bees, and the orchid flowers contain all but one of these 15 among the 27 compounds in their floral bouquet (fig. 23.5). A decrease in total volatiles following pollination leads to reduced copulation attempts; but Schiestl and Ayasse (2001) also showed a specific *increase* in farnesyl hexanoate in the postpollination flowers, which acted as a repellent to the bees. This compound functions normally as part of the brood-cell lining, and is therefore only used by females that have already mated. Used as a repellent by already pollinated flowers, it directs visitors to apparently virgin females and hence to unpollinated flowers. Intriguingly, as with *O. speculum*, there is also a postmating change in the female bees' mate-attracting odor, which can guide males to virgin females. And there is also much reduced attractiveness in cross-pollinated flowers compared with self-pollinated flowers (fig. 23.6), so that the latter will still have some chance of receiving further and "better" visits.

5. Continuing the theme of elaborate scent specificities, Schiestl and Ayasse (2002) showed that two related orchids, *O. fusca* and *O. bilunulata*, attract different species of *Andrena* simply by differing in the alkene components that they offer as part of their floral bouquet.

6. *Ophrys apifera*, the bee orchid, occurring across much of Europe and into North Africa, is pollinated in the south by *Eucera* and *Tetralonia* bees, but in its northern range is substantially self-pollinated. In this and other *Ophrys* species, selfing is achieved when after a few days the pollinia of an unpollinated flower fall out of the anther and swing around on their viscid threads, readily being blown against the stigmatic surface.

Southern Hemisphere Examples

1. The Australian hammer orchids, such as *Drakea*, are unusual in including a motile structure in the copulatory relationship (Stoutamire 1974). The flower has a very long stalk, and the glossy dark red tip of the labellum mimics the flightless female thynnine wasps that climb up such stalks and emit pheromones that

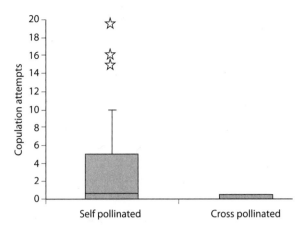

Figure 23.6 Decreased attractiveness in cross-pollinated flowers as compared with self-pollinated flowers, in *Ophrys sphegodes*; this may be attributable to an increase in the repellent compound farnesyl hexanoate in the cross-pollinated individuals (see text and fig. 6.7). (Redrawn from Schiestl and Ayasse 2001), with stars showing three outlying values.)

attract flying males. The flower thus provides a dummy with visual, olfactory, and tactile cues (Peakall 1990). This dummy insect is tightly grasped by a visiting naive male thynnine, active before any females have emerged; he attempts to fly off with it, since his own wingless females are carried by him while being mated. This makes the dummy female swing upward about a hinge point further down the labellum, and the visiting male wasp is "hammered" against the anthers a few times, receiving the pollinia on his back (fig. 23.7). Several related *Drakea* species occur, each pollinated by a different wasp. Similar mechanisms also occur in other Australian genera, including the elbow orchids (*Spiculea*) pollinated by thynnine wasps, the many species of *Caladenia* orchids visited by small bees or by thynnines, and some bird orchids (*Chiloglottis*), again visited by thynnines. In this last example, the females mitigate the cost of deception of the male wasps because their scents are more attractive to males when not near orchids (Wong et al. 2004).

2. The tongue orchids *Cryptostylis*, with narrow reddish petals bearing wart-like lumps, are pollinated by ichneumon wasps (*Lissopimpla excelsa*) that alight with their abdomen tips toward the column and not infrequently deposit a substantial sperm package there as they "mate" with the flower. Orchids with this extreme pollinator behavior show the highest pollination success, possibly because the solitary haplodiploid wasps concerned may (once their sperm is depleted) produce unusually high frequencies of unfertilized and hence male eggs, enhancing future pollination events (Gaskett et al. 2008). At least five species of *Cryptostylis* utilize the same wasp, but they are mutually incompatible so hybridization does not occur.

3. In southern Africa, a few species of *Disa* orchids are pollinated by pseudocopulating *Podalonia* (sphecid) and *Hemipepsis* (pompilid) wasps (Steiner et al. 1994).

4. Turning to more unusual hymenopteran pollinators, sawflies in the genus *Lophyrotoma* attempt to mate with duck orchids (*Caleana*) (Cady 1965); and winged male ants (*Myrmecia*) pollinate the fringed hare orchid *Leporella* while attempting to mate with it (Peakall et al. 1987, 1991; Peakall 1989).

5. Perhaps most unusually, fungus gnats appear to be sexually attracted to the greenhood orchid *Pterostylis rufa* by a small brown insect-like structure on its labellum; the whole labellum is motile and rapidly snaps upward when this structure is touched, trapping the visitor with its back to the column while pollinia are attached (Beardsell and Bernhardt 1983). Blanco and Barboza (2005) also reported pseudocopulatory pollination by "sexually aroused" fungus gnats in *Lepanthes* orchids, where the fly grabs a small appendix on the labellum with his genital claspers and appears to ejaculate into it.

General Points

The males that are being deceived in pseudocopulatory interactions are seemingly always (and certainly in the *Ophrys* examples tested) less attracted to the orchid than to their own females, indicating reliance on naive early-season males. There may not be too much cost to the deceived visitor in most such cases, other than some wastage of foraging time (not an issue if the females are not yet available, and assuming males acquire the minimum food needed to keep alive); the exception occurs where significant sperm deposits are made. Hence the system may be closer to commensalism than exploitation. There is often a low or very low frequency of success (table 23.2) with less than 20% successful pollination commonly recorded. But each success does mean thousands of pollen grains

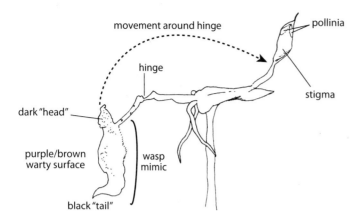

Figure 23.7 The hammer orchid, *Drakea*, visited by male thynnine wasps. See text for details. (Drawn from photographs.)

TABLE 23.2
Percentage of Pollinated Flowers in Pseudocopulatory Orchids

Orchid species	Pollinator	Percent pollinated
Ophrys insectifera	Argogorytes wasps	42%*
Ophrys lutea	Andrena bees	3%–80%[†]
Ophrys sphegodes	Andrena bees	15%*
Drakaea	Zaspilothynnus wasps	0%–58%

Note: Measured as * pollinia removed, [†] seed capsules produced.

transferred, almost always to a conspecific flower; and the recorded values may not differ greatly from rates of success for orchids as a whole.

Schlüter and Schiestl (2008) reviewed the molecular basis of both color and odor mimicries in these systems, and pointed out new possibilities for testing adaptation using gene-silencing and genetic manipulations. From current knowledge, it seems likely that evolution of a specific scent attractant came first in most of these relationships and visual mimicry later, given the prevalence of scent attraction in other kinds of orchid pollination relationships (Bergström 1978; Harborne 1993). The visual deception is often only partial; orchids do not manage to produce stripes to mimic bee and wasp abdomens, for example. Hence the overall specificity involved is also only partial, as evidenced by the many naturally occurring *Ophrys* hybrids.

As a footnote here, *Gilliesia* (Alliaceae) was reported as an insect mimic (Rudall et al. 2002) with secondary bilaterality and unusual appendages that appear very like legs, the flowers also being nectarless; this perhaps is a rare example of pollination by pseudocopulatory deceit outside the orchids.

Mimicry of Other Sites Attractive to Insects

There are rare examples of flowers mimicking other features attractive to potential visitors. One case is that of sleeping sites for male bees (with no nest to retreat to, males often sleep or rest overnight singly or in groups in specific sites, as described in chapter 9). Hence the orchid *Serapias vomeracea* apparently imitates the holes normally used by sleeping solitary male bees (Dafni et al. 1981).

3. Overview

Why are flowers so often deceptive? Most obviously, deceptive mimicries save resources, and could potentially therefore aid any plant in balancing its economic budget while achieving some reproductive output. But the whole range of deceptive mimicries may represent particularly good strategies for plants that are usually rare and nongregarious, such that they cannot provide a worthwhile resource for flower-constant pollinators (Cohen and Shmida 1993). This would cer-

tainly help explain why they are so common in orchids. Deception—whether by empty flowers, flower mimicry, or brood site mimicry—may also be especially worthwhile for spikes of flowers, as a way of discouraging visitors from continuing to visit more flowers on the same plant (Nilsson 1992), so avoiding excessive geitonogamy.

Chapter 24

FLOWER VISITORS AS CHEATS AND
THE PLANTS' RESPONSES

In the last chapter we saw many examples of plants cheating in the plant-pollinator interaction. However, the reverse can also be true—it should at least sometimes pay visitors to gather food from a flower yet resist being manipulated into carrying pollen around, by avoiding the anthers or indeed by eating all the pollen before they move on. This chapter examines some of the kinds of cheating shown by flower visitors and considers what plants can do to avoid the costs of being cheated.

1. Animals That Cheat: Floral Theft

Many visitors to a flower are classed as illegitimate, meaning that they interact with the flower in the "wrong" fashion, entering from the rear or sideways between the petals, or being the wrong size or shape so that they enter in the "right" way but make no contact with anthers or stigma; it is then fairly evident that they are not a successful or coevolved pollinator. In many cases the proportion of effective legitimate visits

to a given flower is remarkably low: for example, in *Asclepias*, only about 30% of visits moved pollen in some observation periods (though up to 80% in others) (Fishbein and Venable 1996). Many related examples were given in chapter 11 where the issue of visitor versus pollinator was a major theme.

An illegitimate visitor may have several ways of getting rewards without effecting pollination. Inouye (1980b) offered a clarification of the terminology of various kinds of floral larceny, as in figure 24.1.

Nectar Theft

Simple Theft or "Pickpocketing"
Nectar theft may occur in different ways. Very small visitors may just crawl into an elongate corolla without contacting the anthers; this not infrequently occurs with small flower beetles, ants, thrips, etc. It is sometimes called simple theft, where no floral damage is incurred (Inouye 1980b, 1983). The same term could apply to very long-tongued moths or butterflies visiting certain bee flowers, where only their tongues enter the relatively broad corolla and no part of their anatomy touches the anthers at all. This kind of theft is most commonly due to a mismatch of morphologies, especially of size. However, it may also sometimes occur because a visitor has learned a way of handling that avoids picking up pollen. For example, bees sometimes insert their tongues into the side or base of a flower tube where the petals are not fused; *Apis* bees behave thus on many *Brassica* flowers or on bluebells, their tongues passing laterally between petals and sepals. Bumblebees also cheat on bluebells by robbing nectar, although this does not reduce the plants' fe-

NECTAR			Activity	Effects	Pollination?
A	Robbery	Primary	Making and using a hole	Nectar depletion AND corolla damage	Some possible increases in pollination success, direct or indirect
B		Secondary	Using an existing hole made by others		
C	Theft		Normal entry, but morphological mismatch precludes pollination	Nectar depletion	
D	Baseworking		Basal entry between petals/sepals	Nectar depletion	
POLLEN					
E	Robbery		Pollen gathered and tissues damaged	Pollen loss	No effective pollination
F	Theft		Pollen gathered without damage	Pollen loss	

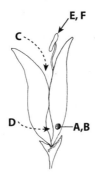

Figure 24.1 An overview of theft types occurring in flowers.

male fitness (Morris 1996). Inouye termed this behavior "base working."

True Primary Robbery, with Floral Damage

Here a visitor effects entry to a flower by making a hole through the side of the corolla—nectar robbery being a term implying the use of force and resulting damage (plate 33A,B). This biting, piercing, or slicing usually occurs at the base of a stout or tubular corolla where there is petal or sepal fusion (plate 33G,H), so that there is no other way in for a relatively short-tongued visitor. Nectar is obtained through the hole, usually without any possibility of touching anthers.

Among insects, major culprits are short-tongued bumblebees (such as *B. terrestris* and *B. lucorum* in Europe, or *B. ternarius* and *B. terricola* in North America), using their mandibles to pierce the corollas of longer-tubed labiates, comfrey, daffodils, columbines, and some legumes like *Vicia*. Carpenter bees are also frequent culprits using their stout mandibles (plate 33A), and ants too can sometimes be robbers.

Robbery can also be achieved by bird beaks, especially in the case of the flower piercer *Diglossa* which may rip or tear whole flowers (Arizmendi et al. 1996). It may be possible for some mammals too: for exam-

ple, the striped squirrel in China is a robber of ginger flowers (Deng et al. 2004).

Nectar robbery can be extremely common, so that more than 90% of all flowers in some tropical plants are attacked by birds such as *Diglossa*, with similar figures in some species attacked by bumblebees (see Maloof and Inouye 2000). Hence a very high proportion of visits to a flower may be illegitimate robbing visits and have little to do with flower pollination.

Primary robbery of this kind can have serious deleterious consequences for a flower and for the plant. First, it may shorten flower life, since damage can trigger the same (often ethylene-mediated) senescence processes mentioned in chapter 21. Second, the robber may leave scent marks that discourage legitimate visitors, or may alter the scent of the flower via damage-related effects. Third, it may reduce the visual attractiveness of the flower. And finally it can lead directly to further nectar losses as outlined in the next section.

Secondary Robbery

The activities of primary nectar robbers in turn allow in other illegitimate cheating visitors, often termed secondary nectar robbers, stealing nectar through the hole left behind in the corolla. *Apis* is very often a secondary robber; its mandibles are not strong enough to cut through any but the more delicate corollas, but it readily adopts secondary cheating wherever it can. For example, it goes to a range of beans and other legumes in this fashion—holes are regularly made in the base of flowers on runnerbeans *Phaseolus*, or on broadbeans *Vicia*, by short-tongued bumblebees such as *B. terrestris* (e.g., Newton and Hill 1983), but *Apis* is then the commonest user of these holes over the next two or three days. Dedej and Delaplane (2005) showed that on *Vaccinium* flowers honeybees clearly received a net energetic advantage from their habit of larceny; the absolute amounts of nectar removed legitimately and illegitimately were similar, but illegitimate visitations could be completed much more quickly. Ants, attracted to a sugar reward that has become unexpectedly available, are also major secondary robbers.

Theft, primary robbery, and secondary robbery may be exhibited by the same flower visitors on different flowers, or even by the same visitor species on one flower if there is a substantial size variation such that larger individuals can reach the nectar legitimately but smaller ones either cannot do so or find it quicker and cheaper to bite a hole or fortuitously encounter an existing hole.

Pollen Theft

Simple Theft

Some small hoverflies, solitary bees, and stingless bees gather pollen directly from anthers without ever going close to the plant's stigmatic surfaces (plate 33C–F), and often do not need to visit more than one flower in one foraging bout. Others of a similar size may collect pollen off the stigmatic surface directly, which may doubly disadvantage the plant by producing stigmatic damage (physical or chemical). Arthropods that rest on flowers or use them as encounter sites (see section 4, *Other Cheats: Floral Exploitation by Hitchhikers and Ambushers*) may also take some pollen in a manner that can have no benefit to the plant. More unusually, a few orchid pollinators become pollinia thieves by circumventing the elaborate deceptions of their flowers (Gregg 1991).

A review by Hargreaves et al. (2009) showed that most pollen thieves were bees, and that thieves on a given plant were effective pollinators of other plants, so the theft was probably opportunistic as a result of

physical mismatch between the flower and the thief rather than a specialist strategy. Theft was especially common in herkogamous or dichogamous plants, that is, where there was either spatial or temporal separation of sex functions within flowers.

Robbery

Pollen robbery involving floral damage is rare, but there are a few documented cases (e.g., McDade and Kinsman 1980) of small bees nibbling through the corolla walls of hummingbird-pollinated flowers and reaching the concealed anthers, before or shortly after normal anthesis.

Some of the more notorious illegitimate flower visitors may indulge in simple theft *and* primary robbery *and* secondary robbery, depending on the resources available in any given vicinity. *Apis* behaves thus, and a wide range of ants (usually too small to pollinate, but highly attracted to any sugary fluids) are also culpable in this respect.

Equally, it should be noted that any one plant may be suffering all these kinds of cheating at once. For example, *Justicia aurea* in Costa Rica received legitimate visits from two hummingbirds, but also nectar robbery from several smaller hummingbirds, secondary nectar robbery from both bees and ants, and pollen theft from small stingless bees (Willmer and Corbet 1981).

Florivory

The eating of flowers themselves needs inclusion here, as it could be regarded as the ultimate theft, usually with unequivocally deleterious effects on the plant and its reproductive success. Many insects cheat a plant by using the flowers that the adults (legitimately) feed from as egg-laying sites, so that their larvae after hatching then eat some part of the flower. For example, many flies use relatively dense flower-heads such as thistles and other composites as oviposition sites. By far the most famous examples here are the fig and yucca *active pollination* systems, covered in chapter 26; although these cannot really be called "cheating," as the plant gains very obvious benefits most of the time.

A further range of larger adult animals eat whole flowers, notably beetles and some birds, and the habit is particularly common in the tropics, with birds and monkeys implicated. However, florivory is best seen as a specialist subcategory of herbivory, so examples are covered more fully in chapter 25.

2. Selective Effects: Protection against Theft

Any or all robbery, whether of nectar or pollen, could obviously have serious effects on floral visitation and pollen transfer by reducing the rewards on offer, but it may affect the plant much more than this, since damage inflicted by robbing visitors (especially those that bite the corolla) can decrease floral longevity and cause premature wilting or abscission. Ants engaged in their exploitative thieving consumption of nectar may also reduce pollinator visit frequency or duration since they can be aggressive to incoming visitors, thus reducing seed set (Galen and Geib 2007; Willmer et al. 2009). At worst, resident ants can produce castration effects, for example by cutting the styles in *Polemonium* (Galen 1983) or partly destroying flower buds in the semi-myrmecophyte *Humboldtia* (Gaume et al. 2005). It should therefore be expected that at-risk flowers will evolve floral defenses against the cheats (Guerrant and Fiedler 1981; Irwin, Adler, and Brody 2004). In fact this issue was explored in considerable detail in the nineteenth century by Kerner (1878/2008), who noted many features of flowers that were keeping thieves out rather than attracting pollinators.

Hence cheating visitors may exert distinct selective pressures on flower design. For example, Galen (1983, 1999b) showed that ant thief activity could have a strong selective effect on floral morphology, leading to compromises in floral design: a balancing of traits for pollinators versus what is in effect a "predator." However, in some cases selection favoring thief and pollinator may work in the same direction; for example, in *Polemonium* where Galen (1999b) showed that ants prefer wide corollas to narrow ones, interacting with selection for the same preference by bumblebees. Irwin (2006) modeled the conflicting effects of pollinators and robbers on *Ipomopsis aggregata* flowers and showed that both were important selective agents, but that the direction and intensity of selection by each varied from year to year.

Defense of the plants and its rewards against theft can be achieved in various ways, but there are broadly three main options (see also Willmer, Nuttman et al. 2009):

1. Physical barriers
2. Chemical deterrents, often using volatile organic compounds (VOCs)
3. Bribes, for example, food or lodging placed some distance from the flowers

A summary of possible defensive characters is shown in figure 24.2.

Physical Protection and Barriers

Perhaps the most obvious possibility to prevent primary nectar theft is to thicken the base of the corolla; this is the same response as seen in many hummingbird-visited plants, where the base of the tube must resist piercing from within by the birds' beaks. Alternatively the plant could thicken the sepals or bracts that surround the base of the corolla, as in the calyx of *Dianthus*, which is rather leathery, or the calyx of *Silene*, which is inflated and may serve the same purpose. The calyx may in turn be protected by bracts, and in some cases the bracts are so arranged that they contain a moat acting as a water barrier that can be highly effective against ants and small herbivores (e.g., Carlson and Harms 2007). Water-filled bracts or calyces (plate 34D) are moderately common in tropical flowers in the families Bignoniaceae, Gesneriaceae, and Solanaceae. Mucilage in the calyx (e.g., in *Commelina* and *Malvaviscus* species) may also be effective. Taken to an extreme, aquatic plants with floating flowers are entirely protected from theft by nonflying insects. But amphibious plants then present an interesting conundrum, and Kerner (1878/2008) cited *Polygonum amphibium* as having no physical defenses for its flowers when these were floating, but (reversibly) developing numerous glandular trichomes on the leaves and especially on the floral stems (which became sticky enough to trap ants) when the habitat dried out so that flowers were accessible to crawling insects.

To deter walking thieves such as ants, spiny or very hairy outer surfaces on stems or on the corolla, calyx, or bracts can be effective, where passage is impeded or tarsi cannot grip through entangling fine hairs (plate 34C). These fine hairs (trichomes) may also be glandular, overlapping with the chemical defenses described in the next section. A fine example is the calyx protecting the long narrow corolla of various *Plumbago* species, covered in delicate secretory hairs in the

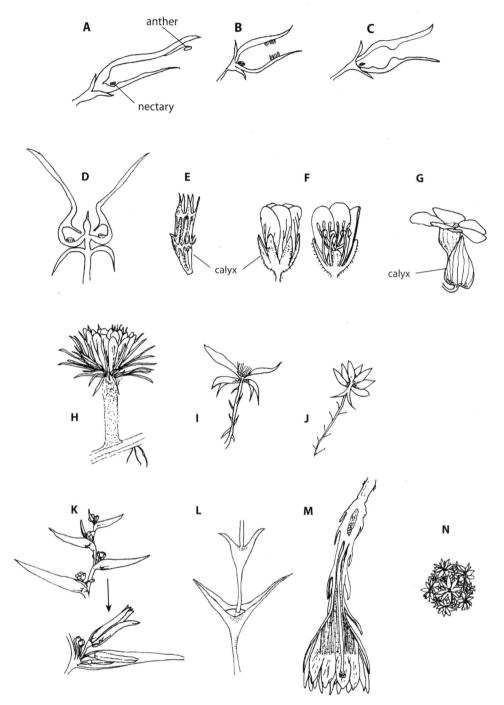

Figure 24.2 An overview of physical defences that may surround and protect flowers: (A) thickened corolla bases; (B) hairs within corolla; (C) constrictions within corolla; (D) lidded nectaries; (E–G) spiky, bristly or inflated calyces; (H–J) stems bearing bristles, spines, or glandular trichomes; (K,L) bracts protecting flowers, sometimes with a water moat within; (M) pendant stems, often with overlapping scales or bristles or trichomes; (N) tight packing of flowers within an inflorescence, as in *Heracleum* and other umbels. (Drawn from photographs or from flowers.)

area where they might otherwise be robbed (plate 6G). Sticky surfaces are also excellent defenses against walking ants, and slippery waxy surfaces can make access impossible (e.g., Harley 1991). Pendant flowers on delicate bending stems are also difficult for ants to access.

Defenses against access by crawling insects may also occur within the flower—rings of fine hairs as in *Stephanotis* (plate 6H), extremely narrow tubes (or areas of constriction on a wider tube), and tissues that form "lids" over the nectaries are all encountered quite commonly in more elongate flowers.

Another possible physical defense might be the sheer density of flowers in an inflorescence, so that small thieving insects (or the mandibles or beaks of larger robbers) cannot get access to the corolla bases. For example, very tight flower heads of *Brassica* species do not get robbed as much as loose heads. This may also apply to clovers, mints, thymes, and also some kinds of Asteraceae; indeed this may be the strongest of all selective forces favoring the evolution of tight inflorescences.

Chemical Protection

The presence of glandular hairs on, or en route to, a flower might be a suitable chemical deterrent against both crawling florivores and possible nectar thieves such as ants. Any contact with or damage to such hairs can potentially release deterrent or irritant chemicals.

Another possible chemical defense would be to make the nectar unattractive to casual or illegitimate visitors. Here an example might be the provision of very dilute nectar in *Ipomopsis*, acceptable to a pollinating hummingbird but deterrent to bumblebee robbers (Irwin, Adler, and Agrawal et al. 2004). A more specific case is provided by the very alkaline character of nectar in a ground-dwelling flower, *Lathraea clandestina* (plate 34E), arising from ammonia in the nectar, and deterrent to birds and perhaps to ants who try to rob the flowers but manageable for pollinating bumblebees (Prŷs-Jones and Willmer 1992). Taking this a stage further, there is the option of secreting toxic nectar, discussed (as a rarity) in chapter 8.

Having specific repellents in the flowers is perhaps the ultimate option, and the best-known cases of this involve repellents targeted at ants. Pollen-based repellents in *Acacia* flowers are covered below, as they relate specifically to ant-guarded plants, but since that work was published it has become clear that other flowers also produce ant repellents. Ness (2006) and Agarwal and Rastogi (2008) found repellence of some but not all ants by *Ferocactus wislizeni* petals and *Luffa* flowers, respectively. Jaffé et al. (2003) studied encounter rates of ants with flowers of Venezuelan plants in situ and found lower ant repellency in forest canopy flowers compared to savanna flowers. Junker et al. (2007) also studied behavioral effects of whole flowers of various ages, and reported ant repellence for 8 out of 18 plant species from Borneo, with greater repellence in canopy flowers than in forest understory flowers. But these studies generally did not allow separation of contact versus volatile ant repellence, and they recorded repellence merely in terms of ant location relative to floral cues. Because multiple ants were tested together, these studies also suffered potential confounding effects from individuals following each other and magnifying apparent effects.

Junker and Blüthgen (2008) used a four-way olfactometric assay instead, and found *Camponotus floridanus* ants to be repelled from 20 of the 30 flowers tested, and *Lasius fuliginosus* from 8 of 26. Willmer, Nuttman et al. (2009) used the unequivocal method of puffing flower-volatile-loaded air over individual ants, with stereotyped alarm behaviors as a proxy for floral repellence. Here *Formica aquilonia* showed clear alarm or aggressive responses to about half of the 67 floral species tested, and *Lasius niger* responded to a smaller subset of these. Behavioral responses were always greatest when flowers were at peak dehiscence and were elicited to pollen alone when this was tested. There was also some trade-off between morphological barriers and the strength of volatile repellence in flowers; no species tested showed high levels of both kinds of defense (fig. 24.3). More exploration of the role of pollen volatiles in controlling potential thief and pollinator behaviors is needed. There may be interesting links here to suggestions discussed in chapter 7 that some pollens may possess chemical properties that limit their use by some bees (Praz et al. 2008b).

Offering Bribes

Where nectar is being stolen, or is under threat of being stolen, one strategy is to offer an alternative sugary fluid somewhere away from the flowers—in other

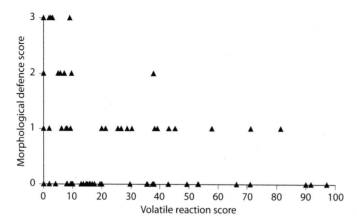

Figure 24.3 An indication of trade-offs between volatile protection against ants in flowers (especially from pollen), and the presence of physical barriers or decoys on the plant. Volatile reactions are scored from behavioral responses of ants as percentages of a maximum; morphological defences are on a 0–3 scale from no visible defence, one deterrent feature, two or more such features, and total exclusion. No plants scored highly in both respects. (Redrawn from Willmer, Stanley et al. 2009.)

words, to use extrafloral nectar as a distraction or bribe. Extrafloral nectaries (EFNs) are extremely common, often in leaf petioles or blades or stems (plate 34H), sometimes on bracts or sepals (plate 34F,G). Bentley (1977) provided an early review, stressing the great range of plant families and genera that possess EFNs and the evidence that they really have a purpose in attracting animals away from flowers (see also Wagner and Kay 2002). A neat demonstration of their effectiveness came from species of *Hibiscus* on an oceanic island, where the ancestral *H. tiliaceus* has sepal EFNs and a derived endemic *H. glaber* does not (Sugiura, Abe et al. 2006). Moth larvae attacked 20% of the latter's buds, compared with just 0.2% of buds on the former; it seems likely that loss of the EFNs on the endemic species reflects an original paucity of herbivores on the island.

In particular, EFNs are attractive to ants, and some ants will become resident on a plant well endowed with extrafloral nectaries because of the sugary reward, without straying onto the flowers. Ants commonly patrol the leathery EFN-bearing buds of *Costus* (plate 34F,G), from which individual large flowers emerge. Since the essentially carnivorous ants will also use their host plant as a source of insect prey, they may also often attack the plant's invertebrate herbivores. For example, on *Caryocar* plants ants were shown to prey upon bugs, flies, caterpillars, and gall wasps, and as a result ant-occupied plants produced more flowers and had higher initial fruit set than those with ants excluded (Oliveira 1997). However, this is a very nonspecific interaction, where more or less any plant with some EFNs will be acceptable for any ant; more spe-

cific and structured ant-plant interactions are covered in detail in the next section. Note, though, that the distinction between the generalized ants taking EFN resources and the resident ants employed as bodyguards is not all that clear-cut, since the presence of nonspecific ants can still reduce robbery and herbivory and improve seed set (as in *Caryocar*, and also on *Passiflora* flowers; Leal et al. 2006) and can cause pollinators to move around more and improve outcrossing (e.g., Altshuler 1999).

Employing Bodyguards

Where *Physical Protection and Barriers* covered physical defense, and *Chemical Protection* and *Offering Bribes* both concerned forms of chemical defense, this category covers *biotic defense*. Keeping enemies at bay is classic work for bouncers and bodyguards, and in rather analogous roles floral guards are surprisingly common, notably in the tropics but also in many temperate plants. By far the commonest groups employed as guards are the ants, generally small and nonflying; thus not mobile enough to be good pollinators yet highly attracted to sugary nectar. Most ants have elongate mandibles, useful for aggression and carnivory, and most are eusocial with workers that can be recruited and organized by pheromonal signals. Hence ants can be extremely useful guards against flower robbers, as long as they too can be kept away from the insides of flowers and do not interfere with pollination.

Ants as guards are much better known as defenses

against florivores and herbivores, so we will delay their detailed consideration for chapter 25. But there is no doubt that they can be highly efficient in reducing theft from flowers.

3. Overall Effects of Theft on Flowers

Some theft of floral resources is very common (Maloof and Inouye 2000), and even in the best-defended plants some losses may still occur. Robbing varies in intensity within seasons, between years, and across sites, as well as between species (Irwin et al. 2001; Irwin and Maloof 2002). How far does its existence and its variation exert effects on pollination success? In addition to the possible negative effects already mentioned, floral thieves may change the nectar dynamics of the plant and hence visitor behavior (Newman and Thomson 2005) or the visitation preferences of legitimate visitors (Krupnick et al. 1999), or may affect the amount of pollen moved by such a visitor. Any of these effects might intuitively be expected to be deleterious overall, with fewer legitimate visits made per unit of resource provided by the plant, leading to lower seed set. This is certainly the case in many plants (e.g., Galen 1983; Irwin and Brody 1999, 2000; and see Irwin 2003), and Burkle et al. (2007) found that negative impacts were especially likely in pollen-limited and self-incompatible plants.

The review by Maloof and Inouye (2000) addressed the balance of obvious negative and less obvious positive effects from floral robbery and noted that there were several possible benefits. For example, in red clover high numbers of robbing short-tongued bees caused increased seed set (Fussell 1992) because the legitimate visitors moved around more to get enough nectar, so promoting outcrossing. Likewise, Richardson (2004) showed that robbery by *Xylocopa* on *Chilopsis* could increase the effectiveness of legitimate *Bombus* visits by making the bumblebees move around more. Maloof (2001) found a similar effect on visitor behavior (bumblebee pollinators and robbers) in *Corydalis*, although there were no effects on seed set here. In fact robbery can sometimes alter the behavior of the normal visitor or pollinator quite subtly, via changes in flower shape or appearance, or via the reward, so that the animal moves more pollen more effectively. For example, Goulson et al. (2007) showed

that bumblebees and honeybees could detect robbed flowers of *Tropaeolum* and tended to inspect but bypass them, increasing their own average reward per flower visited while also moving around the flower patch more than would otherwise have occurred.

In at least a few cases the robbers may have another beneficial effect, by encouraging ants to move in to exploit the holes as secondary thieves. The presence of the ants may then benefit the plant by reducing herbivore numbers, as documented for *Linaria* flowers in the Rockies, where seed-eating beetle attacks were greater on the unrobbed and relatively ant-free flowers (Newman and Thomson 2005).

A further possible benefit is that a robber may also effect some pollination itself, for example by robbing for nectar first but then visiting properly for pollen, or by receiving pollen on its body while biting the corolla; Koeman-Kwak (1973) showed that short-tongued bumblebees pollinated some *Pedicularis* despite acting as robbers. In *Anthyllis* flowers about 45% of visits were by legitimately visiting *Anthophora* bees and a further 45% by short-tongued robbing bumblebees, yet the robbed flowers had a higher seed set than the unrobbed ones because the bumblebees did inadvertently contact anthers while making their corolla holes (Navarro 2000).

For these various reasons, it may be not uncommon for thieves to have a degree of positive effect on plants, and some of them may even be best regarded as mutualists, depending on the relative abundance of thieves and legitimate visitors at a given time and place. But we should retain a sense of proportion here, as net benefits are probably relatively rare; Irwin (2003) concluded that negative effects from reduced visitation and increased departure usually far outweighed any bonus accruing from factors such as reduced geitonogamy. Perhaps not surprisingly then, plants may have subtle ways of responding to and partly compensating for the negative effects of robbery. Irwin et al. (2008) noted that some "tolerance" of floral larceny could be achieved by "banking" some extra flowers on larger inflorescences, in both *Polemonium viscosum* and *Ipomopsis aggregata*, although at high thief density the effect became rather limited. As another example, nectar robbery reduced the duration of the male phase in the protandrous *Impatiens capensis* (Temeles and Pan 2002), presumably because after robbery the chances of all the remaining pollen being removed were lower than the chances of receiving enough incoming pollen in the female phase to fertilize the 3–5 ovules in this plant.

This is a suitable place to note that cheats are always likely to be rarer than legitimate visitors in any one system, of necessity; otherwise a natural or coevolved community would potentially collapse. Hence, for example, nectar-robbing hummingbirds are always far less abundant than "proper" hummingbird visitors in any one natural community.

4. Other Cheats: Floral Exploitation by Hitchhikers and Ambushers

One other possible kind of cheating in visitor-flower relationships concerns an animal making use of the flower not directly as food but as habitat; either as a convenient microclimate to rest in, or as a meeting place or bus station, as discussed in chapter 9. Any such interactions may in turn influence the efficiency of pollination.

Flowers as Places to Encounter Prey

Flower visiting is not a risk-free occupation; indeed there are substantial predation risks to pollinators, precisely because predators can reliably predict that they will encounter particular kinds of prey on particular flowers. This in turn engenders some risks to the plant, whose pollination may be compromised; and these effects have almost certainly been underestimated in the past (Dukas 2001a). In this situation, the predators become flower cheats.

Flower predators can be conveniently split into two types, the *ambush predators* who sit on or close to flowers to await prey, and some nonambush types. Ambush predators include many generalist predaceous bugs, some larger ants, and some beetles. There are also specialist sphecid wasps in the genus *Philanthus* (commonly known as bee-wolves because they prey on flower-visiting bees) which can cause local reduction in bumblebee densities (Dukas 2001a, 2005) and associated reduction in local pollination of specialist bumblebee plants such as *Aconitum*. Perhaps most spectacular among the ambushers are various kinds of

preying mantids and crab-spiders that are often exceptional floral mimics and thus become cryptic as they await incoming prey (plate 32G,H). Several bugs and spiders do take some nectar while resident on their flowers, and some may include pollen in their diet; mantids do so especially when only just hatched (Beckman and Hurd 2003). However, as these predators are by definition fairly stationary, they must be acting purely as pollen thieves and not as potential pollinators.

Crab-spiders respond to just the same advertisement signals as do the pollinators when choosing flowers to wait on (Heiling et al. 2004), that is, visual cues, the presence of high reward, and above all olfactory signals. Usually they are cryptic, and naive flower visitors do not detect them in time to avoid the flower, although experienced *Apis* do avoid flowers with spiders on them (Dukas 2001a,b), particularly recognizing the cues given by spider forelimb shapes (Goncalves-Souza et al. 2008). Reader et al. (2006) found that *Apis* were better at avoiding spider-occupied flowers than was the solitary bee *Eucera*, and confirmed that *Apis* would also avoid flowers with honeybee corpses present indicating the past or current presence of a predator, although this effect did vary with the particular flower studied. Dukas and Morse (2005) found that, while honeybees were more affected than bumblebees on milkweeds, neither was deterred to the extent of reducing pollinia removal, especially when bee abundance was high. Heiling et al. (2006) showed that, remarkably, honeybees were induced to *prefer* a chrysanthemum flower that had a crab-spider sitting off center on it and creating an enlarged area of color contrast, compared with a vacant flower. Hence the effect of crab-spider predation varies greatly not only with visitor type and abundance, but also with flower type, mainly dependent on how efficiently predator crypsis maintains flower attractiveness and even with potentially increasing attraction if the predator gets its own behavior just right. Given appropriate circumstances, invertebrate ambush predators can clearly have significant effects on pollination outcomes and eventual seed set (e.g., Suttle 2003).

On a larger scale, vertebrate ambushers include the birds known as bee eaters (Meropidae). While taking a wide range of insect prey, these birds have a particularly high intake of honeybees in North America, whereas the European bee eater prefers bumblebees, with a 33% success rate per attack. Other birds that will take flower visitors include flycatchers, swifts and swallows, shrikes, and drongos, while small to medium-sized raptors are fairly regular attackers of hummingbirds. Bats active in the evenings inevitably go for flower-visiting moths; and there are occasional instances of bats themselves being taken at flowers by snakes (Hopkins and Hopkins 1982).

Among the *nonambush predators* with different strategies for capture the most important are probably the orb-web spiders, using their carefully sited webs to snare flower visitors that are moving between plants. Some species of *Nephila* have golden-colored webs that seem to attract unusual numbers of bees. Robber flies and dragonflies may also take flower visitors as they move to and from their flower patches. Perhaps more surprising are the effects of bird insectivores, notably swallows and swifts; Meehan et al. (2005) documented reduced seed set occurring in sweet clover (*Melilotus officinalis*) close to cliff-swallow nesting sites.

There are also parasitoids and parasites that use flowers to encounter hosts, in particular a range of parasitic wasps that use insects as their hosts and exploit flowers as sites to locate them. Many of these also take some nectar from flowers, and Patt et al. (1997) showed that floral architecture had substantial effects on parasitoid feeding success (with more open flowers on umbels best suited to attract parasitoids, thought to be part of a possible pest-control strategy for crops). Conopid flies are probably the worst parasitoids for bumblebees, which regularly suffer around 10%–15% infestation, and if not lethal this can certainly reduce lifespan (Otterstatter et al. 2002). Phorid flies also attack *Bombus* and *Apis* and may steal pollen from them. Mites are often exceptionally abundant as ectoparasites on the outer surfaces of bees and can become so numerous that they incapacitate the insect; *Varroa* mites (currently a major concern as pests of honeybees; chapter 29) also affect learning and behavior (Kralj et al. 2007).

Given this range of possible enemies to be encountered on and around flowers, predation can sometimes be frequent enough to influence floral visitors and pollination outcomes. Existing records show that around 3%–10% of bees do not return from any one foraging trip (fig. 24.4), with rather higher rates in honeybees than in bumblebees. More specifically, bumblebees had a 14% chance of crab-spider attack when visiting *Asclepias* (also shown in fig. 24.4; from Morse 1981b).

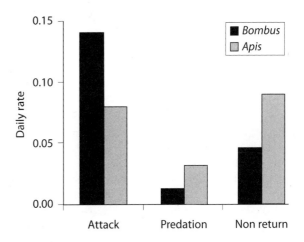

Figure 24.4 Daily rates of attack and of predation, and of bees not returning from foraging bouts as recorded at the nests; these are higher for honeybees than for bumblebees excepting the rate of attack at flowers, which can reach 14% for bumblebees. (Modified from Dukas 2001a, based on several sources.)

It seems that bees (and perhaps most other flower-visiting insects, with rather similar visual systems) are poor at spotting motionless cryptic predators such as crab-spiders and mantids, although they may detect the predator odor or scents left by previously damaged prey.

The visitors should not be passive in all this, however. Avoidance of plants where risks are higher would be an obvious response. Social bees appear to do this and also to become more risk-averse when well fed. There is good evidence of specific avoidance learning, so that a bee that has been subject to disturbance or attack on a flower will not go back to that patch or that flower, and more experienced foragers pick up cues of danger more readily than naive foragers. Not surprisingly, then, bee-wolves catch a higher proportion of bees with relatively unworn wings, that is, younger and less experienced individuals. Bees also learn of dangers at the colony level, so that the numbers emerging to forage will drop markedly (from 90 per minute to 4) when a bee eater is active in the vicinity. Butterflies also appear to learn to avoid danger spots, so that *Heliconius* (which normally come back repeatedly to the same feeding sites) do not return if briefly captured and released.

Alternative strategies to reduce predation include varying the time of foraging in ways that shift the active period away from that of the predator, or more drastically switching plants (with very clear consequences for the original host plant). And the visitor's morphology may also come into play as a defense; longer legs or longer tongues might limit the risk of being caught, and an even longer tongue allows a pollinator to swing hover some distance from the flower and so be extremely hard to catch (Wasserthal 1993; see chapter 14).

Birds as a group are reported to maintain lower fat reserves under high predation risk, which increases their maneuverability, and this general response probably applies to hummingbirds. It may be that some insects do the same thing, in terms of food loads carried; it would be useful to reduce the load if predation occurs at the flower, but perhaps to increase it (and so decrease the number of trips needed) if predation occurs in flight, although this has not been explicitly tested.

In the longer term, being aposematic would be another possible response to high predation risk, and the presence of groups of similarly colored bumblebees (implying Müllerian mimicry) has been interpreted in this light (reviewed by P. Williams 2007).

Do the flowers help? Having traits that help to deter predators and reduce predation risk to pollinators might make good adaptive sense for a plant. The defensive strategies against herbivores are extremely well known, with chemical defenses in leaves especially obvious, and these could readily be extended to supply a defensive chemistry in flowers (see Detzel and Wink 1993; Carisey and Bruce 1997) and specifically in nectar (Adler 2000; Adler et al. 2001; Irwin et al. 2004) or pollen (Roulston and Cane 2000). An alternative to chemical defenses would be a floral morphology that did not provide an easy cryptic background or offer any suitable hiding place near the nectar or the anthers. Perhaps concealed nectar has a benefit in being hard for predators to find, so that they are less likely to choose such a flower to ambush from. A short floral lifespan (less than a day) would also be useful, so that the predator would have to move to new flowers regularly and reduce its hunting period (and perhaps spiders would have to resite their webs more often).

We have mentioned various ways in which visitors might avoid predators and ways that flowers might help to reduce predation risks for visitors, but all of that begs the question as to whether the predators are always deleterious for the flowers. It would seem that predators on flowers attacking incoming pollinators would inevitably be bad news for pollination success, and that may often be true, especially when casualty proportions are high. But predators can at least sometimes also be partly beneficial, and in at least two very different ways:

1. They reduce the effects of herbivorous insects on the plant (Romero et al. 2004), with knock-on benefits of a reduced loss of photosynthate and perhaps reduced florivory
2. They may startle but not attack the incoming flower visitors, so keeping potential pollinators moving around more and effecting more outcrossing.

Flowers as Places to Encounter Mates

The possibility of encountering a mate is almost certainly an additional attraction of flowers for nonresident predators such as conopids, but it may also be relevant in quite a range of more traditional pollinators. Male bees very frequently wander about flower patches seeking incoming females (see plate 29G) and may indeed hold and patrol territories around especially rewarding flower patches as a way to attract good mates. Mating on flowers is thus rather common (plates 1C and 5C). However, as discussed in chapter 18 these males are rarely cheats and can themselves be rather effective as pollinators. Hummingbirds and some perching birds show the same habits of seeking mates around flowers and again may feed and pollinate as they do so (see chapter 15).

Flowers as Places to Hitch a Lift

Technically this is termed *phoresy*. Many flowers provide habitats and breeding sites for flower mites, which may also feed on nectar and/or pollen, and these mites can disperse and find new habitats only by traveling on the body of a flower visitor. This may be a bee, a moth or butterfly, or a bat (large bats have been recorded carrying up to 360 mites at a time). But probably the best-known examples are the mites that live in hummingbird nostrils and beaks, and hitch around between flowers via the birds' tongues, leaping on to a tongue inserted in a flower and thus getting carried to another flower. There can be quite specific associations between flower species, bird visitor, and mite species (Colwell 1986), although some mites are more polyphagous. The more specialist mites appear to recognize their host plant by scent and to use cues from the nectar as a trigger to disembark (Heyneman et al. 1991); and their rate of hitchhiking increases in successive instars.

In some cases the mites consume quite a high proportion of the floral nectar, for example up to 50% in long-lived *Moussonia* flowers (Lara and Ornelas 2002) and a similar proportion even in very short-lived *Heliconia* species (da Cruz et al. 2007). While this is likely to be classified as theft, and potentially bad for the plant, it may at the same time influence pollinator movements between flowers to give some benefit to the plant (and indeed the mite).

Bee-mites such as *Varroa* are parasitic on bee larvae once in the nest, also transmitting many viruses, so their phoretic habit is potentially seriously deleterious for a bee host. As they use phoresy to disperse between flowers and thus to find new hosts, and also to encoun-

ter mates, these last three categories of floral cheats begin to overlap somewhat confusingly!

5. Overview

While both plants and visitors have many ways of cheating, it seems that the diversity and deviousness of cheating by the plants, covered in the last chapter, are substantially greater than the surreptitious stealing and ambushing that goes on in the animals, where even obvious cheats are often in fact giving an indirect benefit to the plant. This may seem a little surprising at first, given the stationary and rather passive character of a plant compared with the sophisticated behavioral repertoires of its visitors; but is probably not surprising when thought of, once again, in terms of the life-dinner principle, and what each participant has at stake.

Chapter 25

THE INTERACTIONS OF POLLINATION AND HERBIVORY

Outline

1. Effects of Florivory on Pollinators
2. Effects of Herbivory on Flowering and Pollination
3. Defenses against Florivory and Herbivory Affecting Flowers
4. Overview

Herbivores in the broadest sense include not just folivores whose diet is green leaves, but also browsers on twigs and bark, seed predators, underground root feeders, florivores, and even nectar robbers. Herbivory is not just due to animals, though: the effects of fungal spores are often very evident on flowers and flowering patterns, and other decomposers that gain entry through any damage to the plant's protective surfaces can also have a massive effect on overall plant growth and hence on resources allocated to flowering. Pollinators and herbivores therefore rarely operate completely independently, and although in the past there was a tendency to recognize only the effects of specific and direct damage to flowers (florivory), more recently there has been analysis of broader herbivore effects, as one part of the wider investigation of non-pollinator agents of selection on floral traits (see Ehrlén 2002; Strauss and Whittall 2006; Wackers et al. 2007). Several levels of interaction between flower visitors and the agents that damage plant tissues can be identified:

1. At the simplest level, many pollinators can discriminate between damaged and undamaged plants (or flowers), and prefer the latter
2. Herbivory depletes a plant's resources, and this will often affect the allocation made to flowering, and to specific floral features that alter pollinator behavior, inducing the more complex and interesting effects that form the centerpiece of this chapter
3. The act of flowering may make a plant both more conspicuous (since it normally involves advertising signals) and more attractive to a herbivore, raising additional conflicts for the plant.

Thus when herbivores and pollinators are both active on plants, there is much scope for differential selection on plant traits, and pollinator-mediated selection can sometimes be overwhelmed by opposing selective forces operating due to herbivory. This can result in increased genetic variation and a compromise phenotype and could potentially promote generalization in the flowers (see chapter 20). This chapter explores the balance between such potentially conflicting selective influences on a flowering plant, from both florivores and more general herbivores, and some ways in which the conflicts can be resolved.

1. Effects of Florivory on Pollinators

Florivory can involve complete removal of flowers by larger feeders, mainly vertebrate grazers but also including large beetles or orthopterans and even occasionally the tree-climbing grapsid land crabs that eat the petals, stamens, and stigmas of some bromeliads (Canela and Sazima 2003). Alternatively, florivory can refer to partial damage to flowers by smaller insects and a few other invertebrates. Quite often the smaller but more numerous insect feeders inflict greater overall damage to plant fecundity (e.g., Amsbery and Maron 2006).

Total removal of flowers not only cuts out any possible reproductive output from the missing blooms but also affects the floral display and the plant's overall appearance, potentially deterring pollinators, altering phenology, and possibly affecting resource allocation to new flowers. Sharaf and Price (2004) documented the effects of flowering stalk removal in *Ipomopsis aggregata*, resulting in increased branching of new inflorescences; neither hummingbirds nor robbing bumblebees showed any preference for browsed or unbrowsed stalks, but the delay to flowering did make the plants miss the peak hummingbird season, giving lower per flower visitation rates and lower fruit set.

More localized damage to an individual flower could occur in several ways, each with different possible effects:

1. Effects on advertisement: most obviously the alteration of shape by removal of petal tissue, potentially affecting pollinator choice (selection or avoidance of that flower), or behaviors such as intrafloral movement after choosing the flower, or constancy (i.e., decisions to visit further flowers)

2. Effects on reward: feeding on nectar or pollen, damaging nectaries or more subtly the protective features affecting nectar concentration, or damaging anthers; possibly affecting pollinator choice, but more likely to affect behavior or constancy

3. Damage to reproductive structures: potentially affecting production or dispensing of pollen, or its deposition, or both, and also subsequent fertilization processes; there are no direct effects on most pollinators unless advertisement signals are also affected.

All of these outcomes could have substantial influences on the plant's pollination and reproductive success. Where the reproductive structures are damaged this is usually obvious, but where advertisements or rewards are altered there has been rather little investigation of the outcomes; in fact, until recently the common occurrence of nectar and pollen feeding by herbivores was underappreciated (Wackers et al. 2007).

One well-known effect relating to advertisement is the avoidance of damaged flowers by visitors, documented for *Erigeron* (Karban and Strauss 1993), for *Isomeris* (Krupnick et al. 1999), for *Rudbeckia* (Hambäck 2001), and for hawkmoths on *Oenothera* (Mothershead and Marquis 2000). Effects are especially noticeable where a complex flower has some part of its symmetrical features removed, or where a labiate-type flower has either its upper or lower lip removed. In *Linaria*, removal of the lower petal reduces visitation and lowers fruit production (Sánchez-Lafuente 2007a), and in *Lavandula* the effects of the removal of the upper flag were described in chapter 5. Damage by beetles to petals of *Lepidium* (Brassicaceae) almost halved the fruit set, although the flowers were still fully functional if hand-pollinated (Leavitt and Robertson 2006). Birds can be affected too, hummingbirds showing reduced visitation rate to damaged *Mimulus* flowers (Pohl et al. 2006).

As an example of effects on rewards, Krupnick et al. (1999) found that when florivores were excluded from *Isomeris* the remaining undamaged flowers often produced more nectar and more anthers; but the generality of this effect and its temporal and spatial patterning across a plant are largely unexplored.

The balance of effects of florivory on a plant's mating system could clearly be highly variable. Penet et al. (2009) pointed out that a florivore-induced decrease in flower number might lead to decreased selfing (less intraplant pollen movement) or to increased selfing (fewer visitors and more facilitated or autonomous selfing); in *Fragaria virginiana* they found that the latter predominated, but other species with different selfing abilities may respond in the other direction. The timing of damage is also critical to the overall outcome, so that florivory at the bud stage produced more reduction in fruit production in *Iris gracilipes* than did damage at the flower or early-fruit-set stages (Oguro and Sakai 2009).

2. Effects of Herbivory on Flowering and Pollination

The effects of herbivory can be either direct or indirect (Mothershead and Marquis 2000). *Direct* effects occur

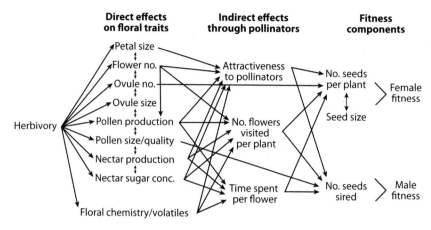

Figure 25.1 An overview of possible effects of herbivore activities on flowering, in terms of both male and female fitness. (Modified from Strauss 1997.)

where female fitness is reduced through lower seed production due to declining resource levels, after loss of leaf area. *Indirect* effects occur when a folivore mediates changes in floral traits that in turn influence pollinator preference or efficiency, so reducing pollen export and/or pollen receipt.

Direct Effects on Flowers

Herbivores almost by definition both damage the plant and deplete its resources. This can involve a whole range of effects at different scales:

1. Removal of large parts of the above-ground biomass, especially when grazing ungulates and other vertebrates are the perpetrators
2. More subtle effects on leaves from small insect herbivores that nibble away tissue (caterpillars, beetles, orthopterans, etc.)
3. Very minor surface damage but loss of photosynthate to insects that pierce and suck out fluids (aphids and other bugs)
4. Damage to below-ground roots, as yet a largely unexplored area of interactions, but just beginning to be documented (Blossey and Hunt-Joshi 2003; Poveda et al. 2003, 2005)

The relative effects of these various types and sites of damage have rarely been compared, although Poveda's group found that the greatest effects on *Sinapis* plants came from leaf herbivores, in part offset by increased nutritional inputs from underground decomposer activity.

Indirect Effects on Flowers

All the levels of damage outlined above can indirectly affect many of the floral features that also alter pollinator behavior and pollination outcomes. Studies to examine specific effects of leaf damage on flowers flourished for a decade or so from the mid-1990s and documented a whole range of effects, although earlier studies varied markedly in how specifically or rigorously they took account of the relative timing of (simulated) herbivory and flower development. It is only in the last few years that issues of timing have been taken into account, and that potential interactions with other selective agents have been added in to the analyses. The additional possibility of effects lasting beyond one season and impacting on lifetime reproductive success (e.g., Ehrlén 2002) has only very recently been fully considered. Nevertheless, it is worth reviewing the various effects on flowers that have been recorded thus far; figure 25.1 summarizes many of the possibilities.

Flowering Phenology and Flower Position
Strauss et al. (1996) found that moderate defoliation in *Raphanus* delayed flowering time, and Frazee and Marquis (1994) recorded similar delays in *Chamaecrista*. Reduced inflorescence height was recorded in damaged wild strawberries, though with no effect on number of open flowers per day (Cole and Ashman 2005). In *Erigeron*, caterpillar herbivory on apical buds led to increased growth of axillary buds and change of architecture but no overall change in flower number (Karban and Strauss 1993). Kliber and Eckert

(2004) found that herbivory in *Aquilegia canadensis* was the main adaptive reason for decline in allocation to pollen in successive flowers up a spike.

Flower Number and Duration

Reduced flower number is a rather common response to herbivory, occurring, for example, in *Oenothera* (Mothershead and Marquis 2000) and *Cucurbita* (Quesada et al. 1995) with simulated herbivory, and in *Rudbeckia* when fed on by spittlebugs, leading to reduced seed set (Hambäck 2001). Karban and Strauss 1993) also found a reduced flower number in *Erigeron* plants damaged by spittlebugs (although not when damaged by caterpillars, as above).

Size of Flowers and Floral Components

Size effects have been recorded in both directions in response to herbivores. Increased petal size occurred in wild strawberries (Cole and Ashman 2005), thus increasing pollinator attraction to damaged plants. But petal size decreased in *Raphanus* (Lehtila and Strauss 1999), and smaller female flower size was recorded in *Cucumis* (Thomson et al. 2004) and in *Oenothera* (Mothershead and Marquis 2000) where both diameter and tube length declined. In *Brassica*, petal size and stamen size were both reduced (Cresswell et al. 2001), and in *Chamaecrista* both ovules and stamens were reduced in size (Frazee and Marquis 1994).

Flower Color and Odor

Genes affecting floral pigments can have pleiotropic effects and affect the same pathways that produce defensive compounds to deter herbivores. Thus herbivores could indirectly affect flower color and lead to nonadaptive floral traits (Armbruster 2002). Herbivores may also function in the maintenance of flower color polymorphisms; for example, in *Claytonia* they exerted selection in the opposite direction to that of pollinators (Frey 2004), with flower color being relatively unconstrained by associations with any other floral trait (Frey 2007).

The demonstration that leaf damage also affected flower odor in *Cucurbita pepo*, with increased fragrance production in male flowers (Theis et al. 2009), may open up new research on odor as another mediator of pollinator-herbivore interactions.

Pollen Abundance

Anther number was reduced in *Isomeris* after herbivory by beetles (Krupnick et al. 1999), leading to substantially lower pollen production. Pollen decreases (in grain size as well as number) were also recorded in *Chamaecrista* (Frazee and Marquis 1994), *Raphanus* (Lehtila and Strauss 1999), *Alstroemeria* (Aizen and Raffaele 1998), *Cucurbita* (Quesada et al. 1995), and *Ipomoea* (Hersch 2006). Avila-Sakar et al. (2003) found increases in pollen grain number in *Cucurbita pepo* but only if leaf damage occurred after meiosis was already complete within the bud; they also showed that the increased pollen production occurred only where simulated damage was concentrated locally rather than dispersed (Avila-Sakar and Stephenson 2006), a variable that has rarely been explored before.

Pollen Performance

This can effectively be measured as the rate of pollen tube growth, and was shown to be reduced in *Cucurbita* (Quesada et al. 1995), *Alstroemeria* (Aizen and Raffaele 1998), and *Lobelia* (Mutikainen and Delph 1996) following leaf herbivory. Delph et al. (1997) reviewed this effect and its likely dependence on the provisioning of pollen grains.

Nectar Quantity or Quality

Undamaged flowers of *Isomeris* produced three times as much nectar as damaged flowers (Krupnick et al. 1999). But effects could also operate through plant defenses, with more alkaloids induced in nectar of *Nicotiana* after herbivore damage to the leaves (Adler et al. 2006).

Interacting Male and Female Effects

Whilst many earlier studies focused on particular plant traits, more recently some have attempted to compare overall male and female effects, given that it might be possible for plants to compensate for resources lost to herbivores by shifting allocation from seeds to pollen, or vice versa. In *Raphanus* the male flower traits were usually more affected than female traits (Lehtila and Strauss 1999), but damaged plants often showed greater success in siring seeds, probably because they increased relative allocation to flowers; the effect disappeared in an environment with more resources available (Strauss et al. 2001). In *Cucumis*, Thomson et al. (2004) found that herbivory (in this case by snails) during flowering produced more plants without any male flowers, so that there was less pollen export; the size of female flowers was also reduced but with no effect on pollen receipt, so that the decreased investment in male function could be seen as adaptive.

Hersch (2006) specifically tested relative effects of damage to male and female functions in *Isomeris*, showing that damaged paternal plants sired fewer offspring, while damaged maternal plants were more selective (i.e., exerted a greater degree of female choice about their partners) than undamaged ones. In orchids, male reproductive traits seemed to be more buffered than female traits against defoliation influences (Pellegrino and Musacchio 2006). However, in gynodioecious wild strawberry (*Fragaria*) the sexes differed in their susceptibility to spittlebug damage, with hermaphrodites more susceptible than females (Cole and Ashman 2005), and the same was true in *Geranium sylvaticum* (Asikainen and Mutikainen 2005); perhaps this helps to maintain females in the populations.

The effects listed above are known for only a limited number of plant species and mainly concern annuals subjected to simulated herbivory under greenhouse conditions. A few studies on perennials such as *Paeonia* (Sanchez-Lafuente 2007b) have shown rather little plasticity in resource allocation between sexual structures and functions, and in the shrub *Cnidoscolus* in Mexico defoliation tended to reduce the number of male flowers but floral visit rates were unaffected (Arceo-Gomez et al. 2009). However, studies on *Acacia* trees in East Africa (where herbivores were mainly grazing mammals) revealed effects on inflorescence size and anther number and hence pollen quantity (fig. 25.2; Otieno et al. 2010). Work on perennial *Helleborus* in Spain (Herrera et al. 2002) also showed important interactions of pollinator and herbivore selection regimes, with marked fitness increases across more than one season for plants that had the "best" combination of pollinator attraction and antiherbivore defensive traits. The most extensive review to date on a perennial is that of Ehrlén (2002) with *Lathyrus*, documenting complex interactions between mollusk damage to shoots, vertebrate damage to mature plants, seed predation, florivory, and pollination, but finding that some effects took more than one season to appear. In that light, much of what has been documented above as short-term effects probably needs revisiting.

These recent studies also highlight how unfortunate it may be that studies looking at specific reproductive traits have been less common recently, even though no clear patterns have yet emerged. Many of the analyses of effects on pollination from herbivores and other countering influences have moved on to a habitat or community scale, often with only seed outputs being

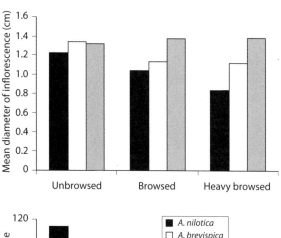

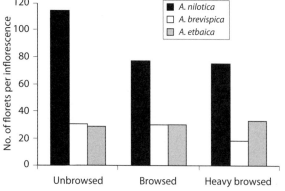

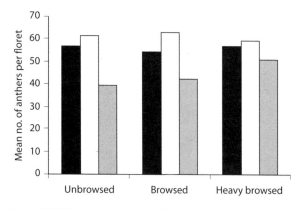

Figure 25.2 Effects of natural and artificially enhanced browsing levels, in three acacia tree species, on aspects of flowering, as compared with protected unbrowsed areas: (A) diameter of inflorescences, (B) florets per inflorescence, and (C) anthers per floret. Note that *A. nilotica* has smaller pollen amounts when browsed, largely due to smaller inflorescences, while *A. etbaica* shows some compensatory effects and produces more anthers per floret when heavily browsed. (Modified from Otieno et al. 2010.)

measured for particular plants; it is therefore almost impossible to know which mechanisms underlie a change of reproductive success in any particular scenario.

Indirect Effects on Pollinators and Their Visits

Damage to plants due to folivores or herbivores in any of the ways outlined in the last sections could potentially influence pollinator behavior, although rather few studies have pursued this. Damage that relates to attraction or to resources available from a plant (or a patch of plants) could reduce visits to the undamaged flowers as well as to the damaged ones. Strauss et al. (1996) recorded fewer pollinator visits to damaged plants of *Raphanus*, with fewer flowers probed and less time spent per plant, and Suarez et al. (2009) documented fewer and shorter flower visits on *Alstroemeria* plants after these had incurred foliar damage and as a result had fewer open flowers. Similar effects occurred in *Isomeris* (Krupnick and Weis 1999; Krupnick et al. 1999), where pollen beetles (*Meligethes*) had damaged flowers in ways that made whole inflorescences less attractive so they received fewer bee visits (53% fewer per plant and 68% fewer per surviving flower) and suffered 65% reduced pollen export.

The results described above for *Acacia* (Otieno et al. 2010) also showed indirect effects on pollinator visit timings, in particular affecting the period of bee visitation (fig. 25.3); the bees visited flowers on the browsed trees only later in the morning, after the more attractive and rewarding flowers on the nonbrowsed (herbivore-excluded) trees were presumably depleted. However, this kind of indirect effect may be rather uncommon. Vázquez and Simberloff (2004) found that nine out of ten species they tested showed no indirect effects of ungulate herbivory on pollinator visitation, pollen deposition, or reproductive output.

Effects on Selection

Selection against herbivory can at least sometimes completely overwhelm pollinator-mediated selection, as mentioned briefly in chapters 11 and 20. For example, Gómez (2003, 2005), working with *Erisymum* in Spain, showed that in areas with ungulates present pollinators had no effect on plant fitness, whereas pro-

tected areas (i.e., with herbivores excluded) showed good correlations of visitation rate (for beetles, bees, and syrphids) with plant fecundity. He also showed (2008) that the direction of selection from pollinators, herbivores, and gall-makers could vary markedly between years and between life-cycle stages.

Irwin and Adler (2006) tested the traits of distylous *Gelsemium* flowers that affected pollinators, flower robbers, and herbivores. They found that leaf and flower alkaloid defense levels were correlated, and that the thrum morphs of the plant received only half as much pollen as the pin morphs that had much weaker chemical defenses, so that traits associated with pollination and with herbivore resistance were not independent.

Matters are made more complicated where the pollinators are also the herbivores, as with many lepidopterans, where the larvae feed on the same plant that the adults pollinate and use for oviposition. The case of *Datura* and its dual interactions with *Manduca* moths presents an ideal study scenario here (Bronstein et al. 2009). For example, increased floral nectar production attracted more moths and more pollination but also led to more oviposition on the plant by the female moths (Adler and Bronstein 2004), giving directly conflicting selection pressures.

3. Defenses against Florivory and Herbivory Affecting Flowers

Physical and Chemical Defenses

Chapter 24 considered possible defenses for flowers against robbery, especially by ants, and a rather similar set of possibilities exists for defense against herbivores. Physical and chemical defenses are very well known for a huge array of plants, either or both being potentially *constitutive* at fixed levels depending on plant age, or *inducible* under the influence of herbivore damage (or indeed of receipt of volatile signals indicating damage to nearby plants).

Whereas physical defenses tend to occur in stems and sometimes calyces, so that enemies never reach the flowers, there are occasions when chemical defenses can occur in the flowers themselves. *Nicotiana* flowers developed more nicotine in the corollas when under attack by leaf herbivores (Euler and Baldwin 1996), and *Raphanus* flowers had higher glucosinolate

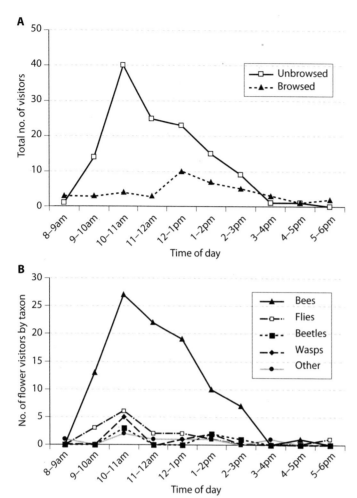

Figure 25.3 Effects of browsing levels on flower visitation in *Acacia nilotica*, where unbrowsed protected trees receive far more visitors (A), primarily due to the large early peak of bee visits seen in (B); on unbrowsed trees bee visits are fewer and later in the day, peaking around 1300. Similar effects occurred for *A. brevispica* and *A. etbaica*. (Modified from Otieno et al. 2010.)

levels when the leaves were damaged (Strauss et al. 2004). A particularly interesting example is the presence of cyanide defenses in the flowers of some *Hakea* species with red flowers pollinated by birds, although in the same genus those species pollinated by insects have no such floral chemical defense and are instead protected by leaf spines (Hanley et al. 2009). Thus the pollination and defense traits effectively covary, and the authors speculated that preexisting antiflorivore defenses may have been more important than selection for particular pollinators in the evolution of floral characters.

Direct effects on flower defenses when a plant undergoes florivory are much less well documented, but *Nemophila* flowers damaged artificially or attacked by caterpillars did gradually develop increased resistance, and suffered less subsequent petal damage (McCall 2006).

Biotic Defenses: Ant Guards and Ant-Plant Mutualisms

The phenomenon of biotic defense using ant guards deserves more attention here, as it often impacts directly on pollination ecology. Many plants are visited and patrolled by ants that feed from their extrafloral nectaries (chapter 24, section 2, *Selective Effects: Protection against Theft*). But there are more specific and interesting relationships where coevolution between particular ants and specific plants has occurred, going well beyond simple bribery scenarios (e.g., Beattie 1985). Best known are the acacia trees, part of a large genus (more than 700 species, although sometimes split into three genera) present as trees and shrubs in all the southern continents. Acacias are often the only trees among dominant grasses in savanna habitats and are well known for being very spiny, providing a defense against large vertebrate herbivores. About 10% of the spiny acacias have a proportion of their thorns greatly expanded, and these swollen *pseudogalls* provide homes for aggressive ants, especially in the genera *Crematogaster* (Africa) and *Pseudomyrmex* (New World). These acacias are an example of **myrmecophytic** plants, and the antguard-myrmecophyte interaction is close to being a real mutualism where the relationship is obligate for the ants.

In the chosen hollowed-out thorns the ants form small colonies each with its own queen; there is not normally a species-specific interaction, and any ant colony is relatively unfussy about which tree it lives in. The workers patrol the tree and bring in nectar from the EFNs on the stems or petioles (outside the flowers) and also small insect herbivores that have attempted to nibble at the leaves. These ant-defended plants sometimes offer EFN nectar with higher amino acid levels and/or reduced sucrose, both of which are known to be preferred by ants (Heil et al. 2005). The ants depend on the plant for shelter but also for virtually all of their food; in a few New World cases extra nutrition is offered by special proteinaceous food bodies at the tips of leaflets (Beltian bodies) that are chewed by the ants.

In return for offering food and shelter, the plant gets protection, as the ants will aggressively attack any herbivore that tries to eat the plant leaves, not only other insects but even mammals (bovids, giraffes, gazelles, even rhinoceroses, in Africa). And when an ant detects an intruding herbivore it not only bites at it itself but liberates an alarm pheromone that brings in many other ants who also attack. At any one time 25% of the ants may be out of their thorns patrolling their host plants, but up to 50% (essentially all the old workers) may come out in response to attack under the influence of the alarm pheromone. Ants may also nip off vegetation such as lianes and vines that may try to scramble over their tree; and they tend to destroy all seedlings growing within a basal ring around their tree's trunk,

perhaps within a 1–2 m diameter, so that this cleared space may also help the acacia trees resist the wildfires that sweep across the habitats they grow in. Clearly both partners—ant and tree—benefit.

There are many other examples of myrmecophytic plants, and no doubt many more to be described, but the well-studied examples beyond the acacias are

1. *Cecropia* from Central America, producing both EFNs and two other rewards for its ants (glycogen-rich Müllerian bodies at petiole bases, and lipid-rich pearl bodies on the undersides of leaves),
2. *Macaranga* from southeast Asia, with EFNs, and
3. *Leonardoxa* from West Africa, again with EFNs.

It was originally assumed that ant-guarded plant species gained by being able to invest less heavily in chemical defenses, that is, they made some saving on other defenses by paying off the ants instead. When herbivores were excluded, certain African acacia trees did have fewer and shorter thorns and fewer EFNs (Huntzinger et al. 2004), implying that the ants were a useful but costly defense. However, Heil et al. (2002) assessed three ant-plant genera and found a good negative trade-off relation (biotic versus chemical) only for *Leonardoxa*, and not for either *Acacia* or *Macaranga*. Thus use of ants is expensive and does not necessarily save on other defenses. In fact the use of ants as bodyguards may have evolved in relation to *different* herbivores; fibrous and tannin-rich leaves may protect against generalized herbivores such as orthopterans, while the ants are particularly good against specialist feeders such as flower-feeding beetles, caterpillars, and bugs. In addition, biotic defenses can sometimes be made more efficient to reduce their costs; for example, *Leonardoxa* can produce an ant attractant specifically in its young leaves, ensuring that ants mainly patrol there and give maximum benefit (Brouat et al. 2000). Extrafloral nectar can be kept less costly by being inducible, both in *Macaranga* (Heil et al. 2001) in response to jasmonate released only when the plant is damaged, and in *Catalpa* plants when ant tended (Ness 2003).

There are, however, some problems with the use of ant defense to protect flowers; how does an ant-patrolled acacia tree achieve pollination and seed dispersal, the two aspects of its reproduction for which it might be expected to rely on other animals having safe and uninterrupted access? In most *Acacia* species the

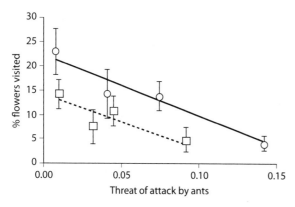

Figure 25.4 The percentage of flowers visited by pollinators in relation to the threat of ant attack on those flowers (a composite of percent occupation by ants, mean number of ants per flower, and per capita effectiveness of ants as deterrents) for *Ferocactus*; squares show data for 2003, and circles for 2004. (Redrawn from Ness 2006.)

seeds are large, pendant, and tough, and are probably bird- or mammal-dispersed; these dispersers are likely to be little bothered by the ants, and several examples of ant-guarded plants having higher seed set than unguarded congeners are now known (e.g., Willmer and Stone 1997a; Wagner 1997, 2000).

But letting in the pollinators is a different matter. Particularly aggressive ants patrolling on a plant are reported to deter some flower visitors (e.g., Tsuji et al. 2004; Gaume et al. 2005; Ness 2006); figure 25.4 shows an example for *Ferocactus* flowers tended by ants. The invasive Argentine ant *Linepithema humile* is a particular problem, and can reduce honeybee visitation by 75% on some Proteaceae (Lach 2008). *Acacia* flowers (plate 8A) have their pollen superficial and very easily accessible and are visited by a wide range of insects, but they have several ways of keeping ants away from the flowers. Willmer and Stone (1997a) showed that a novel system of chemical exclusion occurred, where ant-deterrent volatiles were emitted from flowers (the emission is now known to be specifically from the pollen; Willmer, Nuttman et al. 2009). Thus ants that do trespass briefly onto young flowers show alarm behavior (plate 20G). The repellence is highly regulated in space and time, commencing as flowers become attractive to pollinators, and declining conveniently in proportion to visitation and pollen removal (fig. 25.5). Pollen-based repellence is particularly strong in the ant-guarded acacias and is specifically targeted at the resident ants. Moreover, in at least some

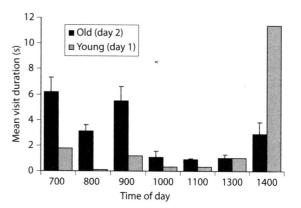

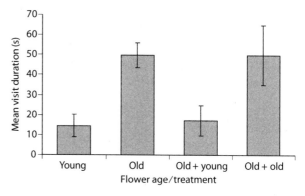

Figure 25.5 (A) Visit duration of resident ants (*Crematogaster*) to young and old *Acacia zanzibarica* flowers, with only very short visits to young dehiscing inflorescences until late in the day when pollen is depleted. (B) Deterrence from young flowers can be transferred to old ones by wiping, so that an old flower wiped with a young one (O+Y) receives only very short visits thereafter. Both show means ± SEs. (Redrawn from Willmer and Stone 1997a.)

ant-guarded species the floral bouquet incorporates an ant alarm pheromone component, E,E-α-farnesene, which while deterrent to some ants is also known to be an attractant for some bees (Blight et al. 1997; Valterova et al. 2007). The volatile emissions provide transient highly focused protection from ants for the sparse and valuable young inflorescences, but still allow the ants to return and protect older postpollination flowers as seed set commences (Willmer, Nuttman et al. 2009). The net result is that the presence of ants can improve fruit set (fig. 25.6), so that ant guards benefit reproduction. The same strategy of pollen-based chemical deterrence of ants is now known to occur in several other Kenyan acacia species (Willmer, Nuttman et al. 2009) and in *Acacia constricta* (Nicklen and Wagner 2006). However, some other ant-guarded acacias use an alter-

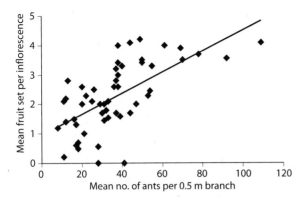

Figure 25.6 The presence of ants can aid fruit set in *Acacia zanzibarica*, because older pollinated flowers receive protection as they set seed, and the potential deterrence of pollinators by aggressive ants is resolved by the short-term volatile deterrent from dehiscing flowers as in figure 25.5. (Redrawn from Willmer and Stone 1997a.)

native strategy of temporal and spatial patterning of their rewards to manipulate ant distributions and so keep them away from young flowers (Raine et al. 2002; Gaume et al. 2005).

Quite how widespread is ant repellence in flowers remains unclear, although chapter 24 reviewed available evidence when considering ants as robbers. There are likely to be patterns linked with floral type, habitat, and phenology, and figure 24.3 (Willmer, Nuttman et al. 2009) showed some apparent trade-off between morphological exclusion features on and around flowers and the strength of volatile repellence emanating from those flowers.

4. Overview

How are the conflicts between pollination and herbivory resolved? In essence, the problem for the flower is one of attracting the right visitors and deterring the wrong ones. Most flowers are producing multiple advertising signals, and since these may attract atten-

tion from many animals the flowers are as a result likely to receive visitors with opposing—beneficial and deleterious—effects. Conflicting selection resulting from the two main kinds of visitor—pollinators and herbivores—is quite common and increasingly recognized (see Strauss and Whittall 2006); zoophilous flowers are almost by definition meant to be visible and attractive and may draw the notice of many non-pollinators to an otherwise inconspicuous or dull foliage plant. Table 25.1 shows a range of examples of potentially conflicting selective pressures (although there are also occasional complementary examples). Other specific cases have been considered by Euler and Baldwin (1996), Armbruster (1997), and—on a plant lifetime basis—Ehrlén (2002).

What can a plant do about this conflict? It may be possible to adjust the temporal signaling from flowers to favor pollinators without unduly attracting florivores or herbivores; an example from chapter 6 is the thistle *Cirsium arvense* which produces maximum fragrance at the times of maximum pollinator activity and low fragrance when folivores are active (Theis et al. 2007), and more examples of this kind must surely await discovery. Or flowers can use defenses and deterrents to manage their visitors in the ways that we have seen. Defenses specifically against florivores are rather similar to those against cheats (see chapter 24); large tough sepals, trichomes on sepals or bracts, and water moats formed by bracts can all be highly effective against small caterpillars and other florivores. As a specific example, some *Dalechampia* species have movable bracts that can close around the flowers at night and prevent up to 90% of nocturnal florivory (Armbruster 1997; Armbruster et al. 1997). Other floral movements are likely to have similar effects, and a whole range of diurnal or postvisitation changes in flower shape, orientation, scent, and color may well be interpretable in this light in future, aimed at making a pollinated flower inconspicuous to enemies as soon as possible after it has served its reproductive purpose.

TABLE 25.1
Examples of Selection by Pollinators and Herbivores on the Same Flower Traits

Trait	Pollinator preference	Concomitant herbivory effects of preferred trait	Plant	Reference
No. of flowers	High no. per spike preferred by bats	More damage by orthopterans	Calyptrogyne ghiesbreghtiana	Cunningham 1995
	High no. per plant preferred by bees	More weevil damage	Fragaria virginiana	Ashman et al. 2004
Flower size	Larger flowers	More weevil damage	Fragaria virginiana	Ashman et al. 2004
	Larger flowers	More florivory	Nemophila menziesii	McCall 2006
Flower shape	Bees prefer open flared corollas	More damage by ants	Polemonium viscosum	Galen and Cuba 2001 Galen and Butchart 2003
	Bees prefer longer petals	Ungulates indirectly select against long petals	Erysimum mediohispanicum	Gómez 2003
	Hummingbirds prefer long corollas	More robbery by bumblebees	Campsidium valdivianum	Urcelay et al. 2006
Flower color	Pale or white preferred over mauve	Pale or white damaged more	Raphanus raphinastrum	Strauss et al. 2004
Plant height	Taller spikes	More damage by ungulates	Erysimum mediohispanicum	Gómez 2003
Calyx shape	Shorter	Attract more seed predators	Castilleja linariaefolia	Cariveau et al. 2004
Nectar volume	Hawkmoths prefer high volumes	Hawkmoth larvae do more leaf damage	Datura stramonium	Adler and Bronstein 2004

1. Generalist flowers and visitors. (A) Beetle on *Spiraea*; (B) hoverflies and butterfly on *Spiraea*; (C) *Rhagonycha* beetles mating on *Heracleum* (SE); (D) small dipterans on *Caltha*; (E) *Apis* on *Skimmia*; (F) *Apis* on *Bupleurum* (GB); (G) *Chloroselas* butterfly and bluebottle flies on a euphorb (DM); (H) generalist flowers of the tropical *Erythroxylon*.

2. Open radial bowl-shaped flowers allowing generalist visitor access. (A) *Magnolia*, not fully open; (B) *Paeonia* hybrid; (C) *Cistus*; (D) *Eucryphia*; (E) *Helianthemum*; (F) *Rhododendron* hybrid; (G) *Hepatica*; (H) *Geranium* with small dipteran visitors.

3. Deeper bowl-shaped flowers. (A) *Ipomoea*; (B) *Campanula* from the front, showing the revolver flower effect with five separate access points to the nectar (see text); (C) *Abutilon*, also with a revolver effect; (D) *Hibiscus* and (E) *Lavatera*, both with petals separated basally; (F) deeper bowls in *Schizostlyis*; (G) *Crocus* (note the frilly conspicuous style); (H) elongate flared bowl in *Datura*.

4. Pendant flowers (A–F) and two examples of complex radial flowers (G–H). (A) *Erythronium*; (B) *Fritillaria*; (C) *Cyclamen*; (D) *Campanula* (sw); (E) *Galanthus*; (F) *Soldanella*; (G) *Tricyrtis* (toad lily) with complex central style ending in three branched stylodia set just above six pendant stamens; (H) *Nymphaea* water lily, showing the two phases: on the right, a day-1 female phase apparently with a pool of nectar centrally, in which visitors may be trapped overnight, and on the left, the later male phase with pollen.

5. Complex radial flowers (A–B), spherical flowers (C–G), and a bilaterally symmetrical example (H). (A) *Passiflora edulis* with frilled lilac-tipped coronal filaments adding to the display beneath the five horizontal anthers and three-part style, with *Bombus terrestris* feeding from the nectar chamber; (B) *Passiflora incarnata* with more elaborate corona; (C) *Eucera caspica* bees mating on *Muscari* flowers (AG); (D) *Andrena lapponica* feeding on blueberry flowers (*Vaccinium*) (AG); (E) *Arbutus* flowers; (F) *Erica*, a tree heather; (G) butterfly (with deceptive "eyes" at its rear wing tips) feeding on an unidentified spherical Costa Rican flower (GB); (H) *Dicentra* flowers with pink bilaterally symmetrical outer petals, deflected sideways at their tips to form nectar guides, with the white anthers and stigma protruding below.

6. Different forms of radial tubular flowers, often with a terminal flare. (A) *Daphne*, where the long calyx extends the tube; (B) *Gentian*; (C) *Primula* flowers with heterostyly, having pin (three, to upper left) and thrum (lower right) morphs; (D) flowers of *Azadirachta*, the neem tree, with a central corolla tube; male and bisexual flowers may occur on the same tree; (E) *Narcissus* (daffodil) with outer petals and central corona; (F) *Nerium oleander* with short, frilled corona; (G) *Plumbago* with the elongate corolla protected by a trichome-bearing calyx; (H) highly scented *Stephanotis* flowers where the long corolla is guarded with a ring of internal hairs.

7. A variety of zygomorphic tubular flowers. (A) *Phlomis* with visiting bumblebee (only large *Bombus* and *Xylocopa* are strong enough to enter the flower legitimately, and sturdy bracts protect the base of the corolla from theft); (B) *Teucrium*; (C) *Salvia*; (A–C) are all Lamiaceae, the latter two with an expanded lower petal; (D) *Aconitum* with dorsal hood containing the petal nectary (Ranunculaceae); (E) *Corydalis* (Scrophulariaceae); (F) *Ancharina*; (G) *Tapinanthus* (a parasitic plant); (H) *Anigozanthos* (kangaroo paw).

8. Brush blossoms and composites. (A) *Acacia brevispica*, with a freshly opened day-1 flower and wilting day-2 flowers; (B) *Albizia* with many neuter florets and a single larger central floret with ovules and copious nectar; (C) *Maerua* with incoming *Xylocopa* bee (DM); (D) *Callistemon* (bottlebrush) with visible nectar in the base of each floret; (E) *Mimosa pudica* with anthers clearly visible on the tip of each filament; (F) the two classic composite flower types: thistle, with honeybee visitor, and daisy; (G) *Lasianthaea*, clearly showing the separate maturing disk florets centrally (GB); (H) *Senecio* with the syrphid *Episyrphus balteatus* feeding on central disk florets.

9. Composites(A–B), flag flowers (C), and keel (legume) flowers (D–H). (A) *Erigeron*, young and old blossoms; (B) *Cicerbita*, a unicolored composite; (C) *Iris* hybrid with enlarged fall sepals and erect petaloid stigmas; (D) *Ulex* (gorse) flower with upper flag and lower keel petals; (E) *Osmia cephalotes* bee visiting a *Genista* flower (note the long tongue inserted to probe for nectar) and tripping the keel petal so that the stigma and anthers are exposed and pollen deposited on the lower abdomen (AG); (F) *Desmodium* flowers after tripping by bee visits, with deposited pollen clearly visible on the stigma to the right; (G) *Wisteria* inflorescences, with sprays of legume-type flowers; (H) the bee *Eucera longicornis* foraging for nectar on the sweet pea *Lathyrus* (AG).

10. Flower spikes (A–E) and orchids (F–H). (A) *Mentha* spike with a butterfly visitor; (B) *Acanthus* spike with protective spiny calyces; (C) *Veronica spicata*, the spike maturing from the base upward; (D) *Echium* flower spike; (E) *Plectranthus scutellariodes*, with just one to two whorls of flowers maturing at a time from the base upwards (ɢʙ); (F) *Habenaria radiate*, fringed orchid; (G) *Cymbidium* hybrid orchid; (H) *Cypripedium* slipper orchid.

11. Inflorescences of various kinds. (A) *Viburnum* with an *Eristalis* hoverfly visitor; (B) *Scilla peruviana* with florets maturing from the outside inward; (C) *Buddleia globosa*; (D) the umbellifer coriander (*Coriandrum*) with the display enhanced by enlarged outer petals on the outer protandrous flowers, the inner flowers being male or sometimes sterile; (E) *Hydrangea* hybrid with enlarged and brightly colored outer florets enhancing the display but usually being sterile; only the small inner flowers produce seed; (F) *Dalechampia*, with enlarged bracts forming the display and a group of small inflorescences centrally; (G) clumped *Rhododendron* flowers, each with a dorsal dark spot acting as a nectar guide; (H) *Lavandula stoechas*, where a pair of large purple bracts act as an advertising flag above the spike of darker corollas.

12. Variety in stamens and styles. (A) *Commelina* showing heteranthy with blue reproductive stamens and showy yellow feeding stamens; (B) the same *Mahonia* flower before (left) and after (right) being probed, the anthers moving inward after contact is made; (C) a hybrid lily (*Lilium*) showing the six "seesaw"-mounted stamens; (D) *Cassia* flower with pronounced heteranthy—two large, curved reproductive anthers that contact a visitor's abdomen and several smaller central feeding anthers; (E) heteranthy in *Dissotis*, with the two types of anthers distinctively colored and located (purple, reproductive; white, feeding); (F) *Clerodendron* flower with asymmetric bilobed style, deflected laterally; (G) extreme style deflection in *Gloriosa superba*, thus siting the triple stigma lobes in the same area as the anthers; (H) highly elongate and asymmetric style in a pendant *Hibiscus* flower, with the anthers emerging from the sides of the stylar column and the stigmatic surfaces themselves being hairy (sw).

13. Different methods of color-based advertising in flowers. (A) Buttercup (*Ranunculus*) flower with shiny upper surfaces and matte lower surfaces; (B) *Schlumbergera* with intense magenta coloration due to betalain pigment; (C) intense and similar colors in petals and sepals of *Epiphyllum*; (D) *Caesalpinia* flowers with intensely colored elongate stamens contrasting with the paler petals; (E) *Ochna* with the stamens forming a main part of the advertisement; (F) *Warszwiczia coccinea*, where the enormous red bracts form the main display and the flowers are rather insignificant as signals (GB); (G) *Salvia* "Black Knight" hybrid, with petals and sepals together providing a strong color signal; (H) *Triplochlamys*, with dark red bracts, largely hidden petals, and a protruding set of strongly colored red stamens and lilac pollen as additional advertising signals.

14. Examples of two contrasting colors acting together in flowers. (A) *Abutilon megapotanicum* with red sepals and yellow petals; (B) *Geranium* with a darker outer ring and paler inner ring of color contrasting with the dark anthers (also with radiating nectar guide lines); (C) *Thunbergia battiescombei* with blue petals peripherally contrasting with translucent yellow petal bases, seen from the front as a "window" effect; (D) *Hibiscus* hybrid with three concentric color rings; (E) *Musa* inflorescence, with dark red bracts and a cluster of small yellow flowers below; (F) *Trillium* flower with white petals alternating with green sepals, and a strongly colored central column bearing the stigmatic lobes; (G) flowering currant, *Ribes*, younger flowers being all pink with contrasting white anthers but ageing to white petals contrasting with the sepals and with the green style gradually elongating; (H) *Dichrostachys*, a mimosoid legume related to acacias but with the lilac upper flowers sterile and strongly contrasting with the functional yellow flowers below.

15. Nectar guides in a variety of flower types. (A) *Streptocarpus* with guide lines converging into the tubular corolla on the ventral "landing platform" petal; (B) *Alstroemeria* with short converging lines on the upper petal; (C) *Iris* with nectar guides on three sepals ("falls"); (D) *Thunbergia grandiflora*, with conspicuous stripes on the lower and lateral petals and a contrasting yellow interior; (E) *Phylloctenium*, with cauliflorous flowers, mainly visited by anthophorine bees and having three strong, converging nectar guide lines; (F) *Podranea* with stripes on all petals and centrally converging stamens; also bearing hairs on the lower corolla interior; (G) *Digitalis* with multiple nectar guide spots, which are denser and larger on the lower petal interior; (H) *Cistus* with large contrasting single spots at the base of each petal. Note that in (B–E) the contrasting central yellow coloration is sometimes taken to be a form of pollen mimicry (chapter 23).

16. Examples of floral color change in different flower parts. (A) *Lonicera japonica* hybrid, with the whole flowers ageing from white through cream to yellow (note the style initially lying below the anthers but ending up well above them); (B) *Lantana* hybrid with central young pale yellow flowers and older peripheral pink flowers; (C) an *Aesculus* species where the banner spot on the upper petal is initially a contrasting yellow but changes to dark pink after visitation; (D) *Pulmonaria* with pink buds turning to purple and then blue; (E) *Koehneria* where the color change occurs in stamens, initially white/cream (upper left flower) but then turning mid-pink and finally dark pink; (F) *Myosotis*, with color change in the coronal scales that protect the corolla tube entrance, initially yellow but with the change to white accelerated by pollination; note that the petals also change from pink to blue, but this is purely age-related; (G) *Parkinsonia* where the most dorsal petal changes from yellow to deep orange soon after visitation; (H) *Lupinus pilosus* where the banner spot changes from white to dark pink after visitation.

17. Buzz-pollinated flowers. (A) *Borago*, with a honeybee (unable to sonicate the anthers for pollen) foraging for nectar; (B) and (C) *Solanum* (front and side views) showing the protruding composite anther cone that is grasped by a sonicating bee; (D) similar anatomy in a tomato flower (*Lycopersicon*, now included in *Solanum*); (E) *Dodecatheon* (shooting star) flowers, again with the protruding anther cone; (F) *Dianella* (flax lily), with six stout protruding anthers; (G) *Vaccinium* (blueberry), with bell-shaped flowers where sonication is not required but is often used by visiting bees; (H) *Arctostaphylos*, where again sonication is optional but can improve pollen gathering.

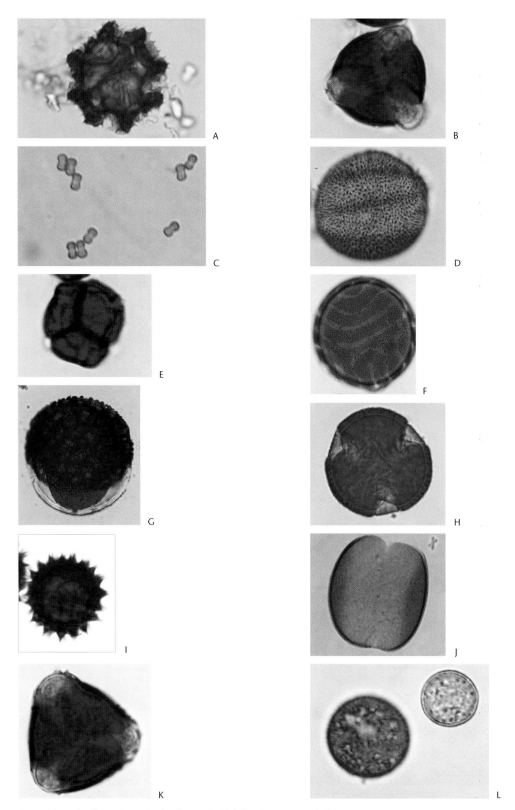

18. A variety of pollen grains stained and viewed with light microscopy, with diameters. Reproduced with grateful thanks to John Chapman, Harrogate and Ripon Beekeepers Association (from whom CD copies are available on request for teaching purposes j.a.chandler.479@btinternet.com). (A) *Taraxacum* 30 μm; (B) *Acer* 37 μm; (C) *Myosotis* 5 μm; (D) *Crocosmia* 70 μm; (E) *Erica* 37 μm; (F) *Crocus* 150 μm (G) *Lilium* 85 μm (H) *Helleborus* 37 μm (I) *Aster* 27 μm (J) *Narcissus* 60 μm (K) *Cotoneaster* 35 μm (L) *Primula* thrum 30μm, pin 20μm

(M) *Carduus* 40 μm (N) *Rosmarinus* 35 μm (O) *Viola* 75 μm (P) *Phlox* 45 μm (Q) *Symphytum* 25 μm (R) *Cucumus* 150 μm (S) *Helianthus* 42 μm (T) *Centaurea* 37 μm (U) *Valeriana* 70 μm (V) *Pinus* 50x75 μm (W) *Alnus* 20 μm (X) *Scrophularia* 32 μm.

19. Nectaries, nectar, and other rewards. (A) Large nectaries at petal bases with visible globules of nectar in *Fritillaria imperialis* flowers; (B) a ring of green fingerlike petal nectaries with narrow slit openings in a *Helleborus* flower (note also the successively dehiscing whorls of anthers); (C) nectar spurs protruding at the back of *Viola* flowers; (D) *Corydalis* flowers with nectar spurs rising at the back of the corolla; (E) long nectar spurs at the rear of *Tropaeolum speciosum*, which is native to Chile, where it is visited by hummingbirds; (F) long, narrow nectar spurs in an *Impatiens* species; (G) the invasive weed *Impatiens glandulifera* with much shorter nectar spurs, notorious for very high nectar production and prone to overflow; (H) a *Lysimachia* flower visited by its specialist pollinator bee *Macropis europeaea*, which collects oil from the orange-colored elaiosome tissues (see fig. 9.1A) as well as receiving pollen on the legs and abdomen (AG).

20. Ants, wasps, and other mainly nonspecialist visitors. (A) A *Hoya* species with many ants gathering nectar (note also the color change occurring in the older flowers); (B) *Camponotus* ants foraging on a *Euphorbia* (DM); (C) *Crematogaster* ants on *Gurania makoyanna* (GB); (D) a typical "wasp" flower, *Scrophularia*; (E) a vespid wasp taking nectar on *Callistemon* (KS); (F) a specialized case: *Hemipepsis*, a pompilid wasp, acting as the specialist pollinator on *Pachycarpus grandiflorus* (SJ); (G) a *Crematogaster* ant straying briefly onto an *Acacia zanzibarica* flower; since this flower has ant-deterrent properties, the ant has the cocked abdomen, rapid antennation, and scuttling action typical of alarm behavior; (H) a snail using its radula to rasp at the spathe of *Calla*, an aroid.

21. Beetles as flower visitors. (A) *Cistus* flower with relatively destructive feeding by four *Oedemara* beetles (IJ); (B) beetles feeding on the spathe of an aroid, *Calla*; (C) a South African *Ixia* flower being pollinated by monkey beetles (hopliine scarabs) (SJ); (D) turquoise *Oedemera nobilis* beetle feeding on *Lavandula*; (E) large chafer beetle feeding on tissues of a euphorb (note that ants are also present taking nectar) (DM); (F) numerous *Meligethes* pollen beetles feeding in a *Hypericum* flower; (G) soldier beetle (probably *Trichodes*) feeding on *Eryngium* flowers (a vespid is also present); (H) destructive feeding by beetles on rosaceous flowers (IJ).

22. Carrion flowers. (A) *Aristolochia* flowers, with the frontal maroon-and-white display of a mature flower, and the pouched corolla tube that lies behind this seen in younger flowers; (B) and (C) small and large species of *Ceropegia*, *C. woodii*, and *C. sandersonii*, each with five petals fused at the tip to form a lantern-like structure with "windows" into the corolla below (note that the petals are hairy, as is the inner corolla tube); (D) *Arisaema* flower; (E) the giant titan lily, or corpse flower, *Amorphophallus*; (F) the dragon lily, *Dracunculus* (sw); (G) the enormous *Rafflesia* flower (the booted socked feet to the right give the scale) (AF); (H) inside the *Rafflesia* flower, showing the white "windows" under the rim (AF).

23. Flies visiting various flowers. (A) hoverfly on *Scabiosa*: honeybee mimic, male *Eristalis*; (B) two different hoverflies on an *Echinops* inflorescence; (C) *Eristalis* female on the spathe of a *Calla* aroid; (D) *Episyrphus* hoverfly on a composite, again revealing the classic hoverfly venation; (E) *Syrphus* hoverfly feeding on *Philadelphus* pollen; (F) the poppy *Papaver sendtneri* with two muscoid flies; (G) a South African long-tongued tabanid fly taking nectar from *Watsonia* (sɪ); (H) *Bombylius major*, a bee-fly, with its long proboscis inserted in a primrose.

24. Butterflies as flower visitors. (A) *Papilio palinurus* on *Lantana*; (B) a peacock (*Inachis io*) on thistle (SE); (C) unidentified butterfly on *Callistemon* in Uganda (KS); (D) close-up of a Red Admiral (*Vanessa atalanta*) feeding on *Buddleia* (SE); (E) lycaenid butterfly on heliotrope flowers (DM); (F) a Painted Lady (*Vanessa cardui*) on *Ceratostigma*; (G) Painted Lady (underside seen) on clover (SE); (H) a swallowtail (*Papilio machaon*) feeding on *Lavandula*.

25. Moth flowers. (A) unidentified moth flower from Madagascar, possibly *Leucosalpa*; (B) *Rothmannia*, also from Madagascar (note that *Stephanotis* in plate 8H is another classic native moth flower of Madagascar); (C) daytime-flying European hummingbird hawkmoth (*Macroglossum stellatarum*) foraging on *Lavandula*; (D–H) are all nocturnal hawkmoth pollinators, mainly African: (D) *Nephele* foraging on *Grewia* (DM); (E) *Hippotion celerio* on *Plumbago* (DM); (F) two brown *Hippotion celerio* and a green *Basiothia medea* on the asclepiad *Pergularia* (DM); (G) *Agrius* on *Turraea* (DM); (H) *Coelonia* moth feeding on *Bonatea* orchids (DM).

26. Typical hummingbird flowers (A–G). (A) *Castilleja* (Indian paintbrush; western USA and southwestern Canada); (B) *Hamelia patens* from Costa Rica; (C) *Mitraria coccinea* (native to Chile); (D) *Fuchsia boliviana* (western South America); (E) *Heliconia* (tropical Americas); (F) *Columnea* (Central and South America); (G) *Passiflora coccinea* (South America); (H) herkogamy in *Aeschynanthus* (Southeast Asian origin, therefore not visited naturally by hummingbirds), young flower below just dehiscing, with style still hidden; older flower above with style extended.

27. Some typical perching bird flowers. (A) *Strelitzia* (bird of paradise) with an unstable horizontal perch formed by the bracts containing sticky exudate, holding clusters of the small bluish flowers dispensing liberal white pollen that gets transferred on the birds' feet; (B) *Chadsia*, visited by sunbirds in Madagascar; (C) *Spathodea* (African tulip tree) with visiting sunbird in Uganda (bees, other insects, and some bats may also visit) (KS); (D) a *Protea*, typically visited by sunbirds in southern Africa; (E) *Watsonia* (see also plate 23G); (F) a sunbird foraging on an African aloe inflorescence (DM); (G) *Techoma capensis* from southern Africa; (H) a bananaquit foraging on *Stachytarpheta* (GB).

28. Flowers associated with bat visitation and with nonflying mammals. (A) *Cobaea*, initially green and with a cabbagelike smell, but later purple and odorless as the anthers dehisce; (B) *Kigelia* (sausage tree) with long pendant inflorescences of cream- to maroon-colored flowers (CN); (C) *Merinthopodium neuranthum* from Costa Rica (GB); (D) *Crescentia*, with large cauliflorous extremely odorous flowers, from Central America (GB); (E) a baobab (*Adansonia*) from southern Africa, some species being more pendant than this one (SJ); (F) a gerbil foraging on *Massonia* (SJ); (G) a rodent feeding at *Protea acaulos* (SJ); (H) a black-and-white Colobus monkey eating *Erythrina* flowers (KS). (E–H all photographed in southern or eastern Africa.)

29. Smaller bees and flowers. (A) a small and short-tongued halictid bee feeding on *Helianthemum*; (B) a leaf-cutter bee, *Megachile pilicrus*, foraging on *Centaurea*, with substantial pollen visible on its abdominal scopa (AG); (C) *Dasypoda hirtipes* on *Leontodon*, with dense pollen packing in the long hairs of the hind-leg scopae (AG); (D) blue-striped *Amegilla* bee feeding on *Lantana* (SW); (E) stingless *Trigona* bee foraging on *Musa* flowers (GB); (F) *Trigona* approaching *Plectranthus* (GB); (G) *Anthophora plumipes* males queuing to mate with a (labelled) foraging female on comfrey (*Symphytum*) (GS); (H) *Andrena fulva* feeding on a *Salix* catkin (AG).

30. Larger bees and flowers. (A) *Systropha* feeding on *Ipomoea*, with pollen on legs and abdomen (DM); (B) a typical metallic green euglossine with exceptionally long tongue and expanded hind tibia for fragrance collection; (C) *Xylocopa* (carpenter bee) foraging on *Stachytarpheta* (GB); (D) *Xylocopa* on pigeon pea (DM); (E) *Xylocopa* on *Disa graminifolia*, a genus with very different pollinators for different species (SJ); (F) *Apis* forcing entry into the constricted corolla of a snapdragon (SE); (G) *Bombus lucorum* on *Eryngium*; (H) *Bombus lucorum* on *Scabiosa*.

31. Bumblebees showing versatility on many different flower types. (A) *Bombus pascuorum* on *Verbena bonariensis*; (B) the long-tongued *B. hortorum* on *Lonicera*; (C) *Bombus terrestris* taking nectar on *Philadelphus*; (D) nectar-feeding on *Echium*; (E) *B. pascuorum* on *Lamiastrum*; (F) European bumblebees attempting to forage on introduced purple *Passiflora*; the bee to the right has inserted its tongue into the tightly enclosed nectar chamber and is temporarily stuck, unable to extricate it due to the size mismatch; (G) bumblebees foraging on introduced *Protea*, normally visited by sunbirds; (H) a *Bombus* visiting and pollinating the nectarless flowers of the orchid *Anacamptis morio*, which attract bees by coflowering with rewarding flowers such as *Allium* (SJ) (S. Johnson, Peter et al. 2003).

32. Different forms of mimicry associated with flowers and flower visitors. (A) *Begonia* flowers showing intersexual mimics, female flowers (left) bearing styles that resemble the pollen-bearing anthers of a male flower (right); (B) *Disa pulchra* (left), a mimic of *Watsonia lepida* (right) in southern Africa (other *Disa* species mimic other genera) (sj); (C) *Linaria* with bright yellow pseudopollen on raised areas of the lower petals; *Impatiens* flower bearing yellow pollen mimics; (D) *Impatients* flower bearing yellow polen mimics; (E) *Iris* flower with the typical "beard" of antherlike yellow tissue; (F) *Parnassia* with glistening pseudonectaries; (G) and (H) show predatory animals waiting cryptically on flowers by mimicking flower structures and colors, both having successfully captured potential pollinating visitors: (G) a white crab spider on white flowers (GB); (H) a green mantid on yellow-green flowers (GB).

33. Examples of theft and robbery on flowers. (A) *Xylocopa* acting as a primary nectar robber; (B) a robber biting the corolla on *Stachytarpheta* (GB); (C) bee stealing pollen on a *Penstemon* flower; (D) syrphids feeding on *Hypericum* pollen; (E) bee taking pollen from a normally bat-pollinated *Crescentia* flower (GB); (F) *Trigona* collecting from *Passiflora* anthers, while many ants search below the petals (GB) (note that in C–F, due to a size mismatch, the bee is operating well away from the stigma); (G) robbery holes pierced in the fleshy nectary-containing base of a red *Passiflora*, giving access for secondary theft by ants (GB); (H) holes in corollas of *Aconitum*, where the nectary hidden in the hood is only accessible legitimately to long-tongued bumblebees.

34. Defenses against herbivores and florivores. (A) hairy surfaces in edelweiss flowers, protecting against small herbivores and giving thermal protection; (B) *Salvia leucantha* with intensely furry calyces and corollas; (C) very long hairs on the calyx of *Merremia aegyptia*, which also has hairs within the corolla precluding access to small insects (GB); (D) *Neoregelia* with many small flowers emerging from a water moat that keeps out ants (GB); (E) the hemiparasitic flower of *Lathraea clandestina*, emerging at ground level and with a water moat within the calyx; (F) *Pachycondyla* ants foraging on extrafloral nectaries (EFNs) on the stout calyx of a *Costus* inflorescence, from which single, large flowers emerge periodically, protected by the ants from florivores (GB); (G) *Ectatomma ruidum* ants foraging on EFNs on another *Costus* inflorescence (GB); (H) an ant feeding on the very obvious petiole EFN on *Inga* (GB).

35. Active pollination, and aspects of pollinator management. (A) A typical *Yucca* flower with six stout anthers that the yucca moths gather pollen from and the stout central stigma on which they specifically deposit their gathered pollen (see text); (B) a typical fig wasp with elongated ovipositor that can extend into the styles within the fig (DM); (C) typical African honeybee hives suspended in an acacia tree; (D) a commercial bumblebee nest as used in tomato and strawberry polytunnels; (E) solitary-bee nests in a UK garden, providing nesting for bees such as *Osmia* and *Andrena* species; here residency is high and two bees are seen returning with fresh supplies; (F) commercial solitary-bee housings in North America (SP).

36. Wind pollination and hand pollination. (A) Typical grass flowers with dangling anthers and copious pollen (sw); (B) a male inflorescence of *Zamia fairchildiana* (GB); (C) the large inflorescence of *Gunnera manicata*; (D) hand pollination of *Vanilla* flowers (DM).

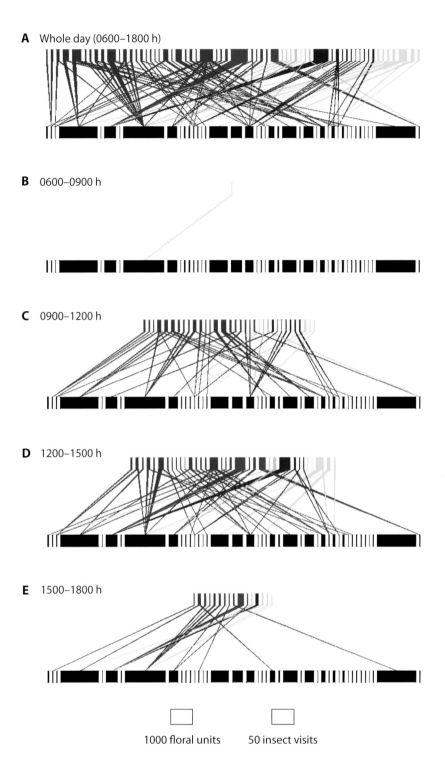

A Whole day (0600–1800 h)

B 0600–0900 h

C 0900–1200 h

D 1200–1500 h

E 1500–1800 h

☐ 1000 floral units   ☐ 50 insect visits

37. The importance of considering time effects in pollination or visitation webs: here the overall daily web is compared with very different webs resulting when daily visits are considered in four separate time-sliced periods (B–E) for a Kenyan savanna community There is a very simple web until 0900, and a maximally complex web at 1200–1500, and it could be very misleading to leave out these temporal patterns. (Reproduced from Baldock et al. 2010.)

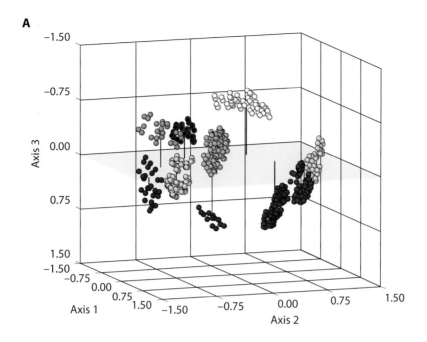

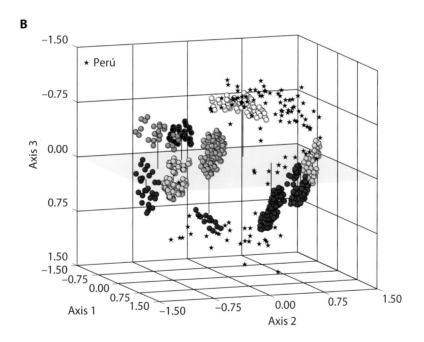

38. (A) Ordinations of idealized pollination syndrome traits in three-dimensional phenotypic space (reproduced from Ollerton et al. 2009). (B) The actual spatial occurrence of flowers in a Peruvian cloud forest community, shown by stars, as an example of the poor fit to expectations found for six communities. More details of methods and axes are given in the text.

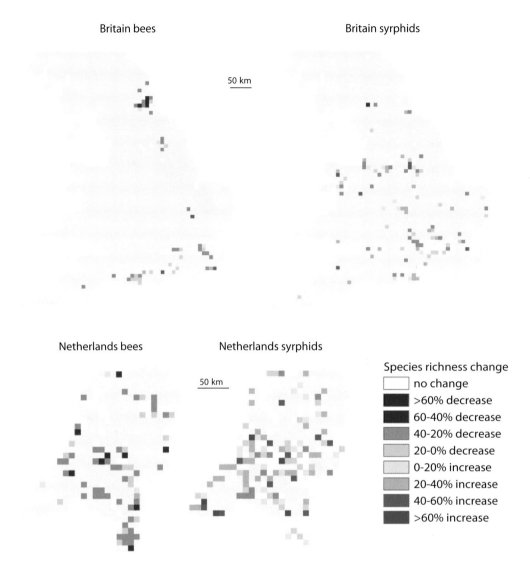

39. Bee population declines (and also changes in hoverfly populations) documented per 10 km × 10 km squares in the United Kingdom and the Netherlands. Blank squares had inadequate records. In some squares declines in excess of 60% have been found. (Reproduced from Biesmeijer et al. 2006.)

Chapter 26

POLLINATION USING FLORIVORES: FROM BROOD SITE MUTUALISM TO ACTIVE POLLINATION

In chapter 23 various one-sided *brood site mimicries* were described where the plant benefits from cheating its visitor, whose progeny usually die because their mother has been deceived into laying on a floral tissue that was mimicking the normal egg-laying site. This chapter deals instead with *brood site mutualisms*, where there may be no mimicry involved and where the pollinators also benefit: their eggs hatch and the progeny survive by feeding on plant tissues, so they are sometimes termed *nursery pollinators*. Here pollination success affects not only plant fitness but also pollinator fitness, and the balance between costs and benefits may be highly variable from place to place and across seasons.

There are at least 13 known nursery pollination systems (Dufay and Anstett 2003), and Sakai (2002) divided this phenomenon into three categories. Two of these are relatively unspecialized, where beetle or lepidopteran larvae develop in decomposing flower heads, or where thrips feed in flowers as pollen parasites; these are dealt with in sections 1, *Unspecialized Examples: Nursery Pollination*, and 2, *Pollen Parasites: Thrips*. But the third category includes highly specialist cases ("ovule parasites"), where a flower visitor laying eggs on a flower does not merely bring about pollination by accident while pursuing its own search for rewards; instead, the female flower visitor pollinates the flowers in a very specific and deliberate fashion as part of her behavioral repertoire while laying her eggs upon them, and the progeny then survive by eating some of the maturing seeds. This is termed "active pollination, also known as the "seed-eating pollination syndrome." Three of these specific cases, where active pollen transfer occurs and a clear mutualism results, are covered in later sections of the chapter.

1. Unspecialized Examples: Nursery Pollination

In some of the less specialist cases of brood site mutualism, the pollinators—having effected normal passive or accidental pollination by visiting the flowers as adults—also lay eggs there, so that their progeny subsequently eat some part of the plants their mothers helped to fertilize. It seems likely that in most of these cases of nursery pollination the habits of breeding in the plants, and eating some plant tissue, were developed after the flower visitor was already acting as a pollinator. For the plant, the benefits and costs depend on the balance of pollination and seed consumption; for the larval animal, there must be a balancing of the food acquired against the costs incurred from encountering plant defenses.

A simple version of this can be seen in *Greya* moths (Thompson and Pellmyr 1992; Thompson and Fernandez 2006) visiting their host plant *Lithophragma* (Saxifragaceae), where the moths pollinate the flowers but then lay eggs so that their larvae become florivorous parasites. In different habitats, and depending on the availability of other pollinators, the moths can be mutualists, commensalists, or antagonists.

Another well-known example involves nocturnal moths in the genera *Hadena* and *Perizoma* laying eggs in the flowers of *Silene* and other Caryophyllaceae (Pettersson 1991; Kephart et al. 2006), the larvae then acting as seed predators. The interaction can be entirely antagonistic but sometimes verges on a mutualism, since the hadenine moths drink nectar and are very effective pollinators but show rather limited seed consumption (Westerbergh 2004). *Hadena bicruris* showed specific responses to the volatiles of its most important host *Silene latifolia* (Dötterl et al. 2006), especially at night, and could avoid laying eggs in fungus-infected plants (where fruit development does not occur; Biere and Honders 2006) even though it still visited the infected plants for nectar.

In each of these cases, other pollinators also visit the same flowers and can affect the overall outcome. For example, in *Silene ciliata* the specific nocturnal moth pollinator is *Hadena consparcatoides* but the flowers are at least as effectively pollinated diurnally by other visitors, probably shifting the nursery pollination interactions more toward a parasitic effect (Gimenez-Benavides et al. 2007). The widespread *Silene-Hadena* interaction is therefore a useful model for exploring how such systems evolve between parasitic and mutualistic status (Kephart et al. 2006).

Other examples are linked to the floral trapping mechanisms discussed in chapter 23, where some aroids such as *Alocasia*, while trapping and then releasing their adult pollinators in the conventional manner of their relatives, do not then condemn the hatching larvae to starvation; instead, the larval flies feed on floral secretions and complete their development through to pupation in a chamber within the spathe.

In yet other cases the pollinators' larvae are "allowed" to feed on male flowers, for example in jackfruit (*Artocarpus*, Moraceae), where pollinating gall-midge flies breed in the male inflorescences (although this case is unusual in that a fungal mycelium on the flowers is the main feeding attractant for the flies; Sakai et al. 2000). Many palms, including the commercial oil palm *Elaeis gunieensis*, use a related system; the pollinators are mainly beetles and lay eggs on the male flowers which drop off the plant a few days later. However, in the palm *Chamaerops*, whose male

flowers are used successfully as a nursery by its pollinating weevil (Dufay and Anstett 2004), the female flowers as they develop into fruit cause the larval weevils to die, so the plants "turn the tables" and cheat the pollinator. The weevils seem unable to distinguish between male and female flowers when ovipositing, so the balance of benefits is somewhat uncertain and may vary with time and site. All these florivorous pollination systems of course require flowers (or inflorescences) to be unisexual and produced in large numbers.

A related but slightly more complex arrangement occurs in some bisexual tropical dipterocarp trees, especially species of *Shorea* in Malaysia, which tend to mass-flower. The flowers are pollinated by thrips, which invade the young buds, feeding there and passing through several generations over 2–4 weeks. Because the flowering is so intense, enough remaining undamaged buds eventually mature, and a cohort of them open at dusk as full flowers; some of the thrips then enter these fresh flowers to feed on petals and pollen. During the following day these flowers absciss and fall off to the forest floor, with some lateral drift as the corollas act like miniature propellers. When further buds open that evening, the thrips fly up to the canopy and invade the new flowers, taking pollen with them which will (at least reasonably often) be from a different tree and so effect outcrossing (Appanah and Chan 1981).

Yet another version of florivory coupled with pollination occurs in *Trollius* globeflowers, which never open fully but are invaded by small anthomyid flies in the genus *Chiastochaeta*, often known as globeflower flies. Both male and female flies enter the flowers, mating and feeding there on pollen and nectar (Pellmyr 1989, 1992b). The females lay their eggs, with three or four fly species each laying in a different site and in a fixed temporal pattern, largely avoiding competition with each other when the larvae then feed on the seeds. Movements relating to either feeding or ovipositing can result in effective pollination. The number of fly eggs left in a flower corresponds closely with the balance of costs and benefits, so presumably the mutualism shows an overall balance in the long term; in fact, Despres et al. (2007) calculated a stable favorable benefit-to-cost ratio of around 3, across 38 different populations.

Use of postpollination food resources is also possible; for example in cocoa (*Theobroma cacao*), which is pollinated by biting midges (see chapter 28), subsequent breeding of the pollinator occurs in some of the decaying seed pods (Dessart 1961; Young 1985).

All these arrangements occur in medium to large perennial plants and ensure for the plant a population of pollinators kept in close proximity for the next batch of flowers. They also seem to involve only very small insects, suggesting that the plant, while trading off some of its own tissues for pollination services, is minimizing the overall costs of doing so.

However, there is one highly unusual case of a plant offering a brood site to a larger animal which is also its pollinator. In the Indian tree *Humboldtia brunonis* (Fabaceae), domatia exist on the trunks and branches that are occupied by various invertebrates, including aggressive ants that provide a protection service as in other myrmecophytes (chapter 25). But some of the domatia are inhabited by adults and brood of the pollinating bee *Braunsapis puangensis*, which visit the host tree's flowers by day and are the best carriers of its pollen; this seems to be the first known case of a pollinator being housed in any plant structure other than flowers, and with no cost to reproductive structures (Shenoy and Borges 2008).

Dufay and Anstett (2003) reviewed the conflicts between plants and pollinators in these nursery systems, highlighting three key questions: why do the plants not kill off the intruding larvae, why do the pollinators continue to visit the deceptive flowers, and why do they pollinate them? The examples above show that the plant's reproductive system is crucial to resolving these issues, and that evolutionary stability often depends on avoiding interspecific competition between pollinator larvae.

2. Pollen Parasites: Thrips

Sakai (2002) identified thrips within flowers as a separate category of brood site mutualists, feeding on the fresh pollen grains still contained within the flower that produced them. Thrips can grow from egg to maturity inside a flower within 2 weeks, so they reproduce explosively on flowers to the point where they can be present in extremely large numbers, but they are so small that they cannot carry much pollen around and rarely move over long distances. However, they do effect pollination of quite a range of plants, especially in tropical forests, as outlined in chapter 9, though at the same time destroying much of the pollen from their hosts.

3. Active Pollination

Figs and Fig Wasps

Figs (*Ficus* species, family Moraceae) are relatively common and take many woody forms, including deciduous and evergreen trees, bushes, vines, and **epiphytes**. At least 800 species are distributed throughout the tropics and subtropics, and they have often been described as ecological "keystone" species. This is mainly because their fruits are a vital food resource for numerous birds and mammals (Bawa 1990; Mabberley 1991), especially bats, primates, and parrots, which often also act as seed dispersers. Figs are eaten by up to 70% of all vertebrates in some forest communities, and they comprise the greater part of the diet in more animal species than any other perennial tropical fruit. Flowering and fruiting are normally asynchronous within populations and between trees, but strongly synchronous within trees. Hence both fruits and flowers are conveniently available in most habitats throughout the year.

Yet even where the density of flowering conspecifics is low, outcrossing levels in figs are high. This turns out to be because of the very tightly coupled fig-pollinator relationship. Fig pollination is perhaps the most extraordinary and complex of all interactions between animals and plants and has attracted attention for decades. It is an obligate mutualism, the figs having no other pollinators and the fig wasps relying totally on the figs. At least 750 *Ficus* species are pollinated by fig wasps from the family Agaonidae (Hymenoptera, Parasitica), each wasp tending to have its own fig (or occasionally two closely related fig species). These agaonid wasps are tiny (1–2 mm) and relatively short-lived insects (just 2–7 days as adults) that are especially abundant in tropical rainforests. They are sexually dimorphic, females being winged and with very long ovipositors (plate 35B), males wingless and with reduced legs, eyes, and antennae. Bronstein (1992), Cook and Rasplus (2003), and Cook, Bean et al. (2004) provided reviews.

The pollination process takes place within what is usually described as the fruit of the fig; but in this instance the term "fruit" is a misnomer; technically it

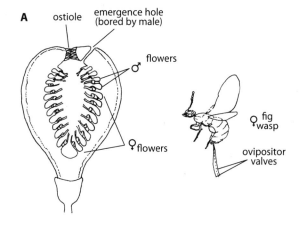

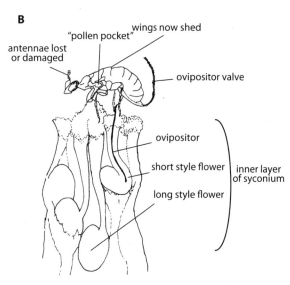

Figure 26.1 (A) Basic structure of a typical *Ficus* (fig) syconium (the "fruit") containing male flowers near the top and female flowers basally, in transverse slice; also showing a female fig wasp from the genus *Blastophaga*. In practice the male and female flowers are not mature synchronously. (B) shows a female wasp egg-laying, her ovipositor only able to reach into the flowers with shorter styles, while her forelegs unload pollen from the pocket and transfer it onto the styles that she grips. (Modified from Galil and Eisikowitch 1968, 1969.)

should be reserved for the later structure that develops from the ovary walls to enclose and protect the maturing seeds. The structure that is termed a fig is technically a *syconium* (fig. 26.1), a hollow receptacle that bears numerous tiny flowers on its inner surface—really a modified inside-out inflorescence. Flowers may be unisexual or neuter, and each flower is reduced

to the simplest possible form, just a single ovary and style, or a tuft of anthers. The syconium has one narrow entrance (the *ostiole*) at its distal end, which is partially closed off by flexible scalelike structures; this is the only access point to the flowers within.

The relationship between the edible fig common all around the Mediterranean (*Ficus carica*) and its main pollinator *Blastophega psenes* will serve as an example of the fig–fig wasp relation, although it is somewhat atypical due its marked seasonality. The fig "fruits" in wild *F. carica* are of three types (fig. 26.2). The *type-1* receptacles appear in late winter and early spring and have many neuter flowers (sterile females) plus a few male flowers close to the entrance. A female wasp enters the fig via the ostiole, which is such a tight fit that the wasp usually loses her wings and sometimes parts of her antennae. She lays her eggs in the neuter flowers (one per flower) and then dies, only occasionally escaping to lay in another fig (Gibernau et al. 1996; Zavodna et al. 2007). The neuter flowers subsequently swell up and become morphologically modified, essentially a form of galling induced by a secretion onto the sterile ovaries from the mother's ovipositor. The offspring complete their development in these swollen ovary galls, male wasps hatching first and seeking out ovaries containing female wasps. They bore into those ovaries and fertilize the females, then die shortly afterward; thus the life cycle of the male fig wasp takes place entirely within the fig (this is linked to his lack of wings and small eyes). The fertilized females subsequently hatch and leave via the ostiole, passing the freshly opened male fig flowers near the entrance on the way and gathering pollen into small pouches on their thoraxes as they leave.

In summer the fig tree produces *type-2* receptacles, ready to receive the pollen-bearing female wasps that emerged from the winter receptacles. These summer figs contain either a mixture of neuter and female flowers or are female only. The flower types are morphologically different: neuters have short styles and the ovary is accessible via a canal, whereas females have a longer, solid style lacking any opening. The arriving female wasps attempt to lay eggs in both types; this is successful in neuters (just as in the winter receptacles) and the same cycle as above results, with new females eventually emerging. But the ovipositor cannot penetrate the full length of the solid styles of female flowers, so any eggs that the female lays there must fail to develop; thus if a female wasp has entered a female-

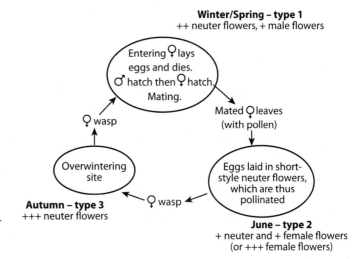

Winter/Spring – type 1
++ neuter flowers, + male flowers

Entering ♀ lays eggs and dies. ♂ hatch then ♀ hatch. Mating.

♀ wasp

Overwintering site

Mated ♀ leaves (with pollen)

Eggs laid in short-style neuter flowers, which are thus pollinated

♀ wasp

Autumn – type 3
+++ neuter flowers

June – type 2
+ neuter and + female flowers
(or +++ female flowers)

Figure 26.2 The annual life cycles of *Ficus carica* and *Blastophaga psenes*, with three different types of fig in different parts of the season. See text for more details.

only syconium she will have no progeny, and the figs presumably must offer no external clues to their sexuality. Note also that in mixed neuter-female syconia in practice it is the style-length variation that effectively limits the number of ovaries supporting larval wasps (in theory this might select for wasps with ever-longer ovipositors, but runaway selection may be braked by the figs' ability to abort syconia too heavily laden with developing wasps). Immediately after each egg-laying attempt, the fig wasp pollinates the flower she has just laid in, and this is the most extraordinary part of the whole story, because she specifically extracts a little pollen from her pollen pouches with her forelegs and places this pollen neatly onto the stigmatic surface—active pollination.

In autumn, small *type-3* receptacles containing neuter flowers only are produced, and are found and entered by the female wasps that emerged from the summer figs. Newly developing females are then ready to emerge in time for the winter-spring type-1 figs and repeat the cycle. In effect, these autumn-winter figs are specially modified flowers whose only real "purpose" is to provide a protected winter breeding place for the pollinator.

Ficus is largely a tropical genus, and in most species flowering and fruiting patterns are relatively aseasonal, so that the whole cycle is usually much less rigid than for the Mediterranean fig just described. Thus the interaction carries on at all stages throughout the year, with different plants having internally synchronized flowering and fruiting, but the population as a whole showing occasional and irregular flowering, which ensures outcrossing. The "normal" pattern for a tropical dioecious fig species is for receptacles containing female-only flowers to occur on one individual and receptacles containing neuters and male flowers to occur on another. Fertilized female wasps emerge from the latter and find either type of tree; those with female flowers are pollinated and set seed but do not become a brood site (given the barrier to egg laying), while eggs that are laid in figs on the neuter-male trees result in offspring that gather pollen from the male flowers as they leave. Monoecious tropical fig species are even simpler, having all three flower types in single receptacles, where the wasps can pollinate the female flowers and lay eggs in the neuter flowers, with pollen subsequently collected from the male flowers.

The traits of the partners in general, and wasp behaviors in particular, are remarkably varied across the fig wasps. Females of at least some species do not just pick up pollen by accident in leaving a fig, but actively collect the pollen and place it in their thoracic pouches, which can carry 2–3000 pollen grains (Galil and Eisikowitch 1969; Kjellberg et al. 2001). Females of some species have both thoracic pollen pouches and coxal corbiculae and use special leg combs and hooks to manipulate the pollen (Ramirez 1969). In some dioecious figs such as *F. Fistulosa*, the female wasps successfully lay eggs and place pollen in both female and neuter syconia, and "pollination" of the neuter flowers appears to have a benefit in inducing growth of the ovule endosperm and so increasing the food supply for the larva (Galil 1973). Males too have complex behaviors. Sometimes (in their brief captive life before dying) they assist females by snipping the anthers from male flowers, and these snipped pollen loads are then collected by the emerging females (Galil and Eisikowitch 1968, 1974). In a few species the males not only snip off the anther tips ready for the females to carry, but also collaborate in making a new exit hole in the syconium, from which the pollen-laden females depart. The timing of this exit opening may be crucial, as in some fig species male activity may be partly controlled by the carbon dioxide levels within the receptacle (Galil et al. 1973); as the flowers mature, accumulating high levels of CO_2 will increase the male emergence rate and mating rate, but once the males have opened an exit hole the CO_2 level drops and this stimulates female emergence and pollen gathering. Such complex behaviors seem surprising, since as a rule these kinds of tiny hymenopteran wasps neither eat pollen nor provision their larvae. But since many figs are entered by only a single ovipositing female, the emergent males are highly likely to be related to the females they are assisting; for example, estimates using microsatellites indicated that about 40% of all broods involved only a single foundress in *Ficus montana*, with (rarely) up to six foundresses having been active (Zavodna et al. 2007), confirming that high levels of inbreeding must occur.

All of this may seem complicated enough, but fig and fig-wasp biology is a fertile area of research, and ever-increasing complexity was revealed once molecular techniques became available to aid species identification. It was already clear that with around 800+ *Ficus* species known, yet only 300 or so obligate agaonid wasps formally described, the specificity with supposed 1:1 matching was suspect, but the wasp species are extremely abundant and very difficult to identify to species level, as indeed are figs. We know now that the apparently specific relationships of figs and agaonid species hide much complexity. Molbo et al. (2003) and Cook and Rasplus (2003) showed the widespread occurrence of wasps other than the one known pollinator species per fig, with these additional "cryptic" wasps occurring in at least 50% of studied figs, some of them probably assisting in pollination. But many of these extra wasps turn out to be parasitic, and nonpollinating. Some are agaonids, ovipositing through the syconium wall and "cheating the system"; others are from unrelated genera in different families

of Parasitica (especially Torymidae), and these lay their eggs in galled flowers within the syconia, their larvae killing those of the true fig wasp.

In specific coevolved relationships, if there has been tight cospeciation of the participants there should be identical or at least closely matching phylogenetic patterns; such patterns were reported in the past for figs and fig wasps (e.g., Herre et al. 1996), but more recent results indicate otherwise (e.g., Erasmus et al. 2007), with rather frequent host jumps, presumably reflecting the complexity of the fig-wasp communities now revealed to be operating within these plants. There have been two separate lines of evolution of oviposition habits, and two separate origins of the cheating fig-wasp habit (Althoff et al. 2006).

Despite this, it remains the case that most figs do have only one effective pollinating fig wasp, especially in the Neotropics (although perhaps less often in Africa, where 2–3 species may be present and effective; Ware and Compton 1992; Michaloud et al. 1996). One major problem for tropical fig wasps should therefore be the location of host trees. Bearing in mind the great diversity of fig species in Neotropical forests and the likely distances between conspecifics, exacerbated by their irregular flowering pattern, how do fig wasps disperse and find the correct host plant at a suitable stage of flowering? It was always assumed that olfactory cues were critical (Janzen 1979; Addicott et al. 1990), and analyses of odors collected from various figs (Grison-Pige et al. 2002) confirmed that individual fig species did indeed have different blends of compounds. Behavioral assays on their putative pollinating fig wasps also revealed different behavioral responses by obligate wasps when presented with either control scents or their "own" fig scent. In at least one case (*Ficus semicordata*), a single volatile, 4-methylanisole, is the key compound ensuring attraction of the one obligate pollinator *Ceratosolen graveli* (Chen et al. 2009).

Work in Borneo suggested that various fig wasps would travel up to 30 km (wind-assisted, flying above the canopy) in search of the correct syconia (Harrison 2003). This kind of directed dispersal is crucial in areas where forest fragmentation is occurring; since fig fruits are such a vital resource to other components of the tropical communities, if the obligate associations of fig to fig wasp were to break down then a cascade of extinctions could follow (see chapter 29). McKey (1989) estimated that a minimum of 300 fig trees would be needed in a given community to ensure survival of the fig-wasp community, and that in turn would require at least 800 acres of typical forest.

Yuccas and Yucca Moths

This is an equally famous obligate mutualism, reviewed by Powell (1992), Pellmyr et al. (1996), and Pellmyr (2003), which again centers around the larvae of pollinators feeding on the internal parts of plants; here it involves a moth rather than a wasp, but the female behavior again involves active pollination that ensures outcrossing in yuccas.

There are about 40–50 species of *Yucca* (Agavaceae) (including the well-known and widely used plants sisal hemp and mezcal), found in North and Central America and the Caribbean, with the most diverse and abundant flora being found in the southwest United States in semidesert habitats. Yuccas are highly variable in size, some having a large treelike form, but generally with narrow and pointed leaves that are often semisucculent. Most species produce long erect flower spikes, the flowers (plate 35A) usually being creamy-white, large, and bell shaped (though sometimes partly closed by the converging tips of the petals), and with a strong scent especially at night. They are protogynous, with six swollen stamens and a central column of three fused styles, overlying a tripartite ovary often with 200–300 ovules.

Yucca moths (in the small family Prodoxidae) are divided into the two genera, *Tegeticula* and *Parategeticula*. The family as a whole contains at least 78 species and is among the oldest of moth groupings, also being among the first that began to develop the nocturnal habit. In fact, yucca moths are fully nocturnal, resting cryptically in yucca flowers by day, while most other prodoxids are diurnal. Importantly, most members of the family are monophagous or narrowly oligophagous (i.e., they have extremely specialized food requirements), the larvae eating particular parts of one or of just a few closely related plants.

Figure 26.3 shows a *Yucca* flower and the basic cycle for the moth *Tegeticula yuccasella* on *Y. filamentosa*, revealing a very stereotyped behavioral sequence. Arriving at a flower, a female moth ascends one of the large individual stamens and, holding position with her proboscis, gathers and compacts pollen using her maxillary palps, which incorporate bristly coiled "ten-

tacles" (fig. 26.4; these apparently have no homologous counterparts in any other insect). The pollen of yuccas is particularly sticky and adheres to the underside of the moth's head. She may repeat her behavior at 2–3 further stamens (accumulating a pollen load between the palps and the trochanters of her forelegs that can equate to 10% of her body weight), before flying off to find further yucca plants. These must be in the female stage, a little earlier in their protogynous cycle, so ensuring crossing. She then inspects the ovaries carefully and can apparently tell if they are of a suitable age or if they already contain moth eggs; choice may be made on the basis of past visitation history due to host-marking pheromonal "footprints" (see chapter 7). Having selected a new flower, the female climbs the central style, reverses down toward the base, and cuts into the ovaries with her ovipositor to lay an egg. Then she climbs back up the styles and actively places part of her pollen load onto the stigmatic surface, usually repeating this process twice more so that she has laid one egg adjacent to each of the sections of the tripartite ovary. This pollen placement is critical for both plant and moth; unpollinated flowers are aborted (Huth and Pellmyr 2000), while pollinated flowers assure reproduction for the plant but also undergo abnormal growth of at least one ovule near to a moth egg, again giving a gall-like effect. This benefits the developing larvae, which feed on the abnormal ovules but leave alone most of the other growing seeds. At maturity the larvae burrow out of the fruit and pupate in the surrounding soil, where diapause may take place. This is important because yuccas do not always flower every year; intraspecific variation in the period of moth diapause ensures survival of at least some moths from each generation, so that the adults meet up with a new flowering yucca.

There are only slight differences in this process for other *Tegeticula* species (Pellmyr 2003), perhaps most notably for *T. maculata* which is a daytime pollinator of the southern Californian yucca *Y. whipplei*. But the genus *Parategeticula* differs more substantially, even though the active pollination process remains unchanged. Unlike the saw-like ovipositor of *Tegeticula*, female *Parategeticula* have blunt ovipositors and cannot lay eggs directly into the ovaries of yucca flowers. Instead a groove is made either in the stalk or in one of the thick petals, and eggs are laid in this shallow structure. The larvae emerge and burrow into the nearby fruit, creating a gall which replaces several seeds. No

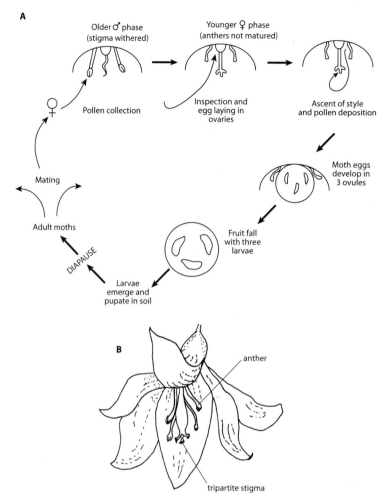

A

Older ♂ phase
(stigma withered)

Younger ♀ phase
(anthers not matured)

Pollen collection

Inspection and
egg laying in
ovaries

Ascent of style
and pollen deposition

Mating

Moth eggs
develop in
3 ovules

Adult moths

Fruit fall
with three
larvae

DIAPAUSE

Larvae
emerge and
pupate in soil

B

anther

tripartite stigma

Figure 26.3 (A) Typical cycle for yucca moth. (B) Structure of a *Yucca* flower. (Redrawn from Proctor et al. 1996 and from photographs.)

further seeds are eaten during development (although some may be eaten prior to pupation), the gall tissue becoming the sole food source during development.

Curiously, some yucca flowers do produce nectar (in fairly small quantities) but visiting yucca moths do not feed on this resource, having a poorly developed sucking proboscis, and an incomplete gut. Other organisms will visit the yucca flowers; honeybees, bumblebees, flies, and beetles have all been recorded as visitors, and the bees may effect some pollination (Powell 1992). However, in the absence of yucca moths, fruit set does not occur, except in one species (*Y. aliofola*) that may be an escaped cultivar and which has also been observed to propagate vegetatively. The

presence of nectar has been interpreted as evidence that the yucca moth–yucca relationship is evolutionarily relatively recent,

As with the fig wasps, there have been recent developments that have added complexity to the story. Again, molecular analyses have somewhat undermined earlier taxonomic assumptions about precise one-to-one relationships of plant and pollinator; for example, the *Tegeticula yuccasella* complex is now known to include at least 11 moth species (four of them nonpollinating) where previously only one was recognized, and it is reasonable to assume that further species await full description. Yucca moths have been found on 17 host yucca species, and to date seven of the moth

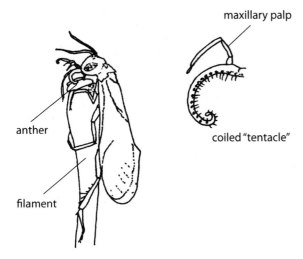

Figure 26.4 Yucca moth on a flower, and a view of the pollen-gathering tentacles. (Redrawn from Proctor et al. 1996.)

species have been found only in association with one yucca species, although other moth species have up to six hosts. Furthermore, and again paralleling the fig-wasp system, some nonpollinating parasitic species of prodoxid moths have been discovered (Pellmyr et al. 1996; Peng et al. 2005), in the "bogus yucca moth" genus *Prodoxus*. These have no pollen-gathering tentacles but breed in the ovaries of yucca flowers that are being pollinated and therefore maintained by the true yucca moths.

There is an additional complexity for yuccas compared with figs, in that the plants are from highly seasonal habitats, and synchrony of flowering and insect emergence, together with the abundance of both in any one year, become critical. This may have encouraged forms of cheating by moths. If an individual moth emerges rather late in its host yucca's flowering period and encounters already well-pollinated flowers, it should lay its own eggs not in the young fresh flowers (which are the most likely to be aborted in favor of the older flowers already invested in by the plant) but into already fertilized and developing ovaries. Late moths should therefore be selected as cheats (Addicott et al. 1990), and even the "good" pollinators often do cheat facultatively, depending on season and on moth abundance. In fact it is increasingly clear that there are *multiple* modes of interaction between yuccas and yucca moths (Addicott and Bao 1999).

There must be a trade-off of pollination gained against seeds lost in yuccas, and some resulting balance of seed production and moth production. Addicott (1986) reported 150–350 ovules per flower in the eight species he studied, which produced between 90 and 200 seeds and normally supported up to ten larvae (but often none at all, and occasionally much higher numbers). The mean ratio of viable seeds to ovules varied between 0.36 and 0.60 in different species, although highly variable between populations. One developing moth larva could eat up to 25 ovules and could damage others sufficiently to prevent seed maturity, but overall reductions in seed production were estimated at only 1%–20%. The balance is complicated by high levels of egg mortality and high levels of fruit abortion (Richter and Weis 1995; see also Pellmyr and Huth 1995) but final fruit set is low in most years, in most habitats, for most yucca species, and it may be that pollinators are usually limiting.

Presumably, though, moths that produce just enough eggs to maintain their populations while dependably pollinating the yucca flower, are selectively favored over moths that overexploit the flowers with eggs so that the flower aborts, or are miserly with their pollen deposition so that seeds fail to develop; thus the symbiosis can be stably maintained through evolutionary time, and fitness gains are balanced.

Other Specialized Examples

In recent years a few more examples of specialized active pollination have been documented. *Glochidium* trees are visited and actively pollinated by *Epicephala* moths (Kato et al. 2003) and up to 500 species in several closely related genera in the Phyllantheae also receive species-specific nocturnal visits from the same moth genus (Kawakita and Kato 2004, 2009), some flowers having all their ovules consumed and others left intact. This relationship evolved at least five times, although active pollination in the moths probably evolved only once (Kawakita and Kato 2009).

In the Sonoran Desert, the senita cacti (*Lophocereus*) are visited by senita moths (*Upiga*) which collect pollen on special abdominal scales and actively pollinate and then oviposit on petals (Fleming and Holland 1998; Holland and Fleming 1999, 2002), thereby producing good fruit set. Unusually, in this case some bees also visit, although they do not improve fruit set (and are excluded from around dawn by flower closure).

4. Overview

Whereas the relatively unspecialized brood-site-associated visitors covered in the first part of this chapter were assumed to have begun as pollinators and then started to lay eggs in their flowers, the more spectacular wasp and moth floral associations in later sections are much more likely to have evolved from larval feeding syndromes, where the adults gradually turned from incidental pollinators into successful specialist pollinators (Pellmyr and Thompson 1992).

The fig and yucca systems represent the best-understood and tightest so-called obligate plant-pollinator associations known, both of them adding active pollination by deliberately targeted pollen placement to a suite of other highly specialist coevolved features. The pollinators ensure the successful seed production of their host plant, and in return they get both food and shelter for their offspring. The trade-offs for each are probably highly variable in time and space, and the success of the mutualism is governed by many interconnected variables, but must work out to mutual benefit overall.

Mutual dependence is complete, but it is now clear that for both fig wasps and yucca moths the mutualisms are not as one-to-one as was originally supposed. Even where the partners are mutually dependent in these specialist associations, there is an inevitable evolutionary conflict, and both parties will be under selective pressure to exploit one another (Pellmyr 1997). Hence these examples of extremely specialized plant-pollinator relationships are fragile and remain rare.

Chapter 27

POLLINATION IN DIFFERENT HABITATS

Thus far pollination has been dealt with in a collective sense, but it will have been apparent that examples from different habitats have often given rather different impressions of the complexity and level of specialization involved. This chapter therefore dissects the issues that are to the fore in current debates in different kinds of ecosystems and habitats, in search of some messages on community structures and how they affect plant-pollinator interactions.

1. Pollination in Deserts and Semiarid Systems

Deserts occur in many low-latitude areas just to the north and south of the equator, between 15° and 40° latitude (fig. 27.1). Most are both hot and dry by day but can be very cold at night as heat is lost through the clear skies; and consistently cold deserts also occur, especially in the center of larger continents and in the rain shadow behind large mountain ranges. Deserts are often visualized as exceptionally hot sunny landscapes with sweeping sand dunes, but in fact a surprising variety of biotopes are included within this general category. Their most important shared characteristic is aridity; the three subdivisions of semiarid, arid, and hyperarid deserts cover one-third of the Earth's land surface, about half of this being true desert as traditionally perceived by the layman.

Less extreme are the semiarid zones such as cool grasslands and hot savannas (see fig. 27.1), also dry for much of the year but highly seasonal, usually with one or two rainy seasons annually.

Habitat Characteristics

Deserts may be defined by their drought status, but their aridity can arise for three distinct reasons, varying in importance for different desert areas. In parts of North and South America desertification arises from a rain-shadow effect, on the leeward side of high mountain ranges (sierras) where the rising air has precipitated all its water. Deserts in the center of large continents are arid principally due to their sheer distance from the sea; the Gobi and Turkestan Deserts are examples. However, the largest and hottest deserts on Earth arise from the third factor, occurring in latitudes 25°–35° N or S where there are dry, stable air masses of high pressure, resistant to invasion by storm systems from north or south. The Saharan and Arabian Deserts, and the Australian Desert, are of this type; these hyperarid deserts, where high-pressure weather dominates, normally receive less than 25 mm of rainfall per year (and within them there are often areas that effectively receive no rain at all). Even when rain does fall it may be as violent convective showers, causing flash-flooding in dry river beds (wadis) and very fierce runoff, with little of the rain becoming available to plants or animals in the area; that which does remain local may be subject to very rapid evaporation due to high temperatures. These deserts usually also have high wind speeds, leading to even more fierce evaporation. Because of cloudless

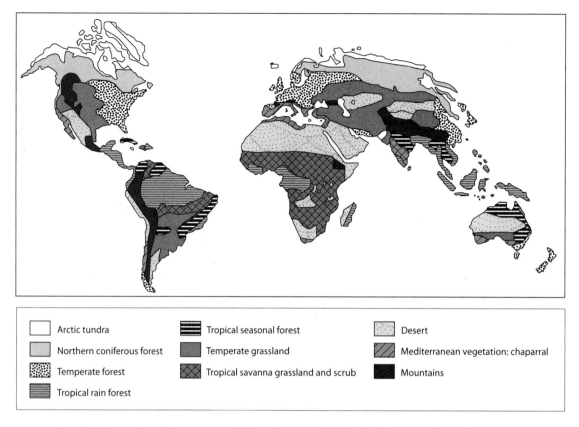

Figure 27.1 An overview of the occurrence of different habitats worldwide. (Modified from Willmer et al. 2005.)

☐ Arctic tundra	▤ Tropical seasonal forest	░ Desert
▦ Northern coniferous forest	▦ Temperate grassland	▨ Mediterranean vegetation: chaparral
▦ Temperate forest	▩ Tropical savanna grassland and scrub	■ Mountains
▤ Tropical rain forest		

skies they can also become very cold at night as heat radiates away from the soil. Their soils are hardly leached at all by any passage of water through them, so are often sandy or stony, leading to deep heat penetration and very poor water retention.

Less extreme areas, the arid and semi-arid deserts, may have up to 600 mm of rain per year, more evenly and predictably distributed, and they are often much cooler than the hyperarid areas. Coastal deserts, such as the Namib, the Atacama, parts of southern Israel, and the Baja California coast, are characterized by cooling fogs, especially after dawn. Cooler inland deserts, such as the Gobi and large areas of Patagonia, also have fogs but experience prolonged periods of winter cold.

Temperatures at and close to ground level in most deserts may be exceptionally high in the daytime, up to 70°C or more, and animals (even insects) cannot function in such conditions. However, within 30–50 mm above the still boundary layer the air is markedly cooler, and in the vicinity of bushes or trees it may fall

to around 40°C, so that both flowers and their potential pollinators are under much less thermal stress.

Most deserts are aseasonal but they are also areas of *unpredictable* rain and therefore are likely to show unpredictable flowering. In some deserts a burst of greenery with very rapid growth and flowering may follow a short rain at any time of year. However, most deserts are not uniform in terms of soil and substratum, and often show patches of unexpectedly high floral diversity, associated with particular gravel deposits or underlying geological features that affect drainage or mineral composition.

Flora and Pollinating Fauna

Desert vegetation inevitably has characteristic specializations. There is usually a crust of just a few millimeters of modified sand densely occupied by a community of microorganisms held together by mucilaginous secretions. Small plants that do survive are mostly evaders in a temporal sense, almost invisible for most of their life as seeds or dried-out prickly husks, undergoing sudden synchronous bursts of growth, flowering, and seed production after occasional rains. Such *ephemeral* plants are inactive when dried, and some can complete their life cycle in only 14 days when rain does occur. On a larger and more obvious scale, "normal" plants occur only at transient water courses and oases. These include a variety of palms, a small range of annual flowers, and the few specialist perennial forms that can keep photosynthesizing even in drought. Cacti are the classic examples, with more than 2000 species in the Americas. There are also small prickly trees such as creosote bush, mesquite, or desert willows, and many eucalypts and wattles in Australian desert fringes; most of these have enormous roots, penetrating to depths of 50 m or more and vastly in excess of the aerial biomass, and some will lose all their leaves and even drop whole branches when they are particularly severely desiccated.

Despite the extreme conditions, some deserts are surprisingly varied in terms of flora and fauna. The deserts of the southwestern United States have a particularly rich flora and bee fauna: the Sonoran Desert houses at least 580 plant species, and there may be 1500 bee species in arid Arizona. But even in these areas of higher diversity the flowering is often very localized and sparse. Earlier views that drought was the main stimulus to desert flowering have been doubted (Fox 1990), and it seems more likely that while drought does hasten completion of the life cycle this is often at the expense of the plant dying before it flowers properly or sets any seed.

With flowering unpredictable, and wind pollination usually rare, many plant species are likely to have to be generalists. It is common to find a range of low and fast-growing composites, which in the depths of a desert (in the hot air at ground level where insects are excluded) are almost always selfing (Hagerup 1932). But the taller desert succulents and larger bushes and trees provide a local microclimate suited to more delicate plants as an understory. Here both insects and birds can function by day (although not often in the midday hours), and bats may be pollinators after dusk.

Desert soils and barren surfaces provide poor habitats for most potential pollinators, but are reasonably accommodating to ground-nesting bees, which become especially important as flower visitors. Indeed, bees are

perhaps the only group to break the rules by being more common and speciose in deserts than in more equable tropics, and in many American deserts it is the more specialist bees that are particularly speciose (Minckley et al. 2000), visiting plants such as creosote bush (*Larrea*) and mesquite (*Prosopis*). However, a surprisingly high proportion of the cacti in these zones are bat pollinated, providing large nocturnal flowers and also sturdy plants that can offer roosting sites.

Semiarid scrubs and savannas tend to be dominated by grasses, prickly shrubs, and some herbs, most of these flowering at or after the onset of rains, but interspersed with small and usually very thorny trees (acacias are typical through most of the southern continents). In many zones there is an abundance of bulbous monocots, flowering in the late winter–spring period; some of these have temperature-dependent petal closure systems that may protect the pollen from overnight moisture, fog, and dew (von Hase et al. 2006). Wind pollination here is reasonably common (Regal 1982), especially in the denser grasslands with limited plant diversity (see chapter 19). Bees are again abundant, with butterflies and moths also much in evidence. Australasia is unusual in having a substantial component of bird pollination in these kinds of habitats; many eucalypts are pollinated by birds, as also are members of Myrtaceae and the well-known shrubs *Banksia* and *Grevillea*, on which honeyeaters and brush-tongued parrots are especially effective. South American savannas also have reasonable numbers of honeycreepers and icterids, but hummingbirds are usually restricted to more wooded areas.

Problems with Triggering and Timing of Flowering

A common trigger for flowering in arid lands is the onset of rainfall, or more rarely the cessation of rainfall where there is a distinct constrained rainy season. For example, five out of six desert plants tested in southern US deserts by Bowers and Dimmitt (1994) had rain as a trigger, the sixth depending on photoperiod. The creosote bush (*Larrea*) had its blooming triggered through spring and summer by any rainfall event of more than 12 mm, and *Ambrosia* required just 9 mm. But *Larrea* also showed an additional strong flowering in early spring as soil temperatures rose whenever there had been winter rain.

As noted in chapter 8, many desert flowers secrete nectar at dawn and/or dusk, so investing water into nectar production at times when humidity is higher and the plant is less water stressed (Bertsch 1983; Nicolson 1993). This release of floral resources at times of high relative humidity and low ambient temperature has probably had long-term implications for the physiology of bees in desert habitats; if nectar is a limiting resource, those bees able to collect it at the low temperatures associated with crepuscular secretion would be favored, leading to selection for increased endothermy. This might explain why deserts are home to the most endothermic bees yet studied (e.g., Stone and Willmer 1989a; Willmer and Stone 1997b).

Problems of Highly Dispersed Flowers

Plants tend to occur very irregularly, often clumped in dry river beds or over gravels, or in small depressions where water can briefly accumulate, or in the shade of boulders where their seeds have been caught. From the perspective of pollinators, they are often highly dispersed and separated by inhospitable terrain.

Some desert bees rely on synchronous emergence with patches of more concentrated blooming by rare and irregularly flowering plants, and are pollen specialists on these species (Minckley et al. 2000). These desert bees can facultatively remain in larval diapause (Danforth 1999), perhaps for up to ten years, and do not begin any activity in years when the desert flora is sparse or absent. Only when their preferred plants such as *Larrea* are triggered into flowering by rains do these specialist bees emerge with marked synchronicity, so presumably their diapause termination is linked to the same rainfall trigger. However, even then only about half of the larvae pupate and emerge, giving a bet-hedging strategy for the population. Species richness of the 22 species of bee that are specialists on *Larrea* appears to be greatest in areas with the least predictable blooming, and in these areas polylectic visitors are less common; perhaps the more generalist bees fail because they cannot achieve synchrony with the plants, being unable to diapause in the years when flowering is poor (Minckley et al. 2000; Minckley and Roulston 2006). Specialist bees also do better on *Prosopis* flowers than introduced *Apis,* with *Chalicodoma* and *Perdita* being the more effective pollinators for this tree (Keys et al. 1995).

This may in part explain why bees are at their most diverse in arid zones (Danforth 1999) and why specialist bees are documented to be more common there than in mesic areas (Michener 2000).

Issues of Energetics, Heat Overload, and Water Balance for Desert Plants and Animals

The need to produce at least reasonably dilute nectar to attract visitors clearly creates water balance problems for a desert or savanna plant. Dawn or dusk flowering may be desirable or even essential, allowing the plants to secrete nectar when water from dew is at its maximum availability, and also ensuring that they are dispensing their nectar into a more humid environment where it will not immediately evaporate to a crystalline state suitable only for flies. Hence use of dusk-flying bats or hawkmoths by larger-flowered plants can be advantageous in some deserts; for example, *Merremia* (Convolvulaceae) opened its flowers and secreted its nectar at dusk in the Sonoran Desert and was pollinated almost exclusively by hawkmoths (Willmott and Búrquez 1996). For day-flowering plants there may be no alternative but to invest scarce water resources into nectar, so that *Echium wildpretii* was interpreted as producing dilute nectar specifically as a water resource to desert bees (Olesen 1988), as were *Calotropis* flowers (Willmer 1988); in both cases the nectar is highly protected by the floral architecture to avoid wasteful evaporation. In the desert willow *Chilopsis*, there is both groove and pool nectar (see chapter 8), partitioning the resources for different visitors and ensuring that some remains protected later into the day (Whitham 1977).

Temperature and water balance problems for desert pollinators were specifically considered by Chappell (1984) for *Centris* bees foraging on paloverde trees; he found that they had to stop flying periodically to avoid lethal overheating. Willmer and Stone (1997b) reviewed the physiological problems of desert bees and found no particular specializations of warm-up rate or thoracic temperature in flight (fig. 27.2), but a strong tendency to matinal, twilight, or bimodal activity patterns (fig. 27.3) to escape the hottest hours.

Honeybees function at least around the fringes of deserts, and have the advantage over most other bees of specific water-collecting behaviors, taking water from up to 2 km from the hive in aliquots of up to

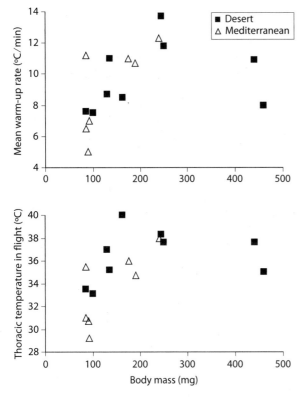

Figure 27.2 Desert bees compared with Mediterranean species, showing warm-up rates and thoracic temperature excesses, with little difference in either parameter at any given body mass; (Modified from Willmer and Stone 1997b.)

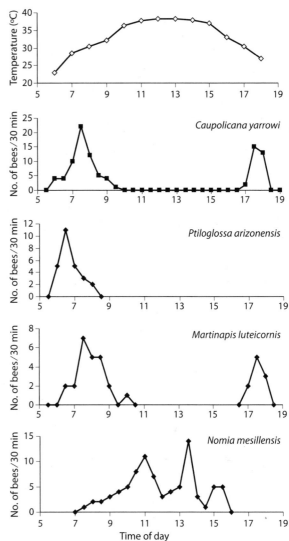

Figure 27.3 Desert bees activity patterns: four species at creosote bush (*Larrea*) in southwestern US desert. (Redrawn from data in Hurd and Linsley 1975.)

40 μl per trip (Visscher et al. 1996). They can also maintain unusually constant and equable hive conditions by communal fanning behaviors.

Increased Reproductive Allocation in Plants

Increased reproductive allocation has been suggested as a likely effect in deserts, with reduced allocation to vegetative plant growth, although direct evidence is scarce. Comparisons of related desert and Mediterranean species of crucifer reported that the desert species showed reduced production of flowers but a higher ratio of reproductive to vegetative biomass; so there is a higher allocation to reproduction overall (Boaz et al. 1994). This is probably a common pattern. The desert crucifers also produced fewer and smaller seeds, but again a higher ratio of seeds to total biomass.

2. Pollination in Mediterranean Ecosystems

Although named after the Mediterranean basin (i.e., southern Europe and North Africa), Mediterranean-type habitats also occur in South Africa and southern Australia, and in the New World in California and parts of Chile (see fig. 27.1). These zones tend to occur toward coastlines (replacing the grasslands that spread further inland) and are at their most obvious in areas around 35°–45° latitude, especially where human influence has been long-standing.

Habitat Characteristics

All the Mediterranean zones exhibit strong seasonality, with mild wet winters and hot dry summers; they therefore tend to have short and highly structured flowering bursts in their springtime months, linked to the very predictable climate that allows plant growth and development to be tightly patterned. These habitats have generally been rather stable through evolutionary time and much less affected by Pleistocene climate cycles (Ice Ages) than cooler temperate zones. They are major centers of plant radiations, especially for small shrubs and herbs rather than trees; the Mediterranean basin has around 30,000 plant species or subspecies, compared with about 6000 in all the rest of Europe. The reasonably continuous perennial Mediterranean vegetation in turn provides amelioration of conditions for a wide diversity of animals. Hence there is high biodiversity and tight packing of flowers and pollinators, with typically high levels of competition. However, these zones are often now particularly subject to substantial anthropogenic stress from increasing fire, overgrazing, tree removal for fuel use, and the demands of human agriculture and urbanization (Blondel and Aronson 1999).

Flora and Pollinating Fauna

Mediterranean soils are often quite shallow and stony and moderately acidic in character. The resulting vegetation is of grasses, perennials, and shrubs, with many bulbous flowering plants, often with scattered stands of relatively small spiny trees; although where the system is allowed to go to a climax vegetation it is often pines that come to dominate above a layer of thorny shrubs. Wherever land is cleared there are also bursts of annual flowers from the long-lasting seed banks. Grazing often helps maintain diversity by preventing the final stages of succession and allowing the annuals to flourish, but native grazers have usually been replaced by abundant sheep, cattle, or goats. Reptiles and birds are common, and insects can be hugely diverse, with orthopterans and ants particularly common but bees and butterflies (and evening moths) also very conspicuous.

Mediterranean areas tend to show a fairly consistent pollinator fauna across different communities and even across continents (Moldenke 1976; Petanidou and Ellis 1993). There is high endemism, and nearly always high bee diversity: southern Africa is the main exception, since bees are relatively uncommon there, but southeastern Europe and California in fact have the highest numbers of bee species in the world (Roubik 1989). This is linked with a predominance of bee-pollinated shrubs, visited by a wide range of solitary bees, but now with *Apis* introduced and often dominating the communities, especially in Europe (Herrera 1988). Even so, many of the shrubs appear specialized to just a few bee species that divide up the flora between them (Moldenke 1976; Simpson 1977; Armstrong 1979; Petanidou and Vokou 1990, Potts et al. 2003a,b). Bees made up 39% and 31%, respectively, of communities in Greece (Petanidou et al. 1995) and in Spain (Herrera 1988). But in southern Australia and in Chile the bees, while still very important, tend to be of different genera and even families from familiar temperate examples, with Colletidae particularly common in Australia (Roubik 1989). Across all these ecosystems the social bees seem to be less important than in other kinds of habitat, perhaps because the flowering season is relatively short and social bees have longer-lived colonies. Furthermore, in all sites the occurrence of parasitic bees is rather low (Petanidou et al. 1995).

All these regions except the actual Mediterranean basin also have about 10% of their species pollinated by birds; in Chile and California the birds (mainly hummingbirds) are not generally visiting the most dominant plants, although in southern Australia and South Africa the honeyeaters, sunbirds, and sugarbirds probably do visit keystone plants. Most of the Mediterranean zones also have a high rate of ant visitation (but usually not pollination) of flowers; Bosch et al. (1997) recorded 58% of all floral visits as due to ants in a northern Spanish site.

The Mediterranean phrygana in Greece is among the most intensively studied habitats (Petanidou and Ellis 1993), where in a 30 ha patch these authors found 132 species of flowering plant and 666 species of insect visitors, of which 225 were solitary bees. Each flower species averaged five kinds of visitor, including flies, lepidopterans, beetles, wasps, and bees.

The South African Cape is the most distinctive of the Mediterranean habitats: it has more fires, a longer flowering season, and especially poor soils (Johnson 1992). The Cape has the most diverse flora in the

TABLE 27.1
Numbers of Plants Observed with Different Pollination Systems in Sub-Saharan Iridaceae

	No. of genera	No. of species
Bee, nectar gathering (zygomorphic)	15	406
Bee, pollen gathering (mainly radial)	8	145
Bee, buzz pollination (radial, porose anthers)	2	8
Long-proboscid fly type 1 (spring flowering, red/mauve)	9	33
Long-proboscid fly type 2 (spring/summer, white/pink)	5	30
Long-proboscid fly type 3 (summer/fall, white/pink)	11	54
Scarab beetle (bright, dark marks, open)	10	62
Scarab and bee (as above but with nectar)	7	59
Moth (pale/dull, long tube)	9	62
Large butterfly (red/yellow, long tube)	6	16
Sunbird (red/orange, elongate)	8	75
Wasp (dull, deep cup)	2	3
Short proboscid fly (yellow, deep cup)	3	9
Dung/carrion fly (dull, putrid odour)	2	7
Generalist	6	24

Source: After Goldblatt and Manning 2006.

world, but a relatively poor representation of pollinating insects, especially of bees: anthophorines are the most common type present (Johnson 1992; McCall and Primack 1992). Johnson and Steiner (2003) noted the relatively few bees overall (roughly 15 times as many plant species as bee species, compared with only 2–3 times as many in north temperate and subtropical systems), but also drew attention to unusually high numbers of long-tongued flies, large satyrid butterflies, hawkmoths, flower-visiting beetles, rodents, and oil-collecting bees. These yield many highly specialized examples of pollination interactions covered in earlier chapters: the *Rediviva* bees taking oils on *Diascia*,

many deceptive orchids using bees and wasps, the specialist flowers visited by long-tongued flies (Bombyliidae, Nemestrinidae, and some Tabanidae), unusual beetle-pollinated flowers, and many geoflorous flowers served by nonflying mammals. There is also an unusually high proportion of wind-pollinated plants. The Iridaceae are instructive of the sheer diversity of plants and of pollination specializations in the Cape area: Goldblatt and Manning (2006) documented 17 different pollination systems (many of them shown in table 27.1), and multiple instances of pollinator shift within the family, and they recorded only 2% of the species in their visitor spectra as being generalist.

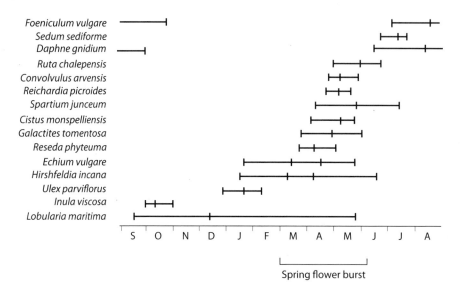

Figure 27.4 Flowering time in key plants from a northeastern Spain community, with vertical lines showing the onset, peak and end of each flowering period; March to May show a substantial community flowering burst. (Redrawn from Pico and Retana 2001.)

Timing and Phenology

Petanidou and Lamborn (2005) provided a useful review of the general characteristics of pollination ecology in Mediterranean communities. Flowering is strongly skewed toward spring (hence March to May in Europe and North Africa, September to December in South Africa, Australia, and Chile), and many plants have summer dormancy. A few flower in autumn, often with no leaves present (hysteranthy), and these often have hydrophobic floral surfaces or a pendant habit to avoid rain wetting their pollen; some *Crocus* and *Cyclamen* species are typical. But a few flowers are often available all year round, mostly from deep-rooted perennials. Figure 27.4 shows a phenological example from northeast Spain. Most plants have a flowering period of 2–3 months (see Kummerov 1983). Typical spring flowers in bulbous species are rather showy, whereas flowers borne on herbs are rather small, diurnal, and with rather low nectar yields (see fig. 8.7), although in the summer period the flowers tend to be even smaller. However, the sequence of flowers over the crucial spring months is such that a wide range of visitors can be supported, as shown for an Israeli site by Potts et al. (2003b). In many sites around the Mediterranean there are apparent "waves" of color in the spring flora, with predominantly red, pink/mauve, and yellow phases (although somewhat overlapping; e.g., Bosch et al. 1997); and it is often the case that the yellow flowers are nectarless with rather dry pollen while the pink/mauve flowers are the best nectar sources.

A few species do break the rules, by flowering at dusk with much larger nectar rewards; *Capparis* (the caper plant) is a classic example, abundant through the Mediterranean basin and attracting both hawkmoths and the nocturnal *Proxylocopa* carpenter bees. *Lobularia maritima* is also unusual in flowering for anything up to 10 months of the year, peaking in autumn (Pico and Retana 2001); it is a common brassicaceous plant of coasts and scrublands around the Mediterranean, mainly visited by flies and a few beetles.

Certain plants may be identified as key components in the community, linking between different phases of flowering and providing the core forage for pollinators. An example of a sequence and key linkages is shown in figure 27.5, modified from Petanidou and Potts (2006).

Higher levels of dioecy in plants are characteristic compared with temperate habitats. About 20% dioecy was recorded for shrubs in maquis and sclerophyll forest in Italy (Aronne and Wilcock 1994), many of the dioecious species having particularly small flowers and small fleshy fruits. Where monoecy occurred it was associated with dry single-seeded fruits, while the hermaphrodites tended to have larger flowers and dry

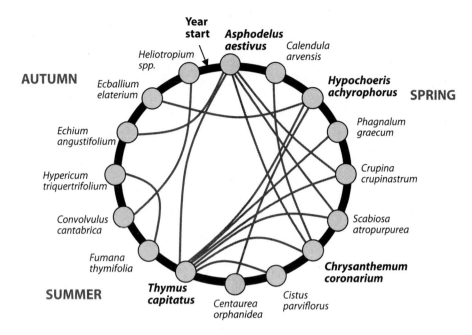

Figure 27.5 Key components in a Mediterranean flowering community. Central thin lines connect two flowering species sharing at least ten visitor species, so that four key plant species (bold) link seasons and provide food continuity for pollinators. (Modified from data in Petanidou and Potts 2006.)

many-seeded fruits, with a lack of vegetative spreading systems. The very common wild thyme (*Thymus vulgaris*) is gynodioecious (small female plus larger hermaphrodite flowers), and is spread mainly vegetatively (which should favor gynodioecy; see Charlesworth and Charlesworth 1978).

Given the structured but rather limited variation in plant phenology, there is surprisingly high variability in insect phenology, a characteristic thought to be linked to the variations of climate (Herrera 1988; Petanidou and Ellis 1993). Specific sequences of solitary bee emergence are particularly obvious in these zones.

Specialization in Mediterranean Systems

Petanidou and Potts (2006) compared three such systems, in Greece, Israel, and Spain. Web connectance values for these were very low (table 27.1), and connectance decreased with increased community size (cf. Olesen and Jordano 2002), being markedly lower for the more speciose Greek and Israeli sites. A very high proportion of monotropic insects was recorded at all three sites (varying from 36% to 56%), while the

proportions of monophilic plants were below 10% in Greece and Spain but higher (36%) in Israel (where only bees were recorded), indicating a marked asymmetry in specialization (an issue discussed in chapter 20). However, when the Greek site was compared using these authors' preferred parameter of "selectivity" (see chapter 20), values were similar and very high for the plants and the animals, with most species of both groups having a selectivity of greater than 0.9 (fig. 27.6). Here, then, both the plants and the insects were selective about their interacting partners, and generalization would definitely *not* appear to be the rule. (Curiously, the authors concluded that because relatively specialized plants and animals were usually associated with moderately generalized partners, giving asymmetry, the communities should be regarded as having "very limited specialization").

3. Pollination in the Humid Tropics

Tropical forests occur on all the southern continents, straddling the equator (see fig. 27.1), notably in West Africa, southeast Asia, and northern Australasia, but the largest areas are found in Neotropical Amazonia.

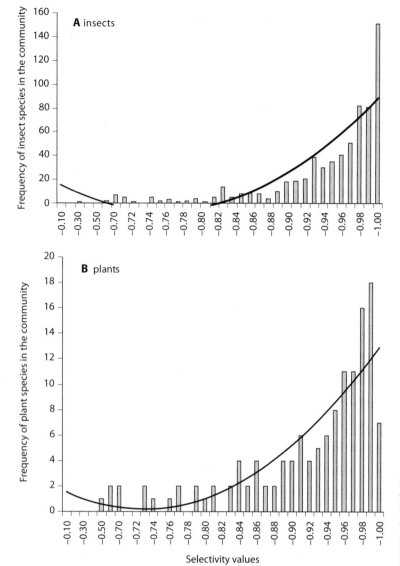

Figure 27.6 Selectivity values for insects and for plants, in a Greek phrygana community; selectivity is the ratio $E-T/E$, where E is the available number of species and T the species used as partners. Note that intervals are smaller at higher S values. (Redrawn from Petanidou and Potts 2006.)

Habitat Characteristics

Where rainfall exceeds 2000 mm per year and is evenly distributed, evergreen rainforest results, almost nonseasonal and consistently warm and wet and windless within the canopy. In areas where there is some variation in rainfall patterns the forest is somewhat more seasonal. Mean monthly temperatures in all these tropical forests may be as high as 24°C–28°C.

The core tropical forests show greater species richness, and indeed greater richness at most taxonomic levels, than any other ecosystems, and often also show the highest primary productivity levels (Gaston and Williams 1996). Neotropical and southeast Asian forests have higher plant species richness than African forests, but structural complexity is similar in all three continents. However, tropical forest soils are diverse in character on different continents and also within large land masses such as Amazonia, and may produce surprising local variability in the overlying forest diversity. Sometimes the soils are deep but well weathered, forming reasonably fertile clays; but where rainfall is

TABLE 27.2
Comparisons of Three Mediterranean Plant-Pollinator Communities

	Greece		Israel	Spain	
	All insects	Bees only	Bees only	All insects	Bees only
No. of plant species visited	132	129	179	26	24
No. of visiting insect species	665	262	340	180	55
No. of interactions	3006	1390	921	415	155
Mean plant phily (range)	23 (1–123)	11 (1–48)	5 (1–43)	16 (1–86)	6 (1–25)
Mean insect tropy (range)	4.5 (1–103)	5.3 (1–103)	2.7 (1–20)	2.3 (1–17)	2.8 (1–14)
Connectance	3.4	4.1	1.5	8.9	11.7

Source: Petanidou and Potts (2006).

particularly high, and where forests lie over sandstone, the soils may get severe weathering and only a thin layer of organic soil persists over meters of whitened quartz.

Flora and Pollinating Fauna

Tropical forests are characterized by a pronounced vertical stratification of vegetation from the tallest emergent and canopy trees (above 50 m) down to woody shrubs, large herbs, and creepers, and then infrequent small herbs, but with vertically growing lianes traversing the strata. The end product may be up to 1000 species of plant per hectare in total. All levels of this vegetation bear flowers and interact with flower visitors, although flowering is often not very obvious near the forest floor. Despite the architectural plant diversity, the overall above-ground animal biomass in rain forests, at least on the visible scale, is surprisingly low. Among the vertebrates, many herbivores, frugivores, and pollinators are arboreal and consequently small. But birds and insects are abundant throughout the three-dimensional architecture of the forest.

The flowers cover the entire range of blossom types, with a very high diversity of floral forms, reflecting in part the greater diversity of plant growth forms in the stratified community. Most flowers are small and not especially conspicuous visually, and a first impression of the forest is often of a rather dull green leafy panorama with flowers hardly to be seen. The larger, deep, and showy flowers present are often rather high up; this is no doubt linked to the fact that the larger pollinators (bats, birds, primates, large moths) are arboreal or are good fliers. The largest flowers known are tropical, although this is probably in part due to the lack of constraints on growth; and the very largest forms tend to grow at floor level where their weight is well supported. Some particularly specialized floral morphologies do occur, although they tend to arise from more variety within particular phenomena rather than from whole new phenomena appearing.

A spectacular variety of flower positions on the plant is notable in many forests. Flowers occur on trunks (cauliflory; see *Phylloctenium*, plate 15E) and on the main branches (ramiflory), long pendulous peduncles dangle from branches (flagelliflory; see *Kigelia*, plate 28B), or a profusion of flowers shoot out of the canopy crown. Occasionally there are even flowers on foliage or flowers from underground roots.

Among the pollinators, there is a very well-documented diversity of pollination mechanisms. Gentry (1982) reported that more than 80% of all woody tropical species are animal-pollinated, and Bawa (1990) put the total figure for all tropical plants at 98%–99%. Table 27.2 shows a breakdown of examples of different

TABLE 27.3
Pollination Types in Neotropical Forests

	Lowland forest canopy	Lowland forest understory	Upland forest canopy	Mean
Wind	-	3	0	2
Vertebrates				
Bats	4	4	2	4
Hummingbirds	2	18	5	8
Insects				
Larger bees	44	22	17	27
Smaller bees	8	17	35	19
Beetles	-	16	4	7
Butterflies	2	5	12	8
Moths	14	7	4	8
Wasps	4	2	3	3
Other insects (mainly flies)	23	8	17	16

Source: Bawa et al. 1985, Kress and Beach 1994, and van Dulmen 2001.
Note: Mean values are derived from five different studies in Costa Rica and Colombia.

pollination types, and mean values for their frequency, from some Neotropical forest sites. Wind pollination is rare, occurring in only about 2%–3% of plants; this is linked to the massive species diversity with conspecifics widely spaced, and to the lack of air movement below the canopy (see chapter 19). Bats visit 4%–5% of both canopy and subcanopy flowers, whereas hummingbirds are visitors to 5%–8% of canopy flowers but approaching 20% of shadier subcanopy and understory flowers. In these same forests, bees may visit at least 50% of the canopy flowers (up to 95% of canopy flowers in some areas) and a slightly smaller proportion of the more shade-tolerant flowers, with larger and medium-sized bees more frequent visitors than small bees at nearly all sites.

Sargent and Vamosi (2008) produced a meta-analysis of pollination guilds in relation to the light levels experienced in the forest, with all but 15 of their 488 plant species being from tropical forests. Their findings are summarized in table 27.3. All groups tended to be most common at low to mid light levels, but this was particularly true for Hymenoptera (effectively the bees) and for Diptera.

In terms of numbers of plant species visited or pol-
linated, bees remain the most important group, and although their diversity is not especially high (Heithaus 1979), some new bee groups do appear, notably euglossines in the Neotropics. Colonial stingless bees are very important too, and may constitute a quarter of all bees in the rainforests of French Guyana (Roubik 1989). In southeast Asian forests, the stingless bees and *Apis* are again particularly common (Corlett 2004), together with a high proportion of beetles; these patterns are especially clear in the lowland dipterocarp forests, with their unique pattern of general flowering episodes. Wasp flower visitors also become more common and conspicuous. Crepuscular flowers (pollinated mainly by bats and sphingid moths) are often very noticeable, and many produce especially strong scents. Birds are of course very important, mainly hummingbirds in the Neotropics, but sunbirds and others elsewhere (chapter 15).

But these are generalizations, and in practice plant type and pollinator distribution vary markedly, above all with height:

1. In the canopy, with flowering trees and additionally some epiphytes and lianes, flower visiting is dom-

Table 27.4
Presence of Major Pollinating Groups, as Percentage of Totals, at Different
Canopy Light Levels in Forests (Mainly Tropical)

Pollinator	Forest floor Light level 1 %	Understory Light level 2 %	Substory * Light level 3 %	Canopy, or gap Light level 4 %	No. of species
Hymenoptera	8.6	44.4	17.9	29.2	247
Coleoptera	7.1	36.0	32.3	25.2	127
Lepidoptera	10.7	30.9	33.3	25.0	84
Diptera	2.7	54.8	17.8	24.7	73
Birds	0	52.6	28.1	19.1	57
Mammals	0	40.0	40.0	20.0	10

Source: From Sargent and Vamosi (2008).
*Includes lianes and epiphytes.

inated by medium and larger bees (see table 27.4), which constitute up to 44% of visitors in Costa Rica. However, other than species with flowers catering for bees and some moths, many canopy trees tend to have small and inconspicuous, often greenish or yellowish flowers, strongly generalist, and often dioecious.

2. The subcanopy trees are more diverse in flower type and pollination type, but bees are still most important. Both trees and lianes have good representation from the Fabaceae, and these are strongly associated with medium and large bees. Also there are more of the smaller bees (Halictidae, Megachilidae, and stingless species) and more beetles and hummingbirds. Bird pollination is less common there outside the Neotropics since perching types largely occur in the canopy and not in the understory (Pettet 1977; Appanah 1981); this must be in part because these birds often forage in flocks, so that they can be accommodated only in the larger trees (Stiles 1981). More beetles appear at lower heights, and larger hawkmoths are prominent in the lowest canopy layers. Some bat pollination occurs sporadically throughout the tree layers. The rest of the vegetation types here, apart from the trees, are not dissimilar in their pollinator-type distribution, with bees still most important, and hummingbirds common on the nonwoody plants in Neotropical areas. Quite strong links appear between families: hummingbirds especially occur on Bromeliaceae (all epiphytes), Gesneriaceae (herbs and epiphytes), Passifloraceae (climbers), and *Heliconia* species (shrubby herbs), especially in gaps

and edges. Understory layers have many species of Rubiaceae, often visited by moths, butterflies, and again hummingbirds. Beetles visit many species of Annonaceae, Lauraceae, and Araceae in the lower understory, while Bombacaceae are often linked with hawkmoths and bats. However, there are also many highly generalist small or tiny social bees from the stingless group (*Trigona,* etc.), often robbing more specialist flowers as well as being part of the generalist community.

3. The forest floor, often with very low light levels, has relatively few herbs, plus a few ferns and orchids and mosses. Where grasses occur they still tend to be insect pollinated (Soderstrom and Calderon 1971).

Much of the description above is derived from Neotropical sites, which have until recently been much better studied. The southeast Asian forests are rather different and particularly odd, being exceptionally stable and often dominated by the single family Dipterocarpaceae, which flower only intermittently and often in bursts (mast flowering), triggered by a drop in temperature (Ashton 1988). One tree can produce several million flowers over just 2 weeks. Medium to large, mainly social, bees are most important, but cannot cope with real flushes when there may be thousands of flowers simultaneously. Some dipterocarps therefore use generalist pollinators, an extreme example being the use of thrips in *Shorea* and some other genera, where a foundress thrip can produce four generations and several thousand individuals

TABLE 27.5
Percentages of Plants with Different Pollination Syndromes in Different
Successional Stages in Costa Rican Rainforest

	Successional age (months from establishment)				
	6	12	24	36	Mature
Wind	38	22	8	6	Lower
Small bee	34	40	40	38	~
Medium bee or wasp	7	18	23	19	Lower
Large bee	0	0	3	6	Higher
Beetle	0	4	0	3	Higher
Small lepidopteran	21	11	10	9	~
Large lepidopteran	0	4	5	3	~
Hummingbird	0	4	10	10	Lower
Bat	0	0	0	3	Higher

Source: From Opler et al. 1980.

in 2–3 weeks and so colonize thousands of flowers (chapter 12).

There is a particularly high diversity of beetle pollination in Australian rainforests, linked with high representation from families such as Annonaceae (Bawa 1990). There is also an unusually high diversity of hawkmoth pollination in Madagascan forests, together with some unique instances of lemur pollination (chapter 17), probably linked to the absence of other long-tongued insect taxa and of primates.

On all continents, where the forests occur at higher altitude they tend to turn into cloud forest, and here the proportion of bird pollination increases (Cruden 1972a; Linhart et al. 1987), probably because the cooler temperatures and low cloud limit insect activity. Bees and beetles certainly tend to become rarer at altitude, although flies, hawkmoths, and other lepidopterans are still commonplace. More plants also become capable of selfing (Sobrevila and Arroyo 1982). Hummingbirds are very dominant on some Neotropical mountains (Linhart et al. 1987), while in Africa many peculiar plants appear in montane forests, often linked with sunbirds; examples include *Leonotis*, *Protea* species, giant heathers (*Erica*), and giant *Lobelia* (Dowsett-Lemaire 1989).

Finally, trends in flowering and pollination can also be seen with secondary successional stages in a rainforest. Opler et al. (1980) reported higher levels of self-compatibility in early stages, with more dioecism and self-incompatibility in later succession; and as the plants in later successional stages were larger so there were also increases in their spacing, in flower size, and in pollinator size (see table 27.5).

Patterns of Reproduction and Flowering in Tropical Rainforests

With the high levels of tropical plant diversity in all forests, there is often a relatively enormous distance between conspecifics; for example, between 30% and 65% of all trees in parts of the Amazon Basin grow at densities of less than one per hectare. This links to differences in plant reproductive strategy, and table 27.6 shows estimates of compatibility patterns and dioecy, based on Bawa (1990), whose study indicated that tropical forests were characterized by moderately high levels of dioecy and rather low levels of self-compatibility. More recent work, including studies from the canopy layer, suggests that levels of dioecy up to 25% in rainforest plants are indeed common (Schatz 1990). Monoecy is rather rare, and selfing also unusual in most strata, with self-incompatible (SI) systems commonplace and probably rather varied in their mechanisms.

Very rapid floral opening is a widespread occurrence (each corolla opening to its full width in just a

TABLE 27.6
Reproductive Systems in Tree Species from Tropical Forests

Forest type	Percent dioecy	Percent self-compatible
Lowland forest, Costa Rica	23	20
Lowland forest, Panama	9	
Island forest, Hawaii	27	
Montane forest, Venezuela	31	62
Montane forest, Jamaica	21	85

Source: From Bawa 1980a. 1990.

few minutes or even seconds); this is often due to sepals or tepals being united in the bud and coming under tension as the flower grows. Sometimes explosive opening is initiated by the arrival of a pollinator (e.g., in some Fabaceae, Loranthaceae, and Proteaceae). This fast-triggered anthesis may help protect the bud from being damaged before it is ready.

Once the bud is open, a brief flowering period is also common, so that the great majority of flowers are attractive and functional for only 1 day (Bawa 1983), and a particular plant may only be in flower for a few days. This is perhaps related to strong competition among plants and among the pollinators, and would make sense in the seasonal tropics especially, where most plants are flowering just as the rains begin, and each takes only a brief "slot" in the prime period of 2–4 weeks (this is particularly obvious in the less diverse seasonal forest of eastern Madagascar (pers. obs.), where many species' flowering periods last just 2–3 days). Synchronous flowering is common and largely related to reducing predation on seeds, but with peak flowering correlated with peak irradiance (usually at the start of a rainy season in the more seasonal forests) (van Schaik et al. 1993).

Flowering patterns in completely nonseasonal forests are more varied, but trees in particular are often noted for big-bang patterns (chapter 21), and masting or mass flowering is common in the Neotropics (Bawa 1983; Bullock et al. 1983). It also occurs in Asian dipterocarp forests, where Appanah (1993) recorded heavy bouts of mass flowering every 2–10 years; these were probably triggered by small temperature dips. They led to almost half the mature plants and over 80% of the canopy trees flowering for a period of 3–4 months, but with some evidence of interspecific sequencing of flowering within this period, as shown in figure 27.7

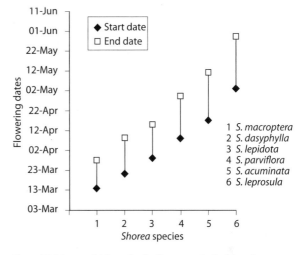

Figure 27.7 Sequential flowering in *Shorea* species in Asian dipterocarp forests. (Modified from Appanah 1993.)

for *Shorea* species. Maycock et al. (2005) showed that masting in Bornean dipterocarp trees helped to increase seed and seedling survival for most species, with low-intensity masting along with many conspecific trees also helping to improve outcrossing.

Problems for Animal Pollination

1. First, a high percentage of plants are obligate outcrossers (Bawa 1974; Bawa and Opler 1975), with Ward et al. (2005) finding greater than 90% outcrossing for 45 species. Yet the vast majority of plants rely on biotic pollination in most of the forests that have been studied in any detail (Bawa et al. 1985; Bawa 1990; Endress 1994). Thus the visiting animals

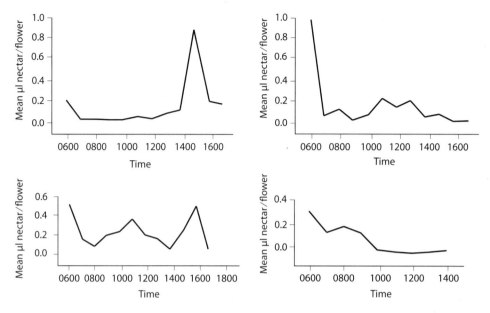

Figure 27.8 Nectar patterns varying on different days and for different trees of *Myrospermum frutescens* in Costa Rican tropical forest. (Redrawn from Frankie and Haber 1983.)

are required to work fairly intensively if the plants are to succeed.

2. There is also a very high floral biodiversity, such that each plant or tree may be hundreds of meters or even kilometers away from its nearest-neighbor conspecific. Yet pollinators can and do move effectively among and between the resulting highly dispersed flowers (Gilbert 1975; Frankie et al. 1976; Augspurger 1980; Appanah 1981, 1990). A review of pollen-mediated gene flow in Neotropical trees by Ward et al. (2005) found pollen dispersal ranging up to 19 km, and with a mean of 200 m. How do tropical forest trees get such long-range pollinator movement?

One strategy is to avoid mass flowering but instead to produce a few flowers each day, over a long period, and attract trap-lining large bees or birds (Janzen 1971; Gilbert 1975). This relies on pollinators that have good spatial memory, and the flowers also need to be particularly attractive; large, showy, often highly scented (except where hummingbirds are used), and with a high reward to more than cover the costs of inter-tree flights.

But that is the minority approach; as discussed above, big-bang mass flowering using photoperiodic cues is more common and can in itself be seen as an adaptation to cope with low population density. Where this occurs, a plant must nevertheless ensure that any one visitor does not go to too many of its flowers: it should be unpredictable, varying its flowering time and the flowers' rewards a little (Perry and Starrett 1980; Frankie and Haber 1983). The timing and quality and quantity of nectar may be crucial, and having empty flowers as cheats might be a bonus (chapter 23). Frankie and Haber (1983) tested this idea with seven species from three families (Bignoniaceae, Caesalpiniaceae, and Fabaceae), all visited mainly by larger bees (*Xylocopa* and euglossines). All had one-day flowers, mostly opening before sunrise. They found that flowering was seasonally synchronous but actually slightly staggered between individual trees in any one area, by up to a few days, so that bees were faced with a mosaic of different reward patterns. The nectar patterns also varied between trees quite substantially (fig. 27.8), and for some species also varied between days for any given tree. Bee visits mapped onto peak nectar availability on a tree-by-tree basis. Even within a tree these authors found an earlier nectar flow in lower branches of *Andira* trees (Fabaceae) compared with the higher branches, the bees (mostly *Gaesischia* and *Centris*) also moving progressively up the tree almost as a circular band around the crown.

In dioecious species this variability can be taken one stage further such that there is different nectar production timing in males and females. In *Mappia racemosa* (Icacinaceae) in Costa Rica, Frankie and Haber (1983) noted that males had nectar available about 1 hour earlier than females, facilitating movement from male to female individuals. A similar pattern occurred in the androdioecious tree *Xerospermum* (Sapindaceae) in Malaysian forests (Appanah 1982).

A further useful possibility if a plant is mass-flowering is for there to be strong aggressive interactions (direct competition, often with territoriality) between visitors or pollinators, which may help reduce within-plant movements by any one individual. Such interactions are particularly common in the ubiquitous meliponine bees of the rainforests (see chapters 18 and 21).

3. Temperature and water balance may cause problems. The high ambient temperatures raise the possibility of overheating in pollinators, and larger endothermic animals when flying may readily overreach their T_b limits and need to cease flight temporarily to cool down. They may also experience excess water loads as they generate metabolic water internally by their own flight activity (chapter 10) and cannot unload this quickly enough into humid air. Birds, bats, and large bees may all suffer in this way, and have a tendency to urinate copiously, often while flying.

In practice, most plants have no shortage of water and produce dilute nectar (15%–25% concentration is common), which stays dilute at the high ambient humidities (chapter 8). Most birds and bats have to drink dilute nectar anyway, because of their tongue structures and lengths, and this high-volume dilute input can exacerbate their water balance problems. But almost all the large bee-pollinated rainforest plants have nectar above 30% sugar concentration (Frankie et al. 1983), which may help to reduce their visitors' water loading.

Specialization in Tropical Forests

It is generally perceived that tropical communities are more specialized at a broad ecological level, and that the same is true in pollination ecology (Momose et al. 1998). Most workers who have analyzed particular taxa have reported such an effect, and the meta-analysis by Sargent and Vamosi (2008) reported that 81.2% of species were visited by just one functional group of pollinators. Certainly in the tropics the classic specialist relations are very well known: pollination by bats, by large long-tongued hawkmoths, by resin- and fragrance-collecting bees, or by long-tongued flies, as well as the obligate relations of figs and fig wasps. High levels of specialization evidently do occur in some families—figs, orchids, etc.—but these are perhaps only a small proportion of the flora. In practice, from ground level everything looks rather sparse and very generalist, with many small white and greenish flowers, and with just a few exciting patches of colorful and unusual flowers at mid-heights or around clearings and paths.

To clarify matters, there have recently been broader analyses of the patterns across different habitats, discussed at length in chapter 20. Ollerton and Cranmer (2002) found no evidence for increased specialization overall in tropical networks, while Olesen and Jordano (2002) reported that the plants showed more specialization, although the insects did not. However, Ollerton et al. (2006) fairly convincingly showed an increase in diversity of pollination types in the tropics compared to all other latitudes, and in forests compared to simpler habitat structures. Work on southeast Asian forests also raised new issues, with communities from India and Malaysia proving structurally more like each other and rather different from most western hemisphere forests (Devy and Davidar 2003).

Species richness and specialization within a community could increase for several reasons, but interactions with pollinators certainly underlie some of the possible explanations (Ollerton et al. 2003). Plants could alter their niche axis by exploiting new kinds of flower visitor; or they could decrease their niche breadth by concentrating on a particular visitor (directly leading to specialization); or they could increase niche overlap with other plants by sharing pollinators. Any of these three mechanisms could be occurring in tropical forest communities, and examples of each have been covered in earlier chapters.

4. Pollination at High Latitude and High Altitude

Essentially this section deals with pollination at low ambient temperatures, whether in the boreal and arctic zones (primarily of the northern hemisphere), or in montane habitats at moderate to high latitudes. Some winter-flowering plants from more temperate zones can also be included here as they share some similar problems.

The circumpolar ecosystems are marked by extreme cold; the records for lowest temperatures in overland climates are held by Siberia in the northern hemisphere (−68°C) and by central Antarctica in the southern hemisphere (−89°C). These zones also show a negative winter energy balance (more heat is lost than is being gained from the sun) and a very short growing season. The cold biomes comprise three main biotic types: the polar zones with continuous ice or snow cover, the tundra zones with minimal low vegetation in the short summer, and the taiga where coniferous forests thrive. Cold biomes could also include areas of cold desert, often found on relatively high-altitude plateaus in the interior of large continents (e.g., the Tibetan plateau).

Temperatures vary seasonally in all these zones; night temperatures are usually below 0°C at all times of year, and daytime temperatures may be only slightly above freezing in the "warmer" summer growing season, which may be further compromised by high winds. Several problems are posed, and solutions found, for flowers to achieve successful pollination in these circumstances.

The Antarctic has just two species of flowering plant, together with an abundant fringe of lichens and algae. But in the terrestrial Arctic fringes, topography and the maritime influence have a substantial effect on local climate, with warmer air flowing down any slopes and giving favorable conditions for plant growth. The plant material itself gives further amelioration, so there may be a temperature excess of at least 20°C above ambient among the diverse small plants that survive.

In tundra the characteristic vegetation is abundant moss and lichen; at best there may also be very sparse and very low-growing trees (especially dwarf willow) and some coarse grasses. No large trees or shrubs can survive because the permafrost precludes taproot penetration, and tall plants with shallow roots would inevitably be blown down. Plant biomass is mostly underground as roots and extensive rhizomes, and low cushion-like plant growth helps to trap the decomposing leaves each autumn and allow local, slow nutrient recycling. Many of the evergreen plants that persist have small leaves with thickened and hairy cuticles, and most have vegetative budding from runners with protected ground-level buds to allow rapid growth in spring.

The coniferous taiga forests stretch in a wide belt across Canada, Scandinavia, and northern Europe, plus most of Russia, from the northernmost treeline down to a gradual merger with more deciduous temperate forests. The climate allows only a short growing period (2–3.5 months), and the summers though fairly warm have limited rainfall. Conifers always dominate, usually in low diversity, so that these regions have a dense layer of decomposing conifer needles up to 100 mm deep, producing a very acid soil but providing a thermally buffered zone; only acid-living shrubs (mainly Ericaceae) thrive in these conditions.

Short Seasons

There may be only a few weeks in the year when temperatures allow plant growth, and flowering must be achieved within this period. This inevitably produces competition within and between species for any available pollinators, and often a tight temporal patterning. For example, in the very short summer in northern Sweden, Stenstrom and Bergmann (1998) found that bees showed strong dependency on pollen-presentation patterns of the very limited available flora, so that *Bombus alpinus* began foraging on unspecialized *Saxifraga* early on, but moved to *Astragulus* and then *Bartsia* (both more specialized and typical of bee-pollinated flowers) as soon as they started flowering; on any one day thereafter the bees had 90%–100% pollen from one of these species in their scopae, and the saxifrages were no longer visited.

In both subpolar and montane habitats the start of the growth season may be asynchronous and patchy, with a mosaic of small habitats arising fairly unpredictably at the snowmelt margins (Kudo 1991, 2006; Stanton et al. 1997; Yamagishi et al. 2005). Flowering may start early in wind-blown areas relatively free of snow, but very late within deep snowfields (fig. 27.9), with a given species such as *Rhododendron aureum* being two months later to begin flowering in the latter sites.

Aizen (2003b) showed the stresses of short seasons that can result for Andean mistletoe (*Tristerix*). The pollinators (hummingbirds) were rare on early flowers, but seed dispersers (marsupials) were abundant in the short summer and thus most effective when the earliest flowers reached seed set; hence flowers had to be produced mainly in winter even though the pollinators were rare.

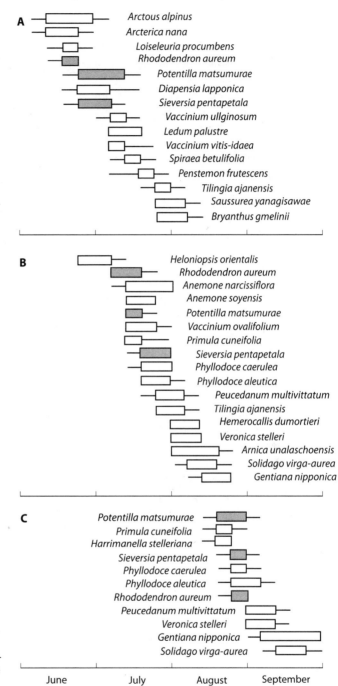

Figure 27.9 Variations in flowering time for entomophilous species along the snow line of (A) a fellfield, (B) an early shallow snowbed, and (C) a late deep snowbed. Three species common to all and with very different phenologies are highlighted in gray. (Redrawn from Kudo 2006.)

Limited Pollinators Available

The fauna of tundra and fringe polar regions is very restricted, and some types of pollinators are infrequent, with many groups missing entirely. Flies (especially the more primitive biting flies and small muscids) are particularly common, and many flowers use them (e.g., Kevan 1972; Pont 1993). Muscids and empids are rather generalized in their flower visiting habits, apart from a few longer-tongued specialized muscids (Elberling and Olesen 1999). With flies abundant, open bowl-shaped white and yellow flowers are common; interestingly Lazaro et al. (2008) found that these flower types were among the more specialist forms in an alpine community, though rather generalist elsewhere.

Endothermic bees are able to cope with low temperatures, hence bee pollination is dominant in the less extreme parts of the northern tundras (Ranta et al. 1981). *Bombus* extends well into the Arctic zone, for example occurring at 81°N on Ellesmere Island in Canada (Hocking 1968; Kevan 1972). Following Bergmann's rule, the bee fauna tends to be of large body size in these colder climates. There is also a trend of increasing tongue length with increasing latitude in bumblebees (Ranta et al. 1981; Pleasants 1983), although this is not matched by any increase of corolla length. Reduced competition from other species probably allows the longer-tongued bees to visit short and medium corollas more successfully (see chapters 18 and 22).

Limited pollinator availability has consequences for plant reproduction. Some plants show a mix of wind and animal pollination perhaps as a fail-safe, with *Salix* species varying between 2% and 52% anemophily in Norway (Peeters and Totland 1999), where the lower values corresponded with longer female catkins and higher nectar content. Low pollinator numbers also lead in some cases to increased influence on overall seed set from rather poor-quality flower visitors such as ants. The alpine skypilot *Polemonium viscosum* in a high tundra population suffered 20% reductions in seed set due to flower damage from ants, and at high density ants could reduce the selection for bumblebee pollination (Galen and Geib 2007).

A little further south but still in the colder ecosystems, Kevan et al. (1993) suggested that the diversity of pollinators in boreal forests was not much reduced overall; but he suggested that specializations though present were less developed, perhaps due to the evolutionary youth of the habitats and the frequent perturbations by fire or by insect pest outbreaks.

Flowers in colder climates show a tendency to be protogynous (chapter 3), because the anthers can persist into old age if the flower has not been pollinated (rather than shriveling at a predetermined age to allow stigma maturation), and thus the flower can still self-fertilize as a last resort; this is often necessary in habitats where pollinators are scarce and crossing is unreliable.

The problem of limited pollinator availability also applies to any plants that flower during the winter in temperate climates. A famous and instructive example here is the gorse *Ulex europaeus*, where some plants only flower in winter and early spring (Bowman et al. 2008), apparently thereby escaping from herbivory. The winter-blooming plants showed similar flower size and pollen production to those flowering in autumn or spring, but had a much longer flowering period, which (combined with their high attractiveness in winter, and almost no competitors) ensured that they received sufficient pollinators on the rare warmer days.

Increased or Decreased Floral Showiness

There is some evidence for enhanced floral longevity in alpine and boreal flowers, giving a longer period of display; for example Utelli and Roy (2000) recorded *Aconitum* flowers lasting 6.4 days at altitude but only 4.2 days at low elevations. Furthermore, while flowers tend to be open and bowl shaped, and thus fairly unspecialized, they are also rather showy, often with quite large glistening white or brightly colored corollas. Where more specialized morphologies occur, as in the bumblebee-pollinated sea pea (*Lathyrus japonicus*), the color is bright and the flowers normally last at least 6 days (Asmussen 1993). These floral traits may represent a strategy to enhance the probability of pollination by taking maximum advantage of the short season and limited pollinator numbers. However, there is apparently a reduced impact of flower size on pollination success (e.g., in *Ranunculus acris*; Totland 2004), probably because seed set is limited by resources rather than by pollen receipt. For *Dryas octopetala*, Lundemo and Totland (2007) found both these effects operating at different elevations in Norway; in particular, sites with lower visitation had significantly longer-lived flowers.

There is also a distinct alternative strategy, of small flowers with high self-compatibility and high levels of autogamy (Arroyo et al. 2006). An example of this is *Campanula uniflora*, growing in Greenland and Iceland, where the most northerly Iceland flowers have switched to cleistogamy, shedding pollen internally to self-fertilize while the buds are still closed (Aegisdottir and Thorhallsdottir 2006). Two Norwegian *Cerastium* species compared by Totland and Schulte-Herbruggen

(2003) also neatly make the points here; *C. cerastoides* is a pioneer species in various disturbed moderate-elevation habitats, while *C. alpinum* is an alpine specialist; both show similar levels of autogamy but the former has small pale flowers and higher self-compatibility, the latter has larger showier flowers and better insect visitation.

Flowers Need to Keep Warm to Speed Ovule Development

Microclimatic variation plays a crucial role in cold habitats, and flowers are able to use the incoming radiation (which can be intense in the brief summer) to achieve a degree of warm-up. This can be in part passive, by choice of growing site; for example *Chaetanthera* flowers growing on more sunlit northern and eastern slopes in the Andes could be 2°C–3°C warmer than those on southern slopes, and received 4–8 times as many visits from butterflies and andrenid bees (Torres-Diaz et al. 2007).

Some flowers have a more active form of radiative gathering, their corolla bowls shaped as parabolic reflectors (see fig. 9.10) to reflect radiation internally onto their own centers and so warm up the ovaries. Several, such as the snow buttercup *Ranunculus adoneus*, take this further and show heliotropism, behaving as solar trackers, with their flowers rotating around each day to keep pointing at the sun and maximizing the radiative input (chapter 9; see Ehleringer and Forseth 1980). In *R. adoneus* this rotation gives a 30%–40% advantage in germination and pollen tube growth compared with artificially tethered flowers (Galen and Stanton 2003), while for *Dryas integrifolia* Krannitz (1996) found that heliotropic flowers achieved warmer gynoecia, more insect visitors, and heavier seeds than conspecifics that were naturally nonheliotropic but merely pointed at the noon sun. In the Austrian Alps, Luzar and Gottsberger (2001) found temperature elevations within five different bowl-shaped plants (two *Ranunculus,* plus *Pulsatilla, Callianthemum,* and *Leucanthemopsis*), with the heliotropism having demonstrable effects on visit frequencies and residence times for *Ranunculus montanus*. Kudo (1995) found a mean temperature elevation of 5.5°C in the heliotropic spring ephemeral *Adonis ramosa* (Ranunculaceae), again with higher visit frequency in the most heliotropic flowers. Most spectacularly, in the high arctic of northern Greenland, heliotropism continued throughout the 24 hours of a midsummer day in *Papaver radicatum* (Mølgaard 1989); furthermore, yellow flowers achieved higher temperatures (6.5°C at the ovary) than white ones (5.1°C), with the yellow morph also increasing in frequency with altitude.

A few plants (especially Asteraceae) from cold high-altitude zones in Ecuador have nodding heads, which serve the dual purposes of protecting the reproductive structures from rain and snow and also gathering reflected radiation from the substratum, leading to raised intrafloral temperatures (Sklenar 1999).

The various mechanisms to increase floral temperatures can also be enhanced by having rather well-insulated flowers, a classic case being the edelweiss of the European Alps (*Leontopodium*, plate 34A), pollinated mainly by flies attracted to its rather odd honey-plus-sweat odor (Erhardt 1993). A number of other arctic and alpine flowers have rather woolly calyces or (less often) petals. Kevan (1990) reported denser pubescence and warmer temperatures in pistillate catkins of *Salix arctica* compared with the (smaller) staminate catkins.

Gentians are interesting alpines in this context, as some species show an ability to close the flowers overnight in response to cooling temperatures, as well as using a second more permanent type of closure in direct response to being pollinated (He et al. 2005). In the latter case the closed flowers are often retained on the plant for some time to add to the display effects, which also supports the point made above about elevated floral longevity.

Pollinators Need to Keep Warm

The issue of the pollinators keeping warm is strongly linked to the previous section, since any flower that is keeping as warm as possible for increased ovule development is also providing a warm spot for a pollinator, and that warmth may become part of its attraction (see chapter 9). This can be critical for small ectothermic flies that would otherwise rarely be warm enough to fly, but even endothermic bees will take advantage of the warmth in flowers by basking there, and thus save a little on their own warm-up costs.

Keeping warm will always be costly for a pollinator in cold weather, however. Flowers growing at higher (hence cooler) latitudes therefore also secrete more nectar than the same species at lower latitudes (Heinrich and Raven 1972), because the energy demands of their pollinators are greater. For example, Hocking (1968) recorded eight out of ten species producing more sugar per day at 82°N than at 57°N in Canada.

The question remains as to whether pollinators functioning in cold habitats have any special thermoregulatory abilities, and as yet there are no indications of this (although the larger body sizes of bees will help). Heinrich and Vogt (1993) found that queen bumblebees in the Arctic did have higher abdominal temperatures than temperate species, but there were no differences for workers or drones, suggesting that the differences may have had more to do with egg maturation than with flower visiting.

Special Problems of Pollination in Mountains

Mountains occur in all continents and in a wide range of forms, both as long mountain chains and as relatively isolated peaks. Elevation has effects on temperature, pressure, and oxygen availability. In mountain ranges daily average temperatures are reduced by about 1°C for every 150 m of altitude, and this effect is roughly similar at all latitudes. Thus even tropical mountain ranges sitting astride the equator can be snow covered and are profoundly cold at night even though highly irradiated by day, these diurnal cycles of extreme heat and cold being accentuated during the dry season. In more temperate latitudes, mountain ranges may show a lesser diurnal temperature range but a very drastic seasonal variation, so that temperate mountain ranges such as the Alps and Rockies have an extremely variable snowline.

Montane habitats suffer from many of the same problems as high-latitude zones, but with more drastic temperature changes between night and day, and high exposure to winds and cloud. Soils are usually very thin and nutrient depleted, so there may be resource limitations on plant reproduction even in summer (Munoz et al. 2005). Many alpine plants are reported to be pollen limited, often severely so (e.g. Kudo and Kasagi 2005 with *Phyllodoce*), but in some cases even when this is demonstrated by increased seed set following hand-pollination the final seed set remains low. For example, in *Chuquiraga* from the Andes, maximum seed set was only 5%–6% (Munoz and Arroyo 2006), suggesting limitation from abiotic resources;

and in *Ranunculus acris* higher-altitude plants did not achieve greater seed set when hand-pollinated, although lower-altitude populations did (Totland 2001), the latter also normally receiving three times higher muscid fly visitation.

Vegetational patterns at altitude parallel those with latitude but may show the same sequences compressed into a much smaller linear dimension. The lower slopes may have deciduous forest, then chaparral or grassland blends into coniferous forest, above which is a distinct treeline; higher again there may be regions of shrubby vegetation including ericaceous plants, sages and junipers, with some thinning grass, then bare rock with lichens. The snowline in winter descends downward through these successive tiers of vegetation. But the actual vegetation on any particular mountain range is very variable with continent and latitude, so that it is hard to generalize beyond this. In addition, mountain ranges have often acted as refugia and as centers of isolation at different times in their history, so that many of them have unique combinations of flora.

At low latitudes, montane floras are often particularly unusual, coping with temperature extremes and drought; many plants are large relative to their lowland relatives, with "giant" herbs appearing on several African mountains, where their foliage may help to ameliorate microclimatic conditions for their own reproductive structures as well as for pollinators. The fauna can be dominated by birds and bees as in the lowland forests, but at the higher elevations the bee fauna diminishes and birds become more important (Cruden 1972). In the high Andean forests, 13 hummingbird species were identified as key pollinators of 29 plants, forming three subcommunities within which there were strong associations of floral nectar supply and the birds' energetic needs in terms of molting, reproduction, and migrations (Gutierrez *et al.* 2004). On Mt. Kenya, close to the equator, insects are rare as flower visitors in the higher vegetation systems, and the large *Lobelia telekii* inflorescences are mainly pollinated by the territorial sunbird *Nectarinia johnstoni* (Evans 1996).

At higher latitudes the montane pollinating fauna tends to be dominated by flies and by bees, although the balance between these varies; in general, once again bees diminish in importance with altitude (Primack 1983; Warren et al. 1988), but tend to increase in size (Malo and Baonza 2002). In the Andes, bee pollination is dominant at most altitudes (Arroyo et al.

1982), but with fly pollination increasing closest to the snowline. In Australian mountains fly pollination is most important, although around 30% of the flora is bee pollinated (Inouye and Pyke 1988). The fly pollinators may not be especially effective though; in Norwegian mountains, Totland (1993) showed that *Ranunculus acris* was visited mainly by higher flies (muscids, anthomyids), and on average a flower got visited only once every 40–120 minutes in daylight, a single fly visit resulting in eventual production of a mean of 5.3 seeds, about 18% of the maximum possible.

By far the majority of studies have focused on the Rockies, where pollinator abundance varies with elevation (Macior 1974). So too does plant species diversity: Moldenke (1975) recorded reduced numbers of bee-pollinated species in Californian mountain habitats, and Pleasants (1977) reported about 20 bumblebee-pollinated species at 3000 m, and only 7 at 4000 m. Hence the average number of plant species in a particular bee-visited guild also decreased. More specifically, Pleasants reported that the pattern of guilds changed at altitude, with larger guilds of short-tongued bumblebee plants (eight species in this guild) compared to long-tongued bumblebee plants (just four). This was probably because short-tongued bees are more speciose anyway and are a resource that can be more finely subdivided; hence the plant species in this guild tend to be more specialist, including some pollen-only flowers. A similar pattern of proportional decreases in longer-corolla flowers was reported at altitude in northern Europe (Ranta et al. 1981). However, there may again be latitudinal difference here, as a specific analysis of altitudinal gradients in Spain (Malo and Baonza 2002) found significant increases in pollinating insect body sizes with altitude, and an associated increase in *Cytisus* flower size and corolla length, while overall pollination success decreased.

Perhaps best known of all montane plants is the alpine sky pilot *Polemonium viscosum*, reviewed in Galen (1996). It is widespread above the treeline in the Rockies, with altitudinal clinal variations in corolla size, color, and fragrance, all tending to increase its showiness at higher elevations. For example, corolla width increased about 12% from just above the treeline to the uppermost tundra-like elevations (Galen et al. 1987). Larger flowers also produced more nectar and more pollen grains. At the highest levels (around 4000 m), 75% of visits and 90% of seed set were

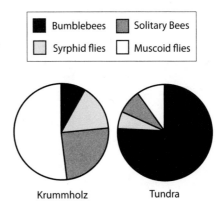

Figure 27.10 Seed set due to bumblebee visits in *Polemonium viscosum* in cold montane zones in the Colorado Rockies, at altitudes of 3520 m (*krummholz*) and at colder 3640 m (tundra). (Redrawn from Galen 1996.)

attributable to bumblebees, while near the treeline smaller bees and many flies were more important, and *Bombus* achieved only 5% of visits and 50% of seed set (fig. 27.10). This suggests that bumblebees are the main selective driver for flower size and rewards, and supports rather efficient pollinator-mediated selection on floral traits (cf. Miller 1981; Armbruster 1985; and see chapter 20).

A similar tendency to be more showy at altitude is reasonably common, despite the poor resources available. This may explain why moderate high-elevation (alpine) habitats also have disproportionately high numbers of blue flowers visited by bees, whether in Europe, the Rockies, or southeast Asia (Weevers 1952).

Resource availability at altitude is further constrained by a reliance on selfing as a safety net, generating a strong need for resource allocation to maintenance and storage. Douglas (1981) studied the balance between vegetative and sexual reproduction with altitude in *Mimulus primuloides* from California, and found vegetative output around 40%–56%, greatest at midelevations (around 2500 m); sexual reproduction, in contrast, was only 2%–4% at low and mid-heights but reached 14% at the uppermost site, where flower production appeared to be genetically fixed at a high rate, again giving a more showy display. Similar findings were reported for *Heloniopsis* (Kawano and Masuda 1980). However, in *Sedum* there was less sexual effort at altitude (30%) than at lower elevations (55%) (Jolls 1980), while a study of three ranunculaceous plants

TABLE 27.7
Altitude Effects in Ranunculaceae Species

	Compatibility	Changes at higher altitude
Trollius ranunculoides	Self-incompatible	More pollen limited; smaller flowers, larger seeds
Anemone rivularis	Self-compatible	Not pollen limited; lower carpel number, larger seeds
Anemone obtusifolia	Self-compatible	Not pollen limited; higher carpel number, smaller seeds

Source: After Zhao et al. 2006.

along an altitudinal gradient in Tibet, by Zhao et al. (2006), found different responses in each (table 27.7), with no consistent trend of decreased allocation to flowering or fruiting at altitude. It would be fair to conclude that strategies vary markedly between plants.

Again there are issues of mosaic populations along the snowmelt line, where flies and larger bees tend to be dominant. In Japanese mountains, at early-flowering sites overwintered queen bumblebees may be particularly important, with worker bees and flies only appearing (but sometimes in very large numbers) a few days or weeks later. This could affect the genetic structure of local populations, with early-flowering plants receiving fewer pollen donors but setting more seeds per fruit (Kudo 2006), while some later-flowering individuals suffered excessive selfing, perhaps from competitively scrabbling flies working on the flowers and moving only very short distances between visits. Similarly, the plants could diverge in their mating system over very short distances due to competitive effects. For two species of *Phyllodoce*, Kudo and Kasagi (2005) showed that at early-flowering sites *P. caerulea* was favored over *P. aleutica* and the latter showed moderate levels of autonomous self-pollination; whereas at later-flowering sites *P. caerulea* was at its phenological limit and produced few flowers, thus being outcompeted for pollinators by *P. aleutica* plants, which were then obligate outcrossers.

Puterbaugh (1998) explored ants as flower visitors in three alpine plant species, showing effective ant pollination in gynodioecious *Paronychia pulvinata*, ant herbivory occurring on *Eritrichium*, and very little effect on *Oreoxis*. He suggested that in each case ants could be influencing the breeding system of the plants.

In fact, ants may be more important at altitude than elsewhere (paralleling the effect at high latitudes already referred to) as other pollinators decline in abundance.

Potential Problems with Global Warming in Colder Climates

Having noted that both the plants and the animals in high-latitude–high-altitude zones may take special precautions to keep warm, and given that the effects of climate change are likely to be especially strong in such zones, it is not hard to predict that pollination interactions might be greatly at risk there. High-latitude zones are susceptible to larger temperature changes, and montane zones are likely to lose snow cover and suffer a very marked seasonal alteration to plant growth regimes. Price and Waser (1998) therefore assessed the effects of warming (using artificial heaters) on alpine meadow communities in the Rockies. All ten species studied showed advanced flowering date, but no effects on duration of flowering or fruiting, or on community structure (Price and Waser 2000). However, at some arctic sites rapid changes of plant distribution and abundance have been recorded at warmed sites (e.g., Hobbie and Chapin 1998), suggesting that different factors are limiting at different sites (notably soil nutrients and soil moisture as affected by the artificial warming). Lambrecht et al. (2007) more specifically reported different effects on flowering rate and total reproductive effort in each of four subalpine species studied, suggesting that overall responses in communities will be very complex. Much longer-term studies, including effects on the insect pollinators, are needed.

5. Pollination on Islands

Islands have always attracted attention for evolutionary studies, and the well-known theory of island **biogeography** (MacArthur and Wilson 1967) made specific predictions about biodiversity and ecological characteristics for island biota that are particularly easy to test with plants. Plant-animal interactions, and plant breeding systems, are likely to differ on islands with limited numbers of species present (Barrett 1996). Islands are also very useful places to study the effects of naturally occurring changes of pollinator on the evolution of floral traits, and island pollination has therefore attracted much attention. Volcanic islands provide even better natural laboratories, where interactions may be disrupted suddenly and regularly over a short evolutionary timespan.

Cox and Elmqvist (2000) specifically discussed pollinator extinction in Pacific islands, pointing out that island floras may be highly vulnerable to invasive introductions because there may be no parasites or herbivores to control the aliens, and because native endemic plants with inevitably small populations and often low genetic diversity may be more susceptible to extinction anyway. Island floras have generally also evolved obligate outcrossing, and are thus unusually reliant on the available pollinators (e.g., Sakai et al. 1995).

Pollinator Paucity and Pollinator Switching

Island biogeography theory predicts that islands will have floral and faunal diversities related to their size and to their distance from other land masses; in general, they are expected to have relatively few species. Numbers of potential pollinating species will inevitably be low, and some groups are likely to be entirely absent. Thus, for example, the Galapagos Islands have just one native bee and two native butterflies; and the number of hummingbird species on Caribbean islands is always low and is roughly related to the size of each island (e.g., Feinsinger et al. 1985).

Pollinator limitation is therefore likely to be common. On Kent Island in the boreal Bay of Fundy (Canada), five out of seven plant species tested were strongly pollinator limited (Wheelwright et al. 2006). In small islands off eastern Spain larger populations of *Daucus* were pollinator limited, the main visitors being *Eristalis* flies migrating in from the mainland (Perez-Banon et al. 2007), and the local plant subspecies had apparently responded to this with extended floral longevity and stigma receptivity, plus a change in nectar production in the male-phase flowers.

Switching to different pollinators, or to more generalist traits, might therefore be quite common. On Puerto Rico, the cactus *Pilosocereus royenii* has apparently relaxed some of the bat syndrome traits found in congeners on other Caribbean islands, and it received very few bat visits, instead being effectively pollinated by carpenter bees (Rivera-Marchand and Ackerman 2006). In the Canary Islands several endemic species now show bird-pollinated traits, seemingly inheriting these traits from mainland relatives, and are visited by passerines (Valido et al. 2004). But ornithophily has probably arisen *de novo* in *Echium wildpretii*, which has two subspecies, the pink-flowered bee-visited *E. wildpretii trichosiphon* on La Palma, and a red bird-visited version (*E. wildpretii wildpretii*) on Tenerife. Dupont et al. (2004) found that the Canarian bird-visited flowers had hexose-rich nectars while those persisting with insect pollination retained sucrose-rich nectars, suggesting a rapid evolution of nectar traits (but see chapter 8 for some problems with the ascription of hexose preferences to birds).

On the most remote oceanic islands, switches sometimes involve the appearance of particularly specialist interactions, so that in Samoa the introduced kapok tree *Ceiba* (normally visited by a wide range of animals including moths and small bats) is exclusively pollinated by the flying fox (a very large pteropid bat), with individuals vigorously defending small territories around the rich resource provided and keeping away all other visitors (Elmqvist et al. 1992).

Switching may extend to the use of unusual taxa as pollinators, especially for certain groups that do well under reduced predation regimes; for example, lizards are often particularly abundant and may become useful as pollinators (Olesen and Valido 2003; chapter 17).

Limited Dispersal

Because islands are small and self-contained, there will normally be lower levels of pollen movement, and a greater potential for inbreeding, so that inbreeding depression becomes a potential problem. This can be aggravated because many plant taxa will have arrived from a single invasion event and so are rather closely related from the start (Silvertown 2004).

But on rare occasions it is also possible to get inbreeding euphoria, as seen in silverswords on Hawaii and tarweeds on Californian offshore islands (Baldwin 2007), where there is extensive radiation following island colonization by a self-incompatible ancestor, with high interfertility in offspring and rapid ecological diversification (see also Price and Wagner 2004).

Selfing as a Solution

Self-compatibility in island plants is commonly more prevalent than in mainland plants (Stebbins 1957; Cox 1989), so that autogamy is rather frequent (e.g., Wheelwright et al. 2006). There may be at least two reasons for this. First, immigration and invasion by self-incompatible species would require at least two individuals to arrive, whereas a single propagule from a self-compatible species could survive. And second, pollinator paucity may make self-compatibility advisable.

For similar reasons, the breaking down of SI systems in island populations has been reported (e.g., Strid 1970; McMullen 1987), including the loss of heterostyly (Barrett and Shore 1987; and Barrett and Husband 1990 for *Turnera* and *Eichhornia* species). Figure 27.11 compares heterostyly, and the effects of its loss, in populations of *Eichhornia* from Brazil and from Jamaica.

High Levels of Dioecy and Mixed Mating Systems

High levels of dioecy appear to be characteristic of island plants (see table 3.7), especially where those islands are particularly isolated; this is presumed to help avoid inbreeding. Charlesworth's arguments on why plants revert to dioecy (chapter 3, section 3, *Benefits of Crossing and Selfing in Plants*) may be relevant here: she maintains that when a plant is free from the constraint of attracting bees, and therefore does not have to contain pollen as a reward, it can use smaller, cheaper, and unisexual flowers.

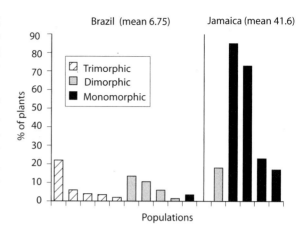

Figure 27.11 Effects of higher selfing rates in island Jamaican populations of *Eichhornia* where trimorphic heterostyly has largely broken down, compared with mainland Brazil where it remains common along with dimorphic and monomorphic plants; there are much higher percentages of abnormal development on the island. (Redrawn from Barrett et al. 1989.)

However, being dioecious could be especially difficult on islands where the fauna is relatively low in species; this situation may favor the presence of a few "supergeneralists" (see below).

Mixed mating systems may also be favored where self-incompatibility breaks down but there are too few pollinators to give enough outcrossing. There have been several attempts at modeling whether mixed mating systems can become stable on islands (see Inoue et al. 1996), especially relating to the likely variability of pollinator availability between years.

Effects of Introductions

Since islands are enclosed limited habitats, introduced species either of flowering plants or of pollinating animals may have more substantial effects on endemic species. For example, introduced honeybees in Mauritius are affecting endemic nectarivorous birds, mainly by lowering the standing crop of nectar on some of their preferred tree species, so that birds such as the local white-eyes stop feeding earlier in the morning (Hansen et al. 2002). Rats may have led to the extinction of a pollinating parrot in Polynesia, and bumblebees are producing potentially deleterious effects in New Zealand (see also chapter 29).

Effects of Volcanic Activity

Many islands are volcanic, and eruptions can produce sudden devastation of an environment followed by gradual recovery, providing unparalleled opportunities to study the evolution of interactions. Abe and Hasegawa (2008) studied *Camellia* pollination by *Zosterops* on a Japanese island following an eruption in 2000, and noted that in the most damaged areas the plants suffered reduced foliage and flower production, and lower pollinator density, but achieved higher pollination rates due to raised pollen transfer efficiency among the fewer flowers. However, fruit abortion was also higher, so that overall fruit set rates and reproductive success were similar in damaged and undamaged areas. These kinds of compensatory mechanisms may be common where the vulcanism does not entirely wipe out plants and pollinators. Furthermore, and on a longer time scale, recovery after such wipeouts may also involve some compensatory effects as new plant growth benefits from increased soil fertility.

Specialization and Generalization on Islands

Adaptive radiations on isolated islands are a classic scenario, where ancestral colonists evolve largely without competition and eventually speciate into various available niches. Coevolutionary radiations of plants and pollinators might therefore be expected to yield unusual levels of specialization, and Olesen and Jordano (2002) did report higher levels of plant specialization on islands, which they presumed to be associated with the impoverished fauna of potential pollinators. However, in other contexts island plants are often asserted to be rather generalist (e.g., Barrett 1996). For example on Chiloe Island in southern Chile Smith-Ramirez et al. (2005) studied 26 plant species over three years, recording 172 "pollinator" (visitor) species and an average of 6.6 pollen vectors per plant species, the animals visiting on average 15.2 plant species. Some specialized bee-pollinated vines and hummingbird-pollinated gesneriads existed, but overall the flora was deemed rather generalist.

The apparent inconsistency here probably relates mainly to the different uses of the term "specialized" (see chapter 20), since island plants may be specialized in the sense of using rather few visitors but unspecialized in that they use whatever visitors are available. Olesen et al. (2002) suggested that islands may therefore often have a few endemic supergeneralists with very broad networking, responding to the low interspecific competition and often high species' abundances found there. Aliens that invade may also act as supergeneralists and weaken the native mutualisms (see Aizen et al. 2008).

Specific Cases

Galapagos Islands

These islands have very low pollinator diversity, with only one bee (*Xylocopa darwinii*) and only two native butterflies. The islands are home to many small, yellow and white, cup-shaped bowl or bellflowers, mostly with very little scent production, many of them primarily autogamous; and to wind-pollinated catkin types (McMullen 1987, 1990, 1993). For example, a member of the Boraginaceae (*Tournefortia rufo-sericea*) has completely atypical small white flowers, visited by ants in daylight and by some beetles and moths at night, but appears to be largely autogamous in practice (McMullen 2007). Likewise, the endemic *Ipomaea habeliana* is effectively pollinated nocturnally by a hawkmoth, but shows substantial autonomous or facilitated autogamy (McMullen 2009). Goosefoots and dropseeds occur widely on beaches, using wind or self-pollination. The prickly pear cactus is visited not by bees as elsewhere but by the cactus finch *Geospiza* (one of Darwin's finches) with an appropriately modified beak.

New Zealand

On these two large islands, the native biota lacks many angiosperm and insect groups found routinely elsewhere, and the native flowers (about 80% endemic) are strongly dominated by rather dull white generalist forms, with flies, small moths, and beetles visiting; there are just a few bee- and bird-pollinated examples (visited mainly by bellbirds and tuis), and no native butterfly flowers. But the biota is now highly modified by introductions (so that a recent Flora lists more aliens than natives), and this has led to a highly modified pollination ecology across the various habitats within the islands.

Very high levels of unisexual flowers occur as natives in New Zealand (Webb et al. 1999) with many gynodioecious and dioecious species recorded, often with male-biased sex ratios (Webb and Lloyd 1980).

The Charlesworth theory may in part explain the high levels of dioecy in the New Zealand floras, where specialist bee pollinators were absent until introduced by the European incomers. Self-compatibility is around 64% in species so far tested (Newstrom and Robertson 2005), not as high as was previously assumed; although as elsewhere it is lower in trees and shrubs (~20%) than in herbs (~80%).

Several endemic plants have declined markedly, and in some cases this is due to pollinator paucity. For example, native mistletoes (*Peraxilla* species) are severely reduced in abundance; this is related to poor pollen receipt (Robertson, Kelly et al. 1999), a service largely provided by native bellbirds. Robertson et al. (2005) reported that short-tongued bees could be reasonably effective as alternative pollinators. But in practice imported bumblebees are posing a threat in New Zealand; this issue of introduced pollinators is covered in more detail in chapter 29.

Hawaii

The Hawaiian islands, with differing sizes and ages of land mass resulting from progressive volcanic activity across a "hot spot," provide an excellent testing ground for theories in ecology generally, and no less so for pollination ecologists.

Once again high levels of dioecy have been recorded (Bawa 1982; Baker and Cox 1984; Sakai et al. 1995), including cryptic dioecy (Mayer and Charlesworth 1991). Mixed mating systems are also rather common (e.g., Sun and Ganders 1988; Weller et al. 1998). As regards pollinators, paucity is again the rule; only six hawkmoths and two butterflies occur, and overall only 15% of insect families are present (Barrett 1996). However, the co-radiations of honeycreepers and lobelioids are well known (see Lunau 2004), each group probably deriving from a single ancestral species, each radiating widely (now with about 30 and 60 species each) and coevolving respectively into closely matched curved-beak birds and curved corollas, ranging from 15 to 85 mm in length.

Again, though, extinctions and introductions have somewhat muddied the waters. It is worth recalling the ieie story from chapter 11, where a plant persists only because pollination has been taken over by an introduced white-eye, after the local bird pollinators went extinct (Cox 1983). The lobelioid cardinal flowers have also suffered, and only 27% of the 273 recorded species and varieties have big enough populations not to be under threat of extinction, a quarter having vanished in the last hundred years and another 20% now being rare or endangered. Two particularly rare species have been recorded as receiving zero visits by any honeycreepers, which are assumed to be the "proper" pollinator; Cory (1984) saw only some hawkmoths, some *Hylaeus* (colletid) bees, and many *Apis*, and he showed that both plants were capable of selfing now. But then Smith et al. (1995) examined museum specimens of one local honeycreeper called the i'iwi (*Vestaria coccinea*) and noted that its bill used to be significantly longer, while contemporary notes revealed that it had fed at the lobelioids. Nowadays it is very rare and feeds mainly on shorter-tubed flowers, especially *Metrosideros*, all of which suggests that the birds have had to shift from the disappearing long-corolla or long-spurred species to the remaining shorter ones, so experiencing directional selection for shorter bills. Many other Hawaiian birds are thought to be close to extinction, and several species of Campanulaceae formerly visited by birds are now extinct, as are many species of bee and moth plants (Cox and Elmqvist 2000).

Madagascar

The large island of Madagascar, off the coast of East Africa but with a geological connection to the Indian mainland, has remarkably high endemism in both fauna and flora. The number of bee species is rather low, with only 1–2 species in most genera, except for one unusual genus *Pachymelus* that has radiated more substantially. There is a high diversity of hawkmoth pollination in forests, with some unusually complex interactions between the sphingids and the angraecoid orchids, and there is also some lemur pollination (see chapter 17). Both these effects are perhaps linked to a relative scarcity of bats (Sussman and Raven 1978). Unfortunately, all these phenomena are under extreme threat from deforestation and soil erosion over large parts of the island.

There are indications that the island's plants do rely on a rather low number of pollinator species and experience low visit rates. For example, Farwig et al. (2004) studied *Commiphora* and found around 1 and 0.2 visits per hour to male and female flowers, respectively, and similar effects occur with *Dalechampia*. This latter genus is particularly well documented in Madagascar, with a range of different rewards and pollination types (chapter 20) and with reversion in at least three instances to more generalized pollen-feeding insect visitors and away from very specialized resin-producing flowers (Armbruster and Baldwin 1998). Another

TABLE 27.8
Campanula Outcross Pollination Patterns on
Japanese Offshore Islands

	Distance to mainland (km)	Outcrossing rates
Mainland	0	0.64–0.79
Islands		
O1–O2	~25	0.63–0.76
T1	~32	0.54
N1–N2	~40	0.46–0.56
K1	~50	0.37
M1	~80	0.16
H1–H3	~200	0.17–0.25

Source: Based on Inoue and Kawahara 1990.

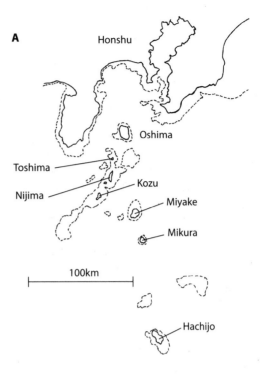

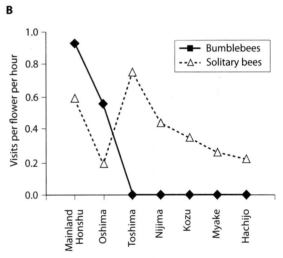

example is provided by 13 endemic species of *Masaola* trees (Bignoniaceae) occurring sympatrically across the island, with low visitation rates but marked specialization of visitors (few species being visited by more than two pollinator groups), and separated by morphological differences, vertical stratification, phenological stagger, and differing flowering displays and durations (Zjhra 2008).

Faroe Islands

These islands provide a useful contrast to most of the others that have been studied because of their extreme isolation and their oceanic climate, with cool wet cloudy conditions even in summer. Not surprisingly, perhaps, self-pollination dominates (Hagerup 1951).

Izu and Ogasawara Islands, Japan

These islands form a chain separated from mainland Honshu, and have recently attracted several studies. *Campanula punctata* occurs as an island version (whether species or variety is still unclear) that has smaller flowers, more flowers per plant, and later flowering (Inoue et al. 1996). The islands furthest from the mainland had the lowest levels of SI in this plant. Outcrossing rates declined from the mainland (by more than 80%) down the island chain to the outlier islands (table 27.8). Pollen-to-ovule ratios also declined, and flowers had a longer initial male phase before their styles elongated. Pollinator diversity decreased progressively along the chain, especially the diversity of

Figure 27.12 (A) and (B) Frequency of visits by *Bombus* and various solitary bees to *Campanula* flowers on Japanese islands at successive distances offshore. (Redrawn from data in Inoue 1993.)

longer-tongued types, so that the *Campanula* flowers were mostly visited by *Bombus* on the mainland but by much smaller and shorter-tongued halictids on the outer islands (fig. 27.12).

Rhododendron species on these islands again had lower SI levels where they were more distant from the mainland (Inoue 1993); and high levels of cryptic dioecy have been documented for several other genera (e.g., Kawakubo 1990).

Florida Keys

Spears (1987) compared one island (Sea Horse Key), 5 miles offshore, with mainland Florida, looking especially at *Opuntia* and at a perennial legume vine *Centrosema*. The island *Opuntia* flowers received only 25%–35% as many bee visits as those on the mainland, and no visits from *Apis*, which was absent from this key. Bees carried pollen for significantly shorter distances on the island and seed set was very low, especially for the legume.

Overview of Island Pollination

Many effects have been documented, largely supporting the effects of island size and of distance to mainland predicted by island biogeography theory. But in addition several mainland-island comparisons show very clearly that plant reproductive strategies and plant-pollinator relationships can change drastically and rather rapidly when the pollinator community changes (Linhart and Feinsinger 1980), and this may also carry a wider message about the feasibility of fast and focused selection in plant-pollinator interactions.

Chapter 28

THE POLLINATION OF CROPS

Outline

1. Food Crop Types Needing Animal Pollination
2. Finding out Who Really Are the Pollinators
3. Case Studies: Examples of Specific Pollinators
4. General Approaches to Encouraging Pollination
5. Problems with Hybrid Crops, Seed Crops, and Crop Breeding
6. Overview

The importance of bee pollination to crops has been well known for at least two millennia, and several ancient civilizations cultivated honeybees or stingless bees in wooden or pottery hives. Likewise, humans have attempted to improve crop productivity for millennia, first by selecting the plants with the more desirable traits but latterly using plant breeding and hybridization techniques, whether at the whole-plant level or more directly at the genetic level. Today there is a critically important applied aspect to pollination ecology, and much room for improvement in our knowledge of what makes a good pollinator for different kinds of crops, of how to manage the crop, or the pollinator population, or the environment, and hence of how to aid pollination and seed set. The solutions are often to be found in the field, and expensive biotechnology should normally be the last resort.

Klein et al. (2007) provided a review of the pollination needs of crops (table 28.1) showing that numerically the large majority of important crops, including those directly eaten by humans, do need animal pollination—although in terms of production volumes such crops are in the minority because most staple cereals are wind pollinated. But note that there are differing absolute and relative needs for pollination. First, some zoophilous crops need cross-pollination to achieve any seed set, while for others it is not an absolute requirement but does result in improved seed or fruit quality, or uniformity, or number (hence the various levels of "dependence" in the table). Growers nearly always wish to improve their yields and quality and so regard animal pollinators as crucial as long as their provision or management is cost effective. Second, in some instances the main requirement of a grower (or more usually now of an agro-industrial company) is managing a crop to produce the next generation of seeds, and here the needs can be quite different and often conflicting with respect to pollination, especially for hybrid crops. Third, management of crops with unusual mating systems is generally difficult; dioecy in particular is problematic because pollinating agents are essential but may not be sufficiently reliable in monocultures or greenhouses, etc. Thus while dioecy is ancestral in certain crops (for example, grapes and hemp) it has been bred out of most cultivars.

Wind-pollinated crops, including the cereals (wheat, rice, maize, barley, rye, oats, sorghum, millet), are not usually in need of any particular management. However, there has been controversy over the pollination of some other key crops such as brassicas; for example, it has only recently been shown how important both bees (Hayter and Cresswell 2006) and hoverflies (Jauker and Wolters 2008) are (relative to wind) for *Brassica napus* (oilseed rape). Furthermore, many wind-pollinated trees are dioecious, and where they are grown commercially the sex ratios need to be managed; an example is the date palm (*Phoenix dactylifera*), usually grown with one male tree to many females, where

Table 28.1
Pollination Needs of the Principal World Crops

	Number of crops	Crops eaten by humans	Production volumes
Requiring animal pollination	87		35%
Essential		13	
Strongly dependent		30	
Moderately dependent		27	
Slightly dependent		21	
Unimportant		7	
Unknown		9	5%
Not animal pollinated	28		60%

the ripe male inflorescences are cut off and one is tied into each female tree.

Some crop plants readily self-pollinate, even where their precultivation ancestors were clearly outcrossing; the pea (*Pisum sativum*) is a classic example. Other crops that can self nevertheless benefit from crossing, both by avoiding the longer-term drawbacks of inbreeding depression and by increasing their practical yield at harvest time through early and well-synchronized seed set. And many plants that can self still produce better quality fruits if cross-pollinated.

1. Food Crop Types Needing Animal Pollination

Roubik (1995) recorded that at least 800 cultivated plants worldwide rely on animal pollinators; about 73% of these are pollinated by various bees, 19% by flies, 6% by bats, 5% by wasps, 5% by beetles, 4% by birds, and 4% by lepidopterans. For Europe and other temperate zones, the figure for all crops relying on animal (insect) pollination may be closer to 84%. Further directories and reviews of crop pollination can be found in Crane and Walker (1984), Free (1993), and Delaplane and Mayer (2000). A listing of key crops is shown in table 28.2.

From such lists it is evident that the term "animal-pollinated crops" nearly always means bee-pollinated crops, especially in temperate growing areas, where *Apis* is often strongly dominant. The US Department of Agriculture (USDA) records only honeybees as pol-

linators, and assumes that they provide 80% of the pollination services in the United States, the rest being recorded simply as "other pollinators." In fact just seven crops in the United States are recorded as pollinated mainly by wild native insects rather than *Apis*—blueberry, cacao, cardamom, cashew, cranberry, mango, and squash. Another 18 major North American crops depend partly on wild species of bee.

Across the temperate zones, crops often or ideally pollinated by bees other than *Apis* include many legumes used both as human food and as animal forage. Red clover is a forage crop needing insect pollination (primarily effected by bumblebees), as also is lucerne (alfalfa), pollinated mainly by *Nomia* and *Megachile* bees in the United States and by *Melitta* in Europe, but which can be pollinated (less well) by *Apis* if necessary. Crops that are pollen-only are almost inevitably linked with bees, but often better served by non-*Apis* types; for example, the solitary eucerine bee *Peponapis* is an important pollinator of cucurbits in the United States, and has a seasonal emergence and flight period closely tied in with the anthesis of pumpkins (Willis and Kevan 1995). Kiwifruit and pomegranate are also pollen-only plants to which bee activity is tightly structured. Tomato flowers are self-fertile and nectarless but buzz-pollinated, so the crops normally need bees to disturb the flowers and move pollen out of the anthers. When tomatoes are grown outside in California, *Anthophora* bees work well, while in the plant's native Peru halictid bees are important (Rick 1950). In greenhouses tomato pollination can be achieved by tapping the plants or using commercial vibrators, but it

TABLE 28.2
Main Pollinators of Principal Crops

Pollinator	Crops
Wind	Wheat, oats, rice, maize, rye, barley, sorghum, olive, walnut, pistachio
Wind and some small insects	Sugar beet, spinach, date
Fly	Taro, cocoa
Beetle	Oil palm
Wasp	Fig
Fly, bee	Carrot, onion, lettuce, currant, pepper, tea, yam
Bee	Peanut, potato, cassava, sugarcane, soybean, beans, cowpea, pigeonpea, chickpea, lentil, cotton, rape, melon, almond, tomato, cabbage, cucurbits, citrus, pear, apple, peach, plum, strawberry, raspberry, apricot, cherry, cardamom, coffee
Bee, fly, bat	Coconut, avocado, mango
Bee, bird, bat	Banana
Bird	Pineapple
Bat	Durian, agave

is now common to introduce nests of bumblebees (Eijnde et al. 1991), or of native *Amegilla* bees in Australia (Hogendoorn et al. 2007). Bumblebees are also often used with strawberries where these are grown under glass or in polytunnels, although the evidence that they do much good is equivocal at least for some habitats. Stingless bees such as *Melipona* are currently under investigation for beneficial greenhouse use with some crops such as sweet pepper (e.g., Cruz et al. 2005). Stingless bees are already recognized as key pollinators of 9 crops (mainly nuts and fruits, including macadamia, mango, and coconut) and partial pollinators of another 20 (Heard 1999).

Some foods found widely in western human diets use non-bee pollinators at least in part. Carrots will regularly be pollinated by short-tongued bees (*Halictus, Lasioglossum, Apis*), although also by hoverflies and some sphecid wasps (Hawthorn et al. 1956). Some crucifers (cabbages and Brussels sprouts) are effectively pollinated by bumblebees, but both *Apis* and blowflies (*Calliphora*) can also do the job less well (Faulkner 1962). Avocado, which is self-fertile but cannot self-pollinate because of timing effects (chapter 3), is now usually pollinated by *Apis* (Afik et al. 2007) but seems to have been pollinated by flies ancestrally. From a worldwide perspective, the most impor-

tant examples of pollination by animals other than bees are mangoes, cacao, papaya, pomegranate, some onions, and some umbels like parsnip and carrot. Bird-pollinated crops (at least in part) may include papaya and okra, and bat pollination is important in durian and some other favored Asian fruits.

It can be readily calculated that despite the exceptions given above about one-third of all human food is derived from bee-pollinated crops. Bees can be staggeringly effective; for example, just 100 *Apis* foragers can ensure acceptable fruit set in 1 ha of a commercial apple orchard in about 5 hours on a good sunny day. Total values of crop pollination in the United States have been suggested as $20–$40 billion annually. The value of honeybees was estimated at $5–$14 billion annually for 63 US crops (Kremen et al. 2002), and in Europe at around €4 billion for 177 crops (Carreck and Williams 1998). But an alternative approach using cost-of-replacement economic models indicated that these were probably substantial undervaluations (Allsopp et al. 2008). Estimates of the total commercial value of pollination services worldwide have been attempted (e.g., Costanza et al. 1997) and have given somewhat controversial figures of $120–$200 billion per annum; a more recent bio-economic survey for 100 key crops (Gallai et al. 2009) produced a figure of

€153 billion, with vegetables and fruits each contributing about €50 billion of this total. Consistent agreed approaches and methods are badly needed, but there can be absolutely no doubt that the financial value of pollination is enormous.

2. Finding out Who Really Are the Pollinators

It should be evident on the basis of earlier chapters that one cannot know what will be a good pollinator simply by looking at a flower. There may be some very clear clues from its appearance, but even then selective breeding may have altered key features (morphology, scent, rewards) and changed the original plant beyond recognition (to an observing human, or indeed to a potential visitor). Yet hard evidence of pollination effectiveness is critical in a commercial world. Ideally, a fairly standard set of tests should be applied:

1. Does the plant need cross-pollination at all, or is seed set or fruit quality improved by pollination? Standard tests will use bagging or cages to exclude pollinators, and hand-pollination with self- or cross-pollen, determining relative yields and qualities.

2. If cross-pollination is essential or at least beneficial, then what actually visits the flowers? This can best be assessed by observation, including observations of visitors' behaviors; trapping adjacent to the flowering crop is perhaps easier, but this may yield many different genera (and even orders) some of which are just herbivores or predators and may not visit the flowers, or if they do so may still not be pollinators.

3. When is pollen available at the particular site? There is a need to assess timing and synchrony of dehiscence from direct observation of anthers, and also rate of removal of pollen if possible.

4. When are particular visitors active? This requires observation throughout the day (sometimes also the night) and all through the flowering period. Only those visitors coming to the flowers when pollen is naturally abundant and undepleted could have much effect. (And only those coming when nectar is abundant and collectable in terms of concentration are likely to be coevolved pollinators.)

5. Which visitors actually gather pollen, and which carry it on their bodies? Visitors can be caught as they leave flowers and examined for pollen quantity and placement (excluding that already packaged into the scopa of bees).

6. Which visitors deposit pollen successfully on a flower's stigmas? This requires observations of numbers of pollen grains on a stigma after single visits by known visitors to previously unvisited (bagged or otherwise protected) flowers.

7. Which visitors deposit pollen at times when the stigma is receptive? This is rather difficult to measure, as estimations of receptivity are not easy (see chapter 7).

8. Which visitors deposit viable pollen? (i.e., do so fairly quickly before it has dried out or lost its germination ability—again rather difficult to measure directly, but it can be estimated from the next two observations).

9. Which visitors are flower-constant enough to deposit sufficient conspecific pollen reliably and repeatedly? This requires observation of visiting patterns among mixed vegetation (e.g., at the edge of the crop and into adjacent hedgerows) and/or assessment of pollen loads as in (5) to detect mixed species and nonconstancy.

10. Which visitors move around enough for the pollen to be from a nonself source? For a tree or bush crop, this may mean movements between plants after just a few flower visits per plant, but in many monoculture crops this is complicated by extremely high relatedness in the source seeds or by clonal effects in some soft fruits.

The list is long and similar to that required for establishment of pollinator effectiveness for less applied studies (see chapter 11). The importance of each stage will vary with the particular crop, but a large number of studies on crop pollination do miss out many of these key components, which can sometimes result in less than sensible advice about the usefulness of particular species as pollinators. However, if most aspects are assessed then enough should be known about the biology of the recommended pollinator to find ways of encouraging its presence and activity.

TABLE 28.3
British Crops by Flower Type Categories

Crop type	Numbers		Pollination type				Main families
	Annual	Perennial	Allophilous	Hemiphilous	Hemi/Euphilous	Euphilous	
Forage crops	8	5	0	0	3	10	Fabaceae, Brassicaceae
Vegetables	9	14	5	6	3	9	Brassicaceae, Apiaceae, Asteraceae
Soft fruits	0	9	0	4	3	2	Rosaceae
Tree fruits	0	7	0	7	0	0	Rosaceae
Oilseeds	4	0	0	0	2	2	Brassicaceae, Fabaceae
Herbs	0	8	1	1	0	6	Lamiaceae

Source: Modified from Corbet 2006.

Clearly, the gathering of the information described above is often beyond the scope of a crop grower, and attempts have been made to simplify protocols or to offer clear guidelines on likely pollination types. Corbet (2006) offered an example of the latter approach. Her typological system (see chapter 20) was used to explore the relations between flower type, crop type, and plant family, and thus the pollination type to be expected, for the commonest crops grown in Britain (see table 28.3).

Using her terminology, most of the herbs and forage crops are deemed euphilous and require pollinators of the eutropous type (larger, longer-tongued, endothermic bees), while most of the fruits are from the family Rosaceae and are hemiphilous, flowering early in the year and requiring hemitropous pollinators (smaller and shorter-tongued bees, hoverflies, some butterflies). Vegetables are the most diverse group, including Apiaceae, Asteraceae, and some brassicas, and with a range of different requirements. Corbet also pointed out that the already well-documented declines in the eutropous insect groups (native bumblebees and some solitary bees) would be more serious for crop growers were it not for the ubiquitous generalist honeybee, which can visit many flower types reasonably effectively.

3. Case Studies: Examples of Specific Pollinators

This section picks out a selection of examples that illustrate particular problems of crop productivity or crop management, from plants grown intensively or on a small and local scale, and in a wide range of habitats.

Alfalfa

Alfalfa (*Medicago sativa*) is the most important forage legume in many temperate countries, grown for immediate pasturing and for hay. It is the main forage crop in the United States and covers the second largest acreage (after soybean). The flower is showy and zygomorphic, with the typical legume tripping system, and it produces large nectar quantities highly acceptable to bees. Selfing is possible in most cultivars, but crossing increases seed production.

Many bees are regular and enthusiastic visitors, and honeybees routinely work the crops. In many cultivars, though, a large majority of *Apis* visits that are purely for nectar collection do not result in tripping. This led to several enormous breeding projects to try and match commercial alfalfa strains to honeybees: clones were bred with improved aroma and easier tripping, to encourage honeybee visits (Erickson 1983), and bee strains that visited mainly for pollen were also selected (Estes et al. 1983), although this led to reduced honey yield and was unpopular with beekeepers.

Leaf-cutting bees proved to be excellent pollinators in Canadian sites in the 1940s, and fields that were too large suffered pollination deficits in their centers because the flowers were too far away from field margins where these bees nested in dead wood and stems. So a variety of solitary bees are now the preferred pollinators. *Nomia* and *Megachile* bees are most commonly used in the United States and Canada, and *Melitta* bees are sometimes used in Europe. *Megachile rotundata* is the favored North American species, and strains of this bee have been selected to fly at lower temperatures in northern parts of the United States (Free 1993). There have also been some attempts to produce alfalfa strains with higher nectar secretion rates to support these bees. At present the value of pollination for alfalfa seed growers is estimated as 35% of the total crop value, and for Canada this represents a figure of around $2 million annually for the services of *Megachile* bees. More details of management of these bees are given in section 4, *General Approaches to Encouraging Pollination*.

Red Clover

Trifolium pratense is regularly used as a forage crop and also as a green manure with the ability to increase nitrogen levels in soils. Its flowers have particularly long corollas, so that it needs long-tongued visitors, mainly the longer-tongued bumblebees. However, the short-tongued *Bombus terrestris* often visits the crop as a major primary nectar thief, with *Apis* then arriving as a secondary thief to exploit the holed corollas. To overcome these problems, attempts have been made to breed clover varieties with shorter corollas, and also to breed *Apis* strains with longer tongues (Free 1993).

Carrot

Carrot is an umbellifer (*Daucus carota*, Apiaceae), and open-pollinated carrot cultivars are highly attractive to a wide variety of insects (bees, flies, wasps, beetles), although commercially the visits by *Apis* are always seen as most important. The flowers are self-fertile, but usually get cross-pollinated as the umbels are strongly protandrous, pollen dehiscing throughout one umbel before any stigmas become receptive. Successive "orders" of umbels open in nonoverlapping waves at 9–13 day intervals (Erickson 1983; and see fig. 2.12), the first three orders producing around 90% of the seed.

However, production of carrot hybrids has proved difficult; the flowers of seed parents are often very different in character, and the resultant hybrids are less attractive to honeybees, commonly with poor overlap

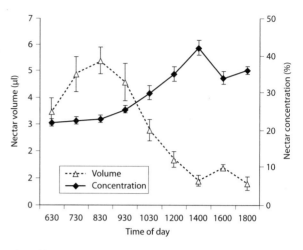

Figure 28.1 Variations in concentration and volume of coffee nectar through the day. (Redrawn from Willmer and Stone 1989.)

of pollen production and stigma receptivity. Moreover, honeybees readily learn to distinguish between cultivars and to prefer those with better scent and more copious nectar flow, giving poor hybrid seed yields.

Coffee

Both highland (*Coffea arabica*) and lowland coffee (*C. canephora*) can self-pollinate, but there is a substantially better yield where native bees are abundant or where there are introduced bee populations (Roubik 2002; Klein et al. 2003a,b,c).

C. canephora was originally reported as self-sterile and predominantly wind pollinated, but later studies (Willmer and Stone 1989) showed that insects are crucial. The flowers (open just before dawn and lasting one day) are conspicuous, creamy white, and heavily scented, the coffee pollen grains 25–35μm and strongly sculpted. Allowing insect access produced an eightfold increase in pollen transfer, with most stigmas receiving more than 50 pollen grains. Floral characteristics, especially the timing and volume of nectar per flower (fig. 28.1), suggested a fairly large and metabolically demanding pollinator.

At sites in Papua New Guinea the leaf-cutter bee *Megachile (Creightonella) frontalis* (large, with a tongue length of 7–10 mm) was the regular visitor, completing 1 nest cell every 1–3 days almost exclusively stocked with coffee pollen. The bees averaged 41 flights per day, of which 6–10 were matinal pollen gathering (the rest for leaves as cell lining); stocking a male cell required 5.8 trips, while a larger female cell required 11.2 trips. One trip would require visits to 29 flowers; thus in one day each bee visited an average of 180–300 flowers and so would be a very effective pollinator. However, bees returning to nests always carried other (leguminous) pollens on their thoraxes, and their crop contents averaged 40%–45% sugar concentration, markedly higher than that of matinal coffee nectar (20%). Thus females were visiting other plants for nectar; *C. canephora* nectar appeared to be nonideal for this bee. Management strategies included encouraging bee nests by maintaining earth drainage banks, retaining alternative nectar sources by leaving leguminous weeds and hedgerow plants in situ, and using insecticides only toward evening once the bees were inactive and the flowers were withering.

In other countries, various different native bees pollinate both this coffee species and the related *C. arabica* when either is introduced. In Costa Rica, Ricketts et al. (2004) showed that native bees coming in from adjacent rainforest were extremely important, the coffee bushes closest to the forest getting twice as many visits and 20% greater yield compared with flowers in the central plantation. Those authors estimated that bee pollination of the coffee was providing around $60,000 annually to the farmer (whereas deforestation and cattle grazing in the same area might raise $24,000). Larger-scale studies in Ecuador showed an 800% increase in revenue to farmers when a fourfold increase in native social bee density occurred (Veddeler et al. 2008). Studies in Indonesia (Priess et al. 2007) found that increasing loss of natural forest would reduce coffee yields by 18% within the next two decades; at present the forests are providing pollination services worth around $46 per hectare. Again, maintaining only a short distance between the crop and natural habitats and thus access to a reserve bank of native pollinators proved the key to crop success. Klein et al. (2003c) showed that overall bee diversity, especially that of solitary bees, was a key factor in enhancing coffee productivity in Indonesia, and it is evident that a whole range of medium-sized bees can be effective pollinators.

Cotton

Cotton production worldwide relies on two main species, long staple (*Gossypium barbadense*) and short staple (*G. hirsutum*). These differ markedly in flower form, the first being yellow with maroon markings, with abundant nectar and orange pollen, the second having cream flowers, low nectar volumes, and yellow pollen. There have been many attempted hybridizations between these since the 1890s, often giving marked hybrid vigor, but the hybrids are difficult to maintain because of high self-fertility in the species. Interspecific hybrids were first grown commercially in India in the 1970s, with seed produced by hand emasculation and pollination, but of such quality in terms of cotton boll size that it proved a worthwhile investment. Intraspecific hybrids are now widespread, aided by the development of male-sterile lines, but these rarely reach the same productivity as the Indian interspecific crosses.

Adequate pollination remains a problem where hybrid seed production is the goal. Bumblebees are useful pollinators in the eastern United States, and various *Anthophora*, *Melissodes*, and *Apis* species are reasonably effective in other parts of the world. Honeybees are usually seen as the answer in large commercial plantings, but unfortunately vary widely in their behavior on different cultivars, often ignoring cotton pollen even when surrounded by acres of it and even grooming it off their bodies. Very intensive import of hives has been necessary to overcome these problems; for example, the use of 900 colonies to get adequate pollination in 1000 acres of cotton in California. This strategy is often regarded as worthwhile because honeybees do generally improve the quality of the lint obtained even if they do not greatly affect seed weight or lint weight (Rhodes 2002).

In practice, the solution to cotton pollination is likely to vary in each growing region, with a range of wild bees and some wasps being effective visitors.

Passion Fruit

Passiflora edulis has a complex flower structure (see fig. 11.1 and plate 5A) and in particular the nectaries are tightly enclosed by a stout lid; relatively few animals can access the nectar, these including larger and strongly built bees (*Xylocopa*, *Ptiloglossa*) and some birds (hummingbirds, bananaquits). Carpenter bees are the dominant and best pollinator in most sites in America and Asia. *P. edulis* was studied as an introduced crop in St. Vincent in the Caribbean, where only the local carpenter bee could feed on the nectar at all, visiting the flowers regularly at times when anthers were dehiscing (Corbet and Willmer 1981). This bee was a good pollen transfer agent, fitting neatly under the pendant anthers to get large pollen loads on its back. However, on return to its nest each bee regurgitated some of the *Passiflora* nectar and evaporated it on its tongue (tongue-lashing; chapter 10) before delivering the more concentrated product into its nest. Each bee also scraped most of the pollen off its back and dumped it outside the nest; it collected most of its pollen from a leguminous weed instead. Clearly there was not a closely coevolved interaction of this plant and this bee, but the net effect was good pollination of the flowers. Here, management included increasing dead log availability as nest sites and encouraging the weed as a pollen source.

In other sites, passion fruit has proved problematic for pollination in East Africa when grown as a large commercial enterprise, although successful nearby in smallholdings where pollinator populations had remained intact. Similar problems occurred in Brazil, where carpenter bees are too rare and hand-pollination has often been used. To improve local populations of carpenter bees, the provision of drilled wooden boards can be beneficial (Gerling et al. 1989; and see Roubik 1995).

Cardamom

This spice (a relative of ginger) is native to the Western Ghats area of India but introduced in many Asian sites. The normal and effective pollinators in India are supposed to be the local *Apis* (*A. cerana* and *A. dorsata*). In Papua New Guinea (Stone and Willmer 1989b) only *A. mellifera* was abundant, and was assumed by the grower to be the pollinator. *Apis* did indeed visit reasonably effectively, but the flowers were getting much better service from a native solitary *Amegilla* bee. This was especially true at less disturbed upland sites, where native bees were much more abundant. Hence the strategy needed was not to import more *Apis* hives but to encourage growth of native bee populations with earth banks and wild hedgerows. In fact, Sinu and Shivanna (2007) indicated that even in India *Apis* are mainly robbers and *Bombus* are the effective pollinators of larger cardamom, so that widespread import of honeybee hives would be particularly inadvisable.

Raspberry

Raspberries (*Rubus idaeus*) can be selfed and do not have an absolute need for insects. But the fruit quality, in terms of numbers of drupelets per fruit, is greatly improved by insect visitors; fruits are less likely to be deformed and are of about 50% greater weight after periods of insect visits. Raspberry flowers are open cup shaped in design, so nectar is freely exposed to a range of visitors, yet the nectar in modern cultivars is produced unusually copiously and stays liquid even on warm days; often 10–50 μl are present, where most temperate flowers have less than 1 μl. Thus insects find the flowers especially rewarding in terms of both nectar reward and pollen availability.

Most early studies of raspberry concentrated on *Apis mellifera*, and some showed very low visit frequencies from other bees. However, bumblebees and some solitary bees do visit raspberry flowers quite commonly. Willmer, Bataw et al. (1994) found that bumblebees were in several respects much better pollinators of raspberry in Scotland: they were more abundant overall (about 60% of all flower visits) and more frequently present in the early mornings when pollen peaked in young flowers; they preferred and selected younger flowers more effectively than honeybees, they carried significantly more pollen on their bodies and deposited significantly more on stigmas per visit, and they visited more flowers per minute than honeybees.

Raspberry is traditionally best grown in areas with relatively cool summers, with the largest acreages in northeast Europe and Canada, so that the importance of non-*Apis* bees as pollinators for this crop may be a rather common phenomenon.

Oil Palm

Oil palm (*Elaeis guineensis*) is unusual in being a monoecious crop, and it is commercially important as a source of oil used in soap, margarine, etc. In its native habitats in West Africa, levels of pollination are high (traditionally assumed to be by wind) and hence fruit set is good. But when it was exported to southeast Asia yields were much lower, and in Malaysia performance was especially poor in newer plantations where wind speeds were higher (Syed 1979), and only by using hand-pollination could a useful yield be obtained. Something seemed wrong here.

Pollination of oil palm in Africa has subsequently been shown to be aided by small beetles (weevils, now named *Elaeiobius*), both male and female inflorescences attracting many hundreds of visits by their scent, produced continuously in males and in short pulsed bursts in females, so that the beetles stay mainly in males and move to females periodically, well coated in pollen. They also disturb the pollen out of the anthers so that it can then become wind-borne. But the only effective flower visitors in Malaysian plantations were a nocturnal moth and some thrips, neither of which responded to the scents or moved pollen effectively. The beetle was therefore introduced to the Malay Peninsula in 1981, where it established well, and has since been taken on to Thailand, New Guinea, etc. Yields have risen to West African levels, with no apparent side effects (Greathead 1983; Hussein et al. 1991), and Malaysia and Indonesia are now the most important world producers of palm oil.

Orchard Fruits

Apples and pears (*Malus, Pyrus*) and plums and cherries (*Prunus*) are all important fruit crops in temperate zones and share features with several important nut crops such as almonds. They have typically open hermaphrodite flowers and are usually self-incompatible; they are grown as vegetatively propagated trees, so that growers commonly need to have two different cultivars in an orchard, one (in the minority) often serving mainly as the pollinizer (reviewed by Free 1993). Bees are widely regarded as the most important pollinators, but early midges and fungus gnats, some higher flies (bibionids, muscids), and also flower beetles (Nitidulidae) occur regularly on the flowers. Use of *Apis* in orchards is common practice, in both Europe (*A. mellifera*) and parts of Asia (*A. mellifera* or *A. cerana*), the growers either having their own hives or renting them from beekeepers or government agencies. Growers often also put out sugar water, which helps balance the supplies of "nectar" and pollen, so that the bees get sufficient nectar rather easily and will then forage more for pollen on the fruit trees. Keeping the trees small by using dwarfing rootstocks is also helpful, as bees then need to go to at least two trees per trip. Honeybees may also cross-transfer pollen within the hive, and so will effectively move pollen from trees they have not themselves visited (deGrandi-Hoffman et al. 1984, 1986).

However, in the United Kingdom it is *Andrena* species and *Osmia rufa* that are the best pollinators in most conditions, the flowers appearing in early to late springtime; only in poor weather do honeybees perform better. *Apis* visit twice as many flowers per minute, but *Andrena* carry more available pollen. In colder areas such as Nova Scotia, *Apis* is rare and *Andrena, Halictus*, and *Lasioglossum* are almost the only pollinators, while in Norway these same three genera are joined by some bumblebees. *Bombus* is recognized to be especially useful on longer-styled apple and plum cultivars; Thomson and Goodell (2001) showed that *Bombus* and *Apis* removed similar amounts of pollen per visit but that *Bombus* deposited far more on stigmas.

In Japan (and sometimes in the United States) hand-pollination is used, the pollen being mixed with inert powders and sprayed onto flowers, and there were early attempts to fully mechanize this process (Ohno 1963). Wild *Osmia* bees are being used now, as a more economical solution. In Japan *O. cornifrons* is favored, but is replaced by *O. lignaria* in US orchards and by *O. cornuta* in Europe (Maccagnani et al. 2007). Species with lower-temperature thresholds are preferred (Bosch and Kemp 2001) as they cope better with the cold spring temperatures at flowering time, and nest boxes are in use that can be closed during crop-spraying periods to protect the bees.

Squash and Other Cucurbits

Squash, pumpkin, and gourd are important crops in more arid areas, especially in southwestern areas of the United States. The solitary gourd bee *Peponapis pruinosa* is the major pollinator of squash; it mainly visits male flowers, whereas *Apis* mainly goes to the female flowers for nectar and moves much less pollen. *Peponapis* also starts foraging much earlier in the morning. For gourds in the Sonoran Desert, *Xenoglossa* squash bees are similarly more effective, full seed set needing 3.3 *Apis* visits but only 1.3 *Xenoglossa* visits. For cucumbers, bumblebees performed better in every respect (earlier starting time, more visits per minute, equal or greater pollen deposition) than did honeybees (Stanghellini et al. 2002). In general, on all these crops, the native bees prove to be better in timing of their visits, better in moving pollen, and better in spreading it more widely. And they show higher fidelity, focusing their attentions only on *Cucurbita* and *Cucumis*, often resting and mating in the flowers. Simulations indicate that native bees would provide sufficient pollination service on at least 90% of watermelon farms in the United States if honeybees were removed from the system (Winfree et al. 2007).

Agave

Agave plants are cultivated as a source for the production of the alcoholic drinks tequila and mescal in Mexico, and also as an important source of fiber; they are pollinated by bats but normally die after reproducing, so the flowering stalks are removed soon after they have formed and commercial reproduction is from vegetative bulbils. However, this has resulted in very low genetic diversity in the Mexican crops (Gil-Vega et al. 2001), which have become very susceptible to fungus and root diseases. Recent breeding has reverted to using pollinators to try and restore some genetic diversity, and some of the resultant new strains are currently under trial.

Cocoa

Theobroma cacao is grown extensively for chocolate production in South and Central America and in West Africa, and it is pollinated by small midges (Young 1985). It can sometimes suffer poor pollination, and problems that were identified at different sites included attacks against flies from various aggressive ants, and overcleanliness in plantations so that there were insufficient clumps of rotting vegetation or small puddles for the midges to breed in (e.g., Ismail and Ibrahim 1986).

Some Crops with Current Problems

Durian (*Durio zibethinus*), a commercially important fruit crop in Thailand and Malaysia, has flowers that open nocturnally and are functional for just a few hours (Honsho et al. 2007). They are pollinated by bats, especially *Eonycteris* (Start and Marshall 1976; though other species of *Durio* in Sarawak are pollinated by nectariniid sunbirds and large bees (Yumoto 2000) with bats playing only a minor role). The duri-

an-pollinating bats feed at mangroves when durians are not in flower, and serious problems arose when mangroves were destroyed for land reclamation, etc. Around the Batu caves near Kuala Lumpur, the mining of limestone was stopped because it would remove the home roosts of essential bats (Pye 1983). However, the durian industry remains under threat from loss of mangroves and also from sport-related or bushmeat killing of bats. Similar problems are occurring in production of the lesser-known bat-pollinated fruits petia and jambu.

Vanilla (*V. planifolia*) is an orchid normally pollinated in its homeland by *Melipona* bees, but it is widely grown in other areas where appropriate stingless bees are absent and often achieves no natural pollination. In Madagascar and elsewhere, where it is a very important cash crop, it is pollinated by hand (plate 36D, but often using boys to climb trees), the rostellum being moved aside and the anther and stigma being pressed together to produce selfing.

Melon farms in the United States have suffered pollination problems when there is too little natural habitat nearby (Kremen et al. 2002), because the most important bee pollinators have tended to become locally extinct and are not adequately substituted by more generalist or less habitat-sensitive species.

Soursop (*Annona muricata*) is a fruit grown quite widely in Brazil and pollinated by small beetles, which also use the flowers as food and as mating sites. In orchards the beetles tend to be too rare, and laborious hand-pollination has often had to be used. In a few sites it has been shown that too much ploughing and excessive use of herbicides have eliminated most of the short grass, on whose roots the pollinator's larvae normally feed, so that replacement of ploughing and spraying with regular grass mowing can restore good pollination (Aguiar et al. 2000).

4. General Approaches to Encouraging Pollination

Moving in Pollinators Artificially

The first approach in the recent past, when poor pollination seemed to be a problem, has nearly always been to import a bee; Bosch and Kemp (2002) discussed the protocol of developing a bee species as a crop pollinator, especially in orchards.

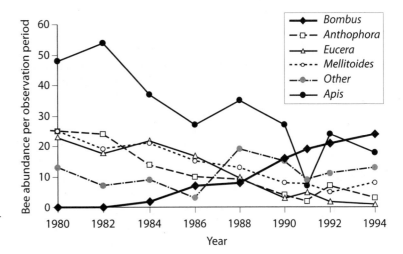

Figure 28.2 Variations in bee abundances in Israel over a 15-year period, with introduced *Bombus* increasing and several key solitary genera decreasing. (Pers. comm. Amots Dafni and Simon Potts.)

Most commonly, import has involved hives of honeybees, since they are seen as such good supergeneralist pollinators, able to tackle most flowers, and are easily managed by experienced beekeepers. They can also be used to spread an inoculum of microbial control agents to flowers and so control certain pests or diseases; for example, carrying streptomycin powder in Canadian orchards to prevent fireblight (Arnason, in Akerberg and Crane 1966). This topic is increasingly being explored for biotechnological agriculture (Al-Mazra'awi et al. 2007; Carreck et al. 2007). For many kinds of orchards use of *Apis* is now common practice, and import of hives into arable fields or areas of mixed horticulture is also becoming widespread. While often useful and effective, and a quick short-term fix, it is clear that *Apis* is not always the best answer for the crop.

The use of other bees is less common, although it is now possible to buy small quantities of various solitary bees in artificial nest tubes in many countries, often for use in gardens (plate 35E). On a larger agricultural scale, import of bumblebee nests has recently become commonplace, especially for greenhouse crops such as tomato. *Bombus terrestris* nests (plate 35D) are now routinely introduced and managed seasonally for improved pollination (Ruijter 1997), and the bees show tight temporal patterns of pollen gathering (e.g., Banda and Paxton 1991; Morandin et al. 2001). Initial work with entire ("queen-right") colonies has now been largely superseded by the use of worker-only colonies, following some problems with escaping queens that were either producing local pop-

ulation explosions (e.g., in Israel in the 1990s, where *Bombus* is not native; Dafni and Shmida 1996; see fig. 28.2) or interacting with and changing the genetic makeup of local bumblebees in Europe. With the new precautions, commercial bumblebee nests are potentially useful for other buzz-pollinated crops too, such as kiwi (*Actinidia*), blueberries, and cranberries, and they are also in regular use with polytunnel strawberries. It may be possible to manage imported bees and their behaviors with the use of bee pheromones (Pankiw et al. 1998), and increasingly there have been attempts to employ genetic selection for *Apis* strains with better pollen-gathering and pollen-hoarding behaviors (Page and Fondrk 1995).

But there is a need for very careful controls in managing bees in these fashions. If hives or nests of non-native species are imported, they can outcompete other insects and in the longer term damage native bee populations (Paton 1993; Gross 2001). Some solitary bees have also proved surprisingly invasive when introduced; for example, *Megachile* and *Anthidium* species have on occasions extended their ranges from introduction sites remarkably rapidly (e.g., Miller et al. 2003). Importation of honeybees or any other exotic pollinators to an area should surely be the strategy of last resort in future. More sustainable approaches are needed, especially given the current threats to honeybee health and productivity (chapter 29).

Most importantly, though, importation of any bees can be inappropriate and simply unnecessary. Kremen et al. (2002) reported that native bee communities can achieve a full pollination service even for demanding crops such as watermelon without the intervention of managed hives, as long as the land is farmed sympathetically and is reasonably adjacent to patches of natural habitat. This free service is destroyed by degrading the natural landscape and reducing bee diversity.

Artificial Pollen Applications

The use of artificial pollen application is another short-term possibility, where local pollinators are unavailable or honeybees are ineffective. Fruit set can sometimes be supplemented by applying extra pollen that bears an electrostatic charge (see chapter 7), whether for wind-pollinated or insect-pollinated crops. This method uses the same techniques as exist for applying powders or tiny droplets in agriculture. Once pollen is collected it can be charged by induction while suspended in liquid, or by friction, or by use of high-voltage electrodes (see Vaknin et al. 2000). Various trials with pollen blowers incorporating a charging nozzle have shown improved pollen deposition on stigmas compared with uncharged pollens. Larch (*Larix*) flowers received three times their normal pollen load and a moderate increase in seed set (Philippe and Baldet 1997), while Bechar et al. (1999) showed increments in fruit set in commercial dates, with similar results reported for almonds and pistachios (Vaknin et al. 2001, 2002).

Encouraging and Managing Native or Specific Pollinators

The general issue of supporting native pollinators is taken up in detail in chapter 29. But there are several ways of helping native (or sometimes introduced) pollinator populations to increase in number, or of manipulating them to concentrate their efforts in particular areas.

1. Supplying or maintaining the substrates that provide *nest sites* can be very helpful. This may involve leaving undisturbed areas of bare well-drained ground for many solitary bees, dry posts or logs for *Xylocopa* (Israel, West Indies), mud slurries for *Chalicodoma* (Arizona), or sandy or earth banks for various ground

nesters. Leaving some carrion for pollinating beetles and flies is useful, and of course leaving alone the roost sites of birds and the roost caves of bats can be absolutely essential in certain ecosystems. More specifically, artificial nest holes can be provided, in the form of stacks of drinking straws, or cardboard tubes, or drilled bricks for many solitary bees (see plate 35F). Even more complicated techniques are used routinely with *Megachile, Osmia*, or *Nomia* for alfalfa crops (United States, Europe, and New Zealand). "Megachileculture" is now a big business routine in parts of North America (e.g., Bosch and Kemp 2005). *Nomia* is used more locally by preference, being native to the Great Basin area of the United States and known as the alkali bee because it nests around the alkaline soils and seeps that surround the occasional natural springs in the desert. It is an excellent pollinator of alfalfa, clover, and mints, and is encouraged to nest in managed bee beds adjacent to all these crops (and also for celery and onion). The bees normally nest shallowly, so farmers dig out a shallow trough and line it, cover it with gravels, sands, topsoil, and some salt, with concrete piping to irrigate above the lining so that water rises upward, as in natural alkali seeps. They inoculate the bee bed with a plug of nest-containing soil from another site (which will include pupae) and provide some shading to control temperatures. But despite careful management there have been problems; in some seasons the bee numbers have been reduced by pesticide spraying (*Nomia* is more susceptible than *Apis*) or by unusually long summer rains leading to mold growth and disease in the beds.

2. Providing specific *forage plants* as part of normal crop rotation patterns is also desirable. Crops such as clovers benefit the soil but also support bees and some longer-tongued flies. Supplying other specific flowers can be useful in some contexts, for example where bees use the crop flowers for nectar but require another plant for pollen, or vice versa (*Xylocopa* in the Caribbean, or leaf-cutter bees on coffee in Papua New Guinea).

3. Providing the specific *host plants* for the larval stages of butterflies, moths, and some beetles are essential. These may be common weeds such as nettles and bedstraws, or occasionally more unusual and rare hosts such as milkweeds or passionflowers.

For both of these last two issues, the plants can be provided as part of the very general strategy of enhancing and increasing hedgerows and field margins (see chapter 29).

Managing the Crops

Turning to the crops themselves, there are further steps that can be useful in encouraging good pollination.

1. *Polyculture* is helpful: avoiding huge areas of monoculture is desirable for many reasons. In particular, extensive areas of wind-pollinated plants not only eliminate all possible bee habitats but also block pollinator movements through the habitat by providing no feeding corridors.

2. The *intercropping* of two or more crops together has benefits; traditionally used for pest control and even for improved yield, suitable intercrops can also favor some pollinators. Certain flowering crops are especially good for encouraging hoverflies (both as pollinators and as aphid controllers) and for supporting native bees.

3. Getting a suitable *crop microenvironment* can also be useful, as insects have to be able to fly and forage at the right times and places for the plant and its flowers. Planting a crop at the best elevation and aspect and with the right spacing (and sometimes the best row direction when sowing) can help provide an ideal microclimate within the crop, so that insects can maintain their thermal and water balance more easily as well as moving between plants most effectively.

4. Care with the development of *hybrid crops* is also necessary; simple morphological mismatches appearing in the flowers are an obvious problem, but many other aspects may change that affect relations with the pollinators, and the levels of genetic diversity can also be problematic (see below).

It is worth emphasizing here that many of the strategies outlined above are already very widely recognized as good practice for longer-term sustainability, and are being incorporated into national and international schemes to improve agri-environmental practice and favor growers who adopt environmentally friendly and pollinator-friendly practices.

5. Problems with Hybrid Crops, Seed Crops, and Crop Breeding

Rather few crops are grown nowadays where they are native; they have been dispersed around the world and tried out as foods and forage in many new countries. Likewise, rather few remain in their native state; they have been bred repeatedly for improved and useful characteristics—greater yield, better taste, ease of handling, etc. Plant breeding these days requires controlled pollination, and hybrids are often crossed manually, which often has to be at particular times of day if dehiscence is at specific times and pollen is short-lived. Chance unwanted crosses must be avoided, and the greenhouses where most plant breeding work is done are insect proofed, with individual flowers often protected additionally with paper or muslin bags. Selfing is often prevented by emasculation of flowers (removing the anthers). For greater efficiency, pollen can be stored and sent to other breeders and other regions (e.g., Shivanna and Johri 1985). Most pollens last well, potentially weeks or months, but grass pollens (i.e., most cereals) are notoriously short-lived, often viable for only a few hours, sometimes only minutes (chapter 7).

Most breeding programs aim to provide uniform seed, giving consistent growth in particular conditions. The best results often come from crossing two different uniform cultivars with hybrid vigor. However, the F_1 progeny are then highly variable and usually cannot be used, so that growers need to buy new seed from their suppliers each year. Furthermore, any plant when grown to provide a seed supply for the following year needs to avoid contamination from nearby different cultivars or from natural populations. For wind-pollinated plants, this is said to need separation of at least 15 m to get contamination below 1% (Bateman 1974a) (but 1% may still seem to be too high). For insect-pollinated crops, the results vary; for crucifers such as turnip and radish, contamination can be reduced below 1% with a separation of 50 m (Bateman 1974b). Cresswell (2005) and Cresswell and Hoyle (2006) reviewed gene dispersal by pollinators and pointed out the many variables that can affect real gene flow from a crop. Contamination for any crop can be reduced by incorporating barrier crops and by discarding the crop from the field perimeters, but a safety-first approach is always preferable. We should recall from chapter 18 that bees, especially social bees, may forage many kilometers from their nests (see Pasquet et al. 2008).

Over many centuries, humans have often selected their crop plants for their autogamous abilities or self-fertility. However, commercial hybrids producing consistent seeds are now crucial for many crops managed on an agro-industrial scale, so that selfing has often had to be bred out again. In fact, where outcrossing crops were needed once more, there were numerous breeding failures initially, and this was often attributed to poor pollination (see Erickson 1983). Modern hybrid seed production may depend on the production and use of male-sterile lines, but these are difficult to work with naturally in the field as bees' foraging behaviors often change when confronted with such anomalies.

Where good cross-pollination is essential in seed production systems, attempts have been made to improve entomophilous pollination in various ways: using complex planting schemes, or importing large numbers of pollinators, or even trying to manipulate the foraging behavior of pollinators, for example by applying chemical attractants to a crop. There have been some successes: with orchard crops such as apples, pears, and various nuts, where many varieties are self-incompatible, it pays off to interplant varieties with similar floral phenotype and phenology. But more often these options designed to improve pollination do not work well, taking too little account of the learning abilities of bees. For example, planting crops in alternating rows to get hybrid seeds has often failed to increase crossing because bees simply learn to work up and down one row; and using attractants fails as the bees quickly learn to negatively associate the new scent with the poor shape or rewards of the flowers they are being encouraged to visit.

Plant breeders therefore are now well aware of the need to take into account the biology of flowers and the needs of pollinators. This can be very hard to achieve, as hybrid seeds often produce flowers that are altered in morphology, in size, in scent, in color, in timing, or in nectar and pollen production. Some, though not all, of these aspects can be restored through selection and appropriate back-crossing. But there are always possible side effects to be wary of. For example, selection to increase nectar reward can decrease interfloral or interplant flights by the visitors and so reduce seed set. And in many instances the carefully prepared commercial hybrids do not compete well with open-pollinated varieties or with surrounding plant species of similar morphology, adding to the ever-increasing stress on monoculture growing practices.

The extent to which F_1 seed production (and therefore parental pollination) matters is highly variable. For the seed merchant it may be crucial in most species, but for the grower the needs vary with the crop. In some plants such as cotton, and in various foods where seed oils are the crucial produce (sunflower, rape, some soya hybrids), the quality and quantity of the seed is the preeminent consideration and floral morphology must be preserved in the F_1 progeny. However, breeders are sometimes also faced with the conflict of needing to produce a large F_1 generation from the parents (to sell as seed to growers), but where seed production from those progeny is unimportant or even undesirable. For example, in forage crops such as alfalfa, or food crops such as carrot, onion, or cucumber, seeds are not wanted and the preservation of normal floral function matters much less.

6. Overview

Why Bees Are Especially Good as Pollinators

Bees are the preferred pollinators for many crops unless the flowers are particularly unusual and/or are co-evolved with something else. There are several key reasons for this. First, flowers provide their only foods, they are specifically adapted to carry pollen, and they requiring more pollen since it provides the larval as well as the adult food; hence they will visit flowers more often than any other animal. Second, they are less susceptible to environmental conditions than many other insects, since nearly all have some degree of endothermy, and many have environmentally well-controlled nests as refuges. Third, they are clever: at finding, learning about, handling, and (in social species) communicating about flowers and at detecting the profitability of plants. They often therefore remain constant to one flower species, whether through one trip, or one day, or a foraging lifetime.

Why Apis *Is Seen as Particularly Good*

Honeybees are bound to be major contributors to most crop pollination. Partly this is because *Apis* species are "medium" bees in many respects: their sizes and colors are both moderate and intraspecifically variable, so they can work in a wide range of conditions over long periods; they are moderately good thermoregulators, which increases this range; and their tongue lengths are medium, so a good range of flowers is accessible. But they are important also because they are highly eusocial: this ensures large numbers, and perhaps makes individuals more prepared to take risks with their costs; it provides conditions in a hive that are very constant (usually 30°C–35°C and 40%–60% relative humidity), giving a safe recovery place; and of course honeybee hives as far as humans are concerned are manageable and movable (whether of the European or African style (plate 35C) or others).

But there are reservations. *Apis* tends to be a thief (secondary) whenever its tongue is too short, so it is problematic for red clover and field bean and many other legumes. It even steals from scopae of other bees if its colony is pollen stressed (Thorp and Briggs 1980). It cannot buzz-pollinate so is no use for blueberry and cranberry, some peppers, and kiwifruit, and is poor at working on tomato; and it reacts unusually and inappropriately to tripping mechanisms on many legumes, so is poor for alfalfa. More generally, it is an especially efficient groomer and pollen storer, so is sometimes described as a pollen waster from the plant's point of view. Finally, it is only moderately endothermic and so is not ideal for very early-flowering crops (e.g., orchard fruits, and some clovers) or for crops growing at altitude or in higher latitudes.

Why Other Bees Are Often Better

Crop managers can get too carried away by hive bees; often native bees have more efficient foraging, a better temporal match of foraging with dehiscence, and special tricks in relation to certain flowers, so they must also be encouraged. In fact it is often more important to avoid excess competition from hive bees and avoid destroying native bee habitats.

In early spring small solitary genera such as *An-drena, Osmia*, and *Halictus* species may be very important and are especially relevant to early-flowering orchard crops. Bumblebees are often especially important for temperate crops; their larger size (and/or size range) and greater hairiness means that they carry significantly more pollen on their bodies and they may deposit significantly more pollen on stigmas per visit, while their variation in tongue length means they can work a wider range of flower types. Bumblebees also commonly visit rather more flowers per minute than honeybees, and they are more likely to move longer distances between plants during any particular trip, so producing better outcrossing. Stronger endothermy also means better activity in the cold weather of early spring, and they are far more frequently present in the early mornings when pollen in many plants is at a peak; in northern Europe and climatically similar zones honeybees are rare in March–May and until at least 0900 on most days.

For all these reasons, the encouragement of maximum bee diversity in every habitat can only work to the benefit of crops, as well as supporting natural biodiversity.

Animal pollination improves the harvest for about 70% of all crops, and conserving habitat for pollinators within the agricultural landscape is therefore key to future food supplies. A synthesis of 23 studies across 16 crops showed that there are strong exponential declines in pollinator richness and visitation rates with distance from such habitat, the declines being somewhat steeper in the tropics and for social bees (Ricketts et al. 2008). On this basis, tropical crops pollinated by social bees may be most at risk in the near future if adjacent natural habitats continue to be lost. A review of global crop production across 45 years (Aizen, Garibaldi et al. 2008) revealed no overall pollination shortage (pollinator-dependent and independent crops having increased their yields at similar rates), but highlighted an increasing acreage devoted to pollinator-dependent crop types, with a rise of more than 300% in the fraction of agriculture that depends on animal pollination (Aizen and Harder 2009). The need for increasing pollinator services by potentially or actually declining pollinator populations therefore poses a real problem for the near future. This theme of potential threats to pollination services is explored further in the final chapter.

Chapter 29

THE GLOBAL POLLINATION CRISIS

Outline

1. Assessing Pollination Services and Pollinator Declines
2. Particular Threats to Pollinators
3. Overall Effects and Measures Needed

For nearly three decades now there have been well-substantiated reports of declines in pollinators worldwide, and the problems were explicitly recognized in the UN Sao Paulo declaration (1998–99), so that pollination disruption is at last being emphasized as a major issue (Kearns et al. 1998). The International Pollinator Initiative (IPI) was set up in 2000 to coordinate worldwide activities. However, the direct evidence for ecosystem-level effects has been slow to appear (Cane and Tepedino 2001; Kevan and Phillips 2001), with uncertainty still over the extent of pollen limitation (Ashman et al. 2004; see chapter 3, section *Pollen Limitation*) or of pollinator community variability (N. M. Williams et al. 2001) in natural communities, with which disturbed habitats could be compared. Therefore long-term data collections are needed to track the changes and to understand their underlying causes with a view to finding sustainable solutions (Roubik 2001; Kearns 2001; Wilcock and Neiland 2002). Pollinating animals may not directly affect most other key ecosystem processes, but as mobile link organisms they can exert important indirect effects via changes in plant community diversity and range (Lundberg and Moberg 2003).

In principle, pollinator loss might lead to plant loss, and vice versa; uncoupling a mutualism by effects on one partner could have knock-on effects at a community level (see Bond 1994, 1995). The impact of pollinator loss on overall pollination service in a community may well depend on the degree of specialization of the interactions, as it is generally assumed that specialists suffer more negative effects from disturbance, and that there is greater resilience to extinctions in more generalized relationships. Vázquez and Simberloff (2002) proposed a *specialization-asymmetry-disturbance* hypothesis, where both the degree of specialization and asymmetry of interactions affect responses to disturbance, although they found no strong support for this from their own data. Hence here again the debate over the extent and importance of specialization is critical, and rather better resolution is urgently needed.

In this chapter the needs for assessment of pollinator decline are assessed, along with some of the key threats to pollinators and to pollination services; good recent reviews on particular themes are also available (e.g., Allen-Wardell et al. 1998; Kearns et al. 1998; Kevan 1999; Klein et al. 2007; Gallai et al. 2009; Aizen and Harder 2009).

1. Assessing Pollination Services and Pollinator Declines

Proper assessments of the current state of pollinator-plant interactions, though crucial for future planning to enhance and sustain pollination services, are surprisingly difficult. Pollination declines cannot necessarily be inferred from pollinator deficits even if these can be adequately recorded (see Thomson 2001b), especially given the prevalence of pollen limitation in so many plants. Current scientific data are usually inadequate, and one of the missions of the IPI has been to establish standard methodologies for documenting pollinator occurrence and abundance, as they vary across time and across environments, and for assessing pollination services. Such core protocols are gradually emerging (see Frankie et al. 1998; Cane et al. 2000; Potts et al. 2005; Campbell and Hanula 2007), but reviews for bees (N. M. Williams et al. 2001) and for flies (Kearns 2001) have indicated that existing studies cannot adequately distinguish between natural fluctuations of populations and human-induced changes. Even when we do have standard protocols, efforts are often hampered by inadequate taxonomic expertise, as the insect fauna of many regions is so poorly documented. Although bees are one of the better-known groups there are far too few experts who can identify to species, and as yet no consistently reliable web-based identification guides for these key groups. However, electronic catalogs are gradually developing (HymenopteraBase, Species 2000, and ITIS Catalogue of Life) and there is a major effort going into image-processing from type specimens and into providing semiautomated identification systems (based on images or on barcoding from molecular data), eventually as on-line facilities.

Repeatable year-on-year samplings for insects that visit flowers at a community level usually rely on netting at flowers (organized in standard grids or transects) or on trapping techniques. Westphal et al. (2008) comprehensively reviewed these approaches for surveying bees and concluded that pan traps were the method with least observer or collector bias. Trapping using Malaise traps or colored water-filled pan traps was compared in three different forest systems by Campbell and Hanula (2007); they found blue pan traps most effective for flower visitors. However, these authors and others noted that the color used in these tended to bias the catch toward bees or flies or other insects, so that use of several colors simultaneously is advisable. Cane et al. (2000) pointed out the depauperate samples of certain taxa found in pan traps compared with traditional net collections, and Wilson et al. (2008) found similar discrepancies for desert sampling of bee communities, although only in certain seasons.

Localized regular sampling for particular bee types has also been attempted, notably for euglossine bees using scented baits (Roubik 2001) and for nocturnal tropical bees using light traps (Roubik and Wolda 2001). Sampling for individual target plant species is easier (and best achieved by intensive observations),

but it is difficult to choose a target that will give representative patterns for the surrounding community. In some cases, Asteraceae are recommended at least for assessing overall bee diversity and abundance, but in other cases the same plant family does not give a good representation for bees (see Eardley et al. 2006). Sampling a single "magnet" plant can often give better results (Minckley et al. 1999), and *Larrea* has been promoted for US desert sites in this respect (Cane et al. 2005), while mass-flowering *Andira* trees in Costa Rica have been used to document very significant reductions in bee numbers since 1972 (Frankie et al. 1997). But subsampling at other plant species is still always needed to see variations across the community (Cane et al. 2005, 2006).

Leaving aside the insect pollinators, efforts to sample other taxa are complicated by their behaviors. Both birds and bats may be migratory, breeding in one site and overwintering in another, and even in one site they may often be roosting at a long distance from the plants that they are mainly feeding on, so that monitoring of populations at different sites can give different impressions. There have been international efforts to monitor hummingbirds across North and Central America (see Allen-Wardell et al. 1998; Sauer et al. 2005), although there are still major problems with data standardization. Likewise there have been good conservation-related surveys of bats in Central America (Medellin 2003) by visiting caves regularly, at least allowing monitoring of abundance and of possible dietary changes.

Another conceivable approach would be regular sampling of fruit and seed sets as a proxy for pollination success, but this is complicated (perhaps impossibly so) by the need to factor out all other possible variables, such as edaphic changes, internal resource constraints for the plant, variable climate and microclimate, and variations in florivory or herbivory.

It can be very useful to look at the problem the other way around and record the losses of known forage plants for particular groups. This has been well documented for bumblebees in the United Kingdom (Carvell et al. 2006), with 76% of the known *Bombus* forage plants decreasing in frequency in the late twentieth century.

However, perhaps the best indicators of the problem come from documented parallel declines in pollinators and in the plants that they pollinate, and this evidence was provided by Biesmeijer et al. (2006) for bees, hoverflies, and entomophilous plants across the United Kingdom and the Netherlands, comparing pre- and post-1980 records. Plate 39 shows the effect for British bees. These authors found the most marked changes in flower specialists and habitat specialists, in univoltine species, and in nonmigrants, with outcrossing plants showing linked declines. Based on these kinds of findings, major initiatives are being set up to record and conserve native bees, especially in Europe and North America. Banaszak (1992) provided an early summary of the necessary measures.

2. Particular Threats to Pollinators

Habitat Degradation and Destruction

Since pollination involves more or less specific interactions, any habitat effects that alter the distribution or abundance of a particular species pose potential risks to the associated partners; and shifts in distribution of one plant or animal in response to habitat change, even without population decline, risk uncoupling relationships whether these be the rare obligate associations or more generalized interactions. Most commonly plants are able to "move" only slowly as their habitats change, so they may be left behind by more mobile animals.

Species that lose their pollinators may persist for some time, perhaps with some asexual reproduction, but are in an evolutionary sense at a dead end. Thus habitat destruction can leave an extinction debt that only becomes apparent years or decades later. While obligate mutualisms with reliance on a single pollinator are very rare and the risks of overdependence are obvious, they do involve some key species, especially figs (chapter 26), whose loss can have a disproportionately large influence on plant and animal community structures.

Habitat loss and disruption can also affect pollination success by disturbing the balance between legitimate visitors and the rarer cheats, nectar robbers, etc., with unpredictable effects.

Hence the persistence of effective pollination can be a good measure of ecosystem health (Aizen and Feinsinger 1994a,b).

Sharp declines in habitat availability have been most conspicuous of late in the tropics, and there are documented cases of effects on the large vertebrate pollinators. Some of the long-nosed bats of Central America have declined in numbers (Medellin 2003), and chiropteran vulnerability to forest breakup is well documented (Meyer et al. 2008).

A global scale meta-analysis (Vamosi et al. 2006) highlighted the particular risk to pollinators and plants in biodiversity hot spots. There is a significant positive relationship between pollen limitation and species richness, especially for obligate outcrossing plants, and for trees relative to shrubs and herbs. Plants in such hot spots, mainly in the tropical forests, are presumably more prone to pollen limitation because of interspecific competition for pollinators, and may experience stronger selection to specialize. They may be more at risk of extinction from the dual risks of pollen limitation and habitat destruction.

Habitat Fragmentation

Loss of habitat in a total sense is rare, but habitat fragmentation is extremely common, as natural landscapes are increasingly subject to deforestation or to more intensive cultivation regimes. Fragmentation creates smaller populations, with greater risk of inbreeding depression and genetic drift, and it amplifies the spatial isolation of these populations. Pollinators are increasingly exposed to invasive competitors, to parasites and diseases, and also to drift from agrochemicals. In isolated habitat fragments plants are also at risk of greater geitonogamy and/or greater heterospecific pollen receipt. We have considered before the estimate that at least 60% of plant species have their seed set limited by pollination at some stages in their lives, that is, they suffer lowered seed production because they receive too few visits. In areas where agriculture or other anthropogenic land use has gradually restricted natural habitats to small isolated fragments, these pollinator-limited species are particularly likely to experience low seed set. Steffan-Dewenter et al. (2002) measured such effects by comparing visits to standard groups of potted plants placed in the center of landscapes within which the area of remaining natural habitat varied from ~1% to 28%, and found that diversity and abundance of solitary bees were strongly correlated with this percentage value, although social bees were less affected.

Some clear examples of fragmentation effects on plant success in specific habitats are also available, for example in prairie wildflowers (Hendrix 1994) and in deciduous woodland herbs (Taki et al. 2007). In the arid southern Mediterranean, populations of cyclamens (*C. balearicum*) have become isolated and now receive very few visits (largely from syrphids) and so have increasingly tended to undergo delayed autogamy (Affre et al. 1995). In calcareous grasslands, *Betonica* plants in isolated fragments were visited less than half as often (by bumblebees) as plants in control areas (Goverde et al. 2002), and with bee behavior altering in ways that could increase inbreeding and reduce pollen dispersal (fig. 29.1). In similar habitats the horseshoe vetch (*Hippocrepis*) required quite large patches and a diverse landscape to attract enough pollinators (Meyer et al. 2007). Hawkmoth-pollinated *Oenothera* flowers suffered greatest pollination limitation in the most disturbed fragmented populations in Missouri (Moody-Weis and Heywood 2001). Perhaps not surprisingly the effects do depend on the habitat and on the pollinators, and in tropical forest remnants orchids visited by euglossine bees that are long-range flyers showed little or no reduction in seed set (Murren 2002), although in other studies Powell and Powell (1987) found that euglossines did not cross open areas of more than 100 m.

Fragmentation is a problem in the longer term mainly because it reduces the habitat for particular species below some critical area, so cutting down the biodiversity in the "islands" of original habitat that are effectively created. But it may also alter the balance between herbivory and pollination and thus alter the plant's responses to damage (Kolb 2008), or it may disrupt the migratory patterns of some pollinators if there are inadequate corridors of reasonably natural habitat between the islands to provide food for the passing migrants and to allow for population movement and gene flow. Habitat fragmentation, and the need to determine suitable minimum areas and suitable maximum spacing between the conserved islands, is therefore a particular concern and focus when it comes to the design of reserves and protected habitats (e.g., Rodrigues and Gaston 2001; Rothley et al. 2003; see also Nebbia and Zalba 2007), and needs special attention where the needs of interacting species operating on different spatial scales are concerned (Tscharntke and Brandl 2004).

Pollination quality generally declines in fragmented populations of tropical trees, as shown for *Samanea* (Cascante et al. 2002) and for figs, where McKey (1989) estimated that a minimum of 800 acres of typi-

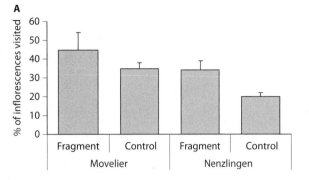

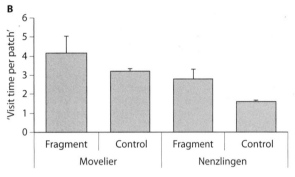

Figure 29.1 Effects of habitat fragmentation on bumblebee behavior when foraging on *Betonica officinalis* at two sites. Bars are shown ± SEs. Bees visit more often in the control patches, but visit a higher proportion of the inflorescences (A), and for longer times (B), in the fragmented areas, potentially increasing selfing and inbreeding. (Redrawn from Goverde et al. 2002.)

cal forest, containing 300 fig trees, would be needed to ensure that a given fig–fig wasp mutualism persisted. In a more general survey, Aizen and Feinsinger (1994a) showed that fragments of less than 5 acres in the dry subtropics of Argentina could not sustain enough wild pollinators; over time, the introduced Africanized *Apis* came to dominate, and at least 16 tree species have declined as a result. Africanized bees may be doing particularly well because they have a preference for pollen over nectar, whereas European bees take more nectar, which enhances honey stores for winter survival in colder climates. The Africanized bees also benefit from higher relative thorax mass and higher specific metabolic rates, features that improve their flight capacity and dispersal powers (Harrison et al. 2006). Although often regarded purely as a problem, they can sometimes be useful in rescuing remnant populations of endemic trees (Dick 2002). However, these generalist pollinators are supplanting the local specialists and taking over (although not necessarily very effectively) the pollination of rarer plants. At least some of these effects have been mainly due to loss of nest sites rather than of flora, since the ground-nesting bee *Dialictus* has suffered much less than *Augochlora* bees that nest in rotting tree stumps, which are largely removed by farmers in the more agricultural areas separating the remnant forest fragments.

In Brazil, fragmentation resulting from deforestation is a major problem, and native bee populations are in decline in at least some areas. Lima-Verde and Freitas (2002) proposed the stingless bee *Melipona quinquefasciata* as a good indicator species, and conservation measures to preserve this bee (notably reduced gathering of firewood and limited agricultural expansion, with retention of larger forest fragments and corridors) should help to conserve the bee fauna in general. Remember too the importance of persistent forest fragments in the tropics in maintaining pollination services for surrounding agricultural areas, an issue highlighted in several studies on coffee pollination (chapter 28).

Turning to temperate and herbaceous plants, a number of specific studies have shown marked effects of habitat fragmentation. Garcia and Chacoff (2007) studied the common hawthorn (*Crataegus monogyna*) and showed reduced pollen tubes per flower as proximity of forest cover declined. Lennartsson (2002), working with *Gentianella*, compared success in large and small habitat fragments, and found reduced cross-pollination and increased extinction in the small fragments, with up to six times fewer visits from bumblebees. As an extreme case Primack and Hall (1990), working on isolated lady slipper orchids (*Cypripedium acaule*), showed that only 2% of flowers were effectively visited by a bumblebee and only 7 successful fruits were produced in a set of 64 plants, over four years. However, on the positive side it seems that where even narrow corridors of remnant land remain between degraded and depauperate zones the animal-plant interactions can persist; for example, Townsend and Levey (2005) specifically demonstrated that habitat corridors could improve pollen transfer between fragments for both butterfly- and bee-pollinated plants. In southern Africa small corridors of grassland between planted pine forests also retain good biodiversity (Bullock and Samways 2005), although those same grassland corridors are now under heavy grazing and human disturbance threat.

Some general patterns are perhaps beginning to emerge. Aizen et al. (2002) reviewed the literature available and could find no obvious difference in the responses to fragmentation of generalist and specialist plants. Results from Florida (Koptur 2006) were similar, and the findings of Aguilar et al. (2006) from a meta-analysis could also be interpreted this way. However, work by Taki and Kevan (2007) found that habitat loss and fragmentation led to increased generalization in the insect communities, but not in the plants. Such findings probably reflect the asymmetry in specialization present in many relationships (Ashworth et al. 2004; chapter 20). However, they have led to the rather worrying assertion that, even if specialist pollinators are wiped out by disturbance, the plants will mostly be buffered against extinction because they also get generalist visitors (as discussed in chapter 20).

Considering just the pollinators, various studies on bee diversity by Steffan-Dewenter and his group (reviewed in Steffan-Dewenter et al. 2006) showed that habitat fragmentation had the greatest impact on solitary species, especially those oligoleges with specialized pollen needs. However, the news is not all bad; other studies found that pockets of native bees persisted surprisingly well in habitats drastically altered by humans (Cane et al. 2006) and even in entirely urban habitats (Frankie et al. 2005). And estimates of pollen dispersal over 5 km in a bird-pollinated shrub (Byrne et al. 2007) indicated that such plants could cope with quite extensive habitat fragmentation.

Edge effects represent an associated issue here; as fragmentation increases, there is an increased ratio of perimeter to interior. Murcia (1995) pointed out that edge effects in forests were critical but very poorly understood, and could be very site specific, with various community features showing increases or decreases in different forests. Around tropical forest edges there may be more light patches and insolated marginal habitats, often invaded by mass-flowering light-tolerant weedy plants that can temporarily promote pollinator abundance. But over time this effect can diminish and there is likely to be a higher ratio of less floriferous but light-tolerant plants.

However, edge effects viewed the other way around also work to the benefit of nearby crops. Agroforestry systems in Sulawesi had higher species diversity (both flowering plants and hymenopterans) the closer they were to the edges of natural forest (Klein et al. 2006). Coffee grown nearer to fragments of native forest gets significant pollination benefits (chapter 28), and grapefruit crops close to subtropical forest in Argentina also experienced greater pollinator diversity (fig. 29.2; Chacoff and Aizen 2006). Chacoff et al. (2008) also pointed out that honeybees were not particularly efficient pollinators of this crop and preserved forest remnants could provide more efficient and resilient pollinators.

Intensive Agriculture

Monocultures and Loss of Hedgerows

Perhaps the most obvious feature of intensification of agriculture is the enlargement of fields to accommodate monoculture cropping, which facilitates use of machinery and reduces costs of tillage, harvesting, and chemical applications. However, this inevitably reduces floral diversity, with the concomitant loss of all the small islands of diversity formerly provided by hedges, field margins, small patches of woodland, and uncultivated land. In the United Kingdom about 40% of hedgerows have been lost since the 1930s. Declines of several key pollinator groups, for example, bumblebees, hoverflies, and solitary bees, have been explicitly linked to this change of land use (Carvell et al. 2004).

Land use around agricultural fields has direct effects on the availability of native pollinators (e.g., Öckinger and Smith 2007). Areas of pastureland within intensive monoculture areas can significantly improve wild bee abundance (Morandin et al. 2007), with 94% of the variation in bumblebee numbers explained by the amount of nearby pasture. Quite small proportions of uncultivated land can be beneficial, and schemes to reinstate arable field margins and hedges have been set in place in recent years. Carvell et al. (2007) evaluated the success of some of these in Europe; field margins deliberately sown with mixtures of pollen- and nectar-producing plants, especially with inclusion of legumes, gave better results (for pollinating species across all tongue lengths) than simply allowing regeneration or sowing mixed grasses.

Agrochemical Effects: Herbicides and Insecticides

Herbicides to control weeds are not necessarily directly toxic to flower visitors, but can have major effects by eliminating key host plants for lepidopterans and key forage plants for bees. They may thereby have major impacts on wild pollinator populations. There are

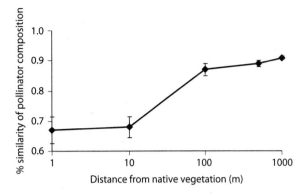

Figure 29.2 Effects of distance from native vegetation on similarity of pollinator species richness for grapefruit crops (*Citrus paradisi*); closer to the edges the pollinator communities are reduced in richness and less similar to the communities deep in the forest. (Redrawn from Chacoff and Aizen 2006.)

suggestions that when herbicides are gathered with pollen they can affect larval growth in bees.

The use of pesticides in agriculture is well documented as causing pollinator declines (Kevan 1975a,b; Johansen 1977; Johansen et al. 1983; Kearns and Inouye 1997; Spira 2001; Gels et al. 2002), especially where spraying time coincides with flowering time. Bees are particularly sensitive to many standard insecticides, and fenitrothion, the organophosphate widely substituted for DDT in the last few decades, is even more toxic to bees. Malathion is also very toxic and especially problematic because it is usually applied in microcapsule form, mimicking pollen and being picked up and transported by bees (Johansen and Mayer 1990). Recent use of phosmet in orchards has also been shown to affect nesting alkali bees in the vicinity (Alston et al. 2007), and there are unproven speculations that nicotinoids may have effects on bee behavior and memory.

The case of the lowbush blueberry in Canada is well known, since Kevan (1975c) noted a massive decrease in yield where adjacent foresters were spraying (first with DDT, then fenitrothion) against spruce budworm; pollinators were getting killed throughout the 1970s. This also affected songbirds, as there were fewer insects for them to feed on and they had taken to damaging the blueberry crop. Growers introduced *Apis* hives to help pollination, but yields only fully recovered once fenitrothion use near blueberry fields was banned (Kevan and Plowright 1995).

Similarly, use of diazinon to control aphids in alfalfa fields in the United States resulted in massive declines in the pollinating alkali bees (chapter 28), which took several years to show recovery (Johansen 1977).

In urban areas, up to 90% of homes may be using sprays indoors, and 50%–80% use insecticides in yards or gardens. Specificity of some is good, but broadspectrum sprays certainly affect beneficial insects and can reduce pollination locally. Even biological controls such as *Bacillus thuringensis* (Bt) sprays, used in organic gardens, are not risk-free—the spray drifting onto butterfly host plants produces quick declines.

The increasingly tight controls on insecticidal spraying (especially against pests that are active at flowering or fruiting time, like the raspberry beetle *Byturus tomentosus*) may mean that in the future native insect pollinator populations will recover, and they could therefore (if we try to conserve them now) come to play a more dominant role in visiting the crops we grow.

Overgrazing

Specific herbivory effects on pollination systems were considered in chapter 25, but at a more general level the moderate vertebrate or insect grazing of grasslands and moorlands is often useful in maintaining plant diversity (e.g., Hartley and Jones 2003) and insect diversity (e.g., Kruess and Tscharntke 2002), and specifically bee abundance (e.g., Steffan-Dewenter and Leschke 2003; Vulliamy et al. 2006). Grazing can increase light availability for some plants, both by removing sheltering foliage for herbs and shrubs and by decreasing the litter thickness around low-growing plants (e.g., Ågren et al. 2006). It may also open up areas of compacted ground suitable for solitary bee nesting and can increase nutrient levels from dung and urine. For these reasons it can have the mild to moderate positive effects on plant diversity and flowering already mentioned. But more damaging effects of intense grazing are recognized as a significant problem in several ecosystems, notably semiarid and Mediterranean habitats; here overgrazing by cattle and goats that takes out large amounts of plant tissue is more likely to be detrimental, both removing flowers and potential flowers and also destroying certain kinds of pollinator habitat and nest site. Livestock can therefore modify pollinator community composition, for example in California, where sheep affected legume pollination and success (Sugden 1985), and in the ka-

roo of South Africa, where the pollinating monkey beetle assemblages changed under more intense grazing pressure (Mayer et al. 2006). Use of late-season grazing rather than continuous grazing may be better management practice for preserving flower abundance (e.g., Vulliamy et al. 2006; Sjodin 2007).

Vázquez and Simberloff (2003), using a path analysis approach, showed that introduced cattle significantly modified some key and frequent plant-pollinator interactions, and the same authors (2004) modeled the various effects of cattle grazing on a single herbaceous plant (*Alstroemeria*) into a path analysis (fig. 29.3; see also Aizen and Vázquez 2006). They found that cattle affected pollen deposition mainly through an effect on the relative density of *Alstroemeria* plants, so reducing conspecific pollen deposition. Vulliamy et al. (2006) analyzed a diverse Mediterranean flora (fig. 29.4) and found that herbs and shrubs were differently affected by grazing, with herb species richness declining only at the highest grazing levels, and with bee diversity responding to herb cover, with a linear relation between bee abundance and grazing level; but they also presented a path analysis which revealed that bee abundance changes were mediated more through changes in nesting availability than through flower cover or floral diversity (fig. 29.5).

Selective Harvesting

Here the effects depend on the abundance of the species that are removed, most commonly trees from forests. Logging of rare valuable species, done with care, may have little effect on the pollination community. For example, monitoring of bird populations in selectively logged Malaysian forests revealed undiminished abundance, diversity, or breeding (Yap et al. 2007), while analyses of bees and syrphids in forests harvested by single-tree selection in Canada found greater abundance and slightly greater diversity in the recently logged areas (Nol et al. 2007). However, logging of dominant trees can produce widespread habitat degradation and will change the light availability and nutrient levels drastically. Cartar (2005) documented the short-term effects of experimental logging in a boreal forest and found negative impacts on plant diversity and on bumblebees, the bees failing to match the changing flower densities and experiencing excess competition in clear-cut areas, but with relatively reduced visitation in control areas where diminished pollination service resulted.

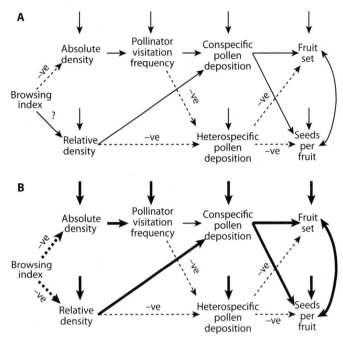

Figure 29.3 Path analysis of the effects of cattle grazing and associated perturbation on pollination and seed set in *Alstroemeria aurea*: (A) predicted and (B) observed effects with the strength of the relation shown by line thickness (with vertical arrows representing unexplained variations). (Redrawn from Aizen and Vázquez 2006.)

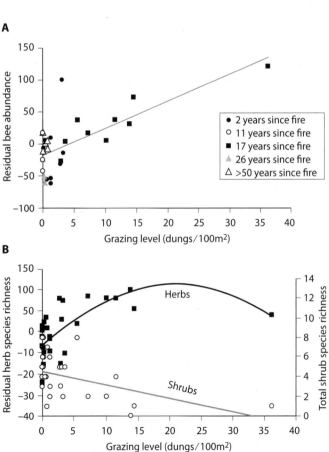

Figure 29.4 Effects of grazing in a mixed Mediterranean flowering community: (A) species richness of herbs and of shrubs in relation to grazing level; (B) general increases in bee abundance with increased grazing levels (which themselves peak at intermediate times since the area suffered fire, an interacting disturbance factor). (Redrawn from Vulliamy et al. 2006.)

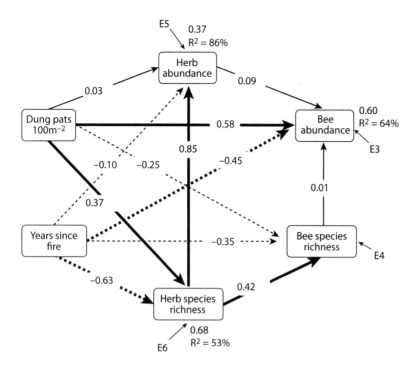

Figure 29.5 A path analysis of the same community as in figure 29.4, showing the strengths of relationships between interacting factors. Bold lines are significant paths, and dotted lines are negative interactions. (Redrawn from Vulliamy et al. 2006.)

Genetic Engineering of Crops

Genes for resistance to herbicide, insecticide, or disease, once in crops that require cross-fertilization, could both in theory and in practice be transferred into compatible weedy relatives, by "escape" of genes from the genetically modified crops (Rieger et al. 2002). There are concerns that this could produce superweeds, although the biotechnologists believe these risks can be avoided.

With the possible advent of a greater variety of genetically engineered crops, it is clearly important to know how all flower visitors (and indeed wind) may act in spreading pollen from cultivation into wild plant communities, as mentioned in chapter 28, and the modeling and measuring of these effects is now a major concern (Hayter and Cresswell 2006; Hoyle and Cresswell 2007).

Increasing Fire

Wildfire is a natural component of many ecosystems (especially Mediterranean zones and grasslands, and some boreal forests) and is important to the reproduction of many plants that require heat to break their seed dormancy. Most fires in areas where human impacts are substantial occur in old woodland undergrowth (with dead brushwood accumulation) and simply serve to open up the ground-cover and stimulate seeds to germinate and renew the normal undergrowth. In fact, managed fire can be very valuable in forests, increasing low herbaceous cover with associated improvements in pollinator diversity (e.g., Campbell et al. 2007). However, excessive frequency or intensity of fire (whether set deliberately, or an accidental consequence of human recreational fires, or even resulting from warmer climates and increased lightning storms) could be severely disruptive to pollination. Fire in the short term creates an open, highly insolated zone with good soil nutrients, and there may be a temporary flourishing of annuals from the seed bank with a high initial floral diversity. It can also affect pollen performance positively, with faster-growing pollen tubes and larger resulting plants (Travers 1999). But in the longer term repeated fire episodes will stop progression to a climax community and simultaneously diminish the seed bank for annuals to grow between the shrubs and small trees that do survive.

Monitoring of zones of different postfire ages in Israel (Potts et al. 2003) therefore showed that bee and plant diversity peaked about two years after a fire, but pollinator composition also varied with postfire age, and overfrequent fires reduced the abundance of some of the key perennials on which bees depended. Fires also had an effect on the availability of key nesting materials for some of the pollinators (Potts, Vulliamy et al. 2005). In Florida pinewoods, the timing of fires also had major effects in the short term on whether flowering occurred when pollinators were available (Pitts-Singer et al. 2002).

In the long run, fires probably have an effect in changing the composition of a particular flora, especially increasing the proportions of fire-tolerant species. They are also interactive over longer time scales with grazing, since heavily grazed areas can act as fire-breaks, and recently burnt areas can act as attractants to grazers during the first postfire green flush (Archibald et al. 2005).

Introduced Animal Species and Pollinators

Introduced mammals are often the most conspicuous threat to pollination systems and communities. Most obviously this would involve livestock, considered above under the heading of grazing. But introductions also include rats, feral cats, and rabbits, and all can be problematic. Cats reduce numbers of birds, lizards, and some small mammals, and this can lead to increases in insect populations, including pest herbivores. Rats may remove key insect groups, again with potential knock-on effects. An introduced predatory tree snake on the island of Guam produced severe reductions in bird pollination for local *Bruguiera* and *Erythrina* trees (Mortensen et al. 2008). Introduced fire ants (*Solenopsis*) can be another unexpected problem, limiting both herbivores and pollinators because they are very aggressive generalist omnivorous foragers (Fleet and Young 2000).

Introduction of pollinators is a different kind of issue, generally involving bees (Goulson 2003) and usually intended to be beneficial. *Apis* is probably the most ubiquitous introduction worldwide, and is characterized as hypercompetitive. It is now hugely dominant in some areas where it has been introduced and can have substantial impacts (reviewed by Butz Huryn 1997). At the resource level the effects seem inevitable for some non-*Apis* bees (e.g., Dupont, Hansen, Valido et al. 2004; Forup and Memmott 2005b). This can in-

clude native bumblebees, which tend to be "living on the edge"; remember (chapter 18) that one honeybee hive can take away the resources of about 100 bumblebee colonies, and hence there are documented effects on bumblebee abundance (Thomson 2004) and size (Goulson and Sparrow 2008). *Apis* may also influence the diversity and abundance of other flower visitors (e.g., Evertz 1995; Brugge et al. 1998), including pollinating birds (Hansen et al. 2002). However, some authors have found little or no overall effect, including Sugden et al. (1996) in Australia, Roubik and Wolda (2001) in a long-term study in Panama, and Steffan-Dewenter and Tscharntke (2000) in central European grasslands. Quite what factors determine impact from honeybees remains unclear, so a precautionary approach is sensible; for example, in New Zealand *Apis* is an exotic introduction and has become an important agricultural pollinator, often outnumbering all other flower visitors, but movement of hives (especially between North and South Island) is tightly controlled and monitored.

In some situations *Bombus* itself has become a problematic introduced species (Dafni 1998; Otterstatter and Thomson 2008). Colonies of *Bombus terrestris* from Europe were taken to Japan and Israel to work in greenhouses, but individuals often escaped and have become naturalized. In Israel in the 1990s *Bombus* numbers increased markedly and other native bees declined (see fig. 28.2) (Dafni and Shmida 1996). In Japan, *B. terrestris* can interfere with native bumblebees partly by its tendency to rob flowers with longer corollas, giving reduced fruit set (Kenta et al. 2007). Even within Europe commercial *B. terrestris* colonies tend to have larger workers with higher nectar-foraging rates and faster gyne production than native colonies and so could outcompete them (Ings et al. 2006). Bumblebees have more recently been introduced deliberately to Mexico and accidentally to Tasmania where they are thriving (Hingston et al. 2002; Hingston 2007). In Argentina the introduced European bee *Bombus ruderatus* has increased in abundance steadily since 1994, at the expense of the native *B. dahlbomi* (Madjidian et al. 2008); the invader is a less effective pollinator of native *Alstroemeria* flowers but has become the better overall pollinator by sheer force of numbers.

The populations and effects of introduced invasive species clearly need to be carefully followed, to avoid competition with natives. It is also critical to monitor introductions to prevent possible spreading of diseases to natives; commercially reared bees may have higher incidence of pathogens, and Colla et al. (2006) reported spillover (of both *Nosema* and *Crithidia*) to natural *Bombus* populations from greenhouses.

Invasive Plant Species and Changing Floras

Invasive alien plants can have marked effects on the biodiversity of plants in particular habitats, and can thereby affect pollination outcomes for natives by rapidly inserting themselves into existing communities. Sometimes the effects are relatively benign: Memmott and Waser (2002) traced the effects of alien plants in a North American pollination web, finding the aliens to be visited by fewer species than the natives, most of the visitors being generalists, but noting that the aliens readily became "well integrated" into the native web (see also Aizen, Morales et al. 2008). However, introduced exotic plants can have an overall negative effect on a pollination community in several ways: by increasing competition for light or for other resources, or by "stealing" pollinators from coflowering natives. The Himalayan balsam (*Impatiens glandulifera*) is a classic example of this latter effect, its superabundant nectar (plate 19G) distracting bees and other flower visitors from native plants in the damp habitats that it invades (Chittka and Schürkens 2001; chapter 22). This same alien was used by Lopezaraiza-Mikel et al. (2007) to test the effect of an invasive plant on a native pollination network (chapter 20); invaded plots had higher visitor species richness and abundance, but the pollen transport in the system was dominated by the alien pollen so that pollination of natives was likely to be compromised. Similar effects have occurred with introduced *Lythrum salicaria* (purple loosestrife), which can reduce pollinator visitation and seed set in native *L. alatum* while probably increasing wasteful interspecific pollen transfer (Brown et al. 2002). Another well-known example is *Rhododendron ponticum*, an invasive in much of northwestern Europe although somewhat declining in its native Spain. Stout et al. (2006) compared its pollination regime in Ireland and Spain and found a range of fairly generalist polylectic visitors in both sites, differing at the species level but with similar visitation rates. In Spain the visitors carried less pollen, and in Ireland the flowers were more nectar depleted, suggesting again that the plant is providing an unusually favored nectar resource for native pollinators where it is an alien.

Taken a stage farther, these kinds of invasions can affect not just pollinator behavior but also wild pollinator population sizes and diversities. For example, alien goldenrods (*Solidago*) are now common in many parts of Europe and North America, and a study in Poland showed that where present they markedly reduced the species richness and diversity of bees, hoverflies, and butterflies (Moron et al. 2009).

All the above examples describe invasive aliens having competitive effects on natives, but Bjerknes et al. (2007) pointed out that facilitative effects are also quite likely in some situations, with aliens supporting an increased pollinator density; such effects could be more likely on a landscape scale. Moragues and Traveset (2005) found just such an effect for *Carpobrotus* introduced in the Balearic Islands, where it competed with *Lotus* but facilitated pollination in *Cistus* and *Anthyllis*. In studies by Jakobsson et al. (2008), this same plant had little impact on natives as its pollen was very rarely transferred to them, while in other Mediterranean sites *Carpobrotus* facilitated visitation to native plants; however, *Opuntia* had the opposite effect in the same communities (Bartomeus et al. 2008). The unpleasantly invasive giant hogweed (*Heracleum mantegazzianum*) also had a surprising slight positive effect on native pollinator abundance on *Mimulus guttatus* (Nielsen et al. 2008). All these cases underline the need to avoid assumptions about negative invasive effects (see Mitchell et al. 2009), although a meta-analysis across 40 available studies showed that negative effects of aliens on visitation to natives were much the commonest outcome, especially where flower shape and color were similar between the interacting species (Morales and Traveset 2009).

It is generally assumed that imported nonnative plants are more likely to establish and spread if they are rather generalist in their pollination requirements (Myers and Bazeley 2003); generalists should therefore be over-represented in existing lists of invasives. However, an analysis by Corbet (2006) for the British flora found no such association between flower type and recorded increases or decreases, for natives or introductions, and she assumed that factors other than pollination system are more critical in determining colonizing ability; larger size and larger seed size are likely candidates (Crawley et al. 1996). Note that this fits with the

Impatiens case already referred to, where pollination is mainly by bumblebees but the plant itself is large and highly successful at explosive seed dispersal.

The converse of the argument above would be that more specialized plants are likely to be more vulnerable to plant invasions and to associated effects on pollinator communities (Johnson and Steiner 2000). Morales and Aizen (2006) produced a pollination web for a temperate Argentinian community with many introduced aliens (both plant and animal) and found that both alien and native plants were visited by similar numbers of species, but that the aliens attracted a different range of visitors and got far more than their "share" of visitors (they were 3.6% of all species but received more than 20% of all individual visits). Hence they may have taken visits away from natives and at the same time facilitated their own spread.

Issues of rare plants and their pollination problems should come into play here. Kwak and Bekker (2006) analyzed visitor patterns for a range of plants in the Netherlands and found that the rare species were particularly likely to be visited by bees in general, and especially bumblebees, although rather few received visits from the ubiquitous *Apis mellifera*.

Diseases and Other Natural Threats to Key Pollinators

There are a number of very real problems for honeybees currently appearing in various parts of the world. Tracheal mites were recorded in the United States in 1984 and are now present in most states. The mite *Varroa* arrived there in 1987 and is a particular concern to beekeepers at present; it now occurs in over 30 states. Acaricides are expensive, and the mite is starting to become resistant anyway. *Nosema* is another parasite that affects 60% of all colonies, and foulbrood occurs in 2%. A new threat is the small hive beetle (*Aethina tumida*), which is native to sub-Saharan Africa where it parasitizes honeybees, but is currently invading North America where it infests bumblebee nests (Spiewok and Neumann 2006). Even if parasites do not always kill bees, they can often affect their foraging behavior and floral constancy (e.g., Otterstatter et al. 2005). And now there is the specter of sudden colony collapse disorder (CCD), which is under intensive investigation (e.g., Anderson and East 2008), and re-

sponsible for loss of around one-third of US hives in 2006–7. It is perhaps most likely to originate from a virus, with Israeli bee paralysis virus strongly implicated by some, but in practice one or more viral infections carried by mites may be interacting with stresses from intensive management and/or from multiple insecticidal or herbicidal influences.

Taken together, these pest and disease issues have so far led to at least $200 million losses being incurred by beekeepers. In the United States, estimates by Southwick and Southwick as far back as 1992 suggested that 50% of managed *Apis* in the northern states would soon be lost to tracheal mites and *Varroa*; and southern hives would be abandoned due to Africanization, giving a net loss of around $2–$5 billion, even without CCD. The Africanization problem arrived in the United States permanently from 1990 and now has led to 80% of hives being given up in many southern areas; this could eventually rise to 100% loss in places. The number of US hives peaked at 5.9 million in the late 1940s, then suffered a major drop due to organochlorine poisoning before recovering fairly strongly; but numbers are now decreasing again, down to 2.7 million in 1995 and much lower again by 2008. With these kinds of bee deficits, alfalfa losses alone might be worth over $300 million; and for all crops, perhaps $5–$6 billion. Hives that used to cost only $40 to hire can now cost $150–$200 or more. Despite this potential "bonus" for the remaining beekeepers, the additional threat of cheap imported honey from Mexico and China is certainly not encouraging them to persist with their work.

Other countries are less affected; indeed the stock of honeybees is probably still increasing globally (Aizen and Harder 2009). Some areas do already have similar problems though, or are aware that they could easily develop them. Hypothetical *Varroa* invasions to the (currently clean) Australian beekeeping industry have been modeled by Cook et al. (2007) and they estimated that costs of $16–$38 million per year (loss of pollination service, reduced yields, and associated production costs) would be avoided only if the mite was kept at bay, to which end the Australian Government is directing very substantial effort.

Clearly, if honeybee populations locally or globally do decline, native bees will be all the more important. Indeed, if other pollinators are encouraged properly, the US crop deficit could probably be reduced to just $1–$2 billion.

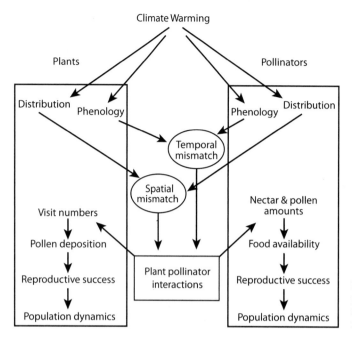

Figure 29.6 Possible effects of climate warming on both plants and pollinators, potentially creating temporal or spatial mismatches. (Redrawn from Hegland, Nielsen et al. 2009.)

Climate Change

The general predicted effects of climate change (warming temperatures, changing precipitation, increased frequency of extreme events) on biodiversity are well documented, as are the specific effects of higher temperatures on aspects of plant growth and reproduction (see Hedhly et al. 2009). The potential effects on species interactions have also been considered for at least three decades, because of the possibility of uncoupling the ecological interactions on which so much community structure depends (see Parmesan 2006; Hegland, Nelsen et al. 2009). Pollination ecology has often been a focus of speculation and modeling here, as attempts to quantify the value of pollination services became an issue for conservationists. The likelihood is of differential effects on pollinating animals (which are relatively mobile and can shift their ranges rather quickly) and on plant reproduction (plants being essentially sessile and shifting range only slowly); hence the threat that some plants, especially those with more specialist needs, might lose their pollinators entirely through spatial or temporal mismatches has been at the core of this debate. Figure 29.6 summarizes possible effects (Hegland, Nielsen et al. 2009).

Problems may be particularly acute for high-latitude and montane species, where early snowmelt and glacial melting lead to longer snow-free periods, but with the associated risk of longer drought that can truncate the growing and flowering period.

Shifts in timing of flowering are already well documented and generally close to linear (averaging around 2.3 days per decade). Flowering is occurring about 8 days earlier now in the vicinity of Boston, Massachusetts (Primack et al. 2004), and similar changes have been recorded in an alpine landscape (Molau et al. 2005) and in temperate UK sites (Fitter and Fitter 2002). More spectacularly, Bowers (2007) estimated an advance of 20–41 days in Sonoran Desert shrubs, with the flowering curve shifted to peak in March rather than May. And in Spain an analysis of more than 200,000 records of plant phenology showed marked effects in the majority of 118 species studied, with growing seasons earlier and longer, increasing by 18 days on average, and with more synchronized onset of several key phases, although with temporal responses differing between plant taxa and causing shifts in interspecies relationships (Gordo and Sanz 2009). Insect-pollinated plants may react more strongly than anemophilous species, and early-season bloomers are also probably the most sensitive (Miller-Rushing et al. 2007). Plant ranges are also shifting by an average of around 6 km northward per decade in the northern hemisphere (Parmesan and Yohe 2003).

Some specific effects on animal phenology and distribution also seem fairly clear. Kerr (2001) documented effects on butterfly species richness and heterogeneity already occurring in Canada that were attributable to climate change, particularly evapotranspiration rates. Northward extensions of butterfly, moth, and bumblebee ranges in Europe and the United Kingdom have been reported. *Apis* bees and pierid butterflies both appear earlier in Spain now (Gordo and Sanz 2006), and bumblebees appear up to 2 weeks earlier in Europe. Notable examples of range spread in the United Kingdom are *Bombus lapidarius,* which used to be almost absent in southern Scotland but had become the single commonest bumblebee by the mid 1990s, and the hawkmoth *Macroglossum*, now occurring as far north as the Orkney Islands (pers. obs.).

Climate change is manifested in the weather but of course underpinned by changes in atmospheric gases. In particular, raised CO_2 concentration is known to affect flowering times in many plants, probably interacting with photoperiod in its actions on developmental pathways. Memmott et al. (2007) incorporated simulated consequences of a doubling of atmospheric CO_2 into a visitation network of 1429 visitors and 428 plants. They found that the phenological shifts that resulted led to floral resource depletion for between 17% and 50% of all the animal visitors. It should be noted that there is as yet no clear evidence (other than modeling) for climate-driven mismatches in pollinator interactions, although Hegland, Nielsen et al. (2009) offered some useful potential scenarios for further research.

3. Overall Effects and Measures Needed

Table 29.1 reviews some of the threats to pollinators and to pollination services and their likely impact on different aspects of the flower-animal interaction at the individual, population, and community levels (see Aizen and Vázquez 2006). The predicted effects are highly varied, as the above discussions should make clear, and we should not assume that all effects are going to be negative. But one fairly safe prediction is that almost all of the kinds of perturbation that are currently worrying biologists will favor those plants that are more perturbation resistant. In many instances, anthro-

pogenic disturbance has increased the dominance of one or a few generalist plants and pollinator species (e.g., Aizen and Feinsinger 1994a), and areas of disturbed habitat now sustain rather similar and convergent pollinator assemblages independent of the kind of disturbance experienced (Morales and Aizen 2006). In large parts of the Neotropics the most successful beneficiary of disturbance has been *Apis mellifera*, but unfortunately of the Africanized aggressive form (Goulson 2003).

Note also that several of the key threats discussed above are interactive. For example, Rao et al. (2001) showed that fragmentation can increase herbivory, and this has knock-on effects for plant communities and hence pollinators. Rymer et al. (2005) found that lower reproductive success in rare rather than common species of *Persoonia* (Proteaceae) in Australia was being exacerbated both by more frequent fires and by introduced honeybees. Probably the best-documented effects to date concern bumblebees, which have been carefully recorded by professionals and amateurs for at least a century. Kosior et al. (2007) summarized the known effects for much of Europe and reported that 80% of the species and subspecies were under threat in at least one country, and 30% throughout their range. Extinctions had peaked in the period 1950–2000, with anthropogenic factors particularly implicated, especially farming practices.

There is an additional cause for concern in that a reduction in pollinators can dramatically change the direction of selection on floral traits, at least in some plants. For example *Mimulus* flowers produced smaller flowers with better selfing abilities when pollinators were excluded (Fishman and Willis 2008), potentially compounding the problems over evolutionary time.

Many of the threats discussed above come down in the end to loss of the flowers on which pollinators depend. This is especially the case for bees, whose reproductive output is entirely dependent on flowers. Müller et al. (2006) attempted calculations of just how many flowers were needed for various solitary bees to raise one offspring, and arrived at figures ranging from 7 to 1100 flowers, or 0.9 to 4.5 inflorescences, varying with bee species and host plant. They estimated that each bee normally only acquired 40% of the pollen in a flower on average, so that these values had to be multiplied by 2.5 to give a real value for the flower numbers needed, thereby suggesting that declines in

TABLE 29.1

Predicted Effects of Various Disturbance Factors on Pollination Systems

Disturbance type	Floral display	Flower morphology	Floral rewards	Plant population size	Plant population density	Plant community diversity	Pollinator diversity	Pollinator abundance
Fragmentation	Inc.	?	Inc.	Dec.	Inc. or dec.	Inc. or dec.	Inc. or dec.	Inc. or dec.
Herbivory	Inc. or dec.	Dec.	Dec.	Dec.	Dec.	Inc. or dec.	Inc. or dec.	Inc. or dec.
Introduced aliens	Dec.	?	Dec.	Dec.	Dec.	Inc. or dec.	Inc. or dec.	Inc. or dec.
Biocides and fertilizers	Dec.	Dec.	Dec.	Dec.	Dec.	Inc. or dec.	Dec.	Dec.
Fire	Inc.	?	Inc.	Inc. or dec.	Inc. or dec.	Inc. or dec.	Inc. or dec.	Inc. or dec.
Harvesting and logging	0 or Inc.	?	0 or inc.	Dec.	Dec.	Inc. or dec.	Inc. or dec.	Inc. or dec.

Source: Based on Aizen and Vazquez 2006.

many bee species could already be attributed to food shortage.

What are the solutions? There is an urgent need to focus on habitat management and use of reserves in terms of keeping floral diversity high and interactions intact, and not just in conserving particular species (Thompson 1997). This is inevitably tricky, requiring preservation of larger areas. In smaller areas that are already at least partly protected it usually makes good sense to remove exotics, to augment key native species, and to manage natural disturbance such as fires and herbivore activity very carefully. Management of the "matrix" landscapes between reserves is also needed, giving a "joined-up" approach (in more than one sense!). Replacing and encouraging the diversity of wider uncropped field margins and hedgerows has real benefits (e.g., Croxton et al. 2002; Carvell et al. 2004; Pywell et al. 2005), as do the set-aside habits of leaving areas of land fallow for several years at a time (Decourtye et al. 2007) or preserving even quite small patches of natural grassland (Öckinger and Smith 2007), in both cases providing habitat and food over a longer flowering period and better-quality water resources with lower nitrogen levels. Naturally grazed meadows, gardens, and general green spaces in urban areas are also part of the picture. Many of the key experts in this field have attempted a joint project to put all these factors together into a conceptual framework that can inform land-use practices (Kremen et al. 2007), also resulting in a first attempt at a quantitative model of pollination services across landscapes (Lonsdorf et al. 2009).

It is encouraging that some restoration projects are already showing good returns. For example, restored hay meadows seemed to rapidly return to the same level of species diversity and interaction as old meadows (Forup and Memmott 2005a), and the same was broadly true for heathlands at least as assessed by pollination network analysis with regard to the main pollinator taxa (Forup et al. 2008). Agri-environment schemes ("ecological compensation areas") in Switzerland promoted diversity and abundance of hoverflies, solitary and social bees, and butterflies (Albrecht et al. 2007). Increased acreages of organic farming are also likely to help native plants and pollinator communities (Gabriel and Tscharntke 2007; Holzschuh et al. 2007). However, on the downside abandoned grasslands undergoing restorative grazing are taking more

than 5 years to recover their natural lepidopteran populations (Poyry et al. 2005), and pollinator biodiversity enhancement from high-quality remnant habitats near farmland has only a very limited spatial range (Kohler et al. 2008).

Overall, a priority in temperate zones has to be to conserve and reinforce the hedgerows and their constituent plants such as dead-nettles, comfrey, clovers, and the herbaceous plants of marginal and fallow land, especially the more specialist long-corolla perennials that tend to have more nectar than annuals (Fussell and Corbet 1992; Corbet 1995; Petanidou and Smets 1995). Table 29.2 gives a listing of particularly useful plants from a western Europe perspective, and similar lists could be readily assembled for other biogeographical regions. Of course hedgerows provide not just food but also shelter and nesting sites for many pollinators, and in their vicinity patches of good butterfly larval food plants such as nettles and bedstraws are easily maintained. It is also very important to conserve the key cornucopia flora that attract the greatest range of insect visitors—in Europe this will include some of the species of *Heracleum, Taraxacum, Daucus, Aegopodium, Cirsium, Anthriscus, Senecio, Achillea, Rubus,* and *Knautia;* and also *Calluna* in upland regions. Many of these are included in the "grass and wildflower" or "nectar and pollen" seed mixtures, available under such schemes as Country Stewardship in the United Kingdom. It has also been shown that a mix of at least two flowering species planted around agricultural plots is significantly more attractive to bees and hoverflies (Pontin et al. 2005; see fig. 29.7).

An understanding of the specific needs of individual endangered pollinator species is also going to be essential, and bee biologists in particular have been putting effort into evaluating the minimum needs of at-risk solitary bees, in terms of the least numbers of their preferred host plants that are needed to supply their pollen budgets (e.g., Müller et al. 2006; Larsson and Franzen 2007).

Extensive planting of nonnative ornamental plants has helped sustain pollinators in the urban environments between natural habitats, for example in several regions of Florida (Koptur 2006). There, surprising substitutions can occur, for example the sausage tree *Kigelia* being pollinated by orioles instead of bats, and squirrels being at least partially effective in visiting the flowers of *Bombax*. Gardeners are generally useful in

TABLE 29.2

Forage Plants of Particular Benefit in Wildlife Seed Mixtures
to Favor Pollinators in Western Europe

Species	Common name	Rank performance
Leucanthemum vulgare	Oxeye daisy	1
Trifolium pratense	Clover	2
Lotus corniculatus	Birdsfoot trefoil	3
Prunella vulgaris	Self-heal	4
Centaurea nigra	Knapweed	5
Hypochaeris radicata	Cat's ear	6
Rhinanthus minor	Yellow rattle	7
Lathyrus pratensis	Meadow vetchling	8
Vicia cracca	Cow vetch	9
Thymus polytrichus	Thyme	10
Anthyllis vulneraria	Kidney vetch	11
Stachys officinalis	Woundwort, betony	12
Succisa pratensis	Devil's bit scabious	13
Knautia arvensis	Field scabious	14
Centaurea scabiosa	Greater knapweed	15
Ajuga reptans	Bugle	
Ballota nigra	Horehound	
Cirsium vulgare	Spear thistle	
Dipsacus fullonum	Teasel	
Echium vulgare	Viper's bugloss	
Glechoma hederacea	Ground ivy	
Lamium album	White deadnettle	
Lamium purpureum	Purple deadnettle	
Odontites vernus	Red bartsia	
Rubus fructicosus	Bramble, blackberry	
Salix cinerea	Gray willow	
Stachys sylvatica	Hedge woundwort	
Symphytum officinale	Comfrey	

Source: Based on Pywell et al. (2003) and Carvell et al. (2006).

Note: The first 15 are given in their rank order of utility and ease of establishment; the others add forage in early spring and at other relative dearth periods.

all relatively prosperous urban areas, especially when they choose their planting carefully. Urban gardens in North America harbor a rich bee biodiversity, although exotic species are over-represented (Matteson et al. 2008). The current tendency to encourage more naturalistic planting and more native plants in gardens is helpful, with avoidance of particularly showy forms such as hybrid double flowers, as these often have reduced nectar or are difficult for pollinators to access for nectar and pollen (and some are even bred to be pollen-free) (see Comba, Corbet, Barron et al. 1999; Comba, Corbet, Hunt et al. 1999; Corbet et al. 2001). Patches that are specifically "bee gardens," "butterfly gardens," and so on can readily be achieved, with appropriate nectariferous plants but also providing nesting sites and nesting materials or host plants. There is, however, some risk of unusual garden plants contributing to genetic variation in nearby native plants

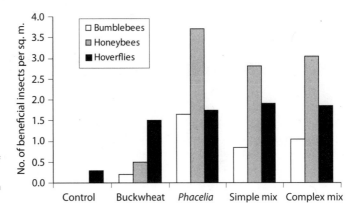

Figure 29.7 The value of planting more than one type of flower to encourage pollinators; *Phacelia* on its own has real benefits, but simple and complex mixtures are often better for hoverflies. (Redrawn from Pontin et al. 2005.)

(Whelan et al. 2006), so that careful management may be needed to maintain the genetic integrity of natural populations.

There are evidently very real problems to be faced, and hopefully resolved, if pollination services are to be preserved for the future. Above all, native pollinators and especially native bees do provide an insurance against the ongoing threats to pollination services from managed honeybees and should—for all possible reasons—be encouraged and cherished.

APPENDIX

Currently Accepted Classification of Major Angiosperm Families

From the Angiosperm Phylogeny Group, 2003, from which a full listing can be obtained

Basal Families
> Amborellaceae
> Chloranthaceae
> Nymphaeaceae
> Austrobaileyales
>> Austrobaileyaceae
>> Schisandraceae
> Ceratophyllales
>> Ceratophyllaceae

Magnoliids
> Canellales
>> Canellaceae
>> Winteraceae
> Laurales
>> Atherospermataceae
>> Calycanthaceae
>> Lauraceae
>> Monimiaceae
> Magnoliales
>> Annonaceae
>> Magnoliaceae
>> Myristicaceae
> Piperales
>> Aristolochiaceae
>> Piperaceae

MONOCOTS
> Acorales
>> Acoraceae
> Alismatales
>> Alismataceae
>> Aponogetonaceae
>> Araceae
>> Hydrocharitaceae
>> Juncaginaceae
>> Potamogetonaceae

> Ruppiaceae
> Zosteraceae
Asparagales
> Alliaceae (=Agapanthaceae)
> Asparagaceae (=Agavaceae)
> Asphodelaceae
> Iridaceae
> Ixioliriaceae
> Orchidaceae
> Tecophilaeaceae
Dioscoreales
> Dioscoreaceae
Liliales
> Alstroemeriaceae
> Colchicaceae
> Liliaceae
> Melanthiaceae
> Smilacaceae
Pandanales
> Cyclanthaceae
> Pandanaceae

Commelinids
Arecales
> Arecaceae
Commelinales
> Commelinaceae
> Pontederiaceae
Poales
> Bromeliaceae
> Cyperaceae
> Juncaceae
> Poaceae
> Typhaceae
Zingiberales
> Cannaceae
> Costaceae

Heliconiaceae
Marantaceae
Musaceae
Strelitziaceae
Zingiberaceae

EUDICOTS
Buxaceae
Proteales
Proteaceae
Ranunculales
Berberidaceae
Lardizabalaceae
Papaveraceae (=Fumariaceae)
Ranunculaceae
Core Eudicots
Berberidopsidaceae
Dilleniaceae
Gunnerales
Gunneraceae
Caryophyllales
Aizoaceae
Amaranthaceae
Cactaceae
Caryophyllaceae
Droseraceae
Frankeniaceae
Nepenthaceae
Nyctaginaceae
Phytolaccaceae
Plumbaginaceae
Polygonaceae
Portulacaceae
Tamaricaceae
Santalales
Loranthaceae
Olacaceae
Santalaceae
Saxifragales
Cercidiphyllaceae
Crassulaceae
Grossulariaceae
Hamamelidaceae
Paeoniaceae
Saxifragaceae
Rosids
Vitaceae
Crossosomatales
Crossosomataceae

Geraniales
Geraniaceae
Melianthaceae (=Francoaceae)
Myrtales
Combretaceae
Lythraceae
Melastomataceae
Myrtaceae
Onagraceae
Eurosids 1
Zygophyllaceae
Celastrales
Celastraceae
Parnassiaceae
Cucurbitales
Begoniaceae
Cucurbitaceae
Fabales
Fabaceae
Polygalaceae
Fagales
Betulaceae
Casuarinaceae
Fagaceae
Juglandaceae
Myricaceae
Nothofagaceae
Malpighiales
Caryocaraceae
Chrysobalanaceae
Clusiaceae
Euphorbiaceae
Hypericaceae
Linaceae
Malpighiaceae
Ochnaceae
Passifloraceae (=Turneraceae)
Podostemaceae
Rhizophoraceae (=Erythroxylaceae)
Salicaceae
Violaceae
Oxalidales
Cunoniaceae
Elaeocarpaceae
Oxalidaceae
Rosales
Cannabaceae
Elaeagnaceae
Moraceae

Rhamnaceae
Rosaceae
Ulmaceae
Urticaceae

Eurosids II
Brassicales
    Brassicaceae
    Caricaceae
    Resedaceae
    Tropaeolaceae
Malvales
    Bixaceae
    Cistaceae
    Dipterocarpaceae
    Malvaceae
    Thymelaeaceae
Sapindales
    Anacardiaceae
    Burseraceae
    Rutaceae
    Sapindaceae

Asterids
Cornales
    Cornaceae (=Nyssaceae)
    Hydrangeaceae
    Loasaceae
Ericales
    Actinidiaceae
    Balsaminaceae
    Clethraceae
    Ericaceae
    Lecythidaceae
    Marcgraviaceae
    Myrsinaceae
    Polemoniaceae
    Primulaceae
    Sapotaceae
    Sarraceniaceae
    Styracaceae
    Theaceae

Euasterids I
Boraginaceae
Icacinaceae

Garryales
    Garryaceae
Gentianales
    Apocynaceae
    Gentianaceae
    Loganiaceae
    Rubiaceae
Lamiales
    Acanthaceae
    Bignoniaceae
    Calceolariaceae
    Gesneriaceae
    Lamiaceae
    Lentibulariaceae
    Oleaceae
    Orobanchaceae
    Plantaginaceae
    Scrophulariaceae
    Verbenaceae
Solanales
    Convolvulaceae
    Solanaceae

Euasterids II
Escalloniaceae
Apiales
    Apiaceae
    Araliaceae
    Pittosporaceae
Aquifoliales
    Aquifoliaceae
Asterales
    Asteraceae
    Campanulaceae (=Lobeliaceae)
    Goodeniaceae
    Menyanthaceae
    Stylidiaceae
Dipsacales
    Adoxaceae
    Caprifoliaceae (=Valerianaceae)

Taxa of uncertain position
Apodanthaceae
Balanophoraceae
Rafflesiaceae

GLOSSARY

abdomen — in insects, the third part of the tripartite body (head, **thorax**, abdomen); in vertebrates, body section between thorax and pelvis.

abiotic pollination — **pollination** not involving a living animal **vector;** involving, instead, wind or water

abort — expel an embryo or seed, or (more loosely) flower or fruit, before it has completed growth or maturation or performed its normal function

abscission — shedding of body part, especially leaf, flower, fruit, or seed

actinomorphic — having **radial symmetry**; divisible into symmetrical halves by more than one axis passing through the center

active constancy — fidelity by a flower visitor to a particular species (or sometimes morph) in a location where several plant species are intermingled and readily available; cf. **passive constancy**

active pollination — **pollination** by pollinators that have specific traits (morphological, behavioral, or both) that facilitate pick-up and transport of **pollen**, its deposition on **stigmas**, or both

adaptationist — view that natural selection is always the prime explanation of evolutionary trait change, even without evidence; the term *adaptationist storytelling* is a critique of careless assumptions that disregard other possible explanations and see causal links without proof of these existing

adaptive radiation — increase in morphological or ecological diversity of a single, usually rapidly diversifying, lineage, with **phenotypes** adjusting in response to change in the environment; often associated with unoccupied environmental niches

advertisement — any feature of flowers that indicates (honestly or by deception) their quality through the use of visual, scent, or other "showy" signals

agamospermy — asexual formation of embryos or seeds without fertilization, technically a type of **apomixis** but almost synonymous with this term for flowering plants; embryo arises either parthenogenetically from an unfertilized **gametophytic** egg or directly from **sporophytic** tissue

alcohol — organic compound composed of a hydroxyl group (-OH) bound to a carbon atom of an alkyl group; examples include ethanol, propanol, cyclohexanol; usually mildly scented, polar, and good solvents

aliphatic compound — organic chemical composed of carbon atoms linked in open chains rather than rings; includes alkanes, alkenes, alkynes, fatty acids

alkaloid — organic compound containing basic nitrogen atoms but otherwise highly variable in structure; subdivided as purines, pyridines, **terpenoids**, and many others; often toxic, defensive secondary chemicals in many plants; examples include nicotine, atropine, morphine, caffeine, quinine

allele — one of several variant forms of a particular gene; results from slightly different DNA sequences at same gene **locus**; for diploid organisms, the **genotype** comprises the particular pair of alleles present at a given locus (two the same in the **homozygous** condition, two different in the **heterozygous** state)

allodapine — member of the subfamily of bees Allodapinae; usually solitary but some genera have varying degrees of sociality

allogamy — **cross-fertilization**; **pollen** being received from or transported to a flower on a separate and genetically distinct plant and fusing with that plant's ovules; also known as *xenogamy*

allometry — scaling effect where change in size necessitates changes in design, so that closely related flowers of very different size can appear divergent; opposite of **isometry**

altruistic — having behavior that benefits another individual without expectation of receiving any reward or benefit in return

ambophily — both wind and insects have **pollen**-transferring effect; an evolutionarily transitional condition between **anemophily** and **zoophily**

amine — organic compound with a basic nitrogen atom; usually derived from ammonia; examples include amino acids and various biogenic **amines**; often with somewhat 'fishy' odor

amino acid — organic compound containing both **amine** group and carboxyl group; linked in chains to form polypeptides and proteins; individually important in metabolism as coenzymes and precursors for other molecules

andrenid—member of large family of solitary bees (Andrenidae); usually relatively short-tongued

androdioecious—(n., androdioecy) having both hermaphrodite flowers and male flowers on different plants of the same species

androecium—functionally male organs of flower, including **stamens** (**filaments** plus **anthers**) producing **pollen**

anemophily—(adj., anemophilous) wind pollination; pollen from one plant being distributed to another in the air flow; main form of **abiotic** pollination

angiosperm—true flowering seed plant, as distinct from nonflowering **gymnosperm** seed plants, such as conifers

annual plant—plant that germinates, flowers, sets seed, and dies within one year (or one season), although some may live longer if prevented from setting seed

antenna—one of a pair of appendages in insects, borne on the head, often segmented, bearing multiple sensilla, especially tactile and chemoreceptive forms

ant guard—ant that spends large part of its life on a plant and acts as a defensive system, deterring **herbivores** from eating foliage; sometimes resident in nests (*domatia*) provided by the plant; sometimes supplied with nutrients via **extrafloral nectaries, food bodies,** or both

anther—upper part of male **stamen**, usually elongate, with visible slit or pore; structure in which **pollen grains** develop and are dispensed during **dehiscence**

anthesis—period when flower is both open and sexually functional; or more specifically, the onset of that period

anthochlor—yellow **flavonoid** floral pigment

anthocyanidin—pigment that is the sugar-free counterpart of an **anthocyanin**

anthocyanin—water-soluble, sugar-containing **flavonoid** pigment contained in plant vacuoles; usually red, blue, or purple; derived from an **anthocyanidin**

anthophilous—literally "flower loving"; used in reference to insects and other animals that are attracted to and usually feed on flowers

anthophorine—member of the subfamily (Anthophorinae) of solitary bees; usually very hairy and long-tongued; also known as *anthophorid*

anthoxanthin—water-soluble **flavonoid** floral pigment; usually white, cream, or yellow

anthropogenic—relating to effects or processes that derives mainly from human activity, including environmental modification, pollutants, and wastes, and climatic change

antioxidant—any molecule that can slow or stop oxidation of other molecules by removing free radical intermediates in oxidative chain reactions; may protect plants from damaging oxidative stress

antiselfing system—system that helps prevent **self-pollination, self-fertilization,** or both in a plant, primarily by physically preventing access of **self-pollen** or by using a **self-incompatibility** (SI) mechanism

aphrodisiac—substance that increases sexual desire; for example, volatiles produced by flowers to entice **pseudocopulatory** pollinators, often by mimicking the visitors' sexual **pheromone(s)**

apical meristem—growing tip of a plant, with undifferentiated tissue, found in buds and root-tips

apomixis—asexual reproduction through seeds; also known as **agamospermy**, although sometimes used more broadly to include asexual reproduction via plantlets or **bulbils**; offspring are genetically identical to parent

aposematism—combination of a warning signal with unprofitability or toxicity of a potential food item, most commonly in the form of "warning coloration" (though odors or sounds or behaviors may also occur)

arboreal—living in or growing in trees

aroid—common name for any member of the Araceae; a large **spadix** forms the inflorescence, often partly enclosed in a leaflike **spathe**

asexual—broad term for reproduction not involving **meiosis** (or **fertilization**); usually achieved by female gametes alone; includes **budding**, parthenogenesis, fission, **agamospermy,** or **apomixis**

assortative mating—mating with a partner like oneself (positive) or very unlike oneself (negative); these behaviors may reduce or increase variation in heritable traits, respectively, and can lead to disruptive or stabilizing natural selection

asymmetry—lack of an axis of symmetry, as opposed to having **radial** or **bilateral symmetry**

attractant—signal (visual, olfactory, etc.) that attracts mobile animals to a source; often used of colors and scents in flowers

autogamy—self-fertilization, the opposite of **allogamy**; in **hermaphrodite** plants it occurs when **pollen** fuses with **ovules** from same individual flower, or flower on same plant (see also **geitonogamy**)

automimic—palatable individual that mimics toxic members of same species

bagging—process of enclosing flower bud in a bag just before it opens to prevent pollinator access; useful commercially for some crops; also for pollination biologists when analyzing effect of subsequent single visits by known animals

Bateman's principle—concept that females invest more energy in reproduction than males, so that females are usually the limiting resource for which males compete

bee-purple—color resulting from mixing the two ends of the bee **visual spectrum** (yellow and ultraviolet); not visible to humans but important for flower discrimination by bees

bellflower—general term for any **pendant** and somewhat flared flower; more specifically, the type of flower in *Campanula* and related genera

Bennettitales—extinct order of seed plants appearing from Triassic to end of **Cretaceous,** characterized by primitive flowerlike structures

benzenoid—organic compound with ring structure of benzene-like character or containing at least one benzene ring; often present in flower scents

bet-hedging—strategy to allow individuals to optimize their **fitness** in unpredictable circumstances by spreading the risk across different options, thus "not putting all one's eggs in one basket"

betalain—intense red or yellow **indole**-derived cell-sap pigment, with five- and six-carbon rings; found especially in the Caryophyllales

biennial—plant that takes two years to complete life cycle and only produces flowers in second year

big bang flowering—form of highly synchronized **mass flowering**; individuals all flower together over a few days

bilabiate—common **zygomorphic,** tubular flower form, with petals forming upper and lower lips; occurs in many **angiosperm** families

bilateral symmetry—anything with two distinct sides; specifically, a form of symmetry between two distinct sides of a structure, such as a flower, with just one midline axis of symmetry

bimodal—any phenomenon with two peaks per day or per year, for example, two daily peaks of pollinator activity or of floral nectar secretion; cf. **unimodal**

biogeography—distribution of animal and plant diversity over surface of the planet in both space and time

biomass—all the biological material of living (and recently living) organisms

biotic pollination—any system in which **pollen** is moved from **anther** to **stigma** by an animal **vector**

bisexual—in reference to flowers, characterized by having both male and female reproductive structures; also termed *perfect* flowers, or **hermaphrodite** flowers

blossom—functional flowering unit, whether one flower or a collection of flowers forming a coherent **inflorescence** (e.g., catkin or daisy); also used informally as a term for flowers on spring-flowering fruit trees

bombyliid—bee fly, in the family Bombyliidae; feeds on **nectar** and **pollen**; often "bee-like," being round and densely furry

boreal—climatic or vegetational region in northern sub-Arctic zones

bract—modified leaf associated with the base of a flower; often involved in advertisement, protection of flower, or both

bromeliad—member of the family Bromeliaceae; mostly from South and Central America; **monocot** plants with large complex **inflorescences**, often **epiphytic**

brood site—egg-laying or breeding site; mimicked by some flowers to entice visitors

brush blossom—**inflorescence** with densely packed tiny flowers giving a flat or dome-like landing surface with protruding brush of **anthers**

bud—undeveloped or embryonic shoot that will become leaf or flower

budding—form of **asexual** reproduction; offspring growing out of rhizomes or runners of mature parent body (but also used in horticulture to denote the grafting of a bud onto another plant)

bulbil—small, above-ground, bulb-like growth, often on stem, able to separate from parent as a form of **asexual** reproduction (see also **viviparous**)

bulb—underground dormant plant, already containing modified leaves and flower buds; a food-storing stage

bumblebee—member of the genus *Bombus*, in the family Apidae; important pollinators occurring mainly in northern hemisphere; social, with copious branching hairs, **pollen baskets** on hind legs; also refers to a few cuckoo bumblebees in the subgenus *Psithyrus*

buzz pollination—technique used by many bees to release **pollen** from **poricidal anthers** by vibrating body against **anther**; also termed *sonication*

caloric reward—in reference to nectar, the total calories provided by a given flower or by a particular nectar intake; readily calculated from the nectar volume and concentration

calyx—outermost whorl of flower, consisting of leaflike structures (**sepals**); usually green and protective; sometimes used for advertisement

cantharophily—pollination by beetles

capitulum—type of flower head characterized by **bracts** that lie flat below the base of the flower and by a short compact cluster of flowers above (as in composites, e.g., daisy)

Carboniferous—geological period roughly 360–295 MYA; characterized by lush swamps of **seed ferns,** horsetails, club mosses, and true ferns, latterly with true conifers appearing

carotene—terpene-based **hydrocarbon** pigment involved in photosynthesis; also produces orange coloration in plant parts including petals; fat soluble but not water soluble

carotenoid—terpene-based pigment in plant **plastids;** comprises two types: **hydrocarbon carotenes** and oxygenated **alcohol** forms, known as **xanthophylls,** both of which produce orange and yellow colors in flowers

carpel—one component of a plant's female organ system (**gynoecium**), which forms innermost whorl of flower; a flower may have one carpel or several distinct or several fused carpels, each made up of **ovary, style**, and **stigma**; upper/visible part of carpel is commonly termed the **pistil**

carrion beetle—beetle that feeds and lays eggs on dead or decaying animal tissue, including members of several families of Coleoptera (especially Silphidae); also known as burying beetles; attracted to certain flowers by scents and colors mimicking dead flesh, thus acting as pollinator

carrion flower—flower that uses deceptive scents and colors to resemble dead or decaying animal carcasses, thus attracting and sometimes trapping **carrion beetles** and **flies** as pollinators

carrion fly—fly that feeds and lays eggs on dead animal tissue, feces, or both; attracted to carrion flowers, which mimic the characteristics of such materials; occur in various families, including the Calliphoridae, Sarcophagidae, and Muscidae; also known as flesh flies and blow flies

cell sap—dilute fluid contained with the large central vacuole of plant cells; important for cell water balance and **turgor**

cellulose—linear **polysaccharide** $(C_6H_{10}O_5)_n$ formed from repeated **glucose** subunits; major component of rigid plant cell walls; most common organic compound on earth, approximately one-third of all plant biomass

central-place foraging—specific behavior where animal forages out from and back to a fixed place (usually a nest)

centripetal—directed to or moving toward a central point

character displacement—accentuation of character differences among similar species in regions where the species co-occur (assumed to be the result of competition); but such differences are minimized or lost where the species' distributions do not overlap

cheating—deception or imposture by one partner in a pollination relationship, either a flower appearing to offer a **reward** where none exists, or an animal acting as **illegitimate visitor** without effecting **pollen** transfer

chelation—linkage of an organic ligand by at least two bonds to a central (usually metallic) atom

chemoreceptor—chemosensor; sensory nerve cell responding to chemical stimulus (taste or odor); located peripherally or within a specialized organ or centrally in the central nervous system (brain plus main nerve tracts) of an animal

chiropterophily—pollination by bats

chloroplast—organelle within plant cell that captures light energy and uses it to carry out photosynthesis

chromatography—analytical technique for separating and identifying components of a mixture in liquid or gas phase

chromosome—coiled and structured length of DNA, containing many genes and regulatory sequences as well as nonfunctional sections; normally wrapped within proteins inside cell nucleus; replicated and redivided into two at cell division

chromosome number—precise number of **chromosomes** typical for a given species; 46 (as 23 pairs) in humans, can be as low as 2 (some ants) or more than 1,000 (some ferns)

clade—taxonomic group defined as a single individual and all of its descendants, all having shared derived **traits** or characters

cleistogamy—automatic **self-pollination** usually without the flower ever opening fully; occurs, for example, in some peanuts and some violets

coadaptation—mutual adaptation between two species, each deriving a **fitness** benefit if the situation is a true **mutualism**; also used to refer to coadapted genes and gene complexes

cob stigma—one type of **stigma** in a **heterostylous** plant; has a corn-cob appearance with coarse sculpturing and is receptive only to cob-type A pollen, which has similarly coarse surface (smaller sculpturing on the alternative **papillate stigmas** receives the less-textured, B type pollen)

coevolution—change in genetic composition of one species in response to genetic change in another; each partner exerts selective pressure on the other; outcome may vary from coevolved change in **amino acid** sequences to covarying macrocharacters in interacting species such as pollinators and flowers, predators and prey, or hosts and parasites

coflowering—temporal overlap in the flowering periods of two or more species

colletid—member of relatively primitive family of bees (Colletidae) with short bifid tongues

colony homeostasis—tightly controlled environmental conditions within nest of a colonial/social animal, with favorable temperatures and humidity, that are often maintained by specific behaviors

color constancy—one aspect of **floral constancy**; characterized by flower visitors returning repeatedly to species or morphs of the same color; (also used to describe an aspect of the visual perception system in animals that allows color to appear the same even in very different lighting conditions)

colorimetric—analytical technique relying on measurement of color change, color spectral composition, or color intensity

color preference—**innate preference** for a particular color, especially of flowers; or learned preference, with color linked to reward status

color receptor—visual receptor with peak response at particular wavelength, so responds mainly to one color range; cones in vertebrates and eyes in insects often have two, three, or four varieties tuned to different colors (hence, dichromy, trichromy, tetrachromy, etc.)

color space — model system for defining colors perceived by a particular animal with known **color receptor** properties, each color being a point in a specific three-dimensional map (the color space) with axes corresponding to the receptor sensitivities

column — in orchids, the fused male and female reproductive organs (technically then a *gynostemium*), characterized by **stamens** and **pistils** combined in an erect single organ in center of flower

comparative analysis — (also *comparative method*); in evolutionary biology, comparisons based on and taking into account the known phylogenetic relationship of species; often used to detect correlated changes in two or more **traits** or to decide if different **clades** of organisms differ significantly in some particular trait

composite flower — member of family Asteraceae (= Compositae); characterized by tightly clustered **inflorescences** usually formed from outer **ray florets** and inner **disk florets** (e.g., daisies with white rays and yellow disks)

cone (plant) — organ in **conifers** that contains reproductive structures between layers of tough scales; male cone less conspicuous and produces **pollen**; female cone larger and produces seeds; technically termed a **strobilus**

cone (vertebrate) — **color receptor** in vertebrate eye

congeneric — in the same genus

conifer — cone-bearing woody seed plants within the **gymnosperms**, technically known as Coniferae or Pinophyta; first appeared in late **Carboniferous** period, approximately 300MYA; around 600–650 living species

connectance — in **pollination webs**, the ratio of interactions actually observed to all possible interactions that could occur

conspecific — of or within the same species

contour intensity — relative proportion of edges (contours) to center in a shape; roughly equivalent to "edginess"; high in a daisy, low in a poppy

convergence — separate and independent evolution of similar **traits** in unrelated species; commonly seen in occurrence of similar flower scents, shapes, or colors in different species that are visited by the same kinds of pollinators; can only be inferred if the plants are unrelated and thus not similar simply because of common descent; **pollination syndromes** are, in effect, suites of convergent character traits

corbicula — **pollen basket** in bees, serving as **scopa**; found on the hind **tibiae** of *Apis, Bombus,* stingless bees, and euglossine bees

cornucopia flowering — pattern characterized by many flowers being produced for many days or weeks on end, giving a distinct seasonal flowering period, as in most temperate plants

corolla — collective term for flower **petals**; located within the **calyx** and normally the most conspicuous part of a flower

corona — outgrowth of corolla in daffodils and similar flowers, appearing as a second and separate inner corolla; frilly in passion flowers and trumpetlike in daffodils

coronal scale — small scalelike appendage arising from upper side of a petal; sometimes providing color contrast (e.g., forget-me-nots), sometimes expanded and fused to form distinct **corona**

crepuscular — being active at twilight, meaning either dusk *or* both dawn and dusk

Cretaceous — geological period roughly 145–65 MYA, when flowering plants first appeared

crop (insect) — expanded area of the foregut; used as storage area

crop (plant) — any plant grown in significant quantities to be harvested as food, fodder, or fuel; also refers to annual yield of such a plant

cross-fertilization — see **allogamy**

cross-pollinator — animal vector that delivers **pollen** from **anthers** of one plant to **stigma** of another individual but **conspecific** plant

crypsis — technique to avoid detection by being hard to see (or sometimes hard to smell or otherwise detect); **cryptic** animals blend into their backgrounds

cuckoo bee — bee that is **kleptoparasitic**, laying its eggs on the **nectar** and **pollen** provisions gathered by another bee of another species, so lacks **scopae** and has no nest of its own; usually closely related to host; evolved independently about 16 times in social bees and about 30 times in solitary bees

cucurbit — plant in family Cucurbitaceae; includes gourds, pumpkins, melons, etc.

cyanidin — natural pigment found in many red berries and red leaves; can be transformed into **anthocyanins**, which are also important as floral pigment, producing red and maroon colors

cycad — seed-plant **gymnosperm**; mainly tropical, with large compound leaves as a crown above stout trunk; reproduction via large **cones**

cycloidea gene — gene involved in symmetry formation in many flowers

cyme — (adj. cymose) **inflorescence** with flat-topped or domed flower clusters where central flower in each cluster opens first (e.g., tomato, dogwood)

danaid — large tropical butterfly (subfamily Danainae) with reduced forelegs and often highly toxic due to sequestration of chemicals from host plants fed on by caterpillar stage

deceit pollination—attraction of pollinators without offering a **reward**, for example, by mimicking rewarding species; common in orchids and Araceae

dehiscence—spontaneous or triggered opening of plant structure, most commonly used in relation to **anthers opening** to release **pollen;** also used to refer to opening sporangium of fern, or opening up of cased fruit or nut; sometimes for reopening of wound

dehiscence furrow—line or groove along which **anther** normally splits open at **dehiscence**

delphinidin—important **anthocyanidin** floral pigment; produces blue and purple colors

dichasial—type of **cymose inflorescence** with two opposite lateral branches developing below each terminal flower

dichogamy—**sequential hermaphroditism**; commonly flower is first male and then female (**protandry**) or, more rarely, vice versa (**protogyny**); usually a means of preventing **self-fertilization**

dichotoma gene—gene involved in symmetry formation in many flowers (cf. *cycloidea*)

dicot—member of dicotyledonous **angiosperms**, which have two embryonic seed leaves (cotyledons) appearing at germination; flower parts usually in multiples of four or five; includes labiates, composites, legumes, (see appendix); informal term, not a strict **monophyletic** group

diel—organized with a 24-hour periodicity

diglyceride—organic compound with two fatty acid chains bonded to one glycerol molecule

dioecious—(n., dioecy, dioecism) species with distinct male and female individuals; plant populations with separate male and female plants

diploid—having two homologous copies of each **chromosome** (2n), one from each parent; common state in most plants (though many are **polyploid**) and almost universal in animals; normally only reducing transiently to **haploid** gamete cells (n) in preparation for **fertilization** and reproduction

dipterocarp—member of large tropical lowland plant family (Dipterocarpaceae) whose trees dominate rainforests of Southeast Asia

directional selection—selective pressure favoring a single **phenotype**, so that **allele** frequency continues to shift in one direction over time

disaccharide—carbohydrate (twelve-carbon) formed from condensation of two **monosaccharides** (e.g., of **glucose** and **fructose** to form sucrose, or of glucose and galactose to form lactose)

disk floret—small tubular **floret** clustered at center of **composite** inflorescence such as a daisy, whose outer parts are usually flattened **ray florets**

distyly—(adj., distylous) form of **heterostyly** characterized by just two morphs, long-styled and short-styled; most familiar in pin and thrum primroses

diurnal—mainly active in the daytime (opposite of nocturnal)

dominant gene—gene that always produces its effect on the phenotype, even if there is only one copy of it in the pair of alleles; effectively masks a nonidentical **recessive** allele

dorsiventral symmetry—form of symmetry more commonly known in flowers as **bilateral**

echo location—biological sonar system used by bats and some other animals for navigation by sensing echoes that are bounced off objects in their path; used by some bats to detect presence and status of a few flowers

ectotherm—(adj., ectothermic) an organism dependent on external heat sources (usually the sun, occasionally geothermal energy) to warm the body

edaphic—affected by soil conditions (rather than climate or other external variables)

EFN—see **extrafloral nectary**

elaiophore—oil-producing (or more generally **lipid**-producing) gland or cell in a plant

electroantennography—(EAG) laboratory system for measuring electrical activity in insects' antennal cells (**chemoreceptors**), hence for detecting responses to flower **volatiles**, **pheromones**, etc.; produces electroantennogram

emasculation—(v., emasculate) removal of genitalia from male animal, or **anthers** and **pollen** (hence, male gametes) from flower

embryo sac—elongated sac inside **ovule** of seed plant within which embryo develops

enantiostyly—**polymorphic asymmetry** in certain flowers having **style** deflected to left or right, producing mirror-image flowers

endosperm—nutritive tissue in **angiosperm** seed, surrounding embryo; formed from fusion of second **haploid** nucleus from **pollen tube** with 2 polar nuclei within **embryo sac** (so endosperm becomes triploid)

endothermy—(adj., endothermic) organism able to generate substantial heat internally from metabolic processes, so warming the body independently of solar radiation

entomophagous—feeding on insects as major component of the diet

entomophily—(adj., entomophilous) pollination by insects

ephemeral—short-lived, transient; in reference to plants, characterized by a life cycle of only a few weeks

epidermis—outermost layer of organism; animal skin; a single-cell layer in plants

epihydrophily—pollination at water surface using water as **vector**

epiphyte—plant growing on the outer surfaces of another plant

ester—organic compound derived from carboxylic acid

plus **alcohol**, glycerol, or phenol; common as floral fragrance components

ethylene—(also ethene); simple alkene, C_2H_4; acts as plant hormone in flower opening, fruit ripening, and abscission; can be triggered by wounding

euglossine—member of bee subfamily Euglossinae; the Neotropical **orchid bees**, males of which collect floral perfumes in hind-leg pouches

eusocial bee—species characterized by overlap of two generations in communal nest, queen with her daughters (workers); includes **honeybees**, **stingless bees**, **bumblebees**, and some others, with varying levels of caste differentiation and complexity

exaptation—trait previously shaped by selection for one function but now used in another role for which it was not directly selected

excreta—composite term for fecal and nitrogenous wastes, which are eliminated together in most insects and birds

explosive pollen presentation—**pollen** presentation by sudden synchronized **dehiscence**, often triggered by arrival of flower visitor (e.g., **tripping** in many legume flowers)

extended flowering—flowers produced in small numbers each day over several weeks or months

extrafloral nectary—(EFN); **nectary** sited outside flower, for example, on stems or bracts; may attract **ant guards**, which defend against herbivores, and/or may decoy ants away from flowers

extrorse anther—anther with **dehiscence furrow** on outer surface, thus presenting **pollen** outward, away from female organs in center of flower

exudate—fluid emerging from plant surface, usually from stigma, and potentially a floral **reward**; also from nonfloral surfaces

facultative—occurring only when needed or triggered; relationship that is nonessential for participants or is a fallback position, for example, selfing occurring only if other routes to reproductive output have failed; opposite of **obligate**

fall petal—lower petal in iris that bears **nectar** grooves and often a raised central "beard" (perhaps mimicking yellow **pollen**)

family—taxonomic group contained within an order and in turn containing many **genera**; names end *-idae* in animals, *-aceae* in plants

female choice—selection (or rejection) of potential mates by females, usually on basis of male traits; in plants usually achieved by physiological mechanisms in the **pistil** (though difficult to distinguish from **pollen competition**)

female function—in flowers, the receipt of **pollen grains** on a **stigma** and the transmission of pollen nuclei to **ovule** followed by **fertilization** and seed production

fertilization—in plants, the fusion of male (**pollen grain**) nucleus with female nucleus in **ovule** to form a **zygote**, restoring the **diploid** state and initiating embryonic seed formation

filament—part of **stamen**; usually a thin stalk supporting terminal **anther**

fitness—relative ability of an individual of a given **genotype** to reproduce itself compared with other genotypes; manifested externally through the **phenotype**, which interacts with the environment, so that a given genotype does not always have the same fitness

flabellum—spoon-like or probe-like tip of **glossa** in many shorter-tongued bees

flag petal—upper, erect, flag-like petal on some legume and iris-type flowers

flavonoid—pigment, dissolved in **cell sap**, formed from a core of coupled six-carbon rings (usually an **anthocyanidin**) plus a sugar (therefore ketone-containing polyphenols); highly variable color range: yellow, red, blue

floral/flower constancy—tendency of visitor to move reliably between flowers of same species (or morph within a species); benefits a plant via **conspecific pollen** transfer and a visitor via improved foraging efficiency

floral display—overall advertising display of flowers on a plant; results from size, spacing, and longevity of individual flowers and their number

floral fidelity—synonym of **floral constancy**; or a composite of constancy and a more innate feeding specialization (**oligolecty**) in a visitor

floral longevity—lifespan of flower in functional (receptive) state; usually hours or days, sometimes weeks

floret—single floral unit within **composite** flower, especially in Asteraceae, which may be either distinct **ray florets** or **disk florets**

florivory—eating flowers; diet composed largely of flowers

flower—complex structure usually with at least four whorls of substructures that collectively form the reproductive organ of an **angiosperm**; usually **hermaphrodite**, producing **pollen** from **anthers** and receiving incoming pollen on **stigma**

fodder pollen—in flowers with dimorphic anthers, **pollen** from the showier anthers taken as food by visitors (more cryptic anthers deposit reproductive pollen elsewhere on visitor's body)

folivory—eating of foliage; a diet largely composed of leaves

food body—floral tissue offered as **reward** to visitor, often high in protein or **lipid** content; also nonfloral, nutrient-rich tissues that attract **ant guards** onto plants

footprint—chemical signal deposited from the **tarsi** on a flower or plant (especially by social bees) as indication to others that it has been visited; a form of **scent mark**

frequency-dependent selection selection of a phenotype whose fitness depends on its frequency relative to other phenotypes in population; may be positive (higher fitness when more common) or negative (e.g., new strains of virus, more successful while still rare and few hosts have acquired immunity)

fructose — simple **monosaccharide** sugar common in nectar; combines with **glucose** to make sucrose

fruit — loosely defined term covering most kinds of **seeds** formed from the swelling **gynoecium** that have some fleshy covering; often confused with *seed* and with *vegetable*

fruit set — roughly synonymous with seed set; used rather loosely

galea — (pl., galeae) paired mouthparts in tongue of bee that surround the **maxillae**

gamete — **haploid** cell resulting from **meiotic division**; male or female participant in subsequent sexual reproduction and **fertilization** that restores **diploid** state with new genetic constitution

gametophyte — haploid multicellular tissue giving rise to and containing **gamete**(s); microgametophytes are **pollen grains**, macrogametophytes are **ovules**

gametophytic self-incompatibility — (GSI) self-incompatibility reaction that occurs within **the style** in response to haploid **genotype** of a **pollen grain** as its **pollen tube** extends

gamopetaly — (adj., gamopetalous) partial fusion of petals to form a tubular **corolla**

gamosepaly — (adj., gamosepalous) partial fusion of sepals to form a tubelike cup for **corolla**

geitonogamy — **pollen** export by **vector** out of one flower but only to another flower on same plant, where fusion with ovules occurs; genetically equivalent to **self-pollination**

gene flow — transfer of alleles of genes from one population to another, or the pattern of this transfer; leads to increased genetic variability

generalist flower — flower that lacks specific traits to attract specific visitors; also, a flower receiving visitors (acting as pollinators) from many taxa

generative nucleus — reproductive nucleus within the **pollen grain** (distinct from the *tube nucleus*); divides to form two *sperm nuclei*

genetic constraint — restriction on selection due to features of the genome itself; occurs, for example, when a heterozygote has greater **fitness** than either homozygote, or when there is a strong **genetic correlation** between traits

genetic correlation — variance that two traits share because of underlying genetic links, usually the result of sharing influences from common genes

genetic drift — change in relative frequency of particular allele in a given population because of random sampling (chance); not driven by adaptive processes; usually has a larger effect in small populations

genome — entire hereditary information in an individual; embedded in DNA, both coding and noncoding sequences; also a term for the full set of chromosomes

genotype — internally coded heritable information in an organism; detailed descriptor of the **genome** (cf. **phenotype**)

genus — (pl., genera); taxonomic group contained within a **family** and in turn normally containing many **species**; first part of binomial name that defines a species, as in *Apis mellifera, Apis* being the genus

germination — in **pollen grains,** the process of emerging from inactive dormant dehydrated state by taking up water and commencing growth into pollen tube

gibberellic acid — (also *gibberellin*); **plant growth substance** (hormone) controlling growth, development, and seed germination

glossa — terminal portion of bee tongue; usually bristly and flexible

glucose — simple **monosaccharide** sugar common in **nectar**

glume — leaflike structure in grass; paired glumes enclose each flower cluster

glycoside — organic molecule with sugar attached, usually producing an inactivated form; glycosidase enzymes produce active form when needed

gnetalean — member of small group of **gymnosperms** (Gnetales) that use animal pollination

grooming — (of pollen) cleansing of passively deposited pollen from its body, by a flower-visiting animal; pollen is either discarded or moved directly to mouth or, in bees, into a carrying area **(scopa)**

gullet flower — tubular and bilateral flower form with landing platform and **nototribic pollen** deposition; especially common in the Lamiaceae, Acathaceae, and Gesneriaceae and in orchids

gymnosperm — seed plant with exposed seeds and lacking true flowers; includes conifers, cycads, etc., but is probably not a **monophyletic** grouping

gynodioecious — (n., gynodioecy) having both **hermaphrodite** flowers and female flowers on different plants of the same species

gynoecium — female organ system of a flower; composed of **pistils** containing **carpels** with **ovules** and a terminal **stigma** that receives **pollen**

halictid — member of relatively primitive family of bees (Halictidae); fairly short-tongued, have **scopae** on hind legs; mostly solitary but includes some species with varying degrees of sociality

handling — movements and behaviors on and within the flower while feeding

hand-pollination — transfer of **pollen** by human intervention for scientific study or in crops whose natural pollinators are scarce or whose yield can be improved artificially

haploid — condition of having just one copy of a **chromosome** (n); usually occurs after **meiotic** cell division and is confined to **gamete** cells in preparation for **fertilization** and restoration of **diploid** (2n) state

hawkmoth — member of one family of large moths (Sphingidae); commonly are **heterothermic** and have an ability to hover

hemolymph — insect equivalent of blood; circulates through open spaces within body

herbaceous plant — nonwoody or non-shrubby plant whose leaves and stems die back in winter

herbivory — eating any or all parts of plants (foliage, seeds, flowers, stems, roots, etc.)

heritability — the proportion of **phenotypic** variation in a population that is attributable to the genetic variation among individuals (the remaining variation being due to environmental factors); heritability values can range from 0 to 1

herkogamy — separation (spatially, temporally, or both) of male and female parts of flower; usually influences visitor behavior and reduces **selfing**

hermaphrodite — organism having both male and female structures and functions; common state of most flowers

heteranthy — (also *heteranthery*) having two kinds of **anther** that are either temporally distinct in their **dehiscence** or are morphologically and functionally distinct (some cryptic and producing reproductive **pollen**, others showy and producing pollen as food for visitors)

heteromorphic — having more than one **morph**; with reference to a species or to a particular structure within a flower, as in **heterostyly**; opposite is *homomorphic* or **monomorphic**

heterospecific — of or from another species; cf. **conspecific**

heterostyly — (adj., heterostylous) having **polymorphic** styles, especially **distyly** with two different style lengths, often termed *pin* and *thrum*; also known as *reciprocal herkogamy*

heterotherm — intermediate between **ectotherm** and **endotherm**; *either* an organism that is normally ectothermic but can generate heat internally for short periods when needed (e.g., many bees, some moths and beetles), *or* one that is normally endothermic but reverts to much lower endogenous heat production overnight or in seasonal torpidity (e.g., small hummingbirds, some small mammals; see **temporal heterotherm**)

heterozygous — having two different **alleles** at a particular genetic **locus**; cf. **homozygous**

hexose — a six-carbon sugar (**monosaccharide**), such as **glucose** or **fructose**

homopteran — member of the insect order or suborder Homoptera (sometimes placed within the large order Hemiptera or true bugs, all with sucking mouthparts); for example, aphids, leafhoppers, scale insects, cicadas, etc.; may be an artificial assemblage, not monophyletic

homostyly — characterized by only one kind of **style**; cf. **heterostyly**

homozygous — having two identical **alleles** at a particular genetic **locus**, that is, on the two homologous **chromosomes**; cf. **heterozygous**

honeycreeper — flower-visiting bird in Hawaii and Neotropics; member of the families Thraupidae or Fringillidae; often important as pollinators

honeydew — sweet secretion from tubular glands (cornicles) on dorsal surface of many **homopterans**, who thus shed excess sugar gained from feeding on plant **phloem**; is used as a food source by other plant visitors, especially ants

honey guide — in flowers, mark or structure indicating presence and site of nectar; also termed **nectar guide**

hoverfly — member of the fly family Syrphidae; such flies often feed on both **pollen** and **nectar** and are therefore important pollinators

hummingbird — member of the family Trochilidae, strictly from the Americas; highly agile, hovering flower visitor

hybrid — offspring resulting from **outbreeding** of different individuals or different species; the latter occurring naturally or by human intervention; usually, but not always, infertile

hybrid vigor — (also *heterosis*, *heterozygote advantage*); enhancement of growth or performance resulting from **outbreeding**, with the hybrid appearing superior to either parent

hydrocarbon — organic compound consisting entirely of hydrogen and carbon atoms; for example, methane, benzene, polyethylene

hydrophily — pollination using water as **vector** (many freshwater plants and seagrasses)

hymenopteran — member of insect order Hymenoptera, which includes ants, wasps, and bees

hypohydrophily — pollination at depth within water, which serves as **vector**

hypopharynx — unpaired mouthpart in flies; together with **labrum** normally forms a short tube (**proboscis**) above the **labium**; often used to regurgitate saliva

ichneumon(id) — parasitoid wasp (family Ichneumonidae) within the order Hymenoptera

illegitimate visitor — flower visitor that does not pollinate; takes nectar without touching **anthers** (or **pollen** without touching **stigma**), often due to physical mismatch with a particular flower type

inbreeding — reproduction between two genetically related individuals; cf. **outcrossing**

inbreeding depression — reduced **fitness** resulting from reproduction between two genetically related individuals because more **recessive**, potentially deleterious, traits can be manifested

incompatibility — in **pollen grain**, the state of being unable to **fertilize** another plant, usually due to **SI** (**self-incompatibility**) mechanisms

indole — organic compound with a six-carbon and a five-carbon ring; flowery scent at low concentration but intense fecal smell if concentrated

inferior ovary — ovary deep within flower; appears to arise *below* **sepals** and **petals**; probably derived condition; occurs in more advanced **angiosperm** groups

inflorescence — massed cluster of flowers, usually forming functional unit in terms of attraction and display

innate behavior — any behavior existing from birth, or inborn; not learned from experience

innate preference — any preference not learned by experience; present without prior training

intersexual mimicry — resemblance between male and female flowers in unisexual plants, usually female resembling male (e.g., appearing to offer **pollen**); strictly speaking, **mimicry** requires that a clear **fitness** benefit must accrue

introrse anther — anther with **dehiscence furrow** on inner surface, thus presenting **pollen** inward toward female organs in center of flower

inversion — in reference to sugars or nectar, the breakdown of **disaccharides** (usually sucrose) to **monosaccharides** (**glucose** and **fructose**)

invertase — enzyme that catalyzes sugar **inversion**

irregular (symmetry) — alternative term for **bilateral** or **zygomorphic** symmetry

isometry — situation where size increase produces simple 1:1 scaling-up of features; opposite of **allometry**

isoprenoid — five-carbon unit (C_5H_8) from which **terpenoids** are formed; the basis of many scents

isozyme — one of several versions of a particular enzyme having slightly differing **amino acid** sequence but same function

Jurassic — geological period approximately 199–145 MYA, between Triassic and **Cretaceous**; characterized by lush vegetation of **conifers**, **cycads**, **seed ferns**, and **Bennettitales**

keel petal — lower central petal of legume flower; two such petals fuse to form a boat-shaped keel containing **anthers** and **stigma**, which are revealed during a flower visit; see also **tripping**

ketone — aliphatic organic compound with carbon-oxygen double bond (C=O); usually volatile and water soluble; common component of floral scents

kleptoparasitic — acting as a cuckoo; laying eggs in another's nest

labellum (fly) — one of pair of pads at tip of **labium**; used for lapping up liquids

labellum (orchid) — enlarged lower petal often providing **landing platform**

labiate — plant in family Lamiaceae (= Labiatae); usually **zygomorphic** with upper hoodlike petal and lower, expanded **landing-platform** petal(s) bearing **nectar guides**

labium — rearmost unpaired mouthpart in insect; forms floor of food canal

labrum — upper-most unpaired mouthpart in insect, forms roof of food canal

landing platform — lower petal (or fused petals) in many flowers forming enlarged surface for visitors to land on; often with nectar guides and with ridges or surface texturing to aid grip

latrorse anther — anther with **dehiscence furrow** on its side(s), thus presenting **pollen** laterally

leaf-cutter bee — member of bee family Megachilidae; solitary bee normally lining its nest cells with small pieces of cut foliage; abdominal pollen carrier

learned association — preference (or sometimes aversion) acquired from learning to link one feature (e.g., flower color or scent) with another (e.g., high reward)

lectin — sugar-binding protein, often involved in cell-cell recognition

leguminous — pertaining to large family Fabaceae, loosely termed *legumes*; includes mainly **keel**-type flowers (but also some **brush blossoms** and other designs)

lek — gathering of males forming an enlarged display that attracts females as potential mates, who then choose between the competing males

lifetime reproductive success — number of surviving offspring of an individual (though hard to define in practice; often confounded by *longevity* and definitions of *surviving*)

lignin — complex biopolymer that is a component of wood; forms parts of secondary plant cell walls, especially of **xylem**

linkage (web) — number of taxa with which a given species interacts (in a constructed visitation or **pollination web**)

linkage disequilibrium — effect occurring when traits do not randomly associate at **meiosis**, either because they are on the same chromosome or are on different chromosomes that retain some association during chromosome assortment

lipid — molecular group that includes fatty acids, oils, waxes, and sterols; hydrophobic; loosely known as *fats*

locule — one of (usually four) pollen sacs within an **anther**, often in two pairs arranged as **thecae**

locus — specific site on **chromosome** of particular gene (normally having two **alleles** present on the paired chromosomes in the **diploid** state)

MADS genes — homeotic genes controlling developmental processes in plants, especially of flowers

male function — in flowers, the production and dispersal of **pollen**

mandible (bird) — lower jaw bone, but commonly used to refer to both upper and lower halves of beak

mandible (insect) — main paired biting mouthpart in insects; often reduced where mouthparts are used for nectar feeding

marginal habitat — habitat with limiting environmental conditions; supports low biotic diversity (or is limiting for certain kinds of organism, which are thus poorly represented)

mass flowering — sudden and highly synchronized flowering in plant population; opposite of **extended flowering**

mate choice — usually of females choosing between males; see **female choice**

maxilla — (pl., maxillae) One of pair of mouthparts forming major tubular part of tongue or **proboscis** in most nectar-feeding insects

maxillary palp — sensory or manipulative palp borne on side of maxilla in many insects

megachilid — member of family Megachilidae; solitary **leaf-cutter bee** with moderately long tongue and ventral abdominal **scopa**

meiotic division — (also **meiosis**); process of **diploid** cell duplicating its chromosomes then splitting into four **haploid** cells, each of which can become a **gamete**

meliponine — member of subfamily Meliponinae; social **stingless bees** within the Apidae

melittid — member of family Melittidae; solitary bees with **scopa** on hind **tarsi**, fairly short bifid tongues, sometimes also are oil collectors

melittophily — pollination by bees, though may be better split into several distinct **pollination syndromes**

meristem — undifferentiated tissue (especially at stem and root tips) from which plant growth and development can occur

mess pollination — visitation by generalists (beetles, some flies, etc.) that scrabble over flower surface often destroying some tissues but also transferring some **pollen**

metapleural gland — gland on dorsal surface of some **hymenopterans**, especially ants, which produces antimicrobial compounds used in nest hygiene but are also damaging to **pollen viability**

microclimate — localized climatic conditions measured on spatial (and temporal) scale appropriate to plant or insect, such as within a flower, under a leaf, etc.

microorganism — informal term for any microscopic life form: protists, bacteria, viruses, etc.

micropyle — narrow apical channel through integument of floral **ovule**; allows entry to **pollen tube** tip and hence delivery of male **gametes**

mimicry — resemblance *not* linked to direct inheritance of similar traits, but achieved convergently; for example, flowers resembling shape, color, or scent of another species; styles mimicking anthers; flowers mimicking animal traits to gain pseudocopulatory pollination; must give a **fitness** benefit to be classed as true mimicry

mimicry ring — several convergently similar unrelated species benefitting from their mutual resemblance, for example, sharing pollinators by exploiting the **search image** formed by a particular visitor

monocarpic — having single reproductive episode in life cycle (**annual** or **biennial**)

monochromatic — light of very narrow band wavelength, appearing as one color; or visual system with only one type of **color receptor**; lacking color discrimination

monocot — member of the monocotyledonous plant group, probably a **monophyletic clade**, having a single cotyledon (seed leaf) at germination; floral structures usually in multiples of three; includes grasses, lilies, arums, palms, onions, orchids; cf. **dicot**

monoecious — (n., monoecy, monoecism) having separate sex flowers that occur on same plant

monolectic — (n., monolecty) using a single type of floral **pollen** for food (referring mainly to bees)

monomorphic — having only one **morph**; cf. **heteromorphic** and **polymorphic**)

monophyletic (taxon) — a taxonomically correct **clade**, defined as containing an ancestor and all of its descendants, and characterized by shared derived characteristics

monosaccharide — simple six-carbon sugar (**glucose**, **fructose**, etc.)

monoterpene — **terpenoid** formed from two **isoprenoid** units (therefore ten-carbon); very common as floral scent component

morph — distinctly different forms within a species; for example, different floral colors, different scents, different style types, etc.

morphogenesis — progression toward adult morphology with unfolding of developmental program

mortality (rate) — measure of number of deaths per unit of population

movement herkogamy — movement of floral parts in response to visitation that alters the relative positions of male and female organs

mucilage — thick and sticky plant exudate, mostly glycoprotein

multiple bang flowering—synchronized **mass flowering** episodes; all or most individuals flower together over just a few days but in several successive bursts

mutation—change in DNA sequence within **genome**, due to radiation, viruses, or other mutagens, or occurring naturally during **meiosis**

mutualism—biological interaction beneficial (in terms of **fitness**) to both partners, may be **facultative** or **obligate**; most pollination interactions are classic mutualisms

mycetophily—pollination by fungus gnats of flowers with parts resembling fungal gills

myophily—pollination by flies

myrmecophyte—(adj., myrmecophytic) plant that lives in a mutualistic obligate association with ants, providing food and/or shelter

nastic response—undirected plant growth movements; cf. *tropism*, a directed movement toward or away from a stimulus

nectar—sugary solution secreted from a **nectary**, normally within flowers, where it forms primary reward for most visitors, but sometimes also extraflorally, where ants and other insects may feed on it

nectar guide—any marks or structures (usually on petals) that indicate presence or position of nectary to a flower visitor

nectar homeostasis—ability in some flowers to maintain **nectar reward** at fairly constant level by resupply after visitors remove some or all of the fluid already secreted

nectar resorption—recovery into floral tissue of unused or surplus nectar sugars; may be more widespread than currently reported

nectar spur—elongated tubular extension, usually formed from **petals** and at rear of a flower, in which nectar can be stored so providing suitable reward for long-tongued visitor

nectar theft—removal of nectar from flower without effecting pollination (illegitimate or cheating visit) by either a small visitor (**illegitimate visitor**) entering **corolla** without touching **anthers**, or a visitor biting or piercing base of corolla from the outside or using such a hole made by others

nectary—**nectar**-producing tissue, more or less discrete, potentially arising from many different floral or extrafloral tissues

Neotropical—(n., Neotropics) from the tropical zones of the New World (North, Central, and South America and Caribbean islands)

nestedness—a measure of order in ecological systems, higher nestedness indicating a more organized system; the degree to which species with few links have a subset of the links also used by other species, rather than a different set of links; thus ecological systems are said to be nested when the species composition of a small assemblage is a subset of a larger assemblage

network—in ecology, the representation of all the biotic pairwise interactions in a given community that describe the structure of the system; often with relatively independent sub-networks (compartments)

noctuid—member of moth family Noctuidae

nonself-pollen—**pollen** from another individual and genetically distinct plant of same species, *or* **pollen** from another species

nototribic—flower with dorsal **anthers** and **style** that deposit **pollen** onto (and receive pollen from) a visitor's back; opposite of **sternotribic**

nucellus—undifferentiated multicellular mass within floral ovule, one cell of which undergoes **meiosis** to give **haploid** cells, in turn forming **embryo sac**

nut—dry seed with hard outer casing that does not open until forced by an animal

nymphalid—butterfly in worldwide family Nymphalidae; has reduced forelegs

obligate—occurring of necessity; a relationship that is essential for both partners, or a phenomenon, such as selfing, that has to occur for any reproductive output in particular plants; cf. **facultative**

oil—liquid form of **lipid**; produced as reward in a few plants

oligolectic—(n., oligolecty) using narrow range of floral **pollens** as food (referring mainly to bees)

ommatidium—functional unit within insect compound eye; composed of about eight elongate cells, the inner margins of each containing a sensory **rhabdomere**

ontogeny—development of organism from fertilized egg to mature form

optimal diet theory—choices made between alternative dietary items that permits **optimal foraging** success

optimal foraging—maximizing energy intake per unit time while seaching for and gathering food; consuming as much as possible while expending least time and energy doing so

orchid bee—member of subfamily Euglossinae within the Apidae; **Neotropical**, long-tongued, having varying degrees of sociality; associated with **trap-lining** foraging and perfume collection from flowers, especially orchids

ordered herkogamy—spatial separation of male and female structures in flower so that visitor meets them sequentially, for example, **stigma** central and **anthers** peripheral, or stigma protruding further than anthers

organic compound—carbon-based chemical (slightly artificial term, excluding carbonates and cyanides, which are considered to be inorganic)

ornithophily—pollination by birds (hummingbirds and perching birds) *osmophore*—specific scent gland in a flower, especially in carrion flowers and in orchids

outbreeding—breeding of individuals that are not (genetically) closely related, thus potentially increasing the number of **heterozygous** progeny

outcrossing—synonym of **outbreeding**, although often used mainly to refer to crosses induced by human intervention (e.g., plant breeders)

ovary—female organ; lower part of **carpel** in flower (below the **style**); comprises one or several **ovules**

oviposition—egg-laying; sometimes occurs in flowers and is then sometimes related to pollination

ovule—component of **ovary** within flower; usually attached to ovary wall internally; each ovule potentially produces one seed

panicle—loose and highly branched type of **inflorescence** where each branch is a **raceme**

papilionate—in reference to flowers, belonging to the Papilionoideae subfamily of the Fabaceae; having keel-type structures and often rather butterfly-like in appearance

papilla—small, often fingerlike protrusion occurring densely on plant surface

papillate stigma—one form of stigma in some heterostylous flowers, which has small surface papillae and reacts with pin-type **pollen**

pappus—hairy calyx in **disk floret** of **composite inflorescence**

parasitoid—insect that lays its eggs in another living organism where they develop and hatch, the host dying as parasitoids reach maturity; approximately intermediate between parasites and predators; common in some flies and wasps

parasocial bee—solitary bee living in physically close, communal group, sometimes sharing nests, but with no interaction across generations

parenchyma—tissue formed from thin-walled, unspecialized cells, making up most of the bulk of a nonwoody plant; includes mesophyll in leaves and pith in stems

passive constancy—constancy by a visitor to a particular flowering species when only one plant species is in flower or where flowering species are strongly aggregated and intraspecific movement is inevitable; cf. **active constancy**

paternity analysis—determination of fatherhood, and hence **gene flow** and genetic structure in a community; usually uses microsatellite analysis

pedicel—flower stalk

pelargonidin—important **anthocyanidin** floral pigment that produces scarlet and orange colors

pendant—in reference to flowers, hanging with corolla mouth pointing downward

PER—see **proboscis extension reflex**

perennial—plant that persists over many years, normally (but not necessarily) flowering and reproducing every year

perianth—outer structures of flower, excluding the reproductive organs

petal—one of inner whorl of corolla components; located inside the **sepals** and outside the reproductive **androecium** and **gynoecium**; petals are usually the main advertising structures

petiole—stalk of a leaf; often used also as term for flower stalk

phalaenophily—pollination by moths (often excluding the hovering hawkmoths)

phenolic—organic chemical with one or more six-carbon phenol unit; includes tannins, **lignin**, and **flavonoids**

phenological displacement—temporal shift of some aspect of life cycle, such as flowering time, when in competition with other species (see **character displacement**)

phenology—periodic timing of life-cycle events in relation to seasonal and annual cycles

phenotype—manifested physical properties of an organism (shape, physiology, behavior) as determined by interaction of its **genotype** and its environment

pheromone—small, volatile organic chemical released from one organism that influences behavior of a **conspecific**, either sexually or in relation to territoriality, trail following, alarm, etc.

phloem—tissue that conducts phloem sap, one of the two main circulating fluids in plants (cf. **xylem**); phloem sap has high sugar content and is main source of **nectar** components

photoperiodic trigger—signal received from day length (thus at a specific time of year, independent of variable weather); causes change in behavior, growth, etc.

photopigment—visual pigment responding to incident light within a **photoreceptor**; usually **rhodopsin** or a modification thereof

photoreceptor—sensory receptor responding to photons of light, resulting in neural impulse sent to the central nervous system (brain plus main nerve tracts) of an animal; eye or subcomponent thereof

photosynthate—sugars produced by photosynthesis

phylogenetic analysis—analysis of morphological or molecular data based on homology to produce a phylogenetic tree or *cladogram*

pin flower—one of two forms in **distylous** flowers; has elongate pin-like **style** set well above the **anthers**; opposite of **thrum flower**

pistil—component of female **gynoecium** in flower; each flower has one or several pistils composed of **style** and **ovary**

pistillate—female flower or phase of a **hermaphrodite** flower) having pistil but not (mature) stamens; cf. **staminate**

plant growth substance—plant equivalent of hormones in animals (often termed *plant hormones* but less specifically defined than in animals)

plastid—intracellular plant organelle, predominant type being **chloroplast**

pleiotropy—two or more traits affected by one chromosomal locus and so genetically correlated; potentially provides nonadaptive effects in one of the linked traits

plumose—featherlike; used to describe, for example, hairs on bees, antennae on moths, or styles in certain plants

pollen—male **gametophyte** of plant; as **pollen grains**; formed within **pollen** sacs of **anthers** and released at **dehiscence**; primary **reward** in flowers originally (now more commonly secondary to **nectar**)

pollen basket—complex **scopa** on hind leg of most social bees; also known as **corbicula**

pollen carryover—**pollen** moved from one source flower to another via a visitor's body, usually decreases with successive flowers visited

pollen clogging—**heterospecific** or infertile **pollen** deposited on **stigma** and physically blocking it from receipt of useful, fertile, **conspecific** pollen

pollen competition—**selection** acting between **pollen grains** on and within **style** for successful germination and faster **pollen tube** growth to achieve **fertilization** of **ovules**; underlying assumption is that an excess of pollen grains lands on a stigma initially

pollen discounting—extent of ineffective **self-pollination**, which reduces number of **pollen grains** left for **cross-pollination**

pollen dispersal—movement of pollen through a flowering community; is affected by amounts carried on visitors' bodies, distance and direction of their travel, and **pollen carryover** between flowers visited

pollen donor—flower or plant that is source of pollen received on a given **stigma**; also sometimes used more loosely to refer to a visitor bringing in pollen

pollen dosing—ability of some **anthers** to control pollen release in spurts, by successively greater opening of **dehiscence furrow**

pollen grain—single pollen unit; contains **haploid** male **gamete** plus a haploid **vegetative nucleus**; has a protective coat

pollenkitt—outermost, oily and pigmented layer of a **pollen grain** coat

pollen limitation—insufficient pollen receipt in a flower, with too few visits by pollen-bearing visitors, leading to reduced **fitness**

pollen placement—deposition of pollen on visitor, ideally in precise location that will ensure subsequent transfer to **stigma** in another **conspecific** flower; more precise in more specialized flowers

pollen press—area between **tarsus** and **tibia** on hind leg of honeybees and bumblebees that compresses collected pollen mass into **pollen basket**

pollen/ovule ratio—(P/O) ratio of **pollen grain** number to number of **ovules** in a flower; high P/O is common in **outcrossing** plants, and especially high in **anemophilous** plants

pollen tube—tube of tissue emerging from pore on **pollen grain** as it **germinates**, the tube then elongating down **style** to deliver male **gamete** to **ovule**

pollen waster—term sometimes used for bees, and *Apis* in particular; their efficient grooming and pollen storing as larval food may leave little pollen remaining on their bodies for effective pollination of additional flowers

pollinarium—(pl., pollinaria) complex of **pollinium** and its accessory stalk and viscid structures in orchids; commonly transferred as one unit

pollination—process of moving nonmotile **pollen** (male **gametes**) from **anthers** of one flower to **stigma** of another **conspecific** flower and thence toward female gamete in **ovule**; outcome is **fertilization** and seed production

pollination effectiveness—measure of real value of a supposed pollination event: for example, the average number of **conspecific outcrossing pollen grains** deposited on a receptive **stigma** at an appropriate time by a given visitor in a single visit

pollination fluid—sticky fluid on female **gymnosperm cones** that assists in **pollen** capture

pollination syndrome—grouping of flower species (often from very different taxonomic groups) that have **convergently** evolved a particular suite of **traits** matched to attraction of and visitation by a particular kind of pollinator

pollination web—mapping of all visits made to all flowers within a given community, showing all interactions that occur; in practice, is often just a "visitation" web, which does not distinguish effective **pollinators** from ineffective/**illegitimate visitors**, and is a potentially misleading term for that reason

pollinator—animal that *both* visits flowers regularly *and* effectively transfers significant quantities of *fertile* **pollen** between **anthers** and **nonself** but conspecific **stigmas**

pollinator limitation—poor levels of pollination because of insufficient numbers of pollinators, especially in marginal or isolated habitats, or in poor weather; or due to plentiful but ineffective flower visitors being inadequate as pollinators

pollinator service—service to plants provided by effective visitors; on a broader scale, the ecosystem service provided by pollinators that may be under threat as pollinator numbers decline

pollinium—(pl., pollinia) complex **polyads** found in orchids and asclepiads, having all **pollen grains** from one **anther** bound together

polyad — multiple **pollen grains** closely adherent and being moved together as one unit; usually multiples of four grains

polycarpic — having multiple flowering episodes in a **perennial** lifestyle

polylectic — using a wide range of floral **pollens** fairly indiscriminately as food (referring mainly to bees)

polymorphic — having more than one **morph** (see also **heteromorphic**)

polypeptide — chain of multiple **amino acids**; component of a protein, which normally has several **polypeptide** chains linked together

polyploid — having more than the normal two (**diploid**) set of **chromosomes**; rather common in plants (which are often tetraploid or hexaploid), estimates varying from 30% to 70% of species, and usually resulting from abnormal cell divisions during **meiosis**; may lead to speciation

polysaccharide — chain or network of multiply linked sugars, for example, **cellulose**, starch

polysporangiate — condition where **anther locules** are subdivided internally so parts can **dehisce** separately

poricidal anther — **anther** with a single pore (or just a few), usually terminal, from which **pollen** is dispensed usually by **sonication** (no elongate **dehiscence furrows** present)

postsecretory change — change in nectar composition or concentration after the initial process of secretion; usually linked to equilibration with environmental conditions

poststigmatic — avoidance of **selfing** by chemical/physiological mechanisms within **style** tissue after **pollen** has landed on **stigma**

prestigmatic — avoidance of **selfing** before **pollen** reaches **stigma**, usually by physical arrangement of male and female parts in flower

proboscis — tongue; especially, a composite term for insect tongue and particularly for its terminal components

proboscis extension reflex — (PER) innate response in flies and other insects, the tongue extending as reflex in response to specific color or scent stimulus

protandry — male phase preceding female phase in flowers, with **anthers dehiscing** before **stigma** is receptive; relatively common; cf. **protogyny**

protogyny — female phase preceding male phase in flowers; cf. **protandry**

pseudocopulation — form of pollination due to **cheating**, flower resembling (visually and often also by odor) female of particular insect species, inducing naive male to attempt mating, during which **pollen** is deposited, or picked up, or both

pseudostamen — any structure mimicking a **stamen**; usually appears to bear yellow **pollen**

pseudotrachea — grooves on pad-like **labella** mouthparts of flies; aid in taking up **nectar** (and sometimes **pollen**)

psychophily — pollination by butterflies

pyrrolizidine alkaloid — highly toxic type of **alkaloid** from certain plants; taken up and used as a defense by specialist insect feeders, especially by some butterflies

quasi-social bee — form of **parasocial** bee; females cooperate on nest building and all retain active ovaries

raceme — (adj., racemose) inflorescence that is unbranched, with row of single flowers along axis

radial symmetry — the ancestral form of floral symmetry, with a circular display having many possible axes of symmetry through center, as in magnolias, poppies, daisies, etc.

ray floret — small floret with strappy petal(s) toward periphery of most **composite** flowers, such as daisies (inner parts normally being tubular **disk florets**)

receptacle — cone-like top of flower stalk, just below flower

receptivity — of stigma to pollen, the ability to receive incoming **conspecific pollen** and then allow it to germinate; often indicated by moist, sticky surface

recessive gene — gene that only produces its effect on the **phenotype** if there are two copies of it in the pair of **alleles**; otherwise it is masked by a **dominant gene**

recurved — structure that curves back on itself

reflexed — structure that is bent back on itself to varying degrees

refractometer — instrument to measure refractive index; used for determining **nectar** concentration because the amount of solute in a solution affects the refraction of light

regular symmetry — alternative term for **radial symmetry**

reproductive assurance — fallback/fail-safe mechanism (often a form of **selfing**) used when **outcrossing** has not occurred

reproductive isolation — prevention of breeding between different species or populations either by prefertilization barriers (physical, behavioral) or postfertilization barriers (biochemical, genetic)

resilin — protein occurring in insects that has exceptional rubberlike properties; allows springlike recovery of shape in flexible tongues and wing bases

resin — sticky and often strongly scented gum-like substance secreted by plant or flower; commonly **terpenoid**; often antibacterial and useful for bees in their nests

resin gland — specific gland that produces resins as **rewards** to bees in a few plants

retina — (adj., retinal) inner, sensitive lining of vertebrate eye that houses a layer of **photoreceptors**

revolver flower — flower with ring of discrete **nectaries** within its base; appears like gun barrel in front view

reward — any useful substance gathered from a flower by a visitor, including not only **nectar** and **pollen** but also oils, resins, perfumes, etc.; forms the currency of the animal/flower exchanges

rhabdom—central sensory region of each unit (**ommatidium**) of insect eye, composed of **rhabdomeres** at inner margin of each ommatidial cell

rhabdomere—sensory region of folded **photoreceptor**-containing membrane within each cell of **ommatidium** of insect compound eye

rhizome—horizontal underground stem that sends out roots and shoots

rhodopsin—visual pigment common to all animal **photoreceptors**/eyes

risk-sensitive/risk-averse foraging—making foraging decisions in variable environments; responding to risk involved by changing to the more stable feeding pattern option (some animals are risk prone or risk resistant and prefer the variable option)

rostellum—modified third **stigma** in an orchid that forms a small projection on top of the **column** and produces a sticky **exudate** that helps stick **pollinia** to visitor

saccus—(pl., sacci); air-filled space in **pollen grains** of many **anemophilous** flowers

sapromyophily—pollination by carrion flies

sawfly—member of order Symphyta (within Hymenoptera but lacking narrow waist of true wasps); often visitors to **generalist flowers**

scent mark—**volatile** chemical signal deposited by animal often to maintain **territoriality** but used on flowers as indicator of recent visitation (see **footprint**)

sclerenchyma—toughened, effectively dead, supporting tissue in woody plants; only lignified secondary walls remain

scopa—(pl., scopae) storage area for **pollen** groomed off a bee's body for transport back to nest; usually on legs, thorax, or ventral abdomen

search image—mental picture (e.g., of flower type) formed by forager; focuses and facilitates location of appropriate food

secondary pollen presentation—transfer of **pollen** from **anthers** to some other surface in flower (often on **stigma** or **petals**) before it is picked up by visitor

seed—embryonic stage of plant; derived from ovule; covered in protective seed coat and with some stored food reserves used during or after dispersal

seed dispersal—movement or transport of seeds away from parent plant using abiotic (wind, water) or biotic (animal) **vectors**

seed fern—several related but all extinct types of seed plant; Devonian to **Cretaceous** periods

selection—process of individuals with advantageous (**adaptive**) traits being more successful than others and so contributing more offspring to the next generation; natural selection is the common form, but **sexual selection** also occurs and artificial selection can be imposed by humans

self-compatibility—ability of **self-pollen** to germinate and successfully **fertilize ovules**

self-fertilization—see **autogamy**

self-incompatibility—(SI) common phenomenon of **self-pollen** grains failing to **germinate** or being blocked in transit down the **style**, preventing **selfing**

selfing—see **self-pollination**

self-pollen—**pollen** from same flower or from another flower on same plant or clone

self-pollination—pollination by **pollen** from same flower or same plant or same clone

semelparous—organism reproducing just once before it dies; equivalent of being **monocarpic** in plants

semisocial bee—form of **parasocial** bee; females cooperate on nest-building and only a few have active ovaries; others serve as workers

senescence—natural ageing of leaf or flower, usually with wilting and then **abscission** from plant

sepal—one of outer whorl of (usually green) floral parts that collectively form **calyx**; often with protective functions

sequential hermaphrodite—flowers having only female or male structures and function at any one time but changing sex across weeks, seasons, or years

sesquiterpene—**terpene** with more than two **isoprenoid** subunits, common in floral odors

sex ratio—ratio of male to female individuals in a population or species; primary sex ratio is at time of fertilization, may be modified through birth and maturity

sexual—reproduction involving **meiosis**, where **haploid gametes** derived from **diploid** parents fuse to form a new diploid zygote genetically distinct from the parents; opposite of **asexual**

sexual dimorphism—morphological differences between males and females

sexual selection—selective pressures operating between two sexes within a species: **female choice** of males as mates, male competition for females; often affecting secondary sexual characters in animals

SI—see **self-incompatibility**

sieve plate—perforated plate between two **phloem** cells, the conducting sieve tubes

SI loci—genetic loci (*S* and *Z* loci) controlling **self-incompatibility** reactions

Silurian—geological period roughly 444–416 MYA, during which the first land plants appeared

size dimorphism—occurrence of two or more different sizes of structures (or of individuals) in a population or species

social facilitation—individuals within a species (usually bees, from one colony) enhancing each other's foraging, for example, by recruiting to higher-**reward** flowers, patches, or species

solar tracking — ability of some flowers to point directly at the sun at all times during daylight, following it through the daily cycle and thereby keeping warmer

solitary bee — bee species with no level of social interaction, each female making and stocking her own nest for her own offspring; the commonest type of bee

sonication — see **buzz-pollination**

spadix — elongate, erect structure within **inflorescences** of Araceae; bears both male and female flowers at its base; housed within the protective **spathe**

spathe — outer sheath-like structure of an Araceae **inflorescence**; protects the **spadix** and often traps visitors within the chamber it encloses

spatial resolution — optical quality of eyes measured as angular separation of two points that can just be detected separately

specialist flower — flower having specific traits (color, shape, scent, **reward**) that attract specific visitors and may exclude others; also, a flower receiving visitors (acting as pollinators) from only a very limited range of related species or genera

speciation — process by which new species arise; often because of habitat fragmentation and reproductive isolation

speciose — in reference to a taxon, rich in number of species

spectral sensitivity — efficiency of detection of light in relation to its wavelength or frequency; used to describe properties of **photoreceptors**

sphingid — member of moth family Sphingidae, hawkmoths; distinct from other moths by being **heterothermic** and able to hover

sphingophily — pollination by hawkmoths (**sphingids**)

spike — **inflorescence** type that is tall and erect with flowers opening usually in sequence along the spike length

spikelet — grouping of small inconspicuous flowers in a grass

spore — reproductive structure used for dispersal but has little or no food storage

sporophyte — **diploid** part of life cycle (cf. **gametophyte**); main lifespan in flowering plants

sporophytic self-incompatibility — reaction that occurs at **stigmatic** surface in response to **genotypes** of **pollen grain** and its parent (**diploid, sporophytic**) genotype

stamen — one of several male organs forming **androecium** of flower; each stamen composed of basal stalk (**filament**) and terminal **pollen**-producing **anther**

staminal tube — **nectar**-containing tube formed between partially fused stamens in the **keel** of many leguminous flowers

staminate — male flower (or phase of a **hermaphrodite** flower) having stamens but no mature or functional **pistil**(s); cf. **pistillate**

staminode — structure formed from embryonically staminal tissue but infertile; often visually attractive and producing **fodder pollen** but no reproductive pollen

standard petal — central, single, erect petal of legume flower, usually the main display; also refers to one of the three erect petals in an iris-type flower

standing crop — average amount of nectar available to natural forager in flower or flowering population; high when visitation rate is low, but in practice much reduced by recurring visitation, therefore *not* equivalent to nectar production measured from bagged, protected flowers

steady-state flowering — flowers produced steadily in small numbers over a very long period, often aseasonally in tropics

sternotribic — flower with ventral **anthers** and **style** that deposit **pollen** onto (and receive pollen from) visitor's underside; opposite of **nototribic** and less common

stigma — upper or terminal portion of **pistil**, above the **style**, and on which **pollen grains** are received

stigmatic exudate — sticky fluid on stigmatic surface that affects **pollen** reception and **germination**; sometimes gathered by flower visitors as **reward**

stigmatic lobe — one of several (commonly two or three) lobes at tip of **stigma**, often opening as surface becomes receptive and closing again later

stingless bee — member of tribe Meliponini, within family Apidae; eusocial bees with no stings but often producing defensive acidic fluid from mouthparts; common flower visitors in tropical/subtropical New World and southern continents

strobilus — (pl., strobili); technical term for **conifer's cone**

style — lower, usually stalk-like, portion of **pistil** that supports the **stigma**

subsocial bee — female bee and her hatched offspring that occur together in one nest, producing overlap of generations, but with no division of labor

sucrose — **disaccharide** (twelve-carbon sugar) produced by condensation of **glucose** with **fructose**

superior ovary — ovary set high in a flower that clearly arises *above* the point of origin of the **sepals** and **petals**

sympatric — organisms with partially or fully overlapping geographical ranges

synandry — several **anthers** uniting into one structure for all or part of their length

synapomorphy — a trait characteristic and defining of a particular **clade**, present in all the members of that group and in their common ancestor

syncarpy — several **carpels** fusing together laterally and contributing to a single **pistil**

syrphid fly — see **hoverfly**

tarsus — (pl., tarsi) terminal section of an insect leg, itself with several segments, often including proximally an enlarged metatarsus that may bear a **scopa** in bees

taxon — (pl., taxa) one of the recognized hierarchical subdivisions of living organisms, ranging from kingdom down to species/subspecies

temperate — climatic and vegetational regions mainly occurring at midlatitudes, between tropics/savannahs and **boreal**/polar regions

temporal herkogamy — change in relative position of **anthers** and **stigma** as flower ages, for example, in **protandrous** and **protogynous** flowers

temporal heterotherm — small endotherm that does not maintain fully raised metabolic rate at night (or sometimes seasonally); state seen in many hummingbirds

tepal — combined **sepal** and **petal**; thus, only one series of **perianth** parts occurs

terpene — organic compound derived from **isoprenoid** units linked as chains or rings

terpenoid — organic compound formed from two or three **isoprenoid** units, modified as esters, **alkaloids**, etc. (e.g., by oxidation) from a terpene; common plant chemicals with diverse, often defensive, functions and are common components of floral scents

territoriality — behavior maintaining a defined area (usually foraging area) for one individual or group that deters entry by others or excludes them by aggression

territorial marking — signaling boundaries of an animal's territories, usually with **scent marks**

tetrachromatic — (n., tetrachromy) visual color discrimination system based on four differently tuned **photoreceptor** types; common in vertebrates

tetrad — with reference to **pollination**, a group of four tightly joined **pollen grains** functioning as a unit, usually due to nonseparation of four microspores during development

theca — (pl., thecae) one of two paired structures composing an **anther**, each theca normally containing two pollen sacs or **locules** and opening at **dehiscence** to release **pollen**

thermogenesis — internal heat generation by raised metabolic rate within body of animal or plant

thorax — in insects, the second body section, located between head and **abdomen** and bearing wings and legs; in vertebrates, upper part of trunk, containing heart and lungs and protected by rib-cage

thrum flower — one of two morphs in **distylous** flowers; has short **style** set well below **corolla** mouth and below **anthers**; cf. **pin flower**

tibia — (pl., tibiae) segment of leg in both vertebrates and insects

tongue-lashing — habit in some bees of regurgitating nectar onto tongue and waving resultant droplet around in air, evaporation then producing head cooling and/or nectar concentration

tongue length — total functional length of tongue, especially in insects where **proboscis** has multiple components; useful measure of floral nectar accessibility

torpid — having lowered metabolic rate at night/seasonally, saving energy in small endotherms; see **temporal heterotherm**

trait — distinct variant of a particular phenotypic character; but often used almost synonymously with *character*

transposon — highly mobile section of DNA in **genome** that can cause mutation by moving position

trap flower — flower with constrictions or hairs that temporarily trap visitors for minutes or hours, enhancing **pollen** deposition or adhesion; common in **carrion flowers**, some orchids

trap-lining — pollinator movements between small numbers of widely spaced **conspecific** flowers along defined routes repeated across hours or days

tree fern — **cycads** and their relatives; mainly wind pollinated but with some animal-mediated **pollen** movement in certain groups

trichome — fine epidermal hairlike growth on plant surface, often glandular

trichromatic — (n., trichromy) visual color discrimination system based on three differently tuned **photoreceptor** types; common in insects, some mammals

trip — flower-visiting episode, usually out from nest, around group of flowers, and back to nest; or period of flower visits between rests, where there is no nest

tripping — process of releasing reproductive organs to spring upward from **keel** of **leguminous** flower; usually triggered by weight of visitor

tristyly — (adj., tristylous) form of **heterostyly** with three style morphs

trochilid — member of the mainly **Neotropical** bird family Trochilidae; hummingbird

turgor — fluid pressure that is maintained in a plant and that keeps cells and vacuoles fully enlarged; cf. **wilting**

ultraviolet — (UV) light wavelengths between 10 and 400 nm; shorter than the human visual spectrum; important component of sunlight and detected by photoreceptors of many insects

umbel — type of **inflorescence** with many short stalks spreading from a common central point; occurs especially in Apiaceae and Alliaceae

unicarpellate — **pistil** (or flower) with only one **carpel**

unimodal — any temporal pattern with only one peak (see **bimodal**)

unisexual flower — flower that is (permanently or temporarily) either male or female; cf. **hermaphrodite**

unrewarding — in reference to flower, having no **reward** of any kind for visitors (may be temporarily empty, or more permanently achieving visits by deception)

valvate stamen — **stamen** that opens by rounded valve rather than a slit or furrow

vector — any mover of **pollen**; wind, water, or animal

vegetative — form of **asexual** reproduction without use of seeds or spores, new tissue arising and differentiating from **meristem** cells

vegetative clone — plant produced by **asexual** (**vegetative**) reproduction from parent, therefore genetically identical to parent

vegetative nucleus — (also *tube nucleus*) one of two **haploid** nuclei in **pollen grain** that passes down **pollen tube** and then disintegrates within the ovary

vegetative trait — plant feature unrelated to reproductive parts and functions

viability — ability to survive and function normally; in reference to pollen, ability to germinate and effect fertilization

viscid — thick and adhesive (of fluids)

visual acuity — ability to distinguish shape, form, etc.; technically a compound of sharpness of focus in photoreceptors and interpretative abilities of the central nervous system of an animal

visual discrimination — ability to recognize similarity and difference by eye, including pattern recognition

visual spectrum — range of wavelengths distinguished by a particular visual system

viviparous — giving rise to live young; often used in reference to **bulbils** in plants

volatile — chemical that readily disperses into vapor phase under normal environmental conditions

water calyx — (also *water jacket*); cup-like base around flower formed by fused **sepals** that collects water and thereby excludes crawling insects (e.g., ants) from **corolla** and its **rewards**

wilting — drooping of plant tissue as it loses **turgor** when insufficiently supplied with water; often followed by **abscission** of leaves or flowers

wing petal — lateral petal of legume flower; two paired, either side of **keel**

worker — of bees, **diploid** female offspring of **eusocial bee** queen that originates from fertilized eggs and carries out most of the work of the colony

xanthophyll — pigment contained in plant **plastid**; normally yellow; oxygenated **carotenoid**

xylem — conducting tissue formed of elongate dead cells in plants that circulates fluid, primarily water (see **phloem**)

zoophily — (adj., zoophilous) pollination by animals

zygomorphic — having **bilateral symmetry**; flowers with one midline axis of symmetry and obviously shaped for animals to land and probe for food

zygote — early embryo formed after **fertilization**; usually **diploid** (but can be **polypoloid** in some plants)

REFERENCES

Aarssen, L. W. (2000) Why are most selfers annuals? A new hypothesis for the fitness benefits of selfing. *Oikos* 89:606–12.

Abe, H., and M. Hasegawa. (2008) Impact of volcanic activity on a plant-pollinator module in an island ecosystem: the example of the association of *Camellia japonica* and *Zosterops japonica*. *Ecol Res* 23:141–50.

Abraham, J. N. (2005) Insect choice and floral size dimorphism: sexual selection or natural selection? *J Insect Behav* 18:743–56.

Acheson, R. M. (1956) The anthocyanins of some Himalayan flowers. *Proc Roy Soc London* B 145:549–54.

Ackerman, J. D. (1981) Pollination biology of *Calypso bulbosa var. occidentalis* (Orchidaceae): a food-deception system. *Madrono* 28:101–10.

———. (1995) Convergence of filiform pollen morphologies in seagrasses: functional mechanisms. *Evol Ecol* 9: 139–53.

———. (1997a) Submarine pollination in the marine angiosperm *Zostera marina*: Part II. Pollen transport in flow fields and capture by stigmas. *Am J Bot* 84:1110–19.

———. (1997b) Submarine pollination in the marine angiosperm *Zostera marina*: Part I. The influence of floral morphology on fluid flow. *Am J Bot* 84:1099–1109.

———. (2000) Abiotic pollen and pollination: ecological, functional, and evolutionary perspectives. *Plant Syst Evol* 222:167–85.

Ackermann, M., and M. Weigend. (2006) Nectar, floral morphology and pollination syndrome in Loasaceae subfam. Loasoideae (Cornales). *Ann Bot* 98:503–14.

Adams, D. E., W. E. Perkins, and J. R. Estes. (1981) Pollination systems in *Paspalum dilatatum* Poir. (Poaceae): an example of insect pollination in a temperate grass. *Am J Bot* 68:389–94.

Adams, W. T., A. R. Griffin, and G. F. Moran. (1992) Using paternity analysis to measure effective pollen dispersal in plant populations. *Am Nat* 140:762–80.

Addicott, J. F. (1986) Variation in the costs and benefits of mutualism: the interaction between yuccas and yucca moths. *Oecologia* 70:486–94.

Addicott, J. F. and T. Bao. (1999) Limiting the cost of mutualism: multiple modes of interaction between yuccas and yucca moths. *Proc Roy Soc London* B 166:197–202.

Addicott, J. F., J. Bronstein, and F. Kjellberg. (1990) Evolution of mutualistic life-cycles; yucca moths and fig wasps. In *Genetics, Evolution and Coordination of Insect Life Cycles*, ed. F. Gilbert, 143–61. London: Springer.

Adler, L. S. (2000) The ecological significance of toxic nectar. *Oikos* 91:409–20.

Adler, L. S., and J. L. Bronstein. (2004) Attracting antagonists: does floral nectar increase herbivory? *Ecology* 85: 1519–26.

Adler, L. S. and R. E. Irwin (2005) Ecological costs and benefits of defenses in nectar. *Ecology* 86:2968–78.

———. (2006) Comparison of pollen transfer dynamics by multiple floral visitors: experiments with pollen and fluorescent dye. *Ann Bot* 97:141–50.

Adler, L. S., R. Karban, and S. Y. Strauss (2001) Direct and indirect effects of alkaloids on plant fitness via herbivory and pollination. *Ecology* 82:2032–44.

Adler, L. S., M. Wink, M. Distl, and A. J. Lentz. (2006) Leaf herbivory and nutrients increase nectar alkaloids. *Ecol Lett* 9:960–67.

Aegisdottir, H. H. and T. S. Thorhallsdottir. (2006) Breeding system evolution in the Arctic: a comparative study of *Campanula uniflora* in Greenland and Iceland. *Arctic Ant Alp Res* 38:305–12.

Affre, L., J. D. Thompson, and J. D. Debussche. (1995) The reproductive biology of the Mediterranean endemic *Cyclamen balearicum* Willk. (Primulaceae). *Bot J Linn Soc* 118:309–30.

Afik, O., A. Dag, and S. Shafir. (2007) Perception of avocado bloom (Lauraceae: *Persea americana*) by the honeybee (Hymenoptera: Apidae: *Apis mellifera*). *Entomol Gen* 30: 135–53.

Agarwal, V. M. and N. Rastogi. (2008) Role of floral repellents in the regulation of flower visits of extra-floral nectar visiting ants in an Indian crop plant. *Mol Ecol Entomol* 33:59–65.

Agelopoulos, N. and J. Pickett. (1998) Headspace analysis in chemical ecology: the effects of different sampling methods on the ratios of volatile compounds present in headspace samples. *J Chem Ecol* 24:1161–72.

Agosta, S. J., and D. H. Janzen. (2005) Body size distributions of large Costa Rican dry forest moths and the underlying relationship between plant and pollinator morphology. *Oikos* 108:183–93.

Agostini, K., M. Sazima, and I. Sazima. (2006) Bird pollination of explosive flowers while foraging for nectar and caterpillars. *Biotropica* 38:674–78.

Agrawal, A. F. (2006) Evolution of sex: why do organisms shuffle their genotypes? *Curr Biol* 16:R696–R704.

Ågren, G. I., K. Danell, T. Elmqvist, L. Ericson, and J. Hjalten. (1999) Sexual dimorphism and biotic interactions. In *Gender and Sexual Dimorphism in Flowering Plants*, ed. M. A. Geber, T. E. Dawson, and L. F. Delph, 217–46. Berlin: Springer-Verlag.

Ågren, J., T. Elmqvist, and A. Tunlid. (1986) Pollination by deceit, floral sex ratios and seed set in dioecious *Rubus chamaemorus* L. *Oecologia* 70:332–38.

Ågren, J., C. Fortunel, and J. Ehrlen. (2006) Selection on floral display in insect-pollinated *Primula farinosa*; effects of vegetation height and litter accumulation. *Oecologia* 150:225–32.

Ågren, J., and D. W. Schemske. (1991) Pollination by deceit in a neotropical monoecious herb, *Begonia involucrata*. *Biotropica* 23:235–41.

Aguiar, J. R., D. M. Bueno, B. M. Freitas, and A. A. Soares. (2000) Tecido nutritivo em flores de gravioleira, *Annona muricata* L. *Ciencia Agronomica* 31:51–55.

Aguilar, I. and D. Briceno. (2002) Sounds in *Melipona costaricensis* (Apidae; Meliponini): effect of sugar concentration and nectar source distance. *Apidologie* 33:375–88.

Aguilar, R., L. Ashworth, L. Galetto, and M. A. Aizen. (2006) Plant reproductive susceptibility to habitat fragmentation: review and synthesis through a meta-analysis. *Ecol Lett* 9:968–80.

Aigner, P. A. (2001) Optimality modelling and fitness trade offs: when should plants become pollinator specialists? *Oikos* 95:177–84.

———. (2004) Floral specialization without trade-offs: optimal corolla flare in contrasting pollination environments. *Ecology* 85:2560–69.

———. (2005) Variation in pollination performance gradients in a *Dudleya* species complex: can generalization promote floral divergence? *Funct Ecol* 19:681–89.

———. (2006) The evolution of specialized floral phenotypes in a fine-grained pollination environment. In *Plant-Pollinator Interactions: from Specialization to Generalization*, ed. N. M. Waser and J. Ollerton, 23–46. Chicago: University of Chicago Press.

Aizen, M. A. (2001) Flower sex ratio, pollinator abundance, and the seasonal pollination dynamics of a protandrous plant. *Ecology* 82:127–44.

———. (2003a) Down-facing flowers, hummingbirds and rain. *Taxon* 52:675–80.

———. (2003b) Influences of animal pollination and seed dispersal on winter flowering in a temperate mistletoe. *Ecology* 84:2613–27.

Aizen, M. A., L. Ashworth, and L. Galetto. (2002) Reproductive success in fragmented habitats: do compatibility systems and pollination specialization matter? *J Veg Sci* 13:885–92.

Aizen, M. A. and A. Basilio. (1998) Sex differential nectar secretion in protandrous *Alstroemeria aurea* (Alstromeriaceae): is production altered by pollen removal and receipt? *Am J Bot* 85:245–52.

Aizen, M. A., and P. Feinsinger. (1994a) Forest fragmentation, pollination and plant reproduction in a Chaco dry forest, Argentina. *Ecology* 75:330–51.

———. (1994b) Habitat fragmentation, native insect pollinators, and feral honey bees in Argentine "Chaco Serrano." *Ecol Appl* 4:378–92.

Aizen, M. A., L. A. Garibaldi, S. A. Cunningham, and A. M. Klein. (2008) Long-term global trends in crop yield and production reveal no current pollinator shortage but increasing pollinator dependency. *Curr Biol* 18:1572–75.

Aizen, M. A. and L. D. Harder. (2007) Expanding the limits of the pollen-limitation concept: effects of pollen quantity and quality. *Ecology* 88:271–81.

———. (2009) The global stock of domesticated honey bees is growing slower than agricultural demand for pollination. *Curr Biol* 19:915–18.

Aizen, M. A., C. L. Morales, and J. M. Morales. (2008) Invasive mutualists erode native pollination networks. *PLoS Biol* 6:396–403.

Aizen, M. A. and E. Raffaele. (1998) Flowering shoot defoliation affects pollen grain size and post-pollination pollen performance in *Alstroemeria aurea*. *Ecology* 79:2133–42.

Aizen, M. A. and D. P. Vázquez. (2006) Flower performance in human-altered habitats. In *Ecology and Evolution of Flowers*, ed. L. D. Harder and S.C.H. Barrett, 159–79. Oxford: Oxford University Press.

Akerberg, E., and E. Crane, eds. (1966) Proc. 2nd Int Symp Pollination, *Bee World* 47.

Alarcón, R., N. M. Waser, and J. Ollerton. (2008) Year-to-year variation in the topology of a plant-pollinator interaction network. *Oikos* 117:1796–807.

Albrecht, M., P. Duelli, C. Müller, D. Kleijn, and B. Schmid. (2007) The Swiss agri-environment scheme enhances pollinator diversity and plant reproductive success in nearby intensively managed farmland. *J Appl Ecol* 44:813–22.

Alexandersson, R., and S. Johnson. (2002) Pollinator-mediated selection on flower-tube length in a hawkmoth-pollinated *Gladiolus* (Iridaceae). *Proc Roy Soc London* B 269:631–36.

Allen, A. M., and S. J. Hiscock (2008) Evolution and phylogeny of self-incompatibility systems in angiosperms. In *Self-Incompatibility in Flowering Plants,* ed. V. E. Franklin-Tong, 73–101. Berlin: Springer-Verlag.

Allen, G. A., and J. A. Antos. (1993) Sex ratio variation in the dioecious shrub *Oemleria cerasiformis. Am Nat* 141: 537–53.

Allen-Wardell, G., P. Bernhardt, R. Bitner, A. Burques, S. Buchman, J. Cane, P. A. Cox, P. Feinsinger, M. Ingram, D. Inouye, C. E. Jones, K. Kennedy, P. Kevan, H. Koopowitz, S. Medellin-Morales, and G. P. Nabhan. (1998) The potential consequences of pollinator declines on the conservation of biodiversity and stability of food crop yields. *Conserv Biol* 12:8–17.

Allison, T. D. (1990) Pollen production and plant density affect pollination and seed production in *Taxus canadensis. Ecology* 71:516–22.

Allsopp, M. H., W. J. de Lange, and R. Veldtman. (2008) Valuing insect pollination services with cost of replacement. *PLoS ONE* 39:e3128.

Al-Mazra'awi, M. S., P. G. Kevan, and L. Shipp. (2007) Development of *Beauveria bassiana* dry formulation for vectoring by honeybees *Apis mellifera* (Hymenoptera: Apidae) to the flowers of crops for pest control. *Biocon Sci Tech* 17:733–41.

Alston, D. G., V. J. Tepedino, B. A. Bradley, T. R. Toler, T. L. Griswold, and S. M. Messinger. (2007) Effects of the insecticide phosmet on solitary bee foraging and nesting in orchards of Capitol Reef National Park, Utah. *Env Entomol* 36:811–16.

Althoff, D. M., K. A. Segraves, J. Leebens-Mack, and O. Pellmyr. (2006) Patterns of speciation in the yucca moths: parallel species radiations within the *Tegeticula yuccasella* species complex. *Syst Biol* 55:398–410.

Altshuler, D. L. (1999) Novel interactions of non-pollinating ants with pollinators and fruit consumers in a tropical forest. *Oecologia* 119:600–6.

Alves-dos-Santos, I., and D. Wittmann. (1999) The proboscis of the long-tongue *Ancycloscelis* bees (Anthophoridae/Apoidea), with remarks on flower visits and pollen collecting with the mouthparts. *J. Kansas Entomol Soc* 72:277–88.

Amasino, R. M. (2005) Vernalization and flowering times. *Curr Opin Biotech* 16:154–58.

Amsbery, L. K., and J. L. Maron. (2006) Effects of herbivory identity on plant fecundity. *Plant Ecol* 187:39–48.

Anand, C., C. Umranikar, P. Shintre, A. Damle, J. Kale, J. Joshi, and M. Watve. (2007) Presence of two types of flowers with respect to nectar sugar in two gregariously flowering species. *J Biosci* 32:769–74.

Anderson, A. M. (1977) Parameters determining attractiveness of stripe patterns in honey bee. *Anim Behav* 25:80–87.

Anderson, B., and Johnson, S. D. (2006) The effects of floral mimics and models on each others' fitness. *Proc Roy Soc London* B 273:969–74.

———. (2008) The geographical mosaic of coevolution in a plant-pollinator mutualism. *Evolution* 62:220–25.

———. (2009) Geographical covariation and local convergence of flower-depth in a guild of fly-pollinated plants. *New Phytol* 182:533–40.

Anderson, D., and I. J. East. (2008) The latest buzz about colony collapse disorder. *Science* 319:724–725.

Anderson, R. C., and S. Schelfhout. (1980) Phenological patterns among tallgrass prairie plants and their implications for pollinator competition. *Am Midl Nat* 104: 253–63.

Andersson, S. (1988) Size-dependent pollination efficiency in *Anchusa officinalis* (Boraginaceae): causes and consequences. *Oecologia* 76:125–30.

———. (1996) Floral display and pollination success in *Senecio jacobae* (Asteraceae): interactive effects of head and corymb size. *Am J Bot* 83:71–75.

———. (2003) Antennal responses to floral scent in the butterflies *Inachis io, Aglais urticae* (Nymphalidae) and *Gonepteryx rhamni* (Pieridae). *Chemoecology* 13:13–22.

———. (2006) Floral scent and butterfly pollinators. In *Biology of Floral Scent*, ed. N. Dudareva and E. Pichersky, 199–217. Boca Raton, FL: CRC Press.

Andersson, S., and H.E.M. Dobson. (2003) Behavioral foraging responses by the butterfly *Heliconius melpomene* to *Lantana camara* floral scent. *J Chem Ecol* 29:2302–18.

Andersson, S., L. A. Nilsson, I. Groth, and G. Bergström. (2002) Floral scents in butterfly-pollinated plants: possible convergence in chemical composition. *Bot J Linn Soc* 140:129–53.

Andersson, S., and B. Widén. (1993) Pollinator-mediated selection on floral traits in a synthetic population of *Senecio integrifolius* (Asteraceae). *Oikos* 66:72–79.

Andrews, E. S., N. Theis, and L. S. Adler. (2007) Pollinator and herbivore attraction to *Cucurbita* floral volatiles. *J Chem Ecol* 33:1682–91.

Antos, J. A., and G. A. Allen. (1994) Biomass allocation among reproductive structures in the dioecious shrub *Oemleria cerasiformes*—a functional interpretation. *J Ecol* 82:21–29.

Appanah, S. (1981) Pollination in Malaysian primary forests. *Malays For* 44:37–42.

———. (1982) Pollination of androdioecious *Xerospermum intermedium* Radlk (Sapindaceae) in a rainforest. *Biol J Linn Soc* 18:11–34.

———. (1990) Plant-pollinator interactions in Malaysian rain forests. In *Reproductive Ecology of Tropical Rainforest Plants*, ed. K. S. Bawa and M. Hadley, 69–84. Paris: UNESCO/Parthenon.

———. (1993) Mass flowering of dipterocarp forests in the aseasonal tropics. *J Biosci* 18:457–74.

Appanah, S., and H. T. Chan. (1981) Thrips: the pollinators of some dipterocarps. *Malays For* 44:234–52.

Arceo-Gomez, G., V. Parra-Tabla, and J. Navarro. (2009) Changes in sexual expression as result of defoliation and environment in a monoecious shrub in Mexico: implications for pollination. *Biotropica* 41:435–41.

Archibald, S., W. J. Bond, W. D. Stock, and D.H.K. Fairbanks. (2005) Shaping the landscape; fire-grazer interactions in an African savanna. *Ecol Appl* 15:96–109.

Arista, M., and P. L. Ortiz. (2007) Differential gender selection on floral size: an experimental approach using *Cistus salvifolius*. *J Ecol* 95:973–82.

Arista, M., P. L. Ortiz, and S. Talavera. (1999) Apical patterns of fruit production in the racemes of *Ceratonia siliqua* (Leguminosae: Caesalpinioideae): role of pollinators. *Am J Bot* 86:1708–16.

Arizmendi, M. C., C. A. Dominguez, and R. Dirzo. (1996) The role of an avian nectar robber and of hummingbird pollinators in the reproduction of two plant species. *Funct Ecol* 10:119–27.

Arizmendi, M. C. and J. F. Ornelas. (1990) Hummingbirds and their floral resources in a tropical dry forest in Mexico. *Biotropica* 22:172–80.

Armbruster, W. S. (1984) The role of resin in angiosperm pollination: ecological and chemical considerations. *Am J Bot* 71:1149–60.

———. (1985) Patterns of character divergence and the evolution of reproductive ecotypes of *Dalechampia scandens* (Euphorbiaceae). *Evolution* 39:733–52.

———. (1986) Reproductive interactions between sympatric *Dalechampia* species: are natural assemblages "random" or organized? *Ecology* 67:522–33.

———. (1991) Multilevel analysis of morphometric data from natural plant populations: insights into ontogenetic, genetic and selective correlations in *Dalechampia scandens*. *Evolution* 45:1229–44.

———. (1992) Early evolution of *Dalechampia* (Euphorbiaceae): insights from phylogeny, biogeography and comparative ecology. *Ann Miss Bot Gdn* 81:302–16.

———. (1993) Evolution of plant pollination systems: hypotheses and tests with the neotropical vine *Dalechampia*. *Evolution* 47:1480–505.

———. (1996) Evolution of floral morphology and function: an integrative approach to adaptation, constraint and compromise in *Dalechampia* (Euphorbiaceae). In *Floral Biology: Studies on Floral Evolution in Animal-Pollinated Systems*, ed. D. G. Lloyd and S.C.H. Barrett, 241–72. New York: Chapman and Hall.

———. (1997) Exaptations link evolution of plant-herbivore and plant-pollinator interactions; a phylogenetic inquiry. *Ecology* 78:1661–72.

———. (2001) Evolution of floral form: Electrostatic forces, pollination and adaptive compromise. *New Phytol* 152:181–83.

———. (2002) Can indirect selection and genetic context contribute to trait diversification? A transition-probability study of blossom-colour evolution in two genera. *J Evol Biol* 15:468–86.

———. (2006) Evolutionary and ecological aspects of specialized pollination; views from the Arctic to the Tropics. In *Plant-Pollinator Interactions: from Specialization to Generalization,* ed. N. M. Waser and J. Ollerton, 260–82. Chicago: University of Chicago Press.

Armbruster, W. S., L. Antonsen, and C. Pelabon. (2005) Phenotypic selection on *Dalechampia* blossoms: honest signalling affects pollination success. *Ecology* 86:3323–33.

Armbruster, W. S., and B. Baldwin. (1998) Switch from specialized to generalized pollination. *Nature* 394:632.

Armbruster, W. S., M. E. Edwards, and E. M. Debevec. (1994) Floral character displacement generates assemblage structure of western Australian triggerplants (*Stylidium*). *Ecology* 75:315–29.

Armbruster, W. S., C. B. Fenster, and M. R. Dudash. (2000) Pollination "principles" revisited: specialization, pollination syndromes and the evolution of flowers. In *The Scandinavian Association for Pollination Ecology Honours Knut Faegri, Mat Naturv Klasse Skrifter Ny Serie* 39:179–200.

Armbruster, W. S., J. J. Howard, T. P. Clausen, E. M. Debevec, J. C. Loquvam, M. Matsuki, B. Cerndolo, and F. Andel. (1997) Do biochemical exaptations link evolution of plant defense and pollination systems? Historical hypotheses and experimental tests with *Dalechampia* vines. *Am Nat* 149:461–84.

Armbruster, W. S., and N. Muchhala. (2009) Associations between floral specialization and species diversity: cause, effect, or correlation? *Evol Ecol* 23:159–79.

Armbruster, W. S., and G. L. Webster. (1979) Pollination of two species of *Dalechampia* (Euphorbiaceae) in Mexico by euglossine bees. *Biotropica* 11:278–83.

Armstrong, J. A. (1979) Biotic pollination mechanisms in the Australian flora: a review. *NZ J Bot* 17:467–508.

Armstrong, J. E. (1992) Lever action anthers and the forcible shedding of pollen in *Torenia* (Scrophulariaceae). *Am J Bot* 79:34–40.

———. (1997) Pollination by deceit in nutmeg (*Myristica insipida*, Myristicaceae): floral displays and beetle activity at male and female trees. *Am J Bot* 84:1266–74.

Aronne, G., and C. C. Wilcock. (1994) Reproductive characteristics and breeding system of shrubs of the Mediterranean region. *Funct Ecol* 8:69–76.

Arroyo, M.T.K., M. S. Munoz, C. Henriquez, I. Till-Bottraus, and F. Perez. (2006) Erratic pollination, high selfing levels and their correlates and consequences in an altitudinally widespread above-tree-line species in the high Andes of Chile. *Acta Oecol* 30:248–57.

Arroyo, M.T.K., R. B. Primack, and J. Armesto. (1982) Community studies in pollination ecology in the high temperate Andes of central Chile. I. Pollination mechanisms and altitudinal variation. *Am J Bot* 69:82–97.

Arroyo, M.T.K., I. Till-Bottraud, C. Torres, C. A. Henriquez, and J. Martinez. (2007) Display size preferences and foraging habitats of high Andean butterflies pollinating *Chaetanthera lycopodioides* (Asteraceae) in the subnival of the central Chilean Andes. *Arctic Ant Alp Res* 39:347–52.

Ashman, T-L. 1994) A dynamic perspective on the physiological cost of reproduction in plants. *Am Nat* 144:300–16.

Ashman, T-L. (2004) Flower longevity. In *Cell Death in Plants*, ed. L. D. Nooden, 349–62. London: Elsevier.

———. (2009) Sniffing out patterns of sexual dimorphism in floral scent. *Funct Ecol* 23:852–62.

Ashman, T-L, and I. Baker. (1992) Variation in floral sex allocation with time of season and currency. *Ecology* 73:1237–43.

Ashman, T-L, M. Bradburn, D. H. Cole, B. H. Blaney, and R. A. Raguso. (2005) The scent of a male: the role of floral volatiles in pollination of a gender dimorphic plant. *Ecology* 86:2099–105.

Ashman, T-L, T. M. Knight, J. A. Steets, P. Amarasekare, M. Burd, D. R. Campbell, M. F. Dudash, M. O. Johnston, S. J. Mazer, R. J. Mitchell, M. T. Morgan, and W. G. Wilson. (2004) Pollen limitation of plant reproduction: ecological and evolutionary causes and consequences. *Ecology* 85:2408–21.

Ashman, T-L, and D. J. Schoen. (1994) How long should flowers live? *Nature* 371:788–91.

———. (1996) Floral longevity: fitness consequences and resource costs. In *Floral Biology: Studies on Floral Evolution in Animal-Pollinated Systems*, ed. D. G. Lloyd and S.C.H. Barrett, 112–39. New York: Chapman and Hall.

———. (1997) The cost of floral longevity in *Clarkia tembloriensis*; an experimental investigation. *Evol Ecol* 11:289–300.

Ashman, T-L, and M. Stanton. (1991) Seasonal variation in pollination dynamics of sexually dimorphic *Sidalcea oregana* ssp *spicata* (Malvaceae). *Ecology* 72:993–1003.

Ashton, P. S. (1988) Dipterocarp biology as a window to the understanding of tropical forest structure. *Ann Rev Ecol Syst* 19:37–370.

Ashton, P. S., T. J. Givnish, and S. Appanah. (1988) Staggered flowering in the Dipterocarpaceae: new insights into floral induction and the evolution of mast flowering in the aseasonal tropics. *Am Nat* 132:44–66.

Ashworth, L., R. Aguilar, L. Galetto, and M. A. Aizen. (2004) Why do pollination generalist and specialist plant species show similar reproductive susceptibility to habitat fragmentation? *J Ecol* 92:717–19.

Asikainen, E., and P. Mutikainen. (2005) Preferences of pollinators and herbivores in gynodioecious *Geranium sylvaticum*. *Ann Bot* 95:879–86.

Asmussen, C. B. (1993) Pollination biology of the sea pea, *Lathyrus japonicus*; floral characters and activity and flight patterns of bumblebees. *Flora* 188:227–37.

Atsatt, P. R., and P. W. Rundel. (1982) Pollinator maintenance vs. fruit production: partitioned reproductive effort in subdioecious *Fuchsia lycioides*. *Ann Miss Bot Gard* 69:199–208.

Attwood, J. T. (1985) Pollination of *Paphiopedilum rothschildianum*; brood site deception. *Nat Geog Res* 1:247–54.

Augspurger, C. K. (1980) Mass flowering of a tropical shrub (*Hybanthus prunifolius*): influence on pollinator attraction and movement. *Evolution* 34:475–88.

———. (1981) Reproductive synchrony of a tropical shrub: experimental studies on effects of pollinators and seed predators in *Hybanthus prunifolius* (Violaceae). *Ecology* 62:775–88.

———. (1983) Phenology, flowering synchrony and fruit set of six neotropical shrubs. *Biotropica* 15:257–67.

Austerlitz, F., C. W. Dick, C. Dutech, E. K. Klein, S. Oddou-Muratorio, P. E. Smouse, and V. L. Sork. (2004) Using genetic markers to estimate the pollen dispersal curve. *Mol Ecol* 13:937–54.

Avila-Sakar, G., S. M. Simmers, and A. G. Stephenson. (2003) The interrelationships among leaf damage, anther development, and pollen production in *Cucurbita pepo* ssp *texana* (Cucurbitaceae). *Int J Plant Sci* 164:395–404.

Avila-Sakar, G., and A. G. Stephenson. (2006) Effects of the spatial pattern of leaf damage on growth and reproduction: whole plants. *Int J Plant Sci* 167:1021–28.

Ayala-Berdon, J., J. E. Schondube, and K. E. Stoner. (2009) Seasonal intake responses in the nectar-feeding bat *Glossophaga soricina*. *J Comp Physiol* B 179:553–62.

Ayasse, M. (2006) Floral scent and pollinator attraction in sexually deceptive orchids. In *Biology of Floral Scent*, ed. N. Dudareva and E. Pichersky, 219–41. Boca Raton, FL: CRC Press.

Bailey, S. F., A. L. Hargreaves, S. D. Hechtenthal, R. A. Laird, T. M. Latty, T. G. Reid, A. C. Teucher, and J. R. Tindall. (2007) Empty flowers as a pollination-enhancement strategy. *Evol Ecol Res* 9:1245–62.

Bain, R. S., A. Rashed, V. J. Cowper, F. S. Gilbert, and T. N. Sherratt. (2007) The key mimetic features of hoverflies through avian eyes. *Proc Roy Soc London* B 274:1949–54.

Baker, H. G. (1975) Sugar concentrations in nectars from hummingbird flowers. *Biotropica* 7:37–41.

———. (1976) "Mistake" pollination as a reproductive system with special reference to the Caricaceae. In *Tropical Trees: Variation, Breeding and Conservation*, ed. J. Burley and B. T. Styles, 161–69. New York: Academic Press.

Baker, H. G. and I. Baker. (1973) Amino acids in nectar and their evolutionary significance. *Nature* 241:543–45.

Baker, H. G. and I. Baker. (1983) Floral nectar sugar constituents in relation to pollinator type. In *Handbook of Experimental Pollination Ecology*, ed. C. E. Jones and R. J. Little, 117–41. New York: Van Nostrand Reinhold.

———. (1986) The occurrence and significance of amino acids in floral nectar. *Plant Syst Evol* 151:175–86.

———. (1990) The predictive value of nectar chemistry to the recognition of pollinator types. *Isr J Bot* 39:159–66.

Baker, H. G., I. Baker, and S. A. Hodges. (1998) Sugar composition of nectars and fruits consumed by birds and bats in the tropics and subtropics. *Biotropica* 30:559–86.

Baker, H. G., I. Baker, and P. A. Opler. (1973) Stigmatic exudates and pollination. In *Pollination and Dispersal*, ed. N.B.M. Brantjes, 47–80. Nijmegen, the Netherlands: University of Nijmegen Press.

Baker, H. G. and P. A. Cox. (1984) Further thoughts on dioecism and islands. *Ann Miss Bot Gdn* 71:244–53.

Baker, H. G. and P. D. Hurd. (1968) Intrafloral ecology. *Ann Rev Entomol* 13:385–414.

Baker, J. D. and R. W. Cruden. (1991) Thrips-mediated self-pollination of two facultatively xenogamous wetland species. *Am J Bot* 78:959–63.

Baldock, K.C.R. (2007) Multi-species interactions in a Kenyan savannah ecosystem. PhD thesis, University of Edinburgh.

Baldwin, B. G. (2007) Adaptive radiation of shrubby tarweeds (*Deinandra*) in the California Islands parallels diversification of the Hawaiian silversword alliance (Compositae, Madiinae). *Am J Bot* 94:237–48.

Balkenius, A. and A. Kelber. (2006) Colour preferences influence odour learning in the hawkmoth *Macroglossum stellatarum*. *Naturwissenschaften* 93:255–58.

Balkenius, A., W. Rosen, and A. Kelber. (2006) The relative importance of olfaction and vision in a diurnal and a nocturnal hawkmoth. *J Comp Physiol* A 192:431–37.

Banaszak, J. (1992) Strategy for conservation of wild bees in an agricultural landscape. *Agric Ecosyst Environ* 40: 179–92.

Banda, H. J. and R. J. Paxton. (1991) Pollination of greenhouse tomatoes by bees. *Acta Hort* 288:194–98.

Banschbach, V. S. and K. D. Waddington. (1994) Risk sensitive foraging in honeybees: no consensus among individuals and no effect of honey bee stores. *Anim Behav* 7:933–41.

Bänziger, H. (1996) The mesmerising wart: the pollination strategy of epiphytic lady slipper orchid *Paphiopedilum villosum* (Lindl.) Stein (Orchidaceae). *Bot J Linn Soc* 121:59–90.

Barkman, T. J., M. Bendiksby, S. H. Lim, K. M. Salleh, J. Nais, D. Madulid, and T. Schumacher. (2008) Accelerated rates of floral evolution at the upper size limit for flowers. *Curr Biol* 18:1508–13.

Barrett, S.C.H. (1988) The evolution, maintenance and loss of self-incompatibility systems. In *Plant Population Biology*, ed. J. Lovett Doust and L. Lovett Doust, 98–124. Oxford: Oxford University Press.

———. (1990) The evolution and adaptive significance of heterostyly. *Trends Ecol Evol* 5:144–48.

———. (1996) The reproductive biology and genetics of island plants. *Phil Trans Roy Soc London* 351:725–33.

———. (2002) The evolution of plant sexual diversity. *Nature Rev Gen* 3:274–84.

Barrett, S.C.H. and L. D. Harder. (2006) David G Lloyd and the evolution of floral biology; from natural history to strategic analysis. In *Ecology and Evolution of Flowers*, ed. L. D. Harder and S.C.H. Barrett, 1–21. Oxford: Oxford University Press.

Barrett, S.C.H., L. D. Harder, and W. W. Cole. (1994) Effects of flower number and position on self-fertilization in experimental populations *of Eichhornia paniculata* (Pontederiaceae). *Funct Ecol* 8:526–35.

Barrett, S.C.H. and B. C. Husband. (1990) Variation in outcrossing rates in *Eichhornia paniculata*: the role of demographic and reproductive factors. *Plant Spec Biol* 5:41–55.

Barrett, S.C.H., L. K. Jesson, and A. M. Baker. (2000) The evolution and function of stylar polymorphisms in flowering plants. *Ann Bot* 85:253–65.

Barrett, S.C.H., D. G. Lloyd, and J. Arroyo. (1996) Stylar polymorphisms and the evolution of heterostyly in *Narcissus* (Amaryllidaceae). In *Floral Biology: Studies on Floral Evolution in Animal-Pollinated Systems*, ed. D. G. Lloyd and S.C.H. Barrett, 339–76. New York: Chapman and Hall.

Barrett, S.C.H., M. T. Morgan, and B. C. Husband. (1989) The dissolution of a complex genetic polymorphism: the evolution of self-fertilization in tristylous *Eichhornia paniculata*. *Evolution* 43:1398–416.

Barrett, S.C.H. and J. S. Shore. (1987) Variation and evolution of breeding systems in the *Turnera ulmifolia* L. complex (Turneraceae). *Evolution* 41:340–54.

Barringer, B. C. (2007) Polyploidy and self-fertilization in flowering plants. *Am J Bot* 94:1527–33.

Barrows, E. M. (1976) Nectar robbing and pollination of *Lantana camara* (Verbenaceae). *Biotropica* 8:132–35.

Barth, F. G. (1985) *Insects and Flowers: the Biology of a Partnership*. Princeton, NJ: Princeton University Press.

Barthell, J. F., and J.M.H. Knops. (1997) Visitation of evening primrose by carpenter bees; evidence of a "mixed" pollination syndrome. *Southwest Nat* 42:86–93.

Bartomeus, I., M. Vila, and L. Santamaria. (2008) Contrasting effects of invasive plants on plant-pollinator networks. *Oecologia* 155:761–70.

Bascompte, J. (2007) Networks in ecology. *Basic Appl Ecol* 8:485–90.

Bascompte, J., P. Jordano, C. J. Melián, and J. M. Olesen.

(2003) The nested assembly of plant-animal mutualistic networks. *Proc Natl Acad Sci USA* 100:9383–87.

Bateman, A. J. (1948) Intrasexual selection in *Drosophila*. *Heredity* 2:349–69.

———. (1974a) Contamination of seed crops. II. Wind pollination. *Heredity* 1:235–46.

———. (1974b) Contamination of seed crops. I. Insect pollination. *J. Genetics* 48:257–75.

Batra, S.W.T. (1984) Solitary bees. *Sci Am* 250:86–93.

———. (1994) *Anthophora pilipes villulosa* SM (Hymenoptera, Anthophoridae), a manageable Japanese bee that visits blueberries and apples during cool rainy spring weather. *Proc Entomol Soc Wash* 96:98–119.

Baude, M., I. Dajoz, and E. Danchin. (2008) Inadvertent social information in foraging bumblebees: effects of flower distribution and implications for pollination. *Anim Behav* 76:1863–73.

Bawa, K. S. (1974) Breeding systems of tree species of a lowland tropical community. *Evolution* 28:85–92.

———. (1980a) Evolution of dioecy in flowering plants. *Ann Rev Ecol Syst* 11:15–39.

———. (1980b) Mimicry of male by female flowers and intersexual competition for pollinators in *Jacaratia dolichaula* (D. Smith) Woodson (Caricaceae). *Evolution* 34: 467–74.

———. (1982) Outcrossing and the incidence of dioecism in island floras. *Am Nat* 119:866–71.

———. (1983) Patterns of flowering in tropical plants. In *Handbook of Experimental Pollination Biology*, ed. C. E. Jones and R. J. Little, 394–410. New York: Van Nostrand Reinhold.

———. (1990) Plant-pollinator interactions in tropical rain forests. *Ann Rev Ecol Syst* 21:399–422.

———. (1994) Pollinators of tropical dioecious angiosperms: a reassessment? No, not yet. *Am J Bot* 81: 456–60.

Bawa, K. S., S. H. Bullock, D. R. Perry, R. E. Coville, and M. H. Grayum. (1985) Reproductive biology of tropical lowland rain forest trees. II. Pollination systems. *Am J Bot* 72:346–56.

Bawa, K. S., and P. A. Opler. (1975) Dioecism in tropical forest trees. *Evolution* 29:167–79.

Bay, D. (1995) Thermogenesis in the Aroids. *Aroideana* 18:32–39.

Beach, J. H. and K. S. Bawa. (1980) The role of pollinators in the evolution of dioecy from distyly. *Evolution* 34: 1138–42.

Beale, G. H., J. R. Price, and V. C. Sturgess. (1941) A survey of anthocyanins. VII. The natural selection of flower colours. *Proc Roy Soc London* B 130:113–26.

Beaman, R. S., P. J. Decker, and J. H. Beaman. (1988) Pollination of *Rafflesia* (Rafflesiaceae). *Am J Bot* 75: 1148–62.

Beardsell, D. V. and P. Bernhardt. (1983) Pollination biology of Australian terrestrial orchids. In *Pollination '82*, ed. E. G. Williams, R. B. Knox, J. H. Gilbert, and P. Bernhardt, 166–83. Parksville, Victoria: University of Melbourne Press.

Beardsell, D. V., S. P. O'Brien, E. G. Williams, R. B. Knox, and D. M. Calder. (1993) Reproductive biology of Australian Myrtaceae. *Aust J Bot* 41:511–26.

Beath, D.D.N. (1996) Pollination of *Amorpholphallus johnsonii*(Araceae) by carrion beetles (*Phaeochrous amplus*) in a Ghanaian rain forest. *J Trop Ecol* 12:409–18.

Beattie, A. J. (1969) Studies in the pollination ecology of *Viola*. 1. The pollen content of stigmatic cavities. *Watsonia* 7:142–56.

———. (1985) *The Evolutionary Ecology of Ant-Plant Mutualisms*. Cambridge: Cambridge University Press.

Beattie, A. J., D. E. Breedlove, and P. R. Ehrlich. (1973) The ecology of the pollinators and predators of *Frasera speciosa*. *Ecology* 54:81–91.

Beattie, A. J., C. L. Turnbull, T. Hough, S. Jobson, and R. B. Knox. (1985) The vulnerability of pollen and fungal spores to ant secretions: evidence and some evolutionary implications. *Am J Bot* 72:606–14.

Beattie, A. J., C. L. Turnbull, R. B. Knox, and E. G. Williams. (1984) Ant inhibition of pollen function: a possible reason why ant pollination is rare. *Am J Bot* 71:421–26.

Bechar, A., I. Shmulevich, D. Eisikowitch, Y. Vaknin, B. Ronen, and S. Gan-Mor. (1999) Simulation and testing of an electrostatic pollination system. *Trans ASAE* 42: 1511–16.

Beck, J. (2007) The importance of amino acids in the adult diet of male tropical rainforest butterflies. *Oecologia* 151:741–47.

Beckman, N., and L. E. Hurd. (2003) Pollen feeding and fitness in praying mantids; the vegetarian side of a tritrophic predator. *Environ Entomol* 32:881–85.

Beekman, M., and F.L.W. Ratnieks. (2000) Long range foraging by the honey bee *Apis mellifera*. *Funct Ecol* 14:490–96.

Beil, M., H. Horn, and A. Schwabe. (2008) Analysis of pollen loads in a wild bee community (Hymenoptera, Apidae)—a method for elucidating habitat use and foraging distances. *Apidologie* 39:456–67.

Bell, G. (1982) *The Masterpiece of Nature; the Evolution and Genetics of Sexuality*. Berkeley, CA: University of California Press.

———. (1985) On the functions of flowers. *Proc Roy Soc London* B 224:223–65.

———. (1986) The evolution of empty flowers. *J Theor Biol* 118:253–58.

Bell, J. M., J. D. Karron, and R. J. Mitchell. (2005) Interspecific competition for pollination lowers seed production and outcrossing in *Mimulus ringens*. *Ecology* 86:762–71.

Bell, P. R. (1992) *Green Plants: Their Origin and Diversity.* Cambridge: Cambridge University Press.

Bellusci, S. and M. D. Marques. (2001) Circadian activity rhythm of the foragers of a eusocial bee (*Scaptotrigona* aff. *depilis*, Hymenoptera, Apidae, Meliponini) outside the nest. *Biol Rhythm Res* 32:117–24.

Benham, B. R. (1969) Insect visitors to *Chamaenerion angustifolium* and their behaviour in relation to pollination. *Entomologist* 102:222–28.

Benitez-Vieyra, S., A. M. Medina, E. Glinos, and A. A. Cocucci. (2006) Pollinator-mediated selection on floral traits and size of floral display in *Cyclopogon elatus*, a sweat bee-pollinated orchid. *Funct Ecol* 20:948–57.

Bentley, B. L. (1977) Extrafloral nectaries and protection by pugnacious bodyguards. *Ann Rev Ecol Syst* 8:407–27.

Bentley, B. and T. Elias, ed. (1983) *Biology of Nectaries.* New York: Columbia University Press.

Bergman, P. and G. Bergström. (1997) Scent-marking, scent origin, and species specificity in male premating behaviour of two Scandinavian bumblebees. *J Chem Ecol* 23:1235–51.

Bergström, G. (1978) Role of volatile chemicals in *Ophrys*-pollinator interactions. In *Biochemical Aspects of Plant-Animal Coevolution*, ed. J. B. Harborne, 207–32. London: Academic Press.

Bergström, G., H.E.M. Dobson, and I. Groth. (1995) Spatial fragrance patterns within the flowers of *Ranunculus acris* (Ranunculaceae). *Plant Syst Evol* 195:221–42.

Bernardello, G. (2007) A systematic survey of floral nectaries. In *Nectaries and Nectar*, ed. S. W. Nicolson, M. Nepi, and E. Pacini, 19–128. Dordrecht: Springer.

Bernhardt, P. (1986) Bee pollination in *Hibbertia fasciculata* (Dilleniaceae). *Plant Syst Evol* 152:231–41.

———. (1996) Anther adaptation in animal pollination. In *The Anther: Form, Function and Phylogeny*, ed. W. G. D'Arcy and R. C. Keating, 192–220. Cambridge: Cambridge University Press.

———. (2000) Convergent evolution and adaptive radiation of beetle-pollinated angiosperms. *Plant Syst Evol* 222: 293–320.

Bernhardt, P., and L. B. Thien. (1987) Self-isolation and insect pollination in primitive angiosperms: New evaluations of older hypotheses. *Plant Syst Evol* 156:159–76.

Berry, P. E., and R. N. Calvo. (1989) Wind pollination, self-incompatibility and altitudinal shifts in pollination systems in the high Andean genus *Espeletia* (Asteraceae). *Am J Bot* 76:1602–14.

Bertin, R. I. (1993) Incidence of monoecy and dichogamy in relation to self-fertilization in angiosperms. *Am J Bot* 80:557–60.

Bertin, R. I. and C. M. Newman. (1993) Dichogamy in angiosperms. *Bot Rev* 59:112–52.

Bertin, R. I. and M. E. Willson. (1980) Effectiveness of diurnal and nocturnal pollination of two milkweeds. *Can J Bot* 58:1744–46.

Bertsch, A. (1983) Nectar production of *Epilobium angustifolium* L. at different air humidities; nectar sugar in individual flowers and the optimal foraging theory. *Oecologia* 59:40–48.

———. (1984) Foraging in male bumblebees (*Bombus lucorum* L.)—maximizing energy or minimizing water load. *Oecologia* 62:325–36.

Bestman, H. J., L. Winkler, and O. von Helversen. (1997) Headspace analysis of volatile flower scent constituents of bat-pollinated plants. *Phytochemistry* 46:1169–72.

Bianchini, M., and E. Pacini. (1996) Explosive anther dehiscence in *Ricinus communis* L. involves cell wall modifications and relative humidity. *Int J Plant Sci* 157: 739–45.

Biere, A., and S. C. Honders. (2006) Coping with third parties in a nursery pollination mutualism: *Hadena bicruris* avoids oviposition on pathogen-infected, less rewarding *Silene latifolia. New Phytol* 169:719–27.

Biernaskie, J. M., and R. V. Cartar. (2004) Using varied nectar production to manipulate pollinators. *Funct Ecol* 18:125–29.

Biernaskie, J. M., R. V. Cartar, and T. A. Hurly. (2002) Risk-averse inflorescence departure in hummingbirds and bumble bees; could plants benefit from variable nectar volumes? *Oikos* 98:98–104.

Biernaskie, J. M., and E. Elle. (2005) Conditional strategies in an animal-pollinated plant: size-dependent adjustment of gender and rewards. *Evol Ecol Res* 7:901–13.

Bierzychudek, P. (1981) *Asclepias, Lantana* and *Epidendrum*: a floral mimicry complex? *Biotropica* 13:1354–58.

———. (1982) The demography of jack-in-the-pulpit, a forest perennial that changes sex. *Ecol Monogr* 52:335–51.

———. (1987) Pollinators increase the cost of sex by avoiding female flowers. *Ecology* 68:444–47.

Biesmeijer, J. C., J.A.P. Richter, M.A.J.P. Smeets, and M. J. Sommeijer (1999) Niche differentiation in nectar-collecting stingless bees: the influence of morphology, floral choice, and interference competition. *Mol Ecol Entomol* 24:380–88.

Biesmeijer, J. C., A.P.M. Roberts, M. Reemer, R. Ohlemuller, M. Edwards, T. Peeters, A. P. Schaffers, S. G. Potts, R. Kleukers, C. D. Thomas, J. Settele, and W. E. Kunin. (2006) Parallel declines in pollinators and insect pollinated plants in Britain and the Netherlands. *Science* 313:351–54.

Biesmeijer, J. C., and E. Toth. (1998) Individual foraging, activity level and longevity in the stingless bee *Melipona beecheii* in Costa Rica (Hymenoptera, Apidae, Meliponinae). *Ins Sociaux* 45:427–43.

Bino, R. J., A. Dafni, and A.D.J. Meeuse. (1984). Entomophily in the dioecious gymnosperm *Ephedra aphylla*, with some notes on *E. campylopoda*. I. Aspects of the entomophilous syndrome. *Proc Kopn Ned Akad Weten* 87:1–24.

Bittrich, V. and M.C.E. Amaral. (1996) Pollination biology of *Symphonia globulifera* (Clusiaceae). *Plant Syst Evol* 200:101–10.

Bjerknes, A. L., Ø. Totland, S. J. Hegland, and A. Nielsen. (2007) Do alien plant invasions really affect pollination success in native plant species? *Biol Cons*138:1–12.

Blanco, M. A., and G. Barboza. (2005) Pseudocopulatory pollination in *Lepanthes* (Orchidaceae, Pleurothallidinae) by fungus gnats. *Ann Bot* 95:763–72.

Blarer, A., T. Keasar, and A. Shmida. (2002) Possible mechanisms for the formation of flower size preferences by foraging bumblebees. *Ethology* 108:341–51.

Bleiweiss, R. (1999) Joint effects of feeding and breeding behaviour on trophic dimorphism in hummingbirds. *Proc Roy Soc London* B 266:2491–97.

———. (2001) Mimicry on the QT(L): genetics of speciation in *Mimulus*. *Evolution* 55:1706–8.

Blight, M. M., M. LeMetayer, M.H.P. Delegue, J. A. Pickett, F. Marion-Poll, and L. J. Wadhams. (1997) Identification of floral volatiles involved in recognition of oilseed rape flowers (*Brassica napus*) by honeybees (*Apis mellifera*). *J Chem Ecol* 23:1715–27.

Bloch, D., N. Werdenberg, and A. Erhardt. (2006) Pollination crisis in the butterfly-pollinated wild carnation *Dianthus carthusianorum*? *New Phytol* 169:699–706.

Blondel, J., and J. Aronson. (1999) *Biology and Wildlife of the Mediterranean Region*. Oxford: Oxford University Press.

Blossey B. and T. R. Hunt-Joshi. (2003) Below-ground herbivory by insects: influence on plants and above-ground herbivores. *Ann Rev Ent* 48:521–47.

Blüthgen, N., F. Menzel, T. Hovestadt, B. Fiala, and N. Blüthgen. (2007) Specialization, constraints and conflicting interests in mutualistic networks. *Curr Biol* 17:341–436.

Boaz, M., U. Plitman, and C. C. Heyn. (1994). Reproductive effort in desert versus Mediterranean crucifers: the allogamous *Erucaria rostrata* and *E. hispanica,* and the autogamous *Erophila minima*. *Oecologia* 100:286–92.

Boch, R. (1956) Die Tänze der Biene bei nahen und fernen Trachtquellen. *Zeits Vergl Physiol* 38:136–67.

Boggs, C. L. (1988) Rates of nectar feeding in butterflies: effects of sex, size, age, and nectar concentration. *Funct Ecol* 2:289–95.

Boisvert, M. J., A. J. Veal, and D. F. Sherry. (2007) Floral reward production is timed by an insect pollinator. *Proc Roy Soc London* B 274:1831–37.

Bolmgren, K., O. Eriksson, and H. P. Linder. (2003) Contrasting flowering phenology and species richness in abiotically and biotically pollinated angiosperms. *Evolution* 57:2001–11.

Bolten, A. B., and P. Feinsinger. (1978) Why do hummingbird flowers secrete dilute nectar? *Biotropica* 10:307–9.

Bond, W. J. (1994) Do mutualisms matter? Assessing the impact of pollinator and disperser disruption on plant extinction. *Phil Trans Roy Soc London B* 344:83–90.

———. (1995) Assessing the risk of plant extinction due to pollinator and disperser failure. In *Extinction Rates*, ed. J. H. Lawton and R. M. May, 131–46. Oxford: Oxford University Press.

Bonod, I., J. C. Sandoz, Y. Loublier, and M. H. Pham-Delegue. (2003) Learning and discrimination of honey odours by the honey bee. *Apidologie* 34:147–59.

Boose, D. L. (1997) Sources of variation in floral nectar production rate in *Epilobium canum* (Onagraceae): implications for natural selection. *Oecologia* 110:493–500.

Boppré, M. (1978) Chemical communication, plant relationships and mimicry in the evolution of danaid butterflies. *Entomol Exp Appl* 24:264–77.

Borchert, R., S. S. Renner, Z. Calle, D. Navarete, A. Tye, L. Gautier, R. Spichiger, and P. Von Hildebrand. (2005) Photoperiodic induction of synchronous flowering near the Equator. *Nature* 433:627–29.

Borg-Karlson, A. K. (1990) Chemical and ethological studies on pollination in the genus *Ophrys* (Orchidaceae). *Phytochem* 29:1359–87.

Borkent, C. J. and Harder LD (2007) Flies (Diptera) as pollinators of two dioecious plants: behaviour and implications for plant mating. *Can Entomol* 139:235–46.

Borkent, C. J. and E. I. Schilinger. (2008) Flower-visiting and mating behaviour of *Eulonchus sapphirinus* (Diptera, Acroceridae). *Can Entomol* 140:250–56.

Borrell, B. J. (2006) Mechanics of nectar feeding in the orchid bee *Euglossa imperialis*: pressure, viscosity and flow. *J Exp Biol* 209:4901–7.

———. (2007) Scaling of nectar foraging in orchid bees. *Am Nat* 169:569–80.

Borrell, B. J. and H. W. Krenn. (2006) Nectar feeding in long-proboscid insects. In *Ecology and Biomechanics; A Mechanical Approach to the Ecology of Animals and Plants*, ed. H. T. Speck and N. Rowe, 185–212. Boca Raton, FL: CRC Press.

Bosch, J. and W. Kemp. (2001) *How to Manage the Blue Orchard Bee*. Beltsville, MD: Sust Agric Network.

———. (2002) Developing and establishing bee species as crop pollinators: the example of *Osmia* spp. (Hymenoptera, Megachilidae) and fruit trees. *Bull Ent Res* 92:3–16.

———. (2005) Alfalfa leafcutting bee population dynamics, flower availability, and pollination rates in two Oregon alfalfa fields. *J Econ Entomol* 98:1077–86.

Bosch, J., J. Retana, and X. Cerda. (1997) Flowering phenology, floral traits and pollinator composition in a herbaceous Mediterranean plant community. *Oecologia* 109: 583–91.

Boulter, S. K., R. L. Kitching, and B. G. Howlett. (2006) Family, visitors and the weather; patterns of flowering in tropical rain forests of northern Australia. *J Ecol* 94: 369–82.

Bower, C. C. (2006) Specific pollinators reveal a cryptic taxon in the bird orchid *Chiloglottis valida* sensu lato (Orchidaceae) in southeastern Australia. *Aust J Bot* 54: 53–64.

Bowers, J. E. (2007) Has climatic warming altered spring flowering date of Sonoran Desert shrubs? *Southwest Nat* 52:347–55.

Bowers, J. E., and M. A. Dimmitt. (1994) Flowering phenology of 6 woody plants in the Northern Sonoran desert. *Bull Torrey Bot Club* 121:215–29.

Bowker, G. E., and H. C. Crenshaw. (2007a) Electrostatic forces in wind-pollination, Part 1: Measurement of the electrostatic charge on pollen. *Atmosph Environ* 41: 1587–95.

———. (2007b) Electrostatic forces in wind-pollination, Part 2: Simulations of pollen capture. *Atmosph Environ* 41:1596–1603.

Bowman, G., M. Tarayre, and A. Atlan. (2008) How is the invasive gorse *Ulex europaeus* pollinated during winter? A lesson from its native range. *Plant Ecol* 197: 197–206.

Bowman, R. N. (1987) Cryptic self-incompatibility and the breeding system of *Clarkia unguiculata* (Onagraceae). *Am J Bot* 74:471–76.

Boyd, E. A. (2004) Breeding system of *Macromeria viridiflora* (Boraginaceae) and geographic variation in pollinator assemblages. *Am J Bot* 91:1809–13.

Boyden, T. C. (1980) Floral mimicry by *Epidendrum ibaguense* (Orchidaceae) in Panama. *Evolution* 34:135–36.

———. (1982) The pollination biology of *Calypso bulbosa* var. *americana* (Orchidaceae): initial deception of bumblebee visitors. *Oecologia* 55:178–84.

Bradshaw, F. J. and S. D. Bradshaw. (2001) Maintenance nitrogen requirement of an obligate nectarivore, the honey possum *Tarsipes rostratus*. *J Comp Physiol B* 171:59–67.

Bradshaw, H. D. and D. W. Schemske. (2003) Allele substitution at a flower color locus produces pollinator shifts in monkey flowers. *Nature* 426:176–78.

Bradshaw, H. D., S. M. Wilbert, K. G. Otto, and D. W. Schemske. (1995) Genetic mapping of floral traits associated with reproductive isolation in monkeyflowers (*Mimulus*). *Nature* 376:762–65.

Brandenburg, A., A. Dell'Olivio, R. Bshary, and C. Kuhlemeier. (2009) The sweetest thing: advances in nectar research. *Curr Opin Plant Biol* 12:1–5.

Branquart, E. and J. L. Hemptinne. (2000) Selectivity in the exploitation of floral resources by hoverflies (Diptera, Syrphinae). *Ecography* 23:732–42.

Brantjes, N.B.M. (1973) Sphingophilous flowers, functions of their scent. In *Pollination and Dispersal*, ed. N.B.M. Brantjes and H. F. Linskens, 27–46. Nijmegen, Netherlands: Publication Deptartment of the Bot University.

———. (1980) Flower morphology of *Aristolochia* species and the consequences for pollination. *Acta Bot Neerl* 25:281–95.

Breed, M. D., E. M. Stocker, L. K. Baumgartner, and S A. Vargas. (2002) Time-place learning and the ecology of recruitment in a stingless bee, *Trigona amalthea* (Hymenoptera, Apidae). *Apidologie* 33:251–58.

Brian, A. D. (1957) Differences in the flowers visited by four species of bumblebees, and their causes. *J Anim Ecol* 26:71–98.

Brice, A. T., K. H. Dahl, and C. R. Grau. (1989) Pollen digestibility by hummingbirds and Psittacines. *Condor* 91: 681–88.

Brink, D. (1980) Reproduction and variation in *Aconitum*. *Am J Bot* 67:263–73.

———. (1982) A bonanza-blank pollinator reward schedule in *Delphinium nelsonii* (Ranunculaceae). *Oecologia* 52: 292–94.

Briscoe, A., and L. Chittka. (2001) The evolution of color vision in insects. *Ann Rev Entomol* 46:471–510.

Brodmann, J., R. Twele, W. Francke, G. Hölzler, Q. H. Zhang, and M. Ayasse. (2008) Orchids mimic green-leaf volatiles to attract prey-hunting wasps for pollination. *Curr Biol* 18:740–44.

Brodmann, J., R. Twele, W. Francke, Y. B. Luo, X. Q. Song, and M. Ayasse. (2008) Orchid mimics honey bee alarm pheromone in order to attract hornets for pollination. *Curr Biol* 19:1368–72.

Brody, A. K. (1997) The effects of pollinators, herbivores and seed predators on flowering phenology. *Ecology* 78: 1624–31.

Bronstein, J. B. (1992) Seed predators as mutualists; ecology and evolution of the fig/pollinator interaction. In *Insect-Plant Interactions*, vol. 4, ed. E. Bernays, 1–44. Boca Raton, FL: CRC Press.

Bronstein, J. L. (1995) The plant-pollinator landscape. In *Mosaic Landscapes and Ecological Processes,* ed. L. Hansson, L. Fahrig, and G. Merriam, 256–88. London: Chapman and Hall.

Bronstein, J. L., T. Huxman, B. Horvath, M. Farabee, and G. Davidowitz. (2009) Reproductive biology of *Datura wrightii*: the benefits of a herbivorous pollinator. *Ann Bot* 103:1435–43.

Brouat, C., D. McKey, J. M. Bessiere, L. Pascal, and M. Hossaert-McKey. (2000) Leaf volatile compounds and the distribution of ant patrolling in an ant-plant protection mutualism: preliminary results on *Leonardoxa* (Fabaceae: Caesalpinoideae) and *Petalomyrmex* (Formicidae: Formicinae). *Acta Oecol* 21:349–57.

Brouillard, R. and O. Dangles. (1993) Flavonoids and flower

colour. In *The Flavonoids*, ed. J. Harborne, 565–88. London: Chapman and Hall.

Brown, B. J., R. J. Mitchell, and S. A. Graham. (2002) Competition for pollination between an invasive species (Purple Loosestrife) and a native congener. *Ecology* 83: 2328–36.

Brown, J. H., and A. Kodric-Brown. (1979) Convergence, competition, and mimicry in a temperate community of hummingbird-pollinated flowers. *Ecology* 60:1022–35.

Broyles, S. B. and R. Wyatt. (1990) Plant parenthood in milkweeds: a direct test of the pollen donation hypothesis. *Plant Spec Biol* 5:131–42.

———. (1995) A re-examination of the pollen-donation hypothesis in an experimental population of *Asclepias exaltata*. *Evolution* 49:89–99.

Brugge, B., E. Van der Spek, and M. M. Kwak. (1998) Honingbijen in natuurgebieden? *De Levende Natuur* 99:71–76.

Brunet, J., and H. R. Sweet. (2006) Impact of insect pollinator group and floral display size on outcrossing rate. *Evolution* 60:234–46.

Buchmann, S. L. (1983) Buzz pollination in angiosperms. In *Handbook of Experimental Pollination Biology*, ed. C. E. Jones and R. J. Little, 73–113. New York: Van Nostrand Reinhold.

———. (1985) Bees use vibration to aid pollen-collection from non-poricidal flowers. *J Kans Entomol Soc* 58: 517–25.

———. (1987) The ecology of oil flowers and their bees. *Ann Rev Ecol Syst* 18:343–69.

Buchmann, S. L. and M. D. Buchmann. (1981) Anthecology of *Mouriri myrtilloides* (Melastomataceae: Memecyleae), an oil flower in Panama. *Biotropica* 13:7–24.

Buchmann, S. L. and J. H. Cane. (1989) Bees assess pollen returns while sonicating *Solanum* flowers. *Oecologia* 81: 289–94.

Buchmann, S. L. and J. P. Hurley. (1978) A biophysical model for buzz pollination in angiosperms. *J Theor Biol* 72:639–57.

Buchmann, S. L. and G. P. Nabhan. (1996) *The Forgotten Pollinators*. Washington, DC: Island Press.

Buchmann, S. L. and M. K. O'Rourke. (1988) Palynological analysis of dietary breadth of honey bee colonies in the Sonoran Desert. Proc XVIII Int Congr Entomol, Vancouver, BC.

Buchmann, S. L., M. K. O'Rourke, and K. J. Niklas. (1989) Aerodynamics of *Ephedra infurca*. III. Selective pollen capture by pollination droplets. *Bot Gaz* 150:122–31.

Buide, M. L. (2006) Pollination ecology of *Silene acutifolia* (Caryophyllaceae): floral traits variation and pollinator attraction. *Ann Bot* 97:289–97.

Bullock, S. H. (1994) Wind pollination of Neotropical deciduous trees. *Biotropica* 17:287–301.

Bullock, S. H. and K. S. Bawa. (1981) Sexual dimorphism and the annual flowering pattern in *Jacaratia dolichaula* (D Smith) Woodson (Caricaceae) in a Costa Rican rain forest. *Ecology* 62:1494–1504.

Bullock, S. H., J. H. Beach, and K. S. Bawa. (1983) Episodic flowering and sexual dimorphism in *Guarea rhopalocarpa* in a Costa Rican rain forest. *Ecology* 64: 851–61.

Bullock, S. H., and A. Pescador. (1983) Wing and proboscis dimensions in a sphingid fauna from western Mexico. *Biotropica* 15:292–94.

Bullock, W. L. and M. J. Samways. (2005) Conservation value of flower-arthropod associations in remnant African grassland corridors in an afforested pine mosaic. *Biodiv and Cons* 14:3093–103.

Burczyk, J., W. T. Adams, D. S. Birkes, and I. J. Chybicki. (2006) Using genetic markers to directly estimate gene flow and reproductive success parameters in plants on the basis of naturally regenerated seedlings. *Genetics* 173: 363–72.

Burd, M. (1994) Bateman's principle and plant reproduction: the role of pollen limitation in fruit and seed set. *Bot Rev* 60:83–139.

Burkle, L. A. and R. E. Irwin. (2009) The effects of nutrient addition on floral characters and pollination in two subalpine plants, *Ipomopsis aggregata* and *Linum lewisii*. *Plant Ecol* 203:83–98.

Burkle, L. A., R. E. Irwin, and D. A. Newman. (2007) Predicting the effects of nectar robbing on plant reproduction: implications of pollen limitation and plant mating system. *Am J Bot* 94:1935–43.

Búrquez, A. (1989) Blue tits, *Parus caeruleus*, as pollinators of the crown imperial, *Fritillaria imperialis*, in Britain. *Oikos* 55:335–40.

Búrquez, A., and S. A. Corbet. (1991) Do flowers reabsorb nectar? *Funct Ecol* 5:369–79.

Butz Huryn, V. M. (1997) Ecological impacts of introduced honeybees. *Q Rev Biol* 72:275–97.

Bynum, M. R. and W. K. Smith. (2001) Floral movements in response to thunderstorms improve reproductive effort in the alpine species *Gentiana algida* (Gentianaceae). *Am J Bot* 88:1088–95.

Byrne, M., C. P. Elliott, C. Yates, and D. J. Coates. (2007) Extensive pollen dispersal in a bird-pollinated shrub, *Calothamnus quadrifidus*, in a fragmented landscape. *Mol Ecol* 16:1303–14.

Cady, L. (1965) The flying duck orchids. *Aust. Plants* 3: 174–77.

Cahill, J. F., E. Elle, G. R. Smith, and B. H. Shore. (2008) Disruption of a below-ground mutualism alters interactions between plants and their floral visitors. *Ecology* 89:1791–801.

Caissard, J. C., S. Meekijjironenroj, S. Baudino, and M. C.

Anstett. (2004) Localization of production and emission of pollinator attractant on whole leaves of *Chamaerops humilis* (Araceae). *Am J Bot* 91:1190–199.

Calvo, R. N. (1993) Evolutionary demography of orchids: intensity and frequency of pollination and the costs of fruiting. *Ecology* 74:1033–42.

Calvo, R. N. and C. C. Horvitz. (1990) Pollinator limitation, cost of reproduction, and fitness in plants: a transition-matrix approach. *Am Nat* 136:499–516.

Cameron, S. A. (1981) Chemical signals in bumblebee foraging. *Behav Ecol Sociobiol* 9:257–60.

Cameron, S. A. and S. Ramirez. (2001) Nest architecture and nesting ecology of the orchid bee *Eulaema meriana* (Apinae: Euglossini). *J Kans Entomol Soc* 74:142–65.

Campan, R. and M. Lehrer. (2002) Discrimination of closed shapes by two species of bee, *Apis mellifera* and *Megachile rotundata*. *J Exp Biol* 205:559–72.

Campbell, D. R. (1989a) Inflorescence size: test of the male function hypothesis. *Am J Bot* 76:730–38.

———. (1989b) Measurements of selection in a hermaphroditic plant: variation in male and female pollination success. *Evolution* 43:318–34.

———. (1991) Comparing pollen dispersal and gene flow in a natural population. *Evolution* 45:1965–68.

———. (1998) Multiple paternity in fruits of *Ipomopsis aggregata*. *Am J Bot* 85:1022–27.

Campbell, D. R. and A. F. Motten. (1985) The mechanism of competition for pollination between two forest herbs. *Ecology* 66:554–63.

Campbell, D. R., N. M. Waser, and E. J. Meléndez-Ackerman. (1997) Analyzing pollinator-mediated selection in a plant hybrid zone: hummingbird visitation patterns on three spatial scales. *Am Nat* 149:295–315.

Campbell, D. R., N. M. Waser, and M. V. Price. (1994) Indirect selection on stigma position in *Ipomopsis aggregata* via a genetically controlled trait. *Evolution* 48:55–68.

Campbell, D. R., N. M. Waser, M. V. Price, E. A. Lynch, and R. J. Mitchell. (1991) Components of phenotypic selection: pollen export and flower corolla width in *Ipomopsis aggregata*. *Evolution* 45:1458–67.

Campbell, J. W. and J. L. Hanula. (2007) Efficiency of Malaise traps and colored pan traps for collecting flower-visiting insects from three forested ecosystems. *J Insect Cons* 11:399–408.

Campbell, J. W., J. L. Hanula, and T. A. Waldrop. (2007) Effects of prescribed fire and fire surrogates on floral-visiting insects of the Blue Ridge province in North Carolina. *Biol Cons* 134:393–404.

Campbell, S. A. and T. J. Close. (1997) Dehydrins: genes, proteins and associations with phenotypic traits. *New Phytol* 137:61–74.

Cane, J. H. (1993a) Reproductive role of sterile pollen in cryptically dioecious species of flowering plants. *Curr Sci* 65:223–25.

———. (1993b) Reproductive role of sterile pollen in *Saurauia* (Actinidiaceae), a cryptically dioecious neotropical tree. *Biotropica* 25:493–95.

Cane, J. H. and S. L. Buchmann. (1989) Novel pollen-harvesting behavior by the bee *Protandrena mexicanorum* (Hymenoptera, Andrenidae). *J Insect Behav* 2:431–36.

Cane, J. H., G. C. Eickwort, F. R. Wesley, and J. Spielholz. (1983) Foraging, grooming and mate seeking behaviors of *Macropis nuda* (Hymenoptera, Melittidae) and use of *Lysimachia ciliata* (Primulaceae) oils in larval provisions and cell linings. *Am Midl Nat* 110:257–64.

Cane, J. H., K. MacKenzie, and D. Schiffhauer. (1993) Honey bees harvest pollen from the porose anthers of cranberries (*Vaccinium macrocarpon*) (Ericaceae). *Am Bee J* 133:293–95.

Cane, J. H., R. L. Minckley, and L. Kervin. (2000) Sampling bees (Hymenoptera, Apidiformes) for pollinator community studies; pitfalls of pan-trapping. *J Kans Entomol Soc* 73:225–31.

Cane, J. H., R. L. Minckley, L. Kervin, and T. H. Roulston. (2005) Temporally persistent patterns of incidence and abundance in a pollinator guild at annual and decadal scales: the bees of *Larrea tridentata*. *Biol J Linn Soc* 85:319–29.

Cane, J. H., R. L. Minckley, L. J. Kervin, T. H. Roulston, and N. M. Williams. (2006) Complex responses within a desert bee guild (Hymenoptera, Apiformes) to urban habitat fragmentation. *Ecol Applic* 16:632–44.

Cane, J. H. and J. A. Payne. (1988) Foraging ecology of the bee *Habropoda laboriosa* (Hymenoptera, Anthophoridae), an oligolege of blueberries (Ericaceae: *Vaccinium*) in the southeastern United States. *Ann Entomol Soc Am* 81:419–27.

Cane, J. H. and S. Sipes. (2006) Characterizing floral specialization by bees: analytical methods and a revised lexicon for oligolecty. In *Plant-Pollinator Interactions: from Specialization to Generalization*, ed. N. M. Waser and J. Ollerton, 99–122. Chicago: University of Chicago Press.

Cane, J. H. and V. J. Tepedino. (2001) Causes and extent of declines among native North American invertebrate pollinators; detection, evidence, and consequences. *Conserv Ecol* 5:1.

Canela, M.B.F. and M. Sazima. (2003) Florivory by the crab *Armases angustipes* (Grapsidae) influences hummingbird visits to *Aechmea pectinata* (Bromeliaceae). *Biotropica* 35:289–94.

Canto, S., C. M. Herrera, M. Medrano, R. Perez, and I. M. Garcia. (2008) Pollinator foraging modifies nectar sugar composition in *Helleborus foetidus* (Ranunculaceae): an experimental test. *Am J Bot* 95:315–20.

Canto, S., R. Perez, M. Medrano, M. C. Castellanos, and C. M. Herrera. (2007) Intra-plant variation in nectar sugar composition in two *Aquilegia* species (Ranunculaceae):

contrasting patterns under field and glass house conditions. *Ann Bot* 99:653–60.

Capaldi, E. A., A. D. Smith, J. L. Osborne, S. E. Fahrbach, S. M. Farris, D. R. Reynolds, A. S. Edward, A. Martin, G. E. Robinson, G. M. Poppy, and J. R. Riley. (2000) Ontogeny of orientation flight in the honeybee revealed by harmonic radar. *Nature* 403:537–40.

Carisey, N. and E. Bruce. (1997) Impact of balsam fir flowering on pollen and foliage biochemistry in relation to spruce budworm growth, development and food utilization. *Entomol Exp Appl* 85:17–31.

Cariveau, D., R. E. Irwin, A. K. Brody, L. S. Garcia-Mayeya, and A. von der Ohe. (2004) Direct and indirect effects of pollinators and seed predators to selection on plant and floral traits. *Oikos* 104:15–26.

Carlson, J. E. (2007) Male-biased nectar production in a protandrous herb matches predictions of sexual selection theory in plants. *Am J Bot* 94:674–82.

Carlson, J. E. and K. E. Harms. (2007) The benefits of bathing buds: water calyces protect flowers from a microlepidopteran herbivore. *Biol Lett* 3:405.

Carlsson, M. A., and B. S. Hanson. (2006) Detection and coding of flower volatiles in nectar-foraging insects. In *Biology of Floral Scent*, ed. N. Dudareva and E. Pichersky, 243–61. Boca Raton, FL: CRC Press.

Carmona-Diaz, G., and J. G. Garcia-Franco. (2009) Reproductive success in the Mexican rewardless *Oncidium cosymbephorum* (Orchidaceae) facilitated by the oil-rewarding *Malpighia glabra* (Malpighiaceae). *Plant Ecol* 203:253–61.

Carpenter, F. L. (1976) Plant-pollinator interactions in Hawaii: pollination energetics of *Metrosideros colina* (Myrtaceae). *Ecology* 57:1125–44.

———. (1978a) Hooks for mammal pollination? *Oecologia* 35:123–32.

———. (1978b) A spectrum of nectar-eater communities. *Am Zool* 18:809–19.

———. (1983) Pollination energetics in avian communities: simple concepts and complex realities. In *Handbook of Experimental Pollination Biology*, ed. C. E. Jones and R. J. Little, 215–34. New York: Van Nostrand Reinhold.

Carreck, N. and I. H. Williams. (1998) The economic value of bees in the UK. *Bee World* 79:115–23.

Carreck, N. L., T. M. Butt, S. J. Clark, L. Ibrahim, E. A. Isger, J. K. Pell, and I. H. Williams. (2007) Honey bees can disseminate a microbial control agent to more than one inflorescence pest of oilseed rape. *Biocontrol Sci Tech* 17:179–91.

Carroll, A. B., S. G. Pallardy, and C. Galen. (2001) Drought stress, plant water status, and floral trait expression in fireweed, *Epilobium angustifolium* (Onagraceae). *Am J Bot* 88:438–46.

Carromero, W. and J. L. Hamrick. (2005) The mating system

of *Verbascum thapsus* (Scrophulariaceae): the effect of plant height. *Int J Plant Sci* 166:979–83.

Cartar, R. V. (2005) Short-term effects of experimental boreal forest logging disturbance on bumble bees, bumble bee-pollinated flowers and the bee-flower match. *Biodiv and Cons* 14:1895–1907.

Cartar, R. V. and L. M. Abrahams. (1996) Risk sensitive foraging in a patch departure context: a test with worker bumblebees. *Am Zool* 36:447–58.

Carter, C., R. Healy, N. M. O'Tool, S.M.S. Naqvi, G. Ren, S. Park, G. A. Beattie, H. T. Horner, and R. W. Thornburg. (2007) Tobacco nectaries express a novel NADPH oxidase implicated in the defense of floral reproductive tissues against microorganisms. *Plant Physiol* 143: 389–99.

Carter, C., S. Shafir, L. Yehonatan, R. G. Palmer, and R. Thornburg. (2006) A novel role for proline in plant floral nectars. *Naturwissenschaften* 93:72–79.

Carter, C., and R. W. Thornburg. (2004) Is the nectar redox cycle a floral defense against microbial attack? *Trends Plant Sci* 9:320–24.

Carthew, S. M. (1993) An assessment of pollinator visitation to *Banksia spinulosa*. *Aust J Ecol* 18:257–68.

———. (1994) Foraging behaviour of marsupial pollinators in a population of *Banksia spinulosa*. *Oikos* 69:133–39.

Carthew, S. M. and R. L. Goldingay. (1997) Non-flying mammals as pollinators. *Trends Ecol Evol* 12:104–8.

Caruso, C. M. (1999) Pollination of *Ipomopsis aggregata* (Polemoniaceae); effects of intra- vs. interspecific competition. *Am J Bot* 86:663–68.

Caruso, C. M., D.L.D. Remington, and K. E. Ostergren. (2005). Variation in resource limitation of reproduction influences natural selection on floral traits of *Asclepias syriaca*. *Oecologia* 146:68–76.

Carvalheiro, L. G., E.R.M. Barbosa, and J. Memmott. (2008) Pollinator networks, alien species and the conservation of rare plants: *Trinia glauca* as a case study. *J Appl Biol* 45:1419–27.

Carvell, C., W. R. Meek, R. F. Pywell, D. Goulson, and M. Nowakowski. (2007) Comparing the efficacy of agri-environment schemes to enhance bumblebee abundance and diversity on arable field margins. *J Appl Ecol* 44:29–40.

Carvell, C., W. R. Meek, R. F. Pywell, and M. Nowakowski. (2004) The response of foraging bumblebees to successional change in newly created arable field margins. *Biol Cons* 118:327–39.

Carvell, C., D. B. Roy, S. M. Smart, R. F. Pywell, C. D. Preston, and D. Goulson. (2006) Declines in forage availability for bumblebees at a national scale. *Biol Cons* 132:481–89.

Cascante, A., M. Quesada, J. A. Lobo, and E. J. Fuchs. (2002) Effects of dry tropical forest fragmentation on the reproductive success and genetic structure of the tree *Samanea saman*. *Cons Biol* 16:137–47.

Case, M. A. and Z. R. Bradford. (2009) Enhancing the trap of lady's slippers: a new technique for discovering pollinators yields new data from *Cypripedium parviflorum* (Orchidaceae). *Bot J Linn Soc* 160:1–10.

Casper, B. B. and T. R. LaPine. (1984) Changes in corolla color and other floral characteristics in *Cryptantha humilis* (Boraginaceae): cues to discourage pollinators? *Evolution* 38:128–41.

Castellanos, M. D., P. Wilson, S. J. Keller, A. D. Wolfe, and J. D. Thomson. (2006) Anther evolution: pollen presentation strategies when pollinators differ. *Am Nat* 167: 288–96.

Castellanos, M. D., P. Wilson, and J. D. Thomson. (2002) Dynamic nectar replenishment in flowers of *Penstemon* (Scrophulariaceae). *Am J Bot* 89:111–18.

———. (2004) "Anti-bee" and "pro-bird" changes during the evolution of hummingbird pollination in *Penstemon*. *Evol Biol J Evol Biol* 17:876–85.

Castro, S., P. Silveira, and L. Navarro. (2008) Effect of pollination on floral longevity and costs of delaying fertilization in the outcrossing *Polygala vayredae* Costa (Polygalaceae). *Ann Bot* 102:1043–48.

Celedon-Neghme, C., W. L. Gonzales, and E. Gianoli. (2007) Costs and benefits of attractive floral traits in the annual species *Madia sativa* (Asteraceae). *Evol Ecol* 21:247–57.

Chacoff, N. P. and M. A. Aizen. (2006) Edge effects on flower-visiting insects in grapefruit plantations bordering premontane subtropical forest. *J Appl Ecol* 43:18–27.

Chacoff, N. P., M. A. Aizen, and V. Aschero. (2008) Proximity to forest edge does not affect crop production despite pollen limitation. *Proc Roy Soc London* B 275:907–13.

Chambers, V. H. (1968) Pollens collected by species of *Andrena* (Hymenoptera, Apidae). *Proc Roy Entomol Soc London A* 43:155–60.

Chappell, M. A. (1984) Temperature regulation and energetics of the solitary bee *Centris pallida* during foraging and intermale mate competition. *Physiol Zool* 57:215–25.

Char, M.B.S. and S. S. Bhat. (1975) Antifungal activity of pollen. *Naturwissenschaften* 62:536.

Charlesworth, B. and D. Charlesworth. (1978) Model for evolution of dioecy and gynodioecy. *Am Nat* 112:975–97.

Charlesworth, D. (1993) Why are unisexual flowers associated with wind pollination and unspecialized pollinators? *Am Nat* 141:481–90.

Charlesworth, D. and B. Charlesworth. (1979) Model for the evolution of distyly. *Am Nat* 114:467–98.

———. (1987) Inbreeding depression and its evolutionary consequences. *Ann Rev Ecol Syst* 18:237–68.

Charlesworth, D. and M. T. Morgan. (1991) Allocation of resources to sex functions in flowering plants. *Phil Trans Roy Soc London* B 332:91–102.

Chavez, R. P. (1974) Observaciones en el pollen de plantas con probable polinizacion quiropterofila. *Anal Exc Nac Cienc Biol Mex* 21:115–43.

Chen, C., Q. Song, M. Proffit, J. M. Bessiere, Z. B. Li, and M. Hossaert-McKey. (2009) Private channel: a single unusual compound assures specific pollinator attraction in *Ficus semicordata*. *Func Ecol* 23:941–50.

Chittka, L. (1997) Bee color vision is optimal for coding flower colors, but flower colors are not optimal for being coded—why? *Isr J Plant Sci* 45:115–27.

———. (1998) Sensorimotor learning in bumblebees: long term retention and reversal training. *J Exp Biol* 201: 515–24.

Chittka, L., A. G. Dyer, F. Bock, and A. Dornhaus. (2003) Bees trade off foraging speed for accuracy. *Nature* 424: 388.

Chittka, L., A. Gumbert, and J. Kunze. (1997) Foraging dynamics of bumble bees: correlations of movements within and between plant species. *Behav Ecol* 8:239–49.

Chittka, L. and P. G. Kevan. (2005) Flower colour as advertisement. In *Practical Pollination Biology*, ed. A. Dafni, P. G. Kevan, and B. C. Husband, 157–96. Ontario, Canada: Enviroquest Ltd.

Chittka, L. and S. Schurkens. (2001) Successful invasion of a floral market—an exotic Asian plant has moved in on Europe's river-banks by bribing pollinators. *Nature* 411: 653.

Chittka, L., J. Spaethe, A. Schmidt, and A. Hickelsberger. (2001) Adaptation, constraint, and chance in the evolution of flower color and pollinator color vision. In *Cognitive Ecology of Pollination*, ed. L. Chittka and J. D. Thomson, 106–26. Cambridge: Cambridge University Press.

Chittka, L. and J. D. Thomson. (1997) Sensorimotor learning and its relevance to task specialization in bumble bees. *Beh Ecol Sociobiol* 41:385–98.

Chittka, L. and J. D. Thomson, ed. (2001) *Cognitive Ecology of Pollination*. New York: Cambridge University Press.

Chittka, L., J. D. Thomson, and N. M. Waser. (1999) Flower constancy, insect psychology and plant evolution. *Naturwissenschaften* 86:361–77.

Chouteau, M., D. Barabe, and M. Gibernau. (2007) Thermogenesis in *Syngonium* (Araceae). *Can. J. Bot.* 85:184–90.

Clark, M. J., and B. C. Husband. (2007) Plasticity and timing of flower closure in response to pollination in *Chamerion angustifolium* (Onagraceae). *Int J Plant Sci* 168:619–25.

Clegg, M. T. and M. L. Durbin. (2003) Tracing floral adaptations from ecology to molecules. *Nat Rev Genet* 4: 206–15.

Clement, S. L., B. C. Hellier, L. R. Elberson, R. T. Staska, and M. A. Evans. (2007) Flies (Diptera: Muscidae: Calliphoridae) are efficient pollinators of *Allium ampeloprasum* L. (Alliaceae) in field cages. *J Econ Entomol* 100: 131–35.

Clements, R. E. and F. L. Long. (1923) *Experimental Pollination. An Outline of the Ecology of Flowers and Insects*, 336. Washington DC: Carnegie Institute.

Cnaani, J., J. D. Thomson, and D. R. Papaj. (2006) Flower choice and learning in foraging bumblebees; effects of nectar volume and concentration. *Ethology* 112:278–85.

Cocucci, A. A. and A. N. Sersic. (1998) Evidence of rodent pollination in *Cajophora coronata* (Loasaceae). *Plant Syst Evol* 211:113–28.

Coelho, J. R. (1991) Heat transfer and body temperature in honey bee (Hymenoptera, Apidae) drones and workers. *Environ Entomol* 20:1627–35.

Coen, E. S. and J. M. Nugent. (1994) Evolution of flowers and inflorescences. *Development (Suppl.)* 1994:107–16.

Cohen, D. and A. Shmida. (1993) The evolution of flower display and reward. *Evol Biol* 27:197–243.

Cole, D. H. and T. L. Ashman. (2005) Sexes show differential tolerance to spittlebug damage and consequences of damage for multi-species interactions. *Am J Bot* 92:1708–13.

Colin, L. J. and C. E. Jones. (1980) Pollen energetics and pollination modes. *Am J Bot* 67:210–15.

Colla, S. R., M. C. Otterstatter, R. J. Gegear, and J. D. Thomson. (2006) Plight of the bumblebee; pathogen spillover from commercial to wild populations. *Biol Cons* 129: 461–67.

Collett, T. S. and M. F. Land. (1975) Visual spatial memory in a hoverfly. *J Comp Physiol* 100:59–84.

Collevatti, R. G., L.A.O. Campos, and J. H. Schoereder. (1997) Foraging behaviour of bee pollinators on the tropical weed *Triumfetta semitriloba*: departure rules from flower patches. *Insectes Soc* 44:345–52.

Collins, B .G. and T. Rebelo. (1987) Pollination biology of the Proteaceae in Australia and southern Africa. *Aust J Ecol* 12:387–421.

Colwell, R. K. (1986) Population structure and sexual selection for host fidelity in the speciation of hummingbird flower mites. In *Evolutionary Processes and Theory*, ed. S. Karlin and E. Nevo, 475–95. New York: Academic Press.

Comba, L., S. A. Corbet, A. Barron, A. Bird, S. Collinge, N. Miyazaki, and M. Powell. (1999) Garden flowers: insect visits and the floral reward of horticulturally-modified variants. *Ann Bot* 83:73–86.

Comba, L., S. A. Corbet, H. Hunt, S. Outram, J. S. Parker, and B. J. Glover. (2000) The role of genes influencing the corolla in pollination of *Antirrhinum majus*. *Plant Cell Environ* 23:639–47.

Comba, L., S. A. Corbet, H. Hunt, and B. Warren. (1999) Flowers, nectar and insect visits: evaluating British plant species for pollinator friendly gardens. *Ann Bot* 83: 369–83.

Connell, J. H. (1975) Some mechanisms producing structure in natural communities: a model and evidence from field experiments. In *Ecology and Evolution of Communities*, ed. M. L. Cody and J. M. Diamond, 460–90. Cambridge, MA: Harvard University Press.

Conner, J. K. (2006) Ecological genetics of floral evolution. In *Ecology and Evolution of Flowers*, ed. L. D. Harder and S.C.H. Barrett, 260–77. Oxford: Oxford University Press.

Conner, J. K., R. Davis, and S. Rush. (1995) The effect of wild radish floral morphology on pollination efficiency by four taxa of pollinators. *Oecologia* 104:234–45.

Conner, J. K. and S. Rush. (1996) Effects of flower size and number on pollinator visitation to wild radish, *Raphanus raphanistrum*. *Oecologia* 105:509–16.

Cook, C. D. K. (1988) Wind pollination in aquatic angiosperms. *Ann Miss Bot Gard* 75:768–77.

Cook, D. C., M. B. Thomas, S. A. Cunningham, D. L. Anderson, and P. J. de Barro. (2007) Predicting the economic impact of an invasive species on an ecosystem service. *Ecol Appl* 17:1832–40.

Cook, J. M., D. Bean, S. A. Powe, and D. J. Dixon. (2004) Evolution of a complex coevolved trait: active pollination in a genus of fig wasps. *J Evol Biol* 17:238–46.

Cook, J. M., and J. Y. Rasplus. (2003) Mutualists with attitude: coevolving fig wasps and figs. *Trends Ecol Evol* 18:241–48.

Cook, S. M., C. S. Awmack, D. A. Murray, and I. H. Williams. (2003) Are honey bees' foraging preferences affected by pollen amino acid composition? *Mol Ecol Entomol* 28:622–27.

Cook, S. M., D. A. Murray, and I. H. Williams. (2004) Do pollen beetles need pollen? The effect of pollen on oviposition, survival and development of a flower-feeding herbivore. *Mol Ecol Entomol* 29:164–73.

Cooley, J. R. (1995) Floral heat rewards and direct benefits to insect pollinators. *Ann Entomol Soc Am* 88:576–79.

Cope, F. (1962) The mechanism of pollen incompatibility in *Theobroma* L. *Heredity* 17:157–82.

Corbet, S. A. (1995) Insects, plants and succession: advantages of long-term set-aside. *Agric Ecosyst Environ* 53: 201–17.

———. (2000) What kinds of flowers do butterflies visit? *Entomol Exp Appl* 96:289–98.

———. (2003) Nectar sugar content: estimating standing crop and secretion rate in the field. *Apidologie* 34:1–10.

———. (2006) A typology of pollination systems: implications for crop management and the conservation of wild plants. In *Plant-Pollinator Interactions: from Specialization to Generalization* ed. N. M. Waser and J. Ollerton, 315–40. Chicago: University of Chicago Press.

Corbet, S. A., J.W.L. Beament, and D. Eisikowitch. (1982) Are electrostatic forces involved in pollen transfer? *Plant Cell Environ* 5:125–29.

Corbet, S. A., J. Bee, K. Dasmahapatra, S. Gale, E. Gorringe, B. La Ferla, T. Moorhouse, A. Trevail, Y. van Bergen, and M. Vorontsova. (2001) Native or exotic? Double or single? Evaluating plants for pollinator-friendly gardens. *Ann Bot* 87:219–32.

Corbet, S. A., I. Cuthill, M. Fallows, T. Harrison, and

G. Hartley. (1981) Why do nectar-foraging bees and wasps work upwards on inflorescences? *Oecologia* 51: 79–83.

Corbet, S. A., M. Fussell, R. Ake, A. Fraser, C. Gunson, A. Savage, and K. Smith. (1993) Temperature and the pollinating activity of social bees. *Mol Ecol Entomol* 18: 17–30.

Corbet, S. A., C. J. C. Kerslake, D. Brown, and N. E. Morland. (1984) Can bees select nectar-rich flowers in a patch? *J Apic Res* 23:293–308.

Corbet, S. A., D. M. Unwin, and O. Prys-Jones.. (1979) Humidity, nectar and insect visits to flowers, with special reference to *Crataegus, Tilia* and *Echium. Mol Ecol Entomol* 4:9–22.

Corbet, S. A., and P. G. Willmer. (1981) Pollination of the yellow passionfruit—nectar, pollen and carpenter bees. *J Agric Sci* 95:655–66.

Corbet, S. A., P. G. Willmer, J.W.L. Beament, D. M. Unwin, and O. Prys-Jones. (1979) Post-secretory determinants of sugar concentration in nectar. *Plant Cell Environ* 2: 293–300.

Corlett, R. T. (2004) Flower visitors and pollination in the Oriental (Indomalayan) region. *Biol Rev* 79:497–532.

Cory, C. (1984) Pollination biology of two species of Hawaiian Lobeliaceae (*Clermontia kakeana* and *Cyanea angustifolia*) and their presumed coevolved relationship with native honeycreepers (Drepanididae). MS thesis, University of California, Davis.

Costanza, R., R. d'Arge, R. de Groot, S. Farber, M. Grasso, B. Hannon, K. Limburg, S. Naeem, R. V. O'Neill, J. Paruelo, R. G. Raskin, P. Sutton, and M. van den Belt. (1997) The value of the world's ecosystem services and natural capital. *Nature* 387:253–60.

Cotton, P. A. (1998) Coevolution in an Amazonian hummingbird-plant community. *Ibis* 140:639–46.

———. (2001) The behavior and interactions of birds visiting *Erythrina fusca* flowers in the Colombian Amazon. *Biotropica* 33:662–69.

Courtney, S. P. and C. J. Hill. (1983) Butterflies as pollinators: reply to Tepedino. *Oikos* 41:145–46.

Cox, P. A. (1982) Vertebrate pollination and the maintenance of dioecism in *Freycinetia. Am Nat* 120:65–80.

———. (1983) Extinction of the Hawaiian avifauna resulted in a change of pollinators for the ieie, *Freycinetia arborea. Oikos* 41:195–99.

———. (1988) Hydrophilous pollination. *Ann Rev Ecol Syst* 19:261–79.

———. (1989) Baker's law, plant breeding systems and island colonization. In *The Evolutionary Ecology of Plants* ed. J. H. Bock and Y. B. Linhart, 209–224. Boulder, CO: Westview Press.

———. (1991) Abiotic pollination: an evolutionary escape for animal-pollinated angiosperms. *Phil Trans Roy Soc London* B 33:217–24.

———. (1993) Water-pollinated plants. *Sci Am* 269:50–56.

Cox, P. A., R. H. Laushmann, and M. H. Ruckelshaus. (1992) Surface and submarine pollination in the seagrass *Zostera marina* L. *Bot J Linn Soc* 109:281–91.

Cox, P. E. and T. Elmqvist. (2000) Pollinator extinction in the Pacific islands. *Cons Biol* 14:1237–39.

Cox, P. E., T. Elmqvist, E. D. Pierson, and W. E. Rainey. (1991) Flying foxes as strong interactors in south Pacific island ecosystems: a conservation hypothesis. *Cons Biol* 5:448–54.

Cozzolino, S. and A. Widmer. (2005) Orchid diversity: an evolutionary consequence of deception? *Trends Ecol Evol* 20:487–94.

Crailsheim, K., L.H.W. Schneider, N. Hrassnigg, G. Buhlmann, U. Brosch, R. Gmeinbauer, and B. Schoffmann. (1992) Pollen consumption and utilization in worker honey bees (*Apis mellifera carnica*)—dependence on individual age and function. *J Insect Physiol* 38:409–19.

Crane, E. (1977) Dead bees under lime trees; sugars poisonous to bees. *Bee World* 58:129–30.

———. (1978) Sugars poisonous to bees. *Bee World* 59: 37–38.

Crane, E. and P. Walker. (1984) *Pollination Directory for World Crops*. London: International Bee Research Association.

Crane, P. R. (1986) Form and function in wind dispersed pollen. In *Pollen and Spores: Form and Function*, ed. S. Blackmore and I. K. Ferguson, 179–202. London: Academic Press.

Crane, P. R., E. M. Friis, and K. R. Pedersen. (1995) The origin and early diversification of angiosperms. *Nature* 374: 27–33.

Crawley, M., P. Jarvey, and A. Purvis. (1996) Comparative ecology of the native and alien floras of the British Isles. *Phil Trans Roy Soc London* B 351:1251–59.

Crepet, W. L., E. M. Friis, and K. C. Nixon. (1991) Fossil evidence for the evolution of biotic pollination. *Phil Trans Roy Soc London* B 333:187–95.

Crespi, B. J. and D. Yanega. (1995) The definition of eusociality. *Behav Ecol* 6:109–15.

Cresswell, J. E. (1990) How and why do nectar-foraging bumblebees initiate movements between inflorescences of wild bergamot *Monarda fistulosa* (Lamiaceae)? *Oecologia* 82:450–60.

———. (2005) Accurate theoretical prediction of pollinator-mediated gene dispersal. *Ecology* 86:574–78.

———. (2006) Models of pollinator-mediated gene dispersal in plants. In *Ecology and Evolution of Flowers*, ed. H. D. Harder and S.C.H. Barrett, 83–101. Oxford: Oxford University Press.

Cresswell, J. E., and C. Galen. (1991) Frequency-dependent selection and adaptive surfaces for floral character combinations—the pollination of *Polemonium viscosum. Am Nat* 138:1342–53.

Cresswell, J. E., C. Hagen, and J. M. Woolnough. (2001) Attributes of flowers of *Brassica napus* L. are affected by defoliation but not by intraspecific competition. *Ann Bot* 88:111–17.

Cresswell, J. E., K. Henning, C. Pennel, M. Lahoubi, M. A. Patrick, P. G. Young, and G. R. Tabor. (2007) Conifer ovulate cones accumulate pollen principally by simple impaction. *Proc Natl Acad Sci USA* 104:18141–44.

Cresswell, J. E. and M. Hoyle. (2006) A mathematical method for estimating patterns of flower-to-flower gene dispersal from a simple field experiment. *Funct Ecol* 20:245–51.

Cripps, C., and R. W. Rust. (1989a) Pollen preferences of seven *Osmia* species (Hymenoptera: Megachilidae). *Environ Entomol* 18:133–38.

———. (1989b) Pollen foraging in a community of *Osmia* bees (Hymenoptera: Megachilidae). *Environ Entomol* 18:582–89.

Croat, T. B. (1980) Flowering behavior of the neotropical genus *Anthurium* (Araceae). *Am J Bot* 67:888–904.

Crone, E. E. (2001) Is survivorship a better fitness surrogate than fecundity? *Evolution* 55:2611–14.

Crone, E. E. and P. Lesica. (2004) Causes of synchronous flowering in *Astragulus scaphoides*, an iteroparous perennial plant. *Ecology* 85:1944–54.

Cronk, Q. and I. Ojeda. (2008) Bird-pollinated flowers in an evolutionary and molecular context. *J Exp Bot* 59:715–27.

Crosswhite, F. S. and C. D. Crosswhite. (1981) Hummingbirds as pollinators of flowers in the red-yellow segment of the color spectrum, with special reference to *Penstemon* and the "open habitat." *Desert Plants* 3:156–70.

Croxton, P.O.J., C. Carvell, J. O. Mountford, and T. H. Sparks. (2002) A comparison of green lanes and field margins as bumblebee habitat in an arable landscape. *Biol Cons* 107:365–74.

Cruden, R. W. (1972) Pollinators in high-elevation ecosystems: relative effectiveness of birds and bees. *Science* 176:1439–40.

———. (1973) Reproductive biology of weedy and cultivated *Mirabilis* (Nyctaginaceae). *Am J Bot* 60:802–9.

———. (1977) Pollen-ovule ratios: a conservative indicator of breeding systems in flowering plants. *Evolution* 31:32–46.

———. (1988) Temporal dioecism: systematic breadth, associated traits, and temporal patterns. *Bot Gaz* 149:1–15.

———. (1997) Implications of evolutionary theory to applied pollination ecology. *Acta Hort* 437:27–51.

———. (2000) Pollen grains: why so many? *Plant Syst Evol* 222:143–65.

Cruden, R. W., and S. M. Hermann-Parker. (1977) Temporal dioecism: an alternative to dioecism? *Evolution* 31:863–66.

———. (1979) Butterfly pollination of *Caesalpinia pulcherrima*, with observations on a psychophilous syndrome. *J Ecol* 67:155–68.

Cruden, R. W., S. M. Hermann, and S. Peterson. (1983) Patterns of nectar production and plant-pollinator coevolution. In *The Biology of Nectaries*, ed. B. Bentley and T. Elias, 82–125. New York: Columbia University Press.

Cruden, R. W. and K. G. Jensen. (1979) Viscin threads, pollination efficiency and low pollen-ovule ratios. *Am J Bot* 66:875–79.

Cruden, R. W., S. Kinsman, R. Stockhouse, and Y. Linhart. (1976) Pollination, fecundity and the distribution of moth flowered plants. *Biotropica* 8:204–10.

Cruden, R. W. and D. L. Lyon. (1985) Patterns of biomass allocation to male and female functions in plants with different mating systems. *Oecologia* 66:299–306.

———. (1989) Facultative xenogamy: examination of a mixed mating system. In *The Evolutionary Ecology of Plants*, ed. H. Bock and Y. B. Linhart, 171–207. Boulder, CO: Westview Press.

Cruden, R. W. and S. Miller-Ward. (1981) Pollen-ovule ratio, pollen size, and the ratio of stigmatic area to the pollen-bearing area of the pollinator—an hypothesis. *Evolution* 35:964–74.

Cruz, D. D., B. M. Freitas, L. A. da Silva, E.M.S. da Silva, and I.G.A. Bomfim. (2005) Pollination efficiency of the stingless bee *Melipona subnitida* on greenhouse sweet pepper. *Pesq Agro Brasil* 40:1197–1201.

Cruzan, M. B. (1990) Pollen-pollen and pollen-style interactions during pollen tube growth in *Erythronium grandiflorum* (Liliaceae). *Am J Bot* 77:116–22.

Cruzan, M. B., P. R. Neal, and M. F. Willson. (1988) Floral display in *Phyla incisa*: consequences for male and female reproductive success. *Evolution* 42:505–15.

Cubas, P., V. Coral, and E. Coen. (1999) An epigenetic mutation responsible for natural variation in floral symmetry. *Nature* 401:157–61.

Culley, T. M., and M. R. Klooster. (2007) The cleistogamous breeding system: a review of its frequency, evolution and ecology in angiosperms. *Bot Rev* 73:1–30.

Culley, T. M., S. G. Weller, and A. K. Sakai. (2002) The evolution of wind pollination in angiosperms. *Trends Ecol Evol* 17:361–69.

Cunningham, J. P., C. J. Moore, M. P. Zalucki, and S. A. West. (2004) Learning, odour preference and flower foraging in moths. *J Exp Biol* 207:87–94.

Cunningham, S. A. (1995) Ecological constraints on fruit

intitiation by *Calyptrogyne ghiesbreghtiana* (Arecaceae): floral herbivory, pollen availability, and visitation by pollinating bats. *Am J Bot* 82:1527–36.

Cunningham, S. A. (1997) The effect of light environment, leaf area and stored carbohydrates on inflorescence production by a rain forest understory palm. *Oecologia* 111:36–44.

Cuthill, I. C., J. C. Partridge, A.T.D. Bennett, S. C. Church, N. S. Hart, and S. Hunt. (2000) Ultraviolet vision in birds. *Adv Stud Behav* 29:159–214.

da Cruz, D. D., V.H.R. de Abreu, and M. van Sluys. (2007) The effect of hummingbird flower mites on nectar availability of two sympatric *Heliconia* species in a Brazilian Atlantic forest. *Ann Bot* 100:581–88.

Dafni, A. (1984) Mimicry and deception in pollination. *Ann Rev Ecol Syst* 15:259–78.

———. (1992) *Pollination Ecology: A Practical Approach.* Oxford: IRL Press.

———. (1996) Autumnal and winter pollination adaptations under Mediterranean conditions. *Bocconea* 5:171–81.

———. (1998) The threat of *Bombus terrestris* spread. *Bee World* 79:113–14.

Dafni, A. and P. Bernhardt. (1990) Pollination of terrestrial orchids of Southern Australia and the Mediterranean region; systematic, ecological and evolutionary implications. *Evol Biol* 24:193–252.

Dafni, A., P. Bernhardt, A. Shmida, Y. Ivri, S. Greenham, C. O'Toole, and L. Losito. (1990) Red bowl-shaped flowers: convergence for beetle pollination in the Mediterranean region. *Isr J Bot* 39:81–92.

Dafni, A. and R. Dukas. (1986) Insect and wind pollination in *Urginea maritima* (Liliaceae). *Plant Syst Ecol* 154:1–10.

Dafni, S., D. Eisikowitch, and Y. Ivri. (1987) Nectar flow and pollinators' efficiency in two co-occurring species of *Capparis* (Capparidaceae) in Israel. *Plant Syst Evol* 157:181–86.

Dafni, A. and D. Firmage. (2000) Pollen viability and longevity; practical, ecological and evolutionary implications. *Plant Syst Evol* 222:113–32.

Dafni, A., M. Hesse, and E. Pacini, ed. (2000) *Pollen and Pollination.* Vienna and New York: Springer.

Dafni, A., and Y. Ivri. (1981a) Floral mimicry between *Orchis israelitica* Baumann and Dafni (Orchidaceae) and *Bellevalia flexuosa* Boiss. (Liliaceae). *Oecologia* 49:229–32.

———. (1981b) The flower biology of *Cephalanthera longifolia* (Orchidaceae) – pollen limitation and facultative floral mimicry. *Plant Syst Evol* 137:229–40.

Dafni, A., Y. Ivri, and N.B.M. Brantjes. (1981) Pollination of *Serapias vomeraceae* (Orchidaceae) by imitation of holes for sleeping solitary male bees (Hymenoptera). *Acta Bot Neerl* 30:69–73.

Dafni, A., P. G. Kevan, and B. C. Husband. (2005) *Practical Pollination Biology.* Ontario, Canada: Enviroquest Ltd.

Dafni, A., M. Lehrer, and P. G. Kevan. (1997) Spatial flower parameters and insect spatial vision. *Biol Rev* 72:239–82.

Dafni, A. and S. G. Potts. (2004) The role of flower inclination, depth and height in the preferences of a pollinating beetle (Coleoptera, Glaphyridae). *J Insect Behav* 17: 823–34.

Dafni, A., and A. Shmida. (1996) The possible ecological implications of the invasion of *Bombus terrestris* (L) (Apidae) at Mt Carmel, Israel. In *The Conservation of Bees*, ed. W. Matheson, S. L. Buchmann, and C. O'Toole, 184–200. New York: Academic Press.

Dafni, A., and S.R.J. Woodell. (1986) Stigmatic exudates and the pollination of *Dactylorhiza fuchsii* (Druce) Soo. *Flora* 178:343–50.

Dai, Y. and S. E. Law. (1995) Modelling the transient electric field produced by a charged pollen cloud entering a flower. *IEEE/IAS Conference* 2:1395–1402.

Dalsgaard, B., A.M.M. Gonzalez, J. M. Olesen, A. Timmermann, L. H. Andersen, and J. Ollerton. (2008) Pollination networks and functional specialization: a test using Lesser Antillean plant-hummingbird assemblages. *Oikos* 117: 789–93.

Damgaard, C., and G. Kjellsson. (2005) Gene flow of oilseed rape (*Brassica napus*) according to isolation distance and buffer zone. *Agric Ecosyst Environ* 108:291–301.

Danforth, B. N. (1999) Emergence dynamics and bet hedging in a desert bee, *Perdita portalis*. *Proc Roy Soc London* B 266:1985–94.

Dar, S., M. D. Arizmendi, and A. Valiente-Banuet. (2006) Diurnal and nocturnal pollination of *Marginatocereus marginatus* (Pachycereeae: Cactaceae) in Central Mexico. *Ann Bot* 97:423–27.

D'Arcy, W. G., and R. C. Keating, ed. (1996) *The Anther: Form, Function and Phylogeny.* Cambridge: Cambridge University Press.

Darwin, C. (1862) *On the Various Contrivances by which British and Foreign Orchids and Fertilized by Insects.* London: John Murray.

———. (1876) *The Effects of Cross and Self Fertilisation in the Vegetable Kingdom.* London: John Murray.

———. (1877) *The Different Forms of Flowers on Plants of the Same Species.* London: John Murray.

Daumer, K. (1956) Reizmetrische Untersuchungen des Farbensehens der Bienen. *Z Vergl Physiol* 38:413–78.

———. (1958) Blumenfarben: wie sie der Bienen sehen. *Z Vergl Physiol* 41:49–110.

Davies, K. L., and M. P. Turner. (2004) Pseudopollen in *Dendrobium unicum* Seidenf. (Orchidaceae): reward or deception? *Ann Bot* 94:129–32.

Davila, Y. C. and G. M. Wardle. (2007) Bee boys and fly girls: do pollinators prefer male or female umbels in protandrous parsnip, *Trachymene incisa* (Apiaceae)? *Aust Ecol* 32:798–807.

Davis, C. C., P. K. Endress, and D. A. Baum. (2008) The evolution of floral gigantism. *Curr Opin Plant Biol* 11: 49–57.

de Cock, A.W.A.M. (1980) Flowering, pollination and fruiting in *Zostera marina* L. *Aquat Bot* 9:202–220.

Decourtye, A., P. Lecompte, J. Pierre, M. P. Chauzat, and P. Thiebeau. (2007) Advantages for farmers and beekeepers of introducing flowering fallows in field crop zones. *Cah Agric* 16:213–18.

Dedej, S., and K. Delaplane. (2005) Net energetic advantage drives honey bees (*Apis mellifera* L) to nectar larceny in *Vaccinium ashei* Reade. *Behav Ecol Sociobiol* 57:398–403.

De Frey, H. M., L. A. Coetzer, and P. J. Robbertse. (1992) A unique anther mucilage in the pollination biology of *Tylosaema esculentum*. *Sex Pl Repro* 5:293–303.

deGrandi-Hoffman, G., R. A. Hoopingarner, and K. K. Baker. (1984) Identification and distribution of cross-pollinating honey bees (Hymenoptera: Apidae) in apple orchards. *Environ Entomol* 13:757–64.

deGrandi-Hoffman, G., R. A. Hoopingarner, and K. Klomparens. (1986) Influence of honey bee (Hymenoptera: Apidae) in-hive pollen transfer on cross-pollination and fruit set in apple. *Environ Entomol* 15:723–25.

de Jong, T. J. and P.G.L. Klinkhamer. (1991) Early flowering in *Cynoglossum officinale* L; constraint or adaptation? *Funct Ecol* 5:750–56.

de Jong, T. J., P.G.L. Klinkhamer, and M. J. van Staalduinen. (1992) The consequences of pollination biology for selection of mass of extended blooming. *Funct Ecol* 6:606–15.

de Jong, T., N. M. Waser, and P.G.L. Klinkhamer. (1993) Geitonogamy: the neglected side of selfing. *Trends Ecol Evol* 8:321–25.

de La Barrera, E. and P. S. Nobel. (2004) Nectar: properties, floral aspects and speculations on origin. *Trends Plant Sci* 9:65–69.

Delaplane, K. S. and D. F. Mayer. (2000) *Crop Pollination by Bees*. Wallingford, CT: CABI Publications.

del Lama, M. A. and R. C. Peruquetti. (2006) Mortality of bees visiting *Caesalpinia peltophoroides* Benth. (Leguminosae) flowers in the state of Sao Paulo Brazil. *Rev Bras Entomol* 50:547–49.

del Moral, R. and L. A. Standley. (1979) Pollination of angiosperms in contrasting coniferous forests. *Am J Bot* 66:26–35.

de Los Mozas Pascual, M., and L. M. Domingo. (1991) Flower constancy in *Heliotaurus ruficollis* (Fabricius 1781), (Coleoptera, Alleculidae). *Elytron* 5:9–12.

Delph, L. F. (1996) Flower size dimorphism in plants with unisexual flowers. In *Floral Biology: Studies on Floral Evolution in Animal-Pollinated Systems*, ed. D. G. Lloyd and S.C.H. Barrett, 217–40. New York; Chapman and Hall.

Delph, L. F., J. L. Gehring, F. M. Frey, A. M. Arntz, and M. Levri. (2004) Genetic constraints on floral evolution in a sexually dimorphic plant revealed by artificial selection. *Evolution* 58:1936–46.

Delph, L. F., M. H. Johannsson, and A. G. Stephenson. (1997) How environmental factors affect pollen performance: ecological and evolutionary perspectives. *Ecology* 78:1632–39.

Delph, L. F., and C. M. Lively. (1992) Pollinator visitation, floral display, and nectar production of the sexual morphs of a gynodioecious shrub. *Oikos* 63:161–70.

del Rio, C. M. (1990) Sugar preferences of hummingbirds: the influence of subtle chemical differences on food choice. *Condor* 92:1022–30.

del Rio, C. M., H. G. Baker, and I. Baker. (1992) Ecological and evolutionary implications of digestive processes: bird preferences and the sugar constituents of floral nectar and fruit pulp. *Cell Mol Sci* 48:544–51.

del Rio, C. M., and A. Burquez. (1986) Nectar production and temperature-dependent pollination in *Mirabilis jalapa*. *Biotropica* 18:28–31.

de Martinis, D., G. Cotti, S. te Lintel Hekker, F.J.M. Harren, and C. Mariani. (2002) Ethylene response to pollen tube growth in *Nicotiana tabacum* flowers. *Planta* 214:806–12.

de Meillon, B., and W. W. Wirth. (1989) A new pollen feeding *Atrichopogon* midge from Madagascar (Diptera: Ceratopogonidae). *Rev Franc Entomol* 11:85–89.

Deng, X. B., P. Y. Ren, J. Y. Gao, and Q. J. Li. (2004) The striped squirrel (*Tamiops swinhoei hainanus*) as a nectar robber of ginger (*Alpinia kwangsiensis*). *Biotropica* 36: 633–36.

Despres, L., S. Ibanez, A. M. Hemborg, and B. Godelle. (2007) Geographical and within-population variation in the globeflower-globeflower fly interaction: the costs and benefits of rearing pollinator's larvae. *Oecologia* 153:69–79.

Despres, L., and N. Jaeger. (1999) Evolution of oviposition strategies and speciation in the globeflower flies *Chiastocheta* spp (Anthomyidae). *J Evol Biol* 12:822–31.

Dessart, P. (1961) Contribution a l'etude des Ceratopogonidae (Diptera: Les *Forcipomyia* pollinisateurs du cacaoyer). *Bull Agric Congo Belge* 52:525–40.

Detzel, A., and M. Wink. (1993) Attraction, deterrence or intoxication of bees (*Apis mellifera*) by plant allelochemicals. *Chemoecology* 4:8–18.

de Vega, C., M. Arista, P. L. Ortiz, C. M. Herrera, and S. Talavera. (2009) The ant pollination system of *Cytinus hypocistis* (Cytinaceae), a Mediterranean root holoparasite. *Ann Bot* 103:1065–75.

deVisser, J.A.G.M., and S. F. Elena. (2007) The evolution of sex: empirical insights into the roles of epistasis and drift. *Nature Rev Gen* 8:139–49.

Devlin, B., and N. C. Ellstrand. (1990) Male and female fertility variation in wild radish, a hermaphrodite. *Am Nat* 136:87–107.

Devoto, M., D. Medan, and N. H. Montaldo. (2005) Patterns of interactions between plants and pollinators along an environmental gradient. *Oikos* 109:461–72.

de Vries, P. J. (1979) Pollen-feeding rainforest *Parides* and *Battus* butterflies in Costa Rica. *Biotropica* 11:237–38.

Devy, M. S. and P. Davidar. (2003) Pollination systems of trees in Kakachi, a mid-elevation wet evergreen forest in Western Ghats, India. *Am J Bot* 90:650–57.

Dick, C. W. (2002) Genetic rescue of remnant tropical trees by an alien pollinator. *Proc Roy Soc London* B 268: 2391–96.

Dicks, L. V., S. A. Corbet, and R. F. Pywell. (2002) Compartmentalization in plant-insect visitor food webs. *J Anim Ecol* 71:32–43.

Dieringer, G. (1992) Pollinator effectiveness and seed set in populations of *Agalinis strictifolia* (Scrophulariaceae). *Am J Bot* 79:1018–23.

Dieringer, G., R. L. Cabrera, M. Lara, L. Loya, and P. Resyes-Castillo. (1999) Beetle pollination and floral thermogenicity in *Magnolia tamauloipana* (Magnoliaceae). *Int J Plant Sci* 160:64–71.

Diez, M. J., S. Talavera, and P. Garcia-Murillo. (1988) Contributions to the palynology of hydrophytic non-entomophilous angiosperms. I. Studies with LM and SEM. *Candollea* 43:147–58.

Diggle, P. K. (1992) Development and the evolution of plant reproductive characters. In *Ecology and Evolution of Plant Reproduction*, ed. R. Wyatt, 326–55. London: Chapman and Hall.

Dilcher, D. L. (1995) Plant reproductive strategies: using the fossil record to unravel current issues in plant reproduction. In *Experimental and Molecular Approaches to Plant Biosystematics*, ed. P. C. Hoch and A. G. Stephenson, 187–98. St. Louis: Miss Botanical Garden.

Dilley, J. D., P. Wilson, and M. R. Mesler. (2000) The radiation of *Calochortus*: generalist flowers moving through a mosaic of potential pollinators. *Oikos* 89:209–22.

Dinkel, T., and K. Lunau. (2001) How drone flies (*Eristalis tenax* L., Syrphidae, Diptera) use floral guides to locate food sources. *J Insect Physiol* 47:1111–18.

Dobat, K., and T. Peikert-Holle. (1985) *Blüten und Fledermäuse (Chiropterophilie)*. Frankfurt: Kramer-Verlag.

Dobson, H.E.M. (1987) Role of flower and pollen aromas in host-plant recognition by solitary bees. *Oecologia* 72: 618–23.

———. (1988) Survey of pollen and pollenkitt lipids—chemical cues to flower visitors? *Am J Bot* 75:170–82.

———. (2006) Relationship between floral fragrance composition and type of pollinator. In *Biology of Floral Scent*, ed. N. Dudareva and E. Pichersky, 147–98. Boca Raton, FL: CRC Press.

Dobson, H. E., J. Arroyo, G. Bergström, and I. Groth. (1996) Interspecific variation in floral fragrances within the genus *Narcissus* (Amaryllidaceae). *Biochem Syst Ecol* 25:685–706.

Dobson, H.E.M. and G. Bergström. (2000) The ecology and evolution of pollen odors. *Plant Syst Evol* 222:63–87.

Dobson, H.E.M., G. Bergström, and I. Groth. (1990) Differences in fragrance chemistry between flower parts of *Rosa rugosa* Thunb. (Rosaceae). *Isr J Bot* 39: 143–56.

Dobson, H.E.M., E. M. Danielseon, and I. D. Van Wesep. (1999) Pollen odor chemicals as modulators of bumble bee foraging on *Rosa rugosa* Thunb. (Rosaceae). *Plant Spec Biol* 14:153–66.

Dobson, H.E.M., I. Groth, and G. Bergström. (1996) Pollen advertisement: chemical contrasts between whole-flower and pollen odors. *Am J Bot* 83:877–85.

Dobson, H.E.M., and Y. S. Peng. (1997) Digestion of pollen components by larvae of the flower-specialist bee *Chelostoma florisomne* (Hymenoptera: Megachilidae). *J Insect Physiol* 43:89–100.

Dobson, H.E.M., R. A. Raguso, J. T. Knudsen, and M. Ayasse. (2005) Scent as an attractant. In *Practical Pollination Biology*, ed. A. Dafni, P. G. Kevan, and B. C. Husband, 197–230. Ontario, Canada: Enviroquest Ltd.

Dodd, M. E., J. Silvertown, and M. W. Chase. (1999) Phylogenetic analysis of trait evolution and species diversity radiation among angiosperm families. *Evolution* 53: 732–44.

Dodson, C. H. (1962) The importance of pollination in the evolution of orchids of Tropical America. *Bull Am Orchid Soc* 31:525–34, 641–49, 731–35.

———. (1965) Studies in orchid pollination: the genus *Coryanthes*. *Am. Orchid Soc. Bull.* 34:680–87.

Dodson, C. H., R. L. Dressler, G. H. Hills, R. M. Adams, and N. H. Williams. (1969) Biologically active compounds in orchid fragrances. *Science* 164:1243–49.

Dodson, C. H., and G. P. Frymire. (1961) Natural pollination of orchids. *Miss Bot Gard Bull* 49:135–52.

Dohzono, I., K. Suzuki, and J. Murata. (2004) Temporal changes in calyx tube length of *Clematis stans* (Ranunculaceae): a strategy for pollination by two bumble bee species with different proboscis lengths. *Am J Bot* 91: 2051–59.

Dominguez, C. A., and R. Dirzo. (1995) Rainfall and flowering synchrony in a tropical shrub: variable selection on the flowering time of *Erythroxylum havanense*. *Evol Ecol* 9:204–16.

Dornhaus, A., A. Brockman, and L. Chittka. (2003) Bumble bees alert to food with pheromone from tergal gland. *J Comp Physiol* A 189:47–51.

Dornhaus, A., and L. Chittka. (1999) Evolutionary origin of bee dances. *Nature* 401:38.

———. (2001) Food alert in bumblebees (*Bombus terrestris*): possible mechanisms and evolutionary implications. *Behav Ecol Sociobiol.* 50:570–76.

———. (2004) Why do honeybees dance? *Behav Ecol Sociobiol* 55:395–401.

Dotterl, S., U. Füssel, A. Jürgens, and G. Aas. (2005) 1,4-dimethoxybenzene, a floral scent compound in willows that attracts an oligolectic bee. *J Chem Ecol* 31:2993–98.

Dotterl, S., A. Jürgens, K. Seifert, T. Laube, B. Weissbecker, and S. Schutz. (2006) Nursery pollination by a moth in *Silene latifolia*: the role of odours in eliciting antennal and behavioural responses. *New Phytol* 169:707–18.

Dotterl, S. and I. Schaeffler. (2007) Flower scent of floral oil-producing *Lysimachia punctata* as attractant for the oil-bee *Macropis fulvipes*. *J Chem Ecol* 33:441–45.

Douglas, D. A. (1981) The balance between vegetative and sexual reproduction of *Mimulus primuloides* (Scrophulariaceae) at different altitudes in California. *J Ecol* 69:295–310.

Dover, J. W. (1989) The use of flowers by butterflies foraging in cereal field margins. *Entomol Gaz* 40:283–91.

Dowsett-Lemaire, F. (1989) Food plants and the annual cycle in a mountain community of sunbirds (*Nectarinia* spp) in northern Malawi. *Tauraco* 1:167–85.

Doyle, J. A.,and M. J. Donohue. (1987) The origin of angiosperms: a cladistic approach. In *Origins of Angiosperms and their Biological Consequences*, ed. E. M. Friis, W. G. Chaloner, and P. R. Crane, 17–49. Cambridge: Cambridge University Press.

Dressler, R. L. (1968) Pollination in euglossine bees. *Evolution* 22:202–10.

———. (1971) Dark pollinia in hummingbird-pollinated orchids, or do hummingbirds suffer from strabismus? *Am Nat* 105:80–83.

———. (1982) Biology of the orchid bees (Euglossini). *Ann Rev Ecol Syst* 13:373–94.

———. (1993) *Phylogeny and Classification of the Orchid Family*. Cambridge: Cambridge University Press.

Dronamraju, K. R. (1960) Selective visits of butterflies to flowers: a possible factor in sympatric speciation. *Nature* 186:178.

Drummond, D. C., and P. M. Hammond. (1991) Insects visiting *Arum dioscoridis* Sm. and *A. orientale* M. Bieb. *Entomol Month Mag* 127:151–56.

Dryer, L. and A. Berghard. (1999) Odorant receptors: a plethora of G-protein-coupled receptors. *Trends Pharm Sci* 20:413–17.

Dudareva, N. and E. Pichersky. (2006) Floral scent metabolic pathways: their regulation and evolution. In *Biology of Floral Scent*, ed. N. Dudareva and E. Pichersky, 55–78. Boca Raton, FL: CRC Press.

Dudash, M. R. (1991) Plant size effects on females and male function in hermaphroditic *Sabatia angularis* (Gentianaceae). *Evolution* 72:1004–12.

Dudley, L. S., S. J. Mazer, and P. Galusky. (2007) The joint evolution of mating system, floral traits and life history in *Clarkia* (Onagraceae): genetic constraint vs. independent evolution. *J Evol Biol* 20:2200–18.

Dufay, M., and M. C. Anstett. (2003) Conflicts between plants and pollinators that reproduce within inflorescences: evolutionary variations on a theme. *Oikos* 100:3–14.

———. (2004) Cheating is not always punished: killer female plants and pollination by deceit in the dwarf palm *Chamaerops humilis*. *J Evol Biol* 17:862–68.

Dukas, R. (1987) Foraging behaviour of three bee species in a natural mimicry system: female flowers which mimic male flowers in *Ecballium elaterium* (Cucurbitaceae). *Oecologia* 74:256–63.

———. (1998) Constraints on information processing and their effects on behaviour. In *Cognitive Ecology*, ed. R. Dukas, 89–128. Chicago: University of Chicago Press.

———. (2001a) Effects of predation risk on pollinators and plants. In *Cognitive Ecology of Pollination*, ed. L. Chittka and J. D. Thomson, 214–36. Cambridge: Cambridge University Press.

———. (2001b) Effects of perceived danger on flower choice by bees. *Ecol Lett* 4:327–33.

———. (2005) Bumble bee predators reduce pollinator density and plant fitness. *Ecology* 86:1401–06.

Dukas, R. and A. C. Kamil. (2001) Limited attention; the constraints underlying search image. *Behav Ecol* 12:192–99.

Dukas, R. and D. H. Morse. (2005) Crab spiders show mixed effects on flower-visiting bees and no effect on plant fitness. *Ecoscience* 12:244–47.

Dukas, R. and L. Real. (1991) Learning foraging tasks by bees: a comparison between social and solitary species. *Anim Behav* 42:269–76.

———. (1993) Learning constraints and floral choice behaviour in bumblebees. *Anim Behav* 46:637–44.

Dukas, R. and P. K. Visscher. (1994) Lifetime learning by foraging honeybees. *Anim Behav* 48:1007–12.

Dunn, D. B. (1956) The breeding system of *Lupinus*, Group Micranthii. *Am Midl Nat* 55:443–72.

Dunn, T. and M. H. Richards. (2003) When to bee social: interactions among environmental constraints, incentive, guarding, and relatedness in a facultatively social carpenter bee. *Behav Ecol* 14:417–23.

Dunne, J. A., R. J. Williams, and N. D. Martinez. (2002) Network structure and biodiversity loss in food webs: robustness increases with connectance. *Ecol Lett* 5:558–67.

Dupont, Y. L., D. M. Hansen, J. T. Rasmussen, and J. M. Olesen. (2004) Evolutionary changes in nectar sugar composition associated with switches between bird and insect pollination; the Canarian bird-flower element revisited. *Funct Ecol* 18:670–76.

Dupont, Y. L., D. M. Hansen, A. Valido, and J. M. Olesen. (2004) Impact of introduced honey bees on native pollination interactions of the endemic *Echium wildpretti* (Boraginaceae) on Tenerife, Canary Islands. *Biol Cons* 118:301–11.

Dupont, Y. L., B. Padron, J. M. Olesen, and T. Petanidou. (2009) Spatio-temporal variation in the structure of pollination networks. *Oikos* 118:1261–69.

du Toit, J. T. (1990) Giraffe feeding on *Acacia* flowers: predation or pollination? *Afr J Ecol* 28:63–68.

Dyer, A. G. and L. Chittka. (2004) Bumblebees (*Bombus terrestris*) sacrifice foraging speed to solve difficult color discrimination tasks. *J Comp Physiol* A 190:759–63.

Dyer, A. G., J. Spaethe, and S. Prack. (2008) Comparative psychophysics of bumblebee and honeybee colour discrimination and object detection. *J Comp Physiol* A 194: 617–27.

Dyer, A. G., H. M. Whitney, S. E. Arnold, B. J. Glover, and L. Chittka. (2006) Bees associate warmth with floral colour. *Nature* 442:525.

Eardley, C., D. Roth, J. Clarke, S. Buchmann, and B. Gemmill. (2006) *Pollinators and Pollination*. African Pollinator Initiative.

Eberhard, S. H., N. Hrassnigg, K. Crailsheim, and H. W. Krenn. (2007) Evidence of a protease in the saliva of the butterfly *Heliconius melpomene* (L.) (Nymphalidae, Lepidoptera). *J Insect Physiol* 53:126–31.

Eckert, C. G., K. E. Samis, and S. Dart. (2006) Reproductive assurance and the evolution of uniparental reproduction in flowering plants. In *Ecology and Evolution of Flowers*, ed. L. D. Harder and S.C.H. Barrett, 183–203. Oxford: Oxford University Press.

Eckhart, V. M. (1991) The effects of floral display on pollinator visitation vary among populations of *Phacelia linearis* (Hydrophyllaceae). *Evol Ecol* 5:370–84.

———. (1992) Spatio-temporal variation in abundance and variation in foraging behavior of the pollinators of gynodioecious *Phacelia linearis* (Hydrophyllaceae). *Oikos* 64:573–86.

Eckhart, V. M., N. S. Rushing, G. M. Hart, and J. D. Hansen. (2006) Frequency-dependent pollinator foraging in polymorphic *Clarkia xantiana* ssp *xantiana* populations: implications for flower colour evolution and pollinator interactions. *Oikos* 112:412–21.

Edwards, J., and J. S. Jordan. (1992) Reversible anther opening in *Lilium philadelphicum* (Liliaceae): a possible means of enhancing male fitness. *Am J Bot* 79: 144–48.

Effmert, U., D. Buss, D. Rohrbeck, and B. Piechulla. (2006) Localization of the synthesis and emission of scent compounds within the flower. In *Biology of Floral Scent*, ed. N. Dudareva and E. Pichersky, 105–24. Boca Raton, FL: CRC Press.

Ehleringer, J. R., and I. N. Forseth. (1980) Solar tracking by plants. *Science* 210:1094–98.

Ehlers, B. K., and J. M. Olesen. (1997) The fruit-wasp route to toxic nectar in *Epipactis* orchids? *Flora* 192:223–29.

Ehlers, B. K., and J. D. Thomson. (2004) Temporal variation in sex allocation in hermaphrodites of gynodioecious *Thymus vulgaris* L. *J Ecol* 92:15–23.

Ehrlén, J. (1997) Risk of grazing and flower number in a perennial plant. *Oikos* 80:428–34.

———. (2002) Assessing the lifetime consequences of plant-animal interactions for the perennial herb *Lathyrus vernus* (Fabaceae). *Persp. Plant Ecol Evol Syst.* 5: 145–63.

Eifler, D. A. (1995) Patterns of plant visitation by nectar-feeding lizards. *Oecologia* 101:228–33.

Eijnde, J. van den, A. de Ruijter, and J. van der Steen. (1991) Method for rearing *Bombus terrestris* continuously and the production of bumblebee colonies for pollination purposes. *Acta Hort* 288:154–55.

Eisikowitch, D. and R. Rotem. (1987) Flower orientation and color change in *Quisqualis indica* and their possible role in pollinator partitioning. *Bot Gaz* 148:175–79.

Elberling, H. and J. M. Olesen. (1999) The structure of a high latitude plant-flower visitor system: the dominance of flies. *Ecography* 22:314–23.

Elle, E., and J. D. Hare. (2002) Environmentally induced variation in floral traits affects the mating system in *Datura wrightii*. *Funct Ecol* 16:79–88.

Elle, E., and T. R. Meagher. (2000) Sex allocation and reproductive success in the andromonoecious perennial *Solanum carolinense* (Solanaceae. II. Paternity and functional gender. *Am Nat* 156:622–36.

Ellington, C. P., K. E. Machin, and T. M. Casey. (1990) Oxygen consumption of bumblebees in forward flight. *Nature* 347:472–73.

Ellis, W. N., and A. C. Ellis-Adam. (1993) The entomophilous flora of NW Europe and its insects. *Bijd tot de Dierkinde* 63:193–220.

Ellstrand, N. C. (1992) Gene flow of pollen: implications for plant conservation genetics. *Oikos* 63:77–86.

Ellstrand, N. C., and D. L. Marshall. (1985) Interpopulation gene flow by pollen in wild radish, *Raphanus sativus*. *Am Nat* 126:606–16.

Ellstrand, N. C., R. Whitkus, and L. H. Reiseburg. (1996) Distribution of spontaneous plant hybrids. *Proc Natl Acad Sci USA* 93:5090–93.

Elmqvist, T., P. A. Cox, W. E. Rainey, and E. D. Pierson. (1992) Restricted pollination on Oceanic islands: pollination of *Ceiba pentandra* by flying foxes in Samoa. *Biotropica* 24:15–23.

El Shafie H.A.F., J.B.B. Mogga, and T. Basedow. (2002) Studies on the possible competition for pollen between the honey bee *Apis mellifera sudanensis* and the imported honey bee *Apis florea* (Hym., Apidae) in North Khartoum (Sudan). *J Appl Entomol* 126:557–62.

Eltz, T., D. W. Roubik, and K. Lunau. (2005) Experience-dependent choices ensure species-specific fragrance collection in male orchid bees. *Behav Ecol Sociobiol* 59: 149–56.

Eltz, T., A. Sager, and K. Lunau. (2005) Juggling with volatiles: exposure of perfumes by displaying male orchid bees. *J Comp Physiol* A 191:575–81.

Eltz, T., W. M. Whitten, D. W. Roubik, and K. E. Linsenmair. (1999) Fragrance collection, storage and accumulation by individual male orchid bees. *J Chem Ecol* 25:157–76.

Eltz, T., Y. Zimmermann, J. Haftmann, R. Twele, W. Francke, J.J.G. Quezada-Euan, and K. Lunau. (2007) Enfleurage, lipid recycling and the origin of perfume collection in orchid bees. *Proc Roy Soc London* B 274:2843–48.

Elvers, I. (1977) Flower-visiting lizards on Madeira. *Bot Notiser* 130:231–34.

Elzinga, J. A., A. Atlan, A. Biere, L. Gigord, A. E. Weis, and G. Bernasconi. (2007) Time after time: flowering phenology and biotic interactions. *Trends Ecol Evol* 22:432–39.

Endler, J. A. (1981) An overview of the relationships between mimicry and crypsis. *Biol J Linn Soc* 16:25–31.

Endress, P. K. (1987) The early evolution of the angiosperm flower. *Trends Ecol Evol* 2:300–4.

———. (1994) *Diversity and Evolutionary Biology of Tropical Flowers*. Cambridge: Cambridge University Press.

———. (1999) Symmetry in flowers: diversity and evolution. *Int J Plant Sci* 160:S3–S23.

Engel, E. C., and R. E. Irwin. (2003) Linking pollinator visitation rate and pollen receipt. *Am J Bot* 90:1612–18.

Englund, R. (1993) Movement patterns of *Cetonia* beetles (Scarabaeidae) among flowering *Viburnum opulus* (Caprifoliaceae). *Oecologia* 94:295–302.

Ennos, R. A. (2008) Spurred on by pollinators. *Heredity* 100:3–4.

Erasmus, J. C., S. van Noort, E. Jousselin, and J. M. Greeff. (2007). Molecular phylogeny of fig wasp pollinators (Agaonidae, Hymenoptera) of *Ficus* section *Galoglychia Zool Scripta* 36:61–78.

Erhardt, A. (1993) Pollination of the edelweiss, *Leontopodium alpinum*. *Bot J Linn Soc* 111:229–40.

Erhardt, A., and H.-P. Rusterholz. (1998) Do peacock butterflies (*Inachis io* L.) detect and prefer nectar amino acids and other nitrogenous compounds? *Oecologia* 117:536–42.

Erickson, E. H. (1975) Surface electric potential on worker honey bees leaving and entering the hive. *J Apic Res* 14:141–47.

———. (1983) Pollination of entomophilous hybrid seed parents. In *Handbook of Experimental Pollination Biology*, ed. C. E. Jones and R. J. Little, 493–535. New York: Van Nostrand Reinhold.

Erickson, E. H., and S. L. Buchmann. (1983) Electrostatics in pollination. In *Handbook of Experimental Pollination Biology*, ed. C. E. Jones and R. J. Little, 173–84. New York: Van Nostrand Reinhold.

Eriksson, R. (1994) The remarkable weevil pollination of the neotropical *Carludovicoideae* (Cyclanthaceae). *Plant Syst Evol* 189:75–81.

Ervik, F., and J. T. Knudsen. (2003) Water lilies and scarabs: faithful partners for 100 million years? *Biol J Linn Soc* 80:539–43.

Estes, J., B. B. Amos, and J. R. Sullivan. (1983) Pollination from two perspectives: the agricultural and biological sciences. In *Handbook of Experimental Pollination Biology*, ed. C. E. Jones and R. J. Little, 536–54. New York: Van Nostrand Reinhold.

Etterson, J. R., and L. F. Galloway. (2002) The influence of light on paternal plants in *Campanula americana* (Campanulaceae): pollen characteristics and offspring traits. *Am J Bot* 89:1899–1906.

Euler, M., and I. T. Baldwin. (1996) The chemistry of defense and apparency in the corollas of *Nicotiana attenuata*. *Oecologia* 107:102–12.

Evans, M. R. (1996) Nectar and flower production of *Lobelia telekii* inflorescences, and their influence on territorial behaviour of the scarlet tufted malachite sunbird (*Nectarinia johnstoni*). *Biol J Linn Soc* 57:89–105.

Evenson, W. E. (1983) Experimental studies of reproductive energy allocation in plants. In *Handbook of Experimental Pollination Biology*, ed. C. E. Jones and R. J. Little, 249–74. New York: Van Nostrand Reinhold.

Evertz, S. (1995) Interspezifische Konkurrenz zwischen Honigbienen (*Apis mellifera*) und solitären Wildbienen (Hymenoptera Apoidae). *Natur Landschaft* 70:165–72.

Evoy, W. H., and B. P. Jones. (1971) Motor patterns of male euglossine bees evoked by floral fragrances. *Anim Behav* 19:583–88.

Faegri, K., and L. van der Pijl. (1979) *The Principles of Pollination Ecology*, 3rd ed. Oxford: Pergamon.

Fahn, A. (1979) Ultrastructure of nectaries in relation to secretion. *Am J Bot* 66:977–85.

Fahn, A., and P. Benouaiche. (1979) Ultrastructure, development and secretion in the nectary of banana flowers. *Ann Bot* 44:84–93.

Farwig, N., E. F. Randrianirina, F. A. Voigt, M. Kraemer, and K. Bohning-Gaese. (2004) Pollination ecology of the dioecious tree *Commiphora guillauminii* in Madagascar. *J Trop Ecol* 20:307–16.

Farzad, M., R. Griesbach, J. Hammond, M. R. Weiss, and H. G. Elmendorf. (2003) Differential expression of three key anthocyanin biosynthetic genes in a colour-changing flower, *Viola cornuta* cv Yesterday, Today and Tomorrow. *Plant Sci.* 165:1333–42.

Faulkner, C. J. (1962) Blow flies as pollinators of *Brassica* crops. *Commercial Grower* 3457:807–9.

Faulkner, G. J. (1976) Honeybee behaviour as affected by plant height and flower colour in Brussels sprouts. *J Apic Res* 15:15–18.

Feber, R. E., and H. Smith. (1995) Butterfly conservation on arable farmlandIn *Ecology and Conservation of Butterflies*, ed. A. S. Pullins, 84–97. London: Chapman and Hall.

Feeny, P. (1976) Plant apparency and chemical defense. *Rec Adv Phytochem* 10:1–40.

Feinsinger, P. (1978) Ecological interactions between plants and hummingbirds in a successional tropical community. *Ecol Monogr* 48:269–87.

———. (1983) Coevolution and pollination. In *Coevolution*, ed. D. J. Futuyma and M. Slatkin, 282–310. Sunderland, MA: Sinauer.

Feinsinger, P., and W. H. Busby. (1987) Pollen carryover: experimental comparisons between morphs of *Palicourea lasiorrachis* (Rubiaceae), a bistylous bird-pollinated tropical treelet. *Oecologia* 73:231–35.

Feinsinger, P., and R. K. Colwell. (1978) Community organization among neotropical nectar-feeding birds. *Am Zool* 18:779–95.

Feinsinger, P., and L. B. Swarm. (1978) How common are ant-repellent nectars? *Biotropica* 10:238–39.

Feinsinger, P., L. A. Swarm, and J. A. Wolfe. (1985) Nectar-feeding birds on Trinidad and Tobago: comparison of diverse and depauperate guilds. *Ecol Monogr* 55:1–28.

Feldman, T. S. (2006) Pollinator aggregative and functional responses to flower density: does pollinator response to patches of plants accelerate at low densities? *Oikos* 115:128–40.

Feldman, T. S., W. F. Morris, and W. G. Wilson, (2004) When can two plant species facilitate each other's pollination? *Oikos* 105:197–207.

Fenster, C. B. (1991a) Gene flow in *Chamaecrista fasciculata* (Leguminosae). 1. Gene dispersal. *Evolution* 45:398–409.

———. (1991b) Gene flow in *Chamaecrista fasciculata* (Leguminosae). 2. Gene establishment. *Evolution* 45:410–22.

———. (1991c) Selection on floral morphology by hummingbirds. *Biotropica* 23:98–101.

Fenster, C. B., W. S. Armbruster, and M. R. Dudash. (2009) Specialization of flowers: is floral orientation an overlooked first step? *New Phytol* 183:502–6.

Fenster, C. B., W. S. Armbruster, P. Wilson, M. R. Dudash, and J. D. Thomson. (2004) Pollination syndromes and floral specialization. *Ann Rev Ecol Syst* 35:375–403.

Fenster, C. B., G. Cheely, M. R. Dudash, and R. T. Reynolds. (2006) Nectar reward and advertisement in hummingbird-pollinated *Silene virginica* (Caryophyllaceae). *Am J Bot* 93:1800–7.

Fenster, C. B., P. K. Diggle, S.C.H. Barrett, and K. Ritland. (1995) The genetics of floral development differentiating 2 species of *Mimulus* (Scrophulariaceae). *Heredity* 74:258–66.

Fenster, C. B., and M. R. Dudash. (2001) Spatiotemporal variation in the role of hummingbirds as pollinators of *Silene virginica*. *Ecology* 82:844–51.

Fenster, C. B., and L. F. Galloway. (2000) Population differentiation in an annual legume: genetic architecture. *Evolution* 54:1157–72.

Fenster, C. B., C. L. Hassler, and M. T. Dudash. (2006) Fluorescent dye particles are good pollen analogs for hummingbird-pollinated *Silene virginica*. *Can J Bot* 74:189–93.

Fenster, C. B., and S. Martén-Rodriguez. (2007) Reproductive assurance and the evolution of pollination specialization. *Int J Plant Sci* 168:215–28.

Ferguson, I. K. (1985) The role of pollen morphology in plant systematics. *An Assos Palinol Leng Esp* 2:5–18.

———. (1990) Significance of some pollen morphological characters of the tribe Amorpheae and the genus *Mucuna* (tribe Phaseoleae) in the biology and systematics of the subfamily Papilionoideae (Leguminosae). *Rev Palaeobot Palynol* 64:129–36.

Ferguson, I. K., and K. J. Pearce. (1986) Observations on the pollen morphology of the genus *Bauhinia* L. (Leguminosae, Caesalpiniaceae), in the neotropics. In *Pollen and Spores: Form and Function*, ed. S. Blackmore and I. K. Ferguson, 283–96. London: Academic Press.

Ferguson, I. K., and J. J. Skvarla. (1982) Pollen morphology in relation to pollinators in Papilionoideae (Leguminosae). *Bot J Linn Soc* 84:183–93.

Fetscher, A. E., and J. R. Kohn. (1999) Stigma behavior in *Mimulus aurantiacus* (Scrophulariaceae). *Am J Bot* 86:1130–35.

Feuerbacher, E., J. H. Fewell, S. P. Roberts, E. F. Smith, and J. F. Harrison. (2003) Effects of load type (pollen or nectar) and load mass on hovering metabolic rate and mechanical power output in the honey bee *Apis mellifera*. *J Exp Biol* 206:1855–65.

Fineblum, W. L., and M. D. Rausher. (1997) Do floral pigmentation genes also influence resistance to enemies? The *W* locus in *Ipomoea purpurea*. *Ecology* 78:1646–54.

Firmage, D. H., and F. R. Cole. (1988) Reproductive success and inflorescence sixe of *Calopogon tuberosus* (Orchidaceae). *Am J Bot* 75:1371–77.

Fishbein, M., and D. L. Venable. (1996) Diversity and temporal change in the effective pollinators of *Asclepias tuberosa*. *Ecology* 77:1061–73.

Fisher, R. A. (1941) Average excess and average effect of a gene substitution. *Ann. Eugen.* 11:53–63.

Fishman, L., and J. H. Willis. (2008) Pollen limitation and natural selection on floral characters in the yellow monkey flower *Mimulus guttatus*. *New Phytol* 177:802–10.

Fitter, A. H., and R.S.R Fitter. (2002) Rapid changes in flowering times in British plants. *Science* 296:1689–91.

Fleet, R. R., and B. L. Young. (2000) Facultative mutualism between imported fire ants (*Solenopsis invicta*) and a legume (*Senna occidentalis*). *Southwest Nat* 453:289–98.

Fleming, P. A., S. D. Hofmeyr, S. W. Nicolson, and J. T. du Toit. (2006) Are giraffes pollinators or predators of *Acacia nigrescens* in Kruger National Park, South Africa? *J Trop Ecol* 22:247–53.

Fleming, P. A., and S. W. Nicolson. (2003) Osmoregulation in an avian nectarivore, the whitebellied sunbird *Nectarinia talatala*: response to extremes of diet concentration. *J Exp Biol* 206:1845–54.

Fleming, T. H., and J. N. Holland. (1998) The evolution of obligate pollination mutualisms: the senita cactus and senita moth. *Oecologia* 114:368–75.

Fleming, T. H., C. T. Sahley, J. N. Holland, J. D. Nason, and J. L. Hamrick. (2001) Sonoran Desert columnar cacti and the evolution of generalized pollination systems. *Ecol Monogr* 71:511–30.

Fleming, T. H., M. D. Tuttle, and M. A. Horner. (1996) Pollination biology and the relative importance of nocturnal and diurnal pollinators in three species of Sonoran Desert columnar cacti. *Southwest Nat* 41:257–69.

Fontaine, C., C. L. Collin, and I. Dajoz. (2008) Generalist foraging of pollinators: diet expansion at high density. *J Ecol* 96:1002–10.

Fontaine, C., I. Dajoz, J. Meriguet, and M. Loreau. (2006) Functional diversity of plant-pollinator interaction webs enhances the persistence of plant communities. *PLoS Biol* 4:129–35.

Forcone, A., L. Galetto, and G. Bernardello. (1997) Floral nectar chemical composition of some species from Patagonia. *Biochem Syst Ecol* 25:395–402.

Ford, D. M., H. R. Hepburn, F. B. Moseley, and R. J. Rigby. (1981) Displacement sensors in the honeybee pollen basket. *J Insect Physiol* 27:339–46.

Ford, M. A., and Q.O.N. Kay. (1985) The genetics of incompatibility in *Sinapis arvensis* L. *Heredity* 54:99–102.

Forrest, J., and J. D. Thomson. (2008) Pollen limitation and cleistogamy in subalpine *Viola praemorsa Botany* 86:511–19.

———. (2009) Background complexity affects colour preferences in bumblebees. *Naturwissenschaften* 96:921–25.

Fortuna, M. A., and J. Bascompte. (2006) Habitat loss and the structure of plant-animal mutualistic networks. *Ecol Lett* 9:278–83.

Forup, M. L., K.S.E. Henson, P. G. Craze, and J. Memmott. (2008) The restoration of ecological interactions: plant-pollinator networks on ancient and restored heathlands. *J Appl Ecol* 45:742–52.

Forup, M. L., and J. Memmott. (2005a) The restoration of plant-pollinator interactions in hay meadows. *Restor Ecol* 13:265–74.

———. (2005b) The relationship between the abundances of bumblebees and honeybees in a native habitat. *Mol Ecol Entomol* 30:47–57.

Fox, G. A. (1990) Drought and the evolution of flowering time in desert annuals. *Am J Bot* 77:1508–18.

Fox, L. R., and P. S. Morrow. (1981) Specialization: species property or local phenomenon? *Science* 211:887–93.

Franchi, G. G., L. Bellani, M. Nepi, and E. Pacini. (1996) Types of carbohydrate reserves in pollen: localization, systematic distribution and ecophysiological significance. *Flora* 191:1–17.

Frankie, G. W. (1975) Tropical forest phenology and pollinator plant coevolution. In *Coevolution of Animals and Plants*, ed. L. E. Gilbert and P. H. Raven, 192–209. Austin: University of Texas Press.

Frankie, G. W., H. G. Baker, and P. A. Opler. (1974) Comparative phenological studies of trees in tropical wet and dry forests in the lowlands of Costa Rica. *J Ecol* 62: 881–913.

Frankie, G. W., and W. A. Haber. (1983) Why bees move among mass-flowering neotropical trees. Page 360–73 in *Handbook of Experimental Pollination Biology*, ed. C. E. Jones and R. J. Little. New York: Van Nostrand Reinhold.

Frankie, G. W., W. A. Haber, P. A. Opler, and K. S. Bawa. (1983) Characteristics and organisation of the large bee pollination system in the Costa Rican dry forest. In *Handbook of Experimental Pollination Biology*, ed. C. E. Jones and R. J. Little, 411–47. New York: Van Nostrand Reinhold.

Frankie, G. W., P. A. Opler, and K. S. Bawa. (1976) Foraging behavior of solitary bees: implications for outcrossing of a neotropical forest tree species. *J Ecol* 64:1049–57.

Frankie, G. W., R. W. Thorp, L. E. Newstrom-Lloyd, M. A. Rizzardi, J. F. Barthell, T. L. Griswold, J. Y. Kim, and S. Kappagoda. (1998) Monitoring solitary bees in modified wildland habitats: implications for bee ecology and conservation. *Environ Entomol* 27:1137–48.

Frankie, G. W., R. W. Thorp, M. Schindler, J. Hernandez, B. Ertter, and M. A. Rizzardi. (2005) Ecological patterns of bees and their host ornamental flowers in two northern Californian cities. *J Kans Entomol Soc* 78:227–46.

Frankie, G. W., and S. B. Vinson. (1977) Scent-marking of passion flowers in Texas by females of *Xylocopa virginica texana* (Hymenoptera, Anthophoridae*). *J Kans Entomol Soc* 50:613–25.

Frankie, G. W., S. B. Vinson, M. A. Rizzardi, T. L. Griswold, S. O'Keefe. and R. Snelling. (1997) Diversity and abundance of bees visiting a mass flowering tree species in a disturbed seasonal dry forest, Costa Rica. *J Kans Entomol Soc* 70:281–96.

Fraser, A. M., W. L. Mechaber, and J. G. Hildebrand. (2003) Electroantennographic and behavioral responses of the sphinx moth *Manduca sexta* to host plant headspace volatiles. *J Chem Ecol* 29:1813–33.

Frazee, J. E., and R. J. Marquis. (1994) Environmental contribution to floral trait variation in *Chamaecrista fasciculata* (Fabaceae: Caesalpinoideae). *Am J Bot* 81:206–15.

Free, J. B. (1963) The flower constancy of honeybees. *J Anim Ecol* 29:385–95.

———. (1968) Dandelion as a competitor to fruit trees for bee visitors. *J Appl Ecol* 5:161–78.

———. (1969) Influence of the odour of a honeybee colony's stores on the behaviour of its foragers. *Nature* 222:778.

———. (1970a) Effect of flower shapes and nectar guides on the behaviour of foraging bees. *Behaviour* 37:269–85.

———. (1970b) The flower constancy of bumble bees. *J Anim Ecol* 39:395–402.

———. (1993) *Insect Pollination of Crops*, 2nd ed. London: Academic Press.

Free, J. B., and I. H. Williams. (1983) Scent-marking of flowers by honeybees. *J Apic Res* 22:86–90.

Freitas, L., and M. Sazima. (2003) Daily blooming patterns and pollination by syrphids in *Sisyrinchium vaginatum* (Iridaceae) in southeastern Brazil. *J Torrey Bot Soc* 130:55–61.

———. (2006) Pollination biology in a tropical high-altitude grassland in Brazil: interactions at the community level. *Ann Miss Bot Gard* 93:465–516.

Frey, F. M. (2004) Opposing natural selection from herbivores and pathogens may maintain floral-color variation in *Claytonia virginica* (Portulaceae). *Evolution* 58:2426–37.

———. (2007) Phenotypic integration and the potential for independent color evolution in a polymorphic spring ephemeral. *Am J Bot* 94:437–44.

Frey-Wissling, A., M. Zimmermann, and A. Mauritzio. (1954) Über den enzymatischen Zuckerumbau in Nektarien. *Experientia* 10:490–91.

Friedman, J., and S.C.H. Barrett. (2008) A phylogenetic analysis of the evolution of wind pollination in the angiosperms. *Int J Plant Sci* 169:49–58.

Friedman, J., and L. D. Harder. (2004) Inflorescence architecture and wind pollination in six grass species. *Funct Ecol* 18:851–60.

Friis, E. M., K. R. Pedersen, and P. R. Crane. (2001) Origin and radiation of angiosperms. In *Palaeobiology II*, ed. D.E.G. Briigs and P. R. Crowther, 97–102. Oxford: Blackwell Science.

Fritz, A. L., and L. A. Nilsson. (1994) How pollinator-mediated mating varies with population size in plants. *Oecologia* 100:451–62.

Frohlich, M. W., and D. S. Parker. (2000) The Mostly Male theory of flower evolutionary origins: from genes to fossils. *Syst Bot* 25:155–70.

Fry, J. D., and M. D. Rauscher. (1997) Selection on a floral color polymorphism in the tall morning glory (*Ipomoea purpurea*): transmission success of the alleles through pollen. *Evolution* 51:66–78.

Fukushi, T. (1989) Learning and discrimination of coloured papers in the walking blowfly *Lucilia cuprina*. *J Comp Physiol* A 166:57–64.

Fulton, M., and S. A. Hodges. (1999) Floral isolation between *Aquilegia formosa* and *Aquilegia pubescens*. *Proc Roy Soc London* B 266:2247–52.

Fumero-Caban, J. J., and E. J. Melendez-Ackerman. (2007) Relative pollination effectiveness of floral visitors of *Pitcairnia angustifolia* (Bromeliaceae). *Am J Bot* 94:419–24.

Furness, C. A. (2007) Why does some pollen lack apertures? A review of inaperturate pollen in eudicots. *Bot J Linn Soc* 155:29–48.

Furness, C. A., and P. J. Rudall. (1999) Inaperturate pollen in Monocotyledons. *Int J Plant Sci* 160:395–414.

———. (2004) Pollen aperture evolution—a crucial factor for eudicot success? *Trends Plant Sci* 9:154–58.

Fussell, M. (1992) Diurnal patterns of bee activity, flowering, and nectar reward per flower in tetraploid clover. *NZ J Agric Res* 35:151–56.

Fussell, M., and S. A. Corbet. (1992) Flower usage by bumble bees: a basis for forage plant management. *J Appl Ecol* 29:451–65.

Futuyma, D. J. (2001) Ecological specialization and generalization. In *Evolutionary Ecology*, ed. C. W. Fox, D. A. Roff, and D. J. Fairbairn, 177–92. Oxford: Oxford University Press.

Gabriel, D., and T. Tscharntke. (2007) Insect-pollinated plants benefit from organic farming. *Agric Ecosyst Environ* 118:43–48.

Galen, C. (1983) The effects of nectar thieving ants on seed set in floral scent morphs of *Polemonium viscosum*. *Oikos* 41:245–49.

———. (1989) Measuring pollinator-mediated selection on morphometric floral traits: bumblebees and the alpine sky pilot, *Polemonium viscosum*. *Evolution* 43:882–90.

———. (1993) Cost of reproduction in *Polemonium viscosum*; phenotypic and genetic approaches. *Evolution* 47:1073–79.

———. (1996) The evolution of floral form: insights from an alpine wildflower, *Polemonium viscosum* (Polemoniaceae). In *Floral Biology: Studies on Floral Evolution in Animal-Pollinated Systems*, ed. D. G. Lloyd and S.C.H. Barrett, 273–93. New York: Chapman and Hall.

———. (1999a) Why do flowers vary? The functional ecology of variation in flower size and form within natural plant populations. *Bioscience* 49:631–40.

———. (1999b) Flowers and enemies: predation by nectar-thieving ants in relation to variation in floral form of an alpine wildflower, *Polemonium viscosum*. *Oikos* 85:426–34.

Galen, C., and B. Butchart. (2003) Ants in your plants: effects of nectar thieves on pollen fertility and seed-siring capacity in the alpine wild flower *Polemonium viscosum*. *Oikos* 101:521–28.

Galen, C., and J. Cuba. (2001) Down the tube: pollinators, predators, and the evolution of flower shape in the alpine sky pilot *Polemonium viscosum*. *Evolution* 55:1963–71.

Galen, C., T. E. Dawson, and M. L. Stanton. (1993) Carpels as leaves: meeting the carbon cost of reproduction in an alpine buttercup. *Oecologia* 95:187–93.

Galen, C., and J. C. Geib. (2007) Density-dependent effects of ants on selection for bumblebee pollination in *Polemonium viscosum*. *Ecology* 88:1202–9.

Galen, C., T. Gregory, and L. F. Galloway. (1989) Cost of self-pollination in a self-incompatible plant, *Polemonium viscosum*. *Am J Bot* 76:1675–80.

Galen, C., and P. G. Kevan. (1980) Scent and color, floral polymorphisms and pollination biology in *Polemonium viscosum* Nutt. *Am Midl Nat* 104:281–89.

Galen, C., and M.E.A. Newport. (1987) Bumble bee behavior and selection on flower size in the sky pilot, *Polemonium viscosum*. *Oecologia* 74:20–23.

———. (1988) Pollination quality, seed set and flower traits in *Polemonium viscosum*: complementary effects of variation in flower scent and size. *Am J Bot* 75:900–5.

———. (1989) Bumblebee pollination and floral morphology: factors influencing pollen dispersal in the alpine sky pilot, *Polemonium viscosum* (Polemoniaceae). *Am J Bot* 76:419–26.

———. (2003) Sunny-side up: flower heliotropism as a source of paternal environmental effects on pollen quality and performance in the snow buttercup *Ranunculus adoneus* (Ranunculaceae). *Am J Bot* 90:724–29.

Galen, C., K. A. Zimmer, and M. E. Newport. (1987) Pollination in floral scent morphs of *Polemonium viscosum*: a mechanism for disruptive selection on flower size. *Evolution* 41:599–606.

Galetto, L., and L. Bernardello. (2003) Nectar sugar composition in angiosperms from Chaco and Patagonia (Argentina); do animal visitors matter? *Plant Syst Evol* 238:69–86.

Galil, J. (1973) Pollination in dioecious figs: pollination of *Ficus fistulosa* by *Ceratosolen hewitti*. *Gard. Bull. Sing.* 26:303–11.

Galil, J., and D. Eisikowitch. (1968) On the pollination ecology of *Ficus sycomorus* in East Africa. *Ecology* 49:259–69.

———. (1969) Further studies on the pollination ecology of *Ficus sycomorus* L. *Tijds Entomol.* 112:1–13.

———. (1974) Further studies on pollination ecology in *Ficus sycomorus* . II. Pocket filling and emptying by *Ceratosolen arabicus* Mayr. *New Phytol* 73:515–28.

Galil, J., W. Ramirez, and D. Eisikowitch. (1973) Pollination of *Ficus costaricana* and *F. hemsleyana* by *Blastophaga esterase* and *B. tonduzi* in Costa Rica (Hymenoptera, Chalcidoidea, Agaonidae). *Tijds Entomol* 116:175–83.

Galizia, C. G., J. Kunze, A. Gumbert, A. K. Borg-Karlson, S. Sachse, C. Markl, and R. Menzel. (2005) Relationship of visual and olfactory signal parameters in a food-deceptive flower mimicry system. *Behav Ecol* 16:159–68.

Gallai, N., J.-M. Salles, J. Settele, and B. E. Vaissière. (2009) Economic valuation of the vulnerability of world agriculture confronted with pollinator decline. *Ecol Econ* 68:810–21.

Galliot, C., M. E. Hoballah, C. Kuhlemeier, and J. Stuurman. (2006) Genetics of flower size and nectar volume in *Petunia* pollination syndromes. *Planta* 225:203–12.

Galliot, C., J. Stuurman, and C. Kuhlemeier. (2006) The genetic dissection of floral pollination syndromes. *Curr Opin Plant Biol* 9:78–82.

Galloni, M., L. Podda, D. Vivarelli, M. Quaranta, and G. Cristofolini. (2008) Visitor diversity and pollinator specialization in Mediterranean legumes. *Flora* 203:94–102.

Gan-Mor, S., Y. Schwartz, A. Bechar, D. Eisikowitch, and G. Manor. (1995) Relevance of electrostatic forces in natural and artificial pollination. *Can Agric Eng* 37:189–94.

Garber, P. A. (1988) Foraging decisions during nectar feeding by tamarin monkeys (*Saguinus mystax* and *Saguinus fuscicollis*, Callitrichidae, Primates) in Amazonian Peru. *Biotropica* 20:100–6.

Garcia, D., and N. P. Chacoff. (2007) Scale-dependent effects of habitat fragmentation on hawthorn pollination, frugivory, and seed predation. *Cons Biol* 21:400–11.

Garcia, M. B., R. J. Antor, and X. Espadaler. (1995) Ant pollination of the palaeo-endemic dioecious *Bordera pyreneica* (Diosporeaceae). *Plant Syst Evol* 198:17–27.

Garcia-Fayos, P., and A. Goldarazena. (2008) The role of thrips in pollination of *Arctostaphylos uva-ursi*. *Int J Plant Sci* 169:776–81.

Gardener, M. C., and M. P. Gillman. (2001) Analyzing variability in nectar amino acids: composition is less variable than concentration. *J Chem Ecol* 27:2545–58.

Gardener, M. C., and M. P. Gillman. (2002) The taste of nectar—a neglected area of pollination ecology. *Oikos* 98:552–57.

Gardener, M. C., R. J. Rowe, and M. P. Gillman. (2003) Tropical bees (*Trigona hockingsi*) show no preference for nectar with amino acids. *Biotropica* 35:119–25.

Gaskett, A. C., E. Conti, and F. P. Schiestl. (2005) Floral odor variation in two heterostylous species of *Primula*. *J Chem Ecol* 31:1223–28.

Gaskett, A. C., C. G. Winnick, and M. E. Herberstein. (2008)

Orchid sexual deceit provokes ejaculation. *Am Nat* 171: E206–12.

Gass, C. L. and W. M. Roberts. (1992) The problem of temporal scale in optimization: three contrasting views of hummingbird visits to flowers. *Am Nat* 140, 829-853

Gaston, K. J., and P. H. Williams. (1996) Spatial patterns in taxonomic diversity. In *Biodiversity, A Biology of Numbers and Difference*, ed. K. J. Gaston, 202–9. Oxford: Blackwell Science.

Gathmann, A., and T. Tscharntke. (2002) Foraging ranges of solitary bees. *J Anim Ecol* 71:757–64.

Gaume, L., M. Zacharias, and R. M. Borges. (2005) Ant-plant conflicts and a novel case of castration parasitism in a myrmecophyte. *Evol Ecol Res* 7:435–52.

Gaur, S., A. Rana, and S.V.S. Chauhan. (2007) Pollen allelopathy: past achievements and future research. *Allelopathy J* 20:115–26.

Geber, M. A., and D. A. Moeller. (2006) Pollinator responses to plant communities and implications for reproductive character evolution. In *Ecology and Evolution of Flowers*, ed. L. D. Harder and S.C.H. Barrett, 102–19.Oxford: Oxford University Press.

Geerts, S., and A. Pauw. (2009) African sunbirds hover to pollinate an invasive hummingbird-pollinated plant. *Oikos* 118:573–79.

Gegear, R. J., and J. G. Burns. (2007) The birds, the bees, and the virtual flowers: can pollinator behaviour drive ecological speciation in flowering plants? *Am Nat* 170: 551–66.

Gegear, R. J., and T. M. Laverty. (2001) The effect of variation among floral traits on the flower constancy of pollinators. In *Cognitive Ecology of Pollination: Animal Behavior and Floral Evolution,* ed. J. D. Thomson and L. Chittka, 1–20. Cambridge: Cambridge University Press.

———. (2005) Flower constancy in bees; a test of the trait variability hypothesis. *Anim Behav* 69:939–49.

Gegear, R. J., J. S. Manson, and J. D. Thomson. (2007) Ecological context influences pollinator deterrence by alkaloids in floral nectar. *Ecol Lett* 10:375–82.

Gegear, R. J., and J. D. Thomson. (2004) Does the flower constancy of bumble bees reflect foraging economics? *Ethology* 110:793–805.

Gels, J. A., D. W. Held, and D. A. Potter. (2002) Hazards of insecticides to the bumble bees *Bombus impatiens* (Hymenoptera, Apidae) foraging on flowering white clover in turf. *J Econ Entomol* 5:722–28.

Gentry, A. H. (1974) Flowering phenology and diversity in tropical Bignoniaceae. *Biotropica* 6:64–68.

———. (1982) Patterns of Neotropical plant species diversity. *Evol Biol* 15:1–85.

Gerling, D., H.H.W. Velthuis, and A. Hefetz. (1989) Bionomics of the large carpenter bees of the genus *Xylocopa*. *Ann Rev Entomol* 34:163–90.

Gess, S. K. (1996) *The Pollen Wasps. Ecology and Natural History of the Masarinae.* Cambridge, MA: Harvard University Press.

Ghazoul, J. (2005) Buzziness as usual? Questioning the global pollination crisis. *Trends Ecol Evol* 20:367–73.

———. (2006) Floral diversity and the facilitation of pollination. *J Ecol* 94:295–304.

Ghim, M. M., and A. Hodos. (2006) Spatial contrast sensitivity of birds. *J Comp Physiol A* 192:523–34.

Gibbs, P. E. (1977) Floral biology of *Talauma ovata* St. Hil. (Magnoliaceae). *Cienca Cultura* 29:1437–41.

Gibbs, P. E., and M. Bianchi. (1993) Post pollination events in species of *Chorisia* (Bombacaceae) and *Tabebuia* (Bignoniaceae) with late-acting self-incompatibility. *Bot Acta* 106:64–71.

Gibernau, M., D. Barabe, P. Cerdan, and A. Dejean. (1999) Beetle pollination of *Philodendron solimoesense* (Araceae) in French Guiana. *Int J Plant Sci* 160:1135–43.

Gibernau, M., M. Hossaert-McKey, M. C. Anstett, and F. Kjellberg. (1996) Consequences of protecting a flower in a fig; a one way trip for pollinators? *J. Biogeog.* 23: 425–32.

Gibson, R. H., I. L. Nelson, G. W. Hopkins, B. J. Hamlett, and J. Memmott. (2006) Pollinator webs, plant communities and the conservation of rare plants: arable weeds as a case study. *J Appl Ecol* 43:246–57.

*Proc Nat Acad Sci USA*Gilbert, F. S. (1981) Foraging ecology of hoverflies: morphology of the mouthparts in relation to feeding on nectar and pollen in some common urban species. *Mol Ecol Entomol* 6:245–62.

———. (1983) The foraging ecology of hoverflies (Diptera, Syrphidae): circular movements on composite flowers. *Behav Ecol Sociobiol* 13:253–57.

———. (1985) Diurnal activity patterns in hoverflies (Diptera, Syrphidae). *Ecol Entomol.* 10:385–92.

Gilbert, F. S., et al. (2001) Individually recognizable scent marks on flowers made by a solitary bee. *Anim Behav* 61:217–29.

Gilbert, F. S., N. Haines, and K. Dickson. (1991) Empty flowers. *Funct Ecol* 5:29–39.

Gilbert, F. S., and M. Jervis. (1998) Functional, evolutionary and ecological aspects of feeding-related mouthpart specializations in parasitoid flies. *Biol J Linn Soc* 63:495–535.

Gilbert, F. S., and J. Owen. (1990) Size, shape, competition, and community structure in hoverflies (Diptera, Syrphidae). *J Anim Ecol* 59:21–39.

Gilbert, L. E. (1972) Pollen feeding and reproductive biology of *Heliconius* butterflies. *Proc Natl Acad Sci USA* 69:1403–7.

———. (1975) Ecological consequences of coevolved mutualisms between butterflies and plants. In *Coevolution of*

Animals and Plants, ed. L. E. Gilbert and P. R. Raven. Austin: University of Texas Press.

Gilissen, L.J.W. (1977) Style-controlled wilting of flower. *Planta* 133:275–80.

Gil-Vega, K., M. Gonzalez Chavira, O. Martinez de la Vega, J. Simpson, and G. Vandermark. (2001) Analysis of genetic diversity in *Agave tequilana* var *azul* using RAPD markers. *Euphytica* 119:335–41.

Gimenes, M., A. A. Benedito-Silva, and M. D. Marques. (1996) Circadian rhythm of pollen and nectar collection by bees on the flowers of *Ludwigia elegans* (Onagraceae). *Biol Rhythms Res* 27:281–90.

Gimenez-Benavides, L., S. Dotterl, A. Jürgens, A. Escudero, and J. M. Iriondo. (2007). Generalist diurnal pollination provides greater fitness in a plant with nocturnal pollination syndrome assessing the effects of a *Silene-Hadena* interaction. *Oikos* 116:1461–72.

Giurfa, M. (1993) The repellent scent-mark of the honeybee *Apis mellifera ligustica* and its role as a communication cue during foraging. *Insectes Soc* 40:59–67.

Giurfa, M., B. Eichmann, and R. Menzel. (1996) Symmetry perception in an insect. *Nature* 382:458–61.

Giurfa, M., and M. Lehrer. (2001) Honeybee vision and floral displays: from detection to close-up recognition. In *Cognitive Ecology of Pollination,* ed. L. Chittka and J. D. Thomson, 106–26. Cambridge: Cambridge University Press.

Giurfa, M., and J. Nuñez. (1992) Honeybees mark with scent and reject recently visited flowers. *Oecologia* 89:113–17.

Giurfa, M., J. Nuñez, L. Chittka, and R. Menzel. (1995) Color preferences of flower-naïve honeybees. *J Comp Physiol* A 177:247–59.

Giurfa, M., M. Vorobyev, P. Kevan, and R. Menzel. (1996) Detection of colored stimuli by honeybees: minimum visual angles and receptor specific contrasts. *J Comp Physiol* A 178:699–709.

Givnish, T. J. (1982) Outcrossing versus ecological constraints in the evolution of dioecy. *Am Nat* 119:849–65.

Glover, B. J. (2007) *Understanding Flowers and Flowering: An Integrated Approach.* Oxford: Oxford University Press.

Glover, B. J., and C. Martin. (1998) The role of petal cell shape and pigmentation in pollination success in *Antirrhinum majus. Heredity* 80:778–84.

Goldblatt, P., P. Bernhardt, and J. C. Manning. (1998) Pollination by monkey beetles (Scarabeidae, Hoplinii) in petaloid geophytes in southern Africa. *Ann Miss Bot Gard* 85:215–30.

Goldblatt, P., P. Bernhardt, P. Vogel, and J. C. Manning. (2004) Pollination by fungus gnats (Diptera: Mycetophilidae) and self-recognition sites in *Tolmeia menziesii* (Saxifragaceae). *Plant Syst Evol* 244:55–67.

Goldblatt, P., and J. C. Manning. (1999) The long-proboscid fly pollination system in *Gladiolus* (Iridaceae). *Ann Miss Bot Gard* 86:758–74.

———. (2000) Long-proboscid fly pollination in southern Africa. *Ann Miss Bot Gard* 87:146–70.

———. (2006) Radiation of pollination systems in the Iridaceae of sub-Saharan Africa. *Ann Bot* 97:317–44.

Goldblatt, P., J. C. Manning, and P. Bernhardt (1995) Pollination biology of *Lapeirousia* subgenus *Lapeirousia* (Iridaceae) in southern Africa; floral divergence and adaptation for long-tongued fly pollination. *Ann Miss Bot Gard* 82:517–34.

Goldingay, R. L., S. M. Carthew, and R. J. Whelan. (1991) The importance of non-flying mammals in pollination. *Oikos* 61:79–81.

Goldingay, R. L., and R. J. Whelan. (1993) The influence of pollinators on fruit positioning in the Australian shrub *Telopea speciosissima* (Proteaceae). *Oikos* 68: 501–9.

Goldsmith, T. H., and G. D. Bernard. (1974) The visual system of insects. In *The Physiology of Insecta II*, ed. M. Rockstein, 156–272. New York: Academic Press.

Gómez, J. M. (2000) Effectiveness of ants as pollinators of *Lobularia maritima*: effects on main sequential fitness components of the host plant. *Oecologia* 122:90–97.

———. (2003) Herbivory reduces the strength of pollinator-mediated selection in the Mediterranean herb *Erysimum mediohispanicum*; consequences for plant specialization. *Am Nat* 162:242–56.

———. (2005) Non-additive effects of ungulates on the interaction between *Erysimum mediohispanicum* and its pollinators. *Oecologia* 143:412–18.

———. (2008) Sequential conflicting selection due to multispecific interactions triggers evolutionary trade-offs in a monocarpic herb. *Evolution* 62:668–79.

Gómez, J. M., J. Bosch, F. Perfectti, J. Fernandez, and M. Abdelaziz. (2007) Pollinator diversity affects plant reproduction and recruitment; the tradeoffs of generalization. *Oecologia* 153:597–605.

Gómez, J. M., J. Bosch, F. Perfectti, J. Fernandez, M. Abdelaziz, and J.P.M. Camacho. (2008) Spatial variation in selection on corolla shape in a generalist plant is promoted by the preference patterns of its local pollinators. *Proc Roy Soc London* B 275:2241–49.

Gómez, J. M., and R. Zamora. (1992) Pollination by ants: consequences of the quantitative effects on a mutualistic system. *Oecologia* 91:410–18.

———. (1996) Wind pollination in high-mountain populations of *Hormathophylla spinosa* (Cruciferae). *Am J Bot* 83:580–85.

Gómez, J. M., and R. Zamora. (1999) Generalization vs. specialization in the pollination system of *Hormathophylla spinosa* (Cruciferae). *Ecology* 80:796–805.

———. (2006) Ecological factors that promote the evolution

of generalization in pollination systems. In *Plant-Pollinator Interactions: from Specialization to Generalization*, ed. N. M. Waser and J. Ollerton, 145–66. Chicago: University of Chicago Press.

Gómez, J. M., R. Zamora, J. A. Hodar, and D. Garcia. (1996) Experimental study of pollination by ants in Mediterranean high mountain and arid habitats. *Oecologia* 105: 236–42.

Goncalves-Souza, T., P. M. Omena, J. C. Souza, and G. Q. Romero. (2008) Trait-mediated effects on flowers: artificial spiders deceive pollinators and decrease plant fitness. *Ecology* 89:2407–13.

Gonzalez, A., C. L. Rowe, P. J. Weeks, D. Whittle, F. S. Gilbert, and C. J. Barnard. (1995) Flower choice by honeybees (*Apis mellifera* L.)—sex phase of flowers and preferences among nectar and pollen foragers. *Oecologia* 101:258–64.

Goodrich, K. R., and R. A. Raguso.A (2009)The olfactory component of floral display in *Asimina* and *Deeringothamnus* (Annonaceae). *New Phytol* 183:457–69.

Goodrich, K. R., M. L. Zjhra, C. A. Ley, and R. A. Raguso. (2006) When flowers smell fermented: the chemistry and ontogeny of yeasty floral scent in pawpaw (*Asimina triloba*, Annonaceae). *Int J Plant Sci* 167:33–46.

Goodwillie, C. (1999) Wind pollination and reproductive assurance in *Linanthus parviflorus* (Polemoniaceae), a self-incompatible annual. *Am J Bot* 86:948–54.

Goodwillie, C., S. Kalisz, and C. G. Eckert. (2005) The evolutionary enigma of mixed mating systems in plants: occurrence, theoretical explanations, and empirical evidence. *Ann Rev Ecol Syst* 36:47–79.

Gordo, O., and J. J. Sanz. (2006) Temporal trends in phenology of the honey bee *Apis mellifera* (L.) and the small white *Pieris rapae* (L.) in the Iberian peninsula (1952–2004). *Evol Entomol* 31:261–68.

———. (2009) Long-term temporal changes of plant phenology in the Western Mediterranean. *Global Change Biol.* 15:1930–48.

Gori, D. F. (1983) Post-pollination phenomena and adaptive floral changes. In *Handbook of Experimental Pollination Biology*, ed. C. E. Jones and R. J. Little, 31–49. New York: Van Nostrand Reinhold.

———. (1989) Floral color change in *Lupinus argenteus* (Fabaceae): why should plants advertise the location of unrewarding flowers to pollinators? *Evolution* 43:870–81.

Gottsberger, G. (1974) The structure and function of the primitive angiosperm flower—a discussion. *Acta Bot Neerl* 23:461–71.

———. (1988) The reproductive biology of primitive angiosperms. *Taxon* 37:630–43.

———. (1989a) Beetle pollination and flowering rhythm of *Annona* species (Annonaceae) in Brazil. *Plant Syst Evol* 167:165–187.

———. (1989b) Comments on flower evolution and beetle pollination in the genera *Annona* and *Rollinia* (Annonaceae). *Plant Syst Evol* 167:189–94.

———. (1999) Pollination and evolution in Neotropical Annonaceae. *Plant Spec Biol* 14:143–52.

Gottsberger, G., T. Arnold, and H. F. Linskens. (1990) Variation in floral nectar amino acids with aging of flowers, pollen concentration, and flower damage. *Isr J Bot* 39: 167–76.

Gottsberger, G., J. Schrauwen, and H. F. Linskens. (1984) Amino acids and sugars in nectar, and their putative evolutionary significance. *Plant Syst Evol* 145:55–77.

Gottsberger, G., and I. Silberbauer-Gottsberger. (1988) Evolution of flower structures and pollination in Neotropical Cassiinae (Caesalpiniaceae). *Phyton* 28:293–320.

Gould, E. (1978) Foraging behavior of Malaysian nectar-feeding bats. *Biotropica* 10:184–93.

Gould, J. L. (1987) Honey bees store learned flower-landing behavior according to time of day. *Anim Behav* 35:1579–80.

———. (1993) Ethological and comparative perspectives on honey bee learning. In *Insect Learning, Ecological and Evolutionary Perspectives*, ed. D. R. Papaj and A. C. Lewis, 18–50 . New York: Chapman and Hall.

Goulson, D. (2000) Are insects flower constant because they use search images to find flowers? *Oikos* 88:547–52.

———. (2003) Effects of introduced bees on native ecosystems. *Ann Rev Ecol Syst* 34:1–26.

Goulson, D., J. W. Chapman, and W.O.H. Hughes. (2001) Discrimination of unrewarding flowers by bees; direct detection of rewards and use of repellent scent marks. *J Insect Behav* 14:669–78.

Goulson, D., and J. S. Cory. (1993) Flower constancy and learning in the foraging behaviour of the green-veined white butterfly, *Pieris napi*. *Mol Ecol Entomol* 18:315–20.

Goulson, D., J. L. Cruise, K. R. Sparrow, A. J. Harris, K. J. Park, M. C. Tinsley, and A. S. Gilburn. (2007) Choosing rewarding flowers; perceptual limitations and innate preferences influence decision making in bumblebees and honeybees. *Behav Ecol Sociobiol* 61: 1523–29.

Goulson, D., G. C. Lye, and B. Darvill. (2008) Diet breadth, coexistence and rarity in bumblebees. *Biodiv Cons* 17: 3269–88.

Goulson, D., K. McGuire, E. E. Munro, S. Adamson, L. Colliar, K. J. Park, M. C. Tinsley, and A. S. Gilburn. (2009) Functional significance of the dark central floret of *Daucus carota* (Apiaceae) L.; is it an insect mimic? *Plant Spec Biol* 24:77–82.

Goulson, D., J. Ollerton, and C. Sluman (1997) Foraging strategies in the small skipper butterfly *Thymelicus flavus*: when to switch? *Anim Behav* 53:1009–16.

Goulson, D., and K. R. Sparrow. (2008) Evidence for competition between honeybees and bumblebees; effects on bumblebee worker size. *J Insect Cons* 13:177–81.

Goulson, D., J. C. Stout, and S. A. Hawson. (1997) Can flower constancy in nectaring butterflies be explained by Darwin's interference hypothesis? *Oecologia* 112:225–31.

Goulson, D., and N. P. Wright. (1998) Flower constancy in the hoverflies *Episyrphus balteatus* (Degeer) and *Syrphus ribesii* (L.) (Syrphidae). *Behav Ecol* 9:213–19.

Goverde, M., K. Schweizer, B. Baur. and A. Erhardt. (2002) Small-scale habitat fragmentation effects on pollinator behaviour: experimental evidence from the bumblebee *Bombus veteranus* on calcareous grasslands. *Biol Cons* 104:293–99.

Goyret, J., P. M. Markwell, and R. A. Raguso. (2008) Context- and scale-dependent effects of floral CO_2 on nectar foraging by *Manduca sexta*. *Proc Natl Acad Sci USA* 105:4565–570.

Goyret, J., M. Pfaff, R. A. Raguso, and A. Kelber. (2008) Why do *Manduca sexta* feed from white flowers? Innate and learnt colour preferences in a hawkmoth. *Naturwissenschaften* 95:569–76.

Goyret, J., and R. A. Raguso. (2006) The role of mechanosensory input in flower handling efficiency and learning by *Manduca sexta*. *J Exp Biol* 209:1585–93.

Grant, B. R. (1996) Pollen digestion by Darwin's finches and its importance for early breeding. *Ecology* 77:489–99.

Grant, K. A. (1966) A hypothesis concerning the prevalence of red coloration in California hummingbird flowers. *Am Nat* 100:85–97.

Grant, V. (1949) Pollination systems as isolating mechanisms in flowering plants. *Evolution* 3:82–97.

———. (1950) The flower constancy of bees. *Bot Rev* 16: 379–98.

———. (1952) Isolation and hybridisation between *Aquilegia formosa* and *A. pubescens*. *Aliso* 2:341–60.

———. (1994) Historical development of ornithophily in the western North American flora. *Proc Natl Acad Sci USA* 91:10407–11.

Grant, V., and K. A. Grant. (1968) *Hummingbirds and Their Flowers*. New York: Columbia University Press.

Grayum, M. H. (1986) Correlations between pollination biology and pollen morphology in the Araceae. In *Pollen and Spores: Form and Function*, ed. S. Blackmore and I. K. Ferguson, 313–27. London: Academic Press.

Greathead, D. J. (1983) The multi-million dollar weevil that pollinates oil palms. *Antenna* 7:105–7.

Greenleaf, S. S., N. M. Williams, R. Winfree, and C. Kremen. (2007) Bee foraging ranges and their relationship to body size. *Oecologia* 153:589–96.

Gregg, K. B. (1991) Defrauding the deceitful orchid: pollen collection by pollinators of *Cleistes divaricata* and *C. bifaria*. *Lindleyana* 6:214–20.

Gribel, R. (1988) Visits of *Caluromys lanatus* (Didelphidae) to flowers of *Pseudobombax tomentosum* (Bombacaceae): a probable case of pollination by marsupials in Central Brazil. *Biotropica* 20:344–47.

Gribel, R., P. E. Gibbs, and A. L. Queiroz. (1999) Flowering phenology and pollination biology of *Ceiba pentandra* (Bombacaceae) in central Amazonia. *J Trop Ecol* 15: 247–63.

Griffin, A. R., A. B. Hingston, and C. P. Ohmart. (2009) Pollinators of *Eucalyptus regnans* (Myrtaceae), the world's tallest flowering plant species. *Aust J Bot* 57:18–25.

Griffiths, D. J. (1950) The liability of seed crops of perennial rye grass (*Lolium perenne*) to contamination by windborne pollen. *J Agric Sci* 40:19–8.

Grimaldi, D. (1999) The co-radiations of pollinating insects and angiosperms in the Cretaceous. *Ann Miss Bot Gard* 86:373–406.

Grime, J. P., J. G. Hodgson, and R. P. Hunt. (1988) *Comparative Plant Ecology*. London: Unwin Hyman.

Grison-Pige, L., J. M. Bessiere, and M. Hossaert-McKey. (2002) Specific attraction of fig-pollinating wasps: role of volatile compounds released by tropical figs. *J Chem Ecol* 28:283–95.

Gronquist, M., A. Bezzerides, A. Attygalle, J. Meinwald, M. Eisner, and T. Eisner. (2001) Attractive and defensive functions of the ultraviolet pigments of a flower (*Hypericum calycinum*). *Proc Natl Acad Sci USA* 98:13745–50.

Gross, C. L. (1992) Floral traits and pollinator constancy: foraging by native bees among three sympatric legumes. *Aust. J Ecol* 17:67–74.

———. (2001) The effect of introduced honeybees on native bee visitation and fruit set in *Dillwynia juniperina* (Fabaceae) in a fragmented ecosystem. *Biol Cons* 102: 89–95.

Grotewold, E. (2006) The genetics and biochemistry of floral pigments. *Ann Rev Plant Biol* 57:761–80.

Grüter, C., M. S. Balbuena, and W. M. Farina. (2008) Informational conflicts created by the waggle dance. *Proc Roy Soc London* B 275:1321–27.

Guerenstein, P. G., E. A. Yepez, J. van Haren, D. G. Williams, and J. G. Hildebrand. (2004) Floral CO_2 emission may indicate food abundance to nectar-feeding moths. *Naturwissenschaften* 91:329–33.

Guerrant, E.O.J. (1989) Early maturity, small flowers, and autogamy: a developmental connection? In *The Evolutionary Ecology of Plants*, ed. J. H. Bock and Y. B. Linhart, 61–84. Boulder, CO: Westview Press.

Guerrant, E. O., and P. L. Fiedler. (1981) Flower defense against nectar pilferage by ants. *Biotropica* 13:25–33.

Guimares, P. R., V. Rico-Gray, P. S. Oliveira, T. J. Izzo, S. F. dos Reis, and J. N. Thompson. (2007) Interactions intimacy affects structure and coevolutionary dynamics in mutualistic networks. *Curr Biol* 17:1797–1803.

Guldberg, L. D., and P. R. Atsatt. (1975) Frequency of reflection and absorption of ultraviolet light in flowering plants. *Am Midl Nat* 93:35–43.

Gumbert, A. (2000) Color choices by bumble bees (*Bombus terrestris*): innate preferences and generalization after learning. *Beh Ecol Sociobiol* 48:36–43.

Gumbert, A., and J. Kunze. (2001) Colour similarity to rewarding model plants affects pollination in a food deceptive orchid, *Orchis boryi*. *Biol J Linn Soc* 72:419–33.

Gumbert, A., J. Kunze, and L. Chittka. (1999) Floral colour diversity in plant communities, bee colour space and a null model. *Proc Roy Soc London* B 272:1711–16.

Guo, Y. H., and C.D.K. Cook. (1989) Pollination efficiency of *Potamogeton pectinatus* L. *Aquat Bot* 34:381–84.

Gutierrez, A., S. V. Rojas-Nossa, and F. G. Stiles. (2004) Annual dynamics of hummingbird-flower interactions in high Andean ecosystems. *Ornithol Neotrop* 15:205–13.

Gyulane, L., T. Peter, and V. Gyulane. (1980) Some results in poppy *Papaver somniferum* breeding. 1. Breeding of winter poppy. *Herba Hung* 19:45–54.

Haber, W. A., G. W. Frankie, H. G. Baker, I. Baker, and S. Koptur. (1981) Ants like flower nectar. *Biotropica* 13: 211–14.

Hackett, D. J., and R. L. Goldingay. (2001) Pollination of *Banksia* spp. by non-flying mammals in north-eastern New South Wales. *Aust J Bot* 49:637–44.

Hadacek, F., and M. Weber. (2002) Club-shaped organs as additional osmophores within the *Sauromatum* inflorescence: odour analysis, ultrastructural changes and pollination aspects. *Plant Biol* 4:367–83.

Hagerup, O. (1932) On pollination in the extremely hot air at Timbuctu. *Dansk Bot Arkiv* 8:1–20.

Hagerup, O. (1951) Pollination in the Faeroes – in spite of rain and poverty of insects. *Biol. Medd. Kong. Dansk Videns Selskab* 18:1–48.

Hainsworth, F. R., and L. L. Wolf. (1976) Nectar characteristics and food selection by hummingbirds. *Oecologia* 25:101–13.

Hainsworth, F. R., E. Precup, and T. Hamill. (1991) Feeding, energy processing rates and egg production in painted lady butterflies. *J Exp Biol* 156:249–65.

Hambäck, P. A. (2001) Direct and indirect effects of herbivory: feeding by spittlebugs affects pollinator visitation rates and seedset of *Rudbeckia hirta*. *Ecoscience* 8:45–50.

Hamrick, J. L., M.J.W. Godt, and S. L. Sherman-Broyles. (1995) Gene flow among plant populations: evidence from genetic markers. In *Experimental and Molecular Approaches to Plant Biosystematics*, 215–32.. St. Louis: Missouri Botanical Garden.

Han, Y., C. Dai, C.-F. Yang, Q. F. Wang, and T. J. Motley. (2008) Anther appendages of *Incarvillea* trigger a pollen-dispensing mechanism. *Ann Bot* 102:473–79.

Hanley, M. E., M. Franco, S. Pichon, B. Darvill, and D. Goulson. (2008) Breeding system, pollinator choice and variation in pollen quality in British herbaceous plants. *Funct Ecol* 22:592–98.

Hanley, M. E., B. B. Lamont, and W. S. Armbruster. (2009) Pollination and plant defence traits co-vary in Western Australian *Hakeas*. *New Phytol* 182:251–60.

Hanna, W. W., and L.E. Towill. (1995) Long term pollen storage. *Plant Breed Rev* 13:179–207.

Hansen, D. M., K. Beer, and C. B. Müller. (2006) Mauritian coloured nectar no longer a mystery: a visual signal for lizard pollinators. *Biol Lett* 2:165–68.

Hansen, D. M., H. C. Kiesbuey, G. G. Jones, and C. B. Müller. (2007) Positive indirect interactions between neighbouring plant species via a lizard pollinator. *Am Nat* 169:534–42.

Hansen, D. M., J. M. Olesen, and C. G. Jones. (2002) Trees, birds and bees in Mauritius: exploitative competition between introduced honeybees and endemic nectarivorous birds. *J Biogeog* 29:721-734.

Hansen, D. M., J. M. Olesen, T. Mione, S. D. Johnson, and C. B. Müller. (2007) Coloured nectar: distribution, ecology and evolution of an enigmatic floral trait. *Biol Rev* 82:83–111.

Hansted, L., H. B. Jakobsen, and C. E. Olsen. (1994) Influence of temperature on the rhythmic emission of volatiles from *Ribes nigrum* flowers in situ. *Plant Cell Env* 17:1069–72.

Harborne, J. B. (1993) *Introduction to Ecological Biochemistry*, 4th ed. London: Academic Press.

Harborne, J. B., and D. M. Smith. (1978) Correlations between anthocyanin chemistry and pollinator ecology in the Polemoniaceae. *Biochem Syst Ecol* 6:127–30.

Harder, L. D. (1986) Effects of nectar concentration and flower depth on flower handling efficiency of bumble bees. *Oecologia* 69:309–15.

———. (1988) Choice of individual flowers by bumblebees: interactions of morphology, time and energy. *Behaviour* 104:60–77.

———. (1990a) Behavioral responses by bumblebees to variation in pollen availability. *Oecologia* 85:41–47.

———. (1990b) Pollen removal by bumblebees and its implications for pollen dispersal. *Ecology* 71:1110–25.

———. (1998) Pollen-size comparisons among animal-pollinated angiosperms with different pollination characteristics. *Biol J Linn Soc* 64:51325.

———. (2000) Pollen dispersal and the floral diversity of Monocotyledons. 7 In *Monocots: Systematics and Evolution*, ed. K. L. Wilson and D. Morrison, 243–5. Melbourne: CSIRO Publishing.

Harder, L. D., and R.M.R. Barclay. (1994) The functional significance of poricidal anthers and buzz pollination: controlled pollen removal from *Dodecatheon*. *Funct Ecol* 8:509–17.

Harder, L. D., and S.C.H. Barrett. (1992) The energy cost of bee pollination for *Pontederia cordata*. *Funct Ecol* 6: 226–33.

———. (1995) Mating cost of large floral displays in hermaphroditic plants. *Nature* 373:512–15.

———. (1996) Pollen dispersal and mating patterns in animal-pollinated plants. In *Floral Biology: Studies on Floral Evolution in Animal-Pollinated Systems*, ed. D. G. Lloyd and S.C.H. Barrett, 140–90. New York: Chapman and Hall.

Harder, L. D., and S.C.H. Barrett, ed. (2006) *Ecology and Evolution of Flowers*. Oxford: Oxford University Press.

Harder, L. D., and M. B. Cruzan. (1990) An evaluation of the physiological and evolutionary influences of inflorescence size and flower depth on nectar production. *Funct Ecol* 4:559–72.

Harder, L. D., and S. D. Johnson. (2005) Adaptive plasticity of floral display size in animal-pollinated plants. *Proc Roy Soc London* B 272:2651–57.

———. (2008) Function and evolution of aggregated pollen in angiosperms. *Int J Plant Sci* 169:59–78.

Harder, L. D., and L. A. Real. (1987) Why are bumble bees risk averse? *Ecology* 68:1104–8.

Harder, L. D. and M. B. Routley. (2006) Pollen and ovule fates and reproductive performance by flowering plants. In *Ecology and Evolution of Flowers*, ed. L. D. Harder and S.C.H. Barrett, 61–80. Oxford: Oxford University Press.

Harder, L. D., and J. D. Thomson. (1989) Evolutionary options for maximising pollen dispersal of animal-pollinated plants. *Am Nat* 133:323–44.

Harder, L. D., N. M. Williams, C. Y. Jordan, and W. A. Nelson. (2001) The effects of floral design and display on pollinator economics and pollen dispersal. In *Cognitive Ecology of Pollination*, ed. L. Chittka and J. D. Thomson, 297–317. Cambridge: Cambridge University Press.

Harder, L. D., and W. G. Wilson. (1994) Floral evolution and male reproductive success—optimal dispensing schedules for pollen dispersal by animal-pollinated plants. *Evol Ecol* 8:542–59.

———. (1998) Theoretical consequences of heterogeneous transport conditions for pollen dispersal by animals. *Ecology* 79:2789–2807.

Hargreaves, A. L., L. D. Harder, and S. D. Johnson. (2009) Consumptive emasculation: the ecological and evolutionary consequences of pollen theft. *Biol Rev* 84:259–76.

Hargreaves, A., S. Johnson, and E. Nol. (2004) Do floral syndromes predict specialization in plant pollination systems? An experimental test in an "ornithophilous" African *Protea*. *Oecologia* 140:295–301.

Harley, R. (1991) The greasy pole syndrome. In *Ant-Plant Interactions,* ed. C. R. Huxley and D. F. Cutler, 430–33. Oxford: Oxford University Press.

Harper, J. L., and W. A. Wood. (1957) Biological flora of the British Isles, *Senecio jacobaea*. *J Ecol* 45:617–37.

Harris, F.C.L., and Beattie AJ (1991) Viability of pollen carried by *Apis mellifera* L., *Trigona carbonaria* Smith and *Vespula germanica* (F.) (Hymenoptera: Apidae, Vespidae). *J Aust Entomol Soc* 30:45–47.

Harris, M. S., and J. R. Pannell. (2008) Roots, shoots and reproduction: sexual dimorphism in size and costs of reproductive allocation in an annual herb. *Proc Roy Soc London* B 275:2595–2602.

Harrison, J. F., J. H. Fewell, K. E. Anderson, and G. M. Loper. (2006) Environmental physiology of the invasion of the Americas by Africanized honeybees. *Integr Comp Biol* 46:1110–22.

Harrison, R. D. (2003) Fig wasp dispersal and the stability of a keystone plant resource in Borneo. *Proc Roy Soc London* B 279:S76–S79.

Hartley, S. E., and T. H. Jones. (2003) Plant diversity and insect herbivores: effects of environmental change in contrasting model systems. *Oikos* 101:6–17.

Haslett, J. R. (1983) A photographic account of pollen digestion by adult hoverflies. *Physiol Entomol* 8:167–71.

Hawkeswood, T. (1989) Notes on *Diphucephala affinis* (Coleoptera, Scarabaeidae) associated with flowers of *Hibbertia* and *Acacia* in Western Australia. *Plant Syst Evol* 168:1–5.

Hawthorn, L. R., G. E. Bohart, and E. H. Toole. (1956) Carrot seed yield and germination as affected by different levels of insect pollination. *Proc Am Soc Hort Sci* 67: 384–89.

Haynes, K. F., J. Z. Zhao, and A. Latif. (1991) Identification of floral compounds from *Abelia grandiflora* that stimulate upwind flight in cabbage looper moths. *J Chem Ecol* 17:637–46.

Hayter, K. E., and J. E. Cresswell. (2006) The influence of pollinator abundance on the dynamics and efficiency of pollination in agricultural *Brassica napus*; implications for landscape-scale gene dispersal. *J Appl Ecol* 43:1196–1202.

He, Y. P., Y. W. Duan, J. Q. Liu, and W. K. Smith. (2005) Floral closure in response to temperature and pollination in *Gentiana straminea* Maxim, (Gentianaceae), an alpine perennial in the Qinghai-Tibetan plateau. *Plant Syst Evol* 256:17–33.

Healy, S. D., and T. A. Hurly. (2001) Foraging and spatial learning in hummingbirds. In *Cognitive Ecology of Pollination*, ed. L. Chittka and J. D. Thomson, 127–47. Cambridge: Cambridge University Press.

Heard, T. A. (1999) The role of stingless bees in crop pollination. *Ann Rev Entomol* 44:183–206.

Hedhly, A., J. I. Hormaza, and M. Herrero. (2009) Global

warming and sexual plant reproduction. *Trends Plant Sci* 14:30–36.

Hegland, S. J., J.-A. Grytnes, and Ø. Totland (2009) The relative importance of positive and negative interactions for pollinator attraction in a plant community. *Ecol Res* 24:929–36.

Hegland, S. J., A. Nielsen, A. Lazaro, A.-L. Bjerknes, and Ø. Totland. (2009) How does climate warming affect plant-pollinator interactions? *Ecol Lett* 12:184–95.

Heil, M., T. Delsinne, A. Hilpert, S. Schurkens, C. Andary, K. E. Linsenmair, M. S. Sousa, and D. McKey. (2002) Reduced chemical defence in ant plants? A critical re-evaluation of a widely accepted hypothesis. *Oikos* 99:457–68.

Heil, M., S. Greiner, H. Meimberg, R. Kruger, J. L. Noyer, G. Heubl, K. E. Linsenmair, and W. Boland. (2004) Evolutionary change from induced to constitutive expression of an indirect plant resistance. *Nature* 430:205–8.

Heil, M., T. Koch, A. Hilpert,, B. Fiala, W. Boland, and K. E. Linsenmair. (2001) Extrafloral nectar production of the ant-associated plant *Macaranga tanarius*, is an induced indirect defensive response elicited by jasmonic acid. *Proc Natl Acad Sci USA* 98:1083–88.

Heil, M., J. Rattke, and W. Boland. (2005) Postsecretory hydrolysis of nectar sucrose and specialization in ant-plant mutualism. *Science* 308:560–63.

Heiling, A. M., K. Cheng, and M. E. Herberstein. (2004) Exploitation of floral signals by crab spiders (*Thomisus spectabilis*, Thomisidae). *Behav Ecol* 15:321–26.

———. (2006) Picking the right spot; crab spiders position themselves on flowers to maximize prey attraction. *Behaviour* 143:957–68.

Heinrich, B. (1975a) Bee flowers: a hypothesis on flower variety and blooming times. *Evolution* 29:325–34.

———. (1975b) The energetics of pollination. *Ann Rev Ecol Syst* 6:139–70.

———. (1976a) The foraging specializations of individual bumblebees. *Ecol Monogr* 46:105–28.

———. (1976b) Resource partitioning among some eusocial insects: bumblebees. *Ecology* 57:874–89.

———. (1979a) *Bumblebee Economics* Cambridge, MA: Harvard University Press.

———. (1979b) "Majoring" and "minoring" by foraging bumblebees, *Bombus vagans*; an experimental analysis. *Ecology* 60:245–55.

———. (1980) Mechanisms of body temperature regulation in honeybees, *Apis mellifera*. 2. Regulation of thoracic temperature at high air temperatures. *J Exp Biol* 85:73–87.

———. (1981) The mechanisms and energetic of honeybee swarm temperature regulation. *J Exp Biol* 91:25–55.

———. (1983a) Insect foraging energetics. In *Handbook of Experimental Pollination Biology*, ed. C. E. Jones and R. J. Little, 187–214. New York: Van Nostrand Reinhold.

———. (1983b) Do bumblebees forage optimally, and does it matter? *Am Zool* 23:273–81.

———. (1993) *The Hot-Blooded Insects: Strategies and Mechanisms of Thermoregulation*. Cambridge, MA: Harvard University Press.

Heinrich, B., and H. Esch. (1997) Honeybee thermoregulation. *Science* 276:1015–16.

Heinrich, B., P. R. Mudge, and P. G. Deringis. (1977) Laboratory analysis of flower constancy in foraging bumblebees, *Bombus ternarius* and *B. terricola*. *Behav Ecol Sociobiol* 2:247–65.

Heinrich, B., and P. H. Raven. (1972) Energetics and pollination ecology. *Science* 176:597–602.

Heinrich, B., and F. D. Vogt. (1993) Abdominal temperature regulation by Arctic bumblebees. *Physiol Zool* 66:257–69.

Heithaus, E. R. (1974) The role of plant-pollinator interactions in determining community structure. *Ann Miss Bot Gard* 61:676–91.

———. (1979) Community structure of neotropical flower visiting bees and wasps: diversity and phenology. *Ecology* 60:190–202.

Heithaus, E. R., T. H. Fleming, and P. A. Opler. (1975) Foraging patterns and resource utilization in seven species of bats in a seasonal tropical forest. *Ecology* 56:841–54.

Heithaus, E. R., P. R. Opler, and H. G. Baker. (1974) Bat activity and pollination of *Bauhinia pauletia*: plant-pollinator coevolution. *Ecology* 55:412–19.

Hemborg, A. M., and W. J. Bond. (2005) Different rewards in female and male flowers can explain the evolution of sexual dimorphism in plants. *Biol J Linn Soc* 85:97–109.

Hemsley, A. J., and I. K. Ferguson. (1985) Pollen morphology of the genus *Erythrina* (Leguminosae: Papilionoideae) in relation to floral structure and pollinators. *Ann Miss Bot Gard* 72:570–90.

Hendel-Rahmanim, K., T. Masci, A. Vainstein, and D. Weiss. (2009) Diurnal regulation of scent emission in rose flowers. *Planta* 226:1491–99.

Hendrix, S. D. (1994) Effects of population size on fertilization, seed production and seed predation in two prairie legumes. *Nth Am Prairie Conf Proc* 13:115–19.

Hepburn, H. R. (1971) Proboscis extension and recoil in Lepidoptera. *J Insect Physiol* 17:637–56.

Herre, E. A., C. A. Machado, E. Bermingham, J. D. Nason, D. M. Windsor, S. S. McCafferty, W. Van Houten, and K. Bachmann. (1996) Molecular phylogenies of figs and their pollinating wasps. *J Biogeogr* 23:521–30.

Herrera, C. M. (1987) Components of pollinator "quality": comparative analysis of a diverse insect assemblage. *Oikos* 50:79–90.

———. (1988) Variation in mutualisms: the spatiotemporal mosaic of a pollinator assemblage. *Biol J Linn Soc* 35:95–125.

———. (1990) Daily patterns of pollinator activity, differential pollinating effectiveness, and floral resource availability, in a summer-flowering Mediterranean shrub. *Oikos* 58:277–88.

———. (1992) Activity pattern and thermal biology of a day-flying hawkmoth (*Macroglossum stellatarum*) under Mediterranean summer conditions. *Mol Ecol Entomol* 17:52–56.

———. (1993) Selection on floral morphology and environmental determinants of fecundity in a hawkmoth-pollinated violet. *Ecol Monogr* 63:251–75.

———. (1995a) Microclimate and individual variation in pollinators: flowering plants are more than their flowers. *Ecology* 76:1516–24.

———. (1995b) Floral biology, microclimate and pollination by ectothermic bees in an early-blooming herb. *Ecology* 76:218–28.

———. (1996) Floral traits and plant adaptation to insect pollinators: a devil's advocate approach. In *Floral Biology: Studies on Floral Evolution in Animal-Pollinated Systems*, ed. D. G. Lloyd and S.C.H. Barrett, 65–87. New York: Chapman and Hall.

———. (2000) Measuring the effects of pollinators and herbivores: evidence for non-additivity in a perennial herb. *Ecology* 81:217–176.

———. (2001a) Pollinators selecting for corolla traits in lavender. *J Evol Biol* 15:574–84.

———. (2001b) The variability of organs differentially involved in pollination, and correlation of traits in Genistae (Leguminosae: Papilionoideae). *Ann Bot* 88:1027–37.

———. (2005) Plant generalization on pollinators: species property or local phenomenon? *Am J Bot* 92:13–20.

Herrera, C. M., M. C. Castellanos, and M. Medrano. (2006) Geographical context of floral evolution: towards an improved research programme in floral diversification. In *Ecology and Evolution of Flowers*, ed. L. D. Harder and S.C.H. Barrett, 278–325. Oxford: Oxford University Press.

Herrera, C. M., C. de Vega, A. Canto, and M. I. Pozo. (2009) Yeasts in floral nectar: a quantitative survey. *Ann Bot* 103:1415–23.

Herrera, C. M., I. M. Garcia, and R. Perez. (2008) Invisible floral larcenies: microbial communities degrade floral nectar of bumblebee-pollinated plants. *Ecology* 89:2369–76.

Herrera, C. M., M. Medrano, P. J. Rey, A. M. Sánchez-Lafuente, M. B. Garciá, J. Guitián, and A. J. Manzaneda. (2002) Interactions of pollinators and herbivores on plant fitness suggests a pathway for correlated evolution of mutualism- and antagonism-related traits. *Proc Natl Acad Sci USA* 99:16823–28.

Herrera, C. M., R. Perez, and C. Alonso. (2006) Extreme intraplant variation in nectar sugar composition in an insect-pollinated perennial herb. *Am J Bot* 93:575–81.

Herrera, C. M. and M. I. Pozo. (2010) Nectar yeasts warm the flowers of a winter-blooming plant. *Proc Roy Soc London B* 277:1827–34.

Herrera, J. (1988) Pollination relationships in southern Spanish Mediterranean shrublands. *J Ecol* 76:274–87.

———. (1993) Selection on floral morphology and environmental determinants of fecundity in a hawkmoth-pollinated violet. *Ecol Monogr* 63:251–75.

———. (2005) Flower size variation in *Rosmarinus officinalis*: individuals, populations and habitats. *Ann Bot* 95:431–37.

———. (2009) Visibility vs. biomass in flowers: exploring corolla allocation in Mediterranean entomophilous plants. *Ann Bot* 103:1119–27.

Herrera, L. G., and C. M. del Rio. (1998) Pollen digestion by New World bats: effects of processing time and feeding habits. *Ecology* 79:2828–38.

Hersch, E. I. (2006) Foliar damage to parental plants interacts to influence mating success of *Ipomoea purpurea*. *Ecology* 87:2026–36.

Heslop-Harrison, J. (1979) Pollen walls as adaptive systems. *Ann Miss Bot Gard* 66:813–29.

Heslop-Harrison, J., and Y. Heslop-Harrison. (1991) Structural and functional variation in pollen intines. In *Pollen and Spores: Patterns of Diversification*, ed. S. Blackmore and S. H. Barnes, 331–44. Oxford: Clarendon Press.

Hess, D. (1990) *Die Blüte*. Stuttgart: Ulmer.

Hesse, M. (2000) Pollen wall stratification and pollination. In *Pollen and Pollination*, ed. A. Dafni, M. Hesse, and E. Pacini, 1–17. Vienna: Springer.

Hesse, M., S. Vogel, and H. Halbritter. (2000) Thread-forming structures in angiosperm anthers: their diverse role in pollination ecology. *Plant Syst Evol* 222:281–92.

Hessing, M. B. (1988) Geitonogamous pollination and is consequences in *Geranium caespitosum*. *Am J Bot* 75:1324–33.

Heuschen, B., A. Gumbert, and K. Lunau. (2005) A generalised mimicry system involving angiosperm flower colour, pollen and bumblebees' innate colour preference. *Plant Syst Evol* 252:121–37.

Heyneman, A. (1983) Optimal sugar concentration of floral nectars: dependence on sugar intake efficiency and foraging costs. *Oecologia* 60:198–213.

Heyneman, A., R. K. Colwell, S. Naeem, D. S. Dobkin, and B. Hallet. (1991) Host plant discrimination: experiments with hummingbird flower mites. In *Plant Animal Interactions: Evolutionary Ecology in Tropical and Temperate Regions*, ed. P. W. Price, T. Lewinsohn, G. W. Fernandes, and W. W. Benson, 455–85. New York: John Wiley & Sons.

Hickman, J. C. (1974) Pollination by ants: a low energy system. *Science* 184:1290–92.

Hickman, J. M., G. L. Lovei, and S. I. Wratten. (1995)

Pollen feeding by adults of the hoverfly *Melanostoma fasciatum* (Diptera, Syrphidae). *NZ J Zool* 22:387–92.

Higginson, A. D., and C. J. Barnard. (2004) Accumulating wing damage affects foraging decisions in honeybees (*Apis mellifera* L.). *Mol Ecol Entomol* 29:52–59.

Higginson, A. D., F. S. Gilbert, T. Reader, and C. J. Barnard. (2007) Reduction of visitation rates by honeybees *(Apis mellifera)* to individual inflorescences of lavender (*Lavandula stoechas*) upon removal of coloured accessory bracts (Hymenoptera, Apidae). *Entomol Gen* 29:165–78.

Hilioti, Z., C. Richards, and K. M. Brown. (2000) Regulation of pollination-induced ethylene and its role in petal abscission of *Pelargonium x hortorum*. *Physiol Plant* 109:322–32.

Hill, H. G., N. H. Williams, and C. H. Dodson. (1972) Floral fragrances and isolating mechanisms in the genus *Catasetum* (Orchidaceae). *Biotropica* 4:61–76.

Hill, P. S. M., J. Hollis, and H. Wells. (2001) Foraging decisions in nectarivores: unexpected interactions between flower constancy and energetic rewards. *Anim Behav* 62:729–37.

Hill, P. S. M., P. H. Wells, and H. Wells. (1997) Spontaneous flower constancy and learning in honey bees as a function of color. *Anim Behav* 54:615–27.

Hingston, A. B. (2007) Is the exotic bumblebee *Bombus terrestris* really invading Tasmanian native vegetation? *J Insect Cons* 10:289–93.

Hingston, A. B., J. Marsden-Smedley, D. A. Driscoll, et al. (2002) Extent of invasion of Tasmanian native vegetation by the exotic bumblebee *Bombus terrestris* (Apoidea, Apidae). *Aust Ecol* 27:162–72.

Hingston, A. B., and P. B. McQuillan. (2000) Are pollination syndromes useful predictors of floral visitors in Tasmania? *Aust Ecol* 25:600–9.

Hingston, A. B., and B. M. Potts. (2005) Pollinator activity can explain variation in outcrossing rates within individual trees. *Aust Ecol* 30:319–24.

Hinton, H. E. (1973) Some recent work on the colors of insects and their likely significance. *Proc Br Entomol Nat Hist Soc* 6:43–54.

Hirthe, G., and S. Porembski. (2003) Pollination of *Nymphaea lotus* (Nymphaeaceae) by rhinoceros beetles and bees in the northeastern Ivory Coast. *Plant Biol* 5:670–75.

Hiscock, S. J., and S. M. McInnis. (2003) Pollen recognition and rejection during the sporophytic self-incompatibility response: Brassica and beyond. Trends Plant Sci 8:606–13.

Hitchcock, J. D. (1959) Poisoning of honey bees by death camas blossoms. *Am Bee J* 99:418–19.

Hoballah, M. E., T. Gubitz, J. Stuurman, L. Broger, M. Barone, T. Mandel, A. Dell'Olivo, M. Arnold, and C. Kuhlemeier. (2007) Single gene-mediated shift in pollinator attraction in *Petunia*. *Plant Cell* 19:779–90.

Hoballah, M. E., J. Stuurman, T.C.J. Turling, P. M. Guerin, S. Connetable, and C. Kuhlemeier. (2005) The composition and timing of flower odour emission by wild *Petunia axillaris* coincide with the antennal perception and nocturnal activity of the pollinator *Manduca sexta*. *Planta* 222:141–50.

Hobbie, S. E., and F. S. Chapin. (1998)The response of tundra plant biomass, above-ground production, nitrogen, and CO_2 flux to experimental warming. *Ecology* 79:1526–44.

Hocking, B. (1968) Insect-flower associations in the high arctic, with special reference to nectar. *Oikos* 19:359–87.

Hocking, B., and C. D. Sharplin. (1965) Flower basking by Arctic insects. *Nature* 206:215.

Hodges, S. A. (1995) The influences of nectar production on hawkmoth behavior, self-pollination, and seed production in *Mirabilis multiflora* (Nyctaginaceae). *Am J Bot* 82:197–204.

———. (1997) Floral nectar spurs and diversification. *Int J Plant Sci* 158:S81–S88.

Hodges, S. A., and J. B. Whittall. (2008) One-sided evolution or two? A reply to Ennos. *Heredity* 100:541–42.

Hogarth, P. J. (1999) *The Biology of Mangroves*. Oxford: Oxford University Press.

Hogendoorn, K., S. Coventry, and M. A. Keller. (2007) Foraging behaviour of a blue-banded bee, *Amegilla chlorocyanea*, in greenhouses: implications for use as tomato pollinators. *Apidologie* 38:86–92.

Holdaway-Clarke, T. L., and P. K. Hepler. (2003) Control of pollen tube growth: role of ion gradients and fluxes. *New Phytol* 159:539–63.

Holderied, M. W., and O. von Helversen. (2006) "Binaural echo disparity" as a potential indicator of object orientation and cue for object recognition in echolocating nectar-feeding bats. *J Exp Biol* 209:3457–68.

Holland, G. J., A. F. Bennett, and R. van der Ree. (2007) Time-budget and feeding behaviour of the squirrel glider (*Petaurus norfolcensis*) in remnant linear habitat. *Wildlife Res* 34:288–95.

Holland, J. N., and T. H. Fleming. (1999) Mutualistic interactions between *Upiga virescens* (Pyralidae), a pollinating seed consumer, and *Lophocereus schottii* (Cactaceae). *Ecology* 80:2074–84.

———. (2002) Co-pollinators and specialization in the pollinating seed-consumer mutualism between senita cacti and senita moths. *Oecologia* 133:534–40.

Holland, R. A., P. Winter, and D. A. Waters. (2005) Sensory systems and spatial memory in the fruit bat *Rousettus aegyptiacus*. *Ethology* 111:715–25.

Hölldobler, B., and E. O. Wilson. (1990) *The Ants*. Cambridge, MA: Harvard University Press.

Holsinger, K. E. (1992) Ecological models of plant mating systems and the evolutionary stability of mixed mating systems. In *Ecology and Evolution of Plant Reproduction*, ed. R. Wyatt, 169–91. London: Chapman and Hall.

Holsinger, K. E., M. A. Feldman, and F. B. Christiansen. (1984) The evolution of self-fertilization in plants. *Am Nat* 124:446–53.

Holsinger, K. E. and J. D. Thomson. (1994) Pollen discounting in *Erythronium grandiflorum*—mass-action estimates from pollen transfer dynamics. *Am Nat* 144:799–812.

Holtsford, T. P., and N. C. Ellstrand. (1992) Genetic and environmental variation in floral traits affecting outcrossing rate in *Clarkia tembloriensis* (Onagraceae). *Evolution* 46:216–25.

Holzschuh, A., I. Steffan-Dewenter, D. Kleijn, and T. Tscharntke. (2007) Diversity of flower-visiting bees in cereal fields: effects of farming system, landscape composition and regional context. *J Appl Ecol* 44:41–49.

Honig, M. A., H. P. Linder, and W. J. Bond. (1992) Efficacy of wind pollination: pollen load size and natural microgametophyte populations in wind-pollinated *Staberoha banksii* (Restionaceae). *Am J Bot* 79:443–48.

Honsho, C., S. Somsri, T. Tetsumura, K. Yamashita, and K. Yonemori. (2007) Effective pollination period in durian (*Durio zibethinus* Murr) and the factors regulating it. *Sci Hort* 111:193–96.

Hopkins, C. H., and M.J.G. Hopkins. (1982) Predation by a snake of a flower-visiting bat at *Parkia nitida* (Leguminosae: Mimosidae). *Brittonia* 34:225–27.

Hopkins, C. Y., A. W. Jevans, and R. Boch. (1969) Occurrence of octadeca-trans-2,cis-9,cis-12-trienoic acid in pollen attractive to the honey bee. *Can J Bot* 47:433–36.

Horovitz, A., and J. Harding. (1972) Genetics of *Lupinus*. V. Intraspecific variability for reproductive traits in *Lupinus nanus*. *Bot Gaz* 133:155–65.

Houston, A. I., and D. C. Krakauer. (1993) Hummingbirds as net rate maximisers. *Oecologia* 94:135–38.

Houston, T. F. (1989) *Leioproctus* bees associated with Western Australian smokebushes (*Conospermum* spp.) and their adaptations for foraging and concealment (Hymenoptera, Colletidae, Paracolletini). *Rec West Austral Mus* 14:275–92.

Houston, T. F., and P. G. Ladd. (2002) Buzz pollination in the Epacridaceae. *Aust J Bot* 50:83–91.

Howell, D. J. (1974) Bats and pollen: Physiological aspects of the syndrome of chiropterophily. *Comp. Biochem Physiol* 48A:263–76.

———. (1979) Flock foraging in nectar-feeding bats: advantages to the bats and to the host plants. *Am Nat* 114:23–49.

Howell, D. J., and B. S. Roth. (1981) Sexual reproduction in agaves; the benefits of bats; the cost of semelparous advertising. *Ecology* 62:1–7.

Howell, G. H., A. T. Slater, and R. B. Knox. (1993) Secondary pollen presentation and its biological significance. *Aust J Bot* 41:417–38.

Hoyle, M., and J. E. Cresswell. (2006) Remobilization of initially deposited pollen grains has negligible impact on gene dispersal in bumblebee-pollinated *Brassica napus*. *Funct Ecol* 20:958–65.

———. (2007) The effect of wind direction on cross-pollination in wind-pollinated GM crops. *Ecol Appl* 17:1234–43.

Hrycan, W. C., and A. R. Davis. (2005) Comparative structure and pollen production of the stamens and pollinator-deceptive staminodes of *Commelina coelestis* and *C. dianthifolia* (Commelinaceae). *Ann Bot* 95:1113–30.

Hu, S. S., D. L. Dilcher, D. M. Jarzen, and D. W. Taylor. (2008) Early steps of angiosperm-pollinator coevolution. *Proc Natl Acad Sci USA* 105:240–45.

Huang, S.-Q., Y. H. Guo, G. W. Robert, Y. H. Shi, and K. Sun. (2001) Mechanism of underwater pollination in *Najas marina* (Najadaceae). *Aquat Bot* 70:67–78.

Huang, S.-Q., Y. Takahashi, and A. Dafni. (2002) Why does the flower stalk of *Pulsatilla cernua* (Ranunculaceae) bend during anthesis? *Am J Bot* 89:1599–1603.

Hubbell, S. P. (2001) *The Unified Neutral Theory of Biodiversity and Biogeography*. Princeton, NJ: Princeton University Press.

Huber, F. K., R. Kaiser, W. Sauter, and F. P. Schiestl. (2005) Floral scent emission and specific pollinator attraction in two species of *Gymnadenia* (Orchidaceae). *Oecologia* 142:564–75.

Hughes, M., M. Möller, T. J. Edwards, D. U. Bellstedt, and M. de Villiers. (2007) The impact of pollination syndrome and habitat on gene flow: a comparative study of two *Streptocarpus* (Gesneriaceae) species. *Am J Bot* 94:1688–95.

Hughes, N. F. (1976) *Palaeobiology of Angiosperm Origins*. New York: Cambridge University Press.

Hull, D. A., and Beattie AJ (1988) Adverse effects of pollen exposed to *Atta texana* and other North American ants: implications for ant pollination. *Oecologia* 75:153–55.

Huntzinger, M., R. Karban, T. P. Young, and T. M. Palmer. (2004) Relaxation of induced indirect defences of acacias following exclusion of mammalian herbivores. *Ecology* 85:609–14.

Hurd, P. D., and E. G. Linsley. (1975) The principal *Larrea* bees of the southwestern United States. *Smithsonian Contrib Zool* 193:1–69.

Hurlbert, S. H. (1970) Flower number, flowering time and reproductive isolation among ten species of *Solidago* (Compositae). *Bull Torrey Bot Club* 97:189–93.

Hurlbert, A. H., S. A. Hosoi, E. J. Temeles, and P. W. Ewals. (1996) Mobility of *Impatiens capensis* flowers: effect on pollen deposition and hummingbird foraging. *Oecologia* 105:243–46.

Husband, B. C., B. Ozimec, S. L. Martin, and L. Pollock. (2008) Mating consequences of polyploid evolution in flowering plants; current trends and insights from synthetic polyploids. *Int J Plant Sci* 169:195–206.

Hussein, M. Y., N. H. Lajis, and J. H. Ali. (1991) Biological and chemical factors associated with the successful introduction of *Elaeidobius kamerunicus* Faust, the oil palm pollinator in Malaysia. *Acta Hort* 288:81–87.

Huth C. J. and O. Pellmyr. (2000) Pollen-mediated selective abortion in yuccas and its consequences for the plant–pollinator mutualism. *Ecology* 81:1100–7.

Huth, H. H., and D. Burkhardt. (1972) Der Spektrale sehrbereich eines Violetta Kolibris. *Naturwissenschaften* 59: 650.

Ichikawa, N., and M. Sasaki. (2003) Importance of social stimuli for the development of learning capability in honeybees. *Appl Entomol & Zool* 38:2203–9.

Igic, B., R. Lande, and J. R. Kohn. (2008) Loss of self-incompatibility and its evolutionary consequences. *Int J Plant Sci* 169:93–104.

Iltis, H. H. (1958) Studies in the Capparidaceae. IV. *Polanisia* Raf. *Brittonia* 10:33–58.

Ings, T. C., N. L. Ward, and L. Chittka. (2006) Can commercially imported bumblebees out-compete their native conspecifics? *J Appl Ecol* 43:940–48.

Inoue, K. (1986) Different effects of sphingid and noctuid moths on the fecundity of *Platanthera metabifolia* (Orchidaceae) in Hokkaido. *Ecol Res* 1:25–36.

———. (1993) Evolution of mutualism in plant-pollinator interactions on islands. *Bioscience* 18:525–36.

Inoue, K., and T. Kawahara. (1990) Allozyme differentiation and genetic structure in island and mainland Japanese populations of *Campanula punctata* (Campanulaceae). *Am J Bot* 77:1440–48.

Inoue, K., M. Maki, and M. Masuda. (1996) Evolution of *Campanula* flowers in relation to insect pollinators on islands. In *Floral Biology: Studies on Floral Evolution in Animal-Pollinated Plants*, ed. D. Lloyd and S. Barrett, 377–400. New York: Chapman and Hall.

Inouye, D. W. (1978) Resource partitioning in bumble bees: experimental studies of foraging behavior. *Ecology* 59: 672–78.

———. (1980a) The effect of proboscis and corolla tube lengths on patterns and rates of flower visitation by bumblebees. *Oecologia* 45:197–201.

———. (1980b) The terminology of floral larceny. *Ecology* 61:1251–52.

———. (1983) The ecology of nectar robbing. In *The Biology of Nectaries*, ed. T. S. Elias and B. L. Bentley, 153–73. New York: Columbia University Press.

Inouye, D. W., N. D. Faure, J. A. Lanum, D. M. Levine, J. B. Meyers, M. S. Roberts, R. C. Tsao, and Y. Wang. (1980) The effects of nonsugar nectar constituents on estimates of nectar energy content. *Ecology* 61:992–96.

Inouye, D. W., D. W. Gill, M. R. Dudash, and C. B. Fenster.

(1994) A model and lexicon for pollen fate. *Am J Bot* 81:1517–30.

Inouye, D. W., and G. H. Pyke. (1988) Pollination biology in the Snowy Mountains of Australia: comparisons with montane Colorado, USA. *Aust J Ecol* 13:191–210.

Internicola, A. I., G. Bernasconi, and L.D.B. Gigord. (2008) Should food-deceptive species flower before or after rewarding species? An experimental test of pollinator visitation behaviour under contrasting phenologies. *J Evol Biol* 21:1358–65.

Irvine, A. K., and Armstrong JE (1988) Beetle pollination in Australian tropical rainforests. *Proc Ecol Soc Aust* 15: 107–13.

———. (1990) Beetle pollination in tropical forests of Australia. In *Reproductive Ecology of Tropical Forest Plants*, ed. K. S. Bawa and M. Hadley, 135–49. Paris: UNESCO/ Parthenon.

Irwin, R. E. (2003) Impact of nectar robbing on estimates of pollen flow: conceptual prediction and empirical outcomes. *Ecology* 84:485–95.

———. (2006) The consequences of direct versus indirect species interactions to selection on traits: pollination and nectar robbing in *Ipomopsis aggregata*. *Am Nat* 167:315–28.

Irwin, R. E., and L. S. Adler. (2006) Correlations among traits associated with herbivore resistance and pollination: implications for pollination and nectar robbing in a distylous plant. *Am J Bot* 93:64–72.

———. (2008) Nectar secondary compounds affect self-pollen transfer: implications for female and male reproduction. *Ecology* 89:2207–17.

Irwin, R. E., L. S. Adler, and A. A. Agrawal. (2004) Community and evolutionary ecology of nectar. *Ecology* 85: 1477–78.

Irwin, R. E., L. S. Adler, and A. K. Brody. (2004) The dual roles of floral traits; pollinator ecology and plant defense. *Ecology* 85:1503–11.

Irwin, R. E., and A. K. Brody. (1999) Nectar-robbing bumblebees reduce the fitness of *Ipomopsis aggregata* (Polemoniaceae). *Ecology* 80:1703–12.

———. (2000) Consequences of nectar robbing for realized male function in a hummingbird-pollinated plant. *Ecology* 81:2637–43.

Irwin, R. E., A. K. Brody, and N. M. Waser. (2001) The impact of floral larceny on individuals, populations and communities. *Oecologia* 129:161–68.

Irwin, R. E., C. Galen, J. J. Rabenold, R. Kaczorowski, and M. L. McCutcheon. (2008) Mechanisms of tolerance of floral larceny in two wildflower species. *Ecology* 89:3093–3104.

Irwin, R. E., and J. E. Maloof. (2002) Variation in nectar robbing over time, space and species. *Oecologia* 133:525–33.

Irwin, R. E., and S. Y. Strauss. (2005) Flower color micro-evolution in wild radish: evolutionary response to pollinator-mediated selection. *Am Nat* 165:225–37.

Ish-Am, G., and D. Eisikowitch. (1991) Possible routes of avocado tree pollination by honeybees. *Acta Hort* 288: 225–33.

Ishida, C., M. Kono, and S. Sakai. (2009) A new pollination system: brood-site pollination by flower bugs in *Macaranga* (Euphorbiaceae). *Ann Bot* 103:39–44.

Ishii, H. S., and L. D. Harder. (2006) The size of individual *Delphinium* flowers and the opportunity for geitonogamous pollination. *Funct Ecol* 20:1115–23.

Ismail, A., and A. G. Ibrahim. (1986) The potential for ceratopogonid midges as pollinators of cacao in Malaysia. In *Biological Control in the Tropics*, ed. M. Y. Hussein and A. G. Ibrahim, 471–84. Selangor, Malaysia: Universiti Pertanian Malaysia.

Itagaki, T., and S. Sakai. (2006) Relationship between floral longevity and sex allocation among flowers within inflorescences in *Aquilegia buergeriana* var. *oxysepala* (Ranunculaceae). *Am J Bot* 93:1320–27.

Ivey, C. T., P. Martinez, and R. Wyatt. (2003) Variation in pollinator effectiveness in swamp milkweed, *Asclepias incarnata* (Apocynaceae). *Am J Bot* 90:214–25.

Iwasa, Y., T. J. de Jong, and P.G.L. Klinkhamer. (1995) Why pollinators visit only a fraction of the open flowers on a plant: the plant's point of view. *J Evol Biol* 8: 439–53.

Jackson, R. R., S. D. Pollard, X. J. Nelson, G. B. Edwards, and A. T. Barrion. (2001) Jumping spiders (Araneae, Salticidae) that feed on nectar. *J Zool* 255:25–29.

Jackson, S., and S. W. Nicolson. (2002) Xylose as a nectar sugar: from biochemistry to ecology. *Comp Biochem Physiol* B 131:613–20.

Jaffé, K., J. V. Hernandez, W. Goitía, A. Osio, F. Osborn, H. Cerda, A. Arab, J. Rincones, R. Gajardo, L. Caraballo, C. Andara, and H. Lopez. (2003) Flower ecology in the neotropics: a flower-ant love-hate relationship. In *Arthropods of Tropical Forests: Spatio-Temporal Dynamics and Resource Use in the Canopy*, ed. Y. Basset, V. Novotny, S. Miller, and R. Kitching, 213–19. Cambridge: Cambridge University Press.

Jakobsen, H. B., and K. Kristjansson. (1994) Influence of temperature and floret age on nectar secretion in *Trifolium repens* L. *Ann Bot London* 74:327–34.

Jakobsson, A., B. Padron, and A. Traveset. (2008) Pollen transfer from invasive *Carpobrotus* spp. to natives - as study of pollinator behaviour and reproduction success. *Biol Cons* 141:136–45.

James, R. L. (1948) Some hummingbird flowers east of the Mississippi. *Castanea* 13:97–109.

Janzen, D. H. (1966) Coevolution of mutualism between ants and acacias in Central America. *Evolution* 20:249–75.

———. (1967) Synchronisation of sexual reproduction of trees within the dry season in Central America. *Evolution* 21:620–37.

———. (1971) Euglossine bees as long-distance pollinators of tropical plants. *Science* 171:2035.

———. (1977) Why don't ants visit flowers? *Biotropica* 9:252.

———. (1979) How to be a fig. *Ann Rev Ecol Syst* 10: 13–51.

Jarau, S., M. Hrncir, M. Ayasse, C. Schulz, W. Franckje, R. Zucchi, and F. G. Barth. (2004) A stingless bee (*Melipona seminigra*) marks food sources with a pheromone from its claw retractor tendons. *J Chem Ecol* 30:793–804.

Jauker, F., and V. Wolters. (2008) Hover flies are efficient pollinators of oilseed rape. *Oecologia* 156:819–23.

Jennersten, O. and D. H. Morse. (1991) The quality of pollination by diurnal and nocturnal insects visiting common milkweed, *Asclepias syriaca*. *Am Midl Nat* 125:18–28.

Jersakova, J. and S. D. Johnson. (2006) Lack of floral nectar reduces self-pollination in a fly-pollinated orchid. *Oecologia* 147:60–68.

Jersakova, J. and S. D. Johnson. (2007) Protandry promotes male pollination success in a moth-pollinated orchid. *Funct Ecol* 21:496–504.

Jervis, M. A. and L. Vilhemsen. (2000) Mouthpart evolution in adults of the basal "symphytan" hymenopteran lineages. *Biol J Linn Soc* 70:121–146.

Jesson, L. K. and S.C.H. Barrett. (2002) Enantiostyly in *Wachendorfia* (Haemodoraceae): the influence of reproductive systems on the maintenance of the polymorphism. *Am J Bot* 89:253–62.

———. (2005) Experimental tests of the function of mirror-image flowers. *Biol J Linn Soc* 85:167–79.

Jetter, R. (2006) Examination of the processes involved in the emission of scent volatiles from flowers. In *Biology of Floral Scent*, ed. N. Dudareva and E. Pichersky, 125–44. Boca Raton, FL: CRC Press.

Jewell, J., J. McKee, and A. J. Richards. (1994) The keel color polymorphism in *Lotus corniculatus* L.: differences in internal flower temperatures. *New Phytol* 128: 363–68.

Jin, B., N. Li, N. Jia, W. Z. Zhou, L. Wang, and C. B. Shang. (2007) Observations on the anatomy of reproductive organs and the pollinators of *Viburnum macrocephalum* f. keteleeri (Caprifoliaceae). *Acta Phyto Sin* 45:753–68.

Johansen, C. A. (1977) Pesticides and pollination. *Ann Rev Ent* 22:177–92.

Johansen, C. A. and D. F. Mayer. (1990) *Pollinator Protection: a Bee and Pesticide Handbook*. Cheshire, CT: Wicwas Press.

Johansen, C. A., D. F. Mayer, J. D. Eves, and C. W. Isious. (1983) Pesticides and bees. *Env Entomol* 12:1513–18.

Johnson, L. K. and S. P. Hubbell. (1974) Aggression and competition among stingless bees: field studies. *Ecology* 55:120–27.

Johnson S. D. (1992) Plant-animal relationships. In *The Ecology of Fynbos: Nutrients, Fire and Diversity* ed. R. M. Cowling, 372–88. Cape Town: Oxford University Press.

———. (1993) Carpenter bee pollination of *Herschelianthe graminifolia* (Orchidaceae) on the Cape Peninsula. *Flora* 188:383–86.

———. (1994) Evidence for Batesian mimicry in a butterfly-pollinated orchid. *Biol J Linn Soc* 53:91–104.

———. (2005) Specialized pollination by spider-hunting wasps in the African orchid *Disa sankeyi*. *Plant Syst Evol* 251:153–60.

———. (2006) Pollinator-driven speciation in plants. In *Ecology and Evolution of Flowers*, ed. L. D. Harder and S.C.H. Barrett 295–310. Oxford: Oxford University Press.

Johnson, S. D., R. Alexandersson, and H. P. Linder .(2003) Experimental and phylogenetic evidence for floral mimicry in a guild of fly-pollinated plants. *Biol J Linn Soc* 80:289–304.

Johnson, S. D. and A. Dafni. (1998) Response of bee-flies to the shape and pattern of model flowers: implications for floral evolution in a Mediterranean herb. *Funct Ecol* 12:289–97.

Johnson, S. D. and T. J. Edwards. (2000) The structure and function of orchid pollinia. In *Pollen and Pollination*, ed. A. Dafni, M. Hesse, and E. Pacini, 243–69. Vienna: Springer.

Johnson, S. D., A. Ellis, and S. Dotterl. (2007) Specialization for pollination by beetles and wasps: the role of lollipop hairs and fragrance in *Satyrium microrrhynchum* (Orchidaceae). *Am J Bot* 94:47–55.

Johnson, S. D., A. L. Hargreaves, and M. Brown. (2006) Dark, bitter-tasting nectar functions as a filter of flower visitors in a bird-pollinated plant. *Ecology* 87:2709–16.

Johnson, S. D., H. P. Linder,and K. E. Steiner. (1998) Phylogeny and radiation of pollination syndromes in *Disa* (Orchidaceae). *Am J Bot* 85:402–11.

Johnson, S. D. and J. J. Midgley. (2001) Pollination by monkey beetles (Scarabaeidae: Hoplinii): do color and dark centres of flowers influence alighting behaviour? *Envir Ent* 30:861–68.

Johnson, S. D. and S. Morita. (2006) Lying to Pinocchio: floral deception in an orchid pollinated by long-proboscid flies. *Bot J Linn Soc* 152:271–78.

Johnson, S. D., P. R. Neal, and L. D. Harder. (2005) Pollen fates and the limits on male reproductive success in an orchid population. *Biol J Linn Soc* 86:175–90.

Johnson, S. D. and S. W. Nicolson. (2008) Evolutionary associations between nectar properties and specificity in bird pollination systems. *Biol Lett* 4:49–52.

Johnson, S. D., A. Pauw, and J. J. Midgeley. (2001) Rodent pollination in the African lily *Massonia depressa* (Hyacinthaceae). *Am J Bot* 88:1768–73.

Johnson, S. D., C. I. Peter, L. A. Nilsson, and J. Agren. (2003) Pollination success in a deceptive orchid is enhanced by co-occurring rewarding magnet plants. *Ecology* 84:2919–27.

Johnson, S. D. and K. E. Steiner.(1995) Long-proboscid fly pollination of two orchids in the Cape Drakensberg mountains, South Africa. *Plant Syst Evol* 195:169–175.

———. (1997) Long-tongued fly pollination and evolution of floral spur length in the *Disa draconis* complex (Orchidaceae). *Evolution* 51:45–53.

———. (2000) Generalisation versus specialisation in plant pollination systems. *Trends Ecol Evol* 15:140–43.

———. (2003) Specialized pollination systems in southern Africa. *S Afr J Sci* 99:345–48.

Jolls, C. L. (1980) Phenotypic patterns of variation in biomass allocation in *Sedum lanceolatum* Torr. at four elevational sites in the Front Range, Rocky Mountains, Colorado. *Bull Torry Bot Club* 107:65–70.

Jones, C. E. and S. L. Buchmann. (1974) Ultraviolet floral patterns as functional orientation cues in hymenopterous pollination systems. *Anim Behav* 22:481–85.

Jones, C. E. and R. J. Little, ed. (1983) *Handbook of Experimental Pollination Biology*. New York: Van Nostrand Reinhold.

Jones, K. N. (2001) Pollinator-mediated assortative mating: causes and consequences. In *Cognitive Ecology of Pollination*, ed. L. Chittka and J. D. Thomson, 259–73. Cambridge: Cambridge University Press.

Jonsson, M., A. Lindkvist, and P. Anderson. (2005) Behavioural responses in three ichnemonid pollen beetle parasitoids to volatile emitted from different phonological stages of oilseed rape. *Ent Exp Appl* 115:363–69.

Jordan, C. Y. and L. D. Harder. (2006) Manipulation of bee behaviour by inflorescence architecture and its consequences for plant mating. *Am Nat* 167:496–509.

Jordano, P. (1987) Patterns of mutualistic interaction in pollination and seed dispersal: connectance, dependence, asymmetries, and coevolution. *Am Nat* 129:657–77.

Jordano, P., J. Bascompte, and J. M. Olesen. (2003) Invariant properties in coevolutionary networks of plant-animal interactions. *Ecol Lett* 6:69–81.

———. (2006) The ecological consequences of complex topology and nested structure in pollination webs. In *Plant-Pollinator Interactions: from Specialization to Generalization*, ed. N. M. Waser and J. Ollerton , 173–99. Chicago: University of Chicago Press.

Jorgensen, T. H. and S. Andersson. (2005) Evolution and maintenance of pollen-colour dimorphisms in *Nigella degenii*: habitat-correlated variation and morph-by-environment interactions. *New Phytol* 168:487–98.

Jorgensen, T. H., T. Petanidou, and S. Andersson. (2006) The potential for selection on pollen color dimorphisms in *Nigella digenii*: morph-specific differences in pollinator visitation, fertilisation success and siring ability. *Evol Ecol* 20:291–306.

Josens, R. B. and W. M. Farina. (2001) Nectar feeding by the hovering hawkmoth *Macroglossum stellatarum*: intake rate as a function of viscosity and concentration of sucrose solutions. *J Comp Physiol J Comp Physiol* A 187: 661–65.

Juillet, N., M. A. Gonzalez, P. A. Page and L.D.B. Gigord. (2007) Pollination of the European food-deceptive *Traunsteinera globosa* (Orchidaceae): the importance of nectar-producing neighbouring plants. *Plant Syst Evol* 265: 12–129.

Junker, R. and N. Blüthgen. (2008) Floral scents repel potentially nectar thieving ants. *Evol Ecol Res* 10:295–308.

Junker, R., A.Y.C. Chung, and N. Blüthgen. (2007) Interactions between flowers, ants and pollinators: additional evidence for floral repellence against ants. *Ecol Res* 22: 665–670.

Jürgens, A. (2006) Comparative floral morphometrics in day-flowering, night-flowering and self-pollinated Caryophylloideae (*Agrostemma, Dianthus, Saponaria, Silene* and *Vaccaria*). *Plant Syst Evol* 257:233–50.

———. (2009) The hidden language of flowering plants: floral odours as a key for understanding angiosperm evolution? *New Phytol* 183:240–43.

Jürgens, A. and S. Dotterl. (2004) Chemical composition of anther volatiles in Ranunculaceae: genera-specific profiles in *Anemone, Aquilegia, Caltha, Pulsatilla, Ranunculus* and *Trollius* species. *Am J Bot* 91:1969–80.

Jürgens, A., S. Dotterl, and U. Meve. (2006) The chemical nature of fetid floral odours in stapeliads (Apocynaceae-Asclepiadoideae-Ceropegieae). *New Phytol* 172:452–68.

Kadmon, R. and A. Shmida. (1992) Departure rules used by bees foraging for nectar: a field test. *Evol Ecol* 6: 142–51.

Kaiser, R. (2006) Flowers and fungi use scents to mimic each other. *Science* 311:806–7.

Kalisz, S., R. H. Ree, and R. D. Sargent. (2006) Linking floral symmetry genes to breeding system evolution. *Trends Plant Sci* 11:568–73.

Kalisz, S., D. W. Vogler, and K. M. Hanley.(2004) Context-dependent autonomous self-fertilization yields reproductive assurance in mixed mating. *Nature* 430:334–37.

Kandori, I. (2002) Diverse visitors with various pollinator importance, and temporal change in the important pollinators of *Geranium thunbergii* (Geraniaceae). *Ecol Res* 17:283–94.

Kandori, I. and N. Ohsaki. (1996) The learning abilities of the cabbage white butterfly, *Pieris rapae*, foraging for flowers. *Res Popul Ecol* 38:111–17.

Kaplan, S. M. and D. L. Mulcahy. (1971) Mode of pollination and floral sexuality in *Thalictrum*. *Evolution* 25:659–68.

Karban, R. and S. Y. Strauss (1993) Effects of herbivores on growth and reproduction of their perennial host, *Erigeron glaucus*. *Ecology* 74:39–46.

Karoly, K. (1994) Inbreeding effects on mating system traits for two species of *Lupinus* (Leguminosae). *Am J Bot* 81: 1538–44.

Karron, J. D., K. G. Holmquist, R. J. Flanagan and R. J. Mitchell (2009) Pollinator visitation patterns strongly influence among-flower variation in selfing rate. *Ann Bot* 103:1379–83.

Kastinger, C. and A. Weber. (2001) Bee-flies (*Bombylius* spp, Bombylidae, Diptera) and the pollination of flowers. *Flora* 196:3–25.

Kato, M. and T. Inoue. (1994) Origin of insect pollination. *Nature* 368:195.

Kato, M., A. Takimura, and A. Kawakita. (2003) An obligate pollination mutualism and reciprocal diversification in the tree genus *Glochidium* (Euphorbiaceae). *Proc Nat Acad Sci USA* 100:5264–67.

Kawaguchi, L. G., K. Ohashi, and Y. Toquenaga. (2007) Contrasting responses of bumblebees to feeding conspecifics on their familiar and unfamiliar flowers. *Proc Roy Soc Lond B* 274:2661–67.

Kawakita, A. and M. Kato. (2004) Evolution of obligate pollination mutualism in New Caledonian *Phyllanthus* (Euphorbiaceae). *Am J Bot* 91:410–15.

———. (2009) Repeated independent evolution of obligate pollination mutualism in the Phyllantheae-*Epicephala* association. *Proc Roy Soc Lond* B 276:417–26.

Kawakubo, N. (1990) Dioecism of the genus *Callicarpa* (Verbenaceae) in the Bonin (Ogasawara) Islands. *Bot Mag Tokyo* 103:57–66.

Kawano, S. and J. Masuda. (1980) The productive and reproductive biology of flowering plants. 7. Resource allocation and reproductive capacity in wild populations of *Helionopsis orientalis* (Thumb) C-Tanaka (Liliaceae). *Oecologia* 45:307–17.

Kay, K. M., C. Voelckel, D. Y. Yang, K. M. Hufford, D. D. Kaska, and S. A. Hodges. (2006) Floral characters and species diversification. In *Ecology and Evolution of Flowers*, ed. L. D. Harder and S.C.H. Barrett, 311–25. Oxford: Oxford University Press.

Kay, Q.O.N. (1976) Preferential pollination of yellow-flowered morphs of *Raphanus raphanistrum* by *Pieris* and *Eristalis* spp. *Nature* 261:230–32.

———. (1978) The role of preferential and assortative pollination in the maintenance of flower colour polymorphisms. In *The Pollination of Flowers by Insects*, ed. A. J. Richards, 175–90. London: Academic Press.

———. (1985) Nectar from willow catkins as a food source for blue tits. *Bird Study* 32:41–45.

Kay, Q.O.N., H. S. Daoud, and C. H. Stirton (1981) Pigment distribution, light reflection and cell structure in petals. *Bot J Linn Soc* 83:57–84.

Kay, Q.O.N, A. J. Lack, F. C. Bamber, and C. R. Davies. (1984) Differences between sexes in floral morphology, nectar production and insect visits in a dioecious species, *Silene dioica*. *New Phytol* 98:515–29.

Kay, Q.O.N and D. P. Stevens. (1986) The frequency, distribution and reproductive biology of dioecious species in the native flora of Britain and Ireland. *Biol J Linn Soc* 92:39–64.

Kearns, C. A. (2001) North American dipteran pollinators: assessing their value and conservation status. *Cons Ecol* 5 (1): Art 5.

Kearns, C. A. and D. W. Inouye (1993) Techniques for Pollination Biologists. Niwot, CO: University Press Colorado.

———. (1997) Pollinators, flowering plants and conservation biology. *Bioscience* 47:297–307.

Kearns, C. A., D. W. Inouye, and N. M. Waser.(1998) Endangered mutualisms: the conservation of plant-pollinator interactions. *Ann Rev Ecol Syst* 29:83–112.

Keasar, T., U. Motro, Y. Shur, and A. Shmida. (1996) Overnight memory retention of foraging skills by bumblebees is imperfect. *Anim Behav* 52:95–104.

Keasar, T., G. Pollak, R. Arnon, D. Cohen,and A. Shmida. (2007) Honesty of signalling and pollinator attraction: the case of flag-like bracts. *Isr J Plant Sci* 54:119–28.

Keijzer, C. J. (1987) The process of anther dehiscence and pollen dispersal. 2. The formation and transfer mechanism of pollenkitt, cell wall development in the locule tissues, and a function of the orbicules in pollen dispersal. *New Phytol* 105:499–507.

———. (1999) Mechanisms of angiosperm anther dehiscence, a historical review. In *Anther and Pollen: from Biology to Biotechnology*, ed. C. Clement, E. Pacini and J. C. Audran, 54–67. Berlin: Springer.

Kelber, A. (1996) Colour learning in the hawkmoth *Macroglossum stellatarum*. *J Exp Biol* 199:1127–31.

———. (2001) Receptor-based models for spontaneous colour choices in flies and butterflies. *Ent Exp Appl* 99:231–44.

———. (2002) Pattern discrimination in a hawkmoth: innate preferences, learning performance, and ecology. *Proc Roy Soc Lond B* 269:2573–77.

———. (2006) Invertebrate colour vision. In *Invertebrate Vision*, ed. E. Warrant and D-D Nilsson, 250–90. Cambridge: Cambridge University Press.

Kelber, A. and A. Balkenius. (2007) Sensory ecology of feeding in the hummingbird hawkmoth *Macroglossum stellatarum* (Lepidoptera, Sphingidae). *Entomol Gen* 29: 97–110.

Kelber, A., A. Balkenius, and E. J. Warren. (2003) Colour vision in diurnal and nocturnal hawkmoths. *Integr Comp Biol* 43:571–79.

Kelber, A. and M. Pfaff. (1997) Spontaneous and learned preferences for visual flower features in a diurnal moth. *Isr J Plant Sci* 45:235–46.

Kelly, M. G. and D. A. Levin. (2000) Directional selection on initial flowering date in *Phlox drummondii* (Polemoniaceae). *Am J Bot* 87:382–91.

Kenrick, J. and R. B. Knox. (1989) Pollen-pistil interactions in Leguminosae (Mimosoideae). In *Advances in Legume Biology* 29:127–56. Mississippi Botanical Gardden: St. Louis, MI.

Kenta, T., N. Inari, T. Nagamitsu, K. Goka, and T. Hiura. (2007) Commercialized European bumblebee can cause pollination disturbance: an experiment on seven native plant species in Japan. *Biol Cons* 134:298–309.

Kephart, S., R. J. Reynolds, M. T. Rutter, C. B. Fenster, and M. R. Dudash. (2006) Pollination and seed predation by moths on *Silene* and allied Caryophyllaceae; evaluating a model system to study the evolution of mutualisms. *New Phytol* 169:667–80.

Kerner, A. J. (1878/2008) *Flowers and their Unbidden Guests*. Charleston, SC: BiblioBazaar.

Kerr, J. T. (2001) Butterfly species richness patterns in Canada: energy, heterogeneity and the potential consequences of climate change. *Cons Ecol* 5 (1): Art 10.

Kessler, A. and R. Halitschke. (2009) Testing the potential for conflicting selection on floral chemical traits by pollinators and herbivores: predictions and a case study. *Func Ecol* 23:901–12.

Kessler, D. and I. T. Baldwin. (2007) Making sense of nectar scents: the effects of nectar secondary metabolites on floral visitors of *Nicotiana attenuata*. *Plant J* 49: 840–54.

Kessler, D., K. Gase and I. T. Baldwin. (2008) Field experiments with transformed plants reveal the sense of floral scents. *Science* 321:1200–202.

Kevan, P. G. (1972) Floral colors in the High Arctic with reference to insect flower relations and pollination. *Can J Bot* 50:2289–316.

———. (1975a) Forest application of the insecticide fenitrothion and its effect on wild bee pollinators of Lowbush Blueberry (*Vaccinium* spp) in southern New Brunswick, Canada. *Biol Cons* 7:301–9.

———. (1975b) Pollination and environmental conservation. *Env Cons* 2:293–98.

———. (1975c) Sun-tracking solar furnaces in high arctic flowers: significance for pollination and insects. *Science* 189:723–26.

———. (1978) Floral coloration, its colorimetric analysis and significance in anthecology. In *The Pollination of Flowers by Insects*, ed. A. J. Richards, 51–78. London: Academic Press.

———. (1979) Vegetation and floral colors revealed by ultraviolet light: interpretational difficulties for functional significance. *Am J Bot* 66:749–51.

———. (1983) Floral colors through the insect eye: what they are and what they mean. In *Handbook of Experimental Pollination Ecology*, ed. C. E. Jones and R. J. Little, 3–30. New York: Van Nostrand Reinhold.

———. (1990) Sexual differences in temperatures of blossoms on a dioecious plant, *Salix arctica*: significance for life in the arctic. *Arct Alp Res* 22:283–89.

Kevan, P. G. and M. Lane. (1985) Flower petal microtexture is a tactile cue for bees. *Proc Nat Acad Sci USA* 82: 4750–52.

Kevan, P. G. and T. P. Phillips. (2001) The economic impacts of pollinator declines: an approach to assessing the consequences. *Conserv Ecol* 5 (1): Art 8.

Kevan, P. G. and R. C. Plowright. (1995) Impact of pesticides on forest pollination. In *Forest Insect Pests in Canada*, ed. J. A. Armstrong and W.G.H. Ives, 607–18. Ottawa: Natural Resources Canada.

Kevan, P. G., E. A. Tikhmenev, and M. Usui. (1993) Insects and plants in the pollination ecology of the boreal zone. *Ecol Res* 8:247–67.

Keys, R. N., S. L. Buchmann, and S. E. Smith. (1995) Pollination effectiveness and pollination efficiency of insects foraging *Prosopis velutina* in South-Eastern Arizona. *J Appl Ecol* 32:519–27.

Kimsey, L. S. (1984) The behavioural and structural aspects of grooming and related activities in euglossine bees (Hymenoptera, Apidae). *J Zool* 204:541–50.

Kindlmann, P. and J. Jersakova. (2006) Effect of floral display on reproductive success in terrestrial orchids. *Folia Geobot* 41:47–60.

Kinet, J. M., R. M. Sachs, and G. Bernier.(1985) *The Physiology of Flowering. III. The Development of Flowers.* Boca Raton, FL: CRC Press.

King, M. J. (1993) Buzz foraging mechanism of bumble bees. *J Apic Res* 32:41–49.

King, M. J. and S. L. Buchmann. (1996) Sonication dispensing of pollen from *Solanum laciniatum* flowers. *Funct Ecol* 10:449–56.

———. (2003) Floral sonication by bees: mesosomal vibration by *Bombus* and *Xylocopa*, but not *Apis* (Hymenoptera, Apidae), ejects pollen from poricidal anthers. *J Kans Ent Soc* 76:295–305.

Kingsolver, J. G. and T. L. Daniel. (1983) Mechanical determinants of nectar feeding strategy in hummingbirds: energetics, tongue morphology and licking behavior. *Oecologia* 60:214–26.

———. (1995) Mechanics of food-handling by nectar-feeding insects. In *Regulatory Mechanisms in Insect Feeding*, ed. R. F. Chapman and G. deBoer, 32–73. New York: Chapman and Hall.

Kinoshita, M., N. Shimada, and K. Arikawa. (1999) Colour vision of the foraging swallowtail butterfly *Papilio xuthus*. *J Exp Biol* 202:95–102.

Kirk, W.D.J. (1984a) Pollen-feeding in thrips (Insecta: Thysanoptera). *J Zool* 204:107–17.

———. (1985) Pollen-feeding and the host specificity and fecundity of flower thrips (Thysanoptera). *Ecol Entomol* 10:281–89.

———. (1987) How much pollen can thrips destroy? *Ecol Entomol* 12:31–40.

———. (1993) Interspecific size and number variation in pollen grains and seeds. *Biol J Linn Soc* 49:239–48.

Kite, G. C. and W.L.A. Hetterschieid. (1997) Inflorescence odours of *Amorphophallus* and *Pseudodracontium* (Araceae). *Phytochemistry* 46:71–75.

Kjellberg, B., S. Karlsson and I. Kertensson. (1982) Effects of heliotropic movements of flowers of *Dryas octopetala* L. on gynoecium temperature and seed development. *Oecologia* 54:10–13.

Kjellberg, F., E. Jousselin, J. L. Bronstein, A. Patel, J. Yokoyama and J. V. Rasplus. (2001) Pollination modes in fig wasps: the predictive power of correlated traits. *Proc Roy Soc Lond* B 268:1113–21.

Klein, A. M., I. Steffan-Dewenter, and T. Tscharntke. (2003a) Pollination of *Coffea canephora* in relation to local and regional agroforestry management. *J Appl Ecol* 40:837–45.

———. (2003b) Bee pollination and fruit set of *Coffea arabica* and *C. canephora* (Rubiaceae). *Am J Bot* 90:153–57.

———. (2003c) Fruit-set of highland coffee increases with the diversity of pollinating bees. *Proc Roy Soc Lond* B 270:955–61.

Klein, A. M., I. Steffan-Dewenter, and T. Tscharntke. (2006) Rain forest promotes trophic interactions and diversity of trap-nesting Hymenoptera in adjacent agroforestry. *J Anim Ecol* 75:315–23.

Klein, A. M., B. S. Vaissiere, J. H. Cane, I. Steffan-Dewenter, S. A. Cunningham, C. Kremen and T. Tscharntke. (2007) Importance of pollinators in changing landscapes for world crops. *Proc Roy Soc Lond* B 274:303–13.

Kleizen, C., J. Midgley, and S. D. Johnson. (2008) Pollination systems of *Colchicum* (Colchicaceae) in Southern Africa: evidence for rodent pollination. *Ann Bot* 102:747–55.

Kliber, A. and C. G. Eckert (2004) Sequential decline in allocation among flowers within inflorescence; proximate mechanisms and adaptive significance. *Ecology* 85:1675–87.

Klinkhamer, P. G. and T. J. de Jong. (1993) Attractiveness to pollinators – a plant's dilemma. *Oikos* 66:180–84.

Klinkhamer, P. G., T. J. de Jong, and J.A.J. Metz. (1994)

Why plants can be too attractive: a discussion of measures to estimate male fitness. *J Ecol* 82:191–94.

Klitgaard, B. B. and I. K. Ferguson. (1992) Pollen morphology of *Browneopsis* (Leguminosae, Caesalpiniodeae), and its evolutionary significance. *Grana* 31:285–90.

Knapp, S., V. Persson, and S. Blackmore. (1998) Pollen morphology and functional dioecy in *Solanum* (Solanaceae). *Plant Syst Evol* 210:13–39.

Knight, T. M., M. W. McCoy, J. M. Chase , K. A. McCoy, and R. D. Holt. (2005) Trophic cascades across ecosystems. *Nature* 437:880–83.

Knight, T. M., J. A. Steets,and T. L. Ashman. (2006) A quantitative synthesis of pollen supplementation experiments highlights the contribution of resource reallocation to estimates of pollen limitation. *Am J Bot* 93:271–77.

Knight, T. M., J. A. Steets, J. C. Vamosi, S. J. Mazer, M. Burd, D. R. Campbell, M. R. Dudash, M. O. Johnston, R. J. Mitchell, and T. L. Ashman. (2005) Pollen limitation of plant reproduction: pattern and process. *Ann Rev Ecol Evol Syst* 36:467–97.

Knoll, F. (1922) Lichtsinn und Blütenbesuch des Falters von *Macroglossum stellatarum*. *Abheil Zool Bot Ges Wien* 12:121–78.

———. (1925) Lichtsinn und Blütenbesuch des Falters von *Deilephila livornica*. *Z Vergl Physiol* 2:329–80.

———. (1926) Die Arum-Blütenstände und ihre Besucher. *Abhand Zool-Bot Gesell Wien* 12:382–481.

Knox, R. B., J. Kenrick, P. Bernhardt, R. Marginson, G. Beresford, I. Baker, and H. G. Baker. (1985) Extrafloral nectaries as adaptations for bird pollination in *Acacia terminalis*. *Am J Bot* 72:1185–96.

Knudsen, J. T. (2002) Variation in floral scent composition within and between populations of *Geonoma macrostachys* (Arecaeeae) in the western Amazon. *Am J Bot* 89:1772–78.

Knudsen, J. T. and J. Gershenzon. (2006) The chemical diversity of floral scent. In *Biology of Floral Scent*, ed. N. Dudareva and E. Pichersky, 27–52. Boca Raton, FL: CRC Press.

Knudsen, J. T.and J. M. Olesen. (1993) Buzz-pollination and patterns in sexual traits in North European Pyrolaceae. *Am J Bot* 80:900–13.

Knudsen, J. T.and L. Tollsten. (1993) Trends in floral scent chemistry in pollination syndromes: floral scent composition in moth-pollinated taxa. *Bot J Linn Soc* 113:263–84.

———. (1995) Floral scent in bat-pollinated plants: a case of convergent evolution. *Bot J Linn Soc* 118:45–57.

Knudsen, J. T., L. Tollsten L and G. Bergström. (1993) Floral scents: a check list of volatile compounds isolated by headspace techniques. *Phytochemistry* 33:253–80.

Knudsen, J. T., L. Tollsten, and F. Ervik. (2001) Flower scent and pollination in selected neotropical palms. *Plant Biol* 3:642–53.

Knudsen, J. T., L. Tollsten, I. Groth, G. Bergström, and R. A. Raguso. (2004) Trends in floral scent chemistry in pollination syndromes: floral scent composition in hummingbird-pollinated taxa. *Bot J Linn Soc* 146:191–99.

Knuth, P. (1898) *Handbuch der Blutenbiologie*. vol 1. Leipzig: Engelmann.

Kochmer, J. P. and S. N. Handel. (1986) Constraints and competition in the evolution of flowering phenology. *Ecol Monogr* 56:303–25.

Koeman-Kwak, M. (1973) The pollination of *Pedicularis palustris* by nectar thieves (short-tongued bumblebees). *Acta Bot Neerl* 22:608–15.

Kohler, F., J. Verhulst, R. van Klink, and D. Kleijn. (2008). At what spatial scale do high-quality habitats enhance the diversity of forbs and pollinators in intensively farmed landscapes? *J Appl Ecol* 45:753–62.

Kohn, J. R. and S.C.H. Barrett. (1994) Pollen discounting and the spread of a selfing variant in tristylous *Eichhornia paniculata*: evidence from experimental populations. *Evolution* 48:1576–94.

Kolb, A. (2008) Habitat fragmentation reduced plant fitness by disturbing pollination and modifying response to herbivory. *Biol Cons* 141:2540–49.

Kono, M. and H. Tobe. (2007) Is *Cycas revoluta* (Cycadaceae) wind- or insect-pollinated? *Am J Bot* 94:847–55.

Koopowitz, H. and T. A. Marchant. (1998) Postpollination nectar reabsorption in the African epiphyte *Aerangis verdickii* (Orchidaceae). *Am J Bot* 85:508–12.

Koptur, S. (1984) Outcrossing and pollinator limitation of fruit set: breeding systems of neotropical *Inga* trees. (Fabaceae: Mimosoideae). *Evolution* 38:1130–43.

Koptur, S. (2006) The conservation of specialized and generalized pollination systems in subtropical ecosystems; a case study. In *Plant-Pollinator Interactions: from Specialization to Generalization*, ed. N. M. Waser and J. Ollerton, 341–61. Chicago: University of Chicago Press.

Kosior, A., W. Celary, P. Olejniczak, J. Fijal, W. Krol, W. Solarz and P. Plonka. (2007) The decline of bumblebees and cuckoo bees (Hymenoptera: Apidae: Bombini) of Western and Central Europe. *Oryx* 41:79–88.

Koul, P., A. K. Koul and I. A. Hamal (1989) Reproductive biology of wild and cultivated carrot (*Daucus carota* L.) *New Phytol* 112:437–43.

Kralj, J., A. Brockmann, S. Fuchs and J. Tautz. (2007) The parasitic mite *Varroa destructor* affects non-associative learning in honeybee foragers, *Apis mellifera* L. *J Comp Physiol* A 193:363–70.

Krannitz, P. G. (1996) Reproductive ecology of *Dryas integrifolia* in the high Arctic semi-desert. *Can J Bot* 74: 1451–60.

Krauss, S. L. (2000) Patterns of mating in *Persoonia mollis* (Proteaceae) revealed by an analysis of paternity using

AFLP: implications for conservation. *Aust J Bot* 48: 349–56.

Kremen, C., N. M. Williams, M. A. Aizen, B. Gemmill-Herren, G. LeBuhn, R. Minckley, L. Packer, S. G. Potts, T. Roulston, I. Steffan-Dewenter, D. P. Vazquez, R. Winfree, L. Adams, E. E. Crone, S. S. Greenleaf, T. H. Keitt, A. M. Klein, J. Regetz and T. H. Ricketts. (2007) Pollination and other ecosystem services produced by mobile organisms: a conceptual framework for the effects of land-use change. *Ecol Lett* 10:299–314.

Kremen, C., N. M. Williams, and R. W. Thorp. (2002) Crop pollination from native bees at risk from agricultural intensification. *Proc Nat Acad Sci USA* 99:16812–16.

Krenn, H. W., B-A Gerebne-Krenn, B. M. Steinwender, and A. Popov. (2008) Flower-visiting Neuroptera: mouthparts and feeding behaviour of *Nemoptera sinuata* (Nemopteridae). *Eur J Entomol* 105:267–77.

Krenn, H. W., J. D. Plant, and N. U. Szucsich. (2005) Mouthparts of flower-visiting insects. *Arthropod Struc and Dev* 34:1–40.

Kress, W. J. and J. H. Beach. (1994) Flowering plant reproductive systems at La Selva Biological Station. In *La Selva: Ecology and Natural History of a Neotropical Rain Forest*, ed. L. A. McDade, K. S. Bawa, H. Hespenheide, and G. Hartshorn, 161–82. Chicago: University of Chicago Press.

Kress, W. J., G. E. Schatz, M. Andrianifahanana, and H S. Morland. (1994) Pollination of *Ravenala madagascariensis* (Strelitziaceae) by lemurs in Madagascar: evidence for an archaic coevolutionary system. *Am J Bot* 81:542–51.

Kriston, I. (1971) Zum Problem des Lernverhaltens von *Apis mellifica* L. gegenüber verschiedenen Duftstoffen. *Zeits Vergl Physiol* 74:169–89.

Krombein, K. V. and B. B. Norden. (1997) Nesting behaviour of *Krombeinictus nordenae* Leclercq, a sphecid wasp with vegetarian larvae (Hymenoptera, Sphecidae, Crabroninae). *Proc Ent Soc Wash* 99:42–49.

Kronenberg, F. and H. C. Heller. (1982) Colonial thermoregulation in honeybees (*Apis mellifera*). *J Comp Physiol* 148:65–76.

Kropf, M. and S. S. Renner. (2005) Pollination success in monochromic yellow populations of the rewardless orchid *Dactylorhiza sambucina*. *Plant Syst Evol* 254:185–98.

Kruess, A. and T. Tscharntke. (2002) Grazing intensity and the diversity of grasshoppers, butterflies, and trap-nesting bees and wasps. *Cons Biol* 16:1570–80.

Krupnick, G. A. and A. E. Weis. (1999) The effect of floral herbivory on male and female reproductive success in *Isomeris arborea*. *Ecology* 80:135–49.

Krupnick, G. A., A. E. Weis. and D. R. Campbell. (1999) The consequences of floral herbivory for pollinator service to *Isomeris arborea*. *Ecology* 80:25–34.

Kudo, G. (1991) Effects of snow-free period on the phenology of alpine plants inhabiting snow patches. *Arct and Alp Res* 23:436–43.

———. (1995) Ecological significance of flower heliotropism in the spring ephemeral *Adonis ramosa* (Ranunculaceae). *Oikos* 72:14–0.

———. (2006) Flowering phenologies of animal-pollinated plants: reproductive strategies and agents of selection. In *Ecology and Evolution of Flowers*, ed. L. D. Harder and S.C.H. Barrett, 139–58. Oxford: Oxford University Press.

Kudo, G. and L. D. Harder. (2005) Floral and inflorescence effects on variation in pollen removal and seed production among six legume species. *Funct Ecol* 19:245–54.

Kudo, G., H. S. Ishii, Y. Hirabayashi, and T. Y. Ida. (2007) A test of the effect of floral color change on pollination effectiveness using artificial inflorescences visited by bumblebees. *Oecologia* 154:119–28.

Kudo, G. and T. Kasagi. (2005) Microscale variations in the mating systems and heterospecific incompatibility mediated by pollination competition in alpine snow-bed plants. *Plant Species Biol* 20:93–103.

Kudo, G., T. Maeda, and K. Narita. (2001) Variation in floral sex allocation and reproductive success within inflorescences of *Corydalis ambigua* (Fumariaceae); pollination efficiency or resource limitation? *J Ecol* 89:48–56.

Kugler, H. (1938) Sind *Veronica chamaedrys* L. und *Circaea lutetiana* L. Schwebefliegenblumen? *Bot Arch* 39:147–65.

———. (1950) Der Blutenbesuch der Schammfliege (*Eristalomyia tenax*). *Z Vergl Physiol* 32:328–47.

Kugler, H. (1956) Über die optische Wirkung von Fliegenblumen auf Fliegen. *Ber Deut Bot Ges* 69:387–98.

———. (1963) UV-Musterungen auf Blüten und ihr Zustandekommen. *Planta* 59:296–29.

———. (1966) UV-Male auf Blüten. *Ber Deut Bot Ges* 79:57–70.

———. (1970) Blütenökologie. Stuttgart: Fischer.

Kuhnholz, S. and T. D. Seeley.(1997) The control of water collection in honey bee colonies. *Behav Ecol Sociobiol* 41:407–22.

Kulahci, I. G., A. Dornhaus, and D. R. Papaj. (2008) Multimodal signals enhance decision making in foraging bumble-bees. *Proc Roy Soc Lond* B 275:797–802.

Kullenberg, B. (1961) Studies on *Ophrys* L. pollination. *Zool Bidr Uppsala* 34:1–340.

Kumano-Nomura, Y and R. Yamaoka. (2009) Beetle visitations, and associations with quantitative variation of attractants in floral odors of *Homaolomena propinqua* (Araceae). *J Plant Res* 12:183–192.

Kummerov, J. (1983) Comparative phenology of Mediterranean type plant communities. In *Mediterranean Plant Ecosystems*, ed. F. J. Kruger, D. L. Mitchell, and J.U.M. Jarvis, 300–317. Berlin: Springer Verlag.

Kunte, K. (2007) Allometry and functional constraints on proboscis length in butterflies. *Funct Ecol* 21:982–87.

Kunz, T. H. and M. B. Fenton, ed. (2003) *Bat Ecology*. Chicago: University of Chicago Press.

Kunze, J. (1991) Structure and function in asclepiad pollination. *Plant Syst Evol* 176:227–53.

Kwak, M. M. (1993) The relative importance of syrphids and bumblebees as pollinators of three plant species. *Proc Exp Appl Entomol* 4:137–43.

Kwak, M. M. and R. M. Bekker. (2006) Ecology of plant reproduction: extinction risks and restoration perspectives of rare plant species. In *Plant-Pollinator Interactions: from Specialization to Generalization*, ed. N. M. Waser and J. Ollerton, 362–86. Chicago: University Chicago Press.

Labandeira, C. C., J. Kvacek and M. B. Mostovski. (2007) Pollination drops, pollen, and insect pollination of Mesozoic gymnosperms. *Taxon* 56:663–95.

Lacey, E. A. and P. W. Sherman. (2005) Redefining eusociality: concepts, goals and levels of analysis. *Ann Zool Fenn* 42:573–77.

Lach, L. (2008) Argentine ants displace floral arthropods in a biodiversity hotspot. *Div and Dist* 14:281–90.

Lack, A. J. (1982) Competition for pollinators in the ecology of *Centaurea scabiosa* L. and *Centaurea nigra*. III. Insect visits and the number of successful pollinations. *New Phytol* 91:321–39.

Lack, A. J. and A. Diaz. (1991) The pollination of *Arum maculatum* L. – a historical review and new observations. *Watsonia* 18:333–42.

Ladley, J. J., D. Kelly, and A. W. Robertson. (1997) Explosive flowering, nectar production, breeding system and pollinators of New Zealand mistletoes (Loranthaceae). *NZ J Bot* 35:345–60.

Laloi, D., O. Bailez, M. M. Blight, B. Roger, H. M. Pham-Delègue, and L. J. Wadhams.(2000) Recognition of complex odors by restrained and free-flying honeybees. *J Chem Ecol* 26:2307–19.

Lambrecht, S. C., M. E. Loik, D. W. Inouye, and J. Harte. (2007) Reproductive and physiological responses to simulated climate warming for four subalpine species. *New Phytol* 173:121–34.

Landeck, I. (2002) Feeding plant spectrum of the hairy flower wasp *Scolia hirta* in Lusatia (Central Europe) with special focus on flower colour, morphology of flowers and inflorescences. *Entomol Gen* 26:107–20.

Lara, C. and J. F. Ornelas. (2002) Effects of nectar theft by flower mites on hummingbird behaviour, and the reproductive success of their host plant *Moussonia deppeana* (Gesneriaceae). *Oikos* 96:470–80.

Larson, B.M.H., P. G. Kevan, and D. W. Inouye. (2001) Flies and flowers: taxonomic diversity of anthophiles and pollinators. *Can Ent* 133:439–65.

Larson, B.M.H. and S.C.H. Barrett. (1999) The ecology of pollen limitation in buzz-pollinated *Rhexia virginica* (Melastomataceae). *J Ecol* 87:371–81

———. (2000) A comparative analysis of pollen limitation in flowering plants. *Biol J Linn Soc* 69:503–20.

Larsson, M. (2005) Higher pollinator effectiveness by specialist than generalist flower-visitors of unspecialized *Knautia arvensis* (Dipsacaceae). *Oecologia* 146:394–403.

Larsson, M. and M. Franzen. (2007) Critical resource levels of pollen for the declining bee *Andrena hattorfiana* (Hymenoptera, Andrenidae). *Biol Cons* 134:405–14.

Lau, T-C and A. G. Stephenson. (1993) Effects of soil nitrogen on pollen production, pollen grain size, and pollen performance in *Cucurbita pepo* (Cucurbitaceae). *Am J Bot* 80:763–68.

———. (1994) Effects of soil phosphorus on pollen production, pollen size, pollen phosphorus content and ability to sire seeds in *Cucurbita pepo* (Cucurbitaceae). *Sexual Pl Repro* 7:215–20.

Laverty, T. M. (1980) The flower-visiting behavior of bumble-bees – floral complexity and learning. *Can J Zool* 58:1324–35.

———. (1994a) Costs to foraging bumblebees of switching plant species. *Can J Zool* 72:43–47.

———. (1994b) Bumble bee learning and flower morphology. *Anim Behav* 47:531–45.

Laverty, T. M. and R. C. Plowright. (1988) Flower handling by bumblebees: a comparison of specialists and generalists. *Anim Behav* 36:733–40.

Law, B. S. (1992) Physiological factors affecting pollen use by Queensland blossom bats (*Syconycteris australis*). *Funct Ecol* 6:257–64.

Law, B. S. and M. Lean. (1999) Common blossom bats (*Syconycteris australis*) as pollinators in fragmented Australian tropical rainforest. *Biol Cons* 91:201–12.

Lawton-Rauh, A. L., E. R. Alvarez-Buylla, and M. D. Purugganan. (2000) Molecular evolution of flower development. *Trends Ecol Evol* 15:144–49.

Lazaro, A., S. J. Hegland, and Ø. Totland. (2008) The relationships between floral traits and specificity of pollination systems in three Scandinavian plant communities. *Oecologia* 157:249–57.

Leadbetter, E. and L. Chittka. (2007) The dynamics of social learning in an insect model, the bumblebee (*Bombus terrestris*). *Behav Ecol Sociobiol* 61:1789–96.

Leadbetter, E. and L. Chittka. (2008) Social transmission of nectar-robbing behaviour in bumble-bees. *Proc Roy Soc Lond* B 275:1669–74.

Leal, I. R., E. Fischer, C. Kost, M. Tabarelli, and R. Wirth. (2006) Ant protection against herbivores and nectar thieves in *Passiflora coccinea* flowers. *Ecosci* 13:431–38.

Leavitt, H. and I. C. Robertson. (2006) Petal herbivory by chrysomelid beetles (*Phyllotreta* sp.) is detrimental to pollination and seed production in *Lepidium papilliferum* (Brassicaceae). *Ecol Entomol* 31:657–60.

LeBuhn, G. and K. Holsinger. (1998) A sensitivity analysis of pollen-dispensing schedules. *Evol Ecol* 12:111–121.

Lee, T. D. (1988) Patterns of fruit and seed production. In *Plant Reproductive Ecology*, ed. J. Lovett-Doust and L. Lovett-Doust, 179–202. Oxford: Oxford University Press.

Lee, T. N. and A. A. Snow. (1998) Pollinator preferences and the persistence of crop genes in wild radish populations (*Raphanus raphanistrum*, Brassicaceae). *Am J Bot* 85: 333–39.

Leebens-Mack, J., O. Pellmyr, and M. Brock. (1998) Host specificity and the genetic structure of two yucca moth species in a yucca hybrid zone. *Evolution* 51:1376–82.

Leege, L. M. and L. M. Wolfe. (2002) Do floral herbivores respond to variation in flower characteristics in *Gelsemium sempervirens* (Loganiaceae), a distylous vine? *Am J Bot* 89:1270–74.

Lehrer, M., G. A. Horridge, S. W. Zhang, and R. Gadagkar. (1995) Shape vision in bees: innate preference for flower-like patterns. *Phil Trans R Soc Lond* B 347:123–37.

Lehtilä, K. and S. Y. Strauss. (1999) Effects of foliar herbivory on male and female reproductive traits of wild radish (*Raphanus raphanistrum*). *Ecology* 80:116–24.

Leins, P. and C. Erbar. (1994) On the mechanisms of secondary pollen presentation in the Campanulales-Asterales-complex. *Bot Acta* 103:87–92.

Leiss, K. A. and P.G.L. Klinkhamer. (2005a) Spatial distribution of nectar production in a natural *Echium vulgare* population: implications for pollinator behaviour. *Basic Appl Ecol* 6:317–24.

Leiss, K. A. and P.G.L. Klinkhamer. (2005b) Genotype by environment interactions in the nectar production of *Echium vulgare*. *Funct Ecol* 19:454–59.

Leiss, K. A., M. Peet, and P.G.L. Klinkhamer. (2009) Does spatial aggregations of total nectar production lead to genetic structure? *Basic Appl Ecol* 10:379–86.

LeMetayer, M., M. H. Pham-Delegue, D. Thiery, and D. Masson. (1993) Influence of host- and non-host plant pollen on the calling and oviposition behaviour of the European sunflower moth *Homoeosoma nebulellum* (Lepidoptera, Pyralidae). *Acta Oecol* 14:619–26.

Lennartsson, T. (2002) Extinction thresholds and disrupted plant-pollinator interactions in fragmented plant populations. *Ecology* 83:3060–72.

Leonhardt, S. D., K. Dworschak, T. Eltz, and N. Bluthgen N (2007) Foraging loads of stingless bees and utilisation of stored nectar for pollen harvesting. *Apidologie* 38:125–35.

Lertzman, K. P. and C. L. Gass. (1983) Alternative models of pollen transfer. In *Handbook of Experimental Pollination Ecology*, ed, C. E. Jones and R. J. Little, 474–89. New York: Van Nostrand Reinhold.

Leseigneur, C.D.C., L. Verburgt, and S. W. Nicolson. (2007) Whitebellied sunbirds (*Nectarinia talatala*, Nectariniidae) do not prefer artificial nectar containing amino acids. *J Comp Physiol* B 177:679–85.

Levin, D. A. (1970) Assortative pollination in *Lythrum*. *Am J Bot* 57:1–5.

———. (1978) The origins of isolating mechanisms in flowering plants. *Evol Biol* 11:185–317.

———. (1985) Reproductive character displacement in *Phlox*. *Evolution* 39:1275–81.

Levin, D. A. and D. E. Berube. (1972) *Phlox* and *Colias*: the efficiency of a pollination system. *Evolution* 26:242–50.

Levin, D. A. and H. W. Kerster. (1967) Natural selection for reproductive isolation in *Phlox*. *Evolution* 21:679–87.

———. (1973) Assortative pollination for stature in *Lythrum salicaria*. *Evolution* 27:144–52.

———. (1974) Gene flow in seed plants. *Evol Biol* 7:139–20.

Levin, M. D. and M. H. Haydak. (1957) Comparative value of different pollens in the nutrition of *Osmia lignaria*. *Bee World* 38:221–26.

Lev-Yadun, S., G. Ne'eman, and U. Shanas. (2009) A sheep in wolf's clothing: do carrion and dung odours of flowers not only attract pollinators but also deter herbivores? *BioEssays* 31:84-88.

Lewinsohn, T. M., P. I. Prado, P. Jordano, J. Bascompte, and J. M. Olesen. (2006) Structure in plant-animal interaction assemblages. *Oikos* 113:174–84.

Lewis, A. C. (1986) Memory constraints and flower choice in *Pieris rapae*. *Science* 232:863–65.

———. (1989) Flower visit consistency in *Pieris rapae*, the cabbage butterfly. *J Anim Ecol* 58:1–13.

Lewis, A. C. (1993) Learning and the evolution of resources: pollinators and flower morphology. In *Insect Learning: Ecological and Evolutionary Perspectives*, ed. D. R. Papaj and A. C. Lewis, 219–42. London: Chapman and Hall.

Lewis, H. and M. E. Lewis. (1955) The genus *Clarkia Univ Calif Pub Bot* 20:241–92.

Lex, T (1954) Duftmale an Bluten. *Z Vergl Physiol* 36:212–34.

Lima-Verde, L. W. and B. M. Freitas. (2002) Occurrence and biogeographic aspects of *Melipona quinquefasciata* in NE Brazil (Hymenoptera, Apidae). *Brazilian J Biol* 62: 479–86.

Lin, Y. and D. Y. Tan. (2007) Enantiostyly in angiosperms and its evolutionary significance. *Acta Phytotax Sinica* 45:901–16.

Lindauer, M. (1948) Uber die Einwirkung von Duft und Geschmacksstoffen sowie anderer Faktoren auf die Tänze der Bienen. *Z Vergl Physiol* 31:348–412.

Lindberg, A. B. and J. M. Olesen. (2001) The fragility of extreme specialization: *Passiflora mixta* and its pollinating hummingbird *Ensifera ensifera*. *J Trop Ecol* 17: 323–29.

Linder, H. P. (1998) Morphology and the evolution of wind pollination. In *Reproductive Biology in Systematics, Conservation and Economic Botany*, ed., S. J. Owens and P. J. Rudall, 123–35. Royal Bot Gard Kew.

Linder, H. P. and J. Midgley. (1996) Anemophilous plants select pollen from their own species from the air. *Oecologia* 108:85–87.

Linhart, Y. B. (1973) Ecological and behavioral determinants of pollen dispersal in hummingbird-pollinated *Heliconia*. *Am Nat* 107:511–23.

Linhart, Y. B., W. H. Busby, J. H. Beach, and P. Feinsinger. (1987) Forager behavior, pollen dispersal, and inbreeding in two species of hummingbird-pollinated plants. *Evolution* 41:679–82.

Linhart, Y. B. and J. A. Mendenhall. (1977) Pollen dispersal by hawkmoths in a *Lindenia rivalis* population in Belize. *Biotropica* 9:143.

Linhart, Y. B. and P. Feinsinger. (1980) Plant-hummingbird interactions: effects of island size and degree of specialization on pollination. *J Ecol* 68:745–60.

Linsley, E. G. and M. A. Cazier. (1972) Diurnal and seasonal behavior patterns among adults of *Protoxaea gloriosa* (Hymenoptera, Osaeidae). *Am Mus Novit* 2509:1–25.

Lippok, B., A. Gardine, P. S. Williamson, and S. S. Renner. (2000) Pollination by flies, bees and beetles, of *Nuphar ozarkana* and *N. advena* (Nymphaceae). *Am J Bot* 87: 9889–902.

Lippok, B. and S. S. Renner. (1997) Pollination of *Nuphar* (Nymphaeaceae) in Europe: flies and bees rather than *Donacia* beetles. *Plant Syst Evol* 207:273–83.

Lisci, M., C. Tanda, and E. Pacini. (1994) Pollination ecophysiology of *Mercurialis annua* L. (Euphorbiaceae), an anemophilous species flowering all year round. *Ann Bot Lond* 74:125–35.

Listabarth, C. (1992) Insect-induced wind pollination in the palm *Chamaedora pinnatifrons* and pollination in the related *Weldlandiella*. *Biodiv Cons* 1:39–50.

———. (1996) Pollination of *Bactris* by *Phyllotrox* and *Epurea*: implications of the palm breeding beetles on pollination at the community level. *Biotropica* 28:69–81.

Little, R. J. (1983) A review of floral food deception mimicries with comments on floral mutualism. In: Handbook of Experimental Pollination Biology, ed. C. E. Jones and R. J. Little, 294–309. New York: Van Nostrand Reinhold.

Liu, F. L., J. Chen, J. Chai, X. Zhang, X. Bai, C. He, and D. W. Roubik. (2007) Adaptive functions of defensive plant phenolics and a non-linear bee response to nectar components. *Funct Ecol* 21:96–100.

Liu, F. L., J. Z. He, and W. J. Fu. (2005) Highly controlled nest homeostasis of honey bees helps deactivate phenolics in nectar. *Naturwissenschaften* 92:297–99.

Liu, F. L., X. W. Zhang, J. P. Chai, and D. Yang. (2006) Pollen phenolics and regulation of pollen foraging in honeybee colony. *Behav Ecol Sociobiol* 59:582–88.

Liu, K. W., Z. J. Liu, L. Q. Huang, L. Q. Li, L. J. Chen, and G. D. Tang. (2006) Self-fertilization strategy in an orchid. *Nature* 441:945.

Lloyd, D. G. (1979) Some reproductive factors affecting self-fertilization in angiosperms. *Am Nat* 113:67–79.

———. (1982) Selection of combined versus separate sexes in seed plants. *Am Nat* 120:571–85.

———. (1992) Evolutionarily stable strategies of reproduction in plants: Who benefits and how? In *Ecology and Evolution of Plant Reproduction*, ed. R. Wyatt, 137–68. New York: Chapman and Hall.

Lloyd, D. G. and S.C.H. Barrett. (1996) *Floral Biology: Studies on Floral Evolution in Animal-Pollinated Systems*. New York: Chapman and Hall.

Lloyd, D. G. and D. J. Schoen. (1992) Self- and cross-fertilization in plants. I. Functional dimensions. *Int J Plant Sci* 153:358–69

Lloyd, D. G. and C. J. Webb. (1986) The avoidance of interference between the presentation of pollen and stigmas in angiosperms. I. Dichogamy. *NZ J Bot* 24:134–62

———. (1992) The selection of heterostyly In: Evolution and Function of Heterostyly, ed. S.C.H. Barrett, 179–207. Berlin: Springer-Verlag.

Lloyd, D. G. and M. S. Wells. (1992) Reproductive biology of a primitive angiosperm, *Pseudowintera colorata* (Winteraceae), and the evolution of pollination systems in the Anthophyta. *Plant Syst Evol* 181:77–95.

Lloyd, S., D. J. Ayre, and R. J. Whelan. (2002) A rapid and accurate assessment of nectar production can reveal patterns of temporal variation in *Banksia ericifolia* (Proteaceae). *Aust J Bot* 50:595–600.

Lobo, J. A., M. Quesada, K. E. Stoner, E. J. Fuchs, Y. Herrerias-Diego, J. Rojas, and G. Saborio. (2003) Factors affecting phenological patterns of Bombacaceous trees in seasonal forests in Costa Rica and Mexico. *Am J Bot* 90:1054–63.

Loew, E. (1895) Einführung in die Blütenbiologie. Ferd. Berlin: Dümmlere Verlag.

Lonsdorf, E., C. Kremen, T. Ricketts, R. Winfree, N. Williams, and S. Greenleaf. (2009) Modelling pollination services across agricultural landscapes. *Ann Bot* 103: 1589–1600.

Lopes, A. V., S. Vogel, and I. C. Machado. (2002) Secretory trichomes, a substitutive floral nectar source in *Lundia* A.DC. (Bignoniaceae), a genus lacking a functional disc. *Ann Bot* 90:169–74.

Lopez, J., T. Rodriguez-Riano, A. Ortega-Oliviencia, J. A.

Devesa JA and T. Ruiz. (1999) Pollination mechanisms and pollen-ovule ratios in some Genisteae (Fabaceae) from southwestern Europe. *Plant Syst Evol* 216:23–47.

Lopezaraiza-Mikel, M. E., R. B. Hayes, M. R. Whalley, and J. Memmott. (2007) The impact of an alien plant on a native plant-pollinator network; an experimental approach. *Ecol Lett* 10:539–50.

Lord, E. M. and S. D. Russell. (2002) The mechanism of pollination and fertilization in plants. *Ann Rev Devt Biol* 18:81–105.

Lord, E. M. and B. D. Webster. (1979) The stigmatic exudates of *Phaseolus vulgaris* L. *Bot Gaz* 140:266–71.

Lotz, C. N., A. M.del Rio, and S. W. Nicolson. (2003) Hummingbirds pay a high cost for a warm drink. *J Comp Physiol* B 173:455–62.

Lotz, C. N. and J. E. Schondube. (2006) Sugar preferences in nectar- and fruit-eating birds: behavioral patterns and physiological causes. *Biotropica* 38:3–15.

Lovett, Doust J. and J. L. Harper. (1980) The resource costs of gender and maternal support in an andromonoecious umbellifer, *Smyrnium olusatrum* L. *New Phytol* 85:251–64.

Lovkam, J. and J. F. Braddock. (1999) Anti-bacterial function in the sexually dimorphic pollinator rewards of *Clusia grandiflora* (Clusiaceae). *Oecologia* 119:534–40.

Lumer, C. and R. D. Schoer.(1980) Pollination of *Blakea austin-smithii* and *Blakea penduliflora* (Melastomataceae) by small rodents in Costa Rica. *Biotropica* 18:363–64.

Lunau, K. (1991) Innate flower recognition in bumblebees (*Bombus terrestris, B. lucorum:* Apidae) - optical signals from stamens as landing reaction releasers. *Ethology* 88:203–14.

———. (1992a) Innate recognition of flowers by bumble bees: orientation of antennae to visual stamen signals. *Can J Zool* 70:2139–44.

———. (1992b) Evolutionary aspects of perfume collection in male euglossine bees (Hymenoptera) and of nest deception in bee-pollinated flowers. *Chemoecology* 3:65–73.

———. (1992c) Limits of colour learning in a flower-visiting hoverfly *Eristalis tenax* L. (Syrphidae, Diptera). *Eur J Neurosci Suppl* 5:103.

———. (1995) Notes on the colour of pollen. *Plant Syst Evol* 198:235–52.

———. (1996a) Signalling functions of floral color patterns for insect flower visitors. *Zool Anz* 235:11–30.

———. (1996b) Unidirectionality of floral colour changes. *Plant Syst Evol* 200:125–40.

———. (2000) The ecology and evolution of visual pollen signals. *Plant Syst Evol* 222:89–111.

———. (2004) Adaptive radiation and coevolution – pollination biology case studies. *Organisms Div Evol* 4:207–24.

Lunau, K., N. Hofmann, and S. Valentin. (2005) Response of the hoverfly *Eristalis tenax* towards floral dot guides with colour transition from red to yellow (Diptera: Syrphidae). *Entomol Gen* 27:249–56.

Lunau, K. and E. J. Maier. (1995) Innate color preferences of flower visitors. *J Comp Physiol* A 177:1–19.

Lunau, K. and S. Wacht. (1994) Optical releasers of the innate proboscis extension in the hoverfly *Eristalis tenax* L. (Syrphidae, Diptera). *J Comp Physiol* A 174:575–79

Lunau, K., S. Wacht, and L. Chittka. (1996) Colour choices of naïve bumble bees and their implications for colour perception. *J Comp Physiol* A 178:477–89.

Lundberg, J. and F. Moberg. (2003) Mobile link organisms and ecosystem functioning; implications for ecosystem resilience and management. *Ecosystems* 6:87–98.

Lundemo, S. and Ø. Totland. (2007) Within-population spatial variation in pollinator visitation rates, pollen limitation on seed set, and flower longevity in an alpine species. *Acta Oecol* 32:262–68.

Lundgren, R. and J. M. Olesen. (2005) The dense and highly connected world of Greenland's plants and their pollinators. *Arct Ant Alp Res* 37:514–20.

Luo, Z., D. Zhang, and S. S. Renner. (2008) Why two kinds of stamen in buzz-pollinated flowers? Experimental support for Darwin's division-of-labour hypothesis. *Func Ecol* 22:794–800.

Luyt, R. and S. D. Johnson. (2002) Post-pollination nectar reabsorption and its implications for fruit quality in an epiphytic orchid. *Biotropica* 34:442–46.

Luzar, N. and G. Gottsberger. (2001) Flower heliotropism and floral heating of five alpine plant species, and the effect on flower visiting in *Ranunculus montanus* in the Austrian Alps. *Arct, Ant and Alp Res* 33:93–99.

Mabberley, D. J. (1991) *Tropical Rain Forest Ecology*. London: Blackie.

MacArthur, R. H. and E. O. Wilson. (1967) *The Theory of Island Biogeography*. Princeton, NJ: Princeton University Press.

Maccagnani, B., G. Burgio, L. Z. Stanisavljevic, and S. Maini. (2007) *Osmia cornuta* management in pear orchards. *Bull Insectol* 60:77–82.

Machado, I. C., S. Vogel, and A. V. Lopes. (2002) Pollination of *Angelonia cornigera* Hook. (Scrophulariaceae) by long-legged oil-collecting bees in NE Brazil. *Plant Biol* 4:352–59.

Macior, L. W. (1968) Pollination adaptation in *Pedicularis canadensis*. *Am J Bot* 55:1031–35.

Macior, L. W. (1970) The pollination ecology of *Pedicularis* in Colorado. *Am J Bot* 57:716–728.

———. (1971) Coevolution of plants and animals – systematic insights from plant-insect interactions. *Taxon* 20:17–28.

———. (1974) Pollination ecology of the Front Range of the Colorado Rocky Mountains. *Melanderia* 14:1–59.

Macior, L. W. (1983) The pollination dynamics of sympatric

species of *Pedicularis* (Scrophulariaceae). *Am J Bot* 70: 844–53.

———. (1994) Pollen-foraging dynamics of subalpine bumblebees (*Bombus* Latr.). *Plant Spec Biol* 9:99–106.

Madjidian, J. A., C. L. Morales, and H. G. Smith. (2008) Displacement of a native by an alien bumblebee: lower pollination efficiency overcome by overwhelmingly higher visitation frequency. *Oecologia* 156:835–45.

Magin, N., R. Classen, and C. Gack.(1989) The morphology of false anthers in *Craterostigma plantagineum* and *Torenia polygonoides* (Scrophulariaceae). *Can J Bot* 67: 1931–37

Mahabale, T. S. (1968) Spores and pollen grains of water plants and their dispersal. *Rev Paolaeobot Palynol* 7:285–96.

Mahy, G., J. de Sloover, and A-L Jacquemart.(1998) The generalist pollination system and reproductive success of *Calluna vulgaris* in the Upper Ardenne. *Can J Bot* 76: 1843–51.

Makholela, T. and J. C. Manning. (2006) First report of moth pollination in *Struthiola ciliata* (Thymelaeaceae) in southern Africa. *S Afr J Bot* 72:597–603.

Makino, T. T. (2008) Bumble bee preference for flowers arranged on a horizontal plane versus inclined planes. *Func Ecol* 22:1027–32.

Makino, T. T., K. Ohashi, and S. Sakai. (2007) How do floral display size and the density of surrounding flowers influence the likelihood of bumblebee visitation to a plant? *Funct Ecol* 21:87–95.

Makino, T. T. and S. Sakai. (2007) Experience changes pollinator responses to floral display size: from size-based to reward-based foraging. *Funct Ecol* 21:854–63.

Malho, R. (1999) The role of calcium and associated proteins in tip growth and orientation. In *Fertilization in Higher Plants*, ed. M. Cresti, G. Cai, and A. S. Moscatelli, 253–70. Heidelberg: Springer..

Mallet, J., J. T. Longino, D. Murawski, A. Murawski, and A. S. de Gamboa. (1987) Handling effects in *Heliconius*: where do all the butterflies go? *J Anim Ecol* 56:377–86.

Malo, J. E. and J. Baonza. (2002) Are there predictable clines in plant-pollinator interactions along altitudinal gradients? The example of *Cytisus scoparius* (L) in the Sierra de Guadarrama (Central Spain). *Divers Distr* 8:365–71.

Maloof, J. E. (2001) The effects of a bumble bee nectar robber on plant reproductive success and pollinator behavior. *Am J Bot* 88:1960–1965.

Maloof, J. E. and D. W. Inouye. (2000) Are nectar robbers cheaters or mutualists? Ecology 81:2651–61.

Mancina, C. A., F. Balseiro, and L. G. Herrera. (2005) Pollen digestion by nectarivorous and frugivorous Antillean bats. *Mamm Biol* 70:282–90.

Manning, A. (1956a) The effect of honey-guides. *Behaviour* 9:114–39.

———. (1956b) Some aspects of the foraging behaviour of bumble-bees. *Behaviour* 9:164–201.

———. (1957) Some evolutionary aspects of the flower constancy of bees. *Proc R Phys Soc Edin* 25:67–71.

Manning, R. (2001) Fatty acids in pollen; a review of their importance for honeybees. *Bee World* 82:60–75.

Mapalad, K. S., D. Leu, and J. C. Nieh. (2008) Bumble bees heat up for high quality pollen. *J Exp Biol* 211:2239–42.

Marazzi, B., E. Conti, and P. K. Endress. (2007) Diversity in anthers and stigmas in the buzz-pollinated genus *Senna* (Leguminosae, Cassiinae). *Int J Plant Sci* 168:371–91.

Marquis, R. J. (1988) Phenological variation in the neotropical understory shrub *Piper arielanum*: causes and consequences. *Ecology* 69:1552–65.

Marshall, A. G. (1983) Bats, flowers, and fruit: evolutionary relationships in the Old World. *Biol J Linn Soc* 20: 115–35.

Marshall, D. L., J. J. Avritt, S. Maliakal-Witt, J. S. Medeiros, and M.G.M. Shaner (2010) The impact of plant and flower age on mating patterns. *Ann Bot* 105:7–22.

Marshall., D. L. and N. C. Ellstrand. (1986) Sexual selection in *Raphanus sativus*: experimental data on non-random fertilization, maternal choice, and consequences of multiple paternity. *Am Nat* 127:446–61.

Marshall., D. L. and N. C. Ellstrand. (1988) Effective mate choice in wild radish: an anatomical to population perspective. *Ann Rev Ecol Syst* 22:37–63.

Marshall., D. L. and M. W. Folsom. (1991) Mate choice in plants: an anatomical to population perspective. *Ann Rev Ecol Syst* 22:37–63.

Marshall., D. L., J. Reynolds, N. J. Abrahamson, H. L. Simpson, M. G. Barnes, J. S. Medeiros, S. Walsh, D. M. Oliveras, and J. J. Avritt. (2007). Do differences in plant and flower age change mating patterns and alter offspring fitness in *Raphanus sativus* (Brassicaceae)? *Am J Bot* 94: 409–18.

Marten-Rodriguez, S., A. Almarales-Castro, and C. B. Fenster. (2009) Evaluation of pollination syndromes in Antillean Gesneriaceae: evidence for bat, hummingbird and generalized flowers. *J Ecol* 97:348–59.

Martin, F. W. (1969) Compounds from the stigmas of ten species. *Am J Bot* 56:1023–27.

Martin, N. H. (2004) Flower size preferences of the honeybee (*Apis mellifera*) foraging on *Mimulus guttatus* (Scrophulariaceae). *Evol Ecol Res* 6:777–82.

Martins, D. and S. D. Johnson. (2007) Hawkmoth pollination of aerangoid orchids in Kenya, with special reference to nectar sugar concentration gradients in the floral spurs. *Am J Bot* 94:650–59.

Matheson, A.,ed. (1994) *Forage for Bees in an Agricultural Landscape*. Cardiff, UK: IBRA.

Mathur, G. and H. Y. Mohan Ram. (1978) Significance of petal colour in thrips-pollinated *Lantana camara* L. *Ann Bot* 42:1473–76.

Matile, P. (2006) Circadian rhythmicity of nectar secretion in *Hoya carnosa*. *Bot Helv* 116:1–7.

Matile, P. and R. Altenburger. (1988) Rhythms of fragrance emission in flowers. *Planta* 174:242–47.

Matsuki, Y., R. Tateno, M. Shibata, and Y. Isagi. (2008) Pollination efficiencies of flower-visiting insects as determined by direct genetic analysis of pollen origin. *Am J Bot* 95:925–30.

Matteson, K. C., J. S. Ascher, and G. A. Langellotto. (2008) Bee richness and abundance in New York city urban gardens. *Ann Entomol Soc Am* 101:140–50.

Matteson, N. A. and L. I. Terry. (1992) Response to color by male and female *Frankliniella occidentalis* during swarming and non-swarming behavior. *Ent Exp Appl* 63:187–201.

Mauck, W. M. and K. J. Burns. (2009) Phylogeny, biogeography and recurrent evolution of divergent bill types in the nectar-stealing flowerpiercers (Thraupini: *Diglossa* and *Diglossopis*). *Biol J Linn Soc* 98:14–28.

Mauss, V. A., A. Müller, and R. Prosi. (2007) Mating, nesting and flower association of the east Mediterranean pollen wasp *Ceramius bureschi* in Greece (Hymenoptera: Vespidae: Masarinae). *Entomol Gen* 29:1–26.

May, P. G. (1985) Nectar uptake rates and optimal nectar concentrations of 2 butterfly species. *Oecologia* 66:381–86.

———. (1988) Determinants of foraging profitability in two nectarivorous butterflies. *Ecol Entomol* 13:171–84.

———. (1992) Flower selection and the dynamics of lipid reserves in two nectarivorous butterflies. *Ecology* 73:2181–91.

Maycock, C. R., R. N. Thewlis, J. Ghazoul, R. Nilus, and D.F.R.P. Burslem. (2005) Reproduction of dipterocarps during low intensity masting events in a Bornean rain forest. *J Veg Sci* 16:635–46.

Mayer, C., G. Soka, and M. Picker. (2006) The importance of monkey beetle (Scarabaeidae: Hoplinii) pollination for Aizoaceae and Asteraceae in grazed and ungrazed areas at Paulshoek, Succulent Karoo, South Africa. *J Insect Cons* 10:323–33.

Mayer, S. S. and D. Charlesworth. (1991) Cryptic dioecy in flowering plants: its occurrence and significance. *Trends Ecol Evol* 6:320–25.

Mayfield, M. M., N. M. Waser, and M. V. Price. (2001a) Exploring the "most effective pollinator principle" with complex flowers: bumblebees and *Ipomopsis aggregata*. *Ann Bot* 88:591–96.

Maynard Smith, J. (1978) *The Evolution of Sex*. Cambridge: Cambridge University Press.

Mazer, S. J., H. Paz, and M. D. Bell. (2004) Life history, floral development and mating system in *Clarkia xantiana* (Onagraceae); do floral and whole plant rates of development evolve independently? *Am J Bot* 91:2041–50.

McCall, A. C. (2006) Natural and artificial floral damage induces resistance in *Nemophila menziesii* (Hydrophyllaceae) flowers. *Oikos* 112:660–66.

McCall, C. and R. B. Primack. (1992) Influence of flower characteristics, weather, time of day, and season on insect visitation rates in three plant communities. *Am J Bot* 79:434–42.

McClure, B. A. and V. Franklin-Tong. (2006) Gametophytic self-incompatibility: understanding the cellular mechanisms involved in 'self' pollen tube inhibition. *Planta* 224:233–45.

McDade, L. A. (1986) Protandry, synchronized flowering and sequential phenotypic unisexuality in neotropical *Pentagonia macrophylla* (Rubiaceae). *Oecologia* 68:218–23.

———. (1992) Pollinator relationships, biogeography and phylogenetics. *Bioscience* 42:21–26.

McDade, L. A and S. Kinsman. (1980) The impact of floral parasitism in two neotropical hummingbird-pollinated plant species. *Evolution* 34:944–58.

McGregor, S. E., S. M. Alcorn, and G. Olin. (1962) Pollination and pollinating agents of the saguaro. *Ecology* 43:259–67.

McKey, D. (1989) Population biology of figs: applications for conservation. *Experientia* 45:661–73.

McLellan, A. R. (1977) Minerals, carbohydrates and amino acids of pollens from some woody and herbaceous plants. *Ann Bot* 41:1225–32.

McMullen, C. K. (1987) Breeding systems of selected Galapagos Islands angiosperms. *Am J Bot* 74:1694–1705.

———. (1990) Reproductive biology of Galapagos Islands angiosperms. *Monogr Syst Bot* 32:35–45.

———. (1993) Flower-visiting insects of the Galapagos Islands. *Pan Pac Ent* 69:95–106.

———. (2007) Pollination biology of the Galapagos endemic, *Tournefortia rufo-sericea* (Boraginaceae). *Bot J Linn Soc* 153:21–31.

———. (2009) Pollination biology of a night-flowering Galapagos endemic, *Ipomoea habeliana* (Convolvulaceae). *Bot J Linn Soc* 160:11–20.

McNeely, C. and M. C. Singer. (2001) Contrasting the roles of learning in butterflies foraging for nectar and oviposition sites. *Anim Behav* 61:847–52.

McWhorter, T. J. and C. M. del Rio. (1999) Food ingestion and water turnover in hummingbirds: how much dietary water is absorbed? *J Exp Biol* 202:2851–58.

McWhorter, T. J., C. M. del Rio, and B. Pinshow. (2003) Food ingestion and water turnover in hummingbirds: how much dietary water is absorbed? *J Exp Biol* 202:2851–58.

Meagher, T. R. (1986) Sex determination in plants. In *Plant Reproductive Ecology*, ed. J. Lovett-Doust and L. Lovett-Doust, 125–38. Oxford: Oxford University Press.

Meagher, T. R. (1991) Analysis of paternity within a natural population of *Chamaelirium luteum*. 1. Identification of most-likely parents. *Am Nat* 128:199–215.

———. (1992) The quantitative genetics of sexual dimorphism in *Silene latifolia* (Caryophyllaceae). I. Genetic variation. *Evolution* 46:445–57.

Medan, D., A. M. Basilio, M. Devoto, N. J. Bartolini, J. P. Torretta. and T. Petanidou. (2006) Measuring generalization and connectance in temperate, year-long active systems. In *Plant-Pollinator Interactions: from Specialization to Generalization*, ed. N. M. Waser NM and J. Ollerton, 245–59. Chicago: University of Chicago Press.

Medan, D., R.P.J. Perazzo, M. Devoto, E. Burgos, M. G. Zimmermann, H. Ceva, and A. M. Delbue. (2007) Analysis and assembling of network structure in mutualistic systems. *J Theor Biol* 246:510–21.

Medel, R., C. Botto-Mahan, and M. Kalin-Aroyo. (2003) Pollinator-mediated selection on the nectar guide phenotype in the Andean monkey flower *Mimulus luteus*. *Ecology* 84:1721–32.

Medel, R., A. Valiente, C. Botto-Mahan, G. Carvallo, F. Perez, N. Pohl, and L. Navarro. (2007) The influence of insects and hummingbirds on the geographical variation of the flower phenotype in *Mimulus luteus*. *Ecography* 30:812–18.

Medellin, R. A. (2003) Diversity and conservation of bats in Mexico: research priorities, strategies, and actions. *Wildlife Soc Bull* 31:87–97.

Medway, Lord. (1972) Phenology of a tropical rain forest in Malaya. *Biol J Linn Soc* 4:117–46.

Meehan, T. D., H. M. Lease, and B. O. Wolf. (2005) Negative indirect effects of an avian insectivore on the fruit set of an insect-pollinated herb. *Oikos* 109:297–304.

Meeuse, A.D.J. (1978) Nectar secretion, floral evolution and the pollination syndrome in early angiosperms. *Proc K Ned Akad Wet Ser* 81:300–326.

Meeuse, B.J.D. (1966) The Voodoo lily. *Sci Am* 215:80–88.

Meeuse, B.J.D. and S. Morris. (1984) *The Sex Life of Flowers*. New York: Facts on File.

Memmott, J. (1999) The structure of a plant-pollinator food web. *Ecol Lett* 2:276–80.

Memmott, J. and N. M. Waser. (2002) Integration of alien plants into a native flower-pollinator visitation web. *Proc Roy Soc Lond* B 269:2395–99.

Memmott, J., N. M. Waser, and M. V. Price. (2004) Tolerance of pollination networks to species extinctions. *Proc Roy Soc Lond* B 271:2605–11.

Memmott, J., P. G. Craze, N. M. Waser, and M. V. Price. (2007) Global warming and the disruption of plant-pollinator interactions. *Ecol Lett* 10:710–17.

Mendonca, L. B. and L. dos Anjos. (2006) Flower morphology, nectar features, and hummingbird visitation to *Palicourea crocea* (Rubiaceae) in the Upper Parana River floodplain, Brazil. *An Acad Bras Ciencas* 78:45–57.

Menzel, R. (2001) Behavioral and neural mechanisms of learning and memory as determinants of flower constancy. In *Cognitive Ecology of Pollination*, ed. L. Chittka and J. D. Thomson, 127–47. Cambridge: Cambridge University Press.

Menzel, R., I. Erber, and T. Masuhr. (1974) Learning and memory in the honeybee. In *Experimental Analysis of Insect Behaviour*, ed. L. Barton Browne, 195–217. Berlin: Springer.

Menzel, R, U. Greggers, and M. Hammer. (1993) Functional organization of appetitive learning and memory in a generalist pollinator, the honey bee. In *Insect Learning*, ed. D. R. Papaj and A. C. Lewis, 79–125. Berlin: Springer.

Menzel, R. and U. Müller. (1996) Learning and memory in honeybees: from behaviour to neural substrates. *Ann Rev Neurobiol* 19:379–404.

Menzel, R. and A. Shmida. (1993) The ecology of flower colours and the natural colour vision of insect pollinators: the Israeli flora as a study case. *Biol Rev* 68:81–120.

Messer, A. C. (1985) Fresh dipterocarp resins gathered by megachilid bees inhibit growth of pollen-associated fungi. *Biotropica* 17:175–76.

Metcalf, R. L. (1990) Chemical ecology of Dacinae fruit flies (Diptera, Tephritidae). *Ann Entomol Soc Am* 83:1017–30.

Mevi-Schütz, J., M. Goverde, and A. Erhardt. (2003) Effects of fertilization and elevated CO_2 on larval food and butterfly nectar amino acid preference in *Coenonympha pamphilus* L. *Behav Ecol Sociobiol* 54:36–43.

Mevi-Schütz, J. and A. Erhardt. (2005) Amino acids in nectar enhance butterfly fecundity: a long awaited link. *Am Nat* 165:411–19.

Meyer, B, V. Gaebele, and I. D. Steffan-Dewenter. (2007) Patch size and landscape effects on pollinators and seed set in horseshoe vetch, *Hippocrepis comosa*, in an agricultural landscape of central Europe. *Entomol Gen* 30:173–85.

Meyer, C.F.J., J. Frund, W. P. Lizaon, and E.K.V. Kalko. (2008) Ecological correlates of vulnerability to fragmentation in Neotropical bats. *J Appl Ecol* 45:381–91.

Michaelson Yeates, T.P.T., A. H. Marshall, I. H. Williams, N. L. Carreck, and J. R. Simpkins. (1997) The use of isoenzyme markers to determine pollen flow and seed paternity mediated by *Apis mellifera* and *Bombus* spp. in *Trifolium repens*, a self-incompatible species. *J Apic Res* 36:57–62.

Michaloud, G., S. Carriere, and M. Kobbi. (1996) Exceptions to the one:one relationship between African fig trees and their fig wasp pollinators: possible evolutionary scenarios. *J Biogeog* 23:513–20.

Michalski, S. G. and W. Durka. (2009) Pollination mode and life form strongly affect the relation between mating system and pollen to ovule ratios. *New Phytol* 183:470–79.

Micheneau, C., J. Fournel, and T. Pailler. (2006) Bird pollination in an angraecoid orchid on Reunion Island. *Ann Bot* 97:965–74.

Micheneau, C., J. Fournel, B. H. Warren, S. Hugel, A. Gauvin-Bialecki, T. Pailler, D. Strasberg, and M. W. Chase. (2010) Orthoptera, a new order of pollinator. *Ann Bot* 105:355–64.

Michener, C. D. (1974) *The Social Behavior of the Bees: a Comparative Study.* Cambridge, MA: Belknap Press.

———. (1979) Biogeography of the bees. *Ann Miss Bot Gard* 66:277–47.

———. (1985) From solitary to eusocial: need there be a series of intervening species? In *Experimental Behavioural Ecology and Sociobiology*, ed. B. Holldobler and M. Lindauer, 293–306. Stuttgart: Gustav Fischer Verlag.

———. (2000) *The Bees of the World.* Baltimore: Johns Hopkins University Press.

Michener, C. D and D. A. Grimaldi. (1988) A *Trigona* from late Cretaceous amber of New Jersey (Hymenoptera: Apidae: Meliponinae). *Am Mus Novit* 2917:1–10.

Midgley, J. J. and W. J. Bond. (1991) How important is biotic pollination and dispersal to the success of the angiosperms? *Phil Trans R Soc Lond* B 333:209–15.

Midgley, J. J. and S. D. Johnson. (1998) Some pollinators do not prefer symmetrically marked or shaped daisy (Asteraceae) flowers. *Evol Ecol* 12:123–26.

Miller, R. B. (1981) Hawkmoths and the geographic patterns of floral variation in *Aquilegia caerulea*. *Evolution* 35: 763–74.

Miller, S. R., R. Gaebel, R. J. Mitchell, and M. Arduser. (2003) Occurrence of two species of old world bees, *Anthidium manicatum* and *A. oblongatum,* in northern Ohio and southern Michigan. *Great Lakes Ent* 35:65–69.

Miller-Rushing A. J., R. Katsuki, R. B. Primack, Y. Ishii, S. D. Lee, and H. Higuchi. (2007) Impact of global warming on a group of related species and their hybrids: cherry tree (Rosaceae) flowering at Mt. Takao, Japan. *Am J Bot* 94:1470–78.

Minckley R. L., J. H. Cane, and L. Kervin. (2000) Origins and ecological consequences of pollen specialization among desert bees. *Proc Roy Soc Lond* B 267:265–71

Minckley R. L., J. H. Cane, L. Kervin, and T. H. Roulston. (1999) Spatial predictability and resource specialization of bees (Hymenoptera, Apoidea) at a superabundant, widespread resource. *Biol J Linn Soc* 67:119–47.

Minckley R. L. and T. H. Roulston. (2006) Incidental mutualisms and pollen specialization among bees. In *Plant-Pollinator Interactions: from Specialization to Generalization*, ed. N. M. Waser and J. Ollerton, 69–98. Chicago: University of Chicago Press.

Mione T. and G. J. Anderson. (1992) Pollen-ovule ratios and breeding system evolution in *Solanum* section *Basarthrum* (Solanaceae). *Am J Bot* 79:279–87.

Mitchell R. J. (2004) Heritability of nectar traits: why do we know so little? *Ecology* 85:1527–33.

Mitchell R. J., R. J. Flanagan, B. J. Brown, N. M. Waser, and J. D. Karron. (2009) New frontiers in competition for pollination. *Ann Bot* 103:1403–13.

Mitchell R. J., J. D. Karron, K. G. Holmquist, and J. M. Bell. (2005) Patterns of multiple paternity in fruits of *Mimulus ringens* (Phrymaceae). *Am J Bot* 92:885–90.

Mitchell R. J. and D. C. Paton. (1990) Effects of nectar volume and concentration on sugar uptake rates of Australian honeyeaters (Meliphagidae). *Oecologia* 83:238–46

Mitchell, R. J. and R. G. Shaw. (1993) Heritability of floral traits for the perennial wild flower *Penstemon centranthifolius* (Scrophulariaceae): clones and crosses. *Heredity* 71:185–92.

Miyake, T. and M. Yafuso M (2003) Floral scents affect reproductive success in fly-pollinated *Alocasia odora* (Araceae). *Am J Bot* 90:370–76.

Miyake, T. and M. Yafuso. (2005) Pollination of *Alocasia cucullata* (Araceae) by two *Colocasiomyia* flies known to be specific pollinators of *Alocasia odora*. *Plant Species Biol* 20:201–8.

Miyake, T. and T. Yahara. (1998) Why does the flower of *Lonicera japonica* open at dusk? *Can J Bot* 76:1806–11.

———. (1999) Theoretical evaluation of pollen transfer by nocturnal and diurnal pollinators: when should a flower open? *Oikos* 86:233–40.

Miyake, T., R. Yamaoka, and T. Yahara. (1998) Floral scents of hawkmoth-pollinated flowers in Japan. *J Plant Res* 111:199–205.

Moeller, D. A. (2004) Facilitative interactions among plants via shared pollinators. *Ecology* 85:3289–301.

Moeller, D. A. and M. A. Geber. (2005) Ecological context of the evolution of self-pollination in *Clarkia xantiana*: population size, plant communities, and reproductive assurance. *Evolution* 59:786–99.

Mogie, M. (1992) *The Evolution of Asexual Reproduction in Plants.* London: Chapman and Hall.

Molau, U., U. Nordenhall, and B. Eriksen.(2005) Onset of flowering and climate variability in an alpine landscape: a 10 year study from Swedish Lapland. *Al J Bot* 92: 422–31.

Molbo, D., C. A. Machado, J. G. Sevenster, L. Keller, and E. A. Herre. (2003) Cryptic species of fig-pollinating wasps: implications for the evolution of the fig-wasp mutualism, sex allocation, and precision of adaptation. *Proc Nat Acad Sci USA* 100:5867–72.

Moldenke, A. R. (1975) Niche specialization and species

diversity along a California transect. *Oecologia* 21:219–42.

Moldenke, A. R. (1976) California pollination ecology and vegetation types. *Phytologia* 34:305–61.

Mølgaard, P. (1989) Temperature relations of yellow and white flowered *Papaver radicatum* in north Greenland. *Arc Alp Res* 21:83–90.

Momose, K., R. Ishii, S. Sakai, and T. Inoue. (1998) Plant reproductive intervals and pollinators in the aseasonal tropics: a new model. *Proc Roy Soc Lond B* 265:2333–39.

Momose, K., T. Nagamitsu, and T. Inoue. (1998) Thrips cross-pollination of *Popowia psiocrpa* (Annonaceae) in a lowland dipterocarp forest. *Biotropica* 30:444–48.

Momose, K., T. Yumoto, T. Nagamitsu, M. Kato, H. Nagamasu, S. Sakai, R. D. Harrison, T. Itioka, A. A. Hamed, and T. Inoue. (1998) Pollination biology of a lowland dipterocarp forest in Sarawak, Malaysia. I. Characteristics of the plant-pollinator community in a lowland dipterocarp forest. *Am J Bot* 85:1477–1501.

Montalvo, A. M. and J. D. Ackerman. (1986) Relative pollinator effectiveness and evolution of floral traits in *Spathiphyllum friedrichsthalii* (Araceae). *Am J Bot* 73:1665–76.

Moody-Weis, J M. and J. S. Heywood. (2001) Pollination limitation to reproductive success in the Missouri Evening Primrose *Oenothera macrocarpa* (Onagraceae). *Am J Bot* 88:1615–22.

Moog, U., B. Fiala, W. Federle, and U. Maschwitz. (2002) Thrips pollination of the dioecious ant plant *Macaranga hullettii* (Euphorbiaceae) in southeast Asia. *Am J Bot* 89:50–59.

Moore, D. (2001) Honey bee circadian clocks: behavioural control from individual workers to whole-colony rhythms. *J Insect Physiol* 47:843–57.

Moore, M. J., C. D. Bell, P. S. Soltis, and D. E. Soltis. (2007) Using plastid genome-scale data to resolve enigmatic relationships among basal angiosperms. *Proc Nat Acad Sci USA* 104:19363–68.

Mora, J. M., V. V. Méndez, and L. D. Gómez. (1999) White-nosed coati *Nasua narica* (Carnivora, Procyonidae) as a potential pollinator of *Ochroma pyramidale* (Bombacaceae). *Rev Biol Trop* 47:719–21.

Moragues, E. and A. Traveset. (2005) Effect of *Carpobrotus* spp. on the pollination success of native plant species of the Balearic Islands. *Biol Cons* 122:611–19.

Morales, M. A. and M. A. Aizen. (2006) Invasive mutualisms and the structure of plant-pollinator interactions in the temperate forest of NW Patagonia, Argentina. *J Ecol* 94:171–80.

Morales, M. A., G. J. Dodge, and D. W. Inouye. (2005) A phenological mid-domain effect in flowering diversity. *Oecologia* 142:83–89.

Morales, C. L. and A. Traveset. (2009). A meta-analysis of impacts of alien vs. native plants on pollinator visitation and reproductive success of coflowering native plants. *Ecol Lett* 12:716–28.

Morandin, L. A., T. M. Laverty, and P. G. Kevan. (2001) Bumble bee (Hymenoptera: Apidae) activity and pollination levels in commercial tomato greenhouses. *J Econ Ent* 94:462–67.

Morandin, L. A., M. L. Winston, V. A. Abbott, and M. T. Franklin. (2007) Can pastureland increase wild bee abundance in agriculturally intense areas? *Basic Appl Ecol* 8:117–24.

Moré, M., A. Sérsic, and A. Cocucci (2006) Specialized use of pollen vectors by *Caesalpinia gilliesii*, a legume species with brush-type flowers. *Biol J Linn Soc* 88:579–92.

Moré, M., A. Sérsic, and A. Cocucci. (2007) Restriction of pollinator assemblage through flower length and width in three long-tongued hawkmoth-pollinated species of *Mandevilla* (Apocynaceae, Apocynoideae). *Ann Miss Bot Gard* 94:485–504.

Morgan, M. T. (2006) Selection on reproductive characters: conceptual foundations and their extension to pollinator interactions. In *Ecology and Evolution of Flowers*, 25–40, ed. L. D. Harder and S.C.H. Barrett. Oxford: Oxford University Press.

Morgan, M. T., D. J. Schoen, and T. M. Bataillon. (1997) The evolution of self-fertilization in perennials. *Am Nat* 150:618–38.

Mori, S. A., J. E. Orchard, and G. T. Prance. (1980) Intrafloral pollen differentiation in the New World Lecythidaceae. *Science* 209:400–403.

Moron, D., M. Lenda, P. Skorka, H. Szentgyorgyi, J. Settele, and M. Woyciechowski. (2009) Wild pollinator communities are negatively affected by invasion of alien goldenrods in grassland landscapes. *Biol Cons* 142:1322–32.

Morrant, D. S., R. Schumann, and S. Petit. (2009) Field methods for sampling and storing nectar from flowers with low nectar volumes. *Ann Bot* 103:533–42.

Morris, W. F. (1996) Mutualism denied? Nectar-robbing bumble bees do not reduce female or male fitness of bluebells. *Ecology* 77:1451–62.

Morris, W. F. (2003) Which mutualists are most essential? Buffering of plant reproduction against the extinction of pollinators. In *The Importance of Species: Perspectives on Expendability and Triage*, ed. P. Kareiva and S. A. Levin, 260–80. Princeton, NJ: Princeton University Press.

Morris, W. F., M. Mangel, and F. R. Adler. (1995) Mechanisms of pollen deposition by insect pollinators. *Evol Ecol* 9:304–17.

Morris, W. F., M. V. Price, N. M. Waser, J. D. Thomson, B. Thomson, and D. A. Stratton. (1994) Systematic increase in pollen carryover and its consequences for geitonogamy in plant populations. *Oikos* 71:431–40.

Morse, D. H. (1977) Resource partitioning in bumble bees: the role of behavioral factors. *Science* 197:678–80.

———. (1981a) Modification of bumblebee foraging: the effect of milkweed pollinia. *Ecology* 62:89–97.

———. (1981b) Prey capture by the crab spider *Misumena vatia* (Thomisidae) on three common native flowers. *Am Midl Nat* 105:358–67.

Morse, D. H. and R. S. Fritz. (1983) Contributions of diurnal and nocturnal insects to the pollination of common milkweed (*Asclepias syriaca* L.) in a pollen-limited system. *Oecologia* 60:190–97.

Mortensen, H. S., Y. L. Dupont, and J. M. Olesen. (2008) A snake in paradise: disturbance of plant reproduction following extirpation of bird flower-visitors in Guam. *Biol Cons* 141:2146–54.

Mosquin, T. (1971) Competition for pollinators as a stimulus for the evolution of flowering time. *Oikos* 22:398–402.

Mothershead, K. and R. J. Marquis. (2000) Fitness impacts of herbivory through indirect effects on plant-pollinator interactions in *Oenothera macrocarpa*. *Ecology* 81:30–40.

Motten, A. F. (1986) Pollination ecology of the spring wildflower community of a temperate deciduous forest. *Ecol Monogr* 5:21–42.

Motten, A. F., D. R. Campbell, D. E. Alexander, and J. L. Miller. (1981) Pollination effectiveness of specialist and generalist visitors to a North Carolina population of *Claytonia virginica*. *Ecology* 62:1278–87.

Motten, A. F. and J. L. Stone. (2000) Heritability of stigma position and the effect of stigma-anther separation on outcrossing in a predominantly self-fertilizing weed, *Datura stramonium* (Solanaceae). *Am J Bot* 87:339–47.

Mound, L. A. and I. Terry. (2001) Thrips pollination of the Central Australian cycad, *Macrozania macdonnellii* (Cycadales). *Int J Plant Sci* 162:147–54.

Muchhala, N. (2006) Nectar bat stows huge tongue in its rib cage. *Nature* 444:701–2.

———. (2007) Adaptive trade-off in floral morphometry mediates specialization for flowers pollinated by bats and hummingbirds. *Am Nat* 169:494–504.

Muchhala, N., A. Caiza, J. C. Vicente, and J. D. Thomson. (2009) A generalized pollination system in the tropics: bats, birds and *Aphelandra acanthus*. *Ann Bot* 103:1481–87.

Muchhala, N. and M. D. Potts. (2007) Character displacement among bat-pollinated flowers of the genus *Burmeistera*: analysis of mechanism, process, and pattern, *Proc Roy Soc Lond* B 274:2731–37.

Muchhala, N. and J. D. Thomson. (2009) Going to great lengths: selection for long corolla tubes in an extremely specialized bat-flower mutualism. *Proc Roy Soc Lond* B 276:2147–52.

Muenchow, G. E. and M. Grebus. (1989) The evolution of dioecy from distyly: re-evaluation of the hypothesis of the loss of long-tongued pollinators. *Am Nat* 133:149–56.

Mulcahy, D. L. (1974) Correlation between speed of pollen tube growth and seedling height in *Zea mays* L. *Nature* 249:491–93.

Mulcahy, D. L., P. S. Curtis, and A. A. Snow. (1983) Pollen competition in a natural population. In *Handbook of Experimental Pollination Biology,* ed. C. E. Jones and R. J. Little, 330–37. New York: Van Nostrand Reinhold.

Mulcahy, D. L. and G. B. Mulcahy. (1987) The effects of pollen competition. *Am Sci* 75:44–50.

Mulcahy, D. L., G. B. Mulcahy, and Searcy. (1992) Evolutionary genetics of pollen competition. In *Ecology and Evolution of Plant Reproduction*, ed. R. Wyatt, 25–36. London: Chapman and Hall.

Müller, A. (1995a) Bumblebee preference for symmetrical flowers. *Proc Nat Acad Sci USA* 92:2288–92.

———. (1995b) Morphological specialisations in central European bees for the uptake of pollen from flowers with anthers hidden in narrow corolla tubes (Hymenoptera, Apoidea). *Entomol Gen* 20:43–57.

———. (1996) Host plant specialization in western palearctic anthidiine bees (Hymenoptera: Apoidea: Megachilidae). *Ecol Monogr* 66:235–64.

———. (2000) Developmental stability and pollination. *Oecologia* 123:149–57.

Müller, A., S. Diener, S. Schnyder, K. Stutz, C. Sedivy, and S. Dorn. (2006) Quantitative pollen requirements of solitary bees: implications for bee conservation and the evolution of bee-flower relationships. *Biol Cons* 130:604–15.

Müller, H. (1873) *Die Befruchtung der Blumen durch Insekten und due gegenseitigen Anpassungen beider: ein Beitrag zur Erkenntniss des ursächlichen Zusammenhanges in der organischen Natur.* Leipzig: Wilhemm Engelmann.

Mulligan, G. A. and P. G. Kevan. (1973) Color, brightness and other floral characteristics attracting insects to the blossoms of some Canadian weeds. *Can J Bot* 51: 1939–52.

Munoz, A. A. and M.T.K. Arroyo. (2006) Pollen limitation and spatial variation of reproductive success in the insect-pollinated shrub *Chuquiraga oppositifolia* (Asteraceae) in the Chilean Andes. *Arct Ant Alp Res* 38:607–13.

Munoz, A., C. Celedon-Neghme, L. A. Cavieres, and M.T.K. Arroyo. (2005) Bottom-up effects of nutrient availability on flower production, pollinator visitation, and seed output in a high-Andean shrub. *Oecologia* 143: 126–35.

Muona, O., G. F. Moran, and J. C. Bell. (1991) Hierarchical patterns of correlated mating in *Acacia melanoxylon*. *Genetics* 127:619–26.

Murawski, D. A. and L. E. Gilbert. (1986) Pollen flow in *Psiguria warscewiczii*: a comparison of *Heliconius* butterflies and hummingbirds. *Oecologia* 68:161–67.

Murcia, C. (1990) Effect of floral morphology and temperature on pollen receipt and removal in *Ipomoea trichocarpa. Ecology* 71:1098–1109.

Murcia, C. (1995) Edge effects in fragmented forests: implications for conservation. *Trends Ecol Evol* 10:58–62.

Murcia, C. and P. Feinsinger. (1996) Interspecific pollen loss by hummingbirds visiting flower mixtures. *Ecology* 77: 550–60.

Murphy, D. D. (1984) Butterflies and their nectar plants: the role of the checkerspot butterfly Euphydryas editha as a pollen vector. *Oikos* 43:113–17.

Murphy, S. D. (1999) Pollen allelopathy. In *Principles and Practices in Plant Ecology*, ed. K.M.M. Dakshini Inderjit and C. L. Foy, 129–48. Boca Raton, FL: CRC Press.

Murren, C. J. (2002) Effects of habitat fragmentation on pollination: pollinators, pollinia viability and reproductive success. *J Ecol* 90:100–107.

Mutikainen, P. and L. F. Delph. (1996) Effects of herbivory on male reproductive success in plants. *Oikos* 75:353–58.

Myers, J. and D. Bazeley. (2003) Ecology and Control of Introduced Plants. Cambridge: Cambridge University Press.

Nagamitsu, T. and T. Inoue. (1997a) Cockroach pollination and breeding system of *Uvaria elmeri* (Annonaceae) in a lowland mixed-dipterocarp forest in Sarawak. *Am J Bot* 84:208–13.

Nagamitsu, T. and T. Inoue. (1997b) Aggressive foraging of social bees as a mechanism of floral resource partitioning in an Asian tropical rainforest. *Oecologia* 110: 432–39.

Nagy, K. A., C. Meienberger, S. D. Bradshaw, and R. D. Wooller. (1995) Field metabolic rate of a small marsupial mammal, the honey possum (*Tarsipes rostratus*). *J Mammalogy* 76:862–66.

Nansen, C. and S. Korie. (2000) Determining the time delay of honey bee (*Apis mellifera*) foraging response to hourly pollen release in a typical pollen flower (*Cistus salvifolius*). *J Apic Res* 39:93–101.

Naug, D. and H. S. Arathi. (2007) Receiver bias for exaggerated signals in honeybees and its implications for the evolution of floral displays. *Biol Lett* 3:635–37.

Navarro, L. (2000) Pollination ecology of *Anthyllis vulneraria* subsp. *vulgaris* (Fabaceae): Nectar robbers as pollinators. *Am J Bot* 87:980–85.

Navarro, L., G. Ayensa, and P. Guitian. (2007) Adaptation of floral traits and mating system to pollinator unpredictability: the case of *Disterigma stereophyllum* (Ericaceae) in southwestern Colombia. *Plant Syst Evol* 266:165–74.

Nazarov, V. V. and G. Gerlach. (1997) The potential seed productivity of orchid flowers and peculiarities of their pollination systems. *Lindleyana* 12:188–204.

Neal, P., A. Dafni, and M. Giurfa. (1998) Floral symmetry and its role in plant pollinator systems: terminology, distribution and hypotheses. *Ann Rev Ecol Syst* 29:345–73.

Nebbia, A. J. and S. M. Zalba. (2007) Designing nature reserves: traditional criteria may act as misleading indicators of quality. *Biodiv Cons* 16:223–33.

Ne'eman, G., A. Dafni,and S. G. Potts. (1999) A new pollination probability index (PPI) for pollen load analysis as a measure of pollination effectiveness in bees. *J Apic Res* 38:19–24.

Ne'eman, G., O. Shavit, L. Shaltiel, and A. Shmida. (2006) Foraging by male and female solitary bees with implications for pollination. *J Insect Behav* 19:383–401.

Neff, J. L. and B. B. Simpson. (1981) Oil-collecting structures in the Anthophoridae (Hymenoptera): morphology, function and use in systematics. *J Kans Ent Soc* 54: 95–123.

———. (1988) Vibratile pollen-harvesting by *Megachile mendica* Cresson (Hymenoptera, Megachilidae). *J Kans Ent Soc* 61:242–44.

Negre, F., C. M. Kish, B. Boatright, K. Shibuya, C. Wagner, D. G. Clark, and N. Dudareva. (2003) Regulation of methylbenzoate emission after pollination in snapdragon and petunia flowers. *Plant Cell* 15:2992–3006.

Neiland, M.R.M. and C. C. Wilcock. (1995) Maximisation of reproductive success by European Orchidaceae under conditions of infrequent pollination. *Protoplasma* 187: 39–48.

Nepi, M. (2007) Nectary structure and ultrastructure. In *Nectaries and Nectar*, ed. S. W. Nicolson, M. Nepi, and E. Pacini, 129–66. Dordrecht: Springer.

Nepi, M., M. Guarnieri, and E. Pacini. (2001) Nectar secretion, reabsorption and sugar composition in male and female flowers of *Cucurbita pepo. Int J Plant Sci* 162: 353–58.

———. (2003) 'Real' and feed pollen in *Lagerstroemia indica*: ecophysiological differences. *Plant Biol* 5:311–14.

Nepi, M. and M. Stpiczynska. (2007) Nectar resorption and translocation in *Cucurbita pepo* L and *Platanthera chlorantha* Custer (Rchb.). *Plant Biol* 9:93–100.

———. (2008) The complexity of nectar: secretion and resorption dynamically regulate nectar features. *Naturwissenschaften* 95:177–84.

Ness, J. H. (2003) *Catalpa bignonioides* alters extrafloral nectar production after herbivory and attracts ant bodyguards. *Oecologia* 134:210–18.

———. (2006) A mutualism's indirect costs; the most aggressive plant bodyguards also deter pollinators. *Oikos* 113:506–14.

Ness, J. H. and I. L. Bronstein. (2004) The effects of invasive ants on prospective ant mutualists. *Biol Invasions* 6: 445–61.

Neumayer, J. and J. Spaethe. (2007) Flower colour, nectar standing crop, and flower visitation of butterflies in an alpine habitat in central Europe. *Entomol Gen* 29:269–84.

Newman, D. A. and J. D. Thomson. (2005) Effects of nectar robbing on nectar dynamics and bumblebee foraging strategies in *Linaria vulgaris* (Scrophulariaceae). *Oikos* 110:309–20.

Newstrom, L. and A. Robertson.(2005) Progress in understanding pollination systems in New Zealand. *NZ J Bot* 43:1–59.

Newton, S. D. and G. D. Hill. (1983) Robbing of field bean flowers by the short-tongued bumble bee *Bombus terrestris* L. *J Apic Res* 22:124–29.

Nicklen, E. F. and D. Wagner. (2006) Conflict resolution in an ant-plant interaction: *Acacia constricta* traits reduce ant costs to reproduction. *Oecologia* 148:81–87.

Nicolson, S. W. (1993) Low nectar concentrations in a dry atmosphere – a study of *Grevillea robusta* (Proteaceae) and *Callistemon viminalis* (Myrtaceae). *S Afr J Sci* 89: 473–77.

———. (1995) Direct demonstration of nectar reabsorption in the flowers of *Grevillea robusta* (Proteaceae). *Funct Ecol* 9:584–88.

———. (1998) The importance of osmosis in nectar secretion and its consumption by insects. *Am Zool* 38:418–25.

———. (2006) Water management in nectar-feeding birds. *Am J Physiol* 291:R828–R829.

———. (2007) Amino acid concentrations in the nectars of Southern African bird-pollinated flowers, especially *Aloe* and *Erythrina*. *J Chem Ecol* 33:1707–20.

Nicolson, S. W.and P. A. Fleming. (2003) Nectar as food for birds; the physiological consequences of drinking dilute nectar solutions. *Plant Syst Evol* 238:139–53.

Nicolson, S. W., M. Nepi, and E. Pacini, ed. (2007) *Nectaries and Nectar*. Dordrecht: Springer.

Nicolson, S. W.and R. T. Thornburg. (2007) Nectar chemistry. In *Nectaries and Nectar*, ed. S. W. Nicolson, M. Nepi, and E. Pacini, 215–64. Dordrecht: Springer.

Nieh, J. C., S. Ramirez, and P. Nogueira-Neto. (2003) Multisource odor-marking of food by a stingless bee, *Melipona mandacaia*. *Behav Ecol Sociobiol* 54:578–86.

Nielsen, A. and J. Bascompte. (2007) Ecological networks, nestedness and sampling effort. *J Ecol* 95:1134–41.

Nielsen, C., C. Heimes, and J. Kollmann. (2008) Little evidence for negative effects of an invasive alien plant on pollinator services. *Biol Invasions* 10:1353–63.

Nierenberg, L. (1972) The mechanism for the maintenance of species integrity in sympatrically occurring equitant *Oncidium* in the Caribbean. *Am Orchid Soc Bull* 41:873–82.

Niesenbaum, R. A. (1999) The effects of pollen load size and donor diversity on pollen performance, selective abortion, and progeny vigor in *Mirabilis jalapa* (Nyctaginaceae). *Am J Bot* 86:261–68.

Niklas, K. J. (1984) The motion of windborne pollen grains around conifer ovulate cones: implications on wind pollination. *Am J Bot* 71:356–74.

———. (1985) The aerodynamics of wind pollination. *Bot Rev* 51:328–86.

———. (1987) Pollen capture and wind-induced movement of compact and diffuse grass panicles: implications for pollination efficiency. *Am J Bot* 84:1542–52.

———. (1992) *Plant Biomechanics*. Chicago: University of Chicago Press.

Niklas, K J.and S. L. Buchmann. (1985) Aerodynamics of wind pollination in *Simmondsia chinensis* (Link) Schneider. *Am J Bot* 72:530–39.

———. (1987) The aerodynamics of pollen capture in two sympatric *Ephedra* species. *Evolution* 41:104–23.

Niklas, K J.and U.K.T. Paw. (1983) Conifer ovulate cone morphology; implications on pollen impaction patterns. *Am J Bot* 70:568–77.

Nilsson, L. A. (1980) The pollination ecology of *Dactylorhiza sambucina* (Orchidaceae). *Bot Notiser* 133:367–85.

———. (1983a) Processes of isolation and introgressive interplay between *Platanthera bifolia* (L) Rich and *P. chlorantha* (Custer) (Orchidaceae). *Bot J Linn Soc* 87:325–36.

———. (1983b) Mimesis of bellflower (*Campanula*) by the red helleborine orchid (*Cephalanthera rubra*). *Nature* 305:799–800.

———. (1988) The evolution of flowers with deep corolla tubes. *Nature* 334:147–49.

———. (1992) Orchid pollination biology. *Trends Ecol Evol* 7:255–59.

———. (1998) Deep flowers for long tongues. *Trends Ecol Evol* 13:259–60.

Nilsson, L. A., L. Johnsson, L. Ralison, and E. Randrianjohany. (1985) Monophily and pollination mechanisms in *Angraecum arachnites* Schltr. (Orchidaceae) in a guild of long-tongued hawk-moths (Sphingidae) in Madagascar. *Biol J Linn Soc* 26:1–19.

Nilsson, L. A., L. Johnsson, L. Ralison, and E. Randrianjohany. (1987) Angraecoid orchids and hawkmoths in central Madagascar: specialized pollination systems and generalist foragers. *Biotropica* 19:310–18.

Nilsson, L. A., E. Rabakonandrianina, B. Pettersson, and R. Grunmeier. (1993) Lemur pollination in the Malagasy rainforest liana *Strongylodon craventae* (Leguminosae). *Evol Trends Plants* 7:49–56.

Nishizawa, T., Y. Watano, D. Kinoshita, T. Kawahara, and K. Ueda. (2005) Pollen movement in a natural population of *Arisaema serratum* (Araceae), a plant with a pitfall-trap flower pollination system. *Am J Bot* 92:1114–23.

Noda, K., B. J. Glover, P. Linstead, and C. Martin. (1994) Flower colour intensity depends on specialized cells shape controlled by a Myb-related transcription factor. *Nature* 369:661–64.

Nol, E., H. Douglas, and W. J. Crins. (2007) Responses of syrphids, elaterids and bees to single-tree selection harvesting in Algonquin Provincial Park, Ontario. *Can Field Nat* 120:15–21.

Norstog, K. (1987) Cycads and the origin of insect pollination. *Am Sci* 75:270–79

Norton, S. A. (1984) Thrips pollination in the lowland forest of New Zealand. *NZ J Ecol* 7:157–64.

Nuttman, C. V., F. M. Semida, S. Zalat, and P. G. Willmer. (2006) Visual cues and foraging choices; bee visits to floral colour phases in *Alkanna orientalis* (Boraginaceae). *Biol J Linn Soc* 87:427–35.

Nuttman, C. V. and P. G. Willmer. (2003) How does insect visitation trigger floral colour change? *Ecol Ent* 28:467–74.

Nuttman, C. V. and P. G. Willmer. (2008) Hoverfly visitation in relation to floral colour change (Diptera : Syrphidae). *Entomol Gen* 31:33–47.

Nyhagen, D. F., C. Kragelund, J. M. Olesen, and C. G. Jones. (2001) Insular interactions between lizards and flowers: flower visitation by an endemic Mauritian gecko. *J Trop Ecol* 17:755–61.

Ober, H. K., R. J. Steidl, and V. M. Dalton. (2005) Resource and spatial-use patterns of an endangered vertebrate pollinator, the lesser long-nosed bat. *J Wildlife Man* 69:1615–22.

Oberrath, R. and K. Böhning-Gaese. (1999) Floral color change and the attraction of insect pollinators in lungwort (*Pulmonaria collina*). *Oecologia* 121:383–91.

Oberrath, R., C. Zanke, and K. Böhning-Gaese. (1995) Triggering and ecological significance of floral color change in lungwort (*Pulmonaria* sp.). *Flora* 190:155–59.

Obeso, J. R. (1992) Pollination ecology and seed set in *Asphodelus albus* (Liliaceae) in Northern Spain. *Flora* 187:219–26.

———. (2002) The costs of reproduction in plants. *New Phytol* 155:321–48.

Öckinger, E. and H. G. Smith. (2007) Semi-natural grasslands as population sources for pollinating insects in agricultural landscapes. *J Appl Ecol* 44:50–59.

O'Donnell, S. and M. J. Lawrence. (1984) The population genetics of the self-incompatibility polymorphism in *Papaver rhoeas*. IV. The estimation of the number of alleles in a population. *Heredity* 53:495–508.

Oguro, M. and S. Sakai. (2009) Floral herbivory at different stages of flower development changes reproduction in *Iris gracilipes* (Iridaceae). *Plant Ecol* 202:2221–34.

Ohashi, K. and T. D. Thomson. (2009) Trapline foraging by pollinators; its ontogeny, economics and possible consequences for plants. *Ann Bot* 103:1365–78.

Ohashi, K. and T. Yahara. (1999) How long to stay on, and how often to visit a flowering plant? A model for foraging strategy when floral displays vary in size. *Oikos* 86:386–92.

———. (2002) Visit larger displays but probe proportionally fewer flowers: counter-intuitive behaviour of nectar-collecting bumble-bees achieves an ideal free distribution. *Funct Ecol* 16:492–503.

Okubo, A. and S. A. Levin. (1989) A theoretical framework for data analysis of wind dispersal of seeds and pollen. *Ecology* 70:329–38.

Okuyama, Y., O. Pellmyr, and M. Kato. (2008) Parallel floral adaptations to pollination by fungus gnats within the genus *Mitella* (Saxifragaceae). *Mol Phyl Evol* 46:560–75.

Olesen, J. M. (1985) The Macronesian bird-flower element and its relation to bird and bee opportunists. *Bot J Linn Soc* 91:395–414.

———. (1988) Floral biology of the Canarian *Echium wildpretii*: a bird-flower or a water resource to desert bees? *Acta Bot Neerl* 37:509–13.

———. (1997) Pollination. In *Methods for risk assessment of transgenic plants. II. Pollination, gene transfer and population impacts,* ed. G. Kjellsson, V. Simonsen, and K. Ammann, 1–10. Basel: Birkhäuser.

Olesen, J. M., J. Bascompte, Y. L. Dupont, and P. Jordano. (2007) The modularity of pollination networks. *Proc Nat Acad Sci USA* 104:19891–96.

Olesen, J. M., J. Bascompte, H. Elberling, and P. Jordano. (2008) Temporal dynamics in a pollination network. *Ecology* 89:1573–82.

Olesen, J. M., Y. L. Dupont, B. K. Ehlers, and D. M. Hansen. (2007) The openness of a flower and its number of flower-visitor species. *Taxon* 56:729–36.

Olesen, J. M., L. I. Eskildsen, and S. Vankatasamy. (2002) Invasion of pollination networks on oceanic islands: importance of invader complexes and endemic super generalists. *Diversity and Dist* 8:181–92.

Olesen, J. M.and P. Jordano. (2002) Geographic patterns in plant-pollinator mutualistic networks. *Ecology* 83:2416–24.

Olesen, J. M. and A. Valido. (2003) Lizards as pollinators and seed dispersers; an island phenomenon. *Trends Ecol Evol* 18:177–81.

Oliveira, P. E., P. E. Gibbs, and A. A. Barbosa. (2004) Moth pollination of woody species in the Cerrados of Central Brazil: a case of so much owed to so few? *Plant Syst Evol* 245:41–54.

Oliveira, P. S. (1997) The ecological function of extrafloral nectaries: herbivore deterrence by visiting ants and reproductive output in *Caryocar brasiliense. Funct Ecol* 11:323–30.

Ollerton, J. (1996) Reconciling ecological processes with phylogenetic patterns: the apparent paradox of plant-pollinator systems. *J Ecol* 84:550–60.

———. (1998) Sunbird surprise for syndromes. *Nature* 394: 726–27.

———. (2006) 'Biological barter': patterns of specialization compared across different mutualisms. In *Plant-Pollinator Interactions: from Specialization to Generalization*,ed. N. M. Waser and J. Ollerton, 411–36. Chicago: University of Chicago Press.

Ollerton, J., R. Alarcon, N. M. Waser, M. V. Price, S. Watts, L. Cranmer, A. Hingston, C. I. Peter, and Rotenberry. (2009) A global test of the pollination syndrome hypothesis. *Ann Bot* 103:1471–80.

Ollerton, J. and L. Cranmer. (2002) Latitudinal trends in plant-pollinator interactions: are tropical plants more specialised? *Oikos* 98:340–50.

Ollerton, J. and A. Diaz. (1999) Evidence for stabilizing selection acting on flowering time in *Arum maculatum* (Araceae): the influence of phylogeny on adaptation. *Oecologia* 199:340–48.

Ollerton, J., J. N. Grace, and K. Smith. (2007) Adaptive floral color change in *Erysimum scoparium* (Brassicaceae) and pollinator behaviour of *Anthophora alluadii* on Tenerife (Hymenoptera, Apidae). *Entomol Gen* 29: 253–68.

Ollerton, J., S. D. Johnson, L. Cranmer, and S. Kellie. (2003) The pollination ecology of an assemblage of grassland asclepiads in South Africa. *Ann Bot* 92:807–34.

Ollerton, J., S. D. Johnson, and A. B. Hingston. (2006) Geographical variation in diversity and specificity of pollination systems. In *Plant-Pollinator Interactions: from Specialization to Generalization*, ed. N. M. Waser and J. Ollerton, 283–308. Chicago: University of Chicago Press.

Ollerton, J., A. Killick, E. Lamborn, S. Watts, and M. Whiston. (2007) Multiple meanings and modes; on the many ways to be a generalist flower. *Taxon* 56:717–28.

Ollerton, J. and A. J. Lack. (1992) Flowering phenology: an example of relaxation of natural selection? *Trends Ecol Evol* 7:274–76.

Ollerton, J. and S. Watts. (2000) Phenotypic space and floral typology: towards an objective assessment of pollination syndromes. In *The Scandinavian Assn for Pollination Ecology Honours Knut Faegri, Mat Naturv Klasse Skrifter Ny Serie* 39:149–59.

Olsen, K. M. (1997) Pollination effectiveness and pollinator importance in a population of *Heterotheca subaxillaris* (Asteraceae). *Oecologia* 109:114–21.

Olsson, M., R. Shine, and E. Ba'K-Olsson. (2000) Lizards as a plant's 'hired help': letting pollinators in and seeds out. *Biol J Linn Soc* 71:191–202.

Omura, H. and K. Honda. (2005) Priority of color over scent during flower visitation by adult *Vanessa indica* butterflies. *Oecologia* 142:588–96.

Omura, H., K. Honda. and N. Hayashi. (2000) Floral scent of *Osmanthus fragrans* discourages foraging behaviour of cabbage butterfly, *Pieris rapae*. *J Chem Ecol* 26, 655–66.

O'Neil, P. (1997) Natural selection on genetically-correlated phenological characters in *Lythrum salicaria* L. (Lythraceae). *Evolution* 51:267–74.

Opler, P. A. (1983) Nectar production in a tropical ecosystem. In *The biology of Nectaries*, ed. B. Bentley and T. Elias, 30–79. New York: Columbia University Press.

Opler, P. A., G. W. Frankie, and H. G. Baker. (1976) Rainfall as a factor in the release, timing, and synchronization of anthesis by tropical trees and shrubs. *J Biogeog* 3: 231–36.

Opler, P. A., H. G. Baker, and G. W. Frankie. (1980) Plant reproductive characteristics during secondary succession in Neotropical lowland forest ecosystems. *Biotropica* 12:40–46.

Ornelas, J. F., C. González, L. Jiménez, C. Lara, and A. J. Martínez. (2004) Reproductive ecology of distylous *Palicourea padifolia* (Rubiaceae) in a tropical montane cloud forest. II. Attracting and rewarding mutualistic and antagonistic visitors. *Am J Bot* 91:1061–69.

Ornelas, J. F. and C. Lara. (2009) Nectar replenishment and pollen receipt interact in their effects on seed production of *Penstemon roseus*. *Oecologia* 160:675–85.

Ornelas, J. F., M. Ordano, and C. Lara. (2007) Nectar removal effects on seed production in *Moussonia deppeana* (Gesneriaceae), a hummingbird-pollinated shrub. *Ecoscience* 14:117–23.

Orth, A. I. and K. D. Waddington. (1997) The movement patterns of carpenter bees *Xylocopa micans* and bumblebees *Bombus pennsylvanicus* on *Pontederia cordata* inflorescences. *J Insect Behav* 10:79–86.

Osborne, J. L., A. P Martin, N. L. Carreck, J. L Swain., M. E. Knight, D. Goulson, R. J. Hale, and R. A. Sanderson. (2008) Bumblebee flight distances in relation to the forage landscape. *J Anim Ecol* 77:406–15.

Otegui, M. and A. Cocucci. (1999) Floral morphology and biology of *Myrsine laetivirens*, structural and evolutionary implications of anemophily in Myrsinaceae. *Nordic J Bot* 19:71–85.

Otieno, M., S. G. Potts, and P. G. Willmer. (2010) Effects of grazing on flowers and pollen in East *African Acacia*. Submitted, *Oecologia*.

Ottaviano, E., M. Sari-Gorla, and D. L. Mulcahy. (1980). Pollen tube growth rates in *Zea mays*: implications for genetic improvement of crops. *Science* 210: 437–38.

Otterstatter, M. C., R. J. Gegear, S. R. Colla, and J. D. Thomson. (2005) Effects of parasitic mites and protozoa on the flower constancy and foraging rate of bumble bees. *Behav Ecol Sociobiol* 58:383–89.

Otterstatter, M. C. and J. D. Thomson. (2008) Does pathogen spillover from commercially reared bumblebees threaten wild pollinators? *PLoS ONE* 3(7):e2771.

Otterstatter, M. C., T. L. Whidden, and R. E. Owen. (2002) Contrasting frequencies of parasitism and host mortality among phorid and conopid parasitoids of bumble-bees. *Ecol Entomol* 27:229–37.

Overdorff, D. J. (1992) Differential patterns in flower feeding by *Eulemur fulvus rufus* and *Eulemur rubriventer* in Madagascar. *Am J Primatol* 28:191–203.

Owens, J. N., T. Takaso, and C. J. Runions. (1998) Pollination in conifers. *Trends Plant Sci* 3:479–85.

Pacini, E. (1994) Cell biology of anther and pollen development. In *Genetic Control of Self Incompatibility and Reproductive Development in Flowers,* ed. E. G. Williams, R. B. Knox, and A. E. Clarke, 289–308. Dordrect: Kluwer.

———. (2000) From anther and pollen ripening to pollen presentation. *Plant Syst Evol* 222:19-43.

Pacini, E. and L. M. Bellani. (1986) *Lagerstroemia indica* L. pollen: form and function. In *Pollen and Spores: Form and Function,* ed. S. Blackmore and I. K. Ferguson, 347–57. London: Academic Press.

Pacini, E. and G. G. Franchi. (1998) Pollen dispersal units, gynoecium and pollination. In *Reproductive Biology,* ed. S. J. Owen and P. J. Rudall, 183–95. Kew: Royal Botanic Gardens.

Pacini, E. and G. G. Franchi. (1999) Types of pollen dispersal units and pollen competition. In *Anther and Pollen: from Biology to Biotechnology,* ed. C. Clement, E. Pacini, and J. C. Audran, 1–11. Berlin: Springer.

Pacini, E., G. G. Franchi, M. Lisci, and M. Nepi. (1997) Pollen viability related to type of pollination in six angiosperm species. *Ann Bot* 80:83–87.

Pacini, E. and M. Nepi. (2007) Nectar production and presentation. In *Nectaries and Nectar,* ed. S. W. Nicolson, M. Nepi, and E. Pacini, 167–214. Dordrecht: Springer.

Pacini, E., M. Nepi, and J. L. Vesprini. (2003) Nectar biodiversity: a short review. *Plant Syst Evol* 238:7–21.

Page, R. E. and K. M. Fondrk. (1995) The effects of colony-level selection on the social organisation of honey bee (*Apis mellifera* L.) colonies: colony level components of pollen hoarding. *Behav Ecol Sociobiol* 36:135–44.

Pais, M. S. and A.C.S. Figueiredo. (1994) Floral nectaries from *Limodorum abortivum* (L) and *Epipactis atropurpurea* Rafin (Orchidaceae) – ultrastructural changes in plastids during the secretory process. *Apidologie* 25:615–26.

Pankiw, T., R. E. Page, and K. M. Fondrk. (1998) Brood pheromone stimulates pollen foraging in honeybees (*Apis mellifera*). *Behav Ecol Sociobiol* 44:193–98.

Pansarin, L. M., E. R. Pansarin, and M. Sazima. (2008) Facultative autogamy in *Cyrtopodium polyphyllum* (Orchidaceae) through a rain-assisted pollination mechanism. *Aust J Bot* 56:363–67.

Parmenter, L. (1958) Flies (Diptera) and their relations with plants. *Nature* 37:115–25.

Parmesan, C. (2006) Ecological and evolutionary responses to recent climate change. *Ann Rev Ecol Syst* 37:637–69.

Parmesan, C. and G. Yohe. (2003) A globally coherent fingerprint of climate change impacts across natural systems. *Nature* 421:37–42.

Parra-Tabla, V., L. Abdala-Roberts, J. C. Rojas, J. Navarro, and L. Salinas-Peba. (2009) Floral longevity and scent respond to pollen manipulation and resource status in the tropical orchid *Myrmecophila christinae*. *Plant Syst Evol* 282:1–11.

Parrish, J.A.D. and F. A. Bazzaz. (1979) Difference in pollination niche relationships in early and late successional plant communities. *Ecology* 60:597–610.

Pasarin, L. M., M. D. Castro, and M. Sazima. (2009) Osmophore and elaiophores of *Grobya amherstiae* (Catasetinae, Orchidaceae) and their relation to pollination. *Bot J Linn Soc* 159:408–15.

Pasquet, R. S., A. Peltier A, M. B. Hufford, E. Oudin, J. Saulnier, L. Paul, J. T. Knudsen, H. R. Herren, and P. Gepts. (2008) Long-distance pollen flow assessment through evaluation of pollinator foraging range suggests trans-gene escape distances. *Proc Nat Acad Sci USA* 105:13456–61.

Patino, S., T. Aalto, A. A. Edwards. and J. Grace. (2002) Is *Rafflesia* an endothermic flower? *New Phytol* 154:429–37.

Patino, S. and J. Grace. (2002) The cooling of convolvulaceous flowers in a tropical environment. *Plant Cell and Envt* 25:41–51.

Patino, S., C. Jeffree, and J. Grace. (2002) The ecological role of orientation in tropical convolvulaceous flowers. *Oecologia* 130:373–79.

Paton, D. C. (1993) Honeybees in the Australian environment: does *Apis mellifera* disrupt or benefit the native biota? *Bioscience* 43:95–103.

Paton, D. C. and H. A. Ford. (1983) The influence of plant characteristics and honeyeater size on levels of pollination in Australian plants. In *Handbook of Experimental Pollination Biology,* ed. C. E. Jones and R. J. Little, 235–48. New York: Van Nostrand Reinhold.

Paton, D. C. and V. Turner. (1985) Pollination of *Banksia ericifolia*: birds, mammals and insects as pollen vectors. *Aust J Bot* 33:271–86.

Patt, J. M., J. C. French, C. Schal, and T. G. Hartmann. (1991) Pollination biology of *Peltandra virginica* (Araceae). 1. Chemical composition and emission pattern of floral odor. *Am J Bot* 78:S67.

Patt, J. M., G. C. Hamilton, and J. H. Lashomb. (1997) Foraging success of parasitoid wasps on flowers: interplay of insect morphology, floral architecture and searching behavior. *Ent Exp Appl* 83:21–30.

Patt, J. M., D. F. Rhoades, and J. A. Corkill. (1988) Analysis of the floral fragrance of *Platanthera stricta*. *Phytochemistry* 27:91–95.

Paul, J. and F. Roces. (2003) Fluid intake rates in ants correlate with their feeding habits. *J Insect Physiol* 49:347–57.

Pauw, A. (1998) Pollen transfer on bird's tongues. *Nature* 394:731–32.

———. (2006) Floral syndromes accurately predict pollination by a specialized oil-collecting bee (*Rediviva peringueyi*, Melittidae) in a guild of South African orchids. *Am J Bot* 93:917–26.

———. (2007) Collapse of a pollination web in small conservation areas. *Ecology* 88:1759–69.

Paxton, R. J., P. F. Kukuk, and J. Tengo. (1999) Effects of familiarity and nestmate number on social interactions in two communal bees, *Andrena scotica* and *Panurgus calcaratus* (Hymenoptera, Andrenidae). *Insectes Soc* 46:109–18.

Paxton, R. J., M. Ayasse, J. Field, and A. Soro. (2002) Complex sociogenetic organization and reproductive skew in a primitively eusocial sweat bee, *Lasioglossum malachurum*, as revealed by microsatellites. *Mol Ecol* 11:2405–16.

Peakall, R. (1989) The unique pollination of *Leporella fimbriata* (Orchidaceae): pollination by pseudocopulating male ants (*Myrmecia urens*, Formicidae). *Plant Syst Evol* 167:137–48.

———. (1990) Responses of male *Zaspilothynnus trilobatus* Turner wasps to females and the sexually-deceptive orchid it pollinates. *Funct Ecol* 4:159–67.

Peakall, R., C. J. Angus, and A. J. Beattie. (1990) The significance of ant and plant traits for ant pollination in *Leporella fimbriata*. *Oecologia* 84:457–60.

Peakall, R. and A. J. Beattie. (1989) Pollination of the orchid *Microtis parviflora* R. by flightless worker ants. *Funct Ecol* 3:515–22.

Peakall, R. and A. J. Beattie. (1996) Ecological and genetic consequences of pollination by sexual deception in the orchid *Caladenia tentaculata*. *Evolution* 50:2207–20.

Peakall, R., A. J. Beattie, and S. H. James. (1987) Pseudocopulation of an orchid by male ants: a test of two hypotheses accounting for the rarity of ant pollination. *Oecologia* 73:511–24.

Peakall, R., S. N. Handel, and A. J. Beattie. (1991) The evidence for, and importance of, ant pollination. In *Ant-Plant Interactions*, ed. S. R. Huxley and D. F. Cutler, 421–29. Oxford: Oxford University Press.

Peat, J. and D. Goulson. (2005) Effects of experience and weather on foraging rate and pollen versus nectar collection in the bumblebee, *Bombus terrestris*. *Behav Ecol Sociobiol* 58:152–56.

Peeters, L. and Ø. Totland. (1999) Wind to insect pollination ratios and floral traits in five alpine *Salix* species. *Can J Bot* 77:556–63.

Pekkarinen, A. (1997) Oligolectic bee species in Northern Europe (Hymenoptera, Apoidea). *Ent Fennica* 8, 205–14.

Pellegrino, G. and A. Musacchio. (2006) Effects of defoliation on reproductive success in two orchids, *Serapias vomeracea* and *Dactylorhiza sambucina*. *Ann Bot Fenn* 43:123–28.

Pellmyr, O. (1985) Flower constancy in individuals of an anthophilous beetle, *Byturus ochraceus* (Scriba) (Coleoptera, Butyridae). *Coleopt Bull* 39:341–45.

———. (1986) Three pollination morphs in *Cimicifuga simplex*: incipient speciation due to inferiority in competition. *Oecologia* 68:304–11.

———. (1988) Bumblebees (Hymenoptera: Apidae) assess pollen availability in *Anemonopsis macrophylla* (Ranunculaceae) through floral shape. *Ann Ent Soc Am* 81:792–97.

———. (1989) The cost of mutualism: interactions between *Trollius europaeus* and its pollinating parasites. *Oecologia* 78:53–59.

———. (1992a) Evolution of insect pollination and angiosperm diversification. *Trends Ecol Evol* 7:46–48.

———. (1992b) The phylogeny of a mutualism: evolution and coadaptation between *Trollius* and its seed-parasitic pollinators. *Biol J Linn Soc* 47:337–65.

———. (1985) Flower constancy in individuals of an anthophilous beetle, *Byturus ochraceus* (Scribs) (Coleoptera, Byturidae). *Coleopt Bull* 39:341–45.

———. (1997) Pollinating seed eaters: why is active pollination so rare? *Ecology* 78:1655–60.

———. (2003) Yuccas, yucca moths and coevolution: a review. *Ann Miss Bot Gard* 90:35–55.

Pellmyr, O. and C. J. Huth. (1995) Differential abortion in the yucca: reply. *Nature* 376, 558.

Pellmyr, O. and J. Leebens-Mack. (2000) Reversal of mutualism as a mechanism for adaptive radiation in yucca moths. *Am Nat* 156:S62–S76.

Pellmyr, O. and J. Patt. (1986) Function of olfactory and visual stimuli in pollination of *Lysichiton americanum* (Araceae) by a staphylinid beetle. *Madrono* 33:47–54.

Pellmyr, O., W. Tang, I. Groth, B. Bergstrom, and L. B. Thien. (1991) Cycad cone and angiosperm floral volatiles: inferences for the evolution of insect pollination. *Biochem Syst Evol* 19:623–27.

Pellmyr, O. and L. B. Thien. (1986) Insect reproduction and floral fragrances: keys to the evolution of the angiosperms? *Taxon* 35:76–85.

Pellmyr, O. and J. N. Thompson. (1992) Multiple occurrences of mutualism in the yucca moth lineage. *Proc Nat Acad Sci USA* 89:2927–29.

———. (1996) Sources of variation in pollinator contribution within a guild: the effects of plants and pollinator factors. *Oecologia* 107:595–604.

Pellmyr, O., J. N. Thompson, J. M. Brown, and R. G. Harrison. (1996) Evolution of pollination and mutualism in the yucca moth lineage. *Am Nat* 148:827–47.

Penet, L., C. L. Collin, and T-L Ashman. (2009) Florivory increases selfing: an experimental study in the wild strawberry, *Fragaria virginiana*. *Plant Biol* 11:38–45.

Peng, Y. Q., D. R. Yang, and Q. Y. Wang. (2005) Quantitative tests of interaction between pollinating and non-pollinating fig wasps on dioecious *Ficus hispida*. *Ecol Ent* 30: 70–77.

Penz, C. M. and H. W. Krenn. (2000) Behavioral adaptations to pollen-feeding in *Heliconius* butterflies (Nymphalidae, Heliconiinae): an experiment using *Lantana* flowers. *J Insect Behav* 13:865–80.

Percival, M. S. (1961) Types of nectar in angiosperms. *New Phytol* 60:235–81.

———. (1965) *Floral Biology*. Oxford: Pergamon Press.

Pereboom, J.J.M. and J. C. Biesmeijer. (2003) Thermal constraints for stingless bee foragers: the importance of body size and coloration. *Oecologia* 137:42–50.

Perez, F., M.T.K. Arroyo, and J. J. Armesto. (2009) Evolution of autonomous selfing accompanies increased specialization in the pollination system of *Schizanthus* (Solanaceae). *Am J Bot* 96:1168–76.

Perez, F., M. T. K. Arroyo, and R. Medel. (2007) Phylogenetic analysis of floral integration in *Schizanthus* (Solanaceae): does pollination truly integrate corolla traits? *J Evol Biol* 20:1730–38.

Perez, S. M. and K. D. Waddington. (1996) Carpenter bee (*Xylocopa micans*) risk indifference and a review of nectarivore risk-sensitivity studies. *Am Zool* 36:435–46.

Perez-Banon, C., T. Petanidou, and M. Marcos-Garcia. (2007) Pollination in small islands by occasional visitors: the case of *Daucus carota* subsp. *commutatus* (Apiaceae) in the Columbretes archipelago, Spain. *Plant Ecol* 192:133–51.

Perez-Mellado, V. and J. L. Casas. (1997) Pollination by a lizard on a Mediterranean island. *Copeia* 1997:593–95.

Pernal, S. F. and R. W. Currie. (2001) The influence of pollen quality on foraging behavior in honeybees (*Apis mellifera* L.) *Behav Ecol Sociobiol* 51:53-68.

Perry, D. R. and A. Starrett. (1980) The pollination ecology and blooming strategy of a neotropical emergent tree, *Dipteryx panamensis*. *Biotropica* 12:307–13.

Petanidou, T. (2005) Sugars in Mediterranean floral nectars: an ecological and evolutionary approach. *J Chem Ecol* 31:1065–88.

Petanidou, T. and W. N. Ellis. (1993) Pollinating fauna of a phryganic ecosystem: composition and diversity. *Biodiv Lett* 1:9–22.

Petanidou, T., W. N. Ellis, N. S. Margaris, and D. Vokou. (1995) Constraints on flowering phenology in a phryganic (east Mediterranean shrub) community. *Am J Bot* 82:607–20.

Petanidou, T., A. S. Kallimanis, J. Tzanopoulos, S. O. Sgardelis, and J. D. Pantis. (2008) Long-term observation of a pollinator network: fluctuation in species and interactions, relative invariance of network structure and implications for estimates of specialization. *Ecol Lett* 11:564–75.

Petanidou, T. and E. Lamborn. (2005) A land for flowers and bees: studying pollination ecology in Mediterranean communities. *Plant Biosys* 139:279–94.

Petanidou, T., V. Goethals, and E. Smets. (2000) Nectary structure of Labiatae in relation to their nectar secretion and characteristics in a Mediterranean shrub community: does flowering time matter? *Plant Syst Evol* 225:103–18.

Petanidou, T. and S. G. Potts. (2006) Mutual use of resources in Mediterranean plant-pollinator communities: how specialized are pollination webs? In *Plant-Pollinator Interactions: from Specialization to Generalization*, ed. N. M. Waser and J. Ollerton, 220–44. Chicago: University of Chicago Press.

Petanidou, T. and E. Smets. (1995) The potential of marginal lands for bees and apiculture: nectar secretion in Mediterranean shrublands. *Apidologie* 26:39–52.

———. (1996) Does temperature stress induce nectar secretion in Mediterranean plants? *New Phytol* 133:513–18.

Petanidou, T., A. Van Laere, W. N. Ellis, and E. Smets. (2006) What shapes amino acid and sugar composition in Mediterranean floral nectars? *Oikos* 115:155–69.

Petanidou, T. and D. Vokou. (1990) Pollination and pollen energetics in Mediterranean ecosystems. *Am J Bot* 77: 986–92.

Peter, C. I., A. P. Dold, N. P. Barker, and B. S. Ripley. (2004) Pollination biology of *Bergeranthus multiceps* (Aizoaceae) with preliminary observations of repeated flower opening and closure. *S Afr J Sci* 100:624–29.

Peter, C. I. and S. D. Johnson. (2006) Anther cap retention prevents self-pollination by elaterid beetles in the South African orchid *Eulophia foliosa*. *Ann Bot* 97:345–55.

——— (2008) Mimics and magnets: the importance of color and ecological facilitation in floral deception. *Ecology* 89:1583–95.

Pettersson, M. W. (1991) Pollination by a guild of fluctuating moth populations: option for unspecialization in *Silene vulgaris*. *J Ecol* 79:591–604.

Pettersson, S., F. Ervik, and J. T. Knudsen. (2004) Floral scent of bat-pollinated species: West Africa vs. the New World. *Biol J Linn Soc* 82:161–68.

Pettet, A. (1977) Seasonal changes in nectar-feeding by birds at Zaria, Nigeria. *Ibis* 119:291–308.

Pettitt, J. M. (1984) Aspects of flowering and pollination in marine angiosperms. *Oceanog and Mar Biol: Ann Rev* 22:315–42.

Pfunder, M. and B. A. Roy. (2006) Fungal pseudoflowers can influence the fecundity of insect-pollinated flowers on *Euphorbia cyparissias*. *Bot Helv* 116:149–58.

Philbrick, C. T. and G. J. Anderson. (1987) Implications of pollen-ovule ratios and pollen size for the reproductive biology of *Potamogeton* and autogamy in aquatic angiosperms. *Syst Bot* 12:98–105.

———. (1992) Pollination biology in the Callitrichaceae. *Syst Bot* 17:282–92.

Philipp, M., J. Böcher, H. R. Siegismund, and L. R. Nielsen. (2006) Structure of a plant-pollinator network on a pahoehoe lava desert of the Galapagos Islands. *Ecography* 29:531–40

Philippe, G. and P. Baldet. (1997) Electrostatic dusting: an efficient technique for pollination in larch. *Ann Sci Forest* 54:301–10.

Pichersky, E., J. P. Noel, and N. Dudareva. (2006) Biosynthesis of plant volatiles: Nature's diversity and ingenuity. *Science* 311:808–11.

Picker, M. D. and J. J. Midgley. (1996) Pollination by monkey beetles (Coleoptera, Scarabaeidae: Hopliini): flower and colour preferences. *Afr Entomol* 4:7–14.

Pico, F. X. and J. Retana. (2001) The flowering pattern of the perennial herb *Lobularia maritima*: an unusual case in the Mediterranean basin. *Acta Oecol* 22:209–17.

Pilson, D. (2000) Herbivory and natural selection on flowering phenology of wild sunflower, *Helianthus annuus*. *Oecologia* 122:72–82.

Piper, J. G., B. Charlesworth, and D. Charlesworth. (1984) A high rate of self-fertilization and increased seed fertility of homostyle primroses. *Nature* 310:50–51.

Pitts-Singer, T. L., J. L. Hanula, and J. L. Walker. (2002) Insect pollinators of three rare plants in a Florida longleaf pine forest. *Flor Entomol* 85:308–16.

Plack, A. (1957) Sexual dimorphism in Labiatae. *Nature* 180:1218–19.

———. (1958) Effect of giberellic acid on corolla size. *Nature* 182:610.

Pleasants, J. M. (1977) The effect of nectar production on neighbourhood size. *Oecologia* 52:104–8.

———. (1980) Competition for bumblebee pollinators in Rocky mountain plant communities. *Ecology* 61:1446–59.

———. (1981) Bumblebee response to variation in nectar availability. *Ecology* 62:1648–61.

———. (1983) Structure of plant and pollinator communities. In *Handbook of Experimental Pollination Biology*, ed. C. E. Jones and R. J. Little, 375–93. New York: Van Nostrand Reinhold.

Pleasants, J. M. and N. M. Waser. (1985) Bumblebee foraging at a 'hummingbird' flower: reward economics and floral choice. *Am Midl Nat* 114:283–91.

Pleasants, J. M. and M. L. Zimmerman. (1983) The distribution of standing crop of nectar: what does it really tell us? *Oecologia* 57:412–14.

———.(1990) The effect of inflorescence size on pollinator visitation of *Delphinium nelsonii* and *Aconitum columbianum*. *Collect Bot* 19:21–39.

Pohl, N., G. Carvallo, C. Botto-Mahan, and R. Medel. (2006) Non-additive effects of flower damage and hummingbird pollination on the fecundity of *Mimulus luteus*. *Oecologia* 149:648–55.

Pohl, M., T. Watolla, and K. Lunau. (2008) Anther-mimicking floral guides exploit a conflict between innate preference and learning in bumblebees (*Bombus terrestris*). *Behav Ecol Sociobiol* 63:295–302.

Pont, A. C. (1993) Observations on anthophilous Muscidae and other Diptera in Abisko National Park, Sweden. *J Nat Hist* 27:631–43.

Pontin, D. R., M. R. Wade, P. Kehrli, and S. D. Wratten. (2005) Attractiveness of single and multiple species flower patches to beneficial insects in agroecosystems. *Ann Appl Biol* 148:39–47.

Poole, R. W. and B. J. Rathcke. (1979) Regularity, randomness and aggregation in flowering phenologies. *Science* 203:470–71.

Possingham, H. P. (1988) A model of resource renewal and depletion: applications to the distribution and abundance of nectar in flowers. *Theor Pop Biol* 33:138–60.

Possingham, H. P., A. I. Houston, and J. M. McNamara. (1990) Risk-averse foraging in bees: a comment on the model of Harder and Real. *Ecology* 71:1622–24.

Potts, S. G., A. Dafni, and G. Ne'eman. (2001) Pollination of a core flowering shrub species in Mediterraneaen phrygana: variation in pollinator diversity, abundance and effectiveness in response to fire. *Oikos* 92:71–80.

Potts, S. G., P. G. Kevan, and J. W. Boone. (2005) Conservation in pollination: collecting, surveying and monitoring. In *Practical Pollination Biology*, ed. A. Dafni, P. G. Kevan, and C. Husband, 401–34. Cambridge, Ontario: Enviroquest.

Potts, S. G., T. Petanidou, S. Roberts, C. O'Toole, A. Hulbert, and P. G. Willmer. (2006) Plant-pollinator diversity and pollination services in a complex Mediterranean landscape. *Biol Conserv* 129:519–29.

Potts, S. G., B. Vulliamy, A. Dafni, G. Ne'eman, C. O'Toole, S. Roberts, and P. G. Willmer. (2003a) Response of plant-pollinator communities following fire: changes in diversity, abundance and reward structure. *Oikos* 101:103–12.

Potts, S. G., B. Vulliamy, A. Dafni, G. Ne'eman, and P. G. Willmer. (2003b) Linking bees and flowers: how do floral communities structure pollinator communities? *Ecology* 84:2628–42.

Potts, S. G., B. Vulliamy, S. Roberts, C. O'Toole, A. Dafni, G. Ne'eman, and P. G. Willmer. (2004) Nectar resource diversity organises flower-visitor community structure. *Ent Exp Appl* 113:103–7.

Potts, S. G., B. Vulliamy, S. Roberts, C. O'Toole, A. Dafni, G. Ne'eman, and P. G. Willmer. (2005) Role of nesting resources in organising diverse bee communities in a Mediterranean landscape. *Ecol Entomol* 30:78–85.

Potts, S. G. and P. G. Willmer. (1997) Abiotic and biotic factors influencing nest-site selection by *Halictus rubicundus*, a ground-nesting halictid bee. *Ecol Entomol* 22: 319–28.

Poveda, K., I. Steffan-Dewenter, S. Scheu, and T. Tscharntke. (2005a) Floral trait expression and plant fitness in response to below- and above-ground plant-animal interactions. *Persp Plant Ecol Evol Syst* 7:77–83.

Powell, A. H. and G.V.N. Powell. (1987) Population dynamics of male euglossine bees in Amazonian forest fragments. *Biotropica* 19:176–79.

Powell, E. A. and C.E. Jones. (1983) Floral mutualism in *Lupinus benthamii* (Fabaceae) and *Delphinium parryi* (Ranunculaceae). In *Handbook of Experimental Pollination Biology*, ed. C. E. Jones and R. J. Little, 310–29. New York: Van Nostrand Reinhold.

Powell, J. A. (1992) Interrelationships of yuccas and yucca moths. *Trends Ecol Evol* 7:10–15.

Poyry, J., S. Lindgren, J. Salminen, and M. Kuussaari. (2005) Responses of butterfly and moth species to restored cattle grazing in semi-natural grasslands. *Biol Cons* 122:465–78.

Prance, G. T. (1980) A note on the pollination of *Nymphaea amazonum* Mart. and Zucc. (Nymphaceae). *Brittonia* 32: 505–7.

Prance, G. T. and J. R. Arias. (1975) A study of the floral biology of *Victoria amazonica* (Poepp.) Sowerby (Nymphaceae). *Acta Amazonica* 5:109–39.

Praz, C. J., A. Müller, and S. Dorn. (2008a) Host recognition in a pollen-specialist bee: evidence for a genetic basis. *Apidologie* 39:547–57.

———. (2008b) Specialized bees fail to develop on non-host-pollen; do plants chemically protect their pollen? *Ecology* 89:795–804.

Preston, J. C. and L. C. Hilernan. (2009) Developmental genetics of floral symmetry evolution. *Trends Plant Sci* 14: 147–54.

Price, J. P. and W. L. Wagner, (2004) Speciation in Hawaiian angiosperm lineages: cause, consequence and mode. *Evolution* 58:2185–200.

Price, M. V. and N. M. Waser. (1982) Experimental studies of pollen carryover: hummingbirds and *Ipomopsis aggregata*. *Oecologia* 54:353–58.

———. (1998) Effects of experimental warming on plant reproductive phenology in a subalpine meadow. *Ecology* 79:1261–71.

———. (2000) Responses of subalpine meadow vegetation to four years of experimental warming. *Ecol Appl* 10: 811–23.

Price, M. V., N. M. Waser, R. E. Irwin, D. R. Campbell, and A. K. Brody. (2005) Temporal and spatial variation in pollination of a montane herb: a seven year study. *Ecology* 86:2106–16.

Priess, J. A., M. Mimler, A. M. Klein, S. Schwarze, T. Tscharntke, and I. Steffan-Dewenter. (2007) Linking deforestation scenarios to pollination services and economic returns in coffee agroforestry systems. *Ecol Appl* 17:407–17.

Primack, R. B. (1980) Phenological variation within natural populations. I. Flowering in New Zealand montane shrubs. *J Ecol* 68:849–62.

———. (1983) Insect pollination in the New Zealand mountain flora. *NZ J Bot* 21:317–33.

———. (1985a) Longevity of individual flowers. *Ann Rev Ecol Syst* 16:15–38.

———. (1985b) Patterns of flowering phenology in communities, individuals, and single flowers. In *Population Structure of Vegetation*, ed. J. White, 571–93. Dordrect: Junk.

Primack, R. B. and P. Hall. (1990) Costs of reproduction in the pink lady's slipper orchid: a four year experimental study. *Am Nat* 136:638–56.

Primack, R. B., C. Imbres, R. Primack, A. Miller-Rushing, and P. Tredici. (2004) Herbarium specimens demonstrate earlier flowering times in response to warming in Boston. *Am J Bot* 91:1260–64.

Primack, R. and J. A. Silander. (1975) Measuring the relative importance of different pollinators to plants. *Nature* 255: 143–44.

Pritchard, J. R. and D. Schluter. (2001) Declining interspecific competition during character displacement: summoning the ghost of competition past. *Evol Ecol Res* 3: 209–20.

Proches, S. and S. D. Johnson. (2009) Beetle pollination of the fruit-scented cones of the South African cycad *Stangeria eriopus*. *Am J Bot* 96:1722–30.

Proctor, J. and S. Proctor. (1978) *Nature's Use of Colour in Plants and their Flowers*. London: Peter Lowe.

Proctor, M. and P. Yeo. (1973) *The Pollination of Flowers*. London: Collins.

Proctor, M., P. Yeo, and A. Lack. (1996) *The Natural History of Pollination*. Portland, OR: Timber Press.

Proença, C.E.B. (1992) Buzz pollination – older and more widespread than we think? *J Trop Ecol* 8:115–20.

Prŷs-Jones, O. E. and S. A. Corbet. (1991) *Bumblebees*. Slough, UK: Richmond Publishing Co.

Prŷs-Jones, O. E. and P. G. Willmer. (1992) The biology of alkaline nectar in the purple toothwort (*Lathraea clandestina*): ground level defences. *Biol J Linn Soc* 45:373–88.

Punt, W., S. Blackmore, S. Nilsson, and A. leThomas. (1994) *Glossary of Pollen and Spore Terminology*. Utrecht: LPP Foundation.

Punzo, F. (2006) Plants whose flowers are utilized by adult of *Pepsis grossa* Fabricius (Hymenoptera, Pompilidae) as a source of nectar. *J Hymen Res* 15:171–76.

Puterbaugh, M. N. (1998) The roles of ants as flower visitors: experimental analysis in three alpine plant species. *Oikos* 83:36–46.

Pye, J. D. (1983) *Moths and bats: pollen and nectar collection at night*. Ilford, UK: Centr Assoc Beekeepers.

Pyke, G. H. (1978) Optimal foraging in hummingbirds: testing the marginal value theorem. *Am Zool* 18:739–52.

———. (1981) Optimal foraging in hummingbirds: rule of movement between inflorescences. *Anim Behav* 39:889–96.

———. (1982) Local geographic distribution of bumblebees near Crested Butte, Colorado: competition and community structure. *Ecology* 63:555–73.

———. (1991) What does it cost a flower to produce nectar? *Nature* 350:58–59.

Pyke, G. H., M. Christy, and R. E. Major. (1996) Territoriality in honeyeaters: reviewing the concept and evaluating available information. *Aust J Zool* 44:297–17.

Pyke, G. H. and N. M. Waser. (1981) The production of dilute nectars by hummingbird and honeyeater flowers. *Biotropica* 13:260–70.

Pywell, R. F., J. M. Bullock, D. B. Roy, L. Warman, K. J. Walker, and P. Rothery. (2003) Plant traits as predictors of performance in ecological restoration. *J Appl Ecol* 40:65–77.

Pywell, R. F., E. A. Warman, C. Carvell, T. H. Sparks, L. V. Dicks, D. Bennett, A. Wright, C.N.R. Critchley, and A. Sherwood. (2005) Providing foraging resources for bumblebees in intensively farmed landscapes. *Biol Cons* 121:479–94.

Qu, R, X. Li, Y. Luo, M. Dong, H. Xu, X. Chen, and A. Dafni. (2007) Wind-dragged corolla enhances self-pollination: a new mechanism of delayed self-pollination. *Ann Bot* 100:1155–64.

Quesada, M., K. Bollman, and A. G. Stephenson. (1995) Leaf damage decreases pollen production and hinders pollen performance in *Cucurbita texana*. *Ecology* 76:437–43.

Rademaker, M.C.J., T. J. de Jong, and P.G.L. Klinkhamer. (1997) Pollen dynamics of bumblebee visitation on *Echium vulgare*. *Funct Ecol* 11:554–63.

Rafinski, J. N. (1979) Geographic variability of flower colour in *Crocus scepusiensis* (Iridaceae). *Plant Syst Evol* 131:107–25.

Raguso, R. A. (2001) Floral scent, olfaction and scent-driven foraging behavior. In *Cognitive Ecology of Pollination*, ed. L. Chittka and J. D. Thomson, 83–105. Cambridge: Cambridge University Press.

———. (2004a) Why do flowers smell? The chemical ecology of fragrance-driven pollination. In *Advances in Chemical Ecology*, ed. R. T. Cardé and J. G. Millar JG, 151–78, Cambridge: Cambridge University Press.

———. (2004b) Flowers as sensory billboards: progress towards an integrated understanding of floral advertisement. *Current Opinion in Plant Biology* 7:434–40.

———. (2004c) Why are some floral nectars scented? *Ecology* 85:1486–94.

———. (2006) Behavioral responses to floral scents; experimental manipulations and the interplay of sensory modalities. In *Biology of Floral Scent*, ed. N. Dudareva and E. Pichersky, 297–318. Boca Raton, FL: CRC Press.

Raguso, R. A. and O. Pellmyr. (1998) Dynamic headspace analysis of floral volatiles: a comparison of methods. *Oikos* 81:238–54.

Raguso, R. A. and E. Pichersky. (1999) A day in the life of a linalool molecule: chemical communication in a plant-pollinator system. Part 1: Linalool biosynthesis in flowering plants. *Plant Species Biol* 14:95–120.

Raguso, R. A. and M. A. Willis. (2004) Synergy between visual and olfactory cues in nectar feeding by wild hawkmoths, *Manduca sexta*. *Anim Behav* 69:407–18.

Raine. N. E. and L. Chittka. (2007a) Pollen foraging: learning a complex motor skill by bumblebees (*Bombus*). *Naturwissenschaften* 94:459–64.

———. (2007b) Flower constancy and memory dynamics in bumblebees (Hymenoptera: Apidae: *Bombus*). *Entomol Gen* 29:179–99.

Raine, N. E., P. G. Willmer, G. N. Stone. (2002) Spatial structuring and floral repellence prevent ant-pollinator conflict in a Mexican ant-acacia. *Ecology* 83:3086–96.

Ram, H. Y. M. and G. Mathur. (1984) Flower colour changes in *Lantana camara*. *J Exp Bot* 35:1656–62.

Ramalho, M., V. L. Imperatriz-Fonseca, A. Kleinert-Giovannini, and M. Cortopassi-Laurino. (1985) Exploitation of floral resources by *Plebeia remota* Holmberg (Apidae, Meliponinae). *Apidologie* 16:307–30.

Rameau, C. and P-H Gouyon. (1991) Resource allocation to growth, reproduction and survival in Gladiolus: the cost of male function. *J Evol Ecol* 4:291–307.

Ramirez, B. W. (1969) Fig wasps: mechanisms of pollen transfer. *Science* 163:580–81.

Ramírez, N. (2003) Floral specialization and pollination: a quantitative analysis of the Leppik and the Faegri and van der Pijl classification systems. *Taxon* 52:687–700.

Ramsey, M. (1988) Differences in pollinator effectiveness of birds and insects visiting *Banksia menziesii* (Proteaceae). *Oecologia* 76:119–24.

———. (1995) Ant pollination of the perennial herb *Blandfordia grandiflora* (Liliaceae). *Oikos* 74:265–72.

Ramsey, M., S. C. Cairns, and G. Vaughton. (1994) Geographical variation in morphological and reproductive characters of coastal and tableland populations of *Blandfordia grandiflora*, (Liliaceae). *Plant Syst Evol* 192:215–30.

Rands, S. A. and H. M. Whitney. (2008) Floral temperature and optimal foraging: is heat a feasible floral reward for pollinators? *PLoS ONE* 3(4):e2007.

Ranta, E. (1984) Proboscis length and the coexistence of bumblebee species. *Oikos* 43:189–96.

Ranta, E. and H. Lundberg. (1980) Resource partitioning in bumblebees: the significance of differences in proboscis length. *Oikos* 35:298–302.

Ranta, E., H. Lundberg, and I. Teras. (1981) Patterns of resource utilization in two Fennoscandian bumblebee communities. *Oikos* 36:1–11.

Ranta, E. and K. Vapsalainen. (1981) Why are there so many species? Spatio-temporal heterogeneity and northern bumblebee communities. *Oikos* 36:28–34.

Rao, M. J., J. Terborgh, and P. Nunez. (2001) Increased herbivory in forest isolates: implications for plant community structure and composition. *Cons Biol* 15:624–33.

Rasheed, S. A. and L. D. Harder. (1997a) Foraging currencies for non-energetic resources: pollen collection by bumblebees. *Anim Behav* 54:911–26.

———. (1997b) Economic motivation for plant species preferences of pollen-collecting bumble bees. *Ecol Entomol* 22:209–19.

Rathcke, B. J. (1988a) Flowering phenologies in a shrub community: competition and constraints. *J Ecol* 76:975–94.

———. (1988b) Interactions for pollination among coflowering shrubs. *Ecology* 69:446-457

———. (1992) Nectar distributions, pollinator behavior, and plant reproductive success. In *Effects of Resource Distribution on Animal-Plant Interactions*, ed. M. D. Hunter, T. Ohgushi,and P. W. Price, 113–38. New York: Academic Press.

———. (2003) Floral longevity and reproductive assurance: seasonal patterns and an experimental test with *Kalmia latifolia* (Ericaceae). *Am J Bot* 90:1329–32.

Rathcke, B. and E. P. Lacey. (1985) Phenological patterns of terrestrial plants. *Ann Rev Ecol Syst* 16:179–214.

Rathcke, B. and L. Real. (1993) Autogamy and inbreeding depression in mountain laurel, *Kalmia latifolia* (Ericaceae). *Am J Bot* 80:143–46.

Ratnayake, R.M.C.S., I.A.U.N. Gunatilleke, D.S.A. Wijesundara, and R.M.K. Saunders. (2007) Pollination ecology and breeding system of *Xylopia championii* (Annonaceae): curculionid beetle pollination, promoted by floral scents and elevated floral temperatures. *Int J Plant Sci* 168:1255–68.

Rausher, M. D. (2008) Evolutionary transitions in floral color. *Int J Plant Sci* 169:7–21.

Raw, A. (1975) Territoriality and scent marking by *Centris* males in Jamaica (Hymenoptera: Anthophoridae). *Behavior* 54:311–21.

Reader, T., A. D. Higginson, C. J. Barnard, and F. S. Gilbert. (2006) The effects of predation risk from crab spiders on bee foraging behaviour. *Behav Ecol* 17:933–39.

Reader, T., I. MacLeod, P. T. Elliott, O. J. Robinson, and A. Manica. (2005) Inter-order interactions between flower-visiting insects: foraging bees avoid flowers previously visited by hoverflies. *J Ins Behav* 18:51–57.

Real, L. (1992) Information processing and the evolutionary ecology of cognitive architecture. *Am Nat* 140:108–45.

Real, L., J. R. Ott, and E. Silverline. (1982) On the tradeoff between the mean and the variance in foraging: effect of spatial distribution and color preference. *Ecology* 63:1617–23.

Real, L. and B. J. Rathke.(1988) Patterns of individual variability in floral resources. *Ecology* 69:728–35.

Reddy, G.V.P. and A. Guerrero. (2004) Interactions of insect pheromones and plant semiochemicals. *Trends Plant Sci* 9:253–61.

Ree, R. H. and M. J. Donoghue. (1999) Inferring rates of change in flower symmetry in asteroid angiosperms. *Syst Biol* 48:633–41.

Regal, P. J. (1982) Pollination by wind and animals: ecology of geographic patterns. *Ann Rev Ecol Syst* 13:497–24.

Regali, A. and P. Rasmont. (1995) New bioassays to evaluate diet in orphan colonies of *Bombus terrestris*. *Apidologie* 26:273–81.

Reinhard, J., M. V. Srinivasa, D. Guez, and S. W. Zhang. (2004) Floral scents induce recall of navigational and visual memories in honeybees. *J Exp Biol* 207:4371–81.

Reith, M., G. Baumann, R. Classen-Bockhoff, and T. Speck. (2007) New insights into the functional morphology of the lever mechanism of *Salvia pratensis* (Lamiaceae). *Ann Bot* 100:393–400.

Rengifo, C, L. Cornejo, and I. Akirov. (2006) One size fits all: corolla compression in *Aphelandra runcinata* (Acanthaceae), an adaptation to short-billed hummingbirds. *J Trop Ecol* 22:613–19.

Renner, S. S. (1983) The widespread occurrence of anther destruction of *Trigona* bees in Melastomataceae. *Biotropica* 15:251–56.

———. (1998) Effects of habitat fragmentation on plant-pollinator interactions in the tropics. In *Dynamics of Tropical Communities*, ed. D. M. Newbery, H.H.T. Prins, and N. D. Brown, 339–60. London: Blackwell Scientific.

———. (2006) Rewardless flowers in the angiosperms and the role of insect cognition in their evolution. In *Plant-Pollinator Interactions: from Specialization to Generali-*

zation, ed. N. M. Waser and J. Ollerton J, 123–44. Chicago: University of Chicago Press.

Renner, S. S. and J. P. Feil. (1993) Pollinators of tropical dioecious angiosperms. *Am J Bot* 80:1100–107.

Renner, S. S. and R. E. Ricklefs. (1995) Dioecy and its correlates in the flowering plants. *Am J Bot* 82:596–606.

Reynolds, R. J. and C. B. Fenster. (2008) Point and interval estimation of pollinator importance: a study using pollination data of *Silene caroliniana*. *Oecologia* 156:325–32.

Rhoades, D F. and J. C. Bergdahl. (1981) Adaptive significance of toxic nectar. *Am Nat* 117:798–803.

Rhodes, J. (2002) Cotton pollination by honey bees. *Aust J Exp Agric* 42:513–18.

Riba-Hernandez, P. and K. E. Stoner KE (2005) Massive destruction of *Symphonia globulifera* (Clusiaceae) flowers by Central American spider monkeys (*Ateles geoffroyi*). *Biotropica* 37:274–78.

Ribbands, C. R. (1949) The foraging method of individual honeybees. *J Anim Ecol* 18:47–66.

Richards, A. J. (1986) *Plant Breeding Systems*. London: Allen and Unwin.

Richards, M. H., E. J. von Wettberg,and A. C. Rutgers. (2003) A novel social polymorphism in a primitively eusocial bee. *Proc Nat Acad Sci USA* 100:7175–80.

Richards, A. J. (2001) Does low biodiversity resulting from modern agricultural practice affect crop pollination and yield? *Ann Bot* 88:165–72.

Richardson, S. C. (2004) Are nectar-robbers mutualists or antagonists? *Oecologia* 139:246–54.

Richardson, T. E. and A. G. Stephenson AG (1992) Effects of parentage and size of pollen load on progeny performance in *Campanula americana*. *Evolution* 46:1731–39.

Richter, K. S. and A. E. Weis AE (1995) Differential abortion in the yucca. *Nature* 376:557–58.

Rick, C. M. (1950) Pollination relations of *Lycopersicon esculentum* in native and foreign regions. *Evolution* 4:110–22.

Ricketts, T. H., G. C. Daily, P. R. Ehrlich, and C. D. Michener. (2004) Economic value of tropical forest to coffee production. *Proc Nat Acad Sci USA* 101:12579–82.

Ricketts, T. H., J. Regetz, I. Steffan-Dewenter, S. A. Cunningham, C. Kremen, A. Bogdanski, B. Gemmill-Herren, S. S. Greenleaf, A. M. Klein, M. M. Mayfield, L. A. Morandin, A. Ochieng, and B. F. Viana. (2008) Landscape effects on crop pollination services: are there general patterns? *Ecol Lett* 11:499–515.

Ricklefs, R. E. and S. S. Renner. (1994) Species richness within families of flowering plants. *Evolution* 48:1619–36.

Rickson, F. R. (1979) Ultrastructural development of the beetle food tissue of *Calycanthus* flowers. *Am J Bot* 66:80–86.

Rieger, M. A., M. Lamond, C. Preston, S. B. Powles,and R. T. Roush. (2002) Pollen-mediated movement of herbicide resistance between commercial canola fields. *Science* 296:2386–88.

Riffell, J. A., R. Alarcon, L. Abrell, G. Davidowitz, J. L. Bronstein, and J. G. Hildebrand. (2008). Behavioral consequences of innate preferences and olfactory learning in hawkmoth-flower interactions. *Proc Nat Acad Sci USA* 105:3404–9.

Rigney, L. P. (1995) Post-fertilization causes of differential success of pollen donors in *Erythronium grandiflorum* (Liliaceae): non-random ovule abortion. *Am J Bot* 82:578–84

Ritland, K. (1991) A genetic approach to measuring pollen discounting in natural plant populations. *Am Nat* 138:1049–57.

Rivera-Marchand, B. and J. D. Ackerman. (2006) Bat pollination breakdown in the Caribbean columnar cactus *Pilosocereus royenii*. *Biotropica* 38:635–42.

Roberts, R. B. and S. R. Vallespir. (1978) Specialisation of hairs bearing pollen and oil on the legs of bees (Apoidea: Hymenoptera). *Ann Ent Soc Am* 71:619–27.

Roberts, S. P., J. F. Harrison, and N. F. Hadley. (1998) Mechanisms of thermal balance in flying *Centris pallida* (Hymenoptera: Anthophoridae). *J Exp Biol* 201:2321–31.

Roberts, W. M. (1995) Hummingbird licking behavior and the energetics of nectar feeding. *Auk* 112:456–63.

———. (1996) Hummingbirds' nectar concentration preferences at low volume: the importance of time scale. *Anim Behav* 52:361–70.

Robertson, A. W. (1992) The relationship between floral display size, pollen carryover and geitonogamy in *Myosotis colensoi* (Kirk) Macbride (Boraginaceae). *Biol J Linn Soc* 46:333–49.

Robertson, A. W., D. Kelly, J. J. Ladley, and A. D. Sparrow. (1999) Effects of pollinator loss on endemic New Zealand mistletoes (Loranthaceae). *Cons Biol* 13:499–508.

Robertson, A. W., J. J. Ladley, and D. Kelly. (2005) Effectiveness of short-tongued bees as pollinators of apparently ornithophilous New Zealand mistletoes. *Austr Ecol* 30:298–309.

Robertson, A. W., C. Mountjoy, B. E. Faulkner, M. V. Roberts, and M. R. Macnair. (1999) Bumble bee selection of *Mimulus guttatus* flowers: the effect of pollen quality and reward depletion. *Ecology* 80:2594–606.

Robertson, C. (1895) The philosophy of flower seasons, and the phenological relations of the entomophilous flora and the anthophilous insect fauna. *Am Nat* 29:97–117.

———. (1924) Phenology of entomophilous flowers. *Ecology* 5:393–407

———. (1929) *Flowers and Insects: Lists of Visitors to four hundred and fifty three flowers*. Carlinville, IL: C Robertson.

Robledo-Arnuncio, J. J., F. Austerlitz, and P. E. Mouse. (2006) A new method of estimating the pollen dispersal curve independently of effective density. *Genetics* 173:1033–45.

Rocca, M. A. and M. Sazima. (2006) The dioecious sphingophilous species *Citharexylum myrianthum* (Verbenaceae): pollination and visitor diversity. *Flora* 201: 440–50.

Rodrigues, A.S.L. and K. J. Gaston. (2001) How large do reserve networks need to be? *Ecol Lett* 4:602–9.

Rodriguez, I., A. Gumbert, N. Hempel de Ibarra, J. Kunze, and M. Giurfa. (2004) Symmetry is in the eye of the 'beeholder'; innate preference for bilateral symmetry in flower-naïve bumblebees. *Naturwissenschaften* 91:374–77.

Rodriguez-Girones, M. A. and L. Santamaria. (2004) Why are so many bird flowers red? *PLoS Biol* 2:1515–19.

———. (2007) Resource competition, character displacement, and the evolution of deep corolla tubes. *Am Nat* 170:455–64.

Rodriguez-Rodriguez, M. C. and A. Valido. (2008) Opportunistic nectar-feeding birds are effective pollinators of bird-flowers from Canary Islands: experimental evidence from *Isoplexis canariensis* (Scrophulariaceae). *Am J Bot* 95:1408–15.

Rodriguez-Pena, N., K. S. Stoner, J. E. Schondube, J. Ayala-Berdon, C. M. Flores-Ortiz, and C. M. del Rio. (2007) Effects of sugar composition and concentration on food selection by Saussure's long-nosed bat (*Leptonycteris curasoae*) and the long-tongued bat (*Glossophaga soricina*). *J Mammalogy* 88:1466–74.

Rogers, H. J. (2006) Programmed cell death in floral organs: how and why do flowers die? *Ann Bot* 97:309–15.

Rose, M-J and W. Barthlott. (1994) Coloured pollen in Cactaceae: a mimetic adaptation to hummingbird pollination. *Bot Acta* 107:402–6.

———. (1995) Pollen-collecting threads in *Heliconia* (Heliconaceae) *Plant Syst Evol* 195:61–65.

Rosenzweig, M. L. (1981) A theory of habitat selection. *Ecology* 62:327–35.

Roshchina, V. V. (2001) Molecular-cellular mechanisms in pollen allelopathy. *Allelopathy J* 8:11–28.

Rothley, K. D., C. N. Berger, C. Gonzalez, E. M. Webster, and D. I. Rubinstein. (2003) Combining strategies to select reserves in fragmented landscapes. *Conserv Biol* 18:1121–31.

Rothschild, M. and N. Marsh. (1978) Some aspects of danaid/plant relationships. *Ent Exp Appl* 24:437–50.

Roubik, D. W. (1989) *Ecology and Natural History of Tropical Bees*. New York: Cambridge University Press.

———. (1992) Loose niches in tropical communities: why are there so few bees and so many trees? In *Effects of Resource Distribution on Animal-Plant Interactions*, ed. M. D. Hunter, T. Ohgushi, and P. W. Price, 327–54. New York: Academic Press.

———. (1993) Tropical pollinators in the canopy and understory: field data and theory for stratum "preferences." *J Ins Behav* 6:659–73.

———. (1995) Pollination of Cultivated Plants in the Tropics. *Agric Serv Bull* 118. Rome: FAO UN.

———. (2001) Ups and downs in pollinator populations: when is there a decline? *Cons Ecol* 5(1): Art 2.

———. (2002) The value of bees to the coffee harvest. *Nature* 417:708.

Roubik, D. W. and S. L. Buchmann. (1984) Nectar selection by *Melipona* and *Apis* (Hymenoptera; Apidae) and the ecology of nectar intake by bee colonies in a tropical forest. *Oecologia* 61:1–10.

Roubik, D. W. and H. Wolda. (2001) Do competing honey bees matter? Dynamics and abundance of native bees before and after honey bee invasion. *Pop Ecol* 43:53–62.

Roubik, D. W., D. Yanega, M. Aluja, S. L. Buchmann, and D. W. Inouye. (1995) On optimal nectar foraging by some tropical bees (Hymenoptera, Apidae). *Apidologie* 26:197–211.

Roulston, T. H. and S. L. Buchmann. (2000) A phylogenetic reconsideration of the pollen starch – pollination correlation. *Evol Ecol Res* 2:627–43.

Roulston, T. H. and J. H. Cane. (2000) Pollen nutritional quality and digestibility for animals *Plant Syst Evol* 222:187–209.

Roulston, T. H., J. H. Cane, and S. L. Buchmann. (2000) What governs protein content of pollen: pollinator preferences, pollen-pistil interactions, or phylogeny? *Ecol Monogr* 70:617–43.

Rudall, P. J., R. M. Bateman, M. F. Fay, and A. Eastman. (2002) Floral anatomy and systematic of Alliaceae with particular reference to *Gillesia*, a presumed insect mimic with strongly zygomorphic flowers. *Am J Bot* 89:1867–83.

Ruijter, A. D. (1997) Commercial bumblebee rearing and its implications. *Acta Hort* 437:261–69.

Rulik, B., S. Wanke, M. Nuss, and C. Neinhaus. (2008) Pollination of *Aristolochia pallida* Wild. (Aristolochiaceae) in the Mediterranean. *Flora* 203:175–84.

Runions, C. J. and M. A. Geber. (2000) Evolution of the self-pollinating flower in *Clarkia xantiana* (Onagraceae). 1. Size and development of floral organs. *Am J Bot* 87: 1439–51.

Runions, C. J. and J. N. Owens. (1995) Pollen scavenging and rain involvement in the pollination mechanism of interior spruce. *Can J Bot* 74:115–24.

Runions, C. J., K. H. Rensing, R. Takaso, and J. N. Owens. (1999) Pollination of *Picea orinetalis* (Pinaceae): saccus morphology governs pollen buoyancy. *Am J Bot* 86:190–97.

Rymer, P. D., R. J. Whelan, D. J. Ayre, P. H. Weston, and K. G. Russell. (2005) Reproductive success and pollinator effectiveness differ in common and rare *Persoonia* species (Proteaceae). *Biol Cons* 123:521–32.

Sahli, H. F. and J. K. Conner. (2006) Characterizing ecological generalization in plant-pollinator systems. *Oecologia* 148:365–72.

———. (2007) Visitation, effectiveness and efficiency of 15 genera of visitors to wild radish, *Raphanus raphanistrum* (Brassicaceae). *Am J Bot* 94:203–9.

Sakai, S. (2002) A review of brood-site pollination mutualism: plants providing breeding sites for their pollinators. *J Plant Res* 115:161–68.

Sakai, S., R. D. Harrison, K. Momose, K. Kuraji, H. Nagamasu H, T. Yasunari, L. Chong, and T. Nakashizuka. (2006) Irregular droughts trigger mass flowering in aseasonal tropical forests in Asia. *Am J Bot* 93:1134–39.

Sakai, S. and T. Inoue. (1999) A new pollination system: dung-beetle pollination discovered in *Orchidantha inouei* (Lowiaceae, Zingiberales) in Sarawak, Malaysia. *Am J Bot* 86:56–61.

Sakai, S., M. Kato, and H. Nagamasu. (2000) *Artocarpus* (Moaceae) gall midge pollination mediated by a male-flower parasitic fungus. *Am J Bot* 87:440–45.

Sakai, S., K. Momose, T. Yumoto, H. Nagamitsu, A. A. Hamid, and T. Nakashizuka. (1999) Plant reproductive phenology over four years including an episode of general flowering in a lowland dipterocarp forest, Sarawak, Malaysia. *Am J Bot* 86:1414–36.

Sakai, S. and S. J. Wright. (2008) Reproductive ecology of 21 existing *Psychotria* species (Rubiaceae); when is heterostyly lost? *Biol J Linn Soc* 93:125–34.

Sakai, A. K., W. L. Wagner, D. M. Ferguson, and D. R. Herbst. (1995) Origins of dioecy in the Hawaiian flora. *Ecology* 76:2517–29.

Saleh, N., K. Ohashi, J. D. Thomson, and L. Chittka. (2006) Facultative use of repellent scent marks in foraging bumblebees: complex versus simple flowers. *Anim Behav* 71:847–54.

Samuelson, G. A. (1994) Pollen consumption and digestion by leaf beetles. In *Novel Aspects of the Biology of Chrysomelidae,* ed. P. H. Jolivet and M. L. Cox, 179–83. Dordrecht: Kluwer Academic Publishers.

Sánchez-Lafuente, A. M. (2007a) Corolla herbivory pollination success and fruit predation in complex flowers; an experimental study with *Linaria lilacina* (Scrophulariaceae). *Ann Bot* 99:355–64.

———. (2007b) Sex allocation under simulated herbivory in the generalist perennial herb *Paeonia broteroi* (Paeoniaceae). *Plant Syst Evol* 265:59–70.

Sánchez-Lafuente, A., J. Guitián, M. Medrano, C. M. Herrera, P. J. Rey, and X. Cerdá. (2005) Plant traits, environmental factors, and pollinator visitation in winter-flowering *Helleborus foetidus* (Ranunculaceae). *Ann Bot* 96:845–52.

Sanders, L. C. and E. M. Lord. (1992) A dynamic role for the stylar matrix in pollen tube extension. *Int Rev Cytol* 140:297–18.

Sanderson, M. J. and M. J. Donoghue. (1994) Shifts in diversification rate with the origin of angiosperms. *Science* 264:1590–93.

Sandoz, J. C., D. Laloi, J. F. Odoux, and Pham-Delegue MH (2000) Olfactory information transfer in the honeybee: compared efficiency of classical conditioning and early exposure. *Anim Behav* 59:1025–34.

Santamaria, L. and M. A. Rodriguez-Girones. (2007) Linkage rules for plant-pollinator networks: trait complementarity or exploitation barriers? *PLoS Biol* 5:354–62.

Sapir, Y., A. Shmida, and G. Ne'eman. (2006) Morning floral heat as a reward to the pollinators of the *Oncocyclus* irises. *Oecologia* 147:53–59.

Sargent, R. D. (2004) Floral symmetry affects speciation rates in angiosperms. *Proc Roy Soc Lond* B 271:603–8.

Sargent, R. D., C. Goodwillie, S. Kalisz, and R. H. Rees. (2007) Phylogenetic evidence for a flower size and number trade-off. *Am J Bot* 94:2059–62.

Sargent, R. D. and S. P. Otto. (2004) A phylogenetic analysis of pollination mode and the evolution of dichogamy in angiosperms. *Evol Ecol Res* 6:1183–99.

———. (2006) The role of local species abundance in the evolution of pollinator attraction in flowering plants. *Am Nat* 167:67–80.

Sargent, R. D. and J .C. Vamosi. (2008) The influence of canopy position, pollinator syndrome, and region on evolutionary transitions in pollinator guild size. *Int J Plant Sci* 169:39–47.

Sarma, K., R. Tandon, K. R. Shivanna, and H.Y.M. Ram. (2007) Snail-pollination in *Volvulopsis nummularium*. *Curr Sci* 93:826–31.

Sauer, J. R., J. E. Hines, and J. Fallon. (2005) *The North American Breeding Bird Survey; results and analysis 1966–2004.* Laurel, MD: USGS Patuxent Wildlife Research Center.

Saxena, P., M. R. Vijayaraghavan, R. K. Sarbohoy, and U. Raizada. (1996) Pollination and gene flow in chillies with *Scirothrips dorsalis* as pollen vectors. Phytomorphology 46:317–27.

Sazima, I., S. Buzato, and M. Sazima. (1995) The saw-billed hermit *Ramphodon naevius* and its flowers in southeastern Brazil. *J fur Ornith* 136:195–206.

Sazima, I., C. Sazima, and Sazima. (2009) A catch-all leguminous tree: *Erythrina velutina* visited and pollinated by vertebrates at an oceanic island. *Aust J Bot* 57:26–30.

Sazima, I. and M. Sazima. (1977) Solitary and group foraging: two flower-visiting patterns of the lesser spear-nosed bat *Phyllostomus discolor*. *Biotropica* 9:213–15.

Sazima, M. and I. Sazima. (1978) Bat pollination of the

passion flower, *Passiflora mucronata*, in southeastern Brazil. *Biotropica* 10:100–9.

———. (1989) Oil-gathering bees visit flowers of eglandular morphs of the oil-producing Malpighiaceae. *Bot Acta* 102:106–11.

Sazima, M., S. Vogel, A. Cocucci, and G. Hausner. (1993) The perfume flowers of *Cyphomandra* (Solanaceae): pollination by euglossine bees, bellows mechanism, osmophores, and volatiles. *Plant Syst Evol* 187:51–88.

Sazima, M., S. Vogel, A. L. do Prado, D. M. de Oliviera, G. Franz, and A. Sazima. (2001) The sweet jelly of *Combretum lanceolatum* flowers (Combretaceae): a cornucopia resource for bird pollination in the Pantanal, western Brazil. *Plant Syst Evol* 227:195–208.

Schaal, B. A. and W. J. Leverich (1980) Pollination and banner markings in *Lupinus texensis* (Leguminosae) *Southwest Nat* 25:280–82.

Schaffer, W. M. and M. V. Schaffer. (1979) The adaptive significance of variations in reproductive habit in the Agavaceae. II. Pollinator foraging behavior and selection for increased reproductive expenditure. *Ecology* 60:1051–69.

Schatz, G. E. (1990) Some aspects of pollination biology in Central American forests. In *Reproductive Ecology of Tropical Forest Plants*, ed. K. S. Bawa and M. Hadley, 69–84. Paris: UNESCO/Parthenon.

Schemske, D. W. (1978) Evolution of reproductive characters in *Impatiens* (Balsaminaceae): the significance of cleistogamy and chasmogamy. *Ecology* 59:596–613.

———. (1980) Floral ecology and hummingbird pollination of *Combretum farinosum* in Costa Rica. *Biotropica* 12:169–81.

———. (1981) Floral convergence and pollinator sharing in two bee-pollinated tropical herbs. *Ecology* 62:946–54.

———. (1984) Population structure and local selection in *Impatiens pallida* (Balsaminaceae), a selfing annual. *Evolution* 38:817–32.

Schemske, D. W. and J. Ågren. (1995) Deceit pollination and selection on female flower size in *Begonia involucrata*: an experimental approach. *Evolution* 49:207–14.

Schemske, D. W., J. Ågren, and J. LeCorff. (1996) Deceit pollination in the monoecious neotropical herb *Begonia oaxacana* (Begoniaceae). In *Floral Biology: Studies on Floral Evolution in Animal-Pollinated Systems*, ed. D. G. Lloyd and S.C.H. Barrett, 292–318. New York: Chapman and Hall.

Schemske, D. W. and P. Bierzychudek. (2001) Evolution of flower color in the desert annual *Linanthus parryae*: Wright revisited. *Evolution* 55:1269–82.

Schemske, D. W. and H .D. Bradshaw. (1999) Pollinator preference and the evolution of floral traits in monkey flowers (*Mimulus*). *Proc Nat Acad Sci USA* 96:11910–15.

Schemske, D. W. and C. C. Horwitz. (1989) Temporal variation in selection on a floral character. *Evolution* 43:461–65.

Schiestl, F. P. (2005) On the success of a swindle: pollination by deception in orchids. *Naturwissenschaften* 92:255–64.

Schiestl, F. P. and M. Ayasse. (2001) Post-pollination emission of a repellent compound in a sexually deceptive orchid: a new mechanism for maximising reproductive success? *Oecologia* 126:531–34.

———. (2002) Do changes in floral odor cause speciation in sexually deceptive orchids? *Plant Syst Evol* 234:111–19.

Schiestl, F. P., M. Ayasse, H. F. Paulus, D. Erdmann, and W. Francke. (1997) Variation of floral scent emission and postpollination changes in individual flowers of *Ophrys sphegodes* subsp. *sphegodes*. *J Chem Ecol* 23:2881–95.

Schiestl, F. P., M. Ayasse, H. F. Paulus, C. Lofstedt, B. S. Hansson, F. Ibarra, and W. Francke.(1999) Orchid pollination by sexual swindle. *Nature* 399:421–22.

Schiestl, F. P. and D. W. Roubik. (2003) Odor compound detection in male euglossine bees. *J Chem Ecol* 29:253–57.

Schiestl, F. P. and P. M. Schlüter. (2009) Floral isolation, specialized pollination, and pollinator behaviour in orchids. *Ann Rev Entomol* 54:425–46.

Schlindwein, C. (1998) Frequent oligolecty characterizing a diverse bee-plant community in a xerophytic bushland of subtropical Brazil. *Studies on Neotropical Fauna and Environment* 33:46–59.

Schlumpberger, B. O., A. A. Cocucci, M. Moré, A. N. Sérsic, and R. A. Raguso. (2009) Extreme variation in floral characters and its consequences for pollinator attraction among populations of an Andean cactus. *Ann Bot* 103:1489–1500.

Schlüter, P. M. and F. P. Schiestl. (2008) Molecular mechanisms of floral mimicry in orchids. *Trends Plant Sci* 13:228–35.

Schmidt, J. O. (1982) Pollen foraging preferences of honeybees. *Southwest Nat* 7:255–59.

———. (1985) Phagostimulants in pollen. *J Apic Res* 24:107–14.

Schmidt, J. O. and S. L. Buchmann. (1985) Pollen digestion and nitrogen utilization by *Apis mellifera* L (Hymenoptera: Apidae). *Comp Biochem Physiol* A 82:499–503.

Schmidt, J. O. and B. E. Johnson. (1984) Pollen feeding preferences of *Apis mellifera*, a polylectic bee. *Southwest Nat* 9:41–47.

Schmidt, J. O., S. C. Thoenes, and M. D. Levin. (1987) Survival of honey bees, *Apis mellifera* (Hymenoptera: Apidae) fed various pollen sources. *Ann Ent Soc Am* 80:176–83.

Schmidt, V. M., R. Zucchi, and F. G. Barth. (2003) A stingless bee marks the feeding site in addition to the scent path (*Scaptotrigona* aff. *depilis*). *Apidologie* 34:237–48.

Schmidt-Lebuhn, A. N., M. Schwerdtfeger, M. Kessler, and

G. Lohaus. (2007). Phylogenetic constraints vs. ecology in the nectar composition of Acanthaceae. *Flora* 202: 62–69.

Schmitt, J. (1983) Flowering plant density and pollinator visitation in *Senecio*. *Oecologia* 606:97–102.

Schmitt, J. and D. W. Ehrhardt. (1990) Enhancement of inbreeding depression by dominance and suppression in *Impatiens capensis*. *Evolution* 44:269–78.

Schmitt, U. and A. Bertsch. (1990) Do foraging bumblebees scent-mark food sources, and does it matter? *Oecologia* 82:137–44.

Schmitt, U., G. Lübke, and W. Francke. (1991) Tarsal secretion marks food sources in bumblebees (Hymenoptera, Apidae). *Chemoecology* 2:35–40.

Schneider, D., M. Wink, F. Sporer, and P. Lounibos. (2002) Cycads: their evolution, toxins, herbivores and insect pollinators. *Naturwissenschaften* 89:281–94.

Schnetter, B. (1972) Experiments on pattern discrimination in honeybees. In *Information Processing in the Visual System of Arthropods*, ed. E. Wehner, 195–201. Berlin: Springer.

Schoen, D. J. (1982) The breeding system of *Gilia achilleifolia*: Variation in floral characteristics and outcrossing rate. *Evolution* 36:352–60.

Schoen, D. J. and J. W. Busch. (2008) On the evolution of self-fertilization in a metapopulation. *Int J Plant Sci* 169: 119–27.

Schoen, D. J. and M. Dubuc. (1990) The evolution of inflorescence size and number: a gamete-packaging strategy in plants. *Am Nat* 135:841–57.

Schoen, D. J., M. O. Johnston, A. L'Heureux, and J. V. Marsolais. (1997) Evolutionary history of the mating system in *Amsinckia* (Boraginaceae). *Evolution* 51:1090–99.

Schoen, D. J. and D. G. Lloyd. (1992) Self- and cross-fertilisation in plants. III. Methods for studying modes and functional aspects of self-fertilization. *Int J Plant Sci* 153:381–93.

Schoen, D. J., M. T. Morgan, and Bataillon. (1996) How does self-pollination evolve? Inferences from floral ecology and molecular genetic variation. *Phil Trans R Soc B* 351:1281–90.

Schoonhoven, L. M., T. Jermy, and J.J.A. van Loon. (2005) *Insect-Plant Biology*. London: Chapman and Hall.

Schremmer, F. (1941) Sinnesphysiologie und Blumenbesuch des Falters von *Plusia gamma* L. *Zool Jahrbuch* 74:361–522.

Schwaegerle, K. E. and D. A. Levin. (1991) Quantitative genetics of fitness traits in a wild population of *Phlox*. *Evolution* 45:169–77.

Schwendemann, A. B., G. Wang, M. L. Mertz, R. T. McWilliams, S. Thatcher, and J. A. Osborn. (2007) Aerodynamics of saccate pollen and its implications for wind pollination. *Am J Bot* 94:1371–81.

Scoble, J. and M. F. Clarke. (2006) Nectar availability and flower choice by eastern spinebills foraging on mountain correa. *Anim Behav* 72:1387–94.

Scogin, R. (1983) Visible floral pigments and pollinators. In *Handbook of Experimental Pollination Biology*, ed. C. E. Jones, and R. J. Little, 160–72. New York: Van Nostrand Reinhold.

Scott, A. C. and T. N. Taylor. (1983) Plant/animal interactions during the Upper Carboniferous. *Bot Rev* 49:259–307.

Scott, P. E., S. L. Buchmann, and M. K. O'Rourke. (1993) Evidence for mutualism between a flower-piercing carpenter bee and ocotillo: use of pollen and nectar by nesting bees. *Ecol Entomol* 18:234–40.

Seavey, S. R. and K. S. Bawa. (1986) Late-acting self-incompatibility in angiosperms. *Bot Rev* 52:195–219.

Sedivy, C., C. J. Praz, A. Müller, A. Widmer, and S. Dorn. (2008) Patterns of host-plant choice in bees in the genus *Chelostoma*: the constraint hypothesis of host range evolution in bees. *Evolution* 62:2487–507.

Seeley, T. D., M. Kleinhenz, B. Bujok, and J. Tautz. (2003) Thorough warm-up before take-off in honey bee swarms. *Naturwissenschaften* 90:256–60.

Seres, A. and N. Ramirez. (1995) Floral biology and pollination of some monocotyledons in a Venezuelan cloud forest. *Ann Miss Bot Gard* 82:61–81.

Sersic, A. N. and A. A. Cocucci. (1996) A remarkable case of ornithophily in *Calceolaria*; food bodies as rewards for a non-nectarivorous bird. *Bot Acta* 109:172–76.

Seymour, R. S., M. Gibernau, and K. Ito. (2003) Thermogenesis and respiration of inflorescences of the dead horse arum *Helicodiceros muscivorus*, a pseudothermoregulatory aroid associated with fly pollination. *Funct Ecol* 17:886–94.

Seymour, R. S. and P. Schultze-Motel. (1998) Physiological temperature regulation by flowers of the sacred lotus. *Phil Trans R Soc Lond* B 353:935–43.

Seymour, R. S. and P.G.D. Matthews. (2006) The role of thermogenesis in the pollination biology of the Amazon water lily *Victoria amazonica*. *Ann Bot* 98:1129–35.

Shafir, S., A. Bechar, and E. U. Weber. (2003) Cognition-mediated coevolution – context-dependent evaluations and sensitivity of pollinators to variability in nectar rewards. *Plant Syst Evol* 238:195–209.

Shafir, S., D. D. Wiegmann, B. H. Smith, and L. Real. (1999) Risk-sensitive foraging: choice behaviour of honeybees in response to variability in volume of reward. *Anim Behav* 57:1055–61.

Sharaf, K. E. and M. V. Price. (2004) Does pollination limit tolerance to browsing in *Ipomopsis aggregata*? *Oecologia* 138:396–404.

Sheffield, C. S., P. G. Kevan, R. F. Smith, S. M. Rigby, and R. E. L. Rogers. (2003) Bee species of Nova Scotia,

Canada, with new records and notes on bionomics and floral relations (Hymenoptera, Apoidea). *J Kans Ent Soc* 76:357–84.

Shelly, T. E. and E. Villalobos. (2000) Buzzing bees (Hymenoptera: Apidae, Halictidae) on *Solanum* (Solanaceae): floral choice and handling time track pollen availability. *Florida Ent* 83:180–87.

Shenoy, M. and R. M. Borges. (2008) A novel mutualism between an ant-plant and its resident pollinator. *Naturwissenschaften* 95:61–65.

Sherry, R. A. and C. Galen. (1998) The mechanism of floral heliotropism in the snow buttercup, *Ranunculus adoneus*. *Plant Cell Envt* 21:983–93.

Shi, Y. Y., W. J. Tao, S. P. Liang, Y. T. Lu, and L. Zhang. (2009) Analysis of the tip-to-base gradient of CaM in pollen tube pulsant growth using in vivo CaM-GFP system. *Plant Cell Rep* 28:1253–64.

Shivanna, K. R. and B. M. Johri. (1985) *The Angiosperm Pollen: Structure and Function*. New Delhi: Wiley.

Shreeve, T. G. (1981) Flight patterns of butterfly species in woodlands. *Oecologia* 51:289–93.

Shuttleworth, A. and S. D. Johnson. (2006) Specialized pollination by large spider-hunting wasps and self-incompatibility in the African milkweed *Pachycarpus asperifolius*. *Int J Plant Sci* 167:1177–86.

———. (2009a) Specialized pollination in the African milkweed *Xysmalobium orbiculare*: a key role for floral scent in the attraction of spider-hunting wasps. *Plant Syst Evol* 280:37–44.

———. (2009b) A key role for floral scent in a wasp-pollination system in *Eucomis* (Hyacinthaceae). *Ann Bot* 103:715–25.

Silander, J. A. and R. B. Primack. (1978) Pollination intensity and seed set in the evening primrose (*Oenothera fruticosa*). *Amer Midl Nat* 100:213–16.

Silvertown, J. (2004) The ghost of competition past in the phylogeny of island endemic plants. *J Ecol* 92:168–73.

Simms, B. L. and M. A. Bucher. (1996) Pleiotropic effects of flower color intensity on herbivore performance in *Ipomoea purpurea*. *Evolution* 50:957–63.

Simon, R., M. W. Holderied, and O. von Helversen. (2006) Size discrimination of hollow hemispheres by echo location in a nectar-feeding bat. *J Exp Biol* 209:3599–609.

Simpson, B. B. (1977) Breeding systems of dominant perennial plants of two disjunct warm desert ecosystems. *Oecologia* 27:203–26.

Simpson, B. B. and J. L. Neff. (1981) Floral rewards: alternatives to pollen and nectar. *Ann Miss Bot Gard* 68: 301–22.

———. (1983) Evolution and diversity of floral rewards. In *Handbook of Experimental Pollination Biology*, ed. C. E. Jones and R. J. Little, 142–59. New York: Van Nostrand Reinhold.

Singaravelan, N., G. Ne'eman, M. Inbar, and I. Izhaki. (2005) Feeding responses of free-flying honeybees to secondary compounds mimicking floral nectars. *J Chem Ecol* 31:2791–804.

Sinu, P. A. and K. R. Shivanna. (2007) Pollination biology of large cardamom (*Amomum subulatum*). *Current Sci* 93: 548–52.

Sjodin, N. E. (2007) Pollinator behavioural responses to grazing intensity. *Biodiv and Cons* 16:2103–21.

Sklenar, P. (1999) Nodding capitula in superparamo Asteraceae: an adaptation to an unpredictable environment. *Biotropica* 31:394–402.

Skutch, A. F. (1971) *A Naturalist in Costa Rica*. Gainesville: University of Florida Press.

Skvarla, J. J., P. H. Raven, W. F. Chissoe, and M. Sharp. (1978) An ultrastructural study of viscin threads in Onagraceae pollen. *Pollen Spores* 20:5–143.

Slaa, E. J., A. Cevaal, and M. J. Sommeijer. (1998) Floral constancy in *Trigona* stingless bees foraging on artificial flower patches; a comparative study. *J Apic Res* 37:191–98.

Slaa, E. J., B. Ruiz, R. Salas, M. Xeiss, and M. J. Sommeijer. (1998) Foraging strategies of stingless bees: flower constancy versus optimal foraging? *Proc Exp Appl Entomol* 9:185–90.

Slaa, E. J., L. A. Sanchez-Chaves, K. S. Malagodie-Braga, and F. E. Hofstede. (2006) Stingless bees in applied pollination: practice and perspective. *Apidologie* 37:293–315.

Slaa, E. J., J. Wassenberg, and J. C. Biesmeijer. (2003) The use of field-based social information in eusocial foragers: local enhancement among nestmates and heterospecifics in stingless bees. *Ecol Entomol* 28:369–79.

Slauson, L. A. (2000) Pollination biology of two chiropterophilous agaves in Arizona. *Am J Bot* 87:825–36.

Small, E. (1988) Pollen-ovule patterns in tribe Trifolieae (Leguminosae). *Plant Syst Evol* 160:195–205.

Small, E., B. Brookes, L. P. Lefkovitch, and D. T. Fairey. (1997) A preliminary analysis of the floral preferences of the Alfalfa Leafcutting Bee, *Megachile rotundata*. *Can Field Nat* 111:445–53.

Smith, A. P. and S. W. Green. (1987) Nitrogen requirements of the sugar glider (*Petaurus breviceps*), and omnivorous marsupial, on a honey-pollen diet. *Physiol Zool* 60:82–92.

Smith, C. A. and W. E. Evenson. (1978) Energy distribution in reproductive structures of *Amaryllis*. *Am J Bot* 65: 714–16.

Smith, C. E., J. T. Stevenes, E. J. Temeles, P. W. Ewald, R. J. Hebert, and R. L. Bonkovsky. (1996) Effect of floral orifice width and shape on hummingbird-flower interactions. *Oecologia* 106:482–92.

Smith, R. A. and M. D. Rausher. (2008) Experimental evidence that selection favors character displacement in the ivyleaf morning glory. *Am Nat* 171:1–9.

Smith, S. D., C. Ane, and D. A. Baum. (2008) The role of pollinator shifts in the floral diversification of *Iochroma* (Solanaceae). *Evolution* 62:793–806.

Smith, T. B., L. A. Freed, J. K. Lepson, and J. H. Carothers. (1995) Evolutionary consequences of extinctions in populations of a Hawaiian honeycreeper. *Conserv Biol* 9: 107–13.

Smith-Ramirez, C., P. Martinez, M. Nunez, C. Gonzalez, and J. J. Armesto. (2005) Diversity, flower visitation frequency and generalism of pollinators in temperate rain forests of Chiloe Island, Chile. *Bot J Linn Soc* 147:399–416.

Smithson, A. (2002) The consequences of rewardlessness in orchids: reward supplementation experiments with *Anacamptis morio* (Orchidaceae). *Am J Bot* 89:1579–87.

Smithson, A. and L.D.B. Gigord. (2003) The evolution of empty flowers revisited. *Am Nat* 161:537–52.

Snell, R. and L. W. Aarssen. (2005) Life history traits in selfing versus outcrossing annuals: exploring the 'time-limitation' hypothesis for the fitness benefits of self-pollination. *BMC Ecology* 5:2.

Snow, A. A. (1986) Pollination dynamics in *Epilobium canum* (Onagraceae): consequences for gametophytic selection. *Am J Bot* 73:139–51.

Snow, A. A. and P. O. Lewis. (1993) Reproductive traits and male fertility in plants: empirical approaches. *Ann Rev Ecol Syst* 24:331–51.

Snow, A. A. and T. P. Spira. (1991) Pollen vigor and the potential for sexual selection in plants. *Nature* 352:796–97.

———. (1993) Individual variation in the vigor of self pollen and selfed progeny in *Hibiscus moscheutos* (Malvaceae). *Am J Bot* 80:160–64.

Snow, A. A., T. P. Spira, R. Simpson, and R. A. Klips. (1996) Ecology of geitonogamous pollination. In *Floral Biology: Studies on Floral Evolution in Animal-Pollinated Systems*, ed. D. G. Lloyd and S.C.H. Barrett, 191–216. New York: Chapman and Hall.

Sobrevila, C. and M.T.K. Arroyo (1982) Breeding systems in a montane tropical cloud forest in Venezuela. *Plant Syst Evol* 140:19–38.

Soderstrom, T. R. and C. E. Calderon. (1971) Insect pollination in tropical rain forest grasses. *Biotropica* 3: 1–16.

Solberg, Y. and G. Remedios. (1980) Chemical composition of pure and bee-collected pollen. *Meld Norges Landbr* 59:2–12.

Soltis, D. E., H. Ma, M. W. Frohlich, P. S. Soltis, V. A. Albert, D. G. Oppenheimer, N. S. Altman, C. dePamphlis, and J. Leebens-Mack. (2007) The floral genome: an evolutionary history of gene duplication and shifting patterns of gene expression. *Trends Plant Sci* 12:358–67.

Soltis, P. S. and D. E. Soltis. (2004) The origin and diversification of angiosperms. *Am J Bot* 91:1614–26.

Somanathan, H. and R. M. Borges. (2001) Nocturnal pollination by the carpenter bee *Xylocopa tenuiscapa* (Apidae) and the effect of floral display on fruit set of *Heterophragma quadriloculare* (Bignoniaceae) in India. *Biotropica* 33:78–89.

Soucy, S. L., T. Giray, and Roubik. (2003) Solitary and group nesting in the orchid bee *Euglossa hyacinthina* (Hymenoptera, Apidae). *Insectes Soc* 50:248–55.

Soucy, S. L. and B. N. Danforth. (2002) Phylogeography of the socially polymorphic sweat bee *Halictus rubicundus* (Hymenoptera: Halictidae). *Evolution* 56:330–41.

Southwick, E. E. (1991) The colony as a thermoregulating superorganism. In *Behaviour and Physiology of Bees*, ed. L. J. Goodman and R. C. Fisher, 28–47. Wallingford, UK: CAB International.

———. (1984) Photosynthate allocation to floral nectar: a neglected energy investment. *Ecology* 65:1775–79.

Southwick, E. E. and L. Southwick. (1992) Economic values of honey bees (Hymenoptera, Apidae) in the United States. *J Econ Ent* 85:621–33.

Southwood, T.R.E. (1961) The evolution of insect host-tree relationship – a new approach. *Verh XI Int Kongr Entomol*: 651–54.

Spaethe, J. and A. Briscoe. (2005) Molecular characterisation and expression of the UV opsin in bumblebees: three ommatidial subtypes in the retina and a new photoreceptor organ in the lamina. *J Exp Biol* 208:2347–61.

Spears, E. E. (1983) A direct measure of pollinator effectiveness. *Oecologia* 57:196–99.

———. (1987) Island and mainland pollination ecology of *Centrosema virginiamum* and *Opuntia stricta*. *J Ecol* 75: 351–62.

Spiewok, S. and P. Neumann. (2006) Infestation of commercial bumblebee (*Bombus impatiens*) field colonies by small hive beetles (*Aethina tumida*). *Ecol Entomol* 31:623–28.

Spira, T. P. (2001) Plant-pollinator interactions: a threatened mutualism with implications for the ecology and management of rare plants. *Natural Areas Journal* 21:78–88.

Sprague, E. F. (1962) Pollination and evolution in *Pedicularis* (Scrophulariaceae). *Aliso* 5:181–209.

Sprayberry, J.D.H. and T. L. Daniel. (2007) Flower tracking in hawkmoths: behaviour and energetics. *J Exp Biol* 210: 37–45.

Sprengel, C. K. (1793) *Das entdeckte Geheimniss der Natur im Bau und in der Befruchtung der Blumen*. Berlin: Friedrich Vieweg.

Srinavasan, M. and S. W. Zhang. (2003) Small brains, smart minds: vision, perception and cognition in honeybees. *IETE J Res* 49:127–34.

Stace, H. M. and Y. J. Fripp. (1977) Raciation in *Epacris impressa*. II. Habitat differences and flowering times. *Aust J Bot* 25:315–23.

Stang, M., P.G.L. Klinkhamer, N. M. Waser, I. Stang, and

and E. van der Meijden. (2009) Size-specific interaction patterns and size matching in a plant-pollinator interaction web. *Ann Bot* 103:1459–69.

Stang, M., P.G.L. Klinkhamer and E. van der Meijden. (2007) Asymmetric specialization and extinction risk in plant-flower visitor webs: a matter of morphology or abundance? *Oecologia* 151:442–53.

Stanghellini, M. S., J. T. Ambrose, and J. R. Schultheis. (2002) Diurnal activity, floral visitation and pollen deposition by honey bees and bumble bees on field-grown cucumber and watermelon. *J Apic Res* 41:27–34.

Stanley, R. G. and H. F. Linskens. (1974) *Pollen Biology, Biochemistry and Management.* Heidelberg: Springer Verlag.

Stanton, M. L. (1994) Male-male competition during pollination in plant populations. *Am Nat* 144: S40–S68.

Stanton, M. L. and C. Galen. (1993) Blue light controls solar tracking by flowers of an alpine plant. *Plant Cell Envt* 16:983–89.

Stanton, M. L., C. Galen, and J. Shore. (1997) Population structure along a steep environmental gradient: consequences of flowering time and habitat variation in the snow buttercup *Ranunculus adoneus. Evolution* 51:79–94.

Stanton, M. L. and L. F. Galloway. (1990) Natural selection and allocation to reproduction in flowering plants. *Lect Mat Life Sci* 22:1–50.

Stanton, M. L., T. M. Palmer, T. P. Young, A. Evans, M. L. Turner. (1999) Sterilization and canopy modification of a swollen thorn acacia tree by a plant-ant. *Nature* 401: 578–81.

Stanton, M. L., A. A. Snow, and S. N. Handel. (1986) Floral evolution: attractiveness to pollinators increases male fitness. *Science* 232:1625–27.

Start, A. N. and A. G. Marshall. (1976) Nectarivorous bats as pollinators of trees in Western Malaysia. In: *Tropical Trees: Variation, Breeding and Conservation,* ed. J. Burley and B. T. Styles, 141–50. London: Academic Press.

Stead, A. D. (1992) Pollination-induced flower senescence: a review. *Plant Growth Reg* 11:13–20.

Stebbins, G. L. (1957) Self fertilization and population variability in the higher plants. *Am Nat* 91:337–54

———. (1970) Adaptive radiation of reproductive characteristics in Angiosperms, I: Pollination mechanisms. *Ann Rev Ecol Syst* 1:307–26.

Stefanescu, C. and A. Traveset A (2009) Factors determining the degree of generalization of flower use by Mediterranean butterflies. *Oikos* 118:1109–17.

Steffan-Dewenter, I., A-M, Klein, V. Gaebele, T. Alfert, and T. Tscharntke. (2006) Bee diversity and plant-pollinator interactions in fragmented landscapes. In *Plant-Pollinator Interactions: from Specialization to Generalization,* ed. N. M. Waser and J. Ollerton, 387–407. Chicago: University of Chicago Press.

Steffan-Dewenter, I. and A. Kuhn. (2003) Honeybee foraging in differentially structured landscapes. *Proc Roy Soc Lond* B 270:569–75.

Steffan-Dewenter, I. and K. Leschke. (2003) Effects of habitat management on vegetation and above-ground nesting bees and wasps of orchard meadows in Central Europe. *Biodiv and Cons* 12:1953–68.

Steffan-Dewenter, I., U. Munzenberg, C. Buerger, C. Thies, and R. Tscharntke. (2002) Scale-dependent effects of landscape context on three pollinator guilds. *Ecology* 83:1421–32.

Steffan-Dewenter, I. and T. Tscharntke. (2000) Resource overlap and possible competition between honey bees and wild bees in central Europe. *Oecologia* 122, 288–96.

Steiner, K. E. (1981) Nectarivory and potential pollination by a neotropical marsupial. *Ann Miss Bot Gard* 68:505–13.

———. (1989) The pollination of *Disperis* (Orchidaceae) by oil collecting bees in southern Africa. *Lindleyana* 4: 164–83.

———. (1998a) Beetle pollination of peacock moraeas (Iridaceae) in South Africa. *Plant Syst Evol* 209:47–65.

———. (1998b) The evolution of beetle pollination in a South African orchid. *Am J Bot* 85:1180–93.

Steiner, K. E. and V. B. Whitehead. (1990) Pollinator adaptation to oil-secreting flowers – *Rediviva* and *Diascia. Evolution* 44:1701–7.

———. (1991) Oil flowers and oil bees: further evidence for pollinator adaptation. *Evolution* 45:1493–1501.

———. (1996) The consequences of specialization for pollination in a rare South African shrub, *Ixianthes rezioides* (Scrophulariaceae). *Plant Syst Evol* 201:1313–38.

———. (2002) Oil secretion and the pollination of *Colpias mollis* (Scrophulariaceae). *Plant Syst Evol* 235:53–66.

Steiner, K. E., V. B. Whitehead, and S. D. Johnson. (1994) Floral and pollinator divergence in two sexually deceptive South African orchids. *Am J Bot* 81:185–94.

Stenstrom, M. and P. Bergmann. (1998) Bumblebees at an alpine site in northern Sweden: temporal development, population size and plant utilization. *Ecography* 21:306–16.

Stephens, D. W. (1993) Learning and behavioural ecology: incomplete information and environmental predictability. In *Insect Learning: Ecological and Evolutionary Perspectives,* ed. D. R. Papai and A. C. Lewis, 195–218. London: Chapman and Hall.

Stephenson, A. G. (1981) Flower and fruit abortion: proximate causes and ultimate functions. *Ann Rev Ecol Syst* 12:253–79.

Stephenson, A. G. and R. I. Bertin. (1983) Male competition, female choice, and sexual selection in plants. In

Pollination Biology, ed. L. Real, 109–49. London: Academic Press.

Stephenson, A. G., T-C Lau, M. Quesada, and J. A. Winsor. (1992) Factors that affect pollen performance. In *Ecology and Evolution of Plant Reproduction*, ed. R. Wyatt, 119–36. London: Chapman and Hall.

Steven, J. C. and D. M. Waller. (2007) Isolation affects reproductive success in low-density but not high-density populations of two wind-pollinated *Thalictrum* species. *Plant Ecol* 190:131–41.

Stiles, F .G. (1975) Ecology, flowering phenology and hummingbird pollination of some Costa Rican *Heliconia* species. *Ecology* 56:285–301.

———. (1976) Taste preferences, color preferences, and flower choice in hummingbirds. *Condor* 78:10–26.

———. (1977) Coadapted competitors: the flowering seasons of hummingbird-pollinated plants in a tropical forest. *Science* 198:1177–78.

———. (1981) Geographical aspects of bird-flower coevolution with particular reference to Central America. *Ann Miss Bot Gard* 68:323–51.

Stiles, F. G. and C. E. Freeman. (1993) Patterns in floral nectar characteristics of some bird-visited plant species from Costa Rica. *Biotropica* 25:191–205.

Stinchcombe, J. R., C. Weinig, M. Ungerer, K. M. Olesen, C. Mays, S. S. Halldorsdottir, M. D. Purugganan, and J. Schmitt. (2004) A latitudinal cline in flowering time in *Arabidopsis thaliana* modulated by the flowering time gene FRIGIDA. *Proc Nat Acad Sci USA* 101:4712–17.

Stone, G. N. (1995) Female foraging responses to harassment in the solitary bee *Anthophora plumipes*. *Anim Behav* 50:405–12.

Stone, G. N., F. S. Gilbert, P. G. Willmer, S. Potts, F. Semida, and S. Zalat. (1999) Windows of opportunity and the temporal structuring of foraging activity in a desert solitary bee. *Ecol Entomol* 24:208–21.

Stone, G. N. and P. G. Willmer. (1989a) Warm-up rates and body temperatures in bees: the importance of body size, thermal regime and phylogeny. *J Exp Biol* 147:303–28.

———. (1989b) Pollination of cardamom in Papua New Guinea. *J Apic Res* 28:228–37.

Stone, G. N., P. G. Willmer, and S. Nee. (1996) Daily partitioning of pollinators in an *Acacia* community. *Proc Roy Soc Lond B* 263:1389–93.

Stone, G. N., P. G. Willmer, and J. A. Rowe. (1998) Partitioning of pollinators during flowering in an African *Acacia* community. *Ecology* 79:2808–27.

Stout, J. C. (2000) Does size matter? Bumblebee behaviour and the pollination of *Cytisus scoparius* L. (Fabaceae). *Apidologie* 31:129–39.

———. (2007) Pollination of invasive *Rhododendron ponticum* (Ericaceae) in Ireland. *Apidologie* 38:198–206.

Stout, J. C., J. A. Allen, and D. Goulson. (2000) Nectar robbing, forager efficiency and seed set: Bumblebees foraging on the self incompatible plant *Linaria vulgaris* (Scrophulariaceae) *Acta Oecol* 21:277–83.

Stout, J. C. and D. Goulson. (2001) The use of conspecific and interspecific scent marks by foraging bumblebees and honeybees. *Anim Behav* 62:183–89.

———. (2002) The influence of nectar secretion rates on the responses of bumblebees (*Bombus* spp.) to previously visited flowers. *Behav Ecol Sociobiol* 52:239–46.

Stout, J. C., D. Goulson, and J. A. Allen. (1998) Repellent scent marking of flowers by a guild of foraging bumblebees (*Bombus* spp.). *Behav Ecol Sociobiol* 43:317–26.

Stout, J. C., J.A.N. Parnell, J. Arroyo, and T. P. Crowe. (2006) Pollination ecology and seed production of *Rhododendron ponticum* in native and exotic habitats. *Biodiv and Cons* 15:755–77.

Stoutamire, W. P. (1974) Australian terrestrial orchids, thynnid wasps and pseudocopulation. *Am Orchid Soc Bull* 43:13–18.

———. (1983) Wasp pollinated species of *Caladenia* (Orchidaceae) in southwestern Australia. *Aust J Bot* 31:383–94.

Stpiczynska, M. (2003) Nectar resorption in the spur of *Platanthera chlorantha* Custer (Rehb.) Orchidaceae: structural and microautoradiographic study. *Plant Syst Evol* 238:119–26.

Strauss, M. and H. Koopowitz. (1973) Floral physiology of *Angraecum*. I. Inheritance of post-pollinator phenomena. *Am Orchid Soc Bull* 42:495.

Strauss, S. Y. (1997) Floral characters link herbivores, pollinators and plant fitness. *Ecology* 78:1640–45.

Strauss, S. Y., J. K. Conner, and K. P. Lehtila. (2001) Effects of foliar herbivory by insects on the fitness of *Raphanus raphinastrum*: damage can increase male fitness. *Am Nat* 158:496–504.

Strauss, S. Y., J. K. Conner, and S. L. Rush. (1996) Foliar herbivory affects floral characters and plant attractiveness to pollinators: implications for male and female plant fitness. *Am Nat* 147:1098–1107.

Strauss, S. Y., R. E. Irwin, and V. M. Lambrix. (2004) Optimal defense theory and flower petal colour predict variation in the secondary chemistry of wild radish. *J Ecol* 92:132–41.

Strauss, S. Y. and P. Murch. (2004) Using pollinators instead of insecticides to compensate for herbivory. *Ecol Entomol* 29:234–39.

Strauss, S. Y., D. H. Siemens, M. B. Decher, and T. Mitchell-Olds. (1999) Ecological costs of plant resistance to herbivores in the currency of pollination. *Evolution* 53:1105–13.

Strauss, S. Y. and J. B. Whittall. (2006) Non-pollinator agents of selection on floral traits. In *Ecology and Evolution of*

Flowers, 120–38, ed. L. D. Harder and S.C.H. Barrett, New York: Oxford University Press.

Streisfeld, M. A. and J. R. Kohn. (2007) Environment- and pollinator-mediated selection on parapatric races of *Mimulus aurantiacus*. *J Evol Biol* 20:122–32.

Strickler, K. (1979) Specialization and foraging efficiency of solitary bees. *Ecology* 60:998–1009.

Strong, D. R., J. H. Lawton, and T.R.E. Southwood. (1984) *Insects on Plants*. Oxford: Blackwell.

Stroo, A. (2000) Pollen morphological evolution in bat-pollinated plants. *Plant Syst Evol* 222:225–42.

Stuurman, J., M. E. Hoballah, L. Broger, J. Moore, C. Basten, and C. Kuhlemeier. (2004) Dissection of floral pollination syndromes in *Petunia*. *Genetics* 168:1585–99.

Suarez, L. H., W. L. Gonzalez, and E. Gianoli (2009) Foliar damage modifies floral attractiveness to pollinators in *Alstroemeria exerens*. *Evol Ecol* 23:545–55.

Sugden, E. A. (1985) Pollinators of *Astragulus monoensis* Barneby (Fabaceae): potential impact of sheep grazing. *Gt Basin Nat* 45:299–312.

Sugden, E. A., R. W. Thorp, and S. L. Buchmann. (1996) Honey bee-native bee competition; focal point for environmental change and apicultural response in Australia. *Bee World* 77:26–44.

Sugiura, N., T. Abe, and S. Makino. (2006) Loss of extrafloral nectary on an oceanic island plant and its consequences for herbivory. *Am J Bot* 93:491–95.

Sugiura, N., S. Miyazaki, and S. Nagaishi. (2006) A supplementary contribution of ants in the pollination of an orchid, *Epipactis thunbergii*, usually pollinated by hoverflies. *Plant Syst Evol* 258:17–26.

Suguira, S. and K. Yamazaki. (2005) Moth pollination of *Metaplexis japonica* (Apocynaceae): pollinaria transfer on the tip of the proboscis. *J Plant Res* 118:257–62.

Sulikowski, D. and D. Burke. (2007) Food-specific spatial memory in an omnivorous bird. *Biol Lett* 3:245–48.

Sun, J. F., Y. B. Gong, S. S. Renner, and S. Q. Huang. (2008) Multifunctional bracts in the dove tree *Davidia involucrata* (Nyssaceae; Cornales): Rain protection and pollinator attraction. *Am Nat* 171:119–24.

Sun, M. and F. R. Ganders. (1988) Mixed mating systems in Hawaiian *Bidens* (Asteraceae). *Evolution* 4:516–27.

Sun, S. G., Y. H. Guo, R. W. Gituru, and S. Q. Huang. (2005) Corolla wilting facilitates delayed autonomous self-pollination in *Pedicularis dunniana* (Orobanchaceae). *Plant Syst Evol* 251:229–37.

Sun, S. G., K. Liao, J. Xia, and Y. H. Guo. (2005) Floral colour change in *Pedicularis monbeigiana* (Orobanchaceae). *Plant Syst Evol* 255:77–85.

Sussman, R. W. and P. H. Raven. (1978) Pollination by lemurs and marsupials: an archaic coevolutionary system. *Science* 200:731–36.

Sutherland, G. D. and C. L. Gass. (1995) Learning and remembering of spatial patterns by hummingbirds. *Anim Behav* 50:1273–86.

Sutherland, J. P., M. S. Sullivan, and G. M. Poppy. (1999) The influence of floral character on the foraging behaviour of the hoverfly *Episyrphus balteatus*. *Entomol Exp Appl* 93:157–64.

Sutherland, S. (1980) Energy limited fruit set in a paniculate agave: a test of the Bateman principle. *Bull Ecol Soc Am* 61:105.

Sutherland, S. D. and R. K. Vickery. (1993) On the relative importance of floral color, shape, and nectar rewards in attracting pollinators to *Mimulus*. *Gt Bas Nat* 53:107–17.

Suttle, K. B. (2003) Pollinators as mediators of top-down effects on plants. *Ecol Lett* 6:688–694.

Swartz-Tachor, R. (1999) Pollination and seed production in two populations of *Cyclamen persicum*. MSc Thesis, Univ Tel Aviv. Cited in Dafni and Firmage (2000).

Swihart, C. A. and S. L. Swihart. (1970) Colour selection and learned feeding preferences in the butterfly *Heliconius charitonius* Linn. *Anim Behav* 18:60–64.

Syed, R. A. (1979) Studies on oil palm pollination by insects. *Bull Ent Res* 69:213–24.

Szusich, N. U. and H. W. Krenn. (2002) Flies and concealed nectar sources: morphological innovations in the proboscis of Bombyliidae (Diptera). *Acta Zool* 83:183–92.

Takács, S., H. Bottomley, I. Andreller, T. Zaradnik, J. Schwarz, R. Bennett, W. Strong, and G. Gries. (2008) Infrared radiation from hot cones on cool conifers attracts seed-feeding insects. *Proc Roy Soc Lond* B 276:649–55.

Takayama, S. and A. Isogai. (2005) Self-incompatibility in plants. *Ann Rev Plant Biol* 56:467–89.

Takebayashi, N., D. Wolf, and L. Delph. (2006) Effect of variation in herkogamy on outcrossing within a population of *Gilia achilleifolia*. *Heredity* 96:159–65.

Taki, H. and P. G. Kevan. (2007) Does habitat loss affect the communities of plants and insects equally in plant-pollinator interactions? Preliminary findings. *Biodiv and Cons* 16:3147–61.

Taki, H., P. G. Kevan, and J. S. Ascher. (2007) Landscape effects of forest loss in a pollination system. *Landscape Ecol* 22:1575–87.

Tan, K. H. and R. Nishida. (2005) Synomone or kairomone? *Bulbophyllum apertum* flower releases raspberry ketone to attract *Bactrocera* fruit flies. *J Chem Ecol* 31:497–507.

Tan, K. H., L. T. Tan, and R. Nishida. (2006) Floral phenylpropanoid cocktail and architecture of *Bulbophyllum vinaceum* orchid in attracting fruit flies for pollination. *J Chem Ecol* 32:2429–41.

Taneyhill, D. (2007) Foraging mechanisms and the currency for models of energy maximization in bumblebees (Hymenoptera: Apide: *Bombus occidentalis*). *Entomol Gen* 30:119–33

Taneyhill, D. E. and J. D. Thomson. (2007) Behavior of inexperienced bumblebees toward spatial clumping of nectar (Hymenoptera, Apidae). *Entomol Gen* 29:149–64.

Tang, L. L. and S. Q. Huang. (2007) Evidence for reduction in floral attractants with increased selfing rates in two heterandrous species. *New Phytol* 175:588–95.

Taylor, R. M. and R. S. Pfannenstiel. (2008) Nectar feeding by wandering spiders on cotton plants. *Envir Entomol* 37:996–1002.

Taylor, T. N. and D. A. Levin. (1975) Pollen morphology of Polemoniaceae in relation to systematics and pollination systems: scanning electron microscopy. *Grana* 15: 91–112.

Tchuenguem Fohou, F. N., A. Pauly, J. Messi, D. Bruckner, L. N. Tinkeu, and E. Basga. (2004) An afrotropical bee specialized in the collection of grass pollen (Poaceae): *Lipotriches notabilis* (Schletterer 1891) (Hymenoptera Apoidea Halictidae). *Ann Soc Ent France* 40:131–43.

Teichert, H, S. Dötterl, B. Zimma, M. Ayasse, and G. Gottsberger. (2008) Perfume-collecting male euglossine bees as pollinators of a basal angiosperm: the case of *Unonopsis stipitata* (Annonaceae). *Plant Biol* 11:29–37.

Temeles, E. J. and W. J. Kress. (2003) Adaptation in a plant-hummingbird association. *Science* 300:630–33.

Temeles, E. J., Y. B. Linhart, M. Masonjones, and H. D. Masonjones. (2002) The role of flower width in hummingbird bill length-flower length relationships. *Biotropica* 34:68–80.

Temeles, E. J. and I. L. Pan. (2002) Effect of nectar robbery on phase duration, nectar volume, and pollination in a protandrous plant. *Int J Plant Sci* 163:803–8.

Temeles, E. J., I. L. Pan, J. L. Brennan, and J. N. Horwitt. (2000) Evidence for ecological causation of sexual dimorphism in a hummingbird. *Science* 289:441–43.

ten Have, A. and E. J. Woltering. (1997) Ethylene biosynthetic genes are differentially expressed during carnation (*Dianthus caryophyllus* L.) flower senescence. *Plant Mol Biol* 34:89–97.

Tepedino, V. J. (1981) Notes on the reproductive biology of *Zigadenus paniculatus*, a toxic range plant. *Gt Bas Nat* 41:427–30.

———. (1983) Pollen carried for long periods on butterflies: some comments. *Oikos* 41:144–45.

Tepedino, V. J. and F. D. Parker. (1982) Interspecific differences in the relative importance of pollen and nectar to bee species foraging on sunflowers. *Environ Entomol* 11: 246–50.

Terry, L. I., G. H. Walter, J. S. Donaldson, E. Snow, P. I. Forster, and P. J. Machin. (2005) Pollination of Australian *Macrozamia* cycads (Zamiacee): effectiveness and behavior of specialist vectors in a dependent mutualism. *Am J Bot* 92:931–40.

Terry, L. I., G. H. Walter, C. Moore, R. Roemer, and C. Hull. (2007) Odor-mediated push-pull pollination in cycads. *Science* 318:70.

Teixeira, S. and G. Bernasconi. (2007) High prevalence of multiple paternity within fruits in natural populations of *Silene latifolia*, as revealed by microsatellite DNA analysis. *Mol Ecol* 16:4370–79.

Thakar, J. D., K. Kunte, A. K. Chauhan, A. V. Watve, and M. G. Watve. (2003) Nectarless flowers: ecological correlates and evolutionary stability. *Oecologia* 136:565–70

Thanikaimoni, G. (1986) Pollen apertures: form and function. In *Pollen and Spores: Form and Function*, ed. S. Blackmore and I. K. Ferguson, 119–36. London: Academic Press.

Theis, N., K. Kesler, and L. S. Adler. (2009) Leaf herbivory increases floral fragrance in male but not female *Cucurbita pepo* subsp. *texana* (Cucurbitaceae) flowers. *Am J Bot* 96:897–903.

Theis, N., M. Lerdau, and R. A. Raguso. (2007) The challenge of attracting pollinators while evading floral herbivores: patterns of fragrance emission in *Cirsium arvense* and *Cirsium repandum* (Asteraceae). *Int J Plant Sci* 168:587–601.

Theis, N. and R. A. Raguso. (2005) The effect of pollination on floral fragrance in thistles. *J Chem Ecol* 31:2581–600.

Theiss, K., S. Kephart, and C. T. Ivey. (2007) Pollinator effectiveness on co-occurring milkweeds (*Asclepias*; Apocynaceae, Asclepiadoideae). *Ann Miss Bot Gard* 94:505–16.

Thiele, J. and Y. Winter. (2001) Primary and secondary strategies of target localization in flower bats: spatial memory and cue-directed search. *Bat Res News* 42:124–25.

Thien, L. B. (1969) Mosquito pollination of *Habenaria obtusata* (Orchidaceae). *Am J Bot* 56:232–37.

Thien, L. N., P. Bernhardt, M. S. Devall, Z. Chen, Y. Kuo, J-H Fan, L-C Yuan, and J. H. Williams. (2009) Pollination biology of basal angiosperms (ANITA Grade). *Am J Bot* 96:166–82.

Thien, L. B., H. Azuma, and S. Kawano. (2000) New perspectives on the pollination biology of basal angiosperms. *Int J Plant Sci* 161:S225–S235.

Tholl, D. and U.S.R. Rose. (2006) Detection and identification of floral scent compounds. In *Biology of Floral Scent*, ed. N. Dudareva and E. Pichersky, 3–25. Boca Raton, FL: CRC Press.

Thom, C. (2003) The tremble dance of honey bees can be caused by hive-external foraging experience. *J Exp Biol* 206:2111–16.

Thom, C., D. C. Gilley, and J. Tautz. (2003) Worker piping in honey bees (*Apis mellifera*): the behavior of piping nectar foragers. *Behav Ecol Sociobiol* 53:199–205.

Thom, C., D. C .Gilley, J. Hooper, and H. E. Esch. (2007) The scent of the waggle dance. *PLoS Biol* 5:1862–67.

Thompson, J. N. (1997) Conserving interaction biodiversity.

In *The Ecological Basis of Conservation*, ed, S.T.A. Pickett, R. S. Ostfeld, M. Shachak, and G. E. Likens, 285–93. New York: Chapman and Hall.

Thompson, J. N. and C. C. Fernandez. (2006) Temporal dynamics of antagonism and mutualism in a geographically variable plant-insect interaction. *Ecology* 87:103–12.

Thompson, J. N. and O. Pellmyr. (1992) Mutualism with pollinating seed parasites amid co-pollinators - constraints on specialization. *Ecology* 73:1780–91.

Thompson, W. R., D. Meinwald, D. Aneshansley, and T. Eisner. (1972) Flavonols: pigments responsible for ultraviolet absorption in nectar guides of flowers. *Science* 177: 528–30.

Thomson, D. (2004) Competitive interactions between the invasive European honey bee and native bumble bees. *Ecology* 85:458–70.

Thomson, J. D. (1980a) Skewed flowering distribution and pollinator attraction. *Ecology* 61:572–79.

———. (1980b) Implications of different sorts of evidence for competition. *Am Nat* 16:719–26.

———. (1981) Spatial and temporal components of resource assessment by flower-feeding insects. *J Anim Ecol* 50:49–59.

———. (1982) Patterns of visitation by animal pollinators. *Oikos* 39:241–50.

———. (1983) Component analysis of community-level interactions in pollination systems. In *Handbook of Experimental Pollination Biology,* ed. C. E. Jones and R. J. Little, 451–60. New York: Van Nostrand Reinhold.

———. (1986) Pollen transport and deposition by bumblebees in *Erythronium*: influences of floral nectar and bee grooming. *J Ecol* 74:329–41.

———. (1988) Effects of variation in inflorescence size and floral rewards on the visitation rates of traplining pollinators of *Aralia hispida*. *Evol Ecol* 2:65–76.

———. (1996) Trapline foraging by bumblebees. 1. Persistence of flight-path memory geometry. *Behav Ecol* 7:158–64.

———. (2001a) How do visitation patterns vary among pollinators in relation to floral display and floral design in a generalist pollination system? *Oecologia* 126:386–94.

———. (2001b) Using pollination deficits to infer pollination declines; can theory guide us? *Cons Ecol* 5 (1): Art 6.

Thomson, J. D. and J. Brunet. (1990) Hypotheses for the evolution of dioecy in seed plants. *Trends Ecol Evol* 5: 11–16.

Thomson, J. D. and K. Goodell. (2001) Pollen removal and deposition by honeybee and bumblebee visitors to apple and almond flowers. *J Appl Ecol* 38:1032–44.

Thomson, J. D. and R. C. Plowright. (1980) Pollen carryover, nectar rewards and pollinator behavior with special reference to *Diervilla lonicera*. *Oecologia* 46:68–74.

Thomson, J. D., L. P. Rigney, K. M. Karoly, and B. A. Thomson. (1994) Pollen viability, vigor, and competitive ability in *Erythronium grandiflorum* (Liliaceae). *Am J Bot* 81:1257–66.

Thomson, J. D. and B. A. Thomson. (1989) Dispersal of *Erythronium grandiflorum* pollen by bumblebees: implications for gene flow and reproductive success. *Evolution* 43:657–61.

———. (1992) Pollen presentation and viability schedules in animal-pollinated plants: consequences for reproductive success. In *Ecology and Evolution of Plant Reproduction: New Approaches,* ed. R. Wyatt R, 1–24. New York: Chapman and Hill.

Thomson, J. D., P. Wilson. (2008) Explaining evolutionary shift between bee and hummingbird pollination: convergence, divergence and directionality. *Int J Plant Sci* 169: 23–38.

Thomson, J. D., P. Wilson, M. Valenzuela, and M. Malzone. (2000). Pollen presentation and pollination syndromes, with special reference to *Penstemon*. *Plant Spec Biol* 15: 11–29.

Thomson, V. P., A. B. Nicotra, and S. A. Cunningham. (2004) Herbivory differentially affects male and female reproductive traits of *Cucumis sativus*. *Plant Biol* 6:621–28.

Thorp, R. W. (1979) Structural, behavioural and physiological adaptations of bees (Apoidea) for collecting pollen. *Ann Missouri Bot Gard* 66:788–812.

———. (2000) The collection of pollen by bees. *Plant Syst Evol* 222:211–23.

Thorp, R. W. and D. L. Briggs. (1980) Bees collecting pollen from other bees. *J Kans Ent Soc* 53:166–70.

Tollsten, L. (1993) A multivariate approach to post-pollination changes in the floral scent of *Platanthera bifolia* (Orchidaceae). *Nordic J Bot* 13:495–99.

Tollsten, L. and G. Bergström. (1989) Variation and post-pollination changes in floral odours released by *Platanthera bifolia* (Orchidaceae). *Nordic J Bot* 9:359–62.

Tollsten, L., J. T. Knudsen, and G. Bergstrom. (1994) Floral scent in generalist *Angelica* (Apiaceae): an adaptive character? *Biochem Syst Ecol* 22:161–73.

Tollsten, L. and D. O. Ovstedal. (1994) Differentiation in floral scent chemistry among populations of *Conopodium majus* (Apiaceae*). Nord J Bot* 14:361–67.

Tomlinson, P. B. (1994) Functional morphology of saccate pollen in conifers, with special reference to Podocarpaceae. *Int J Plant Sci* 155:699–715.

Tooker, J. F. and L. M. Hanks. (2000) Flowering plant hosts of adult Hymenoptera parasitoids of central Illinois. *Ann Ent Soc Am* 93:580–88.

Torres, C. and L. Galetto. (2002) Are nectar sugar composition and corolla tube length related to the diversity of insects that visit Asteraceae flowers? *Plant Biol* 4:360–66.

Torres-Diaz, C., L. A. Cavieres, C. Munoz-Ramirez, and

M.T.K. Arroyoz. (2007) Consequences of microclimate variation on insect pollinator visitation in two species of *Chaetanthera* (Asteraceae) in the central Chilean Andes. *Rev Chilena Hist Nat* 80:455–68.

Totland, Ø. (1993) Pollination in alpine Norway: flowering phenology, insect visitors, and visitation rates in two plant communities. *Can J Bot* 71:1072–79.

———. (2001) Environment-dependent pollen limitation and selection on floral traits in an alpine species. *Ecology* 82:2233–44.

———. (2004) No evidence for a role of pollinator discrimination in causing selection on flower size through female reproduction. *Oikos* 106:558–64.

Totland, Ø. and B. Schulte-Herbruggen. (2003) Breeding system, insect flower visitation, and floral traits of two alpine *Cerastium* species in Norway. *Arct Antarct Alp Res* 35:242–47.

Townsend, P. A. and D. J. Levey. (2005) An experimental test of whether habitat corridors affect pollen transfer. *Ecology* 86:466–75.

Travers, S. E. (1999) Pollen performance of plants in recently burned and unburned environments. *Ecology* 80:2427–34.

Traveset, A. and E. Sáez. (1997) Pollination of *Euphorbia dendroides* by lizards and insects: spatio-temporal variation in patterns of flower visitation. *Oecologia* 111:241–48.

Tripp, E. A. and P. S. Manos. (2008) Is floral specialization a dead-end? Pollination transitions in *Ruellia* (Acanthaceae). *Evolution* 62:1712–36.

Tsahar, E., Z. Ara, J. P. Joy, I. Izhaki, and C. M. del Rio. (2006) Do nectar- and fruit-eating birds have lower nitrogen requirements than omnivores? An allometric test. *Auk* 123:1004–12.

Tschapka, M. (2003) Pollination of the understory palm *Calyptrogyne ghiesbreghtiana* by hovering and perching bats. *Biol J Linn Soc* 80:281–88.

———. (2004) Energy density patterns of nectar resources permit coexistence within a guild of Neotropical flower-visiting bats. *J Zool* 263:7–21.

Tschapka, M. and O. von Helversen. (2007) Phenology, nectar production and visitation behavior of bats on the flowers of the bromeliad *Werauhia gladiolifera* in a Costa Rican lowland rain forest. *J Trop Ecol* 23:385–95.

Tscharntke, T. and R. Brandl. (2004) Plant-insect-interactions in fragmented landscapes. *Ann Rev Ent* 49:405–30.

Tsou, C-H. (1997) Embryology of the Theaceae: anther and ovule development of *Camellia, Franklinia* and *Schima. Am J Bot* 84:369–81.

Tsuji, K, A. Hasyim, Harlion and K. Nakamura. (2004) Asian weaver ants, *Oecophylla smaragdina*, and their repelling of pollinators. *Ecol Res* 19:669–73.

Tucker, V. A. (1974) Energetics of natural flight in birds. In *Avian Energetics*, ed. R. A. Paynter. *Nuttall Ornith Club* 15:298–333.

Turner, V. (1982) Marsupials as pollinators in Australia. In *Pollination and Evolution*, ed. J. A. Powell and R. J. Richards, 55–66. Sydney: Royal Botanic Gardens.

———. (1984) *Banksia* pollen as a source of protein in the diet of two Australian marsupials, *Cercartetus nanus* and *Tarsipes rostratus. Oikos* 43:53–61.

Udovic, D. (1981) Determinants of fruit set in *Yucca whipplei*: reproductive expenditure vs. pollinator availability. *Oecologia* 48:389–99.

Uemachi, T., Y. Kato, and T. Nishio. (2004) Comparison of decorative and non-decorative flowers in *Hydrangea macrophylla* (Thunb.) *Ser Sci Hort* 102:325–34.

Underwood, B. A. (1991) Thermoregulation and energetic decision-making by the honeybees *Apis cerana, Apis dorsata* and *Apis laboriosa. J Exp Biol* 157:19–34.

Urcelay, C., C. L. Morales, and V. R. Chalcoff. (2006) Relationship between corolla length and floral larceny in the South American hummingbird-pollinated *Campsidium valdivianum. Ann Bot Fenn* 43:205–11.

Ushimaru, A., I. Dohzono, Y. Takami, and F. Hyodo. (2009) Flower orientation enhances pollen transfer in bilaterally symmetrical flowers. *Oecologia* 160:667–74.

Ushimaru, A., S. Kikuchi, R. Yonekura, A. Maruyama, N. Yanagisawa, M. Kagami, M. Nakagawa, S. Mahoro, Y. Kohmatsu, A. Hatada, S. Kitamura, and K. Nakata. (2006) The influence of floral symmetry and pollination systems on flower size variation. *Nordic J Bot* 24:593–98.

Ushimaru, A., T. Watanabe and K. Nakata. (2007) Colored floral organs influence pollinator behavior and pollen transfer in *Commelina communis* (Commelinaceae). *Am J Bot* 94:249–58.

Utelli, A-B and B. A. Roy. (2000) Pollinator abundance and behavior on *Aconitum lycoctonum* (Ranunculaceae): an analysis of the quantity and quality components of pollination. *Oikos* 89:461–70.

Vaknin, Y., S. Gan-Mor, A. Bechar, B. Ronen, and D. Eisikowitch. (2000) The role of electrostatic forces in pollination. *Plant Syst Evol* 222:133–42.

———. (2001) Are flowers morphologically adapted to take advantage of electrostatic forces in pollination? *New Phytol* 152:301–6.

———. (2002) Effects of supplementary pollination on cropping success and fruit quality in pistachio. *Plant Breeding* 121:451–55.

Valdovinos, F. S., R. Ramos-Jiliberto, J. D. Flores, C. Espinoza, and G. Lopez. (2009) Structure and dynamics of pollination networks: the role of alien plants. *Oikos* 118:1190–1200.

Valido, S., Y. L. Dupont, and J. M. Olesen. (2004) Bird-flower interactions in the Macronesian islands. *J Biogeog* 31:1945–53.

Vallejo-Marin, M., J. S. Manson, J. D. Thomson, and S.C.H. Barrett. (2009) Division of labor within flowers: heteranthery, a floral strategy to reconcile contrasting pollen fates. *J Evol Biol* 22:828–39.

Vallejo-Marin, M. and M. D. Rausher. (2007) The role of male flowers in andromonoecious species: energetic costs and siring success in *Solanum carolinense* L. Evolution 61:404–12.

Vallius, E. (2000) Position-dependent reproductive success of flowers in *Dactylorhiza maculata* (Orchidaceae). *Funct Ecol* 14:573–79.

Valterova, I., J. Kunze, A. Gumbert, A. Luxova, I. Liblikas, B. Kalinova, and A. K. Borg-Karlson. (2007) Male bumble bee pheromonal components in the scent of deceit-pollinated orchids: unrecognized pollinator cues? *Arth-Plant Interac* 1:137–45.

Valtuena, F. J., A. Ortega-Olivencia, and T. Rodriguez-Riano. (2007) Nectar production in *Anagyris foetida* (Fabaceae): two types of concentration in flowers with hanging droplet. *Int J Plant Sci* 168:627–38.

Vamosi, J. C. and S. P. Otto. (2002) When looks can kill: the evolution of sexually dimorphic floral display and the extinction of dioecious plants. *Proc Roy Soc Lond* B 269:1187–94.

Vamosi, J. C., T. M. Knight, J. A. Steets, S. J. Mazer, M. Burd, and T. L. Ashman. (2006) Pollination decays in biodiversity hotspots. *Proc Nat Acad Sci USA* 103:956–61.

van der Moezel, P. G., J. C. Delfs, J. S. Pate, W. A. Loneragan, and D. T. Bell. (1987) Pollen selection by honeybees in shrublands of the Northern Sandplains of Western Australia. *J Apic Res* 26:224–32.

van der Pijl, L. (1936) Fledermause und Blumen. *Flora Jena* 131:1–40.

———. (1954) *Xylocopa* and flowers in the Tropics. *Proc K Ned Akad Wet* 57:541–62.

———. (1961) Ecological aspects of flower evolution. II. Zoophilous flower classes. *Evolution* 15:44–59.

van der Pijl, L. and C. H. Dodson. (1966) Orchid Flowers, their Pollination and Evolution. Coral Gables, FL: University of Miami Press.

van Doorn, W. G. (1997) Effects of pollination on floral attraction and longevity. *J Exp Bot* 48:1615–22.

———. (2002) Does ethylene treatment mimic the effects of pollination on floral lifespan and attractiveness? *Ann Bot* 89:375–83.

Van Dulmen, A. (2001) Pollination and phenology of flowers in the canopy of two contrasting rain forest types in Amazonia. *Colombia Plant Ecol* 153:73–85.

Vane-Wright, R. I. (1980) On the definition of mimicry. *Biol J Linn Soc* 13:1–6.

van Kleunen, M. and J. Burczyk. (2008) Selection on floral traits through male fertility in a natural plant population. *Evol Ecol* 22:39–54.

van Kleunen, M., A. Meier, M. Saxenhofer, and M. Fischer. (2008) Support for the predictions of the pollinator-mediated stabilizing selection hypothesis. *J Plant Ecol* 1:173–78.

van Kleunen, M., I. Nänni, J. S. Donaldson, and J. C. Manning. (2007) The role of beetle marks and flower colour on visitation by monkey beetles (Hoplinii) in the Greater Cape floral region, South Africa. *Ann Bot* 100:1483–89.

van Schaik, A. P., J. W. Terborgh, and S. J. Wright. (1993) The phenology of tropical forests: adaptive significance and consequences for primary consumers. *Ann Rev Ecol Syst* 24:353–77.

Van Tets, I. G. (1997) Extraction of nutrients from *Protea* pollen by African rodents. *Belg J Zool* 127:59–65.

van Tussenbroek, B. I., J.G.R. Wong, and J. Marquez-Guzman. (2008) Synchronized anthesis and predation on pollen in the marine angiosperm *Thalssia testudinum* (Hydrocharitaceae). *Mar Ecol Prog Ser* 354:119–24.

Varassin, I. G., J. R. Trigo, and M. Sazima. (2001) The role of nectar production, flower pigments and odour in the pollination of four species of *Passiflora* (Passifloraceae) in south-eastern Brazil. *Bot J Linn Soc* 136:139–52.

Vareschi, E. (1971) Duftunterscheidung bei der Honigbiene - Einzelzell-Ableitungen und Verhalktensreaktionen. *Z Vergl Physiol* 75:143–73.

Vasek, F. C., V. Weng, R. J. Beaver, and C. K. Huszar. (1987) Effects of mineral nutrition on components of reproduction in *Clarkia unguiculata*. *Aliso* 11:599–618.

Vaughton, G. (1991) Variation between years in pollen and nutrient limitation of fruit set in *Banksia spinulosa*. *J Ecol* 79:389–400.

———. (1992) Effectiveness of nectarivorous birds and honeybees as pollinators of *Banksia spinulosa* (Proteaceae). *Aust J Ecol* 17:43–50.

Vaughton, G. and M. Ramsey. (1991) Floral biology and inefficient pollen removal in *Banksia spinulosa* var *neoangelica* (Proteaceae). *Aust J Bot* 39:167–77.

Vázquez, D. P. (2005) Degree distribution in plant-animal mutualistic networks: forbidden links or random interactions? *Oikos* 108:421–26.

Vázquez, D. P. and M. A. Aizen. (2003) Null model analyses of specialization in plant-pollinator interactions. *Ecology* 84:2493–2501.

———. (2004) Asymmetric specialization: a pervasive feature of plant-pollinator interactions. *Ecology* 85:1251–57.

———. (2006) Community-wide patterns of specialization in plant-pollinator interactions revealed by null models. In *Plant-Pollinator Interactions: from Specialization to Generalization*, ed. N. M. Waser and J. Ollerton, 200–219. Chicago: University of Chicago Press.

Vázquez, D. P., W. F. Morris, and P. Jordano. (2005) Interaction frequency as a surrogate for the total effect of animal mutualists on plants. *Ecol Lett* 8:1088–94.

Vázquez, D. P. and D. Simberloff. (2002) Ecological specialization and susceptibility to disturbance: conjectures and refutations. *Am Nat* 159:606–23.

———. (2003) Changes in interaction biodiversity induced by an introduced ungulate. *Ecol Lett* 6:1077–83.

———. (2004) Indirect effects of introduced ungulates on pollination and plant reproduction. *Ecol Monogr* 74:281–308.

Veddeler, D., A. M. Klein, and T. Tscharntke. (2006) Contrasting responses of bee communities to coffee flowering at different spatial scales. *Oikos* 112:594–601.

Veddeler, D., R. Olschewski, T. Tscharntke, and A. M. Klein. (2008) The contribution of non-managed social bees to coffee production: new economic insights based on farm-scale data. *Agrofor Syst* 73:109–14.

Venables, B.A.B. and E. M. Barrows. (1985) Skippers: pollinators or nectar thieves? *J Lepid Soc* 39:299–312.

Vesprini, J. L., M. Nepi, and E. Pacini. (1999) Nectary structure, nectar secretion patterns and nectar composition in two *Helleborus* species. *Plant Biol* 1:560–68.

Vesprini, J. L., M. Nepi, F. Ciaspolini, and E. Pacini. (2008) Holocrine secretion and cytoplasmic content of *Helleborus foetidus* L. (Ranunculaceae) nectar. *Plant Biol* 10:268–71.

Vesprini, J. L. and E. Pacini. (2005) Temperature dependent floral longevity in two *Helleborus* species. *Plant Syst Evol* 252:63–70.

Vickery, R. K. (2008) How does *Mimulus verbenaceus* (Phrymaceae) set seed in the absence of pollinators? *Evol Biol* 35:199–207.

Visscher, P. K., K. Crailsheim, and G. Sherman. (1996) How do honey bees (*Apis mellifera*) fuel their water foraging flights? *J Insect Physiol* 42:1089–94

Vlásaková, B., B. Kalinová, M.G.H. Gustafsson, and H. Teichert. (2008) Cockroaches as pollinators of *Clusia* aff. *sellowiana* (Clusiaceae) on inselbergs in French Guiana. *Ann Bot* 102:295–304.

Vogel, S. (1958) Fledermausblumen in Sudamerika. *Ost Bot Zeits* 104:491–530.

———. (1968-9) Chiropterophilie in der neotropischen Flora, neue Mitteilungen. Flora Abt 157:562–602; 158:185–222; 158:289–323.

———. (1954) Blütenbiologische Typen als Elemente der Sippengliederung. Fischer, Jena

———. (1958) Fledermausblumen in Sudamerika. *Österr Bot Z* 104:491–530

———. (1961) Die Bestäubung der Kesselfallen-Blüten von *Ceropegia*. *Beitr Biol Pfl* 35:159–237.

———. (1966) Scent organs of orchid flowers and their relation to insect pollination. *Proc 5th World Orchid Conf* 253–59.

———. (1969) Flowers offering fatty oil instead of nectar. Abstr. 11th Int Bot Congr., Seattle, 229.

———. (1978a) Evolutionary shifts from reward to deception in pollen flowers. In *The Pollination of Flowers by Insects*, ed. A. J. Richards, 89–96. London: Academic Press.

———. (1978b) Pilzmuckenblümen als Pilzmimeten, erster Teil. *Flora* 167:329–98.

———. (1990) *The role of scent glands in pollination*. Washington D.C.: Smithsonian Institute Library.

Vogel, S., A. V. Lopes, and I. C. Machado. (2005) Bat pollination in the NE Brazilian endemic *Mimosa lewisii*: an unusual case and first report for the genus. *Taxon* 54:693–700.

Vogler, D. W. and S. Kalisz. (2001) Sex among the flowers: the distribution of plant mating systems. *Evolution* 55:202–4.

Voigt, C. C., D. H. Kelm, and G. H. Visser. (2006) Field metabolic rates of phytophagous bats: do pollination strategies of plants make life of nectar feeders spin faster? *J Comp Physiol* B 176:213–22.

Voigt, C. C. and Y. Winter. (1999) Energetic cost of hovering flight in nectar-feeding bats (Phyllostomidae: Glossophaginae) and its scaling in moths, birds and bats. *J Comp Physiol* B 169:38–48.

von Aufsess, A. (1960) Geruchliche Nahorientierung der Biene bei entomophilen und ornithophilen Bluten. *Z Vergl Physiol* 43:469–98.

von Frisch, K. (1950) *Bees, their Vision, Chemical Senses and Language*. Ithaca, NY: Cornell University Press.

———. (1954) *The Dancing Bees*. London: Methuen.

———. (1967) *Dance Language and Orientation of Bees*. Cambridge, MA: Belknap Press.

von Hase, A., R. M. Cowling, and A. G. Ellis. (2006) Petal movement in cape wildflowers protects pollen from exposure to moisture. *Plant Ecol* 184:75–87.

von Helversen, D. and O. von Helversen. (1999) Acoustic guide in bat-pollinated flower. *Nature* 398:759–60.

von Helversen, D., M. Holderied, and O. von Helversen. (2003) Echoes of bat-pollinated bell-shaped flowers: conspicuous for nectar feeding bats? *J Exp Biol* 206:1025–34.

von Helversen, D. and O. von Helversen. (2003) Object recognition by echolocation; a nectar-feeding bat exploiting the flowers of a rain forest vine. *J Comp Physiol* A 189:327–36.

von Helversen, O. (1972) Zur spektralen Unterschiedsempfindlichkeit der Honigbiene. *J Comp Physiol* 80:439–72.

———. (1993) Adaptations of flowers to pollination by glossophagine bats. In: *Animal-Plant Interactions in Tropical Environments*, ed. W. Barthlott et al., 41–59. Bonn: Museum Alexander Koenig.

von Helversen, O., L. Winkler, and H. J. Bestman. (2000) Sulphur-containing 'perfumes' attract flower visiting bats. *J Comp Physiol* A 186:143–53.

Vonhof, J. and L. D. Harder. (1995) Size-number trade offs

and pollen production by Papilionaceous legumes. *Am J Bot* 82:230–38.

Vorobyev, M., A. Gumbert, J. Kunze, M. Giurfa, and R. Menzel. (1997) Flowers through insect eyes. *Isr J Plant Sci* 45:93–101.

Vroege, P. W. and P. Stelleman. (1990) Insect and wind pollination in *Salix repens* L. and *Salix caprea* L. *Israel J Bot* 39:125–32.

Vulliamy, B, S. G. Potts, and P. G. Willmer. (2006) The effects of cattle grazing on plant-pollinator communities in a fragmented Mediterranean landscape. *Oikos* 114:529–43.

Wackers, F. L., J. Romeis, and P. van Rijn. (2007) Nectar and pollen feeding by insect herbivores and implications for multitrophic interactions. *Ann Rev Entomol* 52:301–23.

Waddington, K. D. (1979) Divergence in inflorescence height: an evolutionary response to pollinator fidelity. *Oecologia* 40:43–50.

———. (1983) Foraging behavior of pollinators. In *Pollination Biology*, ed. L. A. Real, 213–39. New York: Academic Press.

Waddington, K, D, D. T. Allen, and B. Heinrich. (1981) Floral preferences of bumblebees (*Bombus edwardsii*) in relation to intermittent versus continuous rewards. *Anim Behav* 29:283–89.

Waddington, K. D. and W. H. Kirchner. (1992). Acoustical and behavioral correlates of profitability of food sources in honey-bee round dances. *Ethology* 92:1–6.

Wagner, D. (1997) The influence of ant nests on *Acacia* seed production, soil chemistry and herbivory. *J Ecol* 85:83–94.

———. (2000) Pollen viability reduction as a potential cost of ant association for *Acacia constricta* (Fabaceae). *Am J Bot* 87:711–15.

Wagner, D. and A. Kay. (2002) Do extrafloral nectaries distract ants from visiting flowers? An experimental test of an overlooked hypothesis. *Ecol Evol Res* 4:293–305.

Wainselboim, A. J., F. Roces, and W. M. Farina. (2002) Honeybees assess changes in nectar flow within a single foraging bout. *Anim Behav* 63:1–6.

Walker, J. B. and K. J. Sytsma. (2007) Staminal evolution in the genus *Salvia* (Lamiaceae); molecular phylogenetic evidence for multiple origins of the stamina lever. *Ann Bot* 100:375–91.

Walther-Hellwig, K. and R. Frankl. (2000) Foraging distances of *Bombus muscorum*, *Bombus lapidarius*, and *Bombus terrestris* (Hymenoptera, Apidae). *J Ins Behav* 13:239–46.

Wälti, M. O., J. K. Muhlemann, A. Widmer, and F. P. Schiestl. (2008) Floral odour and reproductive isolation in two species of *Silene*. *J Evol Biol* 21:111–21.

Wang, D. Y., T. Oakley, J. Moser, L. C. Shimmin, S. Yim, R. L. Honeycutt, H. Tsao, and W. H. Li. (2004) Molecular evolution of bat color genes. *Mol Biol Evol* 21:295–302.

Wang, Y., D. Zhang, S. S. Renner, and Z. Chen. (2005) Self-pollination by sliding pollen in *Caulokaempferia coenobialis* (Zingiberaceae). *Int J Plant Sci* 166:753–59

Ward, M., C. W. Dick, R. Gribel, and A. J. Lowe. (2005) To self or not to self... A review of outcrossing and pollen-mediated gene flow in neotropical trees. *Heredity* 95: 246–54.

Ware, A. B. and S. G. Compton. (1992) Breakdown of pollinator specificity in an African fig tree. *Biotropica* 24: 544–49.

Warren, J. and P. James. (2008) Do flowers wave to attract pollinators? A case study with *Silene maritima*. *J Evol Biol* 21:1024–29.

Warren, J. and S. Mackenzie. (2001) Why are all colour combinations not equally represented as flower-colour polymorphisms? *New Phytol* 151:237–41.

Warren, S. D., K. T. Harper, and G. M. Booth. (1988) Elevational distribution of insect pollinators. *Am Midl Nat* 120:325–30.

Waser, N. M. (1978a) Competition for hummingbird pollination and sequential flowering in two Colorado wildflowers. *Ecology* 59:934–44.

———. (1978b) Interspecific pollen transfer and competition between co-occurring plant species. *Oecologia* 36:223–36.

———. (1982) A comparison of the distances flown by different visitors to flowers of the same species. *Oecologia* 55:251–57.

———. (1983a) Competition for pollination and floral character differences among sympatric plant species: a review of the evidence. In *Handbook of Experimental Pollination Biology*, ed. C. E. Jones and R. J. Little, 277–93. New York: Van Nostrand Reinhold.

———. (1983b) The adaptive nature of floral traits: ideas and evidence. In *Pollination Biology*, ed. L. A. Real, 241–85. New York: Academic Press.

———. (1986) Flower constancy: definition, cause and measurement. *Am Nat* 127:593–603.

———. (1988) Comparative pollen and dye transfer by pollinators of *Delphinium nelsonii*. *Func Ecol* 2:41–48.

———. (1998) Pollination, angiosperm speciation, and the nature of species boundaries. *Oikos* 82:198–201.

———. (2001) Pollinator behavior and plant speciation: looking beyond the 'ethological isolation' paradigm. In *Cognitive Ecology of Pollination*, ed. L. Chittka and J. D. Thomson, 318–35. Cambridge: Cambridge University Press.

———. (2006) Specialization and generalization in plant pollinator interactions: a historical perspective. In *Plant-Pollinator Interactions: from Specialization to Generalization*, ed. N. M. Waser and J. Ollerton, 3–17. Chicago: University of Chicago Press.

Waser, N. M., L. Chittka, M. V. Price, N. M. Williams, and J. Ollerton. (1996) Generalization in pollination systems and why it matters. *Ecology* 77:1043–60.

Waser, N. M. and J. Ollerton, ed. (2006) *Plant-Pollinator Interactions: from Specialization to Generalization*. Chicago: University of Chicago Press.

Waser, N. M. and M. V. Price. (1982) A comparison of pollen and fluorescent dye carry-over by natural pollinators of *Ipomopsis aggregata*. Ecology 63:1168–72.

———. (1983a) Optimal and actual outcrossing in plants, and the nature of plant-pollinator interactions. In *Handbook of Experimental Pollination Ecology*, ed. C. E. Jones and R. J. Little, 341–59. New York: Van Nostrand Reinhold.

———. (1984) Experimental studies of pollen carryover: effects of floral variability in *Ipomopsis aggregata*. *Oecologia* 62:262–68.

———. (1991) Reproductive costs of self-pollination in *Ipomopsis aggregata* (Polemoniaceae); are ovules usurped? *Am J Bot* 78:1036–43.

Waser, N. M. and L. A. Real. (1979) Effective mutualism between sequentially flowering plant species. *Nature* 281:670–72.

Wasserthal, L. T. (1993) Swing-hovering combined with long tongue in hawkmoths, an antipredator adaptation during flower visits. In *Animal-plant Interactions in Tropical Environments*, ed. W. Barthlott, C. M. Naumann, K. Schmidt-Loske, and K. L. Schumann, 77–86. Bonn, Germany: Zoologisches Forschungsinstitut.

Watmough, R. H. (1983) Mortality, sex-ratio and fecundity in natural populations of large carpenter bees (*Xycolopa* spp.). *J Anim Ecol* 52:111–25.

Wcislo, W. T. and J. H. Cane. (1996) Floral resource utilization by solitary bees (Hymenoptera, Apoidea) and exploitation of their stored foods by natural enemies. *Ann Rev Ent* 41:257–86

Webb C. J. and D. G. Lloyd (1980) Sex ratios in New Zealand apioid Umbelliferae. *NZ J Bot* 18:121–26.

———. (1986) The avoidance of interference between the presentation of pollen and stigmas in angiosperms. II. Herkogamy. *NZ J Bot* 24:163–78.

Webb, C. J., D. G. Lloyd, and L. F. Delph. (1999) Gender dimorphism in indigenous New Zealand seed plants. *NZ J Bot* 37:119–30.

Weber, J. J. and C. Goodwillie. (2007) Timing of self-compatibility; flower longevity, and potential for male outcross success in *Leptosiphon jepsonii* (Polemoniaceae). *Am J Bot* 94:1338–43.

Weber, M. and H. Halbritter. (2007) Exploding pollen in *Montrichardia arborescens* (Araceae). *Plant Syst Evol* 263:51–57.

Weberling, F. (1989) *The Morphology of Flowers and Inflorescences*. Cambridge: Cambridge University Press.

———. (2007) The problem of generalized flowers: morphological aspects. *Taxon* 56:707–16.

Weevers, T. (1952) Flower colours and their frequency. *Acta Bot Neerl* 1:81–92.

Wehner, R. (1971) The generalization of directional stimuli in the honeybee, Apis mellifera. *J Insect Physiol* 17:1579–91.

———. (1981) Spatial vision in arthropods. In *Handbook of Sensory Physiology*, vol. 7, 6C, ed. H. J. Autrum, 287–616. Berlin: Springer-Verlag.

Wei, C. A., S. L. Rafalko, and F. C. Dyer. (2002) Deciding to learn: modulation of learning flights in honeybees, Apis mellifera. *J Comp Physiol* A 188:725–37.

Wei, X., S. J. Johnson, and A. M. Hammond. (1998) Sugar feeding strategy of adult velvetbean caterpillars (Lepidoptera, Noctuidae). *Environ Entomol* 17:280–88.

Weigend, M. and M. Gottschling. (2006) Evolution of funnel-revolver flowers and ornithophily in *Nasa* (Loasaceae). *Plant Biol* 8:120–42.

Weis, I. M. and L. A. Hermanutz. (1993) Pollination dynamics of arctic dwarf birch (*Betula glandulosa*: Betulaceae) and its role in the loss of seed production. *Am J Bot* 80:1021–27.

Weiss, M. (1991) Floral colour changes as cues for pollinators. *Nature* 354:227–29.

———. (1995a) Floral color change: a widespread functional convergence. *Am J Bot* 82:167–85.

———. (1995b) Associative color learning in a nymphalid butterfly. *Ecol Entomol* 20:298–301.

———. (1997) Innate colour preferences and flexible colour learning in the pipevine swallowtail. *Anim Behav* 53:1043–52.

Weiss, M. R. (2001) Vision and learning in some neglected pollinators; beetles, flies, moths, and butterflies. In *Cognitive Ecology of Pollination*, ed. L. Chittka and J. D. Thomson, 171–90. Cambridge: Cambridge University Press.

Weiss, M. R. and D. R. Papaj. (2003) Colour learning in two behavioural contexts: how much can a butterfly keep in mind? *Anim Behav* 65:425–34.

Welch, K. C., B. H. Bakken, C. M. del Rio, and R. K. Suarez. (2006) Hummingbirds fuel hovering flight with newly ingested sugar. *Phys Bioch Zool* 79:1082–87.

Weller, S. G., A. K. Sakai, T. M. Culley, D. R. Campbell, and A. K. Dunbar-Wallis. (2006) Predicting the pathway to wind pollination: heritabilities and genetic correlations of inflorescence traits associated with wind pollination in *Schiedea salicaria* (Caryophyllaceae). *J Evol Biol* 19:331–42.

Weller, S. G., A. K. Sakai, T. M. Culley, D. R. Campbell, P. Ngo, and A. K. Dunbar-Wallis. (2008) Sexually dimorphic inflorescence traits in a wind-pollinated species: heritabilities and genetic correlations in *Schiedea adamantis* (Caryophyllaceae). *Am J Bot* 94:1716–25.

Weller, S. G., A. K. Sakai, A. E. Rankin, A. Golonka, B. Kutcher, and K. E. Ashby. (1998) Dioecy and the evolution of pollination systems in *Schiedea* and *Alsinidendron* (Caryophyllaceae: Alsinoideae) in the Hawaiian Islands. *Am J Bot* 85:1377–88.

Wells, H., P. S. Hill, and P. H. Wells. (1992) Nectarivore foraging ecology: rewards differing in sugar type. *Ecol Entomol* 17:280–88.

Wenzler, M., D. Holscher, T. Y. Oerther, and B. Schneider. (2008) Nectar formation and floral nectary anatomy of *Anigozanthus flavidus*: a combined magnetic resonance imaging and spectroscopy study. *J Exp Bot* 59:3425–34.

West, E. L. and T. M. Laverty. (1998) Effect of floral symmetry on flower choice and foraging behaviour of bumblebees. *Can J Zool* 76:730–39.

Wester, P. and R. Classen-Bockhoff. (2006) Hummingbird pollination in *Salvia haenkei* (Lamiaceae) lacking the typical lever mechanism. *Plant Syst Evol* 257:133–46.

Westerbergh, A. (2004) An interaction between a specialized seed predator moth and its dioecious host plant shifting from parasitism to mutualism. *Oikos* 105:564–74.

Westerkamp, C. (1989) Von Pollenhaufen. Nudelspritzen und Pseudo-Staubbattern: Blutenstaub aus zweiter Hand. *Palmengarten* 53:146–49.

———. (1990) Bird flowers - hovering versus perching exploitation. *Bot Acta* 103:366–71.

———. (1996) Pollen in bee-flower relations – some considerations on melittophily. *Bot Acta* 109:325–32.

———. (1997a) Keel blossoms: Bee flowers with adaptations against bees. *Flora* 192:125–32.

———. (1997b) Flowers and bees are competitors – not partners. Towards a new understanding of complexity in specialized bee flowers. *Acta Hort* 437:71–74.

———. (1999) Keel flowers of the Polygalaceae and Fabaceae: a functional comparison. *Bot J Linn Soc* 129:207–21.

Westerkamp, C. and R. Classen-Bockhoff. (2007) Bilabiate flowers: the ultimate response to bees? *Ann Bot* 100:361–74.

Westerkamp, C., A. A. Soares, and L. P. D. Neto. (2006) Male and female booths with separate entrances in the tiny flowers of *Guazuma ulmifolia* (Malvaceae, Byttuerioideae). I. Structural integration. *Flora* 201:389–95.

Westphal, C. R. Bommarco, G. Carré, E. Lamborn, N. Morison, T. Petanidou, S. G. Potts, et al. (2008) Measuring bee diversity in different European habitats and biogeographical regions. *Ecol Monogr* 78:653–71.

Westrich, P. and K. Schmidt. (1987) Pollen analysis, an auxiliary tool to study the collecting behavior of solitary bees (Hymenoptera, Apoidea). *Apidologie* 18:199–213.

Whalen, M. D. (1978) Reproductive character displacement and floral diversity in *Solanum* (Section *Androceras*). *Syst Bot* 3:77–86.

Wheelwright, N. T., E. E. Dukeshire, J. B. Fontaine, S. H. Gutow, D. A. Moeller, J. G. Schuetz, T. M. Smith, S. L. Rodgers, and A. G. Zink. (2006). Pollinator limitation, autogamy and minimal inbreeding depression in insect-pollinated plants on a boreal island. *Am Midl Nat* 155:19–38.

Whelan, R. J., D. G. Roberts, P. R. England, and D. J. Ayre. (2006) The potential for genetic contamination vs. augmentation by native plants in urban gardens. *Biol Cons* 128:493–500.

Whitaker, D. L., L. A. Webster, and J. Edwards. (2007). The biomechanics of *Cornus canadensis* stamens are ideal for catapulting pollen vertically. *Funct Ecol* 21:219–25.

White, D., B. W. Cribb, and T. A. Heard. (2001) Flower constancy of the stingless bee *Trigona carbonaria* Smith (Hymenoptera, Apidae, Meliponini). *Aust J Entomol* 40:61–64.

Whitehead, D. R. (1983) Wind pollination: some ecological and evolutionary perspectives. In *Pollination Biology*, ed. L. Real, 97–108. New York; Academic Press.

Whitehead, M. R. and R. Peakall. (2009) Integrating floral scent, pollination ecology and population genetics. *Func Ecol* 23:863–74.

Whitham, T. G. (1977) Coevolution of foraging in *Bombus* and nectar-dispensing in *Chilopsis*: a last dreg theory. *Science* 197:593–95.

Whitney, H. N., L. Chittka, T. J. A. Bruce, and B. J. Glover. (2009) Conical epidermal cells allow bees to grip flowers and increase foraging efficiency. *Curr Biol* 19:948–53.

Whitney, H. N., A. Dyer, L. Chittka, S. A. Rands, B. J. Glover. (2008) The interaction of temperature and sucrose concentration on foraging preferences in bumblebees. *Naturwissenschaften* 95:845–50.

Whitney, H. M., M. Kolle, P. Andrew, L. Chittka, U. Steiner, and B. J. Glover. (2009) Floral iridescence, produced by diffractive optics, acts as a cue for animal pollinators. *Science* 323:130–33.

Whittall, J. B. and S. A. Hodges. (2007) Pollinator shifts drive increasingly long nectar spurs in columbine flowers. *Nature* 447:706–12.

Whitten, W. M., N. H. Williams, W. S. Armbruster, M. A. Battiste, L. Strekowski, and N. Lindquist. (1986) Carvone oxide: an example of convergent evolution in euglossine-pollinated plants. *Syst Bot* 11:222–28.

Wiegmann, B. M., D. K. Yeates, J .L. Thorne, and H. Kishino. (2003) Time flies, a new molecular time-scale for Brachyceran fly evolution without a clock. *Syst Biol* 52:745–56.

Wiegmann, D. D., D. A. Wiegmann, and F. A. Waldron. (2003) Effects of a reward downshift on the consummatory behavior and flower choices of bumblebee foragers. *Physiol Behav* 79:561–66.

Wiens, D. (1978) Mimicry in plants. *Evol Biol* 11:365–403.

Wiens, D., J. P. Rourke, B. B. Casper, E. A. Ricjart, T. R. LaPine, C. J. Peterson, and A. Channing. (1983) Nonflying mammal pollination of southern African *Proteas*: a non-coevolved system. *Ann Miss Bot Gdn* 70:1–31.

Wiens, D., A. Zitzmann, M-A Lachance, M. Yegles, F. Pragst, F. M. Wurst, D. von Holst, S. L. Guan, and R. Spanagel. (2008) Chronic intake of fermented floral nectar by wild treeshrews. *Proc Nat Acad Sci USA* 105:10426–31.

Wiklund, C., T. Eriksson, and H. Lundberg. (1979) The wood white butterfly *Leptidea sinapis* and its nectar plants: a case of mutualism or parasitism? *Oikos* 33:358–62.

Wilcock, C. and R. Neiland. (2002) Pollinator failure in plants: why it happens and when it matters. *Trends Plant Sci* 7:270–77.

Willemstein, S. C. (1987) *An Evolutionary Basis for Pollination Ecology*. Netherlands: EJ Brill/Leiden University Press.

Williams, C. S. (1998) The identity of the previous visitor influences flower rejection by nectar-collecting bees. *Anim Behav* 56:673–81.

Williams, G. A., P. Adam, and L. A. Mound. (2001) Thrips (Thysanoptera) pollination in Australian subtropical rainforests, with particular reference to the pollination of *Wilkiea huegeliana* (Monimiaceae). *J Nat Hist* 35:1–21.

Williams, N. H. (1982) The biology of orchids and euglossine bees. In *Orchid Biology, Reviews and Perspectives*. Ithaca, NY: Cornell University Press.

———. (2003) Use of novel pollen species by specialist and generalist solitary bees (Hymenoptera: Megachilidae). *Oecologia* 134:228–37.

Williams, N. H. and W. M. Whitten. (1999) Molecular phylogeny and floral fragrances of male euglossine-pollinated orchids: a study of *Stanhopea* (Orchidaceae). *Plant Sp Biol* 14:129–36.

Williams, N. M., R. L. Minckley, and F. A. Silveira. (2001) Variation in native bee faunas and its implications for detecting community change. *Cons Ecol* 5:57–89.

Williams, P. (2007) The distribution of bumblebee colour patterns worldwide: possible significance for thermoregulation, crypsis and warning mimicry. *Biol J Linn Soc* 92:97–118.

Williamson, G. B. (1982) Plant mimicry: evolutionary constraints. *Biol J Linn Soc* 18:49–58.

Willis, D. S. and P. G. Kevan. (1995) Foraging dynamics of *Peponapis pruinosa* (Hymenoptera, Anthophoridae) on pumpkin (*Cucurbita pepo*) in southern Ontario. *Can Ent* 127:167–75.

Willmer, P. G. (1980) The effects of insect visitors on nectar constituents in temperate plants. *Oecologia* 47: 270–77.

———. (1983) Thermal constraints on activity patterns in nectar-feeding insects. *Ecol Entomol* 8:455–69.

———. (1986) Foraging patterns and water balance: problems of optimisation for a xerophilic bee, *Chalicodoma sicula*. *J Anim Ecol* 55:941–62.

———. (1985) Size effects on hygrothermal balance and foraging patterns in a sphecid wasp, *Cerceris arenaria*. *Ecol Entomol* 10:469–79.

———. (1988) The role of insect water balance in pollination ecology: *Xylocopa* and *Calotropis*. *Oecologia* 76:430–38.

Willmer, P. G., A. A. M. Bataw, and J. P. Hughes. (1994) The superiority of bumblebees to honeybees as pollinators – insect visitors to raspberry flowers. *Ecol Entomol* 19: 271–84.

Willmer, P. G. and S. A. Corbet. (1981) Temporal and microclimatic partitioning of the floral resources of *Justicia aurea* amongst a concourse of pollen vectors and nectar robbers. *Oecologia* 51:67–78.

Willmer, P. G., F. S. Gilbert, J. Ghazoul, S. Zalat, and F. Semida. (1994) A novel form of territoriality: daily paternal investment in an anthophorid bee. *Anim Behav* 48: 535–49.

Willmer, P. G., J. P. Hughes, J.A.T. Woodford, and S. J. Gordon. (1996) The effects of crop microclimate and associated physiological constraints on seasonal and diurnal distribution patterns of raspberry beetle (*Byturus tomentosus*) on the host plant (*Rubus idaeus*). *Ecol Entomol* 21:87–97.

Willmer, P. G., C. V. Nuttman, N. E. Raine, G. N. Stone, J. G. Pattrick, K. Henson K, P. Stillman, L. McIlroy, S. G. Potts, and J. T. Knudsen. (2009) Floral volatiles controlling ant behaviour. *Func Ecol* 23:888–900.

Willmer, P. G., D. A. Stanley, K. Steijven, I. M. Matthews, and C. V. Nuttman. (2009) Bidirectional flower color and shape changes allow a second opportunity for pollination. *Curr Biol* 19:919–23.

Willmer, P. G. and G. N. Stone. (1989) Pollination of lowland coffee (*Coffea canephora*); costs and benefits of leaf cutter bees in Papua New Guinea. *Ent Exp Appl* 50:113–24.

———. (1997a) Ant deterrence in *Acacia* flowers: how aggressive ant-guards assist seed set. *Nature* 388:165–67.

———. (1997b) Temperature and water relations in desert bees. *J Therm Biol* 22:453–65.

———. (2004) Behavioral, ecological and physiological determinants of the activity patterns of bees. *Adv Study Behav* 34:347–466.

Willmott, A. P. and A. Búrquez. (1996) The pollination of *Merremia palmeri* (Convolvulaceae): can hawkmoths be trusted? *Am J Bot* 83:1050–56.

Willson, M. F. and J. Ågren. (1989) Differential floral rewards and pollination by deceit in unisexual flowers. *Oikos* 55:23–29.

Wilms, J. and T. Eltz. (2008) Foraging scent marks of

bumblebees: footprint cues rather than pheromone signals. *Naturwissenschaften* 95:149–53.

Wilson, E. O. (1971). *The Insect Societies*. Cambridge, MA: Harvard University Press.

Wilson, J. S., T. Griswold, and O. J. Messinger. (2008) Sampling bee communities (Hymenoptera, Apiformes) in a desert landscape: are pan traps sufficient? *J Kans Ent Soc* 81:288–300.

Wilson, P., M. C. Castellanos, J. N. Hogue, J. D. Thomson, and W. S. Armbruster. (2004) A multivariate search for pollination syndromes among penstemons. *Oikos* 104: 345–61.

Wilson, P., M. C. Castellanos, A. D. Wolfe, and J. D. Thomson. (2006) Shifts between bee and bird pollination in penstemons. In *Plant-Pollinator Interactions: from Specialization to Generalization*, ed. N. M. Waser and J. Ollerton, 47–68. Chicago: University of Chicago Press.

Wilson, P. and M. Stine. (1996) Floral constancy in bumble bees: handling efficiency or perceptual conditioning? *Oecologia* 106:493–99.

Wilson, P. and J. D. Thomson. (1991) Heterogeneity among floral visitors leads to discordance between removal and deposition of pollen. *Ecology* 72:1503–7.

———. (1996) How do flowers diverge? In: Floral Biology: Studies on Floral Evolution in Animal-Pollinated Systems, ed. L. G. Lloyd and S.C.H. Barrett, 88–111. New York: Chapman and Hall.

Wilson, P., A. D. Wolfe, W. S. Armbruster, and J. D. Thomson. (2007) Constrained lability in floral evolution: counting convergent origins of hummingbird pollination in *Penstemon* and *Keckiella*. *New Phytol* 176:883–90.

Winfree, R., N. M. Williams,, J. Dushoff, and C. Kremen. (2007) Native bees provide insurance against ongoing honeybee losses. *Ecol Lett* 10:1105–13.

Winter, Y. and O. von Helversen. (2001) Bats as pollinators: foraging energetics and floral adaptations. In *Cognitive Ecology of Pollination*, ed. L. Chittka and J. D. Thomson, 148–70. Cambridge: Cambridge University Press.

Witjes, S. and T. Eltz. (2009) Hydrocarbon footprints as a record of bumblebees flower visitation. *J Chem Ecol* 35:1320–25.

Woittiez, R. D. and M.T.M. Willemse. (1979) Sticking of pollen on stigmas: the factors and a model. *Phytomorph* 29:57–63.

Wolda, H. (1983) "Long-term" stability of tropical insect populations. *Res Pop Ecol Suppl* 3:112–26.

Wolf, L. L. and F. R. Hainsworth. (1971) Time and energy budgets of territorial hummingbirds. *Ecology* 52:980–88.

Wolf, L. L., F. R. Hainsworth, and F. B. Gill. (1975) Foraging efficiencies and time budgets in nectar-feeding birds. *Ecology* 56:117–28.

Wolf, T., C. P. Ellington, and I. S. Begley. (1999) Foraging costs in bumblebees: field conditions cause large individual differences. *Insectes Sociaux* 46:291–95.

Wolfe, L. M. (2001) Associations among multiple floral polymorphisms in *Linum pubescens* (Linaceae), a heterostylous plant. *Int J Plant Sci* 162:335–42.

Wolfe, L. M. and S. C. H. Barrett. (1989) Patterns of pollen removal and deposition in tristylous *Pontederia cordata* L (Pontederiaceae). *Biol J Linn Soc* 36:317–29.

Wolfe, L. M. and D. R. Sowell. (2006) Do pollinator syndromes partition the pollinator community? A test using four sympatric morning glory species. *Int J Plant Sci* 167:1169–75.

Wolff, D. (2006) Nectar sugar composition and volumes of 47 species of Gentianales from a Southern Ecuadorian montane forest. *Ann Bot* 97:767–77.

Wolff, D., M. Braun, and S. Liede. (2003) Nocturnal versus diurnal pollination success in *Isertia laevis* (Rubiaceae); a sphingophilous plant visited by hummingbirds. *Plant Biol* 5:71–78.

Wolff, D., T. Witt T, A. Jürgens, and G. Gottsberger. (2006) Nectar dynamics and reproductive success in *Saponaria officinalis* (Caryophyllaceae) in southern Germany. *Flora* 201:353–64.

Wong, B. B. M., C. Salzmann, and F. P. Schiestl. (2004) Pollinator attractiveness increases with distance from flowering orchids. *Proc Roy Soc Lond* B 271:S212–S214.

Wooller, R. D., K. C. Richardson, and C. M. Pagendham. (1988) The digestion of pollen by some Australian birds. *Austr J Zool* 36:357–62.

Wooller, R. D. and S. Wooller. (2003) The role of non-flying mammals in the pollination of *Banksia nutans*. *Aust J Bot* 51:503–7.

Worden, B. D. and D. R. Papaj. (2005) Flower choice copying in bumblebees. *Biol Lett* 1:504–7.

Worley, A. C. and S.C.H. Barrett. (2000) Evolution of floral display in *Eichhornia paniculata* (Pontederiaceae): direct and correlated Reponses to selection on flower size and number. *Evolution* 54:1533–45.

Wright, G. A., A. Lutmerding, N. Dudareva, and B. H. Smith. (2005) Intensity and the ratios of compounds in the scent of snapdragon flowers affect scent discrimination by honeybees (*Apis mellifera*). *J Comp Phys* A 191:105–14.

Wright, G. A. and Schiestl. (2009) The evolution of floral scent: the influence of olfactory learning by insect pollinators on the honest signalling of floral rewards. *Func Ecol* 23:841–51.

Wright, S. I., R. W. Ness, J. P. Foxe, and S.C.H. Barrett. (2008) Genomic consequences of outcrossing and selfing in plants. *Int J Plant Sci* 169:105–118.

Wright, S. J. and O. Calderon. (1995) Phylogenetic patterns among tropical flowering phenologies. *J Ecol* 83:937–48.

Wyatt, R. (1981) Ant-pollination of the granite endemic *Diamorpha smallii* (Crassulaceae). *Am J Bot* 68:1212–17.

Wyatt, R. and S. B. Broyles. (1994) Ecology and evolution of reproduction in milkweeds. *Ann Rev Ecol Syst* 25:423–41.

Wyatt R., Broyles S. B. and Derda G. S. (1992) Environmental influences on nectar production in milkweeds (*Asclepias syriaca* and *A. exaltata*). *Am J Bot* 79:636–42.

Wyatt, R. and A. Stoneburner. (1981) Patterns of ant-mediated pollen dispersal in *Diamorpha smallii* (Crassulaceae). *Syst Bot* 6:1–7.

Yaku, A., G. H. Walter, and A. J. Najar-Rodriguez. (2007) Thrips see red – flower colour and the host relationships of a polyphagous anthophilic thrips. *Ecol Entomol* 32: 527–35.

Yamagishi, H., A. D. Taber, and M. Ohara. (2005) Effect of snowmelt timing on the genetic structure of an *Erythronium grandiflorum* population in an alpine environment. *Ecol Res* 20:199–204.

Yap, C. A. M., N. C. Sodhi NC and K. S. H. Peh. (2007) Phenology of tropical birds in Peninsular Malaysia; effects of selective logging and food resources. *Auk* 124: 945–61.

Yeates, D. K. and B. M. Wiegmann. (1999) Congruence and controversy: toward a higher-level phylogeny of the Diptera. *Ann Rev Entomol* 44:397–428.

Yeboah, Gyan K. and S. R. J. Woodell. (1987a) Flowering phenology, flower colour and mode of reproduction of *Prunus spinosa* L. (blackthorn), *Crataegus monogyna* Jacq., (hawthorn), *Rosa canina* L. (dog rose) and *Rubus fruticosus* L. (bramble) in Oxfordshire, England. *Funct Ecol* 1:261–68.

———. (1987b) Analysis of insect pollen loads and pollination efficiency of some common insect visitors of four species of woody Rosaceae. *Funct Ecol* 1:269–74.

Yeo, P. F. (1993) Secondary Pollen Presentation. New York: Springer Verlag.

Yokoi, T. and K. Fujisaki. (2009) Recognition of scent marks in solitary bees to avoid previously visited flowers. *Ecol Res* 24:803–9.

Young, A. M. (1985) Studies of cecidomyid midges (Diptera, Cecidomyiidae) as cocoa pollinators (*Theobroma cacao* L.) in Central America. *Proc Ent Soc Wash* 87:49–79.

Young, H. J. (1986) Beetle pollination of *Dieffenbachia longispatha* (Araceae). *Am J Bot* 73:931–44.

Young, H. J. (1988) Differential importance of beetle species pollinating *Dieffenbachia longispatha* (Araceae). *Ecology* 69:832–44.

Young, H. J., D. W. Dunning, and K. W. von Hasseln. (2007) Foraging behaviour affects pollen removal and deposition in *Impatiens capensis* (Balsaminaceae). *Am J Bot* 94:1267–71.

Young, H. J. and M. L. Stanton. (1990a) Influence of environmental quality on pollen competitive ability in wild radish. *Science* 29:1631–33.

———. (1990b) Influences of floral variation on pollen removal and seed production in wild radish. *Ecology* 71:536–47.

Yu, Q and S. Q. Huang. (2006) Flexible stigma presentation assists context-dependent pollination in a wild columbine. *New Phytol* 169:237–41.

Yuan, L. C., Y. B. Luo, L. B. Thien, J. H. Fan, J. L. Xu, and Z. D. Chen. (2007) Pollination of *Schisandra henryi* (Schisandraceae) by female pollen-eating *Megommata* species (Cecidomyiidae, Diptera) in south-central China. *Ann Bot* 99:451–60.

Yumoto, Y. (2000) Bird-pollination of three *Durio* species (Bombacaceae) in a tropical rainforest in Sarawak, Malaysia. *Am J Bot* 87:1181–88.

Yumoto, T., K. Momose, and H. Nagamasu. (1999) A new pollination syndrome – squirrel pollination in a tropical rainforest in Lambir Hills National Park, Sarawak, Malaysia. *Tropics* 9:147–51.

Zamora, R. (1999) Conditional outcomes of interactions: the pollinator-prey conflict of an insectivorous plant. *Ecology* 80:786–95.

Zavodna, M., S. M. Knapp, S. G. Compton, P. Arens, B. Vosman, P. J. van Dijk, P. M. Gilmartin, and J.M.M. Van Damme. (2007) Reconstruction of fig wasp mating structure: how many mothers share a fig? *Ecol Entomol* 32: 485–91.

Zerega, N. J. C., L. A. Mound, and G. D. Weiblen. (2004) Pollination in the New Guinea endemic *Antiaropsis decipiens* (Moraceae) is mediated by a new species of thrips, *Thrips antiaropsidis* sp nov (Thysanoptera, Thripidae). *Int J Plant Sci* 165:1017–26.

Zhang, L., S. C. H. Barrett, J-Y Gao, J. Chen, W. W. Cole, Y. Liu, Z. L. Bai, and Q-J Li. (2005) Predicting mating patterns from pollination syndromes: the case of 'sapromyiophily' in *Tacca chantrieri* (Taccaceae). *Am J Bot* 92: 517–24.

Zhao, Z. G., G. Z. Du, S. H. Zhou, M. T. Wang, and Q. J. Ren. (2006) Variations with altitude in reproductive traits and resource allocation of three Tibetan species of Ranunculaceae. *Aust J Bot* 54:691–700.

Zimmerman, J. K. and T. M. Aide. (1989) Patterns of fruit production in a neotropical orchid: pollinator vs. resource limitation. *Am J Bot* 76:67–73.

Zimmerman, J. K., D. W. Roubik, and J. D. Ackerman. (1989) Asynchronous phenologies of a neotropical orchid and its euglossine bee pollinators. *Ecology* 70: 1192–95.

Zimmerman, M. (1980) Reproduction in *Polemonium*: competition for pollinators. Ecology 61:497–501.

———. (1982) Optimal foraging: random movement by pollen collecting bumblebees. *Oecologia* 53:394–98.

Zimmerman, M. and G. H. Pyke. (1988) Experimental manipulations of *Polemonium foliosissimum*: effects on subsequent nectar production, seed production and growth. *J Ecol* 76:777–89.

Zjhra, M. L. (2008) Facilitating sympatric species coexistence via pollinator partitioning in endemic tropical trees of Madagascar. *Plant Syst Evol* 271:157–76.

Zych, M. (2002) Pollination biology of *Heracleum sphondylium* L. (Apiaceae). The advantages of being white and compact. *Acta Soc Bot Pol* 71:163–70.

———. (2007) On flower visitors and true pollinators; the case of *Heracleum sphondylium* L. (Apiaceae). *Plant Syst Evol* 263:159–79.

SUBJECT INDEX

INDEX OF ANIMAL GENERA

INDEX OF PLANT GENERA